Manufacturing Techniques for Materials

Engineering and Engineered

Manufacturing Techniques for Materials

Engineering and Engineered

Edited by
T. S. Srivatsan
T. S. Sudarshan
K. Manigandan

CRC Press
Taylor & Francis Group
Boca Raton London New York

CRC Press is an imprint of the
Taylor & Francis Group, an **informa** business

CRC Press
Taylor & Francis Group
6000 Broken Sound Parkway NW, Suite 300
Boca Raton, FL 33487-2742

Printed on acid-free paper

International Standard Book Number-13: 978-1-138-09926-5 (Hardback)

Visit the Taylor & Francis Web site at
http://www.taylorandfrancis.com

and the CRC Press Web site at
http://www.crcpress.com

Contents

Preface

The science and engineering specific to manufacturing is an interesting and challenging field that shows both promise and potential for its vast breadth and depth due to its viability for applications in a spectrum of industries that include both performance-critical and non-performance-critical. This book attempts to provide a cohesively convincing and compelling overview of recent trends and emerging developments along with the advantages and possible commercialization of the techniques presented, in the domain specific to the manufacturing of materials. This has become increasingly essential to keep pace with the noticeable strides being made in the domain of materials development and resultant technological applications. Each chapter, in this volume attempts to provide a qualitative description of the technology and practice specific to the manufacturing technique or techniques used and the viable outcomes.

This book consists of two sections. Each section, i.e., Section A and Section B, contains well laid-out technical chapters. To make every effort to meet the needs and requirements of different readers, each chapter has been written and presented by the author to ensure that it offers a clean, clear, and convincing presentation and discussion of the intricacies pertaining to their discussion of the subject matter in a cohesively convincing and compelling manner.

In the first section of the book (Section A) the focus is on "Emerging Techniques, Trends, and Advances in Manufacturing." In this section, the first five chapters are devoted to additive manufacturing (AM). Chapter 1 introduces the interested reader to aspects pertinent to topological optimization of the additive manufacturing technique with specific reference to applications for both internal pattern and support. Chapter 2 provides an in-depth analysis, in a lucid and convincing manner, of the key design features that are essential to address security changes in AM. Chapter 3 is devoted to applications and dwells upon the use of a robotized laser powder feed for the fabrication of metallic metamaterials. Chapter 4 provides a comprehensive coverage and convincing overview of appropriate use of the approach of laser AM for synthesizing in situ metal matrix composites (MMCs) and functionally graded materials (FGMs). In Chapter 5, the presentation and discussion focus on the relative merits and applications of electron beam prototyping using wires and coupled with the ability of using a beam of electrons to achieve both surface modification and even additive layer manufacturing. Chapter 6 presents all the intricacies, approaches, and merits of the techniques used for the processing and fabrication of bioinspired materials and the implications resulting from potentially viable applications for these materials. Chapters 7 and 8 focus on current trends and approaches in the domain specific to manufacturing. Use of the approach of energy value stream mapping as a tool to develop green manufacturing is the theme of presentation and adequate discussion in Chapter 7. Chapter 8 presents a comprehensive overview of all aspects pertinent to Lean and sustainable concepts in manufacturing and provides meaningful discussion at appropriate locations throughout. This chapter is a thorough review of the work done and published in the technical literature, and the authors have made a truly useful attempt to synchronize the observations and findings scattered throughout the literature and present them as one cohesive and complete manuscript. Chapter 9 focuses on the presentation, elaboration, and discussion of the novel features, noticeable outcomes, and noteworthy advantages that can be obtained from using unconventional approaches to the processing of emerging and advanced materials.

Along the lines of aspects pertinent to the importance of surfaces in manufacturing, the key highlights, advantages, and even applications resulting from laser surface engineering of light metal aluminum in the form of three aluminum alloys are the focus of the presentation and exhaustive discussion in Chapter 10. The use of pulsed electron beam processing for light metals with specific reference to the advantages and applications made possible by the treatment, which would ensure the selection and use of light metals for a plethora of applications, is the central theme of presentation and discussion in Chapter 11. In Chapter 12, the theme and focus is on presenting and understanding developments with specific reference to advances in the fabrication of FGMs. The authors present and discuss a spectrum of processing methods used while concurrently highlighting both the advantages and limitations of each method. They also present an overview of the possible applications of FGMs in structural applications. In Chapter 13, the focus is on the use of nanotechnology for viable applications in the field of medicine. The authors provide an overview of the use of nanotechnology to synthesize biomaterials having a high degree of complexity, tissue specificity, and desired mechanical, physical, and biochemical properties. They also highlight how surfaces of scaffolds can be modified using newly developed nano-techniques with the intent of enhancing cell–biomaterial interaction. Overall, the notable advances in nanotechnology with specific reference to cell biology are well presented and discussed. In their comprehensive overview, they dwell upon the most relevant roles of nanotechnology in the various applications.

The second section of this book (Section B), entitled "Conventional Techniques, Approaches, and Applications in Manufacturing," consists of 10 exhaustive chapters. The first chapter in this section (i.e., Chapter 14) is purely introductory in nature and provides the interested reader with the key highlights with specific reference to the conventional techniques that have and are continuing to be used in the manufacturing domain as it relates to both materials and products. This chapter can safely be categorized as a healthy refresher to the knowledgeable reader and learned engineer/technologist while concurrently providing the novice and inquisitive learner an introductory highlight of the techniques available and useable for manufacturing. In Chapter 15, key aspects that would go a long way in not only enabling but also enhancing our understanding of the kinetics of solidification during cooling of molten metal is well presented and thoroughly discussed from both a scientific and engineering perspective with a valued number of examples. Chapter 16 presents the theme for presentation and adequate discussion of all the intricacies and factors that need to be considered during an analysis of solidification kinetics during continuous casting. This chapter can be both educative and enriching from the standpoint of analysis and rationalization of the findings. In Chapter 17, the authors present their views along with valued discussion from the standpoint of developing FGMs through fused deposition modeling assisted investment casting. In Chapter 18, the focus of the presentation and convincing description is on the domain of mechanical deformation, i.e., hot working. In this chapter, the knowledgeable authors shed light on both the role and importance of fragmentation at the nanoscale level in synergism with microstructural optimization for enabling conditions conducive for hot working of both metals and their alloy counterparts. In Chapter 19, the focus is on powder metallurgy processing, specifically the innovative techniques, approaches, and applications resulting from powder processing of Alnico magnets. The presentation and discussions contained in this chapter should be very useful to all those involved in engineering the development of magnetic materials. Chapters 20 and 21 focus on the techniques, trends, and advances in the domain specific to conventional machining of metals and composite materials, and noteworthy advances and applications of nonconventional machining practices for metals and their composite counterparts. These two chapters put together should be a useful source of neatly contained and presented information for all those in need of quick knowledge in the domain encompassing machining of materials to include both metals and composites. Chapter 22 is a cohesive and convincing presentation of all intricacies pertinent to use of the techniques of friction stir welding and friction stir processing for the joining of materials and structures. Besides the intricacies of each of these two techniques, the relative advantages and applications are also well presented and discussed by the authors. For the case of joining using the technique of welding, the contributing authors provide a comprehensive and convincing overview of the intricacies pertinent to the use of laser welding of the nitinol shape memory alloy in Chapter 23. Besides the current approaches in widespread use, the authors try to present and discuss the potential future opportunities and possibilities with specific reference to shape memory alloys.

Overall, this archival monograph on aspects related to viable manufacturing techniques for materials spanning the families of both engineering and engineered does provide a background that should enable an interested reader to both understand and comprehend the immediate past, the prevailing present, and the possible future trends and/or approaches in the domain specific to manufacturing with an emphasis on innovations and applicability of use for a wide range of materials. Thus, based entirely on the contents included in this bound volume, it can very well serve as a single reference book or even as a textbook for:

a. Students spanning seniors in the undergraduate program of study in the fields of materials science and engineering, mechanical engineering, and manufacturing engineering.
b. Beginning graduate students pursuing graduate degrees in materials science and engineering, mechanical engineering, and manufacturing engineering.
c. Researchers spread through both research laboratories and industries striving to specialize and excel on aspects related to research on materials and resultant product, and
d. Engineers striving and seeking for technologically novel and viable innovations for a spectrum of both performance-critical and non-performance-critical applications.

We certainly anticipate this volume to be of much interest to scientists, engineers, technologists, and even entrepreneurs.

T. S. Srivatsan
T. S. Sudarshan
K. Manigandan

Acknowledgments

The editors gratefully acknowledge the understanding and valued support they received from the authors of the different chapters contained in this bound volume. Efforts made by the contributing authors to both present and discuss the different topics included in this volume, thereby enhancing its scientific and technological content, are greatly appreciated. The useful comments, corrections, recommendations, and suggestions made by the referee(s) on the different chapters have certainly helped in enhancing the technical content and merit of the final version of each chapter.

Our publisher, CRC Press, has been very supportive and patient through the entire process initiating with the conception of this intellectual project. We extend an abundance of thanks, valued appreciation, and gratitude to the editorial staff. We specifically mention Ms. Allison Shatkin [senior acquisitions editor, Books (Materials, Chemical, & Petroleum Engineering)] and Ms. Teresita Munoz (editorial assistant—Engineering) for their sustained interest, involvement, attention, and almost always willing assistance arising from an understanding of both the situation and need coupled with their enthusiastic disposition toward extending help to both the editors and the contributing authors. This certainly helped to a sizeable extent in ensuring the timely execution of the numerous intricacies related to smooth completion of this volume from the moment of its approval and up until its compilation and publication. At moments of need, we the editors have found the presence of Ms. Teresita Munoz at CRC Press coupled with her willing and voluntary participation to be a strong pillar of support, almost always energetic and enthusiastic in her helpful disposition.

Most importantly and worthy of recording is that the timely compilation and publication of this bound volume would not have been possible without: (1) the understanding, cooperation, assistance, and importantly the patience extended by several of the contributing authors; and (2) the positive contributions of the peer-reviewer(s).

Editors

Dr. T. S. Srivatsan is a professor of materials science and engineering in the Department of Mechanical Engineering, The University of Akron. He received his bachelor of engineering degree in mechanical engineering in 1980 from Bangalore University, Bangalore, India, and subsequently his graduate degrees (MS degree in aerospace engineering in 1981 and PhD degree in mechanical engineering in 1984) from Georgia Institute of Technology. Dr. Srivatsan joined the faculty at the Department of Mechanical Engineering, the University of Akron in August 1987. Since joining, he has instructed undergraduate and graduate courses in the areas of advanced materials and manufacturing processes, mechanical behavior of materials, fatigue of engineering materials and structures, fracture mechanics, introduction to materials science and engineering, mechanical measurements, design of mechanical systems, and mechanical engineering laboratory. His research areas currently span the fatigue and fracture behavior of advanced materials to include monolithic(s), intermetallic, nanomaterials, and metal–matrix composites; processing techniques for advanced materials and nanostructure materials; interrelationship between processing and mechanical behavior; electron microscopy; failure analysis; and mechanical design. He serves as a coeditor of the *International Journal on Materials and Manufacturing Processes* and on the editorial advisory board of several journals in the domain of materials science and engineering. His research has enabled him to deliver over 210 technical presentations in national and international meetings and symposia; technical/professional societies; and research and educational institutions. He has authored/edited/coedited 63 books in areas of cross-pollinating mechanical design; processing and fabrication of advanced materials; deformation, fatigue, and fracture of ordered intermetallic materials; machining of composites; failure analysis; and technology of rapid solidification processing of materials. He has authored or coauthored over 700 technical publications in archival international journals, chapters in books, proceedings of national and international conferences, reviews of books, and technical reports. In recognition of his efforts, contributions, and their impact on furthering science, technology, and education, he has been elected a fellow of the Society of American Society for Materials, International; a fellow of the American Society of Mechanical Engineers; and a fellow of the American Association for the Advancement of Science. He has also been recognized as an Outstanding Young Alumnus of Georgia Institute of Technology; Outstanding Research Faculty, College of Engineering, The University of Akron; and Outstanding Research Faculty at The University of Akron; and received the Distinguished Service Award given in recognition of his services to the technical and engineering community by the Minerals, Metals and Materials Society. He offers his knowledge in research services to the U.S. Air Force and U.S. Navy, National Research Laboratories, and industries related to aerospace, automotive, power-generation, leisure-related products, and applied medical sciences.

Dr. T. S. Sudarshan obtained his BTech degree in metallurgy from the Indian Institute of Technology (IIT, Madras) and later completed his MS and PhD degrees in materials engineering science from Virginia Polytechnic Institute and State University (Virginia Tech), USA. Dr. Sudarshan is currently the president and CEO of Materials Modification Inc., which is at the forefront of research, development, and commercialization of advanced materials utilizing novel processing techniques. He has demonstrated technological leadership for well over three decades, and has worked extensively throughout his career in the areas spanning nanotechnology and surface engineering for which he is very well known throughout the world. Through his leadership, he has raised over US$60 million in funding for the primary purpose of very high-risk, high payoff advanced technology related programs in several non-traditional areas. Dr. Sudarshan has published well over 175 peer-reviewed papers and has edited 32 books in the areas spanning surface modification technologies and advanced materials. He is currently the editor of two international journals: *Surface Engineering* and *Materials and Manufacturing Processes*, and holds over 25 patents. He has chaired numerous committees for all federal agencies, chaired many committees of the Ohio Third Frontier, and served as a member of the National Materials Advisory Board and trustee of the American Society for Materials, International (ASM Int.). In recognition of his efforts, contributions, and impact on furthering science, technology, and their far-reaching applications, he has been elected as a fellow of the ASM Int.; a fellow of the International Federation for Heat Treatment and Surface Engineering, and a fellow of the Institute of Materials. He was also awarded the Distinguished Alumni of IIT Madras.

Dr. K. Manigandan obtained his BEng degree (2009) in mechanical engineering from SASTRA University (Thirumalaisamudram, Thanjavur, Tamil Nadu, India) and later completed his MS (2011) and Ph.D. (2014) degrees in mechanical engineering from The University of Akron (Akron, Ohio, USA). Dr. Manigandan is presently a research assistant professor in the Department of Mechanical Engineering at The University of Akron. He is the author of over 50 technical publications in archival journals, proceedings of national and international conferences, and final technical reports. His research interests, involvement, and expertise span the domain of mechanical behavior, which include failure and fracture analysis, of materials spanning the broad spectrum of metals, their alloy counterparts, metal–matrix composites, ceramics, ceramic matrix composites, and emerging materials for use in high-temperature applications. His current research interest and involvement include the use of innovative testing techniques to understand the mechanical behavior of novel materials, spanning both engineering (metals) and engineered (composites), for use in both performance-critical and non-performance-critical applications in the industries including space and aerospace.

Contributors

Behzad Aghabarari
Bioengineering Research Group
Nanotechnology and Advanced Materials Department
Materials and Energy Research Center
Tehran, Iran

Li Ai
Department of Mechanical Engineering
Southern Methodist University
Dallas, Texas

Meysam Akbari
Research Center for Advanced Manufacturing
Southern Methodist University
Dallas, Texas

Iver E. Anderson
Ames Laboratory
Division of Materials Sciences and Engineering
U.S. Department of Energy
Ames, Iowa

Maria Anityasari
Department of Industrial Engineering
Institut Teknologi Sepuluh Nopember
Surabaya, Indonesia

Ramanathan Arunachalam
Mechanical and Industrial Engineering Department
Sultan Qaboos University
Muscat, Oman

G. S. Avadhani
Department of Materials Engineering
Indian Institute of Science
Bangalore, India

Rajarshi Banerjee
Department of Materials Science and Engineering
University of North Texas
Denton, Texas

Sourabh Biswas
School of Mechanical and Aerospace Engineering
Oklahoma State University
Stillwater, Oklahoma

Aidin Bordbar-Khiabani
Bioengineering Research Group
Nanotechnology and Advanced Materials Department
Materials and Energy Research Center
Tehran, Iran

Tushar Borkar
Mechanical Engineering Department
Cleveland State University
Cleveland, Ohio

Fei Chen
Mechanical and Aerospace Engineering Department
Tandon School of Engineering
New York University
Brooklyn, New York

Xizhang Chen
Department of Mechanical and Electrical Engineering
Wenzou University
Wenzou, China

Udisubakti Ciptomulyono
Department of Industrial Engineering
Institut Teknologi Sepuluh Nopember
Surabaya, Indonesia

Steve Constantinides
Magnetics & Materials LLC
Rochester, New York

Sivasrinivasu Devadula
Department of Mechanical Engineering
Indian Institute of Technology Madras
Madras, India

Yaoyu Ding
Research Center for Advanced Manufacturing
Southern Methodist University
Dallas, Texas

Mohsen Eshraghi
Department of Mechanical Engineering
California State University
Los Angeles, California

Sergio D. Felicelli
Department of Mechanical Engineering
The University of Akron
Akron, Ohio

Xin-Lin Gao
Department of Mechanical Engineering
Southern Methodist University
Dallas, Texas

Nicolas Gardan
Micado (CAD/CAE Technical Centre DINCCS)
Pôle de Haute technologie
Charleville-Mézières, France

Mazaher Gholipourmalekabadi
Cellular and Molecular Research Center
and
Department of Tissue Engineering and Regenerative Medicine
Faculty of Advanced Technologies in Medicine
Iran University of Medical Sciences
Tehran, Iran

Ankit Gupta
School of Engineering
Indian Institute of Technology Mandi
Himachal Pradesh, India

Manoj Gupta
Department of Mechanical Engineering
National University of Singapore
Singapore, Singapore

Nikhil Gupta
Composite Materials and Mechanics Laboratory
Mechanical and Aerospace Engineering Department
Tandon School of Engineering
New York University
Brooklyn, New York

Sandip P. Harimkar
School of Mechanical and Aerospace Engineering
Oklahoma State University
Stillwater, Oklahoma

Sri Hartini
Department of Industrial Engineering
Diponegoro University
Semarang, Indonesia

Subramanian Jayalakshmi
Department of Aeronautical Engineering
Kumaraguru College of Technology (KCT)
Coimbatore, India

Jayamani Jayaraj
Corrosion Science and Technology Division
Indira Gandhi Centre for Atomic Research
Kalpakkam, India

Pradeep K. Jha
Department of Mechanical and Industrial Engineering
Indian Institute of Technology Roorkee
Uttarakhand, India

Mohammad Taghi Joghataei
Cellular and Molecular Research Center
and
Department of Tissue Engineering and Regenerative Medicine
and
Neuroscience Department
Faculty of Advanced Technologies in Medicine
Iran University of Medical Sciences
Tehran, Iran

Sathish Kannan
Department of Mechanical Engineering
American University of Sharjah
Sharjah, United Arab Emirates

Thangaraju Deepan Bharathi Kannan
Department of Production Engineering
National Institute of Technology, Tiruchirappalli
Tamil Nadu, India

Aaron G. Kassen
Ames Laboratory
Division of Materials Sciences and Engineering
U.S. Department of Energy
Ames, Iowa

Sergey Konovalov
Department of Metals Technology and Aviation Materials
Samara National Research University
Samara, Russia

Radovan Kovacevic
Research Center for Advanced Manufacturing
Southern Methodist University
Dallas, Texas

Lakshminath Kundanati
Laboratory of Bio Inspired and Graphene Nanomechanics
Department of Civil, Environmental and Mechanical Engineering
University of Trento
Mesiano, Italy

K. Manigandan
Department of Mechanical Engineering
The University of Akron
Akron, Ohio

Ambrish Maurya
Department of Mechanical and Industrial Engineering
Indian Institute of Technology Roorkee
Uttarakhand, India

Masuod Mozafari
Cellular and Molecular Research Center
and
Department of Tissue Engineering and Regenerative Medicine
Faculty of Advanced Technologies in Medicine
Iran University of Medical Sciences
Tehran, Iran

Sukhomay Pal
Department of Mechanical Engineering
Indian Institute of Technology Guwahati
Assam, India

Biswajit Parida
Department of Mechanical Engineering
Indian Institute of Technology Guwahati
Assam, India

Y. V. R. K. Prasad
Department of Materials Engineering
Indian Institute of Science
Bangalore, India

Nicola M. Pugno
Laboratory of Bio-Inspired and Graphene Nanomechanics
Department of Civil, Environmental and Mechanical Engineering
University of Trento
Mesiano, Italy

and

School of Engineering and Materials Science
Queen Mary University of London
London, United Kingdom

and

Ket Lab
Edoardo Amaldi Foundation
Italian Space Agency
Rome, Italy

Sayyad Zahid Qamar
Mechanical and Industrial Engineering Department
Sultan Qaboos University
Muscat, Oman

Thillaigovindan Ramesh
Department of Mechanical Engineering
National Institute of Technology, Tiruchirappalli
Tamil Nadu, India

Puthuveettil Sreedharan Robi
Department of Mechanical Engineering
Indian Institute of Technology Guwahati
Assam, India

Ali Samadikuchaksaraei
Cellular and Molecular Research Center
and
Department of Tissue Engineering and Regenerative Medicine
Faculty of Advanced Technologies in Medicine
Iran University of Medical Sciences
Tehran, Iran

Paulraj Sathiya
Department of Production Engineering
National Institute of Technology, Tiruchirappalli
Tamil Nadu, India

Sambasivam Seshan
Department of Mechanical Engineering
Indian Institute of Science
Bangalore, India

Rajashekhara Shabadi
Physical Metallurgy and Materials Engineering
University of Science and Technology of Lille
Lille, France

Khaled Shahin
Division of Engineering
New York University-Abu Dhabi
Saadiyat Island, United Arab Emirates

Vinay Sharma
Department of Production Engineering
BIT, Mesra (Ranchi)
Jharkhand, India

K. R. Y. Simha
Department of Mechanical Engineering
Indian Institute of Science
Bangalore, India

Ramachandra Arvind Singh
Department of Mechanical Engineering
Kumaraguru College of Technology
Coimbatore, India

Rupinder Singh
Department of Production Engineering
Guru Nanak Dev Engineering College
Ludhiana, India

T. S. Srivatsan
Division of Materials Science and Engineering
Department of Mechanical Engineering
The University of Akron
Akron, Ohio

T. S. Sudarshan
Materials Modification Inc.
Fairfax, Virginia

Mohammad Talha
School of Engineering
Indian Institute of Technology Mandi
Himachal Pradesh, India

Rajasekaran Thanigaivelan
Department of Mechanical Engineering
Mahendra Engineering College
Tamilnadu, India

Aleksandra M. Urbanska
Department of Medicine
Irving Cancer Research Center
Columbia University
New York, New York

Neha Verma
Department of Production Engineering
BIT, Mesra (Ranchi)
Jharkhand, India

Marek St. Weglowski
Instytut Spawalnictwa (Institute of Welding)
Gliwice, Poland

Emma M. White
Ames Laboratory
Division of Materials Sciences and Engineering
U.S. Department of Energy
Ames, Iowa

Section A

Emerging Techniques, Trends, and Advances in Manufacturing

1 Topological Optimization and Additive Manufacturing

Applications for Internal Patterns and Support

Nicolas Gardan

CONTENTS

1.1 INTRODUCTION

Additive manufacturing (AM) is widely used in industrial product development. The main advantage of the additive manufacturing concept is the ability to create almost any possible shape. This capacity is governed by the built-up layer-by-layer process. There are several available technologies based on this additive machining concept [1]: stereolithography, photo-masking, selective laser sintering (SLS), fused deposition modeling (FDP), 3D printing, and so on. Researchers work principally on the influence of part orientation, slicing strategy, and matching internal patterns to improve cost, product quality, built time, and so on. Numerical topological optimization is a technical break that allows the modeling of really innovative shapes, based on trade knowledge [2]. The union of the two technologies, AM and numerical topological optimization, seems to be very promising, more particularly for steel machining (for a real return on investment [ROI] on mass, for example).

Since the appearance of rapid prototyping, different technologies emerged. The manufacturing by layers gets a common characteristic and the AM or 3D printing offers several materials. For manufacturing, the complex geometries need to be held to maintain the support material. This material holds the external and internal surfaces in position. In most cases, the support material is cleaned during finishing or confined into the model. The consumable cost is often expensive and the support material is lost. The environment impact is important because of the resins used.

In the last decade, the use of structural optimization has rapidly increased. The upstream phases of the design process represent 5% of the involved time of a product development, but engage 75% of the global development costs [3]. The integration of optimization in the early phases of a project is thus very important. The use of numerical simulation to optimize products has become essential to test different forms and materials as well as to better understand the involved physical phenomena. The main difficulty of using computational optimization is to manage the loops between computer-aided design (CAD) and computer-aided engineering (CAE). Thus, any change in geometry induced by the analysis can greatly increase the delay. Methods for shape optimization automate this chain and find an optimal solution with the inclusion of the specifications. Besides the possibility to test original solutions, the use of numerical optimization can address the problem of CAE integration in the early stages of the design process. It is then necessary to establish a methodology for capitalization and knowledge management.

There are three main categories of shape optimization of mechanical structures [4]:

- "Parametric shape optimization: the shapes are parameterized by a reduced number of variables (thickness, diameters, and dimensions)." This class of optimization does not allow exploration of other possible shapes, but it allows finding (calculating) the optimum dimensions of parametric forms (existing forms of the model).
- "Optimization of geometric shapes that, from an initial shape, vary the position of the boundaries of form." This optimization by the variation of the boundaries allows finding optimized contours structures without changing the initial topology.
- "Topological shape optimization: obtain, without any explicit or implicit restriction, the best shape possible even if topology changes." This third category of optimization is an appropriate method for the design phase of a new part, because it can explore new concepts and solutions in areas of "no comfort" for engineers (see a basic example on Figure 1.1).

In AM, we are particularly interested in topological shape optimization in the thermomechanical process. In multidisciplinary optimization, thermal optimization is widely used. The heat convection

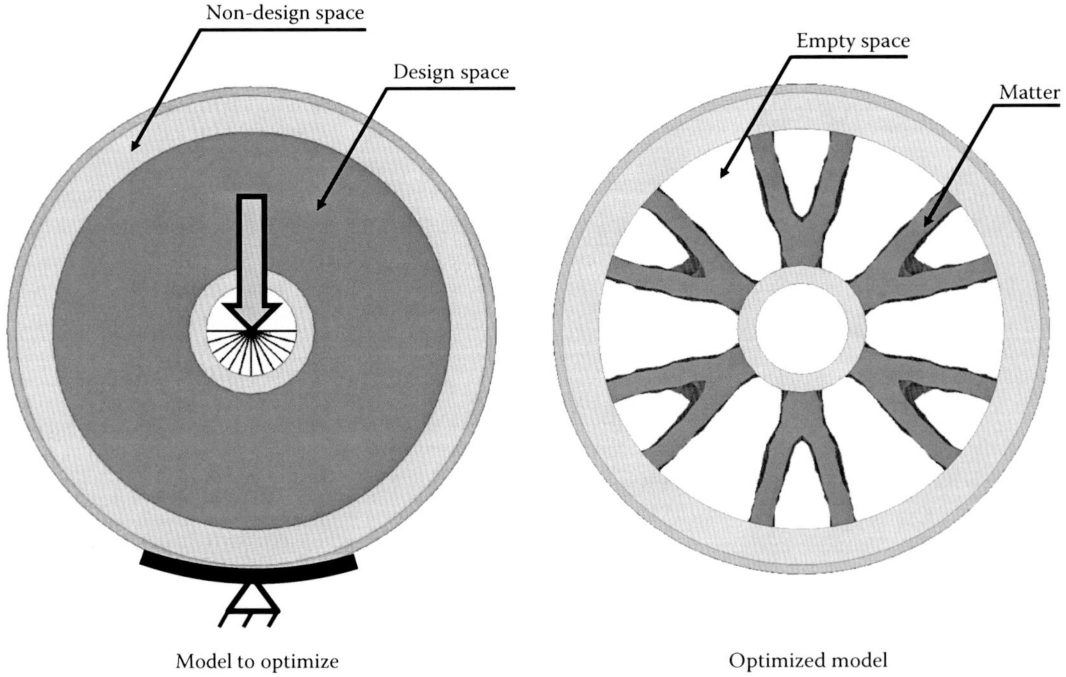

FIGURE 1.1 Simple example of topological optimization.

can significantly influence optimal design configurations [5,6] obtained with structural optimization. Three topology optimization methods are widely applied to thermal problems:

- The level-set method [7,8]
- The evolutionary structural optimization (ESO) method
- The solid isotropic material with penalization (SIMP) method (see below for more details)

Level-set methods are very useful structural optimization methods to avoid complex geometry (like the grayscale problem of density methods such as SIMP) [7]. In a level-set approach, material domain, void domain, and the boundary between them are clearly defined. Many level-set methods for heat transfer have been successfully carried out [9]. It is a very interesting method for forge or foundry problem [10–12], but it is quite difficult to work on microstructures.

To obtain periodic microstructures useful for heat transfer [13], many researches used the ESO method [14–16]. ESO methods represent the structural topology with discrete design variables. The obtained geometry is so clearly defined (by solid mesh or void). Such methods are simply based on the removal of inefficient materials and can be clearly restrictive to obtain innovative shapes.

Methodologies based on the conventional density approach have been successfully proposed by several researchers [17,18]. When such approaches are applied to heat transfer, complex geometry can disrupt the analysis. That is why a density approach such as the SIMP method has to be managed with trade constraints.

Another big challenge of structural optimization is working on the dynamic response of a system [19,20].

1.2 RELATED WORKS

The marriage between AM, which can build almost any shape, and topology optimization seems obvious [21,22]. Indeed, topology optimization will provide innovative forms but requires an adaptation process from traditional manufacturing (a "remodeling" is typically required). The objective of this chapter is to present the development of a methodology that will serve as a basis to develop a product that will be positioned upstream of the rapid prototyping machine. This software and the associated methodology are intended to be added on all types of AM machines. The material and mass savings obtained through digital optimization can be applied to plastics, metals, and so on. However, one of the major interests of optimization in general and more specifically topological optimization is to save mass on products. It is therefore natural to mainly target AM of steel products. In the context of AM center NUM3D, we have access to an SLS machine type. However, the approach can be applied to another AM steel process such as EBM (electron beam melting), DMLS (direct metal laser sintering), and so on. These different machining processes are brought together under the term ALM (additive layer manufacturing metal application). Figure 1.2 shows the positioning of the tool in the AM process.

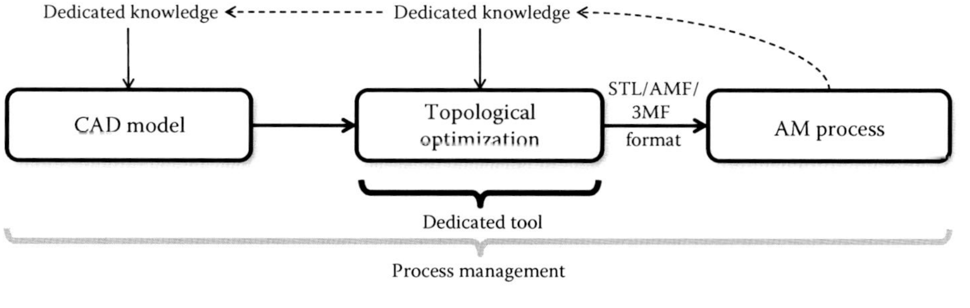

FIGURE 1.2 Process of the project.

The Design for Manufacturing (DFM) approach [23] is used through different mechanical tests in order to integrate the requirement data in the numerical analysis as soon as possible. Several manufacturing characteristics have been integrated to interpret the mechanical behavior of the structure, such as the layered manufacturing or new materials and new 3D printing techniques such as wood or complementary manufacturing systems [24]. The study presents an alternative geometry about existing structures through numerical simulation, mainly numerical optimization and Design for Manufacturing.

1.2.1 Optimization in AM

The use of optimization in AM [25] is generally done in the context of optimization of the build direction [26], parameter optimization trades, optimization construction layers algorithm, and so on. The optimization of the quantity of material used is an important goal. This optimization can match both the product material and the support material. Figure 1.3 shows the case of using a topology

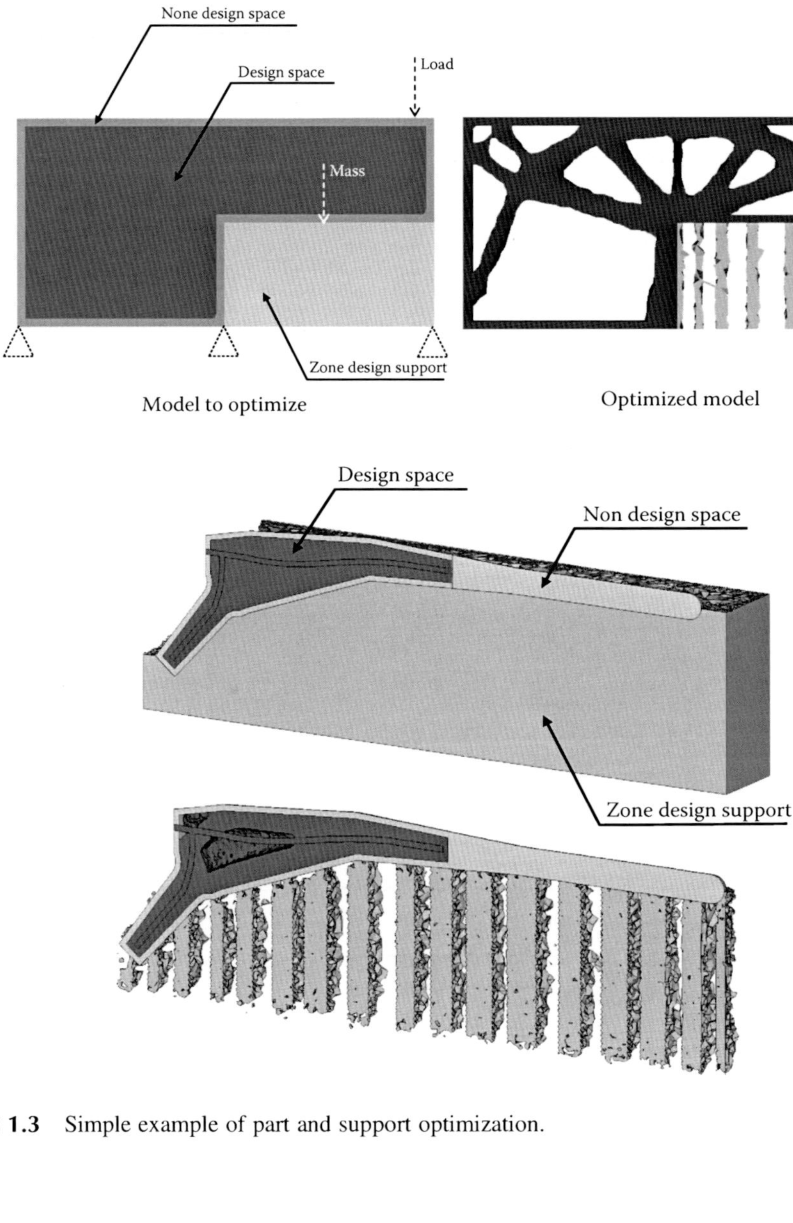

FIGURE 1.3 Simple example of part and support optimization.

optimization on both the part and the support used (two optimizations are performed separately: optimization in the "design" zone is the area that can be optimized; that in the "non-design" zone cannot be changed).

AM machines generally offer the possibility of reducing the mass by using honeycomb shapes, lattices, and so on. The concept of lattice structure (see Figure 1.4) tries to solve the problem of the degradation of the optimization solution with the construction of cells with intermediate densities in order to preserve the geometry of the homogenization solution. To keep the solution density, a unit cell is created in our model and the section of each beam is increased in proportion.

It appears that lattice structures are particularly attractive for their thermal properties [27] and also in the biomedical domain with a good interface definition between bones and prosthesis [28]. With a microstructure definition, it is very important to deal with constraints of additive process to ensure the manufacturability of the structure [29]. In order to improve the mechanic behavior in its industrial use, the research of a considerable number of scientists and engineers is focused on new forms of unit cells [30] in the definition of a lattice structure and its mechanical properties.

These algorithms model the simplified form without taking into account the specifications of mechanical strength. They are mostly applied for internal gain of matter. Actually, there are many researches on the influence of cellular structures. The influence of circular and rectangular shapes on the polyamide was studied through compression tests [31]. The study shows the influence of two types of geometric shapes based on their use. The circular structure is able to absorb 43.5% more energy than a rectangular structure (very useful in high deformation rates for fast dynamics such as crashes, explosions, etc.). Sugimura [32] investigated the use of lattice structures, including rapid prototyping, to lighten sandwich panels while maintaining their mechanical strength. The study enabled one to determine the directions of the lattice anisotropy that influences the mechanical behavior of the entire panel. The lattice modeling can be adjusted according to the specifications of mechanical strength. Other studies develop specific structures such as curved [25], honeycomb [33], and cell shape "tetrachirales" [34] or "hexachirales" [35]. However, these studies did not use the opportunity to integrate the notion of mechanical strength. The topological optimization through numerical simulation can solve this problem. Rochus et al. [36] showed interest in integrating topological optimization and highlighted some difficulties such as

- The difficulty of managing the drainage system of the support part
- The size of the CAD file and the difficulty of implementation

Rezaie et al. [37] developed a methodology that allows the production of the topological optimized part by a low-cost FDM. This methodology uses the classic optimization process to optimize the mass of the part (including the skin of the part).

1.2.2 Topological Problem Specification

Topology optimization problem can be defined as the search for the best allocation or distribution of material in a given design space [38]. There are numerous methods for solving a topological optimization problem: derivative based and level-set method, topological gradient method, homogenization method, ESO, non-gradient methods, and so on. More information on all these methods can be found in Ref. [39].

We are particularly interested in the homogenization method [40–42] and more precisely in the SIMP method. Actually, the SIMP method has been applied successfully in several domains to obtain an innovative structural design and was implanted in most topological optimization commercial software. The aim of the homogenization method is to transform a shape optimization problem into a problem of material density optimization (shape optimization by the homogenization method). This density has a value between 0 and 1 (0 = no material, 1 = material). For example, a density value of 0.2 will define a very porous material. A material density material is thus a

FIGURE 1.4 Density result after topological optimization (left) and its correspondence in terms of lattice structure (right).

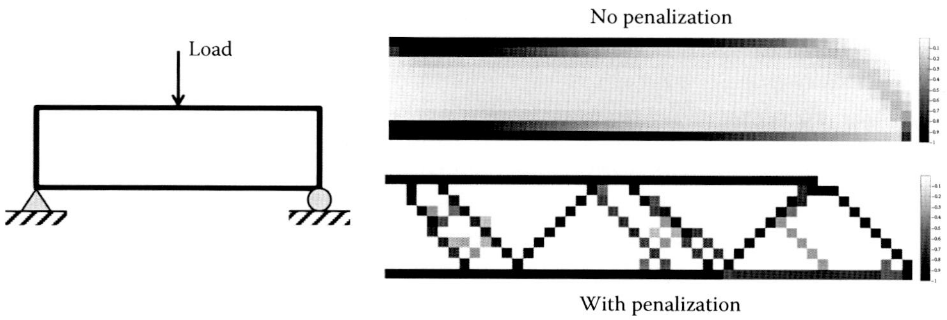

FIGURE 1.5 Penalization comparison with the SIMP model.

continuous problem (the density variable runs to the [0,1] interval) opposite to the classic discrete optimization problem [43]. Basically, we replace the homogenized material by a fictive heterogeneous material. The homogenization algorithm calculates composite shapes, and in mechanical skills, "classic shapes" are needed [44]. To help algorithm to frankly choose between empty material and full material, material characteristics are penalized (evolution of the homogenization method to the SIMP method). Indeed, the penalized method forces the density to take a value of 0 or 1. We can see in Figure 1.5 the difference between using penalization and not using it on the classic plane stress problem called the MBB (Messerschmidt–Bolkow–Blohm) beam problem (we take half of the MBB with symmetric conditions). This result is obtained thanks to Python code developed by Andreassen et al. [45].

According to Garcia-Lopez et al. [46], the homogenization method requires a large number of variables, making the implementation computationally expensive. One of the well-known problems of the SIMP method is that some intermediate densities have no physical meaning.

For a complete study between capabilities of different existing methods, see Ref. [47]. This penalization point is very important in the SLS AM machining context. One example is the one who will drive the minimum thickness.

The aim of the topological optimization is to find the best distribution of material, that is, determine the subdomain ω of a reference domain Ω ($\Omega \in \mathbb{R}^3$) filled with the material. From a mathematical point of view, a topological optimization problem can be written as shown in Equation 1.1. We seek to minimize the objective function f within certain constraints to define χ.

$$\min_{\omega \subset \Omega} f(\omega) : [C] \rightarrow \omega \in \chi \tag{1.1}$$

In practice, the objective function may be represented by weight, volume, deformation energy, and so on; design variables may be represented by dimensions (thickness etc.), type of material, and so on; and constraints may be represented by displacement, mass, frequencies, and so on.

1.2.3 Synthesis

The majority of research mainly focused on the implementation of new forms (honeycomb lattice, etc.) but uses little power topology optimization in numerical simulation that allows automatic modeling of the shape. The lattice functions are nevertheless very interesting because they are geometrically easy to integrate to the model. However, experts have encountered the difficulty of qualifying lattice structures. It is indeed difficult to verify, even with tomographs, whether a lattice element did not break during manufacture. Knowledge management is however necessary to obtain innovative form in the trade context. It is important to note the necessity to manage the drainage system of the support part. As we have seen previously, using the SIMP method for topological optimization dedicated to rapid prototyping seems to be interesting. It requires an adaptation of the modeling process and thus some work on knowledge modeling and its numerical integration. A methodology of application is also required. These points are discussed in the following sections.

1.3 METHODOLOGY

A major interest of AM is to build parts or areas of parts that are not manufacturable by conventional methods (numerical control machines [NCM], plastic injection, etc.). In the context of this research, the goal is to optimize the quantity of material used. Optimization can be used in two cases in AM:

- All the parts can be optimized (inner and outer design and non-design space).
- The outer skin (or part of it) cannot be modified (due to functional/design specifications).

In the first case, we use the AM to obtain an innovative shape. This concept is already considerably used in industry as well as in research.

The different presented processes are found through test inquiries and through CAD/CAE DINCCS research and development center experience. Indeed, we argue that an inductive approach is further interesting in research. Having industrial backup is very useful to capitalize knowledge and to test the different methodologies. In return, industrial applications become very relevant because of the association of researches. The first step of our methodology is to identify and define design spaces (see Figure 1.6). A Boolean operation in CAD software is needed to delimit the different zones. In our application, knowledge is linked with Rhinoceros software to help the designer in different important factors (like thickness of the skin, which is associated with the specific AM process). The knowledge is dependent on two major factors:

- The calculus set: the part is probably subjected to a specific set like loads, use impact (modal analysis for example), and so on. This item is not necessary, if the part to be manufactured is just for design view (touching function, assembly integration, etc.); the only set will not fall down on its own mass.
- The AM process corresponds to previous detailed knowledge capitalization. It depends on the AM machine.

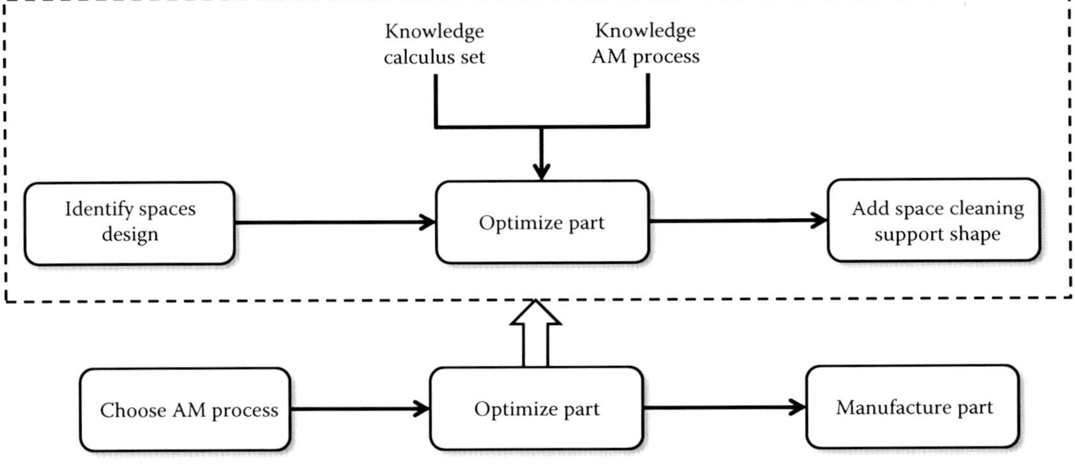

FIGURE 1.6 View of global methodology.

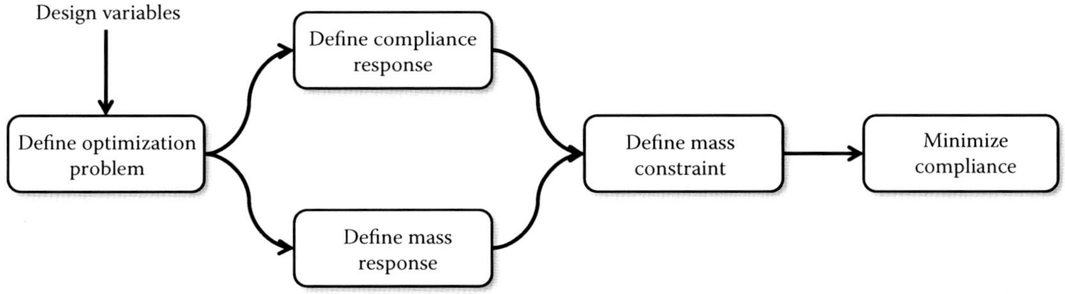

FIGURE 1.7 View of optimization methodology.

When the part is optimized (see details below), we add space shape for the support matter cleaning. This shape is predesigned in CAD software and positioned in empty spaces (e.g., manually). We obtain an STL file that is put in the AM machine process software.

The optimized step is also defined as a sub-methodological process (see Figure 1.7). The first step is to define design variables like the penalization factor as we explained before. This penalization factor is defined according to the minimal thickness obtained by test. We then define two specific responses:

- Compliance response. The compliance is the strain energy of the structure and can be considered as a reciprocal measure for the stiffness of the structure. A global measure of the displacements is the compliance of the structure under the prescribed boundary conditions (see Equation 1.2 with K being the stiffness matrix and U being the displacements).
- Fraction of mass response. The fraction of mass response is the material fraction of the designable material mass. It corresponds to a global response with values between 0 and 1. This allows the user to specify intuitive question like "I want to gain 30% of mass," with the value transcribed as 0.3 in our program.

The next step is to minimize the compliance. Generally, in optimization, the compliance is used to evaluate the stiffness. Minimizing the compliance means having a stiffer structure. The lower the compliance is, the higher the stiffness of the structure will be. Thus, the problem statement involves the objective functional of the strain energy that has to be minimized.

$$C = \frac{1}{2} U^T K U \tag{1.2}$$

1.3.1 Knowledge-Based System

The integration of knowledge in numerical simulation is directly linked to the AM process knowledge and the optimization method used in the finite element solver.

A topology optimization problem relative to an AM process can be defined by the following:

- Design spaces: a design space corresponds to the interior of the objects and a non-design space corresponds to the skin of the object (or any other area that should not be modified such as the apertures for cleaning). These areas are identified in the CAD model.
- Design variables: it is the set of parameters of the design space related to the AM process to define the initialization problem of topological optimization. We find here the penalization factor, the pattern repetition, and so on.

- Responses: responses correspond to structural responses, calculated in a finite element analysis, or combinations of these responses to be used as objective and constraint functions in a structural optimization. Available responses could be, for example, static displacement, mass, volume, temperature, and natural frequency.
- Constraints: constraints are based on responses by marking them with specific values.
- Objective: the objective function is, as we have seen before, the minimization (or the maximization) of the problem, here a specific response (for instance, the aim is to manage one response by objective function).

A knowledge-based system (KBS) has been developed to manage the AM process and material characterization for the topological optimization integration [48]. The KBS uses production rules and constraints to represent the declarative knowledge. It is the most usual representation to model heuristic know-how. We are particularly interested in the scenario model [49], that is, the decomposition of a problem in a series of tasks. We base ourselves on the fact that each time that an expert resolves a problem, he runs a scenario in an intuitive way. Schank and Abelson [50] set up this structure by affirming that there are thousands of scenarios in the human memory. The different scenarios have been collected by tracking the work of experts based on a specific ontology.

Our structure is made up of a knowledge base, including the following:

- A base of scenarios: as we said before, experts translate their step of topological optimization and AM process into diagrams of sequences of tasks. They consequently use a scenario of optimization and AM process.
- A base of rules: it contains the production rules and uses the base of constraints for this purpose. These rules are associative, which means that each formalized rule must contain at the same time the context of application of the rule and the condition of processing of the rule. The rules are rules of production such as If (Conditions) and Then (Conclusions). The "Conditions" part is the process of the rules and the "Conclusions" part describes the actions to be started in the event of release.
- A base of constraints: it contains all the mechanical laws and the message interaction. The base of constraints contains the set of constraints having a relationship with the considered field. For example, a constraint can be like Wall Thickness > 1 mm to point out the limit of manufacturing to avoid collapsing.

The KBS allows the user to describe the boundary conditions, loads, materials, and so on. The CAD system and the CAE system (mesh generator and analysis manager) are fully knowledge driven. They can be used by either a CAD expert or a mechanical engineer and ensure good usage and high-quality results.

The difficulty of the AM and topological optimization coupling is based on two problems:

- Topological optimization problems use FEM resolution and require a fine material characterization (new optimization algorithm can integrate nonlinear material behavior).
- Constraints defined in topological optimization needs to be linked to the AM process to correctly parameterize the solver. For example, if the AM process knowledge is not defined in the topological optimization solver, one can return a very thin wall, which cannot be manufactured (see a test-case example in Figure 1.8).

1.3.2 Trade Integration

The methodology was validated on a laboratory SLS (DTM) machine. To deal with the use of constraints, we started a series of tests to characterize the materials mechanically and dimensionally.

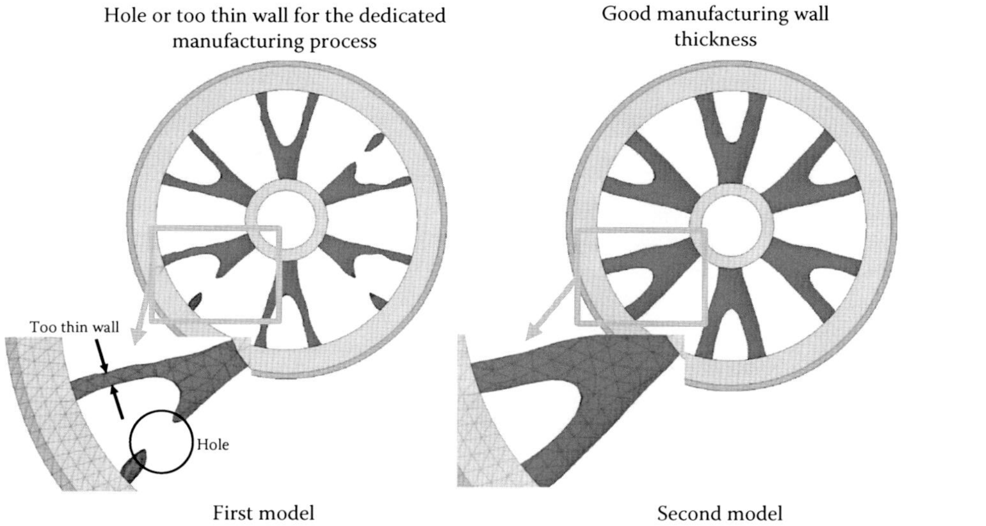

FIGURE 1.8 Some problems for manufacturing process.

This stage will permit us to identify the mechanical limits of the materials provided by the manufacturer without any data.

The experimental approach is based on the use of standardized specimens manufactured by AM. The experimental process to recover AM knowledge is based on two types of specimens:

- Iso campus norm specimens manufactured by the AM process: the aim is to recover material behavior (typically elastoplastic law) to use it in simulation. We use specific tools for the measurement, for instance, a 3D Deformation Aramis camera.
- Dedicated shape specimens also manufactured by the AM process: the aim is to capitalize knowledge management such as determining the influences of parameters: thickness/height limit, pocket depth allowed for powder evacuation, and so on.

Our approach involves the study of three very important factors for topological optimization:

- The minimum thickness printable and cleanable without part deterioration. We seek to maximize the minimum thickness of the wire cloth (final material) without loss of geometric and morphological qualities of the part.
- The minimum diameter printable and cleanable without mechanical cleaning: the objective is to measure the best channel's dimensions for cleaning the internal structure of the piece (allow powder evacuation).
- The maximum height, specifically the ratio between the projected length and height of the part that may cause matter to fall down.

The tested material is a rapid steel metallic powder that is coated with a polymer. The SLS process involves many knowledge capitalizations because the part is involved in the laser phase, the debinding phase, the infiltration phase, and so on. Before the infiltration phase, the part is very breakable and requires specific dimension management. Many features have been studied: holes, slots, ribs, pads, pockets, and so on. They are used in different context. For example, the rib is studied in freeway and in an embedded way. We define 14 test parts that include this feature (as seen in Figure 1.9).

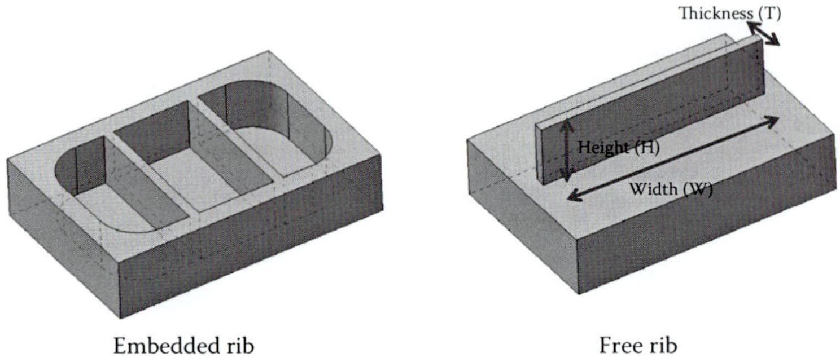

FIGURE 1.9 Two kinds of tested rib features.

TABLE 1.1
Results for Rib

Parameter	Operator	Value	Result
Thickness	<	1 mm	Nonfeasible (matter collapsing)
Thickness	=	2 mm	Deformation for height > 10 mm
Thickness	=	3 mm	Deformation for height > 40 mm
Width	>	15 mm	Cleaning constraint

Note: To help the global strength, pins can be brought back.

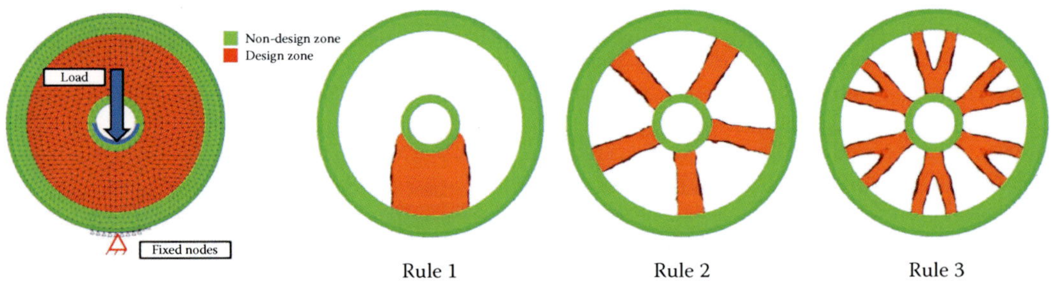

FIGURE 1.10 Impact of geometric rules on topological optimization.

Table 1.1 shows the results for free rib, which is used in optimization parameters.

To trade rules related to the manufacturing process, it is necessary to associate geometric rules that can provide different shapes (symmetry, repetition, etc.) (see Figure 1.10).

Mesh size and quality are also a big issue that requires the deployment of specific rules (see Figure 1.11). These rules must be usable in a generic model approach for integration into processes.

1.4 APPLICATION

To validate our methodology and prepare software integration, we develop a tool called DFX@ Process (see Figure 1.12) to model processes and trade rules. This tool allows the link between commercial tools for CAD/CAE (for instance, Catia V5 and OptiStruct in our case) and the KBS.

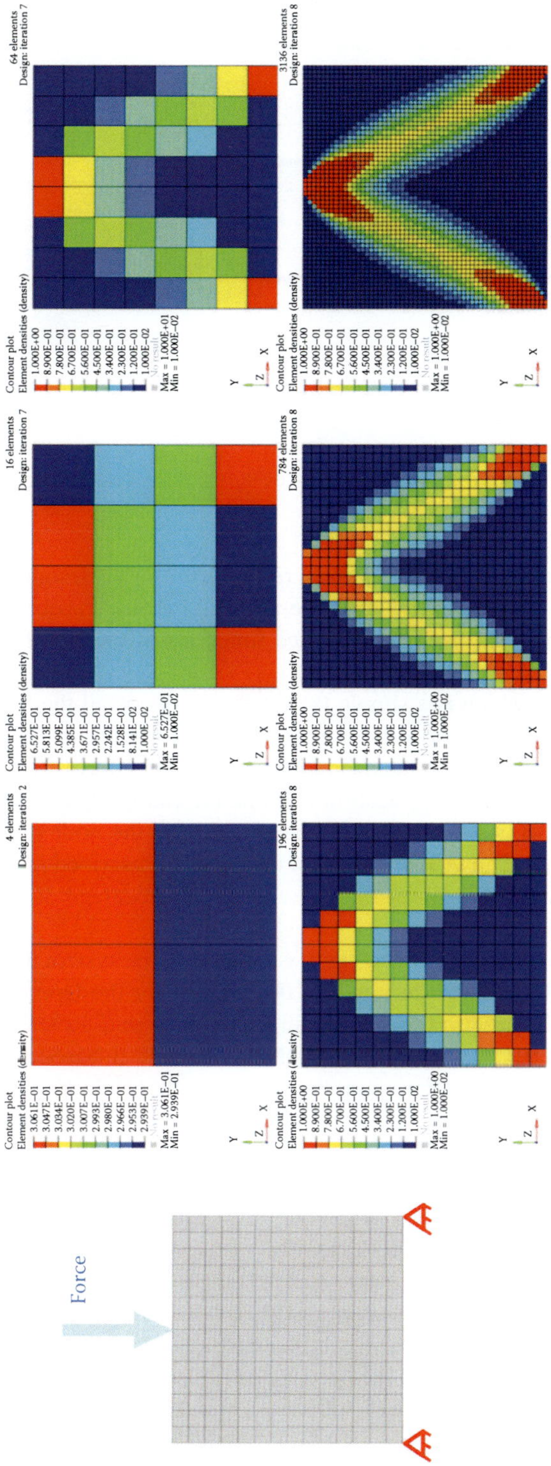

FIGURE 1.11 Impact of mesh on topological optimization.

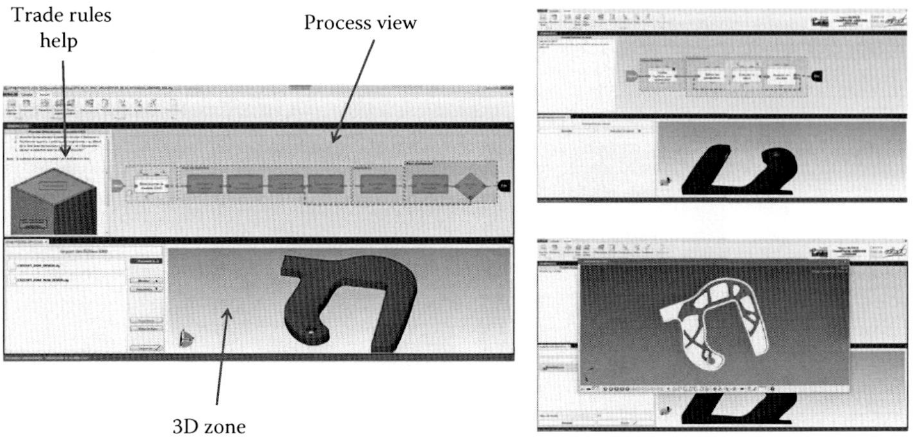

FIGURE 1.12 DFX@Process view.

The program that supports the methodology is developed in C++/VTK and is dedicated to the preparation of optimization model, namely:

- Load the design and non-design CAD model (step format). The design part matches the skin of the part and the non-design part matches the allowed modified space.
- Load the cleaning channel. One of the limited functions of the system is that the user cannot load more than one channel for the moment.
- Load the material file, which is a simple txt file that contains all the basic information for the simulation. For the moment, the system only integrates linear simulation.
- Define the calculus load and boundary condition. The user has to enter the force value and select two faces: one for force application and one for clamping application. A multiple point constraint (RBE2 type) is automatically created based on the node of these two faces.
- Define the optimization knowledge based on the process described in the previous paragraph, namely, the minimum allowed wall thickness and the percentage of weight gain.

Once these parameters are set, the user can verify the model (the system makes a basic verification before running), check the fem ascii file (this file is the entry for the OptiStruct solver; it has been automatically created by the system by creating in the background a mesh [with hypermesh] and linking it with user parameters; note that, for instance, the mesh is based on the global size of the part), and run the optimization. After the optimization, the result is reloaded by the system to view it. The exported model is on STL format. If the user approves the optimization, he can load the STL result on his AM machine.

We deal with numerous industrial and academic examples that enable us to validate our methodologies. Figure 1.13 presents three examples (not only processed on an SLS) of using topological optimization in the internal patterns of products:

a. Stuckmate is a device attached to the front mudguard of a motorbike as well as the forks for extra strength. It has a retractable cord that extends 6 m to help when there are long slippery uphills or massive step-ups where the motorbike needs to be pulled. The device has to be robust, reliable, and lightweight (i.e., it can resist up to 300 kg and should have a weight of 650 g). Several parts were manufactured during the design in AM. From the beginning, topological optimization has been used to make the parts (especially the interior ones) lighter (see Figure 1.13 for an example of the drum). The design fulfills both AM and plastic injection. As shown in the figure, the internal shapes of the drum are very organic and allowed a mass gain of more than 30%.

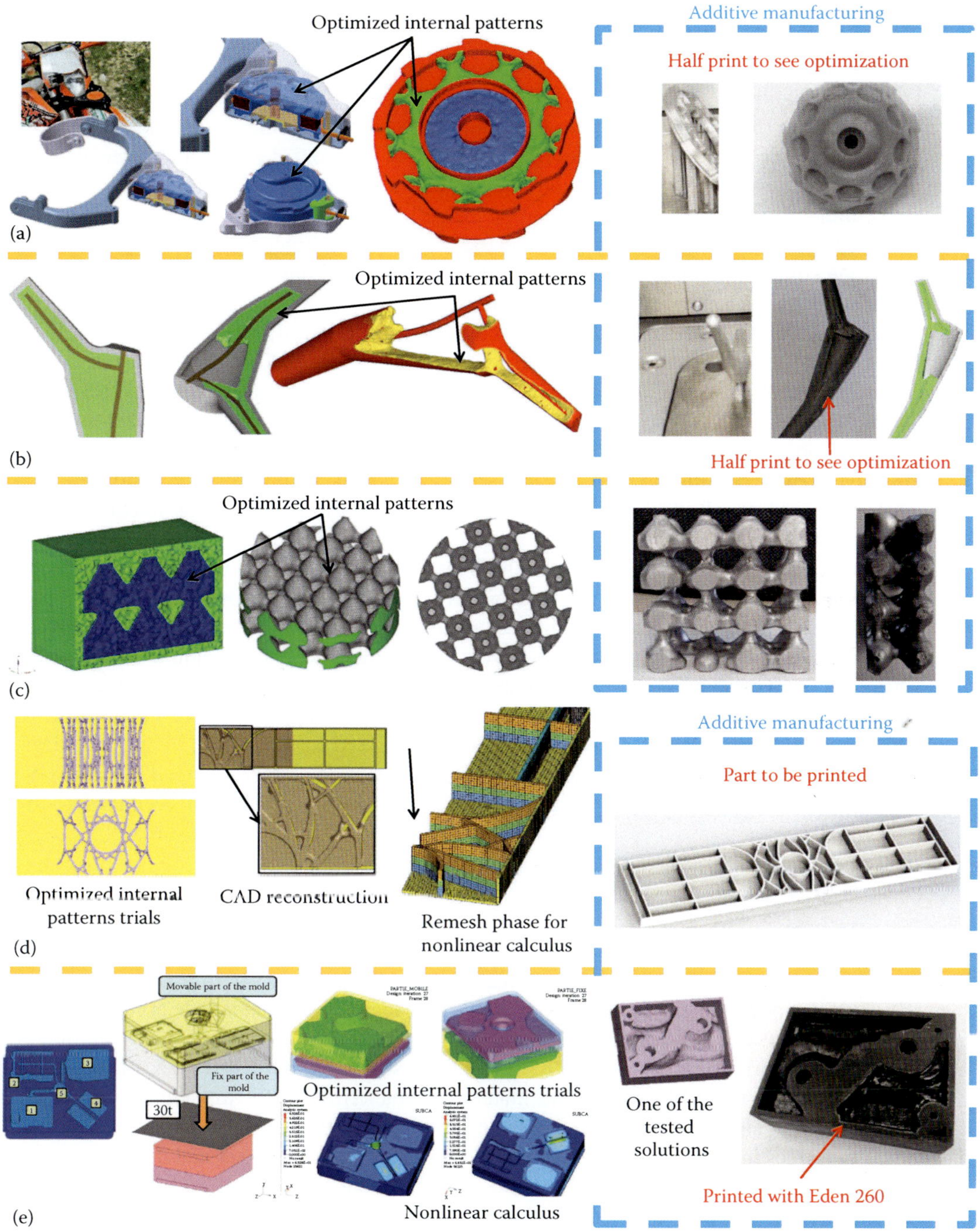

FIGURE 1.13 Industrial cases.

b. This case involves a prosthetic implant used in a hip replacement surgical procedure studied for one of our clients (simplified for confidentiality purposes). There are a large number of hip implant devices on the market. Many different shapes exist but each style falls into one of the four basic material categories (metal on plastic, metal on metal, ceramic on plastic, and ceramic on ceramic). Because of the history of our region (large foundry and forge industrial impact), we are interested in the metal-on-metal material and more particularly on titanium.

This type of prosthesis is built via a forge process using titanium material. The hip has been manufactured on the STS machine in two versions: a full version for physical testing and a half version for academic purposes. The weight gain is almost around 25% in comparison to 30% of the simulation.

c. This research (in collaboration with EPF and UTT) focuses on the methodology of periodic open-cell foams obtained by topological optimization (we considered heat transfer boundary) and by AM to realize a plaster mold used in aluminum casting. The metal foam provides structural bi-material to study its thermal behavior for energy storage and to achieve an adapted geometry with the help of rapid tooling by three-dimensional printing (3DP). The resulting complex geometries are particularly suitable for plaster 3D printing by binder jetting to manufacture a rapid tooling for aluminum casting. Typically, specific lattices with thermal specifications have been created thanks to topological optimization.
d. This case relates to a customer specializing in railway products. They develop a plastic cover to protect electric cables manufactured especially for subways. These covers require a reinforcement to allow the passage of maintenance agents and to avoid the crushing of the cables. Topological optimization has been used to propose several shapes, and with the help of AM, it was possible to manufacture several products with a relative equivalent material compared to plastic injection (just to test assembly function, not traction or compression test).
e. This case is an industrial example for the Vigilock Company with the collaboration of the Crestic Lab of Reims Champagne-Ardenne University. Vigilock works on the development of protection for computer workstations against physical access in the absence of their assigned user. The aim is to develop an optimized mold to manufacture a small series of Vigilock Core products (patented in France and abroad, Vigilock Core ensures that sensitive workstations are locked as soon as their regular user is away). The mold is produced with AM in Verowhite with Eden 260. The mold is then used in injection molding machines to produce a small series of Vigilock Core products (the mold supports around 50 parts before shading off). To manufacture several molds, topological optimization has been used to make the mold lighter.

1.5 SUMMARY AND CONCLUSIONS

The aim of our research in topological optimization for AM is not to develop new mathematical methods of resolution but to use the existing methods and adapt them from a trade point of view. We are particularly interested in the optimization of the inner parts. The aim is to optimize the volume of material to be used and the global mass. A KBS and the associated tool (called DFX@Process) were developed to manage the AM process and material characterization for topological optimization integration. Weight gain is very interesting in the optimization of the material used in the inner parts. The difficulty resides in evacuating the powder for the powder bath processes and in determining the minimum thickness of the outer skin. At the moment, evacuation pipes are prepared upstream and automation methods are being developed. One of the advantages of using topological optimization for the inner parts concerns the possibility of qualifying the organic forms thanks to a tomograph, whereas it is currently difficult for lattices.

We are currently focusing on the use of topological optimization for support. The difficulty with the supports is that they can have several functions based on the manufacturing process. The supports can thus have a holding function to avoid deformations (in SLM for example) to manage thermal (in EBM for example) or electrical effects. Another difficulty lies in exchanging the AM machine to replace the supports created through organic forms obtained thanks to topological optimization, according to the manufacturer's rules. We also set up an observatory based on interviews to enable us to trace the experts' business problems and integrate them into our work.

1.6 AREAS FOR FURTHER RESEARCH

The use of numerical optimization and numerical simulation in AM offers a huge field of research. The application of homogenization methods such as SIMP on porosity is intriguing, but integrating trade constraints has been proven to be difficult. Level set seems to be a good research method especially for controlling thickness. One of the big problems encountered with topological optimization is that it is time-consuming, especially for complex geometries. A possible solution is being examined in our technical research center, with the development of a methodology based on principal stress line computation coupled with composite materials. The trajectories of internal forces are inspired from the capacities of bones, which grow in response to mechanical stress and loading. The architecture of bones has to be strong and lightweight: the structure is then a couple between specific directions oriented in the direction of greatest stress and foam material. This kind of structure is very interesting for industry especially with the possibility of producing it with the help of AM.

ACKNOWLEDGMENT

This project is partly based on a collaboration with UTT and EPF (Xiao Lu Gong, Julien Gardan, and Abdelatif Merabtine). We also thank the Champagne-Ardenne region and the General Council. These studies were part of the OptiFabAdd project (www.optifabadd.com), co-financed by BPIFrance.

REFERENCES

1. Wohlers, T., 2015. *Additive Manufacturing and 3D Printing—State of the Industry*. Wohlers Report 2015 Annual worldwide progress report.
2. Thakur, A., Banerjee, A.G., Gupta, S.K., 2009. A survey of CAD model simplification techniques for physics-based simulation applications. *Computer-Aided Design* 41, 65–80.
3. Sharpe, J.E.E., 1995. Computer tools for integrated conceptual design. *Design Studies* 16, 471–488.
4. Allaire, G., Jouve, F., 1999. Structural optimization by the homogenization method, in: Argoul, P., Frémond, M., Nguyen, Q.S. (Eds.), *IUTAM Symposium on Variations of Domain and Free-Boundary Problems in Solid Mechanics, Solid Mechanics and Its Applications*. Springer Netherlands, pp. 293–300.
5. Iga, A., Nishiwaki, S., Izui, K., Yoshimura, M., 2009. Topology optimization for thermal conductors considering design-dependent effects, including heat conduction and convection. *International Journal of Heat and Mass Transfer* 52(11–12), 2721–2732.
6. Alexandersen, J., 2014. Topology optimisation for natural convection problems. *International Journal for Numerical Methods in Fluids* 76(10), 699–721.
7. Yaji, K., Yamada, T., Kubo, S., Izui, K., Nishiwaki, S., 2015. A topology optimization method for a coupled thermal–fluid problem using level set boundary expressions. *International Journal of Heat and Mass Transfer* 81, 878–888.
8. Feppon, F., Michailidis, G., Sidebottom, M.A., Allaire, G., Krick, B.A., Vermaak, N., 2016. Introducing a level-set based shape and topology optimization method for the wear of composite materials with geometric constraints. *Structural and Multidisciplinary Optimization* 55(2), 547–568.
9. Yamada, T., Yamasaki, S., Nishiwaki, S., Izui, K., Yoshimura, M., 2011. Design of compliant thermal actuators using structural optimization based on the level set method. *Journal of Computing and Information Science in Engineering* 11(1), 11005–11008.
10. Maury, A., Shape optimization for contact and plasticity problems thanks to the level set method. PhD Thesis, Paris 6, 2016.
11. Allaire, G., Jouve, F., Michailidis, G., 2016. Molding direction constraints in structural optimization via a level-set method, in: *Variational Analysis and Aerospace Engineering*. Springer, pp. 1–39.
12. Allaire, G., Dapogny, C., Faure, A., Michailidis, G., 2017. Shape optimization of a layer by layer mechanical constraint for additive manufacturing. *Comptes Rendus Mathematique* 355(6), 699–717.
13. Alexandersen, J., 2015. Topology optimisation of manufacturable microstructural details without length scale separation using a spectral coarse basis preconditioner. *Computer Methods in Applied Mechanics and Engineering*, 290, 156–182.

14. Yan, X., Huang, X., Zha, Y., Xie, Y.M., 2014. Concurrent topology optimization of structures and their composite microstructures. *Computers & Structures* 133, 103–110.
15. Huang, X., Xie, Y.M., Jia, B., Li, Q., Zhou, S.W., 2012. Evolutionary topology optimization of periodic composites for extremal magnetic permeability and electrical permittivity. *Structural and Multidisciplinary Optimization* 46(3), 385–398.
16. Da, D.C., Cui, X.Y., Long, K., Li, G.Y., 2017. Concurrent topological design of composite structures and the underlying multi-phase materials. *Computers & Structures* 179, 1–14.
17. Koga, A.A., Lopes, E.C.C., Villa Nova, H.F., de Lima, C.R., Silva, E.C.N., 2013. Development of heat sink device by using topology optimization. *International Journal of Heat and Mass Transfer* 64, 759–772.
18. Matsumori, T., Kondoh, T., Kawamoto, A., Nomura, T., 2013. Topology optimization for fluid–thermal interaction problems under constant input power. *Structural and Multidisciplinary Optimization* 47(4), 571–581.
19. Dou, S., 2015. Structural optimization for nonlinear dynamic response. *Royal Society of London. Philosophical Transactions A. Mathematical, Physical and Engineering Sciences* 373(2051).
20. Dou, S., 2016. Optimization of hardening/softening behavior of plane frame structures using nonlinear normal modes. *Computers & Structures* 164, pp. 63–74.
21. Langelaar, M., 2016. Topology optimization of 3D self-supporting structures for additive manufacturing. *Additive Manufacturing* 12(Part A), 60–70.
22. Saadlaoui, Y., Milan, J.-L., Rossi, J.-M., Chabrand, P., 2017. Topology optimization and additive manufacturing: Comparison of conception methods using industrial codes. *Journal of Manufacturing Systems 43 (Part* 1), 178–186.
23. Zhao, Z., Shah, J.J., 2002. A normative DFM framework based on benefit–cost analysis. *Design Engineering Technical Conference.*
24. Gardan, J., Roucoules, L., 2011. Characterization of beech wood pulp towards sustainable rapid prototyping. *International Journal of Rapid Manufacturing* 2(4).
25. Galantucci, L.M., Lavecchia, F., Percoco, G., 2008. Study of compression properties of topologically optimized FDM made structured parts. *CIRP Annals—Manufacturing Technology* 57, 243–246.
26. Phatak, A.M., Pande, S.S., 2012. Optimum part orientation in rapid prototyping using genetic algorithm. *Journal of Manufacturing Systems* 31, 395–402.
27. Evans, A.G., Hutchinson, J.W., Fleck, N.A., Ashby, M.F., Wadley, H.N.G., 2001. The topological design of multifunctional cellular metals. *Progress in Materials Science* 46(3–4), 309–327.
28. Xiao, D.-M. Yang, Y.Q., Su, X.-B., Wang, D., Luo, Z.-Y., 2012. Topology optimization of microstructure and selective laser melting fabrication for metallic biomaterial scaffolds. *Transactions of Nonferrous Metals Society of China* 22(10), 2554–2561.
29. Queheillalt, D.T., Wadley, H.N.G., 2005. Pyramidal lattice truss structures with hollow trusses. *Materials Science and Engineering: A* 397(1–2), 132–137.
30. Chen, Q., Pugno, N., Zhao, K., Li, Z., 2014. Mechanical properties of a hollow-cylindrical-joint honeycomb. *Composite Structures* 109, 68–74.
31. Vesenjak, M., Krstulović-Opara, L., Ren, Z., Domazet, Ž., 2010. Cell shape effect evaluation of polyamide cellular structures. *Polymer Testing* 29, 991–994.
32. Sugimura, Y., 2004. Mechanical response of single-layer tetrahedral trusses under shear loading. *Mechanics of Materials* 36, 715–721.
33. Abramovitch, H., Burgard, M., Edery-Azulay, L., Evans, K.E., Hoffmeister, M., Miller, W., Scarpa, F., Smith, C.W., and Tee, K.F., 2010. Smart tetrachiral and hexachiral honeycomb: Sensing and impact detection. *Composites Science and Technology* 70(7), 1072–1079.
34. Miller, W., Smith, C.W., Scarpa, F., Evans, K.E., 2010. Flatwise buckling optimization of hexachiral and tetrachiral honeycombs. *Composites Science and Technology* 70(7), 1049–1056.
35. Prall, D., Lakes, R.S., 1997. Properties of a chiral honeycomb with a Poisson's ratio of −1. *International Journal of Mechanical Sciences* 39(3), 305–314.
36. Rochus, P., Plesseria, J.-Y., Van Elsen, M., Kruth, J.-P., Carrus, R., Dormal, T., 2007. New applications of rapid prototyping and rapid manufacturing (RP/RM) technologies for space instrumentation. *Acta Astronautica* 61(1–6), 352–359.
37. Rezaie, R., Badrossamay, M., Ghaie, A., Moosavi, H, 2013. Topology optimization for fused deposition modeling process. *Procedia CIRP* 6 (2013), 521–526.
38. Calvel, S., 2004. *Conception d'organes automobiles par optimisation topologique*. Université Paul Sabatier Toulouse III.
39. Rozvany, G.I.N., 2009. A critical review of established methods of structural topology optimization. *Structural and Multidisciplinary Optimization* 37, 217–237.

40. Bendsøe, M.P., Kikuchi, N., 1988. Generating optimal topologies in structural design using a homogenization method. *Computer Methods in Applied Mechanics and Engineering* 71, 197–224.
41. Mota Soares, C.A., Bendsoe, M.P., Choi, K.K., Herskovits, J., 2008. Structural optimization. *Computers & Structures* 86(13–14), 1385.
42. Christiansen, A., 2013. Topology optimization using an explicit interface representation. *Structural and Multidisciplinary Optimization*, 1–13.
43. Allaire, G., Jouve, F., 1999. Structural optimization by the homogenization method, in: Argoul, P., Frémond, M., Nguyen, Q.S. (Eds.), *IUTAM Symposium on Variations of Domain and Free-Boundary Problems in Solid Mechanics, Solid Mechanics and Its Applications*. Springer Netherlands, pp. 293–300.
44. Andreassen, E., 2104. How to determine composite material properties using numerical homogenization. *Computational Materials Science* 83, 488–495.
45. Andreassen, E., Clausen, A., Schevenels, M., Lazarov, B.S., Sigmund, O., 2011. Efficient topology optimization in MATLAB using 88 lines of code. *Structural and Multidisciplinary Optimization* 43, 1–16.
46. Garcia-Lopez, N.P., Sanchez-Silva, M., Medaglia, A.L., Chateauneuf, A., 2011. A hybrid topology optimization methodology combining simulated annealing and SIMP. *Computers and Structures* 89, 1512–1522.
47. Rozvany, G.I.N., 2009. A critical review of established methods of structural topology optimization. *Structural and Multidisciplinary Optimization* 37, 217–237.
48. Gardan, N., Schneider, A., 2015. Topological optimization of internal patterns and support in additive manufacturing. *Journal of Manufacturing Systems* 37(1), 417–425.
49. Gardan, N., Gardan, Y., 2003. An application of knowledge based modelling using scripts. *Expert Systems with Applications* 25, 555–568.
50. Schank, R.C., Abelson, R.P., 1977. *Scripts, Plans, Goals and Understanding: An Inquiry into Human Knowledge Structures*. L. Erlbaum, Hillsdale. NJ.

2 Design Features to Address Security Challenges in Additive Manufacturing

Nikhil Gupta, Fei Chen, and Khaled Shahin

CONTENTS

2.1 INTRODUCTION

Additive manufacturing, also commonly referred to as 3D printing, refers to a computer-controlled layer-by-layer material deposition process to create a component. Traditional manufacturing methods are largely subtractive methods that rely on material removal from a large feedstock to create the desired shape (ASTM International 2013). A variety of methods such as welding and lithography have existed for a long time; they are examples of traditional additive manufacturing methods. However, a recent surge in research and development efforts has resulted in the availability of a large suite of additive manufacturing methods that can work at micro- and even nanoscale precision to create macroscale parts. In 2016, the additive manufacturing industry grew to about $6 billion with a 26% annual growth rate, and it is expected to grow at the same rate in the near future (McCue 2016; Wohlers 2016). The additive manufacturing field is expected to grow into a $20 billion market by the year 2020 according to *Wohlers Reports 2015* (Guo and Leu 2013; Melchels et al. 2012; Takaichi et al. 2013; Wohlers 2016).

Some of the commonly known additive manufacturing techniques include Fused Deposition Modeling (FDM), Stereolithography (SLA), Selective Laser Sintering/Melting (SLS/SLM), Direct Metal Laser Sintering (DMLS), Electron Beam Melting (EBM), Polymer Jetting (PolyJet), and Laminated Object Manufacturing (LOM) (ASTM International 2015; Gibson et al. 2015). These techniques can be categorized into seven major groups based on the process types (ASTM International 2015):

1. Binder jetting
2. Directed energy deposition
3. Material extrusion
4. Material jetting
5. Powder bed fusion
6. Sheet lamination
7. Vat photopolymerization

Many of these methods are now commercially viable and 3D printers are available based on these technologies. Material extrusion– and powder bed fusion–based methods are being widely used now for industrial production of parts.

Cost of the different types of 3D printers can vary dramatically based on the printer's capability, build volume, material options, and resolution. For example, a desktop FDM 3D printer can cost as little as $200, while an industrial-level one can cost around $30,000 to $50,000, and an industrial-level DMLS printer can cost more than $500,000. The investment in a 3D printer needs to be in correlation with the machine operating life, manufacturing rate, and the product value, so that the product precision and manufacturing cost can be optimized.

Material extrusion processes like FDM use thermoplastic filament materials such as acrylonitrile butadiene styrene (ABS) or polylactic acid (PLA). Thermoplastics become soft and viscous when heated above their glass transition temperature and can be extruded out of a fine nozzle. In addition, other materials such as carbon fiber filament and metallic filament have been used for making products with tailored properties (Hahnlen and Dapino 2010; Manfredi et al. 2014; Ning et al. 2015; Parthasarathy et al. 2011; Quan et al. 2015; Tekinalp et al. 2014; Willis et al. 2012). FDM methods work by depositing the extruded filament on a build plate in the desired shape to create the component. Powder bed fusion methods create a thin layer of fine powder on a build plate and then use a laser to sinter the powder in the desired shape. The next layer of powder is deposited on the previous one and the sintering is conducted again. The process continues until the entire component is built. The solid part is then retrieved from the powder bed. Powder bed fusion methods can be used for a wide range of materials that can be melted and resolidified, including polymers, metals, and ceramics. The aerospace and automotive industries have been adopting various powder bed fusion methods such as SLS/SLM, DMLS, and EBM to manufacture parts with metallic materials such as aluminum alloy (AlSi10Mg), stainless steel, titanium alloy (TiAl6V4), nickel alloy, and cobalt–chrome alloy (Leuders et al. 2013; Murr et al. 2010; Van Bael et al. 2011). Inconel 718 superalloy has also been used for manufacturing industrial parts used in the oil and gas, chemical, and aerospace industries (Jia and Gu 2014). Additionally, precious metal powders of silver, gold, and platinum are used for manufacturing jewelry and medical and electronic products via the DMLS process (Hancox and McDaniel 2009; O'Connor 2014). Processes like vat photopolymerization or material jetting use photosensitive materials, which are cured by an ultraviolet light source layer by layer to produce the 3D object. Additive manufacturing has been used for some very large scale object printing, for example, an entire office building and a car (Lever 2016; Lim et al. 2012; Saunders 2017; Starr 2016).

Many advantages for adopting additive manufacturing include possible high degree of customization of each part; manufacturing of parts with complex shapes; reduction of manufacturing costs by elimination of dies, molds, and other tooling; and reducing the shipping cost due to on-site manufacturing (Gao et al. 2015; Gibson et al. 2015; Sachs et al. 1990). Artificial hip joints and crowns for teeth are being produced from computed tomography scan data using additive manufacturing to create parts customized for each patient (Sing et al. 2016). An example of industrial-scale additive manufacturing is the burner tips printed by Siemens from steel powder for use in gas turbines as replacement parts (Navrotsky et al. 2015).

2.1.1 Additive Manufacturing Process Chain

The major steps of a typical additive manufacturing process chain are shown in Figure 2.1. The steps include

1. Preliminary computer-aided design (CAD) model development
2. Finite element analysis, computational fluid dynamics, and other analyses to optimize the design
3. Conversion of CAD file to STL (StereoLithography) format
4. Slicing of the model into 2D slices
5. Generation of toolpath in the form of G-code
6. Printing of 3D object on a printer using G-code
7. Testing, imaging, and nondestructive evaluation of the printed objects for quality and reliability

The additive manufacturing process chain starts with the CAD modeling using software programs such as SolidWorks, Blender, or AutoCAD Inventor. It is a standard practice to refine the design by finite element analysis, computational fluid dynamics, or multi-physics analysis to obtain optimized properties and performance of the component. The final CAD model is used for further processing through the additive manufacturing chain. The 3D modeling programs can export the file into STL format, which contains a triangulated representation of the surface geometry of the part. These triangulated surfaces, or facets, are described by the unit normal vector and three vertices ordered by the right-hand rule using the Cartesian coordinate system. The STL file is then imported in slicing software that divides the geometry into a stack of 2D slices. Next, a toolpath is generated to build these slices on the 2D plane of the 3D printer build plate. The toolpath file, called G-code, also contains the machine parameters such as print head speed and temperature. Finally, the G-code is

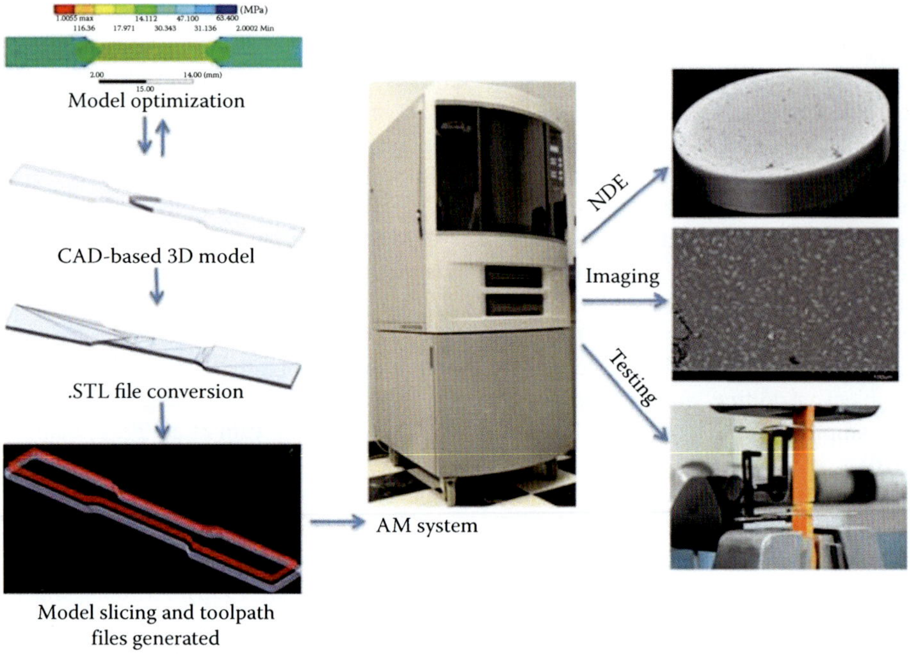

FIGURE 2.1 Typical process chain of additive manufacturing.

sent to the 3D printer to print the part. The printed part goes through a series of inspections and tests to ensure quality. The tests can include both destructive and nondestructive tests. In the case of destructive tests, multiple replicas of the part must be printed to facilitate the testing.

2.1.2 Security Risks in Cyber–Physical Systems

The modern production chain is often distributed over several countries, even continents, and heavily depends on cloud-based resources for cost-effectiveness, availability of talent, and acceleration of work by dividing it in various time zones (Chen et al. 2017). Vulnerabilities may exist in several different places in such distributed systems, which essentially are connected by the Internet (Bradshaw et al. 2010; Chhetri et al. 2016a,b, 2017; Do et al. 2016; Faruque et al. 2016; Krotofil et al. 2014; Nagaraju Kilari 2012; Sturm et al. 2014; Vincent et al. 2015; Wells et al. 2014; Zeltmann et al. 2016). The same vulnerabilities also apply to additive manufacturing systems (Campbell and Ivanova 2013; Chhetri et al. 2016b; Kurfess and Cass 2014; Sturm et al. 2014; Yampolskiy et al. 2015). Companies involved in the development of high-value industrial parts using additive manufacturing take appropriate cybersecurity precautions for protection of sensitive information. However, breaches in cybersecurity occur even for corporations or government resources that are expected to be most heavily protected (Garg et al. 2003; Perlroth 2016; Peterson 2014). Such a scenario poses some special considerations for the additive manufacturing process chain because the stolen CAD files can be used to produce a component that is as good as the original component using the same type of printer and material the company is using for manufacturing that component (Gupta et al. 2017). The CAD file can be encrypted and protected by passwords, but all these possibilities come with their limitations and tools are available to break those security features. Given this scenario, it is highly desired to understand the possible cybersecurity risks and their impact on the product and the company, and develop strategies to counter these threats.

For cyber–physical systems, a threat model needs to be developed based on factors such as cybersecurity tools in place, product line, external threats, and threats from within the organization

(Burmester et al. 2012; Genge et al. 2012). A specific or a targeted threat can have a better success in defense at low cost whereas a generalized threat model requires a more expensive defense mechanism against all possibilities.

Attacks on the additive manufacturing process chain can include malicious changes in CAD and STL files, changes in databases for CAD and finite element analysis software programs, and changes in the printer settings, among numerous other possibilities (Campbell and Ivanova 2013; Deborah and Kelly 2016; Do et al. 2016; Wells et al. 2014; Yampolskiy et al. 2016; Zeltmann et al. 2016). Compared to other cyber–physical systems such as smart grids and smart transportation systems, additive manufacturing is unique in the fact that it produces physical components. These components are usually high-value parts whose failure can have serious consequences. Hence, it is important to understand the vulnerabilities of the additive manufacturing process chain and establish procedures to deter the threats.

Recent advancements in availability and affordability of sensors, data acquisition systems, and networked machines have resulted in the development of numerous cyber–physical systems (Baheti and Gill 2011). The design of security for cyber–physical systems includes feedback between the cyber and physical environments, distributed management and control, quantification of data uncertainty, real-time performance monitoring, and geographic distribution of the system (Lehmhus et al. 2016; Neuman 2009; Wang et al. 2010). Examples include the modern power grid where power generated by wind and solar farms is balanced by the traditional generation methods in response to the demand in real time (Amin 2002; Ericsson 2010). This system requires distributed management and control to increase energy efficiency by production optimization and distributed automation (Baheti and Gill 2011; Sridhar et al. 2012). Protection of such systems requires understanding the vulnerabilities at each connected node and a comprehensive analysis of risk scenarios. In these cases, software actions result in hardware operations and an attack can cause physical damage to infrastructure. Similar risks exist in additive manufacturing process chain where damage to a 3D printer can be caused by a software malfunction, for example, and overheating with a laser can melt the build plate, rails, and electronics present in the machine.

2.1.3 Threat Models in Additive Manufacturing

The vulnerabilities of the additive manufacturing process chain are analyzed with respect to the steps presented in Figure 2.1. The threat model can be broadly divided into three categories (Gupta et al. 2017; Yampolskiy et al. 2015; Zeltmann et al. 2016):

1. Sabotage: An external or internal player can gain access to the system and modify the files in one or more ways.
2. Intellectual property (IP) theft: An external or internal player may make a copy of the original CAD model files and then print the component using a 3D printer to generate unauthorized products.
3. Criminal intent: A hacker can gain unauthorized access to the users' data and withhold it for ransom. The legitimate IP owners can be offered access to the data in exchange for money.

A more detailed description of some of the threats is presented in Yampolskiy et al. (2015).

2.1.3.1 Sabotage

Among the numerous possibilities that exist for sabotage, only a few examples are presented here for illustration. Gaining access to delete the files is the first possibility that does not need any knowledge of additive manufacturing systems or the process chain. However, the standard practice of keeping multiple backups of all cloud resources has reduced the possibility of completely losing all the information. Other subtler changes to the design may go undetected and may be more damaging.

CAD and finite element analysis software programs are connected to the Internet to access license managers and may be the potential targets of attacks, where an outsider can change the CAD model design, the material property values, or dimensions of the model (Turner et al. 2015). In cases where large assemblies are created, small changes in dimensions may cause parts to be incompatible. Similar changes can be made in FEA software, where the database can be compromised to change the material properties, dimensions, or display of results. Attacks on connected 3D printers can also cause problems such as introducing small defects that can remain undetected in a normal testing procedure but reduce the component life in the long run, introducing defects in only a few components in a large production batch or damaging the printer by making it run outside of its normal capabilities. Such kinds of attacks can lead to weak and unreliable components, which can threaten human lives when used in safety-critical systems (Yampolskiy et al. 2015). Failure of only a few components in service can result in recall of the entire batch, which can be expensive for the company.

2.1.3.2 IP Theft

Attacks may focus on stealing the IP for unauthorized reproduction of parts. The worm "ACAD/Medre.A" was designed to infect and steal CAD drawings (Deborah and Kelly 2016; Michael 2013). Stolen files may then be sold to other parties for manufacturing the part. In some cases, a sophisticated attacker can modify the basic design to produce a competing component for the same application, which will not infringe the IP of the original owner.

2.1.3.3 Criminal Intent

Malware and ransomware are used to hide or encrypt files, and then demand payment to retrieve the encryption key. In some cases, the ransom may not be demanded and the files may be sold to other interested parties. The total ransomware market size is estimated to be $5 billion in 2016 and grew 350% compared to the previous year. Although the share of additive manufacturing in this market is currently negligible, the number of instances of ransomware may increase in this sector as the usage of additive manufacturing increases. In some cases, the intent may only involve causing harm to the organization by deleting valuable files.

2.2 SECURITY IN ADDITIVE MANUFACTURING

Protection of the additive manufacturing process chain involves securing software, firmware, hardware, and various files. Two methodologies are described below that are applicable to the additive manufacturing systems. These methodologies can be applied concurrently. In the first methodology, cybersecurity tools are applied to the network. The second methodology, which is the main focus of this chapter, describes a novel strategy of introducing design features in CAD files so that the high-quality parts are printed only under a very specific set of conditions. Stolen files would not print in high quality owing to the lack of knowledge of the processing parameters required to print them.

2.2.1 Network Security Tools

Security tools such as anti-virus software, anti-malware software, network access control, and firewalls can be employed in securing additive manufacturing process. At the hardware level, side channel monitoring such as acoustic emission or heat dissipation incorporated inside the 3D printer can help in monitoring the flow and report any alarming situation such as working environment modification or equipment damage (Chhetri et al. 2016b; Vincent et al. 2015; Yampolskiy et al. 2014).

For example, when testing a software system, reasonable input values are used for most cases while bad input values are only used sometimes based on the assumption that users only do something wrong occasionally. In contrast, security testing assumes that the attacker is intentionally doing something wrong, and thus the concentration should be on testing with unexpected input. When applied to an additive manufacturing cyber–physical system, this means that for testing an additive manufacturing software system, unexpected inputs from attackers should be emphasized rather than assuming that a mistake is accidentally done by the user (Kemmerer 2003). Standards and regulations should be developed for additive manufacturing security testing besides terminology and principals, so that a reference manual can be provided to those who wish to perform a security testing on the system and find the origin, purpose, and effect of the attack (Huang et al. 2013; Zaeh and Ott 2011). This testing can also help in identifying the most vulnerable parts of the process chain. Development of standard practice manuals can also help in establishing accountability for the loss or damage. In addition, patching and maintenance are also important for maintaining a secure system. The software used for generating the toolpath file is routinely updated for fixing bugs or adding more features. The updates should also focus on system vulnerabilities that keep evolving as the nature of threats changes. Though patching a system could be time-consuming, better automatic configuration control and updating are required for securing the additive manufacturing process (Sturm et al. 2014). Security is necessary but must be combined with user-friendliness, efficiency, ease of use, and other desirable system properties.

Compromises at the toolpath generation level are hard to detect because the machine itself cannot easily distinguish between the file that has the correct design and the file that has a compromised design. Only an operator may be able to detect some of the changes in the design or operating conditions. Besides communication of CAD, STL, or toolpath, files are prone to being intercepted while in transmission from one computer to another or to the 3D printer (Gupta et al. 2017). Finally, the last step is to use the additive manufacturing machine to manufacture the product, which is subjected to firmware attacks (Moore et al. 2017). As most additive manufacturing machines have open USB ports for loading files for printing and for installing the firmware and software, this channel can be used for introducing malicious codes.

2.2.2 Security Tools Embedded in CAD Models

Several possibilities have been explored to determine the authenticity of parts produced by additive manufacturing files. Embedding tracking and identification marks, watermarking, and using materials that have tracer elements inside to identify genuine parts from counterfeits are some of the possibilities (Ivanova et al. 2014). A novel security methodology for additive manufacturing is presented here, in which the security features are embedded in the CAD model as design features to act as another layer of defense. When all other cybersecurity tools are breached and access to the CAD, STL, or toolpath file is gained, those files can be used to print the part using the same type of printer as the genuine part. Security features embedded in the CAD designs can help in ensuring that the hackers are not able to print parts in high quality.

The next section will present a methodology in detail that describes the use of design features that can be embedded in CAD files. The presence of these features should not compromise the quality of the product. In addition, it is expected that the detection of these features would not be easy by unsuspecting people. Since these features are based on design principles, they should be implemented in correspondence with the product size, design, and application.

2.3 CAD MODELING STRATEGIES FOR ADDITIVE MANUFACTURING

The additive manufacturing process chain presented in Figure 2.1 shows that the original CAD model goes through a chain of conversions to different formats. When the model undergoes multiple levels of conversions to different formats, sometimes loss of information may take place (Akella et al. 2010; Michael 2013, Sturm et al. 2014; Tata et al. 1998). For example, the information may be lost while converting from CAD to STL format and between slices in the slicing step. In addition, parameters such as print speed, time delay in depositing successive layers, and printer resolution in x, y, and z directions need to be considered throughout this process chain in order to obtain the desired quality of the final product (Asadi-Eydivand et al. 2016). Although these considerations make it challenging to obtain a high-quality product, they also present opportunities to develop features that can be used for security and identification of genuine products. The hypothesis is that in the presence of intentionally introduced features, the model should print in high quality only under a given set of process flow and printing conditions. All other conditions should provide a defective final component that will have an inferior service life or a different printed structure. The embedded features should not compromise the quality of the genuine product. The examples presented here are simple geometries for illustrative purposes. However, industrial component designs are often complex and integrating the proposed security features may be easier in complex geometries.

A few CAD-based security strategies, along with their design methodologies and realization process, are introduced in the following sections. These security features can cause disruption to the final product quality and geometry if the prescribed manufacturing process is not followed exactly. In 3D modeling software such as SolidWorks, a split function can be used for creating multiple bodies from an existing part file with a surface of zero thickness, which can be either kept within the part file or saved into separate files. This work demonstrates use of such surfaces in 3D CAD models for security purposes during additive manufacturing. Parts of simple geometries are designed to demonstrate the clear effects of the features embedded in the geometry.

2.3.1 Spline Split Feature

In this case, a split feature designed in the shape of a spline (curve) is used to create a massless separation inside a single body. The spline split feature cuts through the middle of a standard tensile test bar in the geometry as shown in Figure 2.2a, where the length of spline (21 mm) is 3.5 times the width of the tensile bar gauge section (6 mm). In the CAD file, two bodies are formed within one single part due to the presence of the spline, with zero distance or volume separating them as shown in Figure 2.2b. Next, the CAD file is exported as an STL file. Several different STL resolutions are possible during file export. The finer resolutions use a greater number of triangles to represent the geometry and result in a larger file size. A representative model exported using the "*Coarse*" resolution setting in SolidWorks is shown in Figure 2.3. The difference in the tessellation along the spline in two parts of the geometry results in mismatch at the corners of triangles located at the spline, as shown in the magnified views.

An investigation into implementation of the spline split feature in FDM printed components is performed on different combinations of STL resolutions and printing orientations. Table 2.1 presents results of exporting a CAD file to STL format at three different export resolution settings. *Coarse* and *Fine* resolutions are preset options in SolidWorks, while the *Custom* setting can provide the

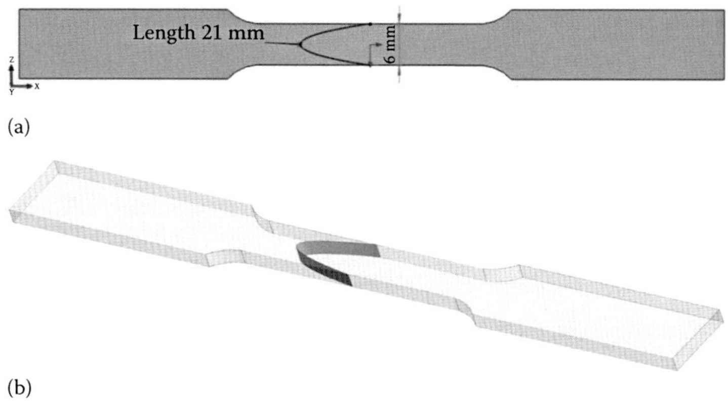

FIGURE 2.2 (a) A tensile test specimen designed with an embedded spline is shown in a CAD solid model. ASTM E8 subsize specimen dimensions are used for this tensile bar design (ASTM International 2016). (b) The spline is shown as a surface in the transparent view.

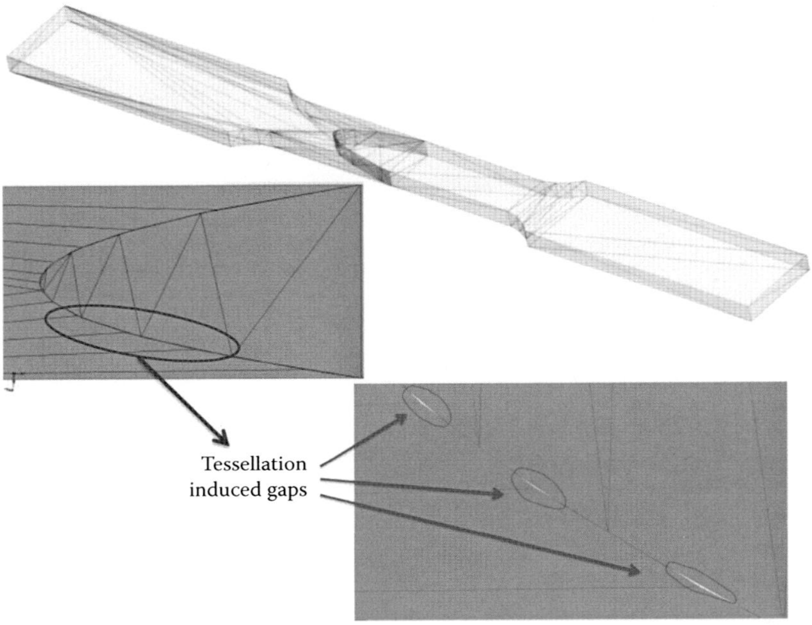

FIGURE 2.3 A tensile test specimen designed with spline split feature exported to STL format, showing gaps induced by tessellation along the spline due to mismatch between vortices of the triangles across the spline.

highest resolution by manually adjusting the *Angle* and *Deviation* permitted to the smallest possible values. The printing orientations of the specimen used in this work are defined in Figure 2.4.

The STL file is imported into slicing software to generate the G-code file containing the 2D toolpaths. The *Preview* function in the slicing software allows visualization and navigation of the 2D toolpaths generated for each layer of the 3D model. Using the slicing properties specified in Table 2.2, the models oriented in *x–y* and *x–z* printing directions show different results. When oriented in the *x–y* direction, the sliced model does not show discontinuity or other signs indicating the existence of the split as shown in Figure 2.5a. Further, models saved in *Fine* and *Custom* STL resolution show similar slicing results under the *x–y* direction. It is therefore expected that tensile bar models will not be printed with discontinuity under the *x–y* direction for all STL resolutions. However, when the model is placed and sliced in the *x–z* direction, discontinuity around the spline feature can be observed for all STL resolutions as shown in Figure 2.6a. The presence of discontinuity in the sliced model makes it likely that the feature will also appear in the printed specimens.

The CAD models are 3D printed using an FDM-based Stratasys Dimension Elite 3D printer. This printer can print a model material (ABS) and a dissolvable support material (SR-10/P400SR Soluble Support Material-acrylic copolymer). Reference models (standard tensile specimen without a split feature) are also printed under both printing orientations to compare with the spline split models. Models printed in the *x–z* and *x–y* orientations are shown in Figure 2.5b and c and Figure 2.6b and c, respectively. Unlike the results observed in sliced models, the printed specimens show that the

TABLE 2.1
Different STL Resolution Settings in SolidWorks

STL Resolution	Coarse	Fine	Custom (Finest)
Angle (°)	30.00	10.00	0.50
Deviation (cm)	1.215×10^{-2}	4.681×10^{-3}	5.062×10^{-4}

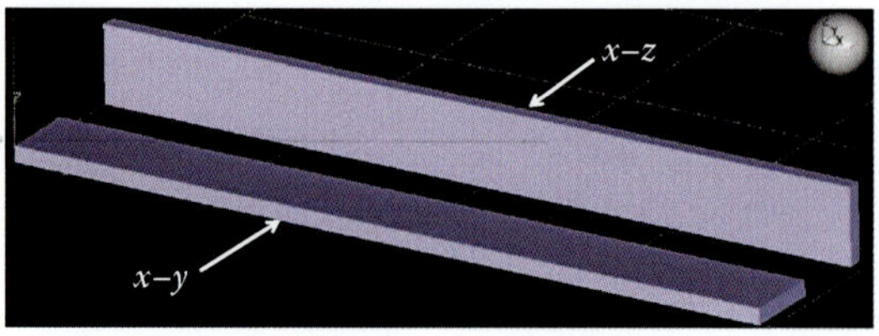

FIGURE 2.4 Printing orientations defined as x–y and x–z used in this work.

TABLE 2.2
Slicing Properties Used in Generating Toolpath for CAD Models Saved in STL Format

Slicing Properties	Value
Layer resolution (cm)	0.01778
Model interior	Solid
Support fill (mode)	Smart
STL units	Millimeters

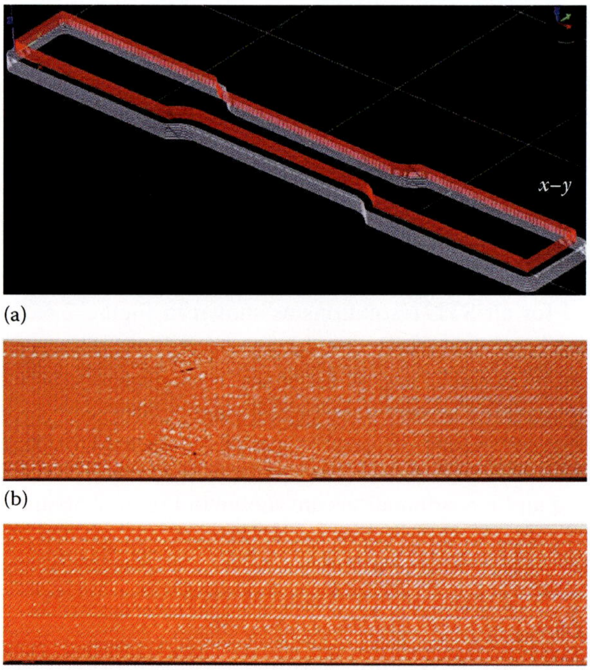

(a)

(b)

(c)

FIGURE 2.5 Tensile test specimens (a) designed with spline split feature is sliced along x–y, (b) 3D printed by Fused Deposition Modeling printer, and (c) a reference intact model (without the spline split) is 3D printed by the same printer. Both models are processed using *Coarse* STL setting and sliced under the x–y orientation.

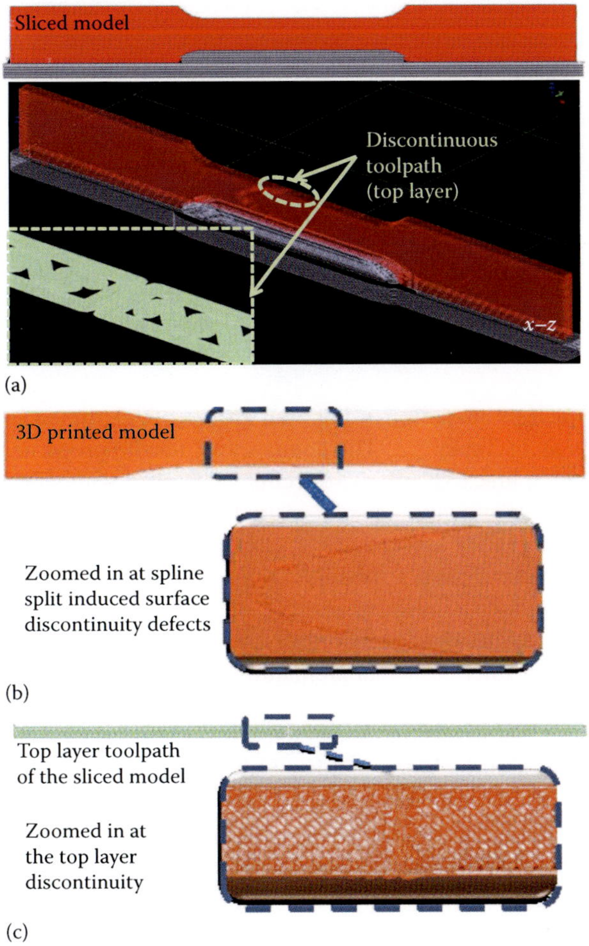

FIGURE 2.6 The tensile test specimen designed with spline split feature (*Coarse* STL): (a) sliced along the x–z direction, showing enlarged details at the discontinuous toolpath, (b) 3D printed in the x–z orientation on an Fused Deposition Modeling printer using thermoplastic filaments; inset shows the spline in the printed model; and (c) discontinuous toolpath visible in the top layer along the thickness of the specimen.

Coarse STL resolution file printed in the x–y orientation has some surface disruption but in a less regular shape compared to the x–z direction. Higher STL resolutions can minimize or even neglect this disruption, leaving the surface texture same as intact samples. However, when printed in the x–z orientation, the spline split feature is printed for all three STL resolutions. Increasing the STL resolution does not seem to help in eliminating the discontinuity along the spline in the samples printed in the x–z orientation.

The results obtained on the FDM printer are then replicated on a material jetting printer (Stratasys Objet30 Pro) using VeroClear material. The Objet30 Pro is capable of printing parts with a minimum layer thickness of 16 μm, as compared to 178 μm for the FDM printer used previously. Similar results are obtained in terms of presence or absence of the spline feature with respect to the STL resolution and print orientation even for the resin printer as shown in Figure 2.7, providing a broader context for these results. It can be concluded from both the ABS thermoplastic specimens and the VeroClear photopolymer specimens that the spline split feature affects the model integrity when printed.

Embedding the spline feature in the standard tensile test specimen allows testing the specimens to obtain quantitative measurement of the mechanical properties in the presence and absence of the split

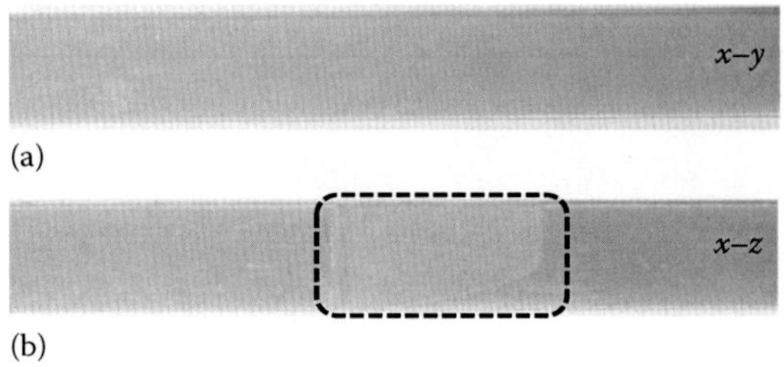

FIGURE 2.7 (a) The spline split feature on a rectangular prismatic bar (101.6 × 9.53 × 2 mm^3) 3D printed by PolyJet technology using VeroClear photopolymer in the *x*–*y* orientation does not show any feature in the printed part while (b) shows a spline when the part is printed in the *x*–*z* orientation.

feature. Therefore, tensile tests are performed on four groups of 3D printed specimens under the same conditions and settings as in Table 2.2. The summary of tensile properties is given in Table 2.3, where "Spline *x*–*y*" and "Spline *x*–*z*" refer to specimen designed with spline split feature printed in *x*–*y* and *x*–*z* orientations, respectively; "Intact *x*–*y*" and "Intact *x*–*z*" refer to the reference specimens with no spline split feature printed in *x*–*y* and *x*–*z* orientations, respectively. For both directions, the ultimate tensile strength and Young's modulus are comparable between intact and spline split samples, but the average failure strain for spline split samples is at least 50% smaller than that for the intact samples. Similarly, the toughness for intact samples is at least twice that of the specimens containing the split. These results show that premature failure can result in specimens containing the spline split as in Figure 2.8.

The example presented in this section was a standard tensile test specimen with simple geometry. Real engineering designs often include complex and multicomponent systems that include many curves, lines, surfaces, and construction lines. Addition of one or more surfaces for security and identification purposes in such complex models is possible with minimal chance of detection. Variations of such features based on the same principle can be developed to use in different designs.

TABLE 2.3
Tensile Properties of Specimens 3D Printed by a Fused Deposition Modeling Printer (the ASTM D638 Standard Is Used)

Property	Spline *x*–*y*	Spline *x*–*z*	Intact *x*–*y*	Intact *x*–*z*
Young's modulus (GPa)	1.89 ± 0.04	2.10 ± 0.05	1.98 ± 0.05	2.05 ± 0.03
Ultimate tensile strength (MPa)	24 ± 1.1	31.5 ± 0.5	30 ± 0.2	32.5 ± 0.3
Failure strain (mm/mm)	0.015 ± 0.001	0.021 ± 0.001	0.029 ± 0.001	0.077 ± 0.041
Toughness (kJ/m^3)	295.4 ± 94.2	453.6 ± 29.5	632.1 ± 33.2	3367.4 ± 902.8

Note: Intact samples are tested for reference.

FIGURE 2.8 A specimen shows that the tensile failure originated at the tip of the spline due to the stress concentration, which resulted in the lower mechanical properties.

These features work independent of cybersecurity tools implemented on the network to guard against duplication from stolen files.

2.3.2 Elliptical Split Feature

In this example, a 2D surface feature is embedded in the component with the aim that the feature will be or will not be printed only under certain conditions. A 3D model is created in the shape of a rectangular prism of dimensions $40 \times 20 \times 5$ mm^3 with an elliptical depression (major radius, 14 mm; minor radius, 6 mm) at the center as illustrated in Figure 2.9a. The split funciton is performed on the same elliptical sketch, generating two bodies within the single part and separating them with zero distance or volume. The CAD model is exported to a *Coarse* STL file as shown in Figure 2.9b, where a number of triangles are used around the ellipse similar to those shown in Figure 2.3, and gaps can be observed around the ellipse in the STL file due to tessellation effects. From the model appearance in Figure 2.9, one cannot easily distinguish whether an ellipse split feature is performed on the model or not.

Using the slicing properties recorded in Table 2.2, the STL file is sliced in the *x–y* and *x–z* orientations. As the model is placed and sliced in the *x–z* orientation as in Figure 2.10, it is noted that many 2D layers of the toolpath show discontinuities when crossing the elliptical depression. These discontinuities are expected to be printed and reflected at the bottom of the rectangular case, aligned with the shape of the ellipse, which will inevitably affect the structural integrity and may cause failure when placed in a load-bearing situation. Figure 2.11 shows the printed parts designed with elliptical split and under the *x–z* direction, which agrees with the observations on the slicing results. On the other hand, when sliced in the *x–y* direction, the model appears as if there were no ellipse split at all, as shown in Figure 2.12. From observation of the toolpath for each layer, it is expected that the model sliced in the *x–y* direction will be printed with no discontinuities or surface defects at the bottom. This again is confirmed by the printed part as shown in Figure 2.13.

The threat model for using this security feature is that a hacker has the capability to retrieve the CAD file or STL file with the intention to manufacture duplicated products, violating the IP holder's rights. The attacker may not have previous knowledge of either the entire model structure or a small key component. Using this method, the true design of the part is disguised by the security feature and, due to the high geometrical complexity, is difficult to determine from the modified model. Such part complexity, which aids in the use of such security features, is fairly common in additive manufacturing and is one of its main advantages. While the hacker expects to be able to print the part as it appears in the CAD file or STL file, without knowledge of the printing parameters that will avoid the intentional defecrs, defective parts would be produced. The elliptical split can be designed in a hidden location and in various sizes, depending on the application, and at a specific tilted angle such that the attacker would have no easy way to tell at a glance if any unusual features have been

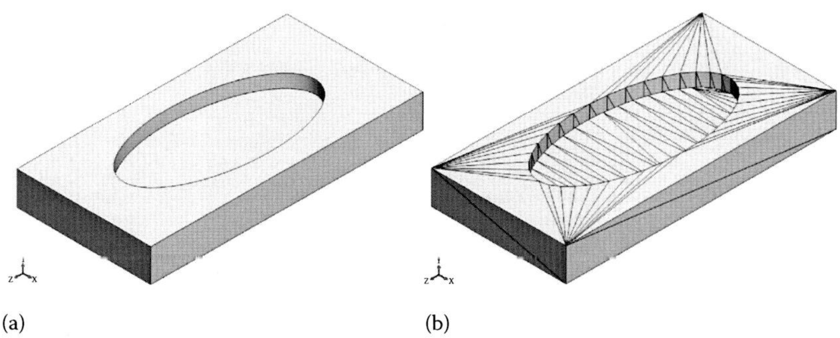

FIGURE 2.9 A rectangular case $40 \times 20 \times 5$ mm^3 (inner ellipse of major radius 14 mm and minor radius 6 mm, at depth 2 mm from the top) designed with the ellipse-based split (a) CAD model, (b) tessellated model at *Coarse* STL resolution.

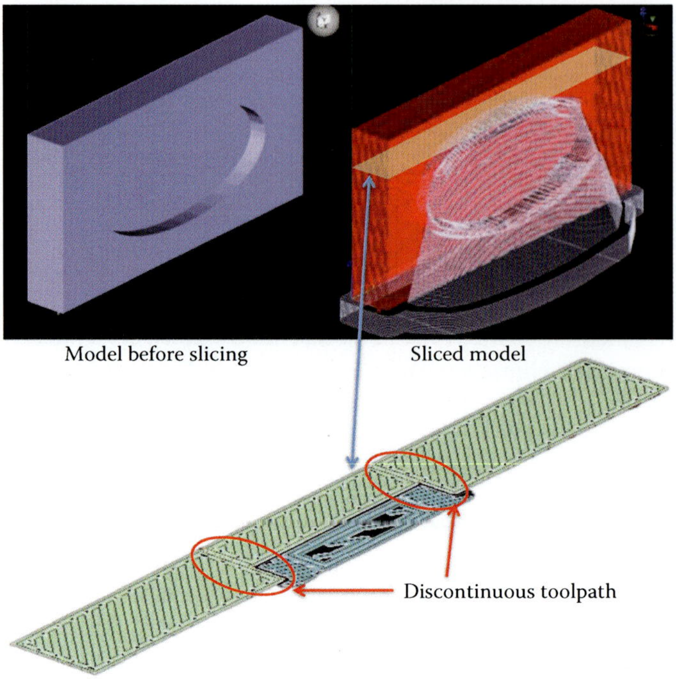

FIGURE 2.10 Model designed with elliptical split when sliced in the x–z direction; discontinuous toolpath can be observed during the insertion of elliptical split (*Coarse* STL resolution).

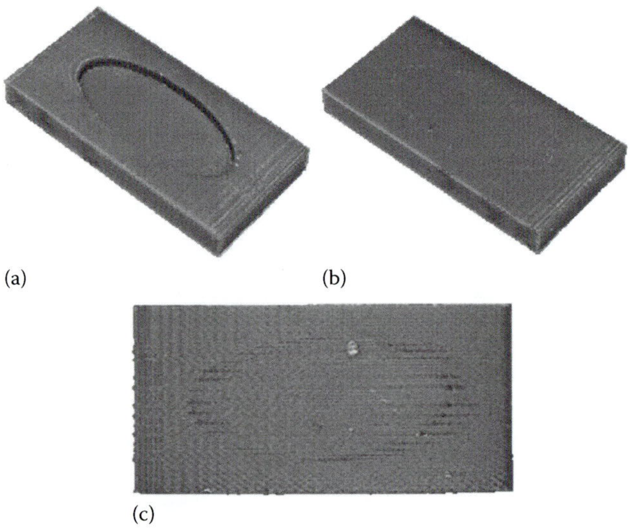

FIGURE 2.11 Model designed with elliptical split when sliced and printed in the x–z direction: (a) top surface, (b) bottom surface, and (c) zoomed in at the bottom showing discontinuity-induced defect details (*Coarse* STL resolution).

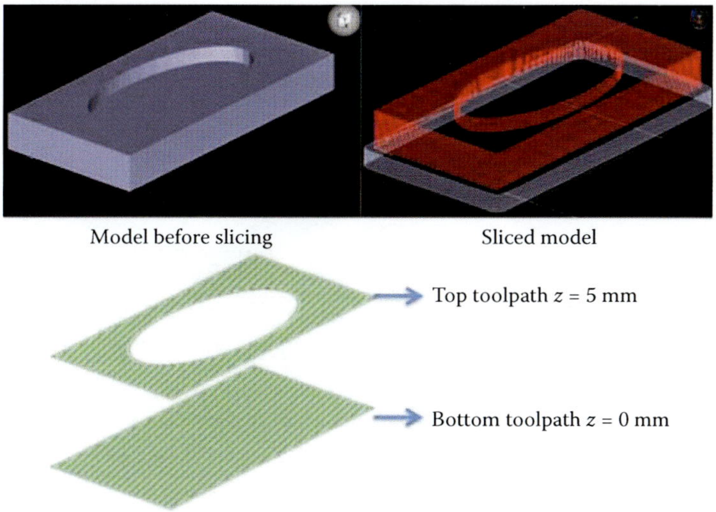

FIGURE 2.12 Model designed with elliptical split when sliced in the *x–y* direction. Toolpath of the bottom shows no discontinuity induced from the elliptical split (*Coarse* STL resolution).

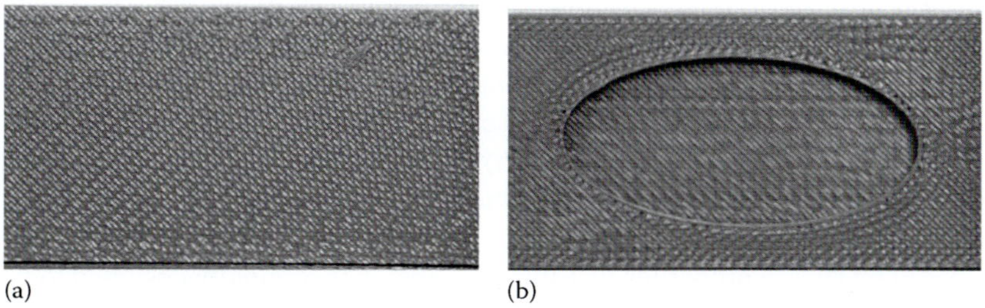

FIGURE 2.13 Model designed with elliptical split when sliced and printed in the *x–y* direction: (a) bottom surface, (b) top surface (*Coarse* STL resolution).

embedded in the model. When such models with embedded features are printed, they would contain the features as defects similar to those shown in Figure 2.11c and experience expected early failure as discussed in Section 2.3.1 and Table 2.3.

2.3.3 Rectangular Split Feature

The previous example explored using a 2D elliptical sketch to split the model and used the inexact rendering of curvature by the STL format to develop a security feature. In the present example, a 2D rectangular sketch is used for the split function, which demonstrates that these features are not reliant on the curvature-tessellation effect.

A 3D CAD model is created in the shape of a rectangular prism of dimensions $40 \times 20 \times 5\ mm^3$, with a rectangular depression of dimensions $28 \times 12 \times 2\ mm^3$ at the center as shown in Figure 2.14a. The sink is only on the top surface of the block, leaving a 3-mm solid bottom. In the second step, a split feature is performed to the same rectangular sink (28 × 12 mm) on the remaining solid bottom but the material is not removed. This operation generates two bodies within the single part, separating them with zero distance or volume. This CAD file is exported using *Coarse* STL resolution setting as shown in Figure 2.14b. The STL view does not show any indication of the presence of split feature in the design because the split lines coincide with model structure lines in the

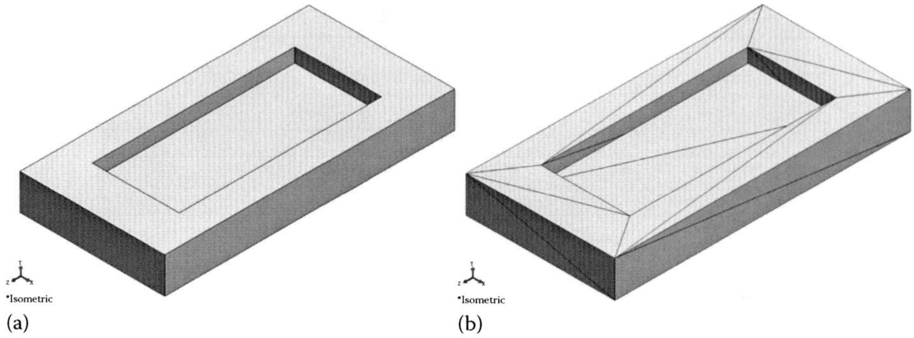

FIGURE 2.14 A rectangular case 40 × 20 × 5 mm^3 (outer rectangle, 40 × 20 mm^3; inner rectangle, 28 × 12 mm^2) designed with rectangular-based (28 × 12 mm^2) split at depth 2 mm from the top: (a) CAD model, (b) tessellated model at *Coarse* STL resolution.

CAD or STL model. In addition, STL file information for the models with and without the split feature such as file size and number of triangles is identical.

Using the slicing properties recorded in Table 2.2, the STL file is sliced in *x–y* and *x–z* orientations represented in Figure 2.4. The slicing results under different printing orientations in the presence of rectangular split feature are different from the elliptical split feature results. The generated toolpath for the rectangular model sliced in the *x–z* orientation is shown in Figure 2.15a, which agrees with the original design of having only a 2-mm-deep sink in the block. However, when sliced in the *x–y* orientation, the generated toolpath also shows an empty volume at the bottom in Figure 2.15b. In the figure, many blue areas (support material toolpath) are found inside the green rectangular area (model material toolpath) close to the bottom of the model. Close inspection of every 2D layer of the model toolpath reveals that support material is used for the inner rectangular area from *z* height 0 to 1.4224 mm (based on slicing properties in Table 2.2), which is originally designed to be printed with the model material.

It can be expected from the generated toolpath that the model will be printed with no material from the top inner rectangular area but support material for most of the bottom inner rectangle area. In FDM technology, the support material is often used around or underneath the model material for overhanging or nested structures, which can be easily removed by scrubbing or dissolving in detergent-based solution. Therefore, it is detrimental to model structure integrity where a part originally designed as solid is produced as empty. Figure 2.16 shows the two blocks placed in front of a mirror to simultaneously show the front and back sides of the printed specimens. It is observed clearly that the *x–y* slicing orientation has resulted in a sink on both sides on this block due to the presence of the split feature. Only a thin layer separates the sinks present in the two sides of the block printed in the *x–y* slicing orientation. Such alteration in the part design during printing can make the component totally useless for a given application.

There are many possibilities of detecting the presence of split feature in the CAD model. The process tree can be investigated to find the presence of such surfaces. However, split feature is a standard feature in CAD software programs and is routinely used for design operations. In a complex model with multiple parts present in an assembly, finding one or two features that are embedded not for design but for security purposes is difficult. These features can also be created in the files after their export to STL format or even coded in the g-code directly before feeding the g-code to the printer. In such cases, CAD model developers can be completely separated from the process of security feature design and embedding process.

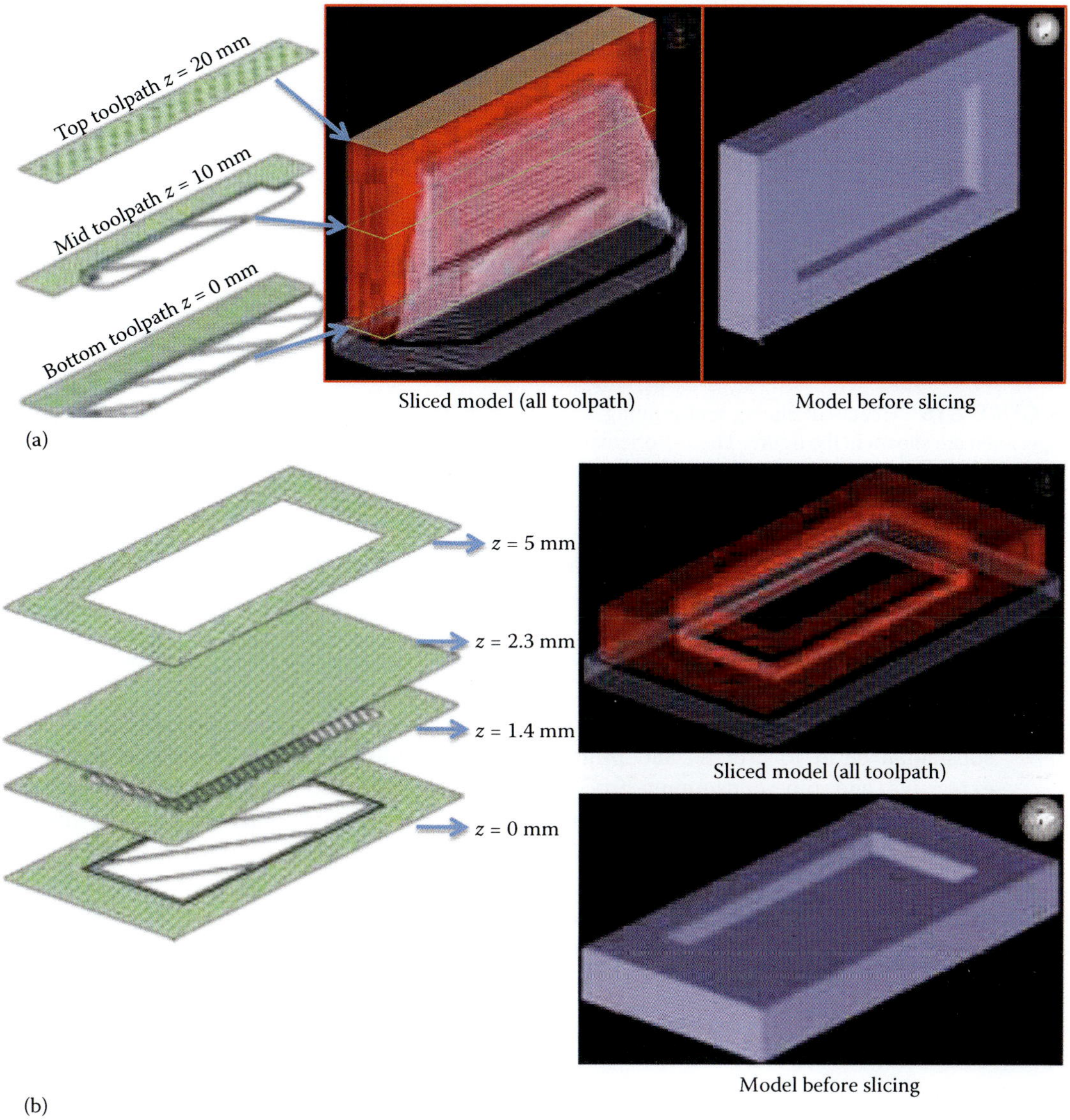

FIGURE 2.15 Model designed with rectangular split sliced (a) in the x–z direction where the middle section toolpath shows that the inner rectangular area will be printed as solid with model material as originally designed and (b) in the x–y direction, showing that support material is printed for the inner rectangular area from the bottom to a 1.4-mm height as opposed to the original design (from $z = 0$ to 3 mm should be solid).

2.3.4 Design-Based Security Features

The examples presented earlier in this section show two possibilities of using the *split* feature to embed surfaces in the CAD models. In both cases, the model does not print in the correct geometry if it is not sliced in the desired orientation. There are numerous other possibilities using routine design features to save files in such a way that only a particular set of conditions will print them correctly.

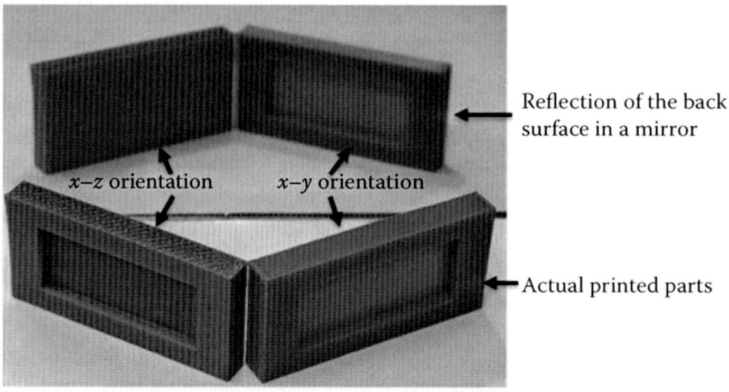

FIGURE 2.16 Effects of slicing and printing orientation on the specimen. The front and back faces of the specimen are shown in the figure. The x–z orientation has provided back side to be solid, which is in compliance with the original design. Meanwhile the x–y orientation has created a rectangular hole on both sides, leaving only a thin surface in between.

The importance of these features can be enormous because they may make printing of the correct part very difficult. These features can serve two purposes:

- If those who stole the file are not able to detect these features, they will most likely print the part in an erroneous configuration.
- If those who stole the files detect the presence of design security features, it will still take substantial effort to process and print the file in multiple configurations. This potentially expensive and time-consuming process can deter some people from making the effort to decode the correct configuration and they move on to find other unprotected files for counterfeiting.

In both these scenarios, having design security features is advantageous. Given the high value of the additive manufacturing parts, network or other security is always required for protection. The potential threats and cybersecurity tools and strategies are discussed in the following sections.

2.4 SECURITY STRATEGIES

Cybersecurity threats are dynamically evolving and cyber-attacks are growing exponentially because of the increasing dependence of systems and machines on information technology. US companies have been reported to lose an average of $15.4 million per year to cybercrime, twice the global average of $7.7 million (Griffiths 2015; Ponemon Institute 2016). According to the report, cybercrimes related to malware, web-based attacks, and social engineering are more common among small organizations. In comparison, attacks associated with Denial of Services (DoS), malicious insiders, malicious codes, and stolen devices are more common for large organizations (Ponemon Institute 2016). Cybercrimes by malicious insiders, DoS attacks, and web-based attacks are among the most expensive cyber-attacks.

Attacks on computer and network systems consist of actions such as remote or local connection, unauthorized file access, and malicious program execution with the intent to compromise the secure operation of the computer and network system. Critical operations in defense, banking, telecommunications, transportation, and electrical power rely on information infrastructure, which makes cyber-attack a significant threat to society (Akella et al. 2010; Ye et al. 2001). In addition, the rapidly growing Internet of Things (IoT) is also susceptible to the same threats. Security goals such as availability, confidentiality, and integrity are achieved by various kinds of techniques for prevention, detection, isolation, assessment, reaction, and vulnerability testing. Hackers intend to enter systems undetected, making it important to review logs on the network, change admin passcodes on a regular basis, and restrict access to particular high-risk user accounts. This section discusses potential cybersecurity risks and available security tools that are implemented on the network or on user computers. The design-based security features discussed in Section 2.3 are intended to provide security in addition to the tools discussed in this section.

2.4.1 Typical Cybersecurity Attacks

A number of common types of attacks are presented in this section, but the list is not exhaustive.

2.4.1.1 Network-Level Attacks

Attacks against distributed systems that consist of connected autonomous computers in a middleware often start at the network layer by approaches including, but not limited to, gathering network-related information such as open Transmission Control Protocol (TCP) ports, exhausting resources such as denial of service attack, or abusing protocols as discussed in several publications (Atighetchi et al. 2003; Murthy and Manoj 2004; Schuba et al. 1997). Therefore, it is critical to increase an application's resistance against cyber-attacks when exposed to environment that is not completely secure (Atighetchi et al. 2003).

2.4.1.2 DoS and Distributed Denial of Service Attack

DoS is the situation in which a system is prevented from functioning as intended either accidentally or intentionally (Schuba et al. 1997). A DoS attack floods the targeted machine or server with an overwhelming amount of traffic to cause disruption or congestion so that it essentially becomes unavailable to its intended users (Schuba et al. 1997; Carl et al. 2006). Many DoS attacks also use unsuspecting home computers as *zombies* to awaken them at a specific time and accomplish a distributed denial of service (DDoS) without being noticed (Carl et al. 2006).

In many DoS attacks, compromising the initial system is just laying down the foundation for later attacks. These zombie systems are compromised but not used until a predetermined time arrives for them to wake up and cause a DoS attack by performing certain steps (Schuba et al. 1997). In a possible scenario, the system may be functioning exactly as it is designed. However, when the DoS attack takes place, the amount of load, scope, or attack parameters imposed on the system exceeds its capability and results in complete shutdown (Carl et al. 2006; Weiss 2012). The major types of DoS attacks can be widely categorized into five groups: the synchronize (SYN) flood; Internet Control Message Protocol (ICMP) flooding via ping requests, or Hypertext Transfer Protocol (HTTP) requests against a web server; teardrop attacks; peer-to-peer attacks; and low-rate DoS attacks (Dalziel 2013; Weiss 2012). Out of the many different types of DoS attacks, the easiest executable one is the single-origin DoS attack (Weiss 2012). This kind of attack sends out a series of network requests from a single machine connected to the Internet to the targeted machine to cause breakdown (Weiss 2012). The single-origin DoS attacks can be effective when targeting undefended victims; however, they can be prevented by blocking the originating IP address at the firewall level or upstream at the Internet service provider level, or blocking the attack with enterprise threat protector products that can identify risks as soon as the attacks begin (Weiss 2012). In addition, web servers can be configured to detect and block the requests and prevent the attacks (Weiss 2012).

A DDoS attack is different from a DoS attack. The DDoS attack uses multiple computers and Internet connections to flood the targeted system at the same time in comparison to the DoS attacks that use one computer and one Internet connection to flood a targeted system (Weiss 2012). In a sophisticated DDoS attack, the attacker appears from many different IP addresses around the world simultaneously, which makes tracing and determining the attack source much more difficult for network administrators. Hence, it is almost impossible to stop the attacks and defend the victim systems by simply blocking one single IP address (Weiss 2012). DDoS attacks have been successfully deployed against online businesses such as Twitter, PayPal, Visa, and MasterCard, causing huge financial loss and severe security concerns (Chen and Perlroth 2013; Weiss 2012). In order to prevent further DDoS attacks once they start, the victim organizations that experienced DDoS attacks should coordinate with the network provider and route traffic originally intended for the victim into the trash to protect other customers (Weiss 2012). Other countermeasures against a DDoS attack like a typical SYN (synchronize) flooding attack make use of the "SYN cookies" either in the server operating system or in a network security device, which can provide a more efficient method for tracking incoming TCP connections and blocking the attack (Eddy 2006; Weiss 2012).

2.4.1.3 Malware

Malware, including computer viruses, worms, and Trojan horses are malicious software that can be used to disrupt the computer or mobile operating environment, retrieve sensitive information, gain access to private computer systems, monitor actions and keystrokes, send confidential data from an untrusted network, and post unwanted advertisements or signs (Milosevic et al. 2017). Malware refers to a variety of forms of intrusive harmful software such as viruses and ransomware. Malware such as spyware can sometimes be embedded in programs that are supplied officially by legitimate companies. For example, around 22 million CDs sold by Sony BMG contained rootkit Trojans, which are silently installed during the installation of the CDs without the user's knowledge or consent when these infected CDs were inserted in CD players. These Trojans creates vulnerabilities that can be exploited by unrelated malwares (Mitchell 2005). They can modify the operating system to interfere with digital rights management and secretly send reports on users' listening habits to help the company gather marketing statistics. This spyware cannot be easily removed or uninstalled without damaging the computer, making the user's computer susceptible to viruses written specifically to target the software (Mitchell 2005). Malware like this can be very damaging in the additive manufacturing field because it can help in stealing not only the final design but also the entire history of creating and refining that design, exposing the design process used by the company. Design software maintains a tree of activities to help revert to a previous version of the design and to deal with individual components in the design assembly. However, such features can be detrimental if the files are stolen as the entire design process can be replicated to build other parts of interest.

2.4.1.4 Computer Worms

A computer worm is a stand-alone malicious computer program that replicates itself and uses a computer network to spread to other computers. Unlike a computer virus, which needs to attach itself to a program or file in order to spread from one computer to another, worms leave infections as they travel. A Trojan horse is a malicious computer program that appears to provide normal functionality, but whose execution usually takes place in the system's background and results in harmful actions (Kemmerer 2003). Severity of this malware ranges widely: some may cause only mildly annoying effects like unwanted advertisements, while others can damage the hardware, software, or private files. Anti-virus software and firewalls are used to protect against malicious activities and to recover from attacks. The infected computers can be isolated and prevented from disseminating further trusted information by imposing an "air gap," by completely disconnecting them from all other networks (Guri et al. 2017).

In the field of additive manufacturing, many problems can be envisioned from virus and malware. The parts produced by additive manufacturing are physical hardware components that are deployed in service. In some cases, a virus implanted on the additive manufacturing machine can wake up at random times to create small defects in the manufactured parts. In a large production run, infection of only a few parts can still cause recall of the entire production batch, making it very expensive for the company. The monetary implications are greater where parts are used in applications where human lives are put into danger or risk of injury is increased (Campbell and Ivanova 2013). Product testing and qualification practices in additive manufacturing need to keep pace with these possibilities and the part qualification process should integrally include certification of the additive manufacturing machine for each batch of the part to minimize such possibilities.

2.4.1.5 Cyber-Attack Levels

Cyber-attackers can be performed by any level of attackers including *script kiddies*, hackers, insiders, or outsiders (Kemmerer 2003). *Script kiddies* usually refer to unsophisticated attackers who simply download and follow online resources or instructions to execute an attack (Kemmerer 2003). The next level is a hacker, who is usually more equipped with technical knowledge to perform sophisticated cyber-attacks. Insiders can be less knowledgeable than hackers but are more dangerous because they have ready access to the resources and can disguise themselves to damage the system or data (Kemmerer 2003). Outsider attacks could be from competitive organizations where one wants to steal private data or information from the other. For these cyber-attacks, it is possible that attackers hide their activities and go undetected by modifying event log or even disabling logging function (Kemmerer 2003). Such attacks in the additive manufacturing process can result in malicious product design modification and monetary losses (Chen et al. 2017). In addition, hacking of additive manufacturing machines can facilitate modifying the G-code on the go, where part print conditions or part design can be changed locally in certain sections (Chen et al. 2017).

2.4.2 Cybersecurity Tools

A number of commonly applied cybersecurity tools are listed below (Staff 2013). Some of these tools are appropriate for the users at any level of network access, while some of these tools are implemented by the administrators on the network. In case the additive manufacturing machines are not connected to the Internet, the cloud-based storage of files and transmitting files via e-mail or messenger programs exposes them to risk.

2.4.2.1 Procedural Approaches

Procedural solutions process jobs at different security levels as per the defined protocols (Kemmerer 2003). An example of this approach is that the private or classified data are processed at one level for each time and wiped off from the system after the processed data from one level are finished and sent to the next level to satisfy security requirements (Kemmerer 2003).

2.4.2.2 Passcodes

From the user security level, the proper choices of strong passcodes to achieve security requirement are recommended, such as a combination of uppercase and lowercase letters, numerals, and special characters as compared to accounts that use no passcodes or weak passcodes (Kemmerer 2003; Soutar et al. 1999). Changing passwords at certain time intervals is also recommended. Passwords should be used for user accounts as well as for individual files of high value (Soutar et al. 1999).

2.4.2.3 Encryption Methods

Encryption methods that scramble data so that it cannot be read by unintended parties are also used for secure communication of files (Kemmerer 2003; Staff 2013). For example, different 3D printer companies have their own encryption methods to generate the toolpath for printers including print head movement algorithm, temperature gradient, and other process parameters unique to the machine, so that the file is only readable when fed to the correct 3D printer. The files can be made more secure by node locking so that only the intended 3D printer can read it.

2.4.2.4 Intrusion Detection System and Intrusion Prevention Systems Tools

Intrusion detection system (IDS) and intrusion prevention system (IPS) tools are used to detect threat activities such as malware, spyware, viruses, worms, and policy violation (Kemmerer 2003). While IDS passively monitors and detects suspicious activity, IPS performs active in-line monitoring and prevents attacks from known or unknown sources (Kemmerer 2003; Staff 2013). IDS tools are automated data analysis tools because manual analysis of the voluminous data is impractical (Kemmerer 2003). Some of the intrusion detection research has explored correlation and fusion of cyber alerts, because identifying attackers and their purposes is rather challenging and false-positive results are expensive and distracting from the real threats (Kemmerer 2003).

2.4.2.5 Firewalls

As a network security system, a firewall monitors, filters, and controls the incoming and outgoing network traffic based on agreed security rules. It is established as a barrier between the trusted network (internal) and untrusted network (external) such as the Internet, to filter and allow data exchange only between authorized sender and receiver, and further prevent unauthorized connections to the computer.

2.4.3 Other Security Tools

Information leakage can happen from a number of sources. Securing against this threat requires an approach that integrates human, machine, and process-related factors. Additional authentication mechanisms, network access controls (NAC), and interference control to the computer system are also encouraged to enhance security (Denning and Elizabeth 1982; Kemmerer 2003; Staff 2013). However, scenarios such as retrieval of information from carelessly disposed of hard drives or printouts require a more systemic approach rather than limited cybersecurity approach.

2.4.3.1 Devices Disposal

Removable storage drives such as CDs, DVDs, and USB flash drives should be carefully handled and disposed of as per the protocol of the organization (Dolgunov 2007; Kemmerer 2003). Use of these devices on multiple computers can make them susceptible to risks presented by viruses and malware. In addition, saving information without encryption can make it very easy to be retrieved by others if the drive is lost or stolen (Dolgunov 2007).

2.4.3.2 Product Marking

IP protection can be realized through product marking using security tags, labels, quick response codes, or barcodes to ensure that each product has its own identity and can be traced down for authenticity (Kemmerer 2003). One of the challenges with additive manufacturing is that the stolen files can produce a part that will be identical to the original part (Chen et al. 2017; Gupta et al. 2017). Innovations are required for embedding the parts with identification marks (Chen et al. 2017).

2.4.3.3 Authentication

An authentication mechanism is used to guarantee that users are who they claim to be (Kemmerer 2003). Traditional authentication and authorization mechanisms provide directory-based service, which authenticate users and grant access based on the organization authorization rules (Staff 2013). Newer technologies use methods like digital certificates and public key infrastructure solutions. In addition, this can be realized by using sophisticated authentication devices that examine and quantify the physiological and behavioral characteristic of each user, convert it into digital form, and compare it to previously stored sample characteristics (Kemmerer 2003). Other biometric authentication devices based on user's fingerprints, handprints, retina patterns, or facial characteristics are also commonly adopted for the authentication process (Kemmerer 2003).

2.4.3.4 Network Access Control

Network access control tools enforce organization security policy by access control mechanisms, allowing network access to only devices that are in accordance with security policy (Staff 2013). It can control the data that specific users can access (Kemmerer 2003). For example, the owner selects a password for each file and distributes those passwords only to relevant people. However, the network access control mechanism is limited in that it cannot prevent unauthorized deliveries of the passcode or revoke access from one user without revoking access from all users (Kemmerer 2003).

2.4.3.5 Inference Control

Inference control mechanism allows access to statistics about groups of individuals while restricting database access to a specific individual's sensitive or private information (Denning and Elizabeth 1982; Kemmerer 2003). It aims to establish a statistical database that discloses no sensitive data or individual identity (Denning and Elizabeth 1982; Kemmerer 2003).

2.4.4 Cybersecurity Threats

In the past, cybersecurity methods primarily focused on preventing access from the outside. However, increasingly sophisticated attacks have originated from people with inside access or by outsiders who steal credentials from legitimate users (Taylor 2015). More and more cybersecurity attacks come in different forms that are never seen before such as the rise of machine-to-machine attacks. Mobile devices such as smartphones are particularly attractive to cybercriminals because of the relatively large user population (Taylor 2015). They are also vulnerable to multiple attack vectors such as malicious applications or unsecured web browsing, where user's private information and data can be attacked and leaked (Taylor 2015). Cyber-attacks such as jailbreaking, ghostware, two-faced malware, and headless worms are among the major cybersecurity threats that emerged in recent years (Li and Clark 2013; Milosevic et al. 2017; Taylor 2015). Some of the threats are listed below:

- *Cloud-based system jailbreaking:* Mobile devices running malwares or compromised applications are perfect targets for attackers to find their ways into both public and private clouds and to access corporate networks (Li and Clark 2013; Taylor 2015).
- *Ghostware*: This is malware designed to break into network systems, steal confidential information, and then cover up its traces. It will be extremely difficult for victim companies to trace the attack or track the amount of compromised data (Taylor 2015).
- *Two-faced malware:* This is malware that appears "safe" when presented to surveillance such as anti-virus, but smoothly transforms into malicious codes once it escapes detection (Taylor 2015).
- *Headless worm*: This is malware that can propagate from device to device, using malicious codes to target "headless devices," such as smartwatches, smartphones, and medical hardware that is freely connected with little security (Milosevic et al. 2017; Taylor 2015). These worms can be targeted toward additive manufacturing machines to sabotage the firmware or parts.

2.4.5 Cybersecurity Tools and Approaches

New protection plans are being proposed to enhance security, such as mobile device management, next-generation firewalls, change management, patch management, traffic analysis, and security information and event management (Defense Systems 2010; Gupta et al. 2010; Kemmerer 2003; Lippmann et al. 2002; Staff 2013). Some of these approaches are listed below:

- *Mobile device management:* use remote monitoring, policy enforcement, and patch installation to strengthen mobile devices security (Staff 2013). The strategy is to remotely block and restrict access to any lost, stolen, or compromised mobile devices and even erase all data if needed (Staff 2013).
- *Next-generation firewalls*: to provide high application visibility and control based on standard firewall inspection capabilities through application-awareness features (Staff 2013).

- *Layered approach*: to create multiple layers of security within the environment to enhance security across the organization (Gupta et al. 2010). In layered approach, each layer of security is made to be autonomous and self-sufficient to detect and block attacks (Gupta et al. 2010). Therefore, communication overhead among different layers and the central decision-making process is eliminated to reduce overall detection time, especially when compared to methods such as decision trees (Gupta et al. 2010).
- *Change management:* to improve network security through changes in monitoring, alerting, backups, and rollbacks (Defense Systems 2015).
- *Firewall management:* to provide detailed information about connectivity and configuration issues or identify any firewall rules inconsistencies to further prevent them (Defense Systems 2015). Firewall management should also offer change tracking to ensure ongoing firewall compliance (Defense Systems 2015).
- *Patch management*: to ensure all users, desktops, and server applications are up to date by continuously applying patches to keep up with newly discovered vulnerabilities (Defense Systems 2015). Automated patch management and update can help save time and keep the system up to date (Lippmann et al. 2002) (Defense Systems 2015).
- *Traffic analysis:* it can help in conducting forensics when exposed to cyber threat (Defense Systems 2015). The traffic analysis tool should find out the history and logs, IP address, and communication history (Defense Systems 2015).
- *Security information and event management*: it is a log and event management tool (Defense Systems 2015).

Deployment of appropriate tools can help in protecting sensitive information. Decision on selecting the right tools for the additive manufacturing process chain should be taken considering the value of the component that is designed by additive manufacturing.

2.5 SUMMARY

The current manufacturing scenario of additive manufacturing is discussed with a special focus on security concerns of this digital-based cyber–physical additive manufacturing system. Vulnerabilities and threat models of additive manufacturing are explored and discussed to understand the potential attacks from different levels. Traditional cyber-security attacks and strategies are introduced and compared with potential attacks in additive manufacturing. The reality is that all computers are vulnerable to compromise, such as to authorized users (insiders) who abuse their privileges and external people with malicious intent. New security methodologies at the CAD level are proposed and described in depth with the help of design examples in this chapter to use as a second line of defense in case the network or cybersecurity is breached and the files are stolen for unauthorized reproduction. The presence of these security features in the CADs will introduce defects in the parts printed using these files. Only a specific set of conditions used to process these files will generate the high-quality part. Such design-based security methodology is extremely useful in the connected cyber–physical system of additive manufacturing.

ACKNOWLEDGMENTS

The authors thank NYU-TSE Dean of Engineering Dr. Sreenivasan and Vice Dean for Research, Innovation, and Entrepreneurship Dr. Kurt Becker for providing institutional fellowship to Fei Chen to work on this project. In addition, NYU Global Seed Grants for Collaborative Research to Gupta and Shahin is acknowledged. Dr. Ramesh Karri of the Electrical and Computer Engineering Department at NYU is thanked for useful discussions. NYU Makerspace and 3D printing facility at NYUAD are acknowledged for printing the parts. Steven E. Zeltmann is thanked for assistance in manuscript preparation.

REFERENCES

Akella, R., H. Tang and B. M. McMillin (2010). Analysis of information flow security in cyber–physical systems. *International Journal of Critical Infrastructure Protection* **3**(3–4): 157–173.

Amin, M. (2002). Security challenges for the electricity infrastructure. *Computer* **35**(4): supl8–supl10.

Asadi-Eydivand, M., M. Solati-Hashjin, A. Farzad and N. A. A. Osman (2016). Effect of technical parameters on porous structure and strength of 3D printed calcium sulfate prototypes. *Robotics and Computer-Integrated Manufacturing* **37**: 57–67.

ASTM International (2013). *ISO/ASTM52921-13, Standard Terminology for Additive Manufacturing—Coordinate Systems and Test Methodologies*. West Conshohocken, PA.

ASTM International (2015). *ASTM ISO/ASTM52900-15 Standard Terminology for Additive Manufacturing—General Principles—Terminology*. West Conshohocken, PA.

ASTM International (2016). *ASTM E8/E8M-16a Standard Test Methods for Tension Testing of Metallic Materials*. West Conshohocken, PA, ASTM International.

Atighetchi, M., P. Pal, F. Webber and C. Jones (2003). Adaptive use of network-centric mechanisms in cyber-defense. Sixth IEEE International Symposium on Object-Oriented Real-Time Distributed Computing, Hakodate, Hokkaido, Japan.

Baheti, R. and H. Gill (2011). Cyber-physical systems. *The Impact of Control Technology* **12**: 161–166.

Boatman, K. (2017). Everything You Need to Know About Firewalls. Retrieved *June* 30, 2017.

Bradshaw, S., A. Bowyer and P. Haufe (2010). The intellectual property implications of low-cost 3D printing. *ScriptEd* **7**(1): 5–31.

Burmester, M., E. Magkos and V. Chrissikopoulos (2012). Modeling security in cyber–physical systems. *International Journal of Critical Infrastructure Protection* **5**(3–4): 118–126.

Campbell, T. A. and O. S. Ivanova (2013). Additive manufacturing as a disruptive technology: Implications of three-dimensional printing. *Technology & Innovation* **15**(1): 67–79.

Carl, G., G. Kesidis, R. R. Brooks and S. Rai (2006). Denial-of-service attack-detection techniques. *IEEE Internet Computing* **10**(1): 82–89.

Chen, B. X. and N. Perlroth. (2013). U.S. accuses 13 hackers in web attacks. Retrieved June 30, 2017, from http://www.nytimes.com/2013/10/04/technology/us-accuses-13-hackers-in-web-attacks.html.

Chen, F., G. Mac and N. Gupta (2017). Security features embedded in computer aided design (CAD) solid models for additive manufacturing. *Materials & Design* **128**: 182–194.

Chhetri, S. R., A. Canedo and M. A. Al Faruque (2016a). KCAD: Kinetic cyber attack detection method for cyber-physical additive manufacturing systems. Proceedings of the 35th International Conference on Computer-Aided Design, Austin, TX, USA, Association for Computing Machinery (ACM).

Chhetri, S. R., S. Faezi, A. Canedo and M. A. Al Faruque (2016b). Thermal side-channel forensics in additive manufacturing systems. 2016 ACM/IEEE 7th International Conference on Cyber-Physical Systems (ICCPS), Vienna, Austria, IEEE.

Chhetri, S. R., S. Faezi and M. A. A. Faruque (2017). Fix the leak! An information leakage aware secured cyber-physical manufacturing system. IEEE/ACM Design Automation and Test in Europe (DATE 2017), Lausanne, Switzerland, IEEE.

Dalziel, H. (2013). Summary of 5 major DOS attack types. Retrieved June 30, 2017, from https://www.concise-courses.com/5-major-types-of-dos-attack/.

Deborah, G. and M. Kelly. (2016). The additive risk of additive manufacturing. Retrieved February 2, 2017, from http://www.industryweek.com/strategic-planning-execution/additive-risk-additive-manufacturing.

Defense Systems (2015). Long live traditional security! Retrieved June 30, 2017, from https://defensesystems.com/articles/2015/08/26/dolisy-comment-traditional-security-dead.aspx.

Denning, R. and D. Elizabeth (1982). *Cryptography and Data Security*. Addison-Wesley Longman Publishing Co., Inc.

Do, Q., B. Martini and K. K. R. Choo (2016). A data exfiltration and remote exploitation attack on consumer 3D printers. *IEEE Transactions on Information Forensics and Security* **11**(10): 2174–2186.

Dolgunov, B. (2007). Enabling optimal security for removable storage devices. Fourth International IEEE Security in Storage Workshop, San Diego, CA, USA, IEEE.

Eddy, W. M. (2006). Defenses against TCP SYN flooding attacks. *The Internet Protocol Journal* **9**(4): 2–16.

Ericsson, G. N. (2010). Cyber security and power system communication—Essential parts of a smart grid infrastructure. *IEEE Transactions on Power Delivery* **25**(3): 1501–1507.

Faruque, A., M. Abdullah, S. R. Chhetri, A. Canedo and J. Wan (2016). Acoustic side-channel attacks on additive manufacturing systems. 2016 ACM/IEEE 7th International Conference on Cyber-Physical Systems (ICCPS), Vienna, Austria IEEE.

Gao, W., Y. Zhang, D. Ramanujan, K. Ramani, Y. Chen, C. B. Williams, C. C. L. Wang, Y. C. Shin, S. Zhang and P. D. Zavattieri (2015). The status, challenges, and future of additive manufacturing in engineering. *Computer-Aided Design* **69**: 65–89.

Garg, A., J. Curtis and H. Halper (2003). Quantifying the financial impact of IT security breaches. *Information Management & Computer Security* **11**(2): 74–83.

Genge, B., C. Siaterlis, I. Nai Fovino and M. Masera (2012). A cyber-physical experimentation environment for the security analysis of networked industrial control systems. *Computers & Electrical Engineering* **38**(5): 1146–1161.

Gibson, I., D. W. Rosen and B. Stucker (2015). *Additive Manufacturing Technologies: 3D Printing, Rapid Prototyping, and Direct Digital Manufacturing*. Springer.

Griffiths, J. (2015). Cybercrime costs the average U.S. firm $15 million a year. CNN. CNNTech.

Guo, N. and M. C. Leu (2013). Additive manufacturing: Technology, applications and research needs. *Frontiers of Mechanical Engineering* **8**(3): 215–243.

Gupta, K. K., B. Nath and R. Kotagiri (2010). Layered approach using conditional random fields for intrusion detection. *IEEE Transactions on Dependable and Secure Computing* **7**(1): 35.

Gupta, N., F. Chen, N. G. Tsoutsos and M. Maniatakos (2017). INVITED: ObfusCADe: Obfuscating additive manufacturing CAD models against counterfeiting. *54th ACM/EDAC/IEEE Design Automation Conference (DAC)*. Austin, TX, USA.

Guri, M., M. Monitz and Y. Elovici (2017). Bridging the air gap between isolated networks and mobile phones in a practical cyber-attack. *ACM Transactions on Intelligent Systems and Technology* **8**(4): 1–25.

Hahnlen, R. and M. J. Dapino (2010). Active metal-matrix composites with embedded smart materials by ultrasonic additive manufacturing. Industrial and Commercial Applications of Smart Structures Technologies, San Diego, California, USA.

Hancox, A. L. and J. A. McDaniel (2009). Additive manufacturing of precious metal dental restorations. *International Journal of Powder Metallurgy* **45**(5): 43.

Huang, S. H., P. Liu, A. Mokasdar and L. Hou (2013). Additive manufacturing and its societal impact: A literature review. *The International Journal of Advanced Manufacturing Technology* **67**(5–8): 1191–1203.

Ivanova, O., A. Elliott, T. Campbell and C. B. Williams (2014). Unclonable security features for additive manufacturing. *Additive Manufacturing* **1–4**: 24–31.

Jia, Q. and D. Gu (2014). Selective laser melting additive manufacturing of Inconel 718 superalloy parts: Densification, microstructure and properties. *Journal of Alloys and Compounds* **585**: 713–721.

Kemmerer, R. A. (2003). Cybersecurity. 25th International Conference on Software Engineering, Portland, OR, USA.

Krotofil, M., A. Cárdenas, J. Larsen and D. Gollmann (2014). Vulnerabilities of cyber-physical systems to stale data—Determining the optimal time to launch attacks. *International Journal of Critical Infrastructure Protection* **7**(4): 213–232.

Kurfess, T. and W. J. Cass (2014). Rethinking additive manufacturing and intellectual property protection. *Research-Technology Management* **57**(5): 35–42.

Lehmhus, D., C. Aumund-Kopp, F. Petzoldt, D. Godlinski, A. Haberkorn, V. Zöllmer and M. Busse (2016). Customized smartness: A survey on links between additive manufacturing and sensor integration. *Procedia Technology* **26**: 284–301.

Leuders, S., M. Thöne, A. Riemer, T. Niendorf, T. Tröster, H. Richard and H. Maier (2013). On the mechanical behaviour of titanium alloy TiAl6V4 manufactured by selective laser melting: Fatigue resistance and crack growth performance. *International Journal of Fatigue* **48**: 300–307.

Lever, R. (2016). Olli, a 3D printed, self-driving minibus, to hit the road in US. Retrieved February 2, 2017, from https://phys.org/news/2016-06-olli-3d-self-driving-minibus-road.html.

Li, Q. and G. Clark (2013). Mobile security: A look ahead. *IEEE Security & Privacy* **11**(1): 78–81.

Lim, S., R. A. Buswell, T. T. Le, S. A. Austin, A. G. F. Gibb and T. Thorpe (2012). Developments in construction-scale additive manufacturing processes. *Automation in Construction* **21**: 262–268.

Lippmann, R., S. Webster and D. Stetson (2002). The effect of identifying vulnerabilities and patching software on the utility of network intrusion detection. The 5th International Symposium of RAID (Recent Advances in Intrusion Detection), Zurich, Switzerland, Springer.

Manfredi, D., F. Calignano, M. Krishnan, R. Canali, E. P. Ambrosio, S. Biamino, D. Ugues, M. Pavese and P. Fino (2014). Additive manufacturing of Al alloys and aluminium matrix composites (AMCs). *Light Metal Alloys Applications* **11**: 3–34.

McCue, T. (2016). Wohlers report 2016: 3D printing industry surpassed $5.1 billion. Retrieved Feb. 2, 2017, from http://www.forbes.com/sites/tjmccue/2016/04/25/wohlers-report-2016-3d-printer-industry-surpassed-5-1-billion/#5b647e607cb1.

Melchels, F. P., M. A. Domingos, T. J. Klein, J. Malda, P. J. Bartolo and D. W. Hutmacher (2012). Additive manufacturing of tissues and organs. *Progress in Polymer Science* **37**(8): 1079–1104.

Michael, K. (2013). AutoCAD malware: Rare but malignant. Retrieved February 2, 2017, from http://www.techrepublic.com/blog/it-security/autocad-malware-rare-but-malignant/.

Milosevic, J., F. Regazzoni and M. Malek (2017). Malware threats and solutions for trustworthy mobile systems design. *Hardware Security and Trust* 149–167.

Mitchell, D. (2005). The rootkit of all evil. Retrieved Feb 4, 2017, from http://www.nytimes.com/2005/11/19/business/media/the-rootkit-of-all-evil.html.

Moore, S. B., W. B. Glisson and M. Yampolskiy (2017). Implications of malicious 3D printer firmware. Proceedings of the 50th Hawaii International Conference on System Sciences, Waikoloa, HI.

Murr, L. E., S. Gaytan, A. Ceylan, E. Martinez, J. Martinez, D. Hernandez, B. Machado, D. Ramirez, F. Medina and S. Collins (2010). Characterization of titanium aluminide alloy components fabricated by additive manufacturing using electron beam melting. *Acta Materialia* **58**(5): 1887–1894.

Murthy, C. S. R. and B. S. Manoj (2004). Transport layer and security protocols for ad hoc wireless networks. *Ad Hoc Wireless Networks: Architectures and Protocols*. Prentice Hall PTR, Prentice Hall PTR.

Murthy, U., O. Bukhres, W. Winn and E. Vanderdez (1998). Firewalls for security in wireless networks. Proceedings of the Thirty-First Hawaii International Conference on System Sciences, Kohala Coast, HI, USA, USA, IEEE

Nagaraju Kilari, D. R. S. (2012). A survey on security threats for cloud computing. *International Journal of Engineering Research & Technology* **1**(7): 1–10.

Navrotsky, V., A. Graichen and H. Brodin (2015). Industrialisation of 3D printing (additive manufacturing) for gas turbine components repair and manufacturing. *VGB PowerTech* **12**: 48–52.

Neuman, C. (2009). Challenges in security for cyber-physical systems. DHS Workshop on Future Directions in Cyber-Physical Systems Security, Newark, NJ, USA.

Ning, F., W. Cong, J. Qiu, J. Wei and S. Wang (2015). Additive manufacturing of carbon fiber reinforced thermoplastic composites using fused deposition modeling. *Composites Part B: Engineering* **80**: 369–378.

O'Connor, C. (2014). How a jewelry company is making $250,000 pieces using 3D printing and Google Earth. Retrieved February 1, 2017, from http://www.forbes.com/sites/clareoconnor/2014/02/28/how-a-jewelry-company-is-making-250000-pieces-using-3d-printing-and-google-earth/#2d0d7da5ed7b.

Parthasarathy, J., B. Starly and S. Raman (2011). A design for the additive manufacture of functionally graded porous structures with tailored mechanical properties for biomedical applications. *Journal of Manufacturing Processes* **13**(2): 160–170.

Perlroth, N. (2016). Yahoo says hackers stole data on 500 million users in 2014. *The New York Times*.

Peterson, A. (2014). The Sony Pictures hack, explained. *The Washington Post*.

Ponemon Institute (2016). 2016 Cost of cyber crime study & the risk of business innovation.

Quan, Z., A. Wu, M. Keefe, X. Qin, J. Yu, J. Suhr, J.-H. Byun, B.-S. Kim and T.-W. Chou (2015). Additive manufacturing of multi-directional preforms for composites: Opportunities and challenges. *Materials Today* **18**(9): 503–512.

Sachs, E., M. Cima and J. Cornie (1990). Three-dimensional printing: Rapid tooling and prototypes directly from a CAD model. *CIRP Annals-Manufacturing Technology* **39**(1): 201–204.

Saunders, S. (2017). Local Motors, Deutsche Bahn launch six-month self-driving Olli vehicle pilot program on Euref Campus in Berlin. Retrieved February 2, 2017, from https://3dprint.com/161739/deutsche-bahn-olli-pilot-program/.

Schuba, C. L., I. V. Krsul, M. G. Kuhn, E. H. Spafford, A. Sundaram and D. Zamboni (1997). Analysis of a denial of service attack on TCP. *IEEE Symposium on Security and Privacy Proceedings*, Oakland, California, USA, IEEE.

Sing, S. L., J. An, W. Y. Yeong and F. E. Wiria (2016). Laser and electron-beam powder-bed additive manufacturing of metallic implants: A review on processes, materials and designs. *Journal of Orthopaedic Research* **34**(3): 369–385.

Soutar, C., D. Roberge, A. Stoianov, R. Gilroy and B. V. Kumar (1999). Biometric encryption. *ICSA Guide to Cryptography*. R. K. Nichols. McGraw-Hill: 649–675.

Sridhar, S., A. Hahn and M. Govindarasu (2012). Cyber–physical system security for the electric power grid. *Proceedings of the IEEE* **100**(1): 210–224.

Staff, F. (2013). 6 network security tools every agency needs. Retrieved February 2, 2017, from http://www.fedtechmagazine.com/article/2013/09/6-network-security-tools-every-agency-needs.

Starr, M. (2016). Dubai unveils world's first 3D-printed office building. Retrieved February 2, 2017, from https://www.cnet.com/news/dubai-unveils-worlds-first-3d-printed-office-building/.

Sturm, L., C. Williams, J. Camelio, J. White and R. Parker (2014). Cyber-physical vunerabilities in additive manufacturing systems. *Context* **7**: 8.

Takaichi, A., T. Nakamoto, N. Joko, N. Nomura, Y. Tsutsumi, S. Migita, H. Doi, S. Kurosu, A. Chiba and N. Wakabayashi (2013). Microstructures and mechanical properties of Co–29Cr–6Mo alloy fabricated by selective laser melting process for dental applications. *Journal of the Mechanical Behavior of Biomedical Materials* **21**: 67–76.

Tata, K., G. Fadel, A. Bagchi and N. Aziz (1998). Efficient slicing for layered manufacturing. *Rapid Prototyping Journal* **4**(4): 151–167.

Taylor, H. (2015). Biggest cybersecurity threats in 2016. CNBC. CNBC Cybersecurity

Tekinalp, H. L., V. Kunc, G. M. Velez-Garcia, C. E. Duty, L. J. Love, A. K. Naskar, C. A. Blue and S. Ozcan (2014). Highly oriented carbon fiber–polymer composites via additive manufacturing. *Composites Science and Technology* **105**: 144–150.

Turner, H., J. White, J. A. Camelio, C. Williams, B. Amos and R. Parker (2015). Bad parts: Are our manufacturing systems at risk of silent cyberattacks? *IEEE Security & Privacy* **13**(3): 40–47.

Van Bael, S., G. Kerckhofs, M. Moesen, G. Pyka, J. Schrooten and J. P. Kruth (2011). Micro-CT-based improvement of geometrical and mechanical controllability of selective laser melted Ti6Al4V porous structures. *Materials Science and Engineering: A* **528**(24): 7423–7431.

Vincent, H., L. Wells, P. Tarazaga and J. Camelio (2015). Trojan detection and side-channel analyses for cyber-security in cyber-physical manufacturing systems. *Procedia Manufacturing* **1**: 77–85.

Wang, E. K., Y. Ye, X. Xu, S.-M. Yiu, L. C. K. Hui and K.-P. Chow (2010). Security issues and challenges for cyber physical system. Proceedings of the 2010 IEEE/ACM International Conference on Green Computing and Communications & International Conference on Cyber, Physical and Social Computing, Hangzhou, China.

Weiss, A. (2012). How to prevent DoS attacks. Retrieved February 2, 2017, from http://www.esecurityplanet.com/network-security/how-to-prevent-dos-attacks.html.

Wells, L. J., J. A. Camelio, C. B. Williams and J. White (2014). Cyber-physical security challenges in manufacturing systems. *Manufacturing Letters* **2**(2): 74–77.

Willis, K., E. Brockmeyer, S. Hudson and I. Poupyrev (2012). Printed optics: 3D printing of embedded optical elements for interactive devices. The 25th annual ACM Symposium on User Interface Software and Technology, Cambridge, Massachusetts, USA, ACM.

Wohlers, T. T. (2016). *Wohlers Report 2016: 3D Printing and Additive Manufacturing State of the Industry*. Wohlers Associates.

Yampolskiy, M., T. R. Andel, J. T. McDonald, W. B. Glisson and A. Yasinsac (2014). Intellectual property protection in additive layer manufacturing: Requirements for secure outsourcing. Proceedings of the 4th Program Protection and Reverse Engineering Workshop, New Orleans, LA, Association for Computing Machinery (ACM).

Yampolskiy, M., L. Schutzle, U. Vaidya and A. Yasinsac (2015). Security challenges of additive manufacturing with metals and alloys. International Conference on Critical Infrastructure Protection, Arlington, Virginia, USA, Springer.

Yampolskiy, M., A. Skjellum, M. Kretzschmar, R. A. Overfelt, K. R. Sloan and A. Yasinsac (2016). Using 3D printers as weapons. *International Journal of Critical Infrastructure Protection* **14**: 58–71.

Ye, N., J. Giordano and J. Feldman (2001). A process control approach to cyber attack detection. *Communications of the ACM* **44**(8): 76–82.

Zaeh, M. and M. Ott (2011). Investigations on heat regulation of additive manufacturing processes for metal structures. *CIRP Annals-Manufacturing Technology* **60**(1): 259–262.

Zeltmann, S. E., N. Gupta, N. G. Tsoutsos, M. Maniatakos, J. Rajendran and R. Karri (2016). Manufacturing and security challenges in 3D printing. *JOM* **68**(7): 1872–1881.

3 Use of a Robotized Laser Powder-Feed Metal Additive Manufacturing System for Fabricating Metallic Metamaterials

Yaoyu Ding, Meysam Akbari, Xin-Lin Gao, Li Ai, and Radovan Kovacevic

CONTENTS

3.1 INTRODUCTION

Metamaterials are materials engineered to have properties that are naturally unavailable, such as a zero/negative Poisson's ratio [1,2], a bi-stable structure with a negative stiffness [3,4], a non-positive coefficient of thermal expansion (CTE) [5–9], and a controllable frequency band gap [10–14]. These abnormal properties hold great potential for developing novel products to be used in wave filtering [15–17], acoustic cloaking [18,19], vibration control [20–23], and energy harvesting [24–27].

A number of manufacturing methods have been developed for fabricating polymeric metamaterials, which include projection micro-stereo-lithography [28], direct ink writing [29,30], electrophoretic deposition [31,32], and two-photon lithography [33]. Compared to polymeric metamaterials, metallic metamaterials can find even broader applications in various industries because of their inherent high stiffness and strength at room and elevated temperatures. In most of the existing fabrication processes for metallic metamaterials, the first step is to decompose a designed structure into small units and then join them through pins [34], adhesive, welding [35], or pressure-fit joints [36]. Such a joining process often requires manual work and can result in concentrations of stress.

As a fast-developing technology, additive manufacturing, based on a layer-by-layer approach, provides a new avenue to fabricate metallic metamaterials that are impossible to produce using traditional manufacturing methods. Currently, the commonly used metal additive manufacturing techniques include selective laser melting (SLM) [37–39], selective laser sintering (SLS) [40], and electron beam melting (EBM) [41,42]. In these processes, metal powders are positioned in the work area, and an energy source (laser beam or electron beam) is used to sinter or melt the powders layer by layer. These two methods have been adopted to print metallic metamaterials such as (a) auxetic cellular structures [43,44], (b) porous metals [45–48], and (c) tensegrity prism structures [49]. The small feature thickness of a layer (20–100 μm) and the tiny beam diameter (~140 μm) allow SLS and EBM to be used to build complex structures with high geometrical accuracy. But the powder bed fashion limits the capacity of fabricating metallic metamaterials with two or more materials such as (a) co-continuous composites (also known as interpenetrating phase composite) [50–54] and (b) functionally graded composites [55–58].

As an alternative approach, the robotized laser powder-feed metal additive manufacturing system has recently been developed for fabricating metallic metamaterials [59,60] at the Research Center for Advanced Manufacturing (RCAM) at Southern Methodist University. The system forms a molten pool on the substrate by using a laser beam and feeds metal powders into the molten pool. The powder-feed module generates an accurate flow of an individual powder or a mixture of two or more metal powders at an adjustable rate. This unique feature makes it prominent for fabricating metallic metamaterials from two or more materials in a continuous and automated deposition process.

To utilize the robotized laser powder-feed metal additive manufacturing system for the purpose of fabrication of metallic metamaterials, the following three issues need to be addressed: (i) the preliminary experiments for different materials, (ii) process planning, and (iii) monitoring and control of the fabrication process. They are described in detail in Sections 3.2.1 through 3.2.3. This will be followed by a discussion on metallic metamaterials with an enlarged positive Poisson's ratio, a negative Poisson's ratio, a bi-stable property, a non-positive CTE, and frequency band gaps, which were additively manufactured using the robotized laser powder-feed metal additive manufacturing system developed at RCAM.

3.2 ROBOTIZED LASER POWDER-FEED METAL ADDITIVE MANUFACTURING SYSTEM

The schematic and photo of the robotized laser powder-feed metal additive manufacturing system are shown in Figure 3.1. It mainly includes (a) a 4-kW fiber laser with a wavelength of 1070 nm, (b) a six-axis robot arm coupled with an additional two-axis tilt and rotatory positioning system, (c) a powder delivery module, and (d) a laser head equipped to the end of the robot arm. During the addition process, two powder feeders developed by RCAM [61,62] feed two different powders independently, as needed. The opto-electronic sensor monitors the powder flow rate at the outlet of the powder feeder. The design of the powder feeder and the opto-electronic sensor are detailed in Section 3.2.3.1. The carrier gas (argon) feeds the powder particles into an annular cone nozzle and injects them into the molten pool formed by the laser beam. The shielding gas (argon) is directed through the nozzle toward the molten pool in order to protect the molten material from contact with the atmosphere. A charge-coupled camera equipped with an infrared filter is installed on the laser head to monitor the molten pool. A firewire 1394 adapter is used to capture the infrared images. The National Instruments (PCI-6221) data control board was used to control the laser powder and powder flow rate by sending analog signals to the laser control box and servomotor of the powder feeder. LabVIEW was used to implement image processing [63,64], control tasks, and end user interface design.

Figure 3.2 shows the procedure for fabricating metamaterials by using the robotized laser powder-feed metal additive manufacturing system. The step of *metamaterial design* determined the types of

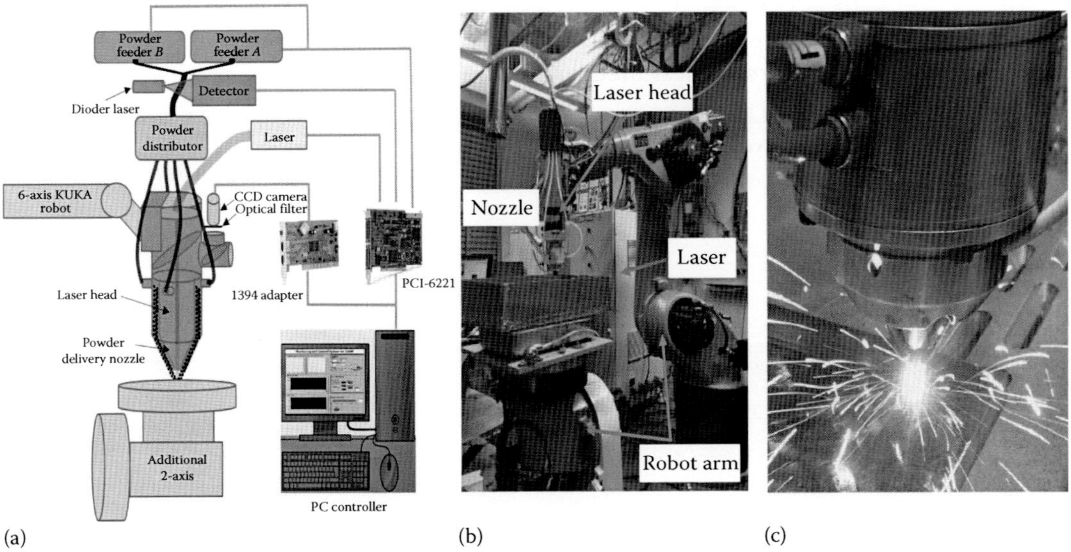

FIGURE 3.1 Setup of the robotized laser powder-feed metal additive manufacturing system: (a) schematic, (b) photo, and (c) a zoomed-in view of the laser head during a fabrication process.

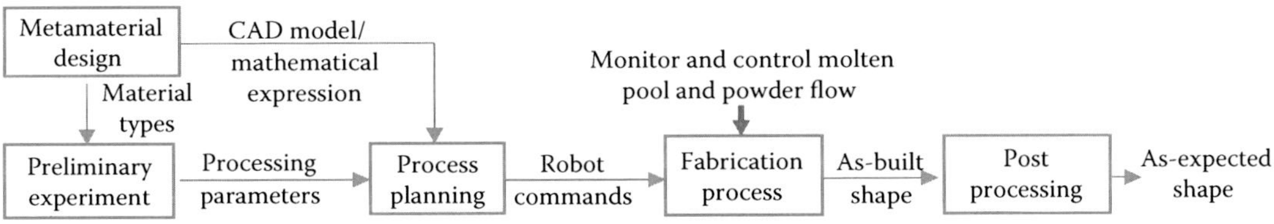

FIGURE 3.2 Procedure for fabricating metamaterials by using the robotized laser powder-feed metal additive manufacturing system.

materials and their distribution described by either *computer-aided design (CAD) models or mathematical expressions*. For different materials, *preliminary experiments* were conducted to find suitable *processing parameters* such as laser power, scanning speed, and powder-feed rate. Then, the *process planning* was conducted to generate the *robot commands* (robot motions and laser/powder/gas operations) by using a *CAD model/mathematical expression* of the metamaterial and the *processing parameters* as the inputs. During the *fabrication process*, both the *powder-feed rate and the molten pool were monitored and controlled* for the purpose of achieving a controllable and repeatable process. Finally, the shape accuracy of the *as-built shapes* could be improved as needed, by adopting a *postprocessing* method, such as (a) machining, (b) milling, and (c) polishing. Sections 3.2.1 through 3.2.3 will detail the steps of preliminary experiments, process planning, and the fabrication process, respectively.

3.2.1 Preliminary Experiments

Several parameters need to be determined for fabricating metamaterials, which include (a) scanning speed, (b) laser power, (c) powder feeding rate, (d) shielding gas flow rate, (e) z increment (i.e., the distance the laser head is lifted along the height direction after finishing one layer), and (f) y offset distance (i.e., the distance between two adjacent tracks in a layer). They are interrelated and need to be tuned by trial and error for building desirable buildups having un-oxidized surfaces and

TABLE 3.1
Processing Parameters for Stainless Steel and Invar

Powder Type	Scanning Speed (mm/s)	Laser Power (W)	Powder Feed Rate (g/s)	Carrier Gas (L/min)	Shielding Gas (L/min)	*Z* Increment (mm)	*y* Offset (mm)
Stainless steel	20	600	0.45	7	12	0.4	1.25
Invar	20	500	0.75	7	12	0.4	1.25

FIGURE 3.3 (a) A single-bead wall and (b) prismatic blocks built by feeding powder of stainless steel with the processing parameters detailed in Table 3.1.

as-expected shapes. Generally, single-bead walls and prismatic blocks were built to find the processing parameters for a certain kind of material. For metamaterials with multiple materials, different sets of processing parameters need to be tuned for the different materials due to the difference in the physical and metallurgical properties. In this chapter, powders of stainless steel and invar are used for the fabrications of the designed metamaterials. Preliminary experiments have been conducted to find desirable sets of processing parameters for stainless steel and invar that are specified in Table 3.1. Figure 3.3 shows a single-bead wall and prismatic blocks built by feeding stainless steel with the processing parameters detailed in Table 3.1.

3.2.2 Process Planning

Process planning aims to calculate the trajectories of the robot arm and schedule the laser/powder/gas operations for metamaterials described by the three-dimensional CAD models or mathematical expressions. Different from the metal additive manufacturing technology having a powder bed fashion that uses a tiny laser beam to scan the layers in a raster fashion [65], the robotized laser powder-feed metal additive manufacturing system has a large beam size (~0.6 mm), which could result in large stair-step effects along the boundaries of the layers if the raster fashion is adopted for filling the layers. A few other filling strategies have been studied, such as (i) zigzag [66], (ii) contour [67], (iii) spiral [68], (iv) continuous [69], and (v) hybrid [70]. The determination of filling pattern for a certain metamaterial is a highly geometry-dependent task. The difficulty of process planning is further compounded by shifting operation of powders during the fabrication processes of metamaterials having multiple materials. No commercial process planning software is capable of planning the fabrication processes of metamaterials by using the robotized laser powder-feed metal additive manufacturing system. RCAM has been developing a MATLAB®-based process planning software for this robotized laser powder-feed metal additive manufacturing system [71–73]. Section 3.3 provides solutions for several representative metamaterials.

3.2.3 Fabrication Process

3.2.3.1 Feeding of Powder

A controllable powder delivery module ensures a stable and consistent fabrication process. The key element for achieving control over powder delivery is to detect powder flow rate in real time. The powder delivery system in RCAM is equipped with two homemade powder feeders and an opto-electronic sensor developed for detecting the powder flow rate. As shown in Figure 3.4, the powder feeder mainly consists of a hopper, a rotating disk, and a suction device. The hopper stores the powder that is fed continuously onto the rotating disk placed under the hopper at a prescribed gap h. Pressurized argon aerates the chamber of the powder feeder. The powder on the disk is continuously sucked out through the suction device due to a pressure difference between the inside of the chamber of the powder feeder and the atmosphere. Given a suitable pressure difference (>10 psi), the powder flow rate is linearly related to angular velocity of the disk that is controlled by the servomotor.

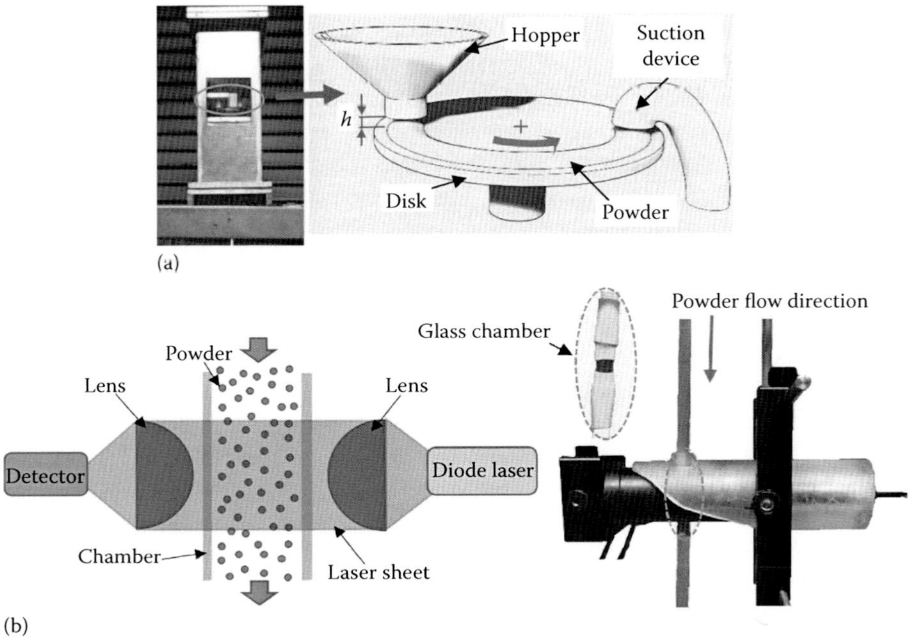

FIGURE 3.4 (a) Powder feeder and (b) optoelectronic sensor. (From Ding, Y., Warton, J., & Kovacevic, R. 2016. Development of sensing and control system for robotized laser-based direct metal addition system. *Additive Manufacturing* 10, 24–35.)

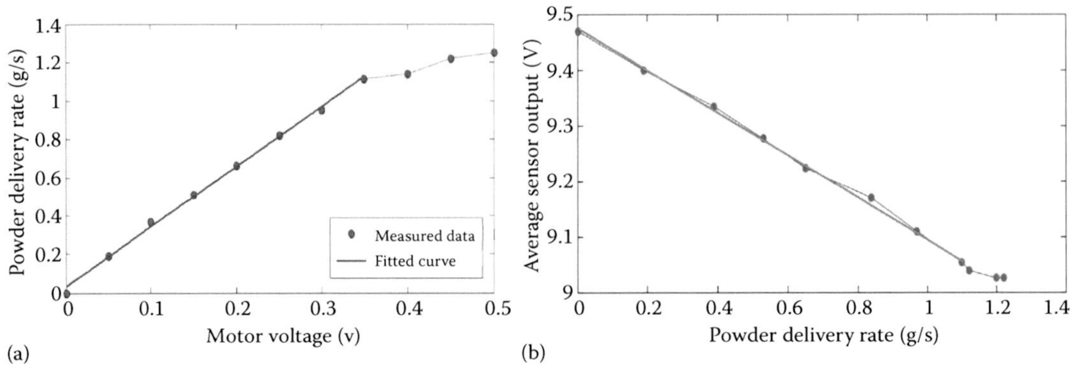

FIGURE 3.5 (a) Relationship between the motor voltage and the powder flow rate and (b) relationship between the powder flow rate and the averaged sensor voltage. (From Ding, Y., Warton, J., & Kovacevic, R. 2016. Development of sensing and control system for robotized laser-based direct metal addition system. *Additive Manufacturing* 10, 24–35.)

By controlling the rotary speed of the servomotor, the developed powder feeder is capable of achieving a high delivery resolution and a high scalability.

The relationship between powder flow rate and motor voltage was calibrated as shown in Figure 3.5a. The maximum powder flow rate for the current design of the powder feeder is approximately 1.2 g/s. The powder feeder exhibits a good linear relationship between the powder flow rate and the motor voltage when the powder flow rate is less than 1.1 g/s. In most of the fabrication processes of metamaterials, the powder flow rate is less than 1.1 g/s.

An optoelectronic sensor has been developed to detect the powder flow rate at the outlet of the powder feeder. As shown in Figure 3.4b, the developed optoelectronic sensor consists of a diode laser with a line generator that emits a thin defocused light sheet with a wavelength of 658 nm and power less than 500 mW, a photo diode, a small rectangular glass chamber (6.4 mm wide, 2.2 mm thick, and 19.5 mm long), and a set of lenses. The corresponding lens collimates the laser beam in the form of a line before passing through the glass chamber. Subsequently, the laser beam that passes through the glass chamber is focused at the photo diode. Because of diffusion, absorption, and reflection of the powder stream passing through the glass chamber, the amount of light detected by the photo diode will change once the density of powder changes. The photo diode is characterized by good linearity between the photo energy it detects and the voltage it gives out. In order to build a relationship between the voltage signal of the sensor and actual powder flow rate, the voltage signal of the sensor for different powder flow rates are acquired and shown in Figure 3.5b. A first-order polynomial is used to match the powder flow rates r (g/s) with the output voltages of the sensor h (V). The fitting result is obtained as

$$h(r) = -0.377 \times r + 9.475. \tag{3.1}$$

3.2.3.2 Closed-Loop Control of Molten Pool Size

During the fabrication process, the molten pool size varies at different positions due to varying cooling boundary conditions. This could result in a nonuniform shape of the buildup. Figure 3.7 shows an "L" shape single-bead wall built without control. High buildups present at the two ends because the heat conduction at the two ends was limited to one side of the built wall and a larger molten pool was formed there. It can also be seen from the horizontal cross-section of the buildup that the wall is thinner at the bottom, essentially because of a heat sink effect of the substrate at the beginning of the process. To build metamaterials having high shape accuracy and uniformity, a

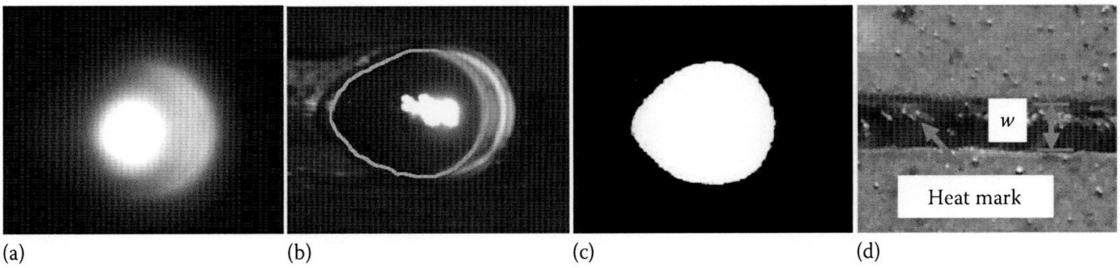

FIGURE 3.6 Captured top full-field images under same scanning conditions: (a) infrared image, (b) video image, (c) the contour of the molten pool on infrared image, and (d) heat mark on the substrate. (From Ding, Y., Warton, J., & Kovacevic, R. 2016. Development of sensing and control system for robotized laser-based direct metal addition system. *Additive Manufacturing* 10, 24–35.)

machine vision–based closed-loop control system was built to monitor and control the size of the molten pool during the fabrication process [61–74]. In this system, an infrared imaging setup was installed at the laser head to monitor the size of the molten pool. It included a high frame rate charge-coupled camera, an infrared filter (>695 nm), an iris, and a set of optical mirrors that were used to guide the light from the molten pool to the charge-coupled camera, as well as an infrared notch filter that was used to block the laser light with a wavelength of 1070 nm. The high frame rate charge-coupled camera was installed at the side of the laser head (see Figure 3.1a). With the help of a set of optical mirrors, a full-field view of the molten pool was acquired. An infrared filter was installed in front of the camera to eliminate the issue of the presence of powder above the molten pool. Based on Planck's law, a particular grayscale contour (isotherm) on the infrared image could represent the actual contour of the molten pool. To determine the isotherm on the infrared image that corresponded with contour of the molten pool, the images from both the video and infrared cameras were overlapped. Figure 3.6a shows the infrared image of the molten pool when scanning the substrate without powder (laser power: 400 W, scanning speed: 20 mm/s). The video image of the molten pool under identical scanning conditions was also captured coaxially by replacing the infrared filter with a band pass optical filter (532 nm) and illuminating the molten pool with a green laser (532 nm) at a power of 5 W. As shown in Figure 3.6b, contour of the molten pool in the video image is clear. By overlapping the video image over the infrared one, the contour of the molten pool on the infrared image can be determined by the isotherm having a specific value of a gray level. In this case, a gray level of 97 on the infrared image represented the contour of the molten pool (see Figure 3.6c).

To achieve a uniform size of molten pool during the fabrication process, an infrared image acquisition system took images of the molten pool in real time. The pixel number inside the molten pool was used as the feedback to adjust the laser power. In this study, a simple proportional-integral-derivative controller was used to build the closed-loop control system. The same "L" shape single-bead walls were built with closed-loop control. As shown in Figure 3.7, a uniform geometry was achieved in both the vertical and horizontal cross sections. There was no buildup at the ends and change-of-direction point. The developed closed-loop control system of the molten pool made a good performance in achieving uniform shapes. This closed-loop control system for molten pool laid a solid foundation for building metamaterials having complex geometries.

3.3 FABRICATION OF METALLIC METAMATERIALS

To verify the feasibility of fabricating metallic metamaterials by the developed robotized laser powder-feed metal additive manufacturing system, five representative metallic metamaterials were fabricated. They were designed for an enlarged positive Poisson's ratio (# 1), a negative Poisson's ratio (# 2), a bi-stable property (# 3), a non-positive CTE (# 4), and frequency band gap (# 5), respectively.

FIGURE 3.7 "L" shape single-bead walls with and without control of the molten pool size: (a) photo, (b) horizontal cross-section of the buildup, and (c) vertical cross-section of the buildup. (From Ding, Y., Warton, J., & Kovacevic, R. 2016. Development of sensing and control system for robotized laser-based direct metal addition system. *Additive Manufacturing* 10, 24–35.)

3.3.1 Design of Metallic Metamaterials

As shown in Figure 3.8 (# 1), a rhomboid shape was chosen as the cell shape to assemble a metallic metamaterial (# 1) with an enlarged Poisson's ratio. The size of the cell shape is specified in Figure 3.8 (# 1, b). The thickness of the cell wall was 1.2 mm. Numerical simulations have shown that this metamaterial displays an average Poisson's ratio of 0.94 in the elastic range, which was significantly enhanced compared to 0.275, the original Poisson's ratio for this kind of stainless steel bar. The metallic metamaterial (# 2) with a negative Poisson's ratio shown in Figure 3.8 was inspired by a cellular material having a negative Poisson's ratio of −0.8 presented in Ref. [75]. Figure 3.8 (# 2,b) shows the size of the unit cell. A novel metallic metamaterial (# 3) for achieving a bi-stable property was designed, as shown in Figure 3.8 (# 3). A ball was initially placed inside a curved rim. By pulling the ball up along a central line, the curved rim expanded sideways. Finally, the entire structure reached a stable state once the ball completely entered the up part of the curved rim. Such a structure is capable of dissipating most of the energy inserted into the system during loading and retaining the deformed shape after unloading. The metallic metamaterial (# 4) [9] was designed to attain a non-positive CTE, where it was numerically shown that this structure can exhibit both a non-positive CTE and a negative Poisson's ratio having suitable geometrical parameters. In this study, the lattice structure with $H_1 = 2$mm, $H_2 = 10$mm, and $\theta = 25°$ was fabricated. The red struts were made from a stainless steel having a higher CTE, while the green ones were made from invar having a lower CTE. Figure 3.8 (# 5) shows a unit cell of a co-continuous metamaterial having a body-centered cubic structure (# 5). This kind of co-continuous metamaterial can exhibit simultaneous wave filtering capability coupled with enhanced mechanical properties [54]. The frequency of complete band gaps can be attained by tailoring the geometrical arrangements and volume fraction of the co-continuous metamaterials. The interface between the two phases is controlled by the equation: $f(x, y, z) = 3(\cos(x) + \cos(y) + \cos(z)) + 4\cos(x)\cos(y)\cos(z) - t = 0$, where x, y, z are the Cartesian coordinates of points inside the cube and t denotes the volume fraction of the composite. In the current design, the space with $f(x, y, z) > 0$ is filled by invar, while the space with $f(x, y, z) \leq 0$ is filled by stainless steel. The value of t was chosen to be −0.7, for which the stainless steel phase achieved a volume fraction of 35%. The length of the cube edge was scaled up from 6.28 to 20 mm.

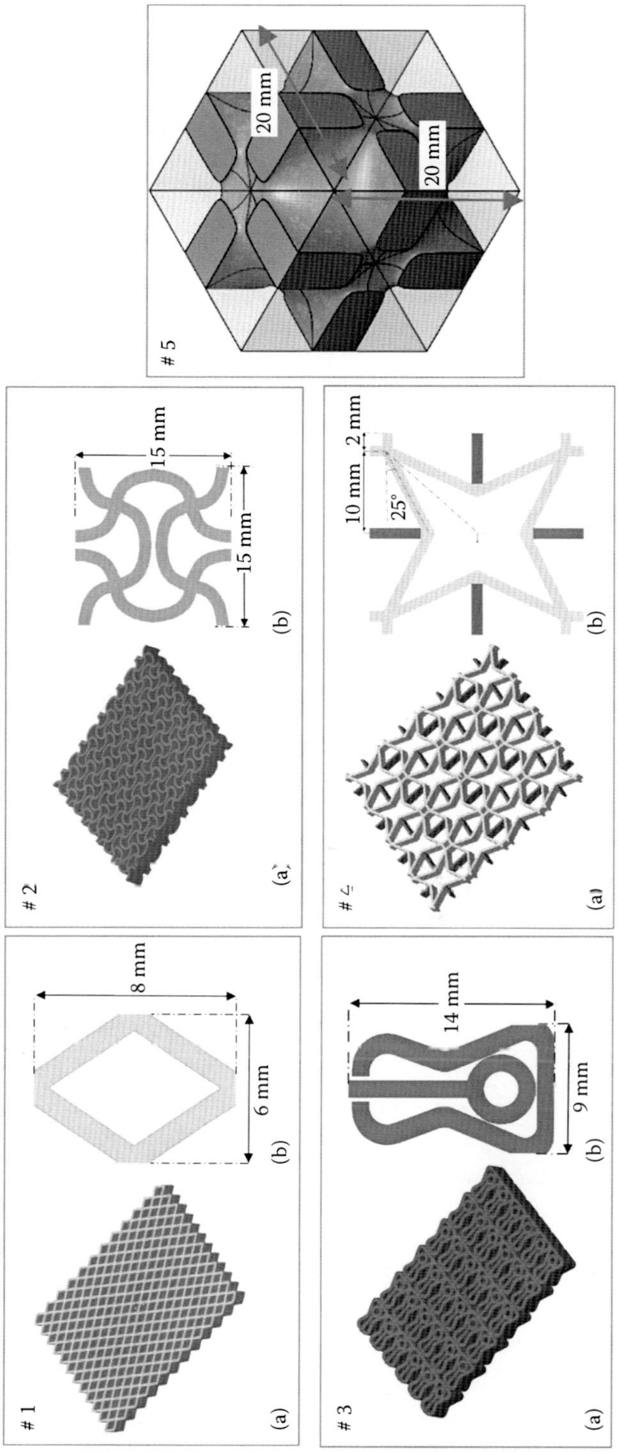

FIGURE 3.8 Five metamaterials designed: (a) model of entire structures and (b) cell shape and size.

3.3.2 Fabrication Process

A 2-D medial axis [76] was computed for metamaterials # 1, # 2, # 3, and # 4 as the scanning trajectories, respectively, as shown in Figure 3.9. Metamaterials # 1, # 2, and # 3 were fabricated by feeding powders of stainless steel with the processing parameters listed in Table 3.1. The laser head moved along the solid lines while the laser power was kept on. When the nozzle came to the path segment denoted by a dashed line, the laser power was turned off. After one layer was built, the laser cladding head was then lifted for an increment of 0.4 mm along the height direction (the z axis). The next layer was built in the same manner as the previous one.

For each layer of metamaterial # 4 that is shown in Figure 3.9d, the struts made from invar were first deposited following the order I-1, I-2, I-3, …, I-8. Then, the powder feeder feeding the stainless steel powders was stopped, and the powder feeder feeding the invar powders started to feed the invar powder. The system waited 5 s after the shifting operation to ensure a stable and pure flow of the powder flow. Then, the system started to build the struts made from stainless steel following the order S-1, S-2, …, S-16. After one layer was built, the feeding of powders was shifted to invar. The same procedure was repeated for the next layer. The printing processes for metamaterials # 2 and # 4 are shown in Figure 3.10.

The slicing and tool-path generation processes for metamaterial # 5 are shown in Figure 3.11. The cross section of the boundary of each of the two material phases varies along the direction of height. The raster pattern was adopted to fill the layers of metamaterial # 5. The parallel red segments were deposited from $y = -10$ mm to 10 mm line by line by feeding invar. Then, the powder delivery system was shifted to feeding the stainless steel powders in a manner similar to that described for metamaterial # 4. The parallel blue segments were deposited in a similar manner. The build times were as follows: (i) 17 min for metamaterial # 1, (ii) 13 min for # 2, (iii) 19 min for # 3, (iv) 22 min for # 4, and (v) 27 min for # 5.

After the printing processes, the samples were milled flat on the top surface (and side surfaces for metamaterial # 5) and cut from the substrate. Figure 3.12 shows the five printed metamaterial samples. The printed shapes provide a good agreement with those given by their CAD models. A good bonding between the two different material phases was obtained in the metamaterial sample # 4 and metamaterial sample # 5. The pores were observed on the surfaces of sample # 5. This can be attributed to the poor wetting and expanding of the molten pool, which can be eliminated by

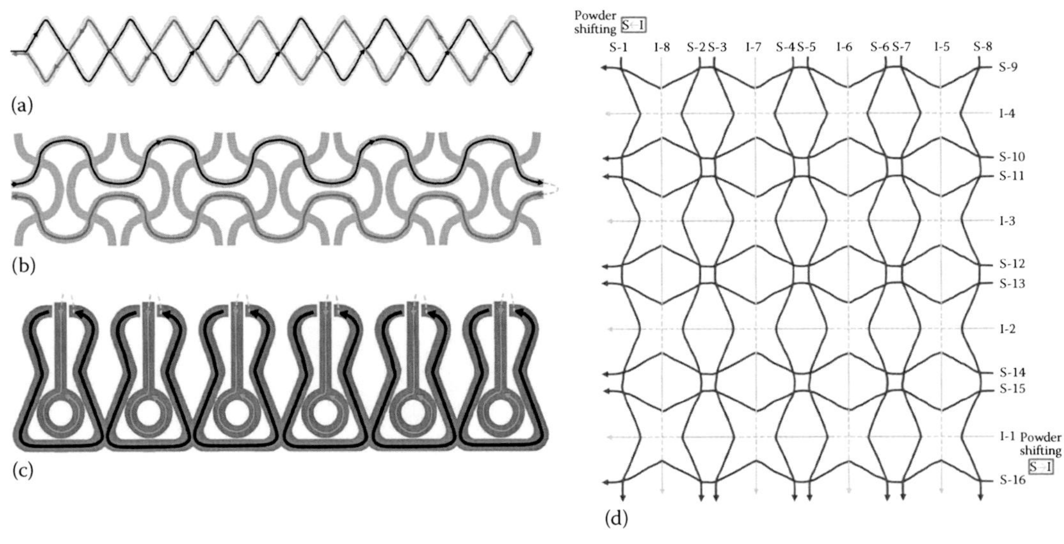

FIGURE 3.9 Process planning of the metallic metamaterials (a) # 1, (b) # 2, (c) # 3, and (d) # 4.

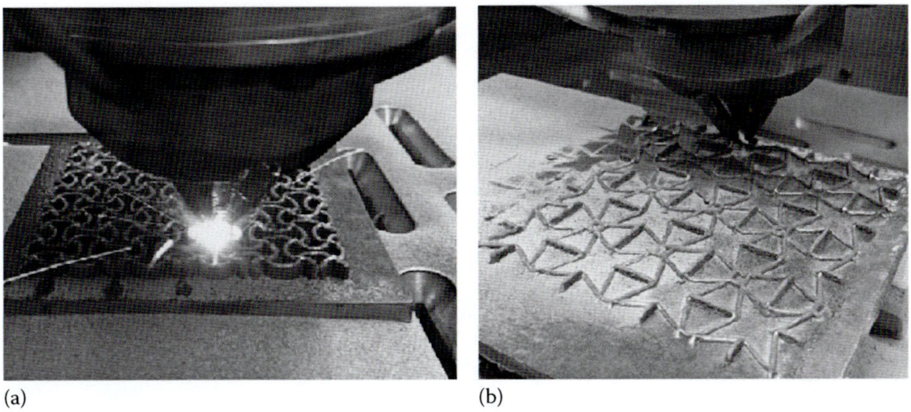

FIGURE 3.10 Fabrication process of (a) metamaterial # 2 and (b) metamaterial # 4.

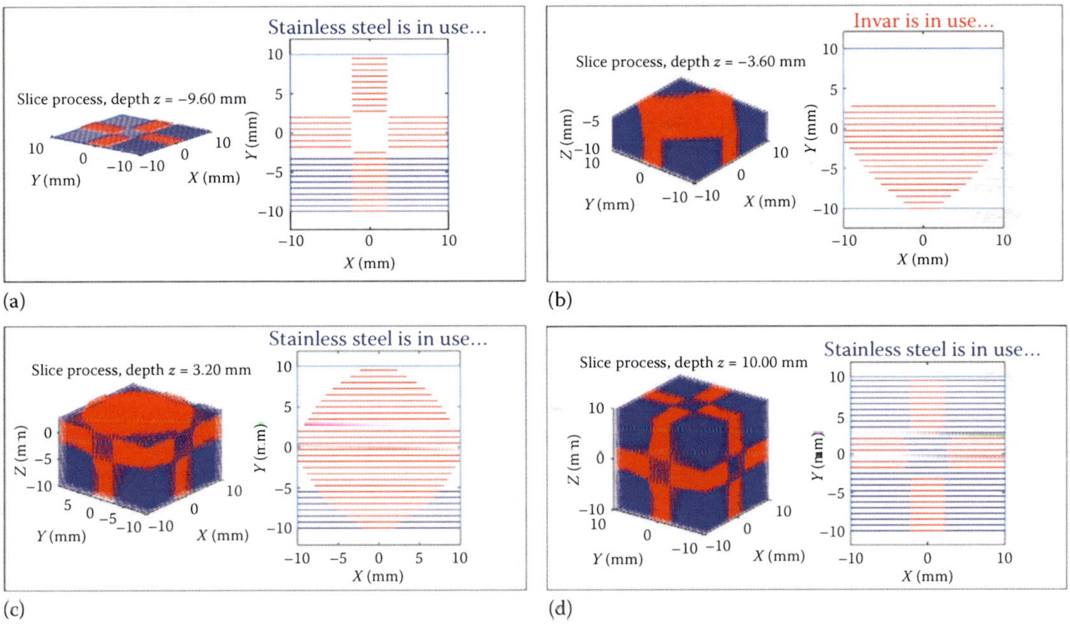

FIGURE 3.11 Process planning for metamaterial # 5 with four depths: (a) depth $z = -9.60$ mm, (b) depth $z = -3.60$ mm, (c) depth $z = 3.20$ mm, and (d) depth $z = 10.00$ mm.

integrating a vibration system that micro-vibrates the molten pool during the printing process [77]. Further mechanical tests of the printed metamaterials will be conducted and reported separately at a later date.

3.4 CONCLUSIONS

1. A robotized laser powder-feed metal additive manufacturing system has been developed for fabricating metallic metamaterials. The key issues addressed in detail here include the preliminary experiments for different materials, process planning for fabrication of metamaterials, and monitoring and control of the fabrication process.

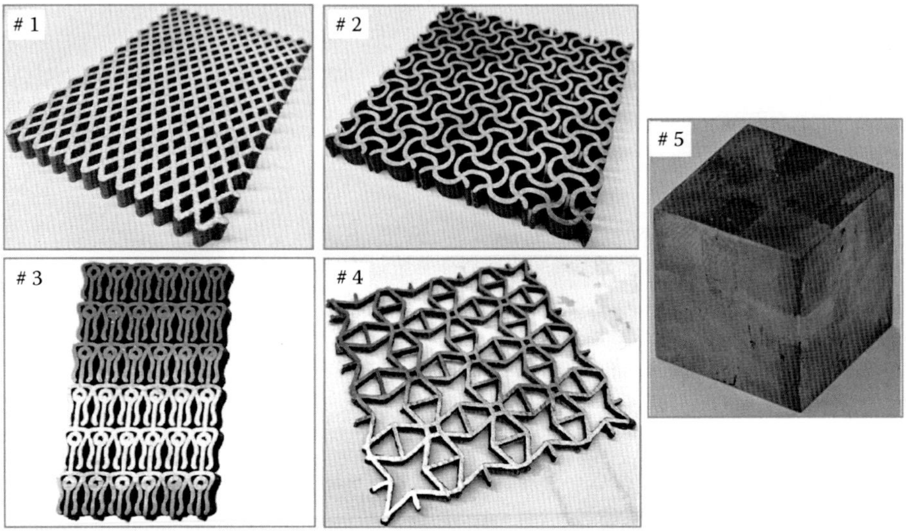

FIGURE 3.12 Printed samples of the five metallic metamaterials.

2. The robotized laser powder-feed metal additive manufacturing system has been successfully applied to fabricate five types of metallic metamaterials with an enlarged positive Poisson's ratio, a negative Poisson's ratio, a bi-stable property, a non-positive CTE, and frequency band gaps, respectively.
3. Challenges of this technology, including the verifications of the corresponding specific properties of the buildups, generalization of path planning software, and elimination of pores in the buildups, need to be further explored.

ACKNOWLEDGMENTS

The authors would like to thank Mr. Andrew Socha at the Research Center for Advanced Manufacturing at SMU for his assistance in this research. The financial support of the US National Science Foundation under grants # IIP-1539853 and CMMI-1463881 is gratefully acknowledged.

REFERENCES

1. Lakes, R. S. 2017. Negative-Poisson's-ratio materials: Auxetic solids. *Annual Review of Materials Research* 47(1).
2. Baughman, R. H., Shacklette, J. M., Zakhidov, A. A., & Stafström S. 1998. Negative Poisson's ratios as a common feature of cubic metals. *Nature* 392(6674), 362–365.
3. Kashdan, L., Conner Seepersad, C., Haberman, M., & Wilson P. S. 2012. Design, fabrication, and evaluation of negative stiffness elements using SLS. *Rapid Prototyping Journal* 18(3), 194–200.
4. Bilal, O. R., Foehr, A., & Daraio C. 2017. Bistable metamaterial for switching and cascading elastic vibrations. *Proceedings of the National Academy of Sciences* 201618314.
5. Mary, T. A., Evans, J. S. O., Vogt, T., & Sleight A. W. 1996. Negative thermal expansion from 0.3 to 1050 Kelvin in ZrW2O8. *Science* 272(5258), 90.
6. Dudek, K. K., Attard, D., Caruana-Gauci, R., Wojciechowski, K. W., & Grima J. N. 2016. Unimode metamaterials exhibiting negative linear compressibility and negative thermal expansion. *Smart Materials and Structures* 25(2), 025009.
7. Wang, Q., Jackson, J. A., Ge, Q., Hopkins, J. B., Spadaccini, C. M., & Fang N. X. 2016. Lightweight mechanical metamaterials with tunable negative thermal expansion. *Physical Review Letters* 117(17), 175901.

8. Ng, C. K., Saxena, K. K., Das, R., & Flores E. S. 2017. On the anisotropic and negative thermal expansion from dual-material re-entrant-type cellular metamaterials. *Journal of Materials Science* 52(2), 899–912.
9. Ai, L., & Gao X. L. 2017. Metamaterials with negative Poisson's ratio and non-positive thermal expansion. *Composite Structures* 162, 70–84.
10. Deymier, P. A. (Ed.). 2013. *Acoustic Metamaterials and Phononic Crystals* (Vol. 173). Springer Science & Business Media.
11. Christensen, J., Kadic, M., Kraft, O., & Wegener M. 2015. Vibrant times for mechanical metamaterials. *MRS Communications* 5(03), 453–462.
12. Krödel, S., Delpero, T., Bergamini, A., Ermanni, P., & Kochmann D. M. 2014. 3D auxetic microlattices with independently controllable acoustic band gaps and quasi-static elastic moduli. *Advanced Engineering Materials* 16(4), 357–363.
13. Matlack, K. H., Bauhofer, A., Krödel, S., Palermo, A., & Daraio C. 2016. Composite 3D-printed metastructures for low-frequency and broadband vibration absorption. *Proceedings of the National Academy of Sciences* 201600171.
14. Xia, B., Chen, N., Xie, L., Qin, Y., & Yu D. 2016. Temperature-controlled tunable acoustic metamaterial with active band gap and negative bulk modulus. *Applied Acoustics* 112, 1–9.
15. Shen, H., Lu, D., VanSaders, B., Kan, J. J., Xu, H., Fullerton, E. E., & Liu Z. 2015. Anomalously weak scattering in metal-semiconductor multilayer hyperbolic metamaterials. *Physical Review X* 5(2), 021021.
16. Pisano, G., Tucker, C., Ade, P. R., Moseley, P., & Ng, M. W. (2015, September). Metal mesh based metamaterials for millimetre wave and THz astronomy applications. In Millimeter Waves and THz Technology Workshop (UCMMT), 2015 8th UK, Europe, China (pp. 1–4). IEEE.
17. Zhang, X., Zhao, G., Xia, L., Wei, D., & Cui H. L. 2016. Band-stop filter characteristics of cross metal structures in terahertz frequency range. In International Symposium on Ultrafast Phenomena and Terahertz Waves (pp. IW1A–5). Optical Society of America.
18. Zigoneanu, L., Popa, B. I., & Cummer S. A. 2014. Three-dimensional broadband omnidirectional acoustic ground cloak. *Nature Materials* 13(4), 352–355.
19. Shen, C., Xu, J., Fang, N. X., & Jing Y. 2014. Anisotropic complementary acoustic metamaterial for canceling out aberrating layers. *Physical Review X* 4(4), 041033.
20. Matlack, K. H., Bauhofer, A., Krödel, S., Palermo, A., & Daraio C. 2016. Composite 3D-printed metastructures for low-frequency and broadband vibration absorption. *Proceedings of the National Academy of Sciences* 201600171.
21. Baravelli, E., & Ruzzene M. 2013. Internally resonating lattices for bandgap generation and low-frequency vibration control. *Journal of Sound and Vibration* 332(25), 6562–6579.
22. Wang, P., Casadei, F., Shan, S., Weaver, J. C., & Bertoldi K. 2014. Harnessing buckling to design tunable locally resonant acoustic metamaterials. *Physical Review Letters* 113(1), 014301.
23. Chen, Y., Li, T., Scarpa, F., & Wang L. 2017. Lattice metamaterials with mechanically tunable Poisson's ratio for vibration control. *Physical Review Applied* 7(2), 024012.
24. Cheng, Y. Z., Fang, C., Zhang, Z., Wang, B., Chen, J., & Gong R. Z. 2016. A compact and polarization-insensitive perfect metamaterial absorber for electromagnetic energy harvesting application. In Progress in Electromagnetic Research Symposium (PIERS) (pp. 1910–1914). IEEE.
25. Zhi Cheng, Y., Wang, Y., Nie, Y., Zhou Gong, R., Xiong, X., & Wang X. 2012. Design, fabrication and measurement of a broadband polarization-insensitive metamaterial absorber based on lumped elements. *Journal of Applied Physics* 111(4), 044902.
26. Carrara, M., Cacan, M. R., Toussaint, J., Leamy, M. J., Ruzzene, M., & Erturk A. 2013. Metamaterial-inspired structures and concepts for elastoacoustic wave energy harvesting. *Smart Materials and Structures* 22(6), 065004.
27. Wang, H., Sivan, V. P., Mitchell, A., Rosengarten, G., Phelan, P., & Wang L. 2015. Highly efficient selective metamaterial absorber for high-temperature solar thermal energy harvesting. *Solar Energy Materials and Solar Cells* 137, 235–242.
28. Zheng, X., Lee, H., Weisgraber, T. H., Shusteff, M., DeOtte, J., Duoss, E. B.,… & Kucheyev S. O. 2014. Ultralight, ultrastiff mechanical metamaterials. *Science* 344(6190), 1373–1377.
29. Lewis, J. A. 2002. Direct-write assembly of ceramics from colloidal inks. *Current Opinion in Solid State and Materials Science* 6(3), 245–250.
30. Smay, J. E., Cesarano, J., & Lewis J. A. 2002. Colloidal inks for directed assembly of 3-D periodic structures. *Langmuir* 18(14), 5429–5437.
31. Besra, L., & Liu M. 2007. A review on fundamentals and applications of electrophoretic deposition (EPD). *Progress in Materials Science* 52(1), 1–61.

32. Pascall, A. J., Qian, F., Wang, G., Worsley, M. A., Li, Y., & Kuntz J. D. 2014. Light-directed electrophoretic deposition: A new additive manufacturing technique for arbitrarily patterned 3D composites. *Advanced Materials* 26(14), 2252–2256.
33. Meza, L. R., Das, S., & Greer J. R. 2014. Strong, lightweight, and recoverable three-dimensional ceramic nanolattices. *Science* 345(6202), 1322–1326.
34. Steeves, C. A., e Lucato, S. L. D. S., He, M., Antinucci, E., Hutchinson, J. W., & Evans A. G. 2007. Concepts for structurally robust materials that combine low thermal expansion with high stiffness. *Journal of the Mechanics and Physics of Solids* 55(9), 1803–1822.
35. Gdoutos, E., Shapiro, A. A., & Daraio C. 2013. Thin and thermally stable periodic metastructures. *Experimental Mechanics* 53(9), 1735–1742.
36. Wei, K., Chen, H., Pei, Y., & Fang D. 2016. Planar lattices with tailorable coefficient of thermal expansion and high stiffness based on dual-material triangle unit. *Journal of the Mechanics and Physics of Solids* 86, 173–191.
37. Wang, X., and Chou, K. 2017. Effects of thermal cycles on the microstructure evolution of Inconel 718 during selective laser melting process. *Additive Manufacturing* 18, 1–14.
38. Wang, X., Keya, T., & Chou K. 2016. Build height effect on the Inconel 718 parts fabricated by selective laser melting. *Procedia Manufacturing*, 5, 1006–1017.
39. Wang, X., & Chou Y. K. 2015. A method to estimate residual stress in metal parts made by Selective Laser Melting. In ASME 2015 *International Mechanical Engineering Congress and Exposition*. American Society of Mechanical Engineers.
40. Bonatti, C., & Mohr D. 2017. Large deformation response of additively-manufactured FCC metamaterials: From octet truss lattices towards continuous shell mesostructures. *International Journal of Plasticity* (92), 122–147.
41. Murr, L. E., Gaytan, S. M., Ramirez, D. A., Martinez, E., Hernandez, J., Amato, K. N.,… & Wicker R. B. 2012. Metal fabrication by additive manufacturing using laser and electron beam melting technologies. *Journal of Materials Science & Technology* 28(1), 1–14.
42. Wang, X., Gong, X., & Chou K. 2015. Scanning speed effect on mechanical properties of Ti–6Al–4V alloy processed by electron beam additive manufacturing. *Procedia Manufacturing* 1, 287–295.
43. Xiong, J., Gu, D., Chen, H., Dai, D., & Shi Q. 2017. Structural optimization of re-entrant negative Poisson's ratio structure fabricated by selective laser melting. *Materials & Design* 120, 307–316.
44. Schwerdtfeger, J., Heinl, P., Singer, R. F., & Körner C. 2010. Auxetic cellular structures through selective electron-beam melting. *Physica Status Solidi (b)* 247(2), 269–272.
45. Jamshidinia, M., Wang, L., Tong, W., & Kovacevic R. 2014. The bio-compatible dental implant designed by using non-stochastic porosity produced by Electron Beam Melting® (EBM). *Journal of Materials Processing Technology* 214(8), 1728–1739.
46. Hernández-Nava, E., Smith, C. J., Derguti, F., Tammas-Williams, S., Léonard, F., Withers, P. J.,… & Goodall R. 2015. The effect of density and feature size on mechanical properties of isostructural metallic foams produced by additive manufacturing. *Acta Materialia* 85, 387–395.
47. van Grunsven, W., Hernandez-Nava, E., Reilly, G. C., & Goodall R. 2014. Fabrication and mechanical characterisation of titanium lattices with graded porosity. *Metals* 4(3), 401–409.
48. Ray, P., Chahine, G., Smith, P., & Kovacevic R. 2011. Optimal design of a golf club using functionally graded porosity. *The 22 International SFF Symposium—An Additive Manufacturing Conference* Aug. 8–10, 2011, Austin, TX.
49. Amendola, A., Hernández-Nava, E., Goodall, R., Todd, I., Skelton, R. E., & Fraternali F. 2015. On the additive manufacturing, post-tensioning and testing of bi-material tensegrity structures. *Composite Structures* 131, 66–71.
50. Zhu, W., Yan, C., Shi, Y., Wen, S., Liu, J., Wei, Q., & Shi Y. 2016. A novel method based on selective laser sintering for preparing high-performance carbon fibres/polyamide12/epoxy ternary composites. *Scientific Reports* 6.
51. Salmoria, G. V., Paggi, R. A., & Beal V. E. 2017. Graded composites of polyamide/carbon nanotubes prepared by laser sintering. *Lasers in Manufacturing and Materials Processing* 4(1), 36–44.
52. Wang, L., Lau, J., Thomas, E. L., & Boyce M. C. 2011. Co-continuous composite materials for stiffness, strength, and energy dissipation. *Advanced Materials* 23(13), 1524–1529.
53. Chen, Y., & Wang L. 2014. Tunable band gaps in bio-inspired periodic composites with nacre-like microstructure. *Journal of Applied Physics* 116(6), 063506.
54. Ai, L., & Gao X. L. 2016. Evaluation of effective elastic properties of 3D printable interpenetrating phase composites using the meshfree radial point interpolation method. *Mechanics of Advanced Materials and Structures* 1–11.

55. Liu, W., & DuPont J. N. 2003. Fabrication of functionally graded TiC/Ti composites by laser engineered net shaping. *Scripta Materialia* 48(9), 1337–1342.
56. Choy, S. Y., Sun, C. N., Leong, K. F., Tan, K. E., & Wei J. 2016. Functionally graded material by additive manufacturing. *Proceedings of the 2nd International Conference on Progress in Additive Manufacturing (Pro-AM 2016).*
57. Sharma, A., Bandari, V., Ito, K., Kohama, K., Ramji, M., & Sai H. 2017. A new process for design and manufacture of tailor-made functionally graded composites through friction stir additive manufacturing. *Journal of Manufacturing Processes* 26, 122–130.
58. Bobbio, L. D., Otis, R. A., Borgonia, J. P., Dillon, R. P., Shapiro, A. A., Liu, Z. K., & Beese A. M. 2017. Additive manufacturing of a functionally graded material from Ti–6Al–4V to invar: Experimental characterization and thermodynamic calculations. *Acta Materialia* (127), 133–142.
59. Kovacevic, R., & Valant M. E. 2006. U.S. Patent No. 7,020,539. Washington, DC: U.S. Patent and Trademark Office.
60. Ding, Y., & Kovacevic R. 2016. Feasibility study on 3-D printing of metallic structural materials with robotized laser-based metal additive manufacturing. *The Journal of the Minerals, Metals & Materials Society* 68(7), 1774–1779.
61. Ding, Y., Warton, J., & Kovacevic R. 2016. Development of sensing and control system for robotized laser-based direct metal addition system. *Additive Manufacturing* 10, 24–35.
62. Yarrapareddy, E., & Kovacevic R. 2008. Synthesis and characterization of laser-based direct metal deposited nano-particles reinforced surface coatings for industrial slurry erosion applications. *Surface and Coatings Technology* 202(10), 1951–1965.
63. Ding, Y., Huang, W., & Kovacevic R. 2016. An on-line shape-matching weld seam tracking system. *Robotics and Computer-Integrated Manufacturing* 42, 103–112.
64. Ding, Y., Zhang, X., & Kovacevic R. 2016. A laser-based machine vision measurement system for laser forming. *Measurement* 82, 345–354.
65. Wang, X., & Chou K. 2017. Electron backscatter diffraction analysis of Inconel 718 parts fabricated by selective laser melting additive manufacturing. *JOM* 69(2), 402–408.
66. Dwivedi, R., & Kovacevic R. 2004. Automated torch path planning using polygon subdivision for solid freeform fabrication based on welding. *Journal of Manufacturing Systems* 23(4), 278–291.
67. Yang, Y., Loh, H. T., Fuh, J. Y. H., & Wang Y. G. 2002. Equidistant path generation for improving scanning efficiency in layered manufacturing. *Rapid Prototyping Journal* 8(1), 30–37.
68. Hongcheck, W., Peter, J., & Stori James A. 2005. A metric-based approach to two-dimensional (2D) tool-path optimization for high-speed machining transactions of ASME. *Journal of Manufacturing Science and Engineering* 127(1), 33–48.
69. Chiu, W. K., Yeung, Y. C., & Yu K. M. 2006. Toolpath generation for layer manufacturing of fractal objects. *Rapid Prototyping Journal* 12(4), 214–221.
70. Zhang, Y. M., Li, P., Chen, Y., & Male A. T. 2002. Automated system for welding-based rapid prototyping. *Mechatronics* 12(1), 37–53.
71. Ding, Y., Dwivedi, R., & Kovacevic R. 2017. Process planning for 8-axis robotized laser-based direct metal deposition system: A case on building revolved part. *Robotics and Computer-Integrated Manufacturing* 44, 67–76.
72. Ding, Y., Akbari, M., & Kovacevic R. 2017 Process planning for laser wire-feed metal additive manufacturing system.*The International Journal of Advanced Manufacturing Technology* 1–11.
73. Akbari, M., Ding, Y., & Kovacevic, R. 2017, June. Process development for a robotized laser wire additive manufacturing. In ASME 2017 12th International Manufacturing Science and Engineering Conference collocated with the JSME/ASME 2017 6th International Conference on Materials and Processing (pp. V002T01A015–V002T01A015). American Society of Mechanical Engineers.
74. Kovacevic, R., Hu, D., & Valant M. E. 2006. U.S. Patent No. 6,995,334. Washington, DC: U.S. Patent and Trademark Office.
75. Clausen, A., Wang, F., Jensen, J. S., Sigmund, O., & Lewis J. A. 2015. Topology optimized architectures with programmable Poisson's ratio over large deformations. *Advanced Materials* 27(37), 5523–5527.
76. Lee, D. T. 1982. Medial axis transformation of a planar shape. *IEEE Transactions on Pattern Analysis and Machine Intelligence* (4), 363–369.
77. Foroozmehr, E., Lin, D., & Kovacevic R. 2009. Application of vibration in the laser powder deposition process. *Journal of Manufacturing Processes* 11(1), 38–44.

4 The Technology and Applicability of Laser Additive Manufacturing for In Situ Metal Matrix Composites and Functionally Graded Materials

Tushar Borkar and Rajarshi Banerjee

CONTENTS

4.1 INTRODUCTION

Additive manufacturing, also known as three-dimensional (3D) printing, has recently emerged as a subject of intense worldwide attention. Additive manufacturing is a freeform fabrication process by which an object is built up by depositing material layer by layer unlike conventional material removal methods. It now constitutes an integral part of the US national network for manufacturing innovation as well as related international efforts in the area of advanced manufacturing. The past three decades have seen significant strides in the development of novel additive manufacturing methods, including but not limited to powder bed fusion (PBF) techniques, such as selective laser melting (SLM) and selective laser sintering (SLS), as well as directed energy deposition (DED) techniques, such as laser engineered net shaping (LENS) or laser metal deposition (LMD), which will be the primary focus of this article.

Advances in the additive manufacturing technology have made it possible to manufacture complex-shaped metal components strong enough for real engineering applications [1–15]. While laser additive manufacturing is becoming more and more important in the context of advanced manufacturing for the future, most of the current efforts are focusing on optimizing the required parameters for processing well-matured alloys from powder feedstock to achieve reproducible properties, comparable to, or better than, their conventionally processed counterparts. However, laser additive manufacturing or processing also opens up a new horizon in terms of processing novel alloys and composites that are difficult to process using conventional techniques. According to the Wohlers Report 2015, the current additive manufacturing market of $4.1 billion is expected to reach $21.2 in 5 years [16]. An additive manufacturing system takes a computer-generated geometric model as its input and builds the geometry by depositing the constituent materials precisely in a layer-by-layer fashion [16–18]. There are more than 10 additive manufacturing techniques that have been developed so far, which include electron-beam melting, SLM, stereolithography, fused deposition modeling, and digital light processing, to name a few [16–24]. Because of the utilization of the layer-by-layer process, additive manufacturing systems are capable of creating geometrically complex prototypes and products efficiently in small to medium quantities. It is thus best suited for applications requiring complex, high-value, time-sensitive, and customized products such as automobile and aerospace parts (i.e., complex designs), broken part replacement (i.e., time-sensitive), and medical implants (i.e., highly customized) such as replacement hip joints. Some examples of the very complex geometries that additive manufacturing is capable of manufacturing are illustrated in Figure 4.1.

LENS is a freeform additive manufacturing process for near-net shaping of nearly fully dense, homogeneous bulk materials. The process begins with a computer-aided design (CAD) file of a 3D

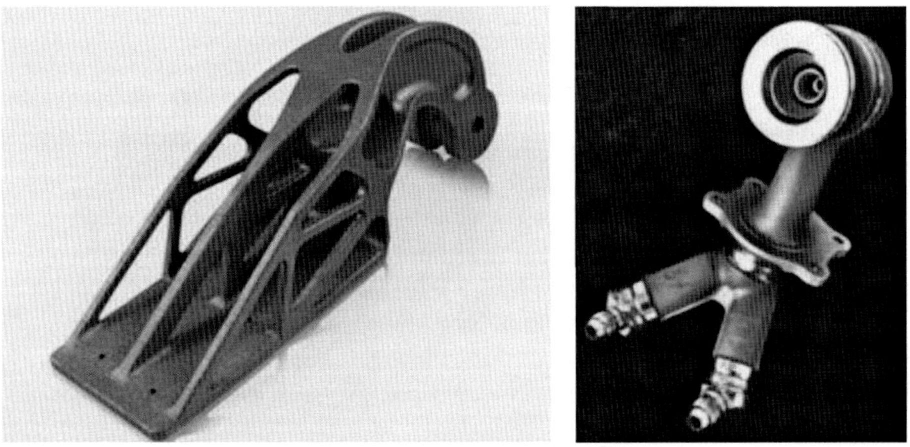

(a) (b)

FIGURE 4.1 Complex geometries' structural components that are redesigned using topology optimization for additive manufacturing: (a) an Airbus A320 nacelle hinge bracket by EADS and Airbus, and (b) a fuel nozzle for the LEAP jet engine by GE aviation. (Reprinted with kind permission from Springer Science+Business Media: *Additive Manufacturing Technologies: 3D Printing, Rapid Prototyping, and Direct Digital* Manufacturing, 2014, 471–472, Gibson et al.)

component, which is sliced into a series of layers with a predetermined layer spacing/thickness on the order of 0.01 in. (0.25 mm). Each layer contains a tool path, which is followed by the multi-axis stage, while pre-alloyed or blends of elemental powders are injected into a melt pool produced by a high-powered laser. The process continues via the sequential deposition of layers to develop the overall 3D shape of the component. Most of the worldwide research and development activities related to additive manufacturing, or 3D printing as it is often referred to, of metallic systems have focused on a rather limited set of alloys, such as stainless steels, certain nickel base superalloys, and conventional titanium alloys, such as Ti–6Al–4V (wt%), typically referred to as Ti–6–4 [12,24]. Furthermore, these alloys have all been processed using pre-alloyed powder (or pre-alloyed wire) feedstock with efforts directed to optimizing feedstock characteristics (morphology, microstructure, composition, impurity content, etc.) and deposition parameters to achieve additively manufactured components with properties comparable to wrought or conventional thermo-mechanically processed materials. While these efforts are critical in order to establish 3D printing/additive manufacturing as viable technologies for future manufacturing, the potential of these technologies is underexploited for the development of both functional and novel materials systems.

4.2 METAL MATRIX COMPOSITES

Metal matrix composites are the new class of advanced materials in which rigid ceramics reinforcements/ceramic fibers/ceramic particulates exhibiting excellent strength as well as elastic modulus are embedded in a ductile metal such as nickel, copper, aluminum, titanium, magnesium, iron, or alloy matrix to overcome the inadequacy of metals and alloys in providing both strength and stiffness to the structure [25–37]. In metal matrix composites, strength and ductility are provided by metal matrix and strength and stiffness are provided by reinforcements [20,32]. Metal matrix composites possess excellent physical as well as mechanical properties, such as high strength in shear and compression, high service temperature capabilities, high specific modulus, fatigue strength, and temperature stability, which make them suitable for structural applications, ground transportation, thermal management devices, and industrial (chemical and transportation), recreational, and infrastructure applications, as well as automotive and aerospace applications [30–32,36,38–41]. There is an increase in demand of metal matrix composites for automotive and aerospace applications due to the availability of inexpensive reinforcements and the development of various processing routes [36]. Metal matrix composites are widely used in electronic packaging applications because of their low coefficient of thermal expansion and higher thermal as well as electrical conductivity [30,31]. Because of excellent physical and mechanical properties, metal matrix composites are candidates for space applications when subjected to extreme conditions, such as vacuum, ionizing, radiation, and plasma [42–44]. Because of ease of fabrication procedures, lower cost, and near-isotropic properties, particulate reinforced metal matrix composites have received increasing attention in many industries [42,45]. Discontinuously reinforced metal matrix composites, also termed conventional or ex situ metal matrix composites, have been fabricated in ways such as mechanical alloying, spray deposition, powder metallurgy, squeeze casting, rheocasting, and compocasting [24–29]. During these techniques, ceramic reinforcements were externally added before composite fabrication into the matrix material (which may be in molten or powder form). The main disadvantages of these conventional ex situ metal matrix composites involve size limitation of reinforcing phase, which is nothing but the starting powder size, interfacial reaction between the reinforcement and matrix, and poor wettability between the reinforcements and the matrix due to surface contamination of reinforcements [36]. The physical and mechanical properties of metal matrix composites are primarily governed by size and volume fraction of reinforcement as well as the nature of the matrix reinforcement interface [36]. The uniform dispersion of fine and thermally stable ceramic particulates in the metal matrix is desirable for achieving optimum mechanical properties of metal matrix composites. This leads to the development of novel in situ metal matrix composites in which precipitates are synthesized in metallic matrix by a chemical reaction between elements or between elements and compounds during the composite fabrication. The in situ metal matrix composites exhibit many advantages over ex situ metal matrix composites [2,3,36,38]:

a. In situ formed reinforcements are thermodynamically stable and lead to less degradation at elevated temperatures.
b. Strong interfacial bonding between the matrix and the reinforcement attributed to clean matrix–reinforcement interface owing to the absence of any interfacial reaction between matrix and reinforcement.
c. Better physical and mechanical properties attributed to homogeneous dispersion of fine-scale reinforcements into matrix.

In situ formation of reinforcement is the promising fabrication route for processing metal matrix composites in terms of both technical and economic considerations. In situ metal matrix composites have better control over physical and mechanical properties because of their greater control on the size and level of reinforcements, as well as the matrix–reinforcement interface. Mechanical and physical properties of metal matrix composites are mainly governed by properties of the matrix, dispersion of the reinforcement, interfacial bonding between matrix and reinforcement, and finally the processing method [3,36]. Various processing routes have been developed as a result of the great potential and widespread applications of these in situ metal matrix composites that involve matrix materials (titanium, nickel, aluminum, and copper) and in situ reinforcements (carbides, nitrides, and borides) [3,36,43]. Laser additive processing of in situ metal matrix composites opens up a new horizon to process multifunctional monolithic metal matrix composites that are difficult to process via traditional manufacturing processing such as laser cladding. These metal matrix composites have been processed either by in situ reaction between elemental blend powders or by in situ reaction between elemental blend powders and reactive gases (nitrogen, oxygen, etc.) during laser additive processing. This section reviews few examples of in situ metal matrix composites processed via LMD processes.

4.2.1 In Situ Reaction between Elemental Blend Powders

4.2.1.1 Nickel–Titanium–Graphite Composites

A new class of Ni–Ti–C–based metal matrix composites has been developed using the LENS process. These composites consist of an in situ formed and homogeneously distributed titanium carbide (TiC) phase reinforcing the nickel matrix. Additionally, by tailoring the C/Ti ratio in these composites, an additional graphitic phase can also be engineered into the microstructure. Serial sectioning followed by 3D reconstruction of the microstructure in a new class of laser-deposited in situ Ni–Ti–C–based metal matrix composites reveals homogeneously distributed primary and eutectic titanium carbide precipitates as well as a graphitic phase encompassing the primary carbides, within a nickel matrix. The morphology and spatial distribution of these phases in three dimensions reveal that the eutectic carbides form a network linked by primary carbides or graphitic nodules at the nodes, suggesting interesting insights into the sequence of phase evolution. These three-phase Ni–TiC–C composites exhibit excellent tribological properties, in terms of an extremely low coefficient of friction while maintaining a relatively high hardness. Backscattered scanning electron microscopy (SEM) images of Ni–Ti–C composites with varying C/Ti ratio are shown in Figure 4.2.

The Ni–10Ti–5C (Figure 4.2a) composite exhibits only fine needle-like eutectic TiC precipitates whereas the Ni–10Ti–10C (Figure 4.2b) composite shows the presence of both fine needle-like eutectic as well as cuboidal primary TiC precipitates. As carbon-to-titanium ratio increases, both Ni–7Ti–20C (Figure 4.2c) and Ni–3Ti–20C (Figure 4.2d) composites show the presence of black graphite phase along with dark gray TiC precipitates. Bright-field transmission electron microscopy (TEM) images and electron diffraction patterns from the Ni–10Ti–10C composite are shown in Figure 4.3.

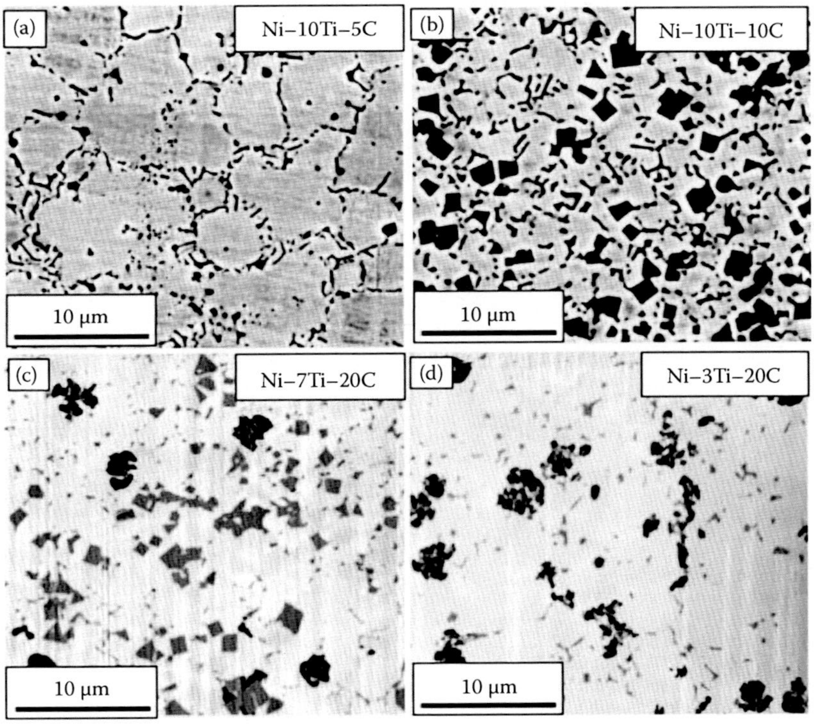

FIGURE 4.2 Backscatter SEM images of LENS-deposited (a) Ni–10Ti–5C, (b) Ni–10Ti–10C, (c) Ni–7Ti–20C, and (d) Ni–3Ti–20C composites. (Modified from Borkar et al. 2014. Laser-deposited in situ TiC-reinforced nickel matrix composites: 3D microstructure and tribological properties. *JOM* 66 (6): 935–942.)

Figure 4.3a shows overall microstructure with both primary and eutectic carbide precipitates. Higher-magnification bright-field images of the primary and eutectic carbides are shown in Figure 4.3b and c, respectively. Selected area diffraction (SAD) patterns from these precipitates are shown as insets in both these figures. These diffraction patterns are consistently indexed as [001] and [112] zone axis of the TiC phase exhibiting NaCl-type rock salt structure. The primary TiC phase exhibits a relatively coarse-faceted structure, while eutectic TiC exhibits a fine grain structure. Figure 4.3d is a high-resolution TEM image showing relatively flat and planer carbide/nickel interface with the lattice planes visible in both phases. The 3D microstructure of Ni–10Ti–10C and Ni–3Ti–20C composites is shown in Figure 4.4, where primary (cuboidal shaped) and eutectic (plate-shaped) TiC reinforcements in a Ni matrix have been reconstructed in 3D via focused ion beam (FIB)–based serial sectioning.

All the Ni–Ti–C composites exhibit very high microhardness as well as excellent tribological properties when compared to pure nickel. The Ni–10Ti–10C composite exhibits highest microhardness because of the presence of primary as well as eutectic TiC precipitates, whereas the Ni–3Ti–20C composite exhibits excellent tribological properties because of the presence of solid lubricious graphite phase along with TiC precipitates. 3D microstructural characterization of Ni–Ti–C composites revealed the following:

a. Different morphologies of primary and eutectic TiC precipitates.
b. Primary graphite phase engulfs TiC precipitates.
c. Primary TiC precipitates act as heterogeneous nucleation sites for the eutectic precipitates during solidification, leading to a carbide network formation.

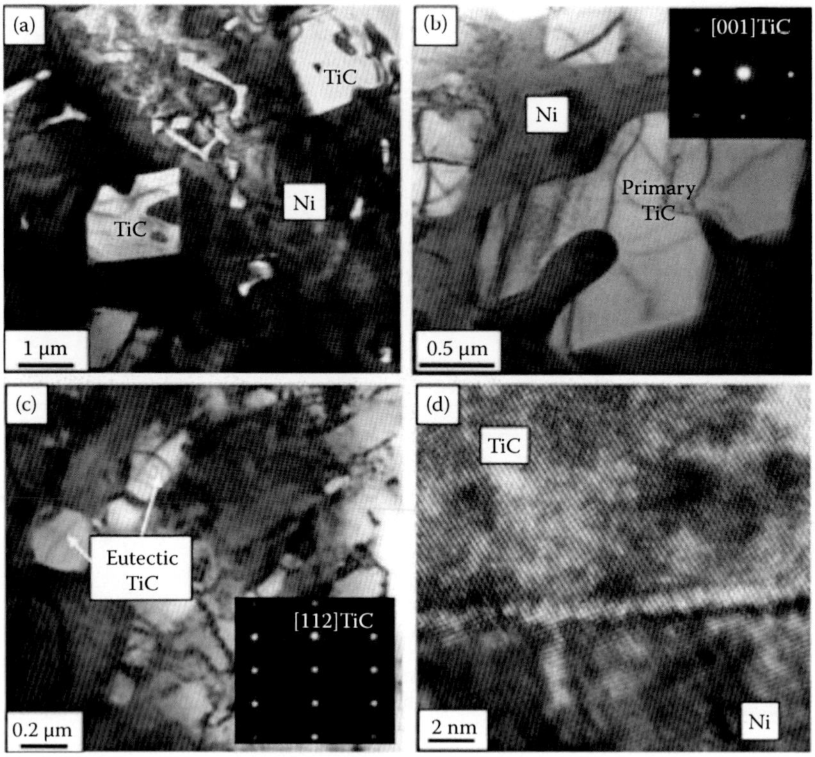

FIGURE 4.3 (a) Bright-field TEM image showing the TiC phase, (b) higher-magnification image showing primary TiC phase with SAD pattern in an inset, (c) higher magnification of eutectic TiC phase with SAD pattern in an inset, and (d) HRTEM image showing interface between Ni matrix and TiC. (Reprinted with permission from Gopagoni et al. 2011. Microstructural evolution in laser deposited nickel–titanium–carbon in situ metal matrix composites, *Journal of Alloys and Compounds* 509: 1255–1260.)

3D characterization leads to novel understanding of the sequence of phase evolution during solidification for these complex metal matrix composites. The connectivity between the carbide precipitates is nearly impossible to visualize based on the 2D SEM images shown in Figure 4.2. The distinction between the cuboidal primary TiC precipitates and the plate or needle-shaped eutectic TiC precipitates is more difficult in Ni–3Ti–20C composites because the primary precipitates are smaller in size as clearly shown in the 3D reconstruction. Figure 4.5 shows friction coefficient versus distance plot LENS-deposited Ni–10Ti–10C, Ni–3Ti–20C, and Ni–7Ti–20C composites and pure nickel.

It is clear that the graphite and TiC phases in the composite were beneficial toward reducing the friction coefficient with respect to the pure nickel sample. While the presence of TiC reduces the coefficient of friction, as observed in case of the Ni–10Ti–10C composite, the presence of the lubricious graphitic phase can play a more dominant role in reducing the friction for these composites. This is evident from the friction curves for the Ni–7Ti–20C and Ni–3Ti–20C composites. The friction coefficient in the case of Ni–7Ti–20C is marginally lower as compared to Ni–10Ti–10C because of the presence of the graphitic phase in the former. However, the most promising composite appears to be the Ni–3Ti–20C composite, which exhibits a drastic reduction in friction coefficient (~0.2) when compared to any of the other composites, mainly due to the presence of a substantial fraction of the graphitic phase as well as TiC precipitates. Thus, these Ni–Ti–C composites, especially the Ni–3Ti–20C composite, appear to be promising materials for high-temperature surface engineering applications requiring high hardness with improved solid lubrication.

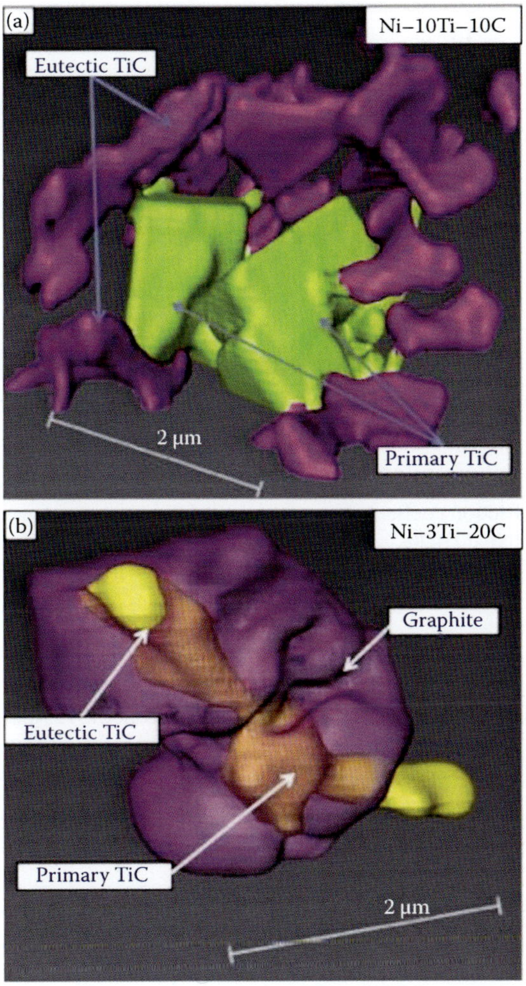

FIGURE 4.4 3D reconstruction of LENS-deposited (a) Ni–10Ti–5C and (b) Ni–3Ti–20 composites. (Modified from Borkar et al. 2014. Laser-deposited in situ TiC-reinforced nickel matrix composites: 3D microstructure and tribological properties. *JOM* 66 (6): 935–942.)

4.2.1.2 Titanium Alloy–Titanium Boride (TiB) Composites

Titanium and its alloys are widely used in biomedical, chemical, and aerospace industries because of their excellent corrosion resistance, mechanical properties, and biocompatibility [3]. However, Ti-based alloys exhibit poor wear resistance and low hardness, which restrict their field of applications. One way of improving the hardness as well as tribological properties of titanium alloys is by reinforcing the soft matrix with hard precipitates, such as titanium borides, carbides, and nitrides [3]. The titanium alloy–TiB composites combine the high strength and stiffness of the borides with the toughness and damage tolerance of the Ti alloy matrix. These composites have been extensively researched since they offer attractive properties, such as increased stiffness, enhanced elevated temperature strength, good creep performance, fatigue resistance, and wear resistance. In situ composites offer additional advantages. Since the boride reinforcement is formed as a consequence of a chemical reaction in such composites, a homogeneous dispersion consisting of refined-scale borides results. Furthermore, the boride phase that forms in these in situ composites is in thermodynamic equilibrium with the matrix. Unlike reinforcement phases added from external sources, in situ

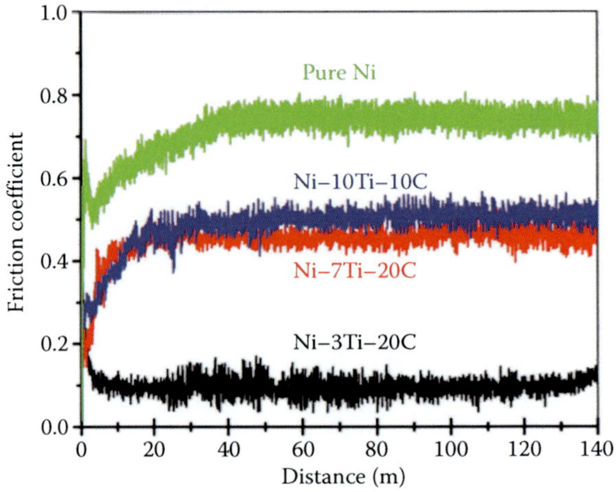

FIGURE 4.5 Steady-state friction coefficient as a function of sliding distance up to 140 m for LENS-deposited pure nickel, Ni–10Ti–10C, Ni–7Ti–20C, and Ni–3Ti–20C composites. (Modified from Borkar et al. 2014. Laser-deposited in situ TiC-reinforced nickel matrix composites: 3D microstructure and tribological properties. *JOM* 66 (6): 935–942.)

composites consist of contaminant-free boride–matrix interfaces, which are significantly stronger. Using conventional ingot metallurgy, it is rather difficult to achieve such a uniform distribution of boride particles because of the low solubility of boron in both α- and β-Ti and rapid particle growth during subsequent thermo-mechanical processing. The Ti–TiB composites have been processed via LENS process from a blend of pure elemental Ti and B powders while Ti alloy–TiB composites were deposited from a blend of pre-alloyed Ti–6Al–4V and elemental boron [46–50].

4.2.1.2.1 Ti–TiB Composites

The Ti–B binary phase diagram is shown in Figure 4.6 [51]. The eutectic in the binary Ti–B phase diagram occurs at a composition of Ti–1.6 wt% B at a temperature of 1540°C. A composition of Ti–2 wt% B lies in the hypereutectic region of this phase diagram. An equilibrium solidification path for an alloy of this composition will be as follows: Liquid → Liquid + Primary TiB → Primary TiB + Eutectic TiB + Eutectic β → Primary TiB + Eutectic TiB + α. A comparison of the microstructures of the arc-melted and LENS-deposited Ti–2B alloys is shown in the backscattered SEM images in Figure 4.7.

The microstructure of the arc-melted composite consists of primary pro-eutectic precipitates of TiB in a eutectic matrix (Figure 4.7a). The primary borides tend to exhibit faceted equiaxed morphology and are typically 20 ± 100 μm in size. The eutectic matrix consists of α-Ti (with relatively negligible solubility for B) and secondary precipitates of TiB (Figure 4.7a). The secondary TiB in the matrix exhibits a highly acicular morphology with diameter of the order of a micron and length ~10–20 μm. As discussed earlier, based on the Ti–B binary phase diagram [51], the formation of such a microstructure is an expected manifestation of equilibrium cooling of an alloy of composition Ti–2B from the liquid phase. In the case of the LENS-deposited Ti–2B alloy, the scale of the microstructure is substantially smaller, as shown in the backscattered SEM image in Figure 4.7b, viewed as the same magnification as Figure 4.7a. At higher magnifications, the backscattered SEM image (Figure 4.7c) exhibits fine precipitates of primary TiB. These primary borides exhibit both equiaxed as well as acicular morphologies. The average size of the equiaxed TiB precipitates are ~0.6 μm while the acicular precipitates typically exhibit an average length of 2.5 μm with an aspect ratio of 7:1. In addition to the primary TiB precipitates (exhibiting the darkest contrast), finer-scale TiB precipitates (exhibiting a lighter gray contrast) are also visible within the α-Ti matrix (Figure 4.7d). The formation of TiB by the reaction of elemental Ti and B is a highly exothermic process, the heat of

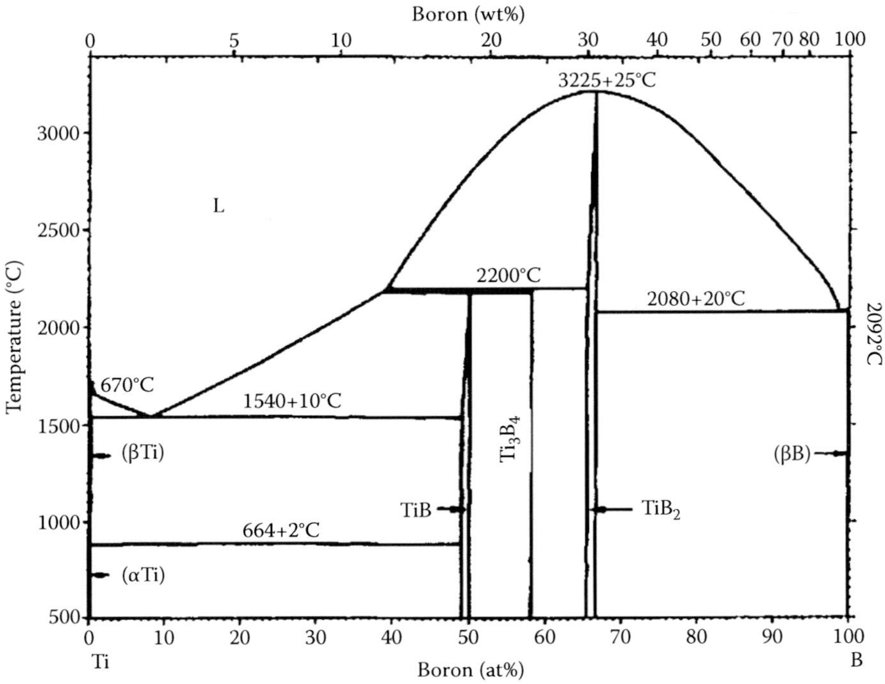

FIGURE 4.6 The binary Ti–B phase diagram. (From ASM. 1992. *ASM Handbook Vol. 3: Alloy Phase Diagrams*. ASM International 9: 2.)

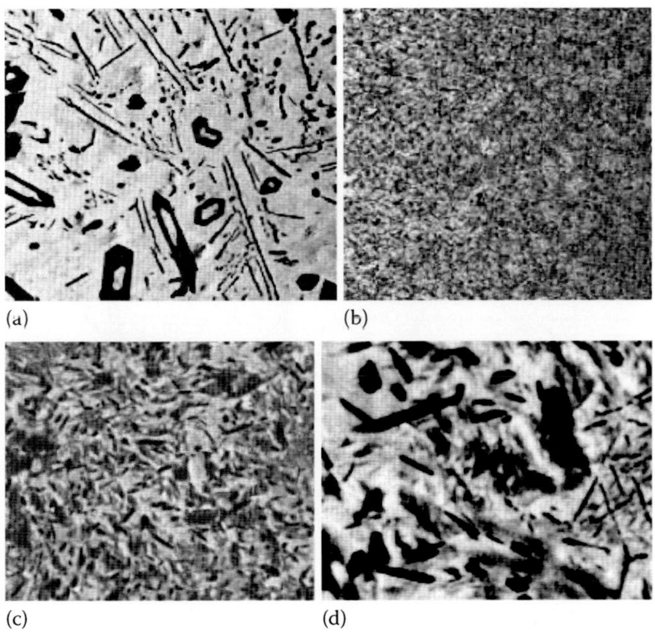

FIGURE 4.7 Comparison of the microstructures in arc-melted and LENS-deposited in situ Ti–2B composites. (a) SEM micrograph of the conventionally arc-melted Ti–2B composite. (b) SEM micrograph of the LENS-deposited Ti–2B composite at the same magnification as (a). (c and d) Higher-magnification SEM images showing the refined nature of the primary TiB precipitates in the LENS-deposited Ti–2B composite. (Reprinted with permission from Banerjee et al. 2002. Laser deposition of in situ Ti–TiB composites. *Advanced Engineering Materials* 4(11): 847–851.)

formation being ~±160 kJ/mol [52]. During the laser deposition of thc Ti–2B alloy, the negative enthalpy of mixing and reaction acts in a highly localized source of heat in the melt pool. This additional heat aids in the mixing of the elemental powders and consequently in the deposition of a homogeneous composite with finely and uniformly distributed TiB reinforcements. The advantages of a negative enthalpy of mixing in the LENS deposition of alloys from elemental powder blends have been discussed in previous papers [53,54]. One of the significant advantages of LENS processing is its ability to produce fully dense, near-net shape components of intricate design in a single step. Thus, combining these advantages of the LENS technology with the use of inexpensive elemental powders as the feedstock offers an attractive processing route for near-net shape high-strength Ti–TiB composites.

4.2.1.2.2 Ti–6Al–4V Alloy–TiB Composites

As mentioned previously, in addition to Ti–TiB composites deposited from elemental powder blends, Ti–6Al–4V with TiB composites have also been deposited using the LENS process. For these depositions, a blend of pre-alloyed Ti–6Al–4V powder and elemental B powder was used. A backscattered SEM image of the as-deposited composite is shown in Figure 4.8a.

The refined and homogeneous dispersion of TiB precipitates in an α/β-Ti alloy matrix appears to be quite promising. In addition, it is interesting to note the refined scale of the α/β microstructure within the matrix, a possible consequence of the rapid solidification rates involved in LENS deposition. A TEM image from the same composite is shown in Figure 4.8b. It is rather interesting to note

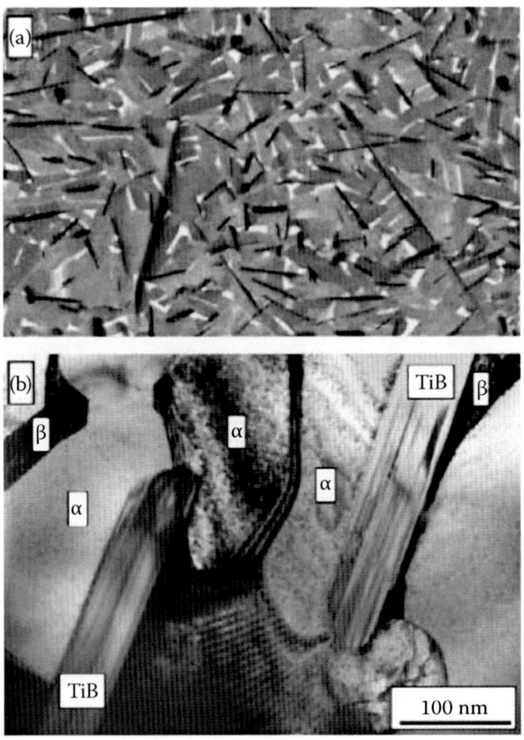

FIGURE 4.8 (a) Backscatter SEM micrograph showing the overall microstructure of the LENS-deposited Ti–6Al–4V–TiB composite with TiB precipitates distributed within an α + β matrix. (b) Bright-field TEM micrograph showing two TiB precipitates in the Ti–6Al–4V matrix. (Reprinted with permission from Genç et al. 2006. Structure of TiB precipitates in laser deposited in situ, Ti–6Al–4V–TiB composites. *Materials Letters* 60 (7): 859–863.)

that while the scale of the microstructure in the LENS-deposited composite is comparable to that processed using the mechanical alloying followed by HIPping route [55], LENS can achieve a full-density near-net shape component in basically a single step with minimal machining for surface finishing.

4.2.1.2.3 Ti–Nb–Zr–Ta Alloy–TiB Composites

The ideal properties for a biomaterial to be used for implant applications are excellent biocompatibility with no adverse tissue reactions, excellent corrosion resistance in body fluid, high mechanical strength and fatigue resistance, low modulus, low density, and good wear resistance [55,56]. Because the β phase in Ti alloys exhibits a significantly lower modulus than the α phase, and the β alloys also satisfy most of the other requirements for an ideal orthopedic alloy, there is a thrust toward the development of lower-modulus β-Ti alloys, which retain a single β phase microstructure after rapid cooling from high temperatures. For major applications such as total joint replacement arthroplasty, higher yield strength is essentially coupled with the requirement of a lower modulus approaching that of human bones [55,56]. Two recently developed promising biomedical alloys, Ti–35Nb–7Zr–5Ta (TNZT) [57] and Ti–29Mo–13Ta–4.6Zr (TNTZ) [58], show significant improvement in these aspects compared to previous generation alloys such as Ti–6Al–4V, stainless steel and cobalt–chromium-based alloys. Despite the numerous beneficial properties of the β-Ti alloys for implant applications, these alloys typically exhibit rather poor wear resistance [59]. Consequently, their application is restricted to those locations on the implant surface where wear resistance is not of critical importance. For example, in a typical hip implant (femoral implant) while the femoral stem is often made of titanium alloys, the poor wear resistance of these alloys prevents their use for femoral head applications. Therefore, more wear-resistant materials are used for the femoral head, such as ceramics (e.g., alumina or zirconia) or cobalt–chromium alloys. While alumina or zirconia ceramic femoral heads offer excellent wear resistance, these ceramics do not have the same level of fracture toughness as their metallic counterparts. This leads to problems such as head fracture when implanted. This has even led to the recall of hip implants using zirconia femoral heads [48]. Furthermore, the use of a ceramic femoral head attached to a metallic femoral stem also leads to an undesirable abrupt ceramic/metal interface in the hip implant. In contrast, by using a metal matrix composite, it might be possible to substantially improve the wear resistance while maintaining acceptable levels of toughness arising from the compliant metal matrix. Because β-Ti alloys, such as Ti–Nb–Zr–Ta, are expected to become the new-generation alloys for implant applications, the present study focuses on the development of composites based on this group of alloys as matrix materials. In recent years, there has been a considerable amount of research devoted to the development of TiB-reinforced Ti alloy composites [46–50,60–62]. These composites combine the high strength and stiffness of the borides with the toughness and damage tolerance of the Ti alloy matrix and offer attractive properties including increased stiffness and substantially enhanced wear resistance [60]. Using the LENS process [46–50,62], it has been successfully demonstrated that in situ TiB-reinforced Ti alloy composites can be fabricated in a single step. These laser-deposited composites exhibit a substantially refined microstructure when compared to their conventionally processed (e.g., ingot processing) counterparts. Furthermore, as discussed previously, because biomedical implants are geometrically quite complex, by employing novel near-net shape processing technologies, it is possible to not only rapidly and efficiently manufacture custom-designed implants but also functionally grade them to exhibit the required site-specific properties. The in situ Ti–Nb–Zr–Ta–TiB composites were processed via a LENS technique from a blend of pure elemental titanium (Ti), niobium (Nb), zirconium (Zr), and tantalum (Ta) powders mixed with titanium diboride (TiB_2) powders [47,48]. The microstructure of the LENS-deposited TNZT+2B alloy at different magnifications is shown in the backscatter SEM images in Figure 4.9.

Figure 4.9a shows a low-magnification backscattered image of the boride composite while Figure 4.9b shows a higher-magnification view of the microstructure consisting of the coarser primary boride precipitates and the finer-scale eutectic boride precipitates dispersed in a matrix. Furthermore, while the finer-scale eutectic borides exhibit a uniform contrast in the backscatter SEM images,

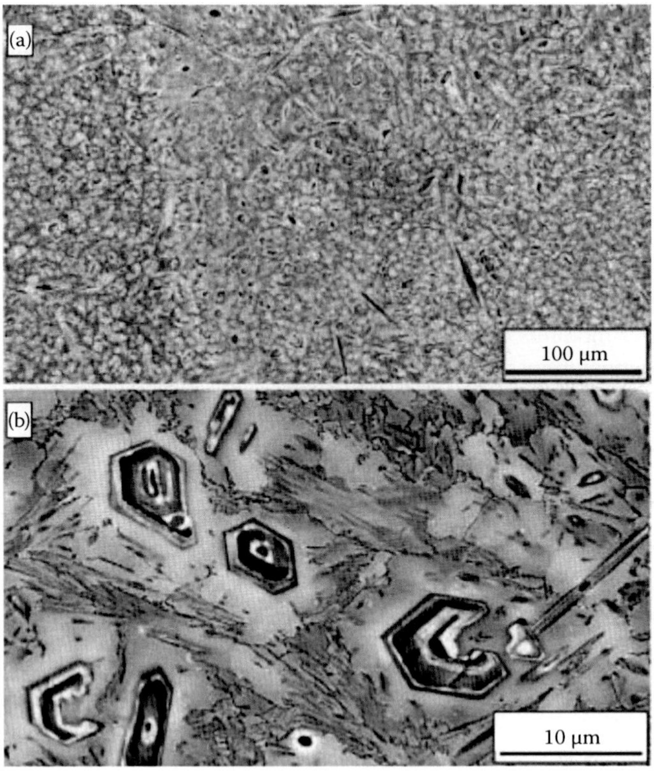

FIGURE 4.9 (a) Lower-magnification and (b) higher-magnification backscatter SEM images of LENS as-deposited TNZT+2B alloy composites showing both coarser primary borides exhibiting contrast within the same boride precipitate as well as finer-scale eutectic borides. (Reprinted with permission from Samuel et al. 2008. Wear resistance of laser-deposited boride reinforced Ti–Nb–Zr–Ta alloy composites for orthopedic implants. *Materials Science and Engineering: C* 28 (3): 414–420.)

the coarser primary boride precipitates exhibit strong contrast variations within the same precipitate, suggesting the possibility of compositional variation within the same boride. In addition, the matrix also exhibits variations in contrast. Thus, while the regions adjoining the primary boride precipitates exhibit a lighter contrast, the regions a further distance exhibit a relatively darker contrast. There is also a clustering of the eutectic boride precipitates within these regions of darker contrast, as clearly visible in Figure 4.9b. Figure 4.10 shows typical friction coefficient curves for the alloys sliding against a stainless steel (SS 440C) counterface ball.

The initial friction coefficients of both Ti–6Al–4V and TNZT+2B are approximately 0.1; however, the friction coefficient of Ti–6Al–4V began to deviate at ~15 m sliding distance. A continual steady increase in the friction coefficient occurs up to ~0.4 to 0.6. These higher and more fluctuating values suggest increased abrasive wear and debris generation. In contrast, the initial friction coefficient of the TNZT+2B alloy remained at a very low steady-state value of ~0.13 for the entire test. Thus, the softer steel counterface eliminated the TiB precipitate pullout. The corresponding SEM images of the wear tracks shown in Figure 4.11 corroborate the friction behavior.

The wear track morphology of Ti–6Al–4V in Figure 4.11a and b shows more extensive plastic deformation, microplowing, and cutting, all typical processes associated with an abrasive wear mechanism. Conversely, the TNZT+2B wear track shown in Figure 4.11c and d exhibits less severity in wear. It is apparent that the surface morphology has undergone a plastic shear deformation mechanism usually associated with less wear volume removal. This is evident by examining some of

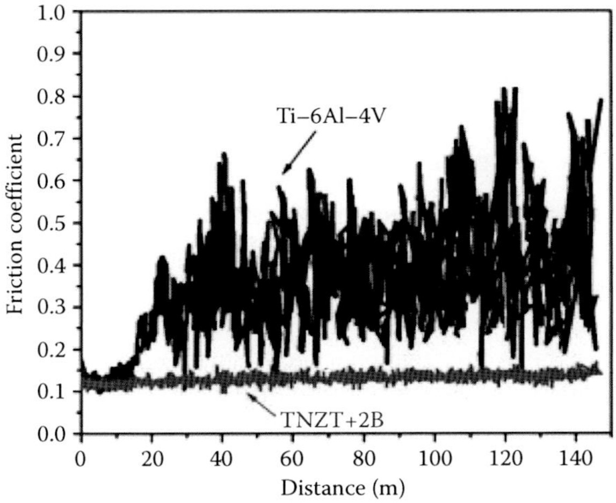

FIGURE 4.10 Wear plots showing the friction behavior as a function of distance of Ti–6Al–4V and TNZT +2B with SS440C stainless steel counterface balls. (Reprinted with permission from Samuel et al. 2008. Wear resistance of laser-deposited boride reinforced Ti–Nb–Zr–Ta alloy composites for orthopedic implants. *Materials Science and Engineering: C* 28 (3): 414–420.)

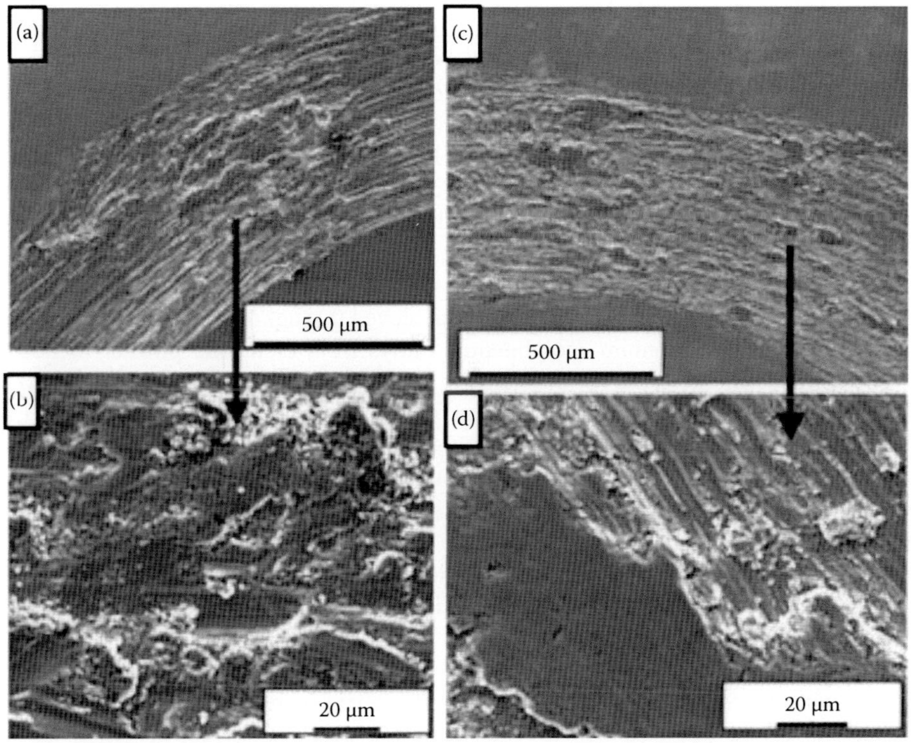

FIGURE 4.11 (a and c) Lower-magnification and (b and d) higher-magnification secondary SEM images of Ti–6Al–4V and TNZT+2B wear tracks after friction studies using SS440 counterface balls. (Reprinted with permission from Samuel et al. 2008. Wear resistance of laser-deposited boride reinforced Ti–Nb–Zr–Ta alloy composites for orthopedic implants. *Materials Science and Engineering: C* 28 (3): 414–420.)

the larger surface depressions with plastic flow process. These results show that much lower friction coefficient and improved wear resistance can be obtained by using a softer counterface material when returning to the TNZT+2B, much lower than observed for current Ti–6Al–4V alloys.

4.2.2 In Situ Reaction between Elemental Blend Powders and Reactive Gases

4.2.2.1 In Situ Nitridation of Titanium–Molybdenum Alloys

Titanium alloys are attractive candidates for structural, marine, aerospace, biomedical (such as in dental and orthopedic, as bone implants), and other industrial applications because of their excellent strength-to-weight ratio, ductility and formability, corrosion resistance, and biocompatible properties. However, titanium alloys suffer from rather poor surface hardness and wear resistance properties. One way of improving the hardness as well as tribological properties of titanium alloys is by reinforcing the soft matrix with hard precipitates, such as titanium nitrides, carbides, and borides. These reinforcements can be introduced either via direct incorporation of such hard compounds in the matrix during processing [63,64] or via in situ reaction with solid or gaseous precursors, resulting in the formation of hard precipitates [46–48], as well as via surface engineering techniques such as nitriding [65–67]. Focusing on nitride-reinforced titanium alloys, the most commonly used technique is surface nitridation of these alloys via heating at elevated temperatures in a flowing nitrogen atmosphere [65–67]. Another approach that has been employed is the direct introduction of δ-TiN and TiB particles during laser deposition of Ti–6Al–4V [63,64]. The forming processing approach, nitridation in a gaseous atmosphere, has been applied to the case of β-Ti–Mo alloys, resulting in the formation of a continuous surface δ-TiN layer and a subsurface microstructure consisting of laths (or plates) of either the same δ-TiN phase or a nitrogen-rich α(Ti, N) solid solution phase, dispersed within the β matrix [66,67]. While such surface nitridation via heating in a nitrogen atmosphere is a simple and inexpensive way to achieve a case-hardened layer, the time required and the depth of penetration are rather limited and it is difficult to introduce hard nitrides (or other nitrogen enriched hard phases) within the bulk of the material. The direct introduction of nitride and boride particles during laser deposition of titanium alloy powders can obviate this problem. However, the quality of such an interface between such externally introduced reinforcements and the alloy matrix can be rather difficult to control, and therefore, ideally, a reinforcement created as a product of an in situ reaction offers the advantage of a thermodynamically stable and clean interface with the matrix. Furthermore, by employing an in situ nitride (or nitrogen-enriched hard α phase) formation reaction, a more uniform and homogeneous distribution of the nitride phase can be potentially achieved throughout the matrix. In situ nitridation during laser deposition of titanium–molybdenum alloys from elemental powder blends has been achieved by introducing the reactive nitrogen gas during the deposition process. Ti–Mo–N alloys have been deposited using the LENS process and resulted in the formation of a hard α(Ti,N) phase, exhibiting a dendritic morphology, distributed within a β(Ti–Mo) matrix with fine-scale transformed α precipitates [3]. Varying the composition of the Ar+N_2 gas employed during laser deposition permits a systematic increase in the nitrogen content of the as-deposited Ti–Mo–N alloy. Interestingly, the addition of nitrogen, which stabilizes the α phase in Ti, changes the solidification pathway and the consequent sequence of phase evolution in these alloys. The nitrogen-enriched hexagonal close packed (hcp) α(Ti,N) phase has higher c/a ratio, exhibits an equiaxed morphology, and tends to form in clusters separated by ribs of the molybdenum (Mo)-rich β phase. The Ti–Mo–N alloys also exhibit a substantial enhancement in microhardness owing to the formation of these α(Ti,N) phase, combining it with the desirable properties of the β-Ti matrix, such as excellent ductility, toughness, and formability. All Ti–Mo–N alloys were deposited inside a glove box of LENS machine with controlled gas atmosphere. Pure Ar (Alloy 1), 25% N_2–75% Ar (Alloy 2), 50% N_2–50% Ar (Alloy 3), and 75% N_2–25% Ar (Alloy 4) were the atmospheres used for fabrication of deposits.

A series of backscattered SEM images recorded from the Ti–Mo–N alloys are shown in Figures 4.12 and 4.13.

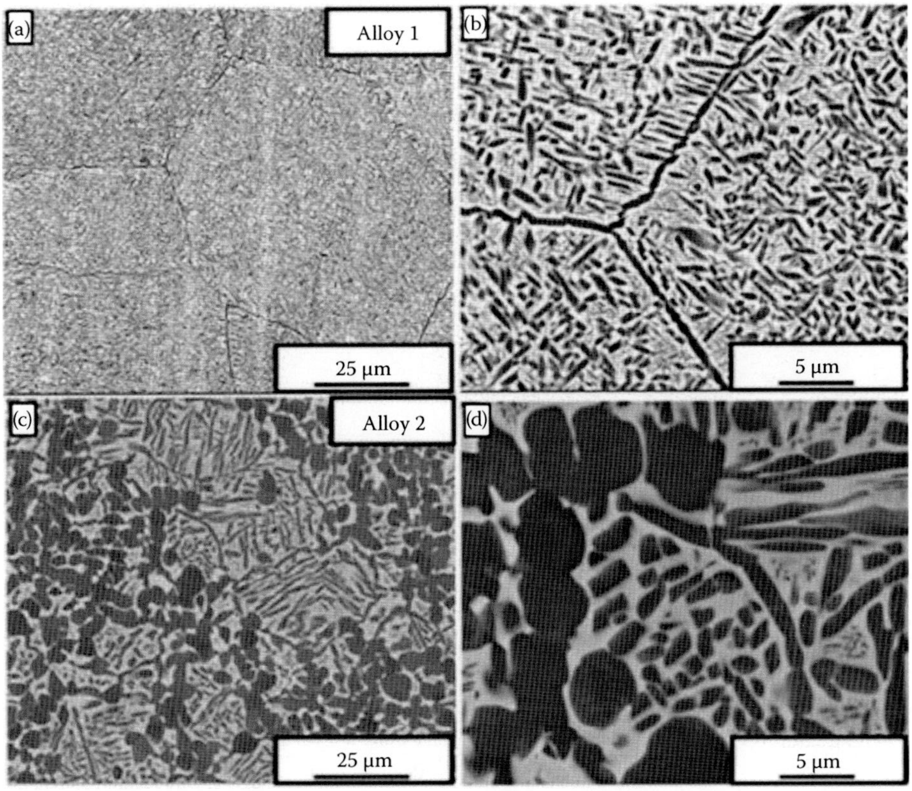

FIGURE 4.12 (a) Low-magnification and (b) high-magnification backscatter SEM image of LENS-deposited Alloy 1. (c) Low-magnification and (d) high magnification backscatter SEM image of LENS-deposited Alloy 2. (Reprinted with permission from Borkar et al. 2012. In situ nitridation of titanium–molybdenum alloys during laser deposition. *Journal of Materials Science* 47 (20): 7157–7166.)

Figure 4.12a and b shows the microstructure for the binary Ti–Mo alloy deposited in 100% argon (Ar) atmosphere. The nominal composition of this alloy measured using energy dispersive spectroscopy (EDS) in the SEM is Ti–10wt%Mo. The microstructure primarily consists of uniformly distributed fine-scale precipitates of a second phase, presumably α precipitates, exhibiting a bimodal size distribution within the β matrix. This bimodal size distribution of α precipitates can be rationalized on the basis of a two-step decomposition of the β matrix, wherein the coarser α precipitates presumably result from the deposition of the layer during LENS processing, while solid-state reheating of the same layer, when the subsequent layer is being deposited on the top, results in the secondary decomposition of β forming the finer-scale α. The inherent rapid cooling rates involved in laser deposition also contribute to such non-equilibrium β decomposition processes. Thus, while the phase evolution follows the sequence liquid to β to β+α during laser deposition of a layer, the inherently rapid cooling rates involved do not result in an equilibrium composition of the β matrix. Therefore, on subsequent reheating of the same layer when a layer on top is being deposited, there is a further decomposition of the retained β matrix to form fine-scale secondary α precipitates. The aspect ratio of this finer-scale α is substantially larger than those of the coarser α precipitates. Figure 4.12c and d show the microstructure of the Ti–Mo–N alloy deposited using a 75% Ar–25% N_2 atmosphere in the LENS glove box as well as the center gas purge and the powder carrier gas. The microstructure clearly exhibits two precipitate phases having rather different morphologies. The coarser precipitates appear to exhibit an equiaxed or globular morphology with curved interfaces separating them from

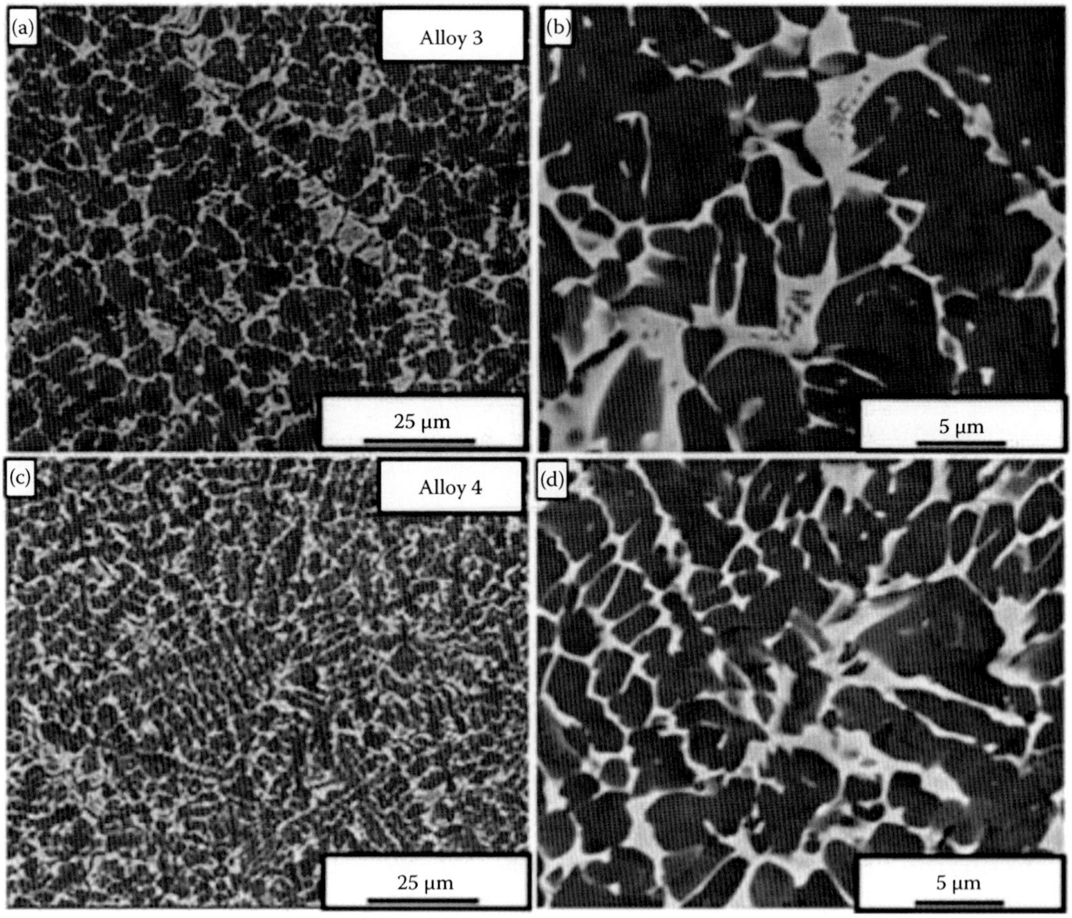

FIGURE 4.13 (a) Low-magnification and (b) high-magnification backscatter SEM image of LENS-deposited Alloy 3. (c) Low-magnification and (d) high-magnification backscatter SEM image of LENS-deposited Alloy 4. (Reprinted with permission from Borkar et al. 2012. In situ nitridation of titanium–molybdenum alloys during laser deposition. *Journal of Materials Science* 47 (20): 7157–7166.)

the β matrix, suggesting that these interfaces are presumably incoherent in nature. In contrast, the finer-scale precipitates exhibit a sharper faceted morphology, suggesting that the precipitate/matrix interface is likely to be semi-coherent. The identity of these two types of precipitates cannot be determined based solely on the backscatter SEM evidence presented in Figure 4.12c and d. However, the morphology and contrast of the finer-scale precipitates suggest that the α precipitates exhibited smaller aspect ratios when compared with those observed in the case of the binary Ti–10Mo alloy deposited under a pure Ar atmosphere (Figure 4.12a and b). As the N_2 content in the Ar+N_2 mixture used in LENS deposition increases, the size and volume fraction of the coarser equiaxed second-phase precipitates increase significantly in the microstructure, as revealed in Figure 4.13.

Backscattered SEM images of Ti–10Mo alloys LENS deposited under a 50% Ar–50%N_2 mixture are shown in Figure 4.13a and b while Figure 4.13c and d corresponds to an alloy deposited under a 25% Ar–75% N_2 mixture. There is also a corresponding decrease in the β volume fraction with increasing N_2 content in the gaseous atmosphere used for deposition. Thus, the primary microstructural influence of introducing nitrogen during laser deposition appears to be the formation of novel equiaxed precipitates within the β matrix together with finer-scale α precipitates, with the volume fraction of the equiaxed precipitates increasing with increasing nitrogen content in the alloy.

The backscatter SEM image of the exact same region in the (25%Ar+75%N_2) alloy where the OIM-EBSD scan was carried out is shown in Figure 4.14a. Figure 4.14b shows the pseudo-colored OIM map of this region wherein both α(Ti,N) and β phases have been attributed colors based on their respective Euler angles. A more detailed analysis of the α(Ti,N) precipitates within equiaxed α clusters and also an intragranular secondary α precipitate are shown in Figure 4.14c and d.

The pseudo-colored OIM map of the first cluster and corresponding α and β pole figures are shown in Figure 4.14c. The region corresponding to this cluster has been marked in the overall OIM map (Figure 4.14b) as well as in the backscatter SEM image (Figure 4.14a). The cluster shown in Figure 4.14c comprises six distinctly oriented α precipitates, marked 1–6 on the OIM map and their corresponding (0001) α poles have been marked on the pole figure. In addition, the $\{11\bar{2}0\}$α poles for these precipitates as well as $\{011\}$β and $\{111\}$β pole figures for the surrounding β grain are shown in Figure 4.14c. It is rather interesting to note that four of the six (0001) α poles (1,2,3,4) lie close to corresponding $\{011\}$β poles but not exactly on these poles. Interestingly, further analysis of the EBSD-OIM data set revealed that the measured misorientation angles between two adjacent α precipitates shown in Figure 4.14c correspond to a rotation ~70° about the common $\{11\bar{2}0\}$ pole between these two precipitates. Furthermore, this common $\{11\bar{2}0\}$ pole between two adjacent precipitates lies close to one of the $\{111\}$β poles of the surrounding grains, but again not exactly parallel to it. All these observations clearly indicate a near-Burgers orientation relationship (i.e., (0001)α//$\{011\}$β and $< 11\bar{2}0 >$ α// $<$ 111 $>$ β) between all the α precipitates within a cluster and the surrounding β grain. However, there is

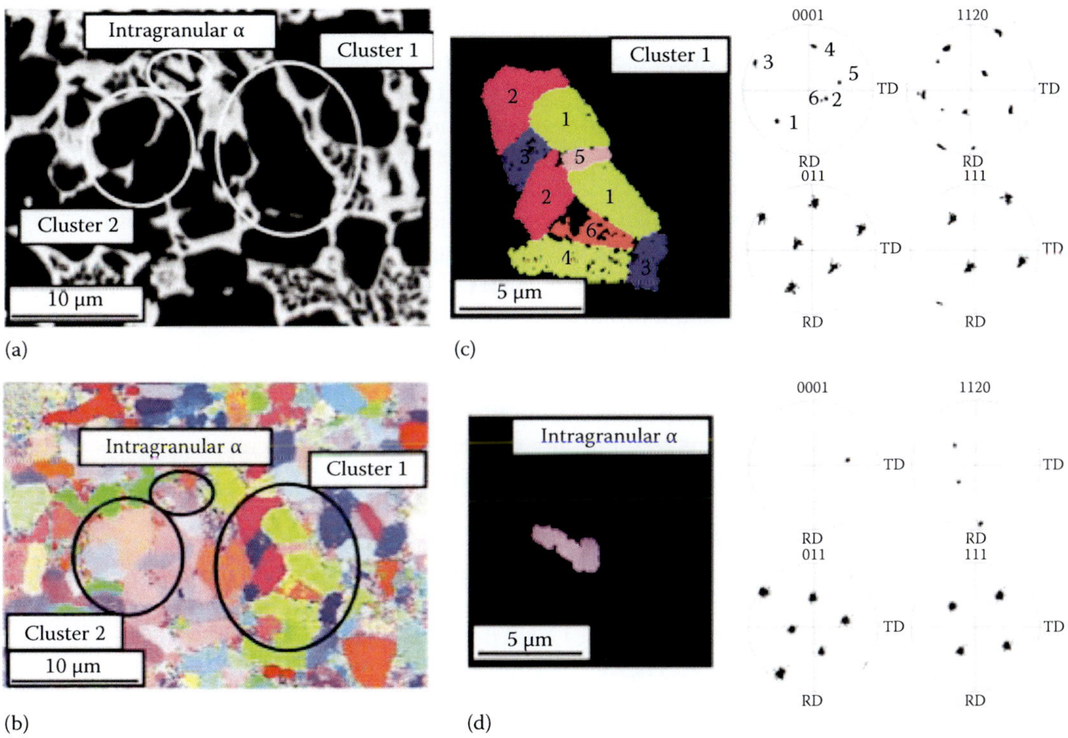

FIGURE 4.14 (a) Backscatter SEM image and (b) corresponding pseudo-colored OIM map of a region in (25%Ar + 75%N_2 alloy where the EBSD scan was carried out. On both the images, equiaxed α clusters and also an intragranular secondary α precipitate have been marked. Pseudo-colored OIM map and corresponding α and β pole figures of (c) cluster and (d) intergranular α. (Reprinted with permission from Borkar et al. 2012. In situ nitridation of titanium–molybdenum alloys during laser deposition. *Journal of Materials Science* 47 (20): 7157–7166.)

a clear departure from the exact Burgers relationship. The α precipitate, marked as 5 in this cluster, does not appear to have even a near-Burgers orientation relationship with the surrounding β grain. EBSD-OIM analysis of an intragranular α precipitate, exhibiting a more faceted lath or plate-like morphology, is shown in Figure 4.14d. The (0001) α and $\{11\bar{2}0\}\alpha$ pole figures for this precipitate as well as the {011}β and {111}β pole figures for the surrounding β grain are also shown. An exact Burgers orientation relationship of the type (0001) α//{011}β and $< 11\bar{2}0 > \alpha // < 111 > \beta$ is evident between this α precipitate and the surrounding β grain. This and other similar observations confirm that the intragranular fine-scale secondary α precipitates exhibit the typical lath or plate-like morphology and have a Burgers orientation relationship with the β matrix.

4.2.2.1.1 Evolution of Microstructure in Laser-Deposited Ti–Mo–N Alloys

Based on the experimental results from the various techniques discussed previously, an overall sequence of microstructural evolution in the Ti–Mo–N alloys can be developed. Thus, during the initial stages of a layer being deposited (via LENS processing), there is a solidification of a mixture of titanium and molybdenum powders fed into a molten pool in an atmosphere enriched in nitrogen. The nitrogen dissolves in this localized Ti–Mo melt pool and the liquid solidifies. During solidification, the entire liquid transforms to the β phase, which is supersaturated with nitrogen at high temperatures. During the cooling of the β phase in the solid state, precipitation of the α phase is initiated below the β transus temperature for this ternary Ti–Mo–N composition. Nitrogen, being a strong α stabilizer, naturally partitions strongly to the precipitating α phase forming the α(Ti,N) phase. Molybdenum (Mo), in contrast, is a strong β stabilizer and hence partitions to the β matrix. As precipitates of α(Ti,N) grow within the β matrix, both during cooling and solid-state reheating of the same layer during subsequent depositions of layers on top, the strain energy increases rather rapidly due to the large misfit between the nitrogen-rich α(Ti,N) phase and the Mo-rich surrounding β matrix. This leads to the loss of coherency at the α(Ti,N)/β-Ti(Mo) interface, resulting in the formation of curved interfaces, as well as a more equiaxed morphology of these precipitates. Furthermore, there is a departure from the Burgers orientation relationship between the α(Ti,N) precipitates and the surrounding β matrix as these precipitates grow and coarsen. Additionally, it should be noted that the Mo-rich β matrix does not achieve equilibrium during deposition of the layer. Therefore, the subsequent solid-state reheating of the same layer also results in the precipitation of finer-scale secondary α within the β matrix. Presumably, this finer-scale secondary α contains a substantially lower amount of nitrogen when compared with the primary α(Ti,N) phase. While this has not been experimentally proven, it can be speculated based on the fact that during the precipitation of the primary α(Ti,N) phase, most of the nitrogen partitions to this phase with only a negligible amount retained in the β matrix. Consequently, the secondary α precipitates exhibit a lower misfit with the β matrix, thus retaining an exact Burgers orientation relationship, as well as a more typical lath or plate-like morphology with faceted interfaces.

4.2.2.1.2 Vickers Microhardness Results

The Vickers microhardness values for all the four alloys, 1 to 4, have been plotted in Figure 4.15. A systematic increase in the microhardness values, with increasing nitrogen content in the alloys can be clearly observed. The binary Ti–Mo alloy, LENS deposited under a pure Ar atmosphere, exhibits an average hardness of ~500 HV. The increase in microhardness of the Ti–Mo–N alloys, with increasing N_2 content in the reactive atmosphere during LENS deposition, can be attributed to the formation of the α-$TiN_{0.3}$ random solid solution hcp phase. For the highest nitrogen containing Alloy 4, the microhardness value is ~800 HV, which is ~60% higher as compared with the LENS-deposited binary Ti–10wt%Mo (Alloy 1).

In situ nitridation of Ti–10wt%Mo alloys has been achieved by the introduction of reactive nitrogen gas during the laser deposition (LENS) of these alloys from elemental powder blends. The nitrogen content in these laser-deposited alloys has been tuned via changing the ratio of argon to nitrogen used in LENS deposition. The enrichment of these alloys with nitrogen results in the

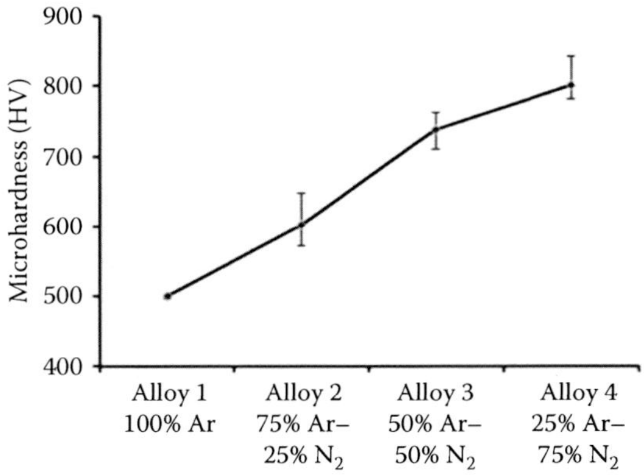

FIGURE 4.15 A plot showing the variation in Vickers microhardness values for LENS-deposited Ti–Mo–N alloys (Alloys 1 to 4). (Reprinted with permission from Borkar et al. 2012. In situ nitridation of titanium–molybdenum alloys during laser deposition. *Journal of Materials Science* 47 (20): 7157–7166.)

formation of primary precipitates of the α(Ti,N) phase within the β matrix. The higher c/a ratio of these hcp α(Ti,N) precipitates, coupled with the Mo enrichment in the β matrix, results in a substantially large misfit, consequently leading to a loss of precipitate/matrix coherency during growth and coarsening. This results in the α(Ti,N) precipitates adopting an equiaxed morphology, and they tend to aggregate into clusters separated by thin Mo-rich β ribs. These α(Ti,N) precipitates increase the microhardness of the alloy to a substantial degree. Additional fine-scale secondary α precipitates, exhibiting a lath or plate-like morphology, are also formed within the retained β matrix of these Ti–Mo–N alloys. The ability to introduce controlled volume fractions of the hard nitrogen-enriched α(Ti,N) phase can be very useful in tailoring the local microhardness and consequently wear resistance of these alloys. Furthermore, using the LENS process, it is possible to grade the nitrogen content within the same alloy and thus process a compositionally graded microstructure with systematically varying properties.

4.2.2.2 In Situ Nitridation of Ti–6Al–4V (Ti64) and Ti–35Nb–7Zr–5Ta (TNZT) Alloys

Titanium (Ti) alloys have long been used in many diverse applications ranging from orthopedic biomedical implants (femoral stems) to aerospace gas turbine engine components (compressor blade roots) [68–70]. While Ti alloys such as the α + β alloy Ti–6Al–4V (Ti64) and the β alloy Ti–35Nb–7Zr–5Ta (TNZT) are well known for their corrosion resistance, adequate fracture toughness, and fatigue strength, they exhibit high wear rates, resulting in galling (adhesive) failure and fretting fatigue [71–76]. Thus, they require surface modification treatments to improve their wear behavior. One such approach is surface nitriding, which produces a high hardness and a chemically inert layer that overcomes the poor wear resistance of Ti alloys [74,75,77–81]. Surface nitriding is a fast, inexpensive, and simple treatment to form a hardened, wear-resistant case layer on the surface. However, the penetration depth of nitrogen into the bulk alloy is limited, thus restricting the thickness of the nitride layer to mitigate wear, and the high processing temperatures can result in alloy softening. The alternative solution to surface nitridation is in situ nitridation of these alloys via the LENS process. Thus, reinforcement ideally created as a product of an in situ gas reaction offers the advantage of a thermodynamically and mechanically stable interface with the matrix. Furthermore, by employing an in situ nitride phase formation reaction, such as a nitrogen-enriched hard α-Ti phase, a more uniform and homogeneous distribution of the nitride phase can be potentially achieved throughout the alloy matrix. To achieve this, in situ nitride-reinforced Ti–Nb–Zr–Ta and Ti–6Al–4V alloys have been fabricated using LENS in the presence of nitrogen gas to study the solidification structure and microhardness properties [82].

4.2.2.2.1 Nitrided Ti–6Al–4V

The comparison of the microstructures of the LENS-deposited Ti64 and nitrided Ti64 samples is shown in the backscatter electron (BSE) micrographs presented in Figure 4.16.

The microstructure of Ti64 appears to exhibit fine-scale α laths in the β matrix (Figure 4.16a and b), typically referred to as a basketweave microstructure in α/β-Ti alloys, containing multiple crystallographic variants of α intersecting within the same prior β grain. Such a basketweave α/β microstructure can be attributed to the high cooling rates involved in LENS deposition due to the substrate acting as an efficient heat sink. These fine-scale α laths have been attributed to martensitic decomposition of the β matrix [78]. However, in the present case, the BSE image shown in Figure 4.16b clearly shows a significant difference in contrast between the darker α laths and the brighter β matrix, indicative of the partitioning of alloying elements between these two phases. Hence, it is unlikely that these fine-scale α laths, observed in the LENS-deposited Ti64 sample, are purely martensitic in nature. In contrast, the BSE images of the nitrided Ti64 sample shown in Figure 4.16c and d exhibit a very different microstructure. Comparing the higher-magnification images for both alloys (Figure 4.16b and d), there are two principal differences. First, Figure 4.16d shows that there is a more pronounced contrast between the α precipitates (darker) and the β matrix (lighter) in nitrided Ti64, indicating more substantial atomic mass difference between these two phases, presumably arising from the partitioning of V into the β matrix, while the lighter Al and N partition into the α precipitates. Second, the morphology of α precipitates in nitrided Ti64 is substantially different compared to base Ti64. The α precipitates in nitrided Ti64 are coarser, exhibit a smaller aspect ratio, and,

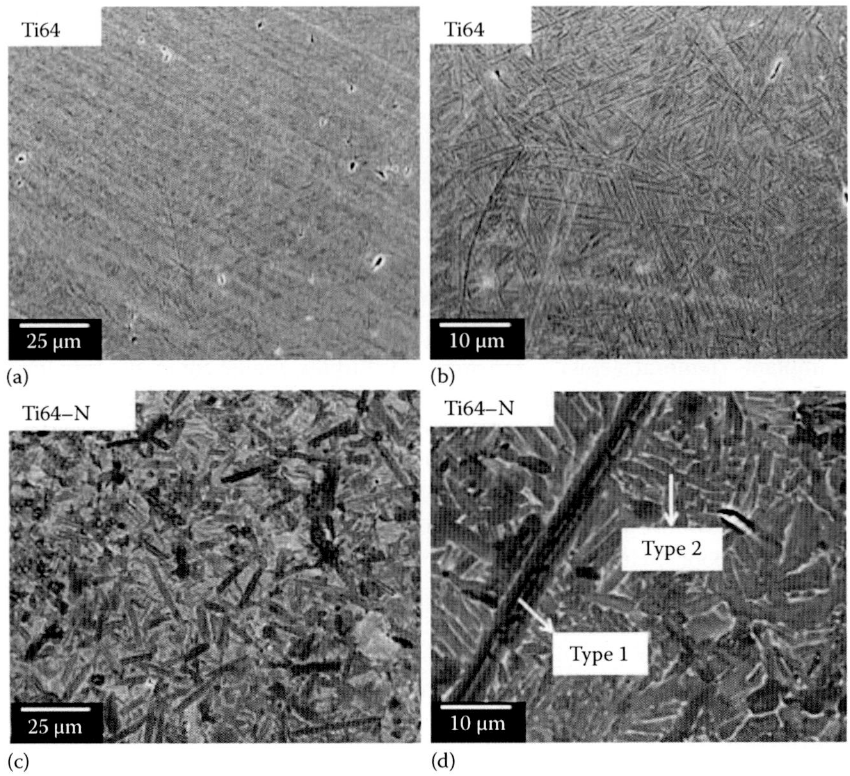

FIGURE 4.16 (a) Low-magnification and (b) high-magnification backscatter SEM images of LENS-deposited Ti64 alloy. (c) Low-magnification and (d) high-magnification backscatter SEM images of LENS-deposited nitrided Ti64 (Ti64–N) alloy. (Reprinted with permission from Mohseni et al. 2015. In situ nitrided titanium alloys: Microstructural evolution during solidification and wear. *Acta Materialia* 83: 61–74.)

in some cases, appear to be more equiaxed in shape. There also appear to be two distinctly different morphologies of α precipitates in nitrided Ti64 (Figure 4.16c and d): a smaller volume fraction of larger elongated needle-like or acicular precipitates exhibiting complex internal contrast variations (referred to as Type 1 in Figure 4.16d), and smaller-scale precipitates exhibiting a lower aspect ratio (referred to as Type 2 in Figure 4.16d).

4.2.2.2.2 Nitrided Ti–35Nb–7Zr–5Ta

The influence of nitrogen on the structure of the LENS-deposited Ti–Nb–Zr–Ta is also evident by comparing the BSE images of base (Figure 4.17a) and nitrided (Figure 4.17b) Ti–Nb–Zr–Ta. The dendritic microstructure typical of single-phase Ti–Nb–Zr–Ta can be seen in Figure 4.17a where β dendrites are separated by darker interdendritic regions that show microsegregation of alloying additions. Such microsegregation is a common feature in alloys with a high concentration of β stabilizers [68]. Figure 4.17b shows a completely different microstructure for nitrided TNZT where the phase with brighter contrast represents the β matrix and the phases exhibiting the darker contrast include TiN/Ti_2N and α (grain boundary and intergranular α precipitates). The higher-magnification inset shown in Figure 4.17b clearly shows these two different precipitate morphologies: finer-scale intragranular α precipitates with platelet morphology, and coarser TiN/Ti_2N exhibiting a dendritic morphology. A cross-sectional scanning transmission electron microscopy–high angle annular dark field (STEM-HAADF) image along with EDS elemental maps of a site-specific sample prepared from the nitrided Ti–Nb–Zr–Ta sample is shown in Figure 4.17c. It is evident from the compositional maps that the intragranular α precipitates (~200 nm thick) are deficient in niobium (Nb) and tantalum (Ta) (β stabilizers), whereas the β matrix is enriched in niobium (Nb) and tantalum (Ta). The zirconium (Zr) map was noisy owing to the lower counts and, hence, has not been included.

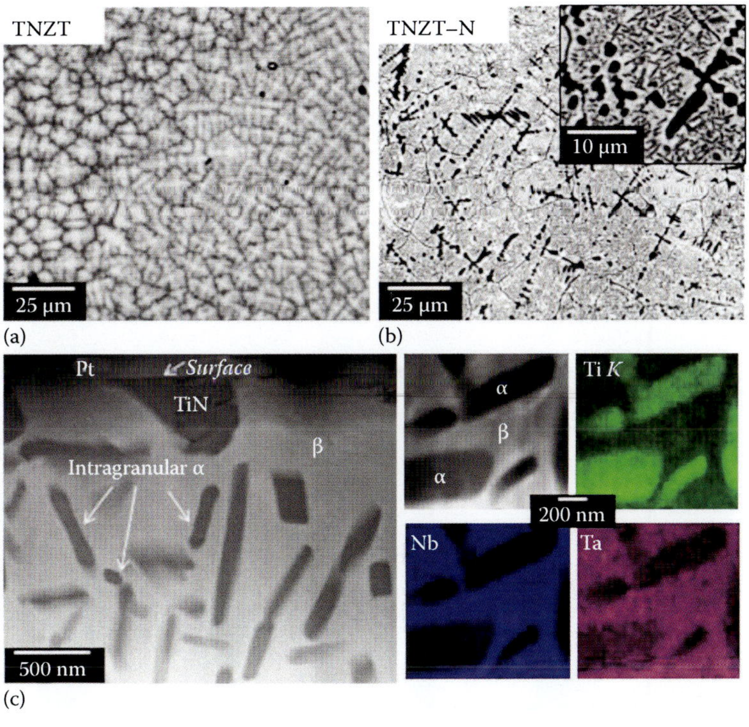

FIGURE 4.17 Low-magnification backscatter SEM images of LENS-deposited (a) TNZT and (b) nitrided TNZT (TNZT-N) alloys. (c) Cross-sectional HAADF-STEM image of nitrided TNZT alloy with corresponding titanium, niobium, and tantalum EDS maps. (Reprinted with permission from Mohseni et al. 2015. In situ nitrided titanium alloys: Microstructural evolution during solidification and wear. *Acta Materialia* 83: 61–74.)

4.2.2.2.2.1 Microhardness The result of microhardness analysis of the LENS-deposited base and nitrided Ti64 and TNZT is summarized in Figure 4.18.

Ti64 has a higher hardness (419 VHN) compared to Ti–Nb–Zr–Ta (288 VHN), since the β phase is softer than α, which makes the Ti64, with α/β structure, harder than Ti–Nb–Zr–Ta. The averaged microhardness of both alloys approximately doubled with nitriding: 1047 and 518 VHN for nitrided Ti64 and Ti–Nb–Zr–Ta, respectively. The increasing hardness with in situ nitriding can be attributed to the formation of TiN/Ti_2N phases in Ti–Nb–Zr–Ta and the nitrogen-enriched α phase, that is, increasing volume of α(Ti,N) precipitates in Ti64.

4.2.2.2.2.2 Evolution of Microstructure in Nitrided Ti-6Al–4V and Ti–35Nb–7Zr–5Ta The Ti–N binary phase diagram [51] in Figure 4.19 can provide some insights toward the probable sequence of phase evolution during in situ nitriding of Ti64 and Ti–Nb–Zr–Ta.

Since Ti64 is an α/β alloy, addition of nitrogen to this system can possibly expand the α phase stability region on the phase diagram. The increased volume fraction of α phase is capable of absorbing more nitrogen, thereby preventing the precipitation of nitride phases. Conversely, the effect of nitrogen can be somewhat complicated in nitrided Ti–Nb–Zr–Ta since the high concentration of β stabilizing elements (niobium [Nb], zirconium [Zr], and tantalum [Ta]) can counteract the α stabilizing effects of nitrogen. Furthermore, the complex interplay of the multiple components in this alloy can possibly lead to changes in the shapes and sizes of the various phase fields on the multicomponent phase diagram in such a manner as to nucleate primary TiN precipitates in the liquid phase followed by nucleation of β grains, as evident by the resulting microstructure in Figure 4.17b. Consequently, the primary TiN precipitates forming in the liquid exhibit a dendritic morphology and can act as heterogeneous nucleation sites for the β grains during solidification, resulting in more equiaxed β grains (Figure 4.17b) as compared to the dendritic β grains observed in LENS-deposited TNZT in the absence of nitrogen (Figure 4.17a). Despite the prior formation of the δ-TiN dendrites, at the elevated temperatures where the β grains nucleate, they have a certain solubility of nitrogen, as evident from the Ti–N binary phase diagram in Figure 4.18. Thus, on subsequent cooling, this retained nitrogen will facilitate the precipitation of α phase at both heterogeneous grain boundary sites and homogeneously within the β grains. Furthermore, as soon as α precipitates within the β matrix, the retained nitrogen within the β partitions to the α phase, resulting in an increase in the c/a ratio of this phase.

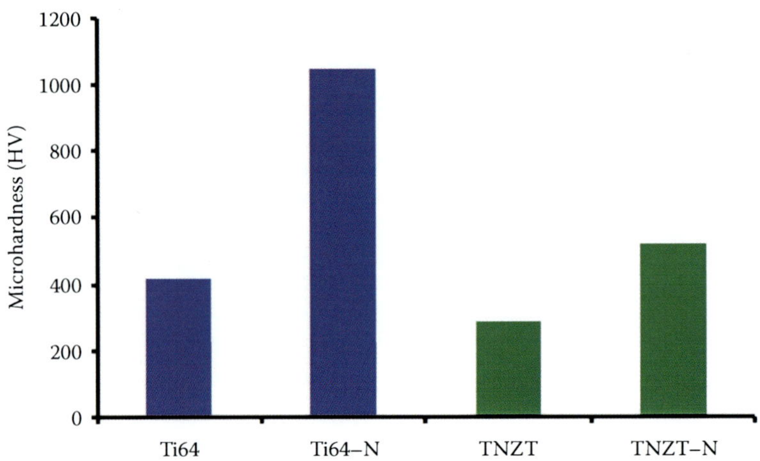

FIGURE 4.18 Plot showing the variation in Vickers microhardness values for LENS-deposited Ti64, Ti64–N, TNZT, and TNZT–N alloys.

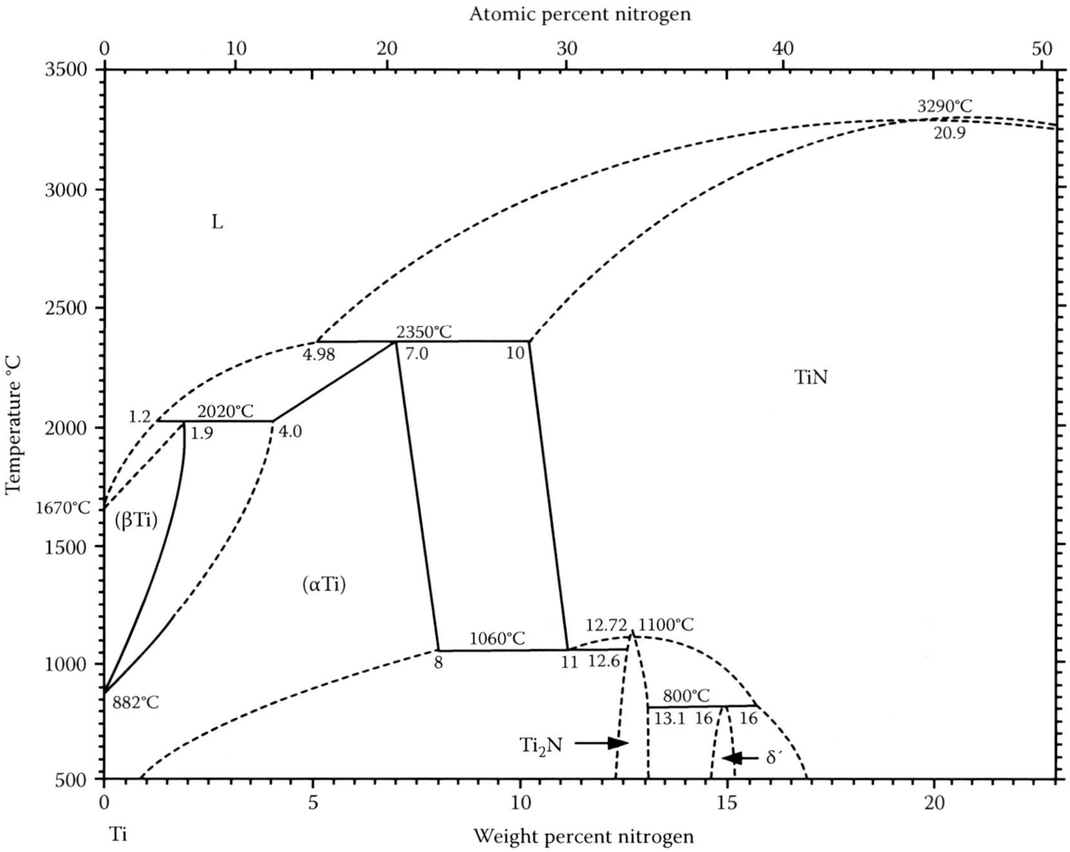

FIGURE 4.19 The binary Ti–N phase diagram. (From ASM. 1992. *ASM Handbook Vol. 3: Alloy Phase Diagrams*. ASM International 9: 2.)

To better understand the possible phase formation sequence, isothermal sections of the Ti–Nb–N ternary system were simulated using PANDAT developed by Computherm LLC and are shown in Figure 4.20a through c, corresponding to temperatures of 2000°C, 1000°C, and 100°C, respectively.

Niobium was chosen as the third component since it is the most abundant element in the alloy after Ti. This ternary phase diagram can provide some valuable insights on the competing nature of the β stabilizing elements and α stabilizing effects of nitrogen on the phase stability in this complex alloy. From Figure 4.20a, it evident that face-centered cubic (FCC) rock salt δ-TiN is a high-temperature phase and exists at 2000°C in conjunction with body-centered cubic (BCC) β phase and liquid for a nominal composition of ~35 wt% Nb and ~10 wt% nitrogen. Figure 4.20b shows for a similar average composition that BCC β, δ-TiN, and Ti_2N are the stable phases at 1000°C, indicating the possible formation of the second nitride, Ti_2N phase. Further lowering the temperature to 100°C results in the precipitation of the hcp α phase in addition to the BCC β, δ-TiN, and Ti_2N phases, as shown in Figure 4.20c. The retention of δ-TiN dendrites, in the LENS-deposited TNZT at room temperature, is explained by the non-equilibrium cooling involved and the sluggish decomposition kinetics of this nitride phase. Furthermore, it should be noted that the PANDAT predictions hold only for the equilibrium phases and are dependent on the accuracy of the thermodynamic models used and their calibration. It can be summarized that the basic variations in the starting microstructures and nature of alloying additions in Ti64 and Ti–Nb–Zr–Ta are responsible for the unambiguous differences that nitrogen addition has in modifying their respective microstructures.

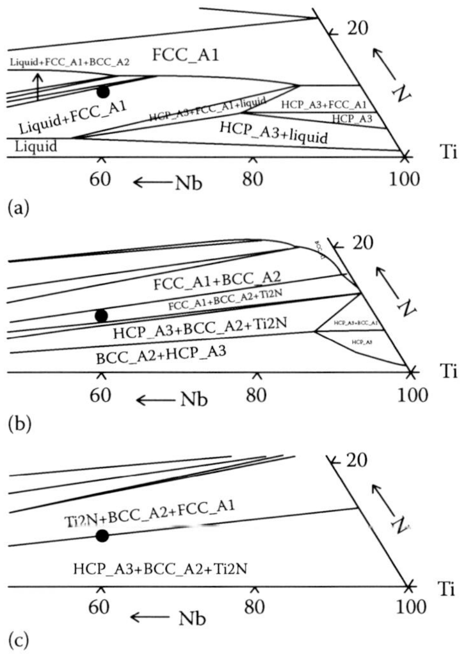

FIGURE 4.20 Isothermal sections of ternary Ti–Nb–N phase diagram showing Ti-rich compositions at (a) 2000°C, (b) 1000°C, and (c) 100°C. Closed circle denotes the approximate ternary composition. (Reprinted with permission from Mohseni et al. 2015. In situ nitrided titanium alloys: Microstructural evolution during solidification and wear. *Acta Materialia* 83: 61–74 [82].)

4.3 FUNCTIONALLY/COMPOSITIONALLY GRADED MATERIALS

The composition and microstructure vary gradually across the length of functionally graded material (FGM), resulting in variation in their properties. The ability to manufacture bulk materials with variation in composition offers great potential and significant advantages over conventional processing. The main feature that distinguishes FGMs from conventional composite materials is the tailoring of graded microstructure and composition in an intentional manner, which enables properties to be selectively enhanced in critical regions. While combinatorial or high-throughput techniques have achieved a high degree of maturity in the pharmaceutical industry, in drug discovery, and in chemistry in general, the development and use of these techniques have been rather limited to the arena of materials. Combinatorial techniques are also now widely used to characterize functional properties in thousands to millions of compositions in a single material library, but the strong dependence of structural properties on microstructure introduces length-scale restrictions that vastly limit this approach for structural materials [83–86]. The most notable example has been in the area of functional materials where the semiconductor industry has exploited multi-target sputter deposition of thin films to rapidly assess the composition–structure–property relationships of a large number of compositions for various device-related applications. While such an approach is well suited for microelectronic devices, since the scale of the thin film being characterized and tested mimics that of eventual use, the extension of the same approach to structural alloys for bulk applications is rather challenging. New approaches to synthesize the library of materials with controlled composition and microstructure gradients are needed, and new experimental techniques that are capable of measuring relevant properties must also be developed. LENS is a solid freeform fabrication process, which involves laser processing of fine powders into fully dense 3D objects directly from a CAD model.

The LENS process is a near-net shape manufacturing process to fabricate complex prototypes leading to cost-effectiveness. In addition, using elemental powder blends in a system with multiple hoppers also allows the possibility of depositing graded composition within the same sample. Using a dual-hopper powder feeder arrangement and blends of elemental powders as feedstock, it is possible to deposit compositionally graded alloys in the LENS system as schematically represented in Figure 4.21a.

To deposit this graded alloy, the first hopper would contain a blend of elemental powders corresponding to one of the end compositions, while the second hopper would contain a blend corresponding to the other end composition. Compositionally graded bars of different geometries can be deposited. For example, an initial rapid screening can be carried out by depositing a cylinder where the

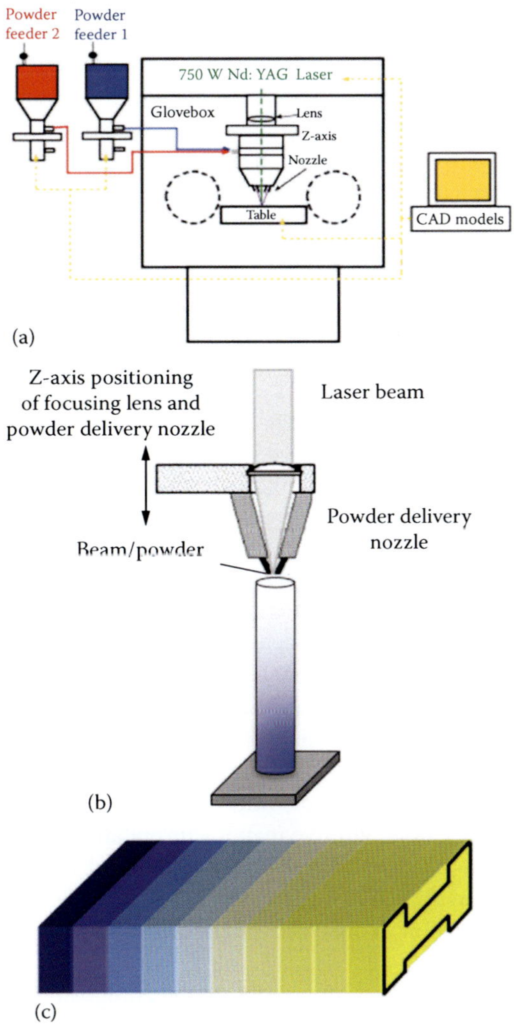

FIGURE 4.21 (a) Schematic layout of the LENS system. (b) Schematic figure showing a compositionally graded alloy deposited in the form of a cylinder. (c) Schematic figure showing a rectangular block-shaped graded alloy deposited with slabs of uniform composition.

compositional gradation is along the axis of the cylinder, as shown in Figure 4.21b. Alternatively, a rectangular bar can be deposited, as shown in Figure 4.21c, where the gradation in composition is lateral across the length of the bar. While the graded cylinder is an ideal geometry for a rapid assessment of the composition–microstructure–microhardness relationships for the range of compositions deposited, the graded rectangular bar is an ideal geometry to investigate the mechanical properties in greater depth by machining mini-tensile specimens from slabs of uniform composition, as shown schematically in Figure 4.21c. In either case, the composition gradation is achieved by controlling the flow rate of powders from each of the powder feeders. Thus, for example, the deposition can be started with only the first powder feeder operating at 100% of a predetermined optimal flow rate. After the first compositional slab has been built, the flow rate of the first feeder is reduced to 90% while the second feeder is then turned on at a flow rate of 10% of the optimal value. After depositing the second compositional slab, the flow rate of the first feeder can be reduced to 70% and that of the second feeder can be increased to 30% and so on. The final section of the graded alloy is deposited with the first powder feeder turned off while the second powder feeder operates at 100% the optimal powder flow rate, resulting in composition of the elemental blend in the second powder feeder. The as-deposited graded alloy is subsequently sectioned and mounted for metallographic characterization. In the case of the graded cylinder (Figure 4.21b), the sectioning would typically be carried out along the axis of the cylinder with the resulting semi-cylindrical section permitting all compositions to be captured within the same metallographic mount for subsequent imaging and microhardness studies. In the case of the graded rectangular bar, postdeposition of this bar will be sectioned into rectangular slabs of nominally uniform composition. Mini-tensile samples can be machined from these rectangular slabs and tested to obtain the tensile properties. Such graded alloys can be used to develop a database relating composition and microstructure to mechanical properties. This database in turn has been used to train and test neural-network models and genetic algorithm–based models for predicting the mechanical properties and also for determining the relative influence of individual composition and microstructural parameters on the properties.

Laser additive manufacturing of compositionally graded materials offers a number of advantages over other bulk rapid alloy prototyping methods [87]. Typically, in LMD, layer-by-layer deposition takes place to fabricate the bulk sample. During deposition of subsequent layers, the consolidated previously deposited material is repeatedly reheated and also partially remelted by the laser beam. Such solid-state reheating will be beneficial for processing of tool steel and related alloys, where strength, toughness, and hardness can be altered by precipitation hardening. Laser additive manufacturing is also beneficial for processing materials that require a high cooling rate, such as oxide dispersion strengthened alloys, or materials containing solute contents above the solubility limit, which are difficult to process via conventional casting process [88,89]. The rapid solidification occurring in laser additive processing because of the high cooling rate (10^3–10^6 K/s) leads to a fine microstructure [1–4]. The raw material used for laser additive manufacturing is either pre-alloyed powders or a mixture of elemental blend powders; laser additive manufacturing is used for processing compositionally graded alloys and widely used for high-throughput as well as alloy development applications. Since laser additive manufacturing allows for a change in composition for a newly depositing layer compared to the previously deposited layer by varying the flow rate of corresponding powders feeders, this technique essentially started using chemical composition as a design parameter in the fabrication of metallic alloys. Hofman et al. [14] have proposed different approaches of processing compositionally graded alloys via laser additive manufacturing processes (Figure 4.22), where the colors represent alloys of different compositions and blended colors indicate that a metallurgical transition has occurred.

An actual build head and a schematic of it are shown in Figure 4.22a and b. Figure 4.22c shows linear transition from alloy A to alloy B following the rule of mixture gradient, where the composition of alloy B will be zero at the bottom and then gradually increases to 100% at the top, and vice versa for alloy A. Step transition in alloy composition is shown in Figure 4.22d where a second alloy will be deposited on the top of the first alloy without any composition gradation. This approach will only

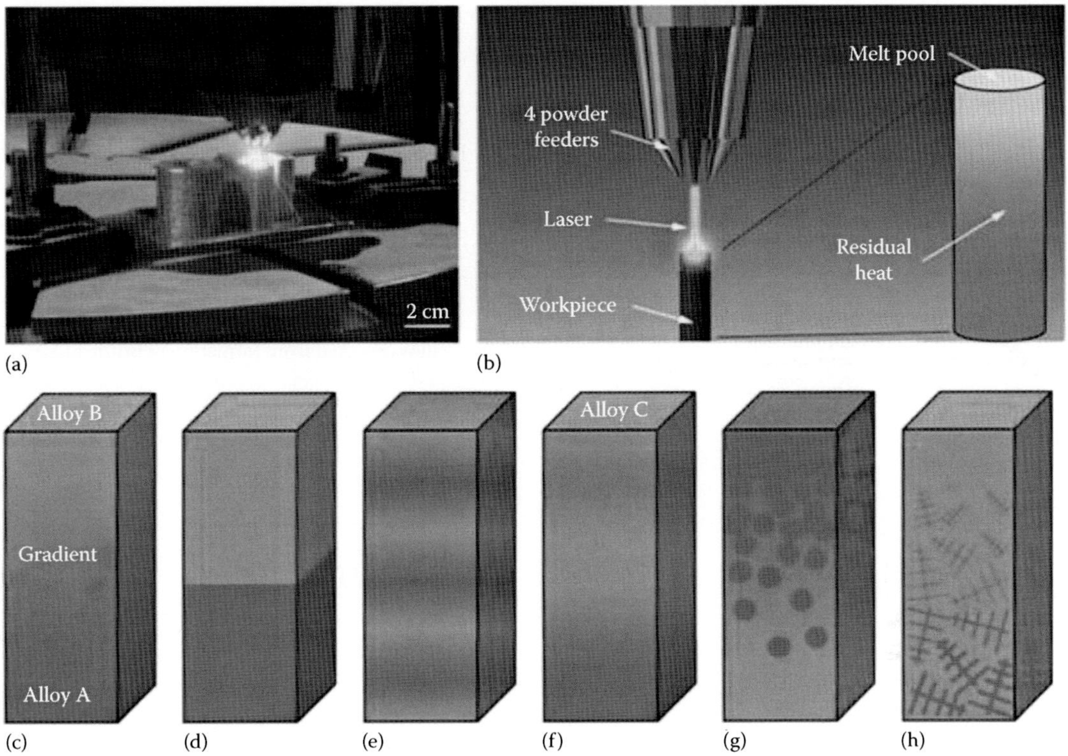

FIGURE 4.22 (a) Gradient alloys being produced at RPM Innovations. (b) A schematic of the build head during the LD process. (c through h) Schematics of some of the different compositionally graded alloys that are possible using the LD process. These include the formation of gradients from one alloy to another and the formation of metal matrix composites (g and h). (Reprinted with permission from Hofman et al. 2014. Compositionally graded metals: A new frontier of additive manufacturing. *Journal of Materials Research* 29 (17): 1899–1910.)

work if the transition layer is free of brittle intermetallic phases. Multiple gradients are shown in Figure 4.22e, where one material has been added strategically throughout the part in order to take advantage of the thermal expansion or magnetism difference between the two deposited layers. Figure 4.22f shows composition transition through three distinct materials. This approach is particularly very useful when gradation between two compositions is very difficult because of the formation of intermetallic brittle phases where a third intermediate material is used to achieve the composition gradient between those materials. Figure 4.22g and h shows two classes of metal matrix composites that can be fabricated via laser additive manufacturing processes. In contrast to traditional alloy development, the FGM allows to vary and process many compositions within the same part. The authors have shown the use of phase diagram to define the gradient path for an additive manufacturing process. As shown in Figure 4.23a, the path could be linear, arbitrary, or binary gradient in order to avoid unwanted brittle phases formed during processing of compositionally graded alloys.

In a linear path, one alloy powder will be mixed linearly to another alloy powder shown in red in Figure 4.23a. If unwanted brittle phases occur with this route, an arbitrary path (shown in green in Figure 4.23a) needs to be taken to circumvent that region. In binary transitions as shown in Figure 4.23a, alloys with minority elements graded to pure metal and then blended with another pure metal in order to reduce complexity occur as a result of their linear transition. Figure 4.23d shows linear (labeled "1"), binary (labeled "2"), and arbitrary (labeled "3") path for gradient transition from Ti–6Al–4V to pure elemental vanadium (V) without encountering intermetallic brittle phases.

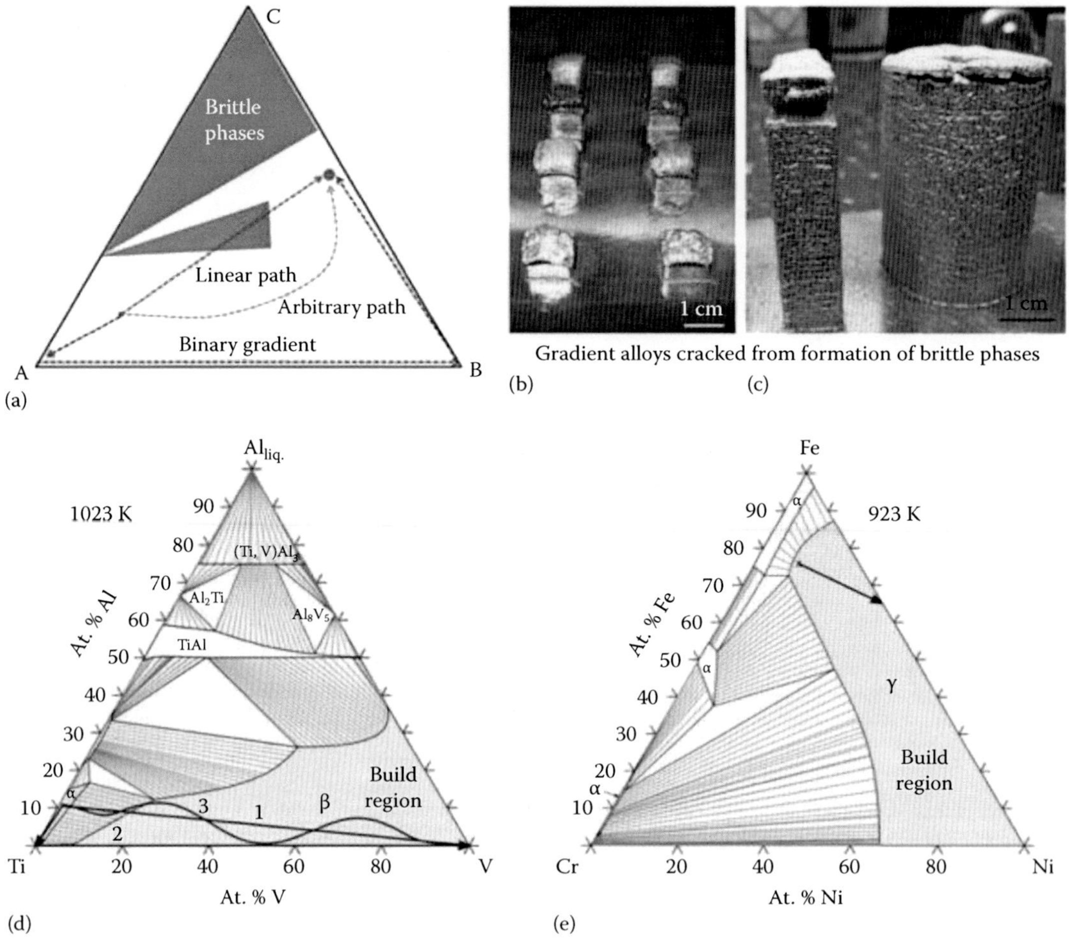

FIGURE 4.23 (a) A schematic of a ternary phase diagram showing possible gradient paths from one alloy to another across a three-element phase space. Several routes are possible, based on factors such as the avoidance of brittle phases. (b and c) Gradient alloys that cracked during fabrication. (b) A failed gradient from Ti to Invar and (c) a failed gradient from Ti to 304L stainless steel. (d) A calculated Al–Ti–V phase diagram showing a gradient from Ti–6Al–4V to pure V showing three different gradient paths: (1) a linear gradient, (2) removing Al and then performing a binary Ti–V gradient, (3) an arbitrary path. The blue region is calculated to be free of brittle intermetallic phases and is composed of hcp and BCC phases. (e) A calculated phase diagram for the Fe–Ni–Cr system showing a gradient from 304L stainless steel to Invar 36. In this diagram, the blue region has been calculated to be all single-phase austenite, which simplifies the build due to the absence of brittle phases. (Reprinted with permission from Hofman et al. 2014. Compositionally graded metals: A new frontier of additive manufacturing. *Journal of Materials Research* 29 (17): 1899–1910.)

The calculated phase diagram for the linear transition from 304L stainless steel to Invar 36 shows the presence of only single-phase austenite (γ) in all gradient compositions (Figure 4.23e). Even though the transition from Ti64 to pure V does not form any brittle phase, the transition from Ti64 to pure Al forms many brittle phases. Figure 4.23b and c show failed gradient in a transition from Ti to Invar 36 and from Ti to 304L steel, respectively, mainly due to thermal residual stresses associated with the additive manufacturing process resulting in cracking of the brittle ordered phases. Figure 4.24 shows potential applications of compositionally graded metal alloys.

The following section reviews few examples of compositionally/functionally graded materials processed via LMD processes.

FIGURE 4.24 Prototypes of gradient alloys fabricated using LD. (a through e) A Ti–6–4 to Nb rocket nozzle where the nozzle is fabricated from a high-temperature refractory metal, Nb, and the body of the rocket is fabricated from low-density Ti–6–4. (f through j) The fabrication of 304L to Inconel 626 valve stems where the valve is high-temperature Inconel 625 and the stem is 304L stainless steel. This part is typically made using a friction weld. (k through n) A 304L to Invar 36 gradient mirror with isogrid backing for increased stiffness. (o) A 304L to Invar 36 gradient alloy without any surface machining, showing the difference in oxide between the two alloys. (p) The alloy from (o) after surface machining and laser welding of a 304L plate, demonstrating weldability of the gradient alloy. (q) A successful gradient from Ti–6–4 to V, then from V to 420 stainless steel. (r) A similar gradient alloy to (q) but with a transition from Ti–6–4 to V, then from V to 304L, showing a crack. (s) After polishing, the crack at a specific composition is visible. (t) SEM micrograph showing that a brittle Fe–V–Cr phase has resulted in the cracking of the gradient alloy. (u) At Ti–6–4 to TiC gradient alloy. (v) Tiled micrographs from the gradient in (u) where the black phase is the TiC particles. This demonstrates the formation of a metal matrix composite using LD. (Reprinted with permission from Hofman et al. 2014. Compositionally graded metals: A new frontier of additive manufacturing. *Journal of Materials Research* 29 (17): 1899–1910.)

4.3.1 Titanium Alloys

4.3.1.1 Titanium–Vanadium and Titanium–Molybdenum Alloys

α/β-Ti alloys, such as Ti–6Al–4V (wt%), have a large number of applications in the aerospace industry. In these Ti-base alloys, vanadium (V) and molybdenum (Mo) are typically added since they are isomorphous β stabilizers. The equilibrium Ti–V and Ti–Mo phase diagrams exhibit monotectoid reactions with a single β phase stable at high temperatures and a phase-separating tendency in the β phase at lower temperatures. At room temperature, the equilibrium phases, for a wide composition range in these binary systems, are α+β. The phase stability and microstructural evolution in Ti alloys containing vanadium (V) and molybdenum (Mo) have been studied quite extensively [90–92]. Williams et al. investigated the effect of ternary additions on the decomposition of the β phase in Ti–V and Ti–Mo alloys [90]. In a more recent paper, the precipitation of α in binary Ti–Mo alloys with Mo contents varying in the range 10–40 wt% has been investigated by Furuhara et al. [91]. All the alloys studied by the authors were solutionized above the β transus temperature and subsequently aged at different temperatures for varying time periods. Their results indicate that for alloys containing 10, 20, and 30 wt% Mo, α precipitation initiates at the β grain boundaries with side plates growing from the boundary into the interior of the grain. The volume fraction of α reduced with increasing Mo content in the alloy. Furthermore, the authors report that for these three alloys, no phase separation was detected within the grains of the β matrix at all aging temperatures. However, for the Ti–40 wt% Mo alloy, aging at low temperatures resulted in the uniform precipitation of α within the β matrix, suggesting a phase separation in the β phase before α precipitation. Ho et al. have studied the structure and properties of a series of binary Ti–Mo alloys, containing 6–20 wt% Mo, from the perspective of biomedical applications [92]. They have discussed the formation of the α″ orthorhombic martensite in these Ti–Mo alloys as a function of Mo content. The possible phase transformations in α/β-Ti alloys when cooled at different rates from above the β transus temperature have been concisely summarized in a recent paper by Ahmed and Rack. This section discusses the deposition of graded binary Ti–V and Ti–Mo alloys using the LENS process starting from a powder feedstock consisting of a blend of elemental titanium (Ti) and vanadium (V) (or molybdenum [Mo]) powders [69,70].

A series of backscattered SEM images from regions of successively increasing V content is shown in Figure 4.25. All the images have been recorded at the same magnification across the compositional gradient.

In Figure 4.25a, corresponding to an average local composition of Ti–1.8% V, the microstructure consists primarily of Widmanstätten α-Ti laths of different orientations with presumably a very small volume fraction of β-Ti at the interfaces between neighboring α laths. The average width of the α laths appears to be ~2 μm. With an increase in the V content, the volume fraction of β increases, as expected based on the binary Ti–V phase diagram. Figure 4.25b shows the microstructure corresponding to an average local composition of Ti–3% V. The phase of lighter contrast in between the α laths is the β phase. In addition to the Widmanstätten α laths, Figure 4.25b also shows a prior β grain boundary decorated with α. Interestingly, when the V content increases to 5%, there is a significant change in the microstructure (Figure 4.25c). Thus, in addition to the increase in the volume fraction of the β phase, there is also a substantial decrease in the average width of the α laths. Furthermore, the resulting microstructure consists of intricately mixed multiple variants of α laths, often referred to as the basketweave microstructure. Interestingly, further increase in the vanadium (V) content (Ti–6.8% V) results in a bimodal mixture of coarser α precipitates and a substantially refined distribution of α laths (Figure 4.25d). During LENS processing, the solid-state annealing of existing layers during the deposition of subsequent layers on top of the existing layers could result in secondary precipitation within the retained β matrix. Thus, the coarser α precipitates is possibly a result of solid-state primary precipitation of α within β during the initial deposition process. The finer-scale α laths are possibly a result of secondary decomposition of the β matrix during postdeposition annealing. An additional point to be noted is the lower aspect ratio of the primary α precipitates as compared

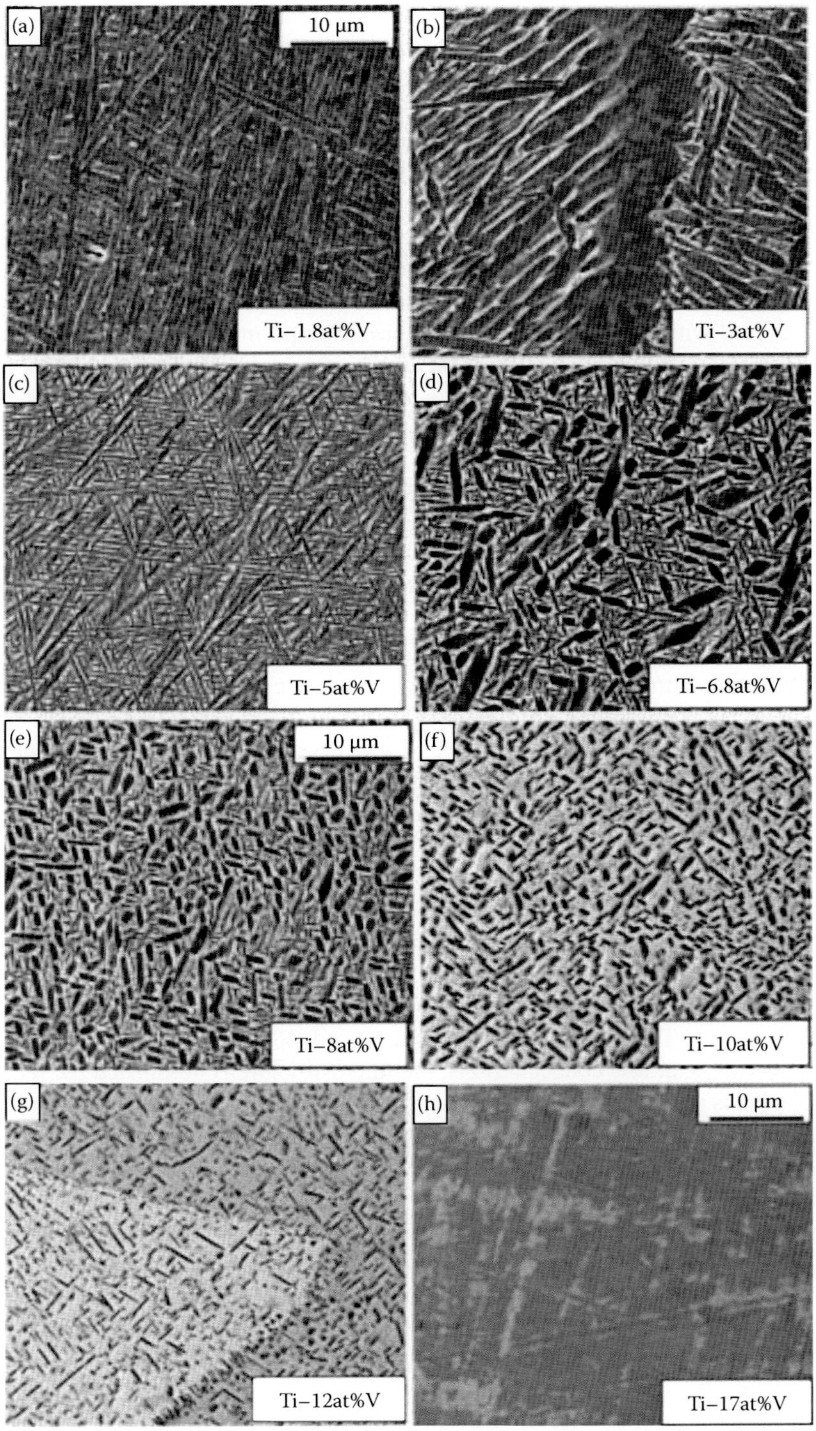

FIGURE 4.25 (a through h) A series of backscatter SEM images from regions with progressively higher V content in the LENS-deposited Ti–*x*V graded alloy. (Reprinted with permission from Collins et al. 2003. Laser deposition of compositionally graded titanium–vanadium and titanium–molybdenum alloys. *Materials Science and Engineering: A* 352 (1): 118–128.)

with that of the secondary α laths. For a composition of Ti–8% V (Figure 4.25e), the microstructural features appear to be similar to those observed for an average alloy composition of Ti–6.8% V with the exception that the average size of the primary α precipitates is relatively smaller. The secondary α laths appear to be of similar size. Further vanadium (V) enrichment to Ti–10% V leads to a uniform distribution of α precipitates of similar size (Figure 4.25f) in contrast to the bimodal distribution of sizes observed in case of the Ti–6.8% V and Ti–8% V compositions. Furthermore, the average size of the α precipitates in the Ti–10% V composition region lies in between the size of primary α precipitates and secondary α laths in the Ti–8% V composition. A substantial decrease in the volume fraction of the α phase can also be observed for this composition (Figure 4.25f). Increase in vanadium (V) content to Ti–12% V (Figure 4.25g) leads to a further decrease in the volume fraction of α precipitates coupled with a further refinement in size. The prior β grain boundary visible in Figure 4.25g also exhibits decoration by discrete α precipitates in contrast to the continuous wetting of the boundary in regions leaner in vanadium (V) content (for example, Figure 4.25b). For a composition of Ti–17% V in the graded deposits (Figure 4.25h), complete stabilization of the β phase was observed.

A series of backscattered SEM images, recorded from the Ti–*x*Mo alloy, are shown in Figure 4.26. The trend in microstructural changes is similar to that observed in case of the Ti–*x*V graded alloy. For low molybdenum (Mo) content (<1%), the microstructure primarily consists of Widmanstätten α laths with a small volume fraction of β at the lath boundaries (Figure 4.26a). For Ti–2.2% Mo, there is a significant reduction in the width of the α laths accompanied by an increase in the volume fraction of β (Figure 4.26b). Interestingly, a small increase in the vanadium (V) content to 2.6% leads to a change in the microstructure, resulting in a bimodal distribution of the α laths within the β matrix (Figure 4.26c). Another interesting observation is that the relatively longer and wider laths appear to be splitting into shorter precipitates. A change in the aspect ratio of these precipitates is apparent. Figure 4.26d shows the microstructure for a composition of Ti–5.4% Mo. The volume fraction of β is significantly higher for this composition and the α phase is homogeneously distributed in the form of refined-scale dispersion of laths within the β matrix. For Ti–7.7% Mo (Figure 4.26e), again a bimodal distribution of a precipitation is observed, similar to the case of Ti–2.6% Mo (Figure 4.26c), but at a substantially more refined scale. Further increase in the Mo content leads to a largely β-Ti alloy with only a small volume fraction of finely distributed α precipitates as shown in Figure 4.26f for the Ti–9.9% Mo composition.

As a function of composition, significant microstructural variations have been observed in the compositionally graded Ti–*x*V and Ti–*x*Mo alloys, laser deposited using the LENS technology. For both alloying elements, at low concentrations, the microstructure is dominated by Widmanstätten α laths forming what is typically referred to as the colony microstructure. A colony of α laths is a group or cluster of laths, formed during the β to α transformation, where all laths belong to the same crystallographic variant. While the colony microstructure can be observed in the Ti–*x*V alloy for compositions up to 3% V (Figure 4.25b), it is seen in the Ti–*x*Mo alloy for compositions typically <1% Mo (Figure 4.26a). With increasing alloying content, there is tendency toward the formation of a more basketweave-like structure with laths exhibiting a higher aspect ratio and smaller average width. This is visible for both V (Figure 4.25c) and Mo (Figure 4.26b) additions with the difference again being in the relative amounts of V and Mo required to achieve similar microstructures (~5% V in case of the Ti–*x*V and ~2% Mo in case of the Ti–*x*Mo alloy). In the basketweave structure, laths of different variants are nucleated in close proximity and subsequently tend to interpenetrate rather than cluster into colonies. The relatively higher amount of V alloying, as compared with Mo alloying, required to form a similar microstructure can be attributed to molybdenum (Mo) being a more effective β stabilizer in titanium alloys as compared with vanadium (V). Further increase in alloying content results in the formation of a bimodal distribution of α precipitates within the β matrix for both V and Mo additions. This is clearly visible in both Figure 4.25d, corresponding to Ti–6.8% V, and Figure 4.26c, corresponding to Ti–2.6% Mo. The longer α laths tend to break up into shorter precipitates with relatively smaller aspect ratios (refer to Figure 4.26c). These larger laths are likely to

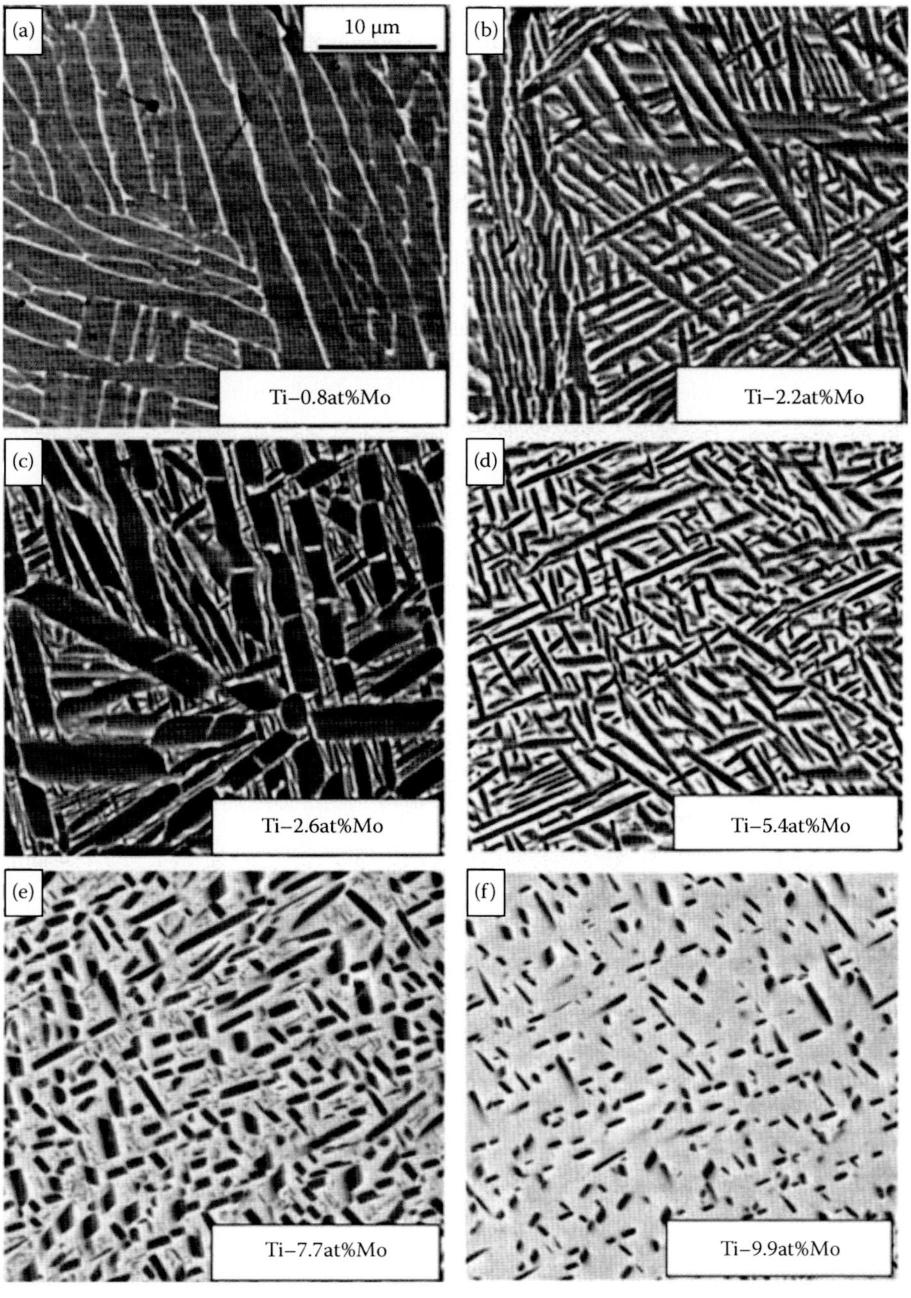

FIGURE 4.26 (a through f) A series of backscatter SEM images from regions with progressively higher Mo content in the LENS-deposited Ti–*x*Mo graded alloy. (Reprinted with permission from Collins et al. 2003. Laser deposition of compositionally graded titanium–vanadium and titanium–molybdenum alloys. *Materials Science and Engineering: A* 352 (1): 118–128.)

have precipitated during the primary $\beta \rightarrow \alpha$ transformation during the deposition of a particular layer and subsequently break up during the reheating of the same layer when newer layers are deposited on top. A closer look at Figure 4.26c suggests that the breaking up of the laths occurs by the nucleation and growth of thin regions of the β phase from the lath/matrix boundary into the interior of the laths.

This observation suggests that during the primary precipitation process, the α laths probably did not achieve their equilibrium composition and morphology, being supersaturated with the alloying addition (V or Mo). Subsequently, during the reheating of the layers after deposition, there is a breakup of the α laths by nucleation and growth of thin β regions along the lath/matrix interface, accompanied by a redistribution of the solute element (either V or Mo). In addition to the primary larger laths, the microstructure consists of a distribution of finer-scale laths in the regions of β retained in between the primary α laths. These finer-scale laths are likely to result from the secondary decomposition of the retained β matrix phase during the reheating process mentioned earlier. Because of the breaking up of the laths, the aspect ratio of the resulting α precipitates is significantly smaller than that of the primary laths. In contrast, the aspect ratio of the finer-scale secondary α laths is much larger than that of the more equiaxed-like α precipitates. Based on the experimental observations, it is possible to propose a sequence of transformations leading to the microstructural development in these graded alloys deposited by LENS. In the initial stages of deposition with relatively low alloying content, the microstructure primary consists of a large volume fraction of α in the formation of Widmanstätten laths (Figure 4.26a). Increase in the alloying content results in an increase in the volume fraction of β. The α laths are unable to thicken to the same extent and are forced to retain larger volumes of inter-lath β phase (Figure 4.26b). The density of α laths (number of laths per unit volume) is larger in the case of Figure 4.26b as compared with Figure 4.26a. A further increase in alloying content, leading to a larger volume fraction of β, is microstructurally manifested with a reduction in the density of primary α laths (larger laths in Figure 4.26c). Few primary laths grow and thicken significantly during the solidification of the layer, still retaining a large volume fraction of β. Subsequently, during reheating of the same layer, two secondary solid-state transformation processes occur. First, there is the precipitation of secondary α laths within the regions of retained β phase. Second, there is a re-precipitation of β at the primary α lath/β matrix interface that grows into the laths, eventually breaking up the laths into more equiaxed-like α precipitates (Figure 4.26c). Both these solid-state processes indicate that the primary α laths as well as the β matrix phase solidified with non-equilibrium compositions, with the α laths being supersaturated (in Mo or V) and the β phase being undersaturated (in Mo or V) at room temperature. The resulting microstructure consists of a bimodal distribution of α precipitates as shown in Figure 4.26c. When the alloying content is increased further, the density of primary α laths increases, with all laths being in the same size range (Figure 4.26d). Comparing Figure 4.26d and Figure 4.26e, the changes occurring owing to the increase in alloying content are similar to those observed from Figure 4.26b and c, only at a substantially finer scale. The transformation processes resulting in the observed changes are also likely to be similar. For even higher alloying contents (~10 at% Mo), the volume fraction of β is very high, and only a small fraction of α, distributed as fine precipitates, is visible in the microstructure (Figure 4.26f).

Using the LENS process, the deposition of graded alloys of Ti–*x*V and Ti–*x*Mo has been successfully demonstrated. While the as-deposited microstructure offers interesting possibilities in terms of microstructure gradients across the compositionally graded alloys, subsequent thermomechanical treatments can be introduced in order to tailor the microstructure for specific applications. The incipient directional solidification in LENS processing results in the formation of columnar grains of β-Ti with dimensions as large as 10 mm along a direction normal to the substrate. Within these large β grains, a substantially refined-scale microstructure develops due to α precipitation under both rapid solidification conditions and subsequent reheating of the deposited layers. Such a microstructure is often desirable for a number of different applications. The ability to produce near-net shape components with graded compositions from elemental powders using the LENS process could potentially be a viable route for manufacturing unitized structures for aerospace applications. Additionally, graded alloys deposited by this method could serve as a very promising tool for combinatorial materials science studies directed toward alloy development for specific applications.

4.3.1.2 Titanium–Niobium–Zirconium–Tantalum (Ti–Nb–Zr–Ta) Alloys

In recent years, there has been a significant thrust directed toward the development of novel implant alloys based on β-Ti. Two recently developed and promising biocompatible β-Ti alloys are Ti–35Nb–7Zr–5Ta and Ti–29Nb–4.6Zr–13Ta. While both these alloy compositions, based on the quaternary Ti–Nb–Zr–Ta system, are promising, there is still a tremendous scope for improvement in terms of alloy design in this and other systems via optimization of alloy composition and thermo-mechanical treatments. Here, a novel combinatorial approach has been used for the development of implant alloys with optimized compositions and microstructures. Using directed laser deposition, compositionally graded alloy samples based on the Ti–Nb–Zr–Ta systems have been fabricated. Subsequently, composition-specific indentation-based hardness and modulus information has been obtained from these samples to construct a database relating the composition and microstructure to the mechanical properties. These databases have been used to train and test fuzzy-logic–based neural-network models for predicting the mechanical properties.

Biomedical alloys have widespread application in joint replacement and other orthopedic surgeries. While the primary properties of interest with respect to biomedical alloys are biocompatibility, corrosion resistance, and wear resistance, it is the lack of their load-bearing capability that limits their applicability [56]. For major applications like total joint replacement arthroplasty, higher yield strength is essentially coupled with the requirement of a lower modulus approaching that of the human bones [55]. Two recently developed promising biomedical alloys, Ti–35Nb–7Zr–5Ta (TNZT) [57] and Ti–12Mo–6Zr–2Fe (TMZF) [93], show significant improvement in these aspects compared to previous generation alloys such as Ti–6Al–4V, stainless steel and cobalt–chromium–based ones. While the mechanical properties and corrosion resistance of Ti–6Al–4V are ideal for implant applications, studies have shown that both V and Al ions may cause long-term health problems. Moreover, the modulus of Ti–6Al–4V (~110 GPa) is substantially higher than that of bone (~10–40 GPa). Finite-element simulations suggest that joint replacements may better simulate the femur in distributing stress if a lower-modulus material is used. Large modulus mismatches cause insufficient loading of bone adjacent to the implant (*stress-shielding* phenomena) and eventual failure of the implant [92]. The high modulus of Ti–6Al–4V is attributable to the high-volume fraction of the α phase in this alloy. The newer biomedical alloys, such as Ti–Nb–Zr–Ta and Ti–Mo–Zr–Fe, exhibit relatively low moduli while maintaining sufficient strength and contain alloying additions that are completely biocompatible. However, the published literature suggests that only a restricted subset of possible alloy compositions in the Ti–Nb–Zr–Ta and Ti–Mo–Zr–Fe complex quaternary systems has been experimentally investigated. In this context, it should be recognized that exploration of such quaternary systems over a wide range of compositions is not only experimentally time consuming but also financially prohibitive. Modeling-assisted combinatorial studies offer a much more attractive alternative. The titanium-based alloys for biomedical applications usually involve complex multi-component chemistries and often exhibit intricate multi-phase microstructures. The property–microstructure–composition relationships in such alloys are relatively poorly understood. There have been some attempts at understanding the influence of alloying additions to titanium on the modulus and yield strength of these alloys [94,95]. Using a discrete variation cluster method, Song et al. [94] have calculated the equilibrium cluster configurations for various Ti alloys and consequently compared their moduli. In another paper, Kuroda et al. [95] have used a d-electron alloy design approach to model the modulus of various Ti alloys for implant applications. In their approach, the bond order and the metal d-orbital energy level have been mapped on a phase stability diagram for various titanium alloy compositions. Based on their locations on this diagram and their moduli and yield strength, the regions of promising model alloy compositions have been determined [95]. While these approaches are good examples of the use of physics-based modeling for alloy design, they are not always appropriate for complex alloy systems involving a number of different alloying additions. In contrast, rules-based modeling approaches, such as fuzzy-logic models, are particularly useful when handling a large number of composition variables, which determine the mechanical properties.

In addition, the benefits of fuzzy-logic modeling as compared to other rules-based approaches are that it requires a relatively small database and the processing steps involved are faster. In the present study, a combinatorial approach, based on laser deposition of compositionally graded alloys, has been used to rapidly develop a database relating composition and microstructure to the mechanical properties. Two types of graded alloys were deposited using a double-powder feeder arrangement via LENS equipment, Ti–20Nb–xTa ($0 < x < 10$) and Ti–xNb–10Zr–5Ta ($20 < x < 35$) (all in wt%) [96]. The database (shown in Table 4.1) consists of composition variables (Nb, Zr, and Ta contents), a microstructural parameter (volume fraction of α), and the corresponding mechanical properties (microhardness and modulus). These experimental results have been used to rapidly populate a database that, in turn, has been used to train and test rules-based fuzzy-logic models for predicting mechanical properties. Thus, by employing such a combination of experimental and modeling techniques via a combinatorial approach, it is possible to accelerate the maturation of existing implant alloys and the development of new ones.

4.3.2 High-Entropy Alloys

High-entropy alloys (HEAs) first came into the limelight in 2004 when the first two publications appeared [97,98], and these materials derived their names from the focus on the configurational entropy [99]. The paradigm shift involved the approach of adding elements in equiatomic proportion to create solid-solution FCC alloys. This method changed the old approach of adding elements to a "parent" element to create alloys. There is no single dominant matrix element. The complexity of the search space is captured with the increase in the number of elements. Thus, for example, it took 9 years since the report of the FCC HEA in 2004 [97] for the first report of an hcp HEA [100]. While single-phase solid solution HEAs are interesting, in most cases, structural materials are incredibly complex systems optimized to provide a demanding balance of properties often under extreme conditions. Traditionally, this complexity has been embodied in the alloy microstructure—and the compositions and processing needed to produce them. HEAs are a new concept to expand the composition space, but the current focus on single-phase microstructures represents a serious limitation [99,101]. Therefore, the new emphasis, specifically of interest for high-temperature structural alloys, is to expand the scope of single-phase alloy exploration in HEAs to include the engineered design of multi-phase alloys. The complexity associated with the vastness of the alloy composition space literally explodes and combines with microstructural complexity to form an interrelated set of daunting challenges, hence the designation of complex concentrated alloys (CCAs) for this new class of metallic alloys.

4.3.2.1 $Al_xCuCrFeNi_2$ HEAs

While most of the published literature on HEAs is focused on single-phase solid solutions, it has been more recently acknowledged that multi-phase microstructures are more likely to exhibit the desired properties for real engineering applications. Hence, some recent efforts have focused on systematic variation of a single element in an HEA composition, to achieve balanced precipitation of a secondary phase within the parent matrix. Previous studies have reported that variation in the aluminum (Al) content affects the microstructure and, consequently, the mechanical properties of HEAs based on the parent FCC solid solution phase CoCrFeNi. Yeh et al. [97] clearly showed that increasing aluminum (Al) in the $Al_xCuCrFeCoNi$ alloy increases the lattice parameter of the parent FCC lattice, inducing strain in the lattice, and, eventually at about $x = 0.8$, results in the precipitation of a BCC/B2 phase within the FCC matrix. Increasing aluminum (Al) further to $x = 2.5$ completely transforms the crystal structure from FCC to BCC/B2. Wang et al. [102] have also reported very similar observations with increasing aluminum (Al) content in $Al_xCoCrFeNi$ alloys. They showed that from $x = 0.5$ to $x = 0.9$, the alloy had a duplex FCC + BCC structure. Karpets et al. [103] evaluated alloy compositions of the type $Al_xCoCrCuFeNi$, which include copper (Cu). They showed the presence of two FCC phases for $x = 0$. The $Al_xCrCuFeNi_2$ system was investigated by Guo et al. [104], where they melted alloys

TABLE 4.1
Database Generated by the Combinatorial Approach and Used for Training and Testing the Fuzzy-Logic Models

Ti (wt%)	Nb (wt%)	Zr (wt%)	Ta (wt%)	VF alpha	*H* (VHN)	*E* (GPa)
66.9	18	10.9	4.3	0	473.3	93.25
61.3	23	10.6	5	0	490.54	122.88
56.2	26	10.5	7.3	0	461.22	101.8
61.3	23	10.6	5	0	470.29	9.42
74.5	19.2	0	6.4	0	359.99	107.67
80.3	19	0	0.6	0	481.69	96.28
76.1	21.4	0	2.5	0	435.01	90.43
78	16	0	6	0	433.72	89.31
70.5	19.3	0	10.2	0	409.58	88.42
68.8	20.1	0	11.1	0	383.18	91.27
74.5	18.7	0	6.8	52.9	453.62	115.96
69.4	20.4	0	10.2	36.2	390.48	119.71
65.4	18.8	10.7	5.2	47.6	423.58	114.86
61.5	22.9	10.2	5.5	35.3	461.43	109.74
76.3	18.2	0	5.5	43.6	398.71	120.6
77.3	20.2	0	2.5	45.5	462.89	118.41
78.5	21.4	0	0.1	41	491.06	123.46
69.1	20	0	10.9	36.2	420.92	107.19
61	24.1	10.7	4.1	36.2	488.73	116.67
66.1	18.2	11	4.8	42.9	449.11	105.65
59.6	25.4	10.8	4.3	21	428.92	99.68
53.1	32.1	10.7	4.2	24	466.3	97.34
68.4	19.8	0	11.8	55.6	426.27	107.78
70.1	20.1	0	9.9	51.2	451.7	110.82
75.3	19.4	0	5.3	47.5	403.57	104.22
75	21.2	0	3.8	45.7	445	113.71
77.9	21.3	0	0.9	42.5	443.27	112.76
79.7	19.7	0	0.7	39.4	368.1	111.94

Source: Adapted from Banerjee et al. 2005. A novel combinatorial approach to the development of beta titanium alloys for orthopedic implants. *Materials Science and Engineering: C* 25 (3): 282–289.

containing differential aluminum (Al) contents and reported the complex eutectic structures forming in the range of compositions, x = 1.2–2.5 in great detail. Despite these previous investigations, a systematic study of the influence of aluminum (Al) content on the microstructure of HEAs or CCAs, transitioning from a FCC parent lattice to a BCC parent lattice, has not been carried out. Furthermore, the transition from a single FCC solid solution or two FCC solid solutions to a two-phase microstructure consisting of ordered $L1_2$ precipitates within the solid-solution phase has not been addressed in detail. The focus of this study is to investigate the changes in microstructure and phase stability, and the associated changes in microhardness properties, in a more systematic manner, using laser additively processed, compositionally graded $Al_xCrCuFeNi_2$ alloys, where x, the atomic/mole fraction of aluminum (Al), was varied from 0 to 1.6 along the compositionally graded alloy [1].

Figure 4.27a shows the x-ray diffraction (XRD) patterns for the alloys with the aluminum (Al) content progressively increasing from $x = 0$ to $x = 1.6$.

As evident from the XRD patterns, for a low content of aluminum (Al) (compositions $x = 0$ and $x = 0.8$ in Figure 4.27a), the dominant phase is FCC based, and as the aluminum (Al) content is

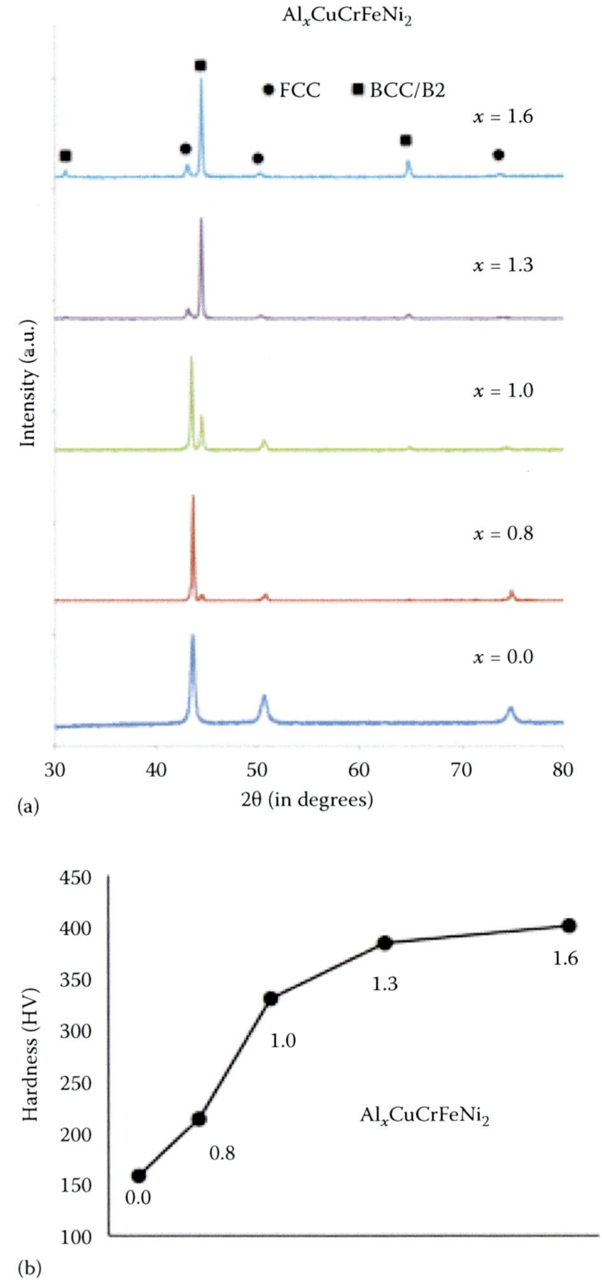

FIGURE 4.27 (a) XRD patterns from different regions of a compositionally graded HEA alloy of composition $Al_xCuCrFeNi_2$ with varying Al content. (b) Microhardness variation across the graded HEA alloy shown in (a) with varying Al content. (Modified from Borkar et al. 2016. A combinatorial assessment of $Al_xCrCuFeNi_2$ ($0<x<1.5$) complex concentrated alloys: Microstructure, microhardness, and magnetic properties. *Acta Materialia* 116: 63–76.)

increased, a BCC-based phase appears. The volume fraction of this BCC-based phase increases with increasing aluminum (Al) content accompanied by a reduction in the volume fraction of the FCC-based phase, and a mixture of both phases is clearly visible in the alloy of composition $x = 1.0$ in Figure 4.27a. The composition corresponding to the highest aluminum (Al) content of $x = 1.6$ contains primarily the BCC-based phase with a small volume fraction of retained FCC-based phase. Furthermore, for compositions of $x = 1.3$ and $x = 1.6$, the {001}B2 superlattice peak is also visible in the XRD patterns, indicating the presence of B2 order in the BCC-based phase for these compositions. The corresponding changes in the microhardness as a function of composition across this graded alloy have been plotted in Figure 4.27b. There is a substantial change in microhardness across this range of compositions, from ~150 VHN for the lowest aluminum (Al)-containing composition to more than 400 VHN for the highest aluminum (Al)-containing composition. It is interesting to note the rapid increase in the microhardness with the increase in the volume fraction of the BCC-based phase as a function of increasing aluminum (Al) content. BSE images showing the change in microstructure as a function of aluminum (Al) content are shown in Figure 4.28.

The dramatic effect of aluminum (Al) is noteworthy considering the single-phase microstructure visible for $x = 0$ changing to a complex two-phase microstructure with increasing volume fraction of the second phase with progressively increasing aluminum (Al) content. The region with $x = 0$ exhibited a typical dendritic microstructure (Figure 4.28a). The cellular/dendritic boundaries between each grain associated with interdendritic compositional segregation. While there is a distinct contrast in between the dendritic and interdendritic regions in Figure 4.28a, this contrast can be attributed to the segregation of copper (Cu) (which tends to partition/segregate readily) into the interdendritic regions, making them appear substantially brighter in BSE images. At the SEM scale of observation, the dendritic structure continued at $x = 0.8$ (Figure 4.28b), but at this composition, secondary phases

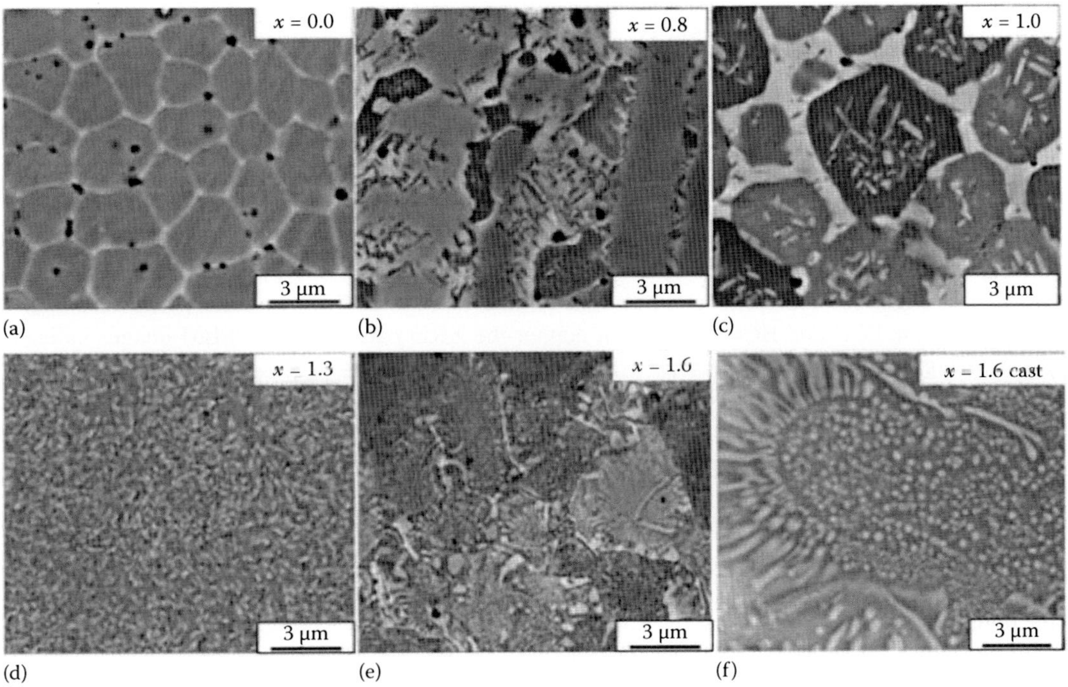

FIGURE 4.28 (a through e) Backscatter SEM images showing the microstructural evolution as a function of Al content in the LENS-deposited graded HEA alloy $Al_xCuCrFeNi_2$. (f) The microstructure for a conventionally cast (arc-melted) alloy of nominal composition $Al_{1.6}CuCrFeNi_2$. (Modified from Borkar et al. 2016. A combinatorial assessment of $Al_xCrCuFeNi_2$ (0<x<1.5) complex concentrated alloys: Microstructure, microhardness, and magnetic properties. *Acta Materialia* 116: 63–76.)

are clearly visible within the microstructure. These secondary phases exhibit two distinct morphologies and size scales, seen more clearly in Figure 4.28b. The regions between the FCC dendrites form coarse precipitates exhibiting a darker contrast that do not appear to exhibit a specific faceted morphology. Fine-scale acicular or needle-like precipitates are another type of precipitate that can be observed for the $x = 0.8$ composition, which also exhibit a darker contrast, but distributed within the dendrites (refer to Figure 4.28b). Based on the XRD results from this region with $x = 0.8$ (Figure 4.27a), it can be speculated that these darker precipitates forming within the dendritic FCC matrix are BCC/B2 precipitates, but a more detailed investigation, at a finer length-scale, using TEM is required in order to understand the details of the microstructure (presented below). At $x = 1.0$, the microstructure (Figure 4.28c) primarily shows equiaxed grains exhibiting a relatively darker contrast, surrounded by a brighter contrast phase wetting the grain boundaries. Additionally, fine-scale needle-like precipitates exhibiting a brighter contrast are visible within the equiaxed grains of the matrix phase. The most drastic change in microstructure is between the compositions of $x = 1.0$ and $x = 1.3$. Furthermore, in order to establish that the LENS-deposited graded alloys exhibited a microstructure similar to that achieved via bulk processing, an arc-melted button was prepared of composition $x = 1.6$ and the microstructure of the as-cast ingot is shown in Figure 4.28f. Comparing the LENS-processed microstructure (Figure 4.28e) with the cast one (Figure 4.28f), it is evident that for nominally similar compositions, the microstructure and phases are similar for both processing routes. The main difference is a substantial refinement in the scale of the microstructure in the case of the LENS-processed alloy that can be attributed to the higher solidification rates encountered during laser deposition (~10^2 to 10^3 K/s). For $x = 0$, or the alloy composition $CuCrFeNi_2$, a bright-field TEM image of the microstructure is shown in Figure 4.29a.

This TEM sample has been prepared in a site-specific manner using a dual-beam FIB instrument, and the specific region chosen for this purpose is indicated in the inset of Figure 4.29a. The foil captures an interdendritic region in the FCC microstructure, which corresponds to the region of darker contrast in the bright-field TEM image. A $[011]_{FCC}$ zone axis diffraction pattern from the FCC matrix is shown in Figure 4.29b. It is evident from this diffraction pattern that there are only reflections from the FCC matrix and no other phase, indicating that in the absence of aluminum (Al) ($x = 0$), ordering within the FCC matrix was not discernible. Additionally, a more detailed investigation of the microstructure for the composition $x = 0.8$ using TEM revealed that matrix grains of the FCC-based phase are in fact ordered and exhibit an $L1_2$-type ordering, as shown in Figure 4.29c. The finer-scale precipitates within the matrix grains, visible in the SEM image shown in Figure 4.28b, have been determined to be ordered B2 precipitates, as shown in Figure 4.29d. Interestingly, the XRD pattern for the same composition, shown in Figure 4.27, exhibits peaks that can be attributed to the FCC and BCC phases, but neither the XRD pattern nor the SEM images indicated any ordering within these phases.

This study presents a novel combinatorial approach for rapid assessment of composition–microstructure–property relationships in single-phase HEAs or multi-phase CCAs. This combinatorial approach, based on laser additive deposition of compositionally graded alloys, has been successfully applied to process graded $Al_xCrCuFeNi_2$ alloys (x varying from 0 to 1.6) using LENS technology. This approach has permitted a detailed assessment of the transition in microstructure along the same alloy gradient, from a predominately FCC solid solution to FCC/$L1_2$ to mixed FCC/$L1_2$ + BCC/B2 and finally to predominantly BCC/B2 as the aluminum (Al) content increases from 0 to 1.6 (molar fraction). This change in microstructure and phase constitution was accompanied by a corresponding progressive increase in microhardness with increasing aluminum (Al) content, clearly indicating that the BCC/B2 microstructure is substantially harder than the FCC/$L1_2$ microstructure.

4.3.2.2 $AlCo_xCr_{1-x}FeNi$ HEAs

The HEA field is flooded with many research articles, which mainly focus on microstructure, and mechanical performance, while systematic work on the understanding of physical properties, especially magnetic properties, is still limited. HEAs based on the AlCoCrFeNi system have been widely

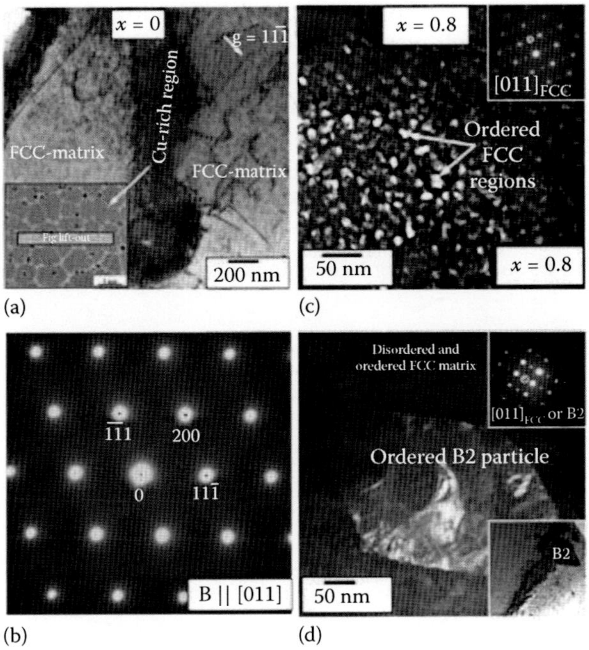

FIGURE 4.29 TEM results from $CrCuFeNi_2$, (a) Bright-field TEM (BFTEM) image showing the FCC matrix (inset shows the SEM images marking the FIB-lift out region in yellow); (b) SADP pattern from a $[011]_{FCC}$ zone axis (ZA). TEM results from $Al_{0.8}CrCuFeNi_2$; (c) DFTEM image showing the ordered (lighted up) and disordered FCC matrix taken from the encircled superlattice spot in SADP pattern from $[011]_{FCC}$ ZA shown in the inset; (d) DFTEM image showing the ordered (lighted up) B_2 particle present as a second phase, taken from the encircled superlattice spot in SADP pattern shown in the inset from $[011]_{BCC}$ ZA. (Modified from Borkar et al. 2016. A combinatorial assessment of $Al_xCrCuFeNi_2$ (0<x<1.5) complex concentrated alloys: Microstructure, microhardness, and magnetic properties. *Acta Materialia* 116: 63–76.)

investigated in the literature, focusing on their microstructure, mechanical, anticorrosive, thermal, electrical, and magnetic properties. However, these previous reports typically investigate the influence of varying aluminum (Al) content in $Al_xCoCrFeNi$-type compositions. Fine-tuning the microstructure and composition can have a substantial impact on the magnetic properties, and from this perspective, the contents of chromium (Cr) and cobalt (Co) in this alloy system becomes critical [105–107]. Therefore, there is a definite need to understand the microstructural evolution and its effect on magnetic properties with varying chromium (Cr) and cobalt (Co) content in $AlCo_xCr_{1-x}FeNi$-type alloys. Additionally, it is worth noting that most of the commercially used hard magnets are based on systems such as AlNiCo or FeCrCo and the soft magnets are based on NiFe, FeSi, or FeCo. Zhang et al. studied the structural and magnetic properties of $FeCoNi(AlSi)_x$ ($0 \le x \le 0.8$) HEAs [108]. They found structural transition from FCC for $x \le 0.2$ to BCC for $x \ge 0.4$ while a mix phase of FCC and BCC was observed in $0 \le x < 0.4$. The structural change from single FCC phase to single BCC phase with a transition duplex FCC/BCC region has also been reported with increasing x in $Al_xCoCrFeNi$ ($0 \le x \le 2$) HEAs [109]. In FeNiCrCuCo HEAs, if copper (Cu) or cobalt (Co) is replaced by aluminum (Al), the microstructures changed from FCC solid solution to BCC solid solution [110]. The FCC–BCC transition with changing composition is normal in HEAs and can be explained by the valence electron concentration and the lattice distortion [111,112]. Phase stability is very important for controlling the mechanical, electrical, and magnetic properties of these HEAs. Thus, the main objective of this investigation is to study the effect of cobalt (Co)/chromium (Cr) ratio on the microstructure and magnetic properties of $AlCo_xCr_{1-x}FeNi$ HEAs [113]. The systematic variation in composition was achieved via laser deposition of a compositionally graded alloy. Such a combinatorial approach permits the investigation of multiple compositional steps in a relatively fast (yet systematic) manner and leads to new insights related to the evolution of microstructure and its consequent influence on the magnetic properties in such complex concentrated multi-component systems.

The composition gradient in the as-deposited $AlCo_xCr_{1-x}FeNi$ graded alloy is schematically shown in Figure 4.30a together with the composition profiles for aluminum (Al), iron (Fe), cobalt (Co), chromium (Cr), and nickel (Ni), based on SEM-EDS measurements.

For convenience of analysis, the entire graded alloy has been divided into six distinct composition steps, A, B, C, D, E, and F, corresponding to average compositions of $x = 1, 0.8, 0.6, 0.4, 0.2$, and 0.

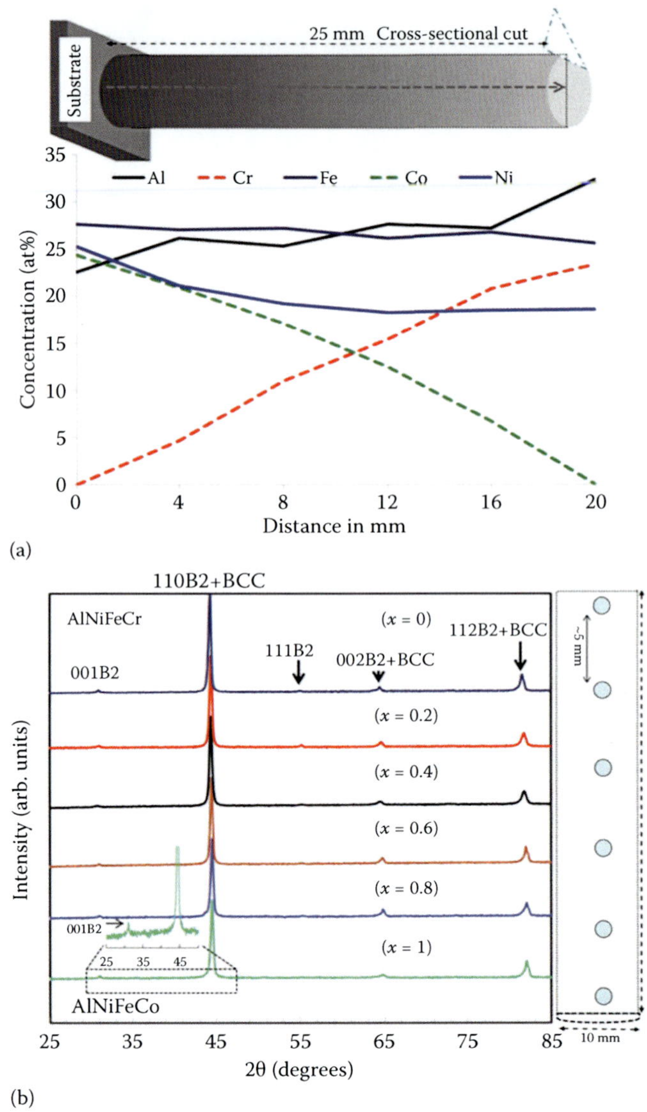

FIGURE 4.30 (a) Composition gradient in the as-deposited $AlCo_xCr_{1-x}FeNi$ graded alloy, based on SEM-EDS measurements. (b) XRD patterns from site-specific locations (compositions) along the LENS-deposited $AlCo_xCr_{1-x}FeNi$ graded alloy. (Reprinted with permission from Borkar et al. In press. A combinatorial approach for assessing the magnetic properties of high entropy alloys: Role of Cr in $AlCo_xCr_{1-x}FeNi$. *Advanced Engineering Materials*.)

The Cr at% varies from 0 to 24 and Co at% varies from 24 to 0; the compositions of the remaining elements (Al, Fe, and Ni) were kept nominally equiatomic. The measured compositions are quite close to the targeted values. XRD patterns from site-specific locations (compositions) along the LENS-deposited $AlCo_xCr_{1-x}FeNi$ graded alloy are shown in Figure 4.30b. The principal peaks in all the XRD patterns have been consistently indexed as the {011}BCC, {002}BCC, and {112}BCC peaks, with a strong {011}BCC texture. Additionally, a peak is observed in the XRD patterns from all the regions, at 2θ of ~30°; this peak can be indexed as the {001} superlattice peak of an ordered B2 phase. Therefore, based on the XRD patterns, it can be concluded that for all the alloy compositions investigated, $x = 1$ to $x = 0$, there is at least one ordered B2 phase, and possibly a mixture of ordered B2 and disordered BCC phases. The microstructure of the six different selected regions ($x = 1$, 0.8, 0.6, 0.4, 0.2, and 0) of the graded $AlCo_xCr_{1-x}FeNi$ alloy has been shown in the montage of backscattered SEM images in Figure 4.31.

At $x = 1$ (i.e., nominal composition AlCoFeNi), the microstructure shows equiaxed grains of a single B2/BCC phase (lower-magnification image on top and higher-magnification image at the bottom). For $x = 0.8$, while the primary grain structure appears to be similar to that of region $x = 1.0$, second-phase precipitation was observed at the B2/BCC grain boundaries. This grain boundary phase exhibits a substantially brighter contrast in the backscatter SEM image, indicating that the average atomic mass of this phase is greater than that of the B2/BCC phase comprising the equiaxed grains. For $x = 0.6$, the equiaxed grains of the B2/BCC phase exhibit a fine substructure; the higher atomic mass grain boundary phase is still present. The region corresponding to $x = 0.4$ exhibits fine-scale decomposition within the B2/BCC grains, similar to that observed in the case of $x = 0.6$. However, the grain boundary phase no longer appears in the case of $x = 0.4$. Regions corresponding to $x = 0.2$ and $x = 0$ exhibit similar microstructures, with the primary B2/BCC grains exhibiting fine-scale intragranular decomposition. However, no other second phase was observed in either of these regions in the higher-magnification backscatter SEM images. Interestingly, the fine-scale intragranular decomposition pattern observed in the case of regions $x = 0.6$, 0.4, 0.2, and 0 appears like a spinodal decomposition within the B2/BCC grains. However, this can be confirmed only via investigations at higher resolution. Furthermore, it should be noted that the tendency for intragranular

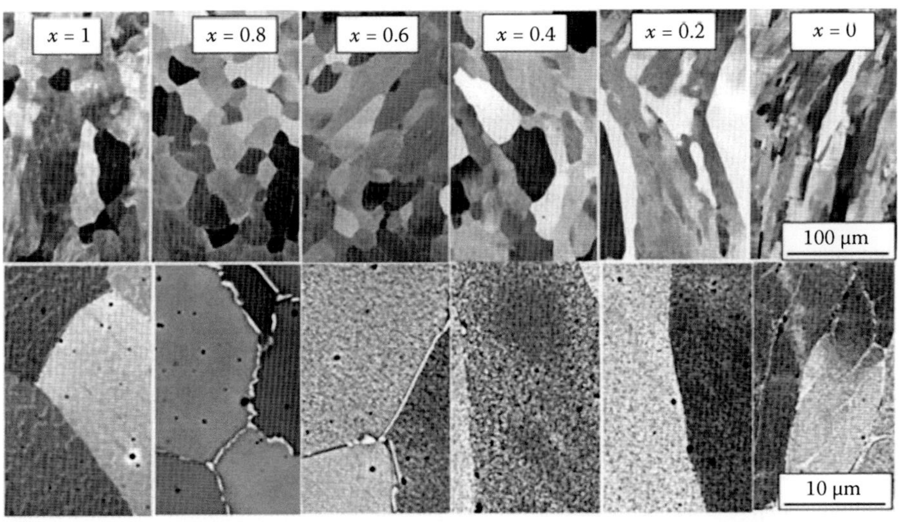

FIGURE 4.31 Backscattered SEM microstructure of the six different selected regions ($x = 1$, 0.8, 0.6, 0.4, 0.2, and 0) of the graded $AlCo_xCr_{1-x}FeNi$ alloy. (Reprinted with permission from Borkar et al. In press. A combinatorial approach for assessing the magnetic properties of high entropy alloys: Role of Cr in $AlCo_xCr_{1-x}FeNi$. *Advanced Engineering Materials*.)

decomposition appears to become more prominent with increasing chromium (Cr) content and reducing cobalt (Co) content in this graded alloy.

Vibrating sample magnetometer results for the LENS-deposited $AlCo_xCr_{1-x}FeNi$ alloys are shown in Figure 4.32.

The hysteresis loops of the $AlCo_xCr_{1-x}FeNi$ alloys are shown in Figure 4.32a. It is evident that the saturation magnetization (M_s) decreases with increasing Cr content and, correspondingly, the decrease in cobalt (Co) content. Cobalt (Co) being the main element that contributes to ferromagnetism and chromium (Cr) being antiferromagnetic in nature, the soft magnetic properties of $AlCo_xCr_{1-x}FeNi$–based alloys are expected to decrease with an increase in chromium (Cr) content and a corresponding decrease in cobalt (Co) content. Figure 4.32a shows the magnetic hysteresis loops obtained at 300 K for $AlCo_xCr_{1-x}FeNi$ ($x = 0, 0.2, 0.4, 0.6, 0.8$, and 1). The variations in M_s and H_c, measured at $T = 300$ K, as a function of composition (x), have been plotted in Figure 4.32b. The saturation magnetization is primarily determined by the composition and atomic-level structures and

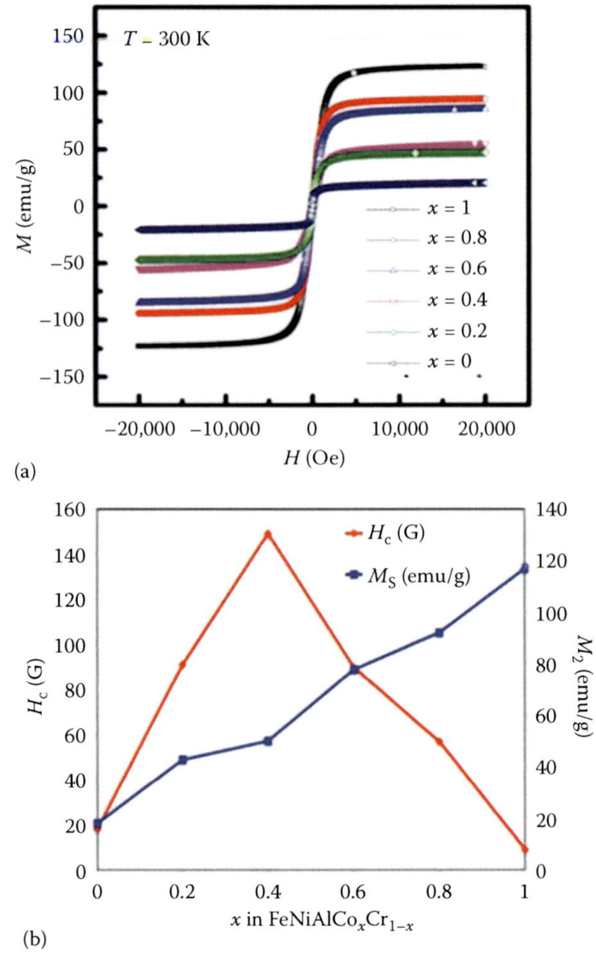

FIGURE 4.32 (a) Magnetizations (M) as a function of applied magnetic field (H) at temperatures of 300 K for $AlCo_xCr_{1-x}FeNi$ ($x = 0, 0.2, 0.4, 0.6, 0.8$ and 1) and (b) show the saturation magnetizations (M_s) and coercivity (H_c) as a function of Co concentration ($X = 0$ to 1) at 300 K. (Modified from Borkar et al. In press. A combinatorial approach for assessing the magnetic properties of high entropy alloys: Role of Cr in $AlCo_xCr_{1-x}FeNi$. *Advanced Engineering Materials*.)

is less sensitive to the microstructural parameters, such as grain size and morphology. Indeed, the saturation magnetization (M_s) raises almost monotonically upon the addition of cobalt (Co) and corresponding reduction in chromium (Cr), from 18.48 emu/g at $x = 0$ (AlNiFeCr) all the way to 117.8 emu/g at $x = 1$ (AlNiFeCo). In contrast, the composition dependence of H_c appears to be nonmonotonic: H_c is low for $x = 0$, but increases dramatically at $x \leq 0.4$ and then falls again for $x >$ 0.4, reaching a low value at $x = 1$. This is mainly due to the increase in cobalt (Co) content and decrease in chromium (Cr) content in these alloys. The addition of chromium (Cr) significantly reduces the magnetization because the magnetic moment of chromium (Cr) is antiparallel to that of Fe/Co/Ni (i.e., antiparallel magnetic coupling), resulting in the reduction/cancellation of the magnetic moment in chromium (Cr)-containing HEAs. Li et al. [114] have reported a similar behavior in the alloy $Fe_{3-x}Cr_xSe_4$ ($x = 0$ to 2), where the coercivity is maximum at room temperature for $x = 0.7$ and then decreases on both sides of x, as observed in the present case. The non-magnetic component aluminum (Al) is found to be nearly non-polarized (or slightly antiparallel with Fe/Co/Ni). Thus, $x = 1.0$ (AlNiFeCo) exhibits the best soft magnetic properties as compared to other compositions in this system.

4.3.2.2.1 *Microstructural Evolution and Its Impact on Magnetization in $AlCo_xCr_{1-x}FeNi$ Alloys*

The combinatorial analysis of the compositionally graded $AlCo_xCr_{1-x}FeNi$ alloy leads to the following observations regarding the microstructural evolution as a function of composition:

1. Over the entire range of compositions, from $x = 0$ to 1, the alloys exhibit a BCC-based microstructure consisting of the ordered B2 phase and a disordered BCC solid solution phase.
2. With decreasing cobalt (Co) content (decreasing x), or increasing chromium (Cr) content, there is a progressive tendency toward phase separation via spinodal decomposition within the BCC-based matrix grains to form a mixed two-phase B2+BCC microstructure. Thus, while $x = 1$ (AlFeNiCo) exhibits grains of a single B2 phase, $x = 0$ (AlFeNiCr) exhibits pronounced spinodal decomposition within the matrix grains, resulting in a patterned two-phase B2+BCC microstructure.
3. Specific compositions along the graded $AlCo_xCr_{1-x}FeNi$ alloy, $x = 0.8$ and $x = 0.6$, exhibit a second grain boundary precipitate phase, which is an FCC solid solution.

The magnetization behavior and magnetic properties of different sections of the $AlCo_xCr_{1-x}FeNi$ graded alloy vary substantially as a function of composition coupled with the microstructure. Thus, increasing the cobalt (Co) content from $x = 0$ (AlCrFeNi) to $x = 1$ (AlCoFeNi) exhibits a monotonically increasing value of M_s, since the saturation magnetization is strongly dependent on the alloy composition with lesser influences from the microstructure. In contrast, the variation in H_c is highly nonmonotonic and appears to be strongly influenced by the microstructure. Thus, while H_c increases with increasing cobalt (Co) content from $x = 0$ to $x = 0.4$, the value drops quite steeply with increasing x from $x = 0.4$ to $x = 1.0$. The decrease in H_c from $x = 0.4$ to $x = 1.0$ can potentially be attributed to the change in microstructure from a pronounced spinodally decomposed B2+BCC microstructure for $x = 0.4$, to a marginally phase-separated B2 microstructure (with composition fluctuations) for $x = 1.0$. Thus, the extent of spinodal decomposition within the matrix grains could potentially influence the pinning of magnetic domains and consequently the coercivity. The well-developed spinodally decomposed B2+BCC microstructure in the case of $x = 0.4$ can exhibit a higher coercivity due to the B2/BCC boundaries acting as pinning sites for the domains, while such boundaries do not exist within the large grains of the single B2 phase observed in the case of $x = 1.0$. Additionally, the substantial difference between the chemical compositions of the B2 and BCC phases at $x = 0.4$ presumably causes a difference in their respective lattice parameters, and the two phases accommodate this mismatch via coherency stresses, which could potentially stabilize the orientation of elementary magnetic domains imposed by an external magnetic field. Furthermore, it should be noted that spinodally decomposed microstructures have been utilized to design most of the modern-day permanent magnets including Fe–22Cr–12Co– and AlNiCo–based ones [106,107,115,116]. The anisotropy of the magnetic properties depends on the form of the elementary domains and can be modified either by annealing in a magnetic field or via cold deformation.

A combinatorial assessment of composition–microstructure–magnetic property relationships in a magnetic high-entropy $AlCo_xCr_{1-x}FeNi$ alloy ($0 \leq x \leq 1$) system has been carried out using compositionally graded alloys fabricated via laser additive manufacturing. At one end, the AlCoFeNi composition ($x = 1$) consisted of equiaxed B2 grains, exhibiting very early stages of phase separation (only compositional partitioning) into Ni–Al–rich and Fe–Co–rich regions within grains of the B2 phase. At the other extreme, the AlCrFeNi composition ($x = 0$) exhibited grains with pronounced spinodal decomposition, resulting in a B2+ BCC microstructure with the degree of spinodal decomposition progressively increasing with Cr content in these $AlCo_xCr_{1-x}FeNi$ alloys. While the saturation magnetization (M_s) monotonically increases six times from $x = 0$ to $x = 1$, the coercivity (H_c) variation is nonmonotonic, increasing seven times from $x = 0$ to $x = 0.4$, and subsequently decreasing 14 times from $x = 0.4$ to $x = 1.0$.

4.3.3 Fe–Si–B–Cu–Nb Soft Magnetic Materials

Iron-based nanocrystalline soft magnetic alloys have attracted great interest in the past 20 years because of their excellent magnetic properties (such as high magnetic permeability, low coercivity, and low magnetostriction), good corrosion, as well as electrical resistance [117–121]. These superior magnetic properties arise from the formation after annealing heat treatment of high density of nanocrystalline Fe_3Si precipitates within the initially amorphous alloy [117,118,121–126]. Because of excellent soft magnetic characteristics, FINEMET-based alloys have been used in various electrical and electronic applications [127]. The role of copper (Cu), niobium (Nb), silicon (Si), and boron (B) elements in achieving such microstructure and its correlation with the magnetic properties has been comprehensively studied [117,118,121,128–131]. It is well known that the addition of copper to the Fe–Si–B alloy results in clustering of this alloying element, with the copper-rich clusters acting as potent nucleation sites for the α-Fe(Si) nanocrystals, while niobium (Nb) restricts growth of these nanocrystals [132–134]. Thus, these alloying additions result in a much finer microstructure with much lower coercivity but somewhat lower saturation magnetization compared to ternary Fe–Si–B–based compositions [135]. This can be attributed in part to the lower concentration of iron (Fe) in FINEMET. The element boron (B), which has very little solubility in the iron (Fe) lattice, strongly affects the volume fraction of the BCC phase. The volume fraction of nanocrystallites increases with increasing silicon (Si)/boron (B) ratio. Boron and silicon are two key elements in FINEMET base alloys, and they are mainly introduced to improve glass-forming ability and electrical resistivity, respectively. The role of alloying elements on the microstructure and magnetic properties of LENS-deposited bulk Fe–Si–B–Cu–Nb soft magnetic alloys needs to be investigated. Soft magnetic material applications, such as magnetostriction transducers, electromagnetic screening, high-power transformers, frequency converters, micromotors, and signal transformers [136–140], can require complex shapes/geometries, which are difficult to produce using conventional processing routes such as melt spun ribbons or sheets. Laser additive manufacturing opens up a new horizon in processing materials exhibiting complex geometries to overcome the issue. Thus, the main objective of the present investigation is to study the effect of Si/B ratio on the microstructure and magnetic properties of laser-deposited compositionally graded Fe–Si–B–Cu–Nb alloys [4].

The compositional gradient in the as-deposited Fe–Si–B–Cu–Nb graded alloy is shown in Figure 4.33a and corresponding compositions of the Fe–Si–B–Cu–Nb alloys are shown in Table 4.2.

The compositions along the gradient have been measured using EDS in the SEM. The Si at% varies from 0.2 to 24.4 and B at% varies from 18.6 to 8, whereas the compositions of the remaining elements were adjusted accordingly. In the LENS system, the powders used were mainly a mixture of

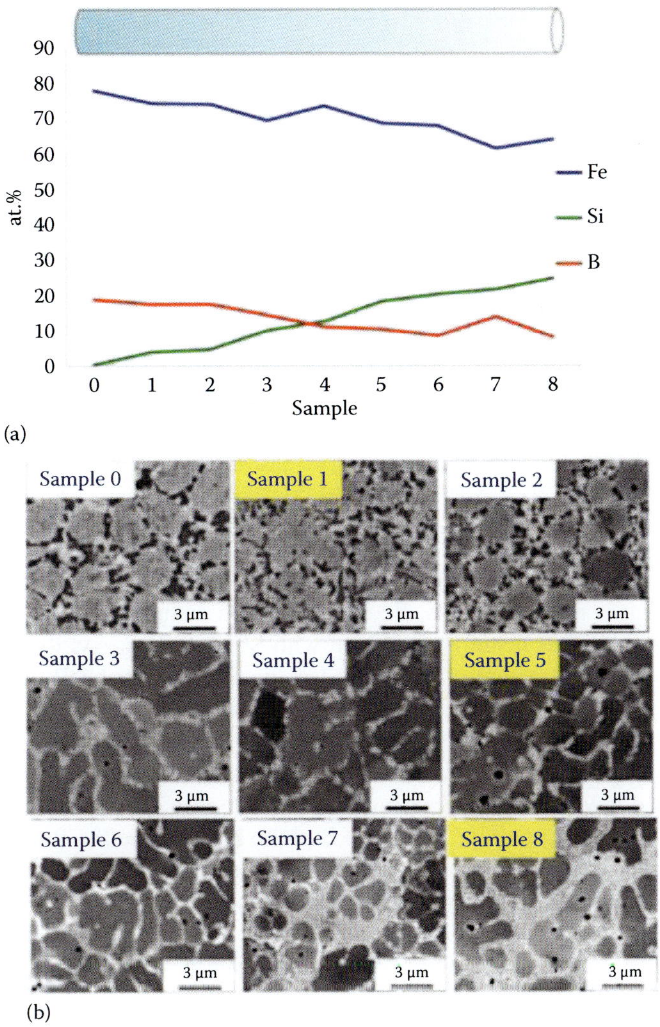

FIGURE 4.33 (a) Schematic representation as composition profile across the LENS-deposited Fe–Si–B–Cu–Nb graded alloy. (b) SEM images of LENS-deposited Fe–Si–B–Cu–Nb graded alloy (sample 0–sample 8). (Reprinted with permission from Borkar et al. 2016. Laser additive processing of functionally-graded Fe–Si–B–Cu–Nb soft magnetic materials. *Materials and Manufacturing Processes*: 1–7.)

elemental powders; therefore, slight variation in the composition, from the intended values, was observed. The microstructures of Fe–Si–B–Cu–Nb alloys processed via LENS with different Si/B contents are shown in Figure 4.33b. These backscatter scanning electron micrographs clearly show the dendritic microstructure for all the compositions with slight variations in the interdendritic products with increasing silicon (Si)/boron (B) ratio. For low silicon (Si)/boron (B) ratio (~ higher boron content) alloys (sample 0–sample 2), strong partitioning of boron (B) and niobium (Nb) was observed. As silicon (Si)/boron (B) ratio increases, partitioning of boron (B) decreases but strong partitioning of niobium (Nb) was still observed. For very high silicon (Si)/boron (B) ratio (~ higher silicon content), niobium-rich eutectic products were observed in interdendritic regions. Samples 1 (Si/B = 0.25), 5 (Si/B = 1.7), and 8 (Si/B = 3.05) were selected for further detailed analysis. Different phase formation during solidification of Fe–Si–B–Cu–Nb alloys processed via LENS was studied by the XRD technique. The XRD pattern (Figure 4.34a) clearly shows the presence of a large volume fraction of α-Fe_3Si phase exhibiting the dendritic morphology.

TABLE 4.2
Compositions of Different Regions of LENS-Deposited Graded Fe–SI–B–Cu–Nb Alloy

Sample	0	**1**	2	3	4	**5**	6	7	**8**
F/rate 1	0	12.5	25	37.5	50	62.5	75	87.5	100
F/rate 2	100	87.5	75	62.5	50	37.5	25	12.5	0
Fe	77.4	74.0	73.6	69.0	73.1	68.3	67.4	61.1	63.6
Si	0.2	3.8	4.5	9.9	12.5	18.0	20.2	21.5	24.4
B	18.6	17.4	17.4	14.3	10.9	10.2	8.4	13.9	8.0
Nb	2.7	3.5	3.1	2.2	2.2	2.2	2.6	2.5	2.7
Cu	1.1	1.3	1.4	1.2	1.2	1.2	1.3	1.0	1.3
Si/B	**0.01**	**0.22**	**0.26**	**0.69**	**1.15**	**1.76**	**2.40**	**1.61**	**3.05**

Source: Adapted from Borkar et al. 2016. Laser additive processing of functionally-graded Fe–Si–B–Cu–Nb soft magnetic materials. *Materials and Manufacturing Processes*: 1–7.

A similar phase was observed in almost all the samples except for minor peaks, which correspond to the secondary phases formed, that is, Fe_3B observed in sample 1 where strong partitioning of boron (B) was observed. This result is consistent with SEM analysis. As the silicon (Si)/boron (B) ratio increases, the volume fraction of the secondary phases, that is, Fe_3B, decreases and only the Fe_3Si phase is observed in samples 5 and 8. Vibrating sample magnetometer results for the LENS-deposited Fe–Si–B–Cu–Nb alloys are shown in Figure 4.34b and Table 4.3. While in general, good saturation magnetization and coercivity values were obtained for all compositions investigated, both parameters decreased in value with increasing silicon (Si)/boron (B) ratio. Sample 1 exhibits higher coercivity mainly due to the presence of the Fe_3B phase because the Fe–B phase has a detrimental effect on soft magnetic properties. Sample 8 exhibits lower saturation magnetization mainly due to the presence of a niobium (Nb)-rich eutectic product in the interdendritic region. Sample 5 exhibits higher saturation magnetization as well as lower coercivity, perhaps the most optimum silicon (Si)/boron (B) ratio within the range of compositions investigated in this study.

LENS-deposited compositionally graded soft magnetic Fe–Si–B–Cu–Nb alloys with varying silicon (Si)/boron (B) ratio exhibit dendritic morphology. XRD analysis clearly shows the presence of Fe_3Si and Fe_3B phases at the dendrite/matrix interface for lower silicon (Si)/boron (B) ratios. As the Si/B ratio increases, only the Fe_3Si phase was observed at the dendrite/matrix interface. Fe–Si–B–Cu–Nb alloys with silicon (Si)/boron (B) ratio ~ 1.7 (very close to the commercial FINEMET composition) exhibited good saturation magnetization and the lowest coercivity compared to the other samples. The ability to produce near-net shape components with graded compositions from elemental powders using the LENS process could potentially be a viable route for manufacturing functionally graded structures for high-power transformers, micromotors, magnetostriction transducers, electromagnetic screening, frequency converters, and signal transformers, which can require complex shapes/geometries.

4.3.4 Dissimilar Metals

Over the last decade, various researchers have developed compositionally graded alloys using dissimilar metals via laser additive manufacturing processes. This section reviews such case studies.

4.3.4.1 Stainless Steel and Titanium

Sahasrabudhe et al. [141] have processed stainless steel to titanium compositionally graded alloy via the LENS process. This stainless steel to titanium bimetallic component offers excellent corrosion

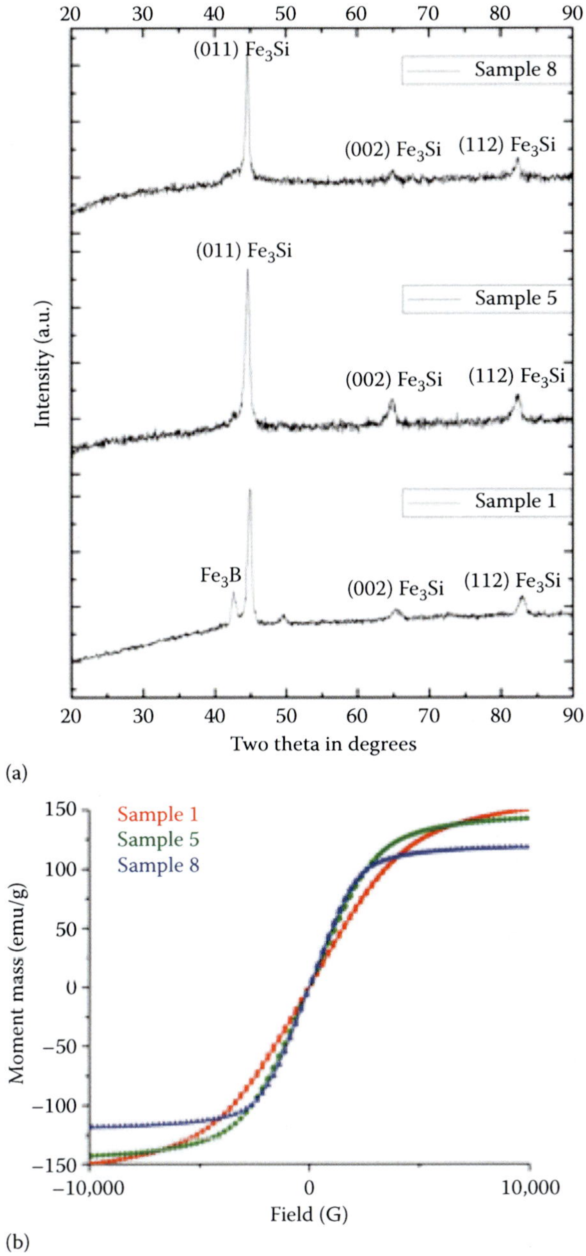

FIGURE 4.34 (a) XRD patterns of LENS processed Fe–Si–B–Cu–Nb graded alloy (Samples 1, 5, and 8) and (b) VSM results of LENS processed Fe–Si–B–Cu–Nb graded alloy (Samples 1, 5, and 8). (Reprinted with permission from Borkar et al. 2016. Laser additive processing of functionally-graded Fe–Si–B–Cu–Nb soft magnetic materials. *Materials and Manufacturing Processes*: 1–7.)

resistance and is cheaper than full Ti64 component. This bimetallic structure has been fabricated using two approaches. In the first approach, Ti64 has been directly deposited on a stainless steel substrate, whereas in the second approach, a NiCr intermediate bonding layer has been deposited between titanium and stainless steel structures. When Ti64 directly deposited on a stainless steel substrate, cracking occurred owing to the formation of brittle intermetallic phases at their interface as

TABLE 4.3
VSM Results of LENS-Deposited Graded Fe–Si–B–Cu–Nb Alloys

Sample	H_c (G)	M_s (emu/g)
Fe–3.8Si–17B–3.5Nb–1.3Cu (Sample 1)	48	150
Fe–18Si–10B–2Nb–1.1Cu (Sample 5)	19	143
Fe–24Si–8B–2.7Nb–1Cu (Sample 8)	16	118

Source: Adapted from Borkar et al. 2016. Laser additive processing of functionally-graded Fe–Si–B–Cu–Nb soft magnetic materials. *Materials and Manufacturing Processes*: 1–7.

well as thermal and residual stresses. The addition of a NiCr intermediate layer significantly improved the bonding between stainless steel and titanium alloys. Figure 4.35a shows crack-free Ti64 bonded to SS410 with the help of an intermediate NiCr layer.

This shows good bonding and smooth transition from stainless steel to NiCr and from NiCr to Ti64 without any brittle intermetallic formation. This study provides guidelines for the processing of compositionally graded dissimilar metals with the help of an intermediate bonding layer in order to avoid formation of brittle intermetallic phases and high residual stresses causing cracking.

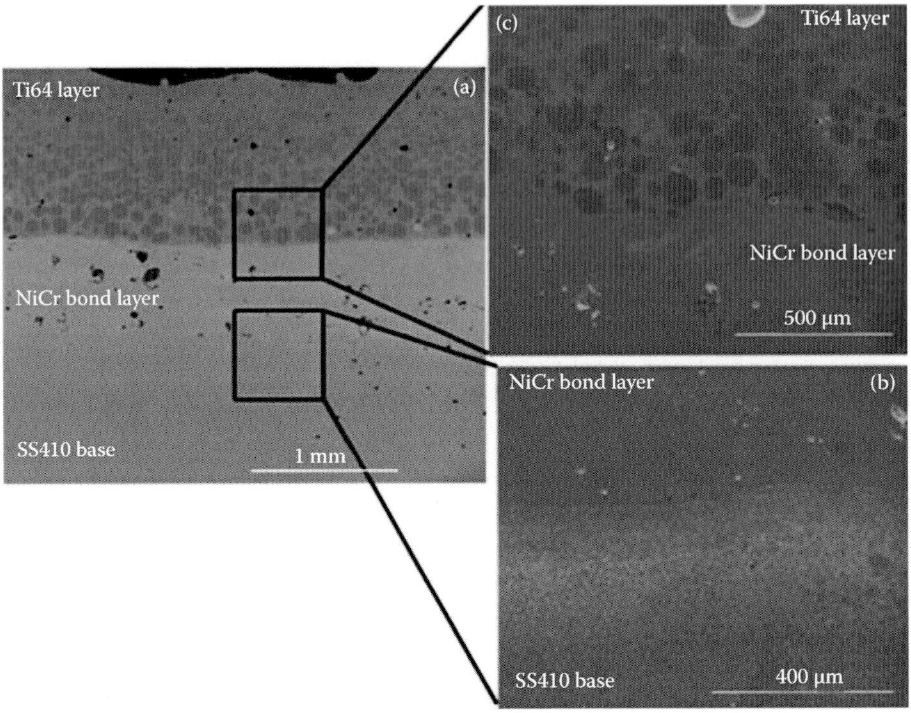

FIGURE 4.35 Cross-sectional SEM image of (a) Ti64 bonded to SS410 using Ni–Cr intermediate layer 750 µm thick, processed using LENS at a power of 350 W for NiCr layer and 425 W for Ti64 layer, in Ar environment with $O_2 < 10$ ppm. (b) Interface of SS410 to NiCr intermediate layer showing a smooth transition. (c) Transition from NiCr to Ti64 showing a sharp interface and globular phases. (Reprinted with permission from Sahasrabudhe et al. 2015. Stainless steel to titanium bimetallic structure using LENS. *Additive Manufacturing* 5: 1–8.)

4.3.4.2 Titanium and Nickel Base Superalloys

Nickel base superalloys are widely used in aerospace application because of their high strength at elevated temperature with long exposure time [142]. Compositionally graded titanium to nickel base superalloys could be more useful in terms of lowering the cost as well as overall mass of the component. Lin et al. [142] have processed compositionally graded Ti to Rene88DT in order to investigate the sequence of the phase evolution in these alloys. The composition gradation has been achieved from 100 wt% Ti to 60 wt% Rene88DT via laser additive manufacturing processes. As Rene88 content increases from 0 to 60 wt% along the compositional gradient, the following sequence of phase evolution takes place: $\alpha \rightarrow \alpha + \beta \rightarrow \alpha + \beta + Ti_2Ni \rightarrow \beta + Ti_2Ni \rightarrow Ti_2Ni + TiNi \rightarrow TiNi$. SEM images of composition gradient from CP Ti to 20% Rene88DT are shown in Figure 4.36.

Microstructure of pure titanium exhibits a large volume fraction of α phase as shown in Figure 4.36a. As Rene88DT content increases (Figure 4.36b), the volume fraction of β phase increases with a corresponding decrease in the volume fraction of α phase. Also, the size and aspect ratio of α laths decrease as compared to α laths observed in pure Ti. The corresponding refinement in α phase is shown as "A" in Figure 4.36b, where multiple variants of α laths exhibiting basketweave morphology are observed. With further increases in Rene88DT content, microstructure shows uniform distribution of fine Ti_2N cubic precipitates within the matrix (Figure 4.36c) as well as those formed at the prior β grain boundary (Figure 4.36d). As Rene88DT content increases further (Figure 4.36e), the volume fraction of Ti_2N precipitates decreases significantly and the morphology of β phase changes from cellular to dendritic. The precipitation of Ti_2N laths as well as Ti_2N particles takes place within the β dendrites with further increases in Rene88DT content in the compositionally graded alloy (Figure 4.36f) and also the stable β dendrites structure was formed at this composition. The eutectoid α-Ti + Ti_2Ni structure was observed at higher magnification (Figure 4.36g) while remaining liquid within interdendritic regions solidified as β-Ti + Ti_2Ni lamellar and dissociate eutectic structure (Figure 4.36h). The microstructure evolution of composition ranging from 30% Rene88DT to 55% Rene88DT is shown in Figure 4.37.

An irregular eutectic structure was observed for the 30% Rene88DT content where granular β-Ti phase uniformly distributed within a continuous Ti_2N matrix (Figure 4.37a). A two-phase dendritic structure of β-Ti + Ti_2Ni (Figure 4.37b) was observed for the composition 60% Ti + 40% Rene88DT along with new two-phase dendritic growth of TiNi + Ti_2Ni (Figure 4.37c). TiNi + Ti_2Ni exhibited spongy morphology (Figure 4.37d) when Rene88DT content increased to 50%. Just before this transition, a eutectic structure of β-Ti + Ti_2Ni was observed within the interdendritic region (inset in Figure 4.37d). As the Rene88DT content increases to 55%, the microstructure exhibits an irregular eutectic structure with Ti_2Ni granules uniformly distributed within a continuous TiNi matrix (Figure 4.37e). Figure 4.38 shows SEM images for the composition 40% Ti + 60% Rene88DT.

The microstructure exhibits an equiaxed TiNi dendrite (Figure 4.38a) along with fishbone-like Ti_2N dendrites (Figure 4.38b) within the interdendritic region. Also, platelet-like white phase and flower-like eutectic growth coupled with continuous TiNi matrix were observed within the interdendritic region (Figure 4.38c).

4.3.4.3 Stainless Steel and Nickel Base Superalloys

Various researchers have probed the compositionally graded stainless steel to nickel base superalloys components [143–148]. Since nickel base superalloys offer high strength as well as corrosion resistance at elevated temperatures and stainless steel aids in cost and mass reduction, these compositionally graded components are particularly suitable for applications such as high-end automobile engine valve stems, nuclear power generation, and oil refineries [143,144]. Compositionally graded stainless steel to nickel base superalloys are a viable solution to join these alloys without any cracking, which typically occurs if these alloys are joined via fusion welding [144,145]. Few examples of compositionally graded different stainless steel to nickel base superalloys are presented here.

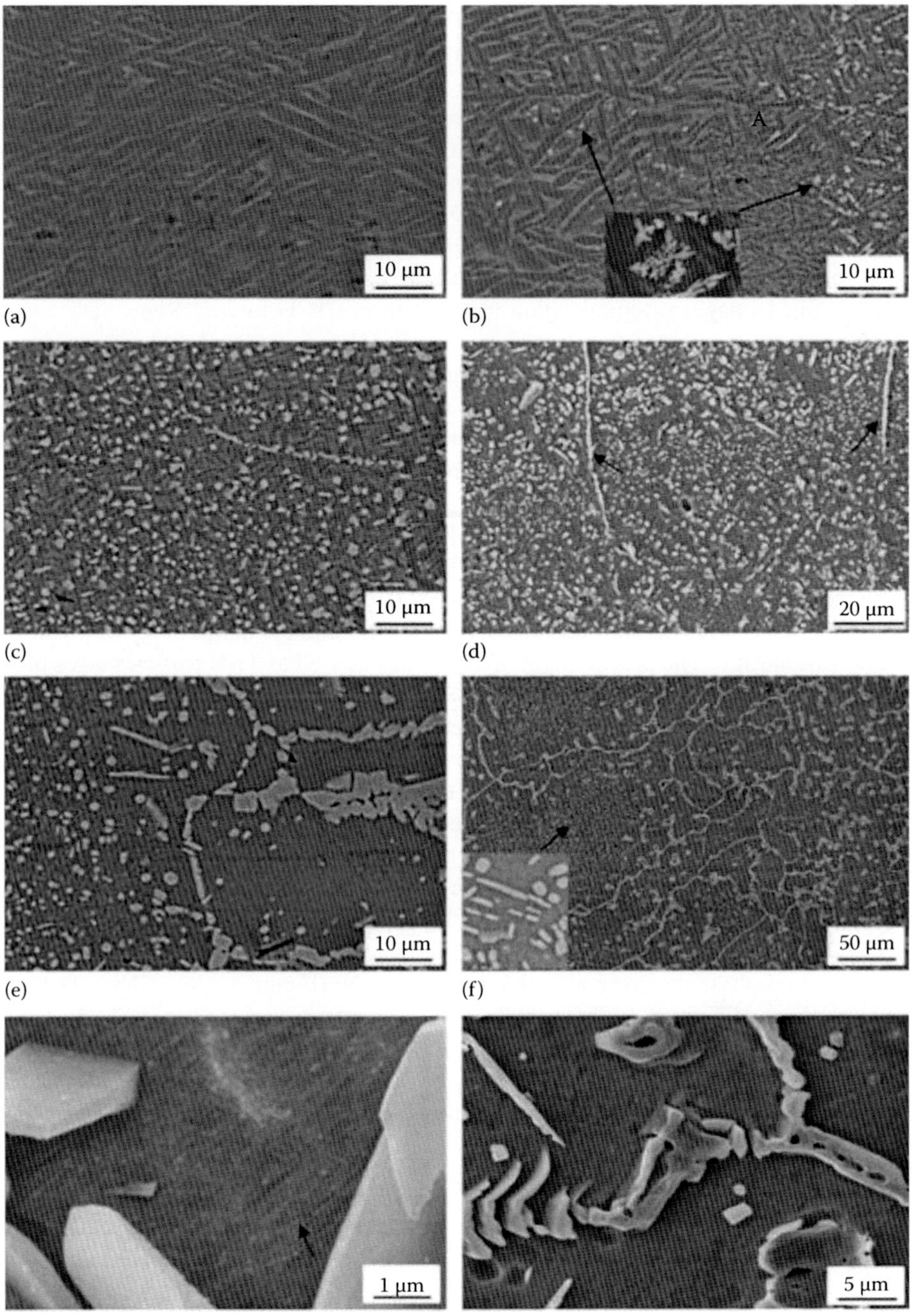

FIGURE 4.36 SEM backscattered and secondary electron images showing the microstructure along the compositional gradient, Rene88DT content increasing from (a) through (c) and (e) and (f); (d) a low magnification showing transition from (b) to (c); (g) shows the lamellar structure; (h) shows the various phases at the interdendritic boundaries. (Reprinted with permission from Lin et al. 2006. Microstructure and phase evolution in laser rapid forming of a functionally graded Ti–Rene88DT alloy. *Acta Materialia* 54 (7): 1901–1915.)

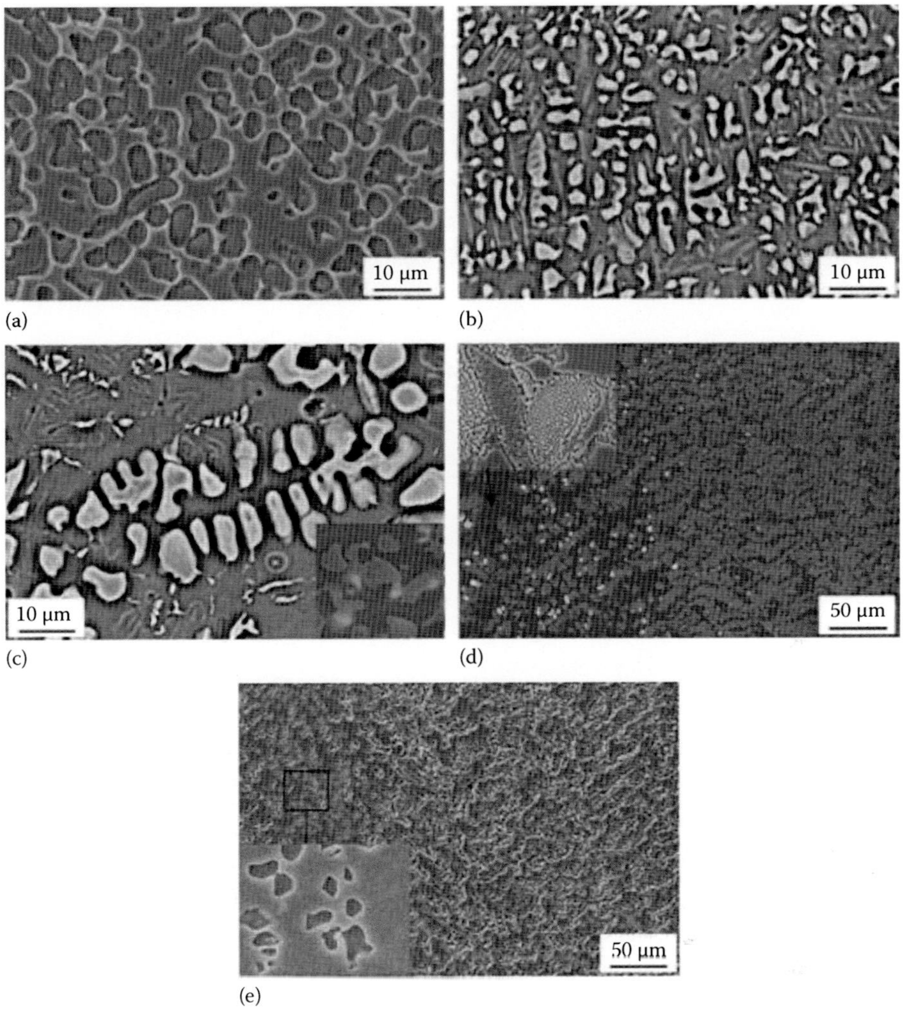

FIGURE 4.37 SEM micrographs showing the microstructure at composition (a) 70%Ti + 30%Rene88DT (secondary electron image), (b and c) 60%Ti + 40%Rene88DT (BSE image), (d) 50%Ti + 50%Rene88DT (BSE image), (e) 45%Ti + 55%Rene88DT (secondary electron image). (Reprinted with permission from Lin et al. 2006. Microstructure and phase evolution in laser rapid forming of a functionally graded Ti–Rene88DT alloy *Acta Materialia* 54 (7): 1901–1915.)

Lin et al. [146] have also fabricated compositionally graded SS316 to Rene88DT alloy via laser additive manufacturing processes. As the Rene88DT content increases along the compositional gradient, microstructure evolution takes place in the following sequence: the dendritic growth of the γ phase, the precipitation of the γ' phase, as well as the formation of the η phase and carbides. Figure 4.39 shows the distribution of the γ' phase with the increasing content of Rene88DT in the overall composition.

For the 40% Rene88DT content, γ' precipitates were observed within the interdendritic region (Figure 4.39a). The formation of γ' precipitates is mainly due to the eutectic reaction of residual liquid. γ' precipitates were observed within the dendritic arm (Figure 4.39b) when the Rene88DT content increased to 60%. Decomposition of metal carbide to (metal$_6$crabide) M_6C also promotes the precipitation of the γ' phase. As Rene88DT content increased from 40% to 100%, there is an increase in volume fraction and an increase in the size of γ' (Figure 4.39a through e).

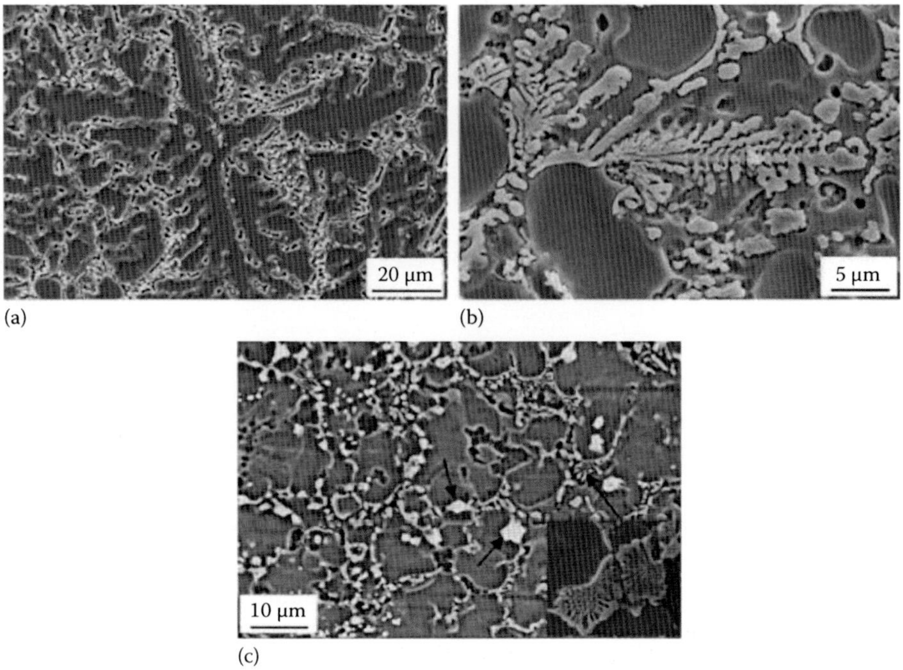

FIGURE 4.38 The microstructure for the composition 40%Ti + 60%Rene88DT: (a and b) secondary electron images; (c) BSE image. (Reprinted with permission from Lin et al. 2006. Microstructure and phase evolution in laser rapid forming of a functionally graded Ti–Rene88DT alloy. *Acta Materialia* 54 (7): 1901–1915.)

Savitha et al. [147] have processed discrete and compositionally graded stainless steel (SS316) to IN625 alloys via LENS equipment. They have processed these graded alloys via two approaches. In the first approach called discrete interface (DI), IN625 directly deposited on SS316 without any composition gradation, whereas in the second approach, IN625 deposited on SS316 in a systematic gradient manner termed compositionally gradient (CG). Figure 4.40 shows the microstructure of DI and CG samples.

The interface between SS316 and IN625 appears to be sharp, without any pores or cracks. The cell structure within IN625 and SS316 is shown in Figure 4.40b and c, respectively. Figure 4.40d shows the microstructure of the CG sample along the compositional gradient where gradual variation in gray level from one side to another was observed with increasing IN625 content. Tensile properties of both DI and CG alloys are comparable to that of end alloys, that is, between tensile properties of SS316 and IN625.

Wu et al. [148] have processed two types of stainless steel 316L to nickel base superalloy Inconel 718 compositionally graded materials. In the first type, composition gradation takes place along the build direction (FGM1), whereas in the second type, composition gradation occurs perpendicular to the build direction (FGM2). In FGM1 where composition gradation is along the build direction, the microstructure primarily exhibited columnar dendrites except at the edge where a columnar to equiaxed transition takes place, whereas in FGM2, the columnar dendritic structure is not as dominant as FGM1. As Inconel 718 content increases along the composition gradation, the increment in primary arm spacing was observed in FGM1 mainly due to the decrease in temperature gradient, whereas primary arm spacing remains constant in FGM2. Shah et al. [144] have investigated the effect of laser additive manufacturing processing parameters on the microstructure and mechanical properties of compositionally graded stainless steel 316L to Inconel 718 alloys. Laser power and powder flow rate have a strong influence on secondary dendritic arm spacing (SDAS), and SDAS

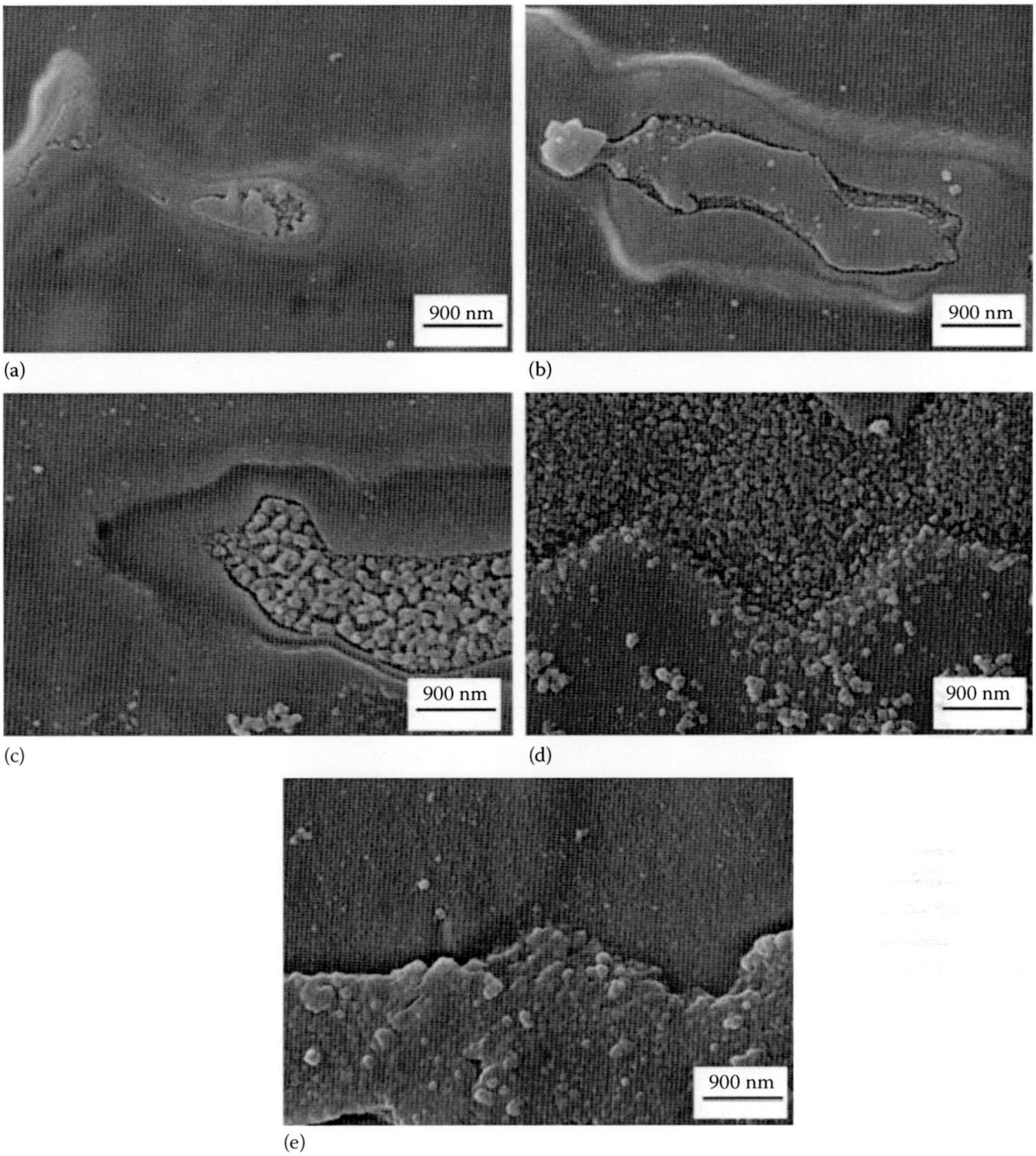

FIGURE 4.39 Showing the distribution of the γ' phase in (a) 40% Rene88DT, (b) 60% Rene88DT, (c) 80% Rene88DT, (d) Rene88DT (4 mm distance from the top), and (e) Rene88DT (1 mm distance from the top). (Reprinted with permission from Lin and Yue 2005. Phase formation and microstructure evolution in laser rapid forming of graded SS316L/Rene88DT alloy. *Materials Science and Engineering: A* 402 (1): 294–306.)

increases with an increase in power and powder flow rate. The powder flow rate has a positive effect in the improvement of mechanical properties (tensile strength), whereas laser powder has an inverse effect on the mechanical properties of compositionally graded alloys.

Carroll et al. [143] have studied FGMs of 304L stainless steel and Inconel 625 processed via the DED process. Figure 4.41 shows SEM images of compositionally graded SS304L and IN625 alloys.

As IN625 content increases from 0% to 17% (Figure 4.41a through c), the microstructure primarily shows austenite and ferrite phases (light areas) and grain boundaries (dark region). Cellular

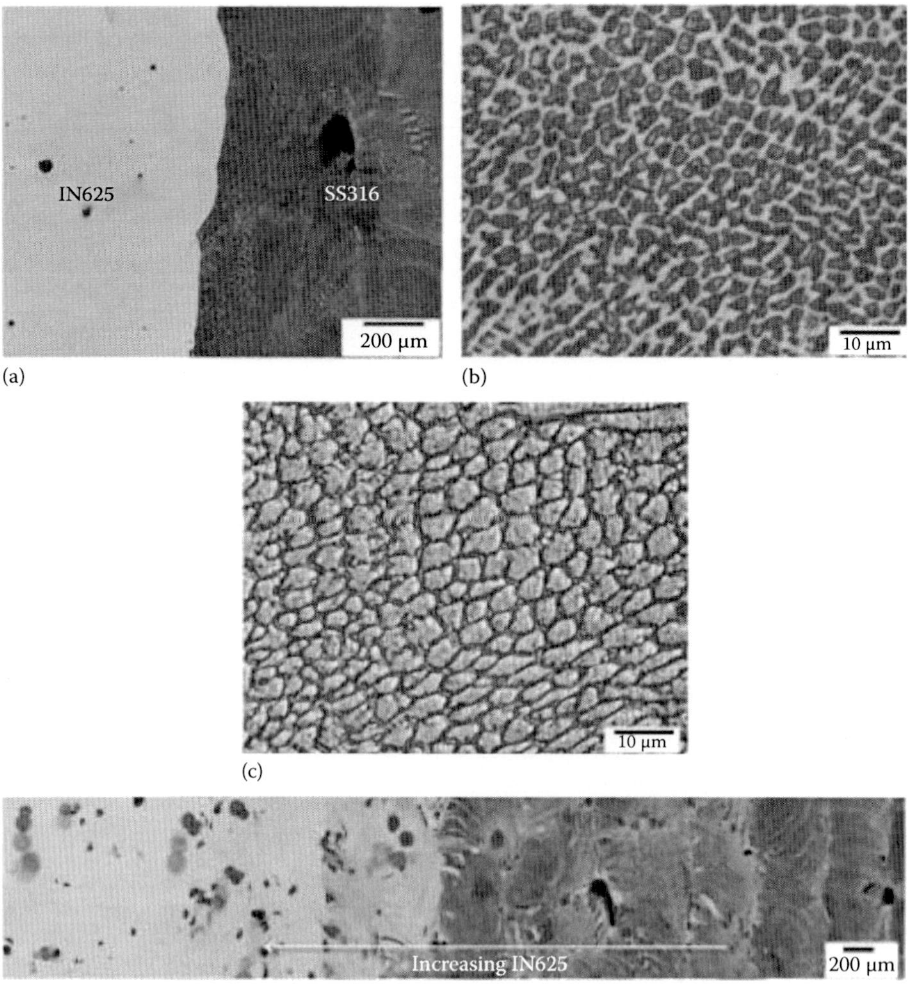

FIGURE 4.40 Optical micrographs of Discrete Interface (DI) specimen. (a) Interface between SS316 and IN625 regions, (b) IN625 side, (c) SS316 side, and (d) transition zone of CG sample. (Reprinted with permission from Savitha et al. 2015. Chemical analysis, structure and mechanical properties of discrete and compositionally graded SS316–IN625 dual materials. *Materials Science and Engineering: A* 647: 344–352.)

dendrites exhibiting high level of contrast were observed in the graded alloy as the IN625 content increased from 20 to 50 wt% (Figure 4.41c through e). Columnar dendrites were observed as the IN625 content increased from 50 to 100 wt% (Figure 4.41f through h).

4.3.4.4 Nickel Base Maraging Steel and Cr–Mo–V Tool Steel

Helene et al. [87] have processed compositionally graded Ni base maraging steel and Cr–Mo–V tool steel via laser additive manufacturing processes. This is an efficient high-throughput method for rapid assessment of various compositions at a shorter time. This investigation suggests that systematic addition of tool steel with maraging steel will be very helpful to fine-tune the mechanical properties of these alloys. As the tool steel content increases along the composition gradient, a significant improvement in tensile strength was observed with a corresponding decrease in strain.

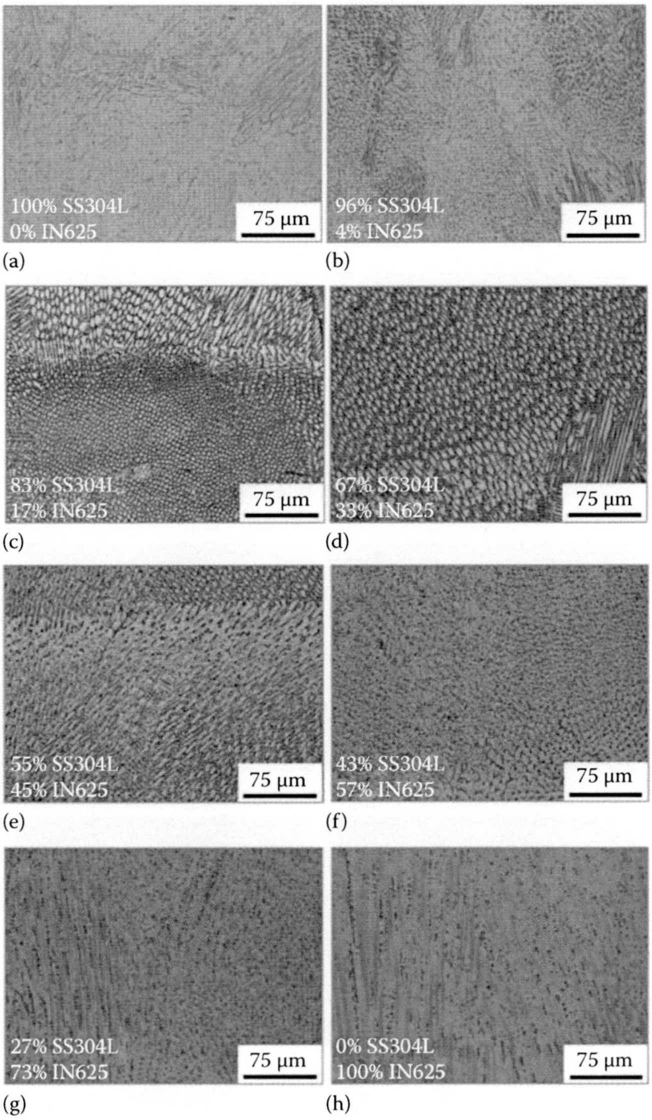

FIGURE 4.41 Microstructure of gradient zone from 100% SS304L (a) to 100% IN625 (h), with weight fractions of SS304L and IN625 indicated. The gradient microstructure resembles that of austenitic SS304L up to a composition of about 15 wt% IN625 (a through c). A high-contrast cellular dendritic structure appears between 15 wt% and approximately 50 wt% IN625 (c through e), and from 50 wt% to 100 wt% IN625 (f through h), the microstructure appears much as IN625, with columnar dendrites. (Reprinted with permission from Carroll et al. 2016. Functionally graded material of 304L stainless steel and inconel 625 fabricated by directed energy deposition: Characterization and thermodynamic modeling. *Acta Materialia* 108: 46–54.)

4.3.5 Metal Matrix Composites

Few researchers have investigated the microstructure and mechanical properties of functionally graded metal matrix composite processes via laser additive manufacturing. Seefeld et al. [149] have processed compositionally graded NiBSi and Cr_3C_2 metal matrix composites via a laser beam cladding process. The microstructure of NiBSi exhibits a fine dendritic structure. As Cr_3C_2 content

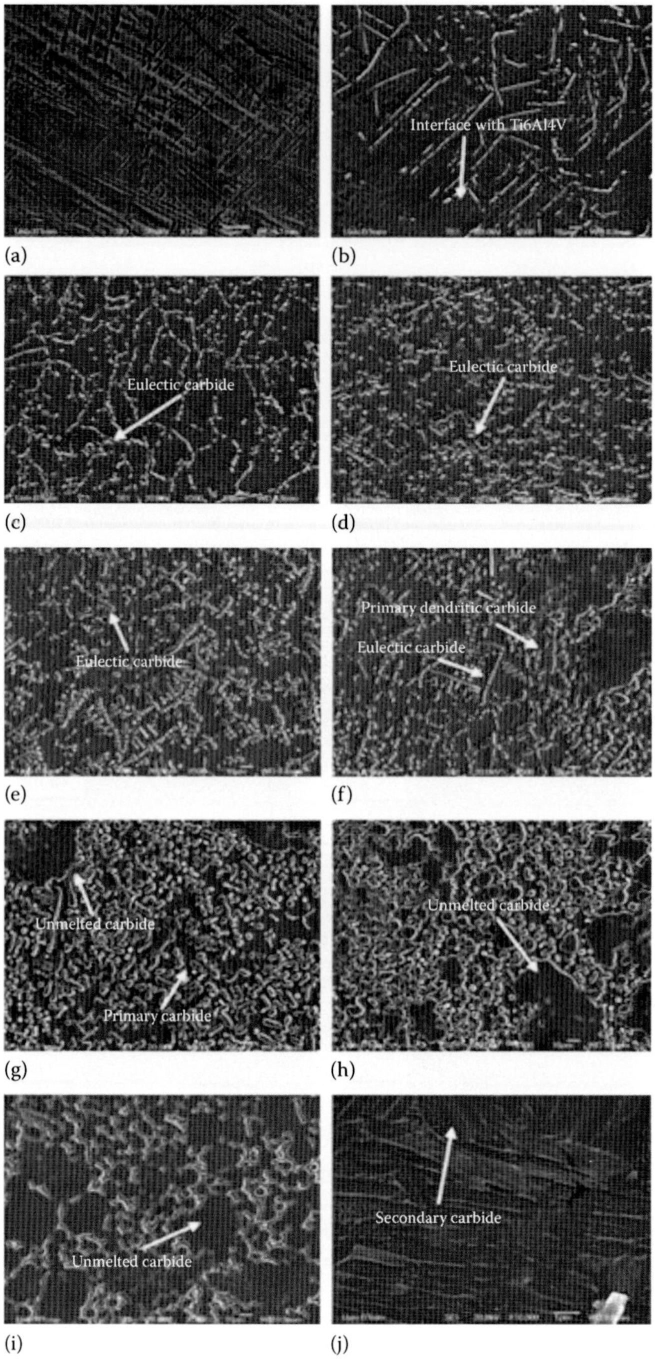

FIGURE 4.42 Microstructure of compositionally graded Ti6Al4V–TiC made by direct laser fabrication using powder with a series of different feed rates and Ti6Al4V wire with a constant feed rate of 1.1 g/min: (a) Ti6Al4V wire feed only; (b) the interface between the composites and the Ti6Al4V; (c) TiC feed rate of 0.14 g/min, with 8% TiC particles in the microstructure; (d) 0.30 g/min, with 15% TiC particles; (e) 0.44 g/min, with 24% TiC particles; (f) 0.57 g/min, with 49% TiC particles; (g) 0.70 g/min, with 74% TiC particles; (h) 0.87 g/min, with 83% TiC particles; (i) 1.03 g/min, with 100% TiC particles; and (j) high-magnification microstructure of the secondary titanium carbides taken from the center of the region where the feed rate of TiC was 0.44 g/min. (Reprinted with permission from Wang et al. 2007. Compositionally graded Ti6Al4V+ TiC made by direct laser fabrication using powder and wire. *Materials & Design* 28 (7): 2040–2046.)

increases in NiBSi–Cr_3C_2 composites, the microstructure primarily exhibits needle-shaped Cr_3C_2 precipitates along with undissolved carbides. The volume fraction of these undissolved carbides increases with increasing Cr_3C_2 in NiBSi–Cr_3C_2 composites. The microhardness of NiBSi alloys increases with the increasing content of Cr_3C_2. The NiBSi alloy exhibits a microhardness of 320 HV, whereas NiBSi–85% Cr_3C_2 exhibits a microhardness of 1200 HV, mainly due to the increase in volume fraction of carbide precipitates.

Wang et al. [150] have processed compositionally graded Ti6Al4V (Ti64) + TiC composites via a direct laser fabrication process using Ti64 wire and TiC powder. Figure 4.42 shows the microstructure of Ti64 + TiC composites with increasing content of TiC. The microstructure of base Ti64 shows the typical α/β microstructure observed in titanium alloys where α-titanium exhibits basketweave morphology.

As TiC content increases, there is an increase in volume fraction for both primary TiC (Figure 4.42c through i) and eutectic TiC (Figure 4.42c through f). Secondary TiC precipitates (Figure 42j) are also observed for Ti64–24%TiC composites due to solid-state precipitation. As TiC content increases above 74%, unmelted TiC particles are observed in Ti64–TiC composites (Figure 4.42g through i). There is an improvement in tribological properties of Ti64 observed due to TiC reinforcement, and Ti64–24vol% TiC exhibits ideal frictional behavior.

Liu et al. [151] have fabricated compositionally graded Ti + TiC composites via the LENS process. The microstructure of Ti + TiC composites exhibits a mixture of primary and eutectic TiC precipitates. As TiC content increases, unmelted TiC particles are observed in the microstructure. The microhardness of Ti + TiC composites increases from 200 HV for pure Ti to 2300 HV for Ti–95vol%TiC composites mainly due to the combined effect of primary, eutectic, and unmelted TiC. Similar results have been observed for compositionally graded Ti–TiC composites processed via a laser powder deposition method [152].

4.4 SUMMARY

- Laser additive manufacturing of in situ metal matrix composites offers various advantages as compared to conventional processing and also has the capability to produce near-net shape components.
- The ability to achieve substantial changes in composition across rather limited lengths makes compositionally/functionally graded alloys highly attractive candidates for investigating the influence of systematic compositional changes on phase transformation and concurrent microstructural evolution in alloys and metal matrix composites. This chapter provided an overview of laser additive manufacturing of in situ metal matrix composites and FGMs.
- Laser additively manufactured in situ nickel titanium graphite composites have shown significant improvement in microhardness as well as tribological properties as compared to pure nickel and are a potential candidate for high-temperature surface engineering applications.
- In situ titanium alloy composites have shown improvement in wear properties and are ideal for biomedical applications.
- Compositionally graded titanium alloys as well as HEAs have shown rapid assessment of those alloys in a very systematic manner. Laser additive manufacturing opens up a new horizon in terms of processing novel alloys and composites that are difficult to process using conventional techniques.

REFERENCES

1. Borkar, T., Gwalani, B., Choudhuri, D., Mikler, C. V., Yannetta, C. J., Chen, X., Ramanujan, R., Styles, M. J., Gibson, M. A., and Banerjee, R. 2016. A combinatorial assessment of $Al_xCrCuFeNi_2$ ($0<x<1.5$) complex concentrated alloys: Microstructure, microhardness, and magnetic properties. *Acta Materialia* 116: 63–76.
2. Borkar, T., Sosa, J., Hwang, J. Y., Scharf, T. W., Tiley, J., Fraser, H., and Banerjee, R. 2014. Laser-deposited in situ TiC-reinforced nickel matrix composites: 3D microstructure and tribological properties. *JOM* 66 (6): 935–942.
3. Borkar, T., Gopagoni, S., Nag, S., Hwang, J. Y., Collins, P. C., and Banerjee, R. 2012. In situ nitridation of titanium–molybdenum alloys during laser deposition. *Journal of Materials Science* 47 (20): 7157–7166.
4. Borkar, T., Conteri, R., Chen, X., Ramanujan, R. V., and Banerjee, R. 2016. Laser additive processing of functionally-graded Fe–Si–B–Cu–Nb soft magnetic materials. *Materials and Manufacturing Processes*: 1–7.
5. Mikler, C. V., Chaudhary, V., Borkar, T., Soni, V., Jaeger, D., Chen, X., Contieri, R., Ramanujan, R. V., and Banerjee, R. 2017. Laser additive manufacturing of magnetic materials. *JOM* 69 (3): 532–543.
6. Das, S., Bourell, D. V., and Babu, S. S. 2016. Metallic materials for 3D printing. *MRS Bulletin* 41 (10): 729–741.
7. William, S. J., List, F. A., Pannala, S., Dehoff, R. R., and Babu, S. S. 2016. The metallurgy and processing science of metal additive manufacturing. *International Materials Reviews* 61 (5): 315–360.
8. Frazier, W. E. 2014. Metal additive manufacturing: A review. *Journal of Materials Engineering and Performance* 23 (6): 1917–1928.
9. Zhao, J.-C. 2006. Combinatorial approaches as effective tools in the study of phase diagrams and composition–structure–property relationships. *Progress in Materials Science* 51 (5): 557–631.
10. Curtarolo, S., Hart, G. L., Nardelli, M. B., Mingo, N., Sanvito, S., and Levy, O. 2013. The high-throughput highway to computational materials design. *Nature Materials* 12 (3): 191–201.
11. Gu, D. D., Meiners, W., Wissenbach, K., and Poprawe, R. 2012. Laser additive manufacturing of metallic components: Materials, processes and mechanisms. *International Materials Reviews* 57 (3): 133–164.
12. Kobryn, P. A., and Semiatin, S. L. 2001. The laser additive manufacture of Ti–6Al–4V. *JOM* 53 (9): 40–42.
13. Liang, Y.-J., Liu, D., and Wang, H.-W. 2014. Microstructure and mechanical behavior of commercial purity Ti/Ti–6Al–2Zr–1Mo–1V structurally graded material fabricated by laser additive manufacturing. *Scripta Materialia* 74: 80–83.
14. Hofman, D. C., Kolodziejska, J., Roberts, S., Otis, R., Dillon, R. P., Suh, J.-O., Liu, J.-K., and Borgonia, J.-P. 2014 Compositionally graded metals: A new frontier of additive manufacturing. *Journal of Materials Research* 29 (17): 1899–1910.
15. Douglas, H. C., Kolodziejska, J., Roberts, S., Otis, R., Dillon, R. P., Suh, J.-O., Liu, J.-K., and Borgonia, J.-P., Shapiro, A. A., Zi-Kui, L., and John-Paul, B. 2014. Developing gradient metal alloys through radial deposition additive manufacturing. *Scientific Reports* 4: 5357.
16. Terry, W. T., and Caffrey, T. 2015. *Wohlers Report 2015: 3D Printing and Additive Manufacturing State of the Industry Annual Worldwide Progress Report*. Wohlers Associates.
17. Gibson, I., Rosen, D., and Stucker, B. 2014. *Additive Manufacturing Technologies: 3D Printing, Rapid Prototyping, and Direct Digital Manufacturing*. Springer.
18. Santos, E. C., Shiomi, M., Osakada, K., and Laoui, T. 2006. Rapid manufacturing of metal components by lasers forming. *International Journal of Machine Tools and Manufacture* 46 (12):1459–1468.
19. Heinl, P., Rottmair, A., Körner, C., and Singer, R. F. 2007. Cellular titanium by selective electron beam melting. *Advanced Engineering Materials* 9 (5): 360–364.
20. Kruth, J. P., Mercelis, P., Van-Vaerenbergh, J., Froyen, L., and Rombouts, M. 2005. Binding mechanisms in selective laser sintering and selective laser melting. *Rapid Prototyping Journal* 11 (1): 26–36.
21. Zein, I., Hutmacher, D. W., Tan, K. C., and Teoh, S. H. 2002. Fused deposition modeling of novel scaffold architectures for tissue engineering applications. *Biomaterials* 23 (4): 1169–1185.
22. Abe, F., Osakada, K., Shiomi, M., Uematsu, K., and Matsumoto, M. 2001. The manufacturing of hard tools from metallic powders by selective laser melting. *Journal of Materials Processing Technology* 111 (1): 210–213.
23. Vandenbroucke, B., and Kruth, J. P. 2007. Selective laser melting of biocompatible metals for rapid manufacturing of medical parts. *Rapid Prototyping Journal* 13 (4): 196–203.

24. Thijs, L., Verhaeghe, F., Craeghs, T., Van Humbeeck, J., and Kruth, J. P. 2010. A study of the microstructural evolution during selective laser melting of Ti–6Al–4V. *Acta Materialia* 58 (9): 3303–3312.
25. Fishman, S. 1986. Interfaces in composites. *JOM* 38: 26–27.
26. Flom, Y., and Arsenault, R. J. 1986. Deformation of SiC/Al Composites. *JOM* 38: 31–34.
27. Mortensen, A., Cornie, J., and Flemings, M. 1988. Solidification processing of metal-matrix composites. *JOM* 40: 12–19.
28. Mortensen, A., Gungor, M. N., Cornie, J. A., and Flemings, M. C. 1986. Alloy microstructures in cast metal matrix composites. *JOM* 38: 30–35.
29. Nardone V. C., and Prewo, K. M. 1986. On the strength of discontinuous silicon carbide reinforced aluminum composites. *Scripta Metallurgica* 20: 43–48.
30. Bakshi, S. R. L. D., and Agarwal, A. 2010. Carbon nanotube reinforced metal matrix composites—A review. *International Materials Reviews* 55: 41–62.
31. Miracle, D. B. 2005. Metal matrix composites—From science to technological significance. *Composites Science and Technology* 65: 2526–2540.
32. Nouari, S., Zafar, I., Abdullah, K., Abbas, H. S., Nasser, A. A., Tahar, L. et. al. 2012. Spark plasma sintering of metals and metal matrix nanocomposites: A review. *Journal of Nanomaterials*: 18–18.
33. Tjong, S. C., and Lau, K. C. 1999. Properties and abrasive wear of TiB_2/Al–4%Cu composites produced by hot isostatic pressing. *Composites Science and Technology* 59: 2005–2013.
34. Tjong, S. C., and Lau, K. C. 2000. Abrasion resistance of stainless-steel composites reinforced with hard TiB_2 particles. *Composites Science and Technology* 60: 1141–1146.
35. Tjong, S. C., and Lau, K. C. 2000. Tribological behaviour of SiC particle-reinforced copper matrix composites. *Materials Letters* 43: 274–280.
36. Tjong, S. C., and Ma, Z. Y. 2000. Microstructural and mechanical characteristics of in situ metal matrix composites. *Materials Science and Engineering: R: Reports* 29: 49–113.
37. Wu, S. Q., Wang, H. Z., and Tjong, S. C. 1996. Mechanical and wear behavior of an Al/Si alloy metal-matrix composite reinforced with aluminosilicate fiber. *Composites Science and Technology* 56: 1261–1270.
38. Gopagoni, S., Hwang, J. Y., Singh, A. R. P., Mensah, B. A., Bunce, N., Tiley, J., Scharf, T., and Banerjee, R. 2011. Microstructural evolution in laser deposited nickel–titanium–carbon in situ metal matrix composites. *Journal of Alloys and Compounds* 509: 1255–1260.
39. Ibrahim, I. A., Mohamed, F. A., and Lavernia, E. J. 1991. Particulate reinforced metal matrix composites—A review. *Journal of Materials Science* 26: 1137–1156.
40. Rawal, S. 2001. Metal-matrix composites for space applications. *JOM* 53: 14–17.
41. Tjong, S. C. 2013. Recent progress in the development and properties of novel metal matrix nanocomposites reinforced with carbon nanotubes and graphene nanosheets. *Materials Science and Engineering: R: Reports* 74: 281–350.
42. Liu, L., Li, W., Tang, Y., Shen, B., and Hu, W. 2009. Friction and wear properties of short carbon fiber reinforced aluminum matrix composites. *Wear* 266: 733–738.
43. Ma, X. C., He, G. Q., He, D. H., Chen, C. S., and Hu, Z. F. 2008. Sliding wear behavior of copper–graphite composite material for use in maglev transportation system. *Wear* 265: 1087–1092.
44. Moustafa, S. F., El-Badry, S. A., Sanad, A. M., and Kieback, B. 2002. Friction and wear of copper–graphite composites made with Cu-coated and uncoated graphite powders. *Wear* 253: 699–710.
45. Tjong, S. C. 2009. *Carbon Nanotube Reinforced Composite*. Wiley.com.
46. Banerjee, R., Collins, P. C., and Fraser, H. L. 2002. Laser deposition of in situ Ti–TiB composites. *Advanced Engineering Materials* 4 (11): 847–851.
47. Samuel, S., Nag, S., Scharf, T. W., and Banerjee, R. 2008. Wear resistance of laser-deposited boride reinforced Ti–Nb–Zr–Ta alloy composites for orthopedic implants. *Materials Science and Engineering* 28 (3): 414–420.
48. Nag, S., Samuel, S., Puthucode, A., and Banerjee, R. 2009. Characterization of novel borides in Ti–Nb–Zr–Ta+ 2B metal-matrix composites. *Materials Characterization* 60 (2): 106–113.
49. Banerjee, R., Collins, P. C., Genc, A., and Fraser, H. L. 2003. Direct laser deposition of in situ Ti–6Al–4V–TiB composites. *Materials Science and Engineering* 358 (1): 343–349.
50. Genc, A., Banerjee, R., Hill, D., and Fraser, H. L. 2006. Structure of TiB precipitates in laser deposited in situ, Ti–6Al–4V–TiB composites. *Materials Letters* 60 (7): 859–863.
51. ASM. 1992. *ASM Handbook Vol. 3: Alloy Phase Diagrams*. ASM International 9: 2.
52. Yamamoto, T., Otsuki, A., Ishihara, K., and Shingu, P. H. 1997. Synthesis of near net shape high density TiB/Ti composite. *Materials Science and Engineering* 239: 647–651.

53. Schwendner, K. I., Banerjee, R., Collins, P. C., Brice, C. A., and Fraser, H. L. 2001. Direct laser deposition of alloys from elemental powder blends. *Scripta Materialia* 45 (10): 1123–1129.
54. Banerjee, R., Collins, P. C., and Fraser, H. L. 2002. Phase evolution in laser-deposited titanium–chromium alloys. *Metallurgical and Materials Transactions* 33 (7): 2129–2138.
55. Long, M., and Rack, H. J. 1998. Titanium alloys in total joint replacement—A materials science perspective. *Biomaterials* 19 (18): 1621–1639.
56. Wang, K. 1996. The use of titanium for medical applications in the USA. *Materials Science and Engineering* 213 (1–2): 134–137.
57. Qazi, J. I., Rack, H. J., and Marquardt, B. 2004. High-strength metastable beta-titanium alloys for biomedical applications. *JOM* 56 (11): 49–51.
58. Niinomi, M., Hanawa, T., and Narushima, T. 2005. Japanese research and development on metallic biomedical, dental, and healthcare materials. *JOM* 57 (4): 18–24.
59. Long, M., and Rack, H. J. 2001. Friction and surface behavior of selected titanium alloys during reciprocating-sliding motion. *Wear* 249 (1): 157–167.
60. Godfrey, T. M. T., Wisbey, A., Goodwin, P. S., Bagnall, K., and Ward-Close, C. M. 2000. Microstructure and tensile properties of mechanically alloyed Ti–6A1–4V with boron additions. *Materials Science and Engineering* 282 (1): 240–250.
61. Fan, Z., Miodownik, A. P., Chandrasekaran, L., and Ward-Close, C. M. 1995. Microstructural investigation of a rapidly solidified Ti 6Al 4V 1B 0.5 Y alloy. *Journal of Materials Science* 30 (7): 1653–1660
62. Fan, Z., and Miodownik, A. P. 1996. Microstructural evolution in rapidly solidified Ti–7.5 Mn–0.5 B alloy. *Acta Materialia* 44 (1): 93–110.
63. Balla, V. K., Bhat, A., Bose, S., and Bandyopadhyay, A. 2012. Laser processed TiN reinforced Ti6Al4V composite coatings. *Journal of the Mechanical Behavior of Biomedical Materials* 6: 9–20.
64. Das, M., Balla, V. K., Basu, D., Manna, I., Sampath Kumar, T. S., and Bandyopadhyay, A. 2012. Laser processing of in situ synthesized TiB–TiN-reinforced Ti6Al4V alloy coatings. *Scripta Materialia* 66 (8): 578–581.
65. Bars, J.-P., David, D., Etchessahar, E., and Debuigne, J. 1983. Titanium α-nitrogen solid solution formed by high temperature nitriding: Diffusion of nitrogen, hardness, and crystallographic parameters. *Metallurgical and Materials Transactions* 14 (8): 1537–1543.
66. Gordin, D. M., Thibon, I., Guillou, A., Cornen, M., and Gloriant, T. 2010. Microstructural characterization of nitrided beta Ti–Mo alloys at 1400°C. *Materials Characterization* 61 (3): 376–380.
67. Gordin, D. M., Guillou, A., Thibon, I., Bohn, M., Ansel, D., and Gloriant, T. 2008. Duplex nitriding treatment of a beta-metastable Ti 94 Mo 6 alloy for biomedical applications. *Journal of Alloys and Compounds* 457 (1): 384–388.
68. Banerjee, R., Banerjee, D., and Williams, J. C. 2013. Perspectives on titanium science and technology. *Acta Materialia* 61 (3): 844–879.
69. Banerjee, R., Collins, P. C., Bhattacharyya, D., Banerjee, S., and Fraser, H. L. 2003. Microstructural evolution in laser deposited compositionally graded α/β titanium–vanadium alloys. *Acta Materialia* 51 (11): 3277–3292.
70. Collins, P. C., Banerjee, R., Banerjee, S., and Fraser, H. L. 2003. Laser deposition of compositionally graded titanium–vanadium and titanium–molybdenum alloys. *Materials Science and Engineering* 352 (1): 118–128.
71. Budinski, K. G. 1991. Tribological properties of titanium alloys. *Wear* 151 (2): 203–217.
72. Blau, P. J., Jolly, B. C., Qu, J., Peter, W. H., and Blue, C. A. 2007. Tribological investigation of titanium-based materials for brakes. *Wear* 263 (7): 1202–1211.
73. Bansal, D. G., Osman, L. E., and Blau, P. J. 2011. Surface engineering to improve the durability and lubricity of Ti–6Al–4V alloy. *Wear* 271 (9): 2006–2015.
74. Ma, G., Singh, A. K., Asokamani, R., and Gogia, A. K. 2009. Ti based biomaterials, the ultimate choice for orthopaedic implants—A review. *Progress in Materials Science* 54 (3): 397–425.
75. Xuanyong, L., Chu, P. K., and Ding, C. 2004. Surface modification of titanium, titanium alloys, and related materials for biomedical applications. *Materials Science and Engineering: R: Reports* 47 (3): 49–121.
76. Mitsuo, N. 1998. Mechanical properties of biomedical titanium alloys. *Materials Science and Engineering* 243 (1): 231–236.
77. Buchanan, R. A., Rigney, E. D., and Williams, J. M. 1987. Ion implantation of surgical Ti–6Al–4V for improved resistance to wear-accelerated corrosion. *Journal of Biomedical Materials Research Part A* 21 (3): 355–366.

78. Hu, C., Xin, H., Watson, L. M., and Baker, T. N. 1997. Analysis of the phases developed by laser nitriding Ti6Al4V alloys. *Acta Materialia* 45 (10): 4311–4322.
79. Czyrska-Filemonowicz, A., Buffat, P. A., Łucki, M., Moskalewicz, T., Rakowski, W., Lekki, J., and Wierzchoń, T. 2005. Transmission electron microscopy and atomic force microscopy characterisation of titanium-base alloys nitrided under glow discharge. *Acta Materialia* 53 (16): 4367–4377.
80. Zhecheva, A., Sha, W., Malinov, S., and Long, A. 2005. Enhancing the microstructure and properties of titanium alloys through nitriding and other surface engineering methods. *Surface and Coatings Technology* 200 (7): 2192–2207.
81. Masaaki, N., Niinomi, M., Akahori, T., Ohtsu, N., Nishimura, H., Toda, H., Fukui, H., and Ogawa, M. 2008. Surface hardening of biomedical Ti–29Nb–13Ta–4.6 Zr and Ti–6Al–4V ELI by gas nitriding. *Materials Science and Engineering* 486 (1):193–201.
82. Mohseni, H., Nandwana, P., Tsoi, A., Banerjee, R., and Scharf, T. W. 2015. In situ nitrided titanium alloys: Microstructural evolution during solidification and wear. *Acta Materialia* 83: 61–74.
83. Zhao, J.-C. 2006. Combinatorial approaches as effective tools in the study of phase diagrams and composition–structure–property relationships. *Progress in Materials Science* 51 (5): 557–631.
84. Deng, Y. P., Guan, Y. F., Fowlkes, J. D., Wen, S. Q., Liu, F. X., Pharr, G. M., Liaw, P. K., Liu, C. T. and Rack, P. D. 2007. A combinatorial thin film sputtering approach for synthesizing and characterizing ternary ZrCuAl metallic glasses. *Intermetallics* 15 (9): 1208–1216.
85. Ludwig, A., Zarnetta, R., Hamann, S., Savan, A., and Thienhaus, S. 2008. Development of multifunctional thin films using high-throughput experimentation methods. *International Journal of Materials Research* 99 (10): 1144–1149.
86. Curtarolo, S., Hart, G. L. W., Nardelli, M. B., Mingo, N., Sanvito, S., and Levy, O. 2013. The high-throughput highway to computational materials design. *Nature Materials* 12 (3): 191–201.
87. Helene, K., Ocylok, S., Weisheit, A., Springer, H., Jägle, E., and Raabe, D. 2017. Combinatorial alloy design by laser additive manufacturing. *Steel Research International.*
88. Raabe, D., Ponge, D., Dmitrieva, O., and Sander, B. 2009. Designing ultrahigh strength steels with good ductility by combining transformation induced plasticity and martensite aging. *Advanced Engineering Materials* 11 (7): 547–555.
89. Pereloma, E. V., Shekhter, A., Miller, M. K., and Ringer, S. P. 2004. Ageing behaviour of an Fe–20Ni–1.8 Mn–1.6 Ti–0.59 Al (wt%) maraging alloy: Clustering, precipitation and hardening. *Acta Materialia* 52 (19): 5589–5602.
90. Williams, J. C., Hickman, B. S., and Leslie, D. H. 1971. The effect of ternary additions on the decompositon of metastable beta-phase titanium alloys. *Metallurgical Transactions* 2 (2): 477–484.
91. Furuhara, T., Makino, T., Idei, Y., Ishigaki, H., Takada, A., and Maki, T. 1998. Morphology and crystallography of α precipitates in β Ti–Mo binary alloys. *Materials Transactions, JIM* 39 (1): 31–39.
92. Ho, W. F., Ju, C. P., and Chern Lin, J. H. 1999. Structure and properties of cast binary Ti–Mo alloys. *Biomaterials* 20 (22): 2115–2122.
93. Wang, K., Gustavson, L., and Dumbleton, J. 1993. The characterization of Ti–12 Mo–6 Zr–2 Fe: A new biocompatible titanium alloy developed for surgical implants. *Beta Titanium Alloys in the 1990's* 49–60.
94. Song, Y., Xu, D. S., Yang, R., Li, D., Wu, W. T., and Guo, Z. X. 1999. Theoretical study of the effects of alloying elements on the strength and modulus of β-type bio-titanium alloys. *Materials Science and Engineering* 260 (1): 269–274.
95. Kuroda, D., Niinomi, M., Morinaga, M., Kato, Y., and Yashiro, T. 1998. Design and mechanical properties of new β type titanium alloys for implant materials. *Materials Science and Engineering* 243 (1): 244–249.
96. Banerjee, R., Nag, S., and Fraser, H. L. 2005. A novel combinatorial approach to the development of beta titanium alloys for orthopaedic implants. *Materials Science and Engineering* 25 (3): 282–289.
97. Yeh, J.-W., Chen, S.-K., Lin, S-J., Gan, J.-Y., Chin, T.-S., Shun, T.-T., Tsau, C.-H., and Chang, S.-Y. 2004. Nanostructured high-entropy alloys with multiple principal elements: Novel alloy design concepts and outcomes. *Advanced Engineering Materials* 6 (5): 299–303.
98. Ta-Kun, C., Shun, T. T., Yeh, J. W., and Wong, M. S. 2004. Nanostructured nitride films of multi-element high-entropy alloys by reactive DC sputtering. *Surface and Coatings Technology* 188: 193–200.
99. Yong, Z., Zuo, T. T., Tang, Z., Gao, M. C., Dahmen, K. A., Liaw, P. K., and Lu, Z. P. 2014. Microstructures and properties of high-entropy alloys. *Progress in Materials Science* 61: 1–93.
100. Akira, T., Kenji, A., Takeshi, W., Kunio, Y., and Wei, Z. 2014. High-entropy alloys with a hexagonal close-packed structure designed by equi-atomic alloy strategy and binary phase diagrams. *JOM* 66 (10): 1984–1992.

101. Miracle, D. B., Miller, J. D., Senkov, O. N., Woodward, C., Uchic, M. D., and Tiley, J. 2014. Exploration and development of high entropy alloys for structural applications. *Entropy* 16 (1): 494–525.
102. Wang, W.-R., Wang, W.-L., Wang, S-C., Tsai, Y-C., Lai, C.-H., and Yeh, J.-W. 2012. Effects of Al addition on the microstructure and mechanical property of $Al_xCoCrFeNi$ high-entropy alloys. *Intermetallics* 26: 44–51.
103. Karpets, M. V., Myslyvchenko, O. M., Makarenko, O. S., and Krapivka, M. O. 2015. Mechanical properties and formation of phases in high-entropy $CrFeNiCuCoAl_x$ alloys. *Powder Metallurgy and Metal Ceramics* 54 (5–6): 344–352.
104. Guo, S., Ng, C., and Liu, C. T. 2013. Anomalous solidification microstructures in Co-free $Al_xCrCuFeNi_2$ high-entropy alloys. *Journal of Alloys and Compounds* 557: 77–81.
105. Tokushichi, M. 1936. Strong permanent magnet with cobalt. U.S. Patent No. 2,027,996.
106. Zhang, L. 2016. Coercivity Enhancement and Gamma Phase Avoidance of Alnico Alloys. Master's Thesis.
107. Chin, T.-S., Wu, T. S., Chang, C. Y., Hsu, T. K., and Chang, Y. H. 1983. Electron microscopy and magnetic properties of Fe–Cr–Co–Si permanent magnet alloys manufactured by rolling-ageing technique. *Journal of Materials Science* 18 (6): 1681–1688.
108. Zhang, Y., Zuo, T., Cheng, Y., and Liaw, P. K. 2013. High-entropy alloys with high saturation magnetization, electrical resistivity, and malleability. *Scientific Reports* 3: 1455.
109. Chou, H.-P., Chang, Y.-S., Chen, S. K., and Yeh, J. W. 2009. Microstructure, thermophysical and electrical properties in $Al_xCoCrFeNi$ ($0 \leq x \leq 2$) high-entropy alloys. *Materials Science and Engineering B* 163: 184–189.
110. Li, C., Li, J. C., Zhao, M., and Jiang, Q. 2009. Effect of alloying elements on microstructure and properties of multiprincipal elements high-entropy alloys. *Journal of Alloys Compounds* 475: 752–757.
111. Guo, S., Ng, C., Lu, J., and Liu, C. T. 2011. Effect of valence electron concentration on stability of fcc or bcc phase in high entropy alloys. *Journal of Applied Physics* 109: 103505.
112. Wang, F. J., Zhang, Y., and Chen, G. L. 2009. Atomic packing efficiency and phase transition in a high entropy alloy. *Journal of Alloys and Compounds* 478: 321–324.
113. Borkar, T., Chaudhary, V., Gwalani, B., Choudhuri, D., Mikler, C.V., Soni, V., Alam, T., Ramanujan, R. V., and Banerjee, R. In press. A combinatorial approach for assessing the magnetic properties of high entropy alloys: Role of Cr in $AlCo_xCr_{1-x}FeNi$. *Advanced Engineering Materials (*Accepted*).*
114. Li, S.-j., Li, D., Liu, W., and Zhang, Z. 2015. High Curie temperature and coercivity performance of $Fe_{3-x}CrxSe_4$ nanostructures. *Nanoscale* 7: 5395–5402.
115. Rao, A. S. 1993. Alnico permanent magnets an overview. *Proceedings of Electrical/Electronics Insulation Conference*, Chicago, IL, 373–383.
116. McCaig, M. 1964. Recent developments in permanent magnetism. *Journal of Applied Physics* 35 (3): 958–965.
117. Smith, C., Katakam, S., Nag, S., Zhang, Y. R., Law, J. Y., Ramanujan, R. V., and Banerjee, R. 2014. Comparison of the crystallization behavior of Fe–Si–B–Cu and Fe–Si–B–Cu–Nb–based amorphous soft magnetic alloys. *Metallurgical and Materials Transactions A* 45 (7): 2998–3009.
118. Zhang, Y. R., and Ramanujan, R. V. 2005. The effect of niobium alloying additions on the crystallization of a Fe–Si–B–Nb alloy. *Journal of Alloys and Compounds* 403 (1): 197–205.
119. Herzer, G. 1997. Nanocrystalline soft magnetic alloys. *Handbook of Magnetic Materials* 10: 415–462.
120. Chen, Y. M., Ohkubo, T., Ohta, M., Yoshizawa, Y., and Hono, K. 2009. Three-dimensional atom probe study of Fe–B-based nanocrystalline soft magnetic materials. *Acta Materialia* 57 (15): 4463–4472.
121. Zhang, Y. R., and Ramanujan, R. V. 2006. Characterization of the effect of alloying additions on the crystallization of an amorphous $Fe_{73.5}Si_{13.5}B_9Nb_3Cu_1$ alloy. *Intermetallics* 14 (6): 710–714.
122. Matsuura, M., Nishijima, M., Takenaka, K., Takeuchi, A., Ofuchi, H., and Makino, A. 2015. Evolution of fcc Cu clusters and their structure changes in the soft magnetic $Fe_{85.2}Si_1B_9P_4Cu_{0.8}$(NANOMET) and FINEMET alloys observed by x-ray absorption fine structure. *Journal of Applied Physics* 117 (17): 17A324.
123. Gheiratmand, T., Hosseini, H. M., Davami, P., Ababei, G., and Song, M. 2015. Mechanism of mechanically induced nanocrystallization of amorphous FINEMET ribbons during milling. *Metallurgical and Materials Transactions A* 46 (6): 2718–2725.
124. Sinha, A. K., Singh, M. N., Upadhyay, A., Satalkar, M., Shah, M., Ghodke, N., Kne, S. N., and Varga, L. K. 2015. A correlation between the magnetic and structural properties of isochronally annealed Cu-free FINEMET alloy with composition Fe72B19.2Si4.8Nb4. *Applied Physics A* 118 (1): 291–299.

125. Gheiratmand, T., Madaah Hosseini, H. R., Davami, P., Gjoka, M., and Song, M. 2015. The effect of mechanical milling on the soft magnetic properties of amorphous FINEMET alloy. *Journal of Magnetism and Magnetic Materials* 381: 322–327.
126. Moya, J. A. 2015. Improving soft magnetic properties in FINEMET-like alloys. A study. *Journal of Alloys and Compounds* 622: 635–639.
127. Muhammad, D., Khan, F. A., Farabi, H. M., and Alam, M. J. 2014. Electrical and magnetic transport properties of nanocrystalline $Fe_{73.5}Si_{13.5}B_9Cu_1Nb_3$ alloy. *Journal of Superconductivity and Novel Magnetism* 27 (6): 1525–1530.
128. Franco, V., Conde, C. F., and Conde, A. 1998. Effect of the Si/B ratio on the magnetic anisotropy distribution of $Fe_{73.5}Si_{22.5-x}B_xCu_1Nb_3$ (x = 7, 9, 16) alloys along nanocrystallization. *Journal of Applied Physics* 84 (9): 5108–5113.
129. Zhi, J., He, K. Y., Cheng, L. Z., and Fu, Y. J. 1996. Influence of the elements Si/B on the structure and magnetic properties of nanocrystalline (Fe, Cu, Nb) $_{77.5}$ $Si_xB_{22.5-x}$ alloys. *Journal of Magnetism and Magnetic Materials* 153 (3): 315–319.
130. Srinivas, M., Majumdar, B., Bysakh, S., Raja, M. M., and Akhtar, D. 2014. Role of Si on structure and soft magnetic properties of $Fe_{87-x}Si_xB_9Nb_3Cu_1$ ribbons. *Journal of Alloys and Compounds* 583: 427–433.
131. Lashgari, H. R., Chu, D., Xie, S., Sun, H., Ferry, M., and Li, S. 2014. Composition dependence of the microstructure and soft magnetic properties of Fe-based amorphous/nanocrystalline alloys: A review study. *Journal of Non-Crystalline Solids* 391: 61–82.
132. Hono, K., Ping, D. H., Ohnuma, M., and Onodera, H. 1999. Cu clustering and Si partitioning in the early crystallization stage of an $Fe_{73.5}Si_{13.5}B_9$ Nb_3Cu_1 amorphous alloy. *Acta Materialia* 47 (3): 997–1006.
133. Ayers, J. D., Harris, V. G., Sprague, J. A., Elam, W. T., and Jones, H. N. 1998. On the formation of nanocrystals in the soft magnetic alloy $Fe_{73.5}Nb_3Cu_1Si_{13.5}B_9$. *Acta Materialia* 46 (6): 1861–1874.
134. Ayers, J. D., Harris, V. G., Sprague, J. A., Elam, W. T., and Jones, H. N. 1997. A model for nucleation of nanocrystals in the soft magnetic alloy $Fe_{73.5}Nb_3Cu_1Si_{13.5}B_9$. *Nanostructured Materials* 9 (1): 391–396.
135. Zhang, Y. R., and Ramanujan, R. V. 2006. Microstructural observations of the crystallization of amorphous Fe–Si–B based magnetic alloys. *Thin Solid Films* 505 (1): 97–102.
136. Zięcbowicz, B., Szewieczek, D., and Dobrzański, L. A. 2007. New possibilities of application of composite materials with soft magnetic properties. *Journal of Achievements in Materials and Manufacturing Engineering* 20 (1–2): 207–210.
137. McHenry, M. E., Willard, M. A., and Laughlin, D. E. 1999. Amorphous and nanocrystalline materials for applications as soft magnets. *Progress in Materials Science* 44 (4): 291–433.
138. Coey, J. D. 2001. Magnetic materials. *Journal of Alloys and Compounds* 326 (1): 2–6.
139. Jiles, D. C. 2003. Recent advances and future directions in magnetic materials. *Acta Materialia* 51 (19): 5907–5939.
140. Jiles, D. C., and Lo, C. C. H. 2003. The role of new materials in the development of magnetic sensors and actuators. *Sensors and Actuators A: Physical* 106 (1): 3–7.
141. Sahasrabudhe, H., Harrison, R., Carpenter, C., and Bandyopadhyay, A. 2015. Stainless steel to titanium bimetallic structure using LENS. *Additive Manufacturing* 5: 1–8.
142. Lin, X., Yue, T. M., Yang, H. O., and Huang, W. D. 2006. Microstructure and phase evolution in laser rapid forming of a functionally graded Ti–Rene88DT alloy. *Acta Materialia* 54 (7): 1901–1915
143. Carroll, B. E., Otis, R. A., Borgonia, J. P., Suh, J., Dillon, R. P., Shapiro, A. A., Hofmann, D. C., Liu, Z-K., and Beese, A. M. 2016. Functionally graded material of 304L stainless steel and inconel 625 fabricated by directed energy deposition: Characterization and thermodynamic modeling. *Acta Materialia* 108: 46–54.
144. Shah, K., Haq, I. u., Khan, A., Shah, S. A., Khan, M., and Pinkerton, A. J. 2014. Parametric study of development of Inconel-steel functionally graded materials by laser direct metal deposition. *Materials & Design (1980–2015)* 54: 531–538.
145. Robinson, J. L., and Scott, M. H. 1980. Liquation cracking during the welding of austenitic stainless steels and nickel alloys. *Philosophical Transactions of the Royal Society of London A: Mathematical, Physical and Engineering Sciences* 295 (1413): 105–117.
146. Lin, X., and Yue, T. M. 2005. Phase formation and microstructure evolution in laser rapid forming of graded SS316L/Rene88DT alloy. *Materials Science and Engineering: A* 402 (1): 294–306.
147. Savitha, U., Jagan Reddy, G., Venkataramana, A., Sambasiva Rao, A., Gokhale, A. A., and Sundararaman, M. 2015. Chemical analysis, structure and mechanical properties of discrete and compositionally graded SS316–IN625 dual materials. *Materials Science and Engineering: A* 647: 344–352.

148. Wu, D., Liang, X., Li, Q., and Jiang, L. 2011. Laser rapid manufacturing of stainless steel 316L /Inconel718 functionally graded materials: Microstructure evolution and mechanical properties. *International Journal of Optics* 2010.
149. Seefeld, T., Theiler, C., Schubert, E., and Sepold, G. 1999. Laser generation of graded metal-carbide components. *Materials Science Forum* 308: 459–466.
150. Wang, F., Mei, J., and Wu, X. 2007. Compositionally graded Ti6Al4V + TiC made by direct laser fabrication using powder and wire. *Materials & Design* 28 (7): 2040–2046.
151. Liu, W. and DuPont, J. N. 2003. Fabrication of functionally graded TiC/Ti composites by laser engineered net shaping. *Scripta Materialia* 48 (9): 1337–1342.
152. Zhang, Y., Zengmin, W., Likai, S., and Mingzhe, X. 2008. Characterization of laser powder deposited Ti–TiC composites and functional gradient materials. *Journal of Materials Processing Technology* 206 (1): 438–444.

5 Electron Beam Rapid Prototyping Using Wires and Modification of the Surface

Marek St. Weglowski

CONTENTS

5.1 INTRODUCTION

In the case of increasing competition in the market, the concept of rapid prototyping (additive manufacturing, additive layer manufacturing) of metallic parts with wire enables shortening the period required to produce a conventional prototype that is no longer necessary in the design phase of a product. It should be noted that the computer-aided design/computer-aided manufacturing (CAD/CAM) systems only allow for preparation of a virtual model. Three technologies of the wire feeding–type processes were developed:

1. Wire and arc–based additive manufacturing
2. Wire and laser–based additive manufacturing
3. Electron beam freeform fabrication (EBF) processes

The layer thickness and deposition rate of the wire feeding process are greater than 1 mm and 10 g/min, respectively. The dimensional accuracy and surface roughness of the wire feeding process are inferior to those of the powder feeding–type rapid prototyping process, while the build speed of the wire feeding process is significantly superior to that of the powder feeding process. Wire arc additive manufacturing processes employ gas metal arc welding, gas tungsten arc welding, and plasma arc welding processes for fusion of the metallic wire [1].

In the initial phase of product development, the most important components of the production cost relating to functions and structures and the materials and technology are generated. The most important stage in the development of a new product is forming the prototype, which is still in the development phase of the product and allows for initial visualization, estimation of utility characteristics, and sometimes even introduced to the market for the purpose of evaluating its characteristics. Prototyping using traditional methods is long and costly. Traditional methods typically require a lot of manual labor coupled with an employee having high professional qualifications, which significantly raises the final cost of the product. Handmade models also disrupt the cycle of electronic information flow between the level of design and the level of actual production [2]. Rapid prototyping technologies are also useful in the production of short runs, which are not cost-effective when applied to processes such as forging or casting. Most metal rapid prototyping processes cannot replace existing processes, but they can offer benefits in certain superalloy applications in terms of cost savings or providing otherwise unavailable processing capabilities (e.g., dual alloy deposition and functionally graded materials). Such capabilities enable innovative design (e.g., for future aerospace applications) [3].

All additive manufacturing technologies (based on powders and wires) have a growth of 34.9% in 2013, the most increasing manufacturing technology in the world. The market volume of machinery and services of additive manufacturing is estimated at 3.7 billion euros, equivalent to 2% of the total machine market. Conservative estimations expect a market volume of 7 billion euros in 2016. In 2020, the estimated volume will reach $11 billion. Overall, there is a total market potential of about 130 billion euros [4].

The additive manufacturing technologies, among others, allow for

1. Reduced down time
2. Overall operation costs
3. Capacity utilization

Also, the supply chain management can be improved, and inventory requirements can be reduced. Thus, customers can be greatly satisfied. Many factors should be taken into account during the development of a business involving additive manufacturing processes, and a comparison with traditional manufacturing processes should be carried out. The following aspects are included [5]:

1. Fixed cost/nonrecurring manufacturing costs
2. The cost of process qualification and component certification
3. Logistical costs
4. The cost of time

A comparative summary between additive manufacturing and traditional manufacturing processes is presented in Table 5.1 [6].

However, to calculate the cost and benefits of additive manufacturing, the different levels of a problem's complexity should be estimated. In Figure 5.1 [7], a typical pyramid of complexity of the problem is shown.

Now that the general processing science of metal additive manufacturing has been explored, this background can be used to compare the technical aspects of existing technologies. A tabulated comparison of SLM (selective laser melting), EBM (electron beam melting), powder-fed systems (laser beam additive manufacturing with powder), and wire-fed systems (laser or electron beams) is provided in Table 5.2 [8].

TABLE 5.1
Comparative Summary between Additive and Traditional Manufacturing Processes

Factor	Additive Manufacturing Technology	Traditional Manufacturing
Cost	Products can be manufactured at comparatively low costs; this is, however, limited to small and medium production batches, rarely in mass production.	These methods, typically, are expensive for small production batches, as costs are involved in casting molds, dyes, tooling, finishing, and other different processes that go into manufacturing the products.
Time	A very short time for manufacturing a single product can be achieved. In additive manufacturing technology, the products are produced directly from the CAD model. This solution helps save time in delivering final products by cutting down on the production development step, supply chain, and dependence on inventory.	Generally, manufacturing times are very long. It depends on the availability of the molds, dyes, inventory, and so on.
Resource consumption	Only optimal quantity required to manufacture the product.	Extremely high.
Product complexity	The complex geometries and products can be manufactured. The products are mainly limited by the design engineer's imagination.	The manufacturing of complex geometries is limited. Many different parts have to be manufactured separately and then assembled.
Material quality and application	The material quality depends on the technology used. Initially, 3D manufactured parts were not used in load-bearing applications (powder bed techniques), but advancements in the technology are rapidly improving the material quality, which has led to their use in some load-bearing applications (deposited techniques with powder or wires).	Because of their excellent quality, the products have always been used for load-carrying applications.
Material wastage	There is little to no wastage of the raw material, as they can be reused. For wire feed system, the waste of material consumption is lower than that for powder techniques.	Involves a lot of material wastage due to post-fabrication finishing processes.
Prototyping	Extremely useful for prototyping and evaluating product concepts. Allows for design changes and feedback.	Not preferred for product prototypes and concepts. Very expensive and time consuming.
Logistic	The logistic issues for powder (bead) technologies are more complicated than those for wire techniques. However, the total logistic aspects are easier than traditional manufacturing.	The logistic issues are very complicated, especially for castings.
Process qualification and component certification	From a practical point of view, the process certification is more difficult than that for traditional methods. The availability of the standards is limited. This is caused by the fact that additive manufacturing technologies are new on the market.	Traditional manufacturing technologies are mature. Thus, a lot of standards and procedures are available.
Space application	3D printing could essentially pave the way for setting up structures off-world, especially on the moon and Mars.	It will be exorbitant to build structures off-world using these techniques.

Source: Joshi, S.C., and Sheikh, A.A. 2015. 3D printing in aerospace and its long-term sustainability. *Virtual Phys Prototyp* 10:175–185.

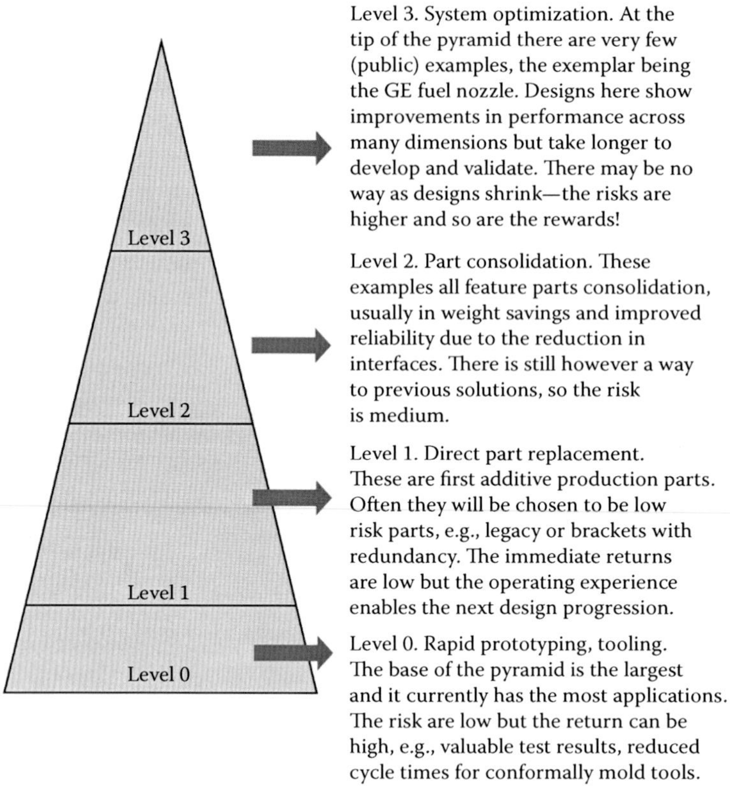

FIGURE 5.1 Additive manufacturing level's pyramid of problem's complexity. (From Canada Makes' Metal additive process guide. http://canadamakes.ca/app/.)

5.2 FUNDAMENTALS OF THE ELECTRON BEAM RAPID PROTOTYPING PROCESS USING WIRE AND TECHNOLOGICAL PARAMETERS

5.2.1 Background

Electron beam rapid prototyping (additive manufacturing, additive layer manufacturing) with wire and an electron beam are unique technologies that allowing for the insertion of additive material in the form of a wire directly into a pool of liquid metal. Additionally, refining occurs as the melting process is carried out in vacuum. Modern devices can precisely adjust to all technological parameters. This is important in order to ensure minimal penetration or minimum volume of molten metal for an electron beam energy density up to 10^{12} W/m^2.

Building up components layer by layer by using a wire is one such technique, which can be adopted for rapid prototyping (Figure 5.2).

Electron beam additive layer manufacturing (with wire) offers numerous advantages:

1. Reduced production and material costs
2. Reduced development and lead times
3. Improved performance

To compare the powder and wire systems, some aspects can be defined. During the powder deposition process, a certain amount of powders cannot be caught by the melt pool. The powders are blown to the surrounding environment, which causes potential hazard to both the operator and the environment. Compared with powder feeding, the wire feeding method offers a higher material usage

TABLE 5.2
Comparison of Defects and Features across Platforms

Defects or Feature	Selective Laser Melting	Electron Beam Melting	Powder Fed	Wire Fed
Feedstock	Powder	Powder	Powder	Wire
Heat source	Laser	E-beam	Laser	Laser or e-beam
Atmosphere	Inert	Vacuum	Inert	Inert/vacuum
Part repair	No	No	Yes	Yes
New parts	Yes	Yes	Yes	Yes
Multi-material	No	No	Possible	Possible
Porosity	Low	Low	Low	Very low
Residual stress	Yes	Low	Yes	Yes
Substrate adherence	Yes	Material dependent	Yes	Yes
Cracking	Yes	Not typical	Yes	Yes
Delamination	Yes	Yes	Yes	Yes
Rapid solidification	Yes	Yes	Yes	Yes
In situ aging	No	Yes	No	No
Overhangs	Yes	Yes	Limited	Limited
Mesh structure	Yes	Yes	No	No
Surface finish	Medium rough	Rough	Medium-poor	Poor but smooth
Built cleanup from process	Loose powder	Sintered powder	Some loose powder	N/A

Source: Sames, W.J., List, F.A., Pannala, S. et al. 2016. The metallurgy and processing science of metal additive manufacturing. *Int Mater Rev* 61:1–47.

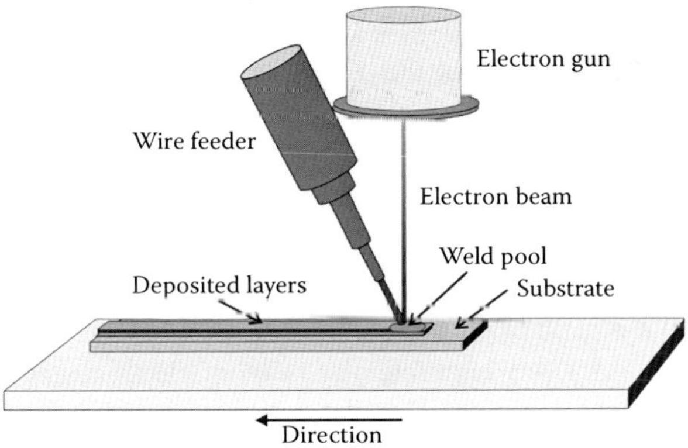

FIGURE 5.2 Scheme of rapid prototyping process with electron beam.

efficiency. Almost all of the materials fed into the melt pool during the process are used to form the deposit. It lowers the risks to the operator and is environment friendly. Moreover, the powder feeding nozzle plays an important role in the powder deposition process. It makes the deposition with powder more complicated than that with wire. Another advantage is that metal wires are easily available and are cheaper than powders, which makes wire deposition very cost-competitive [9].

An earlier investigation [10] revealed that some additive layer manufacturing techniques can be adopted for a variety of different missions in space. One application is the on-orbit construction of large space structures, on the order of tens of meters to a kilometer in size. The second one is a small, multifunctional system that could be used by astronauts on long-duration human exploration missions for the purpose of manufacturing spare parts. A third example of an additive layer manufacturing application in space is a miniaturized automated system for structural health monitoring and repair [10]. Although the filler material in additive layer manufacturing techniques can be applied both as powder and as a wire, from a practical point of view, the wire form is preferred for in-space application because of operational and safety issues associated with the management of metallic powder in a microgravity environment [11]. The concentrated beam of electrons, which has greater than 90% power efficiency and nearly 100% coupling efficiency with electrically conductive materials, is an attractive energy beam source for additive layer manufacturing. The vacuum in space (up to 10^{-17} mbar) can be used as the process environment, thereby eliminating the use of the vacuum chamber and pumping system typical for ground-based electron beam freeform fabrication. Another problem is x-ray radiation at high accelerating voltage. This hazard can be minimized by using low beam power and performing the operation on the outside of the spacecraft where the spacecraft structure does provide shielding for astronauts and the sensitive equipment [11].

To date, there are several studies that discuss the applications of additive layer manufacturing based on an electron beam source. Taminger and Hafley [12] from NASA Langley Research Center focused on electron beam freeform fabrication technology. They presented the results achieved under typical conditions ($1g$) as well as under reduced gravity by using a flight-weight electron beam freeform fabrication (EBF3) system. Their achievements will be especially attractive for space conditions. Moreover, Taminger and Hafley [13] revealed that the microstructures and mechanical properties obtained in titanium alloy have demonstrated the potential for achieving a wide microstructural range with properties comparable to those of wrought products. There is a trade-off between high deposition rates and fine-grained microstructures for materials deposited using the electron beam additive manufacturing process. Mitzner et al. [14] have shown that electron beam with modulation allows for the traditional long columnar grains typical of electron beam freeform fabrication modified to a much more equiaxed structure. Fresh grains were found to nucleate in the steady-state region of the build, where epitaxial growth normally predominates. The size of the β grain was drastically reduced for both modulation waveform conditions, from 1164 to 734 μm. The refinement of the α phase due to modulation was indicated by a reduction of α colony intercept length from 3.2 to 2.2 μm. The width of the α lath was also refined by half from 1.0 to 0.5 μm. Gonzales et al. [15] revealed that powder cored tubular wires successfully replenished the loss of aluminum in EBF3 deposits. Moreover, iron–boron enhancements to the powder core significantly refined the structure of the α lath and hindered the epitaxial growth of the prior β grains in multiple-layer EBF3 deposits of Ti–6Al–4V alloy.

5.2.2 Characteristics of the Process

Electron beam processes use a very high intensity beam of electromagnetic energy as the heating source for additive layer manufacturing. The energy density is approximately 10^{10} to 10^{13} W/m^2 versus 5×10^6 to 5×10^8 W/m^2 for typical arc rapid prototyping processes. Conversion of kinetic energy of the electrons into heat occurs as these particles strike the workpiece and deposition (filler) material, leading to both melting and vaporization. An electron beam can operate in keyhole mode (not useful in additive manufacturing techniques) and weld pool mode (applied in additive manufacturing techniques). The electron beam process provided a low fusion zone with a narrow heat-affected zone and minimal distortion [16].

Shielding for the electron beam process is provided by the vacuum (typically, 10^{-3} to 10^{-5} atm), which is required to allow the beam of electrons to flow to the workpiece unimpeded by collisions with molecules in air or another gaseous atmosphere. While the high energy beam of electrons in electron beam processes is readily absorbed and the kinetic energy is converted to heat by all metallic materials, this is not so for an intense beam of photons in laser beam processes. Some materials reflect photons, or light, depending on the reflectancy of the particular material and the wavelength of the photons or laser light. As a result, since the efficiency of electron absorption is high, the transfer efficiency of electron beam additive manufacturing is also high, approaching 90%.

Electron beams are produced in a gun by thermionically extracting them from a heated filament (cathode) and accelerating them across a high potential achieved using one or more annular anodes along a column of high vacuum. The stream of accelerated electrons is focused into a beam of high energy density using a series of electromagnetic coils and lenses. The electrons then pass from the column to a work chamber to the workpiece and filler material to produce the deposited layer. More details on the physics of electron beam processes can be found elsewhere [17].

In the electron beam additive manufacturing process, the filler material (deposit material) in the wire form is feeding continuously in the deposit area and undergoes melting with high efficiency, and as a result, a single deposit layer can be created. The material simultaneously undergoes melting with the workpiece (previous layer of deposited material) at a low depth. Thus, high-quality metallurgical joining with the workpiece can be achieved. Deposit material is heated and undergoes crystallization in vacuum at a cooling rate of 10^6°C/s. This results in high metallurgical cleanliness and an ultrafine-grained single-layer microstructure. Because the electron beam additive manufacturing process is carried out in vacuum, the refining process allows for an improvement in the metallurgical cleanliness and deoxidation of the deposit material and elimination of inclusions. Given the high energy density, the heat source in the restricted area acts on the filler material as well as the workpiece. Thus, the residual stress and distortion can be reduced. It is possible to produce in one run a single layer with a thickness of 0.1–5 mm and a width of 5–8 mm in a straight single track and with more than 30 mm of swinging movement of the electron beam head or cladding surface. This depends on the machine type, the dimension of deposit material, and the technique. The electron beam additive manufacturing process can be carried out depending on size, shape, and mass of the final element. The movement of the electron beam head of an element can be applied.

However, even if the additive manufacturing process seems to be very straightforward, prediction of mechanical and metallurgical properties as well as microstructure is complex, partly because of its building in layer nature which generates a complex thermal history dictating the mechanical properties, and partly because of the number of parameters involved during the additive manufacturing process itself. Therefore, it must be pointed out that it is most important to understand the physical process of metal transfer during additive manufacturing [18].

The kinetics of the electron beam deposition process using a wire is different from the kinetics of the deposition process for arc welding (e.g., gas metal arc). In the case of arc welding, melted metal droplets from a wire electrode are transferred to the liquid metal by gravity and the forces occur in an electric arc. By changing the current parameters of the arc welding process, a precise regulation of metal transfer can be achieved. However, in the case of electron beam additive manufacturing technology, a drop of liquid metal is mainly affected by the gravity force F_g, the beam pressure force F_e, the surface tension of the filler wire F_s, and the metal-vapor jet force F_v [18]. The forces acting on the liquid metal droplet are shown in Figure 5.3.

If the value of the motive force exceeds the preventive force, the droplet detaches from the tip of the wire. The transfer of metal droplets can be divided into four stages: droplet growth, oscillation of droplet growth, maximum droplet size, and droplet transfer. If the transition height is too large, a drop formed at the tip of the wire cannot contact the surface of the weld pool. On the other hand, instability of the wire and changes in both the value and direction of forces caused by the metal vapor provoke oscillation of the droplet. During electron beam deposition, the metal-vapor jet force can cause the droplets to swing upward because of the changes in both depth and size of the instantaneous keyhole in the molten pool. Periodically, when metal-vapor jet force decreases, droplets move downward in the direction of the weld pool. It should be noted that due to temperature differences between the weld pool and the metal drop, the Leidenfrost phenomenon can occur. Moreover, Zhao et al. [18] revealed that if $H \approx 0$ (the wire nearly contacts the molten pool), instead of droplet transfer of the liquid metal, a molten metal bridge transfer can be achieved (Figure 5.4). When the melting rate is almost equal to the wire feed rate, a stable molten-metal bridge between the wire and molten pool can be obtained. During this mode, little fluctuation in the molten pool is generated. This transfer mode is denoted as the molten-metal bridge transfer. It should be noted that the heat contributing to wire melting includes thermal conduction as well as radiation from the molten pool and the electron beam. Therefore, the molten-metal bridge transfer is a stable transfer mode over a wide range of processing parameters. The molten-metal bridge transfer mode is the most useful transfer during the electron beam additive manufacturing process (Figure 5.4).

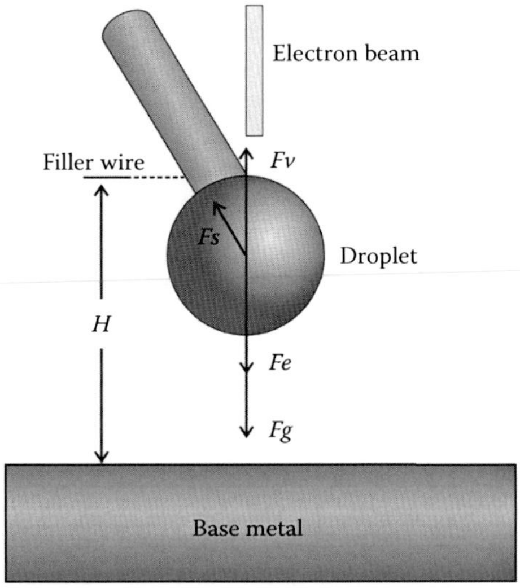

FIGURE 5.3 Schematic of the forces affecting the droplet during laser beam, electron beam, and partial GTA (Gas Tungsten Arc). *H*, transition height. (From Zhao, J., Zhang, B., Li, A. et al. 2015. Effects of metal-vapor jet force on the physical behavior of melting wire transfer in electron beam additive manufacturing. *J Mater Process Technol* 220:243–250.)

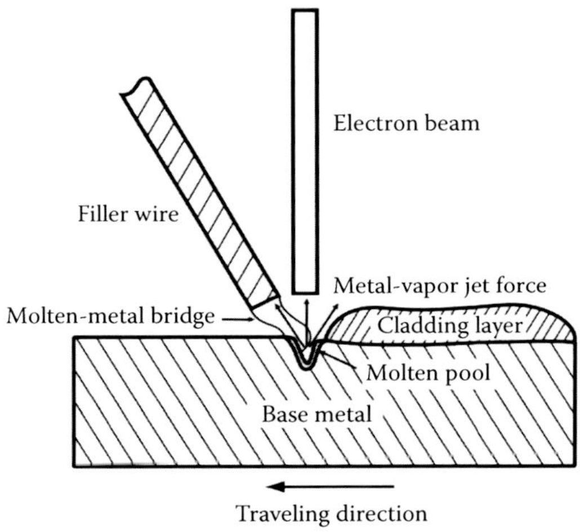

FIGURE 5.4 Molten metal bridge transfer during electron beam additive manufacturing. (From Zhao, J., Zhang, B., Li, A. et al. 2015. Effects of metal-vapor jet force on the physical behavior of melting wire transfer in electron beam additive manufacturing. *J Mater Process Technol* 220:243–250.)

5.2.3 Technological Parameters of the Process

The basic parameters of electron beam additive manufacturing include accelerating voltage, beam current, traveling speed, beam diameter on the workpiece surface, electromagnetic lens focal length, electromagnetic lens current, vacuum in the working chamber, the wire feed rate, geometrical setup of wire, and electron beam and deposited layer, including the following: feeding angle, wire stick-out, longitudinal tip position (Figure 5.5), and feeding position [19].

From the electron beam additive manufacturing technology point of view, the *accelerating voltage* (U) is not the most important parameter involved. For a majority of welding technology operations, the accelerating voltage is kept constant at U_{max}, which depends on the type of high-voltage generator and electron beam gun, and is set at 60 or 150 kV [20]. For the same beam current, an increase in accelerating voltage causes a decrease of the electron beam diameter and an increase of the penetration depth. However, for additive manufacturing, the penetration depth is not an important factor; thus, the accelerating voltage should be reduced. An increase in accelerating voltage causes an increase in the width of the single layer and a decrease in the thickness of the layer.

The *beam current* (I), together with the accelerating voltage, determines the beam power and, above all, affects the resulting deposit layer. An increase in beam current causes an increase in the width of the single layer. Simultaneously, the increase in beam current causes an increase in penetration depth. Hence, it should be matched carefully.

Traveling speed. With the movement of the workpiece to the electron beam, or vice versa, the weld pool (composed of deposit material and substrate) moves through the material and the deposit layer is produced. As a result of this movement, the power beam is transferred into the volume of the material such that for each deposit operation, a typical geometry of heat is set up. In electron beam deposition, in comparison with other fusion processes, the heated volume is particularly small. The heat input from a constant beam power is dependent on the material and not on the width of the weld beam or the depth of fusion. With an increase in traveling speed, the heat input decreases as a result of lower loss of heat conduction. For the electron beam additive manufacturing, the increase in traveling speed caused a decrease of the width as well as thickness of the single layer [21]. For additive

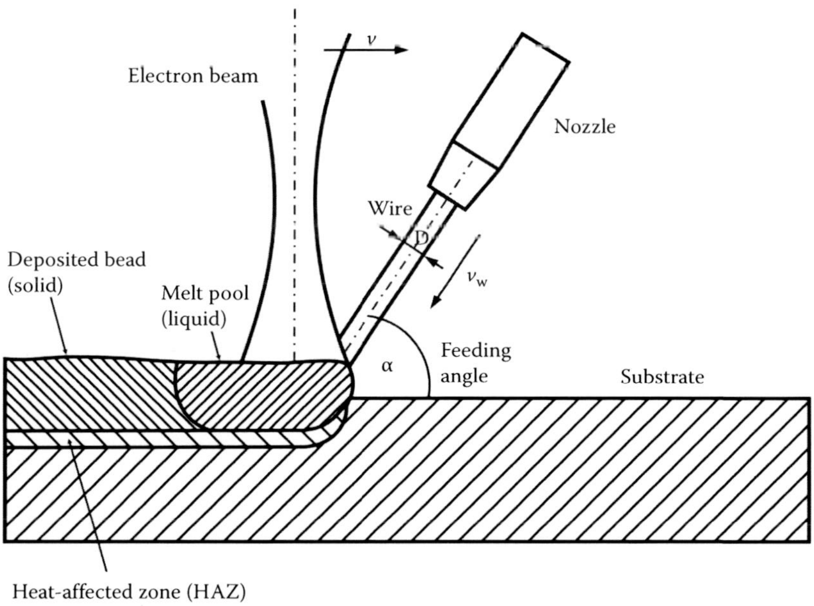

FIGURE 5.5 Setup parameters regarding the wire tip in relation to the deposited pool and the nozzle. v_w, wire feed rate; v, traveling speed. (From Brandl, E., Michailov, V., Viehweger, B. et al. 2011. Deposition of Ti–6Al–4V laser and wire, part I: Microstructural properties of single beads. *Surf Coat Tech* 206:1120–1129.)

manufacturing at a constant power beam and wire feed rate, the deposition height decreased with an increase in traveling speed. The changes in traveling speed have a significant effect on the height of the deposit compared to the power.

All of the electron beam additive manufacturing parameters mentioned so far affect the deposit process quite strongly and in characteristic ways. The beam power, together with the focal point, affects the forces acting in the weld pool as well as the filler material. The dynamic processes, including both solidification and cooling, are affected by the traveling speed. The traveling speed, together with beam power, determines the volume of fusion zone (penetration depth) and the width and thickness of the deposit layer as well as the transport of filler metal to the weld pool.

Beam deflection. The electron beam is made up of negative charge carriers and can thus be deflected in different ways from its normal axis by both electrical and magnetic fields. The static as well as dynamic deflection system can be applied. If the direct current is passed through the coils of a deflection system, then the electron beam will be deflected statically through a certain angle. The required angle of deflection is set by adjusting the direct current to the coils. Generally speaking, the maximum deflection angle is ±5° and is limited more by the increasing distortion occurring at the point of impingement of the beam than by the field of view of an optical viewing system. By periodically varying alternating currents, the electron beam can be deflected dynamically, free of inertia, in all directions, shapes, and frequencies required by electron beam additive manufacturing. Beam oscillation at small amplitudes and high frequencies is an effective and generally useful method that beneficially affects the fusion and solidification process with the objective of improving the quality of the deposited layer. An oscillating electron beam in general increases the size of the fusion zone and allows gas porosity to rise and escape from the weld pool.

Working pressure. Vacuum is an absolute necessity for generating and collimating an electron beam. It is clear that the presence of gas molecules in the path of the beam would result in some loss of beam power by absorption, coupled with a degree of electron scattering. Scattering is, in effect, a defocusing action that reduces the power density of the beam. On the path of the electron gun to the workpiece, a vacuum cannot be easily dispensed with, but for other reasons, an adequate pressure is required. The residual gases in the vicinity of the electron beam should not initiate any chemical reaction, such as oxidation at the point of deposit layer and heated areas of the workpiece, and should not affect the power density of the electron beam. From a practical point of view, the vacuum level should be on the order of 5×10^{-4} mbar or lower especially for titanium alloys [22].

Lens current (focal position). As mentioned before, the electron beam diverges on leaving the electron gun and is collimated and focused by an electromagnetic lens to achieve a typically high power density. The appropriate focus position can be determined using technological trials because of numerous other factors involved. With respect to beam focusing, the terms *normal* and *optimum focus* are often used with no firm definition or agreed upon terms. In electron beam processes, it is most useful to use the term *normal focusing* of an electron beam to mean setting the focus to use the beam on the surface of the workpiece (thus also referred to as surface focus) such that the smallest possible focal spot diameter is achieved [21].

Wire feed rate. The weight of the deposits increased as the wire feed rate increased. When the wire feed rate was set at a high level, the wire could not be fully melted. Since there was no way for the wire to escape, it tends to push backward into the wire feeder and "cracking" sounds are heard [9]. It should be noted that the wire feed rate depends on beam power and traveling speed. The wire feed rate should be at the same level as the melting rate. If the difference is high, the deposit process is unstable. Earlier investigations revealed that for front feeding (Figure 5.6), a good clad can be obtained when the wire is placed at the leading edge of the melt pool [22]. At a constant traveling speed, an increase in wire feed causes an increase in the height-to-width ratio as well as the effective growth rate [23]. It should be noted that the solid and tubular wires can be used.

Feeding directions. The technique of feeding filler material into the weld pool depends on the application and geometry of the final element. Generally speaking, the most popular is front feeding (Figure 5.6). However, for electron beam additive manufacturing technology, especially for more

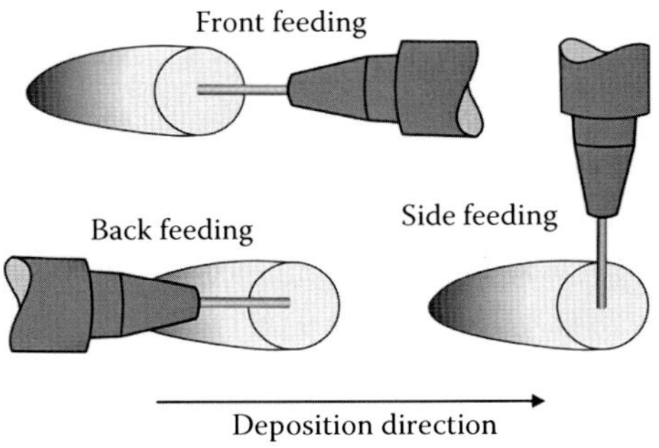

FIGURE 5.6 Different feeding directions. (From Zhao, J., Zhang, B., Li, A. et al. 2015. Effects of metal-vapor jet force on the physical behavior of melting wire transfer in electron beam additive manufacturing. *J Mater Process Technol* 220:243–250.)

complicated parts, side feeding is a promising technique [24]. Positioning the wire in the leading edge of the melt pool in front feeding provides the best results in terms of surface finish, suggesting that the melt pool was least disturbed in this case. The wire did not reflect the laser radiation, since it is not directly in contact with the laser beam. The heat of the melt pool is the main energy source for melting the wire. Marangoni flow type, controlled by thermal gradient, enhances the heat flux at the pool rim. By placing the wire outside the melt pool in front feeding, while placing the wire at the leading edge, an increase in the feeding angle does not cause any appreciable change in the thermal gradient. In general, front feeding and side feeding provide a smoother surface compared with back feeding. During back feeding, the wire is fed on top of the deposit. Thus, the material seems to be pushed away from the deposit. Therefore, a wavy surface with bulbs on the side of the deposit is formed for back feeding specimens. For front and side feeding, the wire is fed from a direction toward the deposit. Therefore, the wire is melted, forms the deposits, and is pushed toward the previous still red-hot deposits. The flow characteristics of melt in the melt pool are better in these configurations. Because of the surface tension effect, a better surface could be formed. Side feeding can provide a smooth surface but can have uneven edges because of the feeding direction. The uneven edges are formed by the wire, which first reacts with the melt pool on one side and more material is deposited on that side. For selecting a suitable feeding orientation, a high deposition rate is a crucial factor. However, it should be noted that consistent feeding is also significant, especially in net-shape component deposition when making the real part. Multiple feeding directions could occur in the same additive manufacturing process [9].

Feeding angle, also known as contact angle, is one of the factors affecting deposition quality, especially in surface deposition and structure build. In the additive manufacturing deposition process, the contact angle is determined not only by the surface tension effect in the molten material but also by the beam size and spatial power density distribution. Acute angles are preferred in the deposition process. Therefore, good dense deposition can be achieved and interrun porosity is not formed. If the deposit angle is greater than 90°, gaps in between overlapping tracks during the deposition would form. The feeding angles recommended in the literature should be in the range of 20°–50° [22]. The surface roughness increased by increasing the feeding angle for front feeding and decreasing the feeding angle for rear feeding [25]. For the additive manufacturing process at the same wire feeding rate, the deposit angle showed significant dependence on the ratio of the power beam to the traveling speed. The angle increases with increasing ratio [9].

Wire stick out. The distance between the end of the nozzle and the tip of the wire strongly influences the stability of the additive manufacturing process. At a short distance, the heat is more effectively removed from the wire to the nozzle, but the stiffness of the wire is higher. For a longer wire stick, the wire is likely to deflect, and heat dissipation to the environment is also higher.

5.2.4 Economic Issues and Applications

Phinazee [26] describes the benefits of using the Sciaky electron beam additive manufacturing process to fabricate a typical Ti–6Al–4V airframe component. The additive manufacturing process was found to have 79% greater material utilization efficiency, and the fabrication cost was reduced from $17,430 to $9810. Kinsella [3] investigated the deposition of IN718 alloy features on a forged engine case. A 30% cost savings was realized using electron beam wire deposition as compared to conventional fabrication methods. However, the societal impact of additive manufacturing from a technical perspective should also be considered [27]. An abundance of evidence was found to support the promises of additive manufacturing in the following areas:

1. Customized healthcare products to improve population health and quality of life
2. Reduced environmental impact for manufacturing sustainability
3. Simplified supply chain to increase efficiency and responsiveness in demand fulfillment

Stecker et al. [28] revealed that the total cost of single parts rapidly decreased with an increase in deposition rate (Figure 5.7). For calculation purposes, the following assumptions were made: deposition size, 45 kg; standard wire spool size, 11 kg; large chamber size; hourly rate, $250/h; and Ti–6Al–4V wire cost, $240/kg. The cost of using the electron beam additive manufacturing process depends on market pricing of raw materials and deposition rate. The cost of raw materials is independent of the process and driven by industry demand and availability. Deposition rates for Ti–6Al–4V have been demonstrated up to 18 kg/h. It should be noted that techniques that allow for higher deposition rates are continually being refined. Based on current experience and power availability with electron beam additive manufacturing systems, deposition rates approaching 23 kg/h can be achieved.

The economic model of the jet engine part made of nickel superalloy was developed and reveals the potential cost savings and benefits associated with the electron beam additive manufacturing (with wire) process. The cost model shows that the additive manufacturing–based process allows 30% savings over traditional forged-and-machined engine cases; the initial forging requires less material, which saves money and time and ultimately creates a lighter case as well. The model also

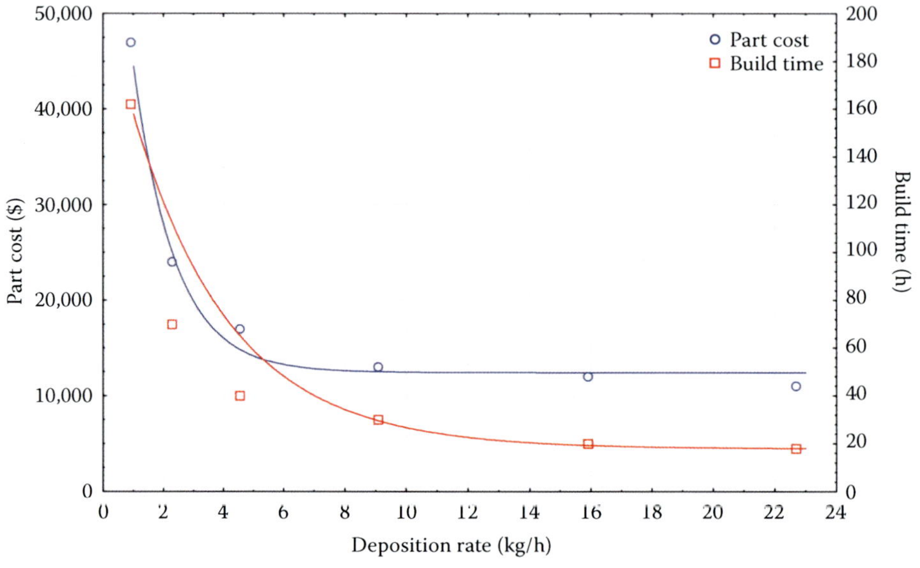

FIGURE 5.7 Cost as a function of deposited rate, with the following assumption: standard wire spool size, 11 kg; hourly rate, $250/h; Ti–6Al–4V wire cost, $240/kg. (From Stecker, S. et al. 2006. Advanced electron beam free form fabrication methods & technology. Session 2. Electron Beam Welding: 35–46.)

FIGURE 5.8 Partially machined half globe (Ti6Al4V). (From Baufeld, B., Widdison, R., and Dutilleu, T. 2017. *Electron Beam Additive Manufacturing: Deposition Strategies and Properties*. 4th IEBW International Electron Beam Conference Aachen.)

FIGURE 5.9 Cylindrical feature added on an existing component. (From Baufeld, B., Widdison, R., and Dutilleu, T. 2017. *Electron Beam Additive Manufacturing: Deposition Strategies and Properties*. 4th IEBW International Electron Beam Conference Aachen.)

shows that while additive manufacturing does not cut costs for a cast engine case, it nonetheless achieves significant weight reduction (up to 10%) by producing thinner case walls [29].

Electron beam additive manufacturing supplies a very efficient, high-power energy source that easily couples with metals, promising one of the highest deposition rates for all metal additive manufacturing technologies. Simultaneously, processing in high vacuum is ideal for reactive alloys. Thus, electron beam additive manufacturing is increasingly more accepted for industrial applications [30,31]. CNC is the standard for most electron beam welding setups, allowing for easy application of an automated build strategy. A variety of complex-shaped components made out of different materials such as steel, aluminum, and titanium are shown in Figures 5.8 through 5.11.

5.3 MODIFICATION OF THE SURFACE

Modification of the surface as the "umbrella" term defines all those electron beam process variants that provide the surface of a component with new properties [34]. Electron beam cladding, hardening, alloying, texturing, and surface modification processes can be recognized. An overview in the form of a summary is shown in Figure 5.12. It should be noticed that the electron beam can be put into effect both very quickly and very precisely at the position desired in each case. Thus, it is no wonder that one

FIGURE 5.10 316L stainless steel Trefoil structure. Machined and polished cross section. (From Baufeld, B., Widdison, R., and Dutilleu, T. 2017. *Electron Beam Additive Manufacturing: Deposition Strategies and Properties*. 4th IEBW International Electron Beam Conference Aachen.)

FIGURE 5.11 Example of electron beam additive manufacturing part after processing and after machining. (From http://www.camvaceng.com/new-technologies/near-net-shape/.)

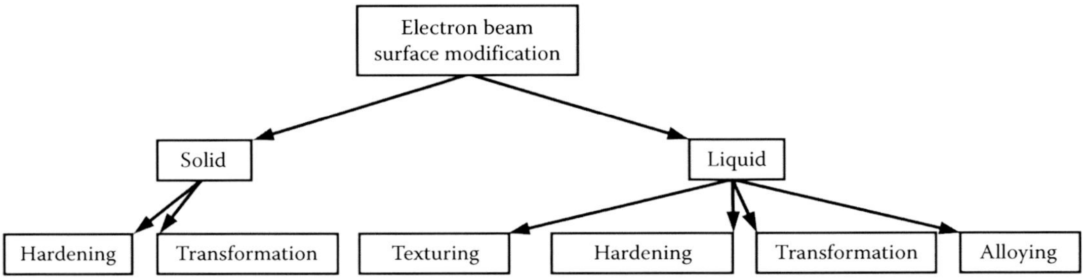

FIGURE 5.12 Electron beam surface modification. (From http://www.ptr-ebeam.com/en/application-areas /surface-treatment.html.)

of the most conspicuous properties of the electron beam surface modification relates to the fact that the desired change can be applied in an absolutely local form precisely to those regions where it is needed.

The electron beam that can be manipulated in an inertia-free operation by means of electromagnetic deflection can subject the surface to linear, punctiform, or planiform loads so that the desired effect in each case is achieved. In this respect, it is possible to modify not only linear or curved tracks but also cylindrical or conical surfaces as well as areas with irregular limits; the control (location and power) of the electron beam is practically unlimited. Moreover, no workpiece-specific facilities whatsoever (such as inductors or similar equipment) are required [35]. In many cases, the electron beam treatment can be carried out as the final processing step, for example, the hardening of an already ground functional surface [34].

5.3.1 Electron Beam Cladding

The cladding process using a concentrated beam of electrons is a unique technology that allows for the deposition of the filler material in the form of a powder or wire directly on the surface of materials. Additionally, because the technological process is carried out in vacuum, refining is concurrently conducted. Modern electron beam devices can precisely adjust parameters, such as beam power, beam diameter, wired and powder feed rate, and traveling speed. All technological parameters are presented in the first part of this chapter. This is particularly important in order to ensure minimal penetration or the minimum volume of molten metal at the electron beam energy density up to 10^{12} W/m^2. Deposition processes may be conducted in a single pass or multiple times, which enables optimization of the chemical composition and properties of the resulting layer.

Gnyusov and Tarasov [36] presented the results of the deposition process using a powder in which an electromagnetic powder feeder provided an efficiency of 40 g/min. Studies included the production of layers deposited on a substrate of alloy steel by cladding two grades of powders:

1. Fe–20%Mn–4%V–4%Mo+15%WC
2. Fe–20%Ni–4%V–4%Mo+15%WC

The particle size was in the 50–350 μm range. To improve the mechanical properties of deposited layers, the annealing process (T = 500°C, 600°C, and 700°C for 1, 2, 3, 5, and 10 h cooling in air) and heat treatment immediately in a vacuum chamber (self-aged coatings) were carried out. The impact of specific technology solutions on wear resistance is shown in Figure 5.13.

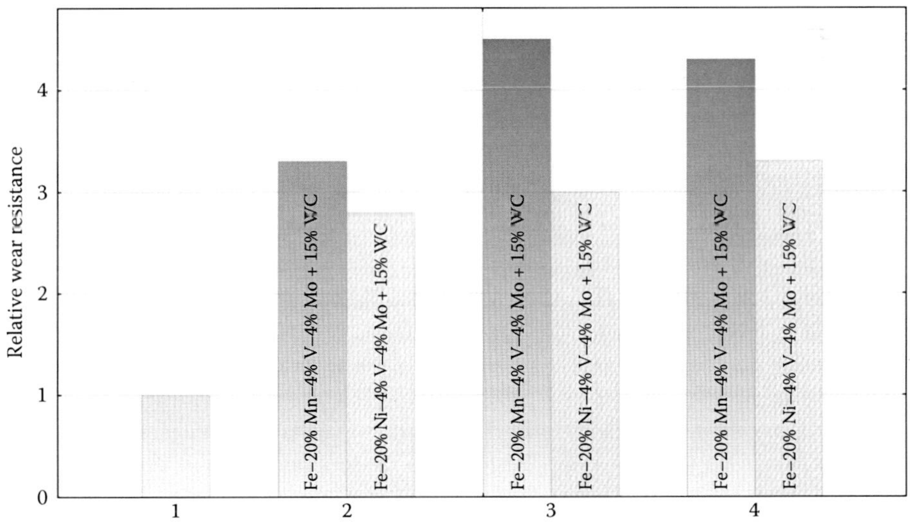

FIGURE 5.13 Relative wear resistance ε of electron beam clad coatings determined by loose particle abrasive test. 1, reference 0.45 wt.% C steel; 2, electron beam coatings; 3, coatings aged at 600°C for 1 h; and 4, self-aged coatings. (From Węglowski, M. St., Blacha S., and Phillips, A. 2016. Electron beam welding—Techniques and trends—Review. *Vacuum* 130:72–92; Gnyusov, S.F., and Tarasov, S. 2013. Structural phase states and heat aging of composite electron-beam clad coatings. *Surf Coat Tech* 232:775–783.)

Also, Morimoto et al. [37] present the results of the deposition process using Cr_3C_2/Ni–Cr powder. The chemical composition of Ni–Cr powder is as follows: Ni, 78.8%; Cr, 19.34%; Fe, 0.34%; C, <0.01%; Mn, <0.99%; Si, 0.47%. Size particle was in the range of 5.5–88 μm. Steel grade S235JR as a substrate was applied. A powder of Cr 17%, B 4%, Si 5%, C 0.9%, Fe <5% Ni residue with a size particle of 53–63 μm as a buffer layer deposit was used. Studies have shown that deposition of the powder allowed a hardness of 791 HV to be reached.

The filler material in powder form allows precise dosing of the additive material and adjustment of layer thickness. The results of electron beam cladding with powders FeB + FeTi and steel grade S235 as substrate are presented elsewhere [38]. The results have shown that the hardness of the cladding layers is 10 times higher than that of the substrate material.

Recently, the electron beam deposition process with filler material in wire form has become popular. Morimoto et al. [37] present the results of surfacing austenitic steel grade X6CrNiMoTi17-12-2 and duplex steel grade X2CrNiMoN22-5-3 with two wires: Fe 79.2%, C 4%, Cr 17.9%, Mo 1.2%, Si <1%, Mn 0.7% and Co 59.2%, C 1.8%, Cr 31%, W 8%. These researchers obtained a significant increase in hardness (up to 550HV0.3) and an increase in corrosion resistance.

In the electron beam cladding with wire, layer thickness and fusion penetration can be correspondingly adapted to the requirements and distortions can be minimized. One special advantage is the minimizing degree of dilution between the base and filler materials. The layer thickness, or number of layers, to be applied can be reduced in comparison to other methods. In contrast to conventionally processed systems, electron beam processed components feature a weld deposit overlay that is close to the final contour. This reduces the cost of finishing.

Richter et al. [39] presented the results of electron beam cladding with wire (CORODUR 300). The raw material was a tempered steel grade 34CrMo4. The investigation revealed that electron beam cladding can be adopted for repair and reconstruction purposes.

It is noteworthy that the deposition process using electron beam in vacuum is particularly important for reactive materials, such as titanium, niobium, and molybdenum. In Ref. [40], the results of cladding with titanium (thickness, 0.1 to 0.3 mm) and austenitic stainless steel grade 304 (thickness, 3.0 mm) as substrate are presented and discussed. The authors revealed that precise control of energy density is the fundamental issue. Good-quality deposit at an energy density of 0.05 to 0.12 kW/mm^2 was obtained.

Jung et al. [41] revealed that the electron beam cladding with wire technology allows the significant improvement the wear resistance of austenitic steels (e.g., X6CrNiMoTi17-12-2) and duplex steels (e.g., X2CrNiMoN22-5-3), without a negative influence on corrosion behavior. Iron and cobalt-additive wires were deposited thermally by electron beam cladding. The wear rate was reduced by a factor of 100 compared to the base materials for electron beam cladding with iron-based wire and by a factor of 10 with cobalt-based wire. Corrosion resistance was preserved for the iron-based cladding layers and slightly increased (by a factor of 3) for the cobalt-based cladding layers.

Electron beam cladding with strip and foil is also possible. However, in order to obtain good cladding layers, the electron beam must have sufficient energy density to penetrate the foil or strip immediately and melt little base materials [42]. La Barbera et al. [43] presented the results of electron beam cladding and alloying of AISI 316 stainless steel (in form of strip) on plain carbon steel. The influence of the electron beam process on corrosion resistance, microstructure, and mechanical properties was demonstrated in their study. The results revealed that an increase in beam energy caused the diffusion zone to become large enough to cause a considerable decrease in the concentration of alloy elements originally contained in the stainless steel, thus allowing a martensitic transformation to occur during cooling. The microstructure of the clad layer consisted of delta-ferrite, austenite, M_6C carbide, and martensite twinned on 1 1 2 crystal faces. The austenite has a heavily dislocated substructure and was characterized by intrinsic stacking faults [44]. This slightly lowered the corrosion resistance of the surface steel but also increased its hardness; other properties, for example, wear, may then be expected to benefit from this new microstructure [43,44].

The other technique is to produce deposited layers by remelting the powder layer previously distributed on the surface of the substrate. This technique is similar to alloying, but the dimensions of the final element after cladding become higher. The results of the melting of the powder layer WC-Co + Ni with a thickness of 3 mm on the substrate steel grade S275JR are presented and discussed by Abe et al. [45]. The remelting process allows a deposit with a hardness above 1400 HV to be obtained. It was found that the cladding layers showed higher hardness values and abrasive resistance with increasing WC-Co mixture ratio; however, corrosion resistance decreased with WC-Co mixture ratio. A coating layer with high abrasive and corrosion resistance was simultaneously achieved by multiple cladding of high WC-Co mixture ratio layers after low WC-Co mixture ratio layers [46].

Electron beam cladding at atmospheric pressure is also possible. Krivezhenko et al. [47] present the results of the powder deposition process. A powder mixture having a composition of 40% B_4C + 10% Fe was used to form the coating. It was established by x-ray diffraction analysis that the major phases in the hardened layer were α-iron, iron boride Fe_2B, and borocementite Fe3B0.6C0.4. The wear resistance of the cladding layers increases twofold when compared to the base material.

Non-vacuum electron beam cladding of pure titanium with Ti–Ta–Nb powder is presented and discussed by the Golkovsky et al. [48]. This method provides high quality of coatings, without pores and cracks. The average microhardness of the coating was about 4000 MPa. The coating structure retained the contours of a dendritic structure and, perhaps, the boundaries of the former β phase grains. At the microlevel, the coating has a martensitic structure. However, elevated oxygen content was found in interdendritic spacing. In order to further reduce the oxygen content, pure starting powders should be used.

It should be noticed that electron beam cladding of vanadium–carbon powder mixtures by high-power electron beam injected to the atmosphere led to the formation of high-quality coatings [49]. The highest microhardness level of cladded layer reached 12 GPa (at thickness of the surface layer in the range 2.5–3.4 mm). The concentration of hard vanadium carbide particles increased with a decrease in layer thickness, which resulted in an increase in the wear resistance level of the modified material.

5.3.2 Hardening

Electron beam hardening is similar to other surface heat treatment processes in that the material to be hardened is heated to the austenitization temperature, held at that temperature for a short period, and then quenched. However, some special characteristics in comparison to other heat treatment technologies can be distinguished. In this process, a short-cycle heat treatment occurred. When compared to other conventional volume and surface-layer heat treatment processes (e.g., induction), the temperature–time (T–t) cycles are characterized by significantly higher austenitizing temperatures, T_A (with T_A close to the melting temperature, T_M), as well as considerably higher heating and cooling rates (10^3 to 10^4 K/s) and short austenitization times (0.1 to 3 to 5 s). Another issue is that in conventional heat treatments, an additional cooling medium is required, whereas in electron beam techniques, cooling occurs from austenitizing temperature via self-quenching; that is, the heat produced in the material by the electron beam flows rapidly into the mass of the material. The advantages of this process when compared to other surface hardening processes, including laser processes, arise in particular from the physical characteristics of the electron beam, which can be transformed with exceptional readiness into technical and technological advantages. The most advantageous characteristics of electron beams are their excellent formability and deflectability, which allow beam deflection frequencies of up to 100 kHz [50]. The other advantages are as follows [51]:

1. Precise control and reproducibility of the energy input with respect to location and time
2. No scaling or oxidation of component surface
3. No preparation of surfaces to be hardened or of regions that have to be left untreated

4. Compatible and easy to integrate with CNC/CAM processing methods
5. High energy efficiency (approximately 75% of the power generated by an electron beam is converted to heat)
6. No waste products generated
7. High process productivity

Taking into account the benefits of electron beam hardening, this process allows for using multi-process technologies [17]. The hardening and tempering for the contour of a controlling camshaft can be applied. Two energy transfer fields interact simultaneously with the rotating cam. Because of stronger load conditions, the depth at the radius must be larger (0.65–0.75 mm) than that at the flanks (0.55–0.65 mm). This process based on CNC control systems in electron beam machines can easily be adopted. The other possibility is applied to electron beam hardening in the case of a calotte carrier. The surface contour is programmed as a rotation-symmetric energy transfer field with a surface contour congruent energy distribution. Thus, the hardening profile is characterized by a constant depth that is independent of the incidence angle of the electron beam. The energy transfer is realized by flash technique. During interaction of the electron beam (≤1.0 s), the component is fitted to the beam before crossing the α/γ transformation temperature (processing time ≤ 0.2 s). This technology guarantees a high productivity (up to 3500 parts per hour) [52]. Songa et al. [53] revealed that the microstructure of the hardened layer of AISI D3 consisted of martensite, a dispersion of fine carbides, and retained austenite, while the transition area consisted of tempered sorbite. Moreover, the microhardness of the hardened layer on the surface increased dramatically, up to 1400HV0.1, compared to the base material. Finally, the hardening response of AISI D3 tool steel to electron beam surface treatment is closely related to the scanning speed of the electron beam. The increase of this speed causes an increase in the hardness level.

5.3.3 Alloying

The electron beam alloying process is based on remelting surface layers with alloying elements that are totally or partially soluble in the substrate. Compared with electron beam hardening, this process is performed at higher power densities and longer heating times. The electron beam alloying process caused deterioration of the surface roughness, improvement of tribological properties, and corrosion resistance when compared to raw surface material.

The alloying process can be conducted by remelting of layers previously manufactured by using other techniques (e.g., electrolytic or thermal spraying) onto a substrate coating or injection into a molten pool [54].

A single-pass or multiple-pass technique for alloying process can be adopted. Hence, optimization of the chemical composition and properties of the resulting layers can be achieved. Moreover, multiple-pass techniques generate a temperature field on the modified surface. Thus, the production of graded materials with a specific gradient property is possible. The modification process is conducted in vacuum; therefore, for the aerospace industry to modify the surface of reactive materials, titanium and its alloys can be adopted. However, in the case of modification of titanium alloys, attention should be paid to the porosity of the modified areas, which leads to deterioration in quality and reduction in functional properties of the newly created areas.

The proper technological parameters allow for the achievement of the modified areas without porosity on the surface. It should be noted that even during welding of titanium alloys in a vacuum, porosity may occur. The tendency to porosity is related to the solubility of hydrogen in the titanium, which is a function of temperature. It should be stressed that the effect of porosity is the amount of heat introduced into the modified area and is directly related to the rate of melting and alloying. Parameters of the process affect nucleation and growth and degas the liquid metal. Limited porosity at low melting speeds is attributed to a sufficiently long time required for degassing, while at high speeds, there is limited time to nucleation of bubbles of gas and degassing the modified region [55,56].

The lamellar microstructure of substrate transforms into a dendritic structure composed of initial α_2 (Ti_3Al) dendrites and an interdendritic phase of the γ (TiAl) during remelting of titanium alloy (Ti–45Al–2Nb–2Mn–1B) using an electron beam at a rate of 16 mm/s [57]. The alloying process with powder (Al8Si20BN) using an electron beam allows for a significant increase in wear resistance of pure titanium alloy [58], and corrosion resistance (H_2SO_4) of the modified layer is not deteriorated when compared to the raw material.

Modification of titanium alloys by using an electron beam at atmospheric pressure at an energy of 1.4 MeV is also possible. Golkovskia et al. [59] obtained a layer with a thickness of 2.0–2.5 mm with tantalum content in the 3.9%–22.4% range when alloying pure titanium with powder Ti–Ta. The microstructure of modified surface layer was composed of α phase(α')+β. The tensile strength was 735 MPa and resistance to corrosion in environment of boiling nitric acid was significantly higher than that of pure Ti (weight loss is 190 times smaller) at 22.4% Ta content. The heating and cooling rates at which new material was produced were different from equilibrium. Thus, the microstructure has high dispersing components, inhomogeneity of chemical composition, and formation of β phase at a lower Ti content of tantalum.

The study of alloying titanium alloy Ti–6Al–4V with powders (TiC, SiC, and TiC + SiC) shows that it is possible to produce composite material with a hardness above 700 HV0.5 (base material hardness was 320HV0.5) on the surface of the titanium alloy [60]. In contrast, alloying with TiN powder [61] allows for obtaining a hardness of 937HV0.5.

It is also possible to increase the hardness of pure titanium by introducing a powder of TiC. The surface layer has the greatest wear resistance of approximately 60% and is characterized by a hardness if 500HV0.1 (titanium 180HV0.1) [62].

The alloying processes for steel were also conducted. The alloying process for steel 10NiCr180 NiCr with powder allows obtaining a layer with a hardness of 1300 HV (base material 220 HV) [63].

There also exist many studies that describe the formation of aluminum-based alloys by electron beam alloying. Petrov [64] presented the results of alloying of hypereutectic alloys AlSi18CuNiMg with Cr–Fe–based powder. The results show a significant increase in microhardness and changes of the structure morphology of the modified zone in comparison to the base material.

Petrov and Dimitroff [65] have conducted an investigation on electron-beam surface alloying of eutectic Al–Si alloys with iron, cobalt, nickel and chromium. Their results show changes of the structure morphology and an increase in hardness and tensile strength. Moreover, the hardness of pure aluminum can be improved by electron beam alloying with titanium and niobium, about 22 times greater in comparison to the base aluminum substrate [66].

Valkov et al. [67] presented the results of electron beam alloying with aluminum–titanium–zirconium and aluminum–titanium–hafnium. The results obtained demonstrate the possibility of formation of hard intermetallic electron beam surface alloying on pure aluminum. Although the microstructure of aluminum–titanium–hafnium modified area is amorphous, the measured microhardness was about 500 HV. The conducted XRD (x-ray diffraction) on the aluminum–titanium–zirconium modified area shows an intermetallic (Ti, Zr)Al_3 phase; the hardness was approximately 260 HV.

5.3.4 Texturing

Electron beam surface texturing process, known also as Surfi-Sculpt developed since 2004 by The Welding Institute [68], is a novel metal surface modification technology, by which customized textured morphology on a metal surface can be produced by controlling process parameters and scanning the route of the electron beam [69]. Surface features are produced by molten displacement generated by repeated swipes of a focused electron beam. The features are characterized by a protrusion and corresponding intrusion as shown in Figure 5.14. There is a critical time delay and feature spacing between each of the swipes in order to achieve the quasi-steady-state temperature conditions and optimized melt displacement [70]. The protrusions of electron beam Surfi-Sculpt can be created

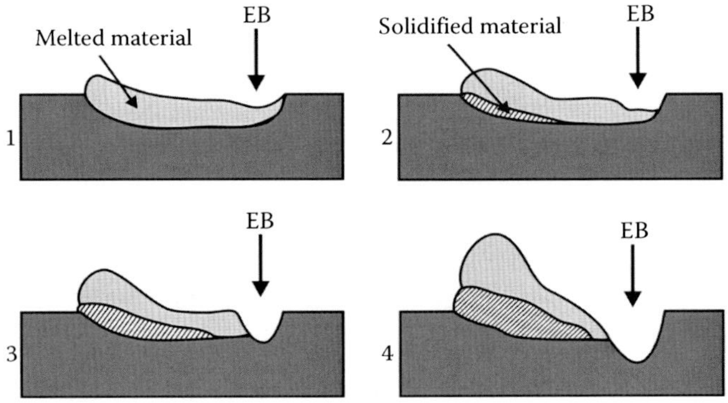

FIGURE 5.14 Process scheme of the electron beam use for surface texturing. (From Xiong, W., Blackman, B., Dear, J.P. et al. 2015. The effect of composite orientation on the mechanical properties of hybrid joints strengthened by surfi-sculpt. *Compos Struct* 134:587–592.)

by repeating a swipe many times in the same position on the workpiece. Molten metal is flown and deposited behind the electron beam. The larger and more complex protrusions could be more easily produced by incorporating several swipes into a motif. Protrusions from microns to tens of millimeters in size can be achieved [71]. Using the flesh and multi-spot techniques, 3500 defined indentations can be simultaneously generated in a process time of 0.15 s [72]. The example of electron beam structures is presented from Figures 5.15 through 5.17.

The waveforms are important keys on the Surfi-Sculpt processing, which could move the electron beam according to a special route under the control of electromagnetic field. Different waveforms may influence differently the developing of protrusions, including shape, time, height, and so on. Wang et al. [69] revealed that the six lines meeting waveform is the best way to obtain protrusions, while proper frequency and beam current are also a key factor to realizing the Surfi-Sculpt process. An increase of the beam current causes an increase in protrusions. However, while the beam current is too high, the decrease of protrusion can be obtained. Hence, the beam current should be matched carefully. Moreover, the microstructure of the protrusion could be divided into four parts: edge zone, central zone, heat-affected zone, and base metal. The hexagon scanning waveform is a very efficient way to produce protrusions [71]. Surface tension plays a dominant role during the electron beam Surfi-Sculpt process. The interaction of vapor pressure and surface tension can be observed. However, microstructures in the protrusion fusion zone and heat-affected zone are different for different cooling rates during the forming process.

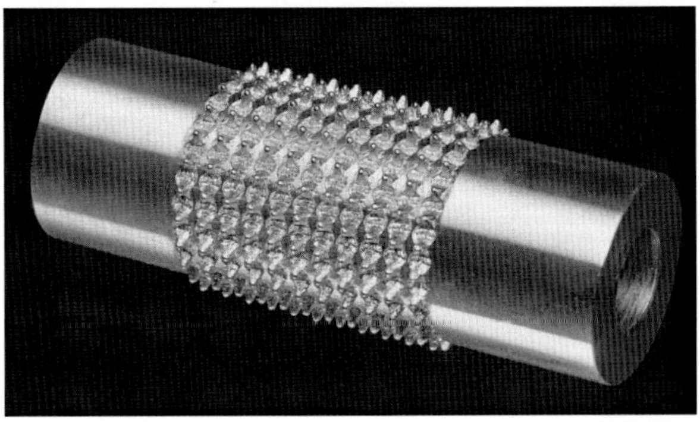

FIGURE 5.15 1.75-mm-high protrusions produced on 30-mm-outer diameter stainless steel tube using TWI's Surfi-Sculpt process. (Image courtesy of TWI Ltd. Materials from TWI company in the field of Surfi-Sculpt technology, 2017.)

FIGURE 5.16 Honeycomb-like structure produced on Ti6Al4V using TWI's Surfi-Sculpt process. (Image courtesy of TWI Ltd. Materials from TWI company in the field of Surfi-Sculpt technology, 2017.)

FIGURE 5.17 2.5-mm-high cones produced on 6-mm-thick Al6082-T6 using TWI's Surfi-Sculpt process. (Image courtesy of TWI Ltd. Materials from TWI company in the field of Surfi-Sculpt technology, 2017.)

Wang et al. [73] revealed that the protrusion, for Ti–6Al–4V, with a greater height and the largest protrusion growth rate can be manufactured more efficiently when using a beam current of 3.5 mA and a scanning frequency of 1500 Hz. The process of transferring the molten substrate into a protrusion was a consequence of the heating and cooling cycle, surface tension gradient, and the viscosity of the molten metal.

Taendl and Enzinger [74] presented results of the electron beam surface structuring of AA6016 aluminum alloy with a thickness of 2.5 mm. Studies include the production of electron beam structures, optimization of the process, and mechanical tests. The beam current had the largest influence on the resulting pin height. Accordingly, all other parameters were held constantly (acceleration voltage was constant at 150 kV) on optimized values, and the pin height was adjusted only by controlling the beam current. Moreover, the mechanical properties of the electron beam–structured aluminum strips dropped to lower levels compared to the base material. In terms of load bearing, this deterioration was mainly attributed to cross-sectional reduction. However, for higher heat input, a strength reduction of the material most presumably due to dissolution of strengthening precipitates was also observed.

The Surfi-Sculpt for advanced applications in many fields can be adopted, for example, for composite to metal joining, also known as Comeld [75], surface coating pretreatment [76], manufacturing of hydrodynamically enhanced surfaces [77], and many others. By accurate control of the electron beam process, a wide variety of well-defined patterns is possible. The process can be used on a range of metals, polymers, ceramics, and glasses and takes only a few seconds to process a square centimeter of surface, depending on the feature complexity [77]. By optimizing parameters of the Surfi-Sculpt process [70], features for rubber-to-metal bonding can be created. Surfi-Sculpt can be tailored to produce a textured surface along the sides of the features, so that the surface forms additional hooks that can attach to a fibrous material. Developments in scanning and beam technology reduce costs and increase production rates so that the possibility of applying the Surfi-Sculpt techniques becomes increasingly feasible.

Xiong et al. [78] presented the results of the effect of composite orientation on the mechanical properties of the joints (Ti–6Al–4V alloy + *carbon fiber*–reinforced polymer SE 84LV). The thickness of the composite adherent was kept constant and the volume content of ±45° ply was increased from 11.1% to 88.9%. The result indicates that Surfi-Sculpt was able to delay the damage initiation and improve the joints' ultimate failure load, failure strain, and absorbed energy. Composite orientation was able to vary the joints' mechanical properties significantly. With optimum composite orientation, the joint damage initiation load was increased by 24.84%, the joint ultimate failure load was increased by 134.5%, and the joint energy absorption was increased by 257.39%.

Tu et al. [79] revealed that the mesh-free method to multi-region problems and to the stress analysis for Comeld joints can be adopted. It is known that a mesh-free method does not need any elements in the domain. The first advantage this method over the finite element method is simplicity and accuracy. Second, optimization of geometry parameters for any design can be carried out easily as there is no remeshing problem.

Surface texturing for coating pretreatment can also be adopted. Galchenko et al. [80] presented the results of the electron beam surface texturing using plasma cathode gun. Surface modification was carried out on Cr18Ni10Ti steel plates intended for subsequent deposition of gas thermal coatings made of nickel chromium alloy. The mode of electron beam treatment of metal substrates was chosen to form a periodic spiked structure on their surface. The formation of an ordered spiked relief on the substrate surface using electron beam modification greatly increases the density and adhesive strength of gas thermal coatings under external loads. The obtained results demonstrate that three-dimensional surface modification of metals and alloys for various industrial and engineering applications can be applied. With electron beam texturing, the shear strength of the coating could be increased by 320% in comparison to substrates without profiles. The reasons were the excellent mechanical interlocking at the bulges and the change of the failure mechanism [81].

A combination of two surface treatment techniques, such as electron beam structuring and electrophoretic deposition, allows one to control the surface topography in metals. The combination of both techniques was presented by the Ramskogler et al. [82] as a promising surface modification approach, which should lead to improved interaction at the interface between bone and implant surface. The versatile electrophoretic deposition technique was used to produce a single layer of chitosan/TiO_2 nanoparticle composite coating. The results revealed that 3D topographies obtained by electron beam structuring can be coated. With this technique, a variety of chitosan composite coatings can be considered in which different types of nanoparticles can be added to affect the surface roughness of implants.

Ramskogler et al. [83] also presented the results of the surface modification process of the Ti–6Al–4V alloy for the purpose of biomedical application. A canal shape with a depth of 1.3 to 9 μm and a width of 68.6 to 119.7 μm was produced. The evaluation of the orientation of the figure showed a pin height of 305 μm and a wall depth of 452 μm. For lower energy input, the pin height was 93 μm. It should be noted that the whole surface of the implant was increased. This allows one to control the surface properties in contact with living tissue. The microstructure of Ti6Al4V after electron beam texturing consisted of the melted zone of martensitic α' as a result of the rapid cooling. The tribological behavior in the melted zone was changed due to the hardness increase. The heat-affected zone showed α' globular, α'' lamellar, and β phase mixed with α' due to the fast cooling. The thickness of these layers was in strong correlation with the energy input during the electron beam surface modification process. Additional heat treatments (poststructuring heat treatments were performed under vacuum with a heating rate of 300 K/min, up to 650°C and 720°C, and then isothermally held for 8 h, followed by quenching in argon atmosphere) showed a stabilization of the martensitic morphology in the melted zone decreasing the hardness.

Moreover, the electron beam Surfi-Sculpt process can be utilized to enhance the functional properties of materials, increasing the surface area for thermal energy exchange in addition to controlling the dynamics of fluid flow over a surface from laminar to turbulent. Heat exchanging applications have demonstrated a combination of benefits including a 50% increase in the measured heat transfer coefficients across a range of fluid flow rates [84].

5.4 SUMMARY

Based on literature review, it should be noted that the electron beam processes including additive manufacturing and surface modification techniques are interesting for the scientific and the industrial community.

Manufacturing near-net-shape elements layer by layer with wire as a deposit material offers a great potential for time and cost savings in comparison to conventional manufacturing technologies such as forging, casting, or machining. The increasing market (namely, aerospace, military, and medical industries) demands for titanium, aluminum, and other materials' serial production parts have promoted wire feed processes. Repeatability, material properties, material usage, possible part size, and building speed have also become issues. This is a novel manufacturing technique that can directly fabricate fully dense large 3D near-net shape elements from metal wires. Of specific interest are the additive layer manufacturing processes with wires that are capable of producing fully dense metallic and hybrid parts in which the resulting parts may be used for loaded structures. These processes are attractive because they eliminate contamination compared to powder processes.

The areas for future development of surface modification processes are presented in Figure 5.18 [85]. As shown, the development will be focused on the techniques, processes, materials, and components, as well as on modeling and simulation. Taking into account the fact that approximately 3000 electron beam machines are installed in the industry and R&D institutions, these techniques can be rapidly implemented in a wide range of industries, such as aerospace, car, and machine building, to name a few.

As regards increasing surface modification and additive manufacturing process efficiency, to ensure the best quality of modified and additive manufacturing materials and to facilitate the work of operators, electron beam equipment manufacturers offer many additional systems, including the following:

1. Automatic beam alignment system
2. Electron-optical monitoring system
3. Automatic seam tracking, control of the process
4. Fast deflection generator and wire as well as powder feeder

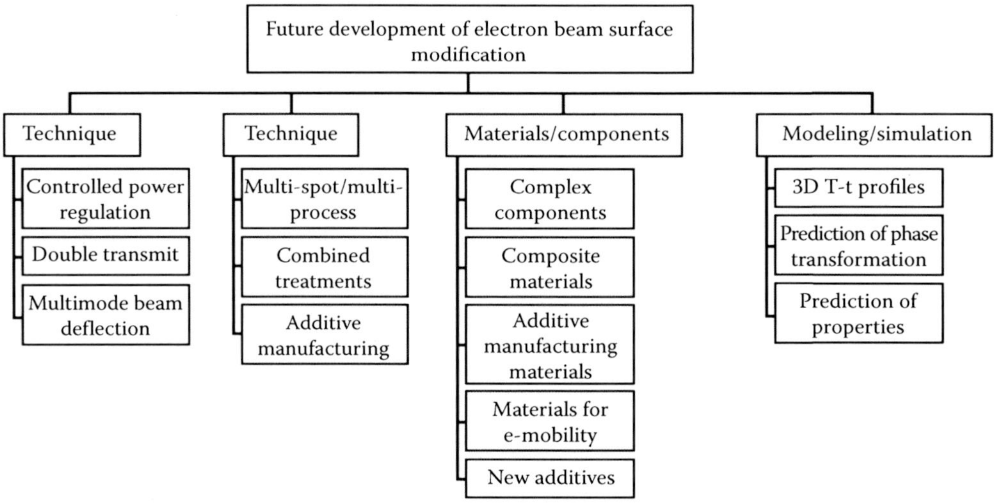

FIGURE 5.18 Future development of electron beam surface modification processes. (From Zenker, R. 2017. *Electron Beam Surface Treatment—History of Development, State of the Art and Prospects.* 4th IEBW International Electron Beam Welding Conference, Aachen, Germany (unpublished).)

Moreover, if the universal electron beam welding machine is considered, the major developments cover the control system and tooling, as well as the following:

Backscattered imaging: In most vacuum chambers, the illumination is quite poor and the field of view is also quite small. The backscattered image technique is a very useful tool where normal viewing is difficult because of the field of view and poor illumination.

Full quality assurance package including pass and fail system: The idea is to monitor key functions of the machine during modification and report any problems in terms of a brief report and more comprehensive data accessible offline. This consists of a data capture card, and in this case, a separate PC independent of the machine system.

Quick change tooling: For a universal electron beam machine, the rotary tilt and Z axis tables can be set up accurately to microns and parts of a degree, meeting ISO standards. Multiple viewing and gun positions including horizontal and vertical position are provided.

One of the most important issues is that the x-ray hazards that occur during electron beam processes are eliminated and the machines are safe to use by operators.

REFERENCES

1. Dong-Gyu, A. 2016. Direct metal additive manufacturing processes and their sustainable applications for green technology: A review. *Int J Pr Eng Man-GT* 3:381–395.
2. Chlebus, E. 2000. *Computer Techniques CAx in Engineering Production.* Warsaw: Wydawnictwa Naukowo Techniczne.
3. Kinsella, M.E. 2008. Additive manufacturing of superalloys for aerospace application. Air Force Research Laboratory Report No AFRL-RX-WP-TP-2008-4318:1–7. www.dtic.mil/cgi-bin/GetTRDoc?AD=ADA489302
4. Schröder, M., Falk, B., and Schmitt, R. 2015. Evaluation of cost structures of additive manufacturing processes using a new business model. *Procedia CIRP* 30:311–316.
5. Frazier, W.W. 2014. Metal additive manufacturing: A review. *J Mater Eng Perform* 23:1917–1928.
6. Joshi, S.C., and Sheikh, A.A. 2015. 3D printing in aerospace and its long-term sustainability. *Virtual Phys Prototyp* 10:175–185.
7. Canada Makes' Metal additive process guide. http://canadamakes.ca/app/.
8. Sames, W.J., List, F.A., Pannala, S. et al. 2016. The metallurgy and processing science of metal additive manufacturing. *Int Mater Rev* 61:1–47.

9. Mok, S.H., Bi, G., Folkes J. et al. 2008. Deposition of Ti–6Al–4V using a high power diode laser and wire, part I: Investigation on the process characteristics. *Surf Coat Tech* 202:3933–3939.
10. Taminger, K.M.B., Harley, R.A., and Dicus, D.L. 2002. Solid freeform fabrication: An enabling technology for future space missions. International Conference on Metal Powder Deposition for Rapid Manufacturing, San Antonio. https://ntrs.nasa.gov/archive/nasa/casi.ntrs.nasa.gov/20030013635.pdf
11. Hafley, R.A., Taminger, K.M.B., and K. Bird. 2007. Electron beam freeform fabrication in the space environment. American Institute of Aeronautics and Astronautics. 45th AIAA Aerospace Sciences Meeting and Exhibit. https://ntrs.nasa.gov/archive/nasa/casi.ntrs.nasa.gov/20070005152.pdf
12. Taminger K.M.B., and Hafley, R.A. 2003. Electron beam freeform fabrication: A rapid metal deposition process. Proceedings of the 3rd Annual Automotive Composite Conference. https://ntrs.nasa.gov/archive/nasa/casi.ntrs.nasa.gov/20040042496.pdf
13. Taminger, K.M.B., and Hafley, R.A. 2006. Electron beam freeform fabrication for cost effective near net shape manufacturing. AVT-139, NATO. https://ntrs.nasa.gov/archive/nasa/casi.ntrs.nasa.gov/20080013538.pdf
14. Mitzner, S., Liu, S. Domack, M. et al. 2012. Grain refinement of freeform fabricated Ti–6Al–4V alloy using beam/arc modulation. Proceeding of the 23st Annual International Solid Freeform Fabrication Symposium: 536–555. https://sffsymposium.engr.utexas.edu/Manuscripts/2012/2012-42-Mitzner.pdf
15. Gonzales, D., Liu, S., Domack M. et al. 2016. Using powder cored tubular wire technology to enhance electron beam freeform fabricated structures. TMS 145th Annual Meeting & Exhibition: 183–189.
16. Messler, R.W. 1999. *Principles of Welding. Processes, Physics, Chemistry and Metallurgy*. New York: John Wiley & Sons.
17. Węglowski, M. St., Blacha S., and Phillips, A. 2016. Electron beam welding—Techniques and trends—Review. *Vacuum* 130:72–92.
18. Zhao, J., Zhang, B., Li, A. et al. 2015. Effects of metal-vapor jet force on the physical behavior of melting wire transfer in electron beam additive manufacturing. *J Mater Process Technol* 220:243–250.
19. Heralic, A. 2009. *Towards Full Automation of Robotized Laser Metal-Wire Deposition*. Chalmers University of Technology Goteborg, Sweden.
20. Schultz, H. 1993. *Electron Beam Welding*. Cambridge: Abington Publishing.
21. *Recommended Practices for Electron Beam Welding and Allied Processes*. American Welding Society Standard No AWS C7.1M/C7.1:2013.
22. Brandl, E., Michailov, V., Viehweger, B. et al. 2011. Deposition of Ti–6Al–4V laser and wire, part I: Microstructural properties of single beads. *Surf Coat Tech* 206:1120–1129.
23. Tang, Q., Pang, S., Chen, B. et al. 2014. A three dimensional model for heat transfer and fluid flow of weld pool during electron beam freeform fabrication of Ti 6Al 4V alloy. *Int J Heat Mass Transf* 78:203–215.
24. Wang, F., Mei J., and Wu, X. 2006. Microstructure study of direct laser fabricated Ti alloys using powder and wire. *Appl Surf Sci* 253:1424–1430.
25. Syed, W.U.H., and Li, L. 2005. Effects of wire feeding direction and location in multiple layer diode laser direct metal deposition. *Appl Surf Sci* 248:518–524.
26. Phinazee, S. 2007. Efficiencies: Saving time and money with electron beam free form fabrication. *Fabricator* 15–20.
27. Huang, S.H., Liu, P., Mokasdar, A. et al. 2013. Additive manufacturing and its societal impact: A literature review. *Int J Adv Manuf Tech* 67:1191–1203.
28. Stecker, S., Lachenberg, K.W., Wang, H. et al. 2006. Advanced electron beam free form fabrication methods & technology. AWS Conference, Session 2: Electron Beam Welding: 35–46. https://app.aws.org/conferences/abstracts/2006/012.pdf
29. Baufeld, B., Widdison, R., and Dutilleu, T. 2017. Electron beam additive manufacturing: Deposition strategies and properties. 4th IEBW International Electron Beam Conference Aachen.
30. Karen, Ms., and Taminger, M. 2006. Electron beam freeform fabrication for cost effective near-net shape manufacturing. NATO Report No AVT-139. https://ntrs.nasa.gov/archive/nasa/casi.ntrs.nasa.gov/20080013538.pdf
31. Taminger, K.M., and Hafley, R.A. Electron beam freeform fabrication for cost effective near-net shape manufacturing. https://ntrs.nasa.gov/archive/nasa/casi.ntrs.nasa.gov/20080013538.pdf
32. http://www.camvaceng.com/new-technologies/near-net-shape/
33. http://www.ptr-ebeam.com/en/application-areas/surface-treatment.html
34. Schulze, K.R. 2012. *Electron Beam Technologies*. Wissen kompakt. Volume 1e. Düsseldorf, DVS Media GmbH.

35. Böhm, S. 2014. The electron beam as a tool for joining technology. DVS Report Volume 299e. DVS Media GmbH, Düsseldorf.
36. Gnyusov, S.F., and Tarasov, S. 2013. Structural phase states and heat aging of composite electron-beam clad coatings. *Surf Coat Tech* 232:775–783.
37. Morimoto, J., Abe, N., Kuriyama F. et al. 2001. Formation of a Cr3C2/Ni–Cr alloy layer by an electron beam cladding method and evaluation of the layer properties. *Vacuum* 62:203–210.
38. Galchenko, N.K., Kolesnikova, K.A., Belyuk, S.I. et al. 2014. Structure and properties of boride coatings synthesized from thermo-reactive powders during electron-beam surfacing. *Adv Mat Res* 880:265–271.
39. Richter, A., Krüssel, Th., and Just C. 2012. Deposition welding with the electron beam as repair technology. International Electron Beam Welding Conference, Aachen.
40. Tomie, M., Abe, N., Yamada, M. et al. 1990. Electron beam cladding of titanium on stainless steel plate. *Transactions of JWRI* 19:51–55.
41. Jung, A., Zenker, R., Gleißner, J. et al. 2014. Elektronenstrahl-Randschichtbehandlung für die Herstellung verschleißbeständiger Auftragschichten auf nichtrostenden Stählen. *Materialwiss Werkst* 45:486–495.
42. Tomie, M., Abe, N., Yamada, M. et al. 1990. Electron beam cladding of titanium on stainless steel plate. *Transactions of JWRI* 19:51–55.
43. La Barbera, A., Mignone, A., Tosto, S. et al. 1991. Electron beam cladding and alloying of AISI 316 on plain carbon steel: Microstructure and electrochemical corrosion behaviour. *Surf Coat Tech* 46:317–329.
44. Tosto, S., Nenci, F., and Jiandong, H. 1994. Microstructure and tensile properties of AISI 316 stainless steel electron-beam cladded on C40 mild steel. *J Mater Sci* 29:5852–5858.
45. Abe, N., Morimoto, J., Tomie, M. et al. 2000. Formation of WC-Co layers by an electron beam cladding method and evaluation of the layer properties. *Vacuum* 59:373–380.
46. Abe, N., and Morimoto, J. 2005. Formation of hard surfacing layers of WC-Co with electron beam cladding method. *J High Temp Soc* 32:137–139.
47. Krivezhenko, D.S., Laptev, L.S., and Zimoglyadova, T.A. 2015. Electron-beam cladding of boron carbide on low-alloyed steel at the air atmosphere. *Appl Mech Mater* 698:369–373.
48. Golkovsky, M.G., Zhuravina, T.V., Bataev, I.A. et al. 2011. Cladding of tantalum and niobium on titanium by electron beam, injected in atmosphere. *Adv Mat Res* 314–316:23–27.
49. Mul, D., Lazurenko, D.B., and Zimoglyadova, T.A. 2014. Electron beam cladding of vanadium and carbon powders on carbon steel in the air atmosphere. *Appl Mech Mater* 682:138–142.
50. Dossett, J., and Totten G.E. 2013. *ASM Handbook, Vol 4A: Steel Heat Treating Fundamentals and Processes.* New York: ASM International.
51. Labeeb, M. 2014. Electron and laser beam hardening. https://www.slideshare.net/labeebmlp/electron-and-laser-beam-hardening
52. Zenker, R. 2009. Modern thermal electron beam processes—Research results and industrial application. *La Metallurgia* 4:1–8.
53. Songa, R.G., Zhanga, K., and Chen, G.N. 2003. Electron beam surface treatment. Part I: Surface hardening of AISI D3 tool steel. *Vacuum* 69:513–516.
54. Adamiec, P., and Dzibinski, J. 2005. *Fabrication and Properties of Surface Layers of Parts for Transportation Machines*. Gliwice: Silesian University of Technology.
55. Mohandas, T., Banerjee, D., and Rao, V.V.K. 1990. Fusion zone microstructure and porosity in electron beam welds of an $\alpha+\beta$ titanium alloy. *Metall Trans A* 30A:789–798.
56. Huang, J., and Warnken, N. 2012. Hydrogen transport and rationalization of porosity formation during welding of titanium alloys. *Metall Trans A* 43A:582–591.
57. Rastkar, A.R., and Shokri, B. 2010. Surface transformation of Ti–45Al–2Nb–2Mn–1B titanium aluminide by electron beam melting. *Surf Coat Tech* 204:1817–1822.
58. Utu, D., and Marginean, G. 2007. Improvement of the wear resistance of titanium alloyed with boron nitride by electron beam irradiation. *Surf Coat Tech* 201:6387–6391.
59. Golkovskia, M.G., Bataev, I.A., Bataev, A.A. et al. 2013. Atmospheric electron-beam surface alloying of titanium with tantalum. *Mat Sci Eng A—Struct* 578:310–317.
60. Oh, J. Ch., and Lee, S. 2004. Correlation of microstructure with hardness and fracture properties of (TiC, SiC)/Ti–6Al–4V surface composites fabricated by high-energy electron-beam irradiation. *Surf Coat Tech* 179:340–348.
61. Yun, E., Lee, K., and Lee, S. 2005. Correlation of microstructure with high-temperature hardness of (TiC, TiN)/Ti–6Al–4V surface composites fabricated by high-energy electron-beam irradiation. *Surf Coat Tech* 191:83–89.

62. Lenivtseva, O.G., Lazurenko, D.V., and Samoylenko V.V. 2014. The structure and wear resistance of the surface layers obtained by the atmospheric electron beam cladding of TiC on titanium substrates. *Appl Mech Mater* 682:14–20.
63. Neagu, D. 2010. Alloying with electron beam of some surfaces on pieces made by non-strengthen materials. *Nonconventional Technol Rev* 10:31–35.
64. Petrov, P. 1997. Electron beam surface remelting and alloying of aluminium alloys. *Vacuum* 48:49–50.
65. Petrov, P., and Dimitroff, D. 1993. Electron beam alloying of aluminium alloys. *Vacuum* 44:857–861.
66. Valkov, S., Petrov, P., Lazarova, R. et al. 2016. Formation and characterization of Al–Ti–Nb alloys by electron beam surface alloying. *Appl Surf Sci* 389:768–774.
67. Valkov, S., Petrov, P., Dechev, D. et al. 2017. Study formation of aluminides in Al–Ti–X (X=Zr, Hf) systems induced by electron-beam surface treatment. 4th IEBW International Electron Beam Welding Conference, Aachen, Germany. 16–18.
68. Dance, B.G.I., and Kellar, E.J.C. 2004. Workpiece structure modification. International patent publication number WO2004028731 A1.
69. Wang, X., Guo, E., Gong, S. et al. 2014. Realization and experimental analysis of electron beam Surfi-Sculpt on Ti–6Al–4V Alloy. *Rare Metal Mat Eng* 43:819–822.
70. Earl, C., Hilton, P., and O'Neill, B. 2012. Parameter influence on Surfi-Sculpt processing efficiency. *Phys Procedia* 39:327–335.
71. Wang, X., Gong, S., Guo, E. et al. 2012. Primary study on electron beam Surfi-sculpt of Ti–6Al–4V. *Adv Mat Res* 418–420:772–776.
72. Zenker, R., Buchwalder, A., Frenkler, N. et al. 2005. Moderne Elektronenstrahltechnologien zum Fügen und zur Randschichtbehandlung. *Vacuum Forchung Praxis* 17:66–72.
73. Wang, X., Ahn, J., Bai, Q. et al. 2015. Effect of forming parameters on electron beam Surfi-Sculpt protrusion for Ti–6Al–4V. *Mater Design* 76:202–206.
74. Taendl, J., and Enzinger, N. 2014. Electron beam surface structuring of AA6016 aluminum alloy. *Weld World* 58:795–803.
75. Tu, W., Wen, P.H., Hogg, P.J. et al. 2011. Optimisation of the protrusion geometry in Comeld™ joints. *Compos Sci Technol* 71:868–876.
76. Thomas, G., Vincent, R., Matthews, G. et al. 2008. Interface topography and residual stress distributions in W coatings for fusion armour applications. *Mater Sci Eng A* 477:35–42.
77. Ferhati, A., Karayiannis, T.G., Lewis, J.S. et al. 2015. Single-phase laminar flow heat transfer from confined electron beam enhanced surfaces. http://bura.brunel.ac.uk/bitstream/2438/11042/2/Fulltext.pdf
78. Xiong, W., Blackman, B., Dear, J.P. et al. 2015. The effect of composite orientation on the mechanical properties of hybrid joints strengthened by surfi-sculpt. *Compos Struct* 134:587–592.
79. Tu, W., Wen, P.H., and Guild F.J. 2010. Multi-region mesh free method for Comeld™ joints. *Comput Mater Sci* 48:481–489.
80. Galchenko, N.K., Kolesnikova, K.A., Semenov, G.V. et al. 2016. AIP Conference Proceedings 1783:020059-1- 020059-5.
81. Hengst, P., Zenker, R., Süß, T. et al. 2017. Electron beam profiling and electron beam remelt-bonding for improving the load-bearing capacity of thermal spray coatings. International Electron Beam Welding Conference Aachen, Germany. 19–26.
82. Ramskogler, C., Cordero, L., Warchomicka, F. et al. 2017. Biocompatible ceramic-biopolymer coatings obtained by electrophoretic deposition on electron beam structured titanium alloy surfaces. *Mat Sci For* 879:1552–1557.
83. Ramskogler, C., Warchomicka F., Mosto, S. et al. 2017. Innovative surface modification of Ti6Al4V alloy by electron beam technique for biomedical application. *Mater Sci Eng C Mater Biol Appl* 78:105–113.
84. Buxton, A.L., Ferhati, A., Glen, R.J.M., Dance, B.G.I., Mullen, D., and Karayiannis, T. 2009. EB surface engineering for high performance heat exchangers. Proc. First International Electona Beam Welding Conference, Chicago, IL, USA.
85. Zenker, R. 2017. Electron beam surface treatment—History of development, state of the art and prospects. 4th IEBW International Electron Beam Welding Conference, Aachen, Germany (unpublished).

6 Advances in the Processing and Fabrication of Bioinspired Materials and Implications by Way of Applications

Lakshminath Kundanati and Nicola M. Pugno

CONTENTS

6.1 INTRODUCTION

In the last few decades, the need for advanced materials has increased significantly, resulting in the development of not only high-performance materials but also smart materials. A significant improvement is achieved with the use of nanostructured materials as they can be useful in building stronger composites because of increased strength at small length scales and offer high surface area for reactivity-based applications and many other advantages. A report by McKinsey & Company has included advanced materials as one of the most disruptive technologies that will transform the life and global economy (Manyika et al. 2013). Engineers are trying to replicate naturally occurring processes to meet this ever-increasing demand for high-performance materials and the need to optimize these processes for energy efficiency. In this regard, biological materials offer unique combinations of mechanical performance with capabilities such as self-healing, self-repairing, and room temperature synthesis. From an engineering perspective, biological materials have an amazing capability of early detection of damage and continuous repair (Taylor 2010). The advantage of looking at natural process offers advantages of process refinement over millions of years of evolution with continuous adaptation to changes in the environment. All natural growth processes are designed to facilitate survival of the organism and usually in situations with limited resources in terms of both quality and quantity (Buehler 2013). In addition, natural materials offer unique capabilities of being multifunctional in the form of gene regulation and signal transduction and allow diffusion. The functionality in such materials is attained using the available simple building block materials in combination with feedback to adapt to the environmental changes. The interplay between hierarchical structure, multiscale mechanics, and physiology of biological materials is observed throughout the spectrum and thus understanding fundamental relations is very challenging because of its highly multidisciplinary nature. For example, biological multifunctionality can be mimicked, likely by using microfibrous-based morphologies to enhance the effects dominated by the surface in comparison to bulk of the material (Lakhtakia 2015).

The primary goal of engineering materials is to perform a function, which is attained by manipulating composition, microstructure, and shape (Bréchet 2013). Some basic functions include high strength and toughness, in combination with low weight. Importantly, biological materials were observed to use the strategy of controlling fracture mode to avoid catastrophic failure (Yao et al. 2015). One such example is wood, with its excellent mechanical properties derived from its cellular structure and variation in its chemical composition (Qin et al. 2014). It is also important to note that interfaces of material constituents play a crucial role in the performance of biological materials (Bréchet 2013). With the rise in biomedical engineering, there are numerous attempts to fabricate materials with the purpose of using them in medical applications such as bone regeneration. Such materials need to be biocompatible, and in some cases, they need to be adsorbed over desired time scales. In this review, several aspects related to the field of bioinspiration were discussed. First to be discussed was the possible impact of bioinspiration in the industrial sector. Second, the relevant fields of applications were highlighted by broadly categorizing them into engineering and biomedical materials. Third, the importance of modeling in processing and fabrication of bioinspired materials was discussed. Finally, various approaches involved in bioinspired fabrication were categorized and discussed in detail to gain understanding of the present state and advancements.

6.2 IMPACT OF BIOINSPIRATION IN THE INDUSTRIAL SECTOR

Technologies and materials inspired by biological materials are expected to have a significant role in transforming economies on a large scale. The effects can be observed in the fields of manufacturing, construction, agriculture, and waste management. During the 2000–2013 period, a fivefold increase was observed in the number of patents and research articles related to bioinspiration (Fermanian Business and Economic Institute 2013). Advances in the field of bioinspiration are expected to have a strong influence in many industrial sectors by the year 2030 (Figure 6.1). A few examples include fabrication of molecular catalysts, piezoelectric nanotubes, surfaces for controlling stem cell interactions, micro-controllable metasurfaces, optical multilayer fibers, tunable lens, soft robotics, and smart fabrics (Brlmblecombe et al. 2010; Carpi et al. 2011; Fisher et al. 2010; Hou et al. 2011; Kholkin et al. 2010; Kolle et al. 2013; Kim et al. 2013; Singh et al. 2012).

6.2.1 Engineering Materials

6.2.1.1 Structural Materials

Structural materials performance improvement is becoming more and more prevalent in recent times because of modern applications. This led to exploration of biological structural materials with high structural performance, which include abalone shell, arthropods, and insect cuticle (Miessner et al.

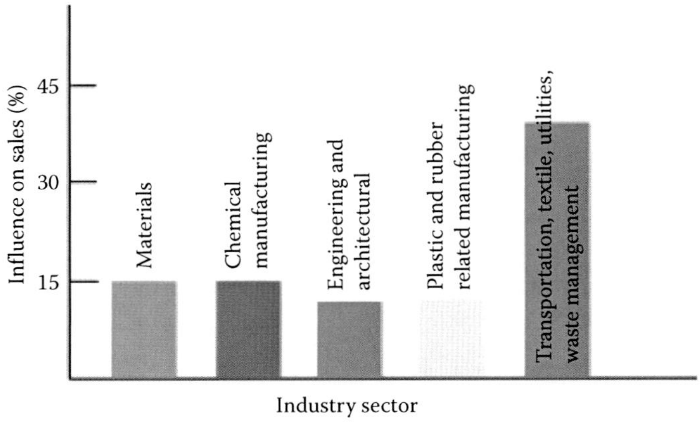

FIGURE 6.1 Projected influence of bioinspiration in industrial sectors by the year 2030 (FBEI, Bioinspiration: An economic progress report).

2001; Ribbans et al. 2016; Wilbrink et al. 2010). The rigid abalone shell–like materials are usually a combination of soft protein matrix and a hard ceramic component, in which the structural arrangement of building blocks is similar to brick mortar arrangement (Weiner and Addadi 1997). These structures led to the fabrication of brick mortar composite materials. Using this concept, scaling relationship models were developed to optimize strength and fracture toughness, which is achieved by modifying the size, strength, and layout of hard components (Wilbrink et al. 2010). Inspired by laminated structures in natural materials, alternating layers of chromium and chromium nitride were deposited to improve wear resistance of tool materials (Zeilinger et al. 2015). Despite numerous studies and extraction of design principles, there is limited applicability of bioinspiration in cases of extreme engineering applications where strength in combination with high-temperature performance is achieved by using elements nickel, iron, and chromium, unlike the biological counterparts (Fratzl 2007).

6.2.1.2 Anti-Wetting and Self-Healing Materials

Anti-wetting and self-healing materials comprise an important subsection of bioinspired engineering materials. Biological materials such as lotus leaf shed off water that carries dirt particles using surface texture at nano- and microscales, which enable macroscale wetting over a large surface area (Sottos and Extrand 2008). Using strategies similar to that of lotus leaf, materials are also developed for preventing ice formation (Lv et al. 2014). Other examples include butterfly wings, water strider legs, cicada wings, and cabbage leaves (Ding et al. 2008; Lepore et al. 2013; Liu 2010; Mo et al. 2015). Bio-fouling is another important phenomenon because of its detrimental effects on water facilities and healthcare systems (Nir and Reches 2016). A recent study demonstrated that gecko skin has antibacterial surface properties, which might serve as inspiration for developing antibacterial artificial microchannels, dental implants, and contact lenses (Watson et al. 2015). The key concept for creating such surfaces is to achieve super-hydrophobicity using size scale, topology, and structure.

The self-healing phenomenon is by far one of the most unique capabilities of biological materials that can be of tremendous use in engineering materials to avoid catastrophic failures. A common example is blood clotting, achieved by using chemical signaling strategy from the site of damage. It is a future possibility, but at present, researchers succeeded in developing a crack healing mechanism. This is achieved by using externally mediated mechanisms that use heat, mechanical, optical or chemical cues, or using a crack-responsive healing material mixed with polymer matrix (Zeilinger et al. 2015). Such systems are challenging to fabricate, as the healing is carried out with the aid of nanocapsules or migrating nanoparticles, which are expected to not alter the properties of polymer matrix much and also remain intact.

6.2.1.3 Materials for Sensing

Extraordinary sensor systems are found in nature that outperforms man-made sensors. For example, the moth species *Bombyx mori* has the capability to detect a single pheromone molecule using its olfactory sensory system (Kaissling 1996). Using inspiration from such a moth organ, an extremely sensitive sensor was developed by using titania nanotubes and a cantilever beam to detect explosive materials such as TNT (Spitzer et al. 2012). Also, miniaturized optical flow sensors were designed, taking inspiration from insects using low-temperature co-fired ceramics that were used as reliable electronic motion detectors (Pudas et al. 2007). Human skin is a good example of a multifunctional sensor that can detect both deformation and heat. A flexible patch antenna capable of structural health monitoring is fabricated, taking inspiration from skin design (Tata et al. 2009).

6.2.1.4 Others Types of Materials for Use in Engineering

Crystalline inorganic materials with desired microstructures find use in applications like microelectronics, optics, and biomedical implants (Aizenberg 2004). The technologies used in fabricating such materials primarily use a bottom-down approach. They involve growing single crystals followed by cutting and polishing them in the desired crystal orientation. This is one such area where inspiration can be drawn from natural materials to grow crystals in the preferred direction, in addition to desired physical and chemical properties. Also, polymer membranes with their unique surface properties find use in many industrial processes such as separation, gas separation, reverse osmosis, microfiltration, nanofiltration, desalination, and batter separation (Yang et al. 2015). Despite challenges, bioinspired membranes are expected to improve membrane technology in the future (Zhao et al. 2014). For example, there is significant research carried out drawing inspiration from dopamine-related chemistry of mussels. Actuation in biological systems has led to the development of a novel chemomechanical hybrid actuation system that enables reversible actuation of pillar structures using variation or gradient in pH levels. The contraction and expansion of a structured hydrogel were used to bring motion in microposts that were embedded in the hydrogel (Zarzar et al. 2011).

6.2.2 Biomedical Materials

Biomedical engineering is an essential aspect of today's world in developing materials and devices for use in healthcare. Two of the subgroups can be broadly classified into tissue engineering and drug delivery systems. Tissue engineering is by far one of the most important emerging fields that has a broad use in the coming future to decrease the need for organ replacement and accelerate development and testing of new drugs (Griffith and Naughton 2002). It involves engineering a scaffold material that is capable of cell interaction similar to the native tissue, to enable their adherence, growth, and division. Drug delivery systems include materials that are used to release drug at a desired target location. With advances in biotechnology and nanotechnology, it is now possible to develop drug carriers in the form of particles and also biological structures such as micelles (Yoo et al. 2011). The crucial part of the delivery process is the design of delivery systems that can carry or attach to the drug molecule and also cross the internal checks of biological systems before reaching the target. There have been numerous attempts to build bioinspired materials for use in both tissue engineering and drug delivery systems. The minimum requirements include biocompatibility and biodegradability, depending on the intended usage of the material. This adds in more complexity in finding materials that can be processed and fabricated to meet the design requirements and simultaneously that are biocompatible and biodegradable. Thus, taking inspiration from biological materials is a natural way to develop such materials. All biological tissues are built with a bottom-up approach using nanoscale materials and controlled hierarchical architecture (Green et al. 2016). One such example is the development of a nanostructural hybrid scaffold fabricated combining chitosan, chondroitin sulfate and hydroxyapatite, which demonstrated promising results for bone repair (Fan et al. 2016). Recent developments in drug delivery systems have led to the development of disk assemblies produced from recombinant Tobacco Mosaic Virus protein as well as protein cages for carrying cancer-treating drugs (Finbloom et al. 2016; Suci et al. 2009).

So far, most of the bioinspired materials were mostly fabricated at laboratory scale and using energy-intensive methods. Thus, it is important to address the issue of scaling up these processes to produce materials in bulk and also make them cost-effective (Wegst et al. 2014). With the advances in technologies, we are able to probe natural materials and also fabricate materials using renewable resources that are eco-friendly and recyclable (Demirel et al. 2015). The parallel developments in microfabrication and nanofabrication techniques have revolutionized the manufacturing field with the ability to control features at very small length scales. Some of these techniques include self-assembly methods, different kinds of lithography, three-dimensional (3D) printing, templating, deposition, and etching techniques. It was observed that the variety of patterns that can be made using a bottom-up approach is limited as compared to top-down approaches (Burgess et al. 2013).

6.3 FABRICATION OF BIOINSPIRED MATERIALS

Bioinspired fabrication recedes the steps of identification and characterization of a biological material with a property or function of interest (Figure 6.2). Material characterization involves understanding its structure–property–function relationship. This step being a crucial one, it can be accompanied with multiscale modeling to understand experimental findings. Modeling adds the additional capability to understand interactions of the building blocks in space and time during their performance, using fundamental interactions at all length scales. Modeling procedures would also need validation from both experimental and theoretical approaches. After determination of structure–property relationships, the material constituents that comply with the possible fabrication route are selected. Selection of materials involves choosing a material depending on the physical attributes in terms of size, shape, and microstructure, and their chemical attributes like surface properties and composition. This is a challenging task because most of the biological materials are built with elements such as carbon, hydrogen, oxygen, nitrogen, calcium, and so on, whereas most of the engineering materials are fabricated using elements such as iron, chromium, aluminum, magnesium, copper, nickel, and so on.

Also, the growth of biological materials is primarily governed by self-assembly unlike the engineering fabrication methods. Both the subroutines of material selection and fabrication process are interrelated because of the constraints that limit either access to a particular material or the capability of a fabrication technique in combination with cost-effectiveness. The final evaluation step after fabrication includes characterization of the produced materials. If the material does not meet the requirements, then a feedback step can be created to fine-tune the materials and fabrication procedure.

6.3.1 Role of Theoretical and Computational Modeling

The key characteristic of biological materials is the display of their bulk behavior that emerges from the behavior at all length scales and at different time scales (Buehler and Genin 2016). To understand the complex behavior of biological materials, it is necessary to systematically develop both theoretical models and computation models. For example, a theoretical study on non–self-similar hierarchical composites has shown that they perform better in resisting fracture and enhance energy storage capacity (An et al. 2014). Also, analytical models were developed to predict the pull-out behavior of bioinspired jigsaw-like sutures that improve the toughness of materials (Malik et al. 2017). Using a different approach, structural elements of biological materials were categorized into eight categories (fibrous, helical, cellular, and tubular, to name a few) to develop basic constitutive equations for all these elements (Naleway et al. 2015). Although it is well known that biological materials draw their mechanical properties from the hierarchical nature, the contribution of structural hierarchy was not well understood. To address this issue, numerical methods were used to

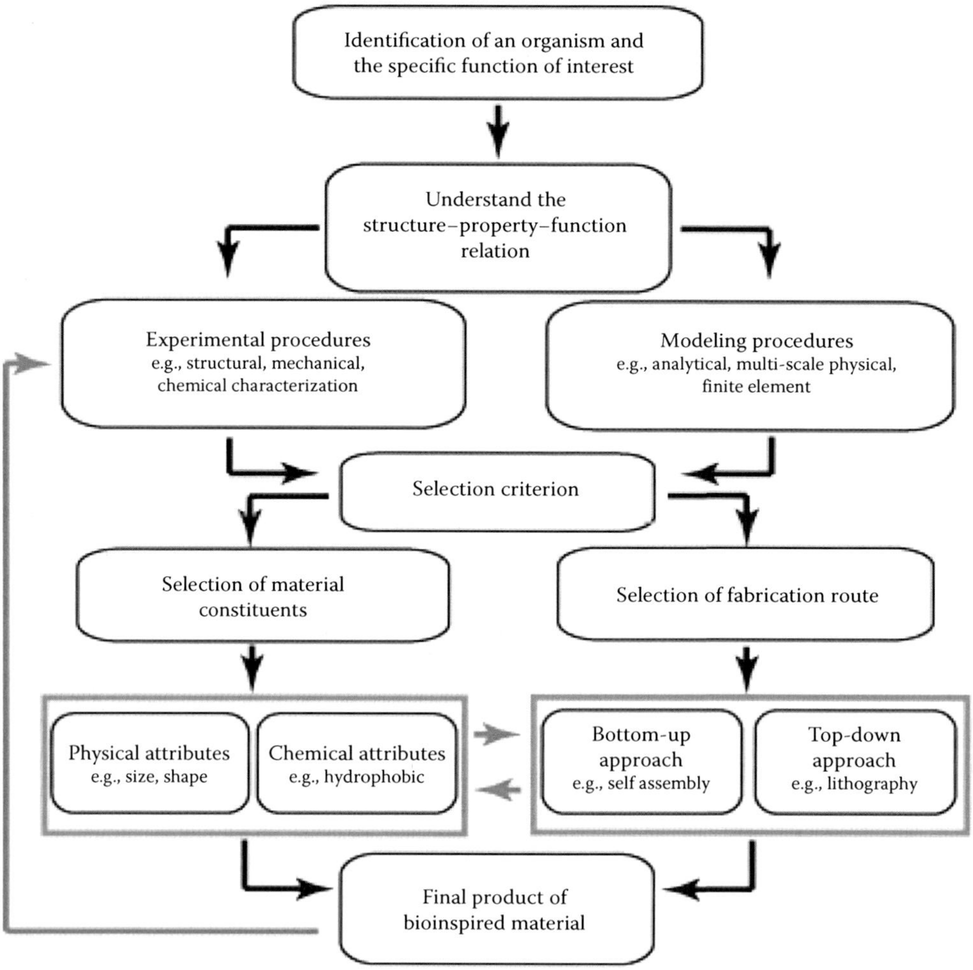

FIGURE 6.2 Schematic of stepwise procedure of a typical bioinspired fabrication.

quantitatively investigate the effects of hierarchy in a heterogeneous fiber-based biological material and demonstrated the role of hierarchy in improvement of properties (Bosia et al. 2013). Similarly, a hierarchical model was developed based on the complex structure of lobster cuticle, which connects from atomic scale to macroscale and can also be adapted to other biological materials (Nikolov et al. 2010). Recently, a novel modeling approach was developed using fiber bundle–based models to understand self-healing mechanisms that enhance toughness (Bosia et al. 2014).

The advances in modeling approaches led to the development of multiscale modeling that incorporates integration of various computational models and combining them with material description (Buehler 2013). Because many of the fundamental computational methods rely on basic interactions at the atomic level, it is possible to describe materials using a bottom-up approach all the way up to macroscale. Using a molecular-based approach that spanned from angstrom scale to mesoscale, superior mechanical properties of silk were explained using only structural elements, as a combination of geometrically confined beta-sheet nanocrystals and highly extensible semi-amorphous regions (Nova et al. 2010). Using integration of mesoscale modeling with block copolymer processing and bioinspired spinning process, a novel material designing approach was described that combines chemistry, processing, and material characterization (Lin et al. 2015). This approach was further used to explain the role of the ratio of hydrophilic and hydrophobic regions in the formation of spider silk. There is also the significant importance of introducing computational approaches that use scalable generative encodings and bioinspired form findings. Using such approach, a novel generative model was developed that created geometries by volumetric gradients that evolve (Richards and Amos 2014). Such methods demonstrate the capability to find performance-oriented forms in structural optimization and in creating designs of multimaterial and functionally graded materials. A novel computational design platform (*Metamesh*) was developed for enabling incorporation of organizational principles at local, regional, and global levels, in combination with physiological information of the biological system (Duro-Royo et al. 2015). This model also demonstrated the capability of extension to human kinetics.

Modeling also has a direct role in fabrication through 3D printing of complex shapes. In particular, creating shapes with volumetric gradient patterns for controlling material structures that are performance oriented could be a challenging task. To address this issue, a novel method was developed to extend architectural research and discover performance-driven designs that fulfill both 2D and 3D topology optimization problems (Richards and Amos 2014). The main idea of this approach was integration of simulation, analysis, and physical testing during the early stages of design. Thus, we see that modeling has a significant role in the fabrication of bioinspired materials.

6.3.2 Fabrication Methods

In this section, we briefly describe the commonly used fabrication methods of bioinspired materials with recent advances and relevant examples. The last subsection was used to describe the advances in fabrication procedures that use a combination of techniques that cannot be categorized under a single fabrication technique.

6.3.2.1 Self-Assembly

Self-assembly is defined as a reversible process that involves assembly of pre-existing components and can also be controlled by designing the components carefully (Whitesides 2002). It eliminates the steps of additional processing as compared to conventional fabrication techniques (Klok and Lecommandoux 2001). Self-assembly can be classified into two main groups depending on whether it is a static or a dynamic process. Static assembly involves components that are either at local or at global equilibrium and do not dissipate energy, such as the formation of molecular crystals. On the contrary, dynamic assembly involves dissipation of energy during interaction of components. In the last two decades, taking inspiration from nature, material scientists looked into nature's principles of self-assembly to create engineered materials using both natural building block materials and synthetic counterparts.

Most of the biological processes are driven by self-assembly right from deoxyribonucleic acid, viruses, proteins, and cell membranes. These processes are an orchestra of molecules organized at nanometer scale under equilibrium conditions to form stable aggregates (Whitesides et al. 1991). Thus, molecular self-assembly has become a crucial route to make bioinspired materials because of its ability to fabricate materials with higher complexity and precision, to achieve multifunctionality (Mendes et al. 2013). At the molecular level, the interactions can be driven by electrostatic interactions, aromatic interactions, hydrophobic effects, and hydrogen bonding. For example, proteins can be seen as self-assembled chains built by the primary building blocks called peptides, and their folding is mitigated by self-assembly with the help of noncovalent interactions in aqueous solution. Protein folding is a very important phenomenon that enables folding of proteins into 3D structures ubiquitous in biological self-assembly. It is one of most critical process in biological systems that led to enormous diversity by selectively influencing particular chemical pathways (Dobson 2002).

Understanding the basics of self-assembly of microscale objects is still very limited and this area needs to be studied further (Boncheva and Whitesides 2005). For example, the process of assembling and folding of microscale polymeric materials is prone to morphological changes as opposed to the assembling that occurs at the atomic scale in a process like crystallization (Cademartiri and Bishop

2014). Although self-assembly is one of the most widely researched areas, it is primarily focused on small-scale and specific applications. It is also limited by the variations in structures that can be achieved because of the constraints in the number and type of interactions that are possible. For example, as regards the fabrication of coiled-coil fibers like spider silk, we are yet to understand the mechanisms of coiling though the assembly of subunits. Another important challenge is the ability to predict synthesis clearly as the length of assembly increases (Quinlan et al. 2015). One such example is biomineralized material like the abalone shell in which the chemical nature of the organic constituent generally determines the path of nucleation and growth of the mineral phase. Although there are significant advances in the understanding of such materials, we are yet to succeed in using the mineralization technique to fabricate large-scale structural materials (Wegst et al. 2014). Thus, we can see the self-assembly process as a challenging route for the fabrication of bioinspired materials.

Self-assembly can be broadly simplified into processes based on the type of interaction between the components, namely, the puzzling approach and the folding approach (Figure 6.3) (Cademartiri and Bishop 2014). The puzzle approach offers the advantage of building any kind of structure and also ease of design because of simple pairwise interactions but has a disadvantage in terms of scaling because of the increased number of multiple interactions. On the contrary, the folding approach offers the advantage of scalability because of limited interactions but has the disadvantage of difficulty in design owing to constrained interactions.

The self-assembly process can also be classified into two broad groups based on the scale of assembly: molecular-level self-assembly and macrolevel self-assembly. Molecular-level assembly involves only molecules or molecules bound by metals and chemicals groups. On the contrary, macro-assembly involves objects bigger than molecules that are guided by the surface chemistry of objects.

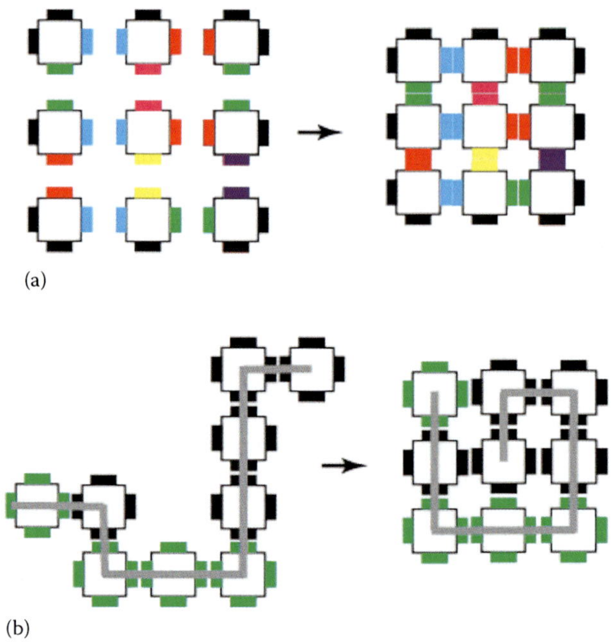

(a)

(b)

FIGURE 6.3 Broad classification of self-assembly. (Adapted from Cademartiri, Ludovico, and Kyle J M Bishop. 2014. Programmable self-assembly. *Nature Materials* 14 (1): 2–9.) (a) Puzzling approach in which multiple interactions are present, and (b) folding approach in which there is only one possible sequence and interaction.

6.3.2.1.1 Molecular Self-Assembly

Molecular assembly can be explained using protein structures and their folding patterns. The basic sequence of amino acids in a protein is called primary structure (Figure 6.4a). A secondary structure of protein is observed when there are local conformational changes in the protein chain leading to two different types of structures, namely, α-helix and β-sheet (Figure 6.4b). The tertiary structure of a protein is a stable 3D structure, formed by bending and twisting of a protein molecule due to interaction between β-sheet and α-helix subunits (Figure 6.4c). When a protein is made of more than one polypeptide chain (protein subunits), a quaternary structure, which is a larger aggregate, is formed as a result of the interaction of these subunits via hydrogen bonding and salt bridging (Figure 6.4d).

A variety of bioinspired nanomaterials can be fabricated using self-assembly of peptide conjugates for use in applications such as biomineralization, bio-imaging, cell targeting, and biocatalysts (Yu et al. 2017). The functional flexibility, compatibility, and biodegradability of peptides make them suitable candidates for design of bioinspired materials. Peptides are also used in biosensors (Liu et al. 2015a) and the challenge is to achieve accurate detection because of numerous interferences. With recent developments, it is possible to fabricate large arrays of peptide nanotubes using the vapor deposition method with fine control over size and length of nanotubes, by controlling the supply of building material blocks. These large–surface area nanotubes offer the potential to be used in energy storage and microfluidic chips (Adler-Abramovich et al. 2009).

It is a great challenge to understand complex biological systems that are influenced by large interlinked networks that primarily include proteins (Vendruscolo et al. 2003). For example, understanding the assembly of amyloid proteins is crucial to tackle diseases like Alzheimer's (Roychaudhuri et al. 2009). Also, in biologically controlled mineralization processes like the formation of skull, metal ion transport proteins play a crucial role in controlling crystallization of particles with superior properties (Bain and Staniland 2015). In a recent study, to gain insight into the complex formation process of mussel byssus fibers of *Mytilus edulis*, in situ tracking of protein assembly was performed (Priemel et al. 2017). This was carried out using 10 different protein precursors

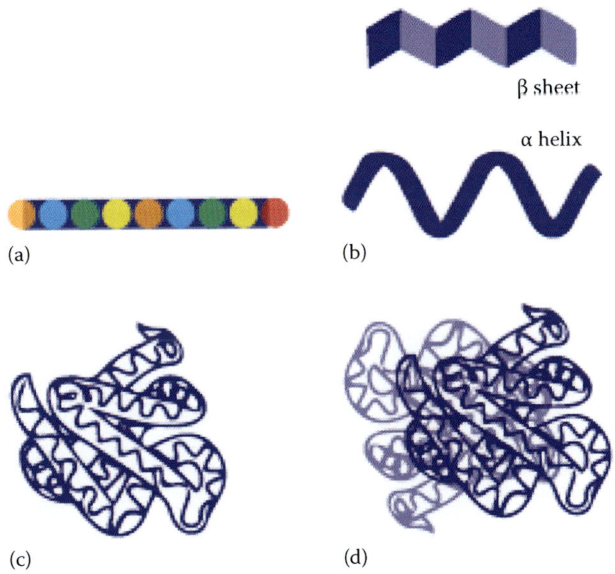

FIGURE 6.4 Structural states of proteins (amino acids shown as spheres of different colors). (a) Primary structure. (b) Secondary structure in terms of either pleated sheet or helical form. (c) Tertiary structure formation due to interactions between two forms of secondary structures. (d) Quaternary structure formation due to interaction of more than one protein chain.

during assembling process and a novel approach that combines histological staining and Raman microspectroscopy.

Transition metals like iron, zinc, copper, and nickel were observed to interact with proteins and other biomolecules influencing biological processes like catalysis, gas transport, and signal transduction (Holm et al. 1996). These interactions were observed to enhance stiffness and adhesive and self-healing properties of biological materials (Degtyar et al. 2014). Using such interactions, a sequential and spatial arrangement of porphyin molecules was achieved with the use of copper, nickel, and palladium metal–ion complexes (Yamada et al. 2014). Specifically, gold nanoparticles were observed to be the most stable and found use in numerous applications spanning fields of material science, catalysis, and biology (Daniel and Astruc 2004). They are used to fabricate 3D architectures with various types of spirals, because of their ability to attach to single-strand deoxyribonucleic acid (Sharma et al. 2009). Thus, nanoparticles can be seen as active elements that control the preference for specific tube conformations through size-dependent steric repulsion effects. Supramolecular assembly of surfactant molecules was used for condensation of inorganic phase at the interfaces, resulting in tailored nanoscale architecture of inorganic composite materials (Aksay et al. 1996). Recently, self-assembly of reconstituted silk fibroin was mediated through application of mechanical forces using a combination of top-down and bottom-up approaches (Tseng et al. 2017). This approach enabled control of morphology at different length scales to obtain complex 3D hierarchical structures. Using tri-block copolymers of miniature size and their self-assembling capabilities, highly regular-sized nanostructures were fabricated into mushroom-shaped supramolecular structures with the aid of crystallization, which further self-organize into films by stacked layers (Stupp 1997). Self-assembly can also be achieved in nature using interactions of molecules with water, depending on whether it is hydrophobic or hydrophilic. Such molecules were observed to aggregate under optimum environmental conditions such as temperature, pressure, and concentration (Gonzalez-Perez and Persson 2016). Such self-assembling methods also provide the flexibility to tailor the material properties such as surface topography and mechanical strength.

6.3.2.1.2 Macro Self-Assembly

Self-assembly of macroscopic objects can be achieved using various interactions such as electrostatic, magnetic, and capillary effects (Boncheva et al. 2005; Bowden et al. 1997; Gartner and Bertozzi 2009). In a recent study, small pieces of acrylamide gels that are functionalized with cyclodextrin rings and small hydrocarbon groups were observed to aggregate by molecular recognition at the surface of gels as shown in Figure 6.5 (Harada et al. 2011). The results of this study suggest that it is possible to assemble macroscopic objects reaching length scales of a few centimeters using a variety of gels. It was even possible to assemble 3D micro-tissues under ambient cell

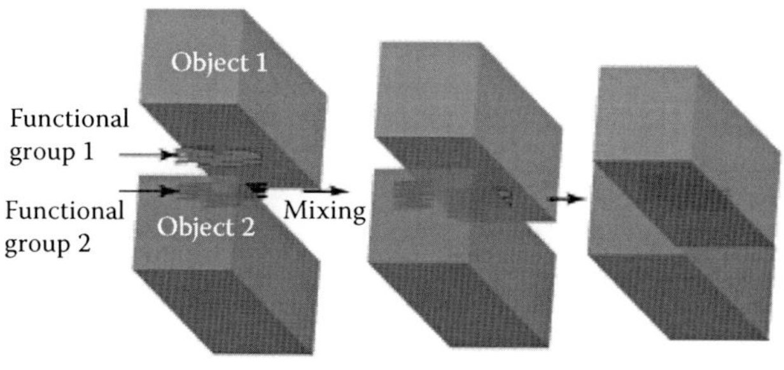

FIGURE 6.5 Schematic of simple macroscopic objects assembling with the aid of different functional groups on the surface. (Adapted from Harada, Akira, Ryosuke Kobayashi, Yoshinori Takashima, Akihito Hashidzume, and Hiroyasu Yamaguchi. 2011. Macroscopic self-assembly through molecular recognition. *Nature Chemistry* 3 (1): 34–37.)

culture conditions with the help of duplex deoxyribonucleic acid to connect the cells (Gartner and Bertozzi 2009). Taking advantage of anisotropic building blocks of microscale, self-shaping materials have been fabricated with inspiration from biological structures. These examples include controlling the orientation, distribution, and interaction of the building blocks that facilitate reinforcing (Studart and Erb 2014).

6.3.2.2 Lithography

Lithography, probably one of the oldest techniques to print patterns, was developed in the seventeenth century for printing on stones. Over the last few decades, with the development of technology, various energy sources were used ranging from light, x-rays, ions, lasers, electron beams, and neutral atoms. The basic steps of lithography remain the same for all types of lithography and they can be broadly classified into optical and nonoptical techniques (Ito and Okazaki 2000). The initial step of lithography includes preparation of a mask and resist material coating on the substrate material on which patterning is desired (Figure 6.6a). The second step involves the use of energy source to focus the energy beam on the coated substrate, mimicking the design of the mask, by selective exposure of certain regions followed by developing (Figure 6.6b). The third step involves etching, in which the exposed coating is either removed or retained, depending on whether it is a negative or positive process (Figure 6.6c). The final step involves stripping the resist coating, leaving the desired pattern for use (Figure 6.6d). These patterns are further used to create a replica with the help of polymeric materials like polydimethylsiloxane. All lithographic techniques need verification of final pattern size

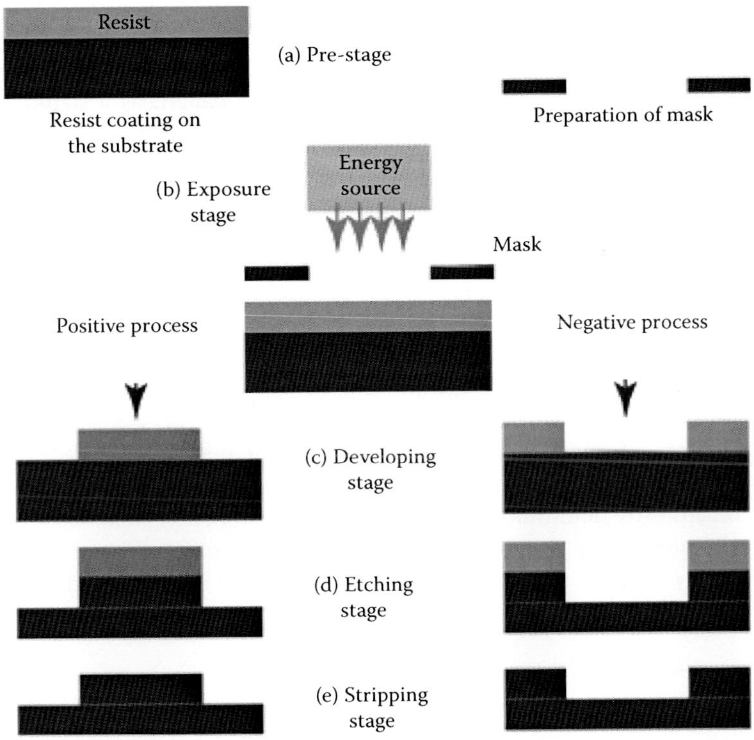

FIGURE 6.6 Schematic of steps involved in standard lithographic technique. (a) Initial preparation of substrate with photoresist coating and also mask design. (b) Exposure of coated surface with an energy source. (c) Developing stage is dependent on the final design, whether it is desired to keep the exposed part intact or remove it. (d) Etching state is where the material removed using etching process. (e) Final stage involves removal of the coated resist material.

and shape, because of variability in the process. This step can be equally challenging as making a pattern, when desired resolutions are in the order of nanoscale, as they require high-end characterization techniques.

Photolithography is one of the oldest and most widely used techniques in the category of lithography. The performance of such process is determined by resolution, fidelity, and throughput. It is necessary to have a clean facility to carry out this process. With advances in technologies that can produce fine focused beams of x-ray and ultraviolet, resolutions reached nanoscale. Depending on the feature size and aspect ratio of the feature, one can choose from a variety of photoresist materials and energy source.

6.3.2.2.1 *Soft Lithography*

Soft lithography can be viewed as an extension of the standard photolithography technique and is primarily used for replica molding in case of micro- and nanofabrication procedures. In brief, a soft elastomeric stamp with patterned microstructures is used to make replicas of structures with a size range from nanometers to micrometers (Xia and Whitesides 1998). It is a low-cost, convenient, and effective method. However, fabricating fibers with a high aspect ratio such as cilia requires higher stability than that of a soft elastomer like polydimethylsiloxane (Grinthal et al. 2012). Using thermo-curable polymers, a novel method was developed that enables fabrication of 3D parts with complex geometries that are difficult to fabricate in conventional processes (Rodrigue et al. 2015).

Micro-contact printing is a type of soft lithography technique that combines both lithography technique and stamp printing. The initial step of this process includes creating a desired pattern on the surface of silicon substrate. The next step uses the principle of spontaneous adsorption of organic thiols on the patterned gold surface to create self-assembled monolayers, which act similar to that of photoresist materials in standard lithographic techniques (Perl et al. 2009). In addition, the technique allows the use of elastomeric materials cured from a master pattern as a stamp material. Taking inspiration from adhesive proteins of mussels, micropatterned polydopamine films were fabricated using micro-contact printing, which also provide the flexibility of tuning morphology by altering the concentration of ink material (Hou et al. 2015). These films can further be functionalized by growing polymer brushes with the aid of self-initiated photografting and photo-polymerization. In a recent study, silica patterns were fabricated under mild conditions using micro-contact printing in combination with layer-by-layer assembly, to control the aspect ratio of patterns (Yang and Choi 2009). Despite recent advances, there are still some challenges in developing strategies for printing arrays of biomolecules (Voskuhl et al. 2014).

Unlike conventional soft lithography, the newly developed bench top lithography method proved the ease of performance on a table top without the requirement of a clean room. In addition, it provides more degrees of freedom in terms of the type of material, geometry, and mechanical properties that can be manipulated from simple structures (Grinthal et al. 2012). It enables fabrication of nontrivial geometries by deforming flexible negative polydimethylsiloxane mold before the final curing, as the technique is not a one-to-one process. One can stretch, bend, twist, and shear this mold to a desired geometry. It also offers the advantage of using both synthetic and biological polymers, low–melting point metals, and even ceramics using atomic layer deposition.

6.3.2.2.2 *Ultraviolet Photolithography*

Ultraviolet photolithography is another form of photolithography in which ultraviolet light source is used as an energy source. Recent advances have led to the development of extreme ultraviolet lithography with light wavelengths reaching 13.5 nm, which is a significant enhancement compared to the existing lithographic techniques. The resolutions of such systems can be pushed to less than 10 nm with the use of tricks like off-axis illumination resolutions and special water immersion lens that can achieve high numerical aperture (Wagner and Harned 2010). This lithography technique has the capability to achieve features with a size of 32 nm using a single exposure, fewer restrictions on design, increased wafer throughput, and reduced costs of manufacturing. In one of the recent studies, a simple and scalable procedure was developed to fabricate dry adhesive structures containing micropillars with the shape of a mushroom, combining roll-to-roll process with a modulated elastic poly(urethane acrylate) resin that is ultraviolet light curable (Yi et al. 2014).

6.3.2.2.3 Laser and Electron Beam Lithography

Laser lithography is another type of photolithography technique in which a laser energy source is used instead of optical source. The use of ultrafast laser processing techniques appears to offer the ability to reach high throughput in combination with good resolution (Malinauskas et al. 2016). With recent advances, femtosecond pulsed laser was used to make patterns on a glass plate that can further be used as a master for creating patterns using polymeric materials (Liu et al. 2012). Direct laser writing is a variation of laser lithography that eliminates the need of recoating or layer-by-layer fabrication and also has the advantage of superior resolution (Mishra 2015). The advances in 3D laser lithography and its compatibility with computer-aided design modeling has enabled one to create negatives of diatoms that were further used to fabricate 3D replicas of diatoms (Belegratis et al. 2014). Using 3D laser lithography that is based on two-photon polymerization of a photoresist surface, a complex 3D pattern was created that mimicked the morphology of aquatic fern (*Salvinia molesta*) leaves (Tricinci et al. 2015). In a recent study, three-beam and four-beam laser interference lithography techniques were used to fabricate 3D hierarchical structures (Hu et al. 2016).

Electron beam lithography uses fine focused electron beam for creating simple patterns on thin films and also complex patterns with high accuracy by monitoring motion using interferometry. It is one of the key microfabrication techniques that allows the use of a wide variety of materials (Altissimo 2010). Advances in technology increased the possibility of achieving resolutions less than 10 nm, and the technique offers flexibility and compatibility in the fabrication of nanostructured patterns (Hu et al. 2004). With the use of aliphatic self-assembling monolayers of a commercially available compound as a resist material, micro- and nanobrushes were fabricated, with extended possible use in biomedical studies (Ballav et al. 2008). In a latest study, electron beam lithography was used as a primary step in building a master mold for developing 3D funnel nanostructured surface topography that is inspired by low frictional surfaces like shark skin (Kirchner et al. 2015).

6.3.2.2.4 Thermal Nano-Imprint Lithography

Imprinting lithography was used for decades but primarily for patterns larger than 1 μm. In recent years, the technique was refined with the use of heating and thermosetting polymers. The refinement led to the fabrication of features close to 25 nm, by heating the mold and the 55-nm poly(propyl methacrylate) film on the substrate above 200°C, thereby taking the polymer above its glass transition temperature (Chou et al. 1995). The mold is then pressed on the film for it to conform to the mold shape and is removed after the temperature drops below transition temperature. The latest advances with the use of thermosetting polymers demonstrated the fabrication of 3D hierarchical structures (Zhang and Low 2006). Although it was demonstrated as a fast and low-cost process, it had the problem of prepatterned structures collapsing in the earlier process. This challenge was overcome by using better temperature control and a better mold that was able to avoid damage to the imprinted structures.

6.3.2.2.5 Lithography Techniques in Conjunction with Other Techniques

Bioinspired materials are also fabricated using lithographic techniques in combination with other fabrication techniques. Standard lithography and other assembly techniques pose limitations in fabricating antireflective surfaces that require a parabolic surface with a wavelength because of the complexity involved in it. This led to the development of interference lithography in combination with thermal reflow followed by pattern transfer (Lee et al. 2010). Lithography is also used in combination with gold coating followed by electro-spinning to create cellulose-based super hydrophobic hierarchical structures on micropillars (Kakunuri et al. 2017). A novel versatile silk protein lithography was developed to fabricate microscale structures with controlled shape and size, using a combination of two proteins, that is, silk fibroin and sericin, through photo-crosslinking (Pal et al. 2016). Such methods provide additional advantages, such as controllable biodegradability of silk and low cost. In recent studies, etching techniques were used in combination with photolithography to fabricate nanostructures mimicking butterfly scales (England et al. 2014), with polymer replica building to fabricate microarray lenses that resemble an insect's compound eye (Bian et al. 2016), with sol-gel method to fabricate single-layer silicon dioxide with antifogging properties (Han et al. 2016) and in alternating sequential steps with polymerization to fabricate artificial hair-like sensors inspired from fish cupulae (McConney et al. 2009). From the above examples, we can see that using lithographic techniques in combination with novel materials and other simple procedures can enhance the speed and versatility of fabrication procedures.

6.3.2.3 3D Printing

3D printing is a manufacturing process that builds a desired physical object layer by layer using a digital model. It is one of the most important disruptive technologies of these times because of the advances made in various fields like computing and related technologies. Global major manufacturers like General Electric, Boeing, and Ford have started using this technology to print some critical components of airplanes, turbines, and automobiles. The market value of this process is expected to reach about $900 million by 2021. By 2030, the role of 3D printing in bioinspired fabrication is expected to reach approximately $1 trillion and reduce the costs involved in wastage and pollution by a quarter of a trillion dollars (Fermanian 2014). 3D printing has great potential in fabrication of bioinspired materials like hydrogel nanocomposites, soft actuating materials, tunable and cell compatible hydrogels, and vascularized tissue constructs (Gou et al. 2014; Kolesky et al. 2014; Roche et al. 2014; Rutz et al. 2015). Recent advances and new capabilities to print layer by layer at small length scale led to the possibility of printing multifunctional nanocomposites. The resolution of industrial 3D printers increased significantly in the last few years, reaching a resolution of 16 μm of layer thickness in stereolithography, 80 μm in stereolithographic sintering, and approximately 170 μm in fused deposition modeling (Chia and Wu 2015). However, challenges are faced in processing, cost, consistency, and issues related to chemical and thermal instability of nanomaterials (Vivek and Mahajan, insights). Earlier, these methods faced the issue of printing a product with a good surface finish, but now the products have a very good surface finish, thereby eliminating an aspect of postprocessing. Advances in polymer chemistry have led to the development of novel printable materials that can withstand temperatures of up to 120°C. Also, using simple mixing to alter composition before printing, cellulose nanofibrils/hydrogel nanocomposites were prepared using an easy direct printing approach that enabled self-healing of hydrogels during drying for the formation of coherent bulk film later, which has a potential use in printed electronics as shown in Figure 6.7 (Wang et al. 2016).

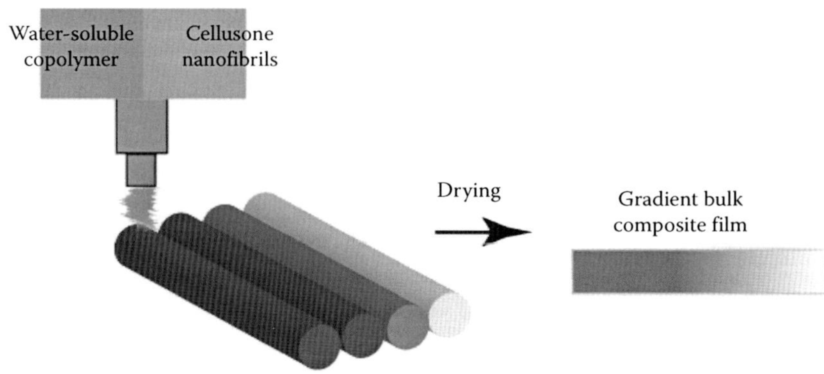

FIGURE 6.7 Schematic of direct filament writing used for fabrication of gradient bulk film.

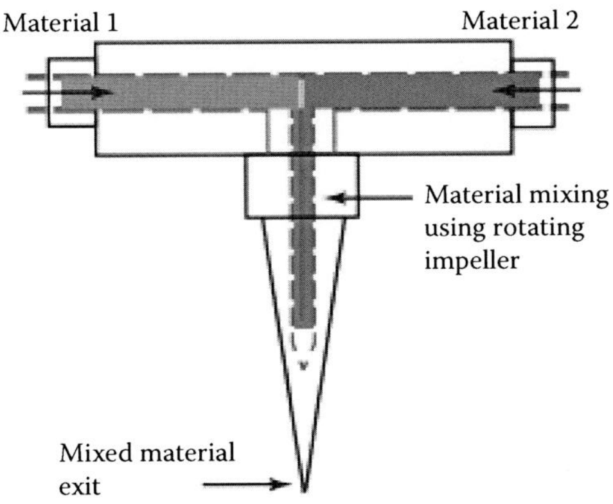

FIGURE 6.8 Design of 3D printing nozzle that allows active mixing of different materials. (Adapted from https://wyss.harvard.edu/new-frontiers-in-3d-printing.)

3D metal printing technology is relatively new compared to its predecessor—polymer printing—and is primarily used at industrial scale using metal powders and metal filaments. This technology has the potential to reduce wastage in the fabrication process by approximately 90%, and the latest advances promise to print an aircraft part in 2 days, which would normally take a few months. It is even possible to print ceramics with tailored structures and properties. For example, cellular ceramics were printed using direct foam writing that enabled the tenability of microstructure, geometry, and stiffness (Muth et al. 2017). The design includes either hexagonal or triangular honeycomb structures that were developed from filamentary struts having closed-cell foam microstructure. Until recently, 3D printers that can produce large-scale objects that are food safe and economical were not available. This led to the development of a novel 3D ceramics printing method using a base material, which enabled printing medium- to large-scale food-safe objects and obtaining a smother surface finish (Herpt 2016). Also, multimaterial 3D printing had limitations of improper mixing due to simple and passive diffuse mixing of fluids containing different materials. This problem was overcome by a method that uses a rotating impeller in the nozzle to actively mix the materials just before printing, as shown in Figure 6.8 (Muth et al. 2017). Ceramic jet printing is another new type of 3D printing technology that builds a model from bottom to top, using a printer head that puts an organic binder on a thin ceramic layer that is already spread on the platform. The model is removed after drying and fired to finally gain maximum strength.

One of latest advances in 3D printing process is the use of magnetic field to orient magnetic particles for improved performance of materials in terms of stiffness and strength. The process is termed 3D magnetic printing, which is robust and scalable and has the capability of controlling the orientation of ceramic particles reaching a size of 90 mm (Martin et al. 2015). To apply magnetic field on nonmagnetic particles, they are usually coated with iron oxide nanoparticles with the help of magnetic labeling techniques. A further advancement includes multimaterial dispensers with the additional ability to control composition locally with the help of a component mixing unit. This novel multimaterial-based 3D printing platform has the capability of fabricating heterogeneous materials with desired microstructures that can closely mimic the biological materials (Figure 6.9) (Kokkinis et al. 2015).

6.3.2.4 Templating

Templating methods are used in the last two decades to fabricate nanomaterials. These methods offer the advantage of ease of operation and implementation in combination with controlling the size, structure, and morphology of nanomaterials (Xie et al. 2016). For example, taking inspiration from insect compound eyes, graphene nanosheets in combination with glycerol were used to fabricate the bioinspired counterparts using a simple template-directed assembly (Wang et al. 2015). These

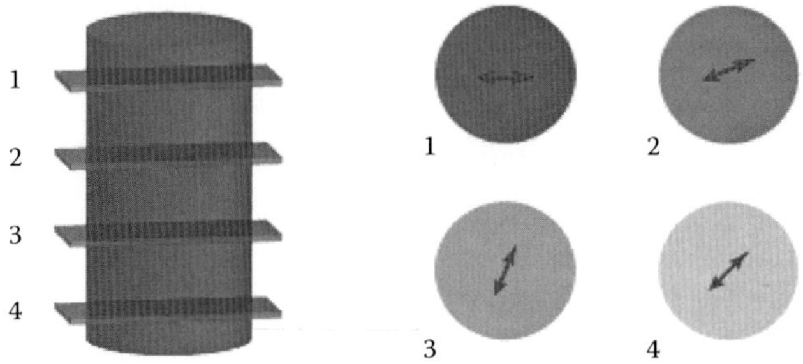

FIGURE 6.9 Design and programming of complex composite with changes in composition (color change) and platelet orientation (arrows) at different levels. (Adapted from Kokkinis, Dimitri, Manuel Schaffner, and André R Studart. 2015. Multimaterial magnetically assisted 3D printing of composite materials. *Nature Communications* 6: 8643.)

bioinspired compound eyes demonstrated the capability of tunable curvature that further enables changing zoom and field of view. A template method based on ice freezing was used to fabricate bioinspired composites that enable detection of microcracks before total failure of the specimen, by increasing the measured voltage, in addition to increased toughness due to crack propagation through the soft graphene layers (Picot et al. 2017). The template method aids in controlling growth direction and the formation of a 3D porous network of long channels with walls made of chemically modified graphene. These porous networks were later infiltrated with polymethyl siloxane and a ceramic material and further heated to have sintered ceramic composite material. An ice templating technique was also used to create porous scaffolds by cooling silicon carbide suspensions to subzero temperatures, which were later infiltrated with poly(methyl methacrylate) to create composites of different architectures (Naglieri et al. 2015). Advances in freezing techniques resulted in bidirectional freezing that enables assembling little building blocks into large aggregate porous structures similar to that of natural nacre (Bai et al. 2015). The novelty of this technique is in controlling growth of ice crystals under dual-temperature control gradients achieved by covering a cold finger with the help of polydimethylsiloxane. Using freeze-drying–based in situ crystallization of chitin/chondroitin sulfate/hydroxyapatite, hybrid nanostructural materials were fabricated for the purpose of bone tissue engineering (Fan et al. 2016). Thus, templating methods provide the advantage of fabricating porous materials and also the directional alignment of constituents to a great extent.

Bio-templating is a type of templating in which the natural material of interest is used for preparing the template; bio-templating is used widely in the fabrication of microscale structured materials. This technique is primarily used to replicate photonic structures of insects like butterfly wings, beetle elytra, and leaves, which enable templating because of their open structural features (Yi et al. 2016). Such a templating procedure was used to fabricate photonic crystals that can be tuned to change color with change in pH (Yang et al. 2013). This was achieved by coating polymethylacrylic acid on the surface of morpho-butterfly wings by surface bonding and polymerization. In a recent study, dried fish skin was used as a template and polydimethylsiloxane was poured to fabricate anisotropic microstructured surfaces that are oleophobic (Cai et al. 2014). Also, the complex venation structural patterns of leaves were used to create templates in an effective and fast way, for use in microfluidic devices (Wu et al. 2016). Using such bio-templating methods, there is a possibility of overcoming the limitations of many manufacturing techniques, specifically in bioinspired fabrication (Chao et al. 2014).

6.3.2.5 Coating and Deposition Techniques

Simple coating and deposition methods can be very effective in fabrication of functionalized surfaces. For example, a nylon fiber is immersed in polymethyl methacrylate solution and drawn using a dip coater to obtain cylindrical film around the fiber, which then forms droplets because of the Rayleigh instability, resulting in a fiber inspired by the design of spider silk, which could effectively be used as a fog collector (Bai et al. 2011). A layer-by-layer deposition technique was used to develop a hybrid film of polyelectrolyte/clay inspired by nacre, which was oleophobic and demonstrated mechanical durability (Xu et al. 2013). Using a simple spin coating technique (Figure 6.10), composite thin films were fabricated comprising alternate layers of graphene oxide and polymethylmethacrylate on a silicon wafer (Beese et al. 2014). A cost-effective and facile method was used to fabricate bioinspired superoleophobic underwater membrane (fish scales) by coating graphene oxide on a commercially stainless steel mesh and making pores in graphene oxide using oxygen plasma (Liu et al. 2015b).

Sequential external deposition was used to fabricate nacre-like material comprising a soft phase (polyelectrolytes) and a stiffer phase (clays), which demonstrated tensile strength similar to that of nacre (Tang et al. 2003). In a similar study, a large-scale nacre-like structure is fabricated by developing an organic layer, using layer-by-layer coating of poly(styrenesulfonate) sodium salt and poly(ethyleneimine), followed by immersion in zirconia sulfate solution for deposition of zirconia (Zlotnikov et al. 2010). Electrodeposition is also used to create rough surfaces on various metals such as steel, aluminum, and magnesium, to achieve superhydrophobic surfaces. For example, multi-functional gold-coated nickel nanocone arrays were fabricated using a simple and cheap electro-deposition method, which were highly stable and superhydrophobic (Mo et al. 2015). Bioinspired composite films were fabricated using electrophoretic deposition of gibbsite nanoplatelets and polyvinyl alcohol that were transparent and flexible (Lin et al. 2010).

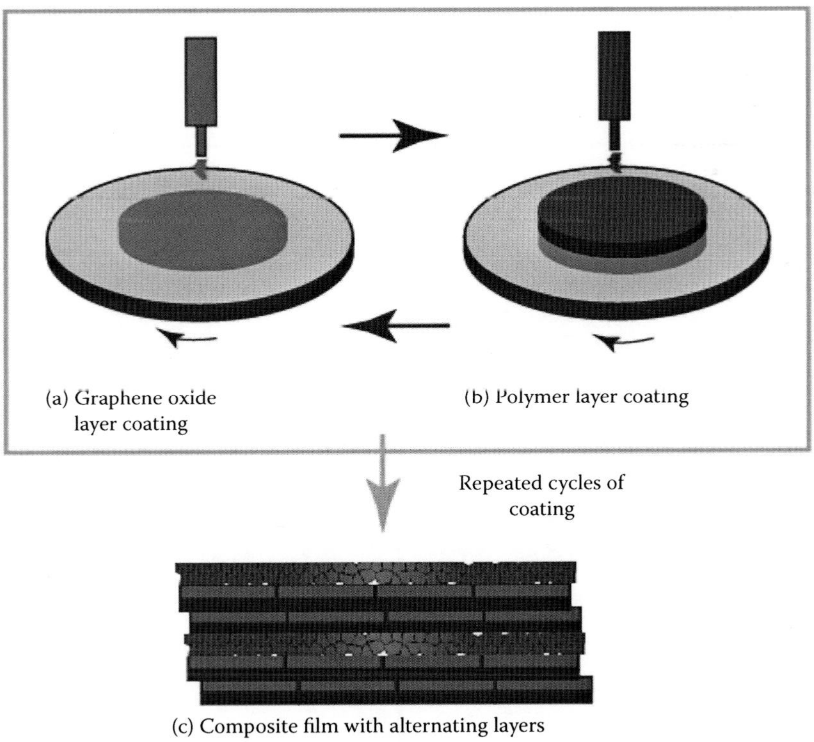

FIGURE 6.10 Schematic showing the spin coating procedure for fabricating composite films. (a) Graphene oxide is spread on the silicon wafer. (b) Polymer is spread on the coated layer of graphene oxide layer. (c) Final composite film comprising alternating layers.

6.3.2.6 Casting and Drawing

Casting process is a manufacturing process in which liquefied metals are poured in a mold to obtain a solid cast. Similarly, films were produced by mixing graphene oxide with polyvinylalcohol as a solution and casting on a polyethylene sheet. The graphene oxide is then reduced using hydroiodic acid to obtain the final nacre-like material (Li et al. 2012). Freeze casting is a variation in which ice formation direction is used to control the solidification direction. As shown in Figure 6.11, the freeze casting technique was used to assemble platelet-like particles to obtain directionally solidified and highly porous material similar to that of nacre (Hunger et al. 2013). Slip casting is a type of casting process that was traditionally used to manufacture ceramic materials. Recently, slip casting process was used to create bioinspired materials made of ceramic, metal, and polymers, in which particles are assembled using a magnetic field to tailor the microstructure (Le Ferrand et al. 2015).

Co-extrusion is a process in which two materials are extruded together through a die to obtain a single material. A unique co-extrusion process was developed that enables multiplication of layers in sequence to fabricate polymers of alternating thin layers up to a thousand in number, with each layer reaching nanoscale thickness (Ponting et al. 2010). Using the above-described technique, a synthetic human eye lens was fabricated with the help of gradation in refractive index of the polymeric thin films (Ji et al. 2012). Extrusion is also used to fabricate a scaffold for bone regeneration by combining hydroxyapatite powders with a binder (Meredith 2014). In a novel attempt, nacre-like fibers were produced using drawing and twisting of two components, resulting in multiscale organization of the fibers and enabling new deformation modalities and improved mechanical performance (Zhang et al. 2016).

6.3.2.7 Sintering and Curing

Sintering is a crucial manufacturing process that is used to fabricate materials in a desired form using ceramics and metals in their powder form. A novel improvised powder metallurgy technique was used to fabricate laminar composites (Kakisawa et al. 2010). The process involved coating glass flake powder with a metal and later sintering the powders after aligning the flakes' orientation (Figure 6.12).

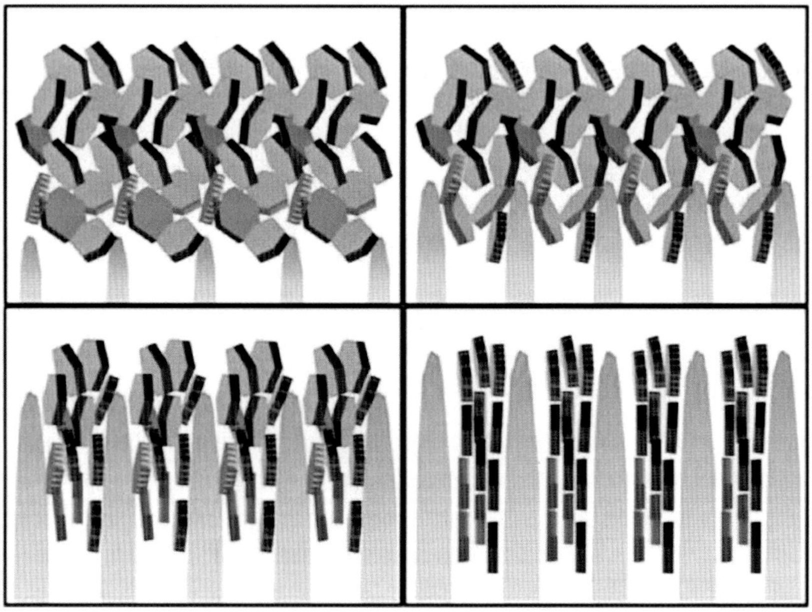

FIGURE 6.11 Schematic of directional freeze casting of ceramic-polymer platelets that assemble in layers during solidification of ice. (Reproduced from Hunger, Philipp M, Amalie E Donius, and Ulrike G K Wegst. 2013. Platelets self-assemble into porous nacre during freeze casting. *Journal of the Mechanical Behavior of Biomedical Materials* 19: 87–93.)

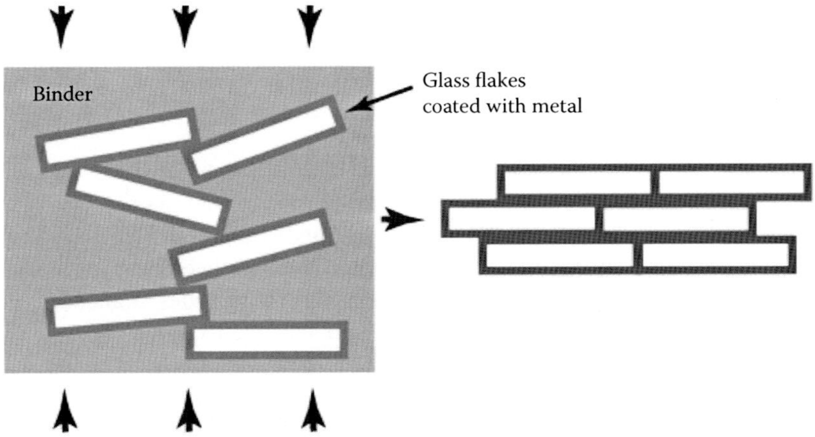

FIGURE 6.12 Schematic of sintering process used in fabrication of laminated composites. (Adapted from Kakisawa, Hideki, Taro Sumitomo, Ryo Inoue, and Yutaka Kagawa. 2010. Fabrication of nature-inspired bulk laminar composites by a powder processing. *Composites Science and Technology* 70 (1): 161–166.)

In another novel study, the commonly available glass fiber–reinforced composite material was cut into the required dimension and stacked to make a bioinspired laminated composite (Cheng et al. 2011). The main pre-impregnated layup with 24 plies was created using the desired sequence of fiber orientation, which is then bagged with film to create vacuum and finally cured at 250°C for 1 h (Figure 6.13). Similarly, a laminated composite was fabricated by using layer-by-layer stacking of yttrium oxide stabilized zirconia sheets with ultraviolet curable adhesive in each step (Tushtev et al. 2014).

The curing process uses a resin to bind the laminates or fibers together to obtain a final composite material. A new class of bioinspired polymers based on functionalized styrene monomers was fabricated by ultraviolet curing of copolymerized vinylbenzyl thymine and vinylbenzyl triethylammonium chloride (Barbarini et al. 2010). Inspired by tree branch joints, aircraft T-joints with increased strain energy until failure were fabricated using carbon epoxy laminates where the skin and stiffener parts were cured together (Burns et al. 2015). Also, novel conductive graphene–epoxy composites were prepared by covering prefabricated graphene film with tin foil, followed by

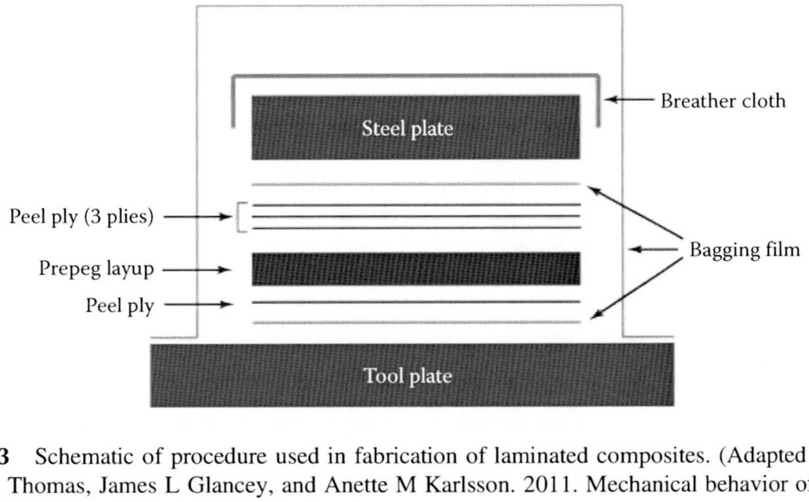

FIGURE 6.13 Schematic of procedure used in fabrication of laminated composites. (Adapted from Cheng, Liang, Adam Thomas, James L Glancey, and Anette M Karlsson. 2011. Mechanical behavior of bio-inspired laminated composites. *Composites Part A: Applied Science and Manufacturing* 42 (2): 211–20.)

autoclaving and then immersing the film into epoxy solution and finally curing them (Ming et al. 2015). Curing processes are a good choice for manufacturing composites on a large scale but has the disadvantage of having low resolution (Libonati and Buehler 2017).

6.3.2.8 Combination of Techniques

It is advantageous to use a combination of techniques to fabricate bioinspired materials, and a few examples are described below. A combination of colloidal crystal self-assembling, lithography, and thermo-pressing techniques was used to fabricate bioinspired compound eye films (Kuo et al. 2015). Superhydrophobic surfaces having high–aspect ratio needles with lateral surface nanogrooves were produced using a simple chemical deposition technique that includes immersing copper foil in ammonia solution (Yao et al. 2010). Using a simple and scalable technique (Figure 6.14), a highly durable superhydrophobic surface was created based on the principles of fingering instabilities that are observed during shear tearing during non-Newtonian flow (Park et al. 2015). Until recently, fabricating close-packed microlenses mimicking insect eyes remained as a challenging task. In a recent study, they were fabricated by making a concave template on silica glass using femtosecond laser and acid etching in the initial step followed by replication on flat plastic films. The films with convex lens on the surface were then subjected to a thermomechanical bending process to obtain a curvilinear-shaped lens (Liu et al. 2012). To mimic the insect body sensory hairs, a simple two-step anodization of an aluminum surface was carried out twice and then followed by cobalt electrodeposition in these pores to obtain pillars (Hein et al. 2013). An armor material was developed using bioinspiration from fish scales, by simple bonding of engraved hard ceramic alumina with the desired shape on a polyurethane surface (Martini and Barthelat 2016).

In a recent study, the fiber deposition technique was used in combination with vacuum-assisted resin injection to fabricate a cortical bone–inspired composite (Libonati et al. 2014). Self-healing carbon fiber–reinforced composites were fabricated by inserting PTFE-coated steel wires, which were later removed to create vascular channels (Figure 6.15). The damage created in the composite is identified by a pressure sensor (Figure 6.15b), followed by a microcontroller that triggers a pump to release the healing resin into these channels, to fill the cracks (Figure 6.15c) (Trask et al. 2014). Thus, it is clear from the above examples that choosing a combination of techniques for the fabrication of bioinspired materials opens up new possibilities.

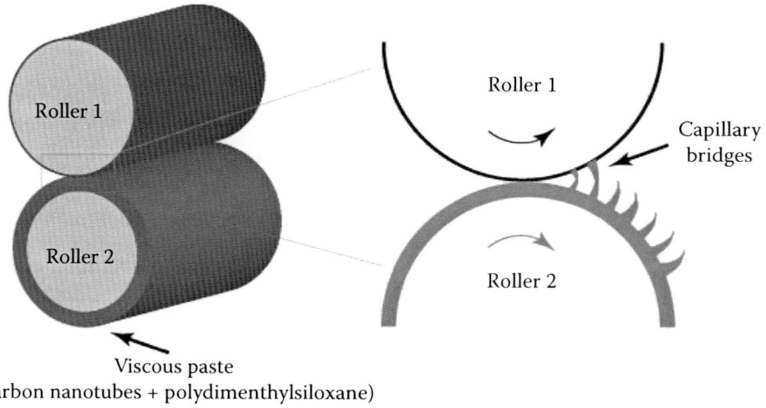

FIGURE 6.14 Schematic showing the roll-to-roll process used to create patterned surface. (Adapted from Park, Sung-Hoon, Sangeui Lee, David Moreira, Prabhakar R Bandaru, Intaek Han, and Dong-Jin Yun. 2015. Bioinspired superhydrophobic surfaces, fabricated through simple and scalable roll-to-roll processing. *Nature Publishing Group*: 1–8.)

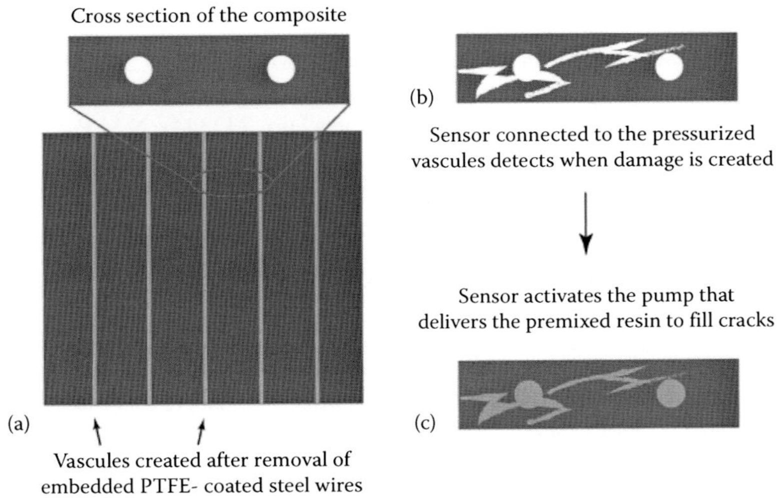

FIGURE 6.15 Schematic of self-healing composites fabrication and working. (a) Carbon fiber–reinforced composites with channels. (b) Cross section showing cracks propagating into the vascules during damage. (c) Resin (blue) is pushed into the cracks through the vascules. (Adapted from Trask, Richard S, Christopher J Norris, and Ian P Bond. 2014. Stimuli-triggered self-healing functionality in advanced fibre-reinforced composites. *Journal of Intelligent Material Systems and Structures* 25 (1): 87–97.)

6.4 CONCLUSION AND FUTURE DIRECTIONS

In this review, we covered a number of processes and methods for fabricating bioinspired materials, although they are not exhaustive. We observe that there is a significant overlap between techniques used in fabrication of bioinspired materials as compared to that of engineering and electronic materials. Thus, major advances in the field of bioinspired material fabrication are expected to occur simultaneously with advances in general manufacturing processes. From the perspective of engineered applications, the bioinspiration route appears to provide immense potential for innovation and efficiency. Further, it is expected to play a significant role in the growth of economies. The key highlights of this review are as follows:

1. The limited numbers of polymers and metals that can be used in most standard fabrication techniques of bioinspired materials pose a challenge. For example, in the fabrication of biomineralized materials, the use of ceramic materials imposes more constraints. As regards sintering requirements, because of the significantly high melting points of ceramics, it becomes extremely difficult to select the mixing materials that can withstand high temperatures in such procedures.
2. Despite significant advances in bioinspired fabrication, we have yet to see its value in terms of commercial production. In most of the studies, large-scale production is not being considered probably because the field is relatively new. The innovative use of multiple fabrication routes in combination could be a possible way to scale the size and speed of fabrication. This is a very crucial aspect from an economic point of view and for the advancement of these fabrication techniques.
3. There are numerous studies that focus on understanding the growth and remodeling aspects of biological materials such as bone and wood. The development of bioinspired materials that mimic such attributes could be a big leap in the field of engineering structural materials.
4. The future of bioinspired materials fabrication will be driven toward the development of multifunctional materials that are similar to biological materials which perform functions such as damage sensing and healing, in combination with suitable mechanical properties.

ACKNOWLEDGMENTS

N.M.P. is supported by the European Research Council (ERC PoC 2015 SILKENE no. 693670) and by the European Commission H2020 under the Graphene Flagship (WP14 "Polymer Composites" no. 696656) and under the FET Proactive ("Neurofibres" no. 732344). N.M.P. is also supported by Fondazione Caritro under "Self-Cleaning Glasses" no. 2016.0278, as L.K.

REFERENCES

Adler-Abramovich, Lihi, Daniel Aronov, Peter Beker, Maya Yevnin, Shiri Stempler, Ludmila Buzhansky, Gil Rosenman, and Ehud Gazit. 2009. Self-assembled arrays of peptide nanotubes by vapour deposition. *Nature Nanotechnology* 4 (12): 849–54.

Aizenberg, Joanna. 2004. Crystallization in patterns: A bio-inspired approach. *Advanced Materials* 16: 1295–302.

Aksay, Ilhan A, Matt Trau, Saravani Manne, Itaru Honma, Nan Yao, L Zhou, Paul Fenter, Peter M Eisenberger, and Sol M Gruner. 1996. Biomimetic pathways for assembling inorganic thin films. *Science* 273 (5277): 892–98.

Altissimo, Matteo. 2010. E-beam lithography for micro-nanofabrication. *Biomicrofluidics* 4 (2): 2–7.

An, Bingbing, Xinluo Zhao, and Dongsheng Zhang. 2014. On the mechanical behavior of bio-inspired materials with non-self-similar hierarchy. *Journal of the Mechanical Behavior of Biomedical Materials* 34: 8–17.

Bai, Hao, Yuan Chen, Benjamin Delattre, Antoni P Tomsia, and Robert O Ritchie. 2015. Bioinspired large-scale aligned porous materials assembled with dual temperature gradients. *Science Advances* 1 (11): e1500849.

Bai, Hao, Jie Ju, Ruize Sun, Yuan Chen, Yongmei Zheng, and Lei Jiang. 2011. Controlled fabrication and water collection ability of bioinspired artificial spider silks. *Advanced Materials* 23 (32): 3708–11.

Bain, Jennifer, and Sarah S Staniland. 2015. Bioinspired nanoreactors for the biomineralisation of metallic-based nanoparticles for nanomedicine. *Physical Chemistry Chemical Physics* 17 (24): 15508–21.

Ballav, Nirmalya, Soeren Schilp, and Michael Zharnikov. 2008. Electron-beam chemical lithography with aliphatic self-assembled monolayers *Angewandte Chemie—International Edition* 47 (8): 1421–24.

Barbarini, Alejandro L, Diana A Estenoz, and Debora M Martino. 2010. Synthesis, characterization and curing of bioinspired polymers based on vinyl benzyl thymine and triethyl ammonium chloride. *Macromolecular Reaction Engineering* 4 (6–7): 453–9.

Beese, Allison M, Zhi An, Sourangsu Sarkar, S Shiva, P Nathamgari, Horacio D Espinosa, and Sonbinh T Nguyen. 2014. Defect-tolerant nanocomposites through bio-inspired stiffness modulation. *Advanced Functional Materials* 24 (19): 2883–91.

Belegratis, Maria R, Volker Schmidt, Dieter Nees, Barbara Stadlober, and Paul Hartmann. 2014. Diatom-inspired templates for 3D Replication: Natural diatoms versus laser written artificial diatoms. *Bioinspiration & Biomimetics* 9: 16004.

Bian, Hao, Yang Wei, Qing Yang, Feng Chen, Fan Zhang, Guangqing Du, Jiale Yong, and Xun Hou. 2016. Direct fabrication of compound-eye microlens array on curved surfaces by a facile femtosecond laser enhanced wet etching process. *Applied Physics Letters* 109 (22): 221109.

Boncheva, Mila, Stefan A Andreev, Lakshminarayan Mahadevan, Adam Winkleman, David R Reichman, Mara G Prentiss, Sue Whitesides, and George M Whitesides. 2005. Magnetic self-assembly of three-dimensional surfaces from planar sheets. *Proceedings of the National Academy of Sciences of the United States of America* 102 (11): 3924–9.

Boncheva, Mila, and George M Whitesides. 2005. Making things by self-assembly. *Materials Research Society Bulletin* 30: 736–42.

Bosia, Federico, Tamer Abdalrahman, and Nicola M Pugno. 2014. Self-healing of hierarchical materials. *Langmuir* 30 (4): 1123–33.

Bosia, Federico, Federico Della Croce, and Nicola M Pugno. 2013. Systematic numerical investigation of the role of hierarchy in heterogeneous bio-inspired materials. *Journal of the Mechanical Behavior of Biomedical Materials* 19: 34–42.

Bowden, Ned, Andreas Terfort, Jeff Carbeck, and George M Whitesides. 1997. Self-assembly of mesoscale objects into ordered two-dimensional arrays. *Science* 276 (5310): 233–5.

Bréchet, Yves J M. 2013. Chapter 1. Architectured materials: An alternative to microstructure control for structural materials design? A possible playground for bio-inspiration? *RSC Smart Materials*, 4: 1–16.

Brimblecombe, Robin, Annette Koo, G Charles Dismukes, Gerhard F Swlegers, and Leone Spiccia. 2010. Solar driven water oxidation by a bioinspired manganese molecular catalyst. *Journal of the American Chemical Society* 132 (9): 2892–4.

Buehler, Markus J. 2013. Materiomics: Multiscale mechanics of biological materials and structures. *Acta Mechanica Solida Sinica* 546: 152.

Buehler, Markus J, and Guy M Genin. 2016. Integrated multiscale biomaterials experiment and modelling: A perspective. *Interface Focus* 6 (1): 20150098.

Burgess, Ian B, Joanna Aizenberg, and Marko Lon. 2013. Creating bio-inspired hierarchical 3D–2D photonic stacks via planar lithography on self-assembled inverse opals. *Bioinspiration & Biomimetics* 8.

Burns, L, Adrian P Mouritz, D Pook, and Stefanie Feih. 2015. Bio-inspired hierarchical design of composite T-joints with improved structural properties. *Composites Part B: Engineering*, 69: 222–31.

Cademartiri, Ludovico, and Kyle J M Bishop. 2014. Programmable self-assembly. *Nature Materials* 14 (1): 2–9.

Cai, Yue, Ling Lin, Zhongxin Xue, Mingjie Liu, Shutao Wang, and Lei Jiang. 2014. Filefish-inspired surface design for anisotropic underwater oleophobicity. *Advanced Functional Materials* 24 (6): 809–16.

Carpi, Federico, Gabriele Frediani, Simona Turco, and Rossi Danilo De. 2011. Bioinspired tunable lens with muscle-like electroactive elastomers. *Advanced Functional Materials* 21 (21): 4152–8.

Chao, Joshua T, Manus J P Biggs, and Abhay S Pandit. 2014. Diatoms: A biotemplating approach to fabricating drug delivery reservoirs. *Expert Opinion on Drug Delivery* 11 (11): 1687–95.

Cheng, Liang, Adam Thomas, James L Glancey, and Anette M Karlsson. 2011. Mechanical behavior of bio-inspired laminated composites. *Composites Part A: Applied Science and Manufacturing* 42 (2): 211–20.

Chia, Helena N, and Benjamin M Wu. 2015. Recent advances in 3D printing of biomaterials. *Journal of Biological Engineering* 9: 4.

Chou, Sy, Pr Krauss, and Pj Renstrom. 1995. Imprint of sub-25 nm vias and trenches in polymers. *Applied Physics Letter* 67: 3114.

Daniel, Marie Christine Mc, and Didier Astruc. 2004. Gold nanoparticles: Assembly, supramolecular chemistry, quantum-size related properties and applications toward biology, catalysis and nanotechnology. *Chemical Reviews* 104: 293–346.

Degtyar, Elena, Matthew J Harrington, Yael Politi, and Peter Fratzl. 2014. The mechanical role of metal ions in biogenic protein-based materials. *Angewandte Chemie—International Edition* 53 (45): 12026–44.

Demirel, Melik C, Murat Cetinkaya, Abdon Pena-Francesch, and Huihun Jung. 2015. Recent advances in nanoscale bioinspired materials. *Macromolecular Bioscience* 15 (3): 300–11.

Ding, Yong, Sheng Xu, Yue Zhang, Aurelia C Wang, Melissa H Wang, Yonghao Xiu, Ching Ping Wong, and Zhong Lin Wang. 2008. Modifying the anti-wetting property of butterfly wings and water strider legs by atomic layer deposition coating: Surface materials versus geometry. *Nanotechnology* 19 (35): 355708.

Dobson, Christopher M. 2002. Protein folding and misfolding. *Nature* 90 (5): 445–53.

Duro-Royo, Jorge, Katia Zolotovsky, Laia Mogas-Soldevila, Swati Varshney, Neri Oxman, Mary C Boyce, and Christine Ortiz. 2015. MetaMesh: A hierarchical computational model for design and fabrication of biomimetic armored surfaces. *CAD Computer Aided Design* 60: 14–27.

England, Grant, Mathias Kolle, Philseok Kim, Mughees Khan, Philip Muñoz, Eric Mazur, and Joanna Aizenberg. 2014. Bioinspired micrograting arrays mimicking the reverse color diffraction elements evolved by the butterfly *Pierella luna*. *Proceedings of the National Academy of Sciences* 111 (44).

Fan, Tiantang, Jingdi Chen, Panpan Pan, Yujue Zhang, Yimin Hu, Xiaocui Liu, Xuetao Shi, and Qiqing Zhang. 2016. Bioinspired double polysaccharides-based nanohybrid scaffold for bone tissue engineering. *Colloids and Surfaces B: Biointerfaces* 147: 217–23.

Fermanian. 2014. Can 3D printing unlock bioinspiration's full potential?

Fermanian Business and Economic Institute. 2013. Bioinspiration: An economic progress report.

Finbloom, Joel A, Kenneth Han, Ioana L Aanei, Emily C Hartman, Daniel T Finley, Michel T Dedeo, Max Fishman, Kenneth H Downing, and Matthew B Francis. 2016. Stable disk assemblies of a tobacco mosaic virus mutant as nanoscale scaffolds for applications in drug delivery. *Bioconjugate Chemistry* 27 (10): 2480–5.

Fisher, Omar Z, Ali Khademhosseini, Robert Langer, and Nicholas A Peppas. 2010. Bioinspired materials for controlling stem cell fate. *Accounts of Chemical Research* 43 (3): 419–28.

Fratzl, Peter. 2007. Biomimetic materials research: What can we really learn from nature's structural materials? *Journal of the Royal Society Interface* 4: 637–42.

Gartner, Zev J, and Carolyn R Bertozzi. 2009. Programmed assembly of 3-dimensional microtissues with defined cellular connectivity. *Proceedings of the National Academy of Sciences of the United States of America* 106 (12): 4606–10.

Gonzalez-Perez, Alfredo, and Kenneth M Persson. 2016. Bioinspired materials for water purification. *Materials* 9 (6).

Gou, Maling, Xin Qu, Wei Zhu, Mingli Xiang, Jun Yang, Kang Zhang, Yuquan Wei, and Shaochen Chen. 2014. Bio-inspired detoxification using 3D-printed hydrogel nanocomposites. *Nature Communications* 5: 3774.

Green, David William, Bessim Ben-Nissan, Kyung-Sik Yoon, Bruce Milthorpe, and H-S Jung. 2016. Bioinspired materials for regenerative medicine: Going beyond the human archetypes. *Journal of Materials Chemistry B* 4 (14): 2396–406.

Griffith, Linda G, and Gail Naughton. 2002. Tissue engineering—Current challenges and expanding opportunities. *Science* 295 (5557): 1009–14.

Grinthal, Alison, Sung H Kang, Alexander K Epstein, Michael Aizenberg, Mughees Khan, and Joanna Aizenberg. 2012. Steering nanofibers: An integrative approach to bio-inspired fiber fabrication and assembly. *Nano Today* 7 (1): 35–52.

Han, Zhiwu, Zhengzhi Mu, Bo Li, Ze Wang, Junqiu Zhang, Shichao Niu, and Luquan Ren. 2016. Active antifogging property of monolayer SiO_2 film with bioinspired multiscale hierarchical pagoda structures. *ACS Nano* 10 (9): 8591–8602.

Harada, Akira, Ryosuke Kobayashi, Yoshinori Takashima, Akihito Hashidzume, and Hiroyasu Yamaguchi. 2011. Macroscopic self-assembly through molecular recognition. *Nature Chemistry* 3 (1): 34–7.

Hein, Matthew A, Mazin M Maqableh, Michael J Delahunt, Mark Tondra, Alison B Flatau, Carol K Shield, and Bethanie J H Stadler. 2013. Fabrication of bioinspired inorganic nanocilia sensors. *IEEE Transactions on Magnetics* 49 (1): 191–6.

Holm, Richard H, Pierre Kennepohl, and Edward I Solomon. 1996. Structural and functional aspects of metal sites in biology. *Chemical Reviews* 96 (7): 2239–314.

Hou, Yanshan, Peng Xiao, Jiawei Zhang, Meiwen Peng, Wei Lu, Youju Huang, Chunfa Ouyang, and Tao Chen. 2015. Fabricating a morphology tunable patterned bio-inspired polydopamine film directly via microcontact printing. *RSC Advances* 5 (75): 60990–2.

Hou, Yidong, Billie L Abrams, Peter C K Vesborg, Mårten E Björketun, Konrad Herbst, Lone Bech, Alessandro M Setti et al. 2011. Bioinspired molecular co-catalysts bonded to a silicon photocathode for solar hydrogen evolution. *Nature Materials* 10 (6): 434–8.

Hu, Wenchuang, Koshala Sarveswaran, Marya Lieberman, and Gary H Bernstein. 2004. Sub-10 nm electron beam lithography using cold development of poly(methylmethacrylate). *Journal of Vacuum Science & Technology B: Microelectronics and Nanometer Structures* 22 (4): 1711–6.

Hu, Yaowei, Zuobin Wang, Zhankun Weng, Miao Yu, and Dapeng Wang. 2016. Bio-inspired hierarchical patterning of silicon by laser interference lithography. *Applied Optics* 55 (12): 3226–32.

Hunger, Philipp M, Amalie E Donius, and Ulrike G K Wegst. 2013. Platelets self-assemble into porous nacre during freeze casting. *Journal of the Mechanical Behavior of Biomedical Materials* 19: 87–93.

Ito, Takashi, and Shinji Okazaki. 2000. Pushing the limits of lithography. *Nature* 406 (6799): 1027–31.

Ji, Shanzuo, Michael Ponting, Richard S Lepkowicz, Armand Rosenberg, Richard Flynn, Guy Beadie, and Eric Baer. 2012. A bio-inspired polymeric gradient refractive index (GRIN) human eye lens. *Optics Express* 20 (24): 26746.

Kaissling, Ke. 1996. Physiology of pheromone reception in insects (an example of moths). *Chemical Senses* 21 (2): 257–68.

Kakisawa, Hideki, Taro Sumitomo, Ryo Inoue, and Yutaka Kagawa. 2010. Fabrication of nature-inspired bulk laminar composites by a powder processing. *Composites Science and Technology* 70 (1): 161–6.

Kakunuri, Manohar, Mudrika Khandelwal, Chandra S Sharma, and Stephen J Eichhorn. 2017. Fabrication of bio-inspired hydrophobic self-assembled electrospun nanofiber based hierarchical structures. *Materials Letters*, 12–15.

Kholkin, Andrei, Nadav Amdursky, Igor Bdikin, Ehud Gazit, and Gil Rosenman. 2010. Strong piezoelectricity in bioinspired peptide nanotubes. *ACS Nano* 4 (2): 610–4.

Kim, Sangbae, Cecilia Laschi, and Barry Trimmer. 2013. Soft robotics: A bioinspired evolution in robotics. *Trends in Biotechnology* 31 (5), 287–94.

Kirchner, Robert, Vitaliy A Guzenko, Michael Rohn, E Sonntag, Michael Mühlberger, I Bergmair, and Helmut Schift. 2015. Bio-inspired 3D funnel structures made by grayscale electron-beam patterning and selective topography equilibration. *Microelectronic Engineering* 141: 107–11.

Klok, Harm-Anton, and Sebastian Lecommandoux. 2001. Supramolecular materials via block copolymer self-assembly. *Advanced Materials* 13 (16): 1217–29.

Kokkinis, Dimitri, Manuel Schaffner, and André R Studart. 2015. Multimaterial magnetically assisted 3D printing of composite materials. *Nature Communications* 6: 8643.

Kolesky, David B, Ryan L Truby, A Sydney Gladman, Travis A Busbee, Kimberly A Homan, and Jennifer A Lewis. 2014. 3D bioprinting of vascularized, heterogeneous cell-laden tissue constructs. *Advanced Materials* 26 (19): 3124–30.

Kolle, Mathias, Alfred Lethbridge, Moritz Kreysing, Jeremy J. Baumberg, Joanna Aizenberg, and Peter Vukusic. 2013. Bio-inspired band-gap tunable elastic optical multilayer fibers. *Advanced Materials* 25 (15): 2239–45.

Kuo, Wen-kai, Guan-fu Kuo, Syuan-you Lin, and Hsin Her Yu. 2015. Fabrication and characterization of artificial miniaturized insect compound eyes for imaging. *Bioinspiration & Biomimetics* 10.

Lakhtakia, Akhlesh. 2015. From bioinspired multifunctionality to mimumes. *Bioinspired, Biomimetic and Nanobiomaterials* 6: 1–11.

Le Ferrand, Hortense, Florian Bouville, Tobias P Niebel, and André R Studart. 2015. Magnetically assisted slip casting of bioinspired heterogeneous composites. *Nature Materials* 14 (September): 1–17.

Lee, Yong Tak, Young Min Song, Sung Jun Jang, and Jae Su Yu. 2010. Bioinspired parabola subwavelength structures for improved broadband antireflection. *Small* 6 (9): 984–7.

Lepore, Emiliano, Mauro Giorcelli, Chiara Saggese, Alberto Tagliaferro, and Nicola Pugno. 2013. Mimicking water striders' legs superhydrophobicity and buoyancy with cabbage leaves and nanotube carpets. *Journal of Materials Research* 28 (7): 976–83.

Li, Yuan Qing, Ting Yu, Tian Yi Yang, Lian Xi Zheng, and Kin Liao. 2012. Bio-inspired nacre-like composite films based on graphene with superior mechanical, electrical, and biocompatible properties. *Advanced Materials* 24 (25): 3426–31.

Libonati, Flavia, Chiara Colombo, and Laura Vergani. 2014. Design and characterization of a biomimetic composite inspired to human bone. *Fatigue and Fracture of Engineering Materials and Structures* 37 (7): 772–81.

Libonati, Flavia, and Markus J Buehler. 2017. Advanced structural materials by bioinspiration. *Advanced Engineering Materials* 19(5): 1600787.

Lin, Shangchao, Seunghwa Ryu, Olena Tokareva, Greta Gronau, Matthew M Jacobsen, Wenwen Huang, Daniel J Rizzo et al. 2015. Predictive modelling-based design and experiments for synthesis and spinning of bioinspired silk fibres. *Nature Communications* 6 (May): 6892.

Lin, Tzung Hua, Wei Han Huang, In Kook Jun, and Peng Jiang. 2010. Bioinspired assembly of surface-roughened nanoplatelets. *Journal of Colloid and Interface Science* 344 (2), 272–8.

Liu, Changchun. 2010. Rapid fabrication of microfluidic chip with three-dimensional structures using natural lotus leaf template. *Microfluidics and Nanofluidics* 9 (4–5): 923–31.

Liu, Hewei, Feng Chen, Qing Yang, Pubo Qu, Shengguan He, Xianhua Wang, Jinhai Si, and Xun Hou. 2012. Fabrication of bioinspired omnidirectional and gapless microlens array for wide field-of-view detections. *Applied Physics Letters* 100 (13): 13–6.

Liu, Qingtao, Jinfeng Wang, and Ben J Boyd. 2015a. Peptide-based biosensors. *Talanta* 136: 114–27.

Liu, Yu Qing, Yong Lai Zhang, Xiu Yan Fu, and Hong Bo Sun. 2015b. Bioinspired Underwater superoleophobic membrane based on a graphene oxide coated wire mesh for efficient oil/water separation. *ACS Applied Materials and Interfaces* 7 (37): 20930–6.

Lv, Jianyong, Yanlin Song, Lei Jiang, and Jianjun Wang. 2014. Bio-inspired strategies for anti-icing. *ACS Nano* 8 (4): 3152–69.

Malik, Idris A, Mohammad Mirkhalaf, and Francois Barthelat. 2017. Bio-inspired 'jigsaw'-like interlocking sutures: Modeling, optimization, 3D printing and testing. *Journal of the Mechanics and Physics of Solids*, 389–99.

Malinauskas, Mangirdas, Albertas Žukauskas, Satoshi Hasegawa, Yoshio Hayasaki, Vygantas Mizeikis, Ričardas Buividas, and Saulius Juodkazis. 2016. Ultrafast laser processing of materials: From science to industry. *Light: Science & Applications* 5 (8): e16133.

Manyika, James, Michael Chui, Jacques Bughin, Richard Dobbs, Peter Bisson, and Alex Marrs. 2013. Disruptive technologies: Advances that will transform life, business, and the global economy. *McKinsey Global Insitute*, no. May: 163. http://www.mckinsey.com/insights/business_technology/disruptive_technologies%5Cnhttp://www.chrysalixevc.com/pdfs/mckinsey_may2013.pdf.

Martin, Joshua J, Brad E Fiore, and Randall M Erb. 2015. Designing bioinspired composite reinforcement architectures via 3D magnetic printing. *Nature Communications* 6: 8641.

Martini, Roberto, and Francois Barthelat. 2016. Stretch-and-release fabrication, testing and optimization of a flexible ceramic armor inspired from fish scales. *Bioinspiration & Biomimetics* 11: 066001.

McConney, Michael E, Kyle D Anderson, Lawrence L Brott, Rajesh R Naik, and Vladimir V Tsukruk. 2009. Bioinspired material approaches to sensing. *Advanced Functional Materials* 19 (16): 2527–44.

Mendes, Ana C, Erkan T Baran, Rui L Reis, and Helena S Azevedo. 2013. Self-assembly in nature: Using the principles of nature to create complex nanobiomaterials. *Wiley Interdisciplinary Reviews: Nanomedicine and Nanobiotechnology* 5 (6): 582–612.

Meredith, James. 2014. High-strength scaffolds for bone regeneration. *Bioinspired, Biomimetic and Nanobiomaterials* 4(1): 48–58.

Miessner, Merle, Martin G Peter, and Julian F Vincent. 2001. Preparation of insect-cuticle-like biomimetic materials. *Biomacromolecules* 2 (2): 369–72.

Ming, Peng, Yuanyuan Zhang, Jianwen Bao, Gang Liu, Zhou Li, Lei Jiang, and Qunfeng Cheng. 2015. Bioinspired highly electrically conductive graphene-epoxy layered composites bioinspired highly electrically conductive graphene–epoxy layered composites. *RSC Advances* 5 (March): 22283–8.

Mishra, Richa. 2015. Direct laser writing as an alternative to photolithography for precise fabrication of microfluidic components for drug delivery system. *19th International Conference on Miniaturized Systems for Chemistry and Life Sciences*, Gyeongju, Korea. 2105–6.

Mo, Xiu, Yunwen Wu, Junhong Zhang, Tao Hang, and Ming Li. 2015. Bioinspired multifunctional Au nanostructures with switchable adhesion. *Langmuir* 31 (39): 10850–8.

Muth, Joseph T, Patrick G Dixon, Logan Woish, Lorna J Gibson, and Jennifer A Lewis. 2017. Architected cellular ceramics with tailored stiffness via direct foam writing. *Proceedings of the National Academy of Sciences*, 201616769.

Naglieri, Valentina, Bernd Gludovatz, Antoni P Tomsia, and Robert O Ritchie. 2015. Developing strength and toughness in bio-inspired silicon carbide hybrid materials containing a compliant phase. *Acta Materialia* 98: 141–51.

Naleway, Steven E, Michael M Porter, Joanna McKittrick, and Marc A Meyers. 2015. Structural design elements in biological materials: Application to bioinspiration. *Advanced Materials* 27 (37): 5455–76.

Nikolov, Svetoslav, Michal Petrov, Liverios Lymperakis, Martin Friák, Christoph Sachs, Helge Otto Fabritius, Dierk Raabe, and Jörg Neugebauer. 2010. Revealing the design principles of high-performance biological composites using ab initio and multiscale simulations: The example of lobster cuticle. *Advanced Materials* 22 (4): 519–26.

Nir, Sivan, and Meital Reches. 2016. Bio-inspired antifouling approaches: The quest towards non-toxic and non-biocidal materials. *Current Opinion in Biotechnology* 39: 48–55.

Nova, Andrea, Sinan Keten, Nicola M Pugno, Alberto Redaelli, and Markus J Buehler. 2010. Molecular and nanostructural mechanisms of deformation, strength and toughness of spider silk fibrils. *Nano Letters* 10 (7): 2626–34.

Pal, Ramendra K, Nicholas E Kurland, Chenyang Jiang, Subhas C Kundu, Ning Zhang, and Vamsi K Yadavalli. 2016. Fabrication of precise shape-defined particles of silk proteins using photolithography. *European Polymer Journal* 85, 421–30.

Park, Sung-Hoon, Sangeui Lee, David Moreira, Prabhakar R Bandaru, Intaek Han, and Dong-Jin Yun. 2015. Bioinspired superhydrophobic surfaces, fabricated through simple and scalable roll-to-roll processing. *Nature Publishing Group* 1–8.

Perl, András, David N. Reinhoudt, and Jurriaan Huskens. 2009. Microcontact printing: Limitations and achievements. *Advanced Materials* 21 (22): 2257–68.

Picot, Olivier T, Victoria G Rocha, Claudio Ferraro, Na Ni, Eleonora D'Elia, Sylvain Meille, Jerome Chevalier et al. 2017. Using graphene networks to build bioinspired self-monitoring ceramics. *Nature Communications* 8: 14425.

Ponting, Michael, Anne Hiltner, and Eric Baer. 2010. Polymer nanostructures by forced assembly: Process, structure, and properties. *Macromolecular Symposia* 294 (1): 19–32.

Priemel, Tobias, Elena Degtyar, Mason N Dean, and Matthew James Harrington. 2017. Rapid self-assembly of complex biomolecular architectures during mussel byssus biofabrication. *Nature Communications* 8 (March): 1–12.

Pudas, Marko, Stephane Viollet, Franck Ruffier, Arvi Kruusing, Stephane Amic, Seppo Leppavuori, and Nicolas Franceschini. 2007. A miniature bio-inspired optic flow sensor based on low temperature co-fired ceramics (LTCC) technology. *Sensors and Actuators, A: Physical* 133 (1): 88–95.

Qin, Zhao, Leon Dimas, David Adler, Graham Bratzel, and Markus J Buehler. 2014. Biological materials by design. *Journal of Physics. Condensed Matter: An Institute of Physics Journal* 26: 73101.

Quinlan, Roy A, Elizabeth H Bromley, and Ehmke Pohl. 2015. A silk purse from a sow's ear—Bioinspired materials based on α-helical coiled coils. *Current Opinion in Cell Biology* 32 (January 2015): 131–7.

Ribbans, Brian, Yujie Li, and Ting Tan. 2016. A bioinspired study on the interlaminar shear resistance of helicoidal fiber structures. *Journal of the Mechanical Behavior of Biomedical Materials* 56: 57–67.

Richards, Daniel, and Martyn Amos. 2014. Designing with gradients: Bio-inspired computation for digital fabrication. *Proceedings of the 34th Annual Conference of the Association for Computer Aided Design in Architecture*. ACADIA/Riverside Architectural Press, pp. 101–11.

Roche, Ellen T, Robert Wohlfarth, Johannes T B Overvelde, Nikolay V Vasilyev, Frank A Pigula, David J Mooney, Katia Bertoldi, and Conor J Walsh. 2014. A bioinspired soft actuated material. *Advanced Materials* 26 (8): 1200–6.

Rodrigue, Hugo, Binayak Bhandari, Wei Wang, and Sung Hoon Ahn. 2015. 3D soft lithography: A fabrication process for thermocurable polymers. *Journal of Materials Processing Technology* 217: 302–9.

Roychaudhuri, Robin, Mingfeng Yang, Minako M Hoshi, and David B Teplow. 2009. Amyloid β-protein assembly and Alzheimer disease. *Journal of Biological Chemistry* 284 (8): 4749–53.

Rutz, Alexandra L, Kelly E Hyland, Adam E Jakus, Wesley R Burghardt, and Ramille N Shah. 2015. A multimaterial bioink method for 3D printing tunable, cell-compatible hydrogels. *Advanced Materials* 27 (9): 1607–14.

Sharma, Jaswinder, Rahul Chhabra, Anchi Cheng, Jonathan Brownell, Yan Liu, and Hao Yan. 2009. Control of self-assembly of DNA tubules through integration of gold nanoparticles. *Science* 323: 112–6.

Singh, Ajay V, Anisur Rahman, N V G Sudhir Kumar, A S Aditi, Massimilano Galluzzi, Simone Bovio, Sara Barozzi, Erica Montani, and Dario Parazzoli. 2012. Bio-inspired approaches to design smart fabrics. *Materials and Design* 36: 829–39.

Sottos, Nancy R, and Chuck Extrand. 2008. Bioinspired materials for self-cleaning and self-healing. *MRS Bulletin* 33: 732–41.

Spitzer, Denis, Thomas Cottineau, Nelly Piazzon, Sébastien Josset, Fabien Schnell, Sergey Nikolayevich Pronkin, Elena Romanovna Savinova, and Valérie Keller. 2012. Bio-inspired nanostructured sensor for the detection of ultralow concentrations of explosives. *Angewandte Chemie—International Edition* 51 (22): 5334–8.

Studart, André R, and Randall M Erb. 2014. Bioinspired materials that self-shape through programmed microstructures. *Soft Matter* 10 (9): 1284–94.

Stupp, Samuel I 1997. Supramolecular materials: Self-organized nanostructures. *Science* 276 (5311): 384–9.

Suci, Peter A, Sebyung Kang, Mark Young, and Trevor Douglas. 2009. A streptavidin-protein cage Janus particle for polarized targeting and modular functionalization. *Journal of the American Chemical Society* 131 (26): 9164–5.

Tang, Zhiyong, Nicholas A Kotov, Sergei Magonov, and Birol Ozturk. 2003. Nanostructured artificial nacre. *Nature Materials* 2 (6): 413–8.

Tata, Uday, S Deshmukh, J C Chiao, Ronald Carter, and H Huang. 2009. Bio-inspired sensor skins for structural health monitoring. *Smart Materials and Structures* 18: 104026.

Taylor, David. 2010. Some of nature's little tricks. *Materials Today* 13 (5): 6–7.

Trask, Richard S, Christopher J Norris, and Ian P Bond. 2014. Stimuli-triggered self-healing functionality in advanced fibre-reinforced composites. *Journal of Intelligent Material Systems and Structures* 25 (1): 87–97.

Tricinci, Omar, Tercio Terencio, Barbara Mazzolai, Nicola M Pugno, Francesco Greco, and Virgilio Mattoli. 2015. 3D micropatterned surface inspired by *Salvinia molesta* via direct laser lithography. *ACS Applied Materials and Interfaces* 7: 25560–7.

Tseng, Peter, Bradley Napier, Siwei Zhao, Alexander N Mitropoulos, Matthew B Applegate, Benedetto Marelli, David L Kaplan, and Fiorenzo G Omenetto. 2017. Directed assembly of bio-inspired hierarchical materials with controlled nano fibrillar architectures. *Nature Nanotechnology*, 114(3): 1–8.

Tushtev, Kamen, Michael Gonsior, Michael Murck, Georg Grathwohl, and Kurosch Rezwan. 2014. A novel bioinspired multilayered polymer-ceramic composite with outstanding crack resistance. *Advanced Engineering Materials* 16 (2): 156–60.

van Herpt, Oliver. 2016. Functional 3D printed ceramics. http://oliviervanherpt.com/functional-3d-printed-ceramics/.

Vendruscolo, Michele, Jesus Zurdo, Cait E MacPhee, and Christopher M Dobson. 2003. Protein folding and misfolding: A paradigm of self-assembly and regulation in complex biological systems. *Philosophical Transactions of the Royal Society A: Mathematical, Physical and Engineering Sciences* 361 (1807): 1205–22.

Voskuhl, Jens, Jenny Brinkmann, and Pascal Jonkheijm. 2014. Advances in contact printing technologies of carbohydrate, peptide and protein arrays. *Current Opinion in Chemical Biology* 18 (1): 1–7.

Wagner, Christian, and Noreen Harned. 2010. EUV lithography: Lithography gets extreme. *Nature Photonics* 4 (1): 24–6.

Wang, Baochun, Alejandro J Benitez, Francisco Lossada, Remi Merindol, and Andreas Walther. 2016. Bioinspired mechanical gradients in cellulose nanofibril/polymer nanopapers. *Angewandte Chemie—International Edition* 55 (20): 5966–70.

Wang, Lanlan, Fang Li, Hongzhong Liu, Weitao Jiang, Dong Niu, Rui Li, Lei Yin, Yongsheng Shi, and Bangdao Chen. 2015. Graphene-based bioinspired compound eyes for programmable focusing and remote actuation. *ACS Applied Materials and Interfaces* 7 (38): 21416–22.

Watson, Gregory S, David W Green, Lin Schwarzkopf, Xin Li, Bronwen W. Cribb, Sverre Myhra, and Jolanta A Watson. 2015. A gecko skin micro/nano structure—A low adhesion, superhydrophobic, anti-wetting, self-cleaning, biocompatible, antibacterial surface. *Acta Biomaterialia* 21: 109–22.

Wegst, Ulrike G K, Hao Bai, Eduardo Siaz, Antoni P Tomsia, and Robert O Ritchie. 2014. Bioinspired structural materials. *Science* 14: 1053–4.

Weiner, Stephen, and Lia Addadi. 1997. Design strategies in mineralized biological materials. *Journal of Materials Chemistry* 7 (5): 689–702.

Whitesides, George M. 2002. Self-assembly at all scales. *Science* 295 (5564): 2418–21.

Whitesides, George M, John P Mathias, and Chrisopher T Seto. 1991. Molecular self-assembly and nanochemistry: A chemical strategy for synthesis of nanostructures. *Science* 254.

Wilbrink, David V, Marcel Utz, Robert O Ritchie, and Matthew R Begley. 2010. Scaling of strength and ductility in bioinspired brick and mortar composites. *Applied Physics Letters* 97 (19): 30–2.

Wu, Wenming, Rosanne M Guijt, Yuliya E Silina, Marcus Koch, and Andreas Manz. 2016. Plant leaves as templates for soft lithography. *RSC Advances* 6 (27): 22469–75.

Xia, Younan, and George M Whitesides. 1998. Soft lithography. *Annual Review of Materials Science*, 12.

Xie, Yadian, Duygu Kocaefe, Chunying Chen, and Yasar Kocaefe. 2016. Review of research on template methods in preparation of nanomaterials. *Journal of Nanomaterials*: 2302595.

Xu, Li Ping, Jitao Peng, Yibiao Liu, Yongqiang Wen, Xueji Zhang, Lei Jiang, and Shutao Wang. 2013. Nacre-inspired design of mechanical stable coating with underwater superoleophobicity. *ACS Nano* 7 (6): 5077–83.

Yamada, Yasuyuki, Takayuki Kubota, Motoki Nishio, and Kentaro Tanaka. 2014. Sequential and spatial organization of metal complexes inside a peptide duplex. *Journal of the American Chemical Society* 136 (17): 6505–9.

Yang, Hao Cheng, Jianquan Luo, Yan Lv, Ping Shen, and Zhi Kang Xu. 2015. Surface engineering of polymer membranes via mussel inspired chemistry. *Journal of Membrane Science* 483 (March): 42–59.

Yang, Qingqing, Shenmin Zhu, Wenhong Peng, Chao Yin, Wanlin Wang, Jiajun Gu, and Wang Zhang. 2013. Bioinspired fabrication of hierarchically structured, ph-tunable photonic crystals with unique transition. *ACS Nano* 7(6): 4911–8.

Yang, Sung Ho, and Insung S Choi. 2009. Bio-inspired silicification on patterned surfaces generated by microcontact printing and layer-by-layer self-assembly. *Chemistry—An Asian Journal* 4 (3): 382–5.

Yao, Haimin, Zhaoqian Xie, Chong He, and Ming Dao. 2015. Fracture mode control: A bio-inspired strategy to combat catastrophic damage. *Scientific Reports* 5: 8011.

Yao, Xi, Qinwen Chen, Liang Xu, Qikai Li, Yanlin Song, Xuefeng Gao, David Quéré, and Lei Jiang. 2010. Bioinspired ribbed nanoneedles with robust superhydrophobicity. *Advanced Functional Materials* 20 (4): 656–62.

Yi, Hoon, Insol Hwang, Jeong Hyeon Lee, Dael Lee, Haneol Lim, Dongha Tahk, Minho Sung et al. 2014. Continuous and scalable fabrication of bioinspired dry adhesives via a roll-to-roll process with modulated ultraviolet-curable resin. *ACS Applied Materials & Interfaces* 6 (16): 14590–9.

Yi, Wu, Ding-Bang Xiong, and Di Zhang. 2016. Biomimetic and bioinspired photonic structures. *Nano Advances* 1 (2): 62–70.

Yoo, Jin-Wook, Darrell J Irvine, Dennis E Discher, and Samir Mitragotri. 2011. Bio-inspired, bioengineered and biomimetic drug delivery carriers. *Nature Reviews. Drug Discovery* 10 (7): 521–35.

Yu, Xiaoqing, Zhenping Wang, Zhiqiang Su, and Gang Wei. 2017. Design, fabrication, and biomedical applications of bioinspired peptide–inorganic nanomaterial hybrids. *Journal of Materials Chemistry B* 5: 1130–42.

Zarzar, Lauren D, Philseok Kim, and Joanna Aizenberg. 2011. Bio-inspired design of submerged hydrogel-actuated polymer microstructures operating in response to pH. *Advanced Materials* 23 (12): 1–5.

Zeilinger, Angelika, Rostislav Daniel, Mario Stefenelli, Bernhard Sartory, Livia Chitu, Manfred Burghammer, Thomas Schöberl, Otmar Kolednik, Jozef Keckes, and Christain Mitterer. 2015. Mechanical property enhancement in laminates through control of morphology and crystal orientation. *Journal of Physics D: Applied Physics* 48 (29): 295303.

Zhang, Fengxiang, and Hong Yee Low. 2006. Ordered three-dimensional hierarchical nanostructures by nanoimprint lithography. *Nanotechnology* 17 (8): 1884–90.

Zhang, Jia, Wenchun Feng, Huangxi Zhang, Zhenlong Wang, Heather A Calcaterra, Bongjun Yeom, Ping An Hu, and Nicholas A Kotov. 2016. Multiscale deformations lead to high toughness and circularly polarized emission in helical nacre-like fibres. *Nature Communications* 7: 10701.

Zhao, Jing, Xueting Zhao, Zhongyi Jiang, Zhen Li, Xiaochen Fan, Junao Zhu, Hong Wu et al. 2014. Biomimetic and bioinspired membranes: Preparation and application. *Progress in Polymer Science* 39 (9): 1668–720.

Zlotnikov, Igor, Irena Gotman, Zaklina Burghard, Joachim Bill, and Elazar Y Gutmanas. 2010. Synthesis and mechanical behavior of bioinspired ZrO_2-organic nacre-like laminar nanocomposites. *Colloids and Surfaces A: Physicochemical and Engineering Aspects* 361 (1–3): 138–42.

7 Energy Value Stream Mapping
A Tool to Develop Green Manufacturing

Neha Verma and Vinay Sharma

CONTENTS

7.1 INTRODUCTION

The manufacturing industry is also the world's premier energy consumer as well as a pattern/aspect of prosperity. With increasing concerns for the environment, countries are being coaxed to make their industries energy efficient. This has led to research that attempts to analyze means and methods to develop energy-efficient machines or reduce energy consumption using existing methods. Improvement in existing setups can be made by eliminating the processes that consume energy and by replacing them with energy-efficient and less costly methods. Recently, research has focused on achieving Lean and green manufacturing, which minimize waste. Lean techniques are focused on reducing lead time and eliminating wastes in all kinds and forms. Green manufacturing is a method for manufacturing that minimizes waste and pollution. Its emphasis is on reducing parts, rationalizing materials, and reusing components, to help make products that are more efficient to build. Green manufacturing involves not just the use of environmental design of products or the use of environmentally friendly raw materials, but also eco-friendly packing, distribution, and destruction or reuse after the lifetime of the product. In the present work, we have focused on using value stream mapping (VSM) as a tool of Lean manufacturing to reduce energy and make the process energy efficient.

7.1.1 Lean Manufacturing

Lean manufacturing is a method for the elimination of waste within a manufacturing process. It basically focused on attaining the right things in the right place at the right time and in the right quantity. Leading manufacturing companies throughout the world are applying Lean manufacturing techniques to save on costs and to minimize waste. The concept of Lean manufacturing originated from the Toyota Production System (TPS). Taiichi Ohno, one of TPS's architects, described its essence as "All we are doing is looking at the time line from the moment the customer gives us an order to the moment we collect the cash. And we are reducing that time line by removing non value-added wastes." Hence, the concept basically focuses on keeping the time interval between taking the order and collecting cash as short as possible by focusing only on the value-added times by eliminating the non–value-adding ones. Seven types of wastes were identified by Ohno:

- Waste of overproduction—Overproduction generates lots of costs and wastage. It also requires time and energy.
- Waste of time on hand (waiting)—Energy usage in machines in standby mode or ready-to-operate mode when they are waiting for the work piece is also a big waste.
- Waste in transportation—Transportation does not add value to a product, but it is a necessary input of energy.
- Waste of processing itself—Designs, manufacturing technologies, and process flows not suited for their intended purpose as well as tools of poor quality entail an energy input that is needlessly high.
- Wastes of stock in hand (inventory)—Unnecessary stocks add to transport cost and storage cost and thus prove to be a waste of energy and costs.
- Waste of movement—Any kind of movement of machinery or product requires a high amount of energy, and if the movement does not add value to the product, it proves to be a waste.
- Waste of making defective products—The energy spent on the product which is rejected also becomes completely vain with the rejected product because in the manufacturing of a new product, it is necessary to give time and equal energy again. Reworking results in wastage of time as well, although it may save the cost of materials.

Thus, Lean manufacturing tries to focus only on the processes that add value to the end product and in the end reduce lead time of the product. The reduction of lead time also leads to saving energy as energy is a function of time and electrical power.

Lean manufacturing can be achieved with the help of a number of tools, including the following:

- Cellular manufacturing: Organizes the entire process for a particular product or similar products into a group (or "cell"), including all the necessary machines, equipment, and operators. Resources within cells are arranged to easily facilitate all operations.
- Just-in-time (JIT): A system where a customer initiates demand, and the demand is then transmitted backward from the final assembly all the way to raw material, thus "pulling" all requirements just when they are required.
- Kanban: A signaling system for implementing JIT production.
- Total preventive maintenance (TPM): Workers carry out regular equipment maintenance to detect any anomalies. The focus is changed from fixing breakdowns to preventing them. Since operators are the closest to the machines, they are included in maintenance and monitoring activities in order to prevent and provide warning of malfunctions.
- Setup time reduction: Continuously try to reduce the setup time on a machine.

- Total quality management (TQM): A system of continuous improvement employing participative management that is centered on the needs of customers. Key components are employee involvement and training, problem-solving teams, statistical methods, long-term goals, and recognition that inefficiencies are produced by the system, not people.
- 5S: Focuses on effective workplace organization and standardized work procedures.
- VSM: It is a collection of all the value-added and non–value-added actions that together brings a product from its raw material stage to its finished stage.

7.1.2 Green Manufacturing

Green manufacturing is a method of manufacturing that minimizes waste and pollution. It retards the depletion of natural resources as well as the extensive amounts of trash that enter landfills. Its emphasis is on reducing parts, rationalizing materials, and reusing components, to help make products more efficient to build. Green manufacturing involves not just the use of environmental design of products, or the use of environmentally friendly raw materials, but also eco-friendly packing, distribution, and destruction or reuse after the lifetime of the product.

This process is the answer to the need of the global community—clean air, water, and earth, while at the same time not having to compromise on quality and quantity of services, products, and technology. Looking at it from a global perspective of population growth, total emissions, rainforest depletion, and all the other maladies afflicting the earth, "it's pretty easy to see that in the next 50 years there are going to be major disasters," and "To keep things the way they are right now, the environmental impact has to be reduced by something on the order of a factor of 10."

Green manufacturing is a relatively new concept that, at present, has brought confusion and misconception about its components, costs, benefits, and implementation. Its future is also uncertain. It is likely, however, that consumers will demand high-quality, low-cost, prompt-produced, and quick-made products. They will also demand that such products be produced by the processes that have an effect on the environment.

Whether green manufacturing is a good business or a bad investment is being argued. On one hand, eliminating waste has the potential for reducing cost and increasing value. On the other hand, because of the investments and negative returns, dividends can be reduced to shareholders. Middle managers are often focused on quality and productivity rather than on environmental or marketing issues. This perspective can hinder development and implementation of green manufacturing.

To ensure success in green construction, corporate cultures need to be developed in which the organization clearly defines the vision of green construction and its objectives for implementation to achieve these objectives, establishes a plan, and accepts the results and costs of achieving these objectives. Along with the support of top management, with representation from engineering, production, purchasing, and human resources, a team's approach has been found to be necessary rather than dependent on plant-level environmental experts or consultants.

Training of a team in total quality management techniques will definitely help. Green manufacturing is likely to be an increasingly important issue because it is an important component of internationally sustainable development. In order to meet the needs of the current generation, sustainable development should be a long-term goal, keeping in view that the future generation would not have to compromise with the needs. To achieve this, interdisciplinary cooperation will be required between politicians, economists, scientists, and engineers.

7.2 LITERATURE REVIEW

A VSM categorizes all processes into value added and non–value added. The ultimate goal of VSM is to identify all types of waste (non–value added) in the value stream and to take steps to eliminate them.

Authors Verma and Sharma (2016) note VSM as one of the best tools for Lean manufacturing. The authors have summarized in their work the calculation of energy as value-adding and non–value-adding energy to optimize value streams holistically.

Murugananthan et al. (2014) gave details of the use of VSM in reducing waste in a manufacturing company. The production process path is visualized by mapping the current state VSM. After tracking the entire process, wastage affecting the cycle time has been identified and its causes are analyzed. A future state VSM is developed and improvement ideas have been suggested. They found that the VSM proved to be a useful technique in minimizing cycle time and in increasing productivity.

Solding and Gullander (2009) presented the concept using simulation for creating dynamic VSM. Creating dynamic value stream maps makes it possible to analyze more complex systems than traditional VSMs that are still able to visualize the result in a language the Lean coordinator can recognize. VSMs are presented in a spreadsheet as they can be modified to make teams consistent.

Fawaz and Rajgopal (2007) identified that Lean manufacturing is applied more frequently in discreet manufacturing rather than in the continuous/process sector and they formulate a simulation model that clearly showed the difference between the past and present scenario, which is acceptable to the managers who are generally skeptical about the application of Lean methods in continuous/process manufacturing.

Egon et al. (2014) described a method to extend VSM to an energy value stream mapping (EVSM) method by maintaining its original character and its inner logic. This chapter includes the time and energy input during the transportation processes into the EVSM and demonstrates the application of EVSM in supply chain management to reduce the energy footprint on a global level. Inclusion of transport into EVSM shows the lead time extending effect as a form of non–value-adding energy consumption.

Keskin et al. (2013) suggested a future-oriented EVSM approach that aims to improve energy efficiency in small- and medium-sized manufacturing companies. EVSM is a graphical technique that allows identifying the level of energy use and, thereby, discovering saving opportunities at each step of different processes either in production or in facility support. To analyze the possible outcomes of improvement options, future scenarios are developed using Bayesian networks. The suggested model can be used not only for diagnostic purposes but also for energy budgeting and saving measures.

Tyagi et al. (2015) introduced the concept of VSM to the product development process stresses and the importance of faster product development for the right edge on the market. The main focus of this article was to exploit Lean thinking concepts in order to manage, improve, and develop the product faster while improving or at least maintaining the level of performance and quality.

Chatterjee et al. (2014) have reported the use of EVSM as a tool to analyze energy consumption in any manufacturing process and demonstrate the use of the EVSM tool to analyze energy consumption in the production of biodiesel.

Nassehi et al. (2012) presented a framework to validate the introduction of energy consumption in the objectives of process planning for CNC machining. The paper considers the critical aspect of energy efficiency in manufacturing and particularly in process planning of products. Computer-aided process planning has been continuously developed for more than 40 years, which has been in the early days since the 1960s.

Gutowski et al. (2006) introduced the specific electrical energy requirements for a wide range of manufacturing processes into a single plot. The analysis is cast in an exergy framework. The following were the results: (1) the specific energy requirements for manufacturing processes are not constant as many life cycle analysis tools assume, (2) the most important variable for estimating this energy requirement is the process rate, and (3) the trend in manufacturing process development is toward more and more energy-intensive processes. The analysis presented here also provides insight into how equipment can be redesigned in order to be more energy efficient.

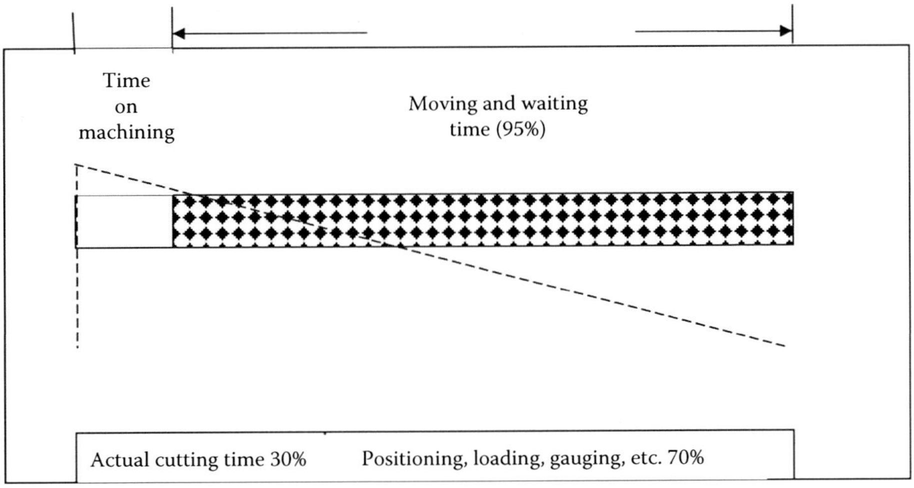

FIGURE 7.1 Machining time breakup.

7.3 OBJECTIVE OF THE RESEARCH WORK

Manufacturing of any product in CNC involved the following steps (either totally or partially):

a. Work holding process
b. Tool holding Process
c. Alignment
d. Coolant on and off
e. Tool changing
f. Work piece orientation changing
g. Tool approach and retracted back
h. Loading and unloading of work piece/pallet fixture
i. Moving and waiting

If the total time on machining is considered 100%, then more than 5% machining lasts real time. If this 5% machining time is considered as 100%, then it is found that 30% of this is actual machining, and the remaining 70% is used for other processes such as positioning, gauging, and so on (Figure 7.1).

A lot of work has been done to make a process green while moving and waiting. A small amount of work has been reported to make a process green during positioning, loading, gauging, and so on. Less time was consumed during positioning, loading, gauging, and so on, but during these processes, it has been found that the coolant flow and cutter movements are not stopped in many cases. The present work concentrates on analyzing the time to make processes green using a new tool called EVSM.

7.4 METHODOLOGY

7.4.1 Value Stream Mapping

A value stream is a collection of all value-added and non–value added actions required to produce a product or a group of products by using the same resources through the main flows, e.i., from raw material straight to customers' hands (Hugh et al. 2002). Value stream maps are a very common technique when you are implementing a Lean system.

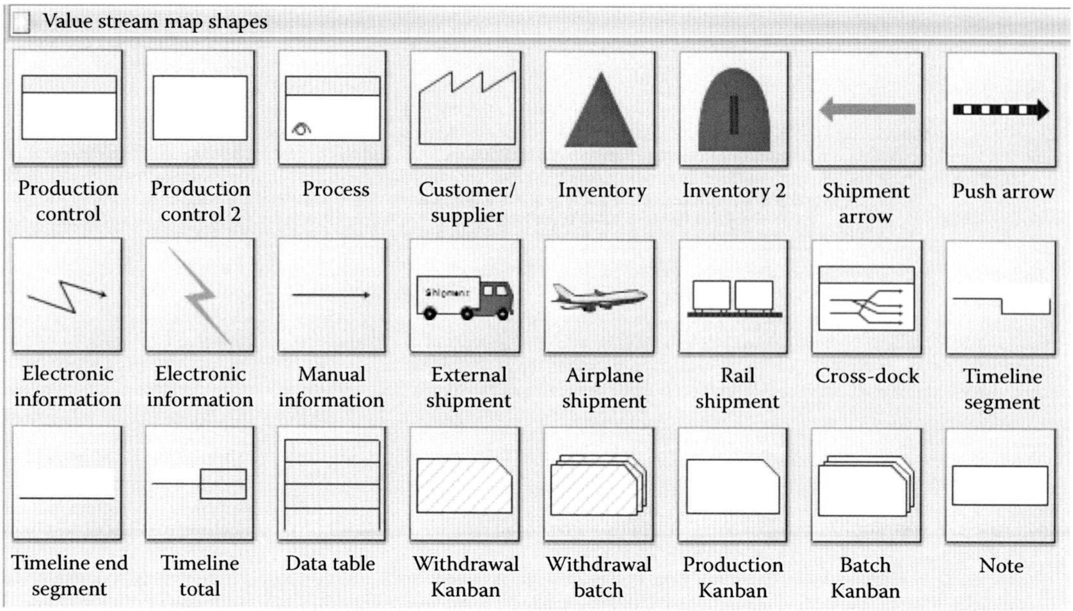

FIGURE 7.2 Standard symbols used for value stream mapping.

- Identify the "present state" and "future state" of the production process.
- Model the information flow, material flow, and lead time from the beginning to the end of the value stream and relative symbols of VSM presented in Figure 7.2.

7.4.2 Case Study

A study was conducted in a small-scale industry—a CNC manufacturing unit located in Jharkhand, which manufactures CNC precision machined components and aluminum and zinc alloy pressure and gravity die castings. The company is accredited and certified by ISO 2001. The study discusses the implementation of VSM that is carried out in the CNC machining center of the industry. The main aim was to reduce the cycle time and cost. The objectives for the implementation of Lean manufacturing tools in this industry are as follows:

- To study the current state VSM by collecting the data
- To identify the problems faced by the industry in terms of non–value-added time
- To propose a future state VSM that can increase value-added time and reduce non–value-added time.

7.4.2.1 Current State Map

Mapping the value stream always starts with customer demand (Shook 1999). To create a current value stream map, do the following steps:

Step 1. Calculate Takt time: It is the rate at which production should run to meet customer demand. The idea is to synchronize the pace of production to that of sales. Takt time for this process is 42 s.

Step 2. Understand customer demand: Customer demand is the monthly or daily demand of customers as per need. Customer demand for this process is 600 pieces per day.

Step 3. Mapping the process flow: This step involves various processes that are in sequence to complete product development and calculation of cycle time, changeover time, and uptime for each.

Step 4. Map the material flow: The flow of material from raw to finish goods is given by the supplier to customers.

Step 5. Map information flow: The information flow is also incorporated to provide demand information. Which is an essential parameter to determine the process in the production system? Various data regarding cycle time, changeover time, uptime, take time?

Step 6. Calculate total product cycle time: After both material and information have been mapped, a timeline is displayed at the bottom of the map showing the processing time for each operation and the transfer delay between operations. The timeline is used to identify the value-adding step as well as waste in the current system. Product lead time (PLT) for our process is 5 days.

Step 7. Detail offline activities: Activities such as placement of orders, material supply, daily scheduling, monthly forecast, etc., are included in this section, which are well executed by way of the transportation, supplier activities, and information flow line.

Step 8. Identify opportunity for improvement: Gather opportunities and write a summary of these observations to further improvement and to draw a future state map to illustrate the change in process. Figure 7.3 shows the current value stream mapping.

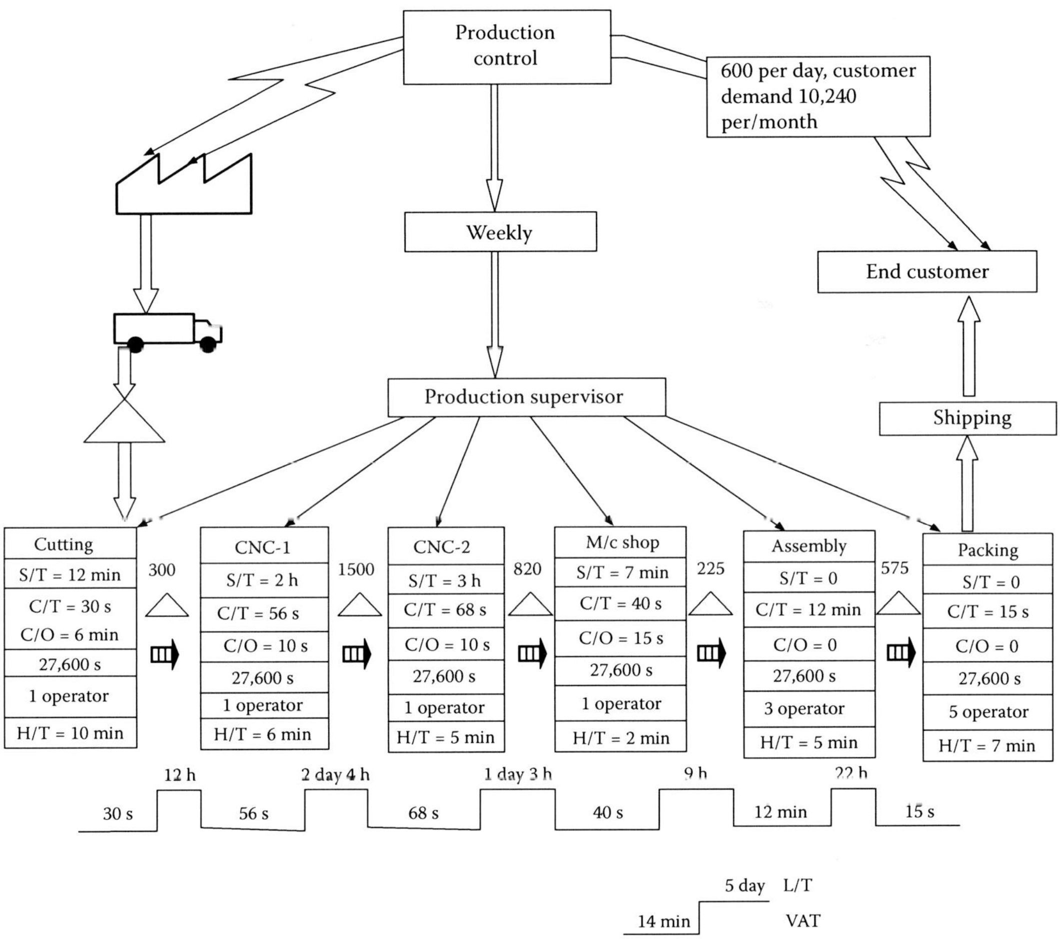

FIGURE 7.3 Current state map.

7.4.2.2 Current State Map Analysis

The current state map is a fancy way of saying "what happens now" or the "as-is" process. The current state map should show all the process steps and sufficient detail on how each step is completed and what happens to the items being processed. This will enable us to spot the causes of problems and thus the means to improving the flow, efficiency, reliability, and flexibility of the process. It can be as detailed or as simple as you need and it can also exist in a number of different versions for consumption by different internal or external groups. We analyzed in our CSM that the PLT for the assembly process is 5 days and the value-added time is 14 min. We have found wastage during changeover from one process to another.

7.4.2.3 Creating a Future State Map

To create a future state map, the following should be observed: PLT, 5 days; value-added time, 14 min; process cycle efficiency (PCE), 0.388%. The following wastes need to be reduced or eliminated first in order for the system to transition to the one proposed in the future value stream map: (1) defects, (2) waiting time, (3) inventory, and (4) material handling.

Data Analysis. Here, calculate the value addition percentage in the different processes and analyze the causes of different problems and suggest their remedial actions. Reduce lead time by improving production.

7.4.2.3.1 Cutting

In this process, cycle time is 30 s, changeover time is 6 min, daily average output is 600 pieces per day, and setup time is 12 min. It can be found that most of the time is spent in handling material between the inventory and the cutting shop. The research concludes that some areas require product improvement as well as inventory reduction.

7.4.2.3.2 CNC

In this process, cycle time is 56 s, changeover time is 10 s, and setup time is 2 h. A high rate of defects was generated by CNC machine, for which the operator had to get rid of the products that were spending additional non–value-added time. It was concluded that the CNC machine that produces a defect needs to be reworked. In addition, inspection time has also been added for this reason.

7.4.2.3.3 VMC

In this process, cycle time is 68 s, changeover time is 10 s, and setup time is 3 h. The CNC operator sometimes idles while waiting for parts to arrive from the VMC machine. The cycle times are different: 56 s in the CNC machine, 30 s in cutting, and 68 s in the VMC machine. There is a problem in this situation. The waiting time will be eliminated by Kanban (pull system).

7.4.2.3.4 Machining Shop

In the machine shop, cycle time is 40 s, changeover time is 15 s, and setup time is 7 min, influenced by the scale formation and dimension and further increases in next process, due to which the time and money are vain when it is admitted in the next process.

As a suggestion, the industry will make proper use of GO–NO-GO gauge for a fairly measured dimension.

7.4.2.3.5 Assembly Shop

At the end of all four lines (cutting shop, CNC shop, VMC shop, machining shop), there is one assembly shop. In the assembly shop, cycle time is 12 min and changeover time is 20 s. Figure 7.4 shows the future state mapping.

7.4.2.4 Analysis of Future Stream Map

For the creation of future state map supermarkets are placed between processes to reduce inventory wastages and to switch the process from "build to stock" to "make to order." According to the orders

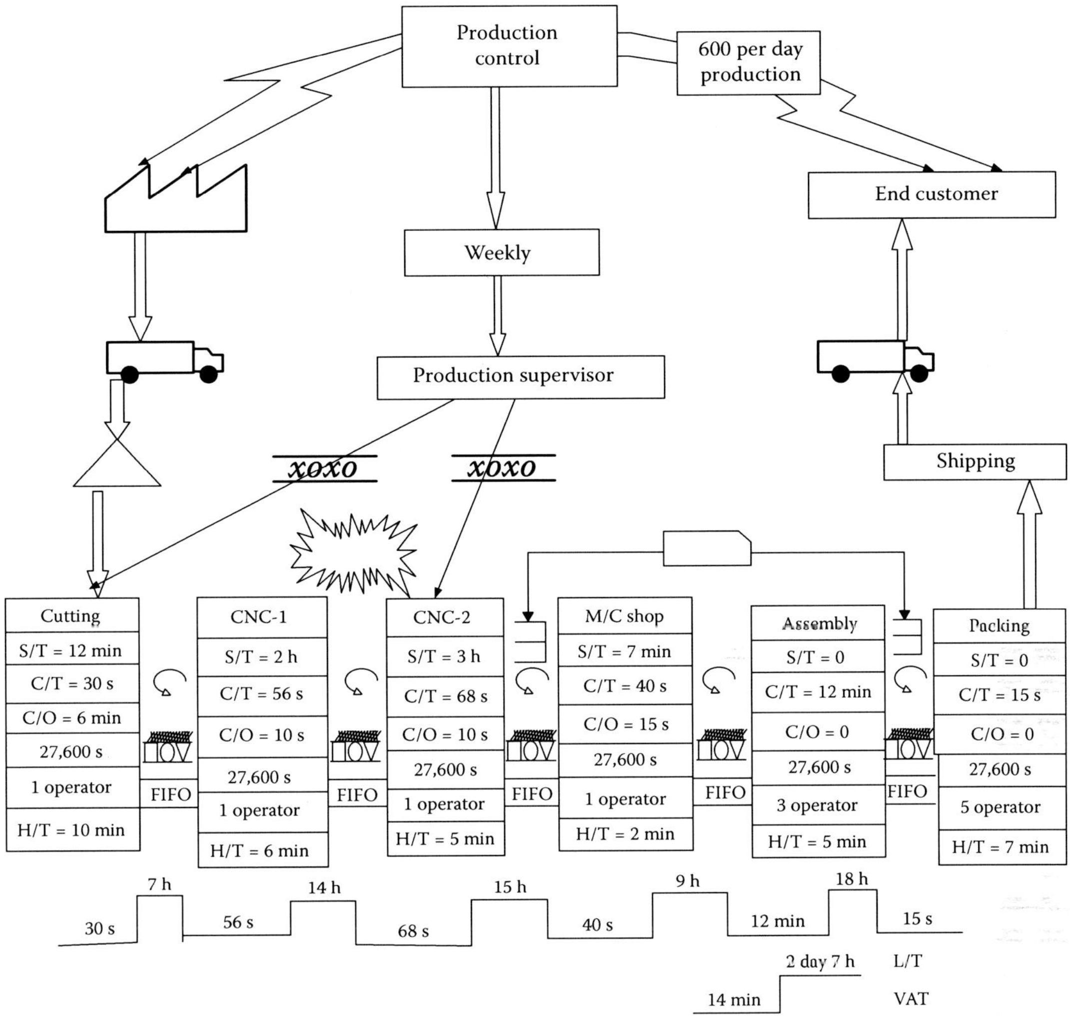

FIGURE 7.4 Future state map.

placed by the customers, the "make to order" process leads to the assembly of parts. It results in reduction in inventories. Between the processing lines, the information and communication flow has been improved by scheduling pacemakers in the process, and process transitioned from push to pull via the Kanban system. We have made considerable improvements in this research. PLT was reduced to 2 days 7 h, and PCE has increased from 0.388% to 0.476%.

7.4.2.5 Identifying the Bottleneck Process

The bottleneck process (presented in Figure 7.5) is the operation with the longest time. From the example, this is the CNC-2 machine at 68 s.

Takt time is the rate a completed product needs to be finished in order to meet customer demand. Calculate available time = working time minus regular "non-direct" time (stand-up meetings, breaks, vacations, sick time, cleaning, etc.). This is simply the work time in the time period selected, regardless of the number of people actually executing the work (William et al. 2011).

$$\textbf{Takt Time} = \textbf{Available Minutes for Production/Required Units of Production}$$

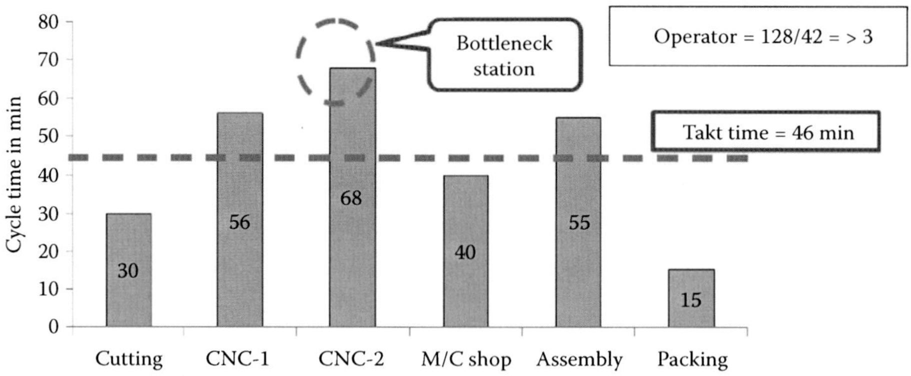

FIGURE 7.5 Bottleneck process.

TABLE 7.1
Comparison of Production in Process (No. of Pieces per Shift)

Process	Current State	Future State
Section	–	–
CNC-1	12 h	7 h
CNC-2	2 day 4 h	14 h
Machine shop	1 day 3 h	15 h
Assembly	9 h	9 h
Packing	22 h	18 h

Calculation for Takt Time: Takt time cannot be measured with a stopwatch. It can only be calculated. To calculate Takt time, think touchdown, or T/D, since we simply divide the net available time by the customer demand. Customer requirements = 600 per day. One shift = 8 h = 480 min. Consider production process to be 85% efficient and there is 480 min in one shift. Now, time required = 27,600/620.

Takt Time = 46s

By the following proposed implementation path of the improvements, the performance of the production process can be improved. This paperwork has been done to improve the overall productivity of a small-scale industry by implementing the Lean methodology. Expected results for some key parameters are presented in Table 7.1.

In the current situation, this results in an average work in progress and lead time before improvement of 5 days by implementing VSM; the non–value-added activities were identified and measures were taken for the improvement of the system. In the future state, lead time after improvement is 2 days 7 h and savings in lead time is 2 day 3 h.

7.4.3 Energy Value Stream Mapping

EVSM is developed based on the standard of value stream methodology. This has been done by adding energy components in addition to the cost in VSM, and the same has been analyzed with respect to time. The EVSM identifies the level of energy utilized and wastage in each step and hence determines the opportunities for energy conservation.

To analyze the possible outcomes of improvement options, future scenarios are also developed using EVSM. The suggested model can be used not only for diagnostic purposes but also for energy budgeting and saving measures.

We will now extend the VSM to energy parameters. The essence of the VSM will be the same, but we will now analyze energy along with time. Work has been done and methods have been proposed to extend VSM to energy VSM. For this chapter, we will limit our study to the method proposed by Egon et al. in 2014. In their method, they used a portable wattmeter to study the energy consumption of the machines in the different processes and obtained an energy signature graph. When plotted against time and studied with the timeline of the machine operations, this energy signature graph tells a lot about value-adding energy and non–value-adding energy. All methods were proposed to draw a stepped energy line similar to the time span we draw in VSMs. Muller and Schilling (2014) suggested to not only draw an energy line but also, unlike other methods, perform a dual assessment of the energy and time input referring to the value-added and non–value-added criteria. The energy required to be spent during the non–value-added time is considered as waste and energy used during the value-added time can be considered as the value of adding energy. Figure 7.6 shows the basic model energy value stream mapping.

This chapter does not just limit its study to production process, it develops a model that implements EVSM to transport processes, along with all second- and third-tier processes that do not contribute to the main process but consume energy. It defines a peripheral model in which there is a

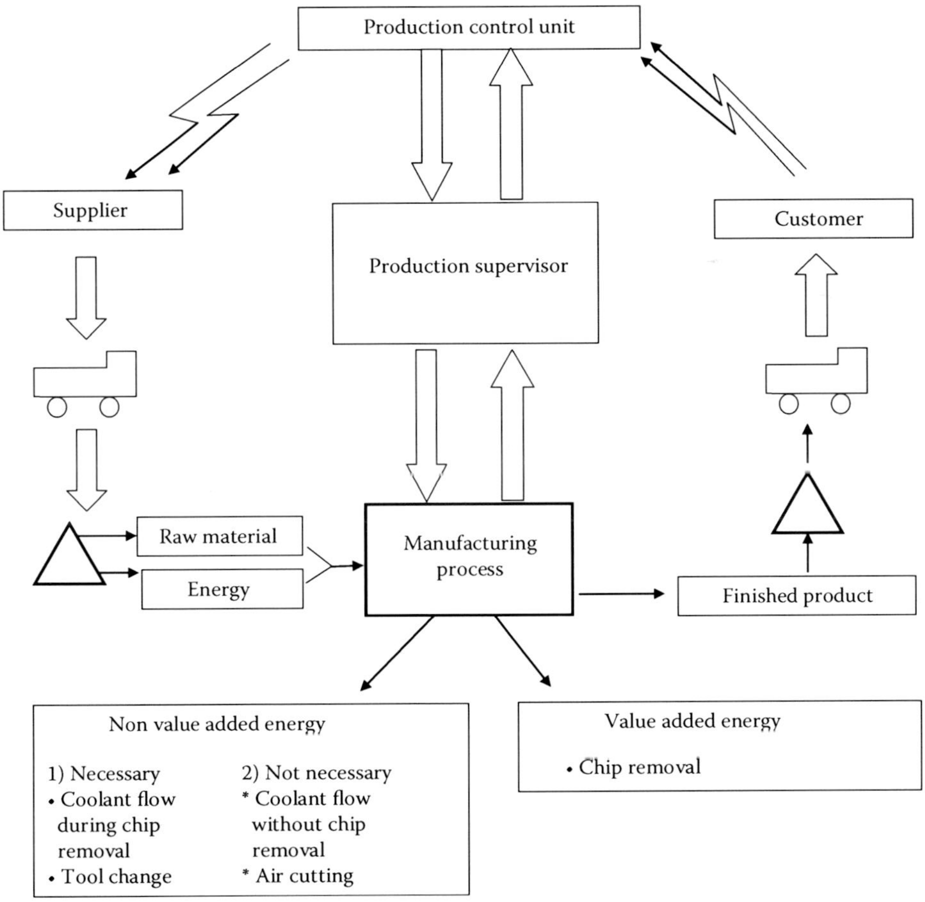

FIGURE 7.6 Basic model of energy value stream mapping.

process of core production and assembly procedure, in which there are three peripherals of processes that are integral to the industry but do not directly contribute as such a product.

A concept of EVSM analysis is introduced, which elaborates machining processes and analyzes the input and output energies along with the production system. The flowchart in Figure 7.6 gives a clear depiction of the energy inputs and acquired outputs. In the figure, the energy flow under the manufacturing process is divided into two: value-added energy and non–value-added energy. The non–value-added energy is further divided into necessary and unnecessary energy usage. Although the necessary energy does not add value (e.g., coolant flow during the chip removal process and tool change), it is important because of the nature of the operation. The unnecessary non–value-added energy usage is focused on non-value processes such as coolant flow during the chip removal process and air cutting process in the machining operations (see in Figure 7.6).

7.4.3.1 Assumptions

The following assumptions are made in order to develop the energy value stream map:

1. Component X is the final component and needs no further assembly.
2. Only the machines with high energy consumption are taken up for energy analysis.
3. Energy consumption during transportation between machines has been ignored.
4. There is no internal loss of energy in the machine by friction, heat, and so on, and energy used by machining is the same as that of power rating of motors.

7.4.3.2 Process Flow

The process flow diagram of component X is shown in Figure 7.7. Component X is produced using all the CNC machines present in the shop floor. The process starts with cutting raw material in the band saw and ends with anodizing operations. The rod raw material from the store is being cut through the band saw and transferred to CNC-1 where various processes such as facing, rough turning, drilling, boring, and threading are performed. Then, it is transferred to CNC-2, where operations such as facing, with taper turning, and V drilling are performed. The next step includes the transfer of components to VMC-1 where operations such as flash hole, detent, and tape hole take place. After which, it is transferred to VMC-2 where hinge pin-hole and locking pin-hole operations are performed. Finally, the component is transferred to the machine shop and anodizing section.

7.4.3.3 Procedure of EVSM

A case study of a small-scale industry was made. The following is the detailed stepwise method followed:

Step 1—Analysis of the process flow of the product
Step 2—Creation of a current state value stream map
Step 3—Analysis of the energy usage to create the energy span similar to the time span in the current state map
Step 4—Analysis of the final energy value stream map created
Step 5—Suggesting improvements in the current process to get closer to green manufacturing and creating a future state value stream map

FIGURE 7.7 Process flow diagram of component X.

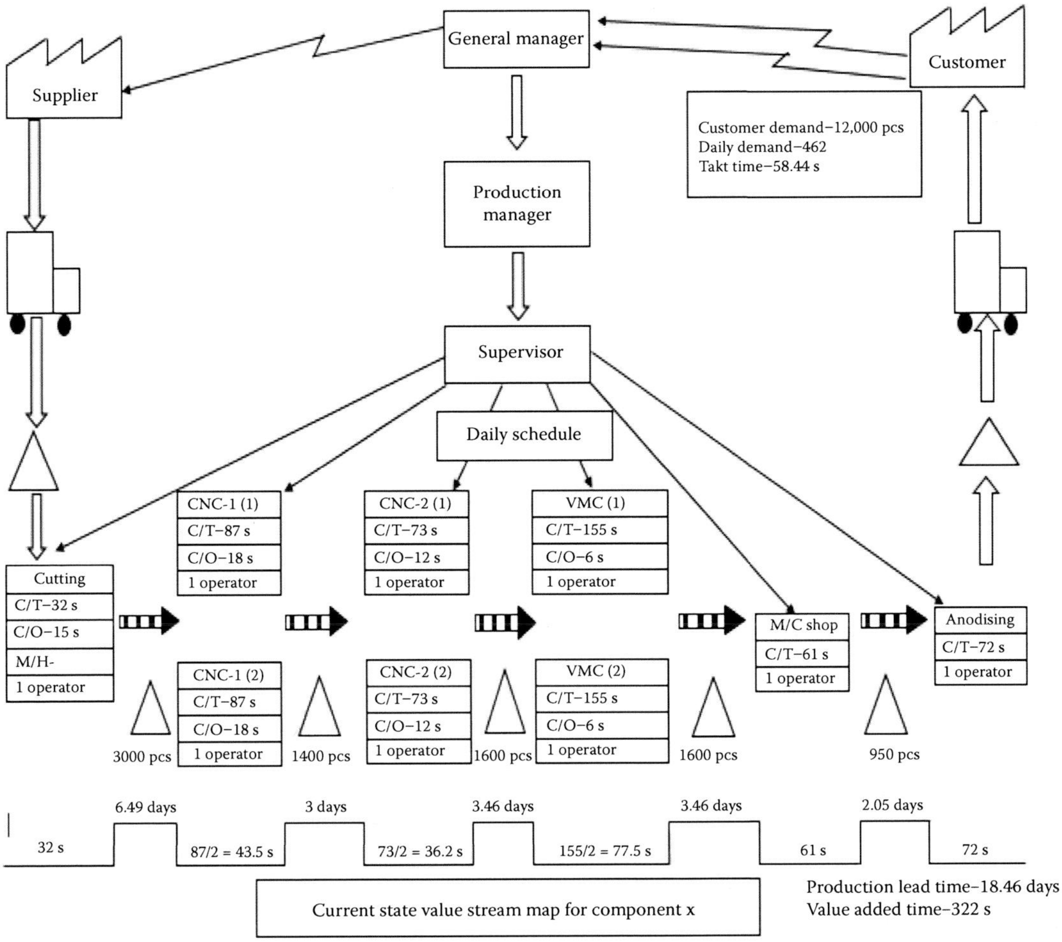

FIGURE 7.8 Current state value stream mapping for component X. (From Verma N., Sharma V. (2015). Lean modelling—A case study for the Indian SME. *International Journal for Technological Research in Engineering (IJTRE)* 2(7) ISSN: 23478–4718.)

7.4.3.3.1 Current State Value Stream Map

The current state value stream map is being prepared for the process flow diagram (Figure 7.7) and is summarized in Figure 7.8 (Verma and Sharma 2015). The standard symbols depicted in Figure 7.2 are used to make the current state value stream mapping. The current energy value stream mapping for component X is developed as shown in Figure 7.9. In order to understand the effect of energy, a future energy value stream map was developed (Figure 7.10).

7.5 RESULTS AND DISCUSSION

Energy usages are calculated for the above value stream map (Figure 7.4) and an energy span is added to the current state map. The energy calculations are done by using the power rating of the motors. The current energy value stream map is shown in Figure 7.5.

The power ratings used for the calculations are stated below:

- Cutting machine: cutting saw motor power, 2.2 kW; coolant motor power, 0.25 kW
- CNC-1: spindle motor power, 6 kW; coolant motor power, 0.25 kW

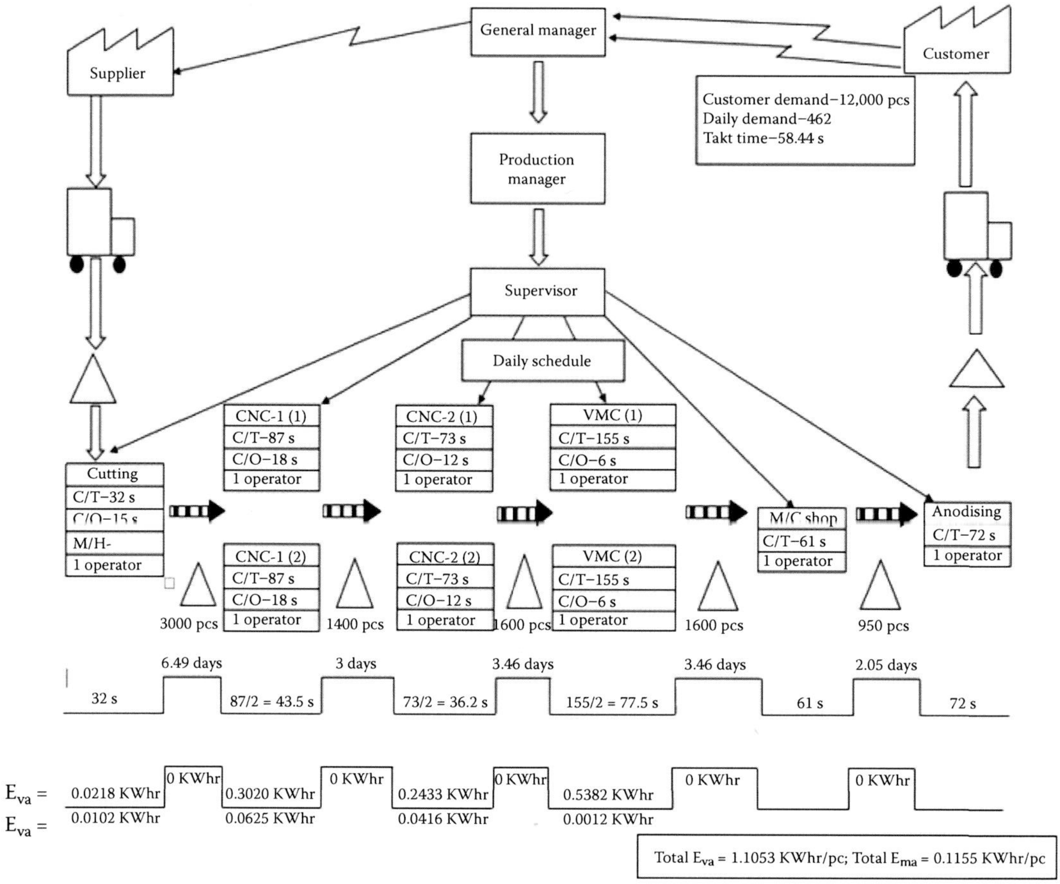

FIGURE 7.9 Energy value stream mapping for component X. (From Verma, N., Sharma V. (2016). Energy value stream mapping a tool to develop green manufacturing. *Procedia Engineering* 149: 526–534.)

- CNC-2: spindle motor power, 6 kW; coolant motor power, 0.25 kW
- VMC: spindle motor power, 6 kW; coolant motor power, 0.37 kW

For calculation of value-adding energy and non–value-adding energy, the cycle time is divided into value-added time when cutting is actually taking place and non–value-added time when there is no cutting operation performed (i.e., tool change time and bed travel time).

For coolant, the energy used during cutting is termed value-added energy, whereas the continuous coolant flow during air cutting and tool change is termed non–value-added energy.

The continuous spindle movement during non-cutting time is also termed non–value-added energy.

Referring to the current state energy value stream map (Figure 7.5), the total non–value-added energy in the energy span per piece of component X is

$$\varepsilon = \text{non}-\text{value-added energy} = 0.1155 \text{ kWh/piece}$$

Monthly demand of component X is 12,000 units (as per data collected).
Thus,

$$\varepsilon_{\text{month}} = \text{total non–value-added energy per month} = 0.1155 \times 12,000 \text{ kWh/month}$$

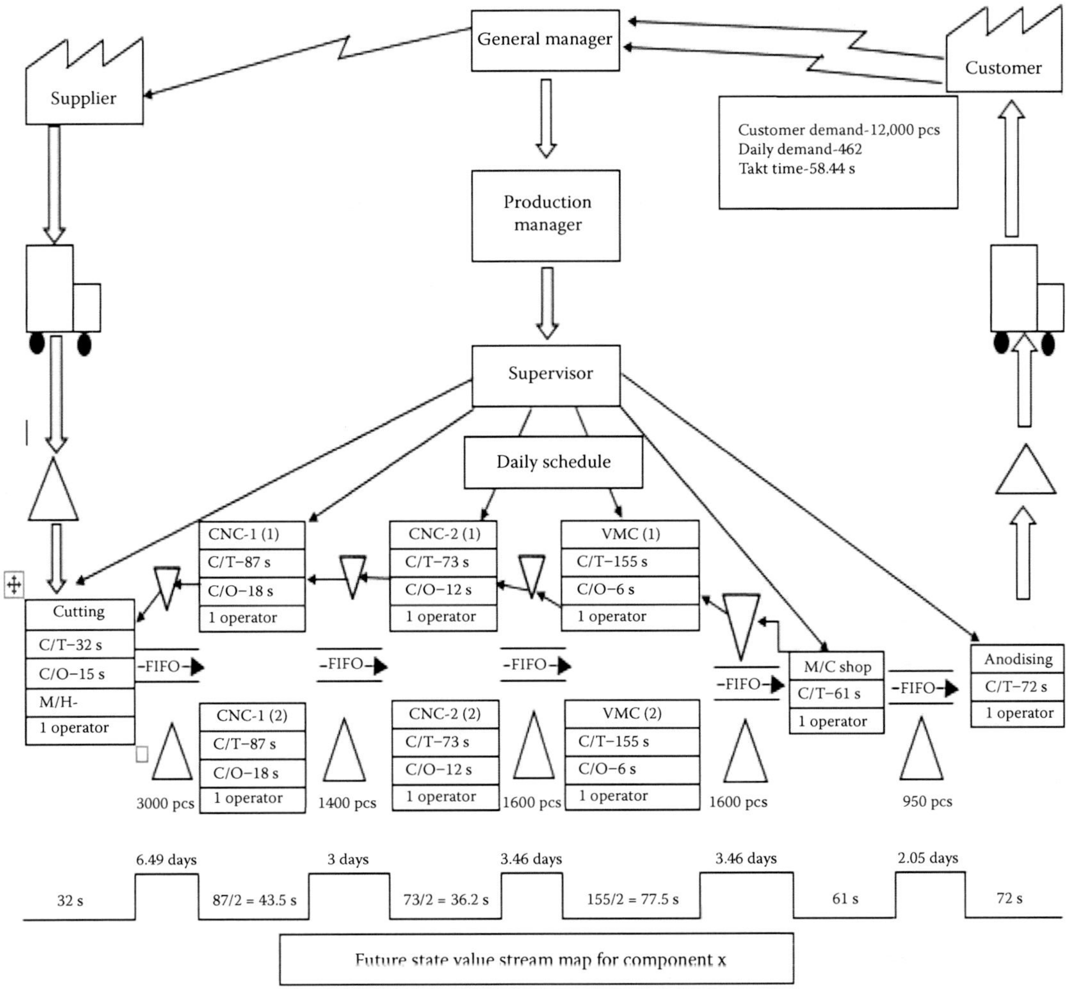

FIGURE 7.10 Future state energy value stream mapping. (From Verma, N., Sharma V. (2016). Energy value stream mapping a tool to develop green manufacturing. *Procedia Engineering* 149: 526–534.)

Assuming the demand of component X to be the same throughout the year, the non–value-added energy in a year is

$$\varepsilon_{\text{year}} = 1386 \times 12 = 16,632 \text{ kWh/year}$$

As per the MSME (micro, small, and medium enterprise) data of MSME 2012–2013, the total number of MSME enterprises in India (registered + unregistered) is 115.01 lakhs.

Assuming all the abovementioned small-scale industries to be of the same stature as the one taken in the case study, the total energy that can be saved in India alone per year is equivalent to

$$E = \text{total energy saved} = \varepsilon_{\text{year}} \times 115.01 = 16,632 \times 115.01 = 1,912,846.32 \text{ kWh}$$

$$= 1912.83 \text{ MWh}$$

Further, the non–value-added energy calculated in the present case study is limited to only the following operations:

1. Coolant flow when no cutting is taking place
2. Spindle rotation during air cutting and tool change operation

This is the kind of energy that can be saved in a small-scale industry if it is producing only one component. The total energy saving potential is large if small modifications are made in the existing machine in operation in the entire industry worldwide.

7.5.1 Future State Energy Value Stream Map

A future energy value stream map for the process is shown in Figure 7.6. From the future EVSM, the following conclusions are drawn for process improvement:

- A pull-type system should be implemented instead of the prevalent push type. This will reduce the excessive inventory being produced, which is then associated with high storage costs for the inventory. Also, a lot of energy is wasted in producing such excessive inventory and the machines are not used judiciously.
- The sizes of the bins on the floor are not fixed. That is, the work in process inventory is not clear at a point of time. The size of the bins should be fixed because the first concept of Lean manufacturing is to make things visible.
- Energy wasted during the process is very large; a small change in the existing setup may lead to higher energy savings.

7.6 CONCLUSION

The results specified in this chapter clearly indicate cost savings, which can be done with small changes in the process flow. Thus, it can be concluded that designing time and energy-efficient EVSM have become very important in the present scenario. With the help of this new VSM, a Lean and green manufacturing system can be developed. This research recommends the path of reducing the impact of the environment and suggests an approach to the implementation of the green system for various manufacturing units. The importance that energy saving holds need not be listed down as the whole world is well aware of it. Based on the direction in which the world is going, it may soon face an energy crisis where the prime focus will not be on saving money or increasing productivity, but on finding alternate sources of energy, and most of the current machinery will become obsolete. What this study focuses on is how one can save energy today and postpone that energy crisis period as much as possible. The world as we know today may not exist in the same form tomorrow. Therefore, the EVSM is an effective tool to identify areas where the energy is wasted, and new designs of machine tools or modifications in the existing setup will help reduce energy wastage.

REFERENCES

Chatterjee R., Sharma V., Mukherjee S. (2014). Value stream mapping based on energy and cost system for biodiesel production. *International Journal of Sustainable Manufacturing*: 95–115.

Egon M., Stock T., Schillig R. (2014). A method to generate energy value streams in production and logistics in respect of time and energy consumption. *Production Engineering Research Development*: 243–251.

Fawaz A., Rajgopal J. (2007). Analyzing the benefits of LAN manufacturing and value stream mapping via simulation—A process sector case study. *International Journal of Production Economics*: 223–236.

Gutowski T., Dahmus J., Thiriez A. (2006). Electrical energy requirements for manufacturing processes. *13th CIRP International Conference on Life Cycle Engineering, Leuven*. CIRP: 1–5.

Hugh L., Manus M., Richard L. (2002). Value stream analysis and mapping for product development. *International Council of the Aeronautical Sciences*: 6103.1–6103.2

Keskin C., Asan U., Kayakutlu G. (2013). Value stream maps for industrial energy efficiency. *Assessment and Simulation Tools for Sustainable Energy Systems*: 357–379.

Murugananthan V., Govindaraj K., Sakthimurugan D. (2014). Process planning through value stream mapping foundry. *International Journal of Innovative Research in Science and Technology*: 1140–1143.

Nassehi N., Asrai I., Dhokia V. (2012). Energy efficiency process planning for CNC machining. *CIRP Journal of Manufacturing Science and Technology*: 127–136.

Shook J. (1999). *Value Stream Mapping to Add Value and Eliminate Muda*. 1.2 Edition Book The Lean Enterprise Institute Inc.

Singh B., Sharma S.K. (2009). Value stream mapping as a versatile tool for lean implementation: An Indian case study of a manufacturing firm. *Measuring Business Excellence* 13(3).

Tyagi S., Alok C., Xianming C. (2015). Value stream mapping to reduce the lead-time of product development process. *International Journal of Production Economics*: 202–212.

Verma N., Sharma V. (2015). Lean modelling—A case study for the Indian SME. *International Journal for Technological Research in Engineering (IJTRE)* 2(7). ISSN: 2347–4718.

Verma, N., Sharma V. (2016). Energy value stream mapping a tool to develop green manufacturing. *Procedia Engineering* 149: 526–534.

William M., Goriwondo S., Alphonce M. (2011). Use of the value stream mapping tool for waste reduction in manufacturing—Case study for bread manufacturing in Zimbabwe. *International Conference on Industrial Engineering and Operations Management*: 22–24.

8 The Relationship between Lean and Sustainable Concepts on Manufacturing Performance

Sri Hartini, Udisubakti Ciptomulyono, and Maria Anityasari

CONTENTS

8.1 INTRODUCTION

The Lean manufacturing concept, which was pioneered by a Japanese automotive company, Toyota, during the 1950s, was famously known as the Toyota Production System (TPS). The primary goal of TPS was to reduce the cost and to improve productivity by eliminating wastes or non–value-added activities (Ghosh 2013; Jasti and Kodali 2014b). Lean manufacturing reorganizes a manufacturing firm into cells and value streams to improve the quality, flexibility, and customer response time of their manufacturing processes (Fullerton et al. 2014). From an operational-level perspective, it can be concluded that Lean manufacturing is very powerful in increasing operational performance on quality (Fullerton and Wempe 2009; Shah and Ward 2003), productivity (Fullerton and Wempe 2009; Singh et al. 2011), reduced inventory (Chong et al. 2012; Losonci and Demeter 2013), reduced cost (Hallgren and Olhager 2009; Pampanelli et al. 2011), reduced defect (Gupta and Kumar 2013; Prashar 2014), and time delivery (Bortolotti et al. 2014). Lean manufacturing is also powerful in increasing systems performance based on cost, quality, delivery, and flexibility (Bortolotti et al. 2014; Boyle and Scherrer-Rathje 2009). However, many researchers believed that conventional Lean manufacturing tools did not consider the environmental and social benefits (Vinodh et al. 2016).

Modern manufacturing systems are expected to be efficient and environment-friendly. The Lean concept focuses on waste and cost reduction, while the sustainability concept focuses on product and process development that is more environment-friendly by considering the economic and social aspects. This means that there are opportunities to incorporate the sustainability concept in extending Lean tools. Some researchers showed the results of applying Lean techniques in developing greener manufacturing strategies and processes (Bergmiller et al. 2009; Chiarini 2014; Hajmohammad et al. 2013).

Most papers that address the connection between Lean and green touch on the efficient use of energy (and resources) and the reduction of waste and pollution (King and Lenox 2001; Rothenberg et al. 2001). Bergmiller and McCright (2009a) identify the correlation between green operations and Lean results. This finding not only suggests that Lean and green systems can co-exist but also provides evidence of synergy. Some researchers developed the Lean concept to achieve sustainable manufacturing, for example, Dombrowski et al. (2012), Aguado et al. 2013, and Faulkner and Badurdeen 2014.

For all these reasons, we identify the interrelationships between Lean and sustainable manufacturing and analysis their effect on sustainability performance, include the environment, economic, and social dimensions.

8.2 EVOLUTION OF SUSTAINABLE MANUFACTURING

Various societal phenomena highly influence the development of the manufacturing system. The many inventions of various machines in the era of the Industrial Revolution has triggered the company to mass production to be more efficient. The conditions during World War II that destroyed Japanese industries encouraged the Toyota Company to develop an efficient production system that focused on maintaining quality. As the population increased (triggering the consumption of resources), so did production. Production increase means the increase of resource consumption, waste, emission, and environmental impact. The impact of production on the environment triggers consumer awareness in protecting of the environment. Such phenomena encourage the creation of environmental protection regulations and costumer demand on environment-friendly products. Therefore, companies are pressured not only to think about profit but also to produce products that are environment-friendly.

Jawahir describes the evolution of the manufacturing system (from traditional to sustainable manufacturing). The traditional system evolved from using the Lean manufacturing concept (with a waste reduction approach [1R]), to using a green manufacturing approach (with the reduce, reuse, recycle [3R] approach), to sustainable manufacturing (with the reduce, reuse, recycle, recovery, redesign, and remanufacturing [6R] approach) (Jawahir 2008).

Lean manufacturing focuses more on economy, while green manufacturing focuses more on the environmental aspect. According to Dornfeld et al. (2013), achieving sustainable manufacturing entails having a concept that focuses more on the environment (namely, green manufacturing) and a concept that focuses more on the economy (namely, Lean manufacturing). If the Lean and green manufacturing concept is equipped with an approach that deals with the social aspect, it will achieve "a manufacturing that minimizes negative impact on the environment, saves energy and natural resources, that is safe for employees, society, and costumers, and is also economic." Many reviews related to Lean and green are needed to realize sustainable manufacturing.

8.2.1 Lean Manufacturing

Lean manufacturing was created post–World War II. During that time, the focus was on mass production to achieve a low production cost per item. Taiichi Ohno and Shigeo Singo were the engineers who developed manufacturing concepts in TPS to overcome this situation. The term "Lean" was first mentioned by John Krafcik in 1988 in a study report in Toyota. It was then popularized by Womack et al. (1992) who reviewed the ability of Toyota to improve the productivity of companies in Japan. After being popularized by Womack et al., the manufacturing system of Toyota has been known as Lean manufacturing. There has been no consensus regarding the definition of Lean manufacturing in general (Pettersen 2009). The definition of Lean manufacturing varies: it can mean philosophy (Shah and Ward 2003, 2007), process (Womack et al. 1992), system (Taj and Berro 2006), practice (Hallgren and Olhager 2009; Simpson and Power 2005), technique, or model (Narasimhan et al. 2006). The scope of Lean manufacturing includes product development, operational management, supply chain management, manufacturing system, market demand, and environmental change (Bhamu and Sangwan 2014). Although the definition of Lean manufacturing varies, in general, Lean manufacturing is a systematic action in reducing invaluable activities by implementing sustainable improvement to increase customer satisfaction. Application of Lean is not limited to the automotive (71%) sector only; it has also found acceptance in a wide range of manufacturing industries: electronics manufacturing (Chong et al. 2012; Doolen and Hacker 2005), aircraft industry (Browning et al. 2009), furniture industry (Almomani et al. 2014; Gurumurthy and Kodali 2011), ceramic industry (Sangwan et al. 2014), and multisector industries (Farris et al. 2009; Fullerton and Wempe 2009; Ghosh 2013; Marodin and Saurin 2015; Taj and Morosan 2011; Thanki and Thakkar 2014; Tortorella et al. 2016).

Lean is categorized into two levels (Hines et al. 2004), namely, Lean thinking and Lean production. Lean thinking lies on a strategic level that centers on customers and can be applied in various functions of companies. Lean thinking is a concept developed from the shop-floor level (i.e., from a manufacturing function) to the function of the whole organization. Lean production lies on an operational level in providing value for customers. Lean production is made up of five elements, namely, Lean manufacturing, Lean product development, supply chain coordination, customer distribution, and Lean enterprise management. Some researchers choose to focus on Lean manufacturing, which is most capable of bringing changes on a shop-floor level, and many manufacturing industries are interested to implement it.

Lean organization consists of five stages (Hines and Taylor 2000).

1. Identification of value and non-value activities from the customers' point of view and not from the companies' perspective
2. Identification of value stream mapping (VSM) and waste finding
3. Reducing waste

4. Manufacturing according to specification demanded by customers and making standardization (pull system)
5. Sustainable improvement to reduce waste (continuous improvement) until zero waste is achieved

8.2.1.1 Types of Waste and Improvement Tools of Lean

In some industries, non–value-added activities can reach more than 90% of total activities and working hours for production. According to Taiichi Ohno in Dombrowski et al. (2012), there are seven types of wastes (Seven Deadly wastes), namely:

- Transportation: product movement that carries risk of waiting and damage, where the value of the product does not increase due to transportation.
- Inventory: waiting of raw material, work-in-process (WIP), or finished product that affects capital without adding value to customers.
- Motion: movement of workers that does not add value to products.
- Waiting: waiting of material because workers have not completed other processes or waiting of workers because material is being processed in another machine.
- Overprocessing: using time or resources that do not add value to customers, including adding detailed specification not required by customers. Customers are not willing to pay for additional work.
- Overproduction: producing more than the needs of customers.
- Defects: defects that require rework cost and are at risk of requiring additional material.

The types of waste most frequently reviewed in studies are inventory (60%), waiting time (51%), defect (41.5%), transportation (29%), motion (27.5%), overprocessing (21%), and overproduction (21%) (Jasti and Kodali 2014a). Some techniques are developed to reduce waste through Lean manufacturing (Table 8.1).

The Lean technique implemented by companies depends on their needs. Jasti and Kodali (2014a) state that Kanban, VSM, and Kaizen are the key techniques in implementing Lean manufacturing, with VSM being the most considered technique.

8.2.1.2 Benefits of Lean Manufacturing

Lean manufacturing is considered profitable by companies (Hines et al. 2004). Fuentes and Diaz (2012) confirm that many companies in various sectors have been using Lean for decades and show improvement in performance and competitiveness (Table 8.2).

The improvement in operational performance achieved by Lean manufacturing, including cost, quality, and flexibility, remains limited (Table 8.2). Lean manufacturing does not consider environmental performance. Changes in environmental law and regulations resulted in stakeholders demanding companies to improve their economic performance while developing greater environmental and social responsibility (Gordon 2001). Collaboration between Lean manufacturing and green manufacturing is necessary to help prevent environmental issues. Lean manufacturing focuses on operational performance and does not consider environmental performance, whereas green manufacturing focuses on the minimization of manufacturing impact on the environment. Integration of Lean and green manufacturing is expected to improve company sustainability.

8.2.2 Green Manufacturing

Green manufacturing is a process or system that minimizes the negative impact of manufacturing on the environment (Dornfeld et al. 2013). Some factors that trigger the implementation of green manufacturing are as follows (Bettley and Burnley 2008):

TABLE 8.1
Techniques in Lean Manufacturing to Reduce Waste

No.	Technique	Description
1	Value stream mapping (VSM)	Value stream mapping is an enterprise improvement tool to help in visualizing the entire production process, representing both material and information flow. Value stream is defined as the collection of all value-added and non–value-added activities that are required to manufacture a product or a group of products that use the same resources through main flows, from raw material to the finished product (Forno et al. 2014; Singh et al. 2011).
2	Kanban system	Kanban is the best-known pull system in which production is triggered by the customer demand at the final stage. In a production environment that is controlled by Kanban, the material flow is from the preceding stage to the succeeding one, but the information flow is downward and from the succeeding stage to the preceding one. At each intermediate stage, production and delivery are triggered by the succeeding stage order (Aghajani et al. 2012).
3	5S	5S is a system to reduce waste and optimize productivity and quality through maintaining an orderly workplace and using visual cues to achieve more consistent operational results. 5S stands for sort (seiri), set in order (seiton), shine (seiso), standardize (seiketsu), and sustain (shitsuke) (Bayo-Moriones et al. 2010).
4	Cellular manufacturing	All the workstations for one family of products are arranged in order in one area or cell, enabling a better process flow, reducing travel within the factory, and making one-piece flow easier (Aghajani et al. 2012; Babu et al. 2000; Biggs 2009).
5	Continuous improvement (CI)	CI is defined as a collection of activities that constitute a process intended to achieve performance improvement (Ali et al. 2013)
6	Total productive maintenance (TPM)	Planned maintenance schedules are devised to keep all equipment running 100% of the time. It means zero breakdown, zero defect, and zero accident (Biggs 2009).
7	Single minute of exchange dies (SMED)	Reducing setup time required to make any changes to equipment in order to switch from producing one product to another allows production by means of a pull system and one-piece flow instead of batch (Biggs 2009; Lacerda et al. 2016)
8	Production leveling (Heijunka)	Create a regular sequence of production of different products where known regular patterns of demand are integrated (Biggs 2009).
9	Visual control (Andon)	All controls and measures (e.g., flow of work, order progress, stock levels, call for replenishment of stock) are done visually, using control boards (including Andon boards), colored cards and markers, and Kanban/two-bin (Biggs 2009).
10	Poka-Yoke (mistake proofing)	Designing tools and so on in such a way that mistakes are impossible (e.g., making a jig so that it is impossible to put a part into it the wrong way around) (Biggs 2009).

- Competitive pressures arising from
  - Recognition of the cost advantages of reducing materials and energy consumption and waste production
  - Cost benefits of taking advantage of the economic incentives of green behavior such as subsidies and reduced taxation
  - The increasing pressures from customers (end consumers and supply chain partners) to demonstrate good environmental stewardship
- The perceived marketing advantages from demonstration of compliance with standards and so on

TABLE 8.2
Impact of Lean Manufacturing on Company Performance

Performance	Author
Reduce defect	(Gupta and Kumar 2013; Prashar 2014)
Reduce cost and processing time but increase quality	Manufacturing (Dora et al. 2013a); service (Bortolotti et al. 2010; Piercy and Rich 2009)
Increased performance on cost, quality, delivery, and flexibility	(Bortolotti et al. 2014; Boyle and Scherrer-Rathje 2009; Taj 2008; Taj and Morosan 2011)
Performance on supply chain	(Jabbour et al. 2014; So and Sun 2010; Taylor 2006)
Reduce waste	(Bergmiller and Mccright 2009a; Shah and Ward 2003)
Minimization on inventory	(Demeter and Matyusz 2011; Singh and Sharma 2009; Singh et al. 2011)
Increase productivity and quality	Industry manufacture (Anand and Kodali 2009; Bhamu et al. 2012; Taj and Berro 2006) and services (Brown 2009)
Increase profitability	(Shen and Han 2006)
Increase customer satisfaction, quality, and responsibility	(Garza-Reyes 2015)
Increase qualitative benefits: behavior, communication, decision making, etc.	(Bhamu and Sangwan 2014)

- Legal obligations, in a climate of increasing regulation as scientific evidence for human influences on climate change, ozone depletion, and so on strengthens
- The demands of investors for security from future liabilities, consumer demands, and expectations of ethical behavior and good corporate citizenship in the wake of scandals such as Enron
- Internal ethical values, reflecting changed values in society as a whole

8.2.3 Sustainable Manufacturing

Although there is no universally accepted definition for the term "sustainable manufacturing," numerous efforts have been made in the recent past, with much more concurrent efforts well underway (Haapala et al. 2013). The US Department of Commerce defines sustainable manufacturing as (USDOC 2011) "the creation of manufactured products that use processes that minimize negative environmental impacts, conserve energy and natural resources, are safe for employees, communities, and consumers and are economically sound."

Sustainable manufacturing includes manufacturing of "sustainable" products and sustainable manufacturing of all products. The former includes manufacturing of renewable energy, energy efficiency, green building, and other "green" and social equity–related products, whereas the latter emphasizes sustainable manufacturing of all products taking into account the full sustainability/total life cycle issues related to the products manufactured (National Council for Advanced Manufacturing [NACFAM] 2009).

Jawahir (2008) stated that a sustainable process leads to (i) improved environmental friendliness, (ii) reduced cost, (iii) reduced power consumption, (iv) reduced wastes, (v) enhanced operational safety, and (vi) improved personnel health. To achieve sustainable manufacturing, Allwood (2009) suggests the following five steps for improvement:

1. Use less material and energy.
2. Substitute input materials (nontoxic for toxic, renewable for nonrenewable).

3. Reduce unwanted outputs (cleaner production, industrial symbiosis).
4. Convert outputs to inputs (recycling and all variants).
5. Change structures of ownership and production (product service systems, supply chain structure).

8.3 THEMATIC ANALYSIS OF THE PUBLISHED LITERATURE

The aim of this chapter is to structure the research field on Lean and sustainable manufacturing in the context of the relationship between the two, including the benefits and their effect on performance, and to point out the most important gaps. This review, which covers academic papers from 2000 to 2016, used the following major research databases: Emerald, ScienceDirect, IEEE, Springer, and Proquest. The database searches yielded hundreds of articles. Each of the articles was examined to ensure that its content was relevant to the aims of our research. Articles were selected based on whether their main contribution revolves around the interrelationships between Lean and sustainable manufacturing and their effect on sustainability performance. This chapter presents a qualitative thematic analysis to elucidate the relationship between Lean and sustainable manufacturing and its impact on performance.

The thematic analysis was reviewed following the approach shown in Figure 8.1. The review began with an analysis of the relation between Lean and sustainable manufacturing, including correlation, synergy, comparison, barrier and driver, critical factor, and assessment. This was followed by an analysis of the effect of integration of Lean and sustainable manufacturing on company performance from a triple bottom line perspective. In particular, the integration of Lean and sustainable concept to develop VSM as a diagnostic tool to improve sustainability was reviewed further.

8.3.1 Relationship between Lean and Sustainable Manufacturing

Some organizations continued to grow on the basis of economic constancy, whereas others struggled because of their lack of understanding of the changed customer mind-sets and cost practices. To overcome this situation and to become more profitable, many manufacturers turned to "Lean manufacturing." The goal of Lean manufacturing is to be highly responsive to customer demand by reducing waste (Bhamu and Sangwan 2014). King and Lenox (2001) found strong evidence that Lean leads to waste and pollution reduction. This evidence is reinforced by Rothenberg et al. (2001), noting that Lean production or just-in-time (JIT) can reduce emission of volatile organic compounds, leading to more efficient solvent use in paints. Also, Simons and Mason (2003) found that Lean has a

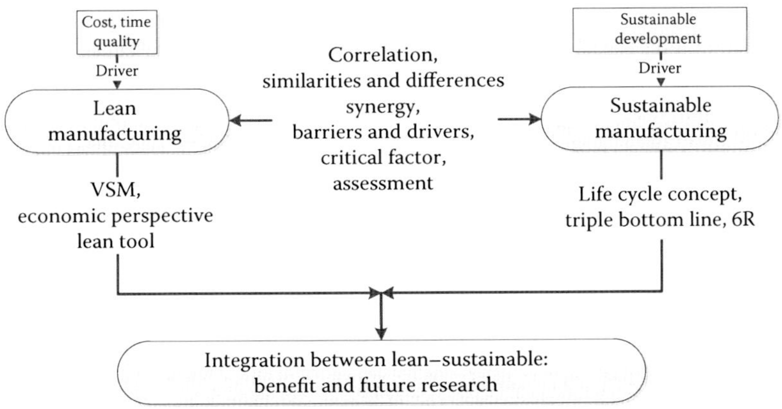

FIGURE 8.1 Flow of thematic analysis.

correlation with emissions reductions. Based on their research, studies on the relationship between Lean and sustainability are growing rapidly.

8.3.1.1 Correlation

Corbett and Klassen (2006) conclude that environmental issues can improve financial returns by opening up to new customers, competitive differentiation (and increasing market share), and reducing cost through waste reduction; focus on environmental improvement may create a more systems-focused approach to management (Corbett and Klassen 2006). Trade-offs between both practices are inevitable (Rothenberg et al. 2001). Clearly, not all Lean processes, procedures, and waste reduction efforts are positively related to environmental performance or pollution reduction, and Lean practices alone will never be enough to address all environmental issues.

The US Environmental Protection Agency (EPA) has successfully analyzed the effect of waste on the environmental impact based on Lean perspective. Conceptually, when a company attempts to reduce waste in a production line using the Lean approach, it indirectly attempts to improve its environmental performance. US EPA states that overproduction caused more raw materials and energy consumed in making the unnecessary products; that extra products may spoil or become obsolete requiring disposal; and that extra hazardous materials used result in extra emissions, waste disposal, worker exposure, etc. Inventory caused more packaging to store work-in-process (WIP); waste from deterioration or damage to stored WIP; more materials needed to replace damaged WIP; and more energy used to heat, cool, and light inventory space. Transportation and motion caused more energy use for transport (Chiarini 2014); caused emissions from transport (Simon and Mason 2003); caused more space required for WIP movement, increasing lighting, heating, and cooling demand and energy consumption; caused more packaging required to protect components during movement, damage, and spills during transport; and caused transportation of hazardous materials that requires special shipping and packaging to prevent risk during accidents. Defects caused more raw materials and energy consumed in making defective products; defective components require recycling or disposal; and defects caused more space required for rework and repair, increasing energy use for heating, cooling, and lighting. Overprocessing caused more parts and raw materials to be consumed per unit of production, and also unnecessary processing increases wastes, energy use, and emissions. Waiting caused potential material spoilage or component damage causing waste, and also wasted energy from heating, cooling, and lighting during production downtime (Chiarini 2014).

Some empirical reviews to determine the effect of Lean manufacturing on environmental performance are discussed in the following subsections.

8.3.1.1.1 Reducing Waste and Pollution

JIT as one of the techniques in Lean manufacturing is capable of reducing supply and waste. It is within the JIT concept to implement sustainable programs to improve efficiency and quality with reduced cost and waste. By reducing supply, the potential damage resulting from storage will decrease, such that waste will potentially decrease as well. However, the amount of waste reduction could not be quantitatively measured because the studies that were carried out used a qualitative method in the form of survey and interview (Florida and Davison 2001). This statement was supported by King and Lenox (2001) who conducted an empirical study regarding the correlation between Lean production and environmental performance. Analysis of the Toxic Release Inventory report of US manufacturing companies for 1991–1996 indicates that companies adopting Lean production were able to manage dangerous material supply to reduce waste and pollution. Companies were also more ready to implement ISO 14000. It is confirmed by Chiarini (2014) who performed a quantitative review on motor component companies in Europe. Results of the study state that 5S could reduce oil leakage and improve waste management, while total productive maintenance could help reduce emission of dust and chemical steam attributed to machinery processes. Pil and

Rothenberg (2003) found that Lean manufacturing could reduce defects and thus reduce the use of materials and emission. From some of the studies, it can be concluded that the implementation of Lean manufacturing techniques can reduce waste and emission.

8.3.1.1.2 Reducing Resource Needs

Rothenberg et al. (2001) reviewed the effect of Lean practice on the environmental and operational performance of car companies in Japan and North America. Air pollution, factory emission, efficiency of water consumption, and efficiency of energy consumption were qualitatively measured. The work culture of Lean manufacturing (i.e., to reduce waste) has encouraged employee participation in reducing consumption of water and energy. Quantitatively, Chiarini (2014) found that cellular manufacturing could reduce electricity consumption. Some of the studies indicate that the implementation of Lean manufacturing can reduce the use of resources.

8.3.1.1.3 Improving Environmental Performance

The implementation of Lean in Boeing Company has the potential to improve the productivity of resources and at the same time improve environmental performance (US EPA 2007). This finding is confirmed by Sawhney et al. (2007), who studied the relation between Lean practice (cellular manufacturing, employee involvement, mistake proofing, product mix, pull system, quick changeover, supplier development, and total productive maintenance) and environmental performance. There is a positive relation between Lean practice and environmental performance, except for the "small lot size" approach. "Small lot" has the potential to increase the frequency of transportation, which results in more emission. This conclusion is confirmed by Yang et al. (2011) with more samples in 22 countries. Lean manufacturing can also improve environmental performance when it is integrated into supply chain management and mediated by environmental practice (Hajmohammad et al. 2013). Some of the studies indicate that the implementation of Lean manufacturing can improve environmental performance.

8.3.1.2 Similarities and Differences

Dües et al. (2013) explored on the similarities and differences between Lean and green manufacturing. Lean and green paradigms overlap as regards waste reduction techniques, people and organization, lead time reduction, supply chain relationship, key performance indicators (KPIs), and service levels; they also share common tools and practices. Lean and sustainable manufacturing aim to reduce waste, and both involve organization to achieve goals; reducing lead time is crucial to reduce waste and environmental impact, supply chain affects achievement of goals, and customer satisfaction level becomes the KPI. The differences between the Lean and green paradigms lie on their focus (e.g., waste, customer satisfaction, product design and manufacturing strategy, end of product life management, KPIs, dominant cost, the principal tool used, and certain practices such as the replenishment frequency) (Table 8.3). Analysis of such differences shows the areas in which Lean and green practices are disconnected. However, it is also recognized that for these attributes, it is also not impossible to combine Lean and green practices.

Deshmukh et al. (2010) try to highlight the different aspects and benefits of Lean manufacturing system (LMS) and its implementation in Indian industries. A model to implement LMS is presented in this chapter, which includes 10 issues, 11 tools, and 14 results. Organizations in the global competitive market as well as those who wish to achieve sustainable development should strive for Lean operations through the application of Lean tools and techniques suitable to their situation.

8.3.1.3 Synergy

Bergmiller and Mccright (2009a) identify the correlation between green operations and Lean results. This study explores the impact of green programs on Lean results. Elements of a green operations

TABLE 8.3
Differences between the Lean and Green Paradigm

No.	Points of Difference	Lean	Green
1	Definition of waste	Waste, known as the Seven Deadly Wastes	Inefficient use of resources that finally becomes solid or liquid waste and CO_2 emission
2	KPI	Cost reduction is done by reducing waste	CO_2 emission
3	Analysis tool	Waste is identified by value stream mapping	Life cycle assessment
4	Focus	Activities that reduce cost and flexibility; customer satisfaction is measured from the aspect of cost and order fulfillment speed (lead time)	Sustainable development and impact of manufacturing on environment
5	Practice	Increase of replenishment frequency to reduce inventory	Reducing replenishment frequency because replenishment triggers transportation that triggers emission
6	Manufacturing strategy	Just-in-time	Remanufacturing
7	Product design	Products are designed to achieve performance maximization and cost minimization	Products are designed by analyzing ecological impact and sustainability during product life cycle
8	End of life	Lean does not concentrate on impact of product use on environment or recovery process of post-use of product	Green concentrates on impact of product during use and recovery strategy at the end of product life by recycle or reuse
9	Customer satisfaction	Customer satisfaction is triggered by cheap price and fast order fulfillment	Customer satisfaction is triggered by their satisfaction in preserving environment
10	Dominant cost	Physical cost	Cost for the future

system are product redesign, process redesign, disassembly, substitution, reduce, recycling, remanufacturing, consume internally, prolong use, returnable packaging, spreading risks, creating markets, waste segregation, and alliances. The Lean results elements consist of quality, cost, delivery, customer satisfaction, and profitability. This paper found that Lean companies that include green practices achieve better Lean results than those companies that do not. Winners and finalists of the Shingo Prize for Manufacturing Excellence (America's preeminent Lean designation) for 2000 through 2005 comprise the sample for this study. Bergmiller et al. (2009c) provide strong evidence of transcendence to green manufacturing by leading Lean manufacturers. The results indicate that the shingo plants were significantly greener in all but 1 of the 26 green manufacturing system measures. The evidence that plants with Lean systems yield higher green results supports the philosophical notion of Lean and green synergy (Bergmiller and McCright 2009b).

8.3.2 Impact of Integration Lean and Sustainable Manufacturing on Performance: Empirical Study

Lean and green have differences and similarities. Focusing on the differences is crucial while taking advantage of the Lean and green synergy when they are implemented together. Some reviews aim to take advantage of the integration of Lean and green. Joint implementation of Lean and green practices and their interaction was researched by Galeazzo et al. (2013). This paper found that reciprocal interdependencies are more likely to be associated with the involvement of external suppliers and that the simultaneous adoption of Lean and green practices ultimately leads to better operational performance.

8.3.2.1 Integration of Lean and Green Strategies to Increase Corporate Performance

Some reviews using case studies in companies were conducted to investigate the effects of integration of Lean and green on system sustainability improvement. Some results indicate that integration of Lean and green can improve sustainability indicators. The examples are discussed in the subsequent subsections.

8.3.2.1.1 Integration between Lean and Green to Reduce Material and Energy Consumption, Waste, and Production Cost (Economic and Environment Dimension)

Pampanelli et al. (2011) suggest an integration model of Lean and green by taking the Kaizen approach to reduce environmental impact on a production cell level. This model integrates the environmental sustainability concept into pure Lean thinking. This model adopts the Kaizen approach to improve the flow of mass and energy in a manufacturing environment that has implemented Lean thinking. This model consists of five stages:

1. Stabilize the value stream. Identify an operational cell that represents a significant use of resources, has a good deployment of Lean tools, and has a stable production flow that justifies application of the Lean and green model.
2. Identify environmental aspects and impacts. Aspect and impact definitions are considered according to ISO 14001:2004.
3. Measure environmental value streams: Identify the actual data on the environmental process. Collect environmental data. Organize the Kaizen event. The improvement metrics used for this Kaizen event are as follows: energy, water, metallic and contaminated waste and other waste, oils and chemicals.
4. Improve environmental value streams.
5. Continuous improvement.

The Lean and green Kaizen model presents a good example of how Lean works using the spirit of Kaizen (bottom-up approach) to involve people to be supporting tools in achieving business sustainability. The study found that a cell was a good initial point to implement Lean and green. A stable implementation level of Lean is vital to the success of Lean and green integration. The result of model implementation can reduce material needs, energy consumption, waste, and cost.

After the success with the pilot project in Monobloc A cell, Pampanelli et al. (2014) implemented the Lean and green Kaizen model in seven other cells in the same company. The result confirms that the Lean and green model can reduce the use of resources, total material cost, and energy.

Aguado et al. (2013) developed a general approach based on Lean and green to help companies achieve efficiency and sustainability. It aimed to add product value to improve competitiveness in the global market. A case study indicated that cost, income, social responsibility, and sustainability could be improved when environmental practice innovation was implemented by changing the traditional production system into a Lean system. Implementation of balance of track and layout improvement reduced processing time. It triggered productivity improvement and cost reduction. Implementation of green was done by replacing steel with aluminum. The effect of material replacement reduced environmental impact and material cost. It can be concluded that the implementation of Lean and green together can reduce cost and improve product quality and environmental performance.

8.3.2.1.2 Integration between Lean and Green to Reduce Production Cost (Economic Dimension)

Elsayed et al. (2013) integrated a Lean and green strategy in a powertrain spare part company. The manufacturing strategy chosen was consulted with experts and project partners. To estimate the impact, the strategy was modeled using discrete event simulation (DES). The Lean strategy simulated involved internal and external setup time reduction, quality level improvement, machine availability, tool life, and batch size reduction. The green strategy involved washing machines, turning, hobbing process, and deburring. The review concluded that integration of Lean and green strategy could reduce cost.

Some reviews regarding integration of Lean and green manufacturing still have limitations in improving social performance. The impact of Lean and sustainable manufacturing on performance (based on empirical studies) is explained in more detail in Tables 8.4 through 8.6.

8.3.2.2 Integration of Lean and Green Tools to Increase Sustainability

8.3.2.2.1 Concept of VSM as a Lean Tool

VSM is a strategic and operational approach in capturing data and in analyzing, planning, and implementing changes on company functions/processes to achieve efficiency (Hines and Rich 1997). Lean practitioners use VSM to identify waste sources in the form of current VSM and waste minimization plan in the form of future VSM. The strength of VSM lies on its ability to illustrate factory floor so as to be communication tool among workers that contains complete observation result of how products are actually made from the beginning to the end (US EPA 2007).

VSM focuses on a certain value stream or a certain product purchased by customers. It prevents chaos on different routes or processes that are adopted from different products or customers. VSM describes whole activities (including value-added activity and non–value-added activity) during the production process, including flow of information, material, decision making, and receipt of material by customers (Rother and Shook 1999). The stages to create VSM are as follows (Hines and Taylor 2000):

1. Preparing customer request record: Request record records data of type of product, number of product demand, time, number of components, number of products in one delivery, frequency of delivery or form of package, and others related to customer request.
2. Identifying flow of information: Flow of information records data from customers, delivery, and processing information until it results in the decision of production order or material purchase order to supplier.
3. Identifying physical flow: Physical flow records data of material reception from supplier including type of material, number of delivery, lead time, period of reception, storage time, number of defects, and others. After the material reaches the production department, the processes involved until the material becomes the final product are recorded. The data include processes, processing time, production size, whether there is waiting time, defect, WIP or bottleneck, and so on.
4. Connecting physical flow and information flow: In this stage, existing work instructions, such as where instruction is delivered and resolving problems that occur on physical flow, are identified.
5. Settlement of mapping and analysis: Mapping is finalized by making an analysis line under the map. The analysis line shows the time efficiency (value-added time ratio).

VSM usually consists of a current state map and a future state map. The current state map identifies waste while the future state map provides information on the technique to be used to reduce waste. After improvement is made, the data will be renewed and reevaluated. The standard structure of VSM is taken from Lasa et al. (2009) (Figure 8.2).

TABLE 8.4
Impact of Lean and Sustainable Manufacturing on Operational and Environment Performance

No.	Reference	Contribution	Criteria	Result
1	Lewis 2000	Impact Lean production and sustainable manufacturing on the overall competitive positions of adopter firms	Manufacturing process, human resource management, supplier input vs. profit, work-in-process, lead time and number of employee	Reduce lead time and work-in-process
2	King and Lenox 2001	Minimizing inventory and adopting quality standards are more likely to have lower emissions of toxic chemicals	ISO 9000 and max inventory vs. total emission and ISO 14000	Integrated Lean production and ISO 9000 have greater source reduction (lower emission) than ISO 14000
3	Rothenberg et al. 2001	Relationship between Lean manufacturing practice and environmental performance	Buffer, work, and HRM vs. water and energy use	Trade-off between Lean and environment performance
4	Pil and Rothenberg 2003	Environment performance as a driver of superior quality	Paint quality vs. resource utilization and emissions	Quality-related tools for environmental issues have corollary implications for quality
5	González-Benito 2005	Environmental proactivity and business performance: an empirical analysis	Environmental proactivity vs. business performance (quality, reliability, and volume flexibility)	A positive and significant effect on operational and on marketing performance
6	Sawhney et al. 2007	En-Lean: a framework to align Lean and green manufacturing in the metal cutting supply chain	Cellular, employee involvement, mistake proofing, product mix, pull system, quick changeover, small lot, supplier development, total productive maintenance vs. environment impact	Positive, except small lot
7	Bergmiller et al. 2009	Lean manufacturers' transcendence to green manufacturing	Green management systems and green waste reduction techniques vs. green business results	Positive relationship between Lean and green
8	Biggs 2009	Exploration of the integration of Lean and environmental improvement	Lean tools vs. environment waste	Lean manufacturing reduces environmental impact
9	Pampanelli et al. 2014	A model based on Lean concepts for managing environmental aspects capable of promoting a better integration of the environmental processes to business needs	Conceptual model: Kaizen vs. environment and operational performance	Lean and green models reduce energy, materials, waste, costs, and operational manufacturing mass

(Continued)

TABLE 8.4 (CONTINUED)
Impact of Lean and Sustainable Manufacturing on Operational and Environment Performance

No.	Reference	Contribution	Criteria	Result
10	Verrier et al. 2014	Combining organizational performance with sustainable development issues: the Lean and green project benchmarking repository	Green initiative vs. green performance	Green initiatives eliminate waste from the environmental perspective
11	Jabbour et al. 2013	The influence of environmental management on operational performance	Environmental management, Lean and human resource vs. operational performance	Positive relationship
12	Hajmohammad et al. 2013	A conceptual model: the magnitude of environmental practices mediates the relationship between Lean and supply management with environmental performance	Lean management vs. environment practice vs. environment performance	The extent of environmental practices mediates the relationship between Lean management and environmental performance
13	Dora et al. 2013b	Analyzes the application of Lean manufacturing, its impact on operational performance, and critical success factors in the food processing SMEs	Lean vs. inventory reduction, productivity improvement, lead or cycle time reduction, quality improvement, on-time delivery	Lean manufacturing improves operational performance, especially productivity and quality
14	Chiarini 2014	Lean production tools can help reduce the environmental impacts of manufacturing companies	Value stream mapping, 5S, cell manufacturing, total productive maintenance, single minute exchange dies vs. total emission	Lean tool reduces emission and improves environment performance
15	Khanchanapong et al. 2014	Manufacturing technologies and Lean practices have unique effects on operational performance dimensions	Manufacturing technology (design, production, admin) and Lean practice (flow, customer, process, supplier, workforce) vs. cost, lead time, quality, and flexibility	Interaction between manufacturing technologies and Lean practices produces positive effects
16	Bortolotti et al. 2014	Examines the relationship between bundles of Lean practices and cumulative performance	Just-in-time (JIT) and total quality management (TQM) versus quality, delivery, flexibility, cost performance	TQM and JIT bundles are directly related to the higher performance
17	Bourlakis et al. 2014	Firm size and sustainable performance in food supply chains: insights from Greek SMEs	Firm size vs. sustainable performance measures (consumption, flexibility, responsiveness, product quality, and total supply chain performance)	The Greek food supply chain members who overperform or underperform in relation to size

Source: Hartini, Sri, and Udisubakti Ciptomulyono. 2015. The relationship between lean and sustainable manufacturing on performance: Literature review. *Procedia Manufacturing*: 38–45.

TABLE 8.5
Impact of Lean and Sustainable Manufacturing on Operational and Economic Performance

No.	Reference	Contribution	Criteria	Result
1	Zhu and Sarkis 2004	Relationships between operational practices and performance among early adopters of green supply chain management practices in Chinese manufacturing enterprises	Green supply chain management vs. organizational performance moderated quality management and JIT	Positive relationships in terms of environmental and economic performance
2	Miller et al. 2010	Integrated Lean tools and sustainability concepts with discrete event simulation modeling to make a positive impact on the environment, society, and its own financial success	Cell vs. lead time vs. environment performance, recycling vs. profit and environment performance	Lean and green manufacturing reduce waste and increase profitability
3	Deif 2011	A system model for the new green manufacturing paradigm that captures various planning activities to migrate from a less green into a greener and more eco-efficient manufacturing	Green manufacturing activities vs. material, labor, emission, disposal, water, energy	Green manufacturing activities reduce material, labor, emission, disposal, water, energy
4	Yang et al. 2011	Empirical evidences with large sample size that environmental management practices (EMPs) become an important mediating variable to resolve the conflicts between Lean manufacturing and environmental performance	Lean manufacturing (JIT, quality, employee involvement vs. EMP vs. market), financial and environment performance	Environment management practices positively affect environmental, market, and financial performance
5	Hong et al. 2012	Firms striving for responsiveness to market and customers also improve environmental performance (EP) and confirm Lean practices as an important mediator to achieve excellent environmental performance; the focal company takes the lead in achieving EP	Response product strategy (RPS) vs. Lean practice (LP), supply chain vs. environment performance and firm performance (FP: sales, market share, rate of return)	Increasing EP is via organizational implementation of LP. This study shows that LP mediates the RPS to progress toward EP
6	Puvanasvaran et al. 2012	The characteristic of the Lean principles in ISO 14001 and to propose linkage between Lean principles and ISO 14001	Linkage between Lean practice (Kaizen, zero defect, Just-in-time, 5S, TPM, Kanban, standardized work) vs. ISO 14000	Lean principles have a positive relationship with ISO 14001 EMS to achieve continual improvement
7	Elsayed et al. 2013	Assessment of Lean and green strategies by simulation of manufacturing systems in discrete production environments	Lean and green strategies vs. production cost	The Lean and green strategy savings in continuous production costs

(Continued)

TABLE 8.5 (CONTINUED)
Impact of Lean and Sustainable Manufacturing on Operational and Economic Performance

No.	Reference	Contribution	Criteria	Result
8	Aguado et al. 2013	Model of efficient and sustainable improvements in a Lean production system through processes of environmental innovation	Model vs. cost, income, and environment impact	Cost, incomes, social responsibility, and sustainability can be improved
9	Azadegan et al. 2013	The results show that environmental complexity positively moderates the effects of Lean operations and Lean purchasing on performance	Lean operation and Lean purchasing (LP) vs. performance	A positive interaction between LP and environmental dynamism on gross margins
10	Fullerton et al. 2014	Lean manufacturing practices are directly related to operations performance and indirectly affect operations performance through Lean MAP (management accounting practices)	Lean MAP vs. operational performance vs. financial performance	Lean manufacturing practices are directly related to operations performance
11	Galeazzo et al. 2013	Discern how Lean and green interact and how they yield maximum synergy in improving both operational and environmental performance	Lean green vs. emissions, hazardous waste disposal, production cost savings, productivity, product quality, volume flexibility	A simultaneous implementation of Lean and green practices can reach higher operational performance

Source: Hartini, Sri, and Udisubakti Ciptomulyono. 2015. The relationship between lean and sustainable manufacturing on performance: Literature review. *Procedia Manufacturing*: 38–45.

TABLE 8.6
Impact of Lean and Sustainable Manufacturing on Operational and Social Performance

No.	Author	Contribution	Criteria	Result
1	Glover et al. 2011	Identifies the factors that most strongly influence the sustainability and commitment to Kaizen events based on a field study	Kaizen event, work area, post-event characteristics vs. work area attitude and commitment	Positive relationship
2	Bockent et al. 2014	A novel value mapping tool was developed to support sustainable business modeling	Value (value captured, missed/destroyed or wasted, and opportunity) and stakeholder (environment, society, customer, and network actors)	Categorization can reduce social and environmental negatives

Source: Hartini, Sri, and Udisubakti Ciptomulyono. 2015. The relationship between lean and sustainable manufacturing on performance: Literature review. *Procedia Manufacturing*: 38–45.

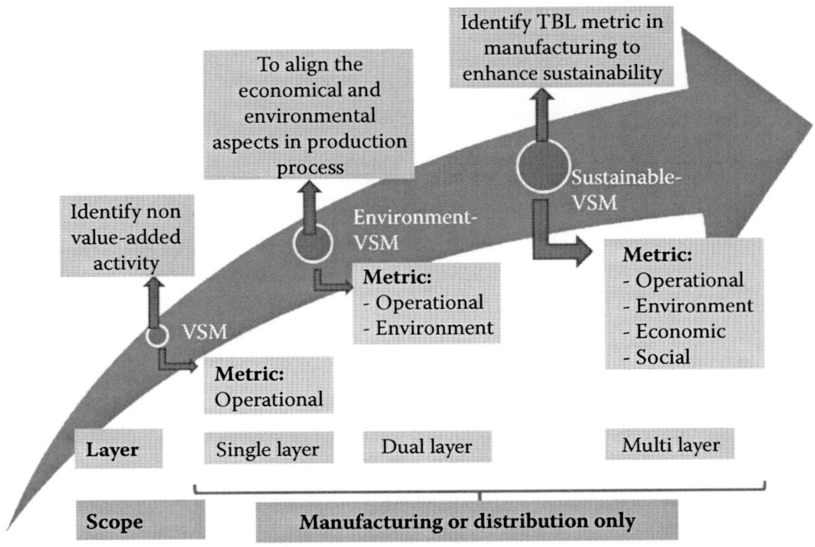

FIGURE 8.2 Evolution of sustainable VSM as an analysis tool in Lean manufacturing.

8.3.2.2.2 Growth of VSM

Some variations of VSM have been developed. VSM has evolved from 1997 to 2016. It is influenced by the needs and characteristics of manufacturing itself or by science advancement and customer demand. Some developments of VSM are as follows:

a. VSM as a Lean technique to identify waste sources: Up until 2000, VSM has been used only as a tool for waste identification (Donatelli and Harris 2001; Hines and Rich 1997; Irani and Zhou 2011; Khaswala and Irani 2001).
b. Tool to redesign production system: VSM is a useful and efficient tool in production system redesign. Although it needs a long period of time to create not only current state mapping but also future state mapping, and although not all alternatives for improvement could be implemented, companies are satisfied by the ability of VSM as a tool rich in important information for production system redesign. It is based on an interview of 12 companies that stably implement Lean manufacturing (Lasa et al. 2008).
c. VSM becomes a roadmap in improving system performance: VSM is a capable tool to transform traditional manufacturing into Lean manufacturing. In VSM, there exist "current" information and expected ideal processes so that management has a roadmap in implementing improvement (Donatelli and Harris 2001). VSM can be combined with simulation to test an improvement plan easily, cheaply, and fast without having to interfere with the production process. Therefore, VSM can give information about indicator performance that can be achieved. As done by Lian and Van Landeghem (2002), who implemented improvement of factory layout and Kanban to reduce lead time, WIP, add value-added ratio, and overcome bottleneck. The same thing was done by Vinodh and Rathod (2010) in implementing VSM in car component industries in India. The result was in the form of VSM that identified waste sources and future state mapping that suggested strategies to improve production line performance. The result was also in the form of reduction of lead time and improvement of machine utility as well as overall equipment effectiveness. Some other researchers also developed VSM combined with simulation to anticipate uncertainty during the production process to improve performance (Abdulmalek and Rajgopal 2007; Belokar et al. 2012;

Braglia et al. 2006; Chen and Meng 2010; Lian and Van Landeghem 2002; Singh et al. 2011; Solding and Gullander 2009).

d. VSM to evaluate and redesign supply chain: Hines and Jones et al. (1999) developed VSM concept to develop supplier network in a distribution. Distribution is important part of supply chain. Coronado and Lyons (2007) use VSM as a tool to redesign supply chain. VSM can illustrate customer demand, supply chain cycle, and inventory flow.
e. VSM combined with other tools: VSM can also be integrated with other concepts to improve the quality of decision making, such as agent-based simulation by Xie and Peng (2012). Braglia et al. (2009) integrated VSM with fuzzy theory to implement VSM in uncertain production systems. VSM was integrated with activity-based costing to improve performance (Abuthakeer et al. 2010).
f. Development of VSM to improve sustainability: Conventional VSM focuses on waste of "time" used in the production process. Environmental indication is the weak side of Lean concept that tends to view waste only as unnecessary material cost excess. From an environmental perspective, it is important to identify the type of waste because different types of waste have different impacts on the environment. Although it was believed that there was a correlation between type of waste and its impact on environment (EPA 2007), VSM does not explicitly identify an environmental metric. It is because environment is not the key indicator in Lean implementation. To obtain a system that is more competitive and responsive toward the environment, some researchers tried to integrate an environmental indicator into VSM. Development of VSM to improve sustainability will be discussed in a separate subsection.

8.3.2.2.3 Extended VSM to Enhance Sustainability

Development of VSM to improve system sustainability was initiated by Simons and Mason (2003). Initially, it only involved an environmental metric, followed by an economic and a social metric.

8.3.2.2.3.1 Addition of an environmental metric to VSM Simons and Mason (2003) suggest sustainable value stream mapping (SVSM) as a means to improve sustainability in manufacturing products by analyzing greenhouse gas emission, CO_2, and value-added ratio. SVSM is implemented for the Transportation Department of England as a distribution model for farmers to retail that maximizes added value and minimizes CO_2 emission for commodities such as lettuce, apples, and cherries. However, the evaluated indicator is only CO_2 emission during transportation and travel time.The suggested performance measurements are "maximize value add %" and "minimize CO_2 %."

The suggestion did not pay attention to time and CO_2 emission during the production process. Neither did the suggestion involve a social metric directly but assumed that increasing economic and environmental value will improve social benefit to achieve sustainability. Unfortunately, the assumption is not supported by proof of how far improvement in economic and environmental performance will improve social performance. Although VSM only adds CO_2 emission calculated based on the transportation system and has not considered the production process, the concept suggested is the beginning of VSM development to improve sustainability. The next advancement in VSM is dominated by the addition of environmental, economic, and social metrics as well as a visualization technique to VSM.

Some later studies that mention as sustainable-VSM are still limited by addition of an environmental metric to VSM. Norton and Fearne (2009) expanded the scope of sustainable VSM by measuring physical waste and wasted resources, as did Simon and Mason (2003). Physical waste includes food waste and packaging material. The performance measurement added "minimize food waste," which is the ratio between supply chain food waste and unit weight of product.

Torres and Gati (2009) developed environment VSM as a tool to map the process by aligning economic vision (profit) with the social and environment aspect (planet and people). Action research

was carried out in alcohol and sugar factories. This focused more on water consumption and loss of water. Water is considered a natural resource with significant social impact. The developed environment VSM could find loss due to inefficient water use. VSM can identify and prioritize the risk of sustainability and formulate systematic steps to analyze the use of resources and create added value in the whole chain. The case study is limited to the identification of water needs, liquid waste produced, and cost needs. The review did not analyze the efficiency of water use.

Kuriger and Chen (2010) added an energy and environmental metric to the standard VSM for manufacturing companies. Special visualization regarding energy and environment in VSM facilitated an understanding of the operational performance, energy efficiency, and environmental impact in value stream. The energy metric identified the type and amount of energy used in each process. The energy metric measured two categories: energy source (electricity/natural gas) and non-process energy use (heating or lighting). The method was applied in manufacturing that involves machinery, heating, painting, assembling, inspection, and delivery. The method can identify the percentage of value-added time, material, and energy used from the beginning of the process until storage. However, the study was not continued until the making of future state-mapping and improvement steps so that suggested method performance was not known. A study evaluating energy was also conducted by Lourenço et al. (2013), Keskin et al. (2013), and Müller et al. (2013).

Li et al. (2012) improved VSM by adding a carbon emission efficiency indicator to the analysis line and a sensitivity analysis approach to manufacturing companies. VSM identifies the carbon emission of each process that will be made as a basis of product carbon efficiency calculation. Improvement of product carbon efficiency is done by improving the process, and a process efficiency sensitivity analysis is performed to identify process obstacles and system improvement potential. The model was tested in printed circuit board (PCB) companies. The result showed that the PCB carbon efficiency increased and the modification made could reduce cost. It also showed that the suggested method can be implemented to improve the process. The formula used to identify carbon emission is the ratio between carbon emission value added and carbon emission total.

Value-added carbon emission concerns carbon emission related to the production process that adds value, whereas total carbon emission also involves process carbon emission that does not add value, such as transportation. The carbon efficiency measurement performed on each process is called process carbon emission. Then, the carbon emission of all processes is added to determine the carbon efficiency of the product.

Ng et al. (2015) developed a carbon value efficiency (CVE) measurement model determined from VSM. The case study was done in a metal-stamped production line and the development done was able to improve carbon efficiency value and lead time efficiency and reduce carbon emission. The case study shows that integration of Lean and green can improve environmental performance. The measurement model is "carbon value efficiency (CVE)," which is determined to be the ratio between the value-added time and carbon $\text{footprint}_{\text{total}}$. Carbon $\text{footprint}_{\text{total}}$ is the sum of carbon footprint value added time (CFP_{VAT}) and carbon footprint nonvalue added time ($CFPN_{VAT}$). CFP_{VAT} is the carbon footprint value-added time determined from the impact of all activities during the process that add value multiplied by the emission factor of the activities, whereas CFP_{NVAT} is calculated with the same method for all processes that do not add value. *N* is the number of processes/activities.

Marimin et al. (2014) and Folinas et al. (2015) developed VSM to map and evaluate environmental performance in the agricultural sector. However, the measured metric was limited to material and energy needs.

US EPA (2007) developed five steps in creating VSM to reduce environmental impact and, at the same time, to include social elements:

1. Using an icon to identify a process with potential EHS (environment, health, and safety) in VSM and involving EHS staff in Lean planning
2. Recording data of environmental performance in VSM and considering the date in developing future state mapping

3. Comparison analysis between the material used and the one needed by products is summarized in the material line
4. Adding VSM with other resources such as consumption of water and energy, waste of water and energy
5. Finding potential improvement of Lean and environment using developing future state mapping

Baxter Healthcare Corp. developed environment VSM to improve performance. With environment VSM, companies are able to lessen water consumption. Although environment VSM involves a social indicator, during the implementation, it tends to evaluate economic and environmental performance more. An EPA toolkit has been implemented by some case studies with companies such as 3M, General Motors, and Lockheed Martin. However, the suggestion does not explain the details of metric visualization in VSM.

8.3.2.2.3.2 Addition of an environmental and a social metric to VSM Faulkner et al. (2012) developed sustainable VSM (Sus-VSM) including economic, environmental, and social performance in manufacturing companies. The measured metric includes environmental and social performance. The environmental indicator measures water, material, and energy consumption, while the social indicator measures work environment and physical work. Physical work measures PLI (physical load index). Work environment involves four categories of risk and noise level. The four categories of risk for workers are electrical systems (S), hazardous materials used (H), pressurized systems (S), and high-speed components (S). Visual symbols are created for each metric to facilitate the identification of area with potential improvement.

In-depth analysis regarding metric and feasibility of Sus-VSM was studied by Faulkner and Badurdeen (2014). A more comprehensive case study was performed by Brown et al. (2014) on three manufacturing systems with different characteristics. The first case involved a manufacturing system with a high volume of a low-variation product (parabolic), the second case involved a low volume of a high-variation product (cathode dispenser), and the third case involved a medium volume of a low-variation product (fin mortar). The result proved that Sus-VSM is able to visualize the production process from a sustainability perspective in various characteristics of a manufacturing system. Sus-VSM provides the facility to measure sustainability performance and to find opportunities to improve performance on an operation level. However, because the study did not reach the stage where future state mapping is developed, it was unable to show the improvement in system performance. Additional case studies are needed to validate the method.

8.3.2.2.3.3 Addition of environmental, social, and economic metrics to VSM Paju et al. (2010) developed a VSM-based process mapping framework introduced as sustainable manufacturing mapping (SMM). This tool includes sustainability indicators, life cycle assessment (LCA), and DES in VSM. LCA can measure environmental indicators and DES considers the dynamics of measured data. The SMM result is in the form of a map as communication tool between shop floor and the management. The SMM output is balance of material and energy in the form of a figure and a graph to achieve strategic goals (future state map). However, the suggestion was still in the form of a framework, namely, a computerization model that identified the production process and that is connected to the indicators to be measured. SMM works based on goals. For example, to reduce production cost and environmental impact, parameters, such as production efficiency, material efficiency, or waste reduction, need to be improved. However, this concept has not yet been tested in companies to determine its feasibility. Vinodh et al. (2016) integrated VSM and LCA to ensure sustainable performance in production. The suggested framework can visualize and assess the performance of the manufacturing process from a sustainability aspect. To measure improvement, the current environmental, economic, and social performances were compared with those after improvement. The case study involved the production process of automotive component parts. The study was able to improve sustainable performance of companies and reduce environmental impact. The study was able to overcome the problem of Faulkner and Badurdeen (2014) as regards visualization of future state mapping. The work framework that integrates LCA and VSM can provide organizational insight with comprehensive assessment regarding energy consumption and added value of processes. Besides its scientific value, the framework helps practitioners identify chances of improvement in reducing environmental impact by combining Lean and sustainability initiatives. The drawback is not having one aggregation value that reflects the sustainability performance of the system identified by VSM. This chapter provided an overview of the development of VSM to enhance sustainable manufacturing (Table 8.7). The evolusion of sustainable VSM as an analysis tool in Lean manufacturing is described in Figure 8.2.

8.4 DISCUSSION

8.4.1 Metrics in Extended VSM

A metric is a parameter that can be measured to indicate the performance of a system. From the extended VSM that has been developed since 2003, there are around 31 metrics and 9 analysis indicators that have been involved (Table 8.8).

Metrics and indicators are classified in a TBL perspective based on economic, environmental, and social dimensions (Figure 8.3). Metrics frequently involved in VSM reviews are written in red. Such metrics are discussed in the subsequent sections.

8.4.1.1 Metric Related to Economic Dimension

The economic metric measures all costs incurred during the production process, either valuable or not (e.g., material cost, worker cost, machinery process cost, and energy cost).

Operational metrics have been widely used in conventional VSM. The metrics are measured to analyze the existence of seven types of system waste. The metrics are cycle time, takt time, changeover time, downtime, setup time, WIP, and transportation time. In triple bottom line analysis with three main pillars, the operational metric has been frequently classified into the economic dimension. The operational metric directly affects system productivity level and indirectly affects economic performance.

8.4.1.2 Metric Related to Environmental Dimension

Many metrics measured in VSM are related to the environmental dimension. However, the most frequently measured ones are energy, material, water consumption, and environmental impact (Hartini et al. 2016).

8.4.1.2.1 Water Consumption Metric

The water consumption metric measures the amount of water used during the manufacturing process as an important aspect that has to be evaluated to improve sustainability from an environment perspective. However, water added to a product (such as liquid chemicals) is included as material consumption.

The metric measured in each process includes

- Water required
- Actual use
- Amount of water lost

Lost water is water not recycled in a wastewater treatment plant (WWTP). If water has been recycled from one process to another, it is not considered "lost water." Coolant and oil in manufacturing processes affecting environmental performance can be measured with the same method.

TABLE 8.7
Development of VSM to Enhance Sustainable Manufacturing

TBL	Writer	Suggestion	Goals	Metric	Limitation
Environment	Simons and Mason 2003	Sustainable VSM (SVSM)	Optimizing value-added time ratio and minimizing CO_2 emission on supply chain network	Added value and CO_2 emission	Limited to CO_2 emission measurement and transportation during distribution Assuming that economy and environment will be profitable socially
	US EPA 2007	EPA Lean and environmental toolkit	Developing a toolkit to evaluate Lean and environmental performance in manufacturing companies	Added value, hazardous waste, water and material consumption	Limited to material, water usage, and hazardous waste, not yet involving energy usage and social metric
	Torres and Gati 2009	Environmental VSM (EVSM)	Developing VSM that involves environmental metric, especially water as a resource with significant social impact	Material and water consumption, amount of waste, liquid waste management cost	Able to find water waste source during production but not yet finding a strategy for system improvement
	Norton and Fearne 2009	Sustainable value chain map (SVCM)	Developing SVCM to improve operational and environmental performance	Time, emission, CO_2, water consumption, amount of food waste, package waste, environmentally friendly package	Expanding the scope of sustainable VSM by adding physical waste and wasted resources
	Kuriger and Chen 2010	Energy and environment VSM (EE-VSM)	Developing VSM to measure efficiency, energy consumption, and environmental impact	Time, energy, material consumption, amount of waste	Limited to material needs and waste where the suggestion is still in conceptual form with validation using simulation
	Dadashzadeh and Wharton 2012	Green value stream map for an IT service	Systematic approach in order of "greening" in information technology department	Time, energy, material consumption, amount of waste, emission, biodiversity	The suggestion is still conceptual
	Li et al. 2012	VSM and carbon emission stream	Improving economic performance and environmental impact with minimization of carbon emission in the form of VSM	Time, energy, water consumption, product carbon efficiency	Developing production performance in the form of product carbon emission with VSM, but limited to environmental performance

(*Continued*)

TABLE 8.7 (CONTINUED)
Development of VSM to Enhance Sustainable Manufacturing

TBL	Writer	Suggestion	Goals	Metric	Limitation
	Lourenço et al. 2013	Multilayer stream mapping (MSM)	VSM to measure and improve energy, time, and cost efficiency	Time, energy, and cost efficiency	Analyzing VSM using efficiency approach for time, energy, and cost needs, but still conceptual
	Keskin et al. 2013	Energy value stream maps (E-VSMs)	VSM to improve energy efficiency of manufacturing companies of UKM scales	Time and energy consumption	Measured performance limited to energy needs
	Müller et al. 2013	Energy value-stream mapping	VSM to improve energy efficiency	Time and energy consumption	Measured performance limited to energy needs
	Marimin et al. 2014	Green value stream	Mapping, evaluating, and improving green productivity index (GPI), including environmental and economic performance	Energy, water, material, waste, transportation, emission, and biodiversity	Minimizing waste by reusing water is the best strategy to improve GPI. However, the chosen strategy has not been validated by case study or simulation
	Folinas et al. 2014	Greening the agrifood supply chain with Lean thinking practices	Systematic approach to measure environmental performance on supply chain of agricultural sector using VSM	Time, water, energy consumption	Limited to energy need in agricultural sector, not yet involving social metric
	Folinas et al. 2015	VSM to greening the canned peach production	Systematic approach to measure environmental performance on supply chain of agricultural sector using VSM	Energy consumption	Limited to energy need in agricultural sector
	Ng et al. 2015	Carbon-value efficiency (CVE)-VSM	Integrating Lean and green practice in VSM and developing index reflecting production system performance called carbon-value efficiency metric	Time, carbon footprint efficiency	Has a production index, but only involves environmental metrics, limited to production process

(Continued)

TABLE 8.7 (CONTINUED)
Development of VSM to Enhance Sustainable Manufacturing

TBL	Writer	Suggestion	Goals	Metric	Limitation
Environment and social	Sparks 2014	SC Sus-VSM: combining Sus-VSM and simulation to SCM	Combining Sus-VSM and simulation to measure performance of supply chain	Time, material, water, and energy consumption, ratio of defect product, ratio of local TK, diversity, hazardous material, TK training	Involving environmental and social metric in some factories but not yet involving economic aspect
	Faulkner and Badurdeen 2014	Sustainable value stream mapping (Sus-VSM)	Developing Sus-VSM to evaluate sustainability performance including all indicators of TBL in manufacturing	Water, material, energy, physical load, work environment	Involving environmental and social metric but not yet economic aspect and limited to scope of manufacturing
	Brown et al. 2014	Sustainable value stream mapping (Sus-VSM)	Testing sustainable VSM (Sus-VSM) by conducting a case study on various types of manufacturing system	Water, material, energy, physical load, work environment	Involving environmental and social metric but not yet economic aspect and limited to scope of manufacturing
Environment, economy, and social	Vinodh et al. 2016	LCA-VSM and end-of-life scenario	Framework integrating VSM and LCA to achieve sustainability performance	Time, material, energy, cost, environmental impact	VSM has measured all indicators, but not yet results in aggregate index reflecting system performance
	Paju et al. 2010	Sustainable manufacturing mapping (SMM)-DES and LCA	Combining LCA, simulation (DES), and value stream mapping (VSM) to be a model to measure sustainability indicators	Energy, material, emission, cost, working hours, number of reclamation	Have involved all economic, environmental, and social indicators, but still limited to conception validated with simulation

TABLE 8.8
Metrics Adding in Extended VSM

No	Year	Author	Operational#							Environment#											Eco#	Sosial#												Indicator Line#								
			1	2	3	4	5	6	7	8	9	10	11	12	13	14	15	16	17	18	19	20	21	22	23	24	25	26	27	28	29	30	31	32	33	34	35	36	37	38	39	40
1	2002	Simons and Mason					*			*																																
2	2007a	US EPA	*	*			*					*	*						*	*				*											*							
3	2007	Norton	*	*			*			*				*	*	*		*		*																						
4	2009	Torres and Gati	*			*											*			*	*													*								
5	2010	Paju et al.	*							*	*		*						*		*		*		*																	
6	2010	Kuriger and Chen	*	*							*		*						*															*	*							
7	2012	Dadashzadeh and Wharton							*	*	*		*						*	*		*																				
8	2012	Li et al.	*							*	*									*														*					*			
9	2013	E.J. Lourenço et al.	*					*			*										*													*			*		*			
10	2013	Keskin et al.	*	*			*	*			*																							*			*					
11	2013	Muller et al.	*				*				*																							*			*					
12	2014	Faulkner and Badurdeen	*	*			*	*			*		*						*	*				*		*	*							*	*	*	*					
13	2014	Sparks and Badurdeen	*					*	*		*		*						*	*								*	*	*	*	*	*	*		*	*			*		
14	2014	Brown et al.	*	*			*	*			*		*						*	*				*		*	*							*	*	*	*					
15	2015	Folinas et al.	*																															*		*	*					
16	2015	Folinas et al.	*																				*														*	*				
17	2016	Vinodh	*	*			*	*			*		*						*	*	*		*	*		*	*							*	*	*	*				*	*
18	2015	Ng et al.	*	*	*																													*								*

Note:

Operational:
1: cycle time
2: change over
3: downtime
4: setup time
5: value added time
6: work in process
7: transportation time

Environment:
8: CO2 emission
9: energy
10: hazard waste
11: material waste
12: Biodegradable Packaging Waste
13: Plastic Packaging Waste
14: waste metal container
15: water waste
16: Food waste
17: material consumption
18: water consumption

Economic:
19: cost

Social:
20: biodiversity
21: number of man hour
22: environment, health, safety
23: absence day
24: physical load index
25: Noise
26: Product defect ratio
27: Local community
28: sickness rate
29: employee training
30: Hazard material
31: Diversity ratio

Analysis Line
32: VAT line
33: Material line
34: Water
35: Energy
36: Steam %
37: Carbon emission
38: GHG emission
39: cost
40: Carbon footprint

Source: Sri Hartini et al. 2017. Extended value stream mapping to enhance sustainability: A literature review. In 3rd International Materials, Industrial and Manufacturing Engineering Conference (MIMEC2017), Vol. 1902(1). AIP Publishing.

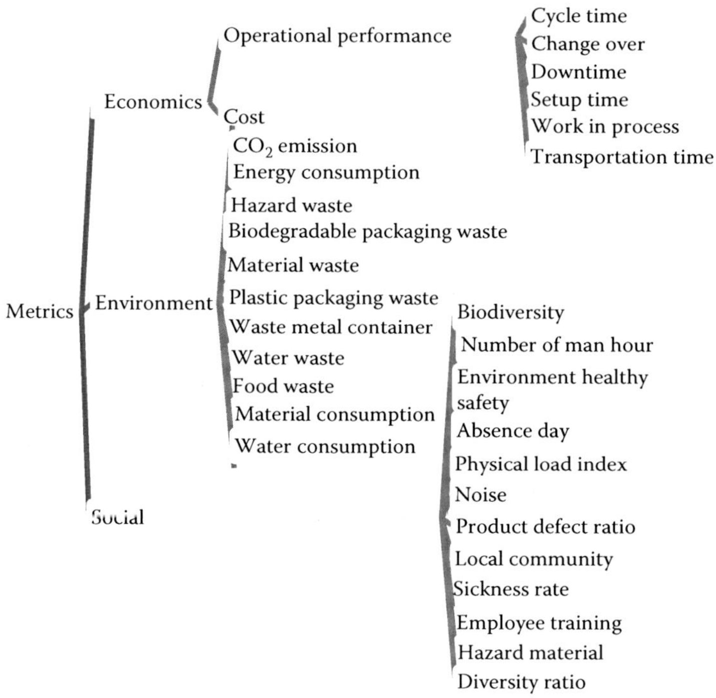

FIGURE 8.3 Metrics for extended-VSM in a triple bottom line perspective.

8.4.1.2.2 Raw Material Usage Metric

Most material waste in manufacturing is in the form of scrap. Recycling of scrap is an alternative to improve sustainability, although it needs additional energy and resources. Therefore, measurement of material use includes used material (added material) and lost material (material removed).

If the system only compares the initial material and the final material without taking pictures of the two indicators above in each process, one cannot identify material use in each process. For example, if the material removed can be identified in a process, the solution to material use efficiency improvement can be easily determined. The value of the material added and removed in each process will be aggregated to investigate total material use.

8.4.1.2.3 Energy Consumption Metric

The energy consumption metric identifies the amount of energy consumed by each process, including energy for transportation during manufacturing and storage. Indirect energy consumption includes lighting, heating, cooling, and other energy not linear to the number of products not involved in sustainable VSM.

8.4.1.3 Metric Related to Social Dimension

Sustainability needs to pay attention to the impact of social performance on stakeholders involved in the product-making process. The worker are important stakeholders (Faulkner and Badurdeen 2014), so that social aspect evaluation usually focuses on health and safety of workers as basis of measurement and monitoring. The social metric is divided into two categories, namely, physical work and work environment.

8.4.1.3.1 Physical Work Metric

This metric begins with observing the economic condition of the work environment. The tools usually used are Rapid Entire Body Assessment and Rapid Upper Limb Assessment, and the NIOSH tool (Faulkner and Badurdeen 2014). The methods are difficult to be stated in VSM. VSM does not need to measure environmental ergonomy comprehensively, but instead identifies more work environment risks to be analyzed further. Therefore, a simple approach is more felt. PLI determines questionnaire-based responses based on the frequency of occurrence against body position change and workload (Hollmann et al. 1999). In VSM, PLI is measured in each process and each operation between processes. When some processes involve some operators, the maximum or the average value of PLI is recommended more in VSM. Processes with a high PLI value can be analyzed further with other tools.

8.4.1.3.2 Work Environment Metric

The work environment metric includes noise level and four categories of risk, namely, electrical systems (E), hazardous chemicals/materials used (H), pressurized systems (P), and high-speed components (S). The metric uses a rating system from 1 to 5, with risk level as in Table 8.9. The noise level is measured by dividing the "time actually spent at sound level" by the "maximum permissible time at sound level" (OSHA 2008).

8.4.2 Indicators in Extended VSM

Among the indicators analyzed in VSM, value-added time ratio and material and energy consumption ratio are the indicators mostly analyzed (Figure 8.4).

8.4.2.1 Value-Added Time Ratio

In conventional VSM, the indicators to be analyzed are limited to the efficiency of process time. Process time efficiency is approached by comparing time needed for valued activities (value-added time) with total time needed during the production process (total time). The analysis line in conventional VSM is shown in Figure 8.5.

8.4.2.2 Cost Analysis

Cost analysis evaluates cost incurred for activities that add value and for those that do not add value. Cost efficiency can be determined by comparing the two. The identified costs are material cost, worker costs, machine costs, and inventory cost.

TABLE 8.9
Risk Scale in Work Environment

Risk Value	Description
–	No potential risk
1	Risk exists, but the impact and the chance of occurrence are low
2	Risk exists, but the impact is low while the chance of occurrence is high or otherwise
3	Risk exists with impact and chance of occurrence that are low
4	Risk exists with medium impact, but the chance of occurrence is high or vice versa
5	Risk exists with high chance of occurrence and impact

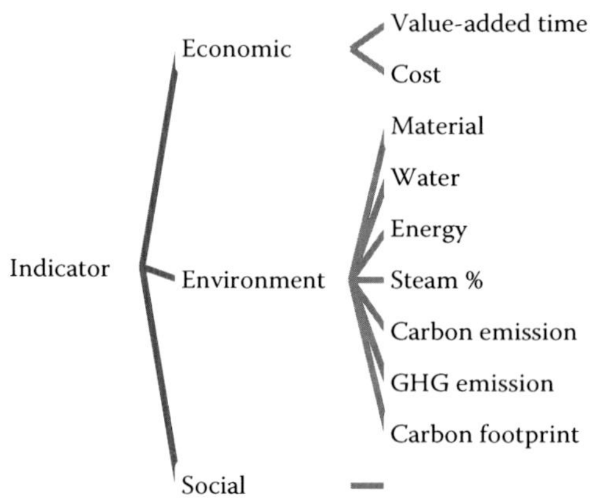

FIGURE 8.4 Classification of indicators in value stream mapping (VSM) based on the triple bottom line (TBL) perspective.

8.4.2.3 Analysis of Material Needs

Analysis of the material needs evaluates the materials to be used and the amount of wasted materials so that the efficiency of material consumption can be determined.

8.4.2.4 Analysis of Energy Needs

Analysis of energy needs evaluates energy need for processes that add value and do not add value. Energy needs that add value include machine processes, whereas energy used for transportation and setup time does not add value.

8.4.2.5 Analysis of Environmental Impact

Environmental impact is calculated in each activity, on both processes that add and do not add value. CVE compares environmental impact in valued activities with total environmental impact during the production process.

There are some aids in calculating environmental impact assessment, such as LCA. LCA is a concept or methodology to estimate and assess the environmental impact contributed by each phase of product life, such as climate change, stratospheric ozone depletion, tropospheric ozone (smog) creation, eutrophication, acidification, toxicological stress on human health and ecosystems, the depletion of resources, water use, land use, and noise (Rebitzer et al. 2004).

The LCA method is usually used to calculate and analyze the environmental impact and/or product damage in all life cycles. LCA is currently the most qualified tool to analyze and measure the consumption of resources and environmental impact produced in all product life cycle phases (Jeong et al. 2015). According to ISO 14040 and 14044 standards, LCA is carried out in four phases, namely, determination of goals and assessment scope, life cycle inventory analysis, assessment of life cycle impact, and assessment result interpretation.

The analysis line in sustainable VSM is multilayered because it includes the analysis line of time, cost, material efficiency, energy, and environmental impact (Figure 8.6).

8.4.3 Needs of Lean and Green Development to Improve Sustainability Rate

Although integration of Lean and green shows a positive effect on system performance, Dües et al. (2013) also recorded the existence of challenges faced by companies during integration and implementation of Lean and green practices. One of the challenges is in the form of resource obstacles. Practitioners felt that it was impractical to implement either Lean or green practices with

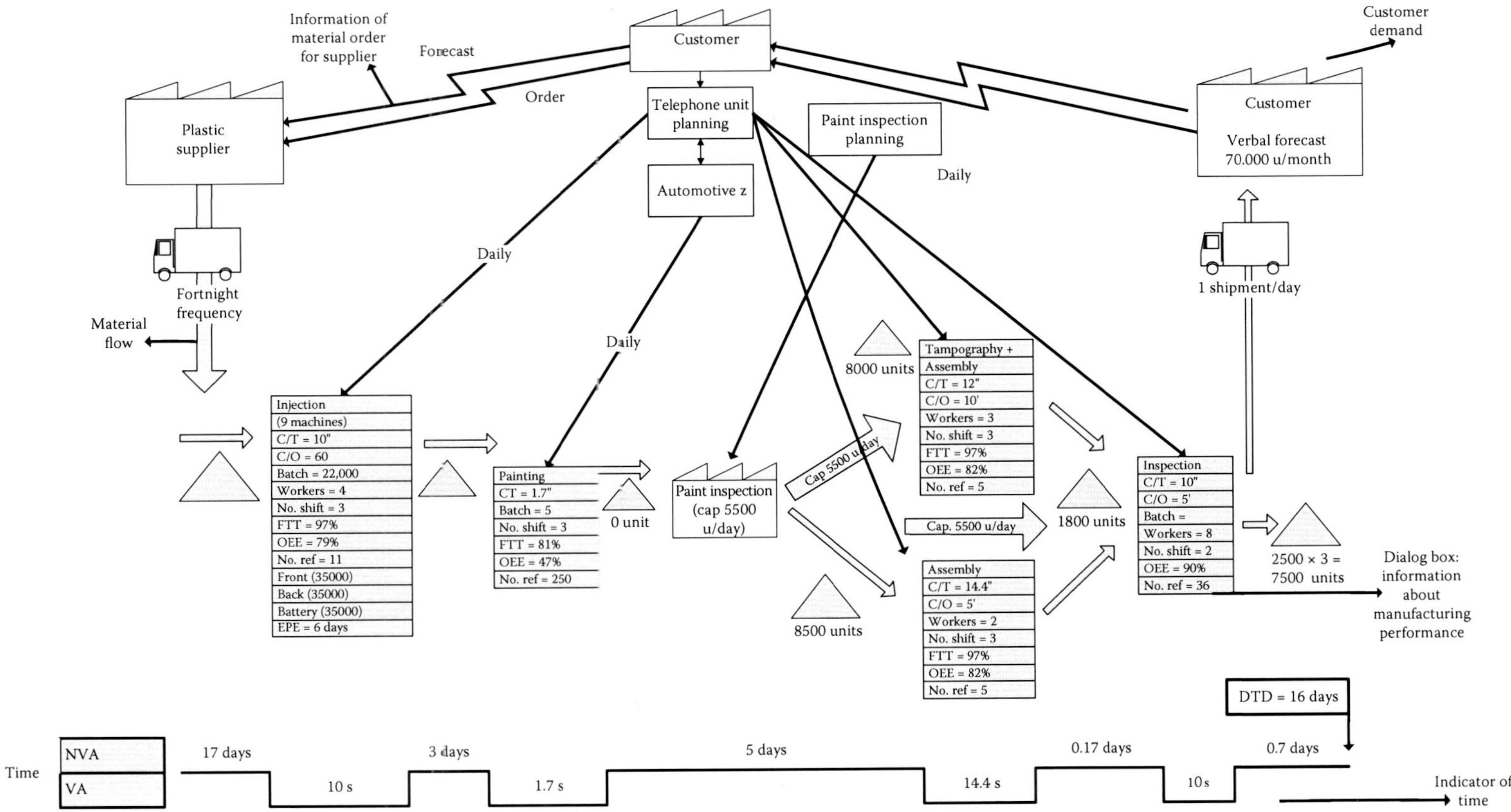

FIGURE 8.5 Conventional VSM: analysis line is a single layer (time indicator). (From Lasa, Serrano Ibon, Rodolfo De Castro, and Carlos Ochoa Laburu. 2009. Extent of the use of lean concepts proposed for a value stream mapping application. *Production Planning & Control* 20(1): 82–98.)

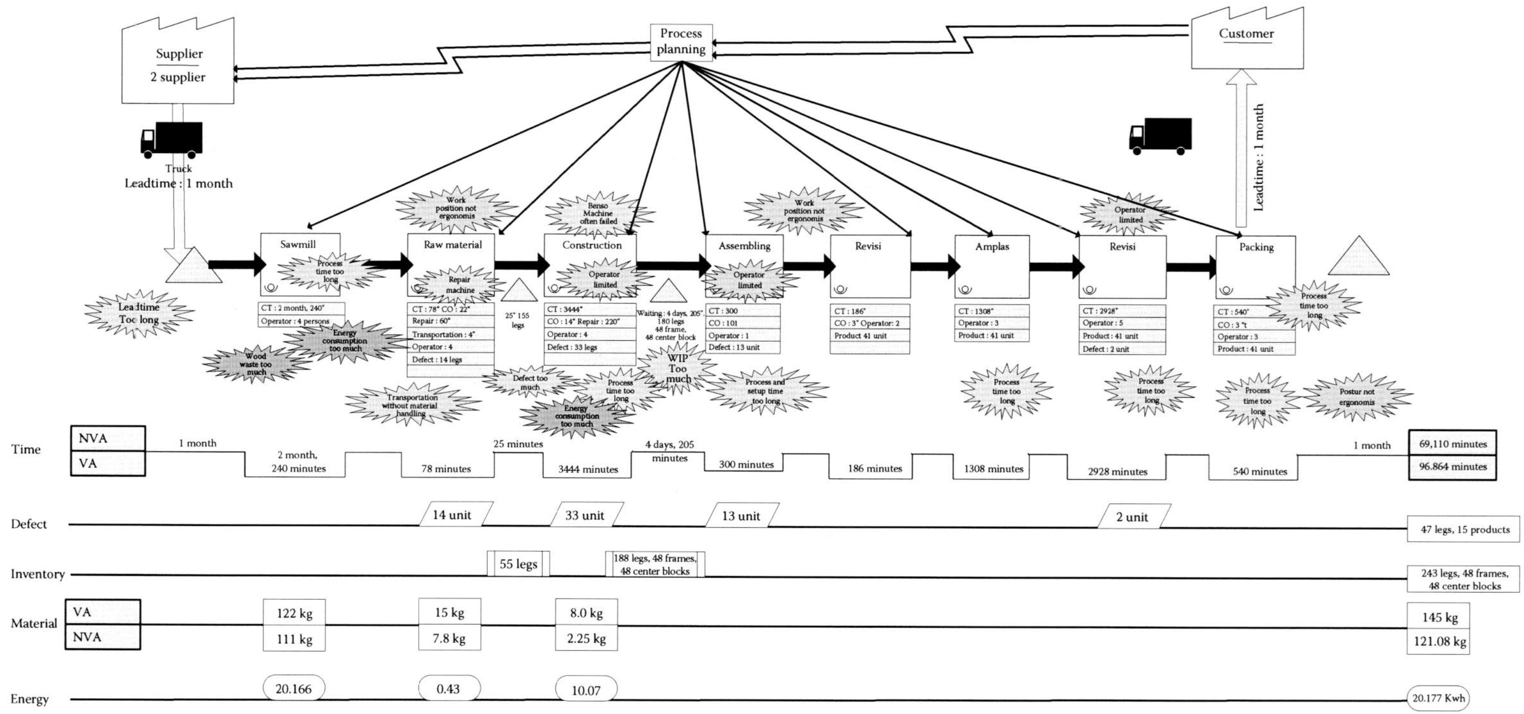

FIGURE 8.6 Sustainable VSM: analysis line is multilayered (indicator: time, defect, inventory, material, energy, level of risk). (From Hartini,Sri, Udisubakti Ciptomulyono, Maria Anityasari, Sriyanto, Darminto Pudjotomo. 2017. Sustainable-value stream mapping to evaluate sustainability performance: Indonesian Furniture Company. In *The 2nd International Conference on Engineering and Technology for Sustainable Development (ICET4SD) 2017*, Yogyakarta, 13–14 September 2017.)

limited resources. Some studies have shown empirical proof that integration of Lean and green can improve sustainability performance, especially in the economic and environmental dimension. However, the model suggested by Pampanelli et al. (2014) is limited to manufacturing at a cell level; it needs to prove the effect of integration of Lean and green on a factory level or supply chain. The model of Elsayed et al. (2013) is still in the simulation model level.

Other limitations faced by the Lean and green integration model lie on the indicator involved and the method used. Involvement of social indicators as one of the sustainable manufacturing pillars remains limited. Not much has been done as regards the mathematical use of the Lean and green integration model (Hartini and Ciptomulyono 2015). Majority of the integrated content remains at the concept level. Dües et al. (2013) state that integrating the differences between Lean and green will prove beneficial and that a study about Lean and green integration to enhance sustainable manufacturing is necessary.

Development of VSM to improve sustainability has been done by researchers. Some limitations of extended VSM are discussed below.

8.4.3.1 Methodology

The VSM review methodology is dominated by case studies (71%); only a small portion involved simulation modeling (10%), whereas the rest are still at the conceptual stage (19%) (Hartini et al. 2016). Mathematical models are limited to scoring of indicators, for example, determination of CVE. Development of VSM using mathematical models for decision making has not been done.

8.4.3.2 Area of Application

Case studies are widely used in automotive industries, as well as in agroindustry and the electronic industry. Because each industry has different characteristics, the development of VSM in other industries has high potential, such as development of sustainable VSM in the food and furniture industry, among others.

8.4.3.3 Scope of the Study

The majority of the discussion is limited to the production process (90%) and the distribution process (10%) VSM only analyzes either of the two processes. A few took pictures of the two processes (Sparks 2014). Sustainable manufacturing highly relates to environmental indicators, and end-of-life strategy highly affects the performance of environmental indicators. Therefore, expanding the scope of sustainable VSM until the end of the product life cycle is highly appealing (Simons and Mason 2003). Only Simons and Mason (2003) developed sustainable VSM that covered pre-manufacturing, manufacturing, distribution usage, and post-usage. However, the metrics evaluated only carbon emissions and transport time.

Therefore, the sustainable VSM developed to date has not been able to analyze total sustainability that evaluates the performance of triple bottom line indicator with scope in accordance with the product life cycle, both inside and outside of the company. It is important for companies that implement reverse logistic strategy or companies with the potential of implementing the 6R strategy (reduce, reuse, recycle, redesign, recovery, and remanufacturing) to know the concept of total sustainability.

8.4.3.4 Content: Metric, Indicator, and Sustainability Rate

Operational metrics that were discussed include cycle time, changeover time, value-added time ratio, and WIP. Environmental metrics that were discussed include energy, material, and water consumption. Social metrics are rarely discussed. Social metrics that were discussed include environmental health and safety. The most efficient metrics evaluated are value-added time ratio and material and energy consumption. The only index that was designed to reflect systems sustainability is carbon value efficiency. This index only considered the environmental aspect and did not involve the economic and social aspects. From a triple bottom line perspective, social metrics are rarely involved in VSM. Moreover, social indicators are not involved in a VSM analysis line.

VSM has succeeded in identifying some current metrics and production performance. However, manufacturing sustainability rate cannot be evaluated although the metric is successfully measured. It is because there is no sustainability rate determination model based on a metric or an indicator in VSM. Consequently, when one indicator is successfully improved, the effect and improvement on sustainability are still difficult to determine. Managerially, it remains difficult for companies to investigate activities to be prioritized in achieving sustainability of VSM produced.

Jayal et al. (2010) developed a framework for evaluating product sustainability covering all stages of the product life cycle. Indicators for each stage (pre-manufacturing, manufacturing, consumption, and post-consumption) are developed in economic, environmental, and social dimensions. This indicator is designed to measure the product sustainability index. Score of indicator are filled by the respondents' perspective so that it is subjective. The model has also not been able to identify the activities that cause the value of sustainable index to be low. To identify such activities, the development of VSM involving all stages in the product life cycle as a basis for determining the total sustainability index is important.

8.5 CONCLUSION

This chapter concludes that there is a positive impact of Lean and sustainable manufacturing on the triple bottom line performance. Many frameworks have been developed and many empirical studies have been carried out to strengthen this synergy. There is a lack of clear and adequate measure for Lean and sustainable manufacturing. Previous research that provides strong evidence that plants with Lean systems yield higher green results supports the philosophy of Lean and sustainable manufacturing synergy. However, research could not determine how large its impact is. When it is not possible to fully quantify performance and the related core characteristics of Lean and sustainable manufacturing through sustainable measures, they need to be represented through the use of models. However, the numerous approaches relevant research gap have been identified and need to explore in the future research, for example integrated Lean and sustainable model to improve performance firm, develop mathematical or simulation modeling for examine impact Lean and sustainable manufacturing for performance firm, and social performance has not been explored widely.

Integrating an analysis tool in Lean manufacturing (VSM) into triple bottom line metrics has been done and has been a valuable effort. Several researchers proposed many approaches to increase the value of sustainable metrics by using these metrics to extend VSM. However, the extended VSM could not provide information about the sustainability index of a system that can be improved using the sustainable VSM. Because of the limitation of the extended VSM to enhance the sustainability system, future research on extending VSM using sustainability metrics is highly appealing. For example, practices regarding development of sustainable VSM show that there is great potential for VSM development to improve system sustainability, by adding a social indicator to the analysis line, expanding the scope of review in accordance with product life cycle, enriching decision making with mathematical models, or applying models to automotive industries.

REFERENCES

Abdulmalek, Fawaz A., and Jayant Rajgopal. 2007. Analyzing the benefits of lean manufacturing and value stream mapping via simulation: A process sector case study. *International Journal of Production Economics* 107(1): 223–36.

Abuthakeer, S. S., P. V. Mohanram, and Gabriel Mohan Kumar. 2010. Activity based costing value steam mapping. *International Journal of Lean Thinking* 1(2): 51–64.

Aghajani, Mojtaba, Abbas Keramati, and Babak Javadi. 2012. Determination of number of Kanban in a cellular manufacturing system with considering rework process. *International Journal of Advance Manufacturing Technology* 63: 1177–89.

Aguado, Sergio, Roberto Alvarez, and Rosario Domingo. 2013. Model of efficient and sustainable improvements in a lean production system through processes of environmental innovation. *Journal of Cleaner Production* 47: 141–8.

Ali, A. J., A. Islam, and L. P. Howe. 2013. A study of sustainability of continuous improvement in the manufacturing industries in Malaysia. Management of Environmental Quality: *An International Journal* 24(3): 408–426.

Allwood, Julian. 2009. *Sustainable Manufacturing: What Would Make a Big Enough Difference?*

Almomani, Mohamed Ali et al. 2014. A proposed integrated model of lean assessment and analytical hierarchy process for a dynamic road map of lean implementation. *International Journal of Advanced Manufacturing Technology* 72(1–4): 161–72.

Anand, G., and Rambabu Kodali. 2009. Selection of lean manufacturing systems using the analytic network process—A case study. *Journal of Manufacturing Technology Management* 20(2): 258–89.

Azadegan, Arash, Pankaj C. Patel, Abouzar Zangoueinezhad, and Kevin Linderman. 2013. The effect of environmental complexity and environmental dynamism on lean practices. *Journal of Operations Management* 31(4): 193–212.

Babu, A. Subash, Keshav N. Nandurkar, and Austin Thomas. 2000. Development of virtual cellular manufacturing systems for SMEs. *Logistics Information Management* 13: 228–42.

Bayo-Moriones, Alberto, Alejandro Bello-Pintado, and Javier Merino-Díaz De Cerio. 2010. 5S use in manufacturing plants: Contextual factors and impact on operating performance. *International Journal of Quality & Reliability Management* 27: 217–30.

Belokar, Rajendra M., Vikas Kumar, and Sandeep Singh Kharb. 2012. An application of value stream mapping in automotive industry: A case study. *International Journal of Innovative Technology and Exploring Engineering* 1(2): 152–7.

Bergmiller, Gary G., and Paul R. McCright. 2009a. Are lean and green programs synergistic? *Proceedings of the 2009 Industrial Engineering Research Conference*: 1–6.

Bergmiller, Gary G., and Paul R. Mccright. 2009b. Parallel models for lean and green operations. *Industrial Engineering Research Conference*: 1138–43.

Bergmiller, Gary G., Paul R. Mccright, and South Florida. 2009. Lean manufacturers' transcendence to green manufacturing. *Industrial Engineering Research Conference*: 1144–8.

Bettley, Alison, and Stephen Burnley. 2008. Towards sustainable operations management integrating sustainability management into operations management strategies and practices. *Handbook of Performability Engineering*: 875–904.

Bhamu, Jaiprakash, J. V. Shailendra Kumar, and Kuldip Singh Sangwan. 2012. Productivity and quality improvement through value stream mapping: A case study of Indian automotive industry. *International Journal of Productivity and Quality Management* 10(3): 288.

Bhamu, Jaiprakash, and Kuldip Singh Sangwan. 2014. Lean manufacturing: Literature review and research issues. *International Journal of Operations & Production Management* 34(7): 876–940.

Biggs, Claire. 2009. *Exploration of the Integration of Lean and Environmental Improvement*. Thesis.

Bocken, N. M. P., S. W. Short, P. Rana, and S. Evans. 2014. A literature and practice review to develop sustainable business model archetypes. *Journal of Cleaner Production* 65: 42–56.

Bortolotti, Thomas, Pamela Danese, Barbara B. Flynn, and Pietro Romano. 2014. Leveraging fitness and lean bundles to build the cumulative performance sand cone model. *International Journal of Production Economics*: 1–15.

Bortolotti, Thomas, Pietro Romano, and Bernardo Nicoletti. 2010. Lean first, then automate: An integrated model for process improvement in pure service-providing companies. *Advances in Production Management Systems: New Challenges; New Approaches*: 579–86.

Bourlakis, Michael et al. 2014. Firm size and sustainable performance in food supply chains: Insights from Greek SMEs. *International Journal of Production Economics* 152: 112–30.

Boyle, Todd A., and Maike Scherrer-Rathje. 2009. An empirical examination of the best practices to ensure manufacturing flexibility: Lean alignment. *Journal of Manufacturing Technology Management* 20(3): 348–66.

Braglia, Marcello et al. 2006. 18th ICPR Paper: A new value stream mapping approach for complex production system. *International Journal of Production Research* 44(18–19): 3929–52.

Braglia, Marcello, Marco Frosolini, and Francesco Zammori. 2009. Uncertainty in value stream mapping analysis uncertainty in value stream mapping analysis. *International Journal of Logistics Research and Applications* 126(6): 435–53.

Brown, Adam. 2009. Quality and continuous improvement in medical device manufacturing. *Emerald Insight* 29(5): 494–519.

Brown, Adam, Joseph Amundson, and Fazleena Badurdeen. 2014. Sustainable value stream mapping (Sus-VSM) in different manufacturing system configurations: Application case studies. *Journal of Cleaner Production* 85: 164–79.
Browning, Tyson R., and Ralph D. Heath. 2009. Reconceptualizing the effects of lean on production costs with evidence from the F-22 program. *Journal of Operations Management* 27(1): 23–44.
Chen, Lixia, and Bo Meng. 2010. The application of value stream mapping based lean production system. *International Journal of Business and Management* 5(6): 203–9.
Chiarini, Andrea. 2014. Sustainable manufacturing-greening processes using specific lean production tools: An empirical observation from european motorcycle component manufacturers. *Journal of Cleaner Production* 85: 226–33.
Chong, Albert, Hooi Cheah, Wai Peng Wong, and Qiang Deng. 2012. Challenges of lean manufacturing implementation: A hierarchical model. In Proceedings of the 2012 International Conference on Industrial Engineering and Operations Management Istanbul, Turkey, 2091–9.
Corbett, Charles J., and Robert D. Klassen. 2006. Extending the horizons: Environmental excellence as key to improving operations. *Manufacturing and Service Operations Management* 8(1): 5–22.
Coronado, Adrian E., and Andrew C. Lyons. 2007. Evaluating operations flexibility in industrial supply chains to support build-to-order initiatives. *Business Process Management Journal* 13(4): 572–87.
Dadashzadeh, Mohammad Dadashzadeh, and T. J. Wharton. 2012. A value stream approach for greening the IT department. *International Journal of Management & Information Systems (IJMIS)* 16(2): 125–36.
Deif, Ahmed M. 2011. A system model for green manufacturing. *Journal of Cleaner Production* 19(14): 1553–9.
Demeter, Krisztina, and Zsolt Matyusz. 2011. The impact of lean practices on inventory turnover. *International Journal of Production Economics* 133(1): 154–63.
Deshmukh, S. G., Nitin Upadhye, and Suresh Garg. 2010. Lean manufacturing for sustainable development. *Global Business and Management Research* 2(1): 125.
Dombrowski, Uwe, Tim Mielke, and Sven Schulze. 2012. Lean production systems as a framework for sustainable manufacturing. In G. Seliger (Ed.), *Sustainable Manufacturing*, Springer-Verlag: Berlin Heidelberg, 17–22.
Donatelli, Anthony J., and Gregory A. Harris. 2001. Combining value stream mapping and discrete event simulation. Proceedings of the Huntsville Simulation Conference by the Society for Modeling and Simulation International, San Diego, CA.
Doolen, Toni L., and Marla E. Hacker. 2005. A review of lean assessment in organizations: An exploratory study of lean practices by electronics manufacturers. *Journal of Manufacturing Systems* 24(1): 55–67.
Dora, Manoj et al. 2013a. Operational performance and critical success factors of lean manufacturing in European food processing SMEs. *Trends in Food Science & Technology* 31(2): 156–64.
Dora, Manoj K,. Maneesh Kumar, Dirk Van Goubergen, and Adrienn Molnar. 2013b. Application of lean practices in small and medium sized food enterprises. *British Food Journal* 116(1): 3103706.
Dornfeld, David et al. 2013. *Green Manufacturing*. Springer: New York Heidelberg Dordrecht London.
Dües, Christina Maria, Kim Hua Tan, and Ming Lim. 2013. Green as the new lean: How to use lean practices as a catalyst to greening your supply chain. *Journal of Cleaner Production* 40: 93–100.
Elsayed, Nancy et al. 2013. Assessment of lean and green strategies by simulation of manufacturing systems in discrete production environments. *CIRP Annals—Manufacturing Technology* 62(1): 475–8.
Farris, Jennifer A., Eileen M. Van Aken, Toni L. Doolen, and June Worley. 2009. Critical success factors for human resource outcomes in kaizen events: An empirical study. *International Journal of Production Economics* 117(1): 42–65.
Faulkner, William, and Fazleena Badurdeen. 2014. Sustainable value stream mapping (Sus-VSM): Methodology to visualize and assess manufacturing sustainability performance. *Journal of Cleaner Production* 85: 8–18.
Faulkner, William, William Templeton, David Gullett, and Fazleena Badurdeen. 2012. Visualizing sustainability performance of manufacturing systems using sustainable value stream mapping (Sus-VSM). In *Proceedings of the 2012 International Conference on Industrial Engineering and Operations Management*, 815–24.
Florida, Richard, and Derek Davison. 2001. Gaining from green management. *California Management Review*.
Folinas, Dimitrios et al. 2014. Greening the agrifood supply chain with lean thinking practices. *International Journal of Agricultural Resources, Governance and Ecology* 10(2): 129.
Folinas, Dimitrios K., Dimitrios Aidonis, and Panayotis Karayannakidis. 2015. Greening the canned peach production. *Journal of Agricultural Informatics* 6(1): 24–39.

Forno, Ana Julia Dal, Fernando Augusto Pereira, Fernando Antonio Forcellini, and Liane M. Kipper. 2014. Value stream mapping: A study about the problems and challenges found in the literature from the past 15 years about application of lean tools. *International Journal of Advanced Manufacturing Technology* 72 (5–8): 779–90.

Fullerton, Rosemary R., Frances A. Kennedy, and Sally K. Widener. 2014a. Lean manufacturing and firm performance: The incremental contribution of lean management accounting practices. *Journal of Operations Management* 32(7): 414–28.

Fullerton, Rosemary R., and William F. Wempe. 2009. Lean manufacturing, non-financial performance measures, and financial performance. *International Journal of Operations & Production Management* 29(3): 214–40.

Galeazzo, Ambra, Andrea Furlan, and Andrea Vinelli. 2013. Lean and green in action: Interdependencies and performance of pollution prevention projects. *Journal of Cleaner Production* 85: 191–200.

Garza-Reyes, Jose Arturo. 2015. Lean and green—A systematic review of the state of the art literature. *Journal of Cleaner Production* 102: 18–29.

Ghosh, Manimay. 2013. Lean manufacturing performance in Indian manufacturing plants. *Journal of Manufacturing Technology Management* 24(1): 113–22.

Glover, Wiljeana J., Jennifer A. Farris, Eileen M. Van Aken, and Toni L. Doolen. 2011. Critical success factors for the sustainability of kaizen event human resource outcomes: An empirical study. *International Journal of Production Economics* 132(2): 197–213.

González-Benito, Javier. 2005. Environmental proactivity and business performance: An empirical analysis environmental proactivity and business performance. *Omega* 33(1): 1–15.

Gordon, Pamela J. 2001. *Lean and Green: Profit for Your Workplace and the Environment*. Berrett-Koehler Publishers.

Gupta, Suraksha, and V. Kumar. 2013. Sustainability as corporate culture of a brand for superior performance. *Journal of World Business* 48(3): 311–20.

Gurumurthy, Anand, and Rambabu Kodali. 2011. Design of lean manufacturing systems using value stream mapping with simulation: A case study. *Journal of Manufacturing Technology Management* 22(4): 444–73.

Haapala, Karl R. et al. 2013. A review of engineering research in sustainable manufacturing. *Journal of Manufacturing Science and Engineering* 135(4): 41013.

Hajmohammad, Sara, Stephan Vachon, Robert D. Klassen, and Iuri Gavronski. 2013. Lean management and supply management: Their role in green practices and performance. *Journal of Cleaner Production* 39: 312–20.

Hallgren, Mattias, and Jan Olhager. 2009. Lean and agile manufacturing: External and internal drivers and performance outcomes. *International Journal of Operations & Production Management* 29(10): 976–99.

Hartini, Sri, and Udisubakti Ciptomulyono. 2015. The relationship between lean and sustainable manufacturing on performance: Literature review. *Procedia Manufacturing*: 38–45.

Hartini, Sri, Udisubakti Ciptomulyono, and Maria Anityasari. 2017. Extended value stream mapping to enhance sustainability: A literature review. In 3rd International Materials, Industrial and Manufacturing Engineering Conference (MIMEC2017), Vol. 1902(1). AIP Publishing.

Hartini, Sri, Udisubakti Ciptomulyono, and Maria Anityasari. 2016. Extended value stream mapping to enhance sustainability: A literature review. In *2016 Annual Conference on Industrial and System Engineering (ACISE)*.

Hartini, Sri, Udisubakti Ciptomulyono, Maria Anityasari, and Sriyanto, Darminto Pudjotomo. 2017. Sustainable-value stream mapping to evaluate sustainability performance: Indonesian Furniture Company. In *The 2nd International Conference on Engineering and Technology for Sustainable Development (ICET4SD) 2017*, Yogyakarta, 13–14 September 2017.

Hines, Peter, and Nicholas Leo Rich. 1997. The seven value stream mapping tools. *International Journal of Operations & Production Management* 17(1): 46–64.

Hines, Peter, Matthias Holweg, and Nick Rich. 2004. Learning to evolve: A review of contemporary lean thinking. *International Journal of Operations & Production Management* 24(10): 994–1011.

Hines, Peter, and David Helliwell Taylor. 2000. *Going Lean*. Lean Enterprise Research Centre.

Hollmann, S., F. Klimmer, K. H. Schmidt, and H. Kylian. 1999. Validation of a questionnaire for assessing physical work load. *Scandinavian Journal of Work, Environment & Health* 25(2): 105–14.

Hong, Paul, James Jungbae Roh, and Greg Rawski. 2012. Benchmarking sustainability practices: Evidence from manufacturing firms. *Benchmarking: An International Journal* 19(4/5): 634–48.

Irani, Shahrukh A., and Jin Zhou. 2011. Value stream mapping of a complete product. *White Paper of Lean Manufacturing Japan* (1): 1–24.

Jabbour, Ana, José Junior, and Charbel Jabbour. 2014. Extending lean manufacturing in supply chains: A successful case in Brazil. *Benchmarking: An International Journal* 21(6): 1070–83.

Jabbour, Charbel José Chiappetta et al. 2013. Environmental management and operational performance in automotive companies in Brazil: The role of human resource management and lean manufacturing. *Journal of Cleaner Production* 47: 129–40.

Jasti, Naga, and Rambabu Kodali. 2014a. A literature review of empirical research methodology in lean manufacturing. *International Journal of Operations & Production Management* 34(8): 1080–122.

Jasti, Naga, and Rambabu Kodali. 2014b. Validity and reliability of lean product development frameworks in Indian manufacturing industry. *Measuring Business Excellence* 18(4): 27–53.

Jawahir, I. S. 2008. Sustainable manufacturing: The driving force for innovative products, processes and systems for next generation manufacturing. *Symposium on Sustainability and Product Development IIT* (859).

Jayal, A. D., F. Badurdeen, O. W. Dillon, and I. S. Jawahir. 2010. Sustainable manufacturing: Modeling and optimization challenges at the product, process and system levels. *CIRP Journal of Manufacturing Science and Technology* 2(3): 144–52.

Jeong, Myeon Gyu, James R. Morrison, and Hyo Won Suh. 2015. Approximate life cycle assessment via case-based reasoning for eco-design. *IEEE Transactions on Automation Science and Engineering* 12(2): 716–28.

Keskin, Cem, Umut Asan, and Gulgun Kayakutlu. 2013. Value stream maps for industrial energy efficiency. In F. Cavallaro (Ed.), *Assessment and Simulation Tools for Sustainable Energy Systems, Green Energy and Technology 129*, London: Springer-Verlag, 357–79.

Khanchanapong, Teerasak et al. 2014. The unique and complementary effects of manufacturing technologies and lean practices on manufacturing operational performance. *International Journal of Production Economics* 153: 191–203.

Khaswala, Zahirabbas N., and Shahrukh A. Irani. 2001. Value Network Mapping (VNM): Visualization and analysis of multiple flows in value stream maps. Proceedings of the Lean Management Solutions Conference St. Louis (September 10–11).

King, Andrew A., and Michael J. Lenox. 2001. Lean and green? An empirical examination of the relationship between lean production and environmental performance. *Production and Operation Management* 10(3): 244–56.

Kuriger, Glenn W., and F. Frank Chen. 2010. Lean and green: A current state view. In *Proceedings of the 2010 Industrial Engineering Research Conference*.

Lacerda, António Pedro et al. 2016. Applying value stream mapping to eliminate waste: A case study of an original equipment manufacturer for the automotive industry. *International Journal of Production Research* 7543(January).

Lasa, Ibon, Carlos Ochoa Laburu, and Rodolfo De Castro. 2008. Evaluation of value stream mapping in manufacturing systems redesigning. *International Journal of Production Research* 46(16): 4409–30.

Lasa, Serrano Ibon, Rodolfo De Castro, and Carlos Ochoa Laburu. 2009. Extent of the use of lean concepts proposed for a value stream mapping application. *Production Planning & Control* 20(1): 82–98.

Lewis, Michael A. 2000. Lean production and sustainable competitive advantage. *International Journal of Operations & Production Management*.

Li, Hongcheng, Huajun Cao, and Xiaoyong Pan. 2012. A carbon emission analysis model for electronics manufacturing process based on value-stream mapping and sensitivity analysis. *International Journal of Computer Integrated Manufacturing* 25(12): 1102–10.

Lian, Yang-hua, and Hendrik Van Landeghem. 2002. An application of simulation and value stream mapping in lean manufacturing. In *Proceedings 14th European Simulation Symposium*, 300–307.

Losonci, Dávid, and Krisztina Demeter. 2013. Lean production and business performance: International empirical results. *Competitiveness Review* 23(3): 218–33.

Lourenço, E. J., A. J. Baptista, J. P. Pereira, and Celia Dias-Ferreira. 2013. Multi-layer stream mapping as a combined approach for industrial processes eco-efficiency assessment. In *20th CIRP International Conference on Life Cycle Engineering*, Singapore.

Marimin et al. 2014. Value chain analysis for green productivity improvement in the natural rubber supply chain: A case study. *Journal of Cleaner Production* 85: 201–11.

Marodin, Giuliano Almeida, and Tarcísio Abreu Saurin. 2015. Classification and relationships between risks that affect lean production implementation: A study in southern Brazil. *Journal of Manufacturing Technology Management* 26: 57–79.

Miller, Geoff, Janice Pawloski, and Charles Robert Standridge. 2010. A case study of lean, sustainable manufacturing. *Journal of Industrial Engineering and Management* 3(1): 11–32.

Müller, Egon, Timo Stock, and Rainer Schillig. 2013. Dual energy signatures enable energy value-stream mapping. In A. Azevedo (Ed.), *Advances in Sustainable and Competitive Manufacturing Systems*, Switzerland: Springer International Publishing, 1603–11.

NACFAM. 2009. National Council for Advanced Manufacturing (NACFAM). 2009, Sustainable Manufacturing. http://nacfam02.dev.web.sba.com/PolicyInitiatives/SustainableManufacturing/tabid/64/Default.aspx.

Narasimhan, Ram, Morgan Swink, and Soo Wook Kim. 2006. Disentangling leanness and agility: An empirical investigation. *Journal of Operations Management* 24: 440–57.

Ng, Ruisheng, Jonathan Sze Choong Low, and Bin Song. 2015. Integrating and implementing lean and green practices based on proposition of carbon-value efficiency metric. *Journal of Cleaner Production* 95: 242–55.

Norton, A., and Andrew Fearne. 2009. Sustainable value stream mapping: A tool for process change and waste reduction. In Keith Waldron (Ed.), *Handbook of Waste Management and Co-Product Recovery in Food Processing*. Cambridge: Woodhead Publishing.

Paju, Marja et al. 2010. Framework and indicator for a sustainable manufacturing mapping methodology. In *Proceedings of the 2010 Winter Simulation Conference*, 3411–22.

Pampanelli, Andrea Brasco, Driveline Brazil, and Joaquim Silveira. 2011. A lean and green kaizen model. In *POMS 21st Annual Conference*, 020–031.

Pampanelli, Andrea Brasco, Pauline Found, and Andrea Moura Bernardes. 2014. A lean & green model for a production cell. *Journal of Cleaner Production* 85: 19–30.

Pettersen, Jostein. 2009. Defining lean production: Some conceptual and practical issues. *The TQM Journal* 21 (2): 127–42.

Piercy, Niall, and Nick Rich. 2009. Lean transformation in the pure service environment: The case of the call service centre. *International Journal of Operations & Production Management* 29(1): 54–76.

Pil, Frits P., and Sandra Rothenberg. 2003. Environmental performance as a driver of superior quality. *Production and Operations Management* 12(3): 404–15.

Prashar, Anupama. 2014. Redesigning an assembly line through lean-kaizen: An Indian case. *TQM Journal* 26 (5): 475.

Puvanasvaran, Perumal, Robert Kerk Swee Tian, Vasu Suresh, and Mohd Razali Muhamad. 2012. Lean principles adoption in Environmental Management System (EMS): A survey on ISO 14001 certified companies in Malaysia. *Journal of Industrial Engineering and Management* 5(2): 406–30.

Rebitzer, G., T. Ekvall, R. Frischknecht, D. Hunkeler, G. Norris, T. Rydberg, W.-P. Schmidt, S. Suh, B. P. Weidema, and D. W. Pennington. 2004. Life cycle assessment. *Environment International* 30(5): 701–20.

Rothenberg, Sandra, Frits K. Pil, and James Maxwell. 2001. Lean, green, and the quest for superior environmental performance. *Production and Operations Management* 10(3). 228–43.

Rother, Mike, and John Shook. 1999. *Learning to See: Value Mapping to Create Value and Eliminate Muda* Lean Enterprise Institute Brookline.

Sangwan, Kuldip Singh, Jaiprakash Bhamu, and Dhwani Mehta. 2014. Development of lean manufacturing implementation drivers for Indian ceramic industry. *International Journal of Productivity and Performance Management* 63(5): 569–87.

Sawhney, Rapinder, Pamuk Teparakul, Aruna Bagchi, and Xueping Li. 2007. En-lean: A framework to align lean and green manufacturing in the metal cutting supply chain. *International Journal of Enterprise Network Management* 1(3): 238.

Shah, Rachna, and Peter T. Ward. 2003. Lean manufacturing: Context, practice bundles, and performance. *Journal of Operations Management* 21(2): 129–49.

Shah, Rachna, and Peter T. Ward. 2007. Defining and developing measures of lean production. *Journal of Operations Management* 25(4): 785–805.

Shen, Synthia Xin, and Chuan Feng Han. 2006. China electrical manufacturing services industry value stream mapping collaboration. *International Journal of Flexible Manufacturing Systems* 18(4): 285–303.

Simons, David, and Robert Mason. 2003. Firms are under pressure to prove their environmental credentials. Now a win–win way of weaving 'green' considerations into business decisions is emerging lean and green: 'Doing more with less.' *International Commerce Review*: 84–91.

Simpson, Dayna F., and Damien J. Power. 2005. Use the supply relationship to develop lean and green suppliers. *Supply Chain Management: An International Journal* 10(1): 60–68.

Singh, Bhim, Suresh K. Garg, and Surrender K. Sharma. 2011. Value stream mapping: Literature review and implications for Indian industry. *International Journal of Advanced Manufacturing Technology* 53(5–8): 799–809.

Singh, Bhim, and S. K. Sharma. 2009. Value stream mapping as a versatile tool for lean implementation: An Indian case study of a manufacturing firm. *Measuring Business Excellence* 13(3): 58–68.
So, Stuart, and Hongyi Sun. 2010. Supplier integration strategy for lean manufacturing adoption in electronic-enabled supply chains. *Supply Chain Management: An International Journal* 15(6): 474–87.
Solding, Petter, and Per Gullander. 2009. Concepts for simulation based value stream mapping. In *Proceedings of the 2009 Winter Simulation Conference (WSC)*, 2231–7.
Sparks, Daniel T. 2014. Combining sustainable value stream mapping and simulation to assess manufacturing supply chain network performance. University of Kentucky.
Taj, Shahram. 2008. Lean manufacturing performance in China: Assessment of 65 manufacturing plants. *Journal of Manufacturing Technology Management* 19(2): 217–34.
Taj, Shahram, and Lismar Berro. 2006. Application of constrained management and lean manufacturing in developing best practices for productivity improvement in an auto-assembly plant. *International Journal of Productivity and Performance Management* 55(3/4): 332–45.
Taj, Shahram, and Cristian Morosan. 2011. The impact of lean operations on the Chinese manufacturing performance. *Journal of Manufacturing Technology Management* 22(2): 223–40.
Taylor, David Helliwell. 2006. Strategic considerations in the development of lean agri-food supply chains: A case study of the UK pork sector. *Supply Chain Management: An International Journal* 11(3): 271–80.
Thanki, S. J., and Jitesh Thakkar. 2014. Status of lean manufacturing practices in Indian industries and government initiatives: A pilot study. *Journal of Manufacturing Technology Management* 25(5): 655–75.
Torres, Alvair Silveira, and Ana Maria Gati. 2009. Environmental Value Stream Mapping (EVSM) as sustainability management tool. In PICMET *'09—2009 Portland International Conference on Management of Engineering & Technology*, 1689–98.
Tortorella, Guilherme Luz et al. 2016. Making the value flow: Application of value stream mapping in a Brazilian public healthcare organisation. *Total Quality Management & Business Excellence* 28(13–14): 1544–58.
USDOC. 2011. *How does Commerce define Sustainable Manufacturing?*, International Trade Administration, U.S. Department of Commerce. http://www.trade.gov/competitiveness/sustainablemanufacturing/how_doc_defines_SM.asp.
US EPA. 2007. *Lean, Energy & Climate Toolkit*. U.S. Environmental Protection Agency (EPA).
Verrier, Brunilde, Bertrand Rose, Emmanuel Caillaud, and Hakim Remita. 2014. Combining organizational performance with sustainable development issues: The lean and green project benchmarking repository. *Journal of Cleaner Production* 85: 83–93.
Vinodh, Sekar, and Gopinath Rathod. 2010. Integration of ECQFD and LCA for sustainable product design. *Journal of Cleaner Production* 18(8): 833–42.
Vinodh, R. Ben Ruben, and P Asokan. 2016. Life cycle assessment integrated value stream mapping framework to ensure sustainable manufacturing: A case study. *Clean Technologies and Environmental Policy* 18(1): 279–95.
Womack, James P., Daniel T. Jones, and Daniel Roos. 1992. The machine that changed the world. *Business Horizons* 35(3): 81–2.
Xie, Yikun, and Qingjin Peng. 2012. Integration of value stream mapping and agent-based modeling for OR improvement. *Business Process Management Journal* 18(4): 585–99.
Yang, Ma Ga (Mark), Paul Hong, and Sachin B. Modi. 2011. Impact of lean manufacturing and environmental management on business performance: An empirical study of manufacturing firms. *International Journal of Production Economics* 129(2): 251–61.
Zhu, Qinghua, and Joseph Sarkis. 2004. Relationships between operational practices and performance among early adopters of green supply chain management practices in chinese manufacturing enterprises. *Journal of Operations Management* 22(3): 265–89.

9 An Overview of Viable Unconventional Processing Methods for Advanced Materials

Subramanian Jayalakshmi, Ramachandra Arvind Singh, Rajashekhara Shabadi, Jayamani Jayaraj, Sambasivam Seshan, and Manoj Gupta

CONTENTS

9.1 INTRODUCTION

Processing of metals is one of the earliest technologies developed by mankind that significantly contributed toward human civilization and rapid industrial growth. Till recently, metals/alloys were processed mainly by conventional casting and forming methods. The demand for newer materials with enhanced properties was a strong driving force for innovation in processing methods. Further, knowledge that properties of conventional metals/alloys can be enhanced multifold by incorporation of high-strength materials, such as ceramics (to form composite materials), propelled the development of synthesizing methods, such as squeeze casting/squeeze infiltration for producing efficient, economical, near-net shape composites.

With time, the advancement in materials science led to the development of nontraditional processing methods that provided good flexibility for the purpose of microstructure/property control.

Materials produced by such nontraditional methods show enhanced performance capability that can be used for a wider range of industrial applications. In this chapter, unconventional processing of a few materials (categorized as "advanced") that are being currently researched is presented and discussed.

9.2 UNCONVENTIONAL PROCESSING TECHNIQUES FOR ADVANCED MATERIALS

Sustained advances in technology have set an ever-growing demand for new materials with high-performance capabilities, for use in critical applications, such as space/aerospace, biomedical, defense, and the nuclear sector. In this context, new materials of interest are the following: (i) metal matrix nanocomposites (MMNCs), (ii) nanocrystalline (NC) materials, (iii) amorphous alloys (AA)/ bulk metallic glasses (BMG), and the recently emerged (iv) high-entropy alloys (HEAs). All these novel advanced materials are made through various unconventional processing techniques. Before the discussion on the unconventional processing techniques used, a brief introduction of advanced materials is provided.

Metal matrix composites and nanocomposites (MMCs/MMNCs) (especially those based on light metals such as aluminum, titanium, and magnesium) are being developed for aircraft and automotive industries owing to the requirement and need for lightweight structures (Ceschini et al. 2016; Gupta and Sharon 2011; Kainer 2006; Miracle 2005). Incorporation of strong and stiff micron-sized ceramic constituents (e.g., silicon carbide, alumina, boron carbide) in light metal matrices provides significant improvement in hardness, stiffness, strength, wear resistance, and high-temperature properties that are not obtainable in conventional alloys (Miracle 2005). The development of nanocomposites incorporating nanosized reinforcements is aimed at improving the ductility of the composites, as micron-sized reinforcements usually lead to low toughness (Gupta and Sharon 2011). Nanoparticles give rise to significant improvement in strength properties owing to the "dispersion strengthening-like" effect, coupled with ductility retention/enhancement, resulting in composites having enhanced toughness (Ceschini et al. 2016). Conventionally, metal matrix composites are prepared by stir casting, squeeze casting, semisolid processing, and powder metallurgy (blend-compact-sinter) methods. In nanocomposites, agglomeration and oxidation/reactivity of nano-reinforcements is of major concern, and an elimination of such issues does require the need for unconventional processing approaches (Ceschini et al. 2016).

Materials with nanometer/sub-micrometer–scale crystalline structure are of interest in materials development. Nanocrystalline materials have an average grain size of less than 100 nm (Lu 1996). A small crystallite structure implies a sizable volume fraction of grain boundaries. Because of this, nanocrystalline materials can exhibit exceptionally high properties (Lu 1996). For example, nanocrystalline materials exhibit increased strength/hardness, improved ductility/toughness, superior soft magnetic properties, and enhanced wear resistance when compared to their microcrystalline counterpart. It has been identified that a grain boundary–based deformation mechanism predominantly manifests during deformation of nanocrystalline material, unlike the dislocation pile-up–based mechanisms occurring in coarse-grained materials (Lu 1996; Suryanarayana 2001). Nanocrystalline materials, at times, can also contain crystalline/quasi-crystalline/amorphous phases when they are made through devitrification of amorphous structure (Lu 1996). Nanocrystalline metallic particles can be obtained using the gas condensation technique. Other methods of synthesizing nanocrystalline materials include the following: (i) mechanical alloying (ball milling), (ii) electrodeposition, (iii) rapid quenching, and (iv) devitrification of amorphous structure (Lu 1996; Suryanarayana 2001). While severe plastic deformation techniques, such as equi-channel angular processing (refer to Section 9.3.3.2), are being widely used, other new severe plastic deformation methods have also been developed and put to use.

Amorphous alloys and bulk metallic glasses are a new class of metallic materials, which are distinctly different from conventional crystalline materials in terms of their structure and properties (Basu and Ranganathan 2003; Inoue 1998; Johnson 1999). Unlike crystalline materials, amorphous alloys/metallic glasses lack long range order (i.e., absence of translational periodicity; Figure 9.1) and consequently have an amorphous structure (Basu and Ranganathan 2003; Inoue 1998; Johnson 1999). They exhibit glass transition and crystallization temperature. Hence, they are considered to be solids with a frozen-in-liquid structure and are thermodynamically metastable. Because of their unique structure, they possess excellent strength (~2 GPa), high elastic limit (2% as against 0.2% in crystalline materials), and superior corrosion resistance (due to the absence of grains and grain boundaries) (Basu and Ranganathan 2003; Inoue 1998; Johnson 1999). In contrast to conventional crystalline alloys where plastic deformation occurs by the motion of dislocations, in metallic glasses, plastic deformation occurs by shear localization (due to the absence of defects and/or dislocations) (Langer 2006). A few interesting applications of amorphous/glassy alloys are as follows: (a) hydrogen membrane separators (e.g., during coal gasification), (b) bipolar plates and catalytic membranes in fuel cells, (c) pressure sensors, (d) micro-gears for motors, (e) magnetic cores for power supplies, and (f) nano-dies (Inoue 1998; Jayalakshmi and Gupta 2015; Langer 2006). In addition, being metallic in nature, amorphous alloys/bulk metallic glasses have a coefficient of thermal expansion close to those of crystalline metals and can provide metal–metal bonding at the interface (less thermal mismatch). Because of this, these are currently being explored as usable reinforcements for metal matrices, as better alternatives to the use of conventional ceramic reinforcements (Jayalakshmi and Gupta 2015).

High-entropy alloys are a newly emerging class of advanced materials that were first reported in 2004. High-entropy alloys are solid solution alloys that contain greater than or equal to five principal elements (Figure 9.2), each with equi-atomic/equi-molar concentration in the range of 5 to 35 at%. This is in contrast to the conventional alloys that usually contain one principal element and few other minor elements (Murthy et al. 2014; Tsai and Yeh 2014; Yeh 2013; Zhang et al. 2014). In high-entropy alloys, the liquid or random solid solution states have significantly higher mixing entropies than those in the conventional alloys. With such constitutional configuration, high-entropy alloys have unique properties, such as (a) very high strength (>1 GPa), (b) high elastic limit, (c) high hardness, (d) strength retention at elevated/cryogenic temperatures, and (e) excellent resistance to corrosion, oxidation, and wear (Murthy et al. 2014; Tsai and Yeh 2014; Yeh 2013; Zhang et al. 2014). Main aspects that give rise to such remarkable properties include (i) high entropy, (ii) severe lattice distortion, (iii) sluggish atomic diffusion, and (iv) the "cocktail effect" (Murthy et al. 2014; Zhang et al. 2014). Prospective applications of high-entropy alloys include the following: (1) thermal barrier coatings, (2) wear-resistant coatings, (3) clads/hulls in nuclear reprocessing systems, (4) cones/nozzles for aircraft engines, and (5) hard tools/molds (Svensson 2014; Zhang et al. 2014). High-entropy alloys are synthesized by the Bridgman solidification, arc melting, thermal spray method, and hot isostatic pressing technique. Unconventional methods, such as (a) spark plasma sintering and (b) laser cladding, are also being used to synthesize these new materials (Svensson 2014; Zhang et al. 2014).

Metal matrix nanocomposites, amorphous alloys/bulk metallic glasses, nanocrystalline materials, and high-entropy alloys have a unique structure and exhibit superior properties that cannot be easily

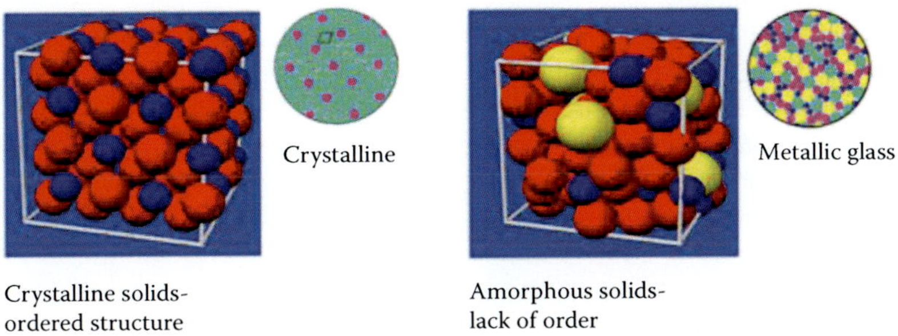

FIGURE 9.1 Difference in atomic arrangement in crystalline and amorphous materials.

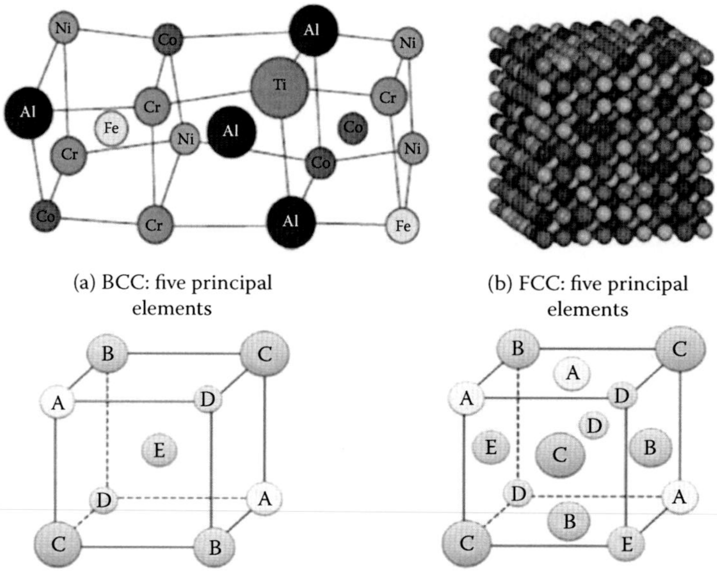

FIGURE 9.2 Solid solution crystalline structures in high-entropy alloys. (Used with permission from Zhang, Y., T. T. Zuo, Z. Tang, M. C. Gao, K. A. Dahmen, P. K. Liaw and Z. P. Lu (2014). Microstructures and properties of high-entropy alloys. *Progress in Materials Science* 61, 1–93.)

realized using conventional processing methods. To highlight this fact, an example is cited here: the preparation of high-entropy alloys by the powder metallurgy method is a challenge as a longer sintering time employed in conventional sintering usually results in the formation and presence of brittle and detrimental microcrystalline intermetallic. With the development of novel rapid sintering techniques such as spark plasma sintering (refer to Section 9.3.1.2), solid-state synthesis of high-entropy alloys without brittle phases has now become a reality.

9.3 PROCESSING METHODS

In this section, unconventional processing techniques that have been developed to synthesize advanced materials are presented. Synthesis of the advanced materials (mentioned above) by one or more of the unconventional processes is stated, and the unique and superior properties of the advanced materials synthesized by the unconventional processes are highlighted. The following are the processing techniques discussed in this section:

I. Solid-state and liquid-state techniques
   a. Microwave sintering
   b. Spark plasma sintering
   c. High-frequency induction sintering
   d. Disintegrated melt deposition technique
II. Surfaces process/coating methods
   a. Friction stir processing
   b. Ultrasonic nanocrystalline surface modification
   c. Laser cladding
III. Severe plastic deformation (bulk deformation) processes
   a. Accumulative roll bonding
   b. Equi-channel angular processing

9.3.1 Solid-State/Liquid-State Processes

9.3.1.1 Microwave Sintering

In the microwave sintering technique, the principle of heating and sintering is fundamentally different from conventional sintering (Gupta and Eugene 2007). Microwave heating is a volumetric heating process involving conversion of electromagnetic energy into thermal energy, which is instantaneous, rapid, and highly efficient. In conventional sintering processes, thermal energy is transferred from outer surface of materials to their inner surface, whereas in microwave sintering, the heat is generated from within the material and radiates outward due to the penetrative power of the microwave (Gupta and Eugene 2005, 2007). In pure microwave heating process, it is observed that higher temperatures exist at the core of samples while their surfaces experience a lower temperature. In the conventional heating processes, it is observed that a higher temperature exists at the surface of the sample whereas the core of the sample experiences a lower temperature (Gupta and Eugene 2007). Such differential heating (core-to-surface or surface-to-core) in a sample causes a variation in microstructure to occur, across its thickness, which would result in poor properties. To eliminate this issue, the bidirectional hybrid microwave heating was developed and is described below.

9.3.1.1.1 Process Description of Bidirectional Microwave Sintering

In the bidirectional hybrid microwave sintering process, microwave susceptors, such as silicon carbide particles/rods, are used to assist in the reduction of thermal gradient during sintering (Gupta and Eugene 2005). Such a type of microwave heating setup is shown in Figure 9.3, which uses a simple household microwave oven, wherein compacted metal/composite powder billets are placed in the inner crucible and silicon carbide powder is placed between the inner and outer crucibles. Silicon carbide powder absorbs microwave readily and heats up quickly, providing heat to the compacted billets (preforms) externally (Gupta and Eugene 2007). On the other hand, the compacted billets absorb the microwave and get heated from inside. Thereby, uniform heat is experienced across the thickness of specimens and the core-to-center thermal variation is minimized (Gupta and Eugene 2005, 2007).

The bidirectional microwave sintering method has proved to be successful for sintering light-metal (both aluminum and magnesium) alloys and their composites. As the method reduces thermal variation, high sintering temperature (620°C–650°C) can be obtained for a short period of time, which is close to the melting point of the specific light metal chosen (Gupta and Eugene 2007).

For the case of nanocomposites, temperature rise close to the melting point of the light metal (Gupta and Eugene 2007) ensures good wettability of the nanoparticles within preforms and eliminates

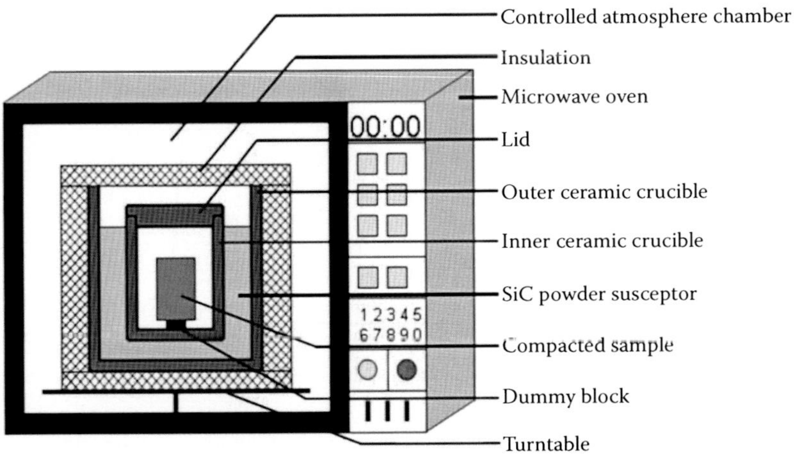

FIGURE 9.3 Schematic of setup showing bidirectional microwave sintering.

porosity at the particle/matrix interface (which is of major concern in conventionally sintered materials). Microwave sintered nanocomposites are dense and have a fine microstructure. The merits of the bidirectional microwave sintering process are (i) high and rapid heating rates, (ii) shorter time duration for sintering (this implies that for materials like magnesium that get oxidized, an inert atmosphere may not be required during sintering), (iii) high sintering temperatures that result in low porosity, and (iv) fine microstructure that results in better mechanical properties (Gupta and Eugene 2005, 2007; Jayalakshmi and Arvind Singh 2015).

9.3.1.1.2 Nanocomposites Sintered by Bidirectional Microwave Sintering Process

Sintering defects, such as circumferential cracks or radial cracks (usually present in conventionally sintered products), are eliminated when the nanocomposites (e.g., magnesium–silicon carbide nanocomposites) were sintered through the microwave process (Eugene and Gupta 2006). These materials showed 26% increase in yield strength, 18% increase in ultimate tensile strength, and 21% improvement in ductility. Significant enhancement in strength properties, ductility, and even thermal stability has been observed for the magnesium–carbon nanotubes (Mg–CNT) composites sintered by this process (Goh et al. 2006a). Uniform reinforcement distribution with good reinforcement-matrix interfacial integrity and significant grain refinement that gives rise to high strength, ductility, and toughness has been reported for magnesium–yttria (Mg–Y_2O_3) nanocomposites (Figure 9.4) (Tun and Gupta 2007). An increase in dislocation density due to coefficient of thermal expansion mismatch and Orowan strengthening enhanced their mechanical properties. Improvement in high-temperature mechanical properties such as strength and ductility (test temperature range: 25°C to 250°C) has been observed for the magnesium–yttria (Mg–Y_2O_3) nanocomposites sintered by the microwave process (Mallick et al. 2010).

Microwave sintered composites exhibit excellent dynamic mechanical properties, for example, magnesium with aluminum reinforced by carbon nanotubes (Mg–Al/CNT) (Habibi et al. 2012). An investigation of the effect of strain rate on the quasi-static and dynamic mechanical properties of these composites has shown a noticeable increase in flow stress under both dynamic tension and dynamic compression conditions when compared to the quasi-static test results, as shown in Figure 9.5a and b

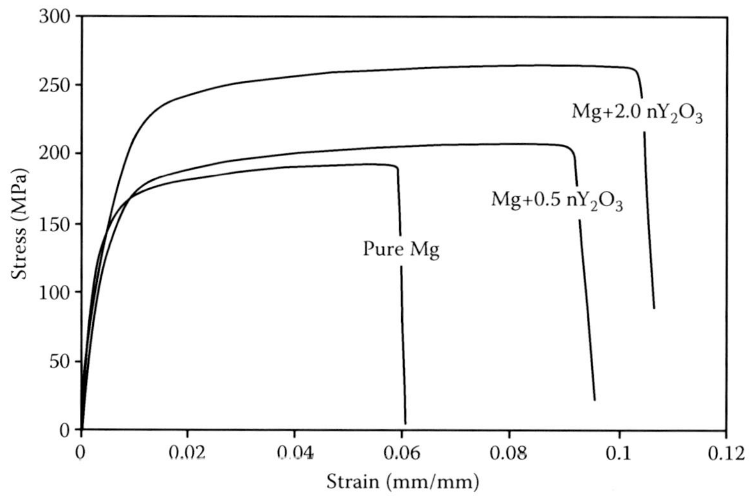

FIGURE 9.4 Engineering stress–strain curves of pure magnesium reinforced with nano-yttria synthesized by bidirectional microwave sintering. (Used with permission from Tun, K. S. and M. Gupta (2007). Improving mechanical properties of magnesium using nano-yttria reinforcement and microwave assisted powder metallurgy method. *Composites Science and Technology* 67 (13): 2657–2664.)

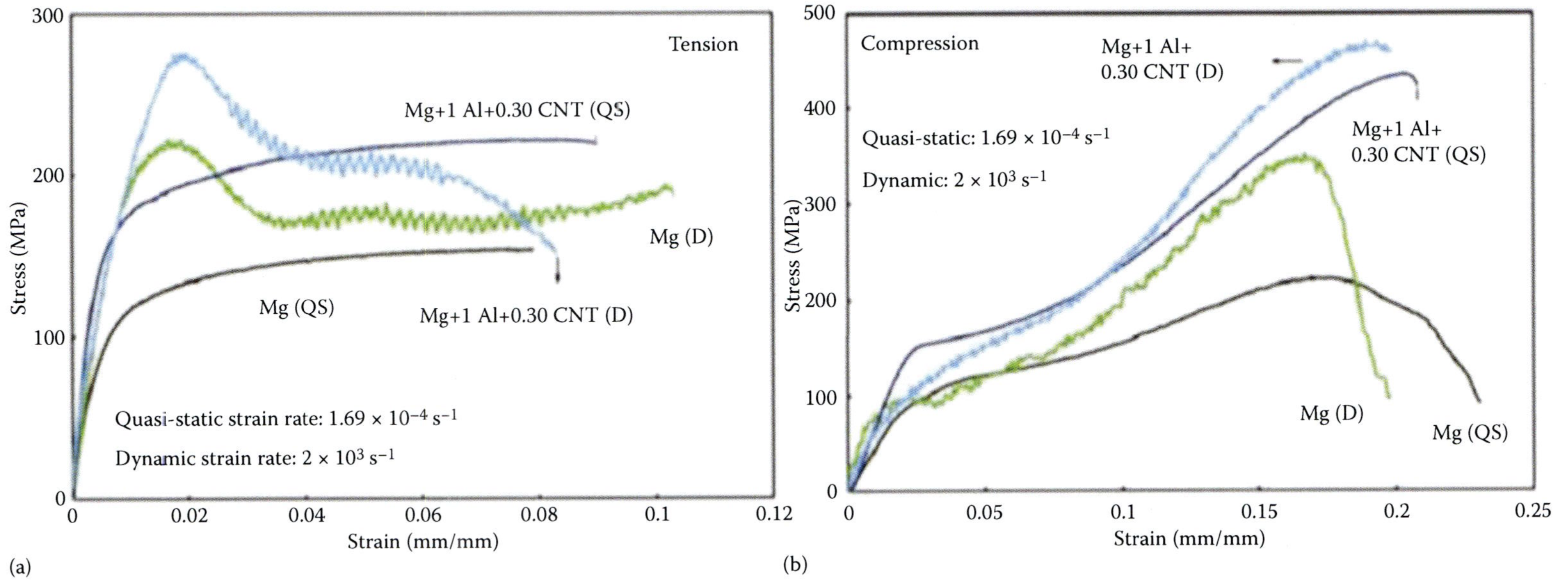

FIGURE 9.5 Engineering stress–strain curves under quasi-static and dynamic (a) tension and (b) compression of magnesium–aluminum alloy with carbon nanotube (Mg–Al+CNT) nanocomposites synthesized by bidirectional microwave sintering. (Used with permission from Habibi, M. K., H. Pouriayevali, A. M. S. Hamouda and M. Gupta (2012). Differentiating the mechanical response of hybridized Mg nano-composites as a function of strain rate. *Materials Science & Engineering A* 545: 51–60.)

(Habibi et al. 2012). These results are significant in terms of crashworthiness, given that magnesium-based materials are the preferred choice for automotive applications.

Microwave sintered magnesium nanocomposite systems with hybridized nano-silicon carbide and alumina (Mg/nSiC+nAl_2O_3), nano-alumina and carbon nanotubes (Mg/nAl_2O_3+CNT), and nano-silicon carbide and carbon nanotubes (Mg/nSiC+CNT) showed enhanced yield strength and ultimate tensile strength with an increase/retention in their ductility (Thakur et al. 2007a,b,c). Composites of magnesium with nano-boron nitride addition (Mg–nBN) have shown significant improvement in ultimate tensile strength with nominal increase in ductility when compared to pure magnesium (Figure 9.6a) (Sankaranarayanan et al. 2013b). Similarly, under compression, an increase in yield strength and ultimate compressive strength has been observed for these materials (Figure 9.6b) (Sankaranarayanan et al. 2013b).

Table 9.1 shows the mechanical properties of magnesium with nano-aluminum nitride (Mg+nAlN) nanocomposites synthesized by microwave process (Sankaranarayanan et al. 2015b). These data are taken from a report on the investigation of microstructure and mechanical properties of these nanocomposites in comparison with pure magnesium (Sankaranarayanan et al. 2015b). Microstructural characterization revealed grain refinement due to uniform distribution of the aluminum nitride nano-particulates (Figure 9.7) (Sankaranarayanan et al. 2015b). A quantitative increase in the tensile yield strength, ultimate tensile strength (UTS), and ductility (tensile fracture strain) of the microwave sintered composites is evident. Among the developed composites, magnesium with 0.8% nano-aluminum nitride had superior strength (30% improvement) and magnesium with 0.2% nano-aluminum nitride had enhanced ductility (80% improvement). Under a compressive load, the nanocomposites showed a response with improved strength properties (compressive yield strength and ultimate compressive strength) without any significant change in compressibility (compressive fracture strain) (Sankaranarayanan et al. 2015b).

9.3.1.1.3 Light Metal Composites with Amorphous Alloy/Bulk Metallic Glass Reinforcements Sintered by Bidirectional Microwave Sintering Process

Reinforcing metallic glass in light metal matrices is a challenge. Metallic glasses are metastable and hence liquid-state processing methods can be adopted only for few of the metallic glasses that have high thermal stability (Lee et al. 2004). Further, traditional sintering methods, such as furnace heating, are time consuming and would cause crystallization of the amorphous phase. Given this issue, any method to introduce metallic glasses as reinforcement in light metals should have the feature of rapid sintering at high sintering temperatures. Bidirectional microwave sintering offers both of these features and hence is suitable for the production of metallic glass reinforced composites. The capability of the microwave process to produce such novel materials was demonstrated for the first time in 2013 (Jayalakshmi et al. 2013, 2014). Aluminum- and magnesium-based composites reinforced with $Ni_{60}Nb_{40}$ (atomic percent) amorphous particles with varying volume fractions were prepared using bidirectional microwave–assisted rapid sintering method (Jayalakshmi et al. 2013, 2014).

$Ni_{60}Nb_{40}$ (atomic percent) amorphous alloy powder prepared by mechanical alloying was reinforced in pure aluminum and magnesium matrices (blend), cold compacted (diameter, 36 mm; height, 50 mm), and microwave sintered (using 100% power level for 750 s) at a sintering temperature of about 550°C. The sintered billets were then hot extruded to 8-mm-diameter rods at 350°C (Jayalakshmi et al. 2013, 2014). Structural analyses of the aluminum reinforced with $Ni_{60}Nb_{40}$ (atomic percent) composites revealed (a) uniform distribution of the reinforcement, (b) absence of interfacial products, and (c) retention of amorphous structure of the reinforcement at all volume fractions (V_f) (Figure 9.8) (Jayalakshmi et al. 2013). When compared to pure aluminum, the composites showed increased microhardness value (130% increase), up to 100% enhancement in compressive strength, and good compressive ductility (Figure 9.9) (Jayalakshmi et al. 2013). Comparison of compressive properties of composites produced by microwave sintering with those

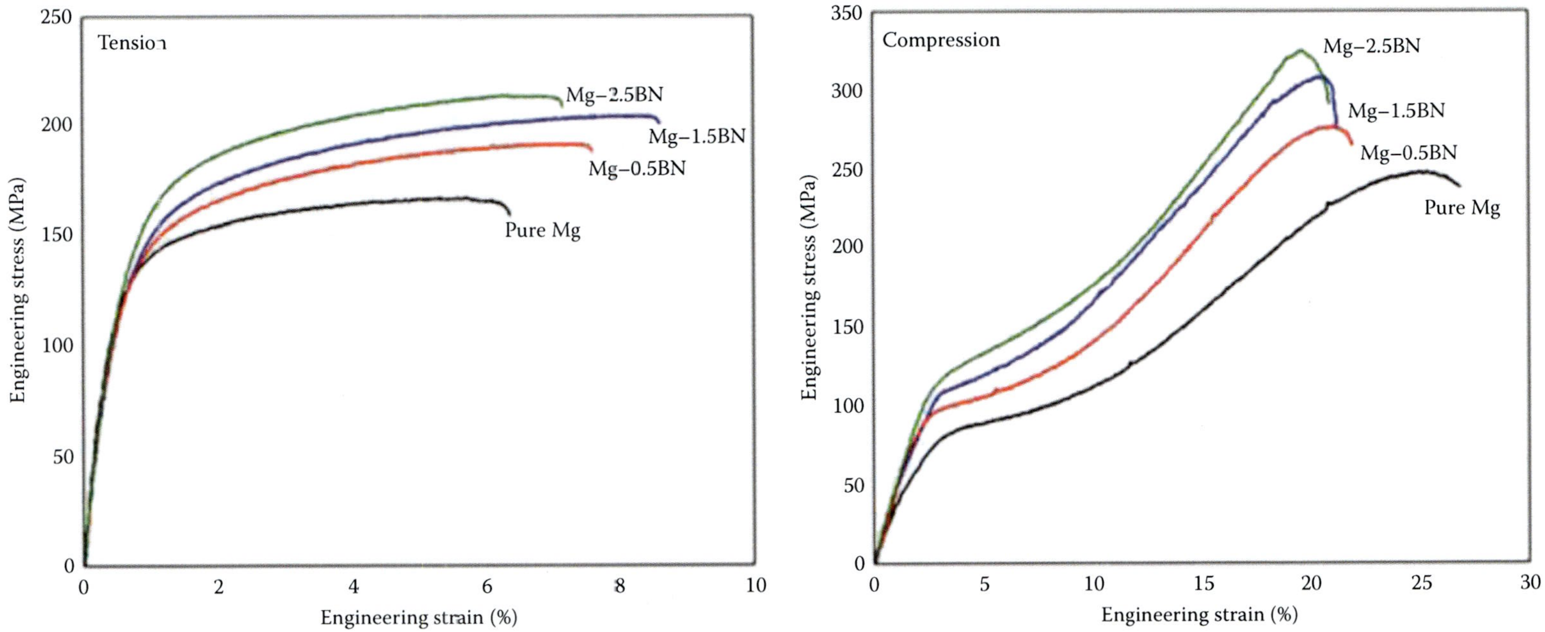

FIGURE 9.6 Tensile and compressive engineering stress–strain curves of pure magnesium reinforced with nano-boron nitride, synthesized by bidirectional microwave sintering. (Used with permission from Sankaranarayanan et al. 2013b.)

TABLE 9.1
Properties of Bidirectional Microwave Sintered Magnesium with Nano-Aluminum Nitride Particle (Mg–nAlN) Composites under Micro-Indentation, Tension, and Compression

		Tensile Properties			Compressive Properties		
Material	Microhardness (Hv)	TYS (MPa)	UTS (MPa)	TFS (%)	CYS (MPa)	UCS (MPa)	CFS (%)
Pure Mg	41 ± 3	96 ± 6	137 ± 9	6.0 ± 3	51 ± 9	268 ± 16	18.9 ± 1.6
Mg–0.2AlN	49 ± 3	102 ± 6	159 ± 8	11.0 ± 2.2	65 ± 1	284 ± 12	19.3 ± 1.5
Mg–0.4AlN	55 ± 3	120 ± 1	164 ± 3	8.4 ± 0.9	72 ± 5	314 ± 20	17.5 ± 0.6
Mg–0.8AlN	53 ± 8	129 ± 5	176 ± 3	6.3 ± 0.4	71 ± 3	307 ± 17	18.3 ± 2.3

Source: Sankaranarayanan, S., M. K. Habibi, S. Jayalakshmi, K. J. Ai, A. Almajid and M. Gupta (2015b). Nano-AlN particle reinforced Mg composites: Microstructural and mechanical properties. *Materials Science & Technology* 31 (9): 1122–1131. Used with permission.

prepared by other methods showed the microwave sintering method to be superior, which was attributed to efficiency of the sintering process (Jayalakshmi et al. 2013). Tensile properties reported were the first of its kind for this class of composites. The capability of the method to produce large

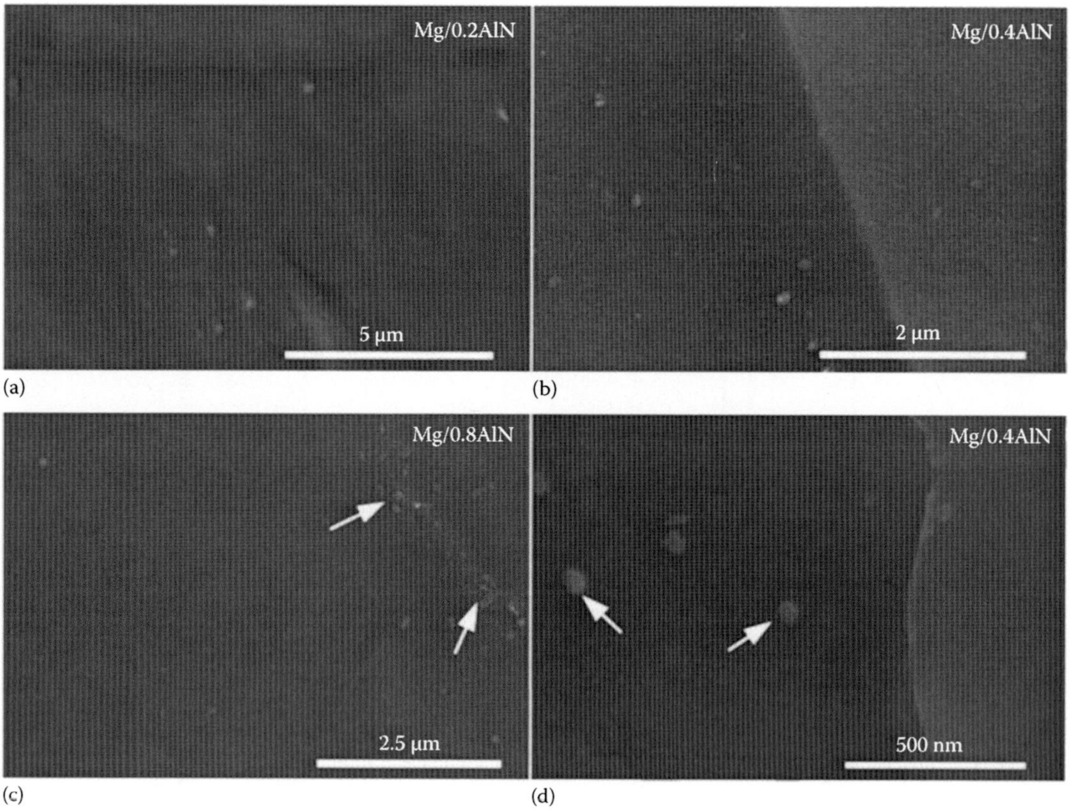

FIGURE 9.7 Scanning electron micrographs showing uniform distribution of nano-aluminum nitride particles in pure magnesium, synthesized by bidirectional microwave sintering process. (Used with permission from Sankaranarayanan, S., M. K. Habibi, S. Jayalakshmi, K. J. Ai, A. Almajid and M. Gupta (2015b). Nano-AlN particle reinforced Mg composites: Microstructural and mechanical properties. *Materials Science & Technology* 31 (9): 1122–1131.)

FIGURE 9.8 Scanning electron micrograph shows clear interface in pure aluminum reinforced with $Ni_{60}Nb_{40}$ (atomic percent) amorphous alloy powder, synthesized by bidirectional microwave sintering.

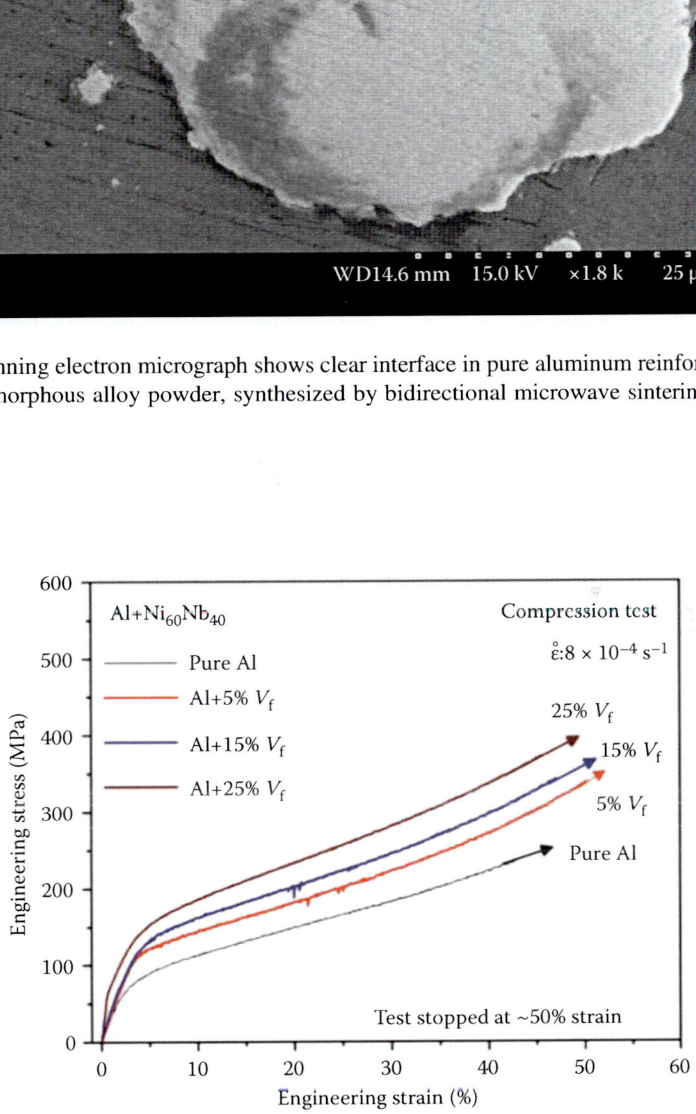

FIGURE 9.9 Compression properties of pure aluminum reinforced with $Ni_{60}Nb_{40}$ (atomic percent) amorphous alloy powder, synthesized by bidirectional microwave sintering. (Used with permission from Jayalakshmi, S., S. Gupta, S. Sankaranarayanan, S. Sahu and M. Gupta (2013). Structural and mechanical properties of $Ni_{60}Nb_{40}$ amorphous alloy particle reinforced Al-based composites produced by microwave assisted rapid sintering. *Materials Science & Engineering A* 581: 119–127.)

size samples was successfully demonstrated. Under tensile loading, a 15%–60% strength increase was observed when compared to pure aluminum. Although the tensile ductility reduced, the reduction was not as drastic as is usually observed in conventional ceramic particle reinforced composites (Jayalakshmi et al. 2013).

Magnesium with $Ni_{60}Nb_{40}$ (atomic percent) composites prepared by microwave sintering showed the following: (a) refinement in grain size, (b) uniform distribution of reinforcement (at low volume fraction), (c) clear reinforcement/matrix interface, and (d) retention of amorphous structure at all volume fractions (Figures 9.10 and 9.11) (Jayalakshmi et al. 2014). X-ray diffraction analyses of the composites revealed a change in crystal orientation; that is, the basal planes were not entirely aligned parallel to the extrusion direction; rather they were oriented at an angle indicating multiple crystallographic orientation (Jayalakshmi et al. 2014). The hardness and compression test properties showed remarkable improvement in properties with increasing volume fraction (V_f), owing to the high hardness, strength, and elastic strain limit of the amorphous reinforcement (Jayalakshmi et al. 2014). The enhancement in properties and retention of amorphous structures in both aluminum and magnesium composites proved that the microwave sintering method, followed by hot extrusion, is an effective route for synthesizing composites with amorphous alloy reinforcements (Jayalakshmi et al. 2014).

Composites of magnesium reinforced with $Ni_{50}Ti_{50}$ (atomic percent) metallic glass produced using microwave sintering showed a uniform distribution and retention of the amorphous structure of the $Ni_{50}Ti_{50}$ (atomic percent) metallic glass reinforcement (Sankaranarayanan et al. 2015c). Under the influence of a mechanical load, a significant improvement in properties was observed. Microhardness improved by 80%, while compression yield strength and ultimate strength improved by more than 70% (Figure 9.12a). Tensile properties of these novel magnesium materials were reported for the first time and showed 95% improvement in yield strength and 50% improvement in ultimate strength. Ductility of the composites did not change at lower volume fractions of the

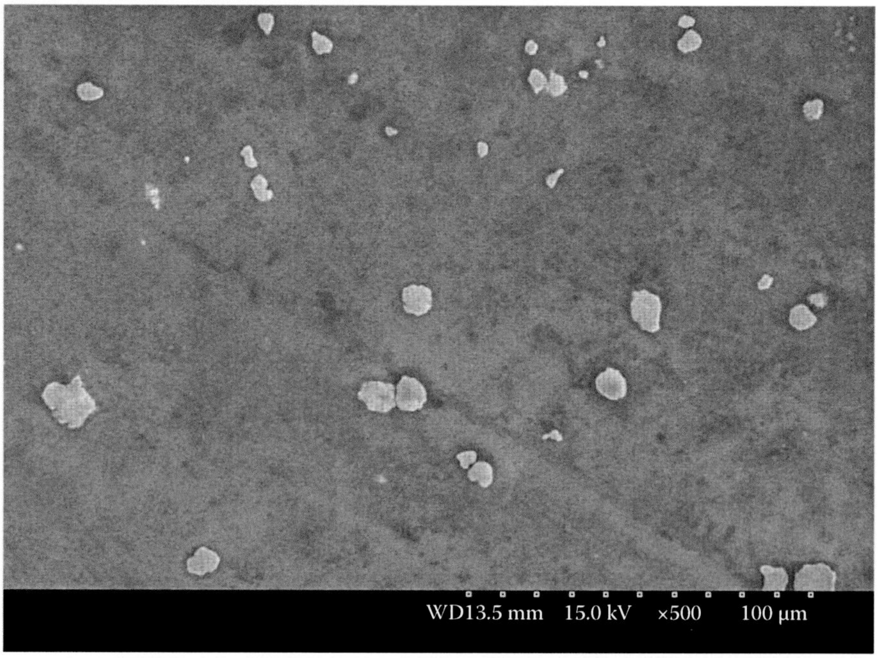

FIGURE 9.10 Scanning electron micrograph showing uniform distribution of $Ni_{60}Nb_{40}$ (atomic percent) amorphous alloy reinforcement in pure magnesium matrix, synthesized by bidirectional microwave sintering.

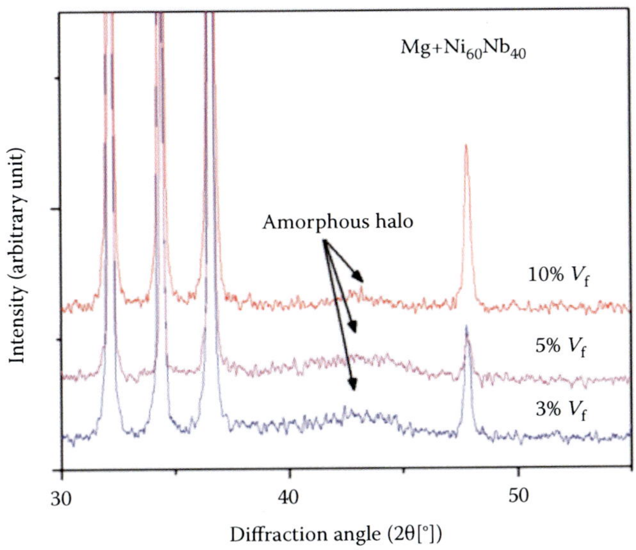

FIGURE 9.11 X-ray diffraction pattern showing retention of the amorphous phase in pure magnesium reinforced with $Ni_{60}Nb_{40}$ (atomic percent) amorphous alloy powder, synthesized by bidirectional microwave sintering. (Used with permission from Jayalakshmi, S., S. Sahu, S. Sankaranarayanan, S. Gupta and M. Gupta (2014). Development of novel Mg–$Ni_{60}Nb_{40}$ amorphous particle reinforced composites with enhanced hardness and compressive response. *Materials & Design* 53: 849–855.)

reinforcement. However, at the high volume fraction of 10% of reinforcement, the ductility decreased (Figure 9.12b). The inherent high-strength properties of the glassy reinforcement and the efficiency of the microwave sintering process contributed toward an enhancement of mechanical properties (Sankaranarayanan et al. 2015c).

Similar results of significant refinement in grain size, retention of amorphous state, retention in ductility, and improvement in compression strength properties were achieved in microwave sintered magnesium composites with $Cu_{50}Ti_{50}$ (atomic percent) amorphous/glassy particles as the reinforcement (Sankaranarayanan et al. 2015a). Microwave sintered magnesium with $Al_{85}Ti_{15}$ composites had equi-axed grains with formation of more twins when compared to the unreinforced counterpart (Bau et al. 2016). The addition of $Al_{85}Ti_{15}$ powder significantly enhanced the damping properties of magnesium. Based on resonant frequency and damping studies, it was seen that the composites showed an increased resistance to damping, with the composites exhibiting a slight reduction in the resonant frequency when compared to pure magnesium. The damping loss rate (vibrational decay of the material, viz., how fast a material stops vibration) increased 3.75 times for the magnesium with $Al_{85}Ti_{15}$ composites (Bau et al. 2016).

9.3.1.1.4 High-Entropy Alloys Sintered by Bidirectional Microwave Sintering Process

Given the rapid sintering rate that can be achieved in bidirectional sintering, the method can be utilized for synthesizing high-entropy alloys. Microwave frequencies of 2450 MHz and 5800 MHz were used to prepare iron-based high-entropy alloys, namely, FeCoNiCuAl, FeCrNiTiAl, and FeCoCrNiAl2.5 high-entropy alloys by direct heating of pressed mixtures of metal powders (Veronesi et al. 2015). In these samples, there was no distortion, that is, change in shape of the powder compacts and dendrite segregation (often seen in high-entropy alloys prepared by arc melting) (Veronesi et al. 2015). A review on the preparation of prospective light metal–based high-entropy alloys by alternative processes including microwave sintering is given by Kumar and Gupta (2016).

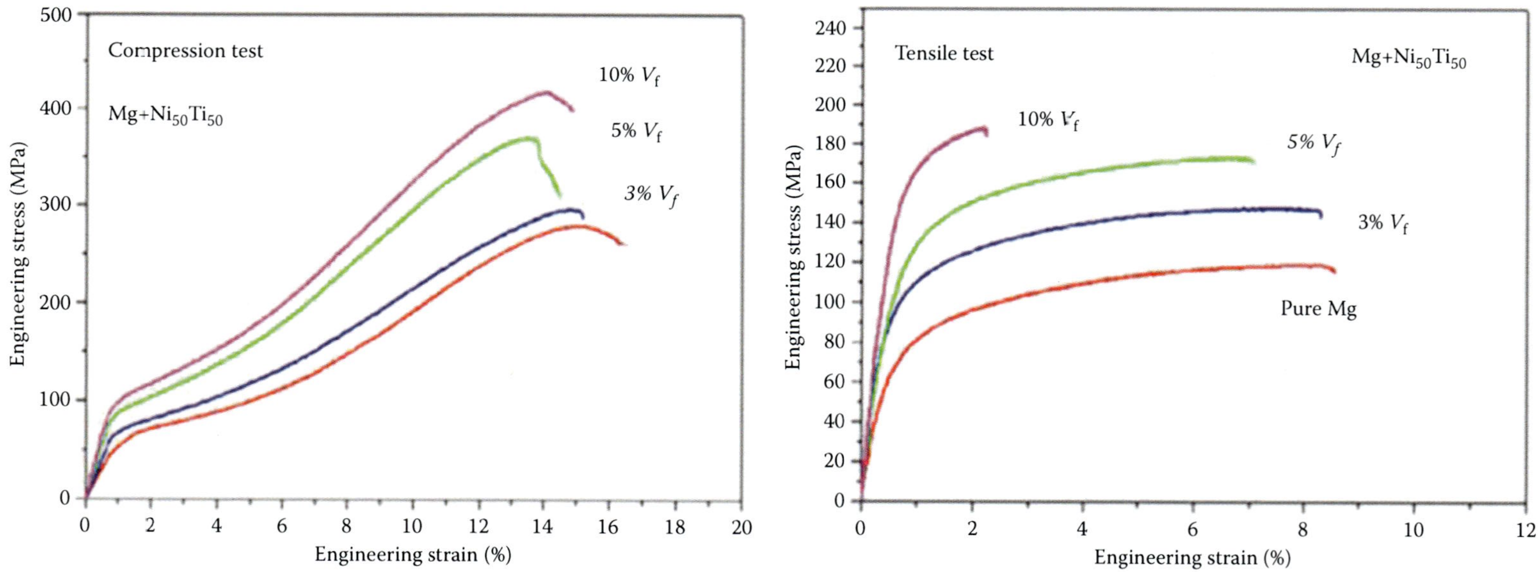

FIGURE 9.12 Compression and tensile properties of bidirectional microwave sintered pure magnesium reinforced with $Ni_{50}Ti_{50}$ metallic glass. (Used with permission from Sankaranarayanan, S., V. Hemanth Shankar, S. Jayalakshmi, N. Q. Bau and M. Gupta (2015). Development of high performance magnesium composites using $Ni_{50}Ti_{50}$ metallic glass reinforcement and microwave sintering approach. *Journal of Alloys and Compounds* 627: 192–199.)

9.3.1.2 Spark Plasma Sintering

When adopting conventional sintering techniques, the main drawback is the occurrence of high porosity in products. Further, when the green compacts are sintered, hot pressed, hot extruded, or hot isostatically pressed, it often results in matrix grain growth (due to long duration of exposure at the sintering temperature) that is detrimental to mechanical properties. Spark plasma sintering is an effective and novel nonconventional sintering method that eliminates this negative aspect. By using spark plasma sintering method, fully dense materials having a refined grain size can be obtained.

In spark plasma sintering, the densification of powder is achieved by directly passing a pulsed direct current through a graphite die containing the compacted metal powder (Saheb et al. 2012). Joule's heating effect and an applied axial load densifies the powder compacts (to near theoretical density) at relatively lower sintering temperatures as compared to conventional sintering techniques. In spark plasma sintering, the heat is generated internally, that is, within the sample, in contrast to the conventional hot pressing methods where the heat is supplied from external heating elements (Saheb et al. 2012). This facilitates very high heating or cooling rates (up to 1000 K/min). Consequently, the sintering process in spark plasma sintering is very rapid (within a few minutes). Such a high process rate in spark plasma sintering ensures densification of powders with nanosize or nanostructure and avoids coarsening, which often occurs in standard densification routes (Saheb et al. 2012).

9.3.1.2.1 Process Description of Spark Plasma Sintering

Figure 9.13 shows the spark plasma sintering process setup (Saheb et al. 2012). Here, sintering is assisted by a uniaxial press. Sintering temperature can be controlled by careful selection of holding time, ramp rate, pulse duration, pulse current, and voltage. The direct current pulse discharge generates spark plasma, spark impact pressure, Joule heating, and the electrical field diffusion effect. In spark plasma sintering, sintering is assisted by the on–off direct current pulse voltage in contrast to the conventional hot pressing as shown in Figure 9.14 (Saheb et al. 2012). The application of pressure aids in the plastic flow of material. The flow of direct current pulse through metal powder particles is shown in Figure 9.15 (Saheb et al. 2012).

The spark plasma sintering process is conducted in four sequential stages as illustrated in Figure 9.16 (Saheb et al. 2012). In the first stage, entrapped gases are released by application of vacuum. Pressure is applied in the second stage for compaction, followed by resistance heating (i.e., sintering) in the third stage and cooling in the fourth stage. As a spark discharge appears at the gap, that is, at the contact points between metal powder particles, the temperature rises locally by several thousand degrees centigrade; momentarily, the rise in temperature causes melting at the surface of powder particles and the formation of necks around the area of contact between particles, resulting in sintering. In addition to high localized temperatures, the presence of pressure and electric current increases the heating rate and drastically reduces the sintering time leading to the consolidation of powders without excessive grain growth (Saheb et al. 2012). Spark plasma sintering is a binder-less process and does not require any pre-compaction step. Materials processed through spark plasma sintering show superior mechanical properties.

9.3.1.2.2 Nanocomposites Sintered by Spark Plasma Sintering

Spark plasma sintering is a rapid sintering method and does not allow sufficient time for grain growth and hence can be effectively used to synthesize nanocomposites. Most of the studies using spark plasma sintering have been carried out in copper–carbon nanotube (Cu–CNT) and in aluminum–carbon nanotube (Al–CNT) systems. The first report on making composites by spark plasma sintering was on the synthesis of copper–carbon nanotube nanocomposites (Rana et al. 2009; Sasaki et al. 2009). These materials were fabricated at 1023 K and 40 MPa with good dispersion and improved density (97%–98.5%). Their sintered microstructure consisted of dual zones of carbon nanotube–free matrix and carbon nanotube–rich grain boundary regions (Rana et al. 2009). Spark plasma sintered nanocomposites (e.g., aluminum–carbon nanotube composites) displayed excellent mechanical

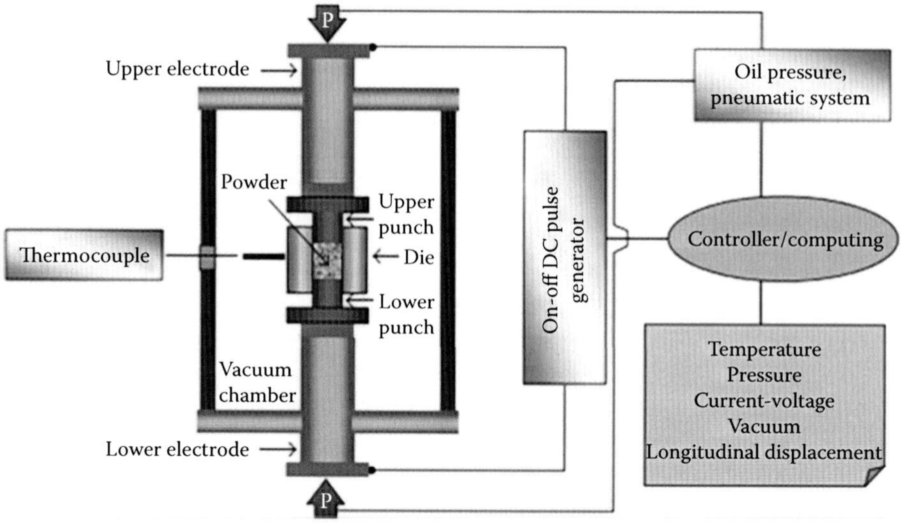

FIGURE 9.13 Schematic showing the spark plasma sintering (SPS) process. (From Saheb, N., Z. Iqbal, A. Khalil, A. S. Hakeem, N. Al Aqeeli, T. L. A. Al-Qutub and R. Kirchner (2012). Spark plasma sintering of metals and metal matrix nanocomposites: A review. *Journal of Nanomaterials*: 1–13. Open access.)

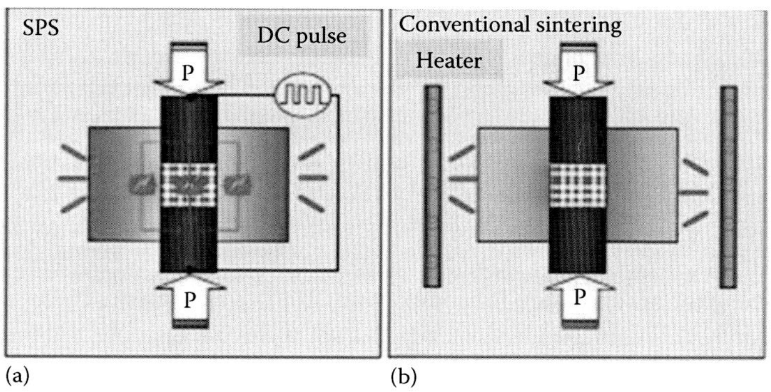

FIGURE 9.14 Comparison of spark plasma sintering (SPS) and conventional sintering. (From Saheb, N., Z. Iqbal, A. Khalil, A. S. Hakeem, N. Al Aqeeli, T. L. A. Al-Qutub and R. Kirchner (2012). Spark plasma sintering of metals and metal matrix nanocomposites: A review. *Journal of Nanomaterials*: 1–13. Open access.)

properties: 129% enhanced strength for 5 vol% carbon nanotubes addition (Khalil et al. 2011). Nanocrystalline aluminum-alloy/silicon carbide (Al alloy/SiC) nanocomposite powders were synthesized using high-energy ball milling and spark plasma sintering at a temperature of 500°C at a heating rate of 300°C/min and a total sintering cycle of 8 min (Bathula et al. 2012a). Substantial enhancement of mechanical properties was reported as a result of sintering by spark plasma sintering. 0.2% compressive proof stress at room temperature for spark plasma sintered aluminum with aluminum diboride (Al–AlB_2) and aluminum with magnesium diboride (Al–MgB_2) (Table 9.2) showed very high strength when compared to those of pure aluminum and commercial 7075-T6 alloy (Kubota et al. 2008). The microstructure and properties of pure aluminum reinforced with 0–2.0 wt% multiwall carbon nanotubes prepared by spark plasma sintering and hot extrusion are reported by Liao et al. (2010). Carbon nanotubes were treated with sodium dodecyl sulfate surfactant (as it decreases the van der Waals force between the carbon nanotubes) in order to facilitate their easy dispersion in aluminum metal. The composites were quite dense with uniform distribution of carbon nanotubes, except at high volume fractions (Figure 9.17) (Liao et al. 2010).

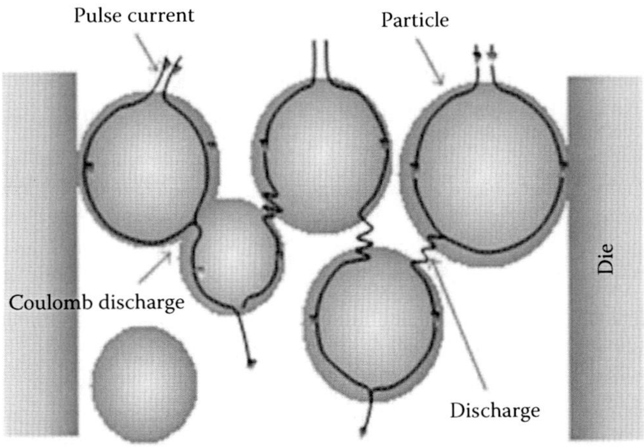

FIGURE 9.15 Direct current pulse current flow through metal powder particles. (From Saheb, N., Z. Iqbal, A. Khalil, A. S. Hakeem, N. Al Aqeeli, T. L. A. Al-Qutub and R. Kirchner (2012). Spark plasma sintering of metals and metal matrix nanocomposites: A review. *Journal of Nanomaterials*: 1–13. Open access.)

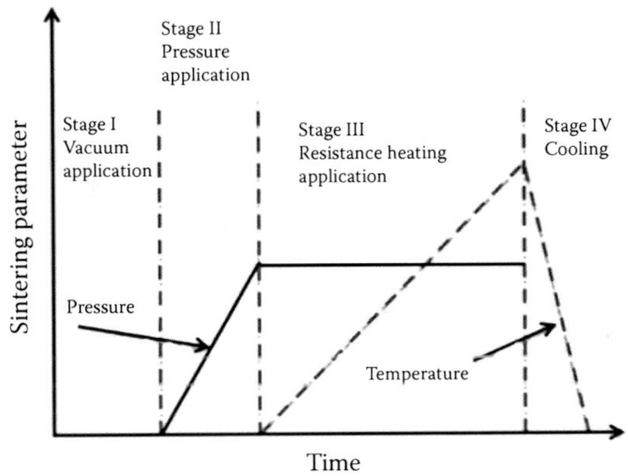

FIGURE 9.16 The four sequential stages in the spark plasma sintering SPS process. (From Saheb, N., Z. Iqbal, A. Khalil, A. S. Hakeem, N. Al Aqeeli, T. L. A. Al-Qutub and R. Kirchner (2012). Spark plasma sintering of metals and metal matrix nanocomposites: A review. *Journal of Nanomaterials*: 1–13. Open access.)

Nanocomposites sintered by spark plasma sintering process have shown exceptional tribological properties. Wear and friction characteristics of pure aluminum–carbon nanotube (Al–CNT) composites synthesized using spark plasma sintering and hot pressing were investigated. It has been reported that the spark plasma sintering method was more effective than the hot pressing method for minimizing the amount of wear and maintaining a stable friction (Kim et al. 2009). Nanostructured aluminum 5083/10 wt% silicon carbide particle nanocomposites by high-energy ball milling followed by spark plasma sintering were produced. These materials exhibited near-theoretical density while retaining nano-grained features (Figure 9.18), indicating less grain coarsening (Bathula et al. 2012b). Nano-indentation studies revealed the composites to exhibit high hardness of 280 Hv with an elastic modulus of 126 GPa. Substantial reduction in wear and friction property was also observed in the nanocomposites when compared to the unreinforced aluminum 5083 alloy (Figure 9.19) (Bathula et al. 2012b).

TABLE 9.2
Compressive Properties of Aluminum with Aluminum Diboride Particles (Al–AlB_2) and Aluminum with Magnesium Diboride Particle (Al–MgB_2) Nanocomposites Synthesized by Spark Plasma Sintering (SPS) Process

Material	0.2% Proof Stress (MPa)	Microhardness (Hv)
7075-T6	505	39
Pure Al MA 0 h SPS	173	158
Pure Al MA 8 h SPS	440	178
Al_{85} $(AlB_2)_{15}$ MA 4 h SPS	520	187
Al_{85} $(AlB_2)_{15}$ MA 8 h SPS	620	215
Al_{85} $(MgB_2)_{15}$ MA 4 h SPS	628	267
Al_{85} $(MgB_2)_{15}$ MA 8 h SPS	846	

Source: Kubota, M., J. Kaneko, and M. Sugamata (2008). Properties of mechanically milled and spark plasma sintered Al–AlB2 and Al–MgB2 nano-composite materials. *Materials Science & Engineering A*, 475 (1–2): 96–100. Used with permission.

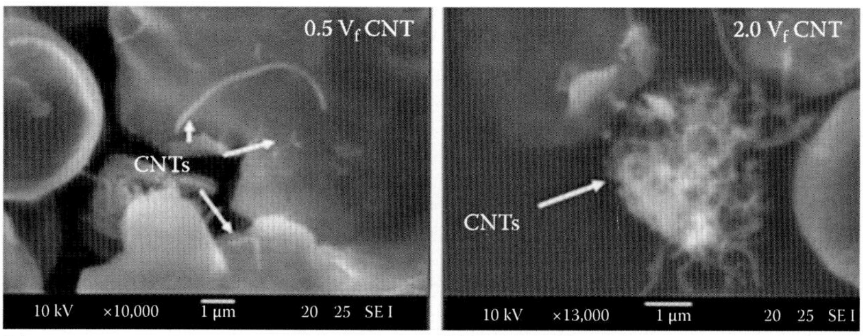

(a) (b)

FIGURE 9.17 Scanning electron micrographs of the mixed aluminum and multiwalled carbon nanotubes powders (a) aluminum–0.5 wt% multiwalled carbon nanotubes and (b) aluminum–2.0 wt% multiwalled carbon nanotubes. (Used with permission from Liao, J. Z., M. J. Tan and I. Sridhar (2010). Spark plasma sintered multiwall carbon nanotube reinforced aluminum matrix composites. *Materials & Design* 31 (1): S96–S100.)

9.3.1.2.3 High-Entropy Alloys Sintered by Spark Plasma Sintering

High-entropy alloys and high-entropy alloy coatings with superior mechanical properties have been successfully synthesized by the spark plasma sintering method. CoCrFeNiAl high-entropy alloy synthesized by mechanical alloying and spark plasma sintering showed the coexistence of body-centered cubic (BCC) and face-centered cubic (FCC) phases (Ji et al. 2014). Their hardness and compressive strength were significantly high: 625 Hv and 1.9 GPa, respectively (Ji et al. 2014). Similarly, CoCrFeNiMnAl high-entropy alloy prepared by mechanical alloying and spark plasma sintering showed excellent hardness of 660 Hv and high compressive strength of 2.2 GPa, with the coexistence of BCC and FCC solid solution phases (Wang et al. 2014). AlCoCrCuFeNi high-entropy alloys with varying aluminum content were prepared using spark plasma sintering method. Hardness of 160 Hv and excellent thermal stability (phase stability and crystallite size) up until 600°C were observed for these materials (Sriharitha et al. 2014).

AlCoCrFeNiTi high-entropy alloy composite with in situ titanium carbide particles were synthesized by mechanical alloying followed by spark plasma sintering method (Moravcik et al. 2016). The material had a high hardness value of 760 Hv. The hardness was retained (600 Hv) even after

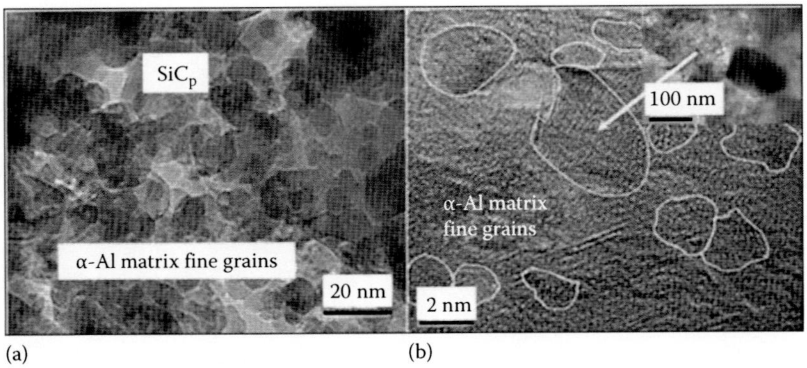

FIGURE 9.18 High-resolution transmission electron microscopic (HRTEM) images of Al 5083 alloy with silicon carbide particles: (a) nanocomposite ball milled for 15 h and (b) spark plasma sintered. Inset in (b) shows the nano-grained microstructure with networked dislocations. (Used with permission from Bathula, S., M. Saravanan and A. Dhar (2012). Nanoindentation and wear characteristics of Al 5083/SiCp nanocomposites synthesized by high energy ball milling and spark plasma sintering. *Journal of Materials Science & Technology* 28 (11): 969–975.)

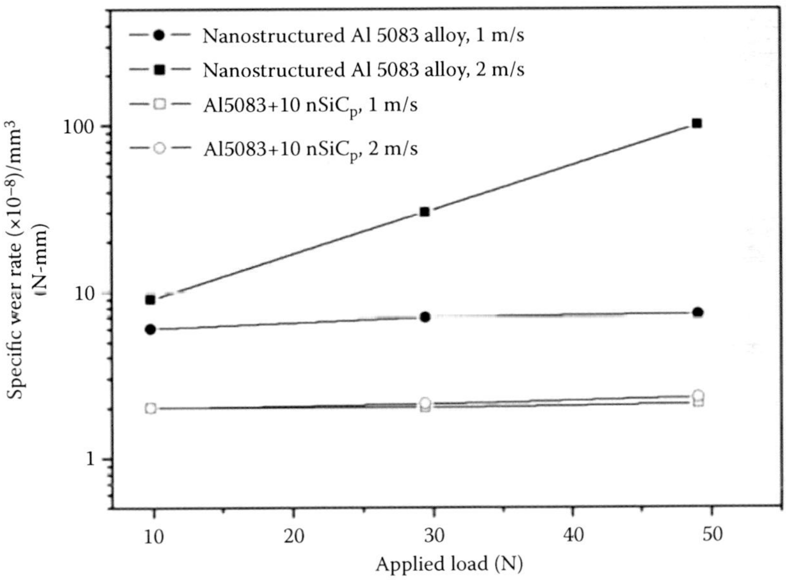

FIGURE 9.19 Specific wear rate as a function of applied load of the Al5083 alloy with silicon carbide particles nanocomposite synthesized by spark plasma sintering process in comparison with the nanostructured aluminum 5083 alloy. (Regenerated from Bathula, S., M. Saravanan and A. Dhar (2012). Nanoindentation and wear characteristics of Al 5083/SiCp nanocomposites synthesized by high energy ball milling and spark plasma sintering. *Journal of Materials Science & Technology* 28 (11): 969–975. Used with permission.)

heat treatment at a temperature of 1100°C, indicating its high thermally stability (Moravcik et al. 2016). The process of spark plasma sintering induced refined microstructure in CoCrFeMnNi high-entropy alloy fabricated via mechanical alloying and spark plasma sintering. These had smaller grain size and exhibited higher strength when compared to the cast high-entropy alloy (Figures 9.20 and 9.21) (Joo et al. 2017).

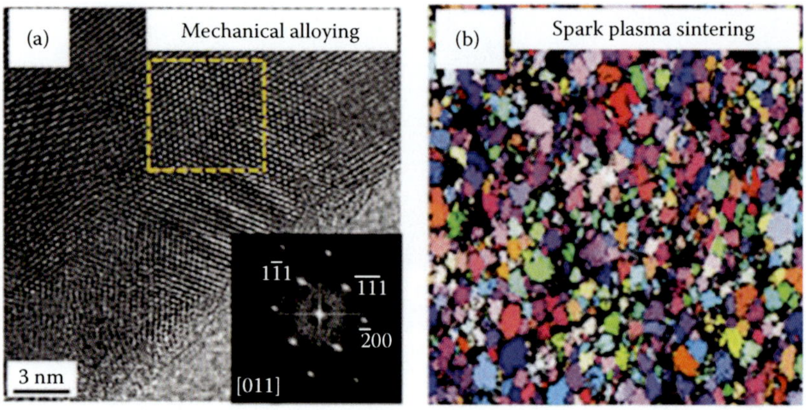

FIGURE 9.20 (a) High-resolution transmission electron microscope image of CoCrFeMnNi high-entropy alloy powder mechanically alloyed for 60 min (60 MA). (b) Electron backscatter diffraction (EBSD) image of CoCrFeMnNi high-entropy alloy mechanically alloyed for 60 min and sintered at 900°C by spark plasma sintering method. (Used with permission from Joo, S. H., H. Kato, M. J. Jang, J. Moon, E. B. Kim, S. J. Hong and H. S. Kim (2017). Structure and properties of ultrafine-grained CoCrFeMnNi high-entropy alloys produced by mechanical alloying and spark plasma sintering. *Journal of Alloys and Compounds* 698: 591–604.)

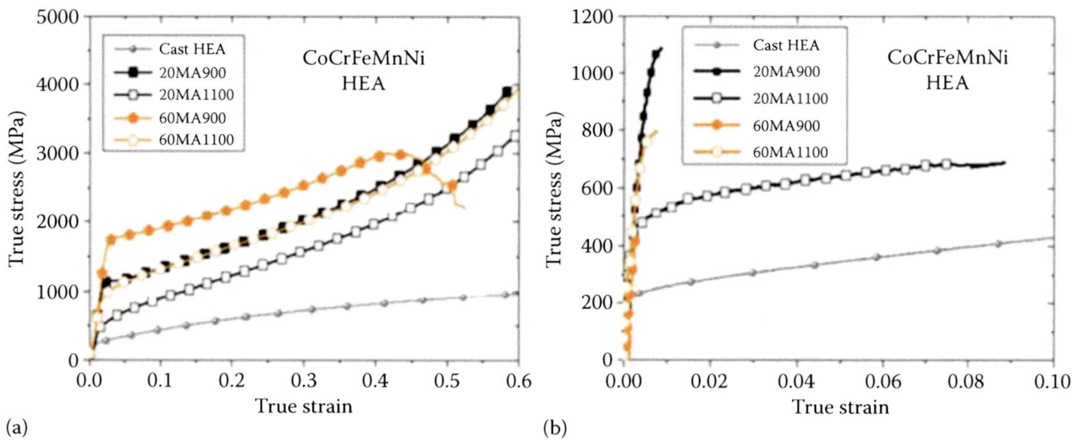

FIGURE 9.21 True stress–true strain curves under (a) compression and (b) tension of CoCrFeMnNi high-entropy alloy sintered by spark plasma sintering method at different duration of mechanical alloying and sintering temperatures. (Used with permission from Joo, S. H., H. Kato, M. J. Jang, J. Moon, E. B. Kim, S. J. Hong and H. S. Kim (2017). Structure and properties of ultrafine-grained CoCrFeMnNi high-entropy alloys produced by mechanical alloying and spark plasma sintering. *Journal of Alloys and Compounds* 698: 591–604.)

Advanced heat-resistant WMoNbV(Ta,Ti) high-entropy alloys based on refractory metals have been developed using mechanical alloying and spark plasma sintering for high-temperature structural applications (Ryu et al. 2016). The refractory high-entropy alloys exhibited superior hardness and compressive strength values coupled with high mechanical stability until 1000°C. The room-temperature strength varied between 2.0 and 3.6 GPa with a fracture strain of 9.7%, whereas the high-temperature strength was about 1 GPa (Ryu et al. 2016). Spark plasma sintering can also be used to develop high-entropy alloy coatings on metal substrates. AlCrFeCoNiCu coatings with varying aluminum content were synthesized using the spark plasma sintering method on AISI 4140 carbon steel (Ji et al. 2016). These high-entropy alloy coatings showed the coexistence of FCC and BCC phases. Their wear and friction behavior indicated that the BCC phase was beneficial in improving the wear resistance and reducing the coefficient of friction (Ji et al. 2016). High-entropy alloy coatings by spark plasma sintering have been suggested for high-temperature applications, for example, NiCoFeCrSiAlTi (Hsu et al. 2016).

9.3.1.3 High-Frequency Induction Heat Sintering

High-frequency induction heat sintering (HFIHS) is an effective sintering process that can successfully consolidate ceramics and metallic powders to near-theoretical density (Khalil and Almajid 2012). The HFIHS process involves rapid sintering of nanostructured hard metals in a short time assisted with high temperature and pressure. The process inhibits grain growth in nanostructured materials.

The configuration of an HFIHS unit is shown in Figure 9.22a and b (Khalil and Almajid 2012). It consists of a uniaxial pressure device, a graphite die, a water-cooled reaction chamber, and regulating systems for induced current, pressure, and temperature (Khalil and Almajid 2012). In the HFIHS process, an intense magnetic field is applied via an electrically conducting pressure die and the die acts as a heating source. Samples also get heated from both inside and outside (Khalil and Almajid 2012). The unit is operated by first evacuating to a vacuum level of 1×10^3 Torr, and then a uniaxial pressure is applied. Subsequently, an induced current (frequency ~50 kHz) is activated and applied until densification and sintering are complete. The four sequential stages involved in the HFIHS process are shown in Figure 9.23 (Khalil and Almajid 2012).

9.3.1.3.1 Nanocomposites Sintered by HFIHS

The HFIHS process has been used to prepare fine crystalline, nearly fully dense nanocomposites of magnesium with hydroxyapatite particles from powders with smaller crystal sizes by optimizing the processing parameters (Khalil and Almajid 2012). Their relative density, microhardness, and compressive strength values increased with an increase in sintering temperature, reaching values as high as 99.7%, 60 Hv, and 194.5 MPa, respectively, at 550°C. The crystal size of the sample sintered at 500°C was 37 nm and was found to be strongly dependent on the heating rate. High heating rates (1200°C/min) produced crystal sizes of 34 nm (Khalil and Almajid 2012).

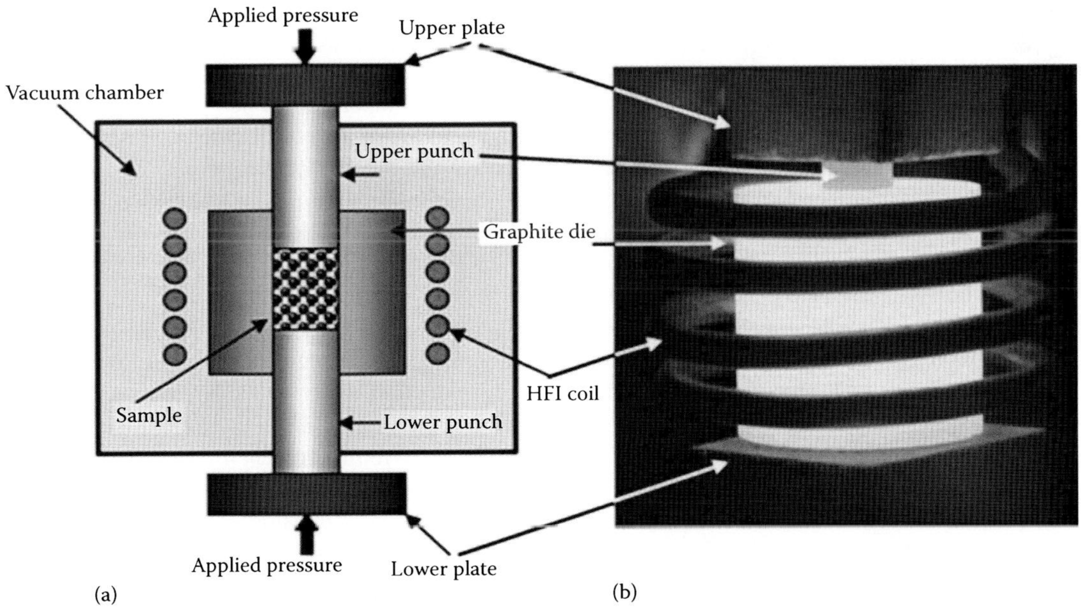

FIGURE 9.22 (a) High-frequency induction-heated sintering apparatus and (b) heated die. (Used with permission from Khalil, K. A. and A. A. Almajid (2012). Effect of HFIHS conditions on the microstructure and mechanical properties of nanostructured magnesium/hydroxyapatite nanocomposites. *Materials & Design* 36: 58–68.)

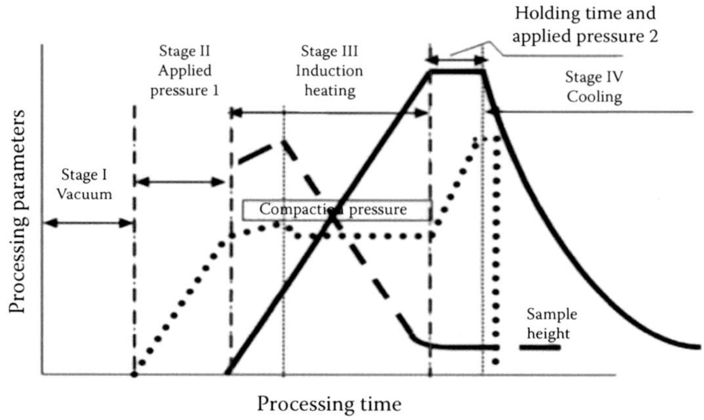

FIGURE 9.23 The four sequential stages of sintering in the HFIHS process. (Used with permission from Khalil, K. A. and A. A. Almajid (2012). Effect of high-frequency induction heat sintering conditions on the microstructure and mechanical properties of nanostructured magnesium/hydroxyapatite nanocomposites. *Materials & Design* 36: 58–68.)

9.3.1.3.2 Amorphous Alloy/Bulk Metallic Glass Reinforced Aluminum and Magnesium Composites Sintered by HFIHS

A brief description of synthesis of aluminum- and magnesium-based composites incorporated with glassy reinforcements by HFIHS is given here (Aljerf et al. 2012; Dudina et al. 2009, 2010). The AZ31 magnesium alloy was incorporated with amorphous/glassy reinforcement, namely, Vitraloy 6 ($Zr_{57}Nb_5Cu_{15.4}Ni_{12.6}Al_{10}$) (Dudina et al. 2009). AZ91 ingots prepared by induction melting under argon were melt-spun into ribbons (thickness, 30 μm). Vitraloy 6 was taken in the form of ribbons (Dudina et al. 2009). The ribbons were cut and milled under argon in a low-speed vibratory mill for 15 h to produce powders. The milled magnesium alloy ribbons were reinforced with 15% volume fraction of milled zirconium glassy powder and the mixture was milled under similar conditions. This composite powder was sintered in vacuum by the HFIHS process (maximum sintering temperature, 713 K; heating rate, 1.6 K/s) at a pressure of 50 MPa, for a duration of 120 s (Dudina et al. 2009). The sintering temperature was selected to be within the super-cooled liquid region of Vitraloy 6, the glassy reinforcement, and close to the solidus temperature of AZ91, the matrix alloy (Dudina et al. 2009). The composite was examined for its structural and mechanical properties. The x-ray diffraction pattern of the composite showed retention of the amorphous structure and crystalline phases corresponding to magnesium and $Mg_{17}Al_{12}$ (Dudina et al. 2009). Microstructure of the composite when observed in a scanning electron microscope (not shown here) revealed clear interface and uniform distribution of the glassy reinforcement particles with no porosity. A significant improvement in hardness and compression strength of the composite is shown in Figure 9.24a (Dudina et al. 2009).

Aluminum alloy A520 with copper-based $Cu_{54}Zr_{36}Ti_{10}$ (atomic percent) glassy alloy (of 15% volume fraction) was prepared using high-frequency induction sintering under pressure (Dudina et al. 2010). The metallic glass ribbons were powdered in a vibratory mill for 24 h and mixed with the cut ribbons of aluminum matrix (Dudina et al. 2010). The alloy-reinforcement mixture was milled for 8 h and processed using high-frequency induction sintering (pressure, 50 MPa) for 120 s, at a temperature of 720 K. The sintering temperature was selected after considering the glass transition temperature (T_g ~ 715 K) and the crystallization temperature (T_x ~ 760 K) of the $Cu_{54}Zr_{36}Ti_{10}$ (atomic percent) amorphous ribbon (Dudina et al. 2010). X-ray diffraction analysis indicated the absence of additional phases and no transformation of the amorphous phase (Dudina et al. 2010). Uniform distribution of the amorphous phase with no preferred orientation was observed. The compression test results showed a significant improvement in strength with a slight reduction in fracture strain (Figure 9.24b) (Dudina et al. 2010). An increase in strength of the composite was

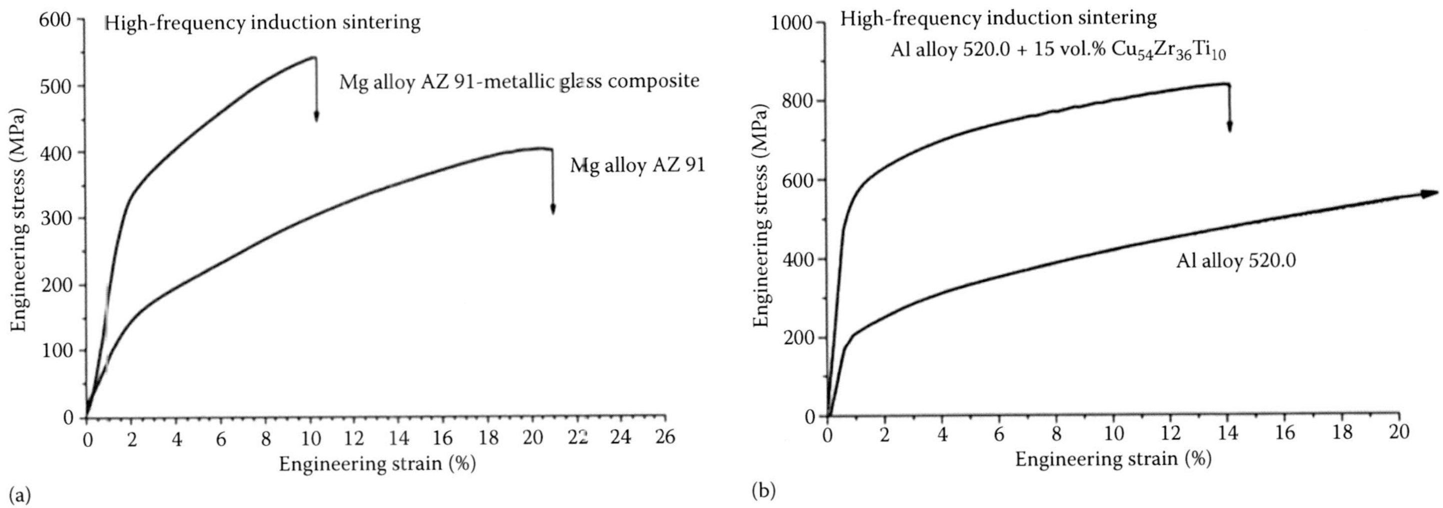

FIGURE 9.24 Comparison of compression stress–strain curves for the HFIHS sintered (a) magnesium alloy with $Zr_{57}Nb_5Cu_{15.4}Ni_{12.6}Al_{10}$ metallic glass composite with that of unreinforced AZ91 alloy (Used with permission from Dudina, D. V., K. Georgarakis, Y. Li, M. Aljerf, A. LeMoulec, A. R. Yavari and A. Inoue (2009). A magnesium alloy matrix composite reinforced with metallic glass. *Composites Science and Technology* 69 (15–16): 2734–2736.) and (b) aluminum alloy with 15 vol% glassy $Cu_{54}Zr_{36}Ti_{1C}$ composite with that of unreinforced cast aluminum alloy 520. (Used with permission from Dudina, D. V., K. Georgarakis, M. Aljerf, Y. Li, M. Braccini, A. R. Yavari and A. Inoue (2010). Cu-based metallic glass particle additions to significantly improve overall compressive properties of an Al alloy. *Composites—A* 41 (10): 1551–1557.)

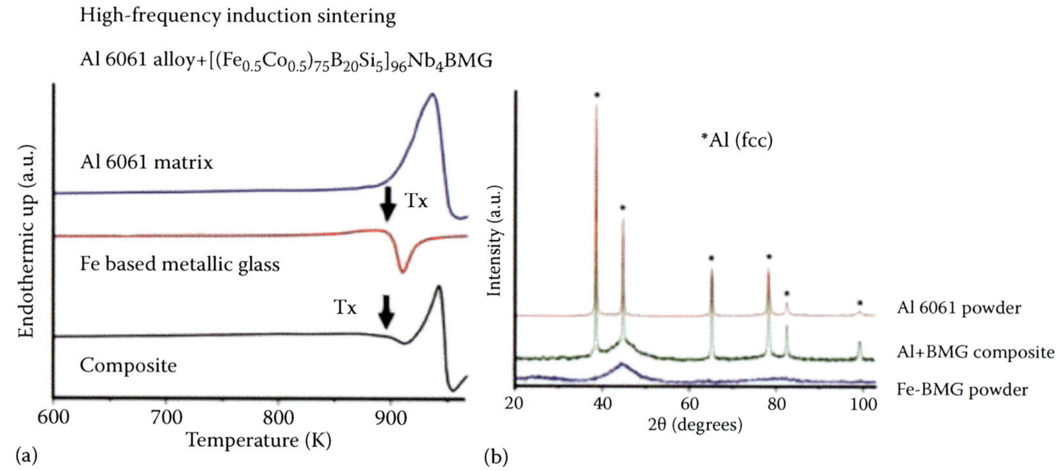

FIGURE 9.25 (a) Differential scanning calorimetry thermograms and (b) x-ray diffraction patterns of aluminum 6061 alloy, iron–cobalt (FeCo)–based metallic glass and Al6061 alloy with $[(Fe_{0.5}Co_{0.5})_{75}B_{20}Si_5]_{96}Nb_4$ composite prepared by HFIHS. (Used with permission from Aljerf, M., K. Georgarakis, D. Louzguine-Luzgin, A. Le Moulec, A. Inoue and A. R. Yavari (2012). Strong and light metal matrix composites with metallic glass particulate reinforcement. *Materials Science & Engineering A* 532: 325–330.)

attributed to (a) the load transfer effect (i.e., high load-carrying capacity of the amorphous reinforcement), (b) dislocation strengthening effect due to the preparation method, (c) Orowan strengthening, (d) grain refinement (due to melt spinning and ball milling of the matrix alloy), and (e) solid solution strengthening due to the possible diffusion of copper from the reinforcement into the aluminum matrix (Dudina et al. 2010).

By the high-frequency induction sintering method assisted with pressure, Al6061 alloy was reinforced with $[(Fe_{0.5}Co_{0.5})_{75}B_{20}Si_5]_{96}Nb_4$ (atomic percent) glassy particles (Aljerf et al. 2012). The milling of the glassy ribbons was done using ceramic balls (time: 6 h), because the glassy ribbons were harder than stainless steel balls. The induction sintering temperature was maintained at 828 K, considering the glass transition temperature and the crystallization temperature of the alloy (T_g: 821 K and T_x: 861 K) and the consolidation was done at 70 MPa pressure. The differential scanning calorimetry curves (Figure 9.25a) and x-ray diffraction patterns (Figure 9.25b) showed retention of amorphous structure in the composites (Aljerf et al. 2012). The stress–strain behavior of the composite indicated an excellent improvement in strength properties (Figure 9.26) (Aljerf et al. 2012).

9.3.1.4 Disintegrated Melt Deposition Technique

The disintegrated melt deposition technique is a liquid-state processing method that has the combined advantages of both gravity die casting and spray forming processes (Ceschini et al. 2016; Gupta and Sharon 2011). Unlike the spray process, disintegrated melt deposition employs higher superheat temperatures and lower impinging gas jet velocity, which provides a high solidification rate. It has bottom-pouring arrangement, which effectively eliminates impurities/oxides (i.e., no requirement for any complex flux additions) (Ceschini et al. 2016; Gupta and Sharon 2011). The method includes both melting and pouring together in a single step, thereby making the process less intensive in terms of time and labor/cost. It is one of the most cost-effective processes to synthesize nanocomposites (Jayalakshmi et al. 2016). The following are the inherent merits of the disintegrated melt deposition method:

- Combined advantages of casting and spray forming processes
- Eliminates the need for separate melting and pouring units
- Removes oxides and slag/dross and least metal wastage

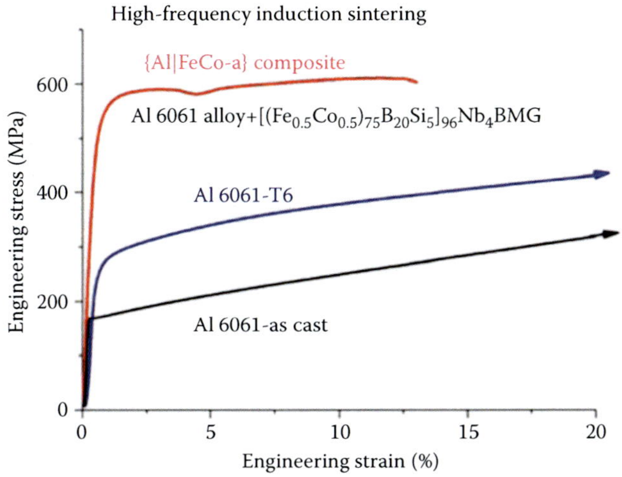

FIGURE 9.26 Compression stress–strain curves of high-frequency induction sintered Al6061 alloy with $[Fe_{0.5}Co_{0.5})_{75}B_{20}Si_5]_{96}Nb_4$ composite compared to those of Al 6061 commercial alloy in the as-cast and T6-tempered conditions. (Used with permission from Aljerf, M., K. Georgarakis, D. Louzguine-Luzgin, A. Le Moulec, A. Inoue and A. R. Yavari (2012). Strong and light metal matrix composites with metallic glass particulate reinforcement. *Materials Science & Engineering A* 532: 325–330.)

- Flexibility of addition/incorporation of alloying elements/nanoparticles
- Eliminates retention/settling of nano-reinforcements in crucible
- High process yield and gives rise to fine-grained materials with minimal porosity
- Capability to be up-scaled from laboratory to industrial level

9.3.1.4.1 Process Description of Disintegrated Melt Deposition Technique

The disintegrated melt deposition process is an innovative processing technology (Figure 9.27) (Ceschini et al. 2016; Gupta and Sharon 2011). For the synthesis of magnesium composites, the magnesium materials and the desired amount of reinforcements are first weighed and placed in a graphite crucible (with an arrangement for bottom-pouring). The materials are superheated to 750°C (1023 K) in a resistance furnace under an inert argon atmosphere. At the superheat temperature, the molten composite melt (i.e., magnesium base matrix, alloying elements, and nanoparticles) is mechanically stirred using a twin-blade impeller (with pitch 45°) to facilitate both the incorporation and uniform distribution of the reinforcement materials in the metallic matrix. The impeller is coated with a water-based zircon wash (Zirtex 25: 86% ZrO_2, 8.8% Y_2O_3, 3.6% SiO_2, 1.2% K_2O and Na_2O, and 0.3% trace inorganic) to prevent iron contamination in the melt. Subsequently, the melt is released through a 10-mm-diameter orifice situated at the base of the crucible. The composite melt is disintegrated by two jets of argon gas oriented normal to the melt stream (as shown in Figure. 27). The argon gas flow rate is maintained at 25 lit/min. The disintegrated composite melt is deposited onto a metallic substrate situated at the base. The magnesium ingot so produced is hot extruded at 350°C (with prior soaking for 1 h at 400°C) at extrusion ratios ranging from 12:1 to 25:1, to obtain rods of 7 to 10 mm diameter for subsequent characterization and testing (Ceschini et al. 2016; Gupta and Sharon 2011).

9.3.1.4.2 Nanocomposites Processed by Disintegrated Melt Deposition Technique

Nanocomposites processed by the disintegrated melt deposition liquid-state processing method have shown exceptional microstructure and mechanical properties. As an example, tensile properties of magnesium and magnesium with carbon nanotubes (Mg–CNT) nanocomposites made by disintegrated

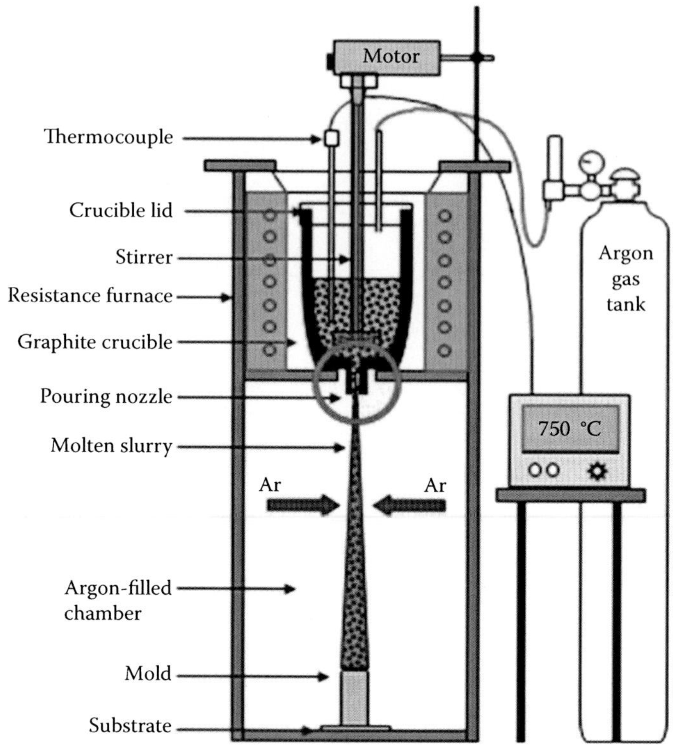

FIGURE 9.27 Schematic showing the disintegrated melt deposition process.

melt deposition showed high yield and ultimate tensile strengths. The ductility of the nanocomposite also improved significantly (Figure 9.28a) (Goh et al. 2006). Similarly, magnesium with nano-yttria ($Mg–Y_2O_3$) nanocomposites prepared by disintegrated melt deposition followed by hot extrusion showed remarkable improvement in yield strength and ultimate tensile strength with the retention of ductility (Goh et al 2007). Nanoparticles hinder dislocation movement and thereby improve the material strength (Figure 9.28b) (Goh et al 2007). A significant improvement in the mechanical properties was observed for ZK60 magnesium alloy reinforced with nano-titanium carbide (TiC) particles (Paramsothy et al. 2011a). Similar observations were made in ZK60A magnesium alloy reinforced with nano-silicon nitride (Si_3N_4) particles (Paramsothy et al. 2011b). In all these nanocomposite systems, the strengthening was attributed to (i) load bearing effect by the nanoparticles, (ii) generation of geometrically necessary dislocations, (iii) Orowan strengthening, and (iv) grain boundary strengthening.

Hybrid composites with micro- and nanoscale reinforcements have been successfully produced by the disintegrated melt deposition technique; for example, pure magnesium-based hybrid nanocomposites with micron-sized titanium and nanosized silicon carbide particles were synthesized by disintegrated melt deposition followed by hot extrusion (Sankaranarayanan et al. 2013c). The addition of nano-reinforcement promoted recrystallization, grain refinement, and improvement in hardness, tensile, and compressive properties. The matrix grain characteristics are shown in Figure 9.29, and the mechanical properties are shown in Figure 9.30 (Sankaranarayanan et al. 2013c). High-temperature properties of nanocomposites produced by the technique of disintegrated melt deposition have been examined and revealed good high-temperature properties, for example, nano-alumina reinforced magnesium metal matrix nanocomposites (Hassan et al. 2008). The nano-alumina reinforced composite showed matrix strain hardening up to 150°C, and beyond that temperature (i.e., 200°C), strain softening was observed (Figure 9.31) (Hassan et al. 2008). Strength retention was seen

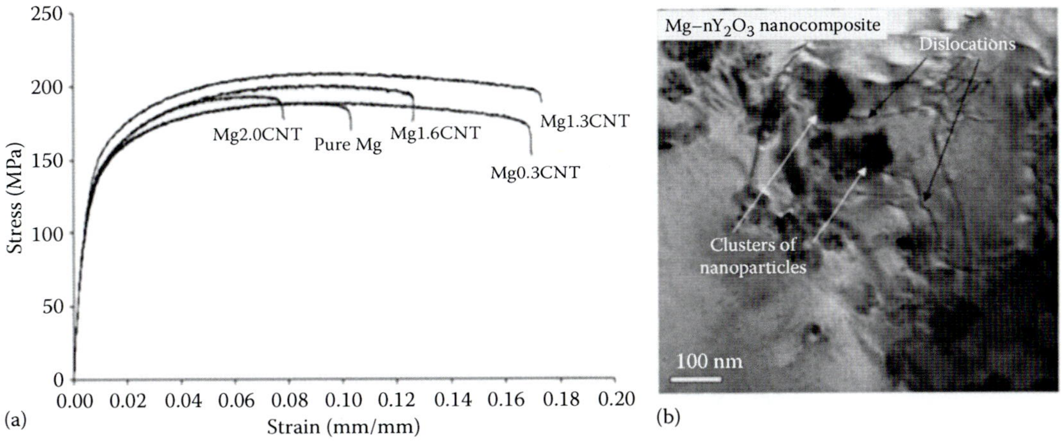

FIGURE 9.28 Representative stress–strain curves of (a) pure magnesium and magnesium with carbon nanotube (Mg–CNT) nanocomposites (Used with permission from Goh, C. S., J. Wei, L. C. Lee and M. Gupta (2006a). Development of novel carbon nanotube reinforced magnesium nanocomposites using the powder metallurgy technique. *Nanotechnology* 17 (1): 7–12.) and (b) nanoparticles acting as obstacles for dislocations. (Used with permission from Goh, C. S., J. Wei, L. C. Lee and M. Gupta (2007). Properties and deformation behaviour of Mg–Y2O3 nanocomposites. *Acta Materialia* 55 (15): 5115–5121.)

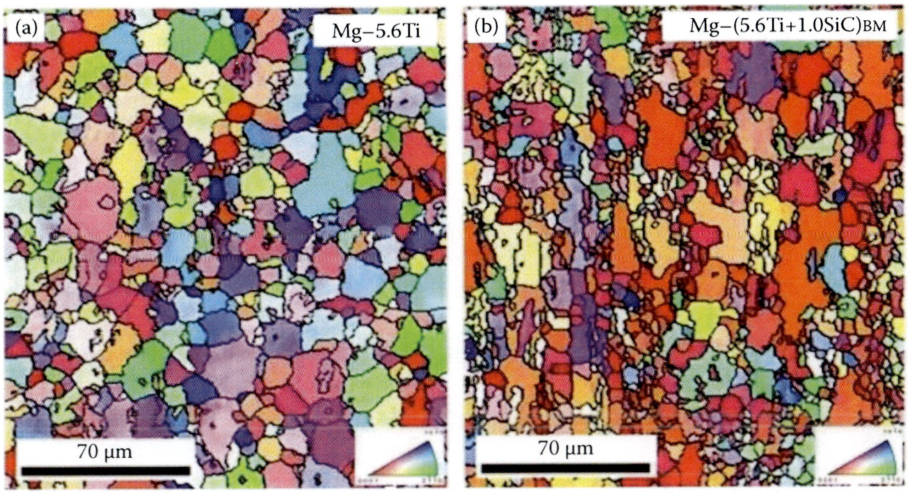

FIGURE 9.29 Electron backscattered diffraction (EBSD) images of (a) magnesium–titanium and (b) magnesium with titanium and nano-silicon carbide composite synthesized using disintegrated melt deposition technique. (Used with permission from Sankaranarayanan, S., R. K. Sabat, S. Jayalakshmi, S. Suwas and M. Gupta (2013c). Effect of hybridizing micron-sized Ti with nano-sized SiC on the microstructural evolution and mechanical response of Mg–5.6Ti composite. *Journal of Alloys and Compounds* 575: 207–217.)

until 150°C, and subsequently it reduced at 200°C. To make composites having superior properties, ball milling of base metal and reinforcements before their processing by disintegrated melt deposition and postprocessing heat treatment was attempted. The effect of heat treatment temperature and heat treatment time on microstructural and mechanical properties of magnesium with titanium and boron carbide nanoparticle (Mg–(5.6Ti+2.5B_4C)) ball-milled composite was investigated

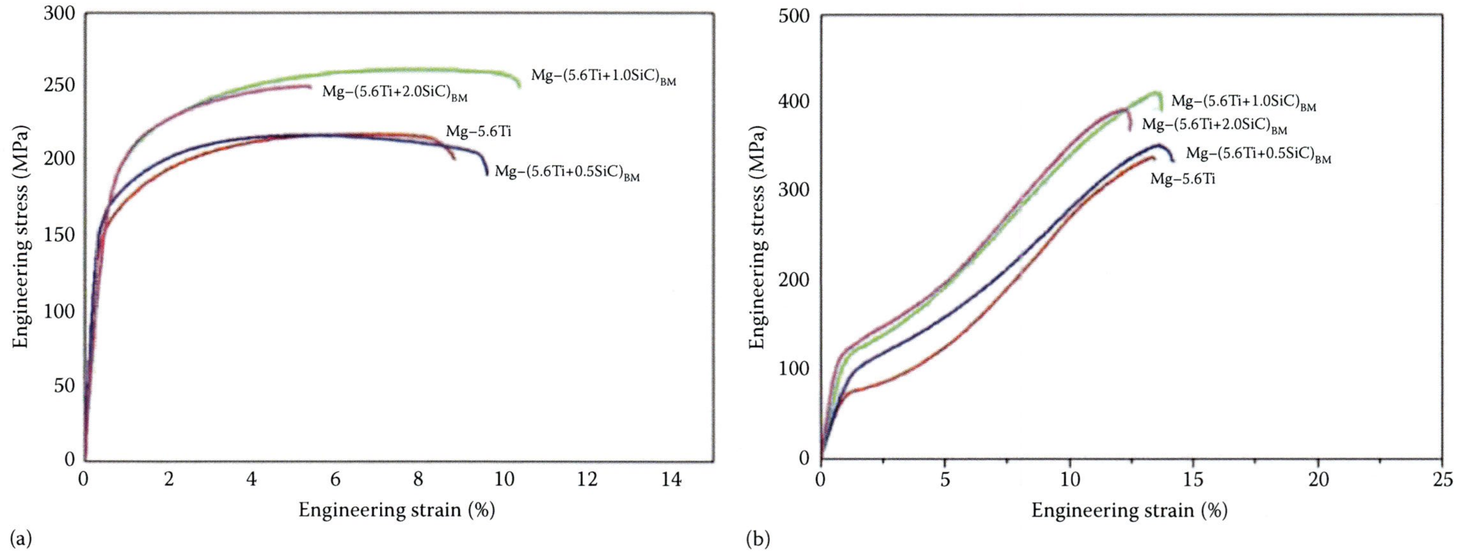

FIGURE 9.30 Representative engineering stress–strain curves of magnesium–titanium and magnesium with titanium and nano-silicon carbide composite synthesized by disintegrated melt deposition technique under (a) tension and (b) compression. (Used with permission from Sankaranarayanan, S., R. K. Sabat, S. Jayalakshmi, S. Suwas and M. Gupta (2013c). Effect of hybridizing micron-sized Ti with nano-sized SiC on the microstructural evolution and mechanical response of Mg–5.6Ti composite. *Journal of Alloys and Compounds* 575: 207–217.)

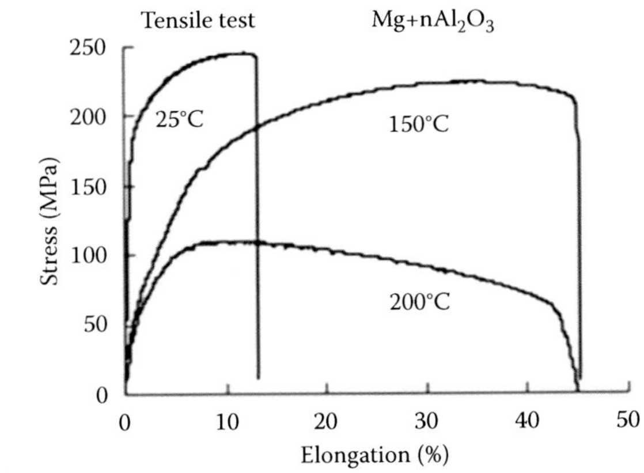

FIGURE 9.31 Representative tensile engineering stress–strain curves of magnesium with nano-alumina composite synthesized by disintegrated melt deposition technique under tension at high temperature. (Used with permission from Hassan, S. F., M. J. Tan and M. Gupta (2008). High-temperature tensile properties of Mg/Al2O3 nanocomposite. *Materials Science & Engineering A* 486: 56–62.)

(Sankaranarayanan et al. 2014). Microstructural studies of the heat-treated sample (200°C for 5 h) revealed strain-free grains due to recrystallization and residual stress relaxation effect. The heat-treated composites showed significant enhancement in ductility of approximately 85% under tensile loads and approximately 60% improvement under compressive loads.

To identify the hot working regions of metal/alloys and composites, processing maps have been constructed from data based on hot compression deformation tests. Hot compressive deformation behavior and microstructural evolution of AZ31B alloy with nano-alumina particle composite using processing maps (temperature range: 250°C–400°C; strain rate range: 0.01 to 1.0 s^{-1}) were studied (Srinivasan et al. 2013). The test samples were produced by the disintegrated melt deposition technique. The optimal region for hot working was observed at a strain rate of 0.01 s^{-1} and at a temperature of 400°C for the magnesium alloy as well as its nanocomposite. Heat treatments can promote ductility of the composites made by disintegrated melt deposition, for example, heat-treated magnesium incorporated with titanium, aluminum, and alumina nanoparticle Mg–Ti–Al–nAl_2O_3 composites (Sankaranarayanan et al. 2013a). These composites were heat treated at 200°C for 5 h and showed significant improvement in ductility (210% or three times increase under tension and 45% increase under compression) with slight or no reduction in yield strength under both tensile loading and compressive loading conditions (Figure 9.32a and b) (Sankaranarayanan et al. 2013a). This was attributed to stress relaxation that occurs at matrix-reinforcement interface, and it was identified that the best combination of strength and ductility can be obtained following heat treatment (Sankaranarayanan et al. 2013a).

Stress-amplitude controlled high cycle fatigue tests on AZ31 alloy with carbon nanotube nanocomposite prepared by disintegrated melt deposition showed superior fatigue resistance, with a 40% improvement in the endurance limit for 10^6 cycles (Srivatsan et al. 2011). From the fracture surface, it was identified that carbon nanotubes enhanced the resistance to crack initiation during cyclic stress-controlled fatigue and suppressed or delayed the crack growth (Srivatsan et al. 2011). In another work, high cycle fatigue test of magnesium with carbon nanotube composites showed that the cycle to failure for the magnesium with carbon nanotube (1.3 wt%) nanocomposite was lower than that of pure magnesium and was attributed to the presence of voids and clusters of carbon nanotubes present in the nanocomposite (Goh et al. 2008). Given that fatigue behavior is highly statistical in nature, these results indicated that behavior of nanocomposites varied significantly with the matrix and reinforcements and should be critically investigated.

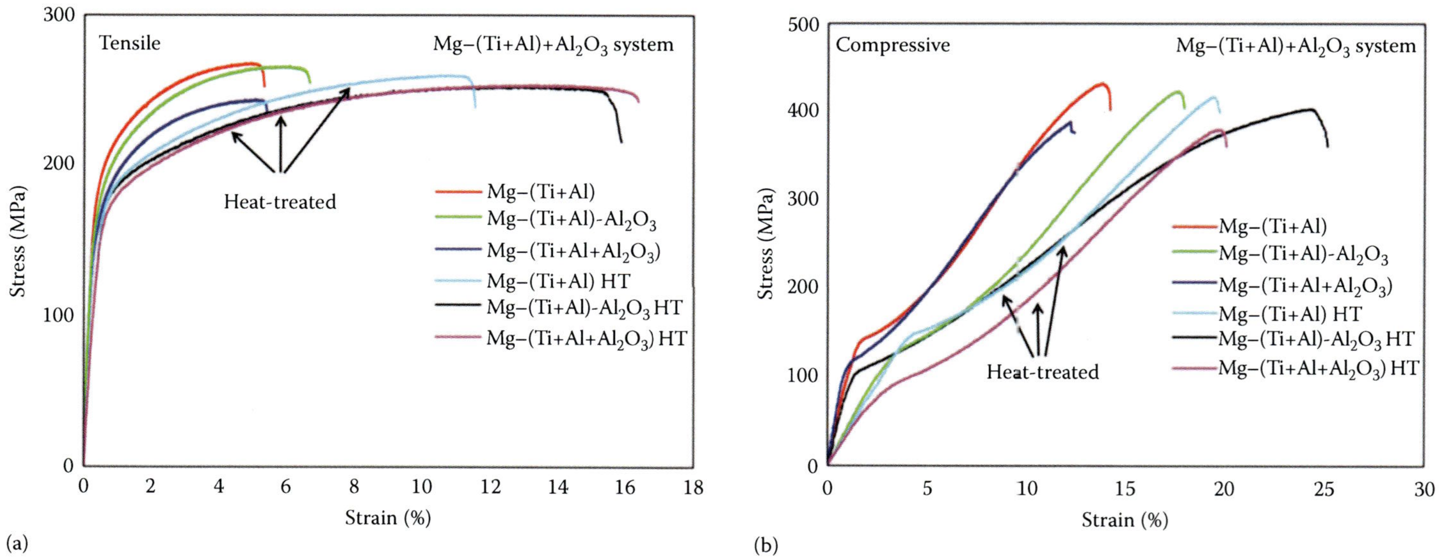

FIGURE 9.32 Representative engineering stress–strain curves of non–heat-treated and heat-treated magnesium with titanium, aluminum and nano-alumina composites synthesized by disintegrated melt deposition technique under (a) tension and (b) compression. (Used with permission from Sankaranarayanan, S., R. K. Sabat, S. Jayalakshmi, S. Suwas and M. Gupta (2013c). Effect of hybridizing micron-sized Ti with nano-sized SiC on the microstructural evolution and mechanical response of Mg–5.6Ti composite. *Journal of Alloys and Compounds* 575: 207–217.)

Magnesium nanocomposites with varying volume fractions (V_f) of 0.22%, 0.66%, and 1.11% nano-alumina particles (size 50 nm) were slid against a steel counterface (speeds: 1–10 m/s; load: 10 N) (Lim et al. 2005). Reduction in wear rate with an increase in volume fraction of reinforcement was observed (Lim et al. 2005). The improvement in wear resistance was attributed to an increase in hardness and strength with an increase in volume fraction of alumina nanoparticles. The highest volume fraction of reinforcement (1.11%) leads to an increased wear resistance of approximately 1.3 times at the lowest speed (1 m/s) and approximately 1.8 times at the highest speed (10 m/s) when compared to the unreinforced metal. At low speeds, wear occurred by abrasion, whereas at high speeds, wear occurred by adhesion mechanism. Wear behavior with respect to sliding speed was investigated for the AZ31 alloy and its nanocomposites containing 1.5 vol% of alumina nanoparticles (Shanthi et al. 2010). At higher speeds (7 and 10 m/s), thermal softening was observed in AZ31 alloy, whereas such an occurrence was absent in the composites. In this same study, it was also observed that with the addition of calcium (Ca) to the nanocomposite, the wear rate decreased. This was attributed to an enhancement of mechanical properties due to the formation of $(MgAl)_2Ca$ intermetallic phase in the nanocomposite.

9.3.1.4.3 Light Metal Bulk Metallic Glass Composites Processed by Disintegrated Melt Deposition Technique

$Mg_{67}Zn_{28}Ca_5$ (atomic percent) bulk metallic glass reinforced with 0.66–1.5 vol% of nano-alumina particulates were synthesized using the disintegrated melt deposition technique (Shanthi et al. 2011). Microstructural characterization showed uniform distribution of alumina particulates in the glass matrix. The reinforced particles had no significant effect on the glass-forming ability of the matrix. Mechanical characterization under compressive loading showed improved hardness, fracture strength, and failure strain with an increase in volume fraction of nano-alumina particulates. The best combination of strength, hardness, and ductility was observed in bulk metallic glass having 1.5 vol% alumina particulates, which showed a fracture strength of 780 MPa and a 2.6% failure strain (Figure 9.33) (Shanthi et al. 2011).

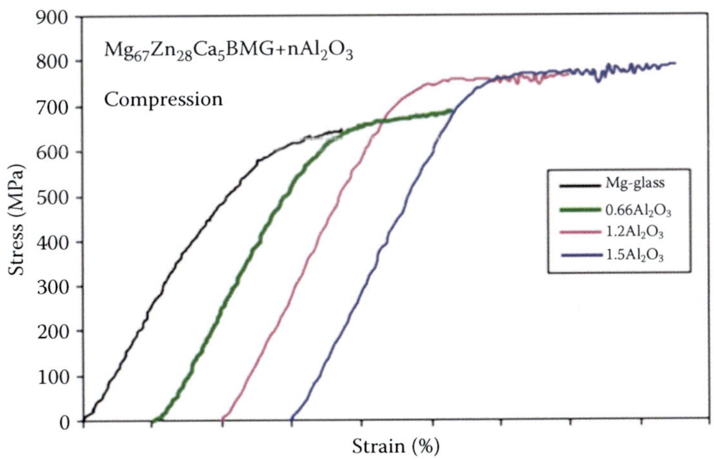

FIGURE 9.33 Compressive stress–strain curves for monolithic $Mg_{67}Zn_{28}Ca_5$ amorphous alloy and $Mg_{67}Zn_{28}Ca_5$ reinforced 0.66 to 1.5 vol% nano-alumina composites. (Used with permission from Shanthi, M., M. Gupta, A. E. W. Jarfors and M. J. Tan (2011). Synthesis, characterization and mechanical properties of nano alumina particulate reinforced magnesium based bulk metallic glass composites. *Materials Science & Engineering A* 528 (18): 6045–6050.)

9.3.2 Surface Processes/Coating Methods

9.3.2.1 Friction Stir Processing

Friction stir processing was developed based on friction stir welding, which is a solid-state joining process. It is an emerging metalworking technique that can provide localized modification and control of microstructure at the near-surface region of metallic components. During friction stir processing, a rotating tool with a shoulder and a pin is plunged into the surface of a work piece (desired base matrix) and moved along the length and breadth of the surface to be processed (Figure 9.34) (Gan et al. 2010; Lee et al. 2006a; Mahoney and Lynch 2016; Nithin Viswanath et al. 2016; Weglowski and Pietras 2011). To produce a composite surface, the surfaces are grooved and filled with the reinforcement powder, which are then processed by the passing friction stir processing tool (Gan et al. 2010; Lee et al. 2006a; Mahoney and Lynch 2016; Nithin Viswanath et al. 2016; Weglowski and Pietras 2011). As the tool rotates, it moves forward to process the region of interest. Friction stir processing causes intense plastic deformation, material mixing, and thermal exposure, resulting in significant microstructural refinement, densification, and homogeneity in the processed zone. This technique has been successfully used (i) to produce fine-grained structures, (ii) to achieve phase modification, (iii) for surface alloying, and (iv) for the synthesis of surface micro/nanocomposites. In most friction stir processed surfaces, the surface grain structure is altered (either by friction stir processing alone or through the addition of reinforcements during friction stir process to produce surface composites). As a consequence, improvement in properties such as greater resistance to wear, fatigue, and corrosion can be realized.

9.3.2.1.1 Process Description of Friction Stir Processing

The process involves a rapidly rotating, nonconsumable tool, which consists of a pin and a shoulder that moves along the surface of a work piece. To friction stir process a sheet, a specially designed

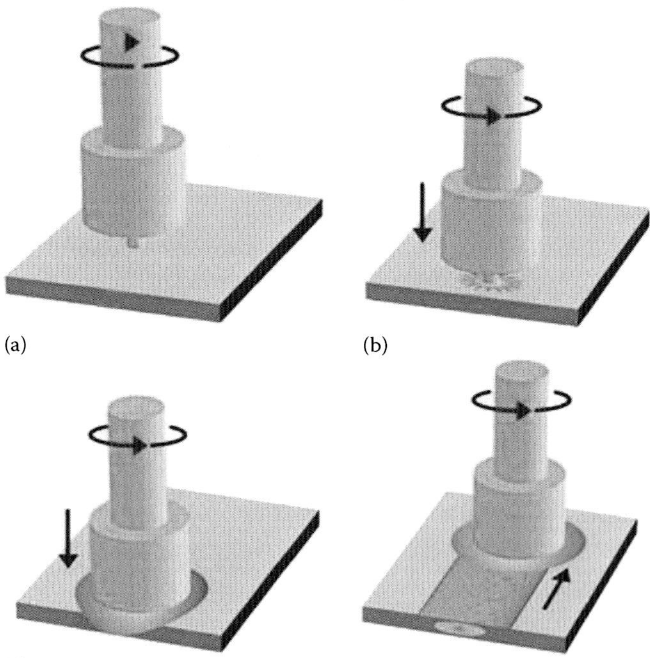

FIGURE 9.34 Schematic showing different steps in friction stir processing. (From Mahoney, M. W. and S. P. Lynch (2016). Friction-stir processing. Rockwell Sci. Company LLC, Thousand Oaks, California, Report. Open access.)

cylindrical tool is used, which, while rotating, is plunged into the selected area that is to be modified. The tool has a small-diameter pin with a concentric larger-diameter shoulder. A schematic illustration of the friction stir process is shown in Figure 9.34, and the steps are as follows: (a) rotating tool before contact with the plate in the plunged position; (b) tool pin makes contact with the plate, creating heat; (c) shoulder makes contact, restricting further penetration while expanding the hot zone; (d) plate moves relative to the rotating tool, creating a fully recrystallized, fine-grained microstructure (Mahoney and Lynch 2016).

The following are the steps involved in making surface micro/nanocomposites by friction stir process. One or more grooves are first cut on the test material surface to be processed and filled with reinforcement particles (Shabadi et al. 2015). The powder-filled grooves can either be covered with another thin sheet of the base material or closed using the passage of a pinless shoulder tool so that it gets entrapped in the material to be processed. Subsequently, a rotating tool is moved over the groove. The rubbing action of the rotating tool against the surface of test material induces high friction that gives rise to high surface temperatures. Such an action generates a stirring effect of the reinforcement with the base metal. The fast cooling that occurs after the rubbing of tool produces surface composites along with micro/nanostructured grains (Lee et al. 2006a) The process is used to incorporate nanoparticles into metal matrix to form surface nanocomposites. However, obtaining uniform dispersion of nanosized reinforcements remains a challenge (Shabadi et al. 2015).

In friction stir processed materials, four distinct microstructural zones can be identified, namely, (i) base metal, (ii) heat-affected zone, (iii) thermo-mechanically affected zone, and (iv) nugget zone (friction stir processing zone) (Mahoney and Lynch 2016). The heat-affected zone is found adjacent to the friction stir processed area and is affected only by the heat generated from the tool. Hence, unlike in the case of conventional welding processes, the volume of the heat-affected zone is small because of lesser heat input, shallow tool penetration, and shorter thermal cycle. In the thermo-mechanically affected zone, because of the application of heat and rotating stir movement of the tool, the area is affected both thermally and mechanically. In this zone, because of mechanical deformation and temperature, grain size is larger when compared to that of the base metal and is slightly lower than that in the heat-affected zone. The nugget zone (friction stir processed zone) is the intended area for friction stir processing, that is, the zone in which surface composite is made by an incorporation of reinforcement into the base matrix via the stirring movement of the tool. Because of high plastic deformation, mixing, and generation of high temperature followed by fast cooling, this area is highly strain hardened and thus grain size is considerably reduced. The fine grain size and the strain hardening increase the hardness of processed materials significantly (Gan et al. 2010; Lee et al. 2006a; Mahoney and Lynch 2016; Nithin Viswanath et al. 2016; Weglowski and Pietras 2011).

9.3.2.1.2 *Composites/Nanocomposites Produced by Friction Stir Processing*

Friction stir processing was initially applied for surface modification of aluminum alloys and to form their surface composites. In recent times, it has been applied to other materials including stainless steels, magnesium, titanium, and copper. Surface composites produced by friction stir processing have superior microstructure (e.g., grain refinement) and enhanced mechanical properties (e.g., hardness, wear resistance) as compared to their base materials.

Silicon carbide particle–reinforced aluminum–magnesium–manganese (Al–6Mg–Mn) composites with dimensions of 150 mm (length) × 60 mm (width) × 6 mm (depth) were produced by friction stir processing (Wang et al. 2009). Good interface bonding between particles and the base metal was obtained with 1.5% volume fraction of silicon carbide particles. Microhardness of the surface composite was reported to be 10% higher than that of the base metal. Nanosized alumina particles (50 nm) were incorporated onto AA6082 alloy surface to form a nanocomposite surface layer (Shafiei-Zarghani et al. 2009). Perfect bonding between the surface composite and the alloy substrate with a defect-free interface was observed (Figure 9.35a). Microhardness of the surface composite showed a value three times higher when compared to the aluminum alloy, and the wear rate reduced to one-third of the aluminum alloy (Figure 9.35b).

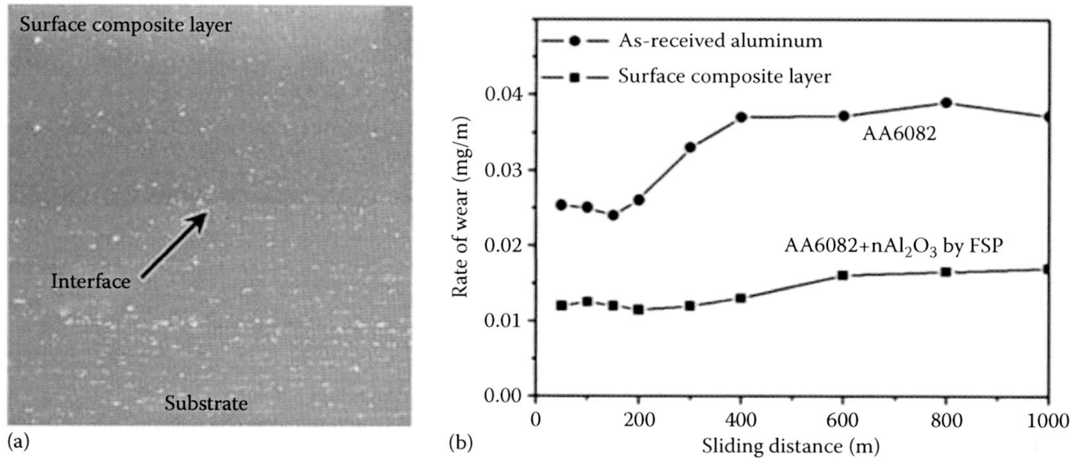

FIGURE 9.35 Friction stir processed nano-alumina particle reinforced A6082 aluminum alloy: (a) scanning electron image showing well-bonded interface (Used with permission from Shafiei-Zarghani, A., S. F. Kashani-Bozorg and A. Zarei-Hanzaki (2009). Microstructures and mechanical properties of Al/Al2O3 surface nano-composite layer produced by friction stir processing. *Materials Science & Engineering A* 500 (1–2): 84–91.) and (b) comparison of wear rate of as-received alloy with the surface nanocomposite layer produced by four friction stir process passes. (Regenerated from Shafiei-Zarghani, A., S. F. Kashani-Bozorg and A. Zarei-Hanzaki (2009). Microstructures and mechanical properties of Al/Al2O3 surface nano-composite layer produced by friction stir processing. *Materials Science & Engineering A* 500 (1–2): 84–91. Used with permission.)

2009 aluminum alloy reinforced with 1.5 vol% and 4.5 vol% carbon nanotubes were fabricated using powder metallurgy followed by four-pass friction stir processing (Liu et al. 2012). The composites thus produced showed homogeneously dispersed carbon nanotubes and refined matrix grains (Figure 9.36a and b). Tensile tests of the composites showed an increase in yield strength of the 1.5 vol% carbon nanotubes in 2009 aluminum alloy when compared to the 2009Al matrix. AZ61 magnesium alloy was incorporated with 5–10 vol% nanosized silicon dioxide to form composites (Lee et al. 2006a). The nanoparticles were uniformly dispersed after four passes, and the average grain size of the composites varied within 0.5 to 2.0 μm. Hardness of the magnesium nanocomposites

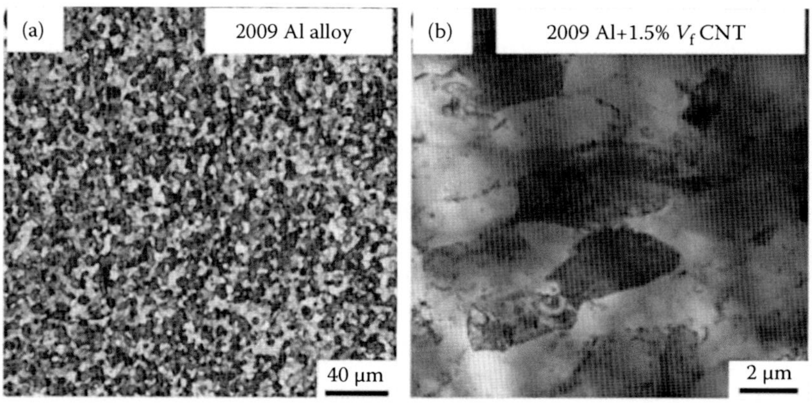

FIGURE 9.36 Microstructure of friction stir processed (a) 2009 aluminum alloy in T4 condition and (b) 1.5 vol% carbon nanotubes in 2009 aluminum composite. (Used with permission from Liu, Z. Y., B. L. Xiao, W. G. Wang and Z. Y. Ma (2012). Elevated temperature tensile properties and thermal expansion of CNT/2009Al composites. *Composites Science and Technology* 72 (15): 1826–1833.)

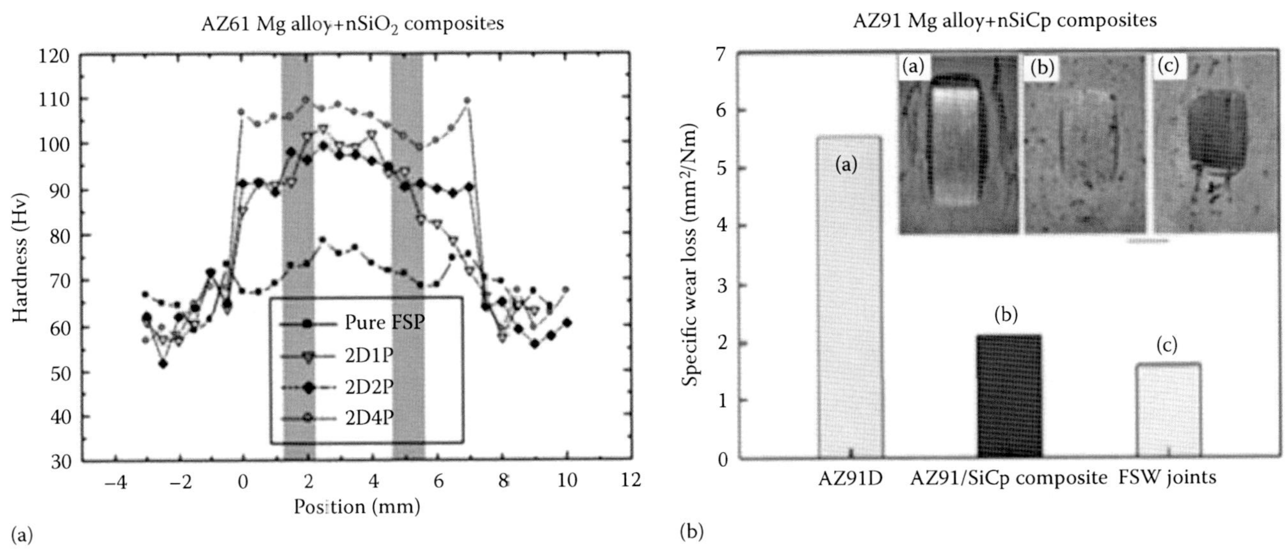

FIGURE 9.37 (a) Improved microhardness of friction stir processed AZ61 alloy with nano-silica composites when compared to the base alloy. (Used with permission from Lee, C. J., J. C. Huang and P. J. Hsieh (2006a). Mg based nano-composites fabricated by friction stir processing. *Scripta Materialia* 54 (7): 1415–1420.) (b) Wear of AZ91 and AZ91 with nano-silicon carbide composite with and without friction stir processing. (Used with permission from Lee, W. B., C. Y. Lee, M. K. Kim, J. I. Yoon, Y. J. Kim, Y. M. Yoen and S. B. Jung (2006b). Microstructures and wear property of friction stir welded AZ91 Mg/SiC particle reinforced composite. *Composites Science and Technology* 66 (11–12): 1513–1520.)

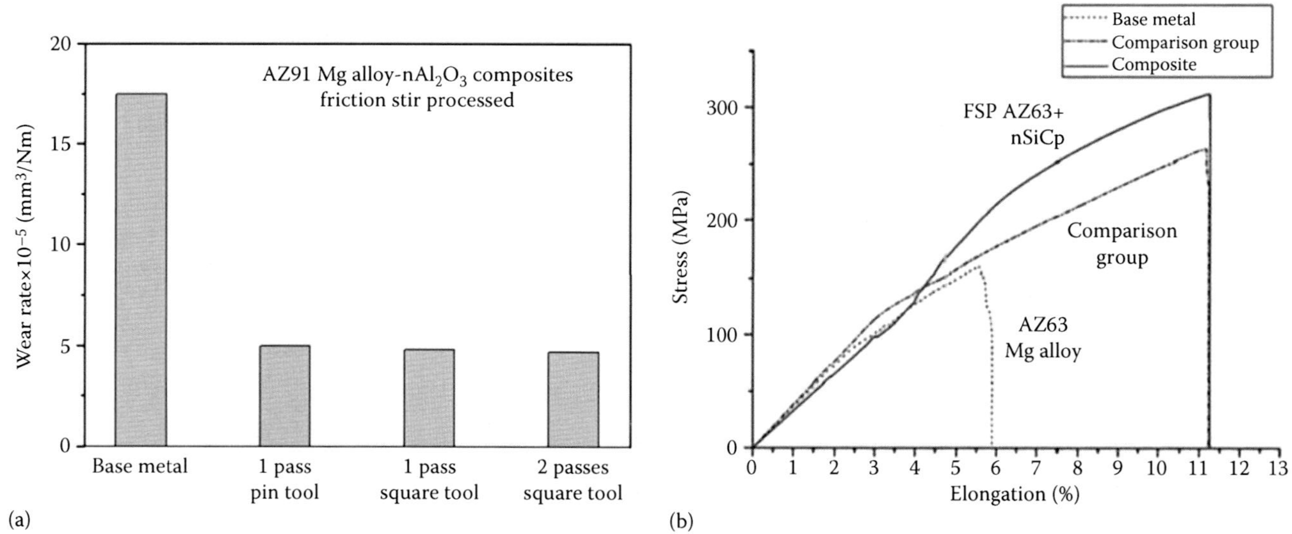

FIGURE 9.38 (a) Wear of AZ91 magnesium alloy and AZ91 reinforced with alumina nanoparticles synthesized by friction stir process. (Regenerated from Faraji, G. and P. Asadi (2011). Characterization of AZ91/alumina nanocomposite produced by FSP. *Materials Science & Engineering A* 528 (6): 2431–2440. Used with permission.) (b) Stress–strain curves of AZ63 base alloy and friction stir processed AZ63 alloy with nano-silicon carbide particles composite. (Used with permission from Sun, K., Q. Y. Shi, Y. J. Sun and G. Q. Chen (2012). Microstructure and mechanical property of nano-SiCp reinforced high strength Mg bulk composites produced by friction stir processing. *Materials Science & Engineering A* 547: 32–37.)

nearly doubled when compared with that of the base material (Figure 9.37a). Wear of the composite was three times lower than that of the base alloy (Figure 9.37b) (Lee et al. 2006b).

AZ91 magnesium alloy reinforced with nano-alumina particles showed a uniform distribution of nanoparticles coupled with significant grain refinement (Faraji and Asadi 2011). Microhardness and wear resistance improved significantly when compared to the unreinforced alloy (Figure 9.38a). In AZ63 surface composites reinforced with nano-silicon carbide particles (size 40 nm) using friction stir process, no interfacial reaction products at the matrix/particle interface were found (Sun et al. 2012). With the particles being present both within and along the grain boundary, the Vickers microhardness values were 109 Hv for the composite and 80 Hv for the unreinforced alloy. The tensile strength of the composite nearly doubled (319 MPa) when compared to the unreinforced alloy (160 MPa) (Figure 9.38b). When AZ31 alloy was reinforced with hybrid nano-reinforcements containing carbon nanotubes and nano-alumina particulates by the friction stir processing technique, wear and friction coefficients of the composites reduced significantly (Lu et al. 2013).

Friction stir processing has also been used for other non-ferrous materials, such as copper (Shabadi et al. 2015), to obtain oxide dispersion strengthened copper composites. Copper (an excellent heat sink material) finds applications in various industrial and home appliances/electronics. However, when in contact with silicon and other materials used in electronics, it experiences large mismatch in coefficient of thermal expansion. Limiting the thermal expansion of copper can be achieved by reinforcing it with yttria particles, which are not only inert but also effective in pinning down grain boundary migration arising due to the heat of copper. Integration and uniform distribution of yttria particles in copper matrix using conventional methods are rather difficult because of a large difference in their densities. Friction stir processing is an efficient method for incorporating yttria particles in copper (Shabadi et al. 2015). The thermal conductivity of friction stir processed copper and copper–yttria composites (derived from thermal diffusivity values) showed that upon friction stir processing, pure copper showed a reduction in coefficient of thermal expansion values due to the fine grain size obtained by friction stir processing (Figure 9.39). At high temperatures, the copper composite showed 27% reduction in coefficient of thermal expansion when compared to pure copper. The uniform distribution of yttria particles in the copper matrix brought forth by friction stir

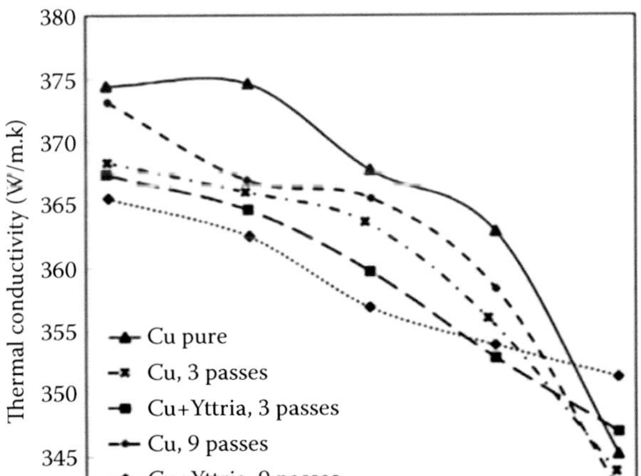

FIGURE 9.39 Thermal conductivity of yttria reinforced copper produced by friction stir processing. (Used with permission from Shabadi, R., M. N. Avettand-Fènoëla, A. Simar, R. Taillarda, P. K. Jain and R. Johnson (2015). Thermal conductivity in yttria dispersed copper. *Materials & Design* 65: 869–877.)

processing stabilized the microstructure and reduced the coefficient of thermal expansion with no loss in thermal conducting ability of copper (Shabadi et al. 2015).

9.3.2.1.3 *Nanocrystalline/Ultra-Fine–Grained Materials Produced by Friction Stir Processing*

Friction stir processing can be used for surface modification of materials to produce fine-grained structures. Figure 9.40a and b shows the microstructure of nickel–aluminum bronze before and after friction stir process (Mahoney and Lynch 2016). Before friction stir processing, the as-cast nickel–aluminum bronze showed a large dendritic structure, which usually results in poor properties. After friction stir processing, the grain size of nickel–aluminum bronze became finer and equi-axed, which resulted in 50% increase in hardness (from 68 to 75 RB, as-cast to 91 RB, friction stir processed). Increase in yield strength was significant (from 214 to 508 MPa). The tensile strength of as-cast nickel–aluminum bronze was 445 MPa, which increased to 776 MPa after friction stir processing (50% increase) (Mahoney and Lynch 2016). Nanocrystalline materials have also been developed by friction stir processing, for example, nanocrystalline structures in copper (substructures formed around the pin tool consisted of crystallites of a few tens of nanometers in size) (Su et al. 2011). By employing two-pass friction stir processing technique, grain size refinement was realized for the AZ61 alloys (Du and Wu 2009). Nanocrystalline AZ61 magnesium alloy with an average grain size of less than 100 nm was obtained. The processed alloy had a microhardness value that was three times higher than that of the AZ61 substrate.

9.3.2.1.4 *Amorphous Alloy/Bulk Metallic Glass Reinforced Aluminum and Magnesium Composites Produced by Stir Processing*

Surface composites with amorphous/metallic glass reinforcement have been developed by friction stir processing technique. Pure aluminum was reinforced with iron-based bulk metallic glass using friction stir processing (Fujii et al. 2011). As the friction stir welding temperature of aluminum is in the range of 300°C to 450°C, amorphous/bulk metallic glass reinforcement with a glass transition temperature higher than that temperature was selected. For this reason, iron-based glassy particles of composition $Fe_{72}B_{14.4}Si_{9.6}Nb_4$ (atomic percent) were used (Fujii et al. 2011). Pure aluminum plate with dimensions of 300 mm × 70 mm × 5 mm was taken as the base matrix. A screw-type probe was used with the tool tilted by 3° during the process. For dispersion of the reinforcements, a gap was intentionally made by placing a shim of 2 mm thickness between the two plates. The reinforcement particles were filled in the gap and friction stir processing was performed initially without the probe so as to prevent scattering/flying of particles and seal the particles on the surface. As the next step, friction stir processing was performed with the probe for single and multiple passes through the

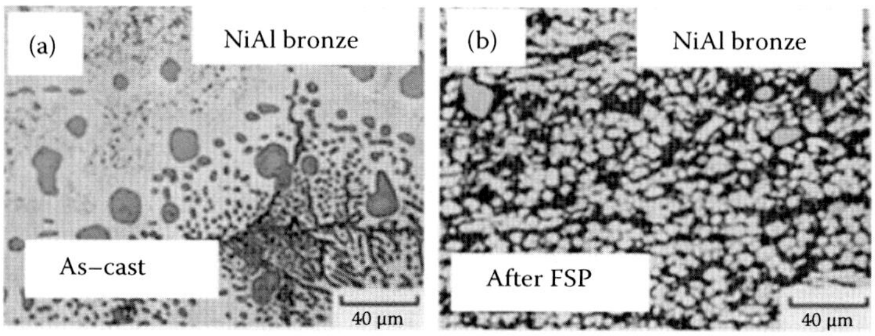

FIGURE 9.40 Microstructure of (a) as-cast nickel–aluminum bronze and (b) friction stir processed surface modified nickel–aluminum bronze showing fine-grained structure. (From Mahoney, M. W. and S. P. Lynch (2016). Friction-stir processing. Rockwell Sci. Company LLC, Thousand Oaks, California, Report. Open access.)

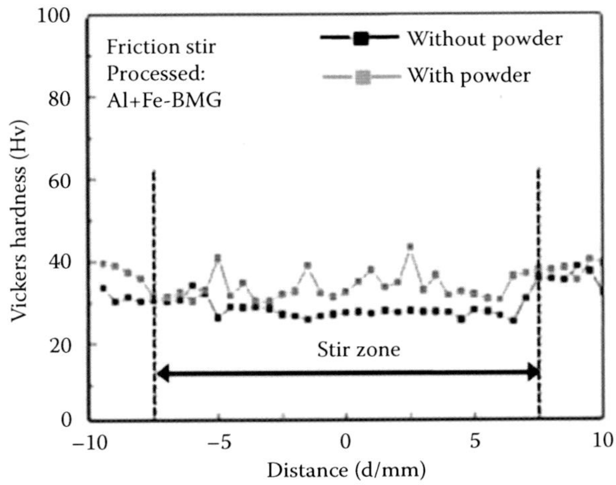

FIGURE 9.41 Hardness distribution on a cross section perpendicular to the friction stir processing direction in pure aluminum reinforced with $Fe_{72}B_{14.4}Si_{9.6}Nb_4$ metallic glass. (Used with permission from Fujii, H., Y. F. Sun, K. Inada, Y. S. Ji, Y. Yokoyama, H. Kimura and A. Inoue (2011). Fabrication of Fe-Based metallic glass particle reinforced Al-based composite materials by friction stir processing. *Materials Transaction* 52 (8): 1634–1640.)

length of the filled area in the base matrix. Structural and mechanical properties were investigated. It was observed that the dispersed iron-based metallic glass particle reduced the coarsening of aluminum grains and improved the hardness (Figure 9.41). With a change in the number of passes and/or speed, reaction between pure aluminum and iron-based metallic glass particles resulted in the formation of $Al_{13}Fe_4$ precipitates that further improved the hardness in the stir zone (Fujii et al. 2011).

9.3.2.1.5 High-Entropy Alloys Produced by Friction Stir Processing

In high-entropy alloys, the evolution of microstructure plays an important role. By applying friction stir processing to AlCoCrFeNi high-entropy alloy, notable grain refinement was achieved (grain size range: 0.35 to 15 μm) when compared to the as-received high-entropy alloy (Komarasamy et al. 2015). Because of the sluggish nature of atomic diffusion in the high-entropy alloy, the friction stir processed region exhibited large variation in microstructure, which was attributed to the accumulated plastic strain during the process. The yield strength increased by a factor of 4 after grain refinement, while higher ductility was maintained (Figure 9.42).

9.3.2.2 Ultrasonic Nanocrystalline Surface Modification

9.3.2.2.1 Process Description of Ultrasonic Nanocrystalline Surface Modification

Ultrasonic nanocrystalline surface modification utilizes ultrasonic vibration to produce nanostructured surfaces. It is a severe plastic deformation method in which the ultrasonic pulse induces severe plastic deformation on the surfaces of materials (Cho et al. 2011a). Ultrasonic nanocrystalline surface modification is based on the instrumental conversion of harmonic oscillations of an acoustically tuned body into resonant impulses of ultrasonic frequency. The acoustically tuned body is resonated by energizing an ultrasonic transducer. This technology modifies material properties in the treatment area and improves fatigue and corrosion resistance, as well as resistance to abrasion and contact failure. It induces compressive residual stress and also refines grain size.

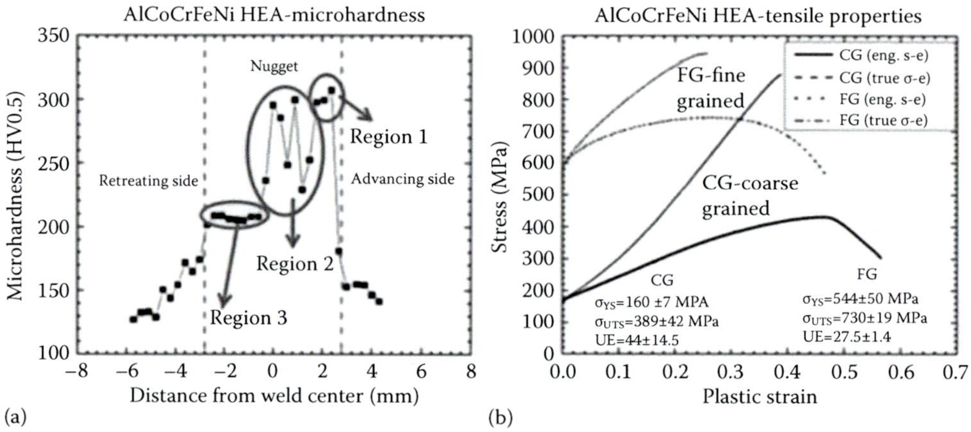

FIGURE 9.42 Friction stir processed AlCoCrFeNi high-entropy alloy showing (a) variation of hardness across the processed region on the transverse cross section and (b) plastic stress–strain curves in both coarse-grained and fine-grained conditions. (Used with permission from Komarasamy, M., N. Kumar, Z. Tang, R. S. Mishra and P. K. Liaw (2015). Effect of microstructure on the deformation mechanism of friction stir-processed Al0.1CoCrFeNi high entropy alloy. *Materials Research Letters* 3 (1): 30–34.)

9.3.2.2.2 Nanocrystalline Materials Processed by Ultrasonic Nanocrystalline Surface Modification

Nanocrystalline materials can be produced using ultrasonic nanocrystalline surface modification. An example is that of the surface modification of SAE 52100 bearing steel using ultrasonic nanocrystalline surface modification. Electron backscattered diffraction analysis confirmed the formation of a nanocrystalline structure (approximately 100 μm thick) in the surface layer (Figure 9.43a)

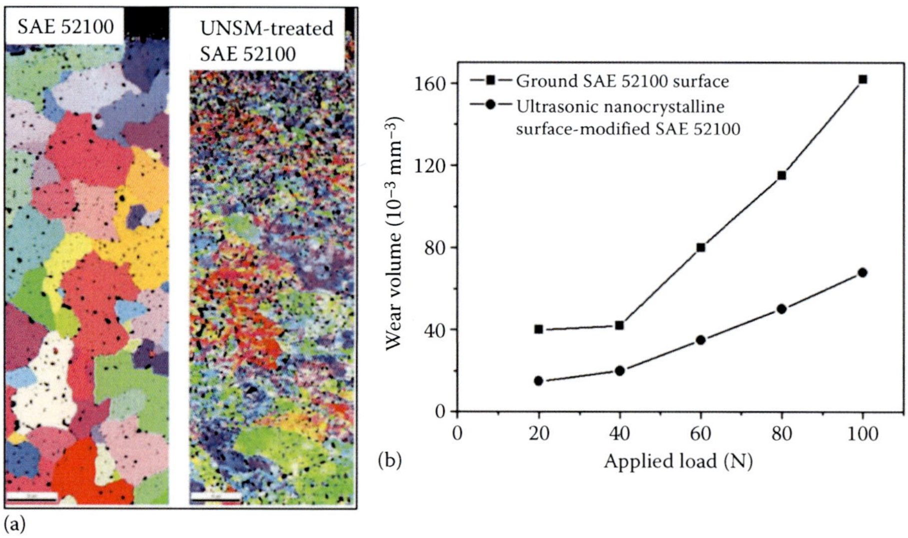

FIGURE 9.43 (a) Electron backscattered diffraction cross-sectional morphologies before and after ultrasonic nanocrystalline surface modification in SAE 52100 bearing steel, and (b) variation of wear with applied load for the ultrasonic nanocrystalline surface–modified and SAE 52100 steel surfaces. (From Cho, I. S., C. S. Lee, A. Amanov, Y. S. Pyoun and I. G. Park (2011). The effect of ultrasonic nanocrystal surface modification on the high frequency fretting beaviour of AISI310 steel. *Journal of Nanoscience & Nanotechnology* 11 (1): 742–746. Open access.)

(Cho et al. 2011a). The grain size was approximately 30 nm in the top surface layer, and it increased with an increase in depth from the ultrasonic nanocrystalline–modified surface. Because of plastic deformation in the surface layer, the coarse-grained structure in the surface layer was refined to nanometer scale without changing chemical composition (Cho et al. 2011a). The intense plastic deformation of the surface layer improved the strength and hardness of the metal surface layer significantly. Wear tests indicated better wear resistance of ultrasonic nanocrystalline surface–modified specimens when compared to the unmodified counterparts (Figure 9.43b) (Cho et al. 2011a).

Similarly, a nanocrystalline surface layer that is approximately 100 μm thick was generated on AZ91D magnesium alloy using the ultrasonic nanocrystalline surface modification technique (Amanov et al. 2012). Tribological properties of the AZ91D magnesium alloy disc specimen against silicon nitride ceramic (Si_3N_4) ball were reported. The ultrasonic nanocrystalline surface–modified specimens showed 23% reduction in friction and approximately 30% reduction in wear rate compared to that of the untreated specimens. When ultrasonic nanocrystalline surface modification treatment was carried on Inconel alloy (Li et al. 2015), an increase in hardness of approximately 60% was observed up to 310 μm depth. Layer-by-layer transmission electron microscopic analysis showed well-refined grains and twins in addition to high dislocation density. Microstructure refinement was attributed to the formation of nano-grains, twin structures, and dislocations.

Ultrasonic nanocrystalline surface modification of materials improves their response under fatigue and fretting conditions. Surface modification of AISI 310 stainless steel by ultrasonic nanocrystalline surface modification resulted in increased life under fatigue condition (Khan et al. 2016). The effect of ultrasonic nanocrystalline surface modification on high-frequency fretting wear behavior of AISI304 steel was investigated (Cho et al. 2011b). A deformed layer of 220 μm (with high dislocation density) was observed at the top layer, having grains of 23 nm in size with mechanical twins. Wear rate was found to decrease by 40% at 800,000 cycles. Enhancement in wear properties of rail steel was observed as a result of ultrasonic nanocrystalline surface modification treatment (Chang et al. 2017). Wear of the ultrasonic nanocrystalline surface–modified rails was lower in comparison with that of the untreated rails (Figure 9.44a and b).

9.3.2.3 Laser Cladding

9.3.2.3.1 Process Description of Laser Cladding

In the laser cladding process, a stream of powder is fed into a laser beam. As the laser beam scans the substrate, it coats/deposits the powder as a layer on the substrate. Desired powders are clad onto surfaces in order to improve their properties (Meriaudeau et al. 1997). This technology is effective in providing high wear-resistant coatings on metallic parts. It is similar to the thermal spray process in which an energy source is used to melt the feedstock (i.e., coating material) that is being applied to a surface. In laser cladding, the laser beam also melts the surface to which the feedstock is being applied. This normally results in a metallurgical bond that has superior bond strength over thermal spray processes (Zhang et al. 2014). The resultant coating is dense with no voids or porosity.

In the laser-cladding process, the laser beam is kept focused onto a very small area and this makes the heat-affected zone of the surface very shallow. This feature minimizes the chance of cracking/distorting and changing the metallurgy of the substrate (Zhang et al. 2014). Further, as the heat-affected zone of the surface is very shallow, it minimizes the dilution of coating with elements diffusing from beneath the surface. Parameters that are controlled during the laser cladding process are (a) processing speed, (b) powder feed rate, (c) laser power, and (d) beam diameter. By judicious combination of these parameters, the process can give rise to dense cladding with good metallurgical bond to surfaces. A schematic of the laser cladding process is shown in Figure 9.45 (Zhang et al. 2014).

9.3.2.3.2 High-Entropy Alloy Coatings by Laser Cladding Process

FeCrCoNiTiAlBC high-entropy alloys were coated (thickness, 1.5 mm) on 45# steel using a laser (Chen et al. 2017). The laser cladding process resulted in the formation of titanium carbide particles,

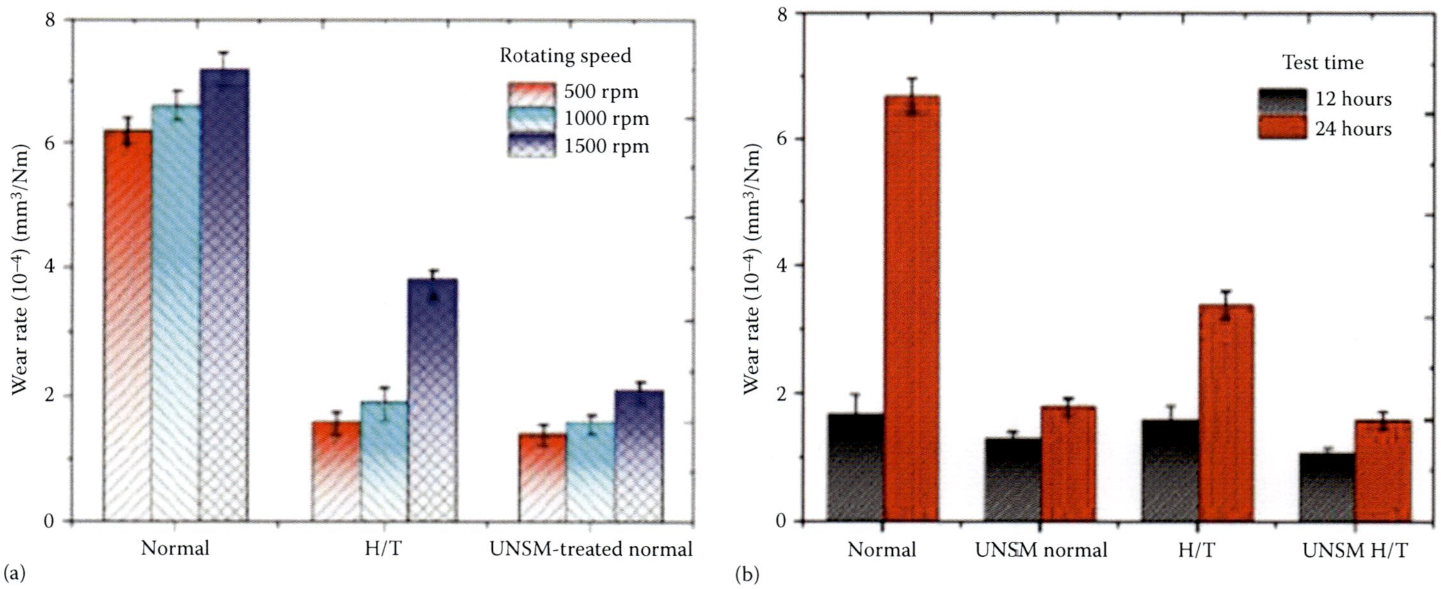

FIGURE 9.44 Variation of wear rate of normal, heat-treated and ultrasonic nanocrystalline surface–modified rail steel: (a) wear rate versus rotating speed, and (b) test time versus rotating speed. (From Chang, S., Y. S. Pyun and A. Amanov (2017). Wear enhancement of wheel-rail interaction by ultrasonic nanocrystalline surface modification technique. *Materials* 10: 188–200. Open access.)

FIGURE 9.45 Schematic of laser cladding process. (Used with permission from Zhang, Y., T. T. Zuo, Z. Tang, M. C. Gao, K. A. Dahmen, P. K. Liaw and Z. P. Lu (2014). Microstructures and properties of high-entropy alloys. *Progress in Materials Science* 61, 1–93.)

which increased with the increase in carbon and boron elements. Microstructural studies revealed that the initial crystal structure of the coating, which was BCC solid solution, changed to FCC solid solution with an increase in the addition of boron and carbon (i.e., with an increase in titanium carbide particle formation) (Chen et al. 2017). The coating hardness was 560 Hv, and the cracking tendency of the coating reduced significantly owing to the formation of titanium carbide particles and the FCC solid solution matrix (Chen et al. 2017).

TiVCrAlSi high-entropy alloy coatings on Ti–6Al–4V alloy surfaces were prepared using laser cladding (Huang et al. 2012). It gave rise to a quality coating with minimal cracks and pores. Microstructural observations showed BCC solid solution along with $(Ti,V)_5Si_3$ intermetallic compound. The hardness of the laser-clad high-entropy alloy coating was significantly higher than that of the titanium alloy surface. Dry sliding wear tests showed significant improvement in wear resistance after laser cladding (Figure 9.46a and b). The enhancement in wear resistance was attributed to the combination of the hard silicide phase (that resists abrasive and adhesive wear) and the relatively ductile/tough BCC matrix (that limits brittle crack propagation) (Huang et al. 2012). The titanium alloy (Ti–6Al–4V) surface showed adhesive and abrasive wear. In contrast, the high-entropy alloy coatings showed mild wear with oxidation as the main mechanism. The friction property of the high-entropy alloy coating was relatively low when compared to that of the uncoated substrate (Figure 9.47) (Huang et al. 2012).

Annealing and oxidation studies were conducted on TiVCrAlSi high-entropy alloy-clad Ti–6Al–4V alloy (Huang et al. 2011). Annealing at 800°C (24 h) in vacuum resulted in the formation of Al8 $(V,Cr)_5$ phase. Laser cladding effectively improved the oxidation resistance of the substrate at 800°C in air, which was because of the formation of a thin and adherent mixed scale consisting of silicon dioxide, chromium oxide, titanium dioxide, alumina, and vanadium pentoxide (SiO_2, Cr_2O_3, TiO_2, Al_2O_3 and V_2O_5) (Figure 9.48) (Huang et al. 2011).

9.3.3 Severe Plastic Deformation (Bulk Deformation) Processes

In severe plastic deformation processes, a very high strain is imposed on bulk solids without inducing any significant change in their overall physical dimension. By these processes, exceptional grain refinement, usually resulting in ultra-fine grains/nanocrystallinity, is achieved; consequently, mechanical properties of materials are enhanced significantly.

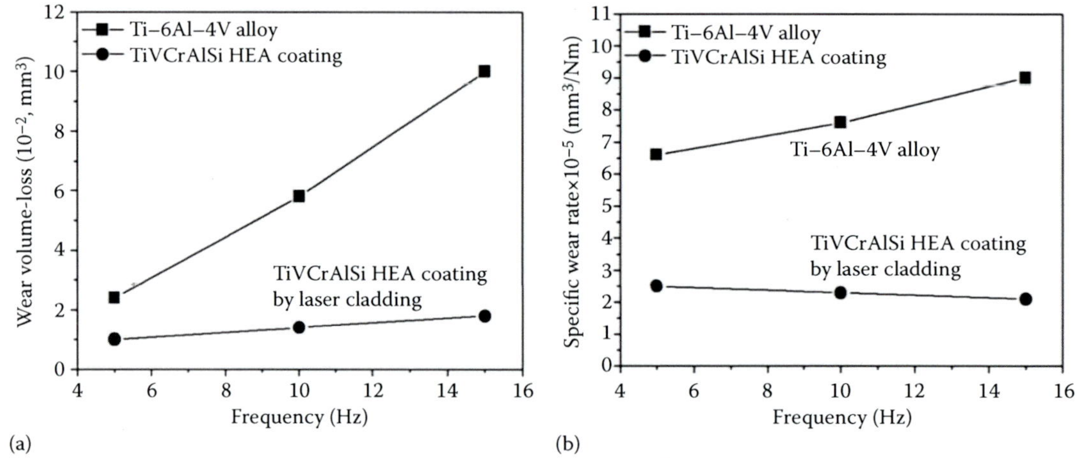

FIGURE 9.46 Comparison of (a) wear volume loss and (b) specific wear rate of laser-clad TiVCrAlSi high-entropy alloy coating on Ti–6Al–4V alloy substrate after dry sliding wear test under different frequencies. (Used with permission from Huang, C., Y. Z. Zhang, R. Vilar and J. Y. Shen (2012). Dry sliding wear behavior of laser clad TiVCrAlSi high entropy alloy coatings on Ti–6Al–4V substrate. *Materials & Design* 41: 338–343.)

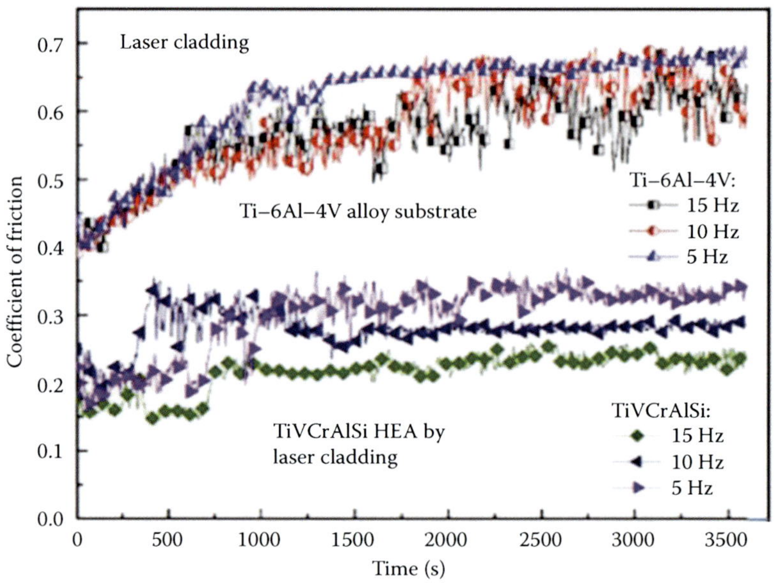

FIGURE 9.47 Friction coefficient as the function of test time for laser-clad TiVCrAlSi high-entropy alloy coating and Ti–6Al–4V alloy under different frequencies during dry sliding wear test. (Used with permission from Huang, C., Y. Z. Zhang, R. Vilar and J. Y. Shen (2012). Dry sliding wear behavior of laser clad TiVCrAlSi high entropy alloy coatings on Ti–6Al–4V substrate. *Materials & Design* 41: 338–343.)

9.3.3.1 Accumulative Roll Bonding

This is a solid-state method that processes materials in the form of sheets. Stacking of materials, followed by roll bonding, is repeated in this process (Saito et al. 1999). In the accumulative roll-bonding process, two sheets of the same material are stacked, heated, and rolled, which eventually bonds the sheets together. Compared to other severe plastic deformation processes, accumulative roll bonding does not require specialized equipment or tooling, as it requires only a conventional rolling

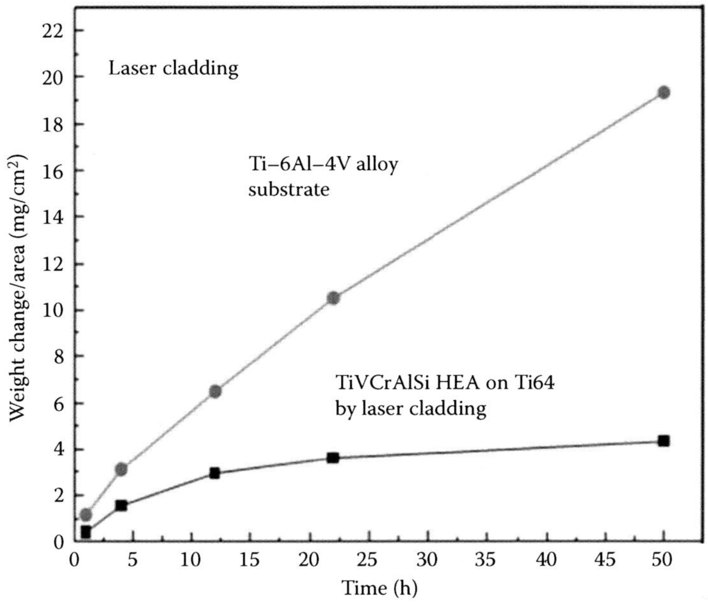

FIGURE 9.48 Oxidation behavior (weight change) of Ti–6Al–4V alloy and TiVCrAlSi high-entropy alloy coated on Ti–6Al–4V (at 800°C). (Used with permission from Huang, C., Y. Z. Zhang, J. Y. Shen and R. Vilar (2011). Thermal stability and oxidation resistance of laser clad TiVCrAlSi high entropy alloy coatings on Ti–6Al–4V alloy. *Surface & Coating Technology* 206 (6): 1389–1395.)

mill. It should be ensured that the surfaces must be well cleaned before rolling to achieve good bonding.

9.3.3.1.1 Process Description of Accumulative Roll Bonding

In the accumulative roll-bonding process, initially, the strip of a selected metal is neatly placed on top of another strip. A prior surface treatment is employed to enhance the bond strength at the interfaces of the two strips. The two layers of material are joined together by rolling in a conventional rolling mill. As the next step, the length of rolled material is sectioned into two halves. These sectioned strips are again surface-treated, stacked, and roll-bonded. The whole process is repeated multiple times (Saito et al. 1999).

It is important to note that the process temperature in accumulative roll bonding should be maintained below recrystallization temperature, because recrystallization annihilates the accumulated strain. On the other hand, low process temperatures should be avoided as they would result in insufficient ductility and low bond strength (Saito et al. 1999). Hence, selection of process temperature is critical. In the accumulative roll-bonding process, reduction in sheet thickness occurs. However, there exists a limit to the reduction in thickness (i.e., threshold deformation) in order to achieve sufficient bonding. This threshold deformation decreases with an increase in temperature. If the homologous temperature of roll bonding is less than 0.5, thickness reduction greater than 50% can be achieved without recrystallization. A schematic of the accumulative roll-bonding process is shown in Figure 9.49 (Saito et al. 1999).

9.3.3.1.2 Composites/Nanocomposites Processed by Accumulative Roll Bonding

To make composites/nanocomposites by accumulative roll-bonding process, reinforcement particles are distributed on metal sheets before processing. By increasing the number of cycles, uniform distribution of particles can be achieved, accompanied by a decrease in porosity and an increase in

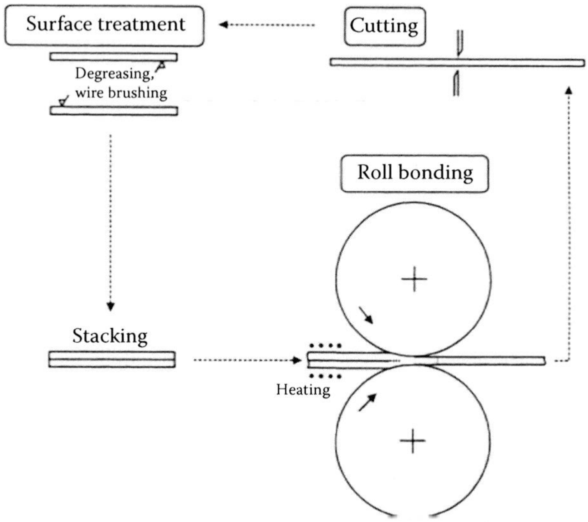

FIGURE 9.49 Schematic of accumulative roll-bonding process. (Used with permission from Saito, Y., H. Utsunomiya, N. Tsuji and T. Ssakai (1999). Novel ultra-high straining process for bulk materials—Development of the accumulative roll-bonding (ARB) process. *Acta Materialia* 47 (2): 579–583.)

bond strength between matrix and reinforcement (Saito et al. 1999). Because of these effects, mechanical properties (viz., yield strength, ultimate strength, and wear resistance) of accumulative roll-bonding processed unreinforced metals and composites are enhanced. Both microsized and nanosized reinforcing particles (e.g., alumina, silicon carbide, silicon dioxide, and boron carbide particles) have been incorporated in metallic matrices through accumulative roll-bonding process.

Aluminum–silicon carbide nanocomposites produced by the accumulative roll-bonding process were investigated for their wear behavior (Darmiani et al. 2013). Pure Al 1050 was taken as the matrix and silicon carbide particles (less than 75 μm) were used as reinforcement. The as-received sheets were annealed for 2 h at 380°C. The roll-bonding process was carried out using laboratory roll mill with a speed of 4 rpm and a load capacity of 20 tons. At each rolling cycle, 0.26 g of silicon carbide particles was dispersed uniformly, and after 5 cycles, 50% reduction was obtained. Next, the roll bonded strips were cut in half, and the same procedure was repeated for five more cycles (Darmiani et al. 2013). Microstructural investigation showed that the accumulative roll-bonding process significantly changed the silicon carbide particle size. It was observed that the size of silicon carbide was less than 100 nm after 10 cycles, resulting in the formation of nanocomposite (Darmiani et al. 2013). The field emission scanning electron micrographs of the aluminum–silicon carbide composite after six cycles with silicon carbide particle sizes of (a) 500 nm and (b) 300 nm are shown in Figure 9.50. Tests showed the microhardness of accumulative roll-bonding processed aluminum-silicon carbide nanocomposite to be much higher than that of the pure aluminum (Darmiani et al. 2013). Sliding wear tests showed low wear (Figure 9.51) and low coefficient of friction for the nanocomposites, which was due to the uniform distribution of silicon carbide particles and the reduction in particle size due to the accumulative roll-bonding process (Darmiani et al. 2013).

9.3.3.1.3 Bulk Metallic Glass Reinforced Aluminum Composites Synthesized by Accumulative Roll Bonding

Recently, accumulative roll-bonding process has been employed to reinforce bulk metallic glass as reinforcements in aluminum matrices. The major advantage of using accumulative roll bonding is that it can give rise to high-density, ultra–fine-grained structured composites at room/low temperature without any protective atmosphere (unlike the high temperature required for casting or sintering

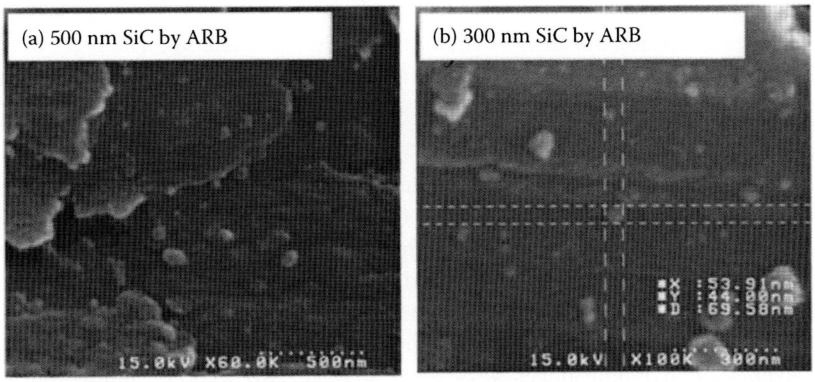

FIGURE 9.50 Aluminum–silicon carbide composite made by accumulative roll-bonding process. Field emission scanning electron micrographs after six cycles show silicon carbide particle sizes of about (a) 500 nm and (b) 300 nm. (Used with permission from Darmiani, E., I. Danaee, M. A. Golozar, M. R. Toroghinejad, A. Ashrafi, and A. Ahmadi (2013). Reciprocating wear resistance of Al–SiC nano-composite fabricated by accumulative roll bonding process. *Materials & Design* 50: 497–502.)

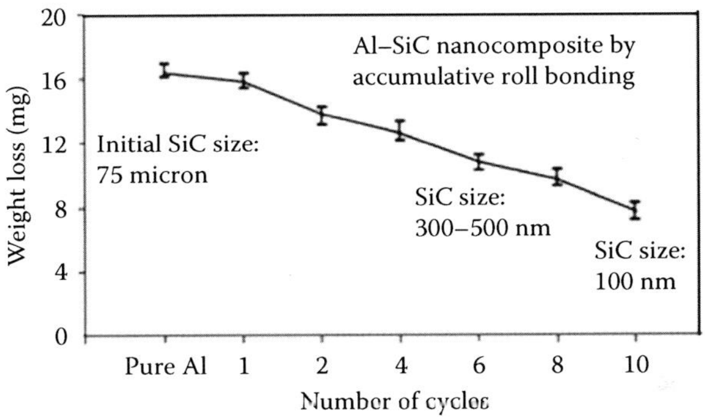

FIGURE 9.51 Variation of weight loss as a function of cycles for pure aluminum and aluminum with nano-silicon carbide composites synthesized by accumulative roll-bonding process. (Regenerated from Darmiani, E., I. Danaee, M. A. Golozar, M. R. Toroghinejad, A. Ashrafi, and A. Ahmadi (2013). Reciprocating wear resistance of Al–SiC nano-composite fabricated by accumulative roll bonding process. *Materials & Design* 50: 497–502. Used with permission.)

methods) (Khoramkhorshid et al. 2016). Especially with bulk metallic glasses as reinforcements, which are metastable (nonequilibrium phases), exposure to increasing temperature and time would result in their devitrification. Aluminum-based composites reinforced with $Al_{84}Gd_6Ni_7Co_3$ glassy powders were produced by accumulative roll bonding (Khoramkhorshid et al. 2016). Microstructural studies and mechanical property evaluation indicated uniform distribution of glassy particles in the aluminum matrix, along with an improvement in tensile strength and microhardness (Khoramkhorshid et al. 2016). Accumulative roll bonding was used to produce a multilayer aluminum-based composite reinforced with Al–Ni–Sm metallic glassy alloy in both amorphous and fully crystallized state to produce aluminum composites with ultra-fine microstructure (Anghelus et al. 2015).

9.3.3.2 Equi-Channel Angular Processing

Equal channel angular process is a severe plastic deformation process, which involves processing/extrusion of metal billet through an angled (typically 90°) channel (Sklenicka et al. 2013). To achieve best results, this process is repeated several times, changing the orientation of the billet with each pass. This produces a uniform shear throughout the bulk of the material. The process is widely used to make fine-grained/ultra–fine-grained materials and has also been used to produce nanocomposites.

9.3.3.2.1 Process Description of Equi-Channel Angular Processing

A schematic of the equi-channel angular processing method is shown in Figure 9.52 (Sklenicka et al. 2013). For the die shown in the schematic figure, the internal channel is bent through an abrupt angle (Φ), and there is an additional angle (Ψ) that represents the outer arc of curvature where the two channels intersect. Initially, the sample (e.g., rod or bar) is forced through the die using a plunger (Sklenicka et al. 2013). The imposed deformation on the sample is the shear, which occurs as the billet passes through the die. Equi-channel angular process retains the same cross-sectional area of samples, even though it introduces very large strains. This feature significantly distinguishes equi-channel angular process from conventional metal working operations (e.g., rolling, extrusion and drawing). As the cross-sectional area remains unchanged, the same billet may be pressed repetitively to introduce exceptionally high strains (Sklenicka et al. 2013).

As repetitive pressing is employed, different slip systems can be activated in materials by simply rotating the sample in different ways during consecutive passes. The four different processing routes by equal channel angular process are summarized schematically in Figure 9.53 (Sklenicka et al. 2013). In Route A, the sample is pressed without rotation. In Route B_A, the sample is rotated by 90° in alternate directions between consecutive passes, whereas in Route B_C, the sample is rotated by 90° in the same direction (either clockwise or counterclockwise) between each pass. In Route C, the sample is rotated by 180° between passes. By employing different routes with multiple passes, homogeneous and equiaxed ultra–fine-grained microstructures can be obtained.

9.3.3.3 Other Severe Plastic Deformation Processes

Other severe plastic deformation processes include cryomilling, high-pressure torsion, twist extrusion, and multiple forging. In cryomilling, which is widely used for processing aluminum nanocrystalline materials, the desired alloy powder is prepared by milling at cryogenic temperatures (Marchi 2006). This induces severe plastic deformation. Milling also reduces grain size significantly

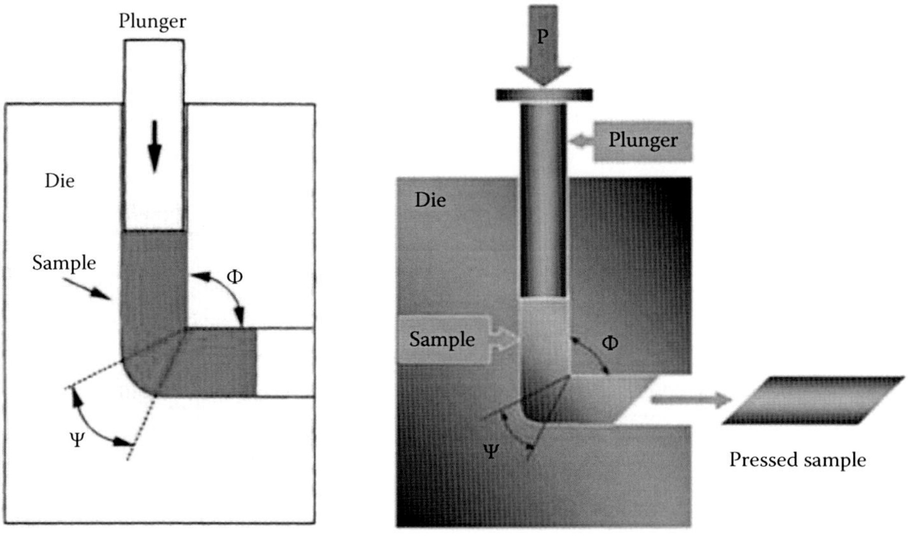

FIGURE 9.52 Equi-channel angular processing method. (From Sklenicka, V., J. Dvorak, M. Svoboda, P. Kral and M. Kvapilova (2013). Equal-channel angular pressing and creep in ultrafine grained aluminium and its alloys. *InTech Open Access Journal* 3–46. Open access.)

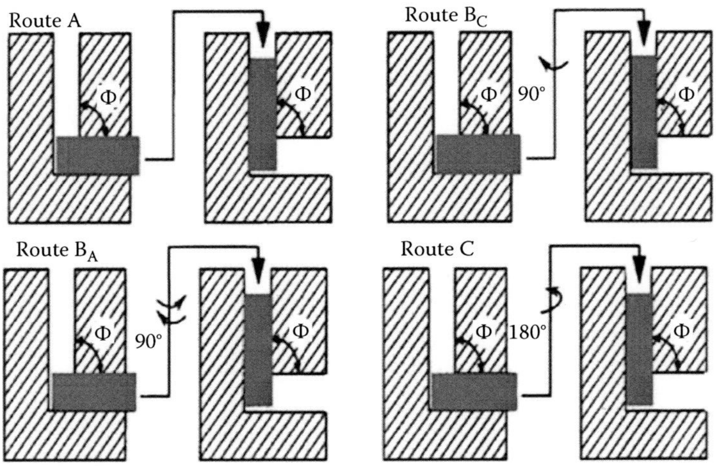

FIGURE 9.53 The four different processing routes by equi-channel angular processing (ECAP). (From Sklenicka, V., J. Dvorak, M. Svoboda, P. Kral and M. Kvapilova (2013). Equal-channel angular pressing and creep in ultrafine grained aluminium and its alloys. *InTech Open Access Journal* 3–46. Open access.)

while the cryogenic temperatures prevent recovery and recrystallization. In addition, any hard oxide present in the starting powder will also be milled thoroughly and uniformly distributed. Following the milling stage, consolidation of powder is performed at high temperature. The bulk ingot so produced is usually hot worked (rolling or extrusion) to form wrought products, such as plates and bars. Optimization of process parameters is important in order to avoid recrystallization or grain growth, which would result in the loss of nanocrystalline structure (Marchi 2006).

The high-pressure torsion process is a severe plastic bulk deformation process. In the high-pressure torsion process method, processing of metals is carried out by subjecting samples to compressive force and torsional strain simultaneously (Edalati and Horita 2016; Zhilyaev and Langdon 2008). Through the high-pressure torsion process, ultra-fine–grained and nanocrystalline materials can be obtained. In this process, a thin disc is placed between two anvils and subjected to torsional straining under high pressure (Todaka et al. 2008). Important parameters in the high-pressure torsion process are the number of turns applied to disc (N) and the magnitude of the imposed pressure (P). Reduction in diffusion rate due to high applied pressures delays the recovery kinetics. This facilitates grain refinement and hardening. Formation of ultra-fine–grained structures depends on deformation temperature and strain rate (torsional rotation speed, R). A detailed review of the high-pressure torsion process can be found in Zhilyaev and Langdon (2008).

A recently developed severe plastic deformation process is the twist extrusion (Beygelzimer et al. 2006). Here, a billet is extruded through a "twist die." Billet's cross section is deformed as follows: first, it becomes twisted to some angle in one direction and then retwisted to the same angle in the opposite direction. This causes high strains to accumulate in the billet, with no change in their form and physical dimensions. Such a process is suitable to produce ultra-fine–grained crystalline and nanocrystalline structures in bulk specimens.

9.4 CONCLUSIONS

In this chapter, the methodologies and salient features of the unconventional processing techniques used to synthesize advanced materials were described and discussed. The effectiveness of the processing methods was highlighted in terms of the obtained microstructure and mechanical properties of the synthesized materials. The following are the conclusions that could be drawn from the overview:

1. Unconventional processing methods involve processing of materials at nonequilibrium conditions such as rapid cooling rates, frictional heating at surfaces, and severe plastic deformation.
2. The unconventional processing methods have wide processing windows; that is, they provide large flexibility and control of process parameters, thereby enabling better manipulation of process–microstructure–properties.
3. Advanced materials that are produced by the unconventional processing methods have unique structure and superior properties when compared to conventional metals/alloys. Advanced materials include composites/nanocomposites, amorphous alloys/metallic glasses, nanocrystalline materials, and high-entropy alloys.
4. The unconventional processing methods discussed in this chapter have been broadly classified into (i) solid-state and liquid-state techniques, (ii) surface process/coating methods, and (iii) severe plastic deformation (bulk deformation) processes.
5. Solid-state and liquid-state techniques include (i) microwave sintering, (ii) spark plasma sintering, (iii) high-frequency induction sintering, and (iv) disintegrated melt deposition technique.
6. The solid-state and liquid-state techniques employ rapid cooling/solidification principles and are suitable to produce bulk materials with amorphous/fine-grained structure.
   - By using bidirectional microwave sintering process, the temperature difference between the core and periphery of powder compacts can be reduced significantly.
   - In spark plasma sintering, Joule heating effect along with an applied axial load give rise to near-theoretical densification of powder compacts.
   - The high-frequency induction sintering process that uses magnetic field–induced current facilitates bidirectional heating and pore-free sintering.
   - In the disintegrated melt deposition technique, the bottom pouring facility and the disintegration of the molten metal during exit ensure rapid heat extraction that results in materials with fine-grained structures.
7. Surface process/coating methods that are used include (i) friction stir processing, (ii) ultrasonic nanocrystalline surface modification, and (iii) laser cladding.
8. The surface processes are suitable to produce advanced materials at surface without affecting the bulk of the material.
   - In friction stir processing, the frictional heat generated between the work piece and the tool is used (i) for modification of surface microstructure and (ii) to produce surface composites.
   - In ultrasonic nanocrystalline surface modification, the high-frequency ultrasonic pulses create severe plastic deformation on surfaces, forming surface nanocrystalline grain structures in materials.
   - Through the laser cladding process, coating of advanced materials can be achieved on metal substrates using a laser beam.
9. Severe plastic deformation (bulk deformation) processes include (i) accumulative roll bonding, (ii) equi-channel angular processing, and (iii) other severe plastic deformation processes.
   - Through the accumulative roll-bonding process, advanced materials are produced in the form of sheets.
   - The equi-channel angular process and other severe plastic deformation processes, such as cryomilling, high-pressure torsion, twist extrusion, and so on, are bulk metal working processes that can produce advanced materials with ultra–fine-grained/nanocrystalline structures.
10. The development of unconventional processing methods has paved the way for synthesizing advanced materials with superior combination of properties and reliability. The unconventional processing techniques described and discussed in this chapter have promising potential to meet the materials demand in current and future technologies.

REFERENCES

Aljerf, M., K. Georgarakis, D. Louzguine-Luzgin, A. Le Moulec, A. Inoue and A. R. Yavari (2012). Strong and light metal matrix composites with metallic glass particulate reinforcement. *Materials Science & Engineering A* 532: 325–330.

Amanov, A., O. V. Penkov, Y. S. Pyun and D. E. Kim (2012). Effects of ultrasonic nanocrystalline surface modification on the tribological properties of AZ91D magnesium alloy. *Tribology International* 54: 106–113.

Anghelus, A., M. N. Avettand-Fènoël, C. Cordier and R. Taillard (2015). Microstructural evolution of aluminium/Al–Ni–Sm glass forming alloy laminates obtained by controlled accumulative roll bonding *Journal of Alloys and Compounds* 631: 209–218.

Basu, J. and S. Ranganathan (2003). Bulk metallic glasses: A new class of engineering materials. *Sadhana-Academy Proceeding in Engineering Sciences* 28: 783–798.

Bathula, S., M. Saravanan and A. Dhar (2012b). Nanoindentation and wear characteristics of Al 5083/SiCp nanocomposites synthesized by high energy ball milling and spark plasma sintering. *Journal of Materials Science & Technology* 28 (11): 969–975.

Bathula, S., R. C. Anandani, A. Dhar, and A. K. Srivastava (2012a). Synthesis and characterization of Al-alloy/SiCp nanocomposites employing high energy ball milling and spark plasma sintering. *Advanced Materials Research* 410: 224–227.

Bau, N. Q., M. L. S. Nai, S. Sankaranarayanan, S. Jayalakshmi, W. L. E. Wong and M. Gupta (2016). Microstructure and damping characteristics of Mg and its composites containing cetastable $Al_{85}Ti_{15}$ particle. *Journal of Composite Materials* 50 (18): 2565–2573.

Beygelzimer, Y., D. Orlov, A. Korshunov, S. Synkov, V. Varyukhin, I. Vedernikova, A. Reshetov, A. Synkov, L. Polyakov and I. Korotchenkova (2006). Features of twist extrusion: Method, structures & material properties. *Solid State Phenomena* 114: 69–78.

Ceschini, L., A. Dahle, M. Gupta, A. Jarfors, S. Jayalakshmi, A. Morri, F. Rotundo, S. Toschi and R. Arvind Singh (2016). *Aluminum and Magnesium Metal Matrix Nanocomposites*. Springer Nature, Singapore.

Chang, S., Y. S. Pyun and A. Amanov (2017). Wear enhancement of wheel–rail interaction by ultrasonic nanocrystalline surface modification technique. *Materials* 10: 188–200.

Chen, S., X. Chen, L. Wang, J. Liang and C. S. Liu (2017). Laser cladding FeCrCoNiTiAl high entropy alloy coatings reinforced with selfgenerated TiC particles. *Journal of Laser Applications* 29 (1): 0120041–0120048.

Cho, I. H., A. Amanov, I. S. Cho and Y. S. Pyun (2011a). Application of ultrasonic nanocrystalline surface modification technology for prolonging the service life of double row angular contact bearings and for reducing friction loss. *11th International Conference on Shot Peening*, Indiana, USA.

Cho, I. S., C. S. Lee, A. Amanov, Y. S. Pyoun and I. G. Park (2011b). The effect of ultrasonic nanocrystal surface modification on the high frequency fretting behaviour of AISI310 steel. *Journal of Nanoscience & Nanotechnology* 11 (1): 742–746.

Darmiani, E., I. Danaee, M. A. Golozar, M. R. Toroghinejad, A. Ashrafi, and A. Ahmadi (2013). Reciprocating wear resistance of Al–SiC nano-composite fabricated by accumulative roll bonding process. *Materials & Design* 50: 497 502.

Du, X. H. and B. L. Wu (2009). Using two-pass frition stir processing to produce nanocrystalline microstructure in AZ61 magnesium alloy. *Science in China Series: Technological Sciences* 52: 1751–1754.

Dudina, D. V., K. Georgarakis, M. Aljerf, Y. Li, M. Braccini, A. R. Yavari and A. Inoue (2010). Cu-based metallic glass particle additions to significantly improve overall compressive properties of an Al alloy. *Composites—A* 41 (10): 1551–1557.

Dudina, D. V., K. Georgarakis, Y. Li, M. Aljerf, A. LeMoulec, A. R. Yavari and A. Inoue (2009). A magnesium alloy matrix composite reinforced with metallic glass. *Composites Science and Technology* 69 (15–16): 2734–2736.

Edalati K. and Z. Horita (2016). A review on high-pressure torsion (HPT) from 1935 to 1988. *Materials Science & Engineering A* 652: 325–352.

Eugene, W. W. L. and M. Gupta (2006). Simultaneously improving strength and ductility of magnesium using nano-size SiC particulates and microwaves. *Advanced Engineering Materials* 8 (8): 735–740.

Faraji, G. and P. Asadi (2011). Characterization of AZ91/alumina nanocomposite produced by FSP. *Materials Science & Engineering A* 528 (6): 2431–2440.

Fujii, H., Y. F. Sun, K. Inada, Y. S. Ji, Y. Yokoyama, H. Kimura and A. Inoue (2011). Fabrication of Fe-based metallic glass particle reinforced Al-based composite materials by friction stir processing. *Materials Transaction* 52 (8): 1634–1640.

Gan, Y. X., D. Solomon and M. Reinbolt (2010). Friction stir processing of particle reinforced composite materials. *Materials* 3: 329–350.

Goh, C. S., J. Wei, L. C. Lee and M. Gupta (2006a). Development of novel carbon nanotube reinforced magnesium nanocomposites using the powder metallurgy technique. *Nanotechnology* 17 (1): 7–12.

Goh, C. S., J. Wei, L. C. Lee and M. Gupta (2006b). Simultaneous enhancement in strength and ductility by reinforcing magnesium with carbon nanotubes. *Materials Science & Engineering A* 423: 153–156.

Goh, C. S., J. Wei, L. C. Lee and M. Gupta (2007). Properties and deformation behaviour of $Mg–Y_2O_3$ nanocomposites. *Acta Materialia* 55 (15): 5115–5121.

Goh, C. S., J. Wei, L. C. Lee and M. Gupta (2008). Ductility improvement and fatigue studies in Mg-CNT nanocomposites. *Composites Science and Technology* 68 (6): 1432–1439.

Gupta, M. and N. M. L. Sharon (2011). *Magnesium, Magnesium Alloys, and Magnesium Composites*. John Wiley & Sons Inc., New Jersey.

Gupta, M. and W. W. L. Eugene (2005). Enhancing overall mechanical performance of metallic materials using two-directional microwave assisted rapid sintering. *Scripta Materialia* 52 (6): 479–483.

Gupta, M. and W. W. L. Eugene. (2007). *Microwaves and Metals*. John Wiley & Sons, Hoboken, NJ, USA.

Habibi, M. K., H. Pouriayevali, A. M. S. Hamouda and M. Gupta (2012). Differentiating the mechanical response of hybridized Mg nano-composites as a function of strain rate. *Materials Science & Engineering A* 545: 51–60.

Hassan, S. F., M. J. Tan and M. Gupta (2008). High-temperature tensile properties of Mg/Al_2O_3 nanocomposite. *Materials Science & Engineering A* 486: 56–62.

Hsu, W. L., H. Murakami, J. W. Yeh, A. C. Yeh and K. Shimoda (2016). A heat-resistant $NiCo_{0.6}Fe_{0.2}Cr_{1.5}SiAlTi_{0.2}$ overlay coating for high-temperature applications. *Journal of Electrochemical Society* 163 (13): C752–C758.

Huang, C., Y. Z. Zhang, J. Y. Shen and R. Vilar (2011). Thermal stability and oxidation resistance of laser clad TiVCrAlSi high entropy alloy coatings on Ti–6Al–4V alloy. *Surface & Coating Technology* 206 (6): 1389–1395.

Huang, C., Y. Z. Zhang, R. Vilar and J. Y. Shen (2012). Dry sliding wear behavior of laser clad TiVCrAlSi high entropy alloy coatings on Ti–6Al–4V substrate. *Materials & Design* 41: 338–343.

Inoue, A. (1998). *Bulk Amorphous Alloys: Preparation and Fundamental Characteristics 4*. Trans Tech Publications, Uetikon-Zürich, Switzerland.

Jayalakshmi, S. and R. Arvind Singh. (2015). *Processing Techniques and Tribological Behavior of Composite Materials*. IGI Global, Pennsylvania, USA, 9.

Jayalakshmi, S., R. Arvind Singh and M. Gupta (2016). Synthesis of light metal nanocomposites: Challenges and opportunities. *Indian Journal of Advanced Chemical Sciences* S1: 83–288.

Jayalakshmi, S. and M. Gupta (2015). *Metallic Amorphous Alloy Reinforcements in Light Metal Matrices, Springer Briefs in Materials*. Springer, New York, USA.

Jayalakshmi, S., S. Gupta, S. Sankaranarayanan, S. Sahu and M. Gupta (2013). Structural and mechanical properties of $Ni_{60}Nb_{40}$ amorphous alloy particle reinforced Al-based composites produced by microwave assisted rapid sintering. *Materials Science & Engineering A* 581: 119–127.

Jayalakshmi, S., S. Sahu, S. Sankaranarayanan, S. Gupta and M. Gupta (2014). Development of novel Mg–$Ni_{60}Nb_{40}$ amorphous particle reinforced composites with enhanced hardness and compressive response. *Materials & Design* 53: 849–855.

Ji, W., Z. Y. Fu, W. Wang, H. Wang, J. Zhang, Y. Wang and F. Zhang (2014). Mechanical alloying synthesis and spark plasma sintering consolidation of CoCrFeNiAl high-entropy alloy. *Journal of Alloys and Compounds* 589, 61–66.

Ji, X., C. Yan, S. H. Alavi and S. P. Harimkar (2016). STLE Annual Meeting & Exhibition, 15–16 May, Las Vegas, Nevada, USA.

Johnson, W. L. (1999). Bulk glass-forming alloys: Science and technology. *MRS Bulletin* 24: 42–56.

Joo, S. H., H. Kato, M. J. Jang, J. Moon, E. B. Kim, S. J. Hong and H. S. Kim (2017). Structure and properties of ultrafine-grained CoCrFeMnNi high-entropy alloys produced by mechanical alloying and spark plasma sintering. *Journal of Alloys and Compounds* 698: 591–604.

Kainer, K. (2006). *Metal Matrix Composites: Custom-Made Materials for Automotive and Aerospace Engineering*. John Wiley & Sons, Weinheim, Germany.

Khalil, A., A. S. Hakeem, and N. Saheb (2011). Optimization of process parameters in spark plasma sintering Al6061 and Al2124 aluminum alloys. *Advanced Materials Research* 328–330: 1517–1522.

Khalil, K. A. and A. A. Almajid (2012). Effect of high-frequency induction heat sintering conditions on the microstructure and mechanical properties of nanostructured magnesium/hydroxyapatite nanocomposites. *Materials & Design* 36: 58–68.

Khan, M. K., Y. J. Liu, Q. Y. Wang, Y. S. Pyun and R. Kaymov (2016). Effect of ultrasonic nanocrystal surface modification on the characterisitcs of AISI310 stainless steel up to very high fatigue. *Fatigue & Fracture of Engineering Materials & Structures* 39 (4): 427–438.

Khoramkhorshid, S., M. Alizadeh, A. H. Taghvaei and S. Scudino (2016). Microstructure and mechanical properties of Al-based metal matrix composites reinforced with Al84Gd6Ni7Co3 glassy particles produced by accumulative roll bonding. *Materials & Design* 90: 137–144.

Kim, I. Y., J. H. Lee, G. S. Lee, S. H. Baik, Y. J. Kim and Y. Z. Lee (2009). Friction and wear characteristics of the carbon nanotube-aluminum composites with different manufacturing conditions. *Wear* 267 (1–4): 593–598.

Komarasamy, M., N. Kumar, Z. Tang, R. S. Mishra and P. K. Liaw (2015). Effect of microstructure on the deformation mechanism of friction stir-processed Al0.1CoCrFeNi high entropy alloy. *Materials Research Letters* 3 (1): 30–34.

Kubota, M., J. Kaneko and M. Sugamata (2008). Properties of mechanically milled and spark plasma sintered Al–AlB_2 and Al–MgB_2 nano-composite materials. *Materials Science & Engineering A* 475 (1–2): 96–100.

Kumar, A. and M. Gupta (2016). An insight into evolution of light weight high entropy alloys: A review. *Metals* 6: 199–226.

Langer, J. (2006). Shear-transformation-zone theory of deformation in metallic glasses. *Scripta Materialia* 54 (3): 375–379.

Lee, C. J., J. C. Huang and P. J. Hsieh (2006a). Mg based nano-composites fabricated by friction stir processing. *Scripta Materialia* 54 (7): 1415–1420.

Lee, M. H., J.-H. Kim, J. S. Park, J. C. Kim, W. T. Kim and D. H. Kim (2004). Fabrication of Ni–Nb–Ta metallic glass reinforced Al-based alloy matrix composites by infiltration casting process. *Scripta Materialia* 50 (11): 1367–1371.

Lee, W. B., C. Y. Lee, M. K. Kim, J. I. Yoon, Y. J. Kim, Y. M. Yoen and S. B. Jung (2006b). Microstructures and wear property of friction stir welded AZ91 Mg/SiC particle reinforced composite. *Composites Science and Technology* 66 (11–12): 1513–1520.

Li, K., Y. He, I. S. Cho, C. S. Lee, I. J. Park and J. I. Song (2015). Effect of ultrasonic nanocrystalline surface modification on the microstructual evolution of Inconel690 alloy. *Materials and Manufacturing Processes* 30 (2): 194–198.

Liao, J. Z., M. J. Tan and I. Sridhar (2010). Spark plasma sintered multi-wall carbon nanotube reinforced aluminum matrix composites. *Materials & Design* 31 (1): S96–S100.

Lim, C. Y. H., D. K. Leo, J. J. S. Ang and M. Gupta (2005). Wear of magnesium composites reinforced with nano-sized alumina particulates. *Wear* 259 (1–6): 620–625.

Liu, Z. Y., B. L. Xiao, W. G. Wang and Z. Y. Ma (2012). Elevated temperature tensile properties and thermal expansion of CNT/2009Al composites. *Composites Science and Technology* 72 (15): 1826–1833.

Lu, D., Y. Jiang and R. Zhou (2013). Wear performance of nano-Al_2O_3 particles and CNTs reinforced magnesium matrix composites by friction stir processing. *Wear* 305 (1–2): 286–290.

Lu, K (1996). Nanocrystalline metals crystallized from amorphous solids: Nanocrystallization, structure, and properties. *Materials Science and Engineering R: Reports* 16 (4): 161–221.

Mahoney, M. W. and S. P. Lynch (2016). Friction-stir processing. Rockwell Sci. Company LLC, Thousand Oaks, California, Report.

Mallick, A., K. S. Tun, V. Srikanth and M. Gupta (2010). Mechanical charateristics of pure Mg and a Mg/Y_2O_3 nanocomposite in the 25–250°C temperature range. *Journal of Materials Science* 45 (11): 3058–3066.

Marchi, C. S. (2006). Nanostructured aluminium for macrostructured world. Research Highlights. University of California-Davis.

Meriaudeau, F., F. Truchetet, D. Grevey and A. B. Vannes (1997). Laser cladding process and image processing. *Journal of Lasers in Engineering* 6: 161–187.

Miracle, D. (2005). Metal matrix composites—From science to technological significance. *Composites Science and Technology* 65 (15–16): 2526–2540.

Moravcik, I., J. Cizek, P. Gavendova, S. Guo and I. Dlouhy (2016). Effect of heat treatment on microstructure and mechanical properties of spark plasma sintered AlCoCrFeNiTi0.5 high entropy alloy. *Materials Letters* 174: 53–56.

Murthy, B. S., J. W. Yeh and S. Ranganathan (2014). *High Entropy Alloys, 1st Edition*. Butterworth-Heinemann.

Nithin Viswanath, V., G. Keerthi Prasad, B. R. Sujith, S. Jayalakshmi, R. Arvind Singh and Kaushik Hebbar (2016). Microstructural studies of friction stir processed aluminium composites with silicon carbide (SiC) and alumina (Al_2O_3) hybrid reinforcements. *International Conference on Breakthrough in Engineering & Advanced Technology (IC-BEAT'16)*, Cochin, India.

Paramsothy, M., J. Chan, R. Kwok and M. Gupta (2011a). Adding TiC nanoparticles to magensium alloy ZK60A for strength/ductility enhancement. *Journal of Nanomaterials* 2011: 1–9.
Paramsothy, M., J. Chan, R. Kwok and M. Gupta (2011b). Enhanced mechanical response of magnesium alloy ZK60A containing Si3N4 nanoparticles. *Composites—A* 42 (12): 2093–3100.
Rana, J. K., D. Sivaprahasam, K. S. Raju, and V. S. Sarma (2009). Microstructure and mechanical properties of nanocrystalline high strength Al–Mg–Si (AA6061) alloy by high energy ball milling and spark plasma sintering. *Materials Science & Engineering A* 527 (1–2): 292–296.
Ryu, H. J., B. C. Kang, O. A. Waseem, J. H. Lee and S. H. Kong (2016). Compositionally complex alloys. ECI Symposium Series. http://dc.engconfintl.org/superalloys_ii/37
Saheb, N., Z. Iqbal, A. Khalil, A. S. Hakeem, N. Al Aqeeli, T. L. A. Al-Qutub and R. Kirchner (2012). Spark plasma sintering of metals and metal matrix nanocomposites: A review. *Journal of Nanomaterials*: 1–13.
Saito, Y., H. Utsunomiya, N. Tsuji and T. Ssakai (1999). Novel ultra-high straining process for bulk materials—Development of the accumulative roll-bonding (ARB) process. *Acta Materialia* 47 (2): 579–583.
Sankaranarayanan, S., N. Agrawal, S. Jayalakshmi, N. Q. Bau and Manoj Gupta (2015a). Synthesis and characterization of novel magnesium materials containing copper–titanium based ($Cu_{50}Ti_{50}$) amorpous alloy particles. Magnesium Technology, TMS Annual Meeting, Florida, USA, 15–19.
Sankaranarayanan, S., M. K. Habibi, S. Jayalakshmi, K. J. Ai, A. Almajid and M. Gupta (2015b). Nano-AlN particle reinforced Mg composites: Microstructural and mechanical properties. *Materials Science & Technology* 31(9). 1122–1131.
Sankaranarayanan, S., V. Hemanth Shankar, S. Jayalakshmi, N. Q. Bau and M. Gupta (2015c). Development of high performance magnesium composites using $Ni_{50}Ti_{50}$ metallic glass reinforcement and microwave sintering approach. *Journal of Alloys and Compounds* 627: 192–199.
Sankaranarayanan, S., S. Jayalakshmi and M. Gupta (2013a). Effect of nano-Al_2O_3 addition and heat treatment on the microstructure and mechanical properties of Mg–(5.6Ti + 3Al) composite. *Materials Characterization* 75: 150–164.
Sankaranarayanan, S., S. Jayalakshmi, K. S. Tun, A. M. S. Hamouda and M. Gupta (2013b). Synthesis and characterization of nano boron nitride reinforced magnesium composites produced by the microwave sintering method. *Materials* 6 (5): 1941–1955.
Sankaranarayanan, S., R. K. Sabat, S. Jayalakshmi, S. Suwas and M. Gupta (2013c). Effect of hybridizing micron-sized Ti with nano-sized SiC on the microstructural evolution and mechanical response of Mg–5.6Ti composite. *Journal of Alloys and Compounds* 575: 207–217.
Sankaranarayanan, S., R. K. Sabat, S. Jayalakshmi, Stefan Walker, S. Suwas and M. Gupta (2014). Using heat treatment effects and EBSD analysis to tailor microstructure of hybrid Mg nanocomposite for enhanced overall mechanical response. *Materials Science and Technology* 30 (11): 1309–1320.
Sasaki, T. T., T. Ohkubo and K. Hono (2009). Microstructure and mechanical properties of bulk nanocrystalline Al–Fe alloy processed by mechanical alloying and spark plasma sintering. *Acta Materialia* 57 (12): 3529–3538.
Shabadi, R., M. N. Avettand-Fènoëla, A. Simar, R. Taillarda, P. K. Jain and R. Johnson (2015). Thermal conductivity in yttria dispersed copper. *Materials & Design* 65: 869–877.
Shafiei-Zarghani, A., S. F. Kashani-Bozorg and A. Zarei-Hanzaki (2009). Microstructures and mechanical properties of Al/Al_2O_3 surface nano-composite layer produced by friction stir processing. *Materials Science & Engineering A* 500 (1–2): 84–91.
Shanthi, M., M. Gupta, A. E. W. Jarfors and M. J. Tan (2011). Synthesis, characterization and mechanical properties of nano alumina particulate reinforced magnesium based bulk metallic glass composites. *Materials Science & Engineering A* 528 (18): 6045–6050.
Shanthi, M., Q. B. Nguyen and M. Gupta (2010). Sliding wear behaviour of calcium containing AZ31B/Al2O3 nanocomposites. *Wear* 269 (5–6): 473–479.
Sklenicka, V., J. Dvorak, M. Svoboda, P. Kral and M. Kvapilova (2013). Equal-channel angular pressing and creep in ultrafine grained aluminium and its alloys. *InTech Open Access Journal* 3–46.
Sriharitha, R., B. S. Murthy, R. S. Kottada (2014). Alloying, thermal stability and strengthening in spark plasma sintered AlxCoCrCuFeNi high entropy alloys. *Journal of Alloys and Compounds* 583: 419–426.
Srinivasan, M., C. Loganathan, R. Narayanasamy, V. Senthilkumar, Q. B. Nguyen and M. Gupta (2013). Study on hot deformation behavior and microstructure evolution of cast-extruded AZ31B magnesium alloy and nanocomposite using processing map. *Materials & Design* 47: 449–455.
Srivatsan, T. S., C. Godbole, M. Paramsothy and M. Gupta (2011). Influence of nano-sized carbon nanotube reinforcements on tensile deformation, cyclic fatigue and final fracture behaviour of a magensium alloy. *Journal Materials Science* 47 (8): 3621–3638.

Su, J. Q., T. W. Nelson, T. R. McNelley and R. S. Mishra (2011). Development of nanocrystalline structure in Cu during friction stir processing (FSP). *Materials Science & Engineering A* 528 (16–17): 5458–5464.

Sun, K., Q. Y. Shi, Y. J. Sun and G. Q. Chen (2012). Microstructure and mechanical property of nano-SiCp reinforced high strength Mg bulk composites produced by friction stir processing. *Materials Science & Engineering A* 547: 32–37.

Suryanarayana, C. (2001). Mechanical alloying and milling. *Progress in Materials Science* 46 (1–2): 1–184.

Svensson, D. O. (2014). High entropy alloys: Breakthrough materials for aero engine applications? Master's Thesis, Chalmers Institute of Tech., Sweden.

Thakur, S. K., K. Balasubramanian and M. Gupta (2007a). Microwave synthesis and characterization of magnesium based composites containing nanosized SiC and hybrid ($SiC+Al_2O_3$) reinforcements. *Transactions of the ASME* 129: 194–199.

Thakur, S. K., G. T. Kwee and M. Gupta (2007b). Development and characterization of magensium composites containing nano-sized silicon carbide and carbon nanotubes as hybrid reinforcements. *Journal of Materials Science* 42 (24): 10040–10046.

Thakur, S. K., T. S. Srivatsan and M. Gupta (2007c). Synthesis and mechanical behavior of carbon nanotube–magnesium composites hybridized with nanoparticles of alumina. *Materials Science & Engineering A* 466 (1–2): 32–37.

Todaka, Y., M. Umemoto, A. Yamazaki, J. Sasaki and K.-i. Tsuchiya (2008). Influence of high-pressure torsion straining conditions on microstructure evolution in commercial purity aluminum. *Materials Transactions* 49 (1): 7–14.

Tsai, M. H. and J. W. Yeh (2014). High-entropy alloys: A critical review. *Materials Research Letters* 2 (3): 107–123.

Tun, K. S. and M. Gupta (2007). Improving mechanical properties of magnesium using nano-yttria reinforcement and microwave assisted powder metallurgy method. *Composites Science and Technology* 67 (13): 2657–2664.

Veronesi, P., R. Rosa, E. Colmbini and C. Leonelli (2015). Microwave-assisted preparation of high entropy alloys. *Technologies* 3(4): 182–197.

Wang. C., W. Ji and Z. Fu (2014). Mechancial alloying and spark plasma sintering of CoCrFeNiMnal high-entropy alloy. *Advanced Powder Technology* 25 (4) 1334–1338.

Wang, W., Q. Shi, P. Liu, H. Li, and T. Li (2009). A novel way to produce bulk SiCp reinforced aluminum metal matrix composites by friction stir processing. *Journal of Materials Processing Technology* 209 (4): 2099–3103.

Weglowski, M. S. and A. Pietras (2011). Friction stir processing—Analysis of the process. *Archives of Metallurgy & Materials* 56 (3): 779–788.

Yeh, J. W. (2013). Alloy design strategies and future trends in high-entropy alloys. *JOM: The Minerals, Metals & Materials Society* 65 (12): 1759–1771.

Zhang, Y., T. T. Zuo, Z. Tang, M. C. Gao, K. A. Dahmen, P. K. Liaw and Z. P. Lu (2014). Microstructures and properties of high-entropy alloys. *Progress in Materials Science* 61: 1–93.

Zhilyaev, A. P. and T. G. Langdon (2008). Using high-pressure torsion for metal processing: Fundamentals and applications. *Progress in Materials Science* 53 (6): 893–979.

10 Kinetics of Laser Surface Engineering of Three Aluminum Alloys

Sourabh Biswas and Sandip P. Harimkar

CONTENTS

10.1 INTRODUCTION

Aluminum (Al) is the third most abundant element and constitutes as high as 8% of the Earth's crust, which is approximately 50% greater than the next most abundant metal, iron. Initially limited by the difficulties in reducing and extracting aluminum because of its abnormally high oxygen affinity, the immense cost of production and inherent silvery luster turned the value of aluminum comparable with that of gold in monetary terms. Since the development of Hall–Heroult process coupled with other innovations in extraction technologies, the cost of aluminum production has declined exponentially. From the first attempt in commercial production of aluminum in 1888 by Alcoa [1], aluminum alloys have progressed significantly in their rather young history.

A very low density (2.7 g/cm^3) gives aluminum excellent specific properties, and its face-centered cubic crystal structure inherently renders the alloys with marked hot forming ability. Also, aluminum alloys have appreciable castability and corrosion resistance. The selection of materials is becoming increasingly competitive in recent times, and consumers often select a material based on a number of properties, performance, and economic factors. The abundant availability and attractive properties have given aluminum a distinct edge over other engineering alloys. The popularity of aluminum can be inferred from the fact that within the first century of commercialized aluminum production, aluminum has replaced copper to become the most produced non-ferrous metal [2,3].

Although the cost of aluminum per unit mass has declined significantly in the past century, it is reported as one of the most popular metal used in research. This popularity can be attributed to its wide spectrum of applications and potential for developing desired properties with varying heat treatments and alloying additions. The wrought aluminum alloy systems successfully developed for commercial applications are tabulated in Table 10.1 [4].

Among the commercially developed wrought aluminum alloys, only 2xxx, 6xxx, and 7xxx (and still developing 8xxx) series alloys are heat treatable by precipitation hardening [5]. An important feature of the precipitation-hardened aluminum alloy systems are the rapidly decreasing solid solubility with decreasing temperature. The sharply sloping solvus line increases the ease of formation of intermetallic phases such as Guinier–Preston (GP) zones in Al–Cu system upon age hardening. This feature offers an excellent control in developing a microstructure with desired mechanical properties, rendering these systems a distinct edge over other commercial alloy systems such as iron-based systems [6,7].

Among all the aluminum alloys, the precipitation-hardened aluminum alloys offer the highest specific strength, one of the most critical parameters for material selection in aerospace, automobile, defense, structural, and sport applications. Also, as fuel economy is becoming increasingly important because of the acute shortage of fossil fuels, reducing weight of vehicles to improve fuel efficiency is gaining more importance. This again underlines the importance of precipitation-hardened aluminum alloys in these applications. Apart from the unmatched strength of the precipitation-hardened aluminum alloys, their non–heat-treatable counterparts also trail in some attributes extremely essential in certain applications. For example, the 5xxx series Al–Mg alloys exhibit dynamic strain aging behavior [8] and often suffer from inadequate surface finish subsequent to forming to meet the high standards for applications in the automobile industry.

Aluminum alloys offer excellent corrosion resistance owing to their high oxygen affinity. Aluminum alloys form an extremely dense (nonporous) and conformal protective oxide layer, promoting their use in structural applications. However, the corrosion resistance of precipitation-hardened aluminum alloy systems, such as 2024 [9], 6061 [10,11], and 7075 [12], is significantly affected by the presence of highly electropositive precipitates [13]. Although precipitation during age hardening is vital for improving the strength of aluminum alloys, it also tends to form internal corrosion cells with the base aluminum alloy matrix as the electronegativity of the matrix is almost always greater than the precipitates. Therefore, to achieve an optimum combination of both enhanced strength and corrosion resistance in the precipitation-hardened aluminum alloys, several surface engineering–based

TABLE 10.1
Principal Alloying Elements in Wrought Al Alloys

Alloy Series	Principal Alloying Element
1xxx	None
2xxx	Cu
3xxx	Mn
4xxx	Si
5xxx	Mg
6xxx	Mg and Si
7xxx	Zn
8xxx	Miscellaneous

Source: *Aluminum*. Vol. 1. 1967. Aluminum/Edited by Kent R. Van Horn. Metals Park (Ohio): American Society for Metals.

technologies have been developed. Brazing, anodizing, plasma electrolytic oxidation, chemical vapor deposition, and laser surface engineering are some of the most successful approaches developed to achieve superior surface properties. While each of these processes has its advantages and limitations, the approaches of laser surface engineering are becoming increasingly important for improving the mechanical and electrochemical properties of precipitation-hardened aluminum alloys. An improvement in surface properties of these alloys is essential to sustain extreme environments often experienced in components such as fuel exhaust nozzles, fuselage bodies, and aircraft structures. In this chapter, various laser surface engineering approaches for surface modifications of the three most common precipitation-hardened aluminum alloys (2024, 6061, and 7075) are examined and discussed. An assessment of the reported data to understand the influence of laser processing parameters on the development of microstructure (grain refinement, compositional segregation, selective vaporization, in situ reactions, and formation of intermetallic phases) and mechanical/electrochemical properties is also presented.

10.2 LASER SURFACE ENGINEERING OF ALUMINUM ALLOY

Surface engineering generally involves modification of the microstructure and composition of the surfaces to improve the surface properties without significantly altering the base/substrate material properties. Lasers are becoming increasingly important in surface engineering of a wide range of materials, including metals, alloys, ceramics, composites, and polymers. Laser surface engineering approaches have been successfully used to improve hardness, strength, toughness, wear resistance, corrosion resistance, heat resistance, and fatigue resistance of metals and their alloy counterparts. For example, the use of laser surface engineering approaches in preventing/minimizing corrosion has been successfully demonstrated in various alloy systems such as aluminum [14], titanium [15], and iron-based alloys [16]. Three major approaches for laser surface engineering of light alloys are laser surface melting, laser surface alloying, and laser composite surfacing. All of these approaches involve laser material interaction that causes surface melting of the alloy. In laser surface melting, a laser is irradiated on the surface, causing melting and rapid resolidification of the alloy. Very high cooling rates of the order of up 10^5–10^{11} K/s during resolidification result in the formation of a refined dendritic microstructure in the surface. The improvement in properties attributed to laser surface melting is primarily due to microstructural refinement and compositional redistribution on the laser-modified surface. In laser surface alloying, the composition of the surfaces is modified by introducing additional desired element during surface melting. Alloying of the surface can be accomplished either by feeding the elemental powder to the melt pool or by pre-placing the elemental powder on the substrate before laser surface melting. The compositional modification in synergism with microstructural refinement that occurs at the surface leads to an improvement in surface properties. Laser composite surfacing involves introducing hard second-phase particles to the laser-melted pool, forming a surface that is essentially a metal matrix composite. The hard second-phase particles are fed to the melt pool or preplaced on the substrate before laser surface melting. Laser composite surfacing can also be accomplished by the occurrence of in situ reactions on the laser-melted surface. The composite structure and microstructural refinement results in combination of surface properties on the surface. Depending on the incident laser power, laser scanning velocity, focus conditions, and thermo-physical properties of the material, the depth of surface modification, up to 1–2 mm, can be readily achieved in a wide range of materials. Laser surface engineering also offers distinct advantages such as non-contact processing, ease of automation, rapid processing, flexible manufacturing, and minimum heat affected zones, as well as ability to produce a range of microstructural/compositional effects. It should be noted that the aforementioned features associated with laser surface engineering have resulted in laser's successful implementation in several other techniques such as shock peening [17,18], drilling [19,20], and selective laser melting [21,22].

All the laser surface engineering approaches such as laser surface melting, laser surface alloying, and laser composite surfacing have been successfully demonstrated for different aluminum alloys. The main objective of laser surface engineering for precipitation-hardened aluminum alloys is to achieve optimum combination of corrosion resistance and hardness/wear resistance at the surface. Laser surface melting and laser surface alloying are particularly important for improving corrosion resistance of relatively stronger precipitation-hardened aluminum alloys such as 2024 and 7075 alloy. For corrosion-resistant 6061 precipitation-hardened aluminum alloy, laser composite surfacing processes have been extensively used to improve surface hardness and wear resistance. Various effects such as microstructure refinement (dendritic microstructure), compositional segregation (copper enrichment at the grain boundaries), compositional modifications (through selective vaporization and surface alloying), and in situ reactions (formation of intermetallic phases of aluminum) have been observed during laser surface engineering of precipitation-hardened aluminum alloys.

10.2.1 Laser Surface Engineering of Aluminum Alloy 2024

The 2024 aluminum alloy, with copper as its primary alloying element, is one of the most popular and earliest developed aluminum alloy. The chemical composition of this alloy is given in Table 10.2 [4]. The 2024 alloy is age hardened, and distinct phases such as GP zone I, GP zone II, θ', and equilibrium θ have been identified to precipitate during the age hardening treatment depending on both treatment temperature and time. Although this alloy exhibits comparatively high tensile strength as observed from Table 10.2 [4], it is reported to be susceptible to pitting corrosion.

The corrosion behavior of the 2024 alloy was investigated in depth by Liao and coworkers [23]. The formation of microstructural galvanic couples in this alloy was studied by performing potentiodynamic corrosion tests in 0.5 M sodium chloride solution. The influence of the microstructural species was examined using scanning electron microscopy (SEM)/optical microscopy and x-ray diffraction (XRD) techniques. The studies showed that corrosion in the 2024 alloy systems occurred primarily in two distinct stages. Initial pitting occurred along the interfaces of highly electropositive precipitates with matrix and was not very severe. However, the later stage of pitting commences when the passive aluminum oxide layer adhering to the surface is removed. The corrosion in the later stage was observed not only at the coarser precipitate–matrix interfaces but also at the interfaces of the matrix and the solute clusters, even though the interfaces have a noticeably lower corrosion potential. The dramatic rise in corrosion rate was attributed to the removal of the corrosion-resistant layer of aluminum oxide (Al_2O_3) and subsequent exposure of the matrix. Also, the solute clusters, most likely the GP zones, are more dispersed than the equilibrium θ phase, thereby significantly increasing a statistical probability of the formation of microstructural corrosion cells. This explains the susceptibility of the aluminum alloy 2024 to corrosion in aggressive environments often encountered by aircraft structures (moisture, acid rain, and environmental emissions). Intergranular corrosion is also a reported issue with this alloy system, as the grain boundaries act as preferential

TABLE 10.2
Typical Chemical Compositions of Common Precipitation-Hardened Al Alloys

	Composition (wt%)							
Alloy	Si	Cu	Mn	Mg	Cr	Zn	Al	Ultimate Tensile Strength (MPa)
2024	—	4.5	0.6	1.5	—	—	Balance	475
6061	0.6	0.25	—	1.0	0.20	—	Balance	310
7075	—	1.6	—	2.5	0.30	5.6	Balance	572

Source: *Aluminum*. Vol. 1. 1967. Aluminum/Edited by Kent R. Van Horn. Metals Park (Ohio): American Society for Metals.

sites for precipitation and hence increase the chance of formation of a corrosion cell at the grain boundaries [24,25]. Most of the efforts on laser surface engineering of aluminum alloy 2024 are focused on achieving an optimum combination of corrosion resistance and surface mechanical properties.

10.2.1.1 Laser Surface Melting of the 2024 Aluminum Alloy

10.2.1.1.1 Corrosion Behavior

The refined dendritic surface microstructure with finer precipitates and compositional effects often generated by laser surface melting is particularly effective in improving corrosion resistance. The corrosion behavior of laser surface–melted AA2024-T351 was investigated by Li and coworkers [26]. The laser surface melting was conducted using a 2-kW CO_2 laser. The laser surface–melted microstructure revealed that rapid solidification during the laser surface melting treatment processing not only reduces the grain size by formation of finer dendrites but also influences the chemistry of precipitation. A resolidification of the surface initiates appreciable segregation of copper to the grain boundaries due to constitutional supercooling. The higher copper concentration at grain boundaries is virtually unaffected as diffusion of copper in aluminum matrix is very sluggish. The diffusion is further restricted as the temperature at the surface falls rapidly due to the high quenching rates associated with laser processing. The insufficient amount of copper in the grain interior reduces the volume fraction of the θ phases, which is susceptible to corrosion. Also, as the copper segregation reduces the grain boundary energy of the alloy, the occurrence of grain boundary corrosion is restricted (i.e., grain boundaries become more cathodic). As observed in Figure 10.1a, the as-received micrograph exhibits considerable pitting identified by the corroded patches appearing dark due to the roughness caused by corrosion. However, the micrograph of the laser surface–melted sample shows a marked difference in the corrosion behavior, as observed in Figure 10.1b, the micrograph of the laser surface–melted sample. This figure reveals preferred corrosion in bulk of the grains while the grain boundary remains shiny and unreacted on account of being more electro-positive than the grain. Thus, the shift from pitting corrosion (for as-received samples) to uniform corrosion (for laser surface–melted samples) as inferred from SEM observation of the microstructure is effective in improving the quality of the surface. Therefore, mechanical properties can be significantly improved by laser surface melting, especially fatigue endurance and wear resistance

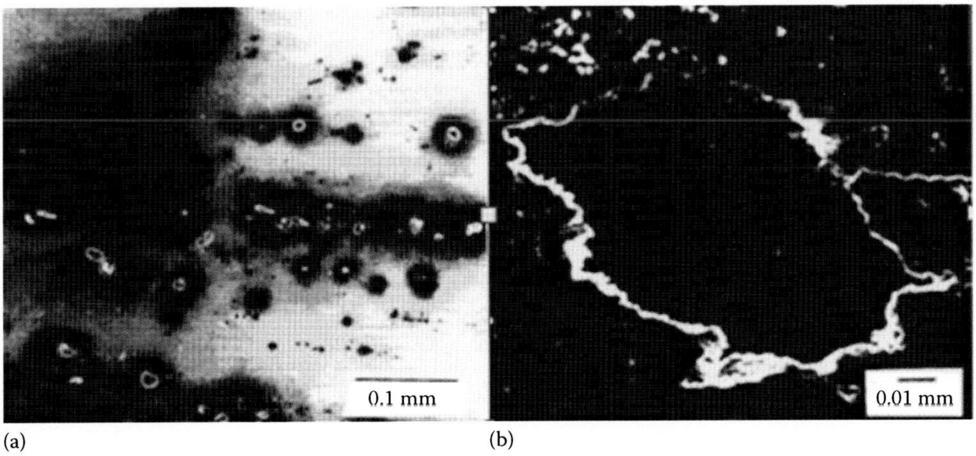

FIGURE 10.1 SEM micrographs of (a) as-received and (b) laser surface–melted 2024 aluminum alloy after corrosion. (Reprinted with permission from Li, R., M. G. S. Ferreira, A. Almeida, R. Vilar, K. G. Watkins, M. A. McMahon, and W. M. Steen. Localized corrosion of laser surface melted 2024-T351 aluminium alloy. *Surface and Coatings Technology* 81, no. 2–3 (1996): 290–296.)

properties, as these properties are sensitive to surface irregularities. This also marks a very uncommon and beneficial feature that the grain boundary is cathodic or far more corrosion resistant than the matrix, thereby minimizing the chances of intergranular corrosion. The Al–Cu system has been reported to show aggravated grain boundary corrosion and embrittlement, and laser surface melting offers an absolute reversal of the situation, further improving its applicability.

The studies also revealed that because of the relatively lower boiling temperature of magnesium, laser surface melting causes selective vaporization of magnesium from the melt pool [26]. Magnesium was undetected from the XRD tests of laser surface–melted samples, indicating a reduction in volume of the magnesium-rich S (Al_2CuMg) intermetallic phase. The selective vaporization improves the corrosion resistance by reducing the surface area of electropositive precipitates due to lower solute (Mg) concentration in the melt. The influence of precipitates on the corrosion behavior of the alloy was studied using free corrosion experiment, which records the variation of open circuit corrosion potential with time. It was observed from the studies that in an acidic or saline environment, the corrosion-resistant oxide undergoes dissolution after a period of time due to essentially enhanced pitting. This results in a significant reduction in the corrosion potential, rendering the surface more anodic. However, the modified microstructure achieved by the laser surface melting process causes a shift in the corrosion mechanism from pitting toward uniform corrosion. The investigation confirmed the overall stability of the laser surface–melted samples to be high since laser surface–melted samples, unlike the as-received samples, do not experience breakdown of the oxide film. The variation of corrosion potential of the laser-treated and base sample can be observed in Figure 10.2 [26]. This can be attributed to the variation of corrosion current density with the mechanism of corrosion occurring in the chosen aluminum alloy. In the as-received sample, the local corrosion current density around the interfaces of the precipitate is higher than that of the bulk material. However, as the interfaces are actual anodic sites, the entire anodic area is just a minor fraction of the bulk area. This creates rapid dissolution around the pits, initiating the breakdown of film in the as-received samples. This phenomenon also supports the observations of Liao and coworkers [23], who indicated the onset of severe pitting subsequent to breakdown of the film due to the initial pitting and activation of the less electronegative solute clusters. The laser surface–melted samples, however, show no breakdown due to the comparatively low corrosion current density and uniform corrosion experienced by the bulk of the grains. The dissolution of oxide film is delayed as

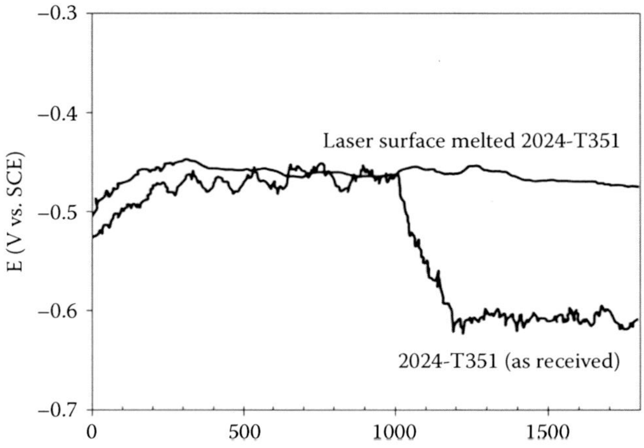

FIGURE 10.2 Potential versus time curve for as-received and laser surface–melted 2024 aluminum alloy. (Reprinted with permission from Li, R., M. G. S. Ferreira, A. Almeida, R. Vilar, K. G. Watkins, M. A. McMahon, and W. M. Steen. Localized corrosion of laser surface melted 2024-T351 aluminium alloy. *Surface and Coatings Technology* 81, no. 2–3 (1996): 290–296.)

the microstructure is stabilized by the lack of precipitates, resulting in a much higher corrosion-resistant surface.

The 2024 aluminum alloy is noted for its complex alloy composition with more than three alloying elements, and hence various precipitate composition and morphology can be achieved by altering the procedure used for heat treatment. The effect of magnesium on the microstructural evolution during laser surface melting and its effect in pitting corrosion of the 2014 and 2024 alloys were investigated by Liu and coworkers [27,28]. The 2014 aluminum alloy had a chemical composition very similar to that of the 2024 alloy, except for much lower magnesium content (2014 has 0.2% to 0.8% Mg while 2024 has 1.2% to 1.8% Mg). Hence, this alloy proved to be effective in identifying the effect of Mg on corrosion behavior. A CO_2 laser with 1.6 kW power, 1 mm spot diameter, 15 to 180 mm/s scanning velocities, and 50% track overlap was used for the purpose of the laser surface melting of the 2024 alloy. The magnesium content in the precipitates was observed to decrease with decreasing laser scanning velocity. Decreasing scanning velocity increases interaction time and thereby increasing vaporization of magnesium due to higher temperature in the melt. Increased vaporization of magnesium lowers the concentration of magnesium in the melt pool, and hence the precipitate composition was affected (the Mg-rich Al_2CuMg phase was replaced by Al_2Cu). The pitting potential studies in 3.5% deaerated sodium chloride solution revealed the Al_2CuMg phase (E_{corr} = 680 mV) to be more electropositive than the Al_2Cu phase (E_{corr} = 560 mV) at identical conditions. Hence, replacement of the Al_2CuMg phase with the Al_2Cu phase is beneficial for the purpose of improved corrosion resistance due to reduction in potential of internal corrosion cells. The purpose of improved migration of copper as compared to magnesium and silicon was also observed in electron probe micro-analyzer mapping. Evidently, the higher atomic volume of copper is more effective in reducing free volume of the grain boundaries, and hence the driving force for copper migration is considerably higher than that for magnesium and silicon, as observed in Figure 10.3 [28]. Thus, reducing scanning velocity and thereby enabling selective vaporization of the magnesium can effectively be used to improve corrosion resistance of the alloy.

10.2.1.1.2 Precipitate Morphology

The use of laser surface melting in a binary eutectic Al–Cu system (reported to exhibit similar Al_xCu_y intermetallic precipitates observed in the 2024 alloy) was investigated by Zimmerman and coworkers [29]. The investigation was performed using a CO_2 laser having an irradiation power of 1 to 1.5 kW. Transmission electron microscopy (TEM) was used to investigate the microstructural evolution at various laser scanning velocities. The analysis revealed that the eutectic microstructure morphology varied significantly with solidification velocity (correlated with scanning velocities of the laser beam) of the laser surface melting. Below a solidification velocity of 20 cm/s, the Al_2Cu precipitates were largely lamellar as observed in Figure 10.4a. The precipitate morphology of laser-treated samples above the threshold solidification velocity of 20 cm/s tended to become wavy, with the amplitudes increasing with an increase in the solidification velocity, as observed in Figure 10.4b. Above a solidification velocity of 50 cm/s, the cooling rate seemed to be too fast to allow precipitate growth, and a highly banded structure as observed in Figure 10.4c, parallel to the solid–liquid interface, was observed. The influence of banded microstructure (developed by friction-stir welding) on crack propagation of the 2024 aluminum alloy was investigated by Sutton and coworkers [30,31]. The investigation revealed that an increased particle density in the precipitate-rich bands did strongly influence the crack-tip field during the early stages of crack growth. The increase in particle density increases the statistical probability of an advancing crack to encounter a particle and thereby triggering crack branching. Hence, this feature can be used as an effective toughening mechanism, especially in improving fatigue endurance. Thus, laser surface melting not only offers the scope to develop finer and more refined microstructures but also controls the morphology of the precipitates by an optimization of the laser scanning velocity. Hence, superior control of the mechanical and surface properties can be achieved as optimization of laser surface melting input parameters is highly effective in developing desired properties required for a specific application. Similar wavy

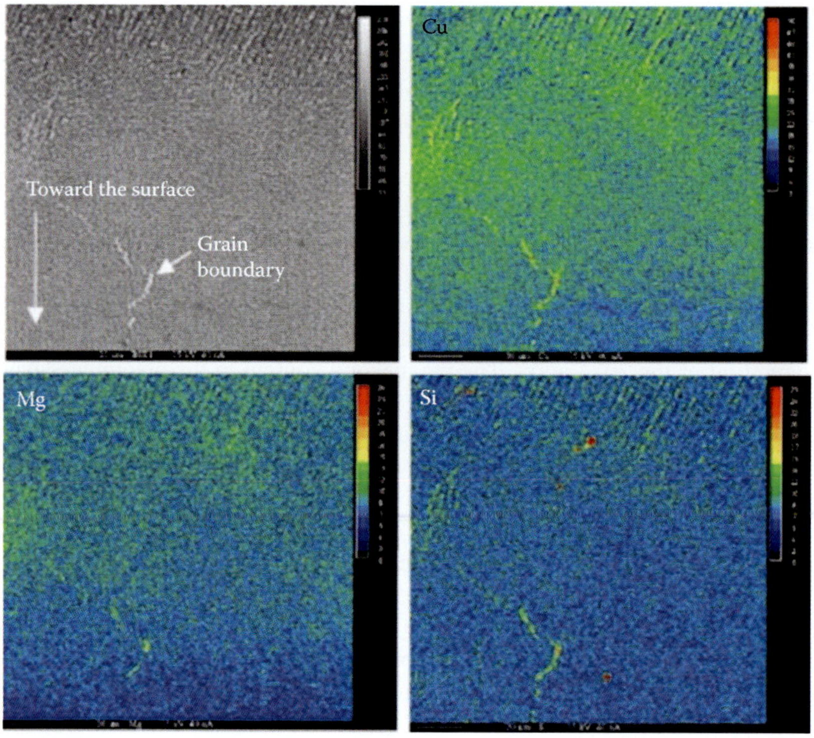

FIGURE 10.3 EPMA mapping of laser surface–melted 2014 alloy. (Reprinted with permission from Liu, Z., P. H. Chong, P. Skeldon, P. A. Hilton, J. T. Spencer, and B. Quayle. Fundamental understanding of the corrosion performance of laser-melted metallic alloys. *Surface and Coatings Technology* 200, no. 18 (2006): 5514–5525.)

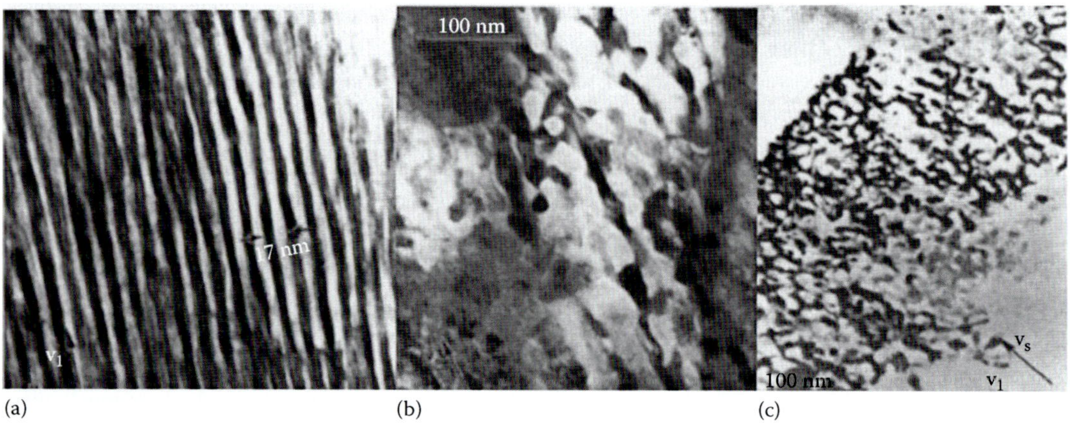

FIGURE 10.4 TEM microstructures of laser surface–melted Al–Cu alloy at solidification rates of: (a) 20 cm/s, (b) 20–50 cm/s, and (c) 50 cm/s. (Reprinted with permission from Zimmermann, M., M. Carrard, and W. Kurz. Rapid solidification of Al–Cu eutectic alloy by laser remelting. *Acta Metallurgica* 37, no. 12 (1989): 3305–3313.)

microstructures in post–laser-melted 2024 aluminum alloy were also mentioned by Embuka and coworkers [32]. The authors studied the effect of heat treatment on laser-melted 2024 aluminum alloy on their pitting behavior. The authors mentioned that precipitation of complex intermetallic phases such as Al_2CuMg and Al(Cu,Fe,Mn) after heat treatment of laser surface–melted alloy decreased pitting resistance, but still exhibited superior resistance compared to the as-received material [32].

10.2.1.2 Laser Composite Surfacing of the 2024 Aluminum Alloy

The 2024 aluminum alloy is extremely sensitive to heat treatment, and a high ratio of peak strength to solutionized strength can be achieved in this alloy. Hence, although laser surface melting of this alloy improves corrosion resistance, it also decreases precipitate density due to remelting of laser-irradiated regions and restricted aging time for precipitation. The reduced precipitate density often results in a drastic reduction of both strength and wear resistance of the laser-modified surface. Ceramic particles such as titanium carbide, silicon carbide, and silicon nitride are noted for their very high hardness and abrasion resistance. Laser composite surfacing of the 2024 aluminum alloy using these hard particles is gaining significant importance to improve the surface hardness. Laser composite surfacing of aluminum alloy with these hard ceramic particles has been reported to exhibit notably higher hardness and wear resistance by other researchers [33,34]. Also, apart from using ceramic particles, hybrid composite surfacing using mixtures of metal and ceramic particles has been attempted. An improvement in wear resistance has been reported to depend on factors, such as surface hardness, size and dispersion of particles, and particle–matrix bonding strength. Hence, research endeavors have not only targeted different hard particles for reinforcement but also identified additives to improve wettability of the particles, thereby improving bonding strength.

10.2.1.2.1 Laser Composite Surfacing with Titanium Carbide Particles

Laser composite surfacing of the 2024 aluminum alloy using titanium carbide powders was investigated by Kadolkar and Dahotre [33]. Their studies revealed that except for very high laser scanning speeds beyond 200 cm/min when diffusion is severely constrained, the particle–matrix bonding is generally good for the 2024 aluminum alloy composite surfaced with titanium carbide. This good particle–matrix bonding can be attributed to the comparatively high copper concentration (4 wt%) in the 2024 alloy, as copper significantly improves the wettability of titanium carbide [35]. The authors reported injecting a mixture of ceramic particles and silicon powder with a weight ratio of 9:1 to the melt pool. The addition of silicon was to enhance the wettability of the particles in the melt zone. The microstructural characterization revealed the existence of three distinct zones: (a) the coating zone, (b) the heat-affected zone, and (c) the base alloy zone, as observed in Figure 10.5a. A marked increase in Knoop hardness from about 100 kg/mm^2 for the substrate to 350–400 kg/mm^2 at the surface was observed due to the ceramic reinforcement (Figure 10.5b).

10.2.1.2.2 In Situ Laser Composite Surfacing

The traditional laser composite surfacing, as discussed, involves injecting hard particles to the laser melt pool. Other novel approaches have also been attempted to develop in situ composites at the surface using laser irradiation, as in situ composites often exhibit superior bonding strength between the matrix and particles. An approach for developing in situ metal matrix composite coating on 2024 alloy was reported by Xu and coworkers [36]. The reinforcement particles were formed in situ due to intermetallic reactions triggered by the injection of iron-coated boron, pure titanium, and pure aluminum powders in the laser melt pool. The input parameters in the investigation were a CO_2 laser power of 1.7 kW and a scanning velocity of 3 mm/s. Surface oxidation was prevented using a coaxial jet and side jet of argon. A distinct relationship of wear resistance with the applied load was observed by the authors (Figure 10.6). When the applied load was low (8.9 N), the wear resistance and hardness of the composite surfaces of the alloy increased with an increase in the volume percentage of titanium diboride (TiB_2), iron aluminide (Al_3Fe), and titanium aluminide (Al_3Ti) dispersions (Samples A to E in figure). This can be explained from the fact that the dispersed particles acted as

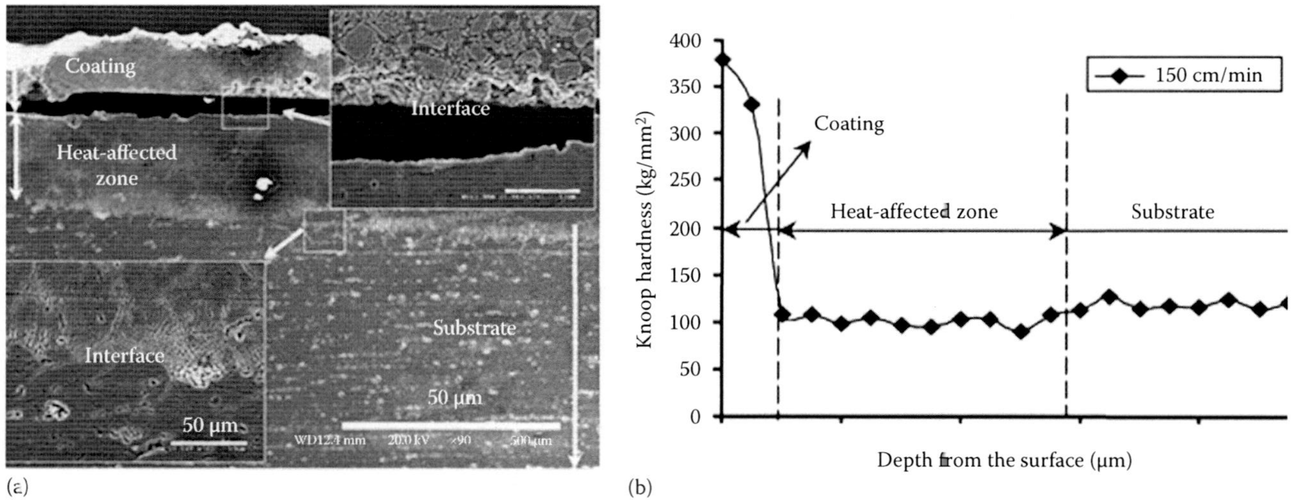

FIGURE 10.5 (a) SEM micrograph of laser surface alloyed TiC-2024 alloy, and (b) Knoop hardness variation with depth. (Reprinted with permission from Kadolkar, P., and N. B. Dahotre. Variation of structure with input energy during laser surface engineering of ceramic coatings on aluminum alloys. *Applied Surface Science* 199, no. 1 (2002): 222–233.)

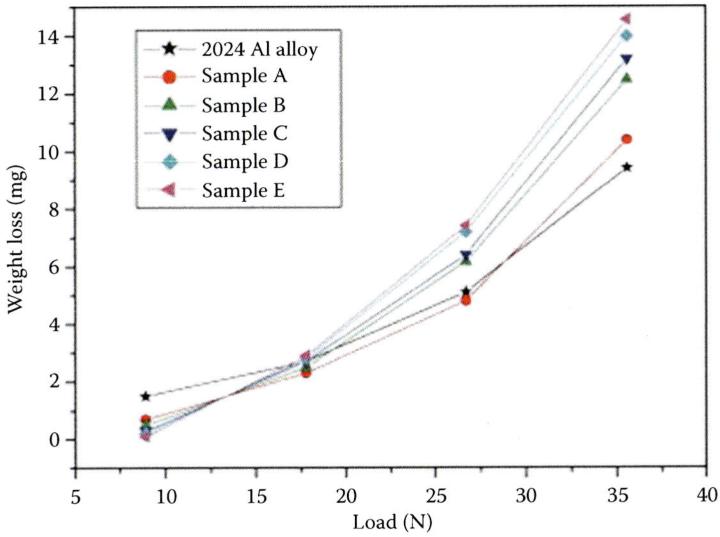

FIGURE 10.6 Variation of weight loss of as-received and laser cladded 2024 aluminum alloy with load. (Reprinted with permission from Xu, J., and W. Liu. Wear characteristic of in situ synthetic TiB 2 particulate-reinforced Al matrix composite formed by laser cladding. *Wear* 260, no. 4 (2006): 486–492.)

higher load-bearing phases, preventing the base Al-matrix from getting loaded by the wear balls. However, as the loads were increased (17.8 N, 26.7 N, and 36.9 N), the wear behavior changed and aggravated wear was observed with a higher volume fraction of the dispersion phase. It was concluded that an increase in load increased particle–matrix strain, thereby causing pulling out of reinforced particles. The wear debris of pulled out hard particles initiates third-body abrasive wear on the material surface, significantly deteriorating the wear resistance of the material. Hence, careful material selection and processing techniques should be used, as the wear resistance depends not only on particle distribution and volume percentage but also on the applied load.

10.2.2 Laser Surface Engineering of Aluminum Alloy 6061

The 6061 aluminum alloy is one of the most popular commercial alloys of the 6xxx series aluminum alloys. The typical chemical composition of the 6061 alloy is given in Table 10.2 [4]. The 6061 alloy is reported to be the preferred material for applications in general-purpose structural, passenger railroad cars and movable radio telescopes [37]. The distinguishable properties of this alloy are its superior corrosion resistance, stress-corrosion cracking resistance, weldability, and formability compared to other precipitation-hardened alloys. The cost of production for the 6061 alloy is relatively lower than that for the 2024 aluminum alloy because of the greater availability of silicon compared to copper. Hence, the popularity of the 6061 aluminum alloy often supersedes the 2024 aluminum alloy, although the maximum peak aged strength achievable is only about two-thirds the strength of the 2024 aluminum alloy, as observed from Table 10.2 [38].

The 6xxx series alloys are unique because this is the only aluminum alloy system that consists of two principal alloying elements: magnesium and silicon. Interestingly, although magnesium and silicon are also the principal alloying elements of the 4xxx and 5xxx alloy systems, respectively, the 4xxx and 5xxx series are non–heat-treatable. The lack of heat treatability is primarily due to the sharper slope of the solvus lines of aluminum–silicon [39] and aluminum–magnesium [40], which renders these alloys unsuitable for heat treatment. However, the unique interaction of magnesium and silicon in aluminum matrix produces Mg_2Si precipitates and enables the 6xxx alloy system to be heat treatable [41]. As the 6061 aluminum alloy has relatively lower strength in the precipitation-hardened aluminum alloy group, most of the laser surface engineering efforts are focused on improving surface mechanical properties.

10.2.2.1 Laser Surface Melting of the 6061 Aluminum Alloy

Laser surface melting of the 6061 alloy is a rather unexplored unlike laser surface alloying and laser composite surfacing of this alloy. This lack of interest in laser surface melting appears to be due to the inherent high corrosion resistance of the 6061 aluminum alloy, making further laser surface melting operations to improve corrosion resistance redundant. However, the results reported limited attempts on laser surface melting of the 6061 aluminum alloy to be promising, and quite a few breakthroughs in this field can be predicted in the future.

The effect of using a high-power Nd:YAG laser on the electrochemical properties of the 6061 aluminum alloy was investigated by Weinman and coworkers [42]. The laser with a pulse energy of 111 J, a pulse time of 1.2 ms, and a defocused spot diameter of 4 mm was used in an air or helium environment. The surface microstructure after laser surface melting showed existence of three zones reminiscent of weld microstructures: (a) the melt zone, (b) the heat-affected zone, and (c) the base material. The SEM characterization showed cracking at the interfaces of inclusions and matrix in the melt zone. Porosity in the melt structure was also observed. In studies of fracture of the laser surface–melted samples, surface cracks appeared to grow preferentially along the pores. The reason for cracking and porosity in the melt matrix was deduced to be governed by (i) the extremely high cooling rates generated during laser processing and (ii) the limited time for the dissolved hydrogen in the melt to escape. The extraordinary cooling rates generate intense thermal stress during processing, and hydrogen is generated due to reduction of water vapor by the aluminum in the melt. The investigations of Delogu and coworkers on the melt zone of the 6061 aluminum alloy created during laser welding operations also confirmed the presence of hydrogen [43]. The solubility of hydrogen in aluminum was reported to be influenced by the alloying elements present in the melt. For example, while copper decreases hydrogen solubility in aluminum, magnesium increases it. Therefore, the presence of magnesium as a premier alloying element in the 6061 aluminum alloy also enhances porosity formation as magnesium increases hydrogen content in the melt.

The influence of laser surface melting on the corrosion resistance of the 6061 aluminum alloy was investigated by Man and coworkers [34]. The corrosion resistance experiments were performed in a de-aerated solution of 3.5% sodium chloride solution, and temperature was maintained at 23°C. The SEM microstructural investigation revealed intense pitting in the base materials, especially at the electropositive Mg_2Si precipitate and matrix interfaces. The corrosion potential showed a mild improvement of 32 mV in laser surface–melted samples due to the highly uniform surface of the laser-treated samples. However, no notable improvement in pitting potential was observed.

The laser surface melting process is not only a highly efficient technique in generating refined chemically redistributed microstructures but also efficient in improving particle bonding and therefore strength. This feature was successfully used in zirconium diboride reinforced 6061 aluminum alloy to improve its surface strength by Zeng and coworkers. The investigation was primarily targeted to study the effect of laser surface melting on surface substructure evolution and their consequences in the mechanical properties. It was observed that nano-ZrB_2 particles preferred to get distributed along the grain boundaries. During the cooling process, significant thermal stress is generated between the particles and the matrix, resulting in generation of dislocations that contributed in further hardening of the laser surface–melted sample [44].

10.2.2.2 Laser Surface Alloying of Aluminum Alloy 6061

The 6061 aluminum alloy, as mentioned earlier, has superior corrosion resistance, weldability, and formability than 2xxx and 7xxx aluminum alloys. But the alloy's popularity is affected by its comparatively lower strength and wear resistance. Hence, laser surface alloying of this alloy has gained immense interest in recent research to address the comparatively lower hardness and wear resistance.

10.2.2.2.1 *Alloying with Nickel and Titanium*

Alloying of the 6061 alloy with metals using laser surface alloying and their effect in wear and corrosion behavior have been successfully demonstrated by several research groups. Wear response of the 6061 aluminum alloy surface alloyed with nickel and titanium powders was investigated by Man and coworkers using a Nd:YAG laser with a laser power of 1.5 kW [45]. The microstructural investigations revealed the formation of very hard intermetallic aluminides and titanides such as Ni_3Al, Al_3Ti, NiTi, and TiAl. The alloy surfaces after laser surface alloying showed marked improvement in wear resistance compared to the original base alloy, with mean wear rates of 3.75×10^{-5} cm^3 h^{-1} and 21.25×10^{-5} cm^3 h^{-1} for the laser surface alloyed samples and base alloy, respectively. The more than fivefold improvement in wear resistance for the laser surface alloyed surfaces was inferred due to a transition in the wear mechanism. The base alloy, being soft and ductile, undergoes extensive plastic deformation and initiates adhesive wear and thereby accelerates the wear process. However, the formation of very hard intermetallic particles in a ductile matrix during the laser surface alloying operation generates an almost-perfect condition for wear resistance. The wear mechanism is primarily abrasive wear as inferred from the remarkably smoother wear tracks. The hardness also showed considerable improvement from 100 HV to 350 HV. The surface modification was achieved for an appreciable thickness of 1.5 mm. It has been reported that efficiency of power absorption increases with decreasing wavelength, resulting in higher depths of alloying [46]. Therefore, the comparatively higher laser melt thickness is most likely due to the shorter wavelength of Nd:YAG lasers (of 1.06 μm) as compared to CO_2 lasers (10.64 μm). Also, studies of aluminide formation have reported these reactions to be intensely exothermic and can also contribute to increasing the size of the laser melt zone [47,48].

10.2.2.2.2 *Alloying with Nickel and Chromium*

Nickel is reported to form tough and very hard aluminides. Thus, surface alloying of aluminum alloys with nickel can potentially produce excellent surface hardness and wear resistance. However, laser surface alloying has been reported to induce surface porosity due to hydrogen evolution associated with reduction of water during laser surface treatment [42,43]. Hence, the influence of input parameters during laser surface alloying of the 6061 alloy with nickel and chromium as alloying elements on cracking and porosity of resolidified region was investigated by Fu and coworkers [49,50]. The studies involved understanding the influence of laser power on the cracking of the laser-treated zone during the processing and thereby optimizing the laser input parameters such as (a) incident power, (b) scanning velocity, and (c) overlapping ratio. The studies were performed using a Nd:YAG laser and a mixture of commercially available nickel and chromium powders in a ratio of 7:3. The powder mixture was sprayed on the substrate using the plasma spraying technique before laser surface alloying. Cracking and porosity were then analyzed using image processing software. The investigation concluded that each parameter has a very critical influence on cracking and porosity of the laser-treated zone. Low incident power generates a very thin and inconsistent surface layer, while very high power showed extensive porosity and microcracking due to the huge thermal distortion generated by high input thermal energy [49]. Similarly, very low scanning velocity enhanced surface roughness, making the materials highly unsuitable for fatigue endurance applications or applications that require very high surface finish, such as automobile bodies. The surface roughness is generated due to the increased time for power absorption, creating depressions on the surface. Again, very high scanning velocities reduced the overall interaction time of the laser on the substrate, therefore restricting time to allow the entrapped gases to escape, creating defects such as pinholes. The laser track overlapping also influences the cracking of the material because greater overlapping generates greater distortion and hence material becomes susceptible to cracking. Lower track overlapping essentially renders the surface to have highly heterogeneous properties, as laser surface treatment generates a semicircular melt region that narrows at the edges. An analysis of these three parameters and cracking concluded the existence of an optimized regime of input parameters where the most favorable results were observed, as shown in Figure 10.7a [49].

additional particle–matrix interfaces lowers the corrosion resistance of the materials, the cavitation resistance circumvents this issue owing to increased surface hardness. During microstructural examination, it was observed that the laser composite surfaced samples had good metallurgical quality and the modified surfaces were free from defects, such as cracks and porosity. The remarkable improvement in silicon nitride–reinforced samples was determined during microstructural studies and erosion mechanism. The relatively lower volume fraction of reinforcement particles in the wear debris was deduced from the extensive plastic deformation and subsequent plastic rupture of the soft and ductile matrix, while the unusually harder reinforcements were not deformed. The nucleation and growth of the voids were observed to have initiated at the particle–matrix interfaces. Hence, a stronger interface was the key to superior cavitation resistance. The microstructures revealed that while silicon carbide had a largely simple polygonal shape, the silicon nitride particles had a highly irregular clustered shape. The larger particle–matrix interfacial area for silicon nitride particles significantly increases the interfacial bonding, resulting in higher erosion resistance.

10.2.2.3.2 Laser Composite Surfacing with Titanium Carbide (TiC)

The effect of laser composite surfacing of the 6061 alloy with titanium carbide was studied by Katipelli and coworkers [54]. The laser composite surfacing was performed using a Nd:YAG laser with a laser power of 1800 W, a laser scanning velocity of 2 mm/s, and a defocused beam diameter of 0.5 mm. The laser parameters were adjusted to considerably high power regimes to involve surface melting on the laser-irradiated region. The laser irradiations caused partial melting of the ceramic powders, thereby enhancing wetting at the particle–matrix interface. The wetting was further improved by using a mixture of titanium carbide particles and silicon powder in a ratio of 9:1 by weight. Silicon is known to reduce the viscosity of molten aluminum, and this contributes to improving wettability [55]. The intermetallic phase formation after laser composite surfacing was subsequently characterized using XRD techniques and was also analyzed using theoretical thermodynamic calculations. Each of the phases detected (titanium carbide, silicon carbide, titanium aluminides) in XRD was predicted from its thermodynamic driving force by free energy calculations and has been presented in Figure 10.8a [54]. However, the aluminum carbide phase was not detected in characterization studies even though it was predicted as one of the possible phases in a thermodynamic study. The absence of this phase was hypothesized because of the scarcity of adequate free silicon in the melt pool. Although only 8% silicon is required for a stable carbothermic reduction of silicon carbide and the amount of silicon added was 10%, the reaction appeared to be unstable since the entire silicon was not active for reactions. Similarly, only traces of titanium aluminide ($TiAl_3$) form only at low concentration of carbon or very high concentration of titanium, as can be observed from Figure 10.8b [54]. The samples after laser composite surfacing exhibited higher roughness than the base material. The wear resistance of the laser-treated material showed appreciable improvement, underlining the efficiency of the laser composite surfacing treatment for the purpose of wear-intensive applications.

10.2.2.3.3 Composite Surfacing with Ceramic and Metal Mixtures

One recurring feature about composite surfacing of aluminum alloys, as discussed from the works of Man and coworkers [34] and Katipelli and coworkers [54], is the low particle–matrix bonding strength. The weak particle–matrix interfaces act as nucleation sites for voids, and therefore, silicon addition for superior wettability was required to achieve superior interfacial bonding during laser composite surfacing. Hence, various combinations of ceramic particles and metallic additions have been attempted to achieve superior interfacial bonding, which presumably is the most critical factor for improving surface properties.

The investigations on hybrid particle–metal addition during laser composite surfacing were reported by Man and coworkers [56]. Laser composite surfacing of aluminum alloy 6061 was conducted using nickel and aluminum (Ni+Al) and nickel and alumina ($Ni+Al_2O_3$) feed powders.

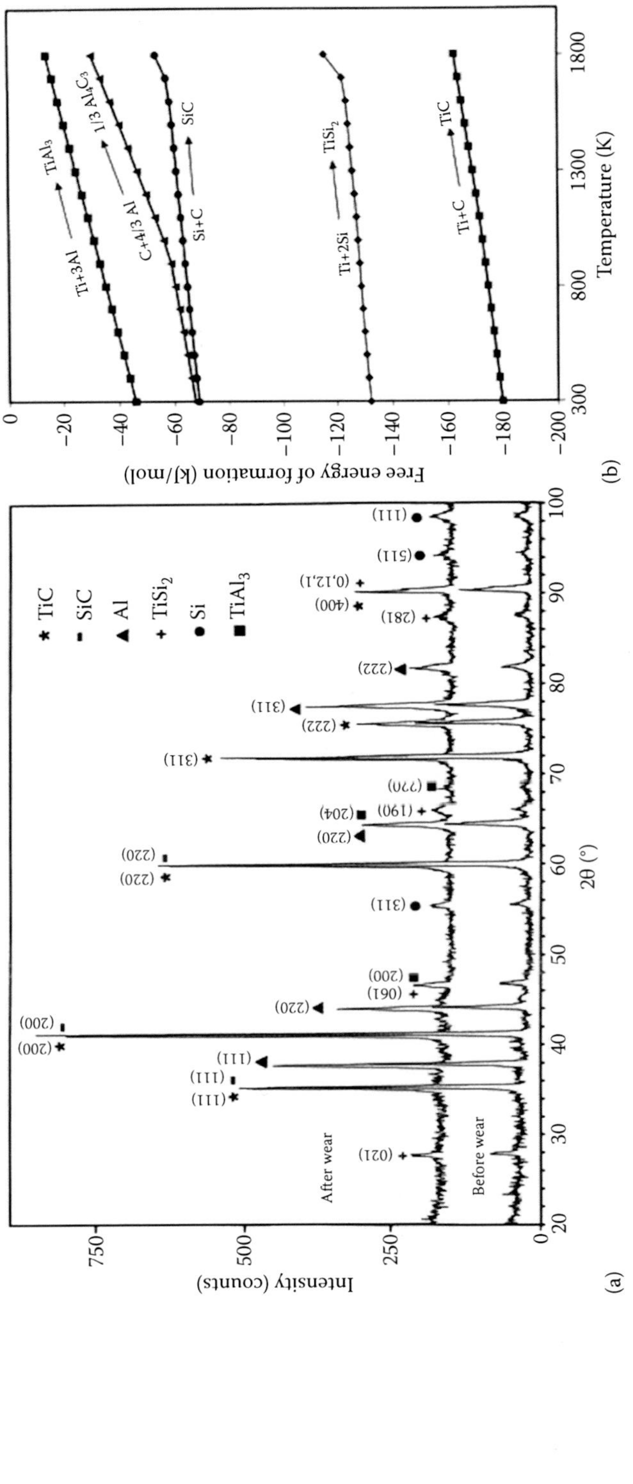

FIGURE 10.8 (a) XRD patterns of TiC reinforced 6061 aluminum alloy, and (b) Gibbs free energy of formation of relevant compounds. (Reprinted with permission from Katipelli, L. R., A. Agarwal, and N. B. Dahotre. Laser surface engineered TiC coating on 6061 Al alloy: Microstructure and wear. *Applied Surface Science* 153, no. 2 (2000): 65–78.)

Also, a 0.3-mm layer of varying compositions of titanium, carbon, tungsten, and tungsten carbide was pasted onto the substrate before laser composite surfacing. A Nd:YAG laser with an incident beam power of 1.6 kW, a spot size of 1.6 mm, a powder feed rate of 24 g/min, a scanning velocity of 20 mm/s, and a 50% track overlap was used for laser composite surfacing. The motivation for the addition of pure elements like titanium, carbon, and tungsten was to facilitate in situ reactions in the melt pool. The samples prepared with nickel and alumina (Ni+Al_2O_3) feed powder and pre-pasted layer of titanium, carbon, and tungsten carbide generated phases such as alumina, titanium carbide, tungsten carbide, and nickel aluminide (Al_3Ni_2) after laser composite surfacing. In samples with identical feed powder composition (nickel and alumina) and pre-pasted layer of tungsten and carbon, presence of alumina, titanium carbide, tungsten carbide, nickel aluminide (Al_3Ni_2 and Al_3Ni), and nickel (Ni) phases was detected. The other set of experiments, where nickel and aluminum (Ni+Al) were used as a feed powder and samples were pre-pasted with tungsten carbide or tungsten and carbon, showed presence of titanium carbide, tungsten carbide, and nickel aluminide (Al_3Ni_2 and Al_3Ni). Mild porosity was also observed in the samples, most likely due to the oxidation of carbon to carbon dioxide. An exceptional improvement in hardness was observed in samples after laser composite surfacing treatment, with hardness of base material increasing from 80 to 800 HV (for nickel and aluminum feed powder, and titanium, carbon, and tungsten pasted layer) and to 1100 HV (for nickel and alumina feed powder, and titanium, carbon, and tungsten carbide pasted layer). The wear response also followed a similar trend, with alumina injected powders showing improved wear resistance compared to aluminum-added samples due to inherent high hardness of alumina. Also, in situ particle formation was explained as a consequence of the high temperatures (about 2350°C) reached due to the laser surface irradiation. The high-temperature results in the dissolution of carbon in titanium, thereby triggering highly exothermic titanium carbide formation reaction. It was also observed that laser composite surfacing using titanium, carbon, and tungsten carbide (instead of uncombined titanium, carbon, and tungsten) resulted in better properties. This seemed to be due to the mildly exothermic nature of tungsten carbide formation reaction compared to titanium carbide reaction. Hence, tungsten carbide formation is triggered only after very high temperature is reached for titanium carbide reaction, thus restricting in situ tungsten carbide formation.

The applicability of laser composite surfacing using molybdenum powder and tungsten carbide as reinforcement particles on the 6061 aluminum alloy was studied by Chong and coworkers [57]. Metals such as nickel and molybdenum are noted for their ease of formation of corresponding high melting metal aluminides. Also, interfacial bonding of particle and matrix is particularly favored when the particle–matrix interface is diffused. Because of the considerably higher melting point of the metal aluminides, the difference in melting point of tungsten carbide particles and aluminides of molybdenum is significantly smaller than the difference in melting point of aluminum and tungsten carbide particles. Therefore, it was predicted that the similar melting points will improve simultaneous melting of the phases to yield continuous and stronger metallurgical bonding among the microstructural phases.

The study was performed using a Nd:YAG laser with an incident power of 1.4 kW, a scanning velocity of 17 mm/s, and a defocused beam diameter of 1 mm. The overlapping on a laser track was maintained at 50%, and surface oxidation was avoided by use of co-axial and side jet of argon gas. The mixture of molybdenum and/or tungsten carbide powders was pre-placed using 4% poly-vinyl alcohol to a thickness of 1 mm on the substrate. The phases characterized in the melt zones after laser composite surfacing were molybdenum aluminides (Al_5Mo, $Al_{22}Mo_5$ and $Al_{17}Mo_4$) in pure molybdenum-alloyed samples. For the samples alloyed with both molybdenum and tungsten carbide, additional phases such as molybdenum carbide (Mo_2C), aluminum carbide (Al_4C_3), tungsten aluminide (WAl_{12}), and tungsten carbide (WC) were identified. A considerably diffused interface, inferring the interface to have extensive interfacial bonding, was observed, as reported in Figure 10.9a [57]. Excellent metallurgical soundness without any cracking and porosity was observed from the microstructural investigation. Also, a drastic reduction in the melt zone thickness was noted in tungsten carbide and molybdenum alloyed samples. While the tungsten carbide and molybdenum

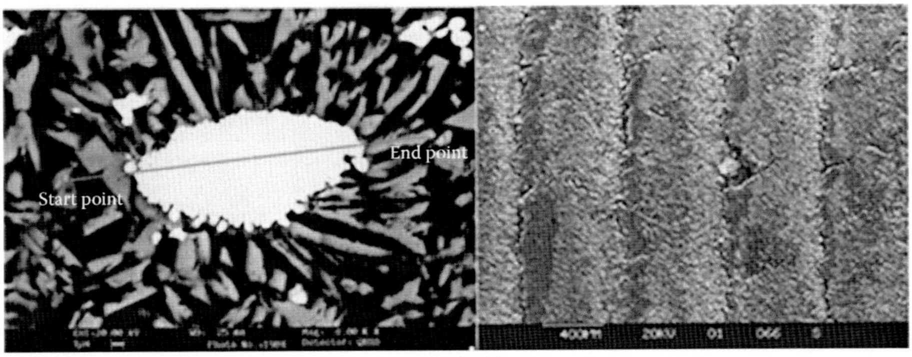

(a) (b)

FIGURE 10.9 (a) SEM micrograph of 6061 alloy laser melted with 40% WC and 60% Mo, and (b) cracking at work surfaces of 6061 alloy laser melted with WC. (Reprinted with permission from Chong, P. H., H. C. Man, and T. M. Yue. Microstructure and wear properties of laser surface-cladded Mo–WC MMC on AA6061 aluminum alloy. *Surface and Coatings Technology* 145, no. 1 (2001): 51–59.)

alloyed samples showed a melt zone thickness of 150–250 μm, a melt zone thickness as high as 430 μm was observed in the pure molybdenum alloyed samples. The analysis of such a substantial influence in melt zone thickness was conjectured due to the result of the significant reduction in thermal conductivity of the 6061 aluminum substrate on addition of tungsten carbide powders. The characterization also revealed that the matrix was primarily an aluminum–silicon alloy system; a plausible explanation might be due to the selective vaporization of low–boiling point magnesium similar to the results of Li and coworkers [26] in the 2024 aluminum alloy. Also, similar to the experience reported by Chuang and coworkers [53] with laser surface alloying of the 6061 aluminum alloy with chromium and cobalt, inhomogeneous distribution of reinforcement phases due to disparity in density was observed. The lower region of the melt zone was observed to have a higher concentration of heavier tungsten carbide particles than the upper region. A relation of the weight ratio of molybdenum and tungsten carbide added to the melt pool and the particle distribution was noted. The homogeneity of particle distribution was observed to increase with an increase in tungsten carbide volume fraction. Hence, apparently the melt zone depth decreases with increasing tungsten carbide particles in the melt, increasing the cooling rate. The higher cooling rate renders lower settling time to the particles, and thus, particle homogeneity increases with increasing tungsten carbide volume. The wear tests on the samples were performed using a pin-on disc abrasive wear apparatus. It was observed that the wear resistance improved with tungsten carbide ratio in the melt. The cause for this improvement in the wear was reasoned due to the superior load-bearing capability of the hard tungsten carbide and intermetallic phases. However, intense plastic deformation of the matrix was observed in the tungsten carbide–rich samples, which triggered microvoid nucleation at the particle–matrix interfaces. Cracking at the worn surfaces was also noted at the overlapped regions of the tracks, due to their lower melt zone depth resulting in lower strength, as observed in Figure 10.9b. Thus, using pure tungsten carbide for composite surfacing is not a recommendable practice, as their lower melt zone depth makes the modified layer more vulnerable to be worn off due to cracking in the overlapped regions.

A summary of different phases formed during laser surface alloying and laser composite surfacing of precipitation-hardened aluminum alloys is presented in Table 10.3.

10.2.3 Laser Surface Engineering of the 7075 Aluminum Alloy

The 7075 aluminum alloy is one of the popular most materials of the 7xxx series of aluminum alloys. The 7075 alloy is particularly popular in aerospace industries for the past 40 years [58,59].

TABLE 10.3
Summary of Phases Reported due to Laser Surface Alloying and Laser Composite Surfacing of Precipitation-Hardened Aluminum Alloys

Alloy	Constituents Added	Phases Detected
2024	Fe-coated B, pure Ti, pure Al [36]	TiB_2, Al_3Fe, and Al_3Ti
6061	Ni and Ti [45]	Ni_3Al, Al_3Ti, NiTi, and TiAl
	Ni and Cr [41,42]	Ni_3Al
	Co and Cr [45]	Al_9Co_2, $Al_{13}Co_4$, Al_7Co
	TiC and Si [46]	TiC, SiC, $TiAl_2$, $TiAl_3$
	Ti, C, TiC, W, WC, Ni, Al, Al_2O_3[48]	TiC, WC, Al_3Ni_2, Al_3Ni, Al_2O_3, Ni
	Mo and WC [49]	Al_5Mo, $Al_{22}Mo_5$, $Al_{17}Mo_4$, Mo_2C, Al_4C_3, WAl_2, WC
7075	SiO_2,Fe_2O_3, CuO, Cr_2O_3, and TiO_2 [58]	α-Al_2O_3, γ-Al_2O_3, corresponding metal aluminides

The typical composition of the 7075 aluminum alloy is given in Table 10.2 [4]. The 7xxx alloys are reported to give the highest strength among the aluminum wrought alloy family [60]. This is one of the most important factors in alloy selection requiring very high strength-to-weight ratio. However, despite their superior strength, the 7075 aluminum alloy is often reported with poor corrosion resistance and stress-corrosion cracking resistance [61]. Therefore, laser surface melting can potentially be one of the most appropriate solutions to address the comparatively higher corrosion susceptibility of the 7075 aluminum alloy. Hence, several scientific investigations have been performed to establish the efficiency of laser surface engineering in this alloy. It should be mentioned that laser surface engineering of the 7075 aluminum alloy is challenging due to the tendency of the alloy to undergo hot shortness [62].

10.2.3.1 Laser Surface Melting of the 7075 Aluminum Alloy

The effect of laser surface melting on corrosion fatigue of the AA 7075-T651 aluminum alloy was investigated by Chan and coworkers [59]. The laser treatment was performed using an excimer laser with energy fluence in the range of 3.3–15.4 J/cm^2, a beam spot size of 0.6 mm, and a track overlapping ratio of 50%. The pulse duration of 25 ns and laser pulse frequency of 10 Hz were used for laser surface melting. The as-received and laser surface–melted samples were subjected to immersion testing [immersed in a solution of 57 g sodium chloride and 10 ml hydrogen peroxide (H_2O_2) in 1 L of distilled water for 48 h] and corrosion fatigue tests in 3.5% sodium chloride solution. The immersion treatment was performed to aggravate pitting and thereby study the effect of corrosion pits in fatigue strength. The immersion studies revealed that while very well defined corrosion pits were observed in as-received samples, pitting was highly constrained in laser-treated samples. The pitting in as-received samples was initiated along the interfaces of second-phase particles (both cathodic Al–Cu–Fe–Zn and anodic Al–Mg–Zn particles). The pitting corrosion advanced along the stringers of these particles to a depth of about 250 μm. It was observed that the elimination of second-phase particles due to remelting in laser surface–melted samples resulted in the formation of continuous passivation film, rendering them better pitting resistance. A very interesting and unusual observation was noted in the fatigue and corrosion fatigue tests. While the fatigue strength of the laser surface–melted samples manifested mild deterioration, the corrosion fatigue strength increased by 30%–50% for the laser surface–melted samples. This behavior was attributed to the highly rough surface associated with laser processing. In addition, surface defects like porosity act as fatigue crack nucleation sites, reducing fatigue strength. However, in corrosion fatigue experiments, the sharp corrosion pits in base materials supersede the effects of surface porosity and rough surface, as the refined grain structure and highly homogeneous surface microstructure significantly improves corrosion resistance of these materials. But an extraordinary rise in corrosion fatigue strength, more than two orders of magnitude higher than base material corrosion fatigue endurance, was observed in shot-peened laser surface–melted sample. The huge improvement can be extrapolated from the fact that shot peening incorporated compressive stress on surface as well reducing surface porosity, considered as the most critical features in fatigue endurance strength.

The 7075 aluminum alloy, as mentioned earlier, is frequently accompanied with poor stress-corrosion cracking resistance. Accordingly, the immense potential of laser surface melting in improving corrosion resistance naturally increases expectations of enhanced stress-corrosion cracking strength from laser-treated materials. The effect of laser surface melting on stress-corrosion cracking of the 7075 aluminum alloy was reported by Yue and coworkers [63]. Laser surface melting was performed on the 7075-T651 alloy using a KrF laser with the pulsed laser beam defocused to a diameter of 0.8 mm, a pulse duration of 25 ns, and a frequency of 20 Hz. The stress-corrosion cracking test was performed by immersing the specimens in a 3.5% sodium chloride solution for 20 and 30 days. The characterization of mechanical and electrochemical properties was conducted using a three-point bend test and electrochemical impedance tests, respectively. The base material showed intense grain boundary corrosion, and about 1-mm crack length was observed in 30-day immersed samples. However, the laser-treated materials' corrosion response was highly restricted and concentrated mainly in the grain interior instead of boundaries. The considerable shift in corrosion rate was explained from copper mapping using electron back-scattered diffraction characterization that showed intense precipitation of coarse precipitates at the grain boundaries in base material [63]. Copper has been reported to improve corrosion resistance in 7xxx alloys [64]. Hence, not only does precipitation of copper increase cathodic phases but also, the depletion of copper from the matrix imparts a synergistic effect to aggravate grain boundary corrosion in base material. However, the highly non-equilibrium solidification almost completely negates migration and eventual precipitation of copper at the grain boundaries in laser surface–melted samples. In contrast to the base material, the TEM study revealed the absence of the highly electropositive η precipitates at grain boundaries in laser-treated samples, thus reducing the preferential corrosion site at the grain boundaries. The influence of the shift of mechanism from grain boundary corrosion (as-received samples) to more uniform corrosion (laser surface–melted samples) in passivation was demonstrated using the electrochemical impedance spectroscopy (EIS) studies. In the EIS studies, more than five times increase in surface resistance was observed for the laser-treated samples. This was attributed to the presence of a uniform passive layer and lack of unaffected cathodic phases in the surface in case of the laser-treated samples [65]. The analogous studies using Nd:YAG lasers in untreated, nitrogen, and air medium by Yue and coworkers also exhibited considerable improvement in electrochemical properties due to similar mechanisms, as can be observed from Figure 10.10a and b [65]. However, since the irradiation power of the Nd:YAG lasers was considerably higher as compared to KrF resulted, significant improvement of melt zone thickness was observed in samples treated with Nd: YAG lasers.

The lasers that are commonly reported to be used for laser surface engineering of aluminum alloy are the Nd:YAG, CO_2, and excimer lasers. However, investigation of the effect of high-power diode laser for laser surface melting of the 7075-T6 alloy (rather unconventional lasers for laser surface engineering applications) was reported by Benedetti and coworkers [66]. The high-power diode lasers are noted for their shorter wavelengths, reported to be as low as 405 nm. Since the power transfer efficiency increases with decreasing wavelengths, more energy per unit area can be achieved with semiconductor lasers. Laser surface melting was performed using high-power diode lasers with a power of 3.3 kW, and the spot size was defocused to 3 × 3 mm. The surface temperature was controlled and varied from 720°C to 780°C by controlling laser power (close loop control). The scanning velocities were varied from 2 to 6 mm/s, and surface oxidation was prevented by using argon as a shield gas. The influence of change in surface temperature on the melt zone thickness was quantified. The experiments revealed that increasing surface temperature from 720°C to 780°C increased melt zone thickness by 58%. However, decreasing the laser scanning velocity from 6 to

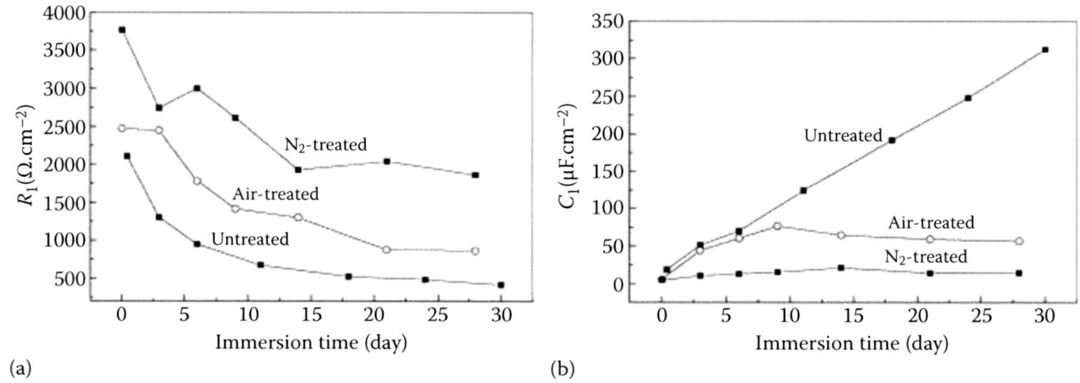

FIGURE 10.10 Variation of (a) surface film resistance and (b) capacitance with immersion time for as-received and laser surface–treated 7075 aluminum alloy specimen. (Reprinted with permission from Yue, T. M., L. J. Yan, and C. P. Chan. Stress corrosion cracking behavior of Nd: YAG laser-treated aluminum alloy 7075. *Applied Surface Science* 252, no. 14 (2006): 5026–5034.)

2 mm/s only mildly increased (by about 6.4%) surface melt zone thickness. This behavior was inferred to be governed by the high thermal conductivity of aluminum, and hence, decreasing scanning velocity essentially increases time for the bulk material to get heated up instead of localized heating of the surface. Microstructural and XRD analysis of the laser-treated surface revealed existence of various intermetallic phases such as $Al_{18}Mg_3Cr$, $MgZn_2$, $Mg(ZnCuAl)_2$, AlCuMg, $Fe_{24}Cu_4Al_{72}$, and Mg_2Si. The microstructures were reminiscent of casting microstructures with microsegregation inside the dendrites as detected in energy-dispersive x-ray spectroscopy characterization. The second-phase particle density was observed to increase significantly along the grain boundaries due to very low solute concentration in the core. Also, existence of a unique partial melted zone was noted, wherein a part of the grains appeared to have no fusion and were an extension of the untreated material's grains. However, the region of the grains closer to the surface melt region resolidified to dendritic structures.

10.2.3.2 Laser Surface Alloying and Laser Composite Surfacing of the 7075 Aluminum Alloy

The inherent high hardness and wear resistance are probably the reasons for the limited interest in laser surface alloying and laser composite surfacing of the 7075 aluminum alloy as compared to the more frequently reported laser surface melting of the 7075 aluminum alloy. However, developing laser composite surfacing of the 7075 aluminum alloy with in situ aluminothermic reactions was reported by Huang and coworkers [67]. In the investigation, mixture of silica and metal oxides such as iron (III) oxide (Fe_2O_3), copper (II) oxide (CuO), chromium oxide (Cr_2O_3), and titanium dioxide (TiO_2) was injected in the laser melt pool. The composite surfacing was performed using a CO_2 laser with a spot diameter of 2 mm, a scanning velocity of 25 mm/s, a laser power of 1800 W, and track overlapping of 50%. The presence of alumina (Al_2O_3) particles as well as intermetallic particles was reported in every laser composite surfaced sample. The microhardness of the laser-treated zone revealed similar values, independent of the metal oxide used in each laser composite surfaced sample, and was in the range of $HV_{0.1}$1500–2300. Evidently, the extraordinary affinity for oxygen of aluminum coupled with intense heating due to laser irradiation triggers aluminothermic reactions in every sample. The nascent reduced metal subsequently reacts with the aluminum of the matrix to form aluminides. The high surface hardness was surmised to be contributed primarily by the hard alumina (α-Al_2O_3 and γ-Al_2O_3) particles, and the corresponding metal aluminides only slightly contributed in the overall surface hardness. The wear response showed similar responses as reported in the laser surface alloying of the 2024 aluminum alloy by Xu and Liu [36]. The wear mechanism was identified to be primarily adhesive wear of the highly deformable base alloy at lower loads. However, abrasive wear dominated the wear mechanism at the higher loads due to abrasion from cracking and removal of hard alumina (Al_2O_3) particles.

10.3 CONCLUSION

Laser surface engineering approaches such as laser surface melting, laser surface alloying, and laser composite surfacing have been extensively used for improving the mechanical and electrochemical properties of precipitation-hardened aluminum alloys, such as (a) 2024, (b) 6061, and (c) 7075. All of these approaches involve rapid surface melting (with or without introduction of additional elements and hard particles) and rapid solidification. The improvement in electrochemical and mechanical properties is attributed to several effects of laser interaction with the alloys. Effects such as grain refinement, elemental segregation, formation of intermetallic phases, and in situ reactions have been observed during laser surface engineering of precipitation-hardened aluminum alloy. For relatively stronger precipitation-hardened aluminum alloys such as the 2024 and 7075 alloys, most of the laser surface engineering efforts were focused on surface melting or alloying to improve the corrosion resistance. It has been observed that laser surface melting of the 2024 aluminum alloy causes segregation of copper to the grain boundaries (making these areas more cathodic) and also selective vaporization of magnesium. These effects not only shift the corrosion mechanism from severe localized pitting (for as-received) to uniform corrosion (laser surface melted) but also reduce the volume of precipitates, which are susceptible to corrosion. For the 7075 aluminum alloy, laser surface melting results in elimination of precipitates (due to remelting) such as Al–Cu–Fe–Zn and Al–Mg–Zn and favors the formation of continuous passivation film. As the 6061 aluminum alloy exhibits superior corrosion resistance but relatively lower strength than the precipitation-hardened aluminum alloy system, most of the laser surface engineering investigations have been focused on improving surface mechanical properties. Laser surface alloying of the 6061 aluminum alloy with metals such as nickel, titanium, chromium, and cobalt results in the formation of intermetallic phases with aluminum. Significant efforts have also been made to introduce hard phases such as silicon carbide, silicon nitride, and titanium carbide directly in the melt during laser composite surfacing. Both laser surface alloying and composite surfacing of the 6061 alloy have demonstrated effectiveness in improving hardness and wear resistance of the surfaces. While the effectiveness of laser surface engineering approaches in improving surface properties of precipitation-hardened aluminum has been demonstrated, significant experimental and computational effects need to be directed toward delineating the roles of grain refinement, compositional redistribution, precipitation, and alloying on corrosion and wear behavior of these alloys. Laser surface engineering of aluminum alloys is still very challenging because of the high reflectivity of the alloys. Also, optimization of laser processing parameters to form defect-free modified surfaces is often involved and time consuming. Development of computational models taking into account thermal, fluid flow, and compositional effects to predict the evolution of microstructure and defects will significantly advance the state of the art in this area of laser surface engineering of precipitation-hardened aluminum alloys.

ACKNOWLEDGMENT

This material is based on work supported by the National Science Foundation under Grant No. CMMI-1149079.

REFERENCES

1. Alcoa. 2017. History/timeline. http://www.alcoa.com/global/en/who-we-are/history/default.asp
2. USGS Minerals Information. 2017. Aluminum statistics and information. https://minerals.usgs.gov/minerals/pubs/commodity/aluminum/

3. C.A.E.I.I. CEIC News Alert. 2017. U.S. Geological Survey: Non-ferrous metals production statistics. http://www.ceicdata.com/en/press/ceic-newslert-us-geological-survey-non-ferrous-metals-production-statistics.
4. Van Horn, K. R. 1967. *Aluminum*. Vol. 1. Metals Park, OH: American Society for Metals.
5. Pohl, M., O. Storz, and T. Glogowski. Effect of intermetallic precipitations on the properties of duplex stainless steel. *Materials Characterization* 58, no. 1 (2007): 65–71.
6. Serajzadeh, S., and H. Sheikh. Investigation into occurring dynamic strain aging in hot rolling of AA5083 using finite elements and stream function method. *Materials Science and Engineering: A* 486, no. 1 (2008): 138–145.
7. Chipman, J. Thermodynamics and phase diagram of the Fe–C system. *Metallurgical and Materials Transactions B* 3.1 (1972): 55–64.
8. Polmear, I. J. Light alloys, metallurgy of the light alloys. *Metallurgy and Materials Science*, Arnold, Great Britain (1995): 168–195.
9. Blanc, C., S. Gastaud, and G. Mankowski. Mechanistic studies of the corrosion of 2024 aluminum alloy in nitrate solutions. *Journal of the Electrochemical Society* 150, no. 8 (2003): B396–B404.
10. Pawel, R. E., G. L. Yoder, D. K. Felde, B. H. Montgomery, and M. T. McFee. The corrosion of 6061 aluminum under heat transfer conditions in the ANS corrosion test loop. *Oxidation of Metals* 36, no. 1–2 (1991): 175–194.
11. Minoda, T., and H. Yoshida. Effect of grain boundary characteristics on intergranular corrosion resistance of 6061 aluminum alloy extrusion. *Metallurgical and Materials Transactions A* 33, no. 9 (2002): 2891–2898.
12. Venugopal, A., R. Panda, S. Manwatkar, K. Sreekumar, L. Ramakrishna, and G. Sundararajan. Effect of microstructure on the localized corrosion and stress corrosion behaviours of plasma-electrolytic-oxidation-treated AA7075 aluminum alloy forging in 3.5 wt.% NaCl solution. *International Journal of Corrosion* 2012 (2012).
13. Mazurkiewicz, B., and A. Piotrowski. The electrochemical behaviour of the Al2Cu intermetallic compound. *Corrosion Science* 23, no. 7 (1983): 697–707.
14. Mabhali, L. A. B., N. Sacks, and S. Pityana. Three body abrasion of laser surface alloyed aluminium AA1200. *Wear* 290 (2012): 1–9.
15. Abboud, J. H., and D. R. F. West. Laser surface alloying of titanium with aluminium. *Journal of Materials Science Letters* 9, no. 3 (1990): 308–310.
16. Lekala, M. B., J. W. Van Der Merwe, and S. L. Pityana. Laser surface alloying of 316L stainless steel with Ru and Ni mixtures. *International Journal of Corrosion* 2012 (2012).
17. Trdan, U., M. Skarba, and J. Grum. Laser shock peening effect on the dislocation transitions and grain refinement of Al–Mg–Si alloy. *Materials Characterization* 97 (2014): 57–68.
18. Wang, J. T., Y. K. Zhang, J. F. Chen, J. Y. Zhou, M. Z. Ge, Y. L. Lu, and X. L. Li. Effects of laser shock peening on stress corrosion behavior of 7075 aluminum alloy laser welded joints. *Materials Science and Engineering: A* 647 (2015): 7–14.
19. Mishra, S., and V. Yadava. Modeling and optimization of laser beam percussion drilling of thin aluminum sheet. *Optics & Laser Technology* 48 (2013): 461–474.
20. Ren, N., L. Jiang, D. Liu, L. Lv, and Q. Wang. Comparison of the simulation and experimental of hole characteristics during nanosecond-pulsed laser drilling of thin titanium sheets. *The International Journal of Advanced Manufacturing Technology* 76, no. 5–8 (2015): 735–743.
21. Thijs, L., K. Kempen, J.-P. Kruth, and J. Van Humbeeck. Fine-structured aluminium products with controllable texture by selective laser melting of pre-alloyed AlSi10Mg powder. *Acta Materialia* 61, no. 5 (2013): 1809–1819.
22. Read, N., W. Wang, K. Essa, and M. M. Attallah. Selective laser melting of AlSi10Mg alloy: Process optimisation and mechanical properties development. *Materials & Design (1980–2015)* 65 (2015): 417–424.
23. Liao, C.-M., J. M. Olive, M. Gao, and R. P. Wei. In-situ monitoring of pitting corrosion in aluminum alloy 2024. *Corrosion* 54, no. 6 (1998): 451–458.
24. Augustin, C., E. Andrieu, C. Blanc, G. Mankowski, and J. Delfosse. Intergranular corrosion of 2024 alloy in chloride solutions. *Journal of the Electrochemical Society* 154, no. 11 (2007): C637–C644.
25. Keddam, M., C. Kuntz, H. Takenouti, D. Schustert, and D. Zuili. Exfoliation corrosion of aluminium alloys examined by electrode impedance. *Electrochimica Acta* 42, no. 1 (1997): 87–97.
26. Li, R., M. G. S. Ferreira, A. Almeida, R. Vilar, K. G. Watkins, M. A. McMahon, and W. M. Steen. Localized corrosion of laser surface melted 2024-T351 aluminium alloy. *Surface and Coatings Technology* 81, no. 2–3 (1996): 290–296.

27. Liu, Z., P. H. Chong, A. N. Butt, P. Skeldon, and G. E. Thompson. Corrosion mechanism of laser-melted AA 2014 and AA 2024 alloys. *Applied Surface Science* 247, no. 1 (2005): 294–299.
28. Liu, Z., P. H. Chong, P. Skeldon, P. A. Hilton, J. T. Spencer, and B. Quayle. Fundamental understanding of the corrosion performance of laser-melted metallic alloys. *Surface and Coatings Technology* 200, no. 18 (2006): 5514–5525.
29. Zimmermann, M., M. Carrard, and W. Kurz. Rapid solidification of Al–Cu eutectic alloy by laser remelting. *Acta Metallurgica* 37, no. 12 (1989): 3305–3313.
30. Sutton, M. A., B. Yang, A. P. Reynolds, and R. Taylor. Microstructural studies of friction stir welds in 2024-T3 aluminum. *Materials Science and Engineering: A* 323, no. 1 (2002): 160–166.
31. Sutton, M. A., B. Yang, A. P. Reynolds, and J. Yan. Banded microstructure in 2024-T351 and 2524-T351 aluminum friction stir welds: Part II. Mechanical characterization. *Materials Science and Engineering: A* 364, no. 1 (2004): 66–74.
32. Embuka, D., A. E. Coy, C. A. Hernandez-Barrios, F. Viejo, and Z. Liu. Thermal stability of excimer laser melted films formed on the AA2024-T351 aluminium alloy: Microstructure and corrosion performance. *Surface and Coatings Technology* 313 (2017): 214–221.
33. Kadolkar, P., and N. B. Dahotre. Variation of structure with input energy during laser surface engineering of ceramic coatings on aluminum alloys. *Applied Surface Science* 199, no. 1 (2002): 222–233.
34. Man, H. C., C. T. Kwok, and T. M. Yue. Cavitation erosion and corrosion behaviour of laser surface alloyed MMC of SiC and Si 3 N 4 on Al alloy AA6061. *Surface and Coatings Technology* 132, no. 1 (2000): 11–20.
35. Eustathopoulos, N., M. G. Nicholas, and B. Drevet, eds. *Wettability at High Temperatures*. Vol. 3. Elsevier, 1999.
36. Xu, J., and W. Liu. Wear characteristic of in situ synthetic TiB 2 particulate-reinforced Al matrix composite formed by laser cladding. *Wear* 260, no. 4 (2006): 486–492.
37. American Society for Metals. *Metals Handbook*. Vol. 3. The Society, 1967.
38. American Society for Metals, and ASM International. 1982. *ASM Handbook*. Materials Park, OH: ASM International.
39. Murray, J. L., and A. J. McAlister. The Al–Si (aluminum–silicon) system. *Journal of Phase Equilibria* 5, no. 1 (1984): 74–84.
40. Murray, J. L. The Al–Mg (aluminum–magnesium) system. *Journal of Phase Equilibria* 3, no. 1 (1982): 60–74.
41. Effenberg, G. 2004. *Light Metal Systems*. Berlin: Springer.
42. Weinman, L. S., C. Kim, T. R. Tucker, and E. A. Metzbower. Characteristics of laser surface melted aluminum alloys. *Applied Optics* 17, no. 6 (1978): 906–910.
43. Delogu, P., and S. Tosto. Auger electron spectroscopy on a laser welded 6061 Al alloy. *Materials Letters* 20, no. 1–2 (1994): 11–17.
44. Zeng, Y., Y. Chao, Z. Luo, Y. Huang, Y. Cai, L. Deng, W. Guo, Y. Lei, T. Lu, and Z. Wang. Surface substructure and properties of ZrB2p/6061Al composite treated by laser surface melting under extreme cooling conditions. *High Temperature Materials and Processes* 36, no. 1 (2017): 69–77.
45. Man, H. C., S. Zhang, and F. T. Cheng. Improving the wear resistance of AA 6061 by laser surface alloying with NiTi. *Materials Letters* 61, no. 19 (2007): 4058–4061.
46. DeWitt, D. P. *Thermal Radiative Properties: Metallic Elements and Alloys*. IFI/Plenum, 1970.
47. Nayak, S., L. Riester, H. M. Meyer, and N. B. Dahotre. Micromechanical properties of a laser-induced iron oxide–aluminum matrix composite coating. *Journal of Materials Research* 18, no. 4 (2003): 833–839.
48. Nayak, S., H. Wang, E. A. Kenik, I. M. Anderson, and N. B. Dahotre. Observation of exothermic reaction during laser-assisted iron oxide coating on aluminum alloy. *Materials Science and Engineering: A* 390, no. 1 (2005): 404–413.
49. Fu, Y., A. W. Batchelor, Y. Gu, K. A. Khor, and H. Xing. Laser alloying of aluminum alloy AA 6061 with Ni and Cr. Part 1. Optimization of processing parameters by x-ray imaging. *Surface and Coatings Technology* 99, no. 3 (1998): 287–294.
50. Fu, Y., and A. W. Batchelor. Laser alloying of aluminum alloy AA 6061 with Ni and Cr. Part II. The effect of laser alloying on the fretting wear resistance. *Surface and Coatings Technology* 102, no. 1 (1998): 119–126.
51. Helander, T., and O. Tolochko. An experimental investigation of possible B2-ordering in the Al–Cr system. *Journal of Phase Equilibria* 20, no. 1 (1999): 57.
52. Venezia, A. M., and C. M. Loxton. Oxidation of Ni3Al at low and high oxygen pressures. *Surface and Interface Analysis* 11, no. 6–7 (1988): 287–290.

53. Chuang, Y.-C., S.-C. Lee, and H.-C. Lin. Wear characteristics of a laser surface alloyed Al–Mg–Si with Co alloy powder. *Materials Transactions* 47, no. 4 (2006): 1140–1144.
54. Katipelli, L. R., A. Agarwal, and N. B. Dahotre. Laser surface engineered TiC coating on 6061 Al alloy: Microstructure and wear. *Applied Surface Science* 153, no. 2 (2000): 65–78.
55. Pech-Canul, M. I., R. N. Katz, M. M. Makhlouf, and S. Pickard. The role of silicon in wetting and pressureless infiltration of SiCp preforms by aluminum alloys. *Journal of Materials Science* 35, no. 9 (2000): 2167–2173.
56. Man, H. C., Y. Q. Yang, and W. B. Lee. Laser induced reaction synthesis of TiC+WC reinforced metal matrix composites coatings on Al 6061. *Surface and Coatings Technology* 185, no. 1 (2004): 74–80.
57. Chong, P. H., H. C. Man, and T. M. Yue. Microstructure and wear properties of laser surface-cladded Mo–WC MMC on AA6061 aluminum alloy. *Surface and Coatings Technology* 145, no. 1 (2001): 51–59.
58. Santner, J. S. A study of fracture in high purity 7075 aluminum alloys. *Metallurgical Transactions A* 9, no. 6 (1978): 769–779.
59. Chan, C. P., T. M. Yue, and H. C. Man. The effect of excimer laser surface treatment on the pitting corrosion fatigue behaviour of aluminium alloy 7075. *Journal of Materials Science* 38, no. 12 (2003): 2689–2702.
60. Polmear, I. J., and M. J. Couper. Design and development of an experimental wrought aluminum alloy for use at elevated temperatures. *Metallurgical and Materials Transactions A* 19, no. 4 (1988): 1027–1035.
61. Maitra, S., and G. C. English. Mechanism of localized corrosion of 7075 alloy plate. *Metallurgical and Materials Transactions A* 12, no. 3 (1981): 535–541.
62. Kaufmann, N., M. Imran, T. M. Wischeropp, C. Emmelmann, S. Siddique, and F. Walther. Influence of process parameters on the quality of aluminium alloy EN AW 7075 using selective laser melting (SLM). *Physics Procedia* 83 (2016): 918–926.
63. Yue, T. M., C. F. Dong, L. J. Yan, and H. C. Man. The effect of laser surface treatment on stress corrosion cracking behaviour of 7075 aluminium alloy. *Materials Letters* 58, no. 5 (2004): 630–635.
64. Sarkar, B., M. Marek, and E. A. Starke. The effect of copper content and heat treatment on the stress corrosion characteristics of Ai–6Zn–2Mg–X Cu alloys. *Metallurgical and Materials Transactions A* 12, no. 11 (1981): 1939–1943.
65. Yue, T. M., L. J. Yan, and C. P. Chan. Stress corrosion cracking behavior of Nd: YAG laser-treated aluminum alloy 7075. *Applied Surface Science* 252, no. 14 (2006): 5026–5034.
66. Benedetti, A., M. Cabeza, G. Castro, I. Feijoo, R. Mosquera, and P. Merino. Surface modification of 7075-T6 aluminium alloy by laser melting. *Surface and Interface Analysis* 44, no. 8 (2012): 977–981.
67. Huang, K., X. Lin, C. Xie, and T. M. Yue. Microstructure and wear behaviour of laser-induced thermite reaction Al_2O_3 ceramic coating on AA7075 aluminum alloy. *Journal of Materials Science and Technology* 23, no. 2 (2007): 201.

11 Overview of Pulsed Electron Beam Treatment of Light Metals

Advantages and Applications

Subramanian Jayalakshmi, Ramachandra Arvind Singh, Sergey Konovalov, Xizhang Chen, and T. S. Srivatsan

CONTENTS

11.1 INTRODUCTION

Aluminum (Al), magnesium (Mg), and titanium (Ti) are classified as lightweight structural metals having a density of 2.7, 1.74, and 4.2 g/cc, respectively. In recent years, commensurate with advances in technology, the light metals are a preferred choice for a variety of structural and functional applications in automotive, aerospace/space, sports, electronics, and biomedical industries. Traditionally, cast iron and steel were the preferential choice for use in structural applications. With an observable advancement in production/extraction methods, these lightweight metals have gradually replaced cast iron and steel for both components and products. For example, in a car, replacement of iron and/or steel engine by magnesium or aluminum can lead to a 22% to 70% reduction in weight. Frame replacement of a car seat made from steel by an alloy of magnesium can reduce weight by as much as 64% (Tharumarajah and Koltun 2007). Since a reduction in density favors a reduction in weight, replacing traditionally used metals by lightweight counterparts will result in an enhancement in specific strength [σ/ρ], specific stiffness [E/ρ] properties, and concurrent improvement in both mileage and fuel efficiency. In Section 11.1.1, properties and applications of aluminum, magnesium, and titanium light metals and their alloy counterparts are presented.

11.1.1 Properties and Applications of Lightweight Structural Materials

11.1.1.1 Aluminum and Its Alloys

Aluminum has a face-centered cubic structure. It has a density of 2.7 g/cc and a melting temperature of 660°C. Apart from being light in weight, the notable advantage of aluminum is that it can be easily alloyed with the appropriate elements to achieve high strength, good fracture toughness, good fatigue resistance, and even good corrosion resistance. The Al–Zn–Mg (7XXX series), Al–Mg–Si (6XXX series), Al–Si (4XXX series), and Al–Cu–Mg (2XXX series) alloys are preferentially chosen and used in the aerospace and automotive industries (Polmear 1995). Further, being able to process these alloys in both cast and wrought forms makes the alloys of aluminum offer increased flexibility coupled with ease of production. The castings of an aluminum alloy can be easily made due to the low melting temperature of aluminum, ease of handling, negligible affinity by way of solubility of the gaseous species (except hydrogen), relatively good fluidity, and an overall good surface finish in the product. The wrought alloys of aluminum can be obtained using various traditional deformation processing methods. They can also be processed using the technique of severe plastic deformation to obtain an alloy with either an ultrafine grain or nanocrystalline microstructure. Another notable advantage of the alloys of aluminum is their ability to age harden/precipitation harden as an outcome of their response to various heat treatments. By selective heat treatments, even a low-strength aluminum alloy can be made to achieve significant improvement in its properties, thereby enhancing the scope of its application. Further, the alloys of aluminum can be joined by most of the methods used for other metals to include the following: (i) welding, (ii) brazing, (iii) soldering, (iv) bolting, (v) riveting, and (vi) adhesive bonding. Aluminum alloys have high specific stiffness [E/ρ], specific strength [σ/ρ], good fracture toughness and high fatigue resistance that make them an appropriate choice for a spectrum of applications in the industries of aerospace, ground transportation, and even rail transportation. However, their surface properties, such as wear and corrosion resistance, need improvement.

Some of the applications of the alloys of aluminum (usually Al–Si cast alloys, the 4XXX series, the 5XXX series, and 6XXX series alloys) in the automobile industry include the following (Jurgen Hirsch 2011):

a. Power train components—engine block and cylinder head, transmission housings, fuel system, liquid lines, and radiators.
b. Chassis and suspension—cradle, axle, wheels, suspension arms, and steering systems.
c. Car body—hoods/bonnets, doors, front structure, wings, crash elements, bumpers, and various interiors.

The applications of Al alloys in the aviation industry include the following (Total Materia 2004b):

i. Transport aircraft—upper skins and spar caps of wings (usually 2XXX series and 7XXX series alloys in the heat-treated condition), trim tabs, servo tabs, control surfaces, flaps and non–load-carrying doors (2XXX series, 6XXX series, and 7XXX series alloys in the heat-treated condition), landing gear structural parts, wheels, forgings in fuselage, and skins
ii. High-performance aircraft—wing skins, canopy supports and frames, fuselage members, and heavily loaded pylons
iii. Supersonic aircrafts—honeycomb-structured wing panels and engine parts made up of 2XXX series and 5XXX series alloys for purpose of high thermal stability and creep resistance because of aerodynamic heating experienced by the parts
iv. Helicopters—rotor blades, main spar members, blade skin, cabin, and fuselage structures (2XXX series and 6XXX series alloys)

11.1.1.2 Magnesium and Its Alloys

Magnesium is the lightest of potentially usable structural metals. It has a density of 1.74 g/cc and a melting temperature of 650°C. Because of its low melting temperature, it is easier to melt magnesium during casting (like aluminum). It is oxidative in nature; hence, proper care must be taken during the casting of magnesium and while handling powders of magnesium. Magnesium has good room temperature properties, good damping capacity, and excellent electromagnetic shielding capability and is abundant in nature (can be obtained from seawater). Magnesium has a hexagonal close-packed (HCP) crystal structure, which makes it difficult to undergo deformation at room temperature, resulting in low ductility and inadequate toughness. This was attributed to a fewer number of "active" slip systems (Gupta and Sharon 2011; Polmear 1995). Further, magnesium is susceptible to degradation when exposed to a marine environment (it has inferior corrosion resistance in a seawater environment). Unlike aluminum, magnesium is rarely used in its pure form. It is often alloyed for use in manufacturing operations in both aerospace and ground transportation industries, specifically automotive. Some of the most commonly used alloying elements present in commercial alloys of magnesium are aluminum, zinc, cerium, silver, thorium, yttrium, and zirconium. Similar to aluminum, the alloys of magnesium can be age hardened/precipitation hardened, which helps control the properties depending on the application. Table 11.1 provides a list of commercially available magnesium alloy designations, as well as their properties and applications. Table 11.2 provides a few examples of application of magnesium alloys in the automotive sector (Kulekci 2008).

In the aerospace sector where high-temperature mechanical stability/creep resistance is an important requirement, Mg–Al–Zn series, Mg–Zn–Zr series, Mg–Zn–RE series, and Mg–Ag–RE–based alloys are the preferential choices for selection and use (Gwynne and Lyon 2007). The AZ92 alloy is used for the brake callipers of an aircraft wheel, where corrosion resistance is important. The AZ92 alloy has also been used for frames, wheels, engine gear boxes, power generation components, and non-primary structural members, to name a few. The alloys WE42 and AZ91E have been identified to be corrosion resistant. In recent years, castings of alloy AZ92A are chosen for use in thrust reverser cascade on turbofan engines. The castings of alloy AZ91E are preferential candidates chosen for the six cylinder pumps in mini-airplanes/glider aircrafts and for rotor hub casings in a Sikorsky helicopter (Gwynne and Lyon 2007). In Sikorsky Blackhawk helicopters, the transmission casing is made up of the ZE41 alloy. The ZE41 magnesium alloy is also used for casings in the Boeing Apache helicopters and Chinook helicopters. The WE43 magnesium alloy is a preferential choice for the transmission casings of Sikorsky S92 helicopters. The EZ33A magnesium alloy is used for gearbox casing of a Rolls Royce engine. The accessory drive gearbox of the F16 fighter aircraft having a General Electric F110 engine is made using the QE22 magnesium alloy (Gwynne and Lyon 2007).

Similar to the alloys of aluminum, the alloys of magnesium have poor surface/subsurface properties, such as low resistance to wear, corrosion, oxidation, and fatigue. This has necessitated the

TABLE 11.1
Composition, Properties, and Applications of Commercial Mg Alloys

Alloy Designation	Alloying Elements (%)	Properties and Applications
AZ91/92	Mg, Al: 9.0, Zn: 0.7–2, Mn: 0.13	Good castability, good mechanical properties at $T < 150°C$
AM60	Mg, Al: 6.0, Mn: 0.15	Greater toughness and ductility than AZ91, slightly lower strength. Preferred for automotive structural applications
AM50	Mg, Al: 5.0, Mn: 0.15	Good strength, ductility, energy absorption properties, and castability
AE44	Mg, Al: 4.0, RE: 4.0 (RE: rare earth)	Automotive alloy Better creep behavior and castability
AE42	Mg, Al: 4.0, RE: 2.0 (RE: rare earth)	Low castability and good creep behavior
AS41	Mg, Al: 4.0, Si: 1.0	Better creep resistance than AZ91 at elevated temperatures, but lower strength
ZE41	Mg, Zn: 4.2, RE: 1.2, Zr: 0.7	RE addition improves creep strength at elevated temperatures
AZ31	Mg, Al: 3.0, Mn: 0.2	Good extrusion alloy, aerospace alloy
QE22	Mg, Ag: 2.2, RE: 2.1, Zr: 0.7	Aerospace alloy
WE42/43	Mg, Y: 4.2, RE: 2.0	Aerospace and biomedical alloy

Sources: Polmear, I. J. (1995). *Light Alloys: Metallurgy of Light Metals.* 3rd Edition, Wiley, USA; Gupta, M. and N. M. L. Sharon (2011). *Magnesium, Magnesium Alloys, and Magnesium Composites.* John Wiley & Sons, New Jersey, USA; Kulekci, M. K. (2008). Magnesium and its alloys applications in automotive industry. *International Journal of Advanced Manufacturing Technology* 39: 851–865.

TABLE 11.2
Examples of Application of Mg Alloys in the Automotive Sector

Engine Parts and Transmission	Interior Parts	Chassis Components	Body Components
• Engine block cores	• Steering wheel	• Rod wheels	• Cast components
• Gear box	• Seat components	• Suspension arms	• Inner bolt lid section
• Intake manifold	• Instrument panel	• Engine cradle	• Cast door inner
• Crankcase components	• Steering column	• Rear support Tailgate (AM50)	• Radiator support
• Cylinder head cover	• Brake, clutch pedal		• Sheet components
• Oil pump housing	• Air bag retainer		• Extruded components
• Oil pump	• Door inner (AM50)		
• Transfer case			
• Support and cover (AZ91D)			
• Cam, bedplate			
• Engine block, oil pan			
• Front cover, engine cradle (Ford)			

Source: Kulekci, M. K. (2008). Magnesium and its alloys applications in automotive industry. *International Journal of Advanced Manufacturing Technology* 39: 851–865.

need for surface modification of both magnesium and its alloy counterparts to facilitate in their selection and use in real-time applications.

11.1.1.3 Titanium and Its Alloys

The alloys of titanium, because of their high specific strength [σ/ρ], excellent corrosion resistance, fatigue resistance, creep resistance, and fracture toughness, are an ideal choice for a plethora of aerospace-related applications. Titanium has an HCP crystal structure. It has a density of 4.2 g/cc and a melting temperature of 1670°C. At 882°C, titanium changes its crystal structure from HCP (α-phase) to BCC (body-centered cubic, β-phase) giving rise to α β, and α–β alloys (Boyer 2010; Polmear 1995). Hence, alloying of titanium is largely dependent on the ability of alloying elements to stabilize either of these phases based on requirement. Because of their properties and suitability for aerospace-related applications, two-thirds of the titanium that is produced is used for the frame and engine of an aircraft in both the civilian and military sector. The SR71 Blackbird is made up of 80% titanium alloys. The aircraft components that are made using titanium alloys include the following: fire wall, landing gears, exhaust ducts (helicopters), and hydraulic systems. In aircraft engines, the titanium alloys are used for rotors, compressor blades, hydraulic system components, heat shields, and nacelles. On account of their resistance to corrosion when exposed to an aggressive environment, the alloys of titanium are used in propeller shafts in desalination plants and in heat exchangers. In the automotive industry, the alloys of titanium are mainly used for the following: (i) exhaust pipes, (ii) mufflers, (iii) intake valves, and (iv) exhaust engine valves, because of their good fatigue strength and heat resistance (Fujii et al. 2003). Another important application of the family of titanium alloys is in biomedical implants. Being bio-compatible and nonmagnetic and having a modulus almost close to that of bone, the alloys of titanium are suitable for dental and orthopedic implants. The alloys of titanium chosen for use for biomedical applications include commercially pure titanium (CP-Ti, α alloy), Ti–6Al–4V (α–β alloy), and Ti–15V–3Cr–3Al–3Sn (β alloy) (Sudhakar and Haque 2015). These titanium alloys are also chosen for use in the aerospace industry. In both aerospace-related and biomedical applications, corrosion resistance and wear resistance of the chosen alloys are important and often the subject of several investigations.

11.2 SURFACE MODIFICATION OF METALS/ALLOYS

The aluminum-, magnesium-, and titanium-based materials are viable candidates for a plethora of emerging applications spanning both performance-critical and non–performance-critical because of their superior properties. Designing and synthesis of new materials is one route to realize high-performance materials. Modification of the surface is another route to achieve the same. The properties of a material are largely dependent on its microstructure. Therefore, the properties of existing metals and their alloy counterparts can be significantly improved by modifying or altering their microstructure. In most cases, the approach of surface modification is less time consuming than the approach taken for making a new material. The bulk microstructure can also be modified by the prudent use of thermomechanical treatments. To improve surface properties, such as corrosion, wear, and fatigue, modification of the bulk microstructure is not always required or essential. For such applications, the microstructure can be optimized, controlled, and/or manipulated using surface processing techniques, wherein the bulk microstructure remains essentially unaltered. Surface modification methods can be classified as being (a) thermomechanical processes, (b) mechanical processes, (c) thermochemical processes, and (d) thermal processes (high-energy surface treatments).

11.2.1 Thermomechanical Processing

The thermomechanical process is a combination of plastic deformation (e.g., rolling, forging, extrusion, etc.) in combination with heating and cooling sequences or cycles (Rudskoy et al. 2016). The combination of thermal treatment and mechanical treatment (plastic deformation) results in the creation and/or rearrangement of dislocations in the microstructure. By thermomechanical processing (TMP), a high dislocation density structure coupled with a reduction in grain size, formation of substructure, and uniformly dispersed secondary-phase particles are obtained. The microstructural changes resulting from TMP aid in improving the strength properties of the material. In TMP, some of the factors that control the plasticity of metals and their alloy counterparts include the following: (a) crystallography of the lattice, (b) chemical composition, (c) presence and distribution of the second-phase particles, (d) temperature of deformation, (e) degree of deformation, (f) dynamic recovery and recrystallization temperature, and (g) strain rate (Total Materia 2004a). TMP was initially used for the family of steels. Recently, the non-ferrous metals, especially the family of aluminum alloys, are being thermomechanically treated to obtain either an ultrafine-grained or a nanocrystalline microstructure, with concomitant improvement in yield strength, toughness, resistance to creep deformation, improved corrosion resistance, and enhanced resistance to stress-corrosion cracking (Total Materia 2004a).

Recently, the friction stir processing (FSP) method was developed and put forth. Unlike the conventional TMP treatment that basically involves bulk structure modification processes, FSP is a thermomechanical treatment undertaken solely for modifying the surface (Venetti 2008). In conventional TMP, the chosen material is often subjected to the conjoint influence of external mechanical deformation (plastic deformation) and thermal treatments (heating/cooling). In contrast, in FSP, the mechanical interaction of a rotating tool on the surface of the chosen material generates high plastic deformation and induces frictional heating of the surface. The temperature resulting from frictional heating is sufficiently high to do the following: (i) refine the microstructure, (ii) achieve phase modification, and (iii) facilitate alloying of the surface (Venetti 2008; Weglowski and Pietras 2011).

11.2.2 Mechanical Processing

In mechanical processing, a cold-worked (i.e., work hardened) material is formed on the surface (Burakowski and Wierzchon 1998). Usually, compressive residual stresses are induced on the surface that is beneficial for enhancing resistance to fatigue, corrosion fatigue, fretting fatigue, and wear. Shot peening and shot blasting are two of the commonly used methods for mechanical surface treatment. In these processes, surface modification of the chosen material is facilitated without any noticeable chemical modification of the surface (Burakowski and Wierzchon 1998; Laouar et al. 2009).

11.2.3 Thermochemical Processing

The thermochemical processing (TCP) treatment is a surface treatment process, which is usually conducted in solid, liquid, or gaseous media using chemically active elements (Czerwinski 2012). During thermochemical treatment, the following mechanisms are facilitated:

i. Decomposition of solid/liquid/gaseous species
ii. Splitting of molecules to form nascent atoms
iii. Absorption of atoms followed by their diffusion into the metallic lattice

The diffused atoms tend to chemically react and/or interact with the substrate microstructure and, in the process, modify the existing phases or form new phases (Czerwinski 2012). Similar to TMP, TCP has been successfully used for the family of steels with processes, such as (a) nitriding/plasma nitriding, (b) carbo-nitriding, (c) carburizing, (d) boriding, (e) boronizing, and (f) aluminizing. In recent years, non-ferrous alloys (mainly the alloys of aluminum) are being treated by TCP, such as (i) coatings (to include chemical conversion coatings), (ii) cladding, and (iii) nitriding. Similarly, the alloys of titanium are subjected to the nitriding process. Basically, the TCP helps in improving the hardness of the surface and resistance to both wear and oxidation, to name a few (Czerwinski 2012).

11.2.4 Thermal Process—High-Energy Surface Processing

High-energy processing is a relatively new method for treating the surface. It can alter the properties of a surface without appreciably changing its dimensions. Common high-energy processing treatments include the following: (i) laser beam treatment, (ii) plasma beam treatment, (iii) ion beam treatment, and (iv) electron beam treatment (Rudskoy et al. 2016). These processes are briefly described below.

a. Laser beam treatment: In this process, a laser beam alters the surface properties by rapid heating in a shallow region near the surface.
b. Plasma beam treatment: This is a reactive treatment process where the selected gas (e.g., argon) is reinforced with additional energy to become ionized and reach a plasma state. The interaction of plasma with the metal substrate initiates reactions and/or modifies the substrate surface.
c. Ion beam treatment: This method uses an electron beam or plasma to impinge ions having sufficient energy. The ions embed into the atomic lattice of the substrate. The mismatch that exists between the ions and the metal surface creates defects at the atomic level that contributes to modifying properties of the surface.
d. Electron beam treatment: In this process, the surface properties are altered by rapid heating using an electron beam followed by rapid cooling, of the order of 10^{6}°C/s (10^{6}°F/s), in a shallow region (~100 μm) near the surface.

Usually, high-energy surface processes are conducted in the continuous mode, for which the energy must be supplied continuously. In recent years, prudent use of pulsed beam treatments has become increasingly popular. For a continuous beam, an accelerating field is required (Wiedemann 2007). However, when the energy source/particles are accelerated by means of radiofrequency fields (rf fields), a "bunched beam" is generated. A beam that contains a finite number of the so generated bunches for a finite length of time is referred to as a "pulsed beam" (Wiedemann 2007). Should the accelerating field and the plasma/electron source operate continuously with time for generating a low current ion beam, it is referred to as "continuous ion beam." As an example, in laser beam treatment, when the laser output power is constant and extends over a prolonged period, it is referred to or known as "continuous wave laser." In contrast, if the laser output power shows a peak value for a

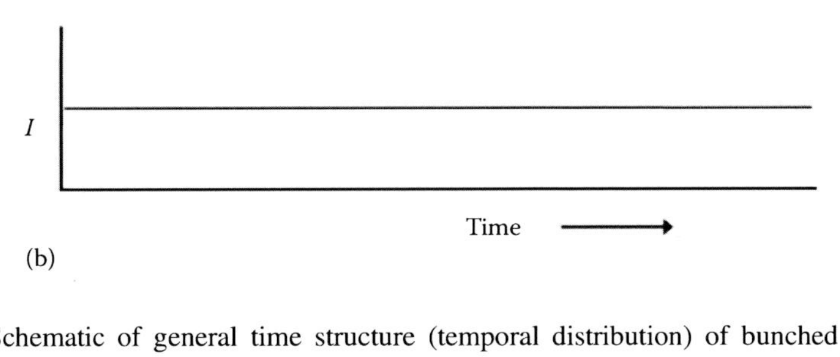

FIGURE 11.1 Schematic of general time structure (temporal distribution) of bunched beams. (a) Pulsed beam. I_{peak} is the "peak current" given by I_{peak} q/τ_μ where q is the charge per microbunch and τ_μ is the microbunch duration, and T_p is the pulse duration. (b) Continuous beam current, having the same magnitude with time. (From Wiedemann, H. (2007). *Particle Accelerator Physics*. Springer, New York.)

short duration of time (at a repeated rate), it is referred to as a "pulsed laser beam" (PLB). In this case, the energy is delivered within a fraction of a second (Wiedemann 2007). For comparison purposes, a pulsed beam provides high current and thus high energy when compared to a continuous beam. A pulsed beam with high energy (high current) induces effective modification of the surface of the chosen material, including a high localized temperature coupled with melting of the surface. A schematic of the general time structure (temporal distribution) of bunched beams (pulsed currents) is shown in Figure 11.1. The smallest unit is known as the "microbunch," which is separated from the next microbunch in the order of wavelength of the accelerating rf field (Wiedemann 2007).

11.3 HIGH-ENERGY SURFACE TREATMENT METHODS

11.3.1 PLB Surface Treatment

Lasers can be a controlled heating source having a high-power density. Lasers are used for metal working purposes since they can generate power densities up to 10^6 W/cm^2 (Brown and Arnold 2010). In laser surface treatments, the two parameters that need to be controlled are (i) applied power density and (ii) dwell time. "Dwell time" is the time duration for which the laser beam interacts with the material that is being treated. Dwell time usually varies from femtoseconds/picoseconds to many tens of seconds. A small dwell time (i.e., short duration) gives rise to a PLB, whereas a longer dwell time (i.e., longer duration) gives rise to a "continuous laser beam." Based on continuous or pulse mode, the depth of interaction of a laser beam with the chosen material can range from a few nanometres to several hundreds of microns/millimeters (Brown and Arnold 2010; Burakowski and Wierzchon 1998; Draper and Mazzoldi 1986).

11.3.1.1 Principle and Workings of the PLB Surface Treatment

In laser surface treatment, an interaction between the laser beam and surface of the material in vacuum/processing gas or a protective atmosphere induces modification of the surface. A schematic of the general setup of a laser surface treatment process is shown in Figure 11.2 (U.S. Laser Corporation 2016). The working principle of PLB surface treatment is briefly described.

In the PLB setup, the laser beam is focused onto the sample surface through an optical transmission system that consists of a string of mirrors or fiber optics. Using beam focusing and/or beam shaping optics, such as (i) lenses, (ii) mirrors, (iii) scanner units, and (iv) beam integrators, the required power density and intensity are established (University of Miskolc 2017). To modify the surface of the chosen material, the laser beam is moved/scanned/passed over the material surface. Consequently, a track pattern is now generated on the surface of the material. The dwell time, that is, interaction time of the laser beam with the material, is determined by both cross section of the beam (i.e., diameter) and the feed rate. Translation stages or robots are now being used to make relative movements of the workpiece material depending on both process type and workpiece geometry. To increase the depth or width of the modified region, the treated regions are exposed to the laser beam for further buildup/extension of the modified region (overlapping tracks). Track widths are in the range of ~0.5 to 2 mm and can vary according to the independent and/or mutually interactive influences of laser interaction time and power density. Usually, multiple passes of the laser beam are required to complete a PLB surface treatment (University of Miskolc 2017).

A judicious control of the process parameters is critical for the success of the laser beam treatment (University of Miskolc 2017; U.S. Laser Corporation 2017). Process parameters that are controlled during a PLB surface treatment include the following: (i) laser power, (ii) laser spot size and shape, (iii) power density (= laser power/area irradiated) distribution on the treated surface, (iv) scanning speed of the laser beam, and (v) dwell time (= scan length/traverse speed) (University of Miskolc 2017; U.S. Laser Corporation 2017). The energy contained in each laser pulse is given by "energy per pulse" and is expressed as P_{Av}/R_{rate}. In this expression, P_{Av} is the average power and R_{rate} is the pulse repetition rate, that is, number of pulses per second (University of Miskolc 2017; U.S. Laser

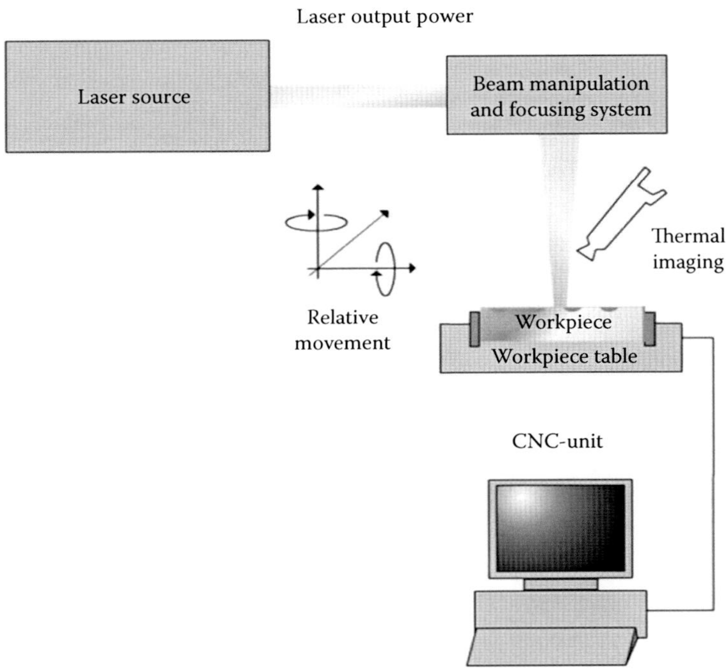

FIGURE 11.2 General setup of a laser surface treatment process. (University of Miskolc (2017). Laser surface treatment. http://emrtk.uni-miskolc.hu/projektek/adveng/home/kurzus/korsz_anyagtech/1_konzultacio_elemei/laser_surface_treat.htm. Last Accessed: May 2017.)

Corporation 2017). The peak power per pulse is given by ratio of energy per pulse to duration of the pulse and is expressed as E/D_{pulse}. In this expression, E is the energy per pulse and D_{pulse} is the pulse duration (University of Miskolc 2017; U.S. Laser Corporation 2017).

Properties of materials (to be laser treated) should be considered to enable the selection of process parameters for purposes of PLB treatment. Some of the material properties that must be considered are the following: (i) density, (ii) thermal conductivity, (iii) specific heat, (iv) absorption coefficient, (v) critical temperature, and (vi) latent heat of phase transformations (University of Miskolc 2017; U.S. Laser Corporation 2017).

Laser surface treatments are undertaken to improve surface properties, such as (a) wear resistance, (b) corrosion resistance, and (c) fatigue resistance of the chosen material. An interaction of the laser beam with the metal surface usually gives rise to phenomena such as (i) surface heating, (ii) surface melting and alloying, and (iii) shock hardening (U.S. Laser Corporation 2017), which help in enhancing the surface properties of the chosen material. The variation of laser power density and beam interaction time at which these phenomena manifest is shown in Figure 11.3 (University of Miskolc 2017). A brief description of these phenomena is provided.

a. Laser surface heating: This process is usually used for annealing treatment of steels (University of Miskolc 2017). In this process, the temperatures generated both at and near the surface of the candidate material due to laser interaction are always less than the melting point of the material.
b. Laser surface melting/remelting: In this process, temperatures generated on the materials surface due to laser interaction exceed the melting point of the chosen material. This helps in realizing better microstructural and surface properties. Upon laser surface melting, microstructural refinement (including fine grain size, reduction in segregation, and the formation and presence of metastable phases), an improvement in hardness, good wear resistance, and acceptable or enhanced corrosion resistance of the material are obtained (University of

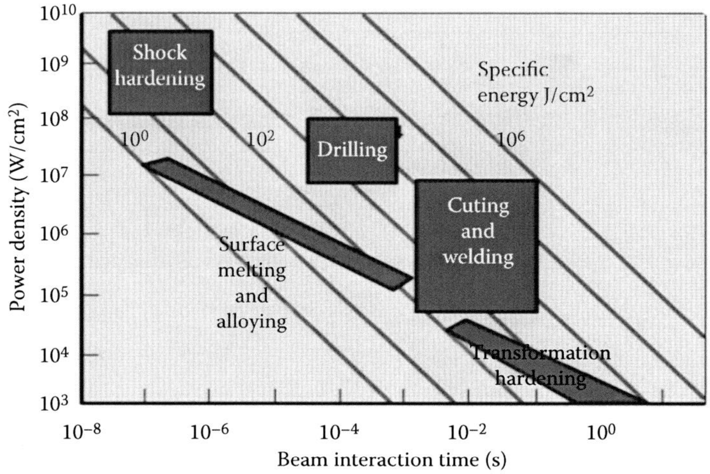

FIGURE 11.3 Operation regime for laser treatment of materials. (University of Miskolc (2017). Laser surface treatment. http://emrtk.uni-miskolc.hu/projektek/adveng/home/kurzus/korsz_anyagtech/1_konzultacio_elemei/laser_surface_treat.htm. Last Accessed: May 2017.).

Miskolc 2017). By laser surface melting process, highly localized surface modification can be achieved without affecting the bulk of the material. The process can produce a material that has fine grains and eliminate surface defects, while concurrently improving surface adhesion for the purpose of applications related to coatings. A drawback of the process is that it generates high residual stresses on the surface, thus causing severe residual stress mismatch between the modified surface layer and the bulk (substrate). Such an occurrence is conducive for the early initiation of fine microscopic cracks (University of Miskolc 2017).

c. Laser surface alloying: Because of the high power of the laser beam, not only does modification of the microstructure take place, but it also alters chemical composition of the surface (University of Miskolc 2017). By the external addition of alloying elements, surface alloys can be synthesized or produced as the process promotes melting, mixing, and alloying of the elements with the base metal. Eventually, an improvement in the surface properties, spanning hardness, wear resistance, and corrosion resistance is achieved.
d. Shock hardening process: In this process, a laser beam having a high-power density is applied for a short interaction time with the metallic surface (University of Miskolc 2017). This results in significant work hardening of the metallic surface. The laser pulse has a power density in the range of 10^8 to 10^9 W/cm^2 and a pulse duration ranging from picoseconds to a few hundred nanoseconds. When this high-power density is applied to a metal surface for a short duration of time, it creates high stress shock waves, which, upon impact with the surface of the candidate metal, induces significant plastic strain on the surface. The high plastic strain results in a high dislocation density that contributes to enhancing the mechanical properties (University of Miskolc 2017).

A few of the disadvantages of the laser beam treatment are the following (University of Miskolc 2017):

i. Nonhomogeneous energy distribution in the laser beam
ii. Narrow temperature field to bring forth the required microstructural changes
iii. Requires an adjustment of workpiece and/or the laser beam to suit different product shapes
iv. Poor absorptivity while interacting with the metallic surfaces

11.3.2 Pulsed Plasma Beam Surface Treatment

Plasma is the fourth state of matter. Phase transitions of matter occur in the following sequence: solid to liquid to gas when sufficient energy is made available. With the addition of energy to the gas phase, the gas becomes ionized and reaches the plasma state. When the plasma meets the surface of a material, the additional energy is transferred from the plasma to the surface, which helps in modifying surface properties of the chosen material (Harry 2010). By judicious use of plasma treatment, a modification of the surfaces of the chosen material or metal surfaces can be achieved.

11.3.2.1 Principle and Workings of Pulsed Plasma Beam Surface Treatment

Plasma is generated using plasma accelerators (flux velocity: 10 to 10^3 km/s, ion kinetic energy: 10 to 10^6 eV) (Morozov 1970–1979). A high-velocity plasma flux can be obtained using various methods. One such method is to irradiate a solid with a laser beam. In plasma accelerators, the plasma is produced through electrical energy by special electric discharges. Plasma accelerator channels contain positive ions and electrons, which give plasma fluxes having an ion current of several million amperes and a particle having an energy of thousands of electron volts (Morozov 1970–1979).

Plasma acceleration occurs because of a drop in the total pressure (sum of ion and electron pressures) and by the action of electromagnetic force originating from an interaction of current in the plasma with the magnetic field (Morozov 1970–1979). Pulsed plasma accelerator systems are basically electromagnetic-type induction accelerators (Morozov 1970–1979). In these accelerators, a plasma bunch is generated and then accelerated at a velocity of the order of 10^8 cm/s (number of particles up to 10^{18}). A schematic diagram of a pulsed plasma system used for surface modification is shown in Figure 11.4 (Özbek and Durman 2015). During surface modification, surfaces of the material are irradiated by the pulse treatment. Eventually, a surface undergoes or experiences rapid melting and resultant solidification (heating and cooling rates: 10^7 to 10^{10} K/s). Because of the high rates of heating and melting, a high-speed diffusion of the ions into the surface takes place, resulting in changes to both the structure and phases present. Rapid cooling rates experienced by the molten material at the surface generate a high dislocation density resulting in the formation of a micro-/nanocrystalline/amorphous surface layer (Byrka et al. 2002; Zhukeshov et al. 2014). Consequently, the plasma irradiation-induced change in microstructure contributes in an observable manner to improving the surface properties of the treated material (Byrka et al. 2002; Özbek 2017; Özbek and Durman 2015; Zhukeshov et al. 2014).

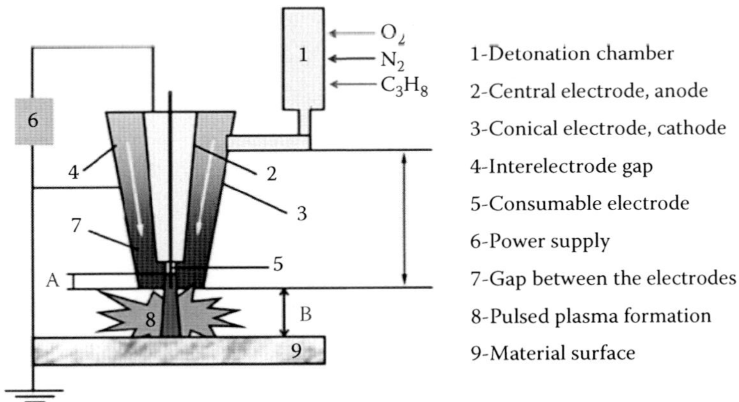

FIGURE 11.4 Schematic diagram of a pulsed plasma system used for surface modification. (From Özbek, Y. Y. and M. Durman (2015). Surface behaviour of AISI 4140 modified with the pulsed-plasma technique. *Materials and Technology* 49 (3): 441–445. Open access.)

11.3.3 Pulsed Ion Beam Surface Treatment

In ion beam treatment of materials, ions (charged particles) are impinged upon surfaces of the chosen material such that they penetrate the surface. Ion impingement is done using either a continuous ion beam or a pulsed ion beam (PIB). In the continuous beam method, ion currents are limited to microamperes (μA). In the pulsed beam method, ion currents can reach up to amperes (A) and the ion beam energy can reach up to mega-electron volts (MeV) (Kondyurin and Bilek 2008).

11.3.3.1 Principle and Workings of PIB Surface Treatment

In PIB surface treatment, surfaces of materials are made to melt to a depth of 5 to 10 μm by exposing them to a high-energy PIB. The bulk region of the material acts as a heat sink and takes away the heat from the melted surface within a few microseconds (i.e., rapid cooling) (Alpha-Omega Power Technologies 2008).

PIB (ion pulse ~500 ns) rapidly heats and melts the surfaces of a material by using 3×10^{13} ions per pulse (Stinnett et al. 1995). The deposition of high energy within a short time (pulse) results in melting of the metallic surfaces having relatively small energy (1–10 J/cm^2). Subsequent to melting of the surface, thermal diffusion into the bulk results in rapid cooling (Figure 11.5) (Stinnett et al. 1995). The rapid heating followed by rapid cooling induces surface modification by an observable change in characteristics of the surface. Typical area treated by a single pulse is about 100 to 1000 cm^2. Depth of surface modification can be varied by varying both the ion species and ion energy. Because of the high cooling rates of 10^9 K/s, nonequilibrium microstructures (i.e., nanocrystalline/amorphous phases) tend to form on the PIB-treated surfaces. This contributes to achieving an improvement in hardness of the surface, and resistance to wear, corrosion, and fatigue (Alpha-Omega Power Technologies 2008).

PIB treatment overcomes certain drawbacks of the PLB treatment, such as (i) poor energy coupling to metals, (ii) inefficient in-depth treatment, (iii) edge effects, and (iv) high cost. In PIB treatment, energy coupling of the ion beam to the surface of the chosen material does not depend on surface preparation (Stinnett et al. 1995). Further, in PIB treatment, the following can be realized: (a) large energy per pulse, (b) in-depth energy deposition, (c) high energy efficiency, and (d) cost-effectiveness. The temperature–depth profile of the PIB treatment when compared to the PLB treatment is shown in Figure 11.6 (Stinnett et al. 1995). Essentially, the PIB treatment is a line-of-sight method because of which the specimen that is being treated should be physically moved and/or rotated to ensure uniform treatment to its entire surface (Kondyurin and Bilek 2008).

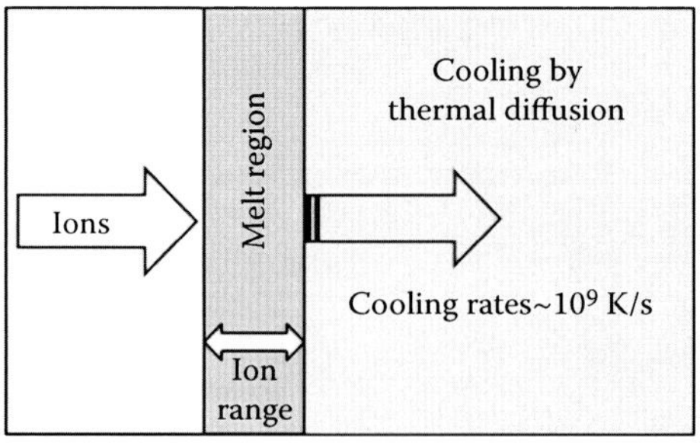

FIGURE 11.5 Pulsed ion beam surface treatment process. (From Stinnett, R. W., R. G. Buchheit, E. L. Neau, M. T. Crawford, K. P. Lamppa and T. J. Renk (1995). Ion beam surface treatment: A new technique for thermally modifying surfaces using intense, pulsed ion beams. *IEEE Xplore*. Open access).

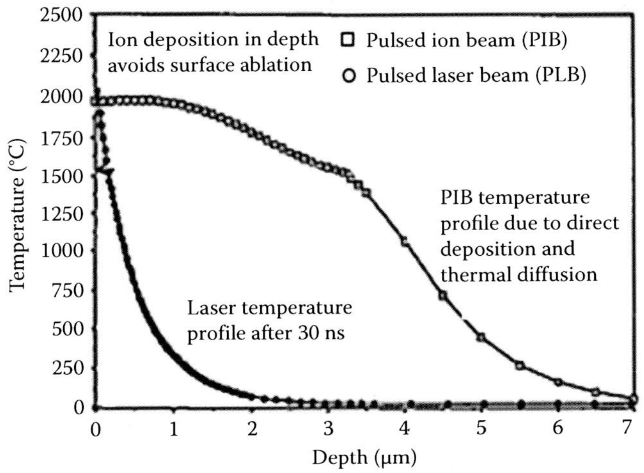

FIGURE 11.6 Temperature–depth profile of PIB treatment as compared to PLB treatment. (From Stinnett, R. W., R. G. Buchheit, E. L. Neau, M. T. Crawford, K. P. Lamppa and T. J. Renk (1995). Ion beam surface treatment: A new technique for thermally modifying surfaces using intense, pulsed ion beams. *IEEE Xplore*. Open access.)

11.3.4 Pulsed Electron Beam (PEB) Surface Treatment

11.3.4.1 Principle and Workings of PEB Surface Treatment

When an electron beam is incident on the surface of a material, it gets reflected/scattered and absorbed (Burakowski and Wierzchon 1998). When the beam strikes a material's surface, the kinetic energy carried by the electrons is transformed into thermal energy that contributes to modifying the surface properties. When the electrons reach and/or strike the surface, they penetrate and deccelerate. During their penetration into the surface, the electrons tend to interact with the atoms, second-phase particles, other microstructural heterogeneities, and the electrons present in the lattice of the material (Burakowski and Wierzchon 1998). As a consequence of this interaction, the electric field in the immediate vicinity of the impinging particles is perturbed and tends to increase their vibrational amplitude, which eventually increases temperature of the surface as a direct consequence of the heat that is generated. The energy of the impinging electrons gradually decreases with depth of peneration, as they gradually lose their energy due to an increased number of collisions within the surface of the chosen material (Burakowski and Wierzchon 1998).

In an electron beam setup, the velocity of electrons is accelerated by an electric field. The kinetic energy of the impinging electrons depends on their velocity. By varying the accelerating voltage, that is, the potential energy between electrodes in a setup, the energy carried by the moving electrons can be varied, as work of the electric field is now transformed into kinetic energy of the moving electrons (Burakowski and Wierzchon 1998). By the application of a magnetic field, an electron beam can be both focused and pulsed. Electron beams can either be used continuously or in the form of a pulse (pulse: 10^{-9} to 10^{-4} s, point pulsed beam diameter: 0.5 nm). Usually, during point pulsed exposure of the electron beam, it remains at a fixed spot on the specimen for a certain duration of time, that is, 10 μs, and gradually moves to another point on the surface of the chosen specimen. The distance the beam moves between the steps is about a few tenths of a millimeter. However, for the purpose of modification of a large surface area, of the order of cm^2, physical movement of the specimen is both essential and required (Burakowski and Wierzchon 1998).

The depth of penetration of the moving electron into the surface of the chosen material is (i) directly dependent on the accelerating voltage and (ii) inversely related to density of the specimen material. This can be expressed mathematically as (Burakowski and Wierzchon 1998)

$$d = k\,10^{-12}\,V^2/\rho \tag{11.1}$$

In this expression, d is the depth of penetration of the moving electron, k is empirical coefficient (2.1 or 2.35), V is the accelerating voltage, and ρ is the density of material chosen for the specimen.

The power carried by a PEB (P_{pb}) is directly dependent on intensity of beam current (I), accelerating voltage (V), pulse duration (τ_p), and pulse frequency (f_p). This can be expressed through the relationship:

$$P_{pb} = IV\tau_p f_p \tag{11.2}$$

11.3.4.2 Large-Area (High-Current) PEB

The major advantage of the PEB process over other high-energy processes is the "large-area high-current pulsed electron beam treatment" (HCPEB) (Karlsruhe Institute of Technology 2014). In an HCPEB setup, there are no cathode filament and focusing/deflecting coils. Instead, it has a plane cathode and solenoid coils. This is shown in Figure 11.7 (Kim et al. 2016). Such an arrangement results in the PEB consuming lower energy at a shorter time for the same performance. For a pulse time range of 60 ns to 2 μs, the beam can produce rapid heating and a cooling rate of 10^7 K/s on the surface of the chosen material. A schematic of the PEB setup is shown in Figure 11.7 (Kim et al. 2016; Uno et al. 2005). The machine irradiates inside a vacuum chamber and the solenoids are biased. A voltage is applied to the anode to attract the generated electrons as they move along the biased solenoids. Near the anode, the electrons experience collision resulting in an unstable state that generates both anions and free electrons (Kim et al. 2016). The cathode is then pulsed with a high voltage that forms an electromagnetic field, which focuses the emission of electrons in the form of a high-current pulsed beam toward the surface of the specimen. In contrast to HCPEB, in the low-energy HCPEB treatment, the energy is deposited in a short time within a shallow or narrow depth on the surfaces of the chosen material (Kim et al. 2016; Uno et al. 2005).

11.3.4.3 Microstructural Modification by PEB Treatment

In PEB treatment, a high-power density electron beam (10^9 to 10^{12} W/cm^2) is generated (Gao et al. 2005). An interaction of the PEB with the surface of the material for a short duration (pulse time: few microseconds) would tend to induce a dynamic temperature field on the surface. This is shown in

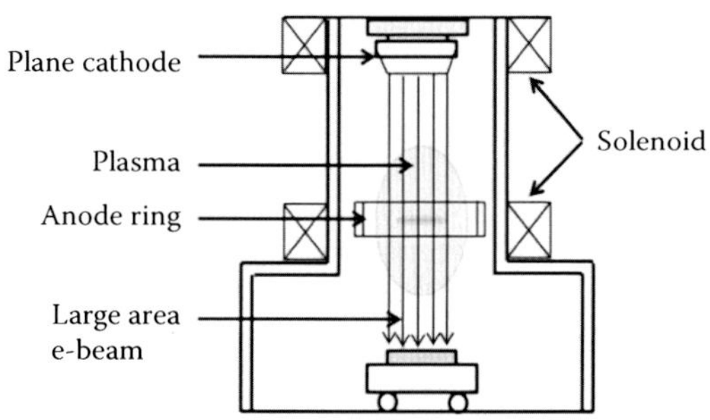

FIGURE 11.7 Schematic of a PEB setup. (Regenerated from Uno, Y., A. Okada, K. Uemura, P. Raharjo, T. Furukawa and K. Karato (2005). High-efficiency finishing process for metal mold by large-area electron beam irradiation. *Precision Engineering* 29 (4): 449–455. Used with permission.)

Figures 11.8 and 11.9 (Hao et al. 2007; Karlsruhe Institute of Technology 2014). This essentially results in the following:

i. Super-fast heating/melting of the surface (10^8 to 10^9 K s^{-1})
ii. Rapid solidification (~10^7 K s^{-1})
iii. Creation of a dynamic stress field that would cause intense deformation (shock waves) on the surface of the chosen material

With PEB treatment of the surface, one or more of the following changes and modification with specific reference to the chosen structure occur at the surfaces of the specimen (Eichmeier and Thumm 2008; Gao et al. 2005; Guo et al. 2013; Li et al. 2014; Walker et al. 2014):

i. Formation of nanostructure/nanograins
ii. Enrichment of the favorable element on the surface
iii. Formation of a nonequilibrium microstructure both at the surface and subsurface
iv. Phase transformation
v. Alloying of the surface

Some of the advantages of PEB treatment include the following: (a) the small size of electrons, (b) small loss in ionization, (c) high depth of penetration, and (d) adiabatic heating of the surface. In contrast, the PIB process has large ions and high ionization loss coupled with a small depth of penetration (Korenev et al. 1996). Depending on the combination of parameters, including power

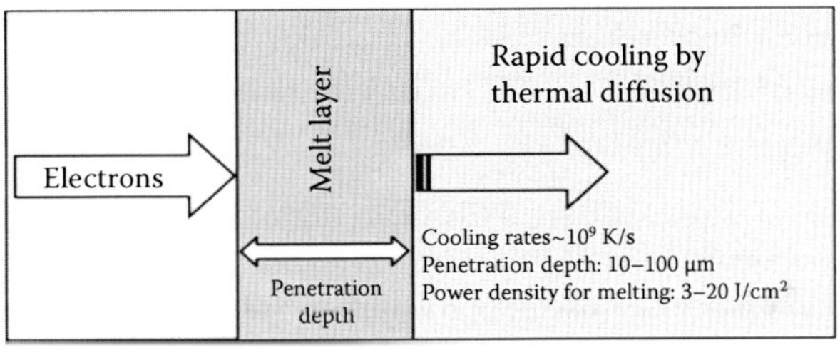

FIGURE 11.8 Schematic of the PEB surface treatment on the surface of the chosen material. (From https://www.ihm.kit.edu/english/300.php (Last Accessed: June 1 2017).)

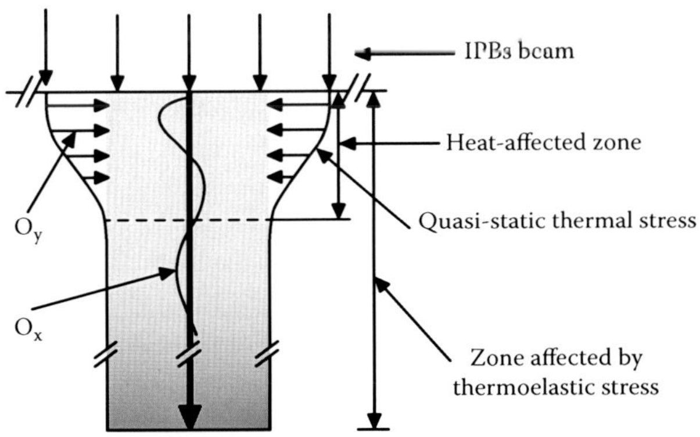

FIGURE 11.9 Schematic diagram of the distribution of temperature and stress fields under pulsed ion beam (PIB) irradiation. The schematic also shows the regions affected by heat and stress. (Used with permission from Hao, S. Z., Y. Qin, X. X. Mei, B. Gao, J. X. Zuo, Q. F. Guan, C. Dong and Q. Y. Zhang (2007). Fundamentals and applications of material modification by intense pulsed beams. *Surface & Coatings Technology* 201: 8588–8595.)

TABLE 11.3
Critical Parameters for PEB Surface Modification for Various Technologial Processes (for an Electron Beam Energy of 100 keV)

Power Density of Electron Beam (W/cm^2)	Pulse Duration	Technological Processes: Hardening Treatment	Technological Processes: Forming Treatment
10^9 to 10^{11}	0.5 to 8.0 μs	Detonation hardening	
7×10^6 to 7×10^9	1.0 to 60 μs	–	Milling, inscribing
2×10^6 to 10^8	10 μs to 10 ms	–	Hollowing, perforation of metals
10^5 to 3×10^6	900 μs to 110 ms	Welding and surface remelting	–
2×10^5 to 2×10^6	1 to 500 ms	Alloying	–
9×10^3 to 4×10^4	1 ms to 1 s	Hardening	–

Source: Burakowski, T., and T. Wierzchon (1998). *Surface Engineering of Metals: Principles, Equipment, Technologies*. CRC Press, 23 Dec 1998. 235 pp.

density and pulse duration, the surfaces of the chosen material can be heated, melted, brought to boiling point, and can also be vaporized.

For real-time applications, a judicious selection of these parameters is critical to facilitate the desired modification to the surface. This is summarized in Table 11.3 (Burakowski and Wierzchon 1998).

11.3.5 Comparison of PEB Surface Treatment with Other Processes Used for Modification of the Surface

As mentioned in the previous sections, the PEB treatment of surfaces offers several advantages over other surface modification techniques. A succinct comparison of the PEB surface treatment with the other surface modification processes used is provided in Table 11.4.

11.4 PEB TREATMENT OF LIGHT METALS

11.4.1 Microstructure and Properties of Aluminum and Its Alloys

In this section, a modification of the surface properties of aluminum and aluminum alloys by PEB surface treatment is presented and briefly discussed. Since pure aluminum has a simple structure without any additional phases or phase transformation, it was the first material to be investigated with specific reference to the PEB treatment. In the PEB treatment, power density of the beam is an important parameter that can be varied for achieving and/or enabling modification of the surface of the chosen material. The beam power density can be controlled by (a) varying the distance between the anode and the target, (b) accelerating the voltage, and (c) magnetic field intensity. It was observed in the PEB treatment of pure aluminum that decreasing the density of the beam and changing the distance between the anode and the target resulted in a reduction in melting experienced by the surface (Hao et al. 2001). When the surface layer was melted, the shock wave produced by the dynamic stresses during the PEB treatment caused the formation of an undulating or wavy front immediately beneath the melted layer. Variation of microhardness with depth revealed high values of the microhardness up

TABLE 11.4
PEB Surface Treatment versus Other Techniques Used for Modification of the Surface

Other Surface Modification Methods	PEB Surface Treatment
1. Chemical Processes and Conversion Coatings	
Major Drawbacks	**Advantage of PEB Treatment**
(a) Use of toxic chemicals (e.g., chromate solutions for magnesium-based materials)	Absence of such issues
(b) Adhesion-related issues with substrates (i.e., interface in coatings and platings)	
2. Friction Stir Processing	
Major Drawbacks	**Advantage of PEB Treatment**
(a) Physical contact of tool with the surfaces, which distorts dimensional finish	No physical contact with surfaces of the chosen material
3. Pulsed Laser Beam Treatment	
Major Drawbacks	**Advantages of PEB Treatment**
(a) Low absorption depths (e.g., 0.02 μm, for laser pulse length ~10 to 50 ns)	(i) High absorption/melt depths (e.g., 2.6 μm, for an electron beam pulse length of 50 ns)
(b) Reflection up to 90% of incident energy (energy loss)	(ii) Only 5% to 10% of incident energy is reflected
(c) Limited to small surface area	(iii) Allows processing on large surface area (up to several 100 cm^2)
4. Pulsed Plasma Beam Treatment	
Major Drawbacks	**Advantage of PEB Treatment**
(a) Erosion of the material used for the electrode materials	Absence of such issues
(b) Deposition of electrode ions on the test sample	
(c) Formation and presence of toxic products	
5. Pulsed Ion Beam Treatment	
Major Drawbacks	**Advantages of PEB Treatment**
(i) Large loss in ionization	(i) Small loss in ionization
(ii) Small penetration depth (0.05 to1.0 mm, at 100 to 1000 keV kinetic energy)	(ii) Large penetration depth (10 to 500 mm, at 100 to 1000 keV kinetic energy)

to a depth of 80 μm, as a direct consequence of the PEB treatment. This is shown in Figure 11.10. The influence of the heat-affected zone (melted zone) extended only to a depth of 10 μm. The structural modifications that occurred in the remaining depth, culminating in a variation of the hardness, were essentially due to the dynamic deformation that occurred during PEB treatment (Hao et al. 2001).

Surface alloying is a major surface modification effect that can be realized using the PEB treatment. This is in addition to microstructural modifications, changes in composition of the phases present in the microstructure, mixing of phases, and the formation of both metastable and nonequilibrium phases that occurs as a direct consequence of the high rate of solidification. Addition of favorable elements to the modified surface layer can create an alloyed surface layer that aids in enhancing both wear and corrosion resistance of the chosen material. Usually, a thin layer of the required material is deposited onto the surface of the chosen specimen. The PEB treatment of this layer does promote mixing of the deposited layer with the base material. For the surface alloying process, the thickness of the deposited

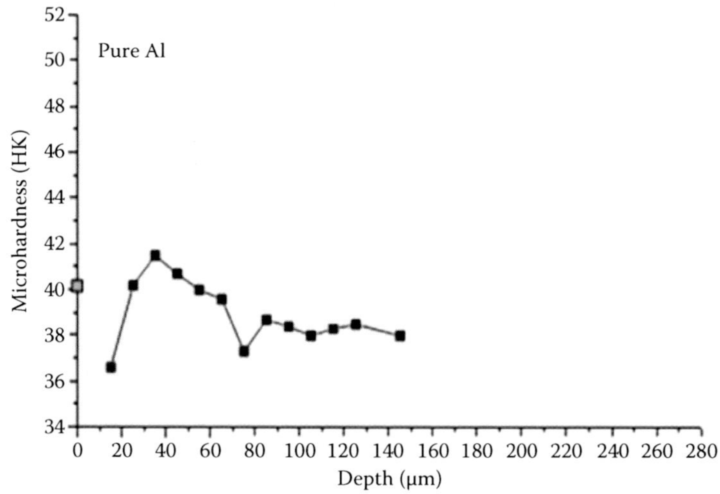

FIGURE 11.10 Variation of microhardness with depth in the PEB-treated pure aluminum. (Used with permission from Hao, S., S. Yao, J. Guan, A. Wu, P. Zhong and C. Dong (2001). Surface treatment of aluminium by high current pulsed electron beam. *Current Applied Physics* 1 (2001): 203–208.)

layer is important. A layer thickness that is greater than the penetration depth of an electron beam will not be effective (Mueller et al. 2005). Modification of the surface of the steel with aluminum can be taken as an example of surface alloying. Steel surfaces usually have a low resistance to corrosion in liquid lead. This aspect is critical in liquid metal–cooled fast breeder reactor systems. To improve the corrosion resistance of steel, an aluminum layer can be created on the steel surface by prudent use of the PEB surface alloying process (Mueller et al. 2005). For the PEB treatment, aluminum foils of varying thickness placed on the steel surface are irradiated using an accelerating voltage of 120 keV, a beam energy density of 0.2 to 0.6 MJ/m^2, and a pulse duration of 20 to 35 μs (Mueller et al. 2005). When the electron beam was bombarded on the aluminum foil, part of the energy was absorbed by the layer and the remaining penetrated into the surface of the chosen steel and caused "localized" melting and mixing of aluminum with steel. Aluminum concentration up to a depth of 10 to 20 μm was easily achieved. This is shown in Figure 11.11a and b (Mueller et al. 2005). The presence of both laminar

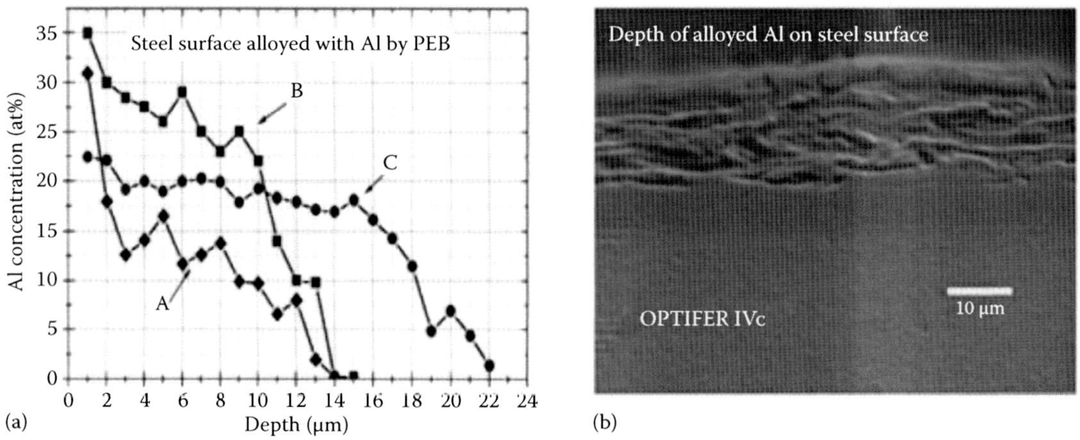

FIGURE 11.11 (a) Variation of aluminum concentration with depth, after surface alloying of aluminum on the steel surface by PEB treatment, and (b) cross section of surface of steel alloyed with aluminum. (Used with permission from Mueller, G., V. Engelko, A. Weisenburger and A. Heinzel (2005). Surface alloying by pulsed intense electron beams. *Vacuum* 77: 469–474.)

and whirling-like features on the surface provided evidence for the dynamic mixing of the process (Mueller et al. 2005). Corrosion studies of steel and PEB-modified steel (aluminum surface-alloyed steel) conducted in liquid lead with dissolved oxygen at 500°C revealed the formation of a stable scale of alumina on the surface, which hindered the penetration of oxygen into the steel and protected the steel from leaching by lead (Figure 11.12) (Mueller et al. 2005).

The diffusion effect of PEB on pure aluminum was studied by spraying carbon powders (<10 μm in size and in ethanol solution) onto the surfaces of pure aluminum (Dong et al. 2003). An electron beam of low energy (10 to 40 keV), high peak current (10^2 to 10^3 A/cm^2), short pulse duration (5 μs), and high efficiency (repeating pulse interval of 10 s) was used for the process. Because of the PEB treatment, the diffusion of carbon into aluminum occurred with a diffusion depth of carbon up to 12 μm (Figure 11.13) (Dong et al. 2003).

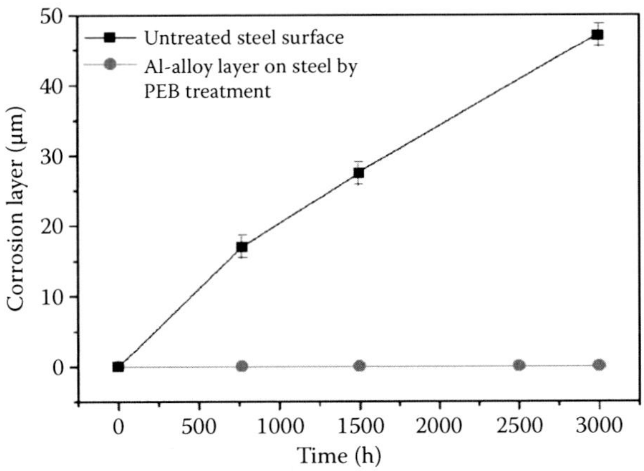

FIGURE 11.12 Oxidation rate of the steel surface upon exposure to liquid lead both before and after PEB treatment. When the surface is alloyed with aluminum by use of the PEB treatment, the steel shows excellent resistance to oxidation. (Regenerated from Mueller, G., V. Engelko, A. Weisenburger and A. Heinzel (2005). Surface alloying by pulsed intense electron beams. *Vacuum* 77: 469–474. Used with permission.)

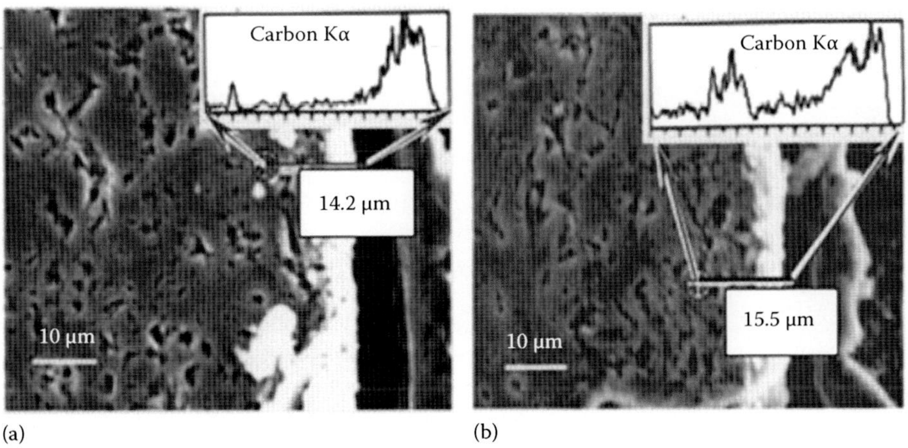

FIGURE 11.13 Diffusion effect of carbon in pure aluminum (depth of C diffusion) after PEB treatment with varying beam energy. (Used with permission from Dong, C., A. Wu, S. Hao, J. Zou, Z. Liu, P. Zhong, A. Zhang, T. Xu, J. Chen, J. Xu, Q. Liu and Z. Zhou (2003). Surface treatment by high current pulsed electron beam. *Surface and Coatings Technology* 163–164: 620–624.)

Homogeneous distribution and refinement of second phases is yet another surface modification that can be achieved on a metallic surface using the PEB treatment. As an example, the aluminum-lead [Al–Pb]-based bearing materials, whose wear properties are greatly influenced by the morphology and distribution of the second phase, were considered for PEB treatment (An et al. 2006). PEB having an energy density of 1.5 to 2.5 J/cm^2 and a pulse duration of 1.5 ns (number of pulses = 15) was bombarded on the surfaces of an Al–Si–Pb alloy (An et al. 2006). Cross-sectional analysis of the PEB-treated samples showed two distinct regions, namely:

i. The molten zone
ii. The heat-affected zone

The heat-affected zone overlapped with the thermal stress-affected zone (identified by its undulating, i.e., wavelike, front). It was observed that the lead particles were homogeneously distributed, with a change in their morphology from a wide strip shape to a thin and fine grain shape (An et al. 2006). The microstructure and microhardness of the PEB-treated sample are shown in Figure 11.14a and b.

Sliding wear and friction behavior of both the untreated and PEB-treated surfaces of the Al–Si–Pb alloy revealed that the energy density and depth of modification play a major role in determining the tribological characteristics (Figure 11.15) (An et al. 2006). The PEB-treated Al–Si–Pb alloy exhibited better wear resistance than the untreated counterpart, particularly for the samples treated at a higher energy density. The wear rate was noticeably decreased when the samples were treated with an increased electron energy density. The heat-affected zone exhibited higher wear resistance. As the wear continued to cause loss of the material beyond the PEB-modified depth (especially at high loads), both the coefficient of friction and wear rate approached the values of the untreated material, providing an indication that the heat-affected zone was highly resistant to wear (An et al. 2006).

Both grain refinement and phase refinement are an important microstructural aspect for ensuring reliable performance of Al–Si alloys, which are widely used in the automobile industry for piston and engine cylinder components. The morphology and distribution of the primary silicon phase are important. The presence of coarse primary silicon reduces the performance durability of the alloys. The presence of fine primary Si-phases can be obtained by using modifiers to include the rare-earth elements, or by choosing alternative manufacturing techniques, such as rapid solidification.

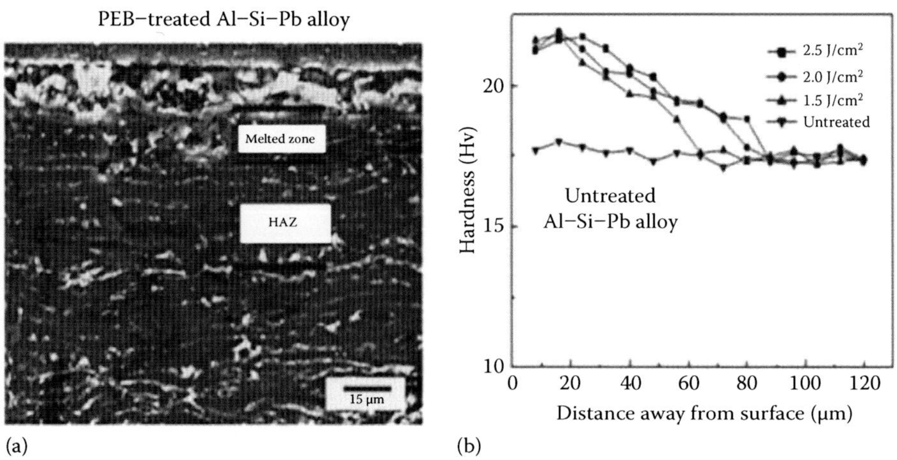

FIGURE 11.14 (a) Scanning electron micrograph of the Al–Si–Pb alloy, and (b) variation of microhardness with depth for the Al–Si–Pb alloy, following PEB treatment. (Used with permission from An, J., X. X. Shen, Y. Lu, Y. B. Liu, R. G. Li, C. M. Chen and M. J. Zhang (2006). Influence of high current pulsed electron beam treatment on the tribological properties of Al–Si–Pb alloy. *Surface & Coatings Technology* 200: 5590–5597.)

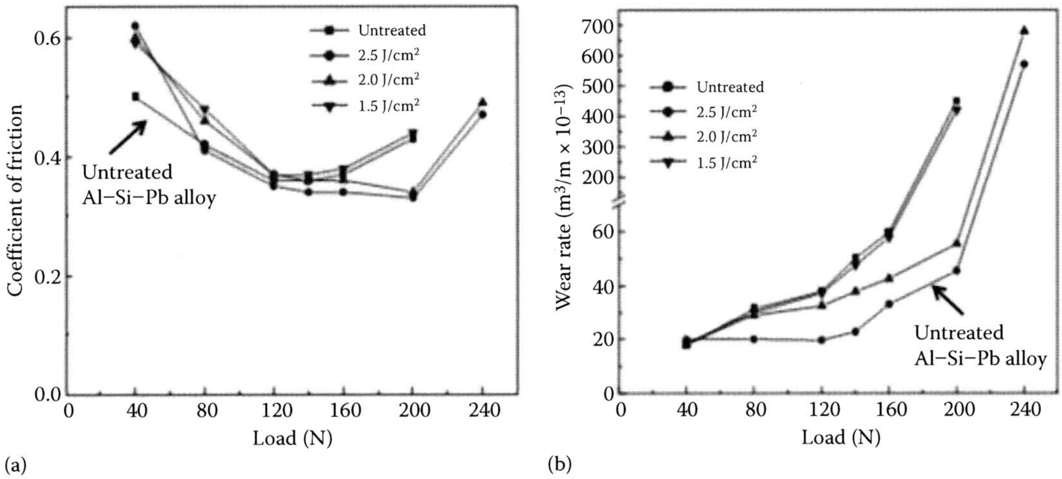

FIGURE 11.15 Variation of (a) coefficient of friction, and (b) wear rate, for the Al–Si–Pb alloy, before and after PEB treatment. (Used with permission from An, J., X. X. Shen, Y. Lu, Y. B. Liu, R. G. Li, C. M. Chen and M. J. Zhang (2006). Influence of high current pulsed electron beam treatment on the tribological properties of Al–Si–Pb alloy. *Surface & Coatings Technology* 200: 5590–5597.)

However, the most important property of piston and engine cylinder components is their resistance to wear. In this context, surface modification of Al–Si can improve its wear resistance property. The PEB treatment can be an effective route for enabling surface modification of the Al–Si alloys with the intent of improving their surface properties and concurrently enhance their performance. Because of PEB irradiation, a high energy is generated at the surfaces of materials within a few microseconds. This would enable fast heating followed by rapid cooling, resulting in the formation of ultrafine-grained/nonequilibrium/metastable phases and a resultant material with exceptional properties, not achievable by use of the conventional treatments. For example, Al–20% Si alloys were PEB-treated with an electron beam at an accelerating voltage of 23 kV and an energy density of 2.5 J/cm^2 having a 1-μs pulse width at a 10-s pulse interval and a target source at 10 cm (Hao et al. 2011b). Upon PEB treatment, the primary silicon particles revealed significant refinement through the formation of a "halo" microstructure. Owing to the high temperatures generated at the surface, remelting of the surface layer occurred, resulting in the formation of supersaturated solid solution of aluminum because of the dissolution of the silicon atoms in the aluminum matrix. The PEB-treated Al–20Si alloy exhibited better compositional homogeneity owing to the interdiffusion of aluminum and silicon atoms (Hao et al. 2011b). Further, it was observed that the aluminum matrix had subgrains that were 50 to100 nm in size and uniformly dispersed nanosilicon particles on the surface of the melted layer (Figure 11.16a) (Hao et al. 2011b). The microhardness of α-Al and the eutectic structure increased after PEB treatment. This is shown in Figure11.16b (Hao et al. 2011b).

PEB treatment of the Al–15Si alloys revealed a similar "halo" microstructure coupled with the formation of nanocrystalline/amorphous Si-phase (Hao et al. 2011a). The sliding wear tests revealed the PEB-modified alloys to exhibit excellent wear resistance when compared one on one with the untreated counterpart (Figure 11.17) (Hao et al. 2011a).

In the context of enhancing the properties of a material and its overall performance, formation of nanostructure/nanograins is an important outcome of surface modification, which can be achieved by PEB treatment. Both surface melting and rapid resolidification determine the formation of nanograins, and this can be controlled by increasing the number of pulses during the PEB treatment. For an Al–17.5Si alloy, such a process was undertaken by increasing the number of pulses from 5, 15, 25, and 100 (Gao et al. 2015). After PEB treatment, the surface melted layer of the chosen Al–Si alloy contained a silicon-rich, an aluminum-rich, and an intermediate zone. Fine, near-spherical shaped

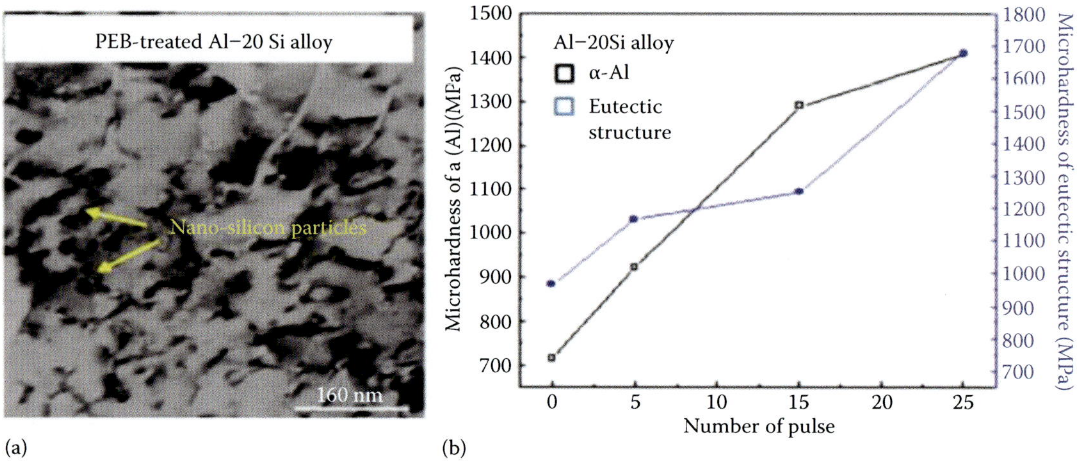

FIGURE 11.16 (a) Uniform dispersion of nanosilicon particles on the surface melted layer of the Al–Si alloy after PEB treatment. (b) Variation of microhardness of the aluminum and eutectic structure of the Al–Si alloy both before and after PEB treatment. (Used with permission from Hao, Y., B. Gao, G. F. Tu, S. W. Li, C. Dong and Z. G. Zhang (2011b). Improved wear resistance of Al–15Si alloy with a high current pulsed electron beam treatment. *Nuclear Instruments and Methods in Physics Research B* 269: 1499–1505.)

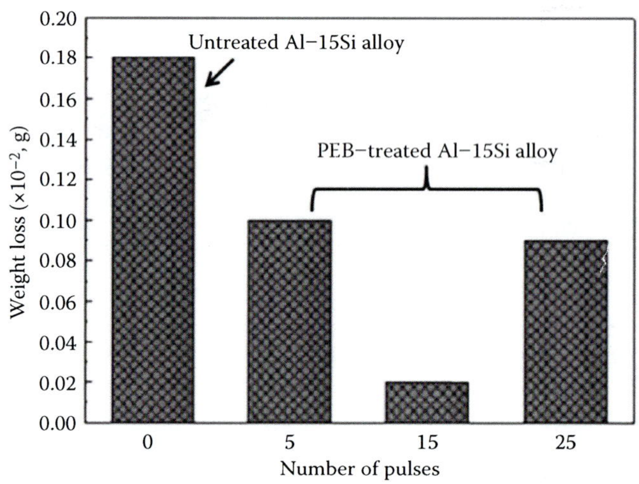

FIGURE 11.17 Weight loss of the untreated and PEB-treated Al–Si alloy after the wear test atoms. (Used with permission from Hao, Y., B. Gao, G. F. Tu, S. W. Li, S. Z. Hao and C. Dong (2011a). Surface modification of Al–20Si alloy by high current pulsed electron beam. *Applied Surface Science* 257: 3913–3919.)

nanosized crystals of silicon were found in the silicon-rich zone surrounded by α-Al. The Al-rich zone exhibited a cellular structure having a diameter of 100 nm. The intermediate zone contained superfine eutectic phases. As the number of pulses increased, the size of the silicon-rich zone increased and gradually diffused into the neighboring zone providing nucleation sites for the formation of ultrafine grains of aluminum. This is shown in Figure 11.18 (Gao et al. 2015).

The surface structure of aluminum alloys significantly controls their fatigue life, as fatigue cracks originate on the surface. In this regard, modifying the structural-phase state of the alloys by exposure to high-intensity PEB can create surface structural modifications that are micrometers thick without affecting the bulk structure. Silumin alloys (aluminum–silicon alloys) have been extensively used for

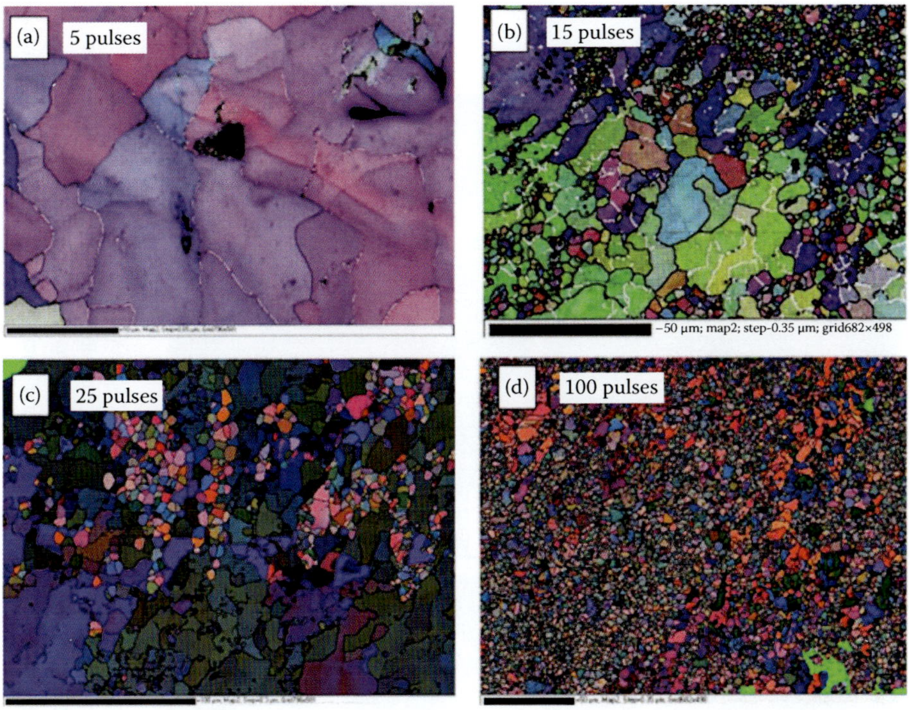

FIGURE 11.18 Evolution of grain size in aluminum-rich zone of the PEB-treated Al–17.5 Si alloy for varying number of pulses. (Used with permission from Gao, B., L. Hu, S. W. Li, Y. Hao, Y. D. Zhang, G. F. Tu and T. Grosdidier (2015). Study on the nanostructure formation mechanism on the hypereutectic Al–17Si alloy induced by pulsed electron beam. *Applied Surface Science* 346: 147–157.)

their fatigue life characteristics before and after PEB treatment (Gromov et al. 2015; Ivanova et al. 2016; Konovalov et al. 2015a,b, 2016). As an example, silumin alloy of composition Al–9.88Si–0.35Fe (Konovalov et al. 2015a) was electron beam treated with an electron energy of 18 KeV, a pulse repetition frequency of 0.3 Hz, a pulse duration of electron beam of 150 μs, and energy densities of electron beam of 15 and 20 J/cm^2. The number of pulses was 3 and 5. A crack was simulated by a half-circle cut with a radius of 10 mm. Fatigue tests were conducted at room temperature, and the samples were loaded by bending at a frequency of 15 Hz under 10 MPa load. The selection of electron beam process parameters plays an important role in determining the fatigue life. When compared to low pulse, a high pulse of 5 improved the fatigue life by a factor of ~3.5 (i.e., the cycle to failure when the number of pulses was 5 was ~517,000 when compared to ~132,000 at 3 pulses). When the number of pulses was increased from 3 to 5, high-speed melting and subsequent high-speed crystallization occurred. Owing to this phenomenon, in the cellular structure, silicon in the form of nanosized interlayers or globular shape inclusions was formed (up to 20 μm thick) as a near-surface layer. Such a surface nanostructure at the subsurface results in an increase in the critical length of the fatigue crack and thereby increases its fatigue life (Konovalov et al. 2015a). Figure 11.19 shows the change in surface structure of electron beam–treated silumin (with number of pulses = 5), before and after fatigue tests (Konovalov et al. 2015b).

Changes in surface structure due to PEB treatment coupled with changes in cell substructure due to fatigue loading influences other surface properties. In this regard, silumin surfaces that were modified by electron beam treatment and fatigue tests were studied for tribological behavior (Konovalov et al. 2016). It was found that the wear coefficient and friction coefficient increased with the number of fatigue cycles (Table 11.5). (Konovalov et al. 2016).

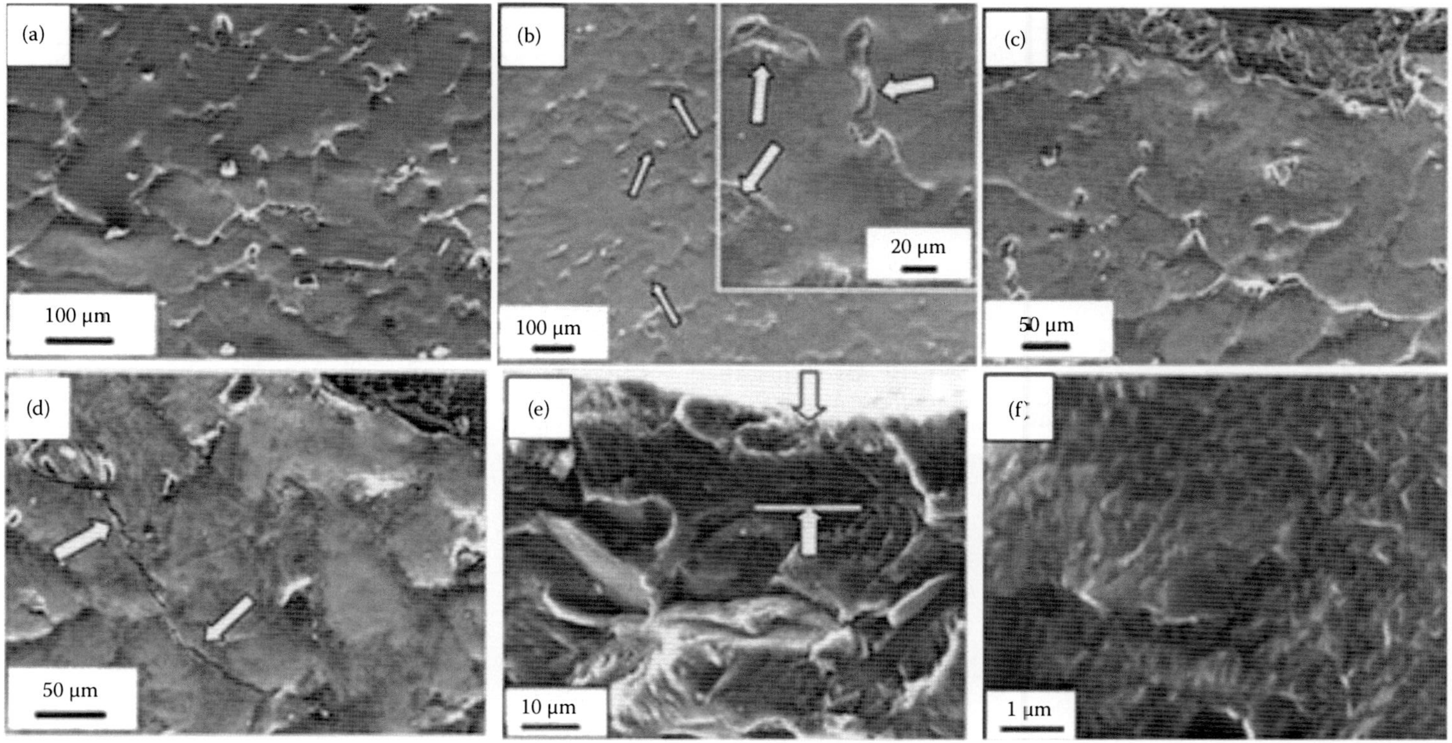

FIGURE 11.19 Surface structure of silumin processed with electron beam with number of pulses = 5, and fractured after 517,000 fatigue cycles. (a and b) Surface microstructure before fatigue tests. (c through f) Surface of post-fatigue tests samples. Arrows indicate (b) silicon particles, (d) formation of microcrack in the process of fatigue tests, and (e) silumin layer melted with electron beam. (From Konovalov, S. V., K. V. Alsaraeva, V. E. Gromov and Y. F. Ivanov (2015). Structure-phase states of silumin surface layer after electron beam and high cycle fatigue. The 12th International Conference on Gas Discharge Plasmas and Their Applications. *Journal of Physics: Conference Series* 652: 012028.)

TABLE 11.5
Change in Tribological Parameters of Silumin (Al–Si alloy) after PEB and Subsequent Fatigue Loading until Failure

Regime of Electron Beam Treatment	Wear Coefficient (10^{-6} mm^3/Nm)	Coefficient of Friction
Initial state without electron beam treatment	8927	0.527
132,000 cycles up to failure without electron beam treatment	13,520	0.520
466,000 cycles up to failure without electron beam treatment	13,920	0.444
20 J/cm^2; 150 μs pulse duration; 1 pulse 2,000 cycles up to failure	6466	0.457
20 J/cm^2; 150 μs pulse duration; 5 pulses 574,000 cycles up to failure	8135	0.480

Source: Konovalov, S. V., K. V. Alsaraeva, V. E. Gromov, Yu. F. Ivanov and O. A. Semina (2016). Fatigue variation of surface properties of silumin subjected to electron-beam treatment. *IOP Conference Series: Materials Science and Engineering* 110: 012012.

11.4.2 Microstructure and Properties of Magnesium and Its Alloys

Magnesium-alloys are light-weight materials having a high strength-to-weight [σ/ρ] ratio. However, due essentially to their poor surface properties, especially low corrosion resistance and inferior wear resistance, their full potential for use in the industries spanning automobile and aerospace has not yet been realized. In this context, PEB treatment is being explored as a viable option to modify the surface characteristics of both pure magnesium and magnesium-base alloys, and a few of these studies are briefly highlighted in this section.

Among the magnesium alloys, AZ91 alloy is the one that is most widely chosen and used. Presently this alloy is commercially available. For purpose of surface modification of this alloy, PEB having electron energy ~27 keV, pulse duration 1 μs, and energy density of 2.2 J/cm^2 per pulse was irradiated on the alloy substrate (Hao et al. 2005). Upon irradiation, 3 layers, namely: (i) melted layer, (ii) heat-affected zone, and (iii) intermediate zone, were formed. In this alloy, the second-phase particles present in the microstructure are coarse, having the composition $Mg_{17}Al_{12}$. Following PEB treatment, the coarse second-phase particles dissolved completely in the melted surface layer after 10 pulses of HCPEB irradiation resulting in the formation of a supersaturated solid solution structure. This is shown in Figure 11.20 (Hao et al. 2005).

The corrosion behavior of high current pulse electron beam treated AZ91 alloy was studied by using the electrochemical polarization method in a 5% sodium chloride (NaCl) solution. The corrosion results revealed that due to surface modification by the PEB treatment, the corrosion potential of the alloy shifted to a positive value (i.e., became more noble). When the number of PEB pulses was increased, the corrosion current decreased significantly by up to four orders when compared to the untreated AZ91alloy, thereby highlighting the improved corrosion resistance as a direct outcome of the PEB treatment (Figure 11.21) (Hao et al. 2005).

The PEB treatment also enhanced the concentration of aluminum on the surface. The improved corrosion resistance was attributed to the following: (i) the formation of homogeneous passive film on the pulse electron beam treated surface, and (ii) the dissolved $Mg_{17}Al_{12}$ phase that formed a supersaturated solid solution on the surface melted layer (Hao et al. 2005).

Alloy AZ31 is a popular choice for use in aerospace-related applications. The surface properties of this magnesium alloy must be improved with the prime intent of ensuring its better utilization. To improve the surface characteristics, surface treatment of the AZ31 alloy with a HCPEB having a

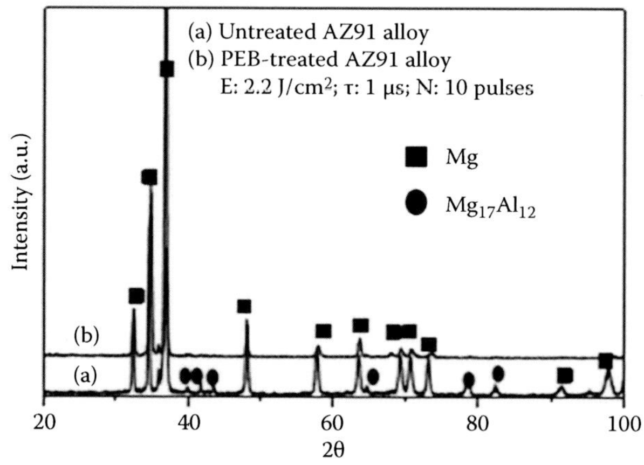

FIGURE 11.20 X-ray diffractograms of the AZ91 Mg alloy both before and after PEB treatment. (Used with permission from Hao, S., B. Gao, A. Wu, J. Zou, Y. Qin, C. Dong, J. An and Q. Guan (2005). Surface modification of steels and magnesium alloy by high current pulsed electron beam. *Nuclear Instruments and Methods in Physics Research B* 240: 646–652.)

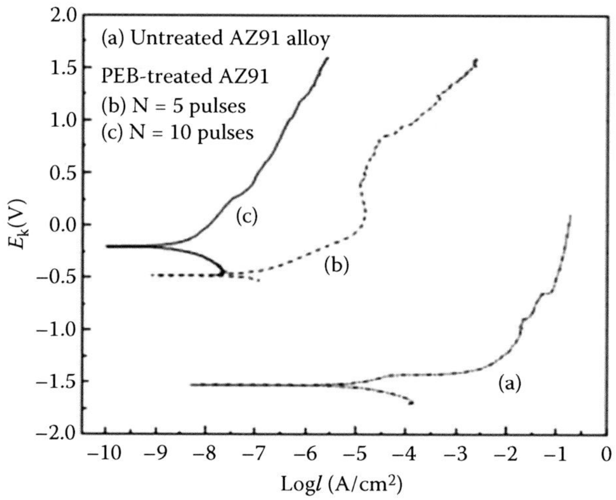

FIGURE 11.21 Polarization curves of the AZ91 alloy before and after PEB treatment. (Used with permission from Hao, S., B. Gao, A. Wu, J. Zou, Y. Qin, C. Dong, J. An and Q. Guan (2005). Surface modification of steels and magnesium alloy by high current pulsed electron beam. *Nuclear Instruments and Methods in Physics Research B* 240: 646–652.)

beam density of 2.5 J/cm^2 and the number of pulses varying from 5 to 15 was conducted (Gao et al. 2005). By using the evaporation mode during treatment, the formation and presence of microscopic craters on the surface was avoided. The surface melted layer was rich in aluminum and secondary phases in the supersaturated solid solution state. The transformed layer included (i) the thermal zone (heat-affected zone), and (ii) the dynamic stress zone, caused by the generation of shock waves as an outcome of the PEB treatment. The values of microhardness increased up until a depth of 500 μm (Figure 11.22) (Gao et al. 2005). An improvement in the surface properties quantified by way of wear and friction behavior (ball-on-flat configuration) revealed a significant decrease (20%) in the coefficient of friction of the PEB-treated surface when compared one-on-one with the untreated AZ31

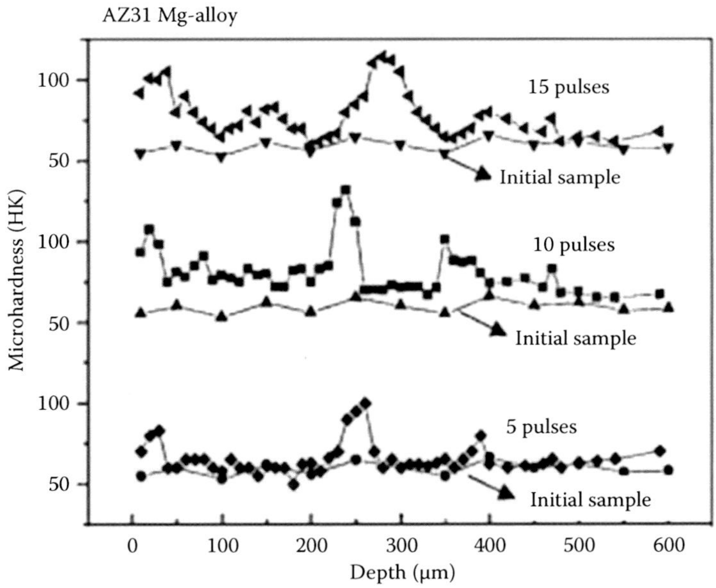

FIGURE 11.22 Variation of microhardness with depth before and after PEB treatment of the AZ31 alloy. (Used with permission from Gao, B., S. Hao, J. Zou, T. Grosdidier, L. Jiang, J. Zhou and C. Dong (2005). High current pulsed electron treatment of AZ31 Mg alloy. *Journal of Vacuum Science & Technology A* 23 (6): 1548–1553.)

alloy (Figure 11.23a) (Gao et al. 2005). The wear resistance of the AZ31 alloy that was treated with PEB improved significantly (i.e., decrease in wear rate) when compared to the untreated alloy. The number of pulses that were used to irradiate the surface played a dominant role in determining the wear behavior. The wear rate revealed a decrease with an increase in the number of pulses (Figure 11.23b) (Gao et al. 2005).

The formation of nanograins is yet another route to improve the wear and corrosion resistance of magnesium-based alloys. PEB treatment can induce the formation of nanograins on the surface due to its high surface heating and rapid solidification rates. When a PEB having an electron energy of

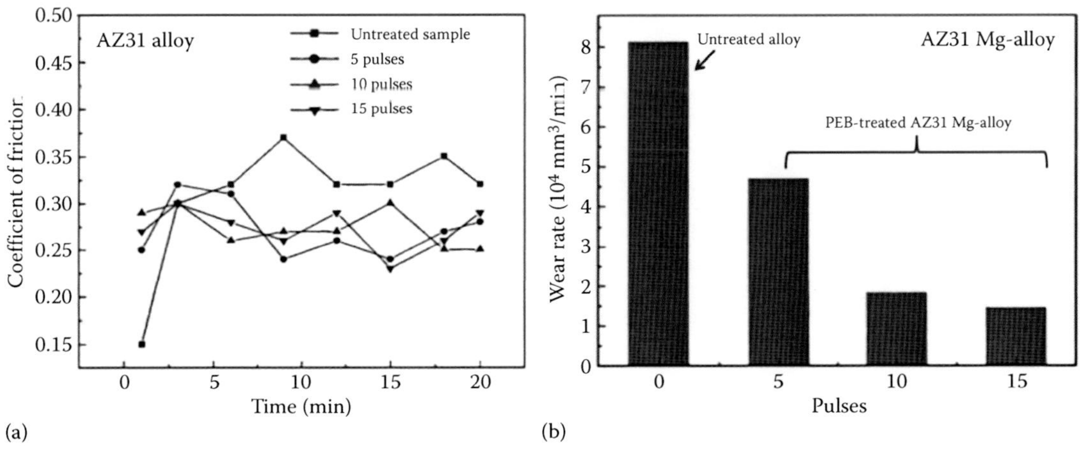

FIGURE 11.23 (a) Variation of friction coefficient as a function of time, and (b) variation of wear rate with the number of pulses, for AZ31 alloy before and after PEB treatment. (Used with permission from Gao, B., S. Hao, J. Zou, T. Grosdidier, L. Jiang, J. Zhou and C. Dong (2005). High current pulsed electron treatment of AZ31 Mg alloy. *Journal of Vacuum Science & Technology A* 23 (6): 1548–1553.)

30keV, pulse duration of 1 μs and an energy density of 3 J/cm^2 per pulse was irradiated on the AZ91 alloy, a thin layer of nanograined magnesium oxide (MgO) formed on the surface, below which substrate of the melted surface layer having a depth of about 10 μm was generated (Gao et al. 2007). The high surface temperature also resulted in the formation of a supersaturated solid solution phase, which otherwise would exist as the coarse intermetallic particles $Mg_{17}Al_{12}$. The dynamic stress from the shock wave produced due to PEB treatment resulted in a high gradient stress field on the subsurface that increased its hardness (Gao et al. 2007). An improvement in surface microstructural characteristics and hardness had an influence on the wear behavior of the PEB-treated sample. The PEB-treated magnesium alloy AZ91 showed a wear track having a width of 270 to 300 μm, in contrast to a wear track of width 440 μm in an untreated alloy (Figure 11.24a and b) (Gao et al. 2007). The PEB-treated sample also revealed a lower wear and lower coefficient of friction when compared one-on-one with the untreated counterpart (Figure 11.25a) (Gao et al. 2007). The PEB treatment also improved the corrosion resistance of this alloy to a remarkable extent (Figure 11.25b) [estimated

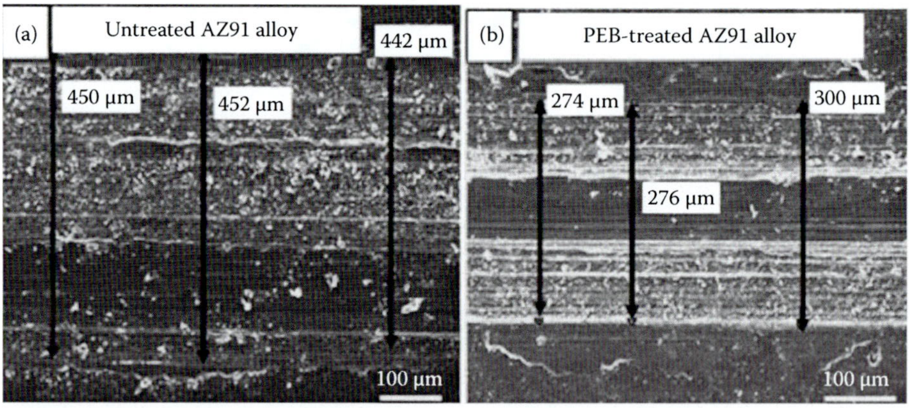

FIGURE 11.24 The wear track on magnesium alloy AZ91, showing (a) untreated and (b) PEB-treated with 10 pulses. (Used with permission from Gao, B., S. Hao, J. Zou, W. Wu, G. Tu and C. Dong (2007). Effect of high current pulsed electron beam treatment on surface microstructure and wear and corrosion resistance of an AZ91HP magnesium alloy. *Surface & Coatings Technology* 201: 6297–6303.)

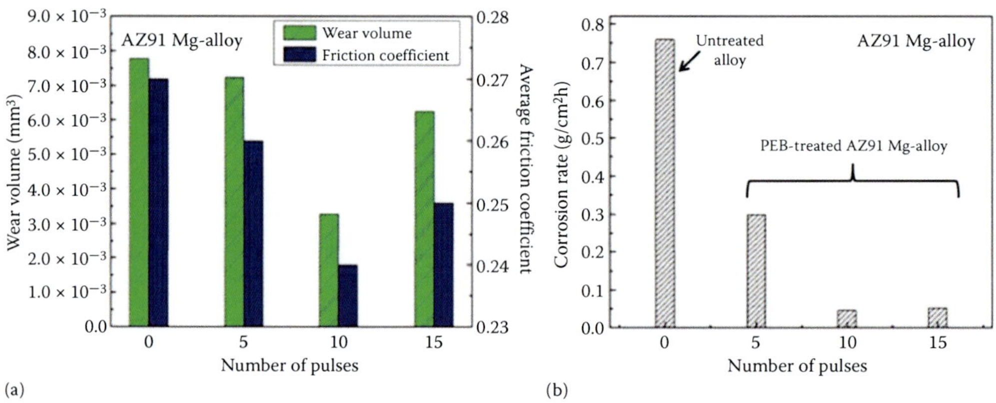

FIGURE 11.25 Variation of (a) wear volume and coefficient of friction, and (b) corrosion rate, with number of pulses for the AZ91 alloy after PEB treatment. (Used with permission from Gao, B., S. Hao, J. Zou, W. Wu, G. Tu and C. Dong (2007). Effect of high current pulsed electron beam treatment on surface microstructure and wear and corrosion resistance of an AZ91HP magnesium alloy. *Surface & Coatings Technology* 201: 6297–6303.)

from weight loss experienced in an immersion test] due to the formation and presence of a homogenous passive protective film on the PEB-treated surface (Gao et al. 2007).

Similar to the formation of nanostructures, a composite nanostructure can also be formed on the surface of a material using the PEB treatment. This is essentially due to its surface melting, mixing and rapid solidification effect. For the magnesium alloy AZ91, a composite nanostructure on the surface was obtained by using a PEB with an accelerating voltage of 27 keV, an energy density of 3 J/cm^2 and a pulse duration of 2.5 μs (Li et al. 2014). In addition to an enrichment of aluminum on the melted layer of the surface along with the formation of a supersaturated solid solution, a composite nanostructure at the surface was formed along with the presence of the $Mg_{17}Al_{12}$ phase (Figure 11.26a) (Li et al. 2014). The composite nanostructure with $Mg_{3.1}Al_{0.9}$ nanograins was formed because of diffusion of the aluminum coupled with the preferential evaporation of magnesium from the surface, due to the rapid solidification process associated with PEB treatment. The PEB surface modified alloy when tested in a solution of sodium chloride (NaCl) using the electrochemical method revealed an actual decrease in corrosion resistance (lower by 2 orders) when compared one-on-one with the untreated alloy (Figure 11.26b) (Li et al. 2014). The modified surface layer coupled with the presence of a composite nanostructure on the surface increased the corrosion resistance of the PEB-treated alloy (Li et al. 2014).

The number of electron pulses irradiating a material surface is an important factor that determines properties of the surface. A high pulse number within the same duration can cause the presence of surface defects, such as microscopic cracks. Such an occurrence was observed in the magnesium alloy AZ91 that PEB-treated with varying number of pulses (3 to 30) (Hao and Li 2016). The fine microscopic cracks were not observed in the subsurface for lower pulses (<10). However, for pulse numbers more than 10, the microscopic cracks were observed. This is shown in Figure 11.27 (Hao and Li 2016). For the PEB-treated surfaces, as thickness of the melted layer of the surface increased, rapid solidification resulted in tensile stresses on the surface that gave rise to fine microscopic cracks. The microhardness of the magnesium alloy AZ91 increased with an increasing number of pulses (Figure 11.28a) (Hao and Li 2016). When tested for corrosion performance, the PEB-treated magnesium alloy AZ91 revealed an increase in corrosion potential and a decrease in

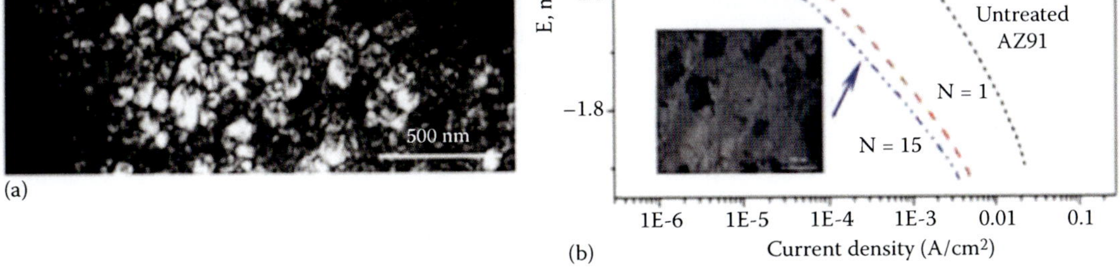

FIGURE 11.26 (a) Transmission electron micrograph (TEM) dark field image of the nano-grained $Mg_{3.1}A_{10.9}$ phase in the PEB-treated AZ91 alloy after 15 pulses. (b) Polarization curves of the untreated and pulsed electron beam (PEB)–treated AZ91 alloy. (Used with permission from Li, M. C., S. Z. Hao, H. Wen and R. F. Huang (2014). Surface composite nanostructures of AZ91 magnesium alloy induced by high current pulsed electron beam treatment. *Applied Surface Science* 303: 350–353.)

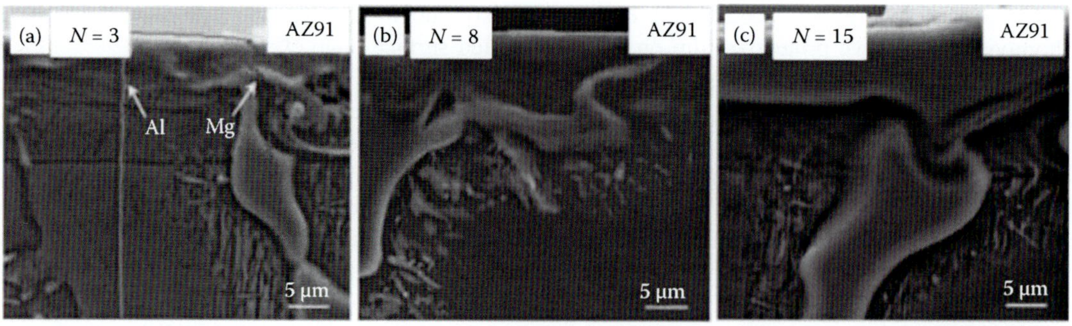

FIGURE 11.27 Scanning electron micrographs of AZ91 alloy after PEB treatment for varying number of pulses: (a) 3 pulses, (b) 8 pulses, and (c) 15 pulses. (Used with permission from Hao, S. and M. Li (2016). Producing nano-grained and Al-enriched surface microstructure on AZ91 magnesium alloy by high current pulsed electron beam treatment. *Nuclear Instruments and Methods in Physics Research B* 375: 1–4.)

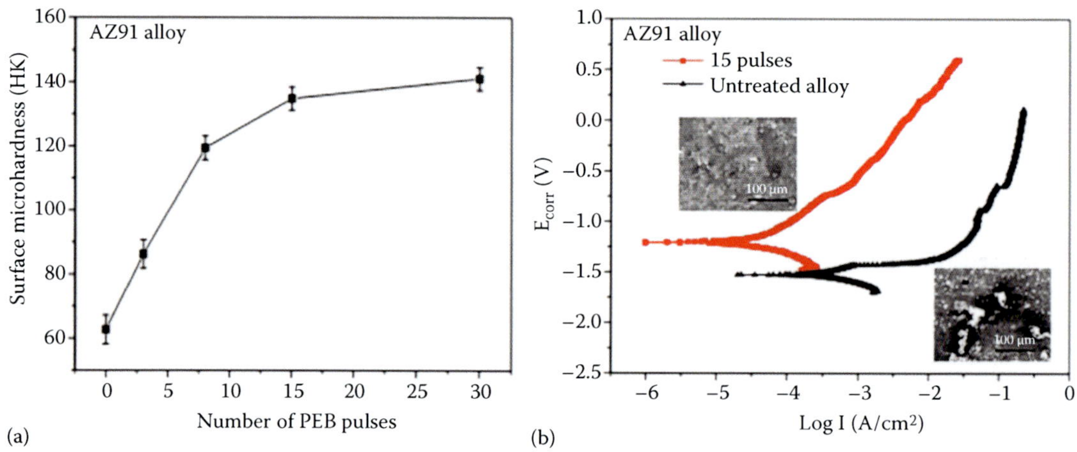

FIGURE 11.28 (a) Variation of surface microhardness with number of pulses, and (b) polarization curves for the magnesium alloy AZ91 both before and after PEB treatment. (Used with permission from Hao, S. and M. Li (2016). Producing nano-grained and Al-enriched surface microstructure on AZ91 magnesium alloy by high current pulsed electron beam treatment. *Nuclear Instruments and Methods in Physics Research B* 375: 1–4.)

corrosion current density, providing an indication of superior corrosion resistance. While the surface of the untreated sample was attacked by the corrosion pits upon exposure to a corrosive environment, the PEB-treated sample did not reveal any such features due to corrosion. The improved corrosion resistance as a direct outcome of the PEB treatment was due to the formation of an enriched aluminum-layer on the surface coupled with ultrafine grains due to rapid solidification (Figure 11.28b) (Hao and Li 2016).

11.4.3 Microstructure and Properties of Titanium and Its Alloys

Titanium alloys have through the years grown in stature to become a preferential choice for a spectrum of applications in the industry spanning aerospace. However, these alloys exhibit poor surface properties and wear resistance. Although their corrosion properties are superior when compared to few other structural alloys, the increasing demand and resultant use of the titanium-base alloys in applications spanning biomedical and nuclear energy require a high level of corrosion resistance in an aggressive environment. The earliest of the PEB treatment of the titanium alloys was studied in 1999 by the Institution of Aviation Materials, Russia (Nochovnaya 1999). Intense PEBs having an accelerating voltage of ~25 to 30 keV, a beam energy density of ~1 to 5 J/cm^2, a pulse duration of ~0.7 to 2.5 μs, and the number of pulses varying between 5 to 100, were used to surface treat the alloys of titanium. When compared one-on-one with an untreated alloy, the PEB-treated alloy exhibited 40 to 50% improvement in fatigue strength, and well over two orders of magnitude better resistance in both an oxidation and erosive environment (Nochovnaya 1999).

The titanium alloy TA15 (Ti–6.5Al–2Zr–1Mo–1V) is a commercially available alpha titanium (α-Ti) alloy. With the view of improving its corrosion properties, the TA15 titanium alloy was treated with a low energy HCPEB at accelerating voltages of 23.4 keV and 27 keV, corresponding to beam energy density of 2.6 J/cm^2 and 3 J/cm^2, and the number of pulses set to 5 and 10 (Zhang et al. 2011b). As would be seen for the PEB-treated material, a wavy front was observed on the surface. This indicates an evaporation of the surface chemical species and resultant formation of a surface melted layer, due to the high surface temperatures arising from high energy of the electron beam (Zhang et al. 2011). The grain size was equiaxed and in the submicron range (~800 nm to 1.2 μm). Further, rapid solidification resulted in the formation of ultrafine lamellar features that indicated a transformation from β (high temperature phase) to α' marten site (Zhang et al. 2011b). The corrosion behavior of the titanium alloy TA15 under potentio-dynamic condition in an aqueous sodium chloride (NaCl) solution revealed significant improvement in corrosion resistance (Figure 11.29) (Zhang et al. 2011b). A decrease in both corrosion current density and corrosion rate, in synergism with an increase in polarisation resistance proved that the surface characteristics of the PEB-treated titanium alloy TA15 was better than the untreated counterpart (Zhang et al. 2011b).

Similarly, when the titanium alloy TA15 was PEB-treated [accelerating voltage ~20 to 40 keV, number of pulses ~3, pulsed time ~15 s, and energy density ~15 J/cm^2], a fine grain microstructure having a grain size of ~168 nm was formed on the surface. The hardness across the PEB-treated depth indicated an increased level of hardness at the surface, which revealed a noticeable decrease with an increase in depth, but still higher than the untreated alloy (Figure 11.30a) (Gao 2013). The high hardness values measured at the surface was essentially due to the melted surface layer, wherein transformation from the high temperature β phase to α' martensite (~90 nm martensite laths, as shown in Figure 11.30b) had taken place (Gao 2013).

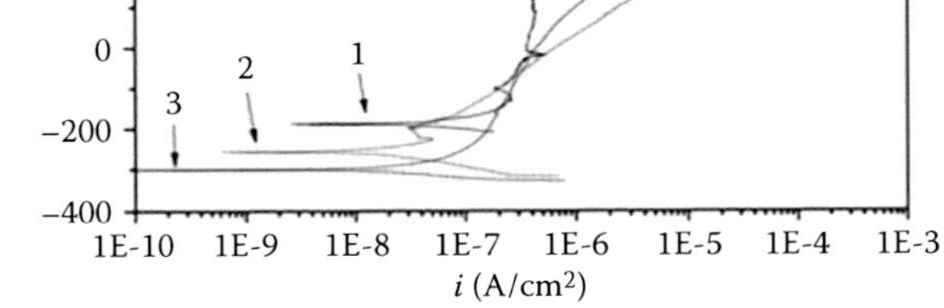

FIGURE 11.29 Polarization curves of the untreated and PEB-treated titanium alloy TA15 alloy. (Used with permission from Zhang, X. D., J. X. Zou, S. Weber, S. Z. Hao, C. Dong and T. Grosdidier (2011). Microstructure and property modifications in a near α Ti alloy induced by pulsed electron beam surface treatment. *Surface & Coatings Technology* 206: 295–304.)

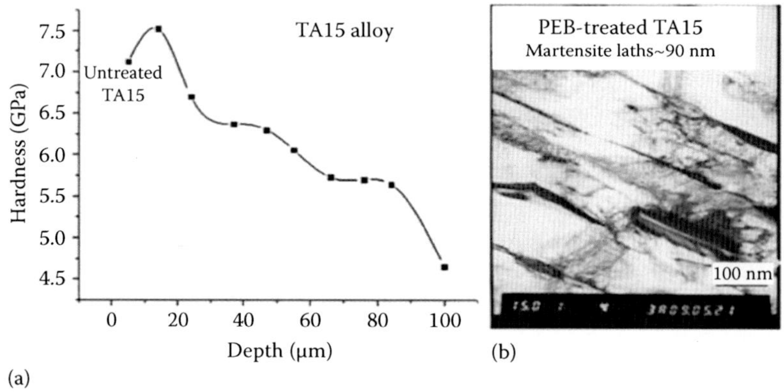

FIGURE 11.30 (a) Variation of hardness with depth for the titanium alloy TA15 alloy. (b) Transmission electron micrograph of the PEB-treated titanium alloy TA15, showing lath-like morphology of the martensite, with an average size of ~90 nm. (Used with permission from Gao, Y. K. (2013). Influence of pulsed electron beam treatment on microstructure and properties of TA15 titanium alloy. *Applied Surface Science* 264: 633–635.)

Although pure titanium exhibits good strength properties, its surface properties, such as wear and corrosion resistance, can be categorized to be poor. Modification of the surfaces of pure titanium is usually carried out to improve its surface characteristics. For this purpose, pure titanium was treated with a PEB (low energy, high current) using an accelerating voltage of 25.2 keV and a pulse time of 15 and 25 (Zhang et al. 2011a). Due to electron beam treatment, the surface of the pure titanium revealed a melted layer. Following rapid solidification, the transformation from predominantly β to α' martensite occurred with ultrafine microstructure. The PEB surface modified titanium showed microhardness values that were ~60% more than the untreated surface, as shown in Figure 11.31 (Zhang et al. 2011a).

Under potentio-dynamic polarization condition, the corrosion behavior of pure titanium in aqueous sodium chloride (NaCl) revealed a remarkable improvement in corrosion resistance. The corrosion potential became more positive (noble), while corrosion current density decreased by well over two-orders of magnitude. This is shown in Figure 11.32 and summarized in Table 11.6 (Zhang et al. 2011a). The formation of an ultrafine structure, redistribution of the impurity elements coupled

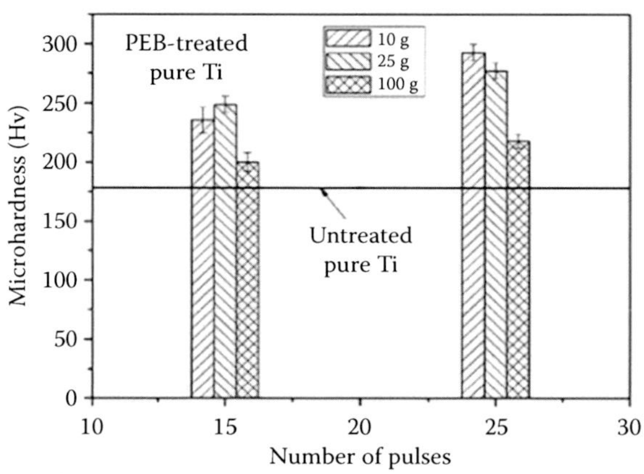

FIGURE 11.31 Surface hardness of pure titanium both before and after PEB treatment. (Used with permission from Zhang, X. D., S. Z. Hao, X. N. Li, C. Dong and T. Grosdidier (2011). Surface modification of pure titanium by pulsed electron beam. *Applied Surface Science* 257: 5899–5902.)

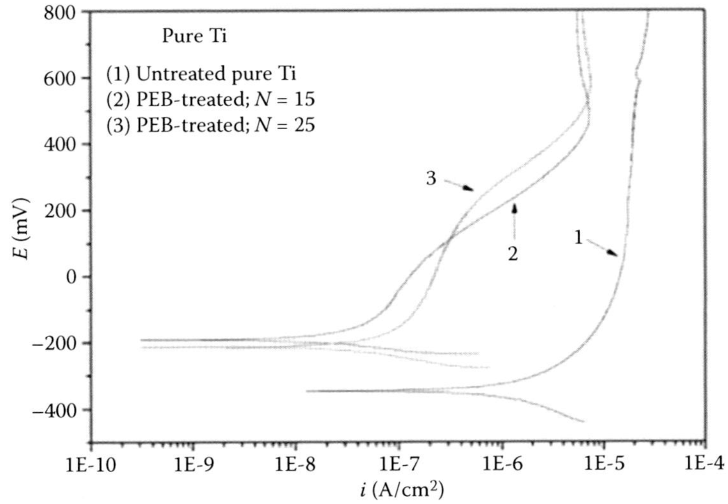

FIGURE 11.32 Potentio-dynamic polarization curves of pure titanium in aqueous sodium chloride (NaCl) solution both before and after PEB treatment. (Used with permission from Zhang, X. D., S. Z. Hao, X. N. Li, C. Dong and T. Grosdidier (2011). Surface modification of pure titanium by pulsed electron beam. *Applied Surface Science* 257: 5899–5902.)

TABLE 11.6
A Comparison of Electrochemical Corrosion Parameters of Untreated and PEB-Treated Pure Titanium with Corrosion Potential (E_{corr}), Corrosion Current (I_{corr}), and Corrosion Resistance (R_p)

Sample	Pulses	E_{corr} (mV)	I_{corr} (μA/cm^2)	R_p (kΩ)
Untreated pure Ti	0	−346.4	2.812	20.49
PEB-treated pure Ti	15	−188.9	0.0454	602.0
	25	−212.1	0.0943	389.5

Source: Zhang, X. D., S. Z. Hao, X. N. Li, C. Dong and T. Grosdidier (2011). Surface modification of pure titanium by pulsed electron beam. *Applied Surface Science* 257: 5899–5902.

with the formation of a protective surface layer, contributed to improving the corrosion resistance (Zhang et al. 2011a).

Surface topography, microstructure and hardness are properties that determine the surface integrity of a material. These properties were investigated for an industrially pure titanium (TA2 titanium) using a low energy HCPEB treatment for varying beam density (10 to 35 J/cm^2), pulse duration (100 to 200 μs) and number of pulses (1 to 40) (Gao 2011). The PEB process parameters, such as: (i) pulse energy density, (ii) pulse duration, and (iii) number of irradiation pulses, determine the roughness of the surface. The surface roughness measurement of the PEB treated TA2 titanium revealed a reduction in surface roughness (<0.1 μm). When the process parameters were carefully selected, R_a values as low as 0.04 μm was achieved (Gao 2011). Due on account to the heat stresses generated during the PEB treatment, the microstructure of the PEB-treated material revealed dislocations, resulting because of plastic deformation induced on the modified surface layer. By changing the process parameters, such as: (i) pulse duration, (ii) number of pulses, and (iii) energy density, the values of hardness can be controlled (Figure 11.33a through c) (Gao 2011).

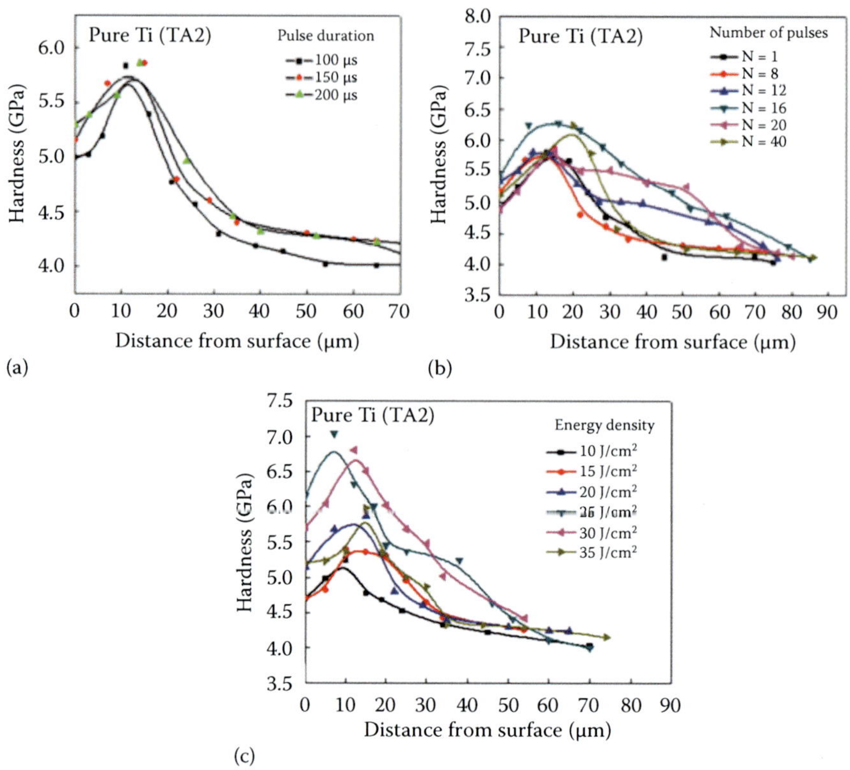

FIGURE 11.33 Variation of nano-hardness with (a) different pulse durations, (b) different pulse times, and (c) number of pulses, for the PEB-treated pure titanium (TA2 titanium). (Used with permission from Gao, Y. K. (2011). Surface modification of TA2 pure titanium by low energy high current pulsed electron beam treatments. *Applied Surface Science* 257: 7455–7460.)

The Ti–6Al–4V alloy is widely chosen and used as an implant material for the human body. Hence, both wear resistance and corrosion resistance of the alloy is an important factor that must be considered. While surface modification techniques such as ion implantation and coatings have been studied, in more recent years the PEB treatment is being explored for surface modification. For example, PEB treatment was conducted on the Ti–6Al–4V alloy using an accelerating voltage of ~33 keV, a pulse duration of ~2.5 μs, a pulse energy density of ~9 J/cm^2 and pulse number varying between 1 to 50 (Guo et al. 2013). The key microstructural features following the PEB treatment revealed phase transformation from a two-phase (α+β) microstructure to a single phase (α'-Ti) microstructure coupled with a significant reduction in the grain size (~100 nm). The corrosion behavior of the PEB-treated alloy in an aqueous solution of sodium chloride (NaCl) revealed much better corrosion resistance when compared one-on-one with the untreated alloy. Further, the number of pulses used to treat the sample played a dominant role in determining the corrosion resistance (Table 11.7) (Guo et al. 2013).

Similar to the Ti–6Al–4V alloy, the Ti–6Al–7Nb alloy is a preferred choice as a bio-implant material for purpose of dental implants and orthopedic implants (Challa et al. 2013). In a study on the Ti–6Al–7Nb alloy, PEB treatment was conducted with the primary intent of improving the surface characteristics while concurrently understanding the corrosion behavior (Kim and Park 2015). For an energy density of 7 to 10 J/cm^2 and an anode-target distance of 30 mm, the surfaces were irradiated with PEB treatment (Kim and Park 2015). The process parameters did play an important role in determining the surface characteristics as was evident from the depth of the resolidified layer following PEB treatment (Figure 11.34) (Kim and Park 2015). Potentio-dynamic polarization behavior

TABLE 11.7
A Comparison of Corrosion Potential and Corrosion Current of Untreated and PEB-Treated Ti–6Al–4V

Number of Pulses (*N*)	Potential (mV)	Current (mA)
0	−709	5.0×10^{-5}
1	−217	5.6×10^{-5}
5	−202	6.4×10^{-5}
20	−310	5.4×10^{-5}
30	−313	4.5×10^{-5}
50	−330	6.6×10^{-5}

Source: Guo, G., G. Tang, X. Ma, M. Sun and G. E. Ozur (2013). Effect of high current pulsed electron beam irradiation on wear and corrosion resistance of Ti6Al4V. *Surface & Coatings Technology* 229: 140–145.

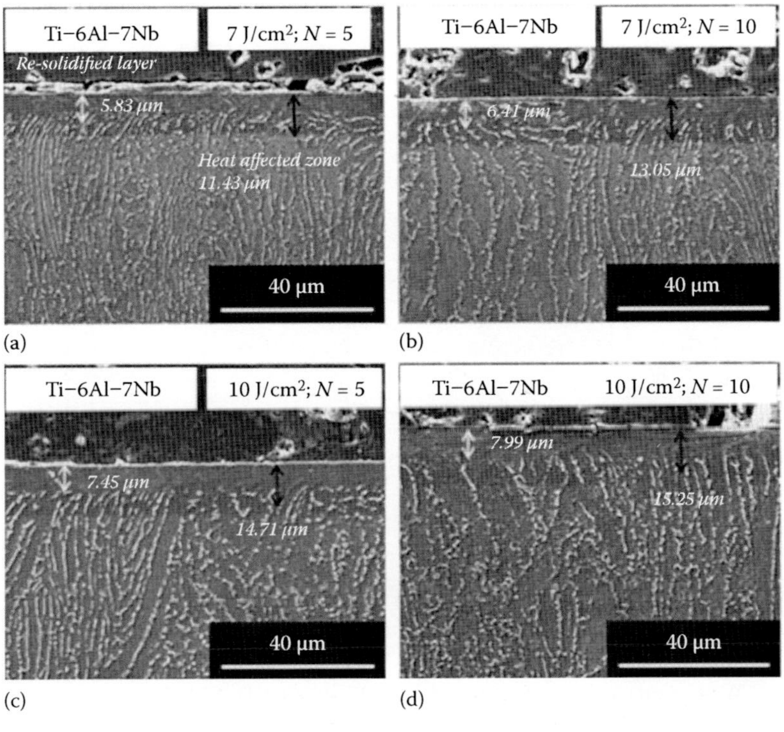

FIGURE 11.34 Cross-sectional scanning electron micrographs of the Ti–6Al–7Nb alloy after PEB treatment with varying energy density and pulse numbers: (a) 7 J/cm^2 and 5 pulses, (b) 7 J/cm^2 and 10 pulses, (c) 10 J/cm^2 and 5 pulses, and (d) 10 J/cm^2 and 10 pulses. (Used with permission from Kim, J. S. and H. W. Park (2015). Influence of large pulsed electron beam (LPEB) on the corrosion resistance of Ti–6Al–7Nb alloys. *Corrosion Science* 90: 153–160.)

in an aqueous solution of sodium chloride (NaCl) revealed the PEB-treated alloy to exhibit excellent corrosion resistance (a more noble corrosion potential, low corrosion current density and a high polarisation resistance) when compared to the untreated alloy (Figure 11.35) (Kim and Park 2015). The formation of re-passivation (protective) surface layer coupled with an increase in surface hydrophobicity (Figure 11.36), as determined by measurement of wettability (contact angle),

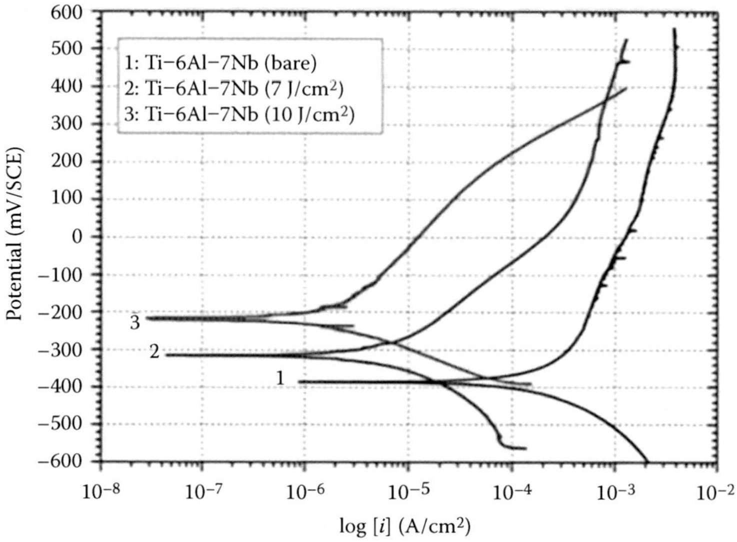

FIGURE 11.35 Potentio-dynamic polarization curves for the Ti–6Al–7Nb alloy both before and after PEB treatment. (Used with permission from Kim, J. S. and H. W. Park (2015). Influence of large pulsed electron beam (LPEB) on the corrosion resistance of Ti–6Al–7Nb alloys. *Corrosion Science* 90: 153–160.)

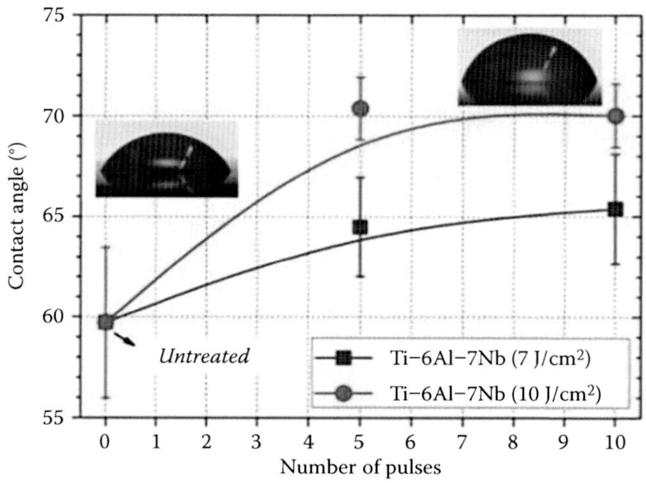

FIGURE 11.36 Water contact angle of untreated and PEB-treated Ti–6Al–7Nb alloy. (Used with permission from Kim, J. S. and H. W. Park (2015). Influence of large pulsed electron beam (LPEB) on the corrosion resistance of Ti–6Al–7Nb alloys. *Corrosion Science* 90: 153–160.)

were considered to be the key reasons for the improved corrosion resistance of the PEB-treated alloy (Kim and Park 2015).

11.5 SUMMARY

The metals and their alloy counterparts based on aluminum, magnesium and titanium are preferential choices for use in a spectrum of applications spanning both performance-critical and non-performance-critical applications due to their light-weight and superior combination of properties. A wider utilization of these materials in the industries spanning aerospace, automotive and biomedical requires a substantial improvement in their surface properties. Surface engineering is an economical and viable method to improve the surface properties of a material, such as hardness of the surface, wear resistance, corrosion resistance, fatigue resistance and oxidation resistance. Amongst the various surface modification methods, high energy processes based on the use of a pulsed beam are promising. Of the pulsed beam techniques, to include laser beam, plasma beam, ion-beam and electron beam, the PEB technique is promising and offers notable advantages over the other methods, which facilitates the treatment of large areas, sizeable depth of penetration, along with a low ionization loss. When a PEB is bombarded on the surface of the chosen metal, the surface experiences the following modifications at varying depths: (i) surface melted layer, (ii) heat-affected zone, and (iii) high stress zone arising due to the shock wave resulting from electron beam bombardment. By a proper selection of the process parameters, such as (a) accelerating voltage, (b) beam energy density, (c) number of pulses, and (d) pulse duration, careful control and/or manipulation of the surface characteristics can be made possible. When PEB treatment of the surfaces of the light metals aluminum, magnesium, and titanium are carried out, one or more microstructural transformations occur:

i. selective enrichment of the element on the surfaces
ii. formation of a protective surface layer
iii. a near-homogenous dispersion of the second phases through the microstructure
iv. a refinement in grain structure (nanocrystalline/ultrafine grains)
v. formation of nonequilibrium phases due to rapid solidification, and
vi. generation of a high dislocation density due to the shock wave.

These microstructural modifications help provide an improvement to the surface properties to include the following: (a) surface hardness, (b) wear resistance, and (c) corrosion resistance. The fatigue resistance and oxidation resistance are also expected to improve. Owing to the superior properties of PEB-treated light metals/alloys over the untreated counterpart, it is envisioned that the PEB treatment will gradually take over to become a future surface modification technique for the purposes of engineering the surfaces of materials for a wide range of critical applications.

REFERENCES

Alpha-Omega Power Technologies (2008). Ion surface treatment summary. www.alphaomegapt.com _pulsedion-8 (Last Accessed: May 1, 2017).

An, J., X. X. Shen, Y. Lu, Y. B. Liu, R. G. Li, C. M. Chen and M. J. Zhang (2006). Influence of high current pulsed electron beam treatment on the tribological properties of Al–Si–Pb alloy. *Surface & Coatings Technology* 200: 5590–5597.

Asian Metal (2017). Titanium. http://metalpedia.asianmetal.com/metal/titanium/applications.shtml (Last Accessed: May 1, 2017).

Boyer, R. R. (2010). Attributes, characteristics and applications of titanium and its alloys. *JOM: The Minerals, Metals & Materials Society* 62 (5): 35–43.

Brown, M. S. and C. B. Arnold (2010). Laser precision microfabrication: Fundamentals of laser–material interaction and application to multiscale surface modification. *Springer Series in Materials Science* 135, Springer-Verlag Berlin Heidelberg.

Burakowski, T., and T. Wierzchon (1998). *Surface Engineering of Metals: Principles, Equipment, Technologies*. CRC Press, 1998.

Byrka, O. V., A. N. Bandura, V. V. Chebotarev, I. E. Garkusha, J. Langner, M. J. Sadowski and V. I. Tereshi (2002). Application of pulsed plasma streams for materials alloying and coatings modification. *Problems of Atomic Science and Technology: Plasma Physics* 7: 173–175.

Challa, V., S. Mali, R. Misra (2013). Reduced toxicity and superior cellular response of preosteoblasts to Ti–6Al–7Nb alloy and comparison with Ti–6Al–4V. *Journal of Biomedical Materials Research A* 101: 2083–2089.

Czerwinski, F. (2012). Heat treatment—Conventional and novel applications: thermochemical treatment of metals. *InTechOpen*.

Dong, C., A. Wu, S. Hao, J. Zou, Z. Liu, P. Zhong, A. Zhang, T. Xu, J. Chen, J. Xu, Q. Liu and Z. Zhou (2003). Surface treatment by high current pulsed electron beam. *Surface and Coatings Technology* 163–164: 620–624.

Draper, C. W. and P. Mazzoldi (1986). *Laser Surface Treatment of Metals*. Martinoff Nijhoff Publishers, Dordrecht/Boston/Lancaster.

Eichmeier, J. A. and M. Thumm (2008). *Vacuum Electronics: Components and Devices*. Springer, Berlin, Heidelberg, Germany.

Fujii, H., K. Takahashi and Y. Yamashita (2003). Application of titanium and alloys for automobile parts. *Nippon Steel Technical Report* 88: 70–75.

Gao, B., L. Hu, S. W. Li, Y. Hao, Y. D. Zhang, G. F. Tu and T. Grosdidier (2015). Study on the nanostructure formation mechanism on the hypereutectic Al–17Si alloy induced by pulsed electron beam. *Applied Surface Science* 346: 147–157.

Gao, B., S. Hao, J. Zou, T. Grosdidier, L. Jiang, J. Zhou and C. Dong (2005). High current pulsed electron treatment of AZ31 Mg alloy. *Journal of Vacuum Science & Technology A* 23 (6): 1548–1553.

Gao, B., S. Hao, J. Zou, W. Wu, G. Tu and C. Dong (2007). Effect of high current pulsed electron beam treatment on surface microstructure and wear and corrosion resistance of an AZ91HP magnesium alloy. *Surface & Coatings Technology* 201: 6297–6303.

Gao, Y. K. (2011). Surface modification of TA2 pure titanium by low energy high current pulsed electron beam treatments. *Applied Surface Science* 257: 7455–7460.

Gao, Y. K. (2013). Influence of pulsed electron beam treatment on microstructure and properties of TA15 titanium alloy. *Applied Surface Science* 264: 633–635.

Guo, G., G. Tang, X. Ma, M. Sun and G. E. Ozur (2013). Effect of high current pulsed electron beam irradiation on wear and corrosion resistance of Ti6Al4V. *Surface & Coatings Technology* 229: 140–145.

Gromov, V. E., Yu. F. Ivanov, A. M. Glezerd, S. V. Konovalov and K. V. Alsaraeva (2015). Structural evolution of silumin treated with a high intensity pulse electron beam and subsequent fatigue loading up to failure. *Bulletin of the Russian Academy of Sciences-Physics* 79 (9): 1169–1172.

Gupta, M. and N. M. L. Sharon (2011). *Magnesium, Magnesium Alloys, and Magnesium Composites*. John Wiley & Sons, New Jersey, USA.

Gwynne, B. and P. Lyon (2007). Magnesium alloys in aerospace applications, past concerns, current solutions. Fifth Triennial International Aircraft Fire & Cabin Safety Research Conference, Atlantic City, New Jersey, USA, October 29–November 1.

Hao, S. and M. Li (2016). Producing nano-grained and Al-enriched surface microstructure on AZ91 magnesium alloy by high current pulsed electron beam treatment. *Nuclear Instruments and Methods in Physics Research B* 375: 1–4.

Hao, S., B. Gao, A. Wu, J. Zou, Y. Qin, C. Dong, J. An and Q. Guan (2005). Surface modification of steels and magnesium alloy by high current pulsed electron beam. *Nuclear Instruments and Methods in Physics Research B* 240: 646–652.

Hao, S., S. Yao, J. Guan, A. Wu, P. Zhong and C. Dong (2001). Surface treatment of aluminium by high current pulsed electron beam. *Current Applied Physics* 1 (2001): 203–208.

Hao, S. Z., Y. Qin, X. X. Mei, B. Gao, J. X. Zuo, Q. F. Guan, C. Dong and Q. Y. Zhang (2007). Fundamentals and applications of material modification by intense pulsed beams. *Surface & Coatings Technology* 201: 8588–8595.

Hao, Y., B. Gao, G. F. Tu, S. W. Li, C. Dong and Z. G. Zhang (2011a). Improved wear resistance of Al–15Si alloy with a high current pulsed electron beam treatment. *Nuclear Instruments and Methods in Physics Research B* 269: 1499–1505.

Hao, Y., B. Gao, G. F. Tu, S. W. Li, S. Z. Hao and C. Dong (2011b). Surface modification of Al–20Si alloy by high current pulsed electron beam. *Applied Surface Science* 257: 3913–3919.

Harry, J. E. (2010). *Introduction to Plasma Technology-Science, Engineering and Applications*. Wiley-VCH Verlag GmBH & Company, Germany.

Ivanova, Yu. F. Ivanov, K. V. Aksenovac, V. E. Gromovc, S. V. Konovalov, and E. A. Petrikova (2016). An increase in fatigue service life of eutectic silumin by electron beam treatment. *Russian Journal of Non-Ferrous Metals* 57 (3): 236–242.

Jurgen H. (2011). Aluminium in innovative light-weight car design. *Materials Transactions* 52 (5) 818–824.

Karlsruhe Institute of Technology (2014). Surface modification of materials using pulsed electron beams. https://www.ihm.kit.edu/english/300.php (Last Accessed: June 1, 2017).

Kim, J. S. and H. W. Park (2015). Influence of large pulsed electron beam (LPEB) on the corrosion resistance of Ti–6Al–7Nb alloys. *Corrosion Science* 90: 153–160.

Kim, J. S., W. J. Lee, and H. W. Park (2016). The state of the art in the electron beam manufacturing processes. *International Journal of Precision Engineering and Manufacturing* 17 (11): 1575–1585.

Kondyurin, A. and M. Bilek (2008). *Ion Beam Treatment of Polymers: Application Aspects from Medicine to Space*. Elsevier, Netherlands.

Konovalov, S. V., K. V. Alsaraeva, V. E. Gromov and Yu. F. Ivanov (2015a). Fatigue life of silumin irradiated by high intensity pulsed electron beam. *IOP Conference Series: Materials Science and Engineering* 91: 012029.

Konovalov, S. V., K. V. Alsaraeva, V. E. Gromov, Yu. F. Ivanov and O. A. Semina (2016). Fatigue variation of surface properties of silumin subjected to electron-beam treatment. *IOP Conference Series: Materials Science and Engineering* 110: 012012.

Konovalov, S. V., K. V. Alsaraeva, V. E. Gromov and Y. F. Ivanov (2015b). Structure-phase states of silumin surface layer after electron beam and high cycle fatigue. The 12th International Conference on Gas Discharge Plasmas and Their Applications. *Journal of Physics: Conference Series* 652: 012028.

Korenev, S. A., A. S. Korenev, I. V. Puzynin, V. N. Samoilov and A. N. Sissakian (1996). The use of pulsed power ion/electron beams for studying of units of electronuclear reactor. *Proceedings of the 2nd International conference on accelerator-driven transmutation technologies and applications*. Kalmar (Sweden); 28 (21) 3–7 Jun 1996.

Kulekci, M. K. (2008). Magnesium and its alloys applications in automotive industry. *International Journal of Advanced Manufacturing Technology* 39: 851–865.

Laouar, L. H. Hamadache, S. Saad and A. Bouchelaghem (2009). Mechanical surface treatment of steel-optimization parameters of regime. *Physics Procedia* 2 (3): 1213–1221.

Li, M. C., S. Z. Hao, H. Wen and R. F. Huang (2014). Surface composite nanostructures of AZ91 magnesium alloy induced by high current pulsed electron beam treatment. *Applied Surface Science* 303: 350–353.

Morozov, A. I. (1970–1979). Plasma accelerators. *The Great Soviet Encyclopedia*, 3rd Edition. (From http://encyclopedia2.thefreedictionary.com/Plasma+Accelerators). Retrieved April 20, 2017.

Mueller, G., V. Engelko, A. Weisenburger and A. Heinzel (2005). Surface alloying by pulsed intense electron beams. *Vacuum* 77: 469–474.

Nochovnaya, N. A. (1999). Surface modification of titanium alloys by intense pulsed electron beams. *3rd International Conference on Interaction of Radiation with Solids* October 6-8, Minsk, Belarus.

Özbek, Y. Y. (2017). Effect of heat treatment and pulse plasma process on surface properties of steels. *Acta Physica Polonica A* 131(1): 182–185.

Özbek, Y. Y. and M. Durman (2015). Surface behaviour of AISI 4140 modified with the pulsed-plasma technique. *Materials and Technology* 49 (3): 441–445.

Polmear, I. J. (1995). *Light Alloys: Metallurgy of Light Metals*. 3rd Edition, Wiley, USA.

Rudskoy, A. I., G. E. Kodzhaspirov, J. Kliber and Ch. Apostolopoulos (2016). Advanced metallic materials and processes. *Materials Physics and Mechanics* 25: 1–8.

Stinnett, R. W., R. G. Buchheit, E. L. Neau, M. T. Crawford, K. P. Lamppa and T. J. Renk (1995). Ion beam surface treatment: A new technique for thermally modifying surfaces using intense, pulsed ion beams. *IEEE Xplore*.

Sudhakar, K. V. and M. E. Haque (2015). Passivity effects of biomedical titanium alloy with chemical etching. *Journal of Mechatronics* 3 (2): 114–116.

Tharumarajah, A. and P. Koltun (2007). Is there an environmental advantage of using magnesium components for light-weighting cars? *Journal of Cleaner Production* 15 (11–12): 1007–1013.

Total Materia (2004a). Thermo-mechanical treatment of aluminium alloys. http://www.totalmateria.com/Article105.htm (Last Accessed: June 15, 2017).

Total Materia (2004b). Aircraft and aerospace applications: Part 1 & 2. http://www.totalmateria.com/Article96.htm (Last Accessed: June 15, 2017).

U.S. Laser Corporation (2016). Laser heat treating. http://www.uslasercorp.com/envoy/heattreating.html (Last Accessed: May 2017).

U.S. Laser Corporation (2017). Pulsed laser beam characterization. Technical Note, (Last Accessed: May, 2017).

University of Miskolc (2017). Laser surface treatment. http://emrtk.uni-miskolc.hu/projektek/adveng/home/kurzus/korsz_anyagtech/1_konzultacio_elemei/laser_surface_treat.htm (Last Accessed: May, 2017).

Uno, Y., A. Okada, K. Uemura, P. Raharjo, T. Furukawa and K. Karato (2005). High-efficiency finishing process for metal mold by large-area electron beam irradiation. *Precision Engineering* 29 (4): 449–455.

Venetti, A. C. (2008). *Progress in Materials Science Research*. Nova Science Publisher, New York.

Walker, J. C., J. W. Murray, M. Nie, R. B. Cook and A. T. Clare (2014). The effect of large-area pulsed electron beam melting on the corrosion and microstructure of a Ti6Al4V alloy. *Applied Surface Science* 311 (2014) 534–540.

Weglowski, M. S. and A. Pietras (2011). Friction stir processing—Analysis of the process. *Archives of Metallurgy and Materials* 56 (3): 779–788.

Wiedemann, H. (2007). *Particle Accelerator Physics*. Springer, New York.

Zhang, X. D., J. X. Zou, S. Weber, S. Z. Hao, C. Dong and T. Grosdidier (2011b). Microstructure and property modifications in a near α Ti alloy induced by pulsed electron beam surface treatment. *Surface & Coatings Technology* 206: 295–304.

Zhang, X. D., S. Z. Hao, X. N. Li, C. Dong and T. Grosdidier (2011a). Surface modification of pure titanium by pulsed electron beam. *Applied Surface Science* 257: 5899–5902.

Zhukeshov, A. M., A. T. Gabdullina, A. U. Amrenova and S. A. Ibraimova (2014). Hardening of structural steel by pulsed plasma treatment. *Journal of Nano and Electronic Physics* 6 (3) 03066.

12 Advances in Fabrication of Functionally Graded Materials
Modeling and Analysis

Ankit Gupta and Mohammad Talha

CONTENTS

12.1 INTRODUCTION

Over the last four decades, the exploitation of composite materials in engineering structures has been progressively diversifying from sports equipment and high-technology automobile, to aerospace and biomedical applications. Composite materials are basically a combination of two or more dissimilar materials that are used together in order to accomplish a new set of characteristics that neither of the constituent materials could achieve individually (Gupta and Talha 2015). It has long been that the multilayered composite materials are extensively used in aerospace, spacecraft, civil engineering, nuclear, and other various industries because of their outstanding features such as the high stiffness-to-weight and strength-to-weight ratio, dimensional stability, design flexibility, and part consolidation (Kant and Mallikarjuna 1989). But at the same time, these advanced composites exhibit delamination, matrix cracking, adhesive bond separation, stress concentration, and damage inspection problems. These problems can be alleviated by replacing conventionally multilayered composites with graded materials. Bever and Duwez (1972) studied the theoretical inferences of graded structure composite materials in 1972. However, because of the inadequate fabrication

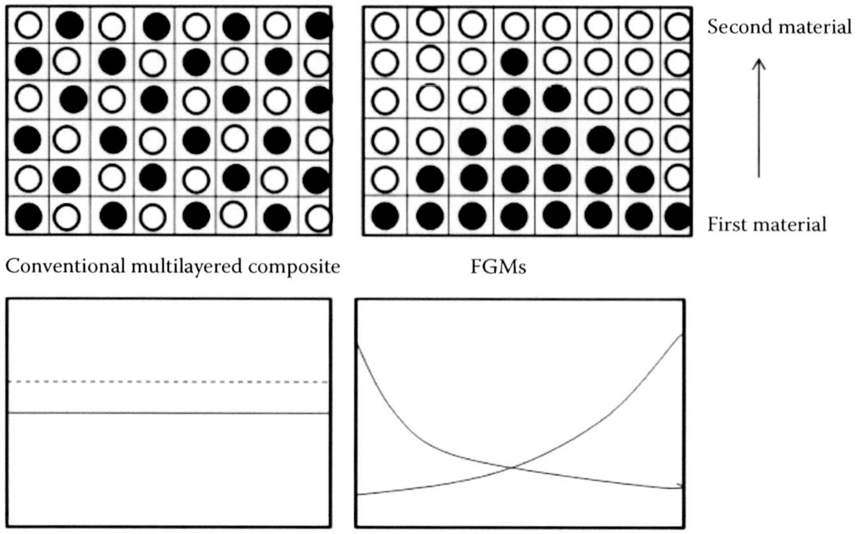

FIGURE 12.1 Variation of properties in conventional multilayered composite and FGMs. (From Gupta, A., and M. Talha. 2015. Recent development in modeling and analysis of functionally graded materials and structures. *Progress in Aerospace Sciences* 79 (November). Elsevier: 1–14. doi:10.1016/j.paerosci.2015.07.001.)

processes available at that time, further development of graded structure materials was delayed (Naebe and Shirvanimoghaddam 2016).

After a decade, in 1984, a group of Japanese material scientists developed a new graded material called "functionally graded material" (FGM) to increase adhesion and to minimize the thermal stresses in metal–ceramic composites developed for reusable rocket engines (Koizumi 1997; Sobczak and Drenchev 2013). FGMs are high-performance, microscopically inhomogeneous advanced composite materials. These materials are usually made of two or more constituent phases with engineered gradients of composition and structure with specific properties in the preferred orientation (Gupta and Talha 2015). These materials do not encompass well-distinguished boundaries or interfaces between their different regions as in the case of conventional multilayered composite materials. Hence, by virtue of this, FGMs reflect reduced mechanical and thermal stress concentration in many structural elements and also eliminates delamination and crack initiation problems. The desired mechanical properties of FGMs, that is, elastic modulus, Poisson's ratio, and material density, can be acquired in a preferred direction through the variation of volume fractions of the constituent materials spatially. Figure 12.1 shows the variation of material properties in conventional composite materials and FGMs (Gupta and Talha 2015). A continuously graded microstructure with typical metal/ceramic FGM is represented in Figure 12.2.

12.1.1 History of FGMs

Although the concept of FGMs was introduced in the late 1980s, spatial variations in the microstructure of materials have been exploited for millions of years by living organisms. The human body contains many examples of FGM parts such as bones, skin, or teeth; another example of naturally occurring FGMs can be found in nature such as in plants and in animal tissues (Singh 2016). Some examples for natural FGMs are shown in Figure 12.3 for illustration. The tooth has been considered as a functionally graded system; enamel and dentin are usually regarded as two homogeneous layers (Figure 12.3a). It requires a high wear resistance outside (enamel) and a ductile inner structure for reasons of fatigue and brittleness. A few reports have indicated that enamel may have graded properties (Giannakopoulos et al. 2010; He and Swain 2009; Niu et al. 2009; Rahbar and Soboyejo 2011).

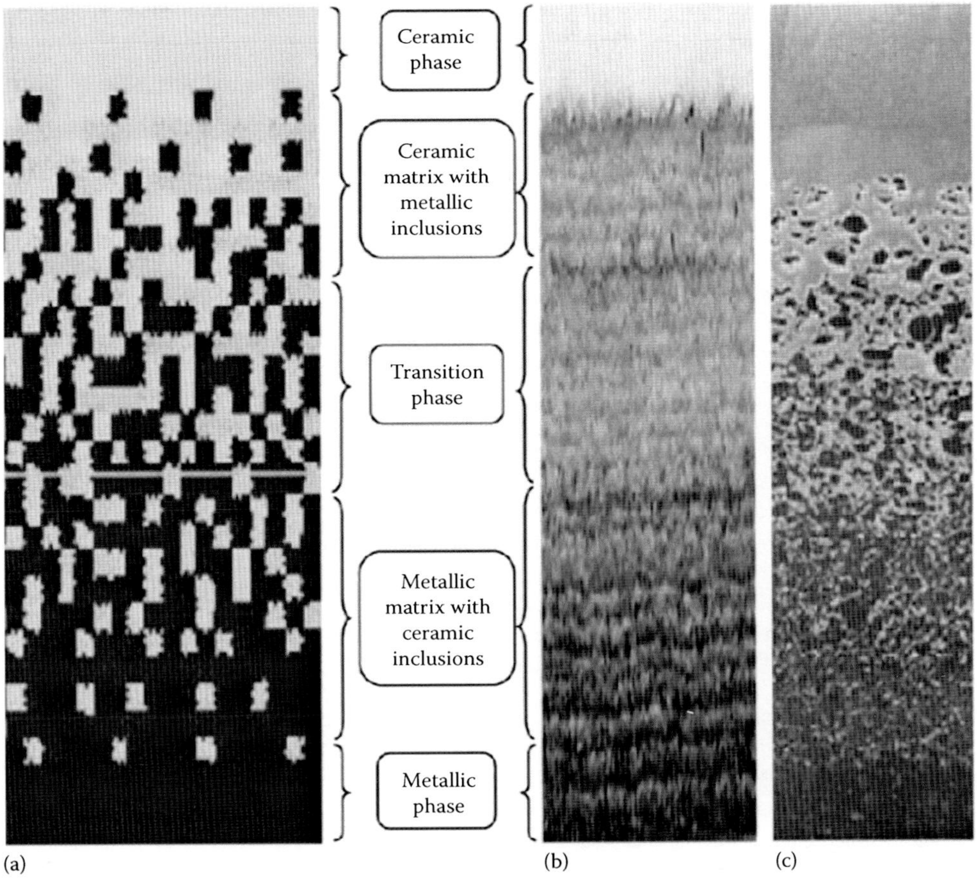

FIGURE 12.2 Gradation of microstructure with metal–ceramic FGMs. (a) Continuous graded microstructure. (b) Enlarged view. (c) Ceramic–Metal FGM. (From Jha, D. K., T. Kant, and R. K. Singh. 2013. A critical review of recent research on functionally graded plates. *Composite Structures* 96. Elsevier Ltd: 833–49. doi:10.1016/j.compstruct.2012.09.001.)

Bamboo is also a natural fiber-reinforced graded structure, composed of hierarchical fiber and parenchyma structures that evolve to resist wind loads (Figure 12.3b). This graded bamboo structure has excellent mechanical properties, such as stiffness, strength, and fracture resistance (Tan et al. 2015). Tissues are also an example of graded material, composed of different extracellular matrix components, cell types, and cell densities, such as the osteochondral tissue (Figure 12.3c) (Melchels et al. 2012). Even human skin is also graded to offer certain toughness and elastic qualities as a function of skin depth and location on the body. Some examples of human-engineered FGM components currently under development are also included in Figure 12.3.

12.1.2 Applications of FGMs

Initially, the FGMs were used in aerospace, but during the last three decades, researchers have shown its application in diversified fields, such as (i) nuclear sector, (ii) optoelectronics, (iii) chemical plants, (iv) solar energy generators, and (v) heat exchangers, to name a few, as depicted in Figure 12.4 (Gupta and Talha 2015). FGMs offer pronounced perspective in applications where the working conditions are severe, including spacecraft heat shields, heat exchanger tubes, biomedical implants, plasma facings in fusion reactors, storage tanks, electrical insulating applications, wear-resistant linings in mining industries, pressure vessels, and general wear and corrosion-resistant

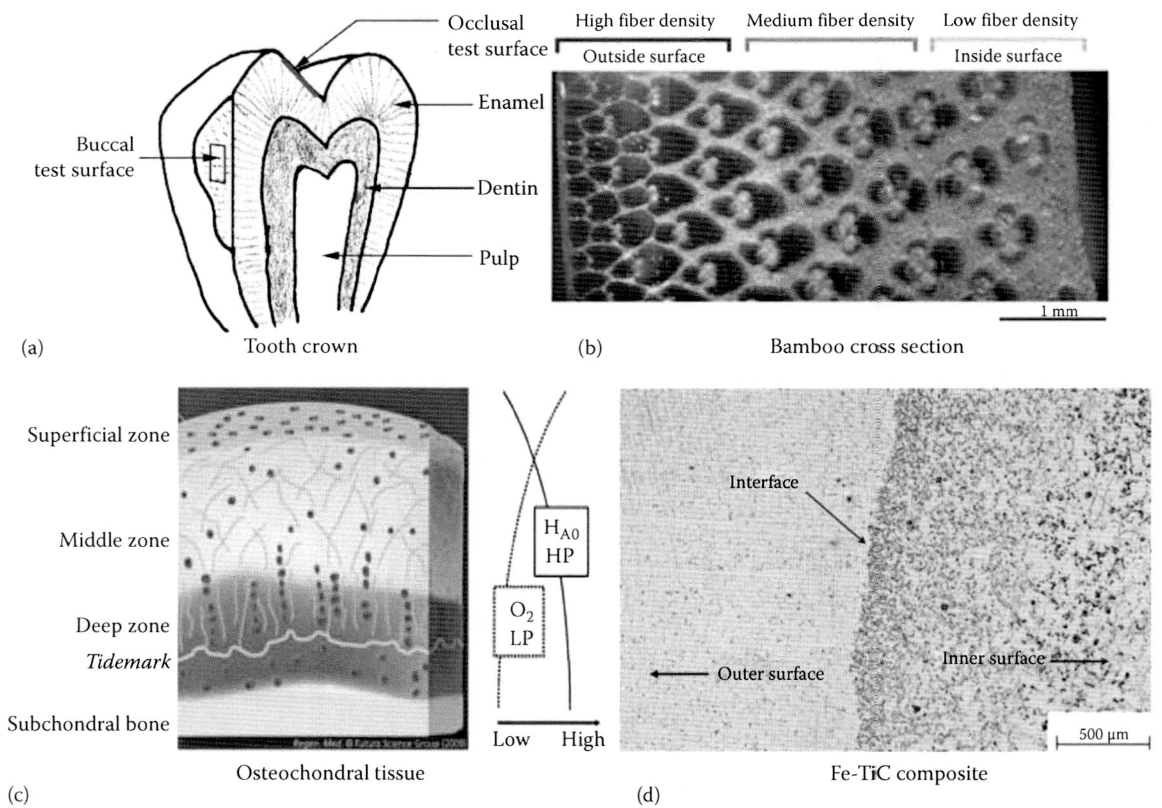

FIGURE 12.3 Natural and engineered. FGMs (a) Tooth crown. (From Rahbar, N., and W. O. Soboyejo. 2011. Design of functionally graded dental multilayers. *Fatigue & Fracture of Engineering Materials & Structures* 34 (11): 887–97. doi:10.1111/j.1460-2695.2011.01581.x.) (b) An optical image of functionally graded bamboo cross section. (From Tan, T., T. Xia, H. Ofolan, J. Dao, Z. Bash, K. Johanson, J. Novotny, M. Ozeki, and M. Smith. 2015. Sustainability in beauty: An innovative proposing-learning model to inspire renewable energy education. *The Journal of Sustainability Education (JSE)* January 17 (May 2010): 1–25.) (c) Mature articular cartilage (osteochondral tissue). (From Stoddart et al. 2009; Melchels, F. P. W., M. A. N. Domingos, T. J. Klein, J. Malda, P. J. Bartolo, and D. W. Hutmacher. 2012. Additive manufacturing of tissues and organs. *Progress in Polymer Science* 37 (8): 1079–104. doi:10.1016/j.progpolymsci.2011.11.007.) (d) Fe–TiC composite. (From Rahimipour, M. R., and M. Sobhani. 2013. Evaluation of centrifugal casting process parameters for in situ fabricated functionally gradient Fe–TiC composite. *Metallurgical and Materials Transactions B: Process Metallurgy and Materials Processing Science* 44 (5): 1120–3. doi:10.1007/s11663-013-9903-z.)

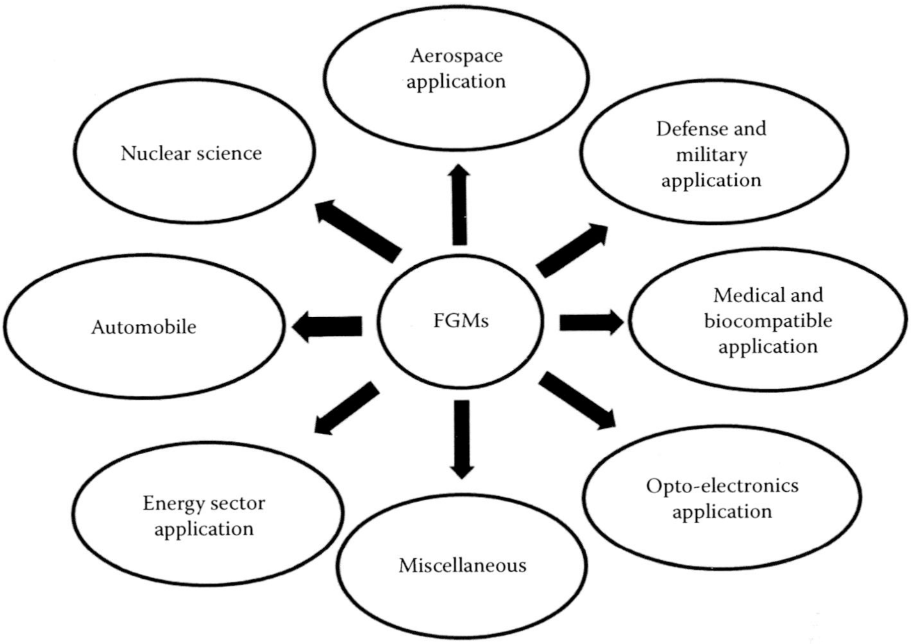

FIGURE 12.4 Applications of functionally graded plate. (From Gupta, A., and M. Talha. 2015. Recent development in modeling and analysis of functionally graded materials and structures. *Progress in Aerospace Sciences* 79 (November). Elsevier: 1–14. doi:10.1016/j.paerosci.2015.07.001.)

coatings in aerospace, automobile, marine, nuclear, and defense industries (Shen 2009; Sofiyev et al. 2016).

Most of the benefits of FGMs depend on the fact that these materials take advantage of the best properties of their constituent phases, mixing them in a graded and tailored way. The gradual change of material properties can be fitted to different applications and working environments, making them preferable in many applications (Bernardo et al. 2016; Birman and Byrd 2007; Reddy 2000). For example, ceramic/metal-based FGM reflects remarkable properties that can be used in a variety of applications. Ceramics typically exhibit high hardness, low density, and brittleness, and withstand high temperature, creep, resistivity toward corrosion, radiation, wear, and shock resistance, whereas metals are ductile in nature, having high tensile strength and high toughness. Therefore, a combination of these two materials is well suited for applications where both toughness and hardness are the primary entities.

Because of the smooth and continuous variation of mechanical and thermal properties, FGMs offer less thermal stress, residual stress, and stress concentration. This attribute allows FGMs to perform exceptionally in thermal environments. For example, in the space shuttle, ceramic tiles are used as thermal protection from heat generated during reentry into the earth's atmosphere. These tiles are laminated to the shuttle's superstructure and are prone to cracking and debonding at the superstructure/tile interface due to a sudden change in thermal expansion coefficients. In other words, the ceramic tile expands by a different amount compared to the substructure it is protecting. This difference in expansion causes stress concentration at the interface of the tile and superstructure, which results in cracking or debonding (Department of Defense n.d.; Reddy 2004). An FGM consists of ceramic on the outer surface and metal on the inner surface, which eliminates the abrupt transition between the coefficients of thermal expansion and, therefore, offers better stability and thermal/corrosion protection and provides load-carrying capability (Zidi et al. 2014).

FGMs have also proved their worth in biomedical applications, particularly in (a) dental implant, (b) knee replacement, and (c) hip joint replacement. In the case of a dental implant, an FGM is

TABLE 12.1
Materials and Fabrication Techniques

Functionally Graded Implants	Fabrication Methods	Literature
Ti/HA	CIP+EF, CIP+HF, SPS, PM	Fujii et al. 2010; Watari et al. 2004
TiN/HA	SPS	Kondo et al. 2004
Ti/Co	CIP+HF	Watari et al. 2004
Ti/ZrO2	CIP+HF, hot pressing	Fujii et al. 2010
HA/ZrO2	SPS, hot pressing	Guo et al. 2003
Porous FGM	PM, PECS, and DLMS	Krishna et al. 2007; Kutty et al. 2004; Traini et al. 2008
Ti/20SiO2, Ti/20ZrO2	CIP+HF	Watari et al. 1997

Sources: Mehrali, M., F. S. Shirazi, M. Mehrali, H. S. C. Metselaar, N. A. B. Kadri, and N. A. A. Osman. 2013. Dental implants from functionally graded materials. *Journal of Biomedical Materials Research—Part A* 101 (10): 3046–57. doi:10.1002/jbm.a.34588; Watari, F., A. Yokoyama, F. Saso, and M. Uo. 1997. Fabrication and properties of functionally graded dental implant. *Composites Part B* 28B: 5–11.

Note: CIP, cold isostatic press: 400–1000 MPa; SPS, spark plasma sintering; EF, electric furnace heating: 1300°C; PM, powder metallurgy; HF, high-frequency induction heating: 1200°C.

usually made of a mixture of titanium and bioactive hydroxyapatite/collagen (HAP/Col) (Mehrali et al. 2013). Tawakol and Bondok (2013) found that the stress concentration in tibia tray can be reduced significantly when benchmark material such as CoCrMo and Ti alloy is replaced by the FGMs. These materials also replace conventional materials such as CoCrMo and Ti alloy in the artificial hip joint replacement(Oshkour et al. 2014), which results in a reduction in stress in bone cement and stem. Table 12.1 shows the various FGMs that are biocompatible in nature and are generally used in implant along with the methods of fabrication.

12.2 PROCESSING TECHNIQUES OF FGMS

Since the inception of FGMs, several processing methods have been developed for FGMs. These processing methods of FGMs are broadly classified into two categories based on constructive processing and mass transport (Groves and Wadley 1997; Gupta and Talha 2015). In the first category, the FGM is constructed layer by layer starting with a predetermined spatial distribution in which the gradients are actually constructed in space. The prime advantage of this technique is the fabrication of FGM having an unconditional and unobstructed number of gradients. In the second category, natural transport phenomena such as the flow of fluid, the diffusion of atomic species, or heat conduction are used to accommodate the gradients within the materials.

There are several methods to accomplish a compositional gradient in the composite material. The gas, liquid, and solid phase of materials can be treated physically or chemically to obtain the desired compositional gradient of the material (Sasaki and Hirai 1991). Chemical vapor deposition (CVD), thermal spray, ion plating, and ion mixing are some examples of gas-based methods used to fabricate the graded materials. In these methods, the gradient composition depends on the phase's reaction ratio in the mixture and production controlling system. In liquid-phase methods such as plasma spray, centrifugal casting, slip casting, and electrodeposition methods are very common methods because of the flexibility of production and the ability to make complex geometry. In the electrodeposition method, composition gradient closely depends on electrochemical factors and proper selection of electrolytic solution (Naebe and Shirvanimoghaddam 2016). The solid-phase methods such as powder metallurgy and sintering diffusion are also employed to fabricate the stepwise FGMs. Some existing techniques for fabrication of FGMs are given in detail and an overview of processing techniques is shown in Tables 12.2 and 12.3.

TABLE 12.2
Fabrication Methods of FGM

Phase	Process	Method	FGM
Gas	Chemical	Chemical vapor deposition/infiltration	SiC/C, SiC/TiC, TiC/C, C/Ceramic
		Surface reaction process	
	Physical	Ion plating, plasma spraying, ion mixing	TiN/Ti, TiC/Ti, ZrO2/Cu, C/Cr
Liquid	Chemical	Electrodeposition, chemical solution deposition,	Ni/Cu, WC/Co, ZrO2/Al2O3
		Electrochemical gradation	PT/PZT, W/Cu
	Physical	Plasma spraying, eutectic reaction	YSZ/NiCrAlY, YSZ/Ni–Cr
		Centrifugal casting	Si/$ZrSi_2$, Al/SiC, W/Al_2O_3
		Slip casting, Gel casting	Ni–Al/Ti–6Al–4V
Solid	Chemical	Self-heating system smearing	TiB_2/Cu, TiB_2/Ni, TiC/Ni, SiC/TiAl
	Physical	Sintering diffusion, Spark plasma sintering,	YSZ/SUS304, YSZ/Mo
		Powder metallurgy	YSZ/Nb, Si_3N_4/Ni
			W/Cu
			ZrB_2–SiC/ZrO_2
			ZrO_2/AlSI316L
			SiC-AA7075

Sources: Sasaki, M., and T. Hirai. 1991. Fabrication and properties of functionally gradient materials. *Journal of the Ceramic Society of Japan* 99 (10): 1002–13; Naebe, M., and K. Shirvanimoghaddam. 2016. Functionally graded materials: A review of fabrication and properties. *Applied Materials Today* 5 (December): 223–45. doi:10.1016/j.apmt.2016.10.001.

TABLE 12.3
Various Processing Techniques of FGMs

Sr. No.	Process	Variability of Transition Function	Versatility in Phase Content	Type of FGM	Versatility in Component Geometry
1	Thermal spraying	Very good	Very good	Bulk, coating	Good
2	Laser cladding	Very good	Very good	Bulk, coating	Very good
3	Sedimentation	Good	Very good	Bulk	Poor
4	Wet powder spraying	Very good	Very good	Bulk	Average
5	Powder stacking	Very good	Very good	Bulk	Average
6	Jet solidification	Very good	Very good	Bulk	Very good
7	GMFC process	Very good	Average	Bulk	Good
8	Filtration/slip casting	Very good	Very good	Bulk	Good
9	Sheet lamination	Very good	Very good	Bulk	Average
10	Electrochemical gradation	Average	Good	Bulk	Good
11	Foaming of polymer	Average	Good	Bulk	Good
12	Slurry dipping	Very good	Very good	Coating	Good
13	Vapor deposition	Very good	Very good	Coating	Average

Sources: Udupa, G., S. S. Rao, and K. V. Gangadharan. 2014. Functionally graded composite materials: An overview. *Procedia Materials Science* 5: 1291–9. doi:10.1016/j.mspro.2014.07.442; Gupta, A., and M. Talha. 2015. Recent development in modeling and analysis of functionally graded materials and structures. *Progress in Aerospace Sciences* 79 (November). Elsevier: 1–14. doi:10.1016/j.paerosci.2015.07.001.

12.2.1 Powder Metallurgy

Fabrication of FGMs using powder metallurgy involves three basic steps of powder preparation: (a) weighing and mixing of powder according to the predesigned spatial distribution, (b) stacking and ramming of the premixed powders (forming operations), and, finally, (c) compacting and forming a solid mass of material (sintering) (Gupta and Talha 2015). Figure 12.5 shows a flowchart of the fabrication process of FGMs using powder metallurgy.

Several techniques have been employed for the preparation of powder such as chemical reactions, electrolytic deposition, grinding, pulverization, atomization, centrifugal disintegration, solid-state reduction, and so on. Processing parameters, such as (i) sintering temperature, (ii) dwelling time, (iii) pressure, and (iv) heat treatment conditions, directly affect the final properties of fabricated FGMs using powder metallurgy (Naebe and Shirvanimoghaddam 2016). Table 12.4 clearly shows the various values of processing parameters considered during the fabrication of FGMs using powder metallurgy.

Xiong et al. (2000) fabricated the W–Mo–Ti–TiAl–Al system graded material using the powder metallurgy method. They found that W, Mo, and W–Mo alloys were basically densified concurrently by sintering at 1473 K for 1 h under a pressure of 30 MPa. It was observed that at low sintering temperatures, the mechanical mixtures of W and Mo were mainly formed in the W–Mo alloys. During sintering, an insufficient solid-solution reaction occurred between Mo and Ti. It was also observed that the residual thermal stresses occurred as a result of the difference in the thermal-expansion coefficient between Ti and TiAl being reduced effectively by inserting a compound mixed powder layer of Ti + TiAl between the Ti and TiAl.

Jin et al. (2005) developed mullite/Mo FGM using powder metallurgy. Mullite and Mo have almost similar thermal expansion coefficients ($\alpha_{Mull} = 5.13 \times 10^{-6}$/K, $\alpha_{Mo} = 5.75 \times 10^{-6}$/K at 1000°C). Because of this special attribute, the residual thermal stress induced during the sintering process is reduced. It was found that the variation of mechanical properties followed the Voigt model, whereas the variation of the thermal properties tended to follow the Reuss model. The fabricated graded materials reflect better thermal shock resistance than the monolithic mullite. In Figure 12.6a, the layer boundaries were observed clearly, which shows that the structures of the graded materials were macroscopically inhomogeneous due to the graded distribution of the composition. Figure 12.6b depicts the microstructural distribution of the concentration of Mo.

12.2.2 Centrifugal Casting

In this process, the reinforcement phase is poured into a molten metal to make a homogeneous mixture. The required gradation of material properties is attained using proper segregation of reinforcement particles through gravitational or centrifugal forces (Saiyathibrahim et al. 2015). There are two different categories of centrifugal casting methods, which are based on the difference between reinforcement particle temperatures and processing temperature. If the processing temperature is higher than the reinforcement particle temperature, then this technique is known as a centrifugal in situ technique and centrifugal forces can be used during the solidification step. Alternatively, if the reinforcement particle is subjected to a higher temperature than the processing temperature, the second phase remains solid in the molten metal, and this is known as the centrifugal solid-particle technique (Naebe and Shirvanimoghaddam 2016; Watanabe et al. 2006). The FGMs obtained from these methods reflect high wear resistance and bulk toughness. The primary advantage of the centrifugal casting method is it improves the density of the metal, increases the actual mechanical properties of the casting by 10% to 15%, and provides a uniform metallurgical structure (Gupta and Talha 2015).

The various steps involved during the fabrication of graded materials (e.g., combination of A/B) using a centrifugal in situ method are as follows: (i) Partial separation of A and B elements in the liquid state occurs due to the density difference. (ii) A compositional gradient is formed before the crystallization of the primary crystal. (iii) The primary crystals in the matrix appear according to the

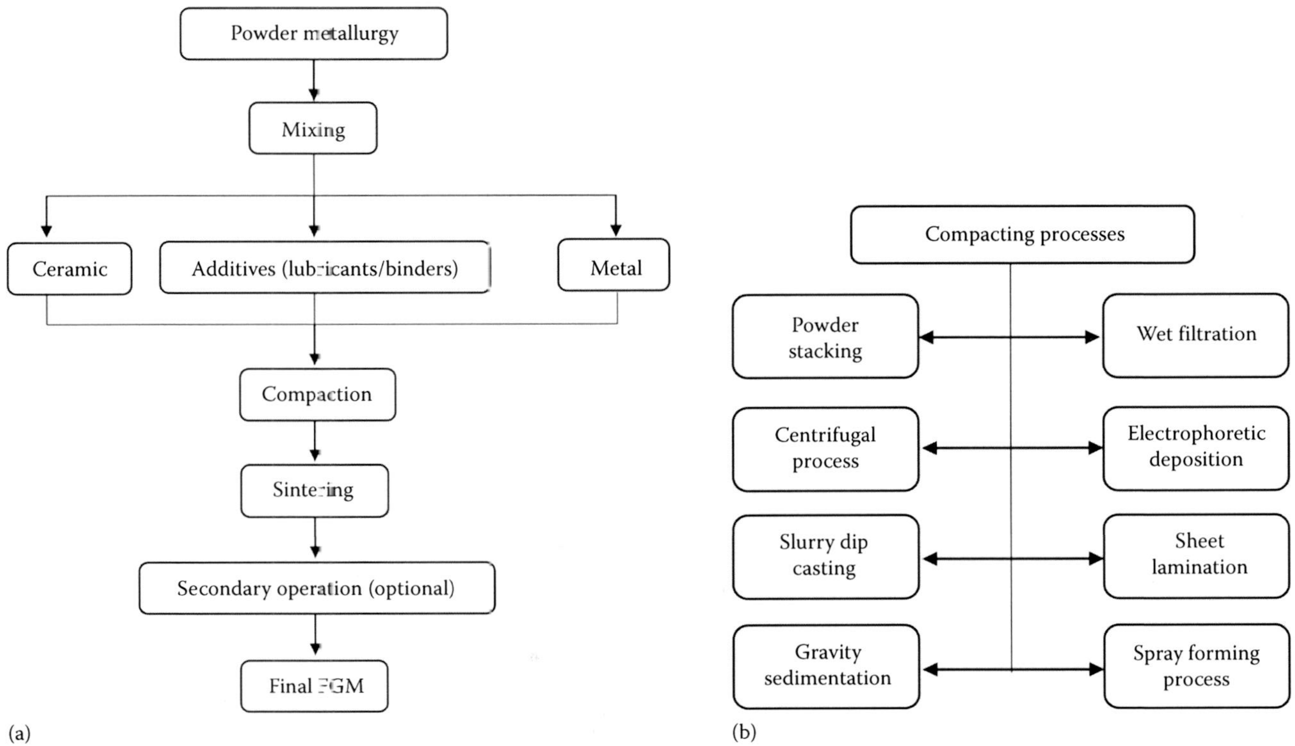

FIGURE 12.5 (a) Steps of fabrication process of FGM using powder metallurgy. (b) Various compacting processes.

TABLE 12.4
Various Process Parameters for FGMs Fabricated Using Powder Metallurgy

Materials	Dwelling Time (min)	Sintering Temperature (°C)	Pressure (MPa)	Literature
Mullite/Mo	10	1500	20	Jin et al. 2005
SiC-AA7075	40	580	–	Übeyli et al. 2014
Hydroxyapatite-Ti	30–90	1100	20	Chenglin et al. 1999
Ti–TiB2	30	1300	5	Ma and Tan 2001
Ti–TiB	–	1350	13.8	Gooch et al. 1999
W–Mo–Ti–TiAl–Al	60	1200	30	Xiong et al. 2000
ZrO2–Ni	60–120	1300–1400	25	Zhu et al. 1996

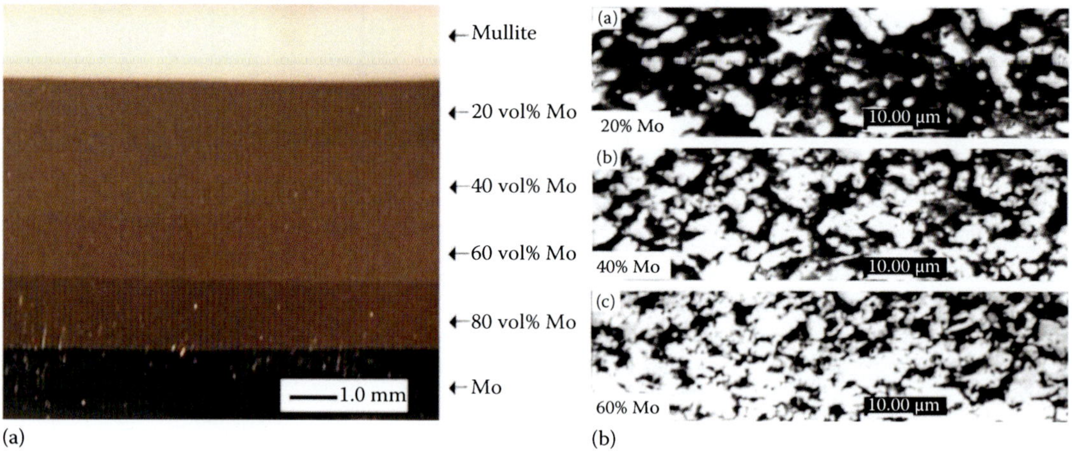

FIGURE 12.6 (a) The optical micrograph of six-layered linear gradient FGM. (b) SEM micrographs of layers. (From Jin, G., M. Takeuchi, S. Honda, T. Nishikawa, and H. Awaji. 2005. Properties of multilayered mullite/Mo functionally graded materials fabricated by powder metallurgy processing. *Materials Chemistry and Physics* 89 (2–3): 238–43. doi:10.1016/j.matchemphys.2004.03.031.)

local chemical composition. (iv) The primary crystals migrate because of density difference, and a further compositional gradient is formed (Watanabe et al. 2005) (Figure 12.7).

Melgarejo et al. (2008) prepared functionally graded Al–Mg/AlB2 metal–matrix composites with Al–2wt.%Mg alloys containing 1, 2, 3, and 4 wt.% boron using centrifugal casting method. The gradient in the volume fraction of the AlB_2 particles along the radial direction was observed, with a denser distribution of diboride particles in the outer regions of the casting due to inertial forces. The hardness of the composite increased along the radial direction and proportionally to the volume fraction of the AlB_2 particles. It was also observed that factors such as the differences in density between the matrix and the dispersed phase, the boron content of the alloy, rotational velocity, size of the diboride particles, the viscosity of the semisolid material, and the local solidification time are important parameters that control the composition gradient.

Watanabe and Oike (2005) used centrifugal in situ method to fabricate the Al–Al_2Cu FGMs. It was observed that compositional gradient can be obtained by the centrifugal in situ method from not only hypoeutectic or hypereutectic alloys but also eutectic alloys. The particle sizes of the Al_2Cu and Al primary crystals were dependent on their locations in the ring thickness. It was also found that because of the difference in the densities between Al and Cu in a liquid state, the graded structure appeared in the eutectic specimen.

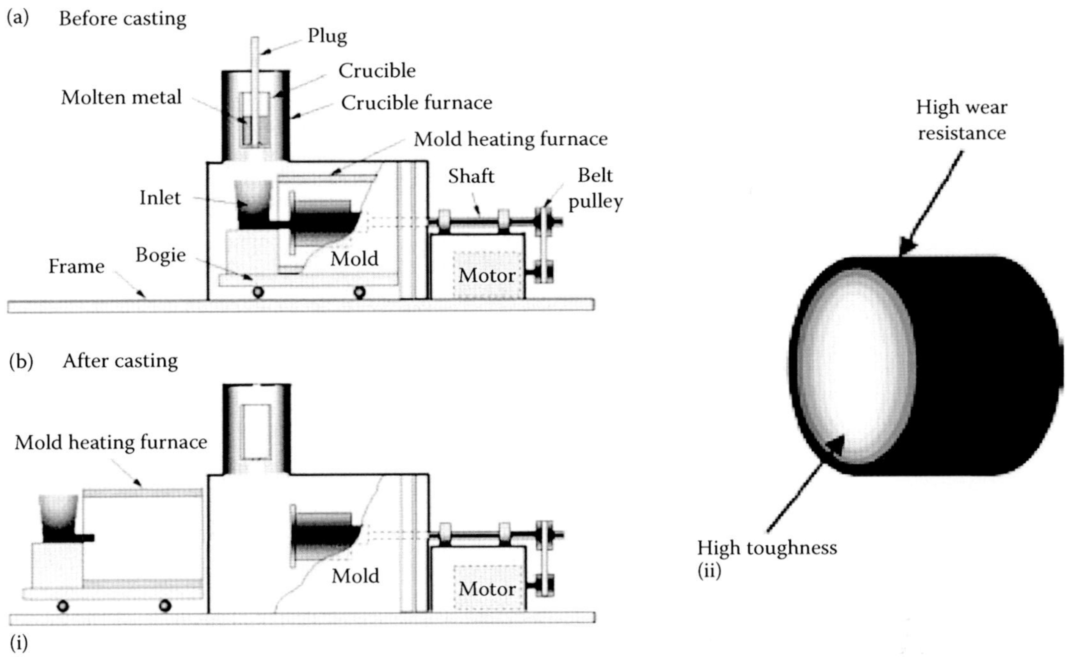

FIGURE 12.7 (a) Schematic of centrifugal casting process. (b) Final graded product FGM fabricated by centrifugal casting. (From Watanabe, Y., and S. Oike. 2005. Formation mechanism of graded composition in Al–Al2Cu functionally graded materials fabricated by a centrifugal in situ method. *Acta Materialia* 53 (6): 1631–41. doi:10.1016/j.actamat.2004.12.013.)

12.2.3 Electrochemical Gradation Process

This method can be employed to fabricate the gradient material having a variety of ceramic or metallic materials as a primary constituent. This process is based on the fact that the reaction in a porous electrode is not restricted to the external surface as in the case of dense electrodes. If the electrolyte occupies the pore space, the electrode reaction will begin inside the pores. An electrode reaction along with the deposition or dissolution of electrode material will change the porosity of the electrode. As the potentials of the different phases are usually position dependent, the electrochemical reaction will have different rates at different positions inside the porous electrode (Jedamzik et al. 2000). It is noteworthy that some processing parameters, such as (i) density, (ii) electrolyte resistivity, (iii) porosity, and (iv) geometry of electrolyte, affect both the shape and morphology of the FGMs (Naebe and Shirvanimoghaddam 2016).

Jedamzik et al. (2000) fabricated W/Cu gradient materials using electrochemical deposition. In the study, tungsten preforms with an open porosity were produced by partially sintering tungsten powders of specific grain size (4 or 10 μm containing about 1% nickel). The preforms were graded in an electrochemical cell using the setup shown in Figure 12.8. For the gradation of tungsten, the anodic dissolution reaction of tungsten in alkaline solution has been used. The reaction consists of the oxidation of tungsten to tungsten trioxide and subsequent dissolution of the tungsten trioxide in the alkaline electrolyte. This method is widely used in the industries because it produces fully dense gradient materials.

12.2.4 Gel Casting

Gel casting is a novel processing technique with a short forming time, high green strength, excellent green machinability, and low-cost machining that is used widely for complex-shaped

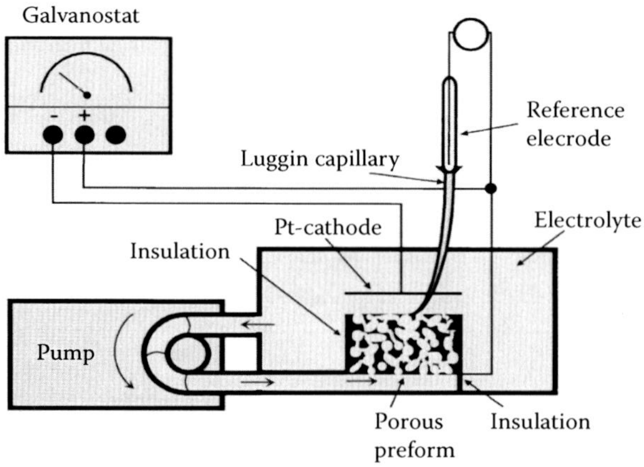

FIGURE 12.8 Experimental setup for electrochemical gradation. (From Jedamzik, R., A. Neubrand, and J. Rödel. 2000. Functionally graded materials by electrochemical processing and infiltration: Application to tungsten/copper composites. *Journal of Materials Science* 35 (2): 477–86. doi:10.1023/A:1004735904984.)

dense/porous ceramic materials (Naebe and Shirvanimoghaddam 2016). This process was first developed by Omatete et al. (1997) from the Ceramic Processing Group at Oak Ridge National Laboratory (USA).

The gel casting process starts with dissolving multifunctional acrylate monomers in organic solvents. These monomers, when polymerized by free-radical initiators, convert into highly cross-linked polymer-solvent gels. When a concentrated slurry of ceramic powder is poured into a mold and polymerized, a green body in the shape of the mold cavity formed (Omatete et al. 1997). Thereafter, the green body is dried under controlled conditions. Binder removal and sintering take place as in other ceramic processes. Details of the gel casting process are shown in Figure 12.9.

Park et al. (2006) characterized the microstructural and mechanical behavior of functionally graded Al_2O_3/ZrO_2 that was fabricated by dip coating with substrates prepared by gel casting. Alumina (Al_2O_3) and 3 mol.% yttria-doped zirconia (TZ-3Y) were used as starting materials for the preparation of each substrate, outer layer, and the graded intermediate layer. The particle size of alumina was 0.32 μm, and those of TZ-3Y powders were 0.024 and 0.5 μm, respectively. Ammonium salt (as a dispersant), acrylamide (as a monomer), methylenebisacrylamide (as a cross-linker), and finally ammonium persulfate and *N,N,N′,N′*-tetramethylethylenediamine (as an initiator and catalyst for the gelation reaction, respectively) were used. Two kinds of graded materials (AZ FGM and ZA FGM) with 11 layers from the substrate to outer layer were fabricated using the gel-casting and dip-coating processes.

Continuous gradation in hardness was observed from the substrate through to the outer layer as shown in Figure 12.10. The coating thickness of AZ FGM and ZA FGM was reported as 300 and 200 μm, respectively. It was also found that in ZA FGM, crack formation was not encountered, whereas in anisotropic, crack growth was reported in AZ FGM.

12.2.5 Vapor Deposition Techniques

There are various types of vapor deposition techniques available in the literature: (a) CVD, (b) physical vapor deposition (PVD), (c) sputter deposition, (d) pulsed laser deposition, and (e) evaporative deposition. In these techniques, the materials in a vapor state are condensed through condensation, chemical reaction, or conversion to form a solid material (Groves and Wadley 1997; Naebe and Shirvanimoghaddam 2016). These vapor deposition methods are used to deposit

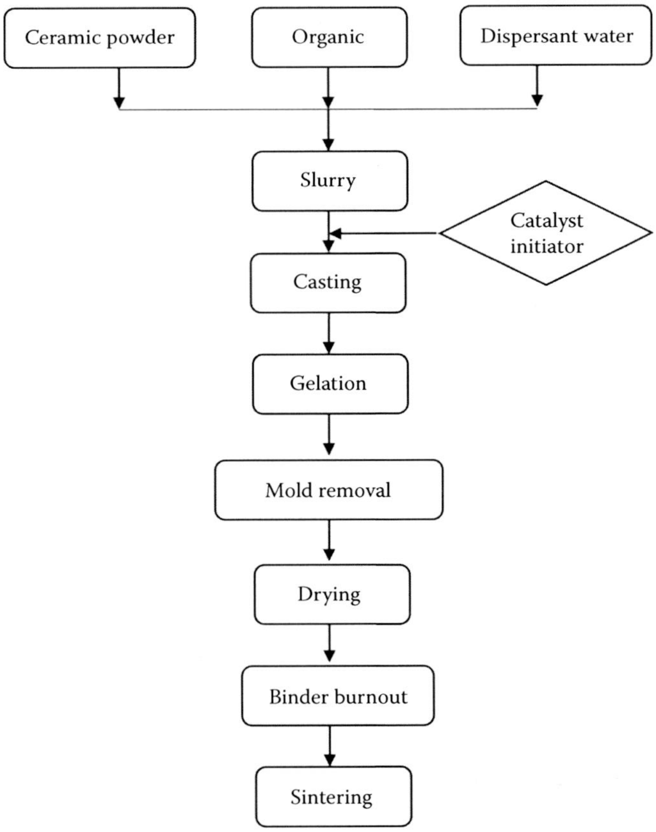

FIGURE 12.9 Flowchart for various steps involved in gel casting. (From Young, A. C., O. O. Omatete, M. A. Janney, and P. A. Menchhofer. 1991. Gelcasting of alumina. *Journal of the American Ceramic Society* 74 (3): 612–8. doi:10.1111/j.1151-2916.1991.tb04068.x.)

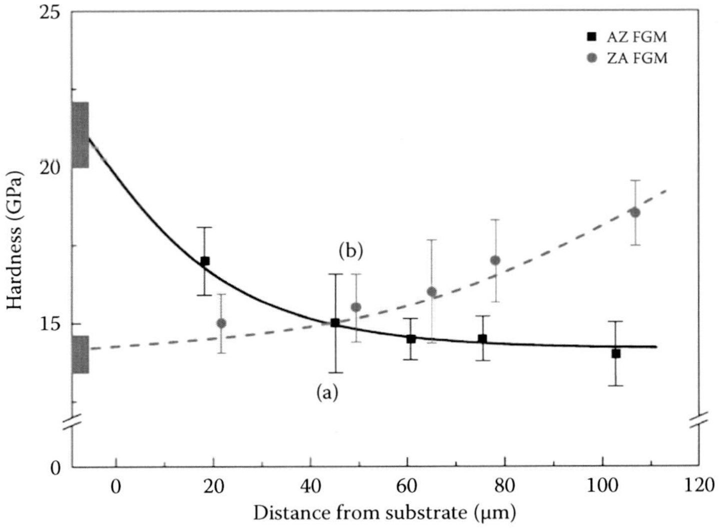

FIGURE 12.10 Hardness values in each region of AZ and ZA functionally graded materials. (From Park, S.-H., J.-H. Kanga, Y.-G. Junga, and U. Paikb. 2006. Fabrication and mechanical characterization of Al2O3/ZrO2 layered composites with graded microstructure. *Multiscale and Functionally Graded Materials*, 273–8.)

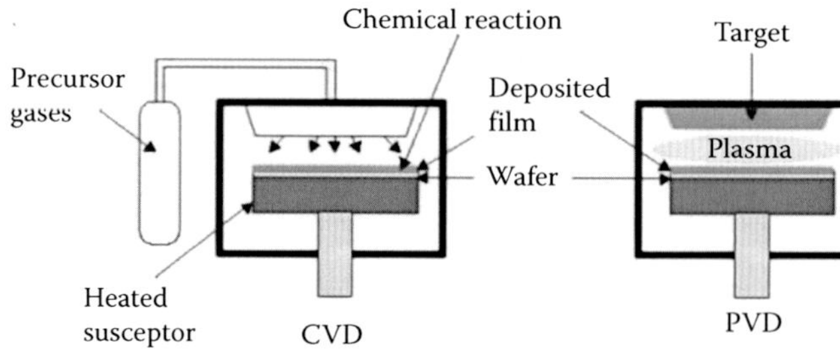

FIGURE 12.11 Comparison of CVD and PVD. (Retrieved from http://thegadgetblog.com/differentiating-physical-vapor-deposition-from-chemical-vapor-deposition/.)

functionally graded surface coatings that give an excellent microstructure for a thin surface coating. In the CVD technique, material deposition processes are carried out using various sources of energy such as (a) heat, (b) plasma, and (c) light.

In the CVD technique, the source material is mixed with a volatile precursor material. This mixture is then injected into a chamber containing the substrate, where the deposition process occurs. When the mixture adheres to the substrate, the precursor will decompose, leaving behind a layer of the source material in the substrate. Thereafter, the product is removed from the chamber through gas flow. Deposition temperature, gas pressure, flow rate, and gas type are the prime parameters that govern the pattern of the targeted chemical composition gradient of FGMs (Hirai 1995). Controlled and continuous change of composition, low-temperature fabrication, and nearly desired shape are the advantages of CVD.

In PVD, deposition processes are performed using evaporation of the source metal onto the substrate (Naebe and Shirvanimoghaddam 2016). Vapor deposition techniques are energy intensive but produce poisonous gases as their by-products. No chemical reactions occur throughout the process. The comparison of CVD and PVD is shown in Figure 12.11.

Sasaki and Hirai (1994) used CVD to fabricate composition gradient material of SiC/C. A $SiC1_4$-CH_4 H_2 system was used and the graphite substrate was heated in a hot wall–type reaction chamber. The temperature of the $SiC1_4$ reservoir was kept constant (293 K), whereas the vapor produced during the process was carried into the furnace by bubbling hydrogen carrier gas. The deposition temperature, gas pressure, and deposition time were selected as 1673–1773 K, 1.3 kPa, and 9 ks, respectively. It was observed that the SiC/C graded materials obtained at 1773 K deposition temperature had pores throughout the film, whereas when the deposition time was 1673 K, pores were restricted near the substrate.

Kawase et al. (1999) fabricated SiC/C FGM using CVD/infiltration techniques. The 3D-woven carbon was used as a preform. The preform was solidified partially by liquid-phase impregnation with phenolic resins. Further solidification of preform was carried out by thermal gradient chemical vapor infiltration of carbon from propane. It is noteworthy that small voids were intentionally left in the partially solidified preform and were filled by CVI after impregnation and carbonization. A consistent compositionally gradient protective layer was prepared by CVD by changing the reactant mixture composition gradually from propane to dimethyldichlorosilane. An outline of the experimental procedure is shown in Figure 12.12.

12.2.6 Slip Casting

Slip casting is a powder-based shaping method that has been used for a long time in the traditional ceramic industry. In this process, "slip" is used, which is a suspension of fine ceramic material powders in a liquid such as water or alcohol with small amounts of secondary materials such as dispersants and binders. Basically, slip casting is a filtration process in which powder suspension (usually a

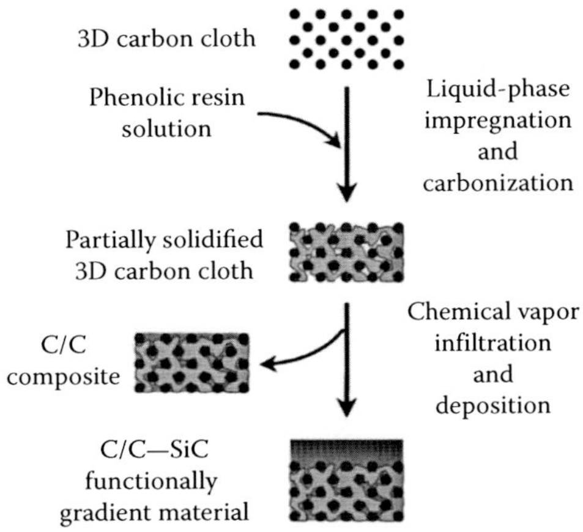

FIGURE 12.12 Procedure of gradient material production using CVI/CVD. (From Kawase, M., T. Tago, M. Kurosawa, H. Utsumi, and K. Hashimoto. 1999. Chemical vapor in infiltration and deposition to produce a silicon carbide-carbon functionally gradient material. *Chemical Engineering Science* 54: 3327–34.)

water-based suspension) is poured into a porous plaster mold. Because of the resulting capillary forces, the liquid is removed from the suspension. When the filtrate is sucked into the plaster mold, the powder particles are forced toward the mold walls. A compositional gradient will be formed by changing grain size of used powder suspension in the casting procedure (Saiyathibrahim et al. 2015).

Yan et al. (2009) fabricated Ni–ZrO_2 FGMs using the slip-casting method. In the fabrication process, Ni and ZrO_2 particles with an average diameter of 1.2 and 0.75 μm, respectively, were considered as raw materials, whereas polyvinylpyrrolidone was used as the deflocculant. The mixed suspension was prepared using the specific volume fraction of raw materials (Ni and ZrO_2) and deflocculant. The mixed suspension was degassed in a vacuum furnace and then solidified in a rubber mold. The cylindrical casting mold was used in the experimental setup as shown in Figure 12.13. During the casting process, the suspension was subjected to various magnetic fields along the vertical direction with various holding times. The gradient field was generated by a Maxwell coil system, consisting of a pair of identical coils with inverse current. It was found that the composition gradient increased with the holding time. It was also observed that the composition gradient decreased as Ni concentration increased from 4 to 15 wt.%, due to the formation of Ni clusters.

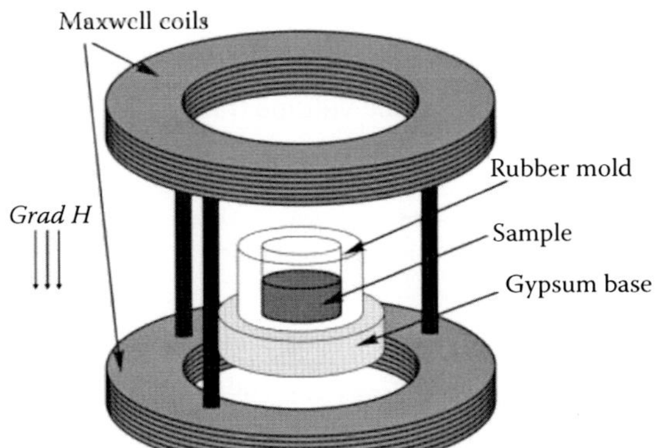

FIGURE 12.13 Setup of slip casting. (From Yan, M., X. Peng, and T. Ma. 2009. Microstructures of Ni–ZrO2 functionally graded materials fabricated via slip casting under gradient magnetic fields. *Journal of Alloys and Compounds* 479 (1–2): 750–4. doi:10.1016/j.jallcom.2009.01.042.)

12.2.7 Tape Casting

Tape casting was initially used for the production of thin, wide ceramic tapes but its applicability has also been observed for the fabrication of stepped FGMs in the last two decades. In this method, raw material (generally ceramic powder) is mixed with solvents and additives to form a free-flowing casting slurry. Then, the forming process takes place in the tape casting plant, where the ceramic slurry is spread continuously over a conveyor belt using a doctor blade to build a thin film. After that, the secondary operations (e.g., drying, sintering, debinding, etc.) are carried out for the final product.

Yeo et al. (1998) fabricated ZrO_2/SUS316 FGMs using tape casting. The starting materials were selected as ceramic (combination of tetragonal zirconia polycrystals and monoclinic zirconia polycrystals) and metal (stainless steel 316 [SUS316]). During fabrication, some additives such as (i) carboxymethylcellulose sodium salt as suspending agent, (ii) polyethylene glycol as plasticizer, (iii) polyvinyl alcohol as binder, (iv) DAXAD-34 as dispersant, and (v) D-SK as antifoaming agent were used. Details of the experimental processor are shown in Figure 12.14. The ZrO_2/SUS316 FGM was fabricated by sintering at 1350°C. It was found that the sintering defects occur during the fabrication due to the fact that different sintering shrinkages of the starting components could be controlled by proper adjustment of the particle size and phase type of ZrO_2. It was also observed that the warping can be controlled by pressing the specimen or increasing the layer number.

12.3 GRADATION OF MATERIAL PROPERTIES OF FGMS

Typically, FGMs are microstructurally inhomogeneous advanced composite materials with a prescribed distribution of volume fraction of its constituent phases. The material properties are generally assumed to follow gradation along the preferred direction in a continuous and smooth manner. This smooth and continuous gradation has to follow some rules to govern both the mechanical and electrical properties of FGMs in a predetermined manner.In the literature, three rules are reported, which are used in the structural analysis of FGMs.

12.3.1 Power Law

This law is widely used in stress analysis of FGM structures. According to this law, the effective material properties $P_{\text{eff}}(z)$ in a specific direction (along z) can be written as

$$P_{\text{eff}}(z) = (P_{\text{t}} - P_{\text{b}})V_{\text{f}}(z) + P_{\text{b}}, \tag{12.1}$$

where P_{t} and P_{b} are properties of the top and the bottom surface of an FGM structure. It is notable that effective material properties depend on the volume fraction $V_{\text{f}}(z)$ of the FGM, which is given by the expression

$$V_{\text{f}} = \left(\frac{2z + h}{2h}\right)^n, 0 \le n < \infty, \tag{12.2}$$

where n is the volume fraction exponent and h is the thickness of plate/shell. It is found from the literature (Bao and Wang 1995; Lee and Erdogan 1995) that during the estimation of effective materials properties using power law, the stress concentration is pronounced at interfaces where the material changes rapidly and continuously. Therefore, to circumvent this issue, Chi and Chung (2002) presented another law called the Sigmoid law in which volume fraction is defined using two power law functions to ensure smooth distribution of stresses among all the interfaces. According to this law, the two power law functions are defined by

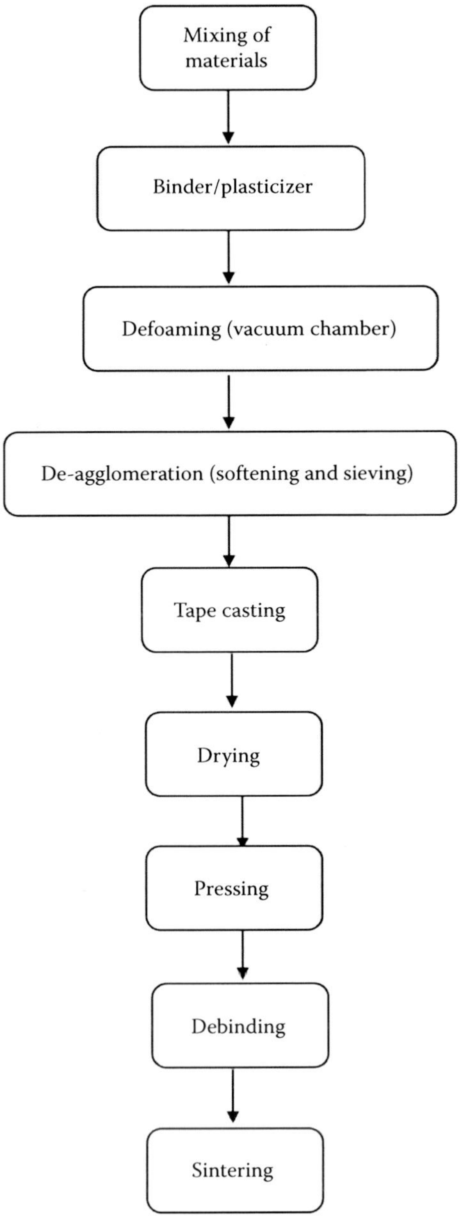

FIGURE 12.14 Schematic diagram of the experimental procedure. (From Yeo, J. G., Y. G. Jung, and S. C. Choi. 1998. Design and microstructure of ZrO2/SUS316 functionally graded materials by tape casting. *Materials Letters* 37 (6): 304–11. doi:10.1016/S0167-577X(98)00111-6.)

$$\left\{V_{\mathrm{fr}}^{1}(z)\right\} = \left\{1 - \frac{1}{2}\left(\frac{(h/2) - z}{(h/2)}\right)^{n}\right\}$$
$$\left\{V_{\mathrm{fr}}^{2}(z)\right\} = \left\{\frac{1}{2}\left(\frac{(h/2) + z}{(h/2)}\right)^{n}\right\}. \tag{12.3}$$

By using the rule of mixture, the material properties of an FGM using the Sigmoid law can be calculated using the following relations:

$$P_{\text{eff}}(z) = P_t V_{\text{fr}}^1(z) + \left(1 - V_{\text{fr}}^1(z)\right) P_{\text{b}} \qquad \text{for } 0 \le z \le h/2$$
$$P_{\text{eff}}(z) = P_t V_{\text{fr}}^2(z) + \left(1 - V_{\text{fr}}^2(z)\right) P_{\text{b}} \qquad \text{for } -h/2 \le z < 0. \tag{12.4}$$

12.3.2 Exponential Law

This law is generally adopted by researchers when they deal with problems in the domain of fracture mechanics. The effective material property of an exponentially graded FGM plate is expressed as (Chakraverty and Pradhan 2014; Gupta and Talha 2016)

$$P_{\text{eff}} = P_{\text{t}} e^{-\delta\left(1-\frac{2z}{h}\right)}, \text{ where } \delta = \frac{1}{2} \ln\left(\frac{P_{\text{t}}}{P_{\text{b}}}\right). \tag{12.5}$$

12.4 HOMOGENIZATION TECHNIQUES OF FGMS

The FGMs are fabricated by mixing two distinct phases of materials, such as a heterogeneous mixture of a metal and a ceramic. Therefore, the precise information about the distribution of constituent particles and shape may not be available. Hence, the effective material properties such as elastic constants and density of the chosen composites are estimated depending on the volume fraction distribution and the approximate shape of the dispersed phase (Jha et al. 2013). The effective properties of macroscopic composite materials can be obtained from the microscopic heterogeneous material structures using the homogenization technique. Several homogenization techniques are reported in the literature; a few of them are discussed in the following section (Hill 1965; Klusemann et al. 2012; Mori and Tanaka 1973; Shen and Wang 2012).

12.4.1 The Voigt Model

This model is generally employed to predict the effective mechanical and electrical properties of a composite material composed of continuous and unidirectional fibers. It provides a theoretical upper and lower bound on properties such as (a) the elastic constants, (b) mass density, (c) thermal conductivity, and (d) electrical conductivity (Gibson et al. 1995; Shen and Wang 2012). The effective material properties (P_{eff}), like Young's modulus (E_{eff}), Poisson's ratio (υ_{eff}), thermal expansion coefficient (α_{eff}), and thermal conductivity (K_{eff}) in a specific direction (along with "z") is determined from the following relations (Gupta et al. 2017):

$$
\begin{aligned}
E_{\text{eff}}(z,T) &= \left[E_{\text{c}}(T) - E_{\text{m}}(T)\right]\left(\frac{2z+h}{2h}\right)^n + E_{\text{m}}(T),\\
\alpha_{\text{eff}}(z,T) &= \left[\alpha_{\text{c}}(T) - \alpha_{\text{m}}(T)\right]\left(\frac{2z+h}{2h}\right)^n + \alpha_{\text{m}}(T),\\
\upsilon_{\text{eff}}(z,T) &= \left[\upsilon_{\text{c}}(T) - \upsilon_{\text{m}}(T)\right]\left(\frac{2z+h}{2h}\right)^n + \upsilon_{\text{m}}(T)\\
\rho_{\text{eff}}(z) &= \left[\rho_{\text{c}} - \rho_{\text{m}}\right]\left(\frac{2z+h}{2h}\right)^n + \rho_{\text{m}},\\
K_{\text{eff}}(z,T) &= \left[K_{\text{c}} - K_{\text{m}}\right]\left(\frac{2z+h}{2h}\right)^n + K_{\text{m}},
\end{aligned}
\tag{12.6}
$$

where subscripts "m" and "c" represent the metallic and ceramic constituents, respectively, and n is the volume fraction index.

12.4.2 The Mori–Tanaka Model

This method is usually used for composites having discontinuous particulate phase regions of the graded microstructure. According to the Mori–Tanaka scheme (Benveniste 1987), the effective Young's modulus E_{eff} and Poisson's ratio υ_{eff} can be expressed by the following relations:

$$\frac{K_{eff}(z) - K_t}{K_b - K_t} = \frac{V_b}{1 + [1 - V_b]\dfrac{3[K_b + K_t]}{[3K_t + 4G_t]}}$$

$$\frac{G_{eff}(z) - G_t}{G_b - G_t} = \frac{G_{eff}(z)}{1 + [1 - V_b]\left(\dfrac{G_b + G_t}{G_t + f_1}\right)}, \tag{12.7}$$

$$\text{where } f_1 = \frac{G_t[9K_t + 8G_t]}{[6K_t + 2G_t]}.$$

From Equation 12.7, the effective material properties are represented as

$$E_{eff} = \frac{9K_{eff}G_{eff}}{3K_f + G_f}, \quad \upsilon_{eff} = \frac{3K_{eff} - 2G_{eff}}{6K_{eff} + 2G_{eff}}. \tag{12.8}$$

12.5 STRUCTURAL KINEMATICS OF PLATES/SHELL

The structural kinematics is always of primary concern since it defines the physical behavior of structures. Prediction of the precise and realistic structural and dynamic response of composite structures depends on their structural kinematics. Various structural kinematics (plate theories) have been developed by researchers for the purpose of analyzing composite plates and shells. In this framework, classical plate theory (CPT) based on the Kirchhoff hypothesis is considered the earliest plate theory, which fails to predict the accurate behavior of a structure due to the negligence of shear deformation (Kirchhoff 1850). Because of the ignorance of transverse shear and normal strain, its applicability is restricted to thin plates (Gupta and Talha 2015; Love 1888; Yang and Shen 2001). To overcome this deficiency, Reissner (1975) and Mindlin (1951) considered the linear shear deformation theory, which is termed as first-order shear deformation theory (FSDT). The FSDT does not account for the traction-free boundary condition at the top and bottom surfaces of the plate. This necessitates the need of a shear correction factor. This correction factor is obtained by comparing the results with an exact elasticity solution and depends on various parameters like loading conditions and geometric configuration. To circumvent the limitations of CPT and FSDT, various higher-order shear deformation theories (HSDTs) have been developed and put forth in the last three decades by assuming quadratic, cubic, or even nonlinear variations along the thickness direction. Although these theories are computationally expensive, they do offer better kinematics to demonstrate the static and dynamic characteristics of a structure. These theories can be broadly classified as polynomial and nonpolynomial HSDTs. In polynomial HSDTs, Taylor's series expansion is used to accommodate the effect of shear deformation. In contrast, shear–strain function is used to accommodate the shear deformation in nonpolynomial HSDTs.

In the context of polynomial HSDTs, Basset (1890) introduced the displacement field in terms of Taylor series expansion along the thickness direction. Levinson (1980), Murthy (1981), and Reddy (1984) developed the renowned theories in which Taylor's series expansion is used in the in-plane displacement component to demonstrate the structural kinematics of a plate. Neves et al. (2013) presented a plate theory having nine unknowns for the FGM plate where the in-plane displacement and transverse displacement components are assumed a cubic and quadratic order of the thickness coordinate, respectively. Talha and Singh (2010) studied the static and vibration responses on functionally graded plates having 13 unknowns. Although these HSDTs are competent to predict the structural behavior with adequate accuracy, they are complex to formulate.

To exterminate the aforesaid intricacy, several shear–strain function–based nonpolynomial HSDTs have been proposed during the last three decades. Touratier (1991) developed sinusoidal strain function–based HSDT for isotropic and laminated composite having five unknowns. Aydogdu (2009) showed logarithmic-based nonpolynomial HSDT with five unknowns. Karama et al. (2009) introduced an exponential HSDT for the static and dynamic analysis of laminated plates having five unknowns. Suganyadevi and Singh (2013) proposed various theories based on trigonometric, logarithmic, exponential-based shear–strain function to investigate the static response of a laminated plate. Hebali et al. (2014) developed a new quasi-3D hyperbolic shear deformation theory having five unknowns. Belabed et al. (2014) developed a five-unknown–based nonpolynomial HSDT with a combination of exponential and hyperbolic function–based shear–strain function. Grover et al. (2013) presented a secant function–based HSDT having five unknowns to investigate the vibration response of advanced composite plate. Neves et al. (2012) proposed a hyperbolic theory having six unknowns for FGM plate, which also includes stretching effect.

In addition to this, recently, few authors split the transverse displacement into bending and shear component to simplify the displacement field and reduce the number of unknowns to some extent. In this framework, Senthilnathan et al. (1987) shortened the third-order shear deformation theory (TSDT) by dividing the transverse displacement into the bending and shear components. Hence, the number of unknowns is reduced by one. Shimpi (2002) used only two unknowns to develop a refined plate theory for isotropic plate by dividing the displacement into the bending and shear component. Thai and Vo (2013) proposed four-unknown HSDTs for the structural response of FGM plate. Gupta and Talha (2016, 2017) proposed hyperbolic and inverse hyperbolic shear–strain function–based HSDTs having four unknowns to investigate the vibration response of FGM plate.

12.6 CONCLUSIONS

FGMs have become a keystone of modern materials research with a widespread array of applications in aerospace, aircraft, medical, civil, and mechanical industries. This comprehensive review of various investigations in FGMs based on their fabrication and applications offers some awareness into the factors that eventually govern their properties. It is observed from the literature review that most of the FGM processing techniques deal with the ceramic–ceramic– and metal–ceramic–based FGMs. This is due to the fact that under severe conditions such as high-temperature and large-temperature gradient environment, the materials should be high-temperature resistant, tough, and lightweight. Among several fabrication techniques of FGMs, powder metallurgy is the most adopted technique, which is also feasible for mass production. However, there is still room for improvement in the techniques to obtain the guaranteed distribution of properties throughout the structure with more reliability and predictability. The research activity in this field is continually increasing, which may become a revolutionary achievement for the benefit of the society in the future.

REFERENCES

Aydogdu, M. 2009. A new shear deformation theory for laminated composite plates. *Composite Structures* 89 (1): 94–101. doi:10.1016/j.compstruct.2008.07.008.

Bao, G., and L. Wang. 1995. Multiple cracking in functionally graded ceramic/metal coating. *International Journal of Solids Structures* 32 (19): 2853–71.

Basset, A. B. 1890. On the extension and flexure of cylindrical and spherical thin elastic shells. *Philosophical Transactions of the Royal Society A: Mathematical, Physical and Engineering Sciences* 181 (0): 433–80. doi:10.1098/rsta.1890.0007.

Belabed, Z., M. S. A. Houari, A. Tounsi, S. R. R. Mahmoud, and O. A. Beg. 2014. An efficient and simple higher order shear and normal deformation theory for functionally graded material (FGM) plates. *Composites Part B: Engineering* 60. Elsevier Ltd: 274–83. doi:10.1016/j.compositesb.2013.12.057.

Benveniste, Y. 1987. A new approach to the application of Mori–Tanaka's theory in composite materials. *Mechanics of Materials* 6 (2): 147–57. doi:10.1016/0167-6636(87)90005-6.

Bernardo, G. M. S., F. R. Damásio, T. A. N. Silva, and M. A. R. Loja. 2016. A study on the structural behaviour of FGM plates static and free vibrations analyses. *Composite Structures* 136: 124–38. doi:10.1016/j.compstruct.2015.09.027.

Bever, M. B., and P. E. Duwez. 1972. Gradients in composite materials. *Materials Science and Engineering* 10: 1–8.

Birman, V., and L. W. Byrd. 2007. Modeling and analysis of functionally graded materials and structures. *Applied Mechanics Reviews* 60 (5): 195. doi:10.1115/1.2777164.

Chakraverty, S., and K. K. Pradhan. 2014. Free vibration of exponential functionally graded rectangular plates in thermal environment with general boundary conditions. *Aerospace Science and Technology* 36 (July): 132–56. doi:10.1016/j.ast.2014.04.005.

Chenglin, C., Z. Jingchuan, Y. Zhongda, and W. Shidong. 1999. Hydroxyapatite–Ti functionally graded biomaterial fabricated by powder metallurgy. *Materials Science and Engineering: A* 271 (1–2): 95–100. doi:10.1016/S0921-5093(99)00152-5.

Chi, S. H., and Y. L. Chung. 2002. Cracking in sigmoid functionally graded coating. *Journal of Mechanics* 18: 41–53.

Department of Defense. n.d. *Metallic Materials and Element for Aerospace Vehicle Structures*. MIL-HDBK-5J. Wright–Patterson Air Force Base.

Fujii, T., K. Tohgo, H. Araki, K. Wakazono, M. Ishikura, and Y. Shimamura. 2010. Fabrication and strength evaluation of biocompatible ceramic–metal composite materials. *Journal of Solid Mechanics and Materials Engineering* 4 (11): 1699–1710. doi:10.1299/jmmp.4.1699.

Giannakopoulos, A. E., A. Kordolemis, and T. Zisis. 2010. Development of strong surfaces using functionally graded composites inspired by natural teeth—A theoretical approach. *Journal of Engineering Materials and Technology* 132 (1): 11009. doi:10.1115/1.3184037.

Gibson, L. J., M. F. Ashby, G. N. Karam, U. Wegst, and H. R. Shercliff. 1995. The mechanical properties of natural materials. II. Microstructures for mechanical efficiency. *Proceedings of the Royal Society A: Mathematical, Physical and Engineering Sciences* 450 (1938). The Royal Society: 141–62. doi:10.1098/rspa.1995.0076.

Gooch, W. A., B. H. C. Chen, M. S. Burkins, R. Palicka, J. J. Rubin, and R. Ravichandran. 1999. Development and ballistic testing of a functionally gradient ceramic/metal applique. *In Functionally Graded Materials*, 308: 614–21. Materials Science Forum. Trans Tech Publications. doi:10.4028/www.scientific.net/MSF.308-311.614.

Grover, N., B. N. Singh, and D. K. Maiti. 2013. Analytical and finite element modeling of laminated composite and sandwich plates: An assessment of a new shear deformation theory for free vibration response. *International Journal of Mechanical Sciences* 67: 89–99. doi:10.1016/j.ijmecsci.2012.12.010.

Groves, J. F., and H. N. G. Wadley. 1997. Functionally graded materials synthesis via low vacuum directed vapor deposition. *Composites Part B* 28: 57–69.

Guo, H., K. Aik, Y. Chiang, and X. Miao. 2003. Laminated and functionally graded hydroxyapatite/yttria stabilized tetragonal zirconia composites fabricated by spark plasma sintering. *Biomaterials* 24: 667–75.

Gupta, A., and M. Talha. 2015. Recent development in modeling and analysis of functionally graded materials and structures. *Progress in Aerospace Sciences* 79 (November). Elsevier: 1–14. doi:10.1016/j.paerosci.2015.07.001.

Gupta, A., and M. Talha. 2016. An assessment of a non-polynomial based higher order shear and normal deformation theory for vibration response of gradient plates with initial geometric imperfections. *Composites Part B* 107: 141–61. doi:10.1016/j.compositesb.2016.09.071.

Gupta, A., and M. Talha. 2017. Large amplitude free flexural vibration analysis of finite element modeled FGM plates using new hyperbolic shear and normal deformation theory. *Aerospace Science and Technology* 1: 1–22. doi:http://dx.doi.org/10.1016/j.ast.2017.04.015.

Gupta, A., M. Talha, and W. Seemann. 2017. Free vibration and flexural response of functionally graded plates resting on Winkler–Pasternak elastic foundations using non-polynomial higher order shear and normal deformation theory. *Mechanics of Advanced Materials and Structures*, 0–0. doi:10.1080/15376494.2017.1285459.

He, L. H., and M. V. Swain. 2009. Enamel—A functionally graded natural coating. *Journal of Dentistry* 37 (8): 596–603. doi:10.1016/j.jdent.2009.03.019.

Hebali, H., A. Tounsi, M. S. A. Houari, A. Bessaim, and E. A. A. Bedia. 2014. New quasi-3D hyperbolic shear deformation theory for the static and free vibration analysis of functionally graded plates. *Journal of Engineering Mechanics* 140 (2): 374–83. doi:10.1061/(ASCE)EM.1943-7889.0000665.

Hill, R. 1965. A self-consistent mechanics of composite materials. *Journal of the Mechanics and Physics of Solids* 13 (4): 213–22. doi:10.1016/0022-5096(65)90010-4.

Hirai, T. 1995. CVD processing. *MRS Bulletin* 20 (1): 45–7.

Jedamzik, R., A. Neubrand, and J. Rödel. 2000. Functionally graded materials by electrochemical processing and infiltration : Application to tungsten/copper composites. *Journal of Materials Science* 35 (2): 477–86. doi:10.1023/A:1004735904984.

Jha, D. K., T. Kant, and R. K. Singh. 2013. A critical review of recent research on functionally graded plates. *Composite Structures* 96. Elsevier Ltd: 833–49. doi:10.1016/j.compstruct.2012.09.001.

Jin, G., M. Takeuchi, S. Honda, T. Nishikawa, and H. Awaji. 2005. Properties of multilayered mullite/Mo functionally graded materials fabricated by powder metallurgy processing. *Materials Chemistry and Physics* 89 (2–3): 238–43. doi:10.1016/j.matchemphys.2004.03.031.

Kant, T., and Mallikarjuna. 1989. Vibrations of unsymmetrically laminated plates analyzed by using a higher order theory with a C0 finite element formulation. *Journal of Sound and Vibration* 134 (1): 1–16.

Karama, M., K. S. Afaq, and S. Mistou. 2009. A new theory for laminated composite plates. *Proceedings of the Institution of Mechanical Engineers, Part L: Journal of Materials: Design and Applications* 223 (2): 53–62. doi:10.1243/14644207JMDA189.

Kawase, M., T. Tago, M. Kurosawa, H. Utsumi, and K. Hashimoto. 1999. Chemical vapor in infiltration and deposition to produce a silicon carbide-carbon functionally gradient material. *Chemical Engineering Science* 54: 3327–34.

Kirchhoff, G. R. 1850. Uber das Gleichgewicht und die Bewegung einer Elastischen Scheibe. *J Reine Angew Math (Crelle's J)* 40: 51–88.

Klusemann, B., H. J. Böhm, and B. Svendsen. 2012. Homogenization methods for multi-phase elastic composites with non-elliptical reinforcements: Comparisons and benchmarks. *European Journal of Mechanics—A/Solids* 34: 21–37. doi:10.1016/j.euromechsol.2011.12.002.

Koizumi, M. 1997. FGM activities in Japan. *Composites Part B: Engineering* 28 (1–2): 1–4. doi:10.1016/S1359-8368(96)00016-9.

Kondo, H., A. Yokoyama, M. Omori, A. Ohkubo, T. Hirai, F. Watari, M. Uo, and T. Kawasaki. 2004. Fabrication of titanium nitride/apatite functionally graded implants by spark plasma sintering. *Materials Transactions* 45 (11): 3156–62. doi:10.2320/matertrans.45.3156.

Krishna, B. V., S. Bose, and A. Bandyopadhyay. 2007. Low stiffness porous Ti structures for load-bearing implants. *Acta Biomaterialia* 3 (6): 997–1006. doi:10.1016/j.actbio.2007.03.008.

Kutty, M. G., S. Bhaduri, and S. B. Bhaduri. 2004. Gradient surface porosity in titanium dental implants: Relation between processing parameters and microstructure. *Journal of Materials Science: Materials in Medicine* 15 (2): 145–50. doi:10.1023/B:JMSM.0000011815.50383.bd.

Lee, Y.-D., and E Erdogan. 1995. Residual/thermal stresses in FGM and laminated thermal barrier coatings. *International Journal of Fracture* 69: 145–65.

Levinson, M. 1980. An accurate simple theory of statics and dynamics of elastic plates. *Mechanics Research Communications* 7 (6): 343–50.

Love, A. E. H. 1888. The small free vibrations and deformation of a thin elastic shell. *Philosophical Transactions of the Royal Society A: Mathematical, Physical and Engineering Sciences* 179 (0). The Royal Society: 491–546. doi:10.1098/rsta.1888.0016.

Ma, J., and G. E. B. Tan. 2001. Processing and characterization of metal ± ceramics functionally gradient materials. *Journal of Materials Processing Technology* 113 (February): 446–9.

Mehrali, M., F. S. Shirazi, M. Mehrali, H. S. C. Metselaar, N. A. B. Kadri, and N. A. A. Osman. 2013. Dental implants from functionally graded materials. *Journal of Biomedical Materials Research—Part A* 101 (10): 3046–57. doi:10.1002/jbm.a.34588.

Melchels, F. P. W., M. A. N. Domingos, T. J. Klein, J. Malda, P. J. Bartolo, and D. W. Hutmacher. 2012. Additive manufacturing of tissues and organs. *Progress in Polymer Science* 37 (8): 1079–104. doi:10.1016/j.progpolymsci.2011.11.007.

Melgarejo, Z. H., O. Marcelo Suárez, and K. Sridharan. 2008. Microstructure and properties of functionally graded Al–Mg–B composites fabricated by centrifugal casting. *Composites Part A: Applied Science and Manufacturing* 39 (7): 1150–8. doi:10.1016/j.compositesa.2008.04.002.

Mindlin, R. D. 1951. Influence of rotatory inertia and shear on flexural motions of isotropic, elastic plates. *ASME Journal of Applied Mechanics* 18: 31–8.

Mori, T., and K. Tanaka. 1973. Average stress in matrix and average elastic energy of materials with misfitting inclusions. *Acta Metallurgica* 21 (5). Pergamon: 571–4. doi:10.1016/0001-6160(73)90064-3.

Murthy, M. V. 1981. An improved transverse shear deformation theory for laminated anisotropic plates. *NASA Technical Paper 1903*, no. November.

Naebe, M., and K. Shirvanimoghaddam. 2016. Functionally graded materials: A review of fabrication and properties. *Applied Materials Today* 5 (December): 223–45. doi:10.1016/j.apmt.2016.10.001.

Neves, A. M. A., A. J. M. Ferreira, E. Carrera, M. Cinefra, C. M. C. Roque, R. M. N. Jorge, and C. M. M. Soares. 2012. A quasi-3D hyperbolic shear deformation theory for the static and free vibration analysis of functionally graded plates. *Composite Structures* 94 (5): 1814–25. doi:10.1016/j.compstruct.2011.12.005.

Neves, A. M. A., A. J. M. Ferreira, E. Carrera, M. Cinefra, C. M. C. Roque, R. M. N. Jorge, and C. M. M. Soares. 2013. Static, free vibration and buckling analysis of isotropic and sandwich functionally graded plates using a quasi-3D higher-order shear deformation theory and a meshless technique. *Composites Part B: Engineering* 44 (1): 657–74. doi:10.1016/j.compositesb.2012.01.089.

Niu, X., N. Rahbar, S. Farias, and W. Soboyejo. 2009. Bio-inspired design of dental multilayers: Experiments and model. *Journal of the Mechanical Behavior of Biomedical Materials* 2 (6): 596–602. doi:10.1016/j.jmbbm.2008.10.009.

Omatete, O. O., M. A. Janney, and S. D. Nunn. 1997. Gelcasting: From laboratory development toward industrial production. *Journal of the European Ceramic Society* 17 (2–3): 407–13. doi:10.1016/S0955-2219(96)00147-1.

Oshkour, A. A., H. Talebi, S. F. Shirazi, M. Bayat, Y. H. Yau, F. Tarlochan, and N. A. A. Osman. 2014. Comparison of various functionally graded femoral prostheses by finite element analysis. *The Scientific World Journal* 1–17.

Park, S.-H., J.-H. Kanga, Y.-G. Junga, and U. Paikb. 2006. Fabrication and mechanical characterization of Al2O3/ZrO2 layered composites with graded microstructure. *Multiscale and Functionally Graded Materials* 273–8.

Rahbar, N., and W. O. Soboyejo. 2011. Design of functionally graded dental multilayers. *Fatigue & Fracture of Engineering Materials & Structures* 34 (11): 887–97. doi:10.1111/j.1460-2695.2011.01581.x.

Rahimipour, M. R., and M. Sobhani. 2013. Evaluation of centrifugal casting process parameters for in situ fabricated functionally gradient Fe–TiC composite. *Metallurgical and Materials Transactions B: Process Metallurgy and Materials Processing Science* 44 (5): 1120–3. doi:10.1007/s11663-013-9903-z.

Reddy, J. N. 1984. A simple higher-order theory for laminated composite plates. *Journal of Applied Mechanics* 51 (4). American Society of Mechanical Engineers: 745. doi:10.1115/1.3167719.

Reddy, J N. 2000. Analysis of functionally graded plates. *International Journal for Numerical Methods in Engineering* 47 (1–3): 663–84.

Reddy, J. N. 2004. *Mechanics of Laminated Composite Plates and Shells : Theory and Analysis*. CRC Press.

Reissner, E. 1975. On transverse bending of plates, including the effect of transverse shear deformation. *International Journal of Solids and Structures* 11 (5). Pergamon: 569–73. doi:10.1016/0020-7683(75)90030-X.

Saiyathibrahim, A., S. S. Mohamed Nazirudeen, and P. Dhanapal. 2015. Processing techniques of functionally graded materials—A review. *Proceedings of the International Conference on Systems, Science, Control, Communication, Engineering and Technology* 1: 98–105.

Sasaki, M., and T. Hirai. 1991. Fabrication and properties of functionally gradient materials. *Journal of the Ceramic Society of Japan* 99 (10): 1002–13.

Sasaki, M., and T. Hirai. 1994. Thermal fatigue resistance of CVD SiC/C functionally gradient material. *Journal of the European Ceramic Society* 14 (3): 257–60. doi:10.1016/0955-2219(94)90094-9.

Senthilnathan, N. R., S. T. Chow, K. H. Lee, and S. P. Lim. 1987. Buckling of shear-deformable plates. *AIAA Journal* 25 (9): 1268–71.

Shen, H. S. 2009. *Functionally Graded Materials: Nonlinear Analysis of Plates and Shells*. Boca Raton, FL: CRC Press.

Shen, H.-S., and Z.-X. Wang. 2012. Assessment of Voigt and Mori–Tanaka models for vibration analysis of functionally graded plates. *Composite Structures* 94: 2197–208. doi:10.1016/j.compstruct.2012.02.018.

Shimpi, R. P. 2002. Refined plate theory and its variants. *AIAA Journal* 40 (1): 137–46. doi:10.2514/2.1622.

Singh, L. 2016. A review on functionally graded ceramic–metal materials. *International Journal of Research in Aeronautical and Mechanical Engineering* 4 (6): 8–14.

Sobczak, J. J., and L. Drenchev. 2013. Metallic functionally graded materials: A specific class of advanced composites. *Journal of Materials Science and Technology* 29 (4). Elsevier Ltd: 297–316. doi:10.1016/j.jmst.2013.02.006.

Sofiyev, A. H., D. Hui, A. A. Valiyev, F. Kadioglu, S. Turkaslan, G. Q. Yuan, V. Kalpakci, and A. Özdemir. 2016. Effects of shear stresses and rotary inertia on the stability and vibration of sandwich cylindrical shells with FGM core surrounded by elastic medium. *Mechanics Based Design of Structures and Machines* 44 (4): 384–404. doi:10.1080/15397734.2015.1083870.

Stoddart, M. J., S. Grad, D. Eglin, and M. Alini. 2009. Cells and biomaterials in cartilage tissue engineering. *Regenerative Medicine* 4: 81–98.

Suganyadevi, S., and B. N. Singh. 2013. A new shear deformation theory for the static analysis of laminated composite and sandwich plates. *Composite Structures* 75: 324–36. doi:http://dx.doi.org/10.1016/j.compstruct.2015.11.049.

Talha, M., and B. N. Singh. 2010. Static response and free vibration analysis of fgm plates using higher order shear deformation theory. *Applied Mathematical Modelling* 34 (12): 3991–4011. doi:10.1016/j.apm.2010.03.034.

Tan, T., T. Xia, H. Ofolan, J. Dao, Z. Bash, K. Johanson, J. Novotny, M. Ozeki, and M. Smith. 2015. Sustainability in beauty: An innovative proposing-learning model to inspire renewable energy education. *The Journal of Sustainability Education (JSE)* January 17 (May 2010): 1–25.

Tawakol, A. E., and N. E. Bondok. 2013. Material selection in the design of the tibia tray component of cemented artificial knee using finite element method. *Materials and Design* 44: 454–60.

Thai, H. T., and T. P. Vo. 2013. A new sinusoidal shear deformation theory for bending, buckling, and vibration of functionally graded plates. *Applied Mathematical Modelling* 37 (5). 3269–81. doi:10.1016/j.apm.2012.08.008.

Touratier, M. 1991. An efficient standard plate theory. *International Journal of Engineering Science* 29 (8): 901–16.

Traini, T., C. Mangano, R. L. Sammons, F. Mangano, A. Macchi, and A. Piattelli. 2008. Direct laser metal sintering as a new approach to fabrication of an isoelastic functionally graded material for manufacture of porous titanium dental implants. *Dental Materials* 24 (11): 1525–33. doi:10.1016/j.dental.2008.03.029.

Übeyli, M., E. Balci, B. Sarikan, and M. K. Öztas. 2014. The ballistic performance of SiC—AA7075 functionally graded composite produced by powder metallurgy. *Materials & Design* 56: 31–36. doi:10.1016/j.matdes.2013.10.092.

Udupa, G., S. S. Rao, and K. V. Gangadharan. 2014. Functionally graded composite materials: An overview. *Procedia Materials Science* 5: 1291–9. doi:10.1016/j.mspro.2014.07.442.

Watanabe, Y., M. Kurahashi, I. S. Kim, S. Miyazaki, S. Kumai, A. Sato, and S.-i. Tanaka. 2006. Fabrication of fiber-reinforced functionally graded materials by a centrifugal in situ method from Al–Cu–Fe ternary alloy. *Composites Part A: Applied Science and Manufacturing* 37 (12): 2186–93. doi:10.1016/j.compositesa.2005.10.003.

Watanabe, Y., and S. Oike. 2005. Formation mechanism of graded composition in Al–Al2Cu functionally graded materials fabricated by a centrifugal in situ method. *Acta Materialia* 53 (6): 1631–41. doi:10.1016/j.actamat.2004.12.013.

Watanabe, Y., R. Sato, I.-S. Kim, S. Miura, and H. Miura. 2005. Functionally graded material fabricated by a centrifugal method from ZK60A magnesium alloy. *Materials Transactions* 46 (5): 944–9. doi:10.2320/matertrans.46.944.

Watari, F., A. Yokoyama, M. Omori, T. Hirai, H. Kondo, M. Uo, and T. Kawasaki. 2004. Biocompatibility of materials and development to functionally graded implant for bio-medical application. *Composites Science and Technology* 64 (6): 893–908. doi:10.1016/j.compscitech.2003.09.005.

Watari, F., A. Yokoyama, F. Saso, and M. Uo. 1997. Fabrication and properties of functionally graded dental implant. *Composites Part B* 28B: 5–11.

Xiong, H., L. Zhang, L. Chen, T. Hirai, and R. Yuan. 2000. Design and fabrication of W–Mo–Ti–TiAl–Al system functionally graded material. *Metallurgical and Materials Transactions A* 31 (9): 2369–76. doi:10.1007/s11661-000-0152-9.

Yan, M., X. Peng, and T. Ma. 2009. Microstructures of Ni–ZrO2 functionally graded materials fabricated via slip casting under gradient magnetic fields. *Journal of Alloys and Compounds* 479 (1–2): 750–4. doi:10.1016/j.jallcom.2009.01.042.

Yang, J., and H.-S. Shen. 2001. Dynamic response of initially stressed functionally graded rectangular thin plates. *Composite Structures* 54 (4): 497–508. doi:10.1016/S0263-8223(01)00122-2.

Yeo, J. G., Y. G. Jung, and S. C. Choi. 1998. Design and microstructure of ZrO2/SUS316 functionally graded materials by tape casting. *Materials Letters* 37 (6): 304–11. doi:10.1016/S0167-577X(98)00111-6.

Young, A. C., O. O. Omatete, M. A. Janney, and P. A. Menchhofer. 1991. Gelcasting of alumina. *Journal of the American Ceramic Society* 74 (3): 612–8. doi:10.1111/j.1151-2916.1991.tb04068.x.

Zhu, J. C., Z. D. Yin, and Z. H. Lai. 1996. Fabrication and microstructure of ZrO2-Ni functional gradient material by powder metallurgy. *Journal of Materials Science* 31 (21): 5829–34. doi:10.1007/BF01160836.

Zidi, M., A. Tounsi, M. S. A. Houari, E. A. A. Bedia, and O. A. Bég. 2014. Bending analysis of FGM plates under hygro-thermo-mechanical loading using a four variable refined plate theory. *Aerospace Science and Technology* 34 (April): 24–34. doi:10.1016/j.ast.2014.02.001.

13 Use of Nanotechnology for Viable Applications in the Field of Medicine

Mazaher Gholipourmalekabadi,
Mohammad Taghi Joghataei, Aleksandra M. Urbanska,
Behzad Aghabarari, Aidin Bordbar-Khiabani,
Ali Samadikuchaksaraei, and Masuod Mozafari

CONTENTS

13.1 NANOTECHNOLOGY—AT A GLANCE

Nanotechnology is an emerging field that has a great potential to have a major effect on human health [1]. Indeed, nanotechnology is a science used in the design, production, characterization, and application of structures, devices, and systems controlling both the size and shape at a length of nanometer scale (0.000000001 m) to improve the character and properties of materials. Technology of nanomaterials has achieved tremendous progress in the last four decades. Nanomaterials—materials with basic structural units, grains, particles, fibers, or other constituent components smaller than 100 nm in at least one dimension [1–3]—have evoked a great amount of attention in improving disease prevention, diagnosis, and treatment [1], and the number of independent research studies based on nanotechnology in the field of drug delivery and tissue engineering is strikingly increasing. The applications of nanotechnology in medical sciences are considered either as "nanomedicine" or "nanobiomedicine." Nanomaterials can be fabricated using either "top-down," i.e., breaking down

TABLE 13.1
Applications and Their Corresponding Scales

Centimeter (cm)	Gravity, friction, and combustion
Millimeter (mm)	Gravity, friction, combustion, and electrostatic
Micrometer (μm)	Electrostatic, van der Walls, and Brownian
Nanometer (nm)	Electrostatic, van der Walls, Brownian, and quantum
Picometer (pm)	Quantum mechanics

matter into more basic building blocks that frequently use chemical or thermal methods [4,5], or "bottom-up," i.e., building complex systems by combining simple atomic-level components [6]. To date, numerous top-down and bottom-up nanofabrication technologies, such as (1) electrospinning, (2) phase separation, (3) self-assembly processes, (4) thin film deposition, (5) chemical vapor deposition, (6) chemical etching, (7) nano-imprinting, (8) photolithography, and (9) electron beam or nanosphere lithographies [1], have been developed to synthesize nanomaterials having ordered or random nanotopographies [1,7,8]. With decreasing material size into the nanoscale, the ratio of surface area to volume is dramatically increased and consequently can lead to improved solubility, multifunctionality, and superior physiochemical properties (i.e., mechanical, electrical, optical, catalytic, magnetic properties, etc.) [9–11]. Table 13.1 shows the effects of size on specific characters of materials.

Nanotechnology has been used in electronics [12,13], information storage capacity [14], pharmaceutics [15], food and agriculture [16], chemistry (as catalysts), drug delivery [17], cell tracking and separation, diagnosis and treatment of disease [18,19], DNA sequencing (using nanopores) [20], DNA delivery into cells [21,22], and tissue engineering [9,14,23,24]. In the following sections, we describe the potential roles of nanotechnology in medicine.

13.2 NANOTECHNOLOGY MEETS MODERN MEDICINE: TISSUE ENGINEERING AND REGENERATIVE MEDICINE

Tissue engineering (also known as regenerative medicine) is a field of research that aims to develop biological substitutes to repair damaged tissues and restore their functionality using a triad of biomaterial scaffolds and soluble factors in combination with cells (i.e., stem cells), [14,23,25–28]. The use of stem and progenitor cells has opened a new frontier in regenerative medicine. Stem cells are undifferentiated cells endowed with a capacity for self-renewal, as well as clonogenic and multilineage differentiation [29,30]. These cells can be differentiated to at least two different cell lines and have a high capacity in tissue engineering applications [29,31]. Embryonic stem cells (ESCs) are pluripotent stem cells derived from the inner cell mass of a blastocyst at an early-stage embryo. Adult stem cells, such as hematopoietic, mesenchymal stem cells, are also multipotent stem cells that present in bone marrow, periphery blood, and other tissues. Stem cell behavior is often correlated with cues that lie in an extracellular microenvironment. It is known that stem cells can respond to genetic signals, such as those imparted by nucleic acids, for the purpose of promoting lineage-specific differentiation. However, before they achieve therapeutic relevancy, proper methods need to be developed to control stem cell differentiation [29,32,33]. Maintenance of stem cells in undifferentiated stages, directing stem cells to differentiate into a special cell line, and prevention of undesired differentiation and delivery of cells into a specific tissue are important objects for using these cells in tissue engineering [34,35]. Many attempts have been made to consider these subjects. For example, ESCs have been cultured in various conditions: feeder layer (usually fetal mouse fibroblasts) [36] or feeder-free condition using Matrigel or laminin [37], conditioned medium or supplement media using specific molecular factors, such as the leukemia inhibitory factor, to be kept in an undifferentiated stage [38]. These methods have some disadvantages: (1) laborious, (2) need enzymatic or mechanical treatment, (3) time consuming and tedious, (4) need an expert operator, and (5) increase the risk of contamination with animal pathogens. Therefore, these limitations made researchers to set up an ideal method for the purpose of solving them. Many strategies such as encapsulation have been adopted to keep stem cells in undifferentiated stages over several passages and deliver cells into a tissue [36,37,39–41]; this will be described in Section 13.2.1.

Langer and Vacanti defined tissue engineering as "an interdisciplinary field that applies the principles of engineering and life sciences toward the development of biological substitutes that restore, maintaining, or improve tissue function, or a whole organ" [9,42]. Every year, world population is remarkably increasing. Thereby, it is expected that the number of patients requiring organ replacement therapy will steadily increase in the future as well. Conventional tissue replacements, such as (1) autografts (tissue or organ transferred into a new position in the body of the same person) and (2) allografts (transplant of an organ or tissue from one individual to another of the same species with a different genotype), have a variety of problems. These include shortage of donor organs and transplant rejection [9,43], which cannot satisfy the high-performance demands necessary for today's patients [12,42]. Furthermore, a two-dimensional (2D) cell culture system (i.e., on a flask glass and a Petri dish) cannot fully simulate the natural tissue microenvironment. For instance, substrate stiffness limits cell spreading, cytoskeleton assembly, as well as directional motility. Use of three-dimensional (3D) tissue scaffolds as a template for purpose of regeneration is the basis of tissue engineering. In general, the 3D scaffolds provide a setting designed to foster self-assembly of components natural to tissue microenvironments and the utero-associated developing embryo environment. In this manner, environmental mechanical forces acting upon cells can be manipulated by engineering mechanical properties of the scaffold. Scaffolds also provide a microenvironment for cells and regulate cell proliferation and differentiation while allowing for a uniform cell population that is essential for the tissue repair process [23,25,27,44–46]. In the body, natural tissues or organs are in a nanometer scale, and cells directly interact with nanostructured extracellular matrices (ECMs). Thus, it is necessary for the engineered implants to interact biologically with cells as ECMs [1].

In tissue engineering, an ideal biomaterial has to fulfill desirable and optimal mechanical and biomaterial properties for a given application. It has to possess adjustable biodegradability and biocompatibility. Ideally, it should mimic the natural environment (natural ECMs) where the materials are to be delivered or implanted [1,9,23,47], and it should facilitate cell adhesion, differentiation, proliferation, and neo tissue generation [23,48]. For this purpose, many techniques have been adopted to modify scaffolds to alter biodegradability, mechanical properties, and their biocompatibility [23,49–52]. Thus far, bioengineers have exploited numerous biomaterials in scaffold fabrication. Interestingly, these materials still are not optimal [1]. Unlike the bulk materials and individual molecules, nanomaterials can be tuned to acquire some unique properties, such as structural, magnetic, and electronic features [53] (Table 13.1). In addition, many proteins exist in ECMs such as collagen, which is the most abundant ECMs protein in the human body and has fibrous structure, with a diameter in the nanometer range at least in one dimension (Figure 13.1) [23,54].

Since nanomaterials can mimic (e.g., topography, physiochemical properties, energy and structure) natural tissues and interact with biological systems at a molecular level [55], they have a key

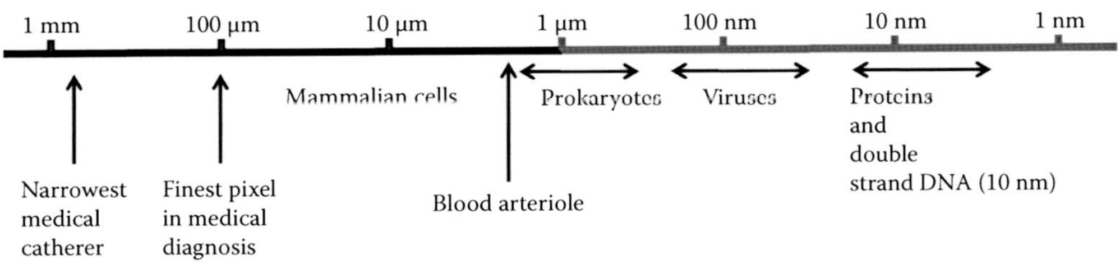

FIGURE 13.1 Comparison of size in different biologic systems.

role in stimulating cell growth and function, as well as in guiding tissue regeneration, and use of these materials may solve aforementioned problems [1,56,57].

On the other hand, nanomaterials can be easily sterilized by filtration with 0.22 μm filters, ultraviolet, ethylene oxide, plasma, etc. The surface of nanomaterials can be modified with newly developed nanotechniques to enhance cell–biomaterial interaction [54,55]. To utilize nanomaterials in tissue engineering, it is critical that used nanoparticles (NPs) have no cytotoxicity effect, keep their physical property under surface modification, and interact with protein or cells without interfering with their behavior and biological activity [56].

The use of nanomaterials in regenerative medicine can be seen widely. For example, from a material property point of view, nanomaterials can be made of metals, ceramics, polymers, organic materials, and even composites, just like conventional or micron-structured materials. Certain metals have superior mechanical properties [23,57], but they are not biodegradable, and this is considered as a disadvantage [23,44]. It is important to state that single-phase materials cannot often provide all of the necessary properties for being used as tissue substitutes (to mimic the desired tissue having suitable strength and modulus property, structure, tissue conductivity and inductivity, biodegradability, etc.). To engineer materials mimicking one or multiple characteristics of the nanofibrous ECM proteins, many technologies, including (1) electrospining, (2) self-assembly, (3) phase separation, and (4) biological effects of nanofibrous architecture, have been developed with varying degrees of success; this will be described under the section 13.4.1 "Electrospinning" [23,58–60].

Nanomaterials include NPs, nanoclusters, nanocrystals, nanotubes, nanofibers, nanowires, nanorods, and nanofilms [1]. Quantum cells (if one dimension is reduced to the nanoscale while the other two remain large), quantum wires (if two dimensions are reduced to the nanoscale), and quantum dots (QDs) (if all three dimensions reach the nanoscale) are some representative NPs [61]. Therefore, composites (heterogeneous combination of two or more materials, in nano- or microscale) have been made to address this issue and to manipulate the scaffold closer to natural tissues [62]. In this chapter, we overview the most relevant applications of nanotechnology in tissue engineering such as cell encapsulation, cell tracking, cell separation, and gene manipulation. Subsequently, we described current opinion on the effect of NPs on the behavior of stem cells, as well as the various techniques for fabrication and modification of tissue engineering scaffolds. In addition, earlier studies on regeneration of some tissues such as bone, cartilage, and nerve are discussed.

13.2.1 Encapsulation Approach

Commonly in tissue engineering, scaffolds made up of biodegradable materials are seeded with cells. After deposition of the ECM of the cells on the scaffold, the resultant construct (scaffold + cells) is transplanted to the defect site. Despite many advances in the tissue engineering area, there are many obstacles to the fabrication of complex organs. Difficulty in regenerating neovascularization and mimicking the architect and physicochemical properties of the complex organ are important barriers in this area. *In vivo*, the cells are fed through blood vessels around them. Therefore, it is important to engineer a functional vascular network for the development of a complex organ. Many efforts have been made to minimize this challenge. For example, researchers cultured endothelial cells on engineered scaffold to enable growth factors related to promotion of vascularization. A major challenge in this approach is that the time required for neovascularization is longer than the cells can stay alive without being fed, which results in the death of the cell. To minimize this limitation, researchers tried to engineer scaffolds containing artificial microvasculature. The challenges of this approach are the engineering of such scaffolds and the functionality of the resulting vasculature. Encapsulation of cells with cytocompatible materials is an interesting strategy in consideration of the above-mentioned challenges. Encapsulation is used for both *in vitro* culture of the cells in 3D conditions and for delivery of the cells to the injured tissues. Capsules could be synthesized with controlled pore structure to restrict the transport of biological molecules. In the cell encapsulation strategy, the scaffold encapsulates the cells during the fabrication process. This method offers a

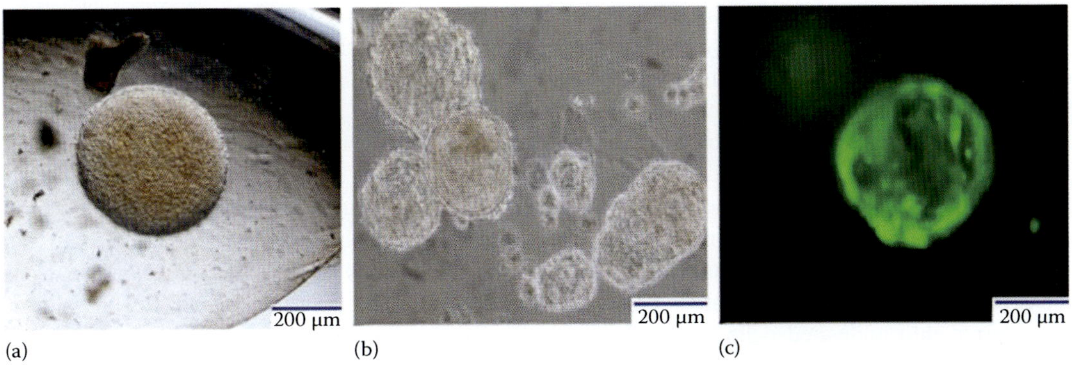

FIGURE 13.2 Morphology and viability of the hESCs encapsulated within alginate hydrogels. (a) hESC aggregates encapsulated for 110 days with no differentiation into the germ layers or cysts being observed. (b) Morphology of the decapsulated hESCs at day 110 cultured in 2D cultures. (c) Encapsulated hESCs remained viable within the aggregates at day 110. (From Siti-Ismail, Norhayati et al., *Biomaterials* 29, no. 29 (2008): 3946–3952.)

variety of advantages. For example, the encapsulated cells can be injected systemically, which delivers the cells to the site of interest. Nanofibrous scaffolds [63–66], hydrogels [67–70], and layer-by-layer structures [71–73] are three major classes of nanomaterials for *in situ* cell delivery and encapsulation. Siti-Ismail et al. [74] encapsulated human ESCs (hESCs) in alginate hydrogels (Figure 13.2). They maintained these cells in an undifferentiated stage for more than 260 days, while their differentiation capacity retained without manipulation the feeder-free media, enzymatic treatment, and mechanical expansion [74].

13.2.2 Cell Tracking and Imaging *In Vivo*

With advances in stem cell application in tissue engineering, development of cell labeling systems *in vivo* has been required. In regenerative medicine tracking of implanted cells is important to monitoring the fate of stem/progenitor cells to include the following: viability, migration, differentiation, and biodistribution of transplanted cells within desire organs, and tissue formation in vivo. several techniques such as positron emission tomography, single-photon emission tomography [75,76], magnetic resonance imaging (MRI) [77–80], x-ray computed microtomography (MicroCT) [81–84], radiotracer [85,86], and ultrasound [87,88] have been developed for *in vivo* imaging of cells. For example, MRI has many advantages—(1) easy accessibility, (2) high spatial resolution, (3) 3D capability, and (4) visualization of implanted cells within the surrounding tissue—which makes it a good choice for *in vivo* cell imaging. For tracking stem cells by MRI, magnetic NPs should be internalized by stem cells *ex vivo* to provide a strong MRI contrast [77–79]. The most important limitation of using MRI is the cell division leads to dilution of agents and gradual loss of MRI cell signal; consequently, it makes it difficult to find an accurate correlation between the detected signal and the transplanted cell number. Various contrast NPs have been used in MRI for cell signal generation, such as iron oxide NPs and superparamagnetic iron oxide NPs (SPIONs) [89–92]. MicroCT is an invasive *in vivo* imaging and cell tracking technique; combined with NP cell labeling, such as iron oxide NPs, and using synchrotron x-ray, MicroCT has been widely used for investigating stem cell migration, differentiation, homing specificity of various stem cells in different stem cell–based therapy, and efficacy of stem cell–related regeneration in small animals [82–84]. Synchrotron x-ray provides high-resolution images (spatial resolution up to 0.3 μm) and a wide range of grayscale values compared to conventional or laboratory x-ray [93,94]. Such methods enable longitudinal studies that can decrease the effect of biological variability in the cohort [95].

Using noninvasive methods for cell labeling has many advantages when compared with invasive methods, such as flow cytometry and intravital microscopy, which are often time-consuming and challenging [85,98,99]. Conventional methods, such as differential interference contrast microscopy and fluorescence microscopy, have some limitations including lack of quantitative data, high background noise from labeling biomolecules, and being time-consuming in observation step, which leads to loss of signals [56,100].

Using NP contrast agents makes us able to sense cell behavior and gives us more knowledge about the process of tissue regeneration *in vivo*. Many contrasting NPs have been used for cell labeling. Having high biocompatibility, stability, and contrasting properties *in vivo* are important points for selecting these agents. On the other hand, these agents must not be diluted over cell division or transferred into undesired cells [101]. For example, for the case of nanosensors, we can monitor protein and metabolite secretion, ion transport, enzyme/cofactor interaction, and cell adhesion [85,102–104]. Nanosensors such as nanotubes have several advantages because of their nanosize and large surface area for use in tissue engineering evaluation [105–107]. A variety of NPs as contrasting agents have been developed for real-time and noninvasive stem cell tracking including iron oxide NPs, SPIONs, colloidal gold, dendrimer composites of iron oxide NPs (also known as magnetodendrimers), polymeric micelles and liposomes, nanotubes, nanowires, nanoshells, and QDs [85,101]. Iron oxide and QDs are inorganic NPs that are widely used in cell tracking [96,97]. These nanosize contrasting agents can be synthesized easily from various materials on a large scale using simple methods.

13.2.2.1 Iron Oxide NPs

Iron oxide NPs can bind to the outer surface of the cell membrane or can be internalized within cells via endocytosis or pinocytosis [102,103]. When they are bound to the cell membrane, iron oxide NPs do not affect cell viability. However, they can interfere with cell surface interactions or detach from the cell surface [104]. Cells can uptake iron oxide NPs via endocytosis, located inside the cytoplasmic vesicles, transferred into lysosomes. In lysosomes, these agents are degraded. Afterward, free irons are released into the cytoplasm [105]. Some modifications have been carried out on iron oxide NPs to reduce cytotoxicity effects coming from the formation of free hydroxyl radicals and reactive oxygen species during NP degradation [106]; to prevent aggregation; and to enhance stability, solubility, contrast, and endocytosis efficiency. For example, superparamagnetic iron oxide NPs have been coated with dextran or other polymers to enhance solubility and stability, and also to prevent aggregation of the superparamagnetic iron oxide NPs [107,108]. Coating superparamagnetic iron oxide NPs with gold provides an inert shell around the NPs that enhances MRI contrast, reduces toxic effects, protects NPs from rapid dissolution, and many other advantages [109–198]. To enhance internalization of the superparamagnetic iron oxide NPs by cells, the surface of this agent has been modified with internalizing ligands, such as HIV-Tat peptides, dendrimers, etc. [91,113,114]. On the other hand, by using superparamagnetic iron oxide NPs for cell labeling, migration of transplanted cells can be easily tracked and monitored using MRI. Also, transplanted stem cells can be retrieved from the implanted site using a magnetic-sorting technique [91,201]. Many researchers have used superparamagnetic iron oxide NPs to track stem/progenitor cells using MRI [111,112,116–121].

13.2.2.2 Carbon Nanotubes

Carbon nanotubes have potential use in biomedical applications, such as (1) drug delivery, (2) delivery of transfection agents, (3) environment sensing, (4) scaffolding for cell-based therapy, (5) *in vivo* imaging, and (6) stem/progenitor cell tracking [85,122–126]. Carbon nanotubes are cylindrical tubes with diameter in the nanometer range, possessing a wide range of thermal, electronic, and structural properties, which are controllable and related to the diameter, length, and twist of the carbon nanotubes [85,127]. Carbon nanotubes can be used as a contrast agent in MRI, optical, and radiotracer [85]. There is contradiction in cytotoxicity effects of carbon nanotubes [128]: Some researchers report the cytotoxicity effects of carbon nanotubes, whereas others have studied the biocompatibility of these agents [129–133]. It has been expressed that several factors are responsible for this contradictory observation [85,134]. To solve this problem and to enhance the safety of carbon nanotubes, this agent has been functionalized using glycopolymers [135].

13.2.2.3 Quantum Dots

QDs are semiconductor nanocrystals (2–5 nm in diameter) with broad absorption and narrow emission spectra. These contrasting agents overcome the limitations of conventional tracking methods and have potential use in (1) cell tracking, (2) labeling, and (3) *in vivo* imaging applications (Figure 13.3) [136,137]. QDs have long fluorescence lifetime (>10 ns) with high fluorescence intensity from high quantum yields. In addition, these agents possess high molar extinction coefficient and are resistant to chemical degradation [137]. Also, we can use QDs with different emission wavelength for multiplex *in vivo* imaging [138–142].

Emission and absorption spectra of QDs are dependent on particle size and can be changed from ultraviolet to infrared. Therefore, with the changing in the size of particles, QDs having the desired and unique emission spectra are acquirable [143,144]. Structurally, QDs are composed of a core (heavy metal, such as CdSe or CdTe) and a surrounding shell (ZnS). The size of the core/shell QDs can be controlled precisely by changing the synthesizing conditions (temperature, durations, and ligands) [145]. QDs have been used widely in stem/progenitor cells labeling [136,141,142,146]. Colhera toxin subunit B- QD conjugates have also been used for long-term cell tracking [147]. Surface modification of QDs with various materials, such as (1) QDs -arginine–glycine–aspartic acid peptide bioconjugates and (2) Streptavidin-coupled QDs, can enhance their properties for the purpose of cell labeling [141,148].

13.2.3 Cell Separation

Separation of pure population of stem cells from heterogeneous cells is an essential part and represents a major contribution in tissue engineering. For example, islet 1+ as cardiac progenitor cells exists in the human heart and mouse heart in very small population and is surrounded by

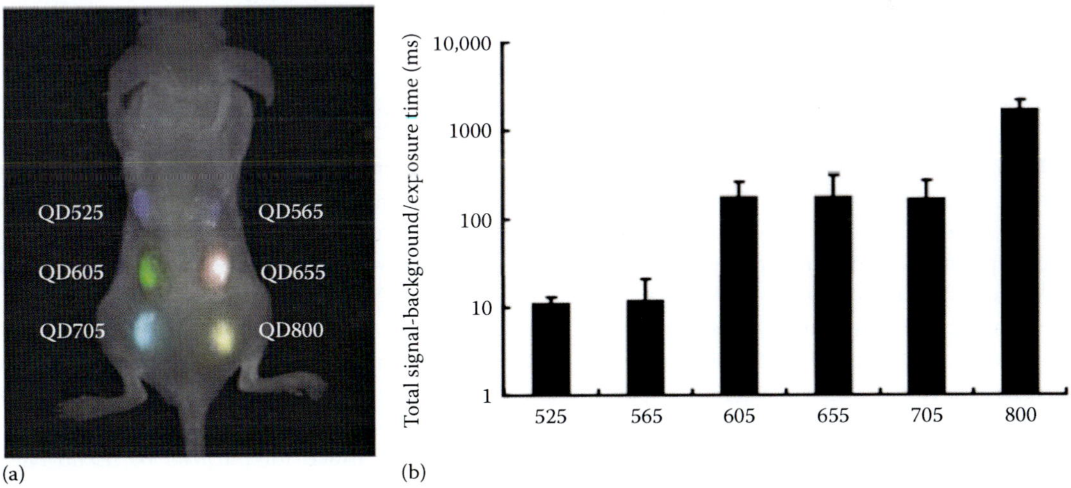

FIGURE 13.3 Multiplex imaging capability of QD in live animals. (a) 1 × 106 ES cells labeled with QD 525, 565, 605, 655, 705, and 800 were subcutaneously injected on the back of the athymic nude mice right after labeling, and the image was taken with a single excitation light source right after injection. (b) Quantification of fluorescent signal intensity defined as total signal-background/exposure time in milliseconds. (From Lin, Shuan et al., *BMC Biotechnology* 7, no. 1 (2007): 67.)

differentiated cells of adult heart tissue [149]. A robust and high efficacy method needs to be developed to separate a pure population of rare islet 1+ cell line. In some cases, stem cells should be differentiated *in vitro* before transplantation, especially in using ESCs. Then, target differentiated cells should be purified and transplanted. Cell-sorting techniques separate cells based on size, density, or affinity (chemical, electrical, or magnetic). Conventional cell separation techniques such as size-based cell sorting, fluorescence-activated cell sorting, and magnetic-activated cell sorting have some disadvantages in terms of cost and time [150]. In modern medicine, miniaturized cell separation systems (microsized/nanosized) have been developed and possess many advantages compared to conventional cell-sorting systems, such as (1) small sample volume, (2) low cost, and (3) portability [150]. Using NPs such as superparamagnetic NPs serves to enhance the resolution. For example, in a study [151], leucocytes (CD 45+ cells) were separated from red blood cells using magnetic-activated cell sorting. After exposure of cells to anti-CD45 conjugated superparamagnetic NPs, CD45+ cells carried about 5000 NPs [151]. Development of a perfect cell-sorting technique for the purpose of separation of rare cells from undesired cells with advantages of high efficacy, low cost, and time is strongly sensed in stem cell biology, and the use of microscale/nanoscale devices enables us to enhance these techniques.

13.2.4 Gene Manipulation of Stem Cells Using NPs

Development of a suitable method for directing stem cell differentiation toward a specific pure cell line is one of the most important points in tissue engineering. All utilized methods for stem cell differentiation, such as (1) specific factor, (2) proteins and small molecules in defined media, (3) vectors for specific gene delivery, and (4) siRNA, lead to alteration in the gene expression profile of the stem cells. As described in the section "Effects of Nanomaterials on Behavior of Stem Cell (13.3)," the NPs can be used in stem cell differentiation. In addition, functionalized NPs can be utilized as efficient gene delivery systems for delivery of specific genes to stem cells for high-level gene expression [152–154]. Kutsuzawa et al. [152] reported functionalized apatite-based gene delivery system by fibronectin and E-cadherin to show a high affinity to needs mouse ESCs. They showed that the system deliver gene was more efficient than commercially available lipofection system [152]. It has been reported that the gold NPs–DNA–polyethylenimine complex, which is patterned on glass (as a solid surface), is a comprising system for the delivery of gene to human mesenchymal stem cells (Figure 13.4) [153].

13.3 EFFECTS OF NANOMATERIALS ON BEHAVIOR OF STEM CELLS

Nanomaterials can be fabricated in the desired shape, size, composition, and various surface properties; they can influence stem cell behavior such as proliferation ability and differentiation direction. It has been shown that mesoporous silica NPs conjugated with fluorescein isothiocyanate enhance actin polymerization and regulate osteogenic differentiation in human mesenchymal stem cells, without affecting stem cell viability and proliferation. Silica NPs are internalized into stem cells by both clathrin and actin-dependent endocytosis, and then escape the endolysosomal vehicle [155,156]. Fibrin polylactide caprolactone NPs are uptaken by stem cells and regulate the gene expression responsible for chondrogenic differentiation. Hence, they can be effectively used for *in situ* cartilage regeneration [155,158].

Nanofabrication techniques provide several types of nanosurface for purpose of tissue engineering. Altering the nanotopography of materials surface surrounding the cells, such as the presence of large, medium, and small nanoscale grooves, pores, pits, ridges, and nodules, can affect cell behavior by several signaling pathways [155,159,160]. A cell response to nanogroove surface in a different manner depends on the cell type, groove depth, and width. Nanoscale groove with dimensions that mimic those *in vivo* affects the behavior of a wide range of cell types, such as (1) fibroblasts [155,161], (2) osteoblasts [155], and (3) mesenchymal stem cells [161]. For example,

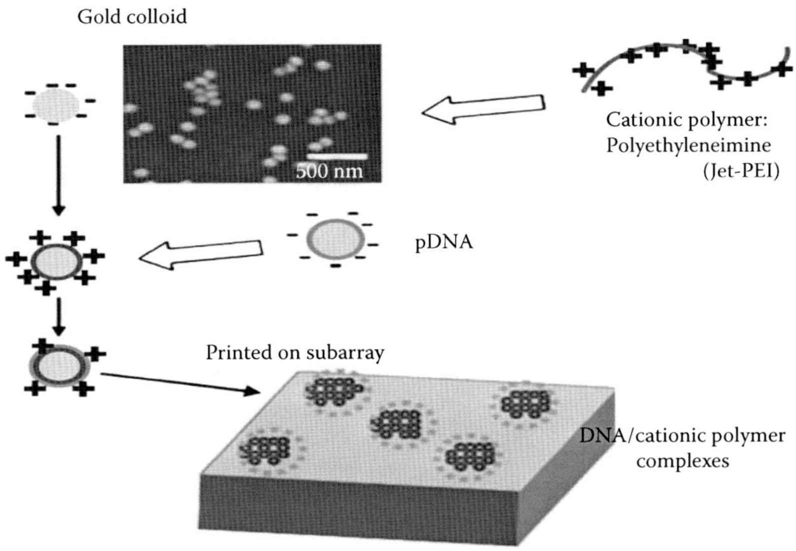

FIGURE 13.4 Reverse transfection with a nanoscaffold. Use of particles of gold colloid (GC) as a nanoscaffold for formation of net positively charged PEI/GC complexes, which can be used in the condensation of DNA and delivery into cells. (From Uchimura, Eiichiro et al., *Journal of Bioscience and Bioengineering* 103, no. 1 (2007): 101–103.)

nanogrooves influenced the behavior of mesenchymal stem cells and osteoblasts after culturing these cells on them [161]. The researchers investigated the effects of nanogroove having a depth of 330 nm and at different widths (10, 25, and 100 nm in width) on osteoblast cells. They showed that a nanogroove having a width of 100 nm did not change cell adhesion property. However, it induced expression of genes involved in skeletal development. It was also found that osteoblast adhesion was reduced on a nanogroove having a width of 10 nm, while osteospecific functions for a nanogroove having a width of 25 nm increased significantly. In addition, mesenchymal stem cells and osteoblasts cultured on a nanoscale groove align their shape and elongation in the direction of the groove surface, and their nuclear is polarized [155,162].

Nanocrystalline diamonds have good biocompatibility, electrical and optical property, high hardness, and low friction coefficient. As the surface of nanocrystalline can mimic the ECM surrounding cells, it can attract cell colonization and support cell viability and proliferation. Thus, it can be applied usefully in tissue engineering [163,164]. For example, nanocrystalline diamonds deposited on silicon substrate developed an adhesion and proliferation rate of the osteogenic and endothelial cells. Nanostructured diamond has also been used in orthopedic transplants [155,163]. Researchers have also shown that titanium oxide has the property of enhancing mesenchymal stem cell viability, proliferation, and differentiation toward the osteogenic cell line. It has also been shown that multilayered titanium films together with chitosan and plasmid DNA induce differentiation of osteoprogenitor cells into mature osteoblast over a long period of time [165–167].

13.4 USE OF NANOMATERIALS IN TISSUE ENGINEERING SCAFFOLDS

In vivo, natural microenvironment around the cells provides a 3D structure that is critical for the normal behavior of cells. With culture of cells in 2D (i.e., culture flask surface), the rate of particular extracellular protein production is reduced, and the morphology of cells is altered. Therefore, using a desired scaffold to provide a 3D supporting surface for cell culture is critical in tissue engineering. An ideal scaffold has to mimic the 3D microenvironment of cells and should (1) be biocompatible,

(2) be degradable, (3) be noncytotoxic, (4) promote cell growth, (5) have a desired mechanical property, and (6) have an interconnected porous network with good attachment property. Developments in nanostructure technology provide a lot of techniques such as (1) electrospinning, (2) phase separation, (3) self-assembly processes, (4) thin film deposition, (5) chemical vapor deposition, (6) chemical etching, (7) nano-imprinting, (8) photolithography, and (9) electron beam or nanosphere lithographies [1,7] for fabrication of 3D nanoscaffolds. These novel nanoscaffolds mimic features of nature ECM and can fulfill all the desired characters of an ideal scaffold, as mentioned above [7,155,169]. Nanostructure scaffold can also act as a growth factor, regulate the cell fate, and trigger differentiation of stem cells toward the tissue of interest, depending on the ingredient of NPs used in nanoscaffold fabrication [170]. In this section, we describe electrospinning, self-assembly, and phase separation as three major techniques for fabrication of nanofibrous scaffolds.

13.4.1 Electrospinning

Use of spinning technology has been started many years ago [23,171]. This technique was used for fabricating household nonwoven textile products [172]. Electrospinning is a technology based on electrostatic interaction that was used for the fabrication of one-dimensional nanostructure fibers [173–180]. Formhals [172] described and patented the operation of electrospinning for the first time in 1934. Up until 1993, this technique had been called electrostatic spinning, but was rarely used [176]. Many researchers focused on optimization of this method for the fabrication of thin nanoscaled fibers from organic polymers and revived this technique in early 1990s. Since then, the term "electrospinning" has been used [177,178]. Figure 13.5 shows three major elements of electrospinning setup: (1) a spinneret (usually a metal needle), (2) a high-voltage power supply (usually in the range of 1–30 kV using direct current power supplies, although the use of alternating current potentials is also possible) [178,179], and (3) a grounded conductor as the collector. In the electrospinning method, polymer solution or melt with a suitable viscosity is dripped from the spinneret to form suspended droplets by force of gravity or mechanical pumping. Solution droplets in the nozzle of the spinneret are electrified upon exposure to a high voltage. As a result, induced charge is distributed over the surface of the droplet [63,181]. Therefore, droplets encounter two forces: (1) repulsion force between surface charge and (2) coulombic force supplied by the power [178,182]. In the level where an electrostatic charge overcomes the surface tension of droplets, the charged jet is ejected. With evaporation of the solvent, the surface charge of the droplet is increased. Consequently, a stream of polymer jet is thinned and forms a slender thread. The charged fluid jet starts a whipping movement and is stretched. As the fluid jet is dried, surface repulsion of the jet exceeds the cohesive force within, which causes a radial force. Eventually, small jets form nanoscale fibers that are randomly oriented on the metal collector screen [181,183,184] (Figure 13.5). Electrospinning is noticeably simple and versatile, but still needs further improvement for use in the fabrication of complex 3D scaffold with interconnected pores [23].

The shape of nanofibers synthesized by electrospinning is usually round in the cross section, although ribbon-like structure with a rectangular cross section has been seen [177,185]. The morphology and the diameter of the nanofibers formed are adjustable depending on (1) the concentration (or viscosity), (2) elasticity, (3) electrical conductivity, (4) the polarity and type of the polymer solution, (5) the strength of the electric field, (6) the feeding rate of the polymer solution, and (7) the distance between the spinneret, and (8) the collector. It has also been published that the humidity and temperature of the chamber in which the electrospinning is carried out are important objects that affect the diameter of synthesized nanofibers [178,186]. It has been reported by many researchers that with a decrease in viscosity of the polymer solution and with the addition of salt, the diameter of synthesized nanofibers is reduced [178,186,187]. Concentration, surface tension, and charge of liquid jet do affect the formation of beads along nanofibers, which is a common problem in nanoscaled fiber fabrication by electrospinning. Researchers have demonstrated that the density of beads in polyethylene oxide (PEO) nanofiber disappeared with the addition of salt (which increases

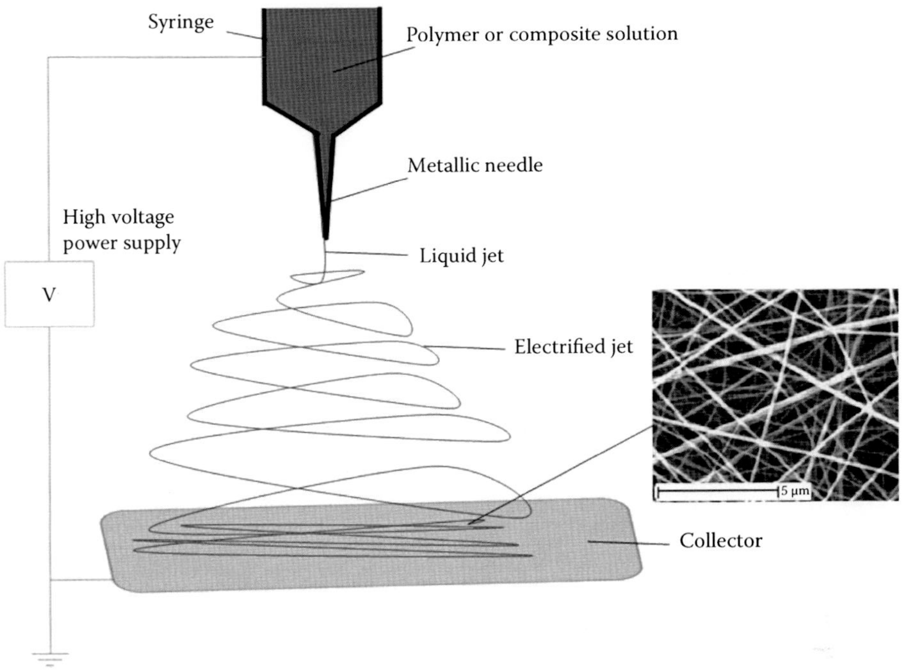

FIGURE 13.5 Schematic illustration shows setup of elements of electrospinning technology. Insert: Nanoscale fibers on a collector. (From Jafari, Javad et al., *Biomedical Materials and Engineering* 21, no. 2 (2011): 99–112.)

the net charge density) and when a more viscous solution and solvents with lower surface tension were utilized [188,189]. In tissue engineering, electrospinning is used for the fabrication of nanofibers from a variety of natural molecules, such as (1) collagen, (2) chitosan, (3) silk fibroin, (4) fibrinogen [58,60,173,190,191], and (5) synthetic polymers [e.g., poly L-lactic acid (PLLA), polylactide-co-glycolide, and poly-ε-caprolactone (PCL)] [60,182,192,193]. Table 13.2 shows some nanofiber materials fabricated by electrospinning used in tissue regeneration.

TABLE 13.2
Synthesized Nanofibrous Scaffolds by Electrospinning Technology for Tissue Engineering Applications

Materials	Average Diameter	Applications in Medicine	References
Natural polymers			
Collagen	250 nm	Fibroblasts	[58,60]
Chitosan and PVA	170–700 nm	Drug carrier for drug delivery application	[191]
Chitosan and PEO	38–62 nm	Osteoblasts and chondrocytes	[180]
Fibroin	50–100 nm	Bone regeneration	[190]
Fibrinogen	320–600 nm	Cardiac fibroblasts	[191]
Synthetic polymers			
PLLA	50–350 nm	Neural tissues	[194]
PLGA	500–800 nm	Soft tissues	[183]
PCL	500–900 nm	Cartilage	[193]

Jafari and coworkers recently blended chitosan and gelatin in five different ratios (70:30, 60:40, 50:50, 40:60, and 30:70) using different electrospinning conditions (flow rate, voltage, viscosity of solution, etc.) to find the best conditions for fabrications of electrospun chitosan–gelatin with the least amount of beads and droplets and the highest uniformity of morphology. They found the values of 15 kV (voltage), 0.2 ml·h^{-1} (flow rate), and a fixed distance of 15 cm as the optimal electrospinning conditions, which produced fibers having a mean diameter of 180 ± 20 nm (Figure 13.6) [58]. They also designed electrospun polyvinyl alcohol (PVA)/chitosan nanofibrous composites to determine the effect of adding chitosan on proliferation of nerve cells and biocompatibility of chitosan-containing scaffold [191].

Studies show that electrospun nanoscaled materials have a profound effect on cell attachment, proliferation, differentiation, and functional properties [23,60,183,194,195]. It has been indicated that electrospun, nonwoven, 3D, and highly porous scaffolds made of poly(lactic-*co*-glycolic) acid (PLGA) influence mesenchymal proliferation over 7 days and migration of mouse fibroblasts [183].

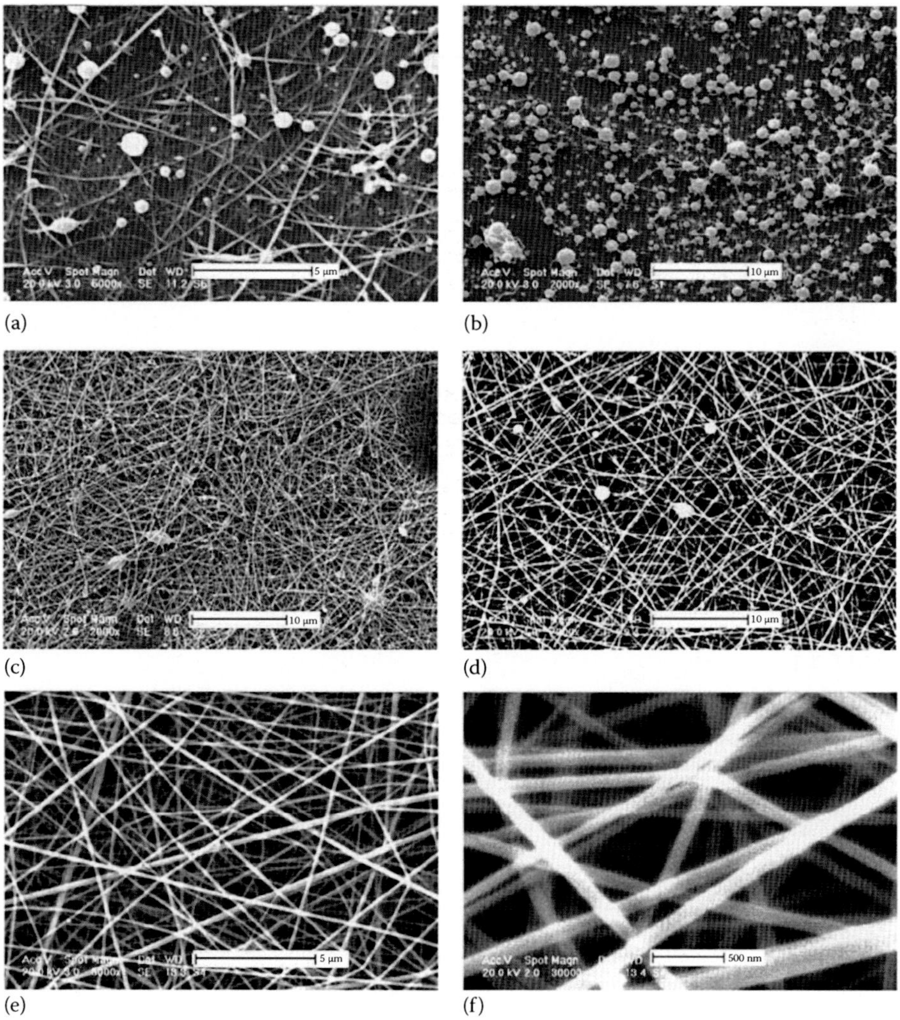

FIGURE 13.6 SEM of the electrospun fibers of chitosan–gelatin in different mass ratios: (a) 70:30, (b) 60:40, (c) 50:50, (d) 40:60, and (e,f) 30:70. Low-viscosity samples (30% chitosan and 70% gelatin) showed the least amount of beads and droplets (e and f). (From Jafari, Javad et al., *Biomedical Materials and Engineering* 21, no. 2 (2011): 99–112.)

It has also been reported that PCL scaffolds depositing collagen influence the migration of mesenchymal stem cells and direct differentiation of these cells into osteoblasts producing calcification [182].

13.4.2 Self-Assembly

Another technique that has been adopted for fabrication of nanofibrous scaffolds is self-assembly. Molecular self-assembly has been utilized to produce supermolecular structures. Materials fabricated by this method are frequently referred to as hydrogels. One of the most desirable properties of hydrogels is the ability to fabricate them in injectable form [196–198]. In this technique, separate components are preprogrammed to assemble themselves into specific patterns or nanostructures spontaneously, without human intervention [199]. Typically, such structures are formed and kept by noncovalent bonds [200]. By using this method, El-sadik [155] is able to synthesize nanoscale fibers that are slender than those synthesized using electrospinning. In nature, they can see the large number of molecular self-assemblies. For example, phospholipids (the basic component of cell membrane) are amphiphilic. They possess both hydrophobic (phosphor in the head of the molecule) and hydrophilic (lipid in tail) moieties. In aqueous solution, phospholipids are self-assembled into highly ordered structures (vesicles, micelles, and tubules) with hydrophobic regions grouped in a core and hydrophilic tails exposed to the surface of water [201]. With modification of the environment (from aqueous to oil, for example), researchers can modify the pattern of their self-assembly. For example, many researchers have synthesized peptide amphiphiles (PAs) that consist of a sequence of collagen (containing 15 amino acids) bonded to a long-chain lipid (mono- or dialkyl ester) to mimic the triple structure of a collagen protein. PA molecules are self-assembled in a suitable solution. Structurally in a typical PA, collagen sequence as the head of a molecule forms a triple helical structure followed by a lipid in its tail, which stabilizes the 3D structure of the peptide [197,202–205]. It has also been demonstrated that mono- or dialkyl lipids and the peptide that did not change the melanoma cell adhesion are used alone. However, their combination (lipid and peptide) through a self-assembly method facilitates cell adhesion [198]. It has also been investigated that under different conditions (charge and shape of PAs), PAs can be self-assembled into channels, sheets, rods, and fibers [63]. Stupp and coworkers have synthesized a novel PA that possesses five structures: (1) a long alkyl lipid that provides hydrophobic tail, (2) four consecutive cysteines for enhanced stability, (3) a region containing three glycines as the flexible hydrophilic head, (4) a region containing serine that is able to interact with calcium ions to promote mineralization, and (5) an arginine–glycine–asparagine motif. In an aqueous environment, PAs are self-assembled, and the resulting nanofibers (cylindrical micelle) inserted vertically have a hydrophobic portion in the interior and a hydrophilic portion on the surface of nanofibers (Figure 13.7) [206].

Researchers have also developed self-assembled ionic complementary oligopeptides to include intermittent blocks of hydrophobic and hydrophilic portions. They have shown that these oligopeptides were self-assembled into stable β-sheets and hydrogels with interwoven nanofibers upon exposure to water and physiological conditions [207]. Self-assembly technique has been used for the fabrication of nanofibers from polyphenylene dendrimers [208,209] and polyglutamines [210,211]. Pores that formed in scaffolds fabricated by self-assembly are not controllable, which is considered to be a disadvantage (Table 13.3).

13.4.3 Phase Separation

Phase separation is another technique applied in the fabrication of 3D highly nanoporous scaffolds having a continuous fibrous network that plays a key role in making scaffolds with the desired mechanical property (approximately 50–500 nm in diameter) [212,213]. In this technique, homogenous multicomponent systems, under certain conditions, are induced thermodynamically to form an unstable multiphase separated system [25,214]. Typically, this process consists of five

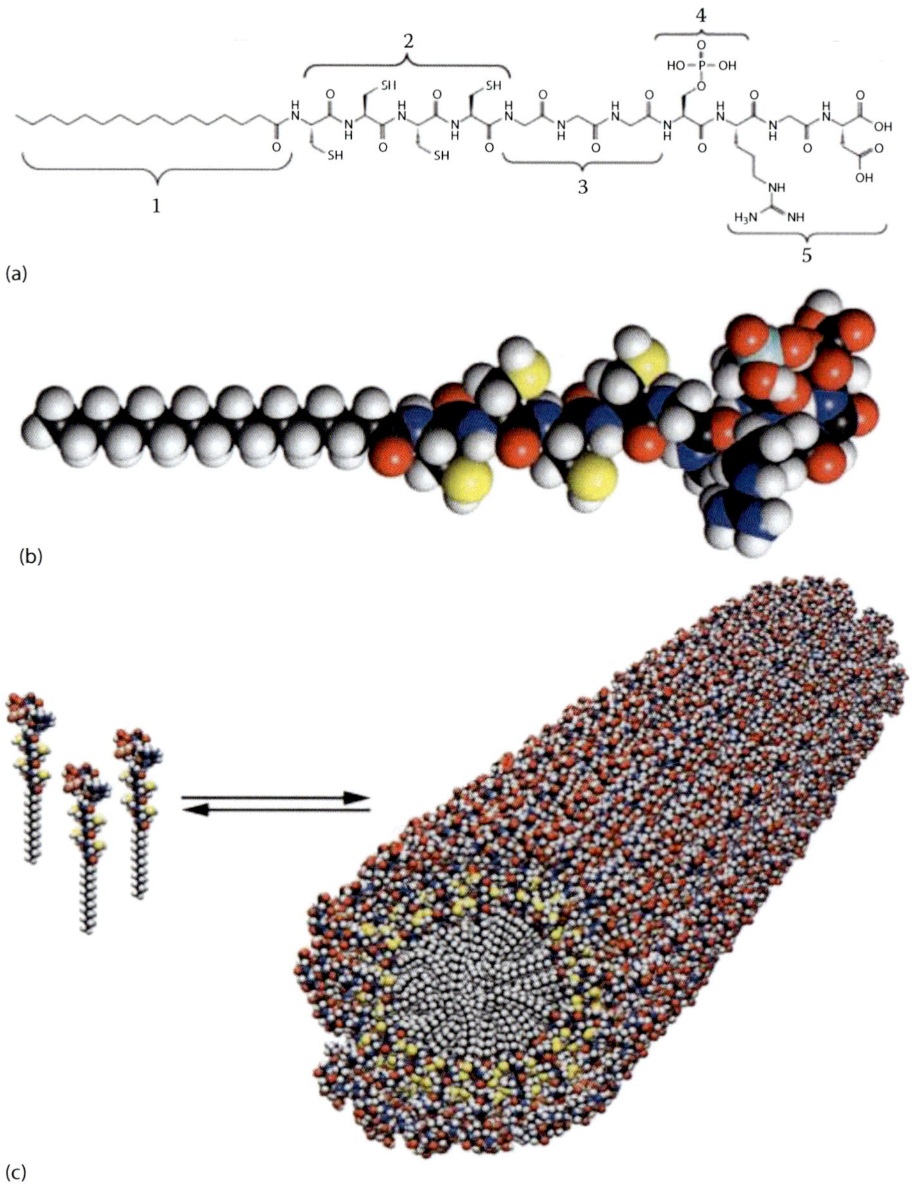

FIGURE 13.7 Chemical (a) and molecular (b) structures of PA (c) Schematic showing the self-assembly of PA molecules into a cylindrical micelle. Five key structural features are shown in (a): (1) a long alkyl lipid that provides hydrophobic tail, (2) four consecutive cysteines to form a cysteine bond to enhance, (3) a region containing three glycines as flexible hydrophilic head, (4) a region containing serine that is able to interact with calcium ions to promote mineralization, and (5) an RGD motif (arginine–glycine–asparagine). (From Hartgerink, Jeffrey D. et al., *Science* 294, no. 5547 (2001): 1684–1688.)

steps—(1) polymer dissolution, (2) gelation (a critical step for the control of porous structure and size), (3) solvent extraction, (4) freezing, and (5) drying [25]—which are time consuming and are a disadvantage. When a homogenous polymer solution is thermally phase-separated, the solution is divided into two phases: a phase having a high density of polymer and another phase with low concentration of polymer. After solvent extraction, polymers form different structures, such as powder and closed or opened pores, depending on (1) process conditions, (2) type of polymer and

TABLE 13.3
Advantages and Disadvantages of Three Major Technologies Used for Fabrication of Nanofibrous Scaffolds

Technique	Advantages	Disadvantages
Electrospinning	Noticeably simple and versatile; can generate large-diameter nanofibers on the upper end of the range of natural; ECM collagen.	Still needs more improvement for use in fabrication of complex 3D scaffold with interconnected pores. Time consuming.
Self-assembly	Materials fabricated by this method are frequently hydrogel, which are suitable for injection. Self-assembly can generate small-diameter nanofibers in the lowest end of the range of natural ECM collagen.	Pores and 3D shapes formed in scaffolds fabricated by self-assembly are not controllable and lack interconnected pores.
Phase separation	Simple and do not require sophisticated equipment [218,219]. The resulting pore sizes and morphology can be controlled by varying the processing parameters [214]. Allows for the design of 3D pore network into the scaffolds. Form continuous fibrous network.	Lack of control over 3D structure and shape (this problem has been resolved by using porogen). Time consuming. Failed to control fiber orientation.

solvent, (3) concentration of polymer in solution, (4) order of produce, and (5) temperature [25]. It is important to state that only open-pore scaffolds are desirable in tissue engineering applications. With freezing and drying, we are able to make interconnected and high-density pores. Several 3D scaffolds have been developed using this method [86,215,216]. For example, a polylactic acid nanofibrous matrix has been fabricated by phase separation [2.5% (w/v) polylactic acid/tetrahydrofuran solution at a gelation temperature of 8°C] [25], which is potentially useful for neural system tissue engineering [217]. However, such a matrix, similar to self-assembled PAs and electrospun materials, does not possess interconnected macropores for the purpose of improving vascular formation and angiogenesis, cell seeding, and better feeding. To solve this problem, a phase separation technique has been developed with building predesigned macroporous/microporous scaffolds [218]. In this method, materials such as sugar, salt, and paraffin have been used as porogens during phase separation. Briefly, 3D macroporous scaffolds are fabricated by porogens. When a polymer solution is cast into the designed scaffold, a phase is separated to form a nanofibrous structure. Eventually, after removal of the solvent, porogen is dissolved using either water or heat. A highly porous 3D scaffold is formed [25]. This method has been commonly employed in the fabrication of 3D scaffolds composed of various rations of polylactic acid with other polymers to obtain spherical (Figure 13.8a and b) [50], helical (Figure 13.8c and d) [7,219,220], and particulate interconnected networks (Figure 13.8e and f) [50]. In comparison to solid-walled polylactic acid scaffolds, it has been shown that macropore nanofibrous polylactic acid (with similar porosity) improved the attachment of osteoblasts twofold [64].

Computer-assisted design and computer-assisted manufacture techniques, solid freeform fabrication (SFF), and rapid prototyping have been employed to optimize better the 3D structure and design of scaffolds [37,64–222]. However, these techniques have some limitations, such as (1) selection of materials, (2) low resolution, and (3) structural heterogeneity of fabricated materials [27]. To overcome this limitation, Chen et al. [223] have developed a reverse SFF technique that is connected with phase separation, which makes it possible to fabricate nanofibrous structured

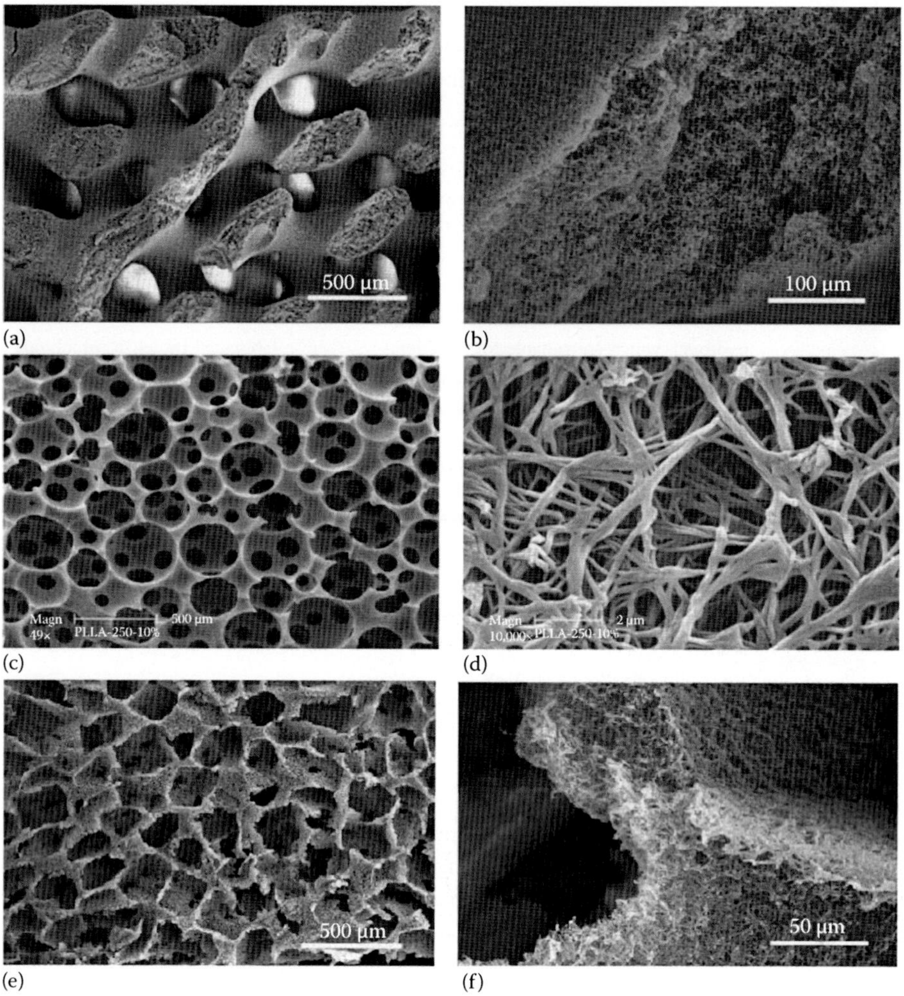

FIGURE 13.8 3D nanofibrous scaffolds with helicoidal tubular (a and b), spherical (c and d), and particulate (e and f) interconnected macropores made of PLLA using the phase separation technique. (a, b, e, and f from Ma, Peter X. et al., *Journal of Biomedical Materials Research* 54, no. 2 (2001): 284–293. c and d from Wei, Guobao, and Peter X. Ma. *Journal of Biomedical Materials Research Part A* 78, no. 2 (2006): 306–315.)

scaffolds using the reversed image of CT scan or histological sections of human tissues and organs (see Figure 13.9).

The advantages and disadvantages of nanotechnology techniques used in the fabrication of tissue engineering 3D scaffolds (described above) are presented in Table 13.3.

13.5 SURFACE MODIFICATIONS

Mammalian cells receive signals through interactions with surrounding environment to control their behavior. Therefore, such interactions of cells with scaffolds play an important role in determination of a success or failure of an implanted material. Physicochemical properties of material surfaces, such as (1) charge, (2) composition, and (3) energy, can remarkably influence cell migration, adhesion, differentiation, proliferation, and behavior. It has been shown that nanotopography affects cell behavior the greatest [224–226]. Nanotopography can also affect the orientation of cells, which

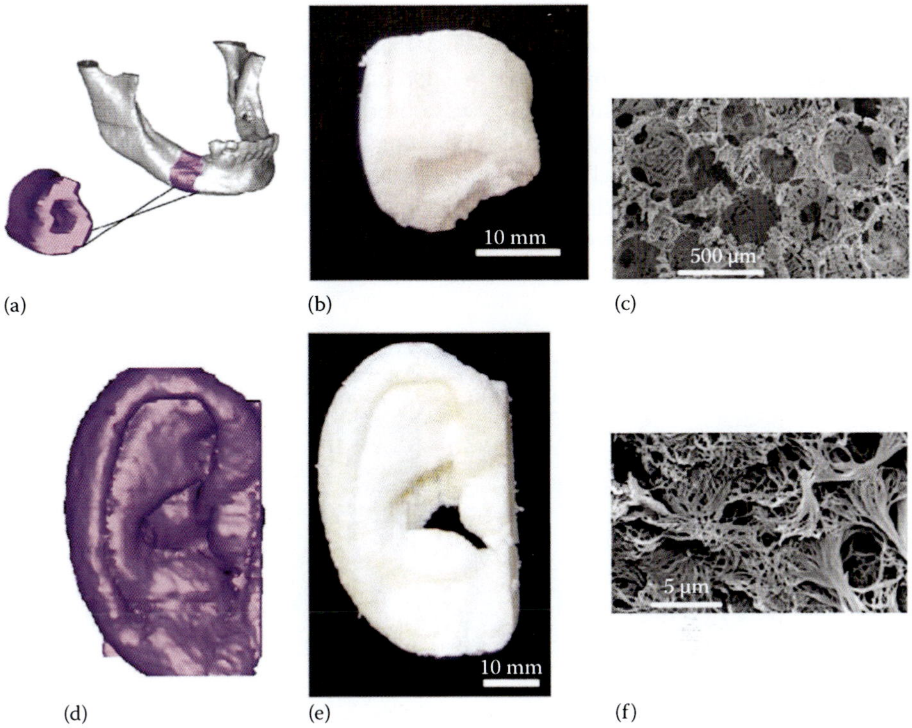

FIGURE 13.9 Nanofibrous scaffolds developed by reverse SFF technique that is connected with phase separation using the reversed image of CT scan or histological sections of human tissues or organs. Human mandible (a) and ear (d) were designed from CT scan and histological sections, respectively; purple sections show the image of the mandible bone fragment and ear to be engineered. Nanofibrous scaffolds of mandible (b) and ear (e) were developed by this method. (c) Interconnected spherical pores of resulting mandible scaffold. (f) Morphology of nanofibrous pores of resulting scaffold. (From Chen, Victor J. et al., *Biomaterials* 27, no. 21 (2006): 3973–3979.)

is crucial for the purpose of achieving functional tissue, especially nerves and tendons [227,228]. Using appropriate chemical compositions, growth factors and adhesion proteins that will make up the matrix of a biomaterial are important in (1) tissue formation, (2) maintenance, and (3) repair process. For example, fibronectin, collagen, and laminin improve the cell attachment property of biomaterials. Therefore, several efforts in the modification of biomaterial surfaces have been done to control cell behavior and improve efficiency of tissue regeneration [229,230]. In addition to enhancing composition by adding growth factors and proteins, some polymers such as PLGA enhance chondrocyte functions [231,232]. It has been reported that coating titanium and tantalum with nanophase apatite enhances bone formation [57,233]. The treatment of coronary and peripheral artery disease using metallic stents has been one of the most revolutionary and rapidly adopted medical interventions of our time [234]. For bare metal stents, coronary in stent restenosis remains a remarkable restriction to the long-term efficacy of coronary artery stent placement. Nanotechnology gives a promising landscape to overcome these impediments. NP-eluting stents have emerged as one of the most promising technologies in the field of cardiovascular nanomedicine. Loaded with anti-inflammatory and antiproliferative agents in the nanoscale, these stents have the potential to reduce post-stent-implantation restenosis significantly [235]. Nakano et al. [236] firstly reported the development of NP-eluting stents. They prepared NP formulations and coated their surfaces with chitosan to achieve electrodeposition. Then, the NPs were coated on a metallic stent surface by a cationic electrodeposition coating technique. Chitosan coating also gave a cationic surface charge to the NPs, and thus this cationicity further helps intracellular uptake because of interaction with negatively charged cellular membranes. *In vivo* study results showed that the NPs were taken up stably and efficiently by cultured vascular smooth muscle cells *in vitro*. In a porcine coronary artery model *in vivo*, substantial fluorescein isothiocyanate (FITC) fluorescence was observed in neointimal and medial layers of the stented segments that had received the FITC-NP-eluting stent until 4 weeks (see Figure 13.10).

Joo et al. [237] developed a novel simple coating process to coat the stent surfaces of PLGA NPs (Figure 13.11). They called this technology as a "ring-shaped surface tension method." It relies on the principle that liquid is held between two very closely spaced surfaces in the form of a meniscus as a result of capillarity. A specially designed ring trails along the immobilized stent surface held along

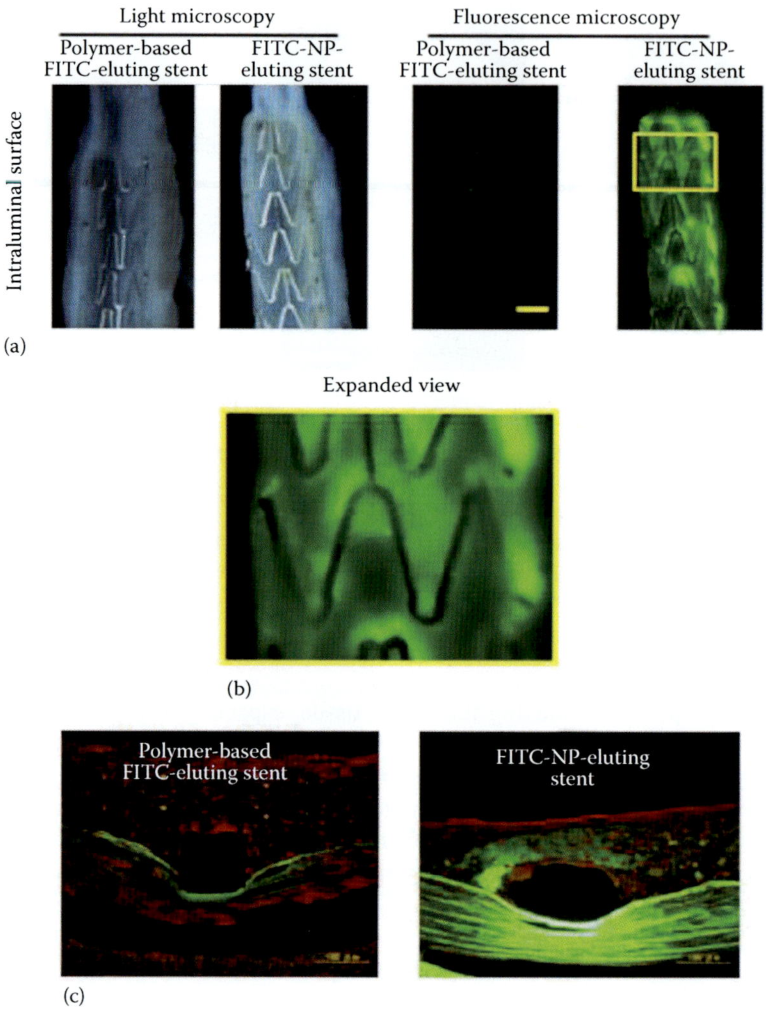

FIGURE 13.10 Localization of FITC fluorescence after deployment of FITC-NP-eluting stent in porcine coronary arteries 4 weeks after stenting. (a) Enface light and fluorescence stereomicroscopic pictures of the intraluminal surface of an isolated stented segment of coronary artery taken from the FITC-NP-eluting and the dip-coated FITC stent sites. Scale bar: 1 mm. (b) Expanded images of boxed area in (a). Expanded images reveal that a number of discrete patterns of fluorescence can be seen, indicating local retention of FITC in the form of NP. (c) Fluorescence microscopic pictures of cross sections from the FITC-NP-eluting stent and polymer-based FITC-eluting stent sites. Microscopic settings (exposure condition, filter, intensity of excitation light, and so forth) are the same between the two pictures. (From Nakano, Kaku et al. *JACC: Cardiovascular Interventions* 2, no. 4 (2009): 277–283.)

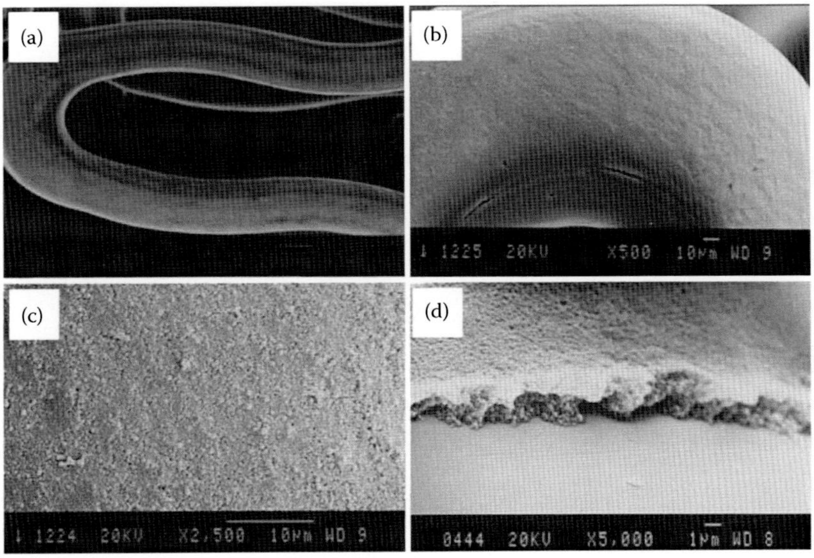

FIGURE 13.11 SEM images of PLGA NP-coated stents. (a) ×110, (b) ×500, (c) ×2,500, and (d) ×5000 (a cross section of a NP layer). (From Joo, Jae-ryang et al., *Bulletin of the Korean Chemical Society* 30 (2009): 1085–1087.)

its axis, just like a ring slides over a finger. The NP suspension was injected between the stent surface and the ring. The deposition occurred at the wedge where the meniscus met the surface when the ring was moved up or down [238]. Uniform deposition of the NPs on the stent surface was observed through the scanning electron microscopy images. The major benefit of this coating process is that the amount of drug on the stent can be modified by various ways for drug-eluting stent purposes [237].

In conclusion, using nanomaterials in surface modification of implantable materials opens up a wide frontier for the purpose of achieving better success in tissue engineering.

13.6 SOME APPLICATIONS OF NANOSTRUCTURED SCAFFOLDS IN TISSUE ENGINEERING

To date, many nanostructured scaffolds have been designed for regeneration of damaged tissue. Table 13.4 shows some nanomaterials applied in the regeneration of different tissues, with different range of success. We now describe a summary of application of various NP-based scaffolds in the regeneration of bone and neural tissue. Structurally, bone can be considered as a real nanocomposite in which stem cells are embedded. The bone matrix consists of two important phases—organic (protein) and non-organic (mineral)—at the nanoscale level (see Table 13.5) [238,239], and osteoblasts are responsible for the creation of new bone when it is damaged [62].

Bone tissue usually heals spontaneously, but severe bone fractures (more than 6 mm in diameter) that lead to loss of its original structure and function can also heal, and this is one of the bigger problems in the science of orthopedics. Bone autografts or allografts have also been widely used. However, some limitations of these grafts, such as (1) limited supply, (2) donor site morbidity, and (3) risk of transmission of infection, reduced their widespread use [257–259]. Cell-based therapy by using stem cells and bioscaffold is trying to address this problem. An ideal scaffold for bone engineering, with the aim of acting as bone ECM, requires a series of material properties, such as (1) biocompatibility, (2) biodegradability, (3) osteoconduction, (4) osteoinduction, (5) nontoxicity, and (6) suitable mechanical property [260,261]. Using nanoscaffolds (nanostructured or nanofibrous)

TABLE 13.4
Examples of Nanomaterials Used in Fabrication of Scaffold for Use in Tissue Engineering

Tissue	Nanomaterials	References
Bone	Ca/P compositions such as hydroxyapatite, TCP, DCPD ($CaHPO_4 \cdot 2H_2O$)	[241] [242] [243] [244]
	PVA	[242]
	PCL	[243] [245] [246]
	Collagen and gelatin	[58] [243] [244] [58,247,248]
	Chitosan	[58] [249]
	Forsterite	[245]
	PLLA	[250] [223]
	Bioactive glass	[251–253]
Cartilage	Nanoporous Ti	[57]
	PLGA	[231,232]
Bladder	PLGA	[254]
	PA	[255]
Nerve	PLGA, PVA, collagen, and chitosan	[191] [256]

TABLE 13.5
Bone Matrix Consists of Two Important Phases—Organic (Protein) and Non-Organic (Mineral)—at the Nanoscale Level

Inorganic Phase	Wt%	Organic Phase	Wt%
Hydroxyapatite	~60	Collagen	~20
Carbonate	~4	Water	~9
Citrate	~0.9	Primary bone cells: osteoblasts, osteocytes, osteoclasts	–
Sodium	~0.7	Noncollagenous proteins (osteocalcin, osteonectin, osteopontin, thrombospondin, morphogenetic proteins, sialoprotein, serum proteins)	~3
Magnesium	~0.5		
Other traces	–	Other traces: polysaccharides, lipids, cytokines	–

Source: Murugan, R., and S. Ramakrishna, *Composites Science and Technology* 65, no. 15 (2005): 2385–2406; LeGeros, Racquel Z. *Hydroxyapatite and Related Materials* 3 (1994); With kind permission from Springer Science+Business Media: *Biomaterials Science and Engineering*, Park, Joon, 2012.

in bone tissue engineering has recently attracted the attention of researchers [57,245,251–257]. Nanoscaled scaffolds improve the important factors involved in cell adhesion, proliferation, differentiation, and migration while providing (1) high surface area-to-volume ratio, (2) high porosity (with a pore size of at least 100 μm in diameter to allow cell migration, vascularization, and tissue ingrowth), and (3) low density [266–268]. To date, several nanomaterials, such as (1) calcium phosphate (CaP), (2) compositions, (3) bioglass, and (4) glass ceramics, have been employed for the fabrication of bioscaffolds with a range of advantages and disadvantages [257,267]. Calcium phosphate (CaP) compositions, such as (1) hydroxyapatite (HAp, $Ca_{10}(PO_4)_6(OH)_2$), (2) octacalcium phosphate, (3) tricalcium phosphate (TCP), (4) dicalcium phosphate dihydrate (DCPD, $CaHPO_4{\cdot}2H_2O$), and (5) dicalcium phosphate bioceramics, are the most well-known bioceramics for use as biological application [257,267,268]. These compositions are equivalent to that of an inorganic phase of the human bone. Hydroxyapatite is a major phase found in bone and one of the frequently used bioceramics in bone reconstruction [269–271]. It is biocompatible [241,269,270], osteoconductive, osteoinductive, and bioactive [271–274]; has neither cytotoxicity nor antigenecity [275,276]; and can be used either alone or in combination with, e.g., collagen, gelatin, silicon carbide, chitosan, alumina, and PCL [58,191,242–246,257,263] for the purpose of fabrication of bone substitutes (Figure 13.12). A major disadvantage of hydroxyapatite is its low degradation rate

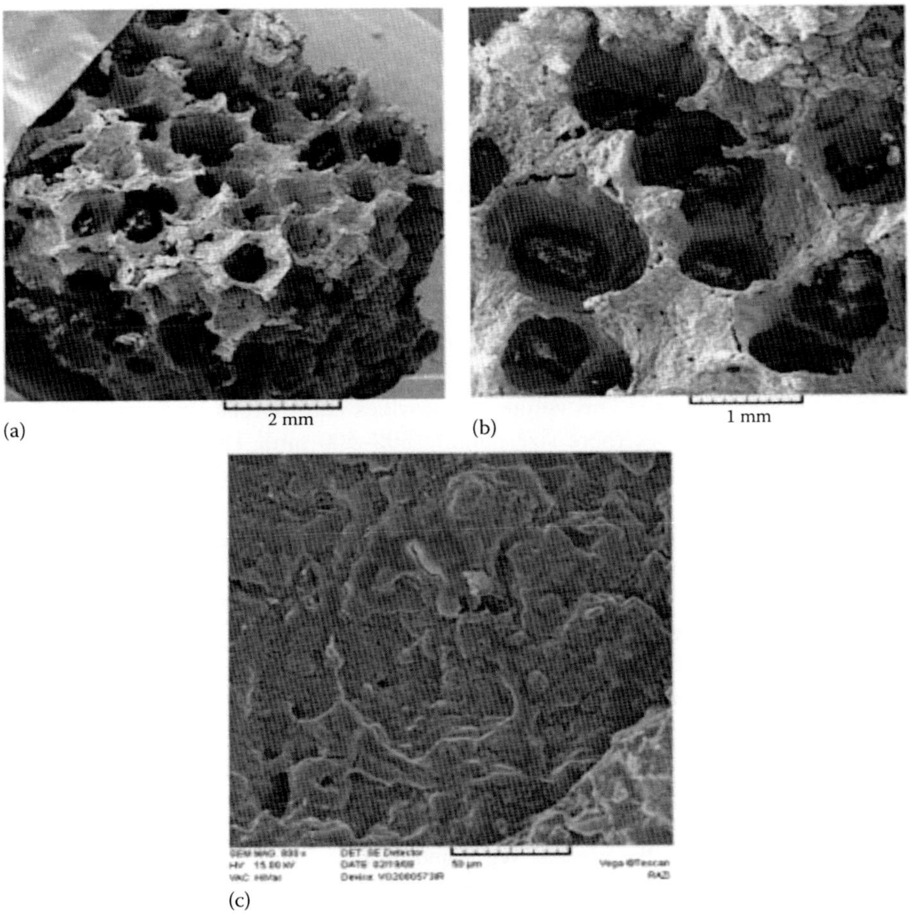

FIGURE 13.12 SEM images of the porous hydroxyapatite–alumina–silicon carbide (HA–Al_2O_3–SiC) composite scaffold (a) Low magnification. (b) Higher magnification of open interconnected micropores. (c) Grain morphology. (From Saki, Mohammad et al., *Yakhteh* 11, no. 1 (2009): 55–60.)

[258,277]. It is important to state that the degradability rate of different calcium phosphate compositions is dependent on the ratio Ca/P, with the highest being that of DCPD, which results in the most extensive bone remodeling around the scaffold [258,278–280].

Baghbani et al. [241] have recently fabricated porous biphasic hydroxyapatite/DCPD composite with various concentrations of DCPD to study the effect of adding DCPD on the mechanical and chemical properties of final composites. It is important to note that after heat treatment, DCPD converted to the TCP phase. Their results revealed that a porous hydroxyapatite/TCP composite generally degrades more rapidly than hydroxyapatite. Increasing the percentage of the DCPD phase enhanced the value of elastic modulus, compressive strength, and density of biphasic hydroxyapatite/DCPD composites. *In vivo* and *in vitro* studies showed that porous hydroxyapatite/TCP composites provide a good 3D ECM for the purpose of bone tissue repair [241]. Fabricated bioscaffolds, based solely on hydroxyapatite, cannot provide the desired mechanical property for bone tissue engineering applications. One of the most commonly used methods for overcoming the mechanical weakness of these materials is forming inorganic–organic composition by introducing a polymeric component into the hydroxyapatite [242]. For example, PVA is one of the most used polymers for the purpose of forming inorganic–organic composition with hydroxyapatite [281,282]. PVA possesses high concentration of hydroxyl pendant groups, which makes it capable of being cross-linked physically without using any chemical cross-linkers [283]. Many studies have reported that incorporating PVA with hydroxyapatite (formation of hydroxyapatite crystals in PVA solution) enhances biodegradability and biocompatibility of the fabricated nanocomposite [282]. The value of the pH of the reactive solution would affect the hydroxyl groups of PVA and would thereby affect both the formation of hydroxyapatite and the incorporation of PVA. Effect of the variation value of pH in synthesized hydroxyapatite nanocomposite structure was investigated by our research group [242]. In our results, a pH value of 9 was an optimum value for best interaction between the organic and inorganic phases in composition, and at the same time allowed acceptable formation of hydroxyapatite crystals within the structure [242].

Johari et al. [246] synthesized porous poly(ε-caprolactone)/nanofluoridated hydroxyapatite (PCL-FHA) scaffolds. They found that viability of seeded cells on PCL-FHA was increased with an increase in porosity. They also showed that the degradation rate of PCL-FHA scaffold is controllable and dependent on the fluorine content of the composite (increased with the content) [246]. Hydroxyapatite–collagen (or gelatin) composite is another promising bioscaffold, which has been used in bone tissue engineering. This composite provides high porous structure (about 85%) with an interconnected pore network (pore size ranging 300–500 μm). Their studies have shown that hydroxyapatite–collagen composites (1) were biocompatible, (2) were osteoconductive, (3) were noncytotoxic, (4) were bioactive, and (5) promoted proliferation of osteoblast-like cells and bone-healing process [244]. Silicon in combination with hydroxyapatite was used for the fabrication of bone substitutes. It has been reported that silicon is essential for the normal growth and development of bone and cartilage, and silica-based materials, such as silicon carbide, remarkably increase the deposition of ECM, biocompatibility, and other cellular activity of composites [263,284–286]. We recently synthesized silicon-substituted hydroxyapatite (Si-HA) composite by incorporating silicon in HA using hydrothermal method, and calcium nitrate [Ca $(NO_3)_2$], ammonium phosphate [$(NH_4)_3PO_4$] or ammonium hydrogen phosphate [$(NH_4)_2HPO_4$], and tetramethyl orthosilicate [Si $(OCH_2CH3)_4$] (TEOS) were used as starting materials. Phosphate (PO_4^{3-}) groups were replaced by the silicate (SiO_4^{4-})) groups resulting in silicon–hydroxyapatite ($Ca_{10}(PO_4)_{6\_x}$ $(SiO_4)x(OH)_{2\_x}$). Replacing phosphate groups with silicon decreased the size of the crystal of silicon–hydroxyapatite but did not alter the shape of the crystal. We found an increase in bioactivity (Figure 13.13) and biocompatibility of silicon–hydroxyapatite compared to pure hydroxyapatite bioceramics [275].

It has been also determined that the addition of poly(ε-caprolactone) (PCLA to hydroxyapatite–gelatin composite improved ultimate stress, stiffness, and elastic modulus [243]. Chitosan, a polysaccharide derived by deacetylation of chitin, has excellent properties for use in medical applications, which include the following: (1) biocompatibility, (2) noncytotoxicity, (3) antifungal stimulator of

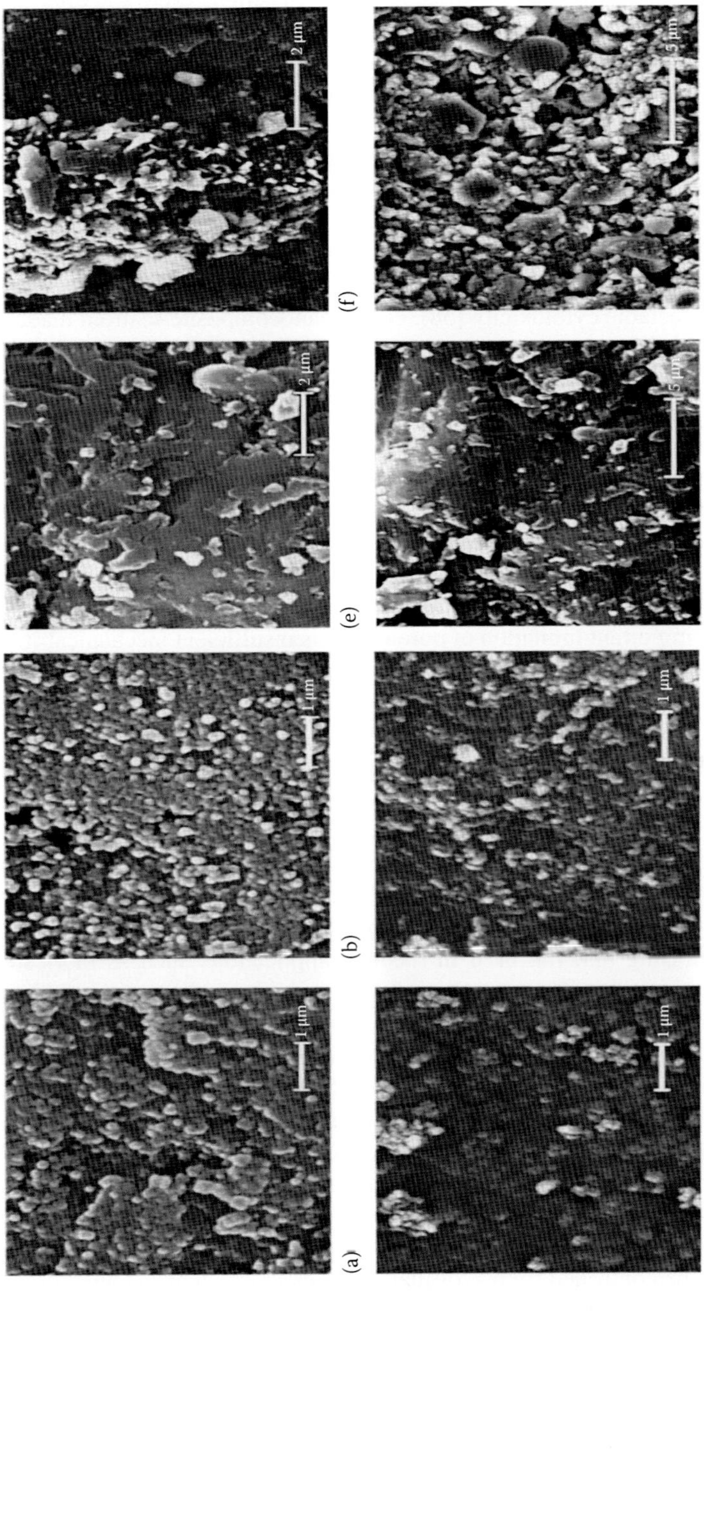

FIGURE 13.13 SEM images of HA (a) and Si-HA (b) before incubation in SBF; HA (c) and Si-HA (d) after 3 days; HA (e) and Si-HA (f) after 8 days; and HA (g) and Si-HA (h) after 14 days incubation in SBF. (From Aminian, Alieh et al., *Ceramics International* 37, no. 4 (2011): 1219–1229.)

the immune system (for recruitment of acute inflammatory cells, such as polymorphonuclear (PMN) and macrophages, which are needed for promoting the healing process), and (4) ability to bind to particular materials such as fats, cholesterol, and protein [249,287]. Gelatin is another natural biomaterial that has widespread application in the fabrication of bioscaffold. It is formed by partial thermal hydrolysis of collagen (the most abundant protein that exist in connective tissue of human), and it possesses some properties that make it attractive for tissue engineering and drug delivery, such as (1) non-immunogenicity, (2) biocompatibility, (3) nontoxicity, and (4) high hemostatic effect [247,248]. Chitosan forms electrostatic interaction with cell membrane and decreases the rate of cell migration. Gelatin can interact electrostatically with chitosan when the two are blended together, thereby decreasing the electrostatic interaction of cell membrane and chitosan, which results in the increase in cell migration [248,289]. We fabricated electrospun chitosan–gelatin nanofibrous scaffold and determined the best conditions for electrospinning (voltage, flow rate, viscosity of samples). For our result conditions, nanofibers were formed having a mean diameter of 180 ± 20 nm. Low-viscosity samples with 30% chitosan and 70% gelatin showed the least amount of beads and droplets coupled with the highest uniformity of morphology. The resultant scaffold did not show high biocompatibility and any detectable cytotoxicity [58]. Novel nanocomposite scaffold made of PCL and forsterite nanopowder was fabricated by our research group. We aimed investigating the effects of forsterite content on mechanical properties, bioactivity, biodegradability, and cytotoxicity of the scaffolds. We found that the degradation rate of PCL nanocomposite scaffold can be modulated by the addition of forsterite nanopowder. Adding forsterite improved bioactivity, mechanical property of scaffold, and proliferation rate of osteoblasts compared to that seen in pure PCL. In addition, the resulting scaffold had no toxicity and allowed cells to grow and proliferate on the surface. All of the results have suggested that a combination of PCL and forsterite nanopowder can be used to form scaffolds suitable for use in bone tissue engineering [245]. PLLA is a biocompatible and biodegradable material and has been used by researchers for the fabrication of bioscaffold for bone tissue engineering [1,250]. An important limitation of pure PLLA is insufficient mechanical strength [296]. To address this issue, researchers have used various types of fillers such as NPs to reinforce these polymers [291,292]. For example, researchers enhanced the mechanical property and mineralization capability of pure PLLA with the addition of octadecylamine-functionalized nanodiamond (ND-ODA). They suggested that these biodegradable composites (ND-ODA/PLLA) can be useful in medical applications, such as bone regenerative engineering [293]. Victor et al. [223] fabricated 3D nanofibrous PLLA matrices' anatomical shape and pore structures (with precise control of pore size and structure) using the reverse SFF technique. They indicated that the fabricated 3D nanofibrous PLLA matrices improved cell proliferation and differentiation capacity toward bone tissue, and had more uniform matrix and mineral production compared to the nonfibrous PLLA [223]. Bioactive glasses are another promising material used in medical applications such as fabrication of bone substitutes and as antiseptic agents. These materials stimulate biological response and bond to tissue, which promotes growth and regeneration of new tissues without the formation of scar. Bioactive glass constructs are more bioactive than calcium phosphate–based composition and conventional biomaterials (e.g., titanium and cobalt–chrome alloys), which makes them a good choice for use in bioactive scaffolds [8,280,294]. Bioactive glasses–based scaffolds can release some ion dissolution products. It has been found that these ions can be consumed by cells and regulate several genes. For example, it has been reported that silicate bioactive glasses can upregulate various families of genes related to bone development through the release of silicon, calcium, phosphorus, and sodium ions during its conversion to a hydroxyapatite-like material [251,252]. The structure of the pore network of bioactive glasses can be controlled by use of many processes, such as (1) alteration of the glass composition, (2) concentration of gelling agents and surfactant, and (3) temperature of foaming [295,296]. Studies on comparison of hydroxyapatite and bioglass indicated that bioactive glasses have more bioactivity and biodegradability properties [297,298].

Cell attachment, proliferation, and differentiation properties of scaffolds for most tissues are important than microstructure. However, the microstructure of scaffolds is an important criterion for use in neural tissue engineering [299–301]. As described in the section "Electrospinning," many nanofibrous biodegradable polymers have been synthesized using the technique of electrospinning. These nanofibrous composites possess a high surface-to-volume ratio, and their shape, size, porosity, and mechanical properties can be precisely controlled (by adjusting the ratio of the components and altering the electrospinning conditions) [192,302,303]. PLGA, PVA, collagen, and chitosan are examples of most used polymers in the area of neural tissue engineering [191,304,305]. It also has been found that nanofibrous biodegradable scaffolds fabricated from these polymers (by electrospinning technique) stimulate axonal regeneration and neural stem cell differentiation [191,306,307]. Alhosseini et al. [191] modified the physicochemical and biological properties of PVA polymer through blending with chitosan, and used the electrospinning method for the fabrication of PVA/chitosan as a neurocompatible scaffold. They found that the addition of chitosan to the PVA scaffolds enhanced the viability and proliferation of nerve cells and increased the biocompatibility of the scaffolds [191].

13.7 CONCLUSION

Nanotechnology has shown many promising applications in medicine. Most recent developments in the field of nanotechnology in medicine are tissue engineering, regenerative medicine, and drug delivery. Another relevant area of interest to modern medicine is the isolation of cells with a particular phenotype using nanotechnology. Cell isolation encompasses presentation of a multiplicity of techniques for which nanotechnology has some revolutionary developments. The advances of nanotechnology in biotechnology enable more accurate studies of cell phenotype, ECM properties, and cell fate after implantation to the living organism that can significantly revolutionize the field of medicine.

REFERENCES

1. Zhang, Lijie, and Thomas J. Webster. "Nanotechnology and nanomaterials: Promises for improved tissue regeneration." *Nano Today* 4, no. 1 (2009): 66–80.
2. Siegel, Richard W., and Gretchen E. Fougere. "Mechanical properties of nanophase metals." *Nanostructured Materials* 6, no. 1 (1995): 205–216.
3. Gao, Jinhao, and Bing Xu. "Applications of nanomaterials inside cells." *Nano Today* 4, no. 1 (2009): 37–51.
4. Peppas, Nicholas A. "Intelligent therapeutics: Biomimetic systems and nanotechnology in drug delivery." *Advanced Drug Delivery Reviews* 56, no. 11 (2004): 1529–1531.
5. Sahoo, Sanjeeb K., and Vinod Labhasetwar. "Nanotech approaches to drug delivery and imaging." *Drug Discovery Today* 8, no. 24 (2003): 1112–1120.
6. Ferrari, Mauro. "Cancer nanotechnology: Opportunities and challenges." *Nature Reviews Cancer* 5, no. 3 (2005): 161–171.
7. Chen, Victor J., and Peter X. Ma. "Nano-fibrous poly (L-lactic acid) scaffolds with interconnected spherical macropores." *Biomaterials* 25, no. 11 (2004): 2065–2073.
8. Nezafati, Nader, Fathollah Moztarzadeh, Saeed Hesaraki, Masoud Mozafari, Ali Samadikuchaksaraei, Leila Hajibaki, and Mazaher Gholipour. "Effect of silver concentration on bioactivity and antibacterial properties of SiO_2-CaO-P_2O_5 sol–gel derived bioactive glass." In *Key Engineering Materials*, vol. 493, pp. 74–79. Trans Tech Publications, Switzerland 2012.
9. Dankers, Patricia Y. W. and E. W. (Bert) Meijer. "Supramolecular biomaterials. A modular approach towards tissue engineering." *Bulletin of the Chemical Society of Japan* 80, no. 11 (2007): 2047–2073.
10. Medintz, Igor L., H. Tetsuo Uyeda, Ellen R. Goldman, and Hedi Mattoussi. "Quantum dot bioconjugates for imaging, labelling and sensing." *Nature Materials* 4, no. 6 (2005): 435–446.
11. El-Sadik, Abir O., Afaf El-Ansary, and Sherif M. Sabry. "Nanoparticle-labeled stem cells: A novel therapeutic vehicle." *Clinical Pharmacology: Advances and Applications* 2, (2010): 9.
12. Bohr, Mark T. "Nanotechnology goals and challenges for electronic applications." *IEEE Transactions on Nanotechnology* 99, no. 1 (2002): 56–62.
13. Peng, Shu, and Kyeongjae Cho. "Chemical control of nanotube electronics." *Nanotechnology* 11, no. 2 (2000): 57.

14. Juang, Jia-Yang, and David B. Bogy. "Nanotechnology advances and applications in information storage." *Microsystem Technologies* 11, nos. 8–10 (2005): 950–957.
15. Uchegbu, Ijeoma F. "Pharmaceutical nanotechnology: Polymeric vesicles for drug and gene delivery." *Expert Opinion on Drug Delivery* 3, no. 5 (2006): 629–640.
16. Morrissey, Susan. "Nanotechnology in food and agriculture." *Chemical & Engineering News* 84, (2006): 31.
17. Farokhzad, Omid C., and Robert Langer. "Impact of nanotechnology on drug delivery." *ACS Nano* 3, no. 1 (2009): 16–20.
18. Wang, Xu, Lily Yang, Zhuo Georgia Chen, and Dong M. Shin. "Application of nanotechnology in cancer therapy and imaging." *CA: A Cancer Journal for Clinicians* 58, no. 2 (2008): 97–110.
19. Jain, Kewal K. "Nanodiagnostics: Application of nanotechnology in molecular diagnostics." *Expert Review of Molecular Diagnostics* 3, no. 2 (2003): 153–161.
20. Clarke, James, Hai-Chen Wu, Lakmal Jayasinghe, Alpesh Patel, Stuart Reid, and Hagan Bayley. "Continuous base identification for single-molecule nanopore DNA sequencing." *Nature Nanotechnology* 4, no. 4 (2009): 265–270.
21. Stupp, Samuel I., and Glenn W. Ciegler. "Organoapatites: Materials for artificial bone. I. Synthesis and microstructure." *Journal of Biomedical Materials Research* 26, no. 2 (1992): 169–183.
22. Stupp, Samuel I., Vassou LeBonheur, K. Walker, Li-Sheng Li, Kevin E. Huggins, M. Keser, and A. Amstutz. "Supramolecular materials: Self-organized nanostructures." *Science* 276, no. 5311 (1997): 384–389.
23. Shin, Heungsoo, Seongbong Jo, and Antonios G. Mikos. "Biomimetic materials for tissue engineering." *Biomaterials* 24, no. 24 (2003): 4353–4364.
24. Liao, Susan, Casey K. Chan, and S. Ramakrishna. "Stem cells and biomimetic materials strategies for tissue engineering." *Materials Science and Engineering: C* 28, no. 8 (2008): 1189–1202.
25. Ma, Peter X. "Biomimetic materials for tissue engineering." *Advanced Drug Delivery Reviews* 60, no. 2 (2008): 184–198.
26. Mooney, David J., Daniel F. Baldwin, Nam P. Suh, Joseph P. Vacanti, and Robert Langer. "Novel approach to fabricate porous sponges of poly (D, L-lactic-co-glycolic acid) without the use of organic solvents." *Biomaterials* 17, no. 14 (1996): 1417–1422.
27. Ma, Peter X. "Scaffolds for tissue fabrication." *Materials Today* 7, no. 5 (2004): 30–40.
28. Rice, Mark A., Brennan T. Dodson, Jeffrey A. Arthur, and Kristi S. Anseth. "Cell-based therapies and tissue engineering." *Otolaryngologic Clinics of North America* 38, no. 2 (2005): 199–214.
29. Samadikuchaksaraei, Ali. "Stem cell therapy for acute myocardial infarction." *Hellenic Journal of Cardiology* 47, no. 2 (2006): 100–111.
30. Weissman, Irving L. "Stem cells: Units of development, units of regeneration, and units in evolution." *Cell* 100, no. 1 (2000): 157–168.
31. Pera, Martin F., and Alan O. Trounson. "Human embryonic stem cells: Prospects for development." *Development* 131, no. 22 (2004): 5515–5525.
32. Andrews, Peter W., Stefan A. Przyborski, and James A. Thomson. "Embryonal carcinoma cells as embryonic stem cells." *Cold Spring Harbor Monograph Series* 40, (2001): 231–266.
33. Donovan, Peter J., and John Gearhart. "The end of the beginning for pluripotent stem cells." *Nature* 414, no. 6859 (2001): 92–97.
34. Sato, Noboru, Laurent Meijer, Leandros Skaltsounis, Paul Greengard, and Ali H. Brivanlou. "Maintenance of pluripotency in human and mouse embryonic stem cells through activation of Wnt signaling by a pharmacological GSK-3-specific inhibitor." *Nature Medicine* 10, no. 1 (2004): 55–63.
35. Bianco, Paolo and Pamela Gehron Robey, "Stem cells in tissue engineering." *Nature* 414, (2001): 118–121.
36. Amit, M., V. Margulets, H. Segev, K. Shariki, I. Laevsky, R. Coleman, and J. Itskovitz-Eldor. "Human feeder layers for human embryonic stem cells 1." *Biology of Reproduction* 68, no. 6 (2003): 2150–2156.
37. Xu, Chunhui, Margaret S. Inokuma, Jerrod Denham, Kathaleen Golds, Pratima Kundu, Joseph D. Gold, and Melissa K. Carpenter. "Feeder-free growth of undifferentiated human embryonic stem cells." *Nature Biotechnology* 19, no. 10 (2001): 971–974.
38. Cheng, Linzhao, Holly Hammond, Zhaohui Ye, Xiangcan Zhan, and Gautam Dravid. "Human adult marrow cells support prolonged expansion of human embryonic stem cells in culture." *Stem Cells* 21, no. 2 (2003): 131–142.
39. Li, Yan, Sandra Powell, Elisa Brunette, Jane Lebkowski, and Ramkumar Mandalam. "Expansion of human embryonic stem cells in defined serum-free medium devoid of animal-derived products." *Biotechnology and Bioengineering* 91, no. 6 (2005): 688–698.
40. Magyar, Josef P., Mohamed Nemir, Elisabeth Ehler, Nicolai Suter, Jean-Claude Perriard, and Hans M. Eppenberger. "Mass production of embryoid bodies in microbeads." *Annals of the New York Academy of Sciences* 944, no. 1 (2001): 135–143.

41. Randle, Wesley L., Jae Min Cha, Yu-Shik Hwang, KL Andrew Chan, Sergei G. Kazarian, Julia M. Polak, and Athanasios Mantalaris. "Integrated 3-dimensional expansion and osteogenic differentiation of murine embryonic stem cells." *Tissue Engineering* 13, no. 12 (2007): 2957–2970.
42. Lanza, Robert, Robert Langer, and Joseph P. Vacanti, eds. *Principles of Tissue Engineering*. Academic Press, Elsevier, North Holland, 2011.
43. Strain, Alastair J., and James M. Neuberger. "A bioartificial liver—State of the art." *Science* 295, no. 5557 (2002): 1005–1009.
44. Liu, Xiaohua, and Peter X. Ma. "Polymeric scaffolds for bone tissue engineering." *Annals of Biomedical Engineering* 32, no. 3. Biomedical Engineering Society (2004): 477–486.
45. Gelain, Fabrizio, Daniele Bottai, Angleo Vescovi, and Shuguang Zhang. "Designer self-assembling peptide nanofiber scaffolds for adult mouse neural stem cell 3-dimensional cultures." *PloS One* 1, no. 1 (2006): e119.
46. Yang, Shoufeng, Kah-Fai Leong, Zhaohui Du, and Chee-Kai Chua. "The design of scaffolds for use in tissue engineering. Part II. Rapid prototyping techniques." *Tissue Engineering* 8, no. 1 (2002): 1–11.
47. Hench, Larry L., and Julia M. Polak. "Third-generation biomedical materials." *Science* 295, no. 5557 (2002): 1014–1017.
48. Oh, Seunghan, Karla S. Brammer, YS Julie Li, Dayu Teng, Adam J. Engler, Shu Chien, and Sungho Jin. "Stem cell fate dictated solely by altered nanotube dimension." *Proceedings of the National Academy of Sciences* 106, no. 7 (2009): 2130–2135.
49. Ma, Peter X., and Robert Langer. "Degradation, structure and properties of fibrous nonwoven poly (glycolic acid) scaffolds for tissue engineering." In *MRS Proceedings*, vol. 394, p. 99. Cambridge University Press, United Kingdom, 1995.
50. Ma, Peter X., Ruiyun Zhang, Guozhi Xiao, and Renny Franceschi. "Engineering new bone tissue in vitro on highly porous poly (α-hydroxyl acids)/hydroxyapatite composite scaffolds." *Journal of Biomedical Materials Research* 54, no. 2 (2001): 284–293.
51. Chen, Victor J., and Peter X. Ma. "The effect of surface area on the degradation rate of nano-fibrous poly (L-lactic acid) foams." *Biomaterials* 27, no. 20 (2006): 3708–3715.
52. Skarja, Gray A., and Kim A. Woodhouse. "In vitro degradation and erosion of degradable, segmented polyurethanes containing an amino acid-based chain extender." *Journal of Biomaterials Science, Polymer Edition* 12, no. 8 (2001): 851–873.
53. Balasundaram, Ganesan, and Thomas J. Webster. "Nanotechnology and biomaterials for orthopedic medical applications." *Nanomedicine* 1, no. 2 (2006): 169–176.
54. Hoshino, Akiyoshi, Kouki Fujioka, Taisuke Oku, Masakazu Suga, Yu F. Sasaki, Toshihiro Ohta, Masato Yasuhara, Kazuo Suzuki, and Kenji Yamamoto. "Physicochemical properties and cellular toxicity of nanocrystal quantum dots depend on their surface modification." *Nano Letters* 4, no. 11 (2004): 2163–2169.
55. Bogunia-Kubik, Katarzyna, and Masanori Sugisaka. "From molecular biology to nanotechnology and nanomedicine." *Biosystems* 65, no. 2 (2002): 123–138.
56. Ye, Zhaoyang, and Ram I. Mahato. "Role of nanomedicines in cell-based therapeutics." *Nanomedicine* 3, no. 1 (2008): 5–8.
57. Liu, Huinan, and Thomas Jay Webster. "Nanomedicine for implants: A review of studies and necessary experimental tools." *Biomaterials* 28, no. 2 (2007): 354–369.
58. Jafari, Javad, Shahriar Hojjati Emami, Ali Samadikuchaksaraei, Mohammad Ali Bahar, and Fazel Gorjipour. "Electrospun chitosan–gelatin nanofiberous scaffold: Fabrication and in vitro evaluation." *Biomedical Materials and Engineering* 21, no. 2 (2011): 99–112.
59. Li, Dan, and Younan Xia. "Electrospinning of nanofibers: Reinventing the wheel?" *Advanced Materials* 16, no. 14 (2004): 1151–1170.
60. Matthews, Jamil A., Gary E. Wnek, David G. Simpson, and Gary L. Bowlin. "Electrospinning of collagen nanofibers." *Biomacromolecules* 3, no. 2 (2002): 232–238.
61. Harrison, Paul, and Alex Valavanis. *Quantum Wells, Wires and Dots: Theoretical and Computational Physics of Semiconductor Nanostructures*. John Wiley & Sons, 2016.
62. Murugan, R., and S. Ramakrishna. "Development of nanocomposites for bone grafting." *Composites Science and Technology* 65, no. 15 (2005): 2385–2406.
63. Smith, L. A., and P. X. Ma. "Nano-fibrous scaffolds for tissue engineering." *Colloids and Surfaces B: Biointerfaces* 39, no. 3 (2004): 125–131.
64. Woo, Kyung Mi, Victor J. Chen, and Peter X. Ma. "Nano-fibrous scaffolding architecture selectively enhances protein adsorption contributing to cell attachment." *Journal of Biomedical Materials Research Part A* 67, no. 2 (2003): 531–537.

65. Liu, Xiaohua, and Peter X. Ma. "The nanofibrous architecture of poly (L-lactic acid)-based functional copolymers." *Biomaterials* 31, no. 2 (2010): 259–269.
66. Hu, Jiang, Kai Feng, Xiaohua Liu, and Peter X. Ma. "Chondrogenic and osteogenic differentiations of human bone marrow-derived mesenchymal stem cells on a nanofibrous scaffold with designed pore network." *Biomaterials* 30, no. 28 (2009): 5061–5067.
67. Tysseling-Mattiace, Vicki M., Vibhu Sahni, Krista L. Niece, Derin Birch, Catherine Czeisler, Michael G. Fehlings, Samuel I. Stupp, and John A. Kessler. "Self-assembling nanofibers inhibit glial scar formation and promote axon elongation after spinal cord injury." *Journal of Neuroscience* 28, no. 14 (2008): 3814–3823.
68. Sargeant, Timothy D., Mustafa O. Guler, Scott M. Oppenheimer, Alvaro Mata, Robert L. Satcher, David C. Dunand, and Samuel I. Stupp. "Hybrid bone implants: Self-assembly of peptide amphiphile nanofibers within porous titanium." *Biomaterials* 29, no. 2 (2008): 161–171.
69. Harrington, Daniel A., Earl Y. Cheng, Mustafa O. Guler, Leslie K. Lee, Jena L. Donovan, Randal C. Claussen, and Samuel I. Stupp. "Branched peptide-amphiphiles as self-assembling coatings for tissue engineering scaffolds." *Journal of Biomedical Materials Research Part A* 78, no. 1 (2006): 157–167.
70. Webber, Matthew J., Jörn Tongers, Marie-Ange Renault, Jerome G. Roncalli, Douglas W. Losordo, and Samuel I. Stupp. "Development of bioactive peptide amphiphiles for therapeutic cell delivery." *Acta Biomaterialia* 6, no. 1 (2010): 3–11.
71. Veerabadran, Nalinkanth G., Poorna L. Goli, Skylar S. Stewart-Clark, Yuri M. Lvov, and David K. Mills. "Nanoencapsulation of stem cells within polyelectrolyte multilayer shells." *Macromolecular Bioscience* 7, no. 7 (2007): 877–882.
72. Teramura, Yuji, Yoshihiro Kaneda, and Hiroo Iwata. "Islet-encapsulation in ultra-thin layer-by-layer membranes of poly (vinyl alcohol) anchored to poly (ethylene glycol)–lipids in the cell membrane." *Biomaterials* 28, no. 32 (2007): 4818–4825.
73. Lvov, Yuri, Alexei A. Antipov, Arif Mamedov, Helmuth Möhwald, and Gleb B. Sukhorukov. "Urease encapsulation in nanoorganized microshells." *Nano Letters* 1, no. 3 (2001): 125–128.
74. Siti-Ismail, Norhayati, Anne E. Bishop, Julia M. Polak, and Athanasios Mantalaris. "The benefit of human embryonic stem cell encapsulation for prolonged feeder-free maintenance." *Biomaterials* 29, no. 29 (2008): 3946–3952.
75. Acton, P. D., and R. Zhou. "Imaging reporter genes for cell tracking with PET and SPECT." *The Quarterly Journal of Nuclear Medicine and Molecular Imaging* 49, no. 4 (2005): 349.
76. Frangioni, John V., and Roger J. Hajjar. "In vivo tracking of stem cells for clinical trials in cardiovascular disease." *Circulation* 110, no. 21 (2004): 3378–3383.
77. Walczak, Piotr, Jian Zhang, Assaf A. Gilad, Dorota A. Kedziorek, Jesus Ruiz-Cabello, Randell G. Young, Mark F. Pittenger, Peter C. M. van Zijl, Judy Huang, and Jeff W. M. Bulte. "Dual-modality monitoring of targeted intraarterial delivery of mesenchymal stem cells after transient ischemia." *Stroke* 39, no. 5 (2008): 1569–1574.
78. Heyn, Chris, John A. Ronald, Lisa T. Mackenzie, Ian C. MacDonald, Ann F. Chambers, Brian K. Rutt, and Paula J. Foster. "In vivo magnetic resonance imaging of single cells in mouse brain with optical validation." *Magnetic Resonance in Medicine* 55, no. 1 (2006): 23–29.
79. Heyn, Chris, John A. Ronald, Soha S. Ramadan, Jonatan A. Snir, Andrea M. Barry, Lisa T. MacKenzie, David J. Mikulis et al. "In vivo MRI of cancer cell fate at the single-cell level in a mouse model of breast cancer metastasis to the brain." *Magnetic Resonance in Medicine* 56, no. 5 (2006): 1001–1010.
80. Himmelreich, Uwe, and Mathias Hoehn. "Stem cell labeling for magnetic resonance imaging." *Minimally Invasive Therapy & Allied Technologies* 17, no. 2 (2008): 132–142.
81. Kinney, J. H., J. T. Ryaby, D. L. Haupt, and Nancy E. Lane. "Three-dimensional in vivo morphometry of trabecular bone in the OVX rat model of osteoporosis." *Technology and Health Care* 6, no. 5 (1998): 339–350.
82. Bayat, S., L. Apostol, E. Boller, T. Brochard, and F. Peyrin. "In vivo imaging of bone micro-architecture in mice with 3D synchrotron radiation micro-tomography." *Nuclear Instruments and Methods in Physics Research Section A: Accelerators, Spectrometers, Detectors and Associated Equipment* 548, no. 1 (2005): 247–252.
83. Boyd, Steven K., Peter Davison, Ralph Müller, and Jürg A. Gasser. "Monitoring individual morphological changes over time in ovariectomized rats by in vivo micro-computed tomography." *Bone* 39, no. 4 (2006): 854–862.
84. Cancedda, R., A. Cedola, A. Giuliani, V. Komlev, S. Lagomarsino, M. Mastrogiacomo, F. Peyrin, and F. Rustichelli. "Bulk and interface investigations of scaffolds and tissue-engineered bones by X-ray microtomography and X-ray microdiffraction." *Biomaterials* 28, no. 15 (2007): 2505–2524.

85. Harrison, Benjamin S., and Anthony Atala. "Carbon nanotube applications for tissue engineering." *Biomaterials* 28, no. 2 (2007): 344–353.
86. Brun-Graeppi, Amanda K. Andriola Silva, Cyrille Richard, Michel Bessodes, Daniel Scherman, and Otto-Wilhelm Merten. "Cell microcarriers and microcapsules of stimuli-responsive polymers." *Journal of Controlled Release* 149, no. 3 (2011): 209–224.
87. Morawski, Anne M., Gregory A. Lanza, and Samuel A. Wickline. "Targeted contrast agents for magnetic resonance imaging and ultrasound." *Current Opinion in Biotechnology* 16, no. 1 (2005): 89–92.
88. Kuliszewski, Michael A., Hiroko Fujii, Christine Liao, Alexandra H. Smith, Aris Xie, Jonathan R. Lindner, and Howard Leong-Poi. "Molecular imaging of endothelial progenitor cell engraftment using contrast-enhanced ultrasound and targeted microbubbles." *Cardiovascular Research* 83, no. 4 (2009): 653–662.
89. Bulte, Jeff W. M., Ali S. Arbab, T. Douglas, and J. A. Frank. "Preparation of magnetically labeled cells for cell tracking by magnetic resonance imaging." *Methods in Enzymology* 386 (2004): 275–299.
90. Bulte, Jeff W. M., and Dara L. Kraitchman. "Iron oxide MR contrast agents for molecular and cellular imaging." *NMR in Biomedicine* 17, no. 7 (2004): 484–499.
91. Lewin, Maite, Nadia Carlesso, Ching-Hsuan Tung, Xiao-Wu Tang, David Cory, David T. Scadden, and Ralph Weissleder. "Tat peptide-derivatized magnetic nanoparticles allow in vivo tracking and recovery of progenitor cells." *Nature Biotechnology* 18, no. 4 (2000): 410–414.
92. Modo, Michel, Diana Cash, Karen Mellodew, Steven C. R. Williams, Scott E. Fraser, Thomas J. Meade, Jack Price, and Helen Hodges. "Tracking transplanted stem cell migration using bifunctional, contrast agent-enhanced, magnetic resonance imaging." *Neuroimage* 17, no. 2 (2002): 803–811.
93. Villa, Chiara, Silvia Erratico, Paola Razini, Fabrizio Fiori, Franco Rustichelli, Yvan Torrente, and Marzia Belicchi. "Stem cell tracking by nanotechnologies." *International Journal of Molecular Sciences* 11, no. 3 (2010): 1070–1081.
94. Ashbridge, Dawn A., Michael S. Thorne, Mark L. Rivers, Julia C. Muccino, and Peggy A. O'Day. "Image optimization and analysis of synchrotron X-ray computed microtomography (CμT) data." *Computers & Geosciences* 29, no. 7 (2003): 823–836.
95. David, Valentin, Norbert Laroche, Benjamin Boudignon, Marie-Hélène Lafage-Proust, Christian Alexandre, Peter Ruegsegger, and Laurence Vico. "Noninvasive in vivo monitoring of bone architecture alterations in hindlimb-unloaded female rats using novel three-dimensional microcomputed tomography." *Journal of Bone and Mineral Research* 18, no. 9 (2003): 1622–1631.
96. Bulte, Jeff W. M., Trevor Douglas, Brian Witwer, Su-Chun Zhang, Erica Strable, Bobbi K. Lewis, Holly Zywicke et al. "Magnetodendrimers allow endosomal magnetic labeling and in vivo tracking of stem cells." *Nature Biotechnology* 19, no. 12 (2001): 1141–1147.
97. Bulte, J. W. M., S. C. Zhang, P. Van Gelderen, V. Herynek, E. K. Jordan, I. D. Duncan, and J. A. Frank. "Neurotransplantation of magnetically labeled oligodendrocyte progenitors: Magnetic resonance tracking of cell migration and myelination." *Proceedings of the National Academy of Sciences* 96, no. 26 (1999): 15256–15261.
98. Anderson, Stasia A., John Glod, Ali S. Arbab, Martha Noel, Parwana Ashari, Howard A. Fine, and Joseph A. Frank. "Noninvasive MR imaging of magnetically labeled stem cells to directly identify neovasculature in a glioma model." *Blood* 105, no. 1 (2005): 420–425.
99. Stuckey, Daniel J., Carolyn A. Carr, Enca Martin-Rendon, Damian J. Tyler, Corinne Willmott, Paul J. Cassidy, Sarah JM Hale et al. "Iron particles for noninvasive monitoring of bone marrow stromal cell engraftment into, and isolation of viable engrafted donor cells from, the heart." *Stem Cells* 24, no. 8 (2006): 1968–1975.
100. Wu, Wuwei, and Alexander D. Q. Li. "Optically switchable nanoparticles for biological imaging." *Nanomedicine* 4, no. 4 (2007): 523–531.
101. Chemaly, Elie R., Ryuichi Yoneyama, John V. Frangioni, and Roger J. Hajjar. "Tracking stem cells in the cardiovascular system." *Trends in Cardiovascular Medicine* 15, no. 8 (2005): 297–302.
102. Rogers, Walter J., Craig H. Meyer, and Christopher M. Kramer. "Technology insight: In vivo cell tracking by use of MRI." *Nature Clinical Practice Cardiovascular Medicine* 3, no. 10 (2006): 554–562.
103. Gupta, Ajay Kumar, and Adam S. G. Curtis. "Lactoferrin and ceruloplasmin derivatized superparamagnetic iron oxide nanoparticles for targeting cell surface receptors." *Biomaterials* 25, no. 15 (2004): 3029–3040.
104. Bulte, M., W. Jeff, and Dara L. Kraitchman. "Monitoring cell therapy using iron oxide MR contrast agents." *Current Pharmaceutical Biotechnology* 5, no. 6 (2004): 567–584.
105. Jing, Xu-hong, Liu Yang, Xiao-jun Duan, Bing Xie, Wei Chen, Zhong Li, and Hong-bo Tan. "In vivo MR imaging tracking of magnetic iron oxide nanoparticle labeled, engineered, autologous bone marrow mesenchymal stem cells following intra-articular injection." *Joint Bone Spine* 75, no. 4 (2008): 432–438.

106. Emerit, J., C. Beaumont, and F. Trivin. "Iron metabolism, free radicals, and oxidative injury." *Biomedicine & Pharmacotherapy* 55, no. 6 (2001): 333–339.
107. Dutz, Silvio, Wilfried Andrä, Rudolf Hergt, Robert Müller, Christiane Oestreich, Christopher Schmidt, Jorg Töpfer, Matthias Zeisberger, and Matthias E. Bellemann. "Influence of dextran coating on the magnetic behaviour of iron oxide nanoparticles." *Journal of Magnetism and Magnetic Materials* 311, no. 1 (2007): 51–54.
108. Lawaczeck, Rüdiger, Michael Menzel, and Hubertus Pietsch. "Superparamagnetic iron oxide particles: contrast media for magnetic resonance imaging." *Applied Organometallic Chemistry* 18, no. 10 (2004): 506–513.
109. Wang, F. H., I. H. Lee, N. Holmström, T. Yoshitake, D. K. Kim, Mamoun Muhammed, J. Frisen, L. Olson, C. Spenger, and J. Kehr. "Magnetic resonance tracking of nanoparticle labelled neural stem cells in a rat's spinal cord." *Nanotechnology* 17, no. 8 (2006): 1911.
110. Niemeyer, Christof M., and Bülent Ceyhan. "DNA-Directed Functionalization of Colloidal Gold with Proteins." *Angewandte Chemie International Edition* 40, no. 19 (2001): 3685–3688.
111. Kim, Daehong, Kwan Soo Hong, and Jihwan Song. "The present status of cell tracking methods in animal models using magnetic resonance imaging technology." *Molecules & Cells (Springer Science & Business Media BV)* 23, no. 2 (2007): 132–137.
112. Küstermann, Ekkehard, Uwe Himmelreich, K. Kandal, T. Geelen, A. Ketkar, Dirk Wiedermann, Cordula Strecker, J. Esser, S. Arnhold, and Mathias Hoehn. "Efficient stem cell labeling for MRI studies." *Contrast Media & Molecular Imaging* 3, no. 1 (2008): 27–37.
113. Gupta, Ajay Kumar, and Mona Gupta. "Synthesis and surface engineering of iron oxide nanoparticles for biomedical applications." *Biomaterials* 26, no. 18 (2005): 3995–4021.
114. Wunderbaldinger, Patrick, Lee Josephson, and Ralph Weissleder. "Tat peptide directs enhanced clearance and hepatic permeability of magnetic nanoparticles." *Bioconjugate Chemistry* 13, no. 2 (2002): 264–268.
116. Sykova, E., and P. Jendelova. "Migration, fate and in vivo imaging of adult stem cells in the CNS." *Cell Death and Differentiation* 14, no. 7 (2007): 1336.
117. Au, Ka-Wing, Song-Yan Liao, Yee-Ki Lee, Wing-Hon Lai, Kwong-Man Ng, Yau-Chi Chan, Mei-Chu Yip et al. "Effects of iron oxide nanoparticles on cardiac differentiation of embryonic stem cells." *Biochemical and Biophysical Research Communications* 379, no. 4 (2009): 898–903.
118. Guzman, Raphael, Nobuko Uchida, Tonya M. Bliss, Dongping He, Karen K. Christopherson, David Stellwagen, Alexandra Capela et al. "Long-term monitoring of transplanted human neural stem cells in developmental and pathological contexts with MRI." *Proceedings of the National Academy of Sciences* 104, no. 24 (2007): 10211–10216.
119. Arai, Takayasu, Theo Kofidis, Jeff W. M. Bulte, Jorg de Bruin, Ross D. Venook, Gerald J. Berry, Michael V. Mcconnell, Thomas Quertermous, Robert C. Robbins, and Phillip C. Yang. "Dual in vivo magnetic resonance evaluation of magnetically labeled mouse embryonic stem cells and cardiac function at 1.5 t." *Magnetic Resonance in Medicine* 55, no. 1 (2006): 203–209.
120. Küstermann, Ekkehard, Wilhelm Roell, Martin Breitbach, Stefan Wecker, Dirk Wiedermann, Christian Buehrle, Armin Welz, Juergen Hescheler, Bernd K. Fleischmann, and Mathias Hoehn. "Stem cell implantation in ischemic mouse heart: A high-resolution magnetic resonance imaging investigation." *NMR in Biomedicine* 18, no. 6 (2005): 362–370.
121. Zhu, Jianhong, Liangfu Zhou, and FengGe XingWu. "Tracking neural stem cells in patients with brain trauma." *The New England Journal of Medicine* 2006, no. 355 (2006): 2376–2378.
122. Gul, Hilal, Weibing Lu, Peng Xu, James Xing, and Jie Chen. "Magnetic carbon nanotube labelling for haematopoietic stem/progenitor cell tracking." *Nanotechnology* 21, no. 15 (2010): 155101.
123. Bianco, Alberto, Kostas Kostarelos, and Maurizio Prato. "Applications of carbon nanotubes in drug delivery." *Current Opinion in Chemical Biology* 9, no. 6 (2005): 674–679.
124. Liu, Zhuang, Kai Chen, Corrine Davis, Sarah Sherlock, Qizhen Cao, Xiaoyuan Chen, and Hongjie Dai. "Drug delivery with carbon nanotubes for in vivo cancer treatment." *Cancer Research* 68, no. 16 (2008): 6652–6660.
125. Pantarotto, Davide, Ravi Singh, David McCarthy, Mathieu Erhardt, Jean-Paul Briand, Maurizio Prato, Kostas Kostarelos, and Alberto Bianco. "Functionalized carbon nanotubes for plasmid DNA gene delivery." *Angewandte Chemie* 116, no. 39 (2004): 5354–5358.
126. Abarrategi, Ander, María C. Gutiérrez, Carolina Moreno-Vicente, María J. Hortigüela, Viviana Ramos, José L. López-Lacomba, María L. Ferrer, and Francisco del Monte. "Multiwall carbon nanotube scaffolds for tissue engineering purposes." *Biomaterials* 29, no. 1 (2008): 94–102.
127. Iijima, Sumio. "Helical microtubules of graphitic carbon." *Nature* 354, no. 6348 (1991): 56.

128. Jia, Guang, Haifang Wang, Lei Yan, Xiang Wang, Rongjuan Pei, Tao Yan, Yuliang Zhao, and Xinbiao Guo. "Cytotoxicity of carbon nanomaterials: Single-wall nanotube, multi-wall nanotube, and fullerene." *Environmental Science & Technology* 39, no. 5 (2005): 1378–1383.
129. MacDonald, Rebecca A., Brendan F. Laurenzi, Gunaranjan Viswanathan, Pulickel M. Ajayan, and Jan P. Stegemann. "Collagen–carbon nanotube composite materials as scaffolds in tissue engineering." *Journal of Biomedical Materials Research Part A* 74, no. 3 (2005): 489–496.
130. Correa-Duarte, Miguel A., Nicholas Wagner, José Rojas-Chapana, Christian Morsczeck, Michael Thie, and Michael Giersig. "Fabrication and biocompatibility of carbon nanotube-based 3D networks as scaffolds for cell seeding and growth." *Nano Letters* 4, no. 11 (2004): 2233–2236.
131. Lovat, Viviana, Davide Pantarotto, Laura Lagostena, Barbara Cacciari, Micaela Grandolfo, Massimo Righi, Giampiero Spalluto, Maurizio Prato, and Laura Ballerini. "Carbon nanotube substrates boost neuronal electrical signaling." *Nano Letters* 5, no. 6 (2005): 1107–1110.
132. Supronowicz, P. R., P. M. Ajayan, K. R. Ullmann, B. P. Arulanandam, D. W. Metzger, and R. Bizios. "Novel current-conducting composite substrates for exposing osteoblasts to alternating current stimulation." *Journal of Biomedical Materials Research* 59, no. 3 (2002): 499–506.
133. Huczko, A., and H. Lange. "Carbon nanotubes: Experimental evidence for a null risk of skin irritation and allergy." *Fullerene Science and Technology* 9, no. 2 (2001): 247–250.
134. Nimmagadda, Aditya, Karen Thurston, Matthias U. Nollert, and Peter S. McFetridge. "Chemical modification of SWNT alters in vitro cell-SWNT interactions." *Journal of Biomedical Materials Research Part A* 76, no. 3 (2006): 614–625.
135. Reynolds, Charles H., Nikoi Annan, Kebede Beshah, Jon H. Huber, Steven H. Shaber, Robert E. Lenkinski, and Jeffrey A. Wortman. "Gadolinium-loaded nanoparticles: New contrast agents for magnetic resonance imaging." *Journal of the American Chemical Society* 122, no. 37 (2000): 8940–8945.
136. Lin, Shuan, Xiaoyan Xie, Manishkumar R. Patel, Yao-Hung Yang, Zongjin Li, Feng Cao, Oliver Gheysens, Yan Zhang, Sanjiv S. Gambhir, Jiang Hong Rao, and Joseph C. Wu. "Quantum dot imaging for embryonic stem cells." *BMC Biotechnology* 7, no. 1 (2007): 67.
137. Maysinger, Dusica, Maik Behrendt, Mélanie Lalancette-Hébert, and Jasna Kriz. "Real-time imaging of astrocyte response to quantum dots: In vivo screening model system for biocompatibility of nanoparticles." *Nano Letters* 7.8, (2007): 2513–2520.
138. Gao, Xiaohu, Lily Yang, John A. Petros, Fray F. Marshall, Jonathan W. Simons, and Shuming Nie. "In vivo molecular and cellular imaging with quantum dots." *Current Opinion in Biotechnology* 16, no. 1 (2005): 63–72.
139. Michalet, X., F. F. Pinaud, L. A. Bentolila, J. M. Tsay, S. J. J. L. Doose, J. J. Li, G. Sundaresan, A. M. Wu, S. S. Gambhir, and S. Weiss. "Quantum dots for live cells, in vivo imaging, and diagnostics." *Science* 307, no. 5709 (2005): 538–544.
140. Rhyner, Matthew N., Andrew M. Smith, Xiaohu Gao, Hui Mao, Lily Yang, and Shuming Nie. "Quantum dots and multifunctional nanoparticles: New contrast agents for tumor imaging." *Nanomedicine* 1, no. 2 (2006): 209–217.
141. Shah, Bhranti S., Paul A. Clark, Eduardo K. Moioli, Michael A. Stroscio, and Jeremy J. Mao. "Labeling of mesenchymal stem cells by bioconjugated quantum dots." *Nano Letters* 7, no. 10 (2007): 3071–3079.
142. Slotkin, Jonathan R., Lina Chakrabarti, Hai Ning Dai, Rosalind SE Carney, Tsutomu Hirata, Barbara S. Bregman, G. Ian Gallicano, Joshua G. Corbin, and Tarik F. Haydar. "In vivo quantum dot labeling of mammalian stem and progenitor cells." *Developmental Dynamics* 236, no. 12 (2007): 3393–3401.
143. Gao, Xiaohu, Warren C. W. Chan, and Shuming Nie. "Quantum-dot nanocrystals for ultrasensitive biological labeling and multicolor optical encoding." *Journal of Biomedical Optics* 7, no. 4 (2002): 532–537.
144. Gao, Xiaohu, and Shuming Nie. "Molecular profiling of single cells and tissue specimens with quantum dots." *Trends in Biotechnology* 21, no. 9 (2003): 371–373.
145. Cui, Bianxiao, Chengbiao Wu, Liang Chen, Alfredo Ramirez, Elaine L. Bearer, Wei-Ping Li, William C. Mobley, and Steven Chu. "One at a time, live tracking of NGF axonal transport using quantum dots." *Proceedings of the National Academy of Sciences* 104, no. 34 (2007): 13666–13671.
146. Rosen, Amy B., Damon J. Kelly, Adam J. T. Schuldt, Jia Lu, Irina A. Potapova, Sergey V. Doronin, Kyle J. Robichaud, Richard B. Robinson, Michael R. Rosen, Peter R. Brink, Glenn R. Gaudette, and Ira S. Cohen. "Finding fluorescent needles in the cardiac haystack: Tracking human mesenchymal stem cells labeled with quantum dots for quantitative in vivo three-dimensional fluorescence analysis." *Stem Cells* 25, no. 8 (2007): 2128–2138.
147. Chakraborty, Subhasish K., James A. J. Fitzpatrick, Julie A. Phillippi, Susan Andreko, Alan S. Waggoner, Marcel P. Bruchez, and Byron Ballou. "Cholera toxin B conjugated quantum dots for live cell labeling." *Nano Letters* 7, no. 9 (2007): 2618–2626.

148. Sekhon, Bhupinder S., and Seema R. Kamboj. "Inorganic nanomedicine—Part 1." *Nanomedicine: Nanotechnology, Biology and Medicine* 6, no. 4 (2010): 516–522.
149. Laugwitz, Karl-Ludwig, Alessandra Moretti, Jason Lam, Peter Gruber, Yinhong Chen, Sarah Woodard, Li-Zhu Lin, Chen-Leng Cai, Min Min Lu, Michael Reth, Oleksandr Platoshyn, Jason X.-J. Yuan, Sylvia Evans, and Kenneth R. Chen "Postnatal isl1+ cardioblasts enter fully differentiated cardiomyocyte lineages." *Nature* 433, no. 7026 (2005): 647–653.
150. Radisic, Milica, Rohin K. Iyer, and Shashi K. Murthy. "Micro-and nanotechnology in cell separation." *International Journal of Nanomedicine* 1, no. 1 (2006): 3.
151. Inglis, David W., Robert Riehn, R. H. Austin, and J. C. Sturm. "Continuous microfluidic immunomagnetic cell separation." *Applied Physics Letters* 85, no. 21 (2004): 5093–5095.
152. Kutsuzawa, K., E. H. Chowdhury, M. Nagaoka, K. Maruyama, Y. Akiyama, and T. Akaike. "Surface functionalization of inorganic nano-crystals with fibronectin and E-cadherin chimera synergistically accelerates trans-gene delivery into embryonic stem cells." *Biochemical and Biophysical Research Communications* 350, no. 3 (2006): 514–520.
153. Uchimura, Eiichiro, Shigeru Yamada, Lorenz Uebersax, Satoshi Fujita, Masato Miyake, and Jun Miyake. "Method for reverse transfection using gold colloid as a nano-scaffold." *Journal of Bioscience and Bioengineering* 103, no. 1 (2007): 101–103.
154. Yang, Fan, Seung-Woo Cho, Sun Mi Son, Said R. Bogatyrev, Deepika Singh, Jordan J. Green, Ying Mei, Sohyun Park, Suk Ho Bhang, Byung-Soo Kim, Robert Langera, and Daniel G. Anderson. "Genetic engineering of human stem cells for enhanced angiogenesis using biodegradable polymeric nanoparticles." *Proceedings of the National Academy of Sciences* 107, no. 8 (2010): 3317–3322.
155. El-Sadik, Abir. "Tissue Regeneration-From Basic Biology to Clinical Application." *Novel Promises of Nanotechnology for Tissue Regeneration*. INTECH Open Access Publisher, 2012.
156. Huang, Dong-Ming, Tsai-Hua Chung, Yann Hung, Fang Lu, Si-Han Wu, Chung-Yuan Mou, Ming Yao, and Yao-Chang Chen. "Internalization of mesoporous silica nanoparticles induces transient but not sufficient osteogenic signals in human mesenchymal stem cells." *Toxicology and Applied Pharmacology* 231, no. 2 (2008): 208–215.
157. Shi, Xuetao, Yingjun Wang, Rohan R. Varshney, Li Ren, Feng Zhang, and Dong-An Wang. "In-vitro osteogenesis of synovium stem cells induced by controlled release of bisphosphate additives from microspherical mesoporous silica composite." *Biomaterials* 30, no. 23 (2009): 3996–4005.
158. Jang, Jun-Hyeog, Oscar Castano, and Hae-Won Kim. "Electrospun materials as potential platforms for bone tissue engineering." *Advanced Drug Delivery Reviews* 61, no. 12 (2009): 1065–1083.
159. Curtis, Adam S. G., Matthew Dalby, and Nikolaj Gadegaard. "Cell signaling arising from nanotopography: Implications for nanomedical devices." *Nanomedicine* 1, no. 1 (2006): 67–72.
160. Dalby, Matthew J., Manus J. P. Biggs, Nikolaj Gadegaard, Gabriela Kalna, Chris D. W. Wilkinson, and Adam S. G. Curtis. "Nanotopographical stimulation of mechanotransduction and changes in interphase centromere positioning." *Journal of Cellular Biochemistry* 100, no. 2 (2007): 326–338.
161. Biggs, M. J. P., R. G. Richards, S. McFarlane, C. D. W. Wilkinson, R. O. C. Oreffo, and M. J. Dalby. "Adhesion formation of primary human osteoblasts and the functional response of mesenchymal stem cells to 330 nm deep microgrooves." *Journal of the Royal Society Interface* 5, no. 27 (2008): 1231–1242.
162. Charest, Joseph L., Lindsay E. Bryant, Andres J. Garcia, and William P. King. "Hot embossing for micropatterned cell substrates." *Biomaterials* 25, no. 19 (2004): 4767–4775.
163. Grausova, L., A. Kromka, L. Bacakova, S. Potocky, M. Vanecek, and V. Lisa. "Bone and vascular endothelial cells in cultures on nanocrystalline diamond films." *Diamond and Related Materials* 17, no. 7 (2008): 1405–1409.
164. Bacakova, L., L. Grausova, J. Vacik, A. Fraczek, S. Blazewicz, A. Kromka, M. Vanecek, and V. Svorcik. "Improved adhesion and growth of human osteoblast-like MG 63 cells on biomaterials modified with carbon nanoparticles." *Diamond and Related Materials* 16, no. 12 (2007): 2133–2140.
165. Kubo, Katsutoshi, Naoki Tsukimura, Fuminori Iwasa, Takeshi Ueno, Lei Saruwatari, Hideki Aita, Wen-An Chiou, and Takahiro Ogawa. "Cellular behavior on TiO_2 nanonodular structures in a micro-to-nanoscale hierarchy model." *Biomaterials* 30, no. 29 (2009): 5319–5329.
166. Webster, Thomas J., Celaletdin Ergun, Robert H. Doremus, Richard W. Siegel, and Rena Bizios. "Enhanced functions of osteoblasts on nanophase ceramics." *Biomaterials* 21, no. 17 (2000): 1803–1810.
167. Kommireddy, Dinesh S., Izumi Ichinose, Yuri M. Lvov, and David K. Mills. "Nanoparticle multilayers: Surface modification for cell attachment and growth." *Journal of Biomedical Nanotechnology* 1, no. 3 (2005): 286–290.

168. Hu, Yan, Kaiyong Cai, Zhong Luo, Rui Zhang, Li Yang, Linhong Deng, and Klaus D. Jandt. "Surface mediated in situ differentiation of mesenchymal stem cells on gene-functionalized titanium films fabricated by layer-by-layer technique." *Biomaterials* 30, no. 21 (2009): 3626–3635.
169. Ma, Zuwei, Masaya Kotaki, Ryuji Inai, and Seeram Ramakrishna. "Potential of nanofiber matrix as tissue-engineering scaffolds." *Tissue Engineering* 11, nos. 1–2 (2005): 101–109.
170. Boudreau, Nancy J., and Peter Lloyd Jones. "Extracellular matrix and integrin signalling: The shape of things to come." *Biochemical Journal* 339, no. 3 (1999): 481–488.
171. Morton, William James. "Method of dispersing fluids." *U.S. Patent 705,691*, issued July 29, 1902.
172. Formhals, A. "Process and apparatus fob pbepabing." *Google Patents* (1934).
173. Wnek, Gary E., Marcus E. Carr, David G. Simpson, and Gary L. Bowlin. "Electrospinning of nanofiber fibrinogen structures." *Nano Letters* 3, no. 2 (2003): 213–216.
174. Xia, Younan, Peidong Yang, Yugang Sun, Yiying Wu, Brian Mayers, Byron Gates, Yadong Yin, Franklin Kim, and Haoquan Yan. "One-dimensional nanostructures: Synthesis, characterization, and applications." *Advanced Materials* 15, no. 5 (2003): 353–389.
175. Frenot, Audrey, and Ioannis S. Chronakis. "Polymer nanofibers assembled by electrospinning." *Current Opinion in Colloid & Interface Science* 8, no. 1 (2003): 64–75.
176. Larrondo, L., and R. St John Manley. "Electrostatic fiber spinning from polymer melts. III. Electrostatic deformation of a pendant drop of polymer melt." *Journal of Polymer Science: Polymer Physics Edition* 19, no. 6 (1981): 933–940.
177. Doshi, Jayesh, and Darrell H. Reneker. "Electrospinning process and applications of electrospun fibers." *Journal of Electrostatics* 35, nos. 2–3 (1995): 151–160.
178. Reneker, Darrell H., and Iksoo Chun. "Nanometre diameter fibres of polymer, produced by electrospinning." *Nanotechnology* 7, no. 3 (1996): 216.
179. Kessick, Royal, John Fenn, and Gary Tepper. "The use of AC potentials in electrospraying and electrospinning processes." *Polymer* 45, no. 9 (2004): 2981–2984.
180. Bhattarai, Shanta Raj, Narayan Bhattarai, Ho Keun Yi, Pyong Han Hwang, Dong Il Cha, and Hak Yong Kim. "Novel biodegradable electrospun membrane: Scaffold for tissue engineering." *Biomaterials* 25, no. 13 (2004): 2595–2602.
181. Bognitzki, Michael, Wolfgang Czado, Thomas Frese, Andreas Schaper, Michael Hellwig, Martin Steinhart, Andreas Greiner, and Joachim H. Wendorff. "Nanostructured fibers via electrospinning." *Advanced Materials* 13, no. 1 (2001): 70–72.
182. Yoshimoto, H., Y. M. Shin, H. Terai, and J. P. Vacanti. "A biodegradable nanofiber scaffold by electrospinning and its potential for bone tissue engineering." *Biomaterials* 24, no. 12 (2003): 2077–2082.
183. Li, Wan-Ju, Cato T. Laurencin, Edward J. Caterson, Rocky S. Tuan, and Frank K. Ko. "Electrospun nanofibrous structure: A novel scaffold for tissue engineering." *Journal of Biomedical Materials Research* 60, no. 4 (2002): 613–621.
184. Deitzel, J. M., W. Kosik, S. H. McKnight, NC Beck Tan, J. M. DeSimone, and S. Crette. "Electrospinning of polymer nanofibers with specific surface chemistry." *Polymer* 43, no. 3 (2002): 1025–1029.
185. Bates, W., C.P. Barnes, Z. Ounaies, and G. E. Wnek "Electrostatic processing of PVDF." *Polymer Preprints—America* 44, no. 2 (2003): 114.
186. McKee, Matthew G., Garth L. Wilkes, Ralph H. Colby, and Timothy E. Long. "Correlations of solution rheology with electrospun fiber formation of linear and branched polyesters." *Macromolecules* 37, no. 5 (2004): 1760–1767.
187. Fridrikh, Sergey V., H. Yu Jian, Michael P. Brenner, and Gregory C. Rutledge. "Controlling the fiber diameter during electrospinning." *Physical Review Letters* 90, no. 14 (2003): 144502.
188. Kim, Moon Suk, Jae Ho Kim, Byoung Hyun Min, Heung Jae Chun, Dong Keun Han, and Hai Bang Lee. "Polymeric scaffolds for regenerative medicine." *Polymer Reviews* 51, no. 1 (2011): 23–52.
189. Deitzel, Joseph M., James Kleinmeyer, D. E. A. Harris, and NC Beck Tan. "The effect of processing variables on the morphology of electrospun nanofibers and textiles." *Polymer* 42, no. 1 (2001): 261–272.
190. Huang, Zheng-Ming, Y.-Z. Zhang, M. Kotaki, and S. Ramakrishna. "A review on polymer nanofibers by electrospinning and their applications in nanocomposites." *Composites Science and Technology* 63, no. 15 (2003): 2223–2253.
191. Alhosseini, Sanaz Naghavi, Fathollah Moztarzadeh, Masoud Mozafari, Shadnaz Asgari, Masumeh Dodel, Ali Samadikuchaksaraei, Saeid Kargozar, and Newsha Jalali. "Synthesis and characterization of electrospun polyvinyl alcohol nanofibrous scaffolds modified by blending with chitosan for neural tissue engineering." *International Journal of Nanomedicine* 7 (2012): 25.

192. Li, Chunmei, Charu Vepari, Hyoung-Joon Jin, Hyeon Joo Kim, and David L. Kaplan. "Electrospun silk-BMP-2 scaffolds for bone tissue engineering." *Biomaterials* 27, no. 16 (2006): 3115–3124.
193. Chew, Sing Yian, Jie Wen, Evelyn K. F. Yim, and Kam W. Leong. "Sustained release of proteins from electrospun biodegradable fibers." *Biomacromolecules* 6, no. 4 (2005): 2017–2024.
194. Yang, F., R. Murugan, S. Wang, and S. Ramakrishna. "Electrospinning of nano/micro scale poly (L-lactic acid) aligned fibers and their potential in neural tissue engineering." *Biomaterials* 26, no. 15 (2005): 2603–2610.
195. Chua, Kian-Ngiap, Wei-Seng Lim, Pengchi Zhang, Hongfang Lu, Jie Wen, Seeram Ramakrishna, Kam W. Leong, and Hai-Quan Mao. "Stable immobilization of rat hepatocyte spheroids on galactosylated nanofiber scaffold." *Biomaterials* 26, no. 15 (2005): 2537–2547.
196. Yu, Ying-Ching, Teika Pakalns, Yoav Dori, James B. McCarthy, Matthew Tirrell, and Gregg B. Fields. "Construction of biologically active protein molecular architecture using self-assembling peptide-amphiphiles." *Methods in Enzymology* 289 (1997): 571–587.
197. Silva, Gabriel A., Catherine Czeisler, Krista L. Niece, Elia Beniash, Daniel A. Harrington, John A. Kessler, and Samuel I. Stupp. "Selective differentiation of neural progenitor cells by high-epitope density nanofibers." *Science* 303, no. 5662 (2004): 1352–1355.
198. Fields, Gregg B., Janelle L. Lauer, Yoav Dori, Pilar Forns, Ying-Ching Yu, and Matthew Tirrell. "Proteinlike molecular architecture: Biomaterial applications for inducing cellular receptor binding and signal transduction." *Peptide Science* 47, no. 2 (1998): 143–151.
199. Whitesides, George M., and Bartosz Grzybowski. "Self-assembly at all scales." *Science* 295, no. 5564 (2002): 2418–2421.
200. Roemer, Dietmar, Heinz H. Buescher, Ronald C. Hill, Janos Pless, Wilfried Bauer, Francis Cardinaux, Annemarie Closse, Daniel Hauser, and Rene Huguenin. "A synthetic enkephalin analogue with prolonged parenteral and oral analgesic activity." *Nature* 268, no. 5620 (1977): 547–549.
201. Schnur, J. M., R. Price, P. Schoen, Paul Yager, J. M. Calvert, J. Georger, and A. Singh. "Lipid-based tubule microstructures." *Thin Solid Films* 152, nos. 1–2 (1987): 181–206.
202. Yu, Ying-Ching, Matthew Tirrell, and Gregg B. Fields. "Minimal lipidation stabilizes protein-like molecular architecture." *Journal of the American Chemical Society* 120, no. 39 (1998): 9979–9987.
203. Yu, Ying-Ching, Vikram Roontga, Vladimir A. Daragan, Kevin H. Mayo, Matthew Tirrell, and Gregg B. Fields. "Structure and dynamics of peptide—amphiphiles incorporating triple-helical proteinlike molecular architecture." *Biochemistry* 38, no. 5 (1999): 1659–1668.
204. Beniash, Elia, Jeffery D. Hartgerink, Hannah Storrie, John C. Stendahl, and Samuel I. Stupp. "Self-assembling peptide amphiphile nanofiber matrices for cell entrapment." *Acta Biomaterialia* 1, no. 4 (2005): 387–397.
205. Niece, Krista L., Jeffrey D. Hartgerink, Jack J. J. M. Donners, and Samuel I. Stupp. "Self-assembly combining two bioactive peptide-amphiphile molecules into nanofibers by electrostatic attraction." *Journal of the American Chemical Society* 125, no. 24 (2003): 7146–7147.
206. Hartgerink, Jeffrey D., Elia Beniash, and Samuel I. Stupp. "Self-assembly and mineralization of peptide-amphiphile nanofibers." *Science* 294, no. 5547 (2001): 1684–1688.
207. Holmes, Todd C., Sonsoles de Lacalle, Xing Su, Guosong Liu, Alexander Rich, and Shuguang Zhang. "Extensive neurite outgrowth and active synapse formation on self-assembling peptide scaffolds." *Proceedings of the National Academy of Sciences* 97, no. 12 (2000): 6728–6733.
208. Liu, Daojun, Steven De Feyter, Mircea Cotlet, Uwe-Martin Wiesler, Tanja Weil, Andreas Herrmann, Klaus Müllen, and Frans C. De Schryver. "Fluorescent self-assembled polyphenylene dendrimer nanofibers." *Macromolecules* 36, no. 22 (2003): 8489–8498.
209. Liu, Daojun, Hua Zhang, P. C. M. Grim, Steven De Feyter, U. -M. Wiesler, A. J. Berresheim, Klaus Müllen, and F. C. De Schryver. "Self-assembly of polyphenylene dendrimers into micrometer long nanofibers: An atomic force microscopy study." *Langmuir* 18, no. 6 (2002): 2385–2391.
210. Perutz, Max F. "Glutamine repeats and neurodegenerative diseases: Molecular aspects." *Trends in Biochemical Sciences* 24, no. 2 (1999): 58–63.
211. Perutz, Max F., and A. H. Windle. "Cause of neural death in neurodegenerative diseases attributable to expansion of glutamine repeats." *Nature* 412, no. 6843 (2001): 143–144.
212. Ma, Peter X., and Ruiyun Zhang. "Synthetic nano-scale fibrous extracellular matrix." *Journal of Biomedical Materials Research* 46, no. 1 (1999): 60–72.
213. Liu, Xiaohua, and Peter X. Ma. "Phase separation, pore structure, and properties of nanofibrous gelatin scaffolds." *Biomaterials* 30, no. 25 (2009): 4094–4103.

214. Van de Witte, P., P. J. Dijkstra, J. W. A. Van den Berg, and J. Feijen. "Phase separation processes in polymer solutions in relation to membrane formation." *Journal of Membrane Science* 117, nos. 1–2 (1996): 1–31.
215. Nam, Yoon Sung, and Tae Gwan Park. "Porous biodegradable polymeric scaffolds prepared by thermally induced phase separation." *Journal of Biomedical Materials Research* 47, no. 1 (1999): 8–17.
216. Wei, Guobao, and Peter X. Ma. "Structure and properties of nano-hydroxyapatite/polymer composite scaffolds for bone tissue engineering." *Biomaterials* 25, no. 19 (2004): 4749–4757.
217. Yang, F., R. Murugan, S. Ramakrishna, X. Wang, Y.-X. Ma, and S. Wang. "Fabrication of nano-structured porous PLLA scaffold intended for nerve tissue engineering." *Biomaterials* 25, no. 10 (2004): 1891–1900.
218. Zhang, Ruiyun, and Peter X. Ma. "Synthetic nano-fibrillar extracellular matrices with predesigned macroporous architectures." *Journal of Biomedical Materials Research* 52, no. 2 (2000): 430–438.
219. Ma, Peter X., and Ji-Won Choi. "Biodegradable polymer scaffolds with well-defined interconnected spherical pore network." *Tissue Engineering* 7, no. 1 (2001): 23–33.
220. Wei, Guobao, and Peter X. Ma. "Macroporous and nanofibrous polymer scaffolds and polymer/bone-like apatite composite scaffolds generated by sugar spheres." *Journal of Biomedical Materials Research Part A* 78, no. 2 (2006): 306–315.
221. Lin, Cheng Yu, Noboru Kikuchi, and Scott J. Hollister. "A novel method for biomaterial scaffold internal architecture design to match bone elastic properties with desired porosity." *Journal of Biomechanics* 37, no. 5 (2004): 623–636.
222. Hutmacher, Dietmar W., Michael Sittinger, and Makarand V. Risbud. "Scaffold-based tissue engineering: Rationale for computer-aided design and solid free-form fabrication systems." *TRENDS in Biotechnology* 22, no. 7 (2004): 354–362.
223. Chen, Victor J., Laura A. Smith, and Peter X. Ma. "Bone regeneration on computer-designed nano-fibrous scaffolds." *Biomaterials* 27, no. 21 (2006): 3973–3979.
224. Engel, Elisabeth, Alexandra Michiardi, Melba Navarro, Damien Lacroix, and Josep A. Planell. "Nano-technology in regenerative medicine: The materials side." *Trends in Biotechnology* 26, no. 1 (2008): 39–47.
225. Hasirci, V., E. Vrana, P. Zorlutuna, A. Ndreu, P. Yilgor, F. B. Basmanav, and E. Aydin. "Nanobiomaterials: A review of the existing science and technology, and new approaches." *Journal of Biomaterials Science, Polymer Edition* 17, no. 11 (2006): 1241–1268.
226. Wilkinson, C. D. W., M. Riehle, M. Wood, J. Gallagher, and A. S. G. Curtis. "The use of materials patterned on a nano-and micro-metric scale in cellular engineering." *Materials Science and Engineering: C* 19, no. 1 (2002): 263–269.
227. Teixeira, Ana I., George A. McKie, John D. Foley, Paul J. Bertics, Paul F. Nealey, and Christopher J. Murphy. "The effect of environmental factors on the response of human corneal epithelial cells to nanoscale substrate topography." *Biomaterials* 27, no. 21 (2006): 3945–3954.
228. Gomez, Natalia, Yi Lu, Shaochen Chen, and Christine E. Schmidt. "Immobilized nerve growth factor and microtopography have distinct effects on polarization versus axon elongation in hippocampal cells in culture." *Biomaterials* 28, no. 2 (2007): 271–284.
229. Massia, Stephen P., and John Stark. "Immobilized RGD peptides on surface grafted dextran promote biospecific cell attachment." *Journal of Biomedical Materials Research Part A* 56, no. 3 (2001): 390–399.
230. VandeVondele, Stephanie, Janos Vörös, and Jeffrey A. Hubbell. "RGD-grafted poly-L-lysine-graft-(polyethylene glycol) copolymers block non-specific protein adsorption while promoting cell adhesion." *Biotechnology and Bioengineering* 82, no. 7 (2003): 784–790.
231. Kay, Sarina, Anil Thapa, Karen M. Haberstroh, and Thomas J. Webster. "Nanostructured polymer/nanophase ceramic composites enhance osteoblast and chondrocyte adhesion." *Tissue Engineering* 8, no. 5 (2002): 753–761.
232. Miller, Derick C., Anil Thapa, Karen M. Haberstroh, and Thomas J. Webster. "Endothelial and vascular smooth muscle cell function on poly (lactic-co-glycolic acid) with nano-structured surface features." *Biomaterials* 25, no. 1 (2004): 53–61.
233. Li, Panjian. "Biomimetic nano-apatite coating capable of promoting bone in growth." *Journal of Biomedical Materials Research Part A* 66, no. 1 (2003): 79–85.
234. Bordbar Khiabani A., Asemani Shahgoli Gh. "Application of nanotechnology in metallic stents for the treatment of cardiovascular diseases." *The First Congress on Sustainable Development in Nanomaterials, Nanostructures and Nanotechnology*, 2016, Tehran, Iran.

235. Puranik, Amey S., Eileen R. Dawson, and Nicholas A. Peppas. "Recent advances in drug eluting stents." *International Journal of Pharmaceutics* 441, no. 1 (2013): 665–679.
236. Nakano, Kaku, Kensuke Egashira, Seigo Masuda, Kouta Funakoshi, Gang Zhao, Satoshi Kimura, Tetsuya Matoba et al. "Formulation of nanoparticle-eluting stents by a cationic electrodeposition coating technology: Efficient nano-drug delivery via bioabsorbable polymeric nanoparticle-eluting stents in porcine coronary arteries." *JACC: Cardiovascular Interventions* 2, no. 4 (2009): 277–283.
237. Joo, Jae-ryang, Hye Yeong Nam, So Hee Nam, Insu Baek, and Jong-Sang Park. "A novel deposition method of PLGA nanoparticles on coronary stents." *Bulletin of the Korean Chemical Society* 30 (2009): 1085–1087.
238. Arsiwala, Ammar, Preshita Desai, and Vandana Patravale. "Recent advances in micro/nanoscale biomedical implants." *Journal of Controlled Release* 189 (2014): 25–45.
239. LeGeros, Racquel Z. "Biological and synthetic apatites." *Hydroxyapatite and Related Materials* 3 (1994): 3–28. CRC Press, Boca Raton.
240. Park, Joon. *Biomaterials Science and Engineering*. Springer Science & Business Media, US, 2012.
241. Baghbani, F., F. Moztarzadeh, A. Gafari Nazari, A.H. Razavi Kamran, F. Tondnevis, N. Nezafati, M. Gholipourmalekabadi, and M. Mozafari. "Biological response of biphasic hydroxyapatite/tricalcium phosphate scaffolds intended for low load-bearing orthopaedic applications." *Advanced Composites Letters* 21, no. 1 (2012): 16–24.
242. Poursamar, Seyed Ali, Mohammad Rabiee, Ali Samadikuchaksaraei, M. Tahriri, Meysam Karimi, and Mahmoud Azami. "Influence of the value of the pH on the preparation of nano hydroxyapatite polyvinyl alcohol composites." *Journal of Ceramic Processing Research* 10, no. 5 (2009): 679–682.
243. Hamlehkhan, Azhang, Masoud Mozafari, Nader Nezafati, Mahmoud Azami, and Ali Samadikuchaksaraei. "Novel bioactive poly (ε-caprolactone)-gelatin-hydroxyapatite nanocomposite scaffolds for bone regeneration." In *Key Engineering Materials*, vol. 493, pp. 909–915. Trans Tech Publications, 2012.
244. Azami, Mahmoud, Shima Tavakol, Ali Samadikuchaksaraei, Mehran Solati Hashjin, Nafiseh Baheiraei, Mehdi Kamali, and Mohammad Reza Nourani. "A porous hydroxyapatite/gelatin nanocomposite scaffold for bone tissue repair: In vitro and in vivo evaluation." *Journal of Biomaterials Science, Polymer Edition* 23, no. 18 (2012): 2353–2368.
245. Diba, M., M. Kharaziha, M. H. Fathi, M. Gholipourmalekabadi, and A. Samadikuchaksaraei. "Preparation and characterization of polycaprolactone/forsterite nanocomposite porous scaffolds designed for bone tissue regeneration." *Composites Science and Technology* 72, no. 6 (2012): 716–723.
246. Johari, N., M. H. Fathi, M. A. Golozar, E. Erfani, and A. Samadikuchaksaraei. "Poly (ε-caprolactone)/nano fluoridated hydroxyapatite scaffolds for bone tissue engineering: In vitro degradation and biocompatibility study." *Journal of Materials Science: Materials in Medicine* 23, no. 3 (2012): 763–770.
247. Sikareepaisan, Panprung, Apichart Suksamrarn, and Pitt Supaphol. "Electrospun gelatin fiber mats containing a herbal—*Centella asiatica*—Extract and release characteristic of asiaticoside." *Nanotechnology* 19, no. 1 (2007): 015102.
248. Xia, Wanyao, Wei Liu, Lei Cui, Yuanchun Liu, Wei Zhong, Deli Liu, Juanjuan Wu, Kienhui Chua, and Yilin Cao. "Tissue engineering of cartilage with the use of chitosan-gelatin complex scaffolds." *Journal of Biomedical Materials Research Part B: Applied Biomaterials* 71, no. 2 (2004): 373–380.
249. Khor, Eugene, and Lee Yong Lim. "Implantable applications of chitin and chitosan." *Biomaterials* 24, no. 13 (2003): 2339–2349.
250. Agrawal, C., and Robert B. Ray. "Biodegradable polymeric scaffolds for musculoskeletal tissue engineering." *Journal of Biomedical Materials Research* 55, no. 2 (2001): 141–150.
251. Doiphode, Nikhil D., Tieshu Huang, Ming C. Leu, Mohamed N. Rahaman, and Delbert E. Day. "Freeze extrusion fabrication of 13–93 bioactive glass scaffolds for bone repair." *Journal of Materials Science: Materials in Medicine* 22, no. 3 (2011): 515–523.
252. Xynos, Ioannis D., Alasdair J. Edgar, Lee D.K. Buttery, Larry L. Hench, and Julia M. Polak. "Gene-expression profiling of human osteoblasts following treatment with the ionic products of Bioglass® 45S5 dissolution." *Journal of Biomedical Materials Research Part A* 55, no. 2 (2001): 151–157.
253. Jones, Julian R., Eileen Gentleman, and Julia Polak. "Bioactive glass scaffolds for bone regeneration." *Elements* 3, no. 6 (2007): 393–399.
254. Pattison, Megan A., Susan Wurster, Thomas J. Webster, and Karen M. Haberstroh. "Three-dimensional, nano-structured PLGA scaffolds for bladder tissue replacement applications." *Biomaterials* 26, no. 15 (2005): 2491–2500.
255. Hartgerink, Jeffrey D., Elia Beniash, and Samuel I. Stupp. "Peptide-amphiphile nanofibers: A versatile scaffold for the preparation of self-assembling materials." *Proceedings of the National Academy of Sciences* 99, no. 8 (2002): 5133–5138.

256. Evans, Gregory R. D., Keith Brandt, Steven Katz, Priscilla Chauvin, Lisa Otto, Melissa Bogle, Bao Wang, Rudolph K. Meszlenyi, Lichun Lu, Antonios G. Mikos, and Charles W. Patrick Jr. "Bioactive poly (L-lactic acid) conduits seeded with Schwann cells for peripheral nerve regeneration." *Biomaterials* 23, no. 3 (2002): 841–848.
257. Saki, Mohammad, M. Kazemzadeh Narbat, Ali Samadikuchaksaraei, Hamed Basir Ghafouri, and Fazel Gorjipour. "Biocompatibility study of a hydroxyapatite-alumina and silicon carbide composite scaffold for bone tissue engineering." *Yakhteh* 11, no. 1 (2009): 55–60.
258. El-Ghannam, Ahmed. "Bone reconstruction: From bioceramics to tissue engineering." *Expert Review of Medical Devices* 2, no. 1 (2005): 87–101.
259. Jones, J. R., and L. L. Hench. "Biomedical materials for new millennium: Perspective on the future." *Materials Science and Technology* 17, no. 8 (2001): 891–900.
260. Pilliar, R. M., M. J. Filiaggi, J. D. Wells, M. D. Grynpas, and R. A. Kandel. "Porous calcium polyphosphate scaffolds for bone substitute applications—In vitro characterization." *Biomaterials* 22, no. 9 (2001): 963–972.
261. Zhang, Yong, and Miqin Zhang. "Three-dimensional macroporous calcium phosphate bioceramics with nested chitosan sponges for load-bearing bone implants." *Journal of Biomedical Materials Research* 61, no. 1 (2002): 1–8.
262. Bhattarai, Narayan, Dennis Edmondson, Omid Veiseh, Frederick A. Matsen, and Miqin Zhang. "Electrospun chitosan-based nanofibers and their cellular compatibility." *Biomaterials* 26, no. 31 (2005): 6176–6184.
263. Aminian, Alieh, Mehran Solati-Hashjin, Ali Samadikuchaksaraei, Farhad Bakhshi, Fazel Gorjipour, Arghavan Farzadi, Fattolah Moztarzadeh, and Martin Schmücker. "Synthesis of silicon-substituted hydroxyapatite by a hydrothermal method with two different phosphorous sources." *Ceramics International* 37, no. 4 (2011): 1219–1229.
264. Greiner, Andreas, and Joachim H. Wendorff. "Electrospinning: A fascinating method for the preparation of ultrathin fibers." *Angewandte Chemie International Edition* 46, no. 30 (2007): 5670–5703.
265. Okii, Norifumi, Shigeru Nishimura, Kaoru Kurisu, Yukio Takeshima, and Tohru Uozumi. "In vivo histological changes occurring in hydroxyapatite cranial reconstruction. Case report." *Neurologia Medico-chirurgica* 41, no. 2 (2001): 100–104.
266. Freyman, T. M., I. V. Yannas, and L. J. Gibson. "Cellular materials as porous scaffolds for tissue engineering." *Progress in Materials Science* 46, no. 3 (2001): 273–282.
267. Wang, Min. "Developing bioactive composite materials for tissue replacement." *Biomaterials* 24, no. 13 (2003): 2133–2151.
268. Ducheyne, Pq, and Q. Qiu. "Bioactive ceramics: The effect of surface reactivity on bone formation and bone cell function." *Biomaterials* 20, no. 23 (1999): 2287–2303.
269. Suchanek, Wojciech, and Masahiro Yoshimura. "Processing and properties of hydroxyapatite-based biomaterials for use as hard tissue replacement implants." *Journal of Materials Research* 13, no. 01 (1998): 94–117.
270. Wozney, John M., and Vicki Rosen. "Bone morphogenetic protein and bone morphogenetic protein gene family in bone formation and repair." *Clinical Orthopaedics and Related Research* 346 (1998): 26–37.
271. Schmidmaier, G., B. Wildemann, P. Schwabe, R. Stange, J. Hoffmann, N. P. Südkamp, N. P. Haas, and M. Raschke. "A new electrochemically graded hydroxyapatite coating for osteosynthetic implants promotes implant osteointegration in a rat model." *Journal of Biomedical Materials Research Part A* 63, no. 2 (2002): 168–172.
272. Lima, Ingrid Russoni de, Gutemberg Gomes Alves, Gustavo Vicentis de Oliveira Fernandes, Eliane Pedra Dias, Glória de Almeida Soares, and José Mauro Granjeiro. "Evaluation of the in vivo biocompatibility of hydroxyapatite granules incorporated with zinc ions." *Materials Research* 13, no. 4 (2010): 563–568.
273. Burg, Karen J. L., Scott Porter, and James F. Kellam. "Biomaterial developments for bone tissue engineering." *Biomaterials* 21, no. 23 (2000): 2347–2359.
274. Reddi, A. H. "Morphogenesis and tissue engineering of bone and cartilage: Inductive signals, stem cells, and biomimetic biomaterials." *Tissue Engineering* 6, no. 4 (2000): 351–359.
275. Kneser, U., D. J. Schaefer, B. Munder, C. Klemt, C. Andree, and G. B. Stark. "Tissue engineering of bone." *Minimally Invasive Therapy & Allied Technologies* 11, no. 3 (2002): 107–116.
276. Boyan, B. D., C. H. Lohmann, J. Romero, and Z. Schwartz. "Bone and cartilage tissue engineering." *Clinics In Plastic Surgery* 4, no. 26 (1999): 629.
277. Asahina, Izumi, Masatoshi Watanabe, Norio Sakurai, Masaji Mori, and Shoji Enomoto. "Repair of bone defect in primate mandible using a bone morphogenetic protein (BMP)-hydroxyapatite–collagen composite." *Journal of Medical and Dental Sciences* 44 (1997): 63–70.

278. Raynaud, S., E. Champion, D. Bernache-Assollant, and P. Thomas. "Calcium phosphate apatites with variable Ca/P atomic ratio I. Synthesis, characterisation and thermal stability of powders." *Biomaterials* 23, no. 4 (2002): 1065–1072.
279. Hamlekhan, A., M. Mozafari, N. Nezafati, M. Azami, and H. Hadipour. "A proposed fabrication method of novel PCL-GEL-HAp nanocomposite scaffolds for bone tissue engineering applications." *Advanced Composites Letters* 19, no. 4 (2010): 123–130.
280. Hench, Larry L. "The story of Bioglass®." *Journal of Materials Science: Materials in Medicine* 17, no. 11 (2006): 967–978.
281. Pon-On, Weeraphat, Siwaporn Meejoo, and I-Ming Tang. "Formation of hydroxyapatite crystallites using organic template of polyvinyl alcohol (PVA) and sodium dodecyl sulfate (SDS)." *Materials Chemistry and Physics* 112, no. 2 (2008): 453–460.
282. Degirmenbasi, Nebahat, Dilhan M. Kalyon, and Elvan Birinci. "Biocomposites of nanohydroxyapatite with collagen and poly (vinyl alcohol)." *Colloids and Surfaces B: Biointerfaces* 48, no. 1 (2006): 42–49.
283. Mansur, Herman S., Rodrigo L. Oréfice, and Alexandra AP Mansur. "Characterization of poly (vinyl alcohol)/poly (ethylene glycol) hydrogels and PVA-derived hybrids by small-angle X-ray scattering and FTIR spectroscopy." *Polymer* 45, no. 21 (2004): 7193–7202.
284. Pietak, Alexis M., Joel W. Reid, Malcom J. Stott, and Michael Sayer. "Silicon substitution in the calcium phosphate bioceramics." *Biomaterials* 28, no. 28 (2007): 4023–4032.
285. Botelho, C. M., M. A. Lopes, I. R. Gibson, S. M. Best, and J. D. Santos. "Structural analysis of Si-substituted hydroxyapatite: Zeta potential and X-ray photoelectron spectroscopy." *Journal of Materials Science: Materials in Medicine* 13, no. 12 (2002): 1123–1127.
286. Tang, Xiao Lian, Xiu Feng Xiao, and Rong Fang Liu. "Structural characterization of silicon-substituted hydroxyapatite synthesized by a hydrothermal method." *Materials Letters* 59, no. 29 (2005): 3841–3846.
287. Kumar, Majeti NV Ravi. "A review of chitin and chitosan applications." *Reactive and functional polymers* 46, no. 1 (2000): 1–27.
288. Di Martino, Alberto, Michael Sittinger, and Makarand V. Risbud. "Chitosan: A versatile biopolymer for orthopaedic tissue-engineering." *Biomaterials* 26, no. 30 (2005): 5983–5990.
289. Mao, Jin Shu, Yuan Lu Cui, Xiang Hui Wang, Yi Sun, Yu Ji Yin, Hui Ming Zhao, and Kang De Yao. "A preliminary study on chitosan and gelatin polyelectrolyte complex cytocompatibility by cell cycle and apoptosis analysis." *Biomaterials* 25, no. 18 (2004): 3973–3981.
290. Mei, Fang, Jisheng Zhong, Xiaoping Yang, Xiangying Ouyang, Shen Zhang, Xiaoyang Hu, Qi Ma, Jigui Lu, Seungkon Ryu, and Xuliang Deng. "Improved biological characteristics of poly (L-lactic acid) electrospun membrane by incorporation of multiwalled carbon nanotubes/hydroxyapatite nanoparticles." *Biomacromolecules* 8.12 (2007): 3729–3735.
291. Agrawal, Sarvesh K., Naomi Sanabria-DeLong, Gregory N. Tew, and Surita R. Bhatia. "Nanoparticle-reinforced associative network hydrogels." *Langmuir* 24, no. 22 (2008): 13148–13154.
292. Teo, W. E., S. Liao, C. Chan, and S. Ramakrishna. "Fabrication and characterization of hierarchically organized nanoparticle-reinforced nanofibrous composite scaffolds." *Acta Biomaterialia* 7, no. 1 (2011): 193–202.
293. Zhang, Qingwei, Vadym N. Mochalin, Ioannis Neitzel, Kavan Hazeli, Junjie Niu, Antonios Kontsos, Jack G. Zhou, Peter I. Lelkes, and Yury Gogotsi. "Mechanical properties and biomineralization of multi-functional nanodiamond-PLLA composites for bone tissue engineering." *Biomaterials* 33, no. 20 (2012): 5067–5075.
294. Martin, Richard A., S. Yue, John V. Hanna, P. D. Lee, Robert J. Newport, Mark E. Smith, and Julian R. Jones. "Characterizing the hierarchical structures of bioactive sol–gel silicate glass and hybrid scaffolds for bone regeneration." *Philosophical Transactions of the Royal Society A* 370, no. 1963 (2012): 1422–1443.
295. Jones, Julian R., and Larry L. Hench. "Factors affecting the structure and properties of bioactive foam scaffolds for tissue engineering." *Journal of Biomedical Materials Research Part B: Applied Biomaterials* 68, no. 1 (2004): 36–44.
296. Jones, J. R., and L. L. Hench. "Effect of surfactant concentration and composition on the structure and properties of sol-gel-derived bioactive glass foam scaffolds for tissue engineering." *Journal of Materials Science* 38, no. 18 (2003): 3783–3790.
297. Fujibayashi, Shunsuke, Masashi Neo, Hyun-Min Kim, Tadashi Kokubo, and Takashi Nakamura. "A comparative study between in vivo bone ingrowth and in vitro apatite formation on Na_2O–CaO–SiO_2 glasses." *Biomaterials* 24, no. 8 (2003): 1349–1356.

298. Oonishi, Hironobu, Shoichi Kushitani, Eiichi Yasukawa, Hiroyoshi Iwaki, Larry L. Hench, June Wilson, Eiji Tsuji, and Tomihito Sugihara. "Particulate bioglass compared with hydroxyapatite as a bone graft substitute." *Clinical Orthopaedics and Related Research* 334 (1997): 316–325.
299. Lee, Jungwoo, Meghan J. Cuddihy, and Nicholas A. Kotov. "Three-dimensional cell culture matrices: State of the art." *Tissue Engineering Part B: Reviews* 14, no. 1 (2008): 61–86.
300. Laura, M. Y., Nic D. Leipzig, and Molly S. Shoichet. "Promoting neuron adhesion and growth." *Materials Today* 11, no. 5 (2008): 36–43.
301. Yang, F., C. Y. Xu, M. Kotaki, S. Wang, and S. Ramakrishna. "Characterization of neural stem cells on electrospun poly (L-lactic acid) nanofibrous scaffold." *Journal of Biomaterials Science, Polymer Edition* 15, no. 12 (2004): 1483–1497.
302. Cheng, Mingyu, Jinguang Deng, Fei Yang, Yandao Gong, Nanming Zhao, and Xiufang Zhang. "Study on physical properties and nerve cell affinity of composite films from chitosan and gelatin solutions." *Biomaterials* 24, no. 17 (2003): 2871–2880.
303. Sanders, Elliot H., Reneé Kloefkorn, Gary L. Bowlin, David G. Simpson, and Gary E. Wnek. "Two-phase electrospinning from a single electrified jet: Microencapsulation of aqueous reservoirs in poly (ethylene-co-vinyl acetate) fibers." *Macromolecules* 36, no. 11 (2003): 3803–3805.
304. Chen, Jyh-Ping, Gwo-Yun Chang, and Jan-Kan Chen. "Electrospun collagen/chitosan nanofibrous membrane as wound dressing." *Colloids and Surfaces A: Physicochemical and Engineering Aspects* 313 (2008): 183–188.
305. Sahoo, S., H. Ouyang, James C.-H. Goh, T. E. Tay, and S. L. Toh. "Characterization of a novel polymeric scaffold for potential application in tendon/ligament tissue engineering." *Tissue Engineering* 12, no. 1 (2006): 91–99.
306. Corey, Joseph M., David Y. Lin, Katherine B. Mycek, Qiaoran Chen, Stanley Samuel, Eva L. Feldman, and David C. Martin. "Aligned electrospun nanofibers specify the direction of dorsal root ganglia neurite growth." *Journal of Biomedical Materials Research Part A* 83, no. 3 (2007): 636–645.
307. Ahmed, Ijaz, Hsing-Yin Liu, Ping C. Mamiya, Abdul S. Ponery, Ashwin N. Babu, Thom Weik, Melvin Schindler, and Sally Meiners. "Three-dimensional nanofibrillar surfaces covalently modified with tenascin-C-derived peptides enhance neuronal growth in vitro." *Journal of Biomedical Materials Research Part A* 76, no. 4 (2006): 851–860.

Section B

Conventional Techniques, Approaches, and Applications in Manufacturing

14 Use of Conventional Manufacturing Techniques for Materials

A Few Highlights

T. S. Srivatsan, K. Manigandan, and T. S. Sudarshan

CONTENTS

14.1 INTRODUCTION

14.1.1 Background

The word *manufacture* first appeared in English way back in 1567. It is derived from the Latin word *manufactus*, which essentially means "made by hand." In 1683, the word *manufacturing* first appeared, while the word *production*, which is often used as an alternative to the word *manufacturing*,

disassemble. Depending on the overall complexity of the product of concern, the assembly cost in manufacturing can be substantial, typically ranging from 10% to 60% of the total cost of the product. Disassembly of the product is an equally important consideration for purpose of activities such as (1) maintenance, (2) servicing, and (3) recycling of the individual components.

Through the years, several methods have been developed, tried, and put forth for purpose of assembly of components. These include the following: (1) use of fasteners; (2) use of adhesives; and (3) joining techniques, such as (a) welding, (b) brazing, and (c) soldering. As is the case for all types of manufacturing-related operations, each of the above assembly operations has its own specific characteristics, time, advantages, limitations, associated costs, and unique design considerations. The individual parts can be assembled either by hand or by the prudent use of variety of automatic equipment and industrial robots. The choice is often governed by the mutually interactive influences of (1) complexity of the product, (2) the number of components to be assembled, (3) the care and protection that is essential to prevent damage to the individual parts, and (4) the cost of labor in comparison with the cost of machinery that is used in automated assembly operations.

14.2 SELECTION OF MANUFACTURING PROCESSES

There is often more than one method that can be chosen to produce a component for a product from a given material. The following categories of manufacturing methods are applicable for most metallic and nonmetallic materials:

1. Casting (Figure 14.1) to essentially include both expensive mold and permanent mold
2. Bulk deforming, also referred to as forming (Figure 14.2) and shaping (Figure 14.3) using the techniques of forging, rolling, extrusion, drawing, sheet-metal forming, molding, and powder metallurgy (PM)
3. Machining (Figure 14.4) to include the techniques of turning, drilling, boring, milling, planning, shaping, broaching, grinding, ultrasonic machining, chemical machining, laser machining, water-jet machining, and high-energy beam machining
4. Joining (Figure 14.5) to include the techniques of welding, brazing, soldering, diffusion bonding, adhesive bonding, and simple mechanical joining
5. Finishing to include the techniques of honing, lapping, polishing, electrochemical polishing, burnishing, deburring, surface treatments, carvings, and even plating
6. Microfabrication and nanofabrication techniques that are capable of producing parts with dimensions at either the micro- or nanolevel; the fabrication of microelectron–mechanical systems and even nanoelectromechanical systems

The selection of a specific manufacturing process or sequence of processes depends essentially on the geometric features of the part being produced to include dimensional tolerance(s), required surface texture, and properties of the material chosen for the workpiece to include its manufacturing properties. This can be particularly challenging when considering the following two situations:

1. The brittle and hard materials chosen cannot be easily shaped and formed without undergoing or experiencing the risk of fracture, unless of course the manufacturing process operation is performed at an elevated temperature, at which temperature the chosen material(s) has adequate ductility that would enable it to be easily cast, machined, and even ground with relative ease due to its ductile nature.
2. Metals that are subject to preshaping treatment at room temperature become less formable during subsequent processing, which, in practice, is necessary to finish the product. This is primarily because the chosen metals have become harder, stronger, and less ductile than they were prior to processing.

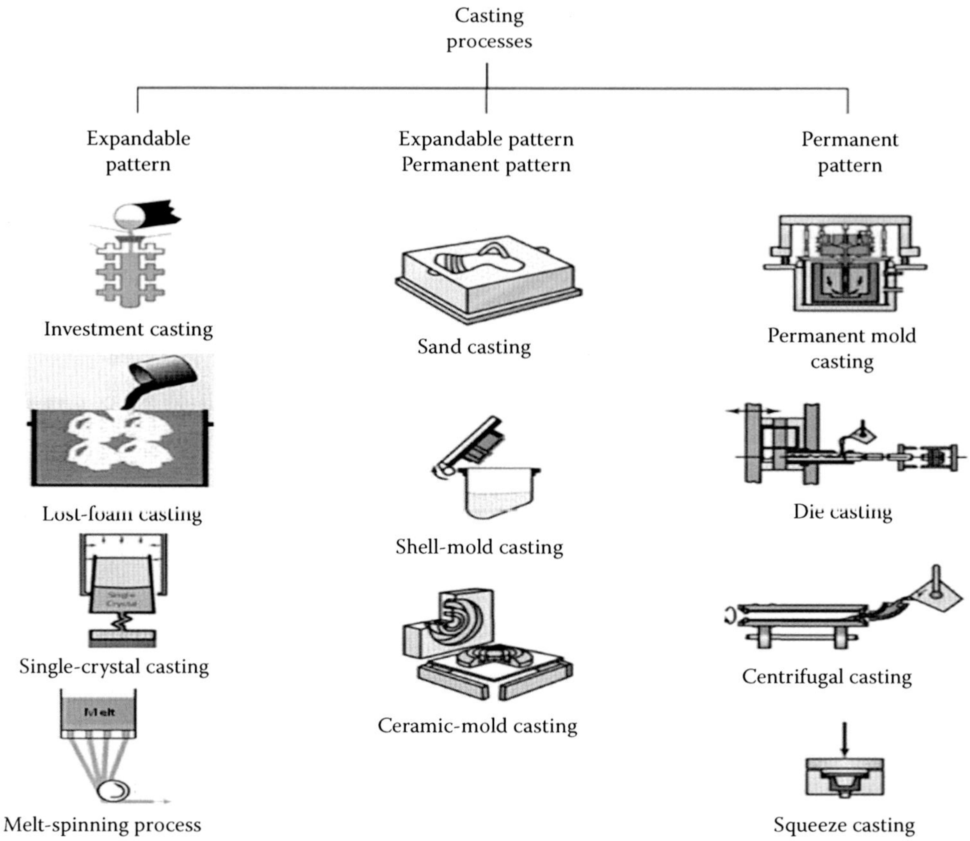

FIGURE 14.1 Schematic illustration of the various casting processes.

With noticeable strides in technology and innovations, there does exist a demand for new and improved approaches to confront the challenges in production while concurrently enabling in a reduction in manufacturing cost. Traditionally, the sheet metal parts have been cut and subsequently fabricated through a prudent use of the common mechanical tools, such as punches and dies. Through the years, these operations have been gradually replaced to the point of noticeable observance by laser cutting. This method eliminates the need for hand tools, which typically have only fixed shapes, and can be both expensive and time-consuming to make.

Several factors do play a role in process selection. This includes the following: (1) size of the part, (2) complexity of shape, (3) the dimensional accuracy required, and (4) surface finish required. For example:

1. Flat parts and those having a thin cross section can be difficult to cast.
2. Complex parts cannot be shaped easily and economically by the available metal-working techniques, such as forging. However, depending on the size of the part and level of complexity, the parts can be precision cast, fabricated, and then assembled from individual pieces. Alternatively, they can even be produced using the PM technique.
3. Dimensional tolerances and surface finish resulting from a hot working operation are often not as fine as those obtained by operations performed at room temperature (25°C), such as cold working. This is primarily because of the dimensional changes coupled with distortion and warping, and oxidation of the surface is favored to occur during the elevated temperature operation.

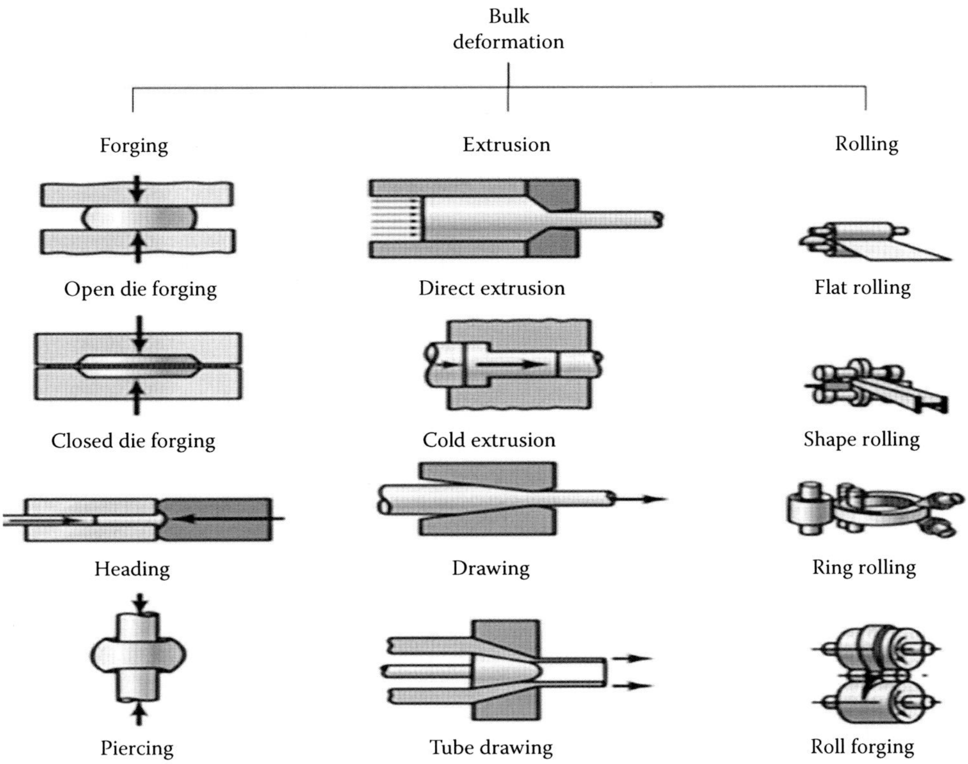

FIGURE 14.2 Schematic illustration of the various bulk-deformation processes.

14.2.1 Net Shape and Near Net-Shape Manufacturing

In synergism, both net shape and near net shape manufacturing can be an important approach by which the desed part is made in one operation and either at or close to the desired dimension and even surface finish. In fact, the essential difference between net-shape and near net-shape is a matter of degree and how close the desired product is to its final dimension and surface finish requirement.

The benefits or advantages of net-shape manufacturing can be best appreciated from the fact that in most cases or products, often more than one additional manufacturing operation is necessary to produce the part. For example, a forged or cast crankshaft will not have the necessary surface finish characteristics, thereby necessitating the need for additional processing by either machining or grinding to produce the required surface finish. These additional operations contribute to a noticeable enhancement in the cost of the product.

14.3 MANUFACTURING PROCESSES

14.3.1 Casting Process

The casting process involves the following steps:

1. Pouring of molten metal into a mold that is patterned after the part that is to be cast
2. Allowing the molten metal to solidify
3. Removing the part from the mold at the end of solidification

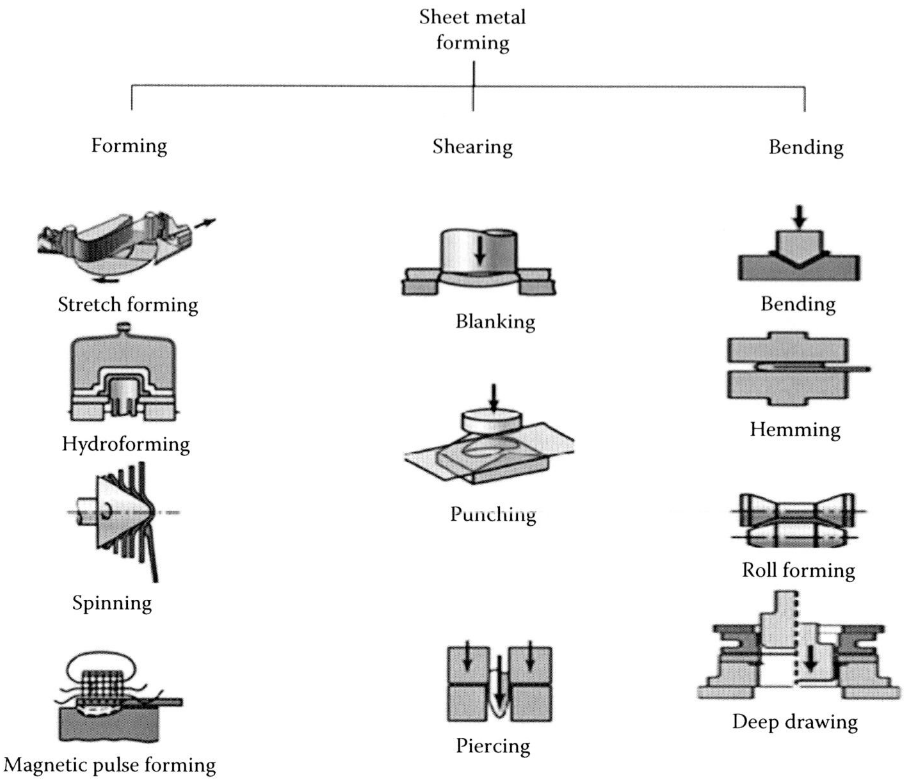

FIGURE 14.3 Schematic illustration of the various sheet-metal forming processes.

In combination with all other manufacturing processes, this technique most certainly necessitates the need for an understanding of the underlying essence to enable in the production of good quality parts at an economical price, and for concurrently establishing the proper techniques for purpose of mold design and casting practice.

Few important considerations that need to be taken in a casting operation are the following (Beeley 1982; Blair et al. 1995; Campbell 2011):

1. Flow of molten metal into the mold cavity coupled with the need for gating systems and pathways to ensure flow of the molten metal to fill the cavity
2. Solidification and resultant cooling of the metal in the mold
3. Influence of mold materials on the casting process

In the last few decades, there have been few to many notable processes and product developments in the domain of metal casting. These have had an observable influence on the structure of foundry manufacturing operations in several aspects and on the nature of the product. In a few cases, the primary emphasis has been on benefits related to cost, while in few to many others, the emphasis has been on an improved quality of the product that is produced coupled with the design capability being achieved and accepted. Environment and health aspects of production have also necessitated the need for modifications to both plant and practice. Progress in the domain of metal casting can be best appreciated considering a short examination of the past growth of the foundry industry and its products, highlighting the cautious steps forward in a noticeable way to the modern system of metal foundry (ASM Handbook 2010; Campbell 2003).

Foundry is of course one of the oldest of the world's industries, and men have for long been produced parts of the desired shape and size by pouring molten metal into molds. After the molten

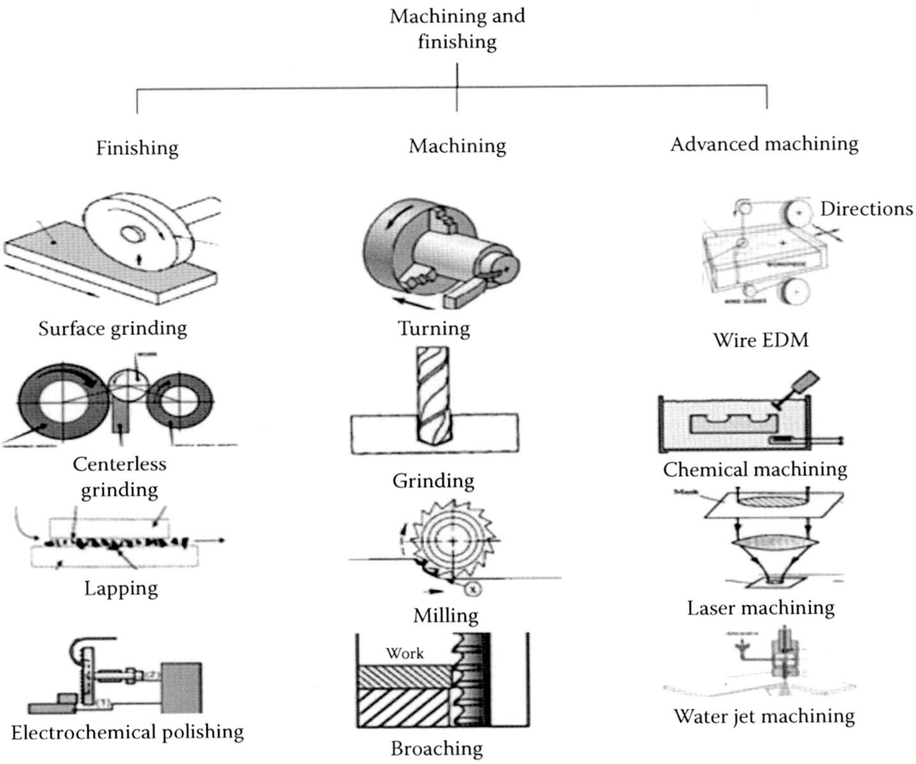

FIGURE 14.4 Schematic illustration of the various machining and finishing processes.

metal is poured into a mold, a sequence of events takes place during the process of solidification and cooling of the metal to ambient temperature (25°C). These events exert an influence on the size, shape, uniformity, and chemical composition of the grains that are formed in the casting, which in turn has an influence on overall properties of the cast product. The significant factor affecting these events are the following: (1) type of metal that is being cast, (2) the thermal properties of both the metal and the mold, (3) shape of the mold, and (4) geometric relationship between volume of casting and surface area of the casting (Campbell 2003). Initially, the copper base alloys were chosen for purposes of casting due to their lower melting temperatures. Subsequently, cast iron was chosen for purposes of casting in molds. Industrial development of the casting processes, both on a sizeable scale and magnitude, occurred following the successful melting of iron with coke around the year 1750. This led to the availability of a method for the bulk product of cast iron, and with the invention of steam power coupled with other developments, formed the basis for the onset of the industrial revolution (ASM Handbook 2010; Campbell 2003; Casting Defects 2002).

Emergence of the coke blast furnace paved the way to produce steel on a large scale following the invention of the Bessemer converter and the "open" hearth furnace in the period immediately after 1850 (ASM Handbook 2010). The first steel castings were made shortly before this period using crucible steel that was made in small quantities by melting wrought iron. This required a temperature of 1600°C and molding materials capable of withstanding the demanding conditions that were required at such high temperatures. In the early days, railway crossings were produced and steel castings eventually became available for a wide range of subsequent applications requiring materials that offered a combination of high strength and ductility. The early steel castings suffered from porosity, shrinkage, and several other defects. However, with the development, emergence and use of deoxidants hailed the beginning of a move toward emphasizing on soundness and integrity

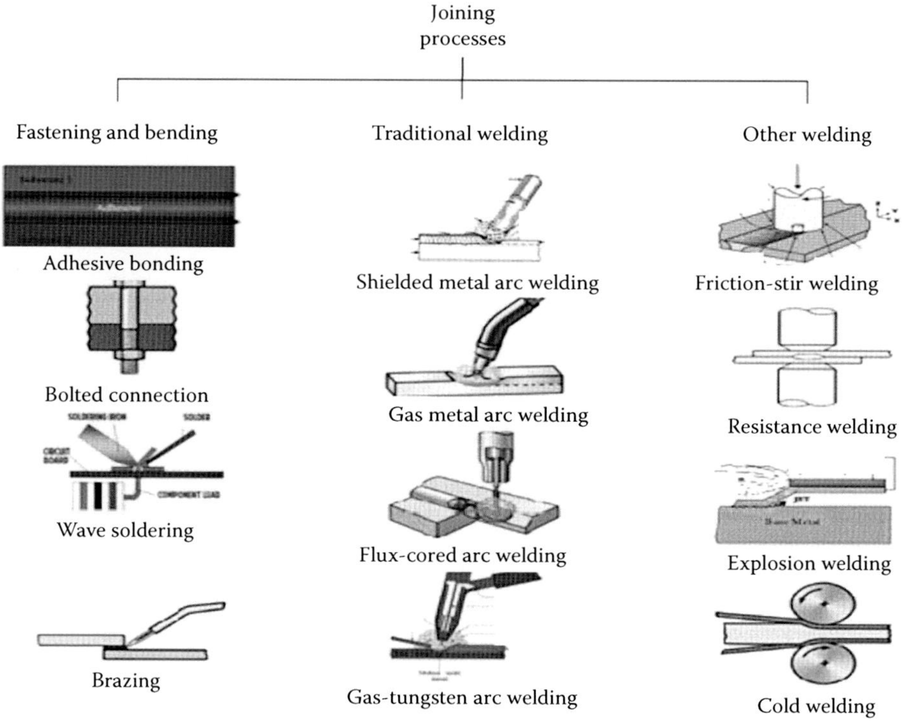

FIGURE 14.5 Schematic illustration of the various joining processes.

bringing in metallurgical control, which is both an essential and required feature for the acceptance of cast products (ASM Handbook 2010; Casting Defects 2002; Stefanescu 2008).

Subsequent developments, though progressive in nature, in the casting industry resulted in a spectrum of castings of the light alloys, die castings of zinc, heat-treatable cast alloy steels, and various cast super-alloys and appeared well into the present century. Cast iron too underwent much further development with the introduction of spheroidal graphite iron or nodular iron as the most important single step (Beeley 1982).

14.3.1.1 Presence of Shrinkage, Defects, and Porosity in Castings

14.3.1.1.1 Shrinkage

Due to their thermal expansion characteristics, metals and even their alloy counterparts usually experience shrinkage or contraction during solidification and while cooling to room temperature (25°C). Shrinkage, which causes dimensional changes, can at times result in warping and cracking, and is the result of the following three events:

1. Contraction of the molten metal as it cools prior to its solidification
2. Contraction of the molten metal during phase change from liquid to solid
3. Contraction of the solidified metal or casting as its temperature drops to ambient temperature (25°C)

The largest shrinkage occurs during the phase change experienced by the material from liquid to solid. Often, this is either reduced or at best even eliminated by the prudent use of risers or pressure-feeding of the molten metal during its gradual cooling in the mold. The amount of contraction experienced during solidification of various metals is summarized in Table 14.1.

TABLE 14.1
Volumetric Solidification Contraction or Expansion for Various Cast Metals

Contraction (%)		Expansion (%)	
Aluminum	7.1	Bismuth	3.3
Zinc	6.5	Silicon	2.9
Aluminum–4.5% copper	6.3	Gray iron	2.5
Gold	5.5		
White iron	4–5.5		
Copper	4.9		
Brass (70–30)	4.5		
Magnesium	4.2		
9% Cu–10% Al	4.0		
Carbon steels	2.5–4.0		
Aluminum–12% silicon	3.8		
Lead	3.2		

The various defects that can result because of manufacturing by casting depends on the competing yet mutually interactive influences of the factors, such as (1) the quality of raw materials used, (2) intricacies of casting design, and (3) an effective control of processing parameters during casting. While a few of the defects tend to be superficial in nature and affect only the appearance, a few others can exert an adverse influence on structural integrity of the part or component that is being produced. A few of the typical defects that develop in cast products are shown in Figures 14.6 and 14.7.

The international committee of the Foundry Technical Association has developed a nomenclature that consists of seven basic categories of casting defects (ASM Handbook 2010; Casting Defects 2002). These are the following:

a. Metallic projections.
b. Cavities.
c. Discontinuities such as cracks, cold tearing, hot tearing, and cold shuts.
d. Defective surfaces such as laps, scars, surface folds, oxide scale, and adhering layer of sand.
e. Incomplete casting such as (1) misruns, (2) runouts, and (3) insufficient volume of metal poured. An incomplete casting can also result due to the following reasons: (1) when the molten metal is at too low a temperature and (2) from pouring the metal too slowly.

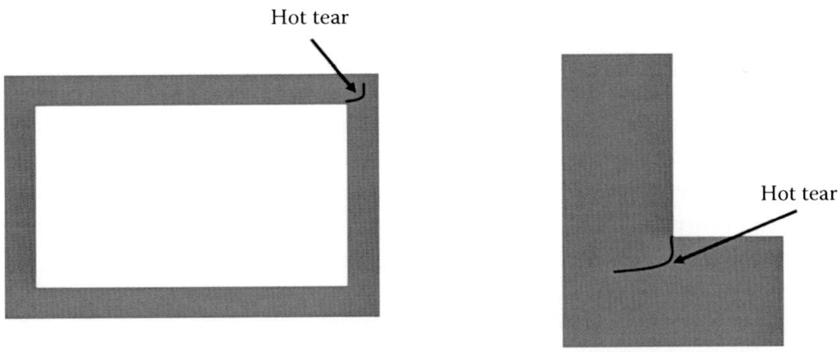

FIGURE 14.6 Examples of hot tear defects in castings. These defects occur because the casting shrank freely during cooling, owing to the presence of constraints in various portions of the molds and cores.

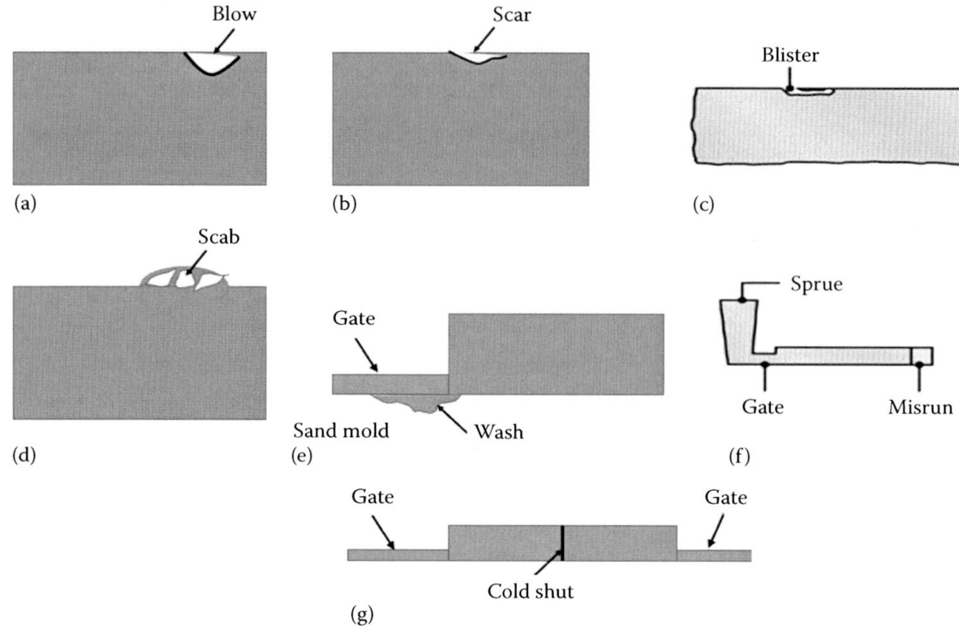

FIGURE 14.7 Examples of common defects in castings: (a) Blow, (b) Scar, (c) Blister, (d) Scab, (e) Wash, (f) Misrun, and (g) Cold shut. These defects can be minimized or eliminated by proper design and preparation of the molds and exercising control over pouring procedure. (Courtesy of J. Datsko).

f. Incorrect dimension or shape that results from improper allowance for shrinkage, irregular contraction, pattern mounting error, a deformed pattern, and a warped casting.
g. Inclusions, generally nonmetallic in nature, which tend to form during melting, solidification, and molding. Generally nonmetallic in nature, their physical presence in the casting can be regarded as being harmful primarily because they act as stress raisers and consequently detrimental to the overall strength of the casting.

14.3.1.1.2 Porosity

Porosity in a casting is often caused by either shrinkage or the presence of both entrapped and dissolved gases, or a combination of both. Regions containing an observable amount of porosity can develop in a casting due to shrinkage experienced by the solidified metal. Often the thin sections in an intricate casting tend to solidify sooner than the thicker sections. This causes the molten metal to flow into the thicker regions that have not yet solidified. At the center, porous regions tend to develop because of contraction as the surfaces of the thicker region begin to solidify first. Microporosity is also favored to occur when the liquid metal solidifies and tends to shrink between the dendrites and between branches of the dendrite. The presence of porosity is often detrimental to strength, ductility, and even surface finish of a casting. The porous regions tend to make the casting permeable. This exerts an influence on overall pressure tightness of the product or cast pressure vessel. The porosity that results because of shrinkage can be either reduced or eliminated by the following methods (ASM Handbook 2010; Campbell 2003; Casting Defects 2002):

a. Providing adequate amount of liquid metal to prevent the formation and presence of cavities due to shrinkage.
b. Use of internal chills or external chills, such as those used in sand castings, is an effective means for reducing shrinkage-related porosity. The primary function of the chill is to increase

the rate of solidification in critical regions of the casting. The internal chills are made from the same material as the casting and are left behind in the casting following solidification.

c. For the case of alloys, the porosity can be either reduced or eliminated by high-temperature gradients, that is, by increasing the cooling rate of the metal. This necessitates the need for a mold material that has high thermal conductivity.
d. Subjecting the casting to *hot isostatic pressing* is an effective method for reducing porosity.

14.3.1.2 Classification of the Casting Processes

The classifications of processes involved in casting are related to the following: (1) mold material, (2) pattern production, (3) molding processes, and (4) method used for feeding the molten metal to the mold. The major categories are discussed next (ASM Handbook 2010; Campbell 2003, 2011).

14.3.1.2.1 Expendable Molds

Expendable molds are typically made of sand, plaster, ceramics, and similar such materials. For purpose of improved properties, the chosen material is mixed with binders or bonding agents. After the casting has solidified, the mold is broken up to remove the casting, hence the name *expendable*. A typical sand mold is 90% sand, 7% clay, and 3% water. The mold is often produced from a pattern. While the mold is expendable, the pattern can be reused to produce several molds. This is referred to the foundry industry as expendable mold, permanent pattern casting.

14.3.1.2.2 Permanent Molds

They are essentially made of metals that maintain their strength at high temperatures. These molds are often used repeatedly and are so designed to facilitate easy removal of the casting while the mold is used for the next casting. Since a metal mold is a better heat conductor than an expendable nonmetallic mold, the solidifying casting experiences a higher rate of cooling, which exerts an influence on its microstructure and the presence of intrinsic microstructural features, such as grain size, of the cast product.

14.3.1.2.3 Composite Molds

These molds are often made up of two or more different materials (such as sand, graphite, and metal), thereby combining the advantages of each material. They have both a permanent and expendable portion and are often used in various casting processes with the primary intent of (1) improving the strength of the mold, (2) controlling the cooling rate, and (3) optimizing the overall economics of the casting operation.

14.3.1.3 Expendable Mold, Permanent Pattern Castings

The major categories of the expendable mold, permanent pattern casting processes are as follows.

14.3.1.3.1 Sand Casting

This can safely be categorized as the traditional method that has been used for casting metals (ASM Handbook 2010; Campbell 2003, 2011). In fact, this is still the most prevalent form of casting. Typical applications of sand castings include the following: (1) large turbine impellers, (2) machine bases, (3) plumbing fixtures, and (4) a wide variety of other products and components. The technique essentially consists of the following steps:

a. Placing the pattern, which has the shape of the part to be cast, in sand to make an imprint
b. Incorporating or introducing a gating system
c. Removing the pattern and filling the mold cavity with molten metal
d. Allowing the metal to gradually cool and, in the process, solidify
e. Breaking away the sand mold
f. Removing the casting

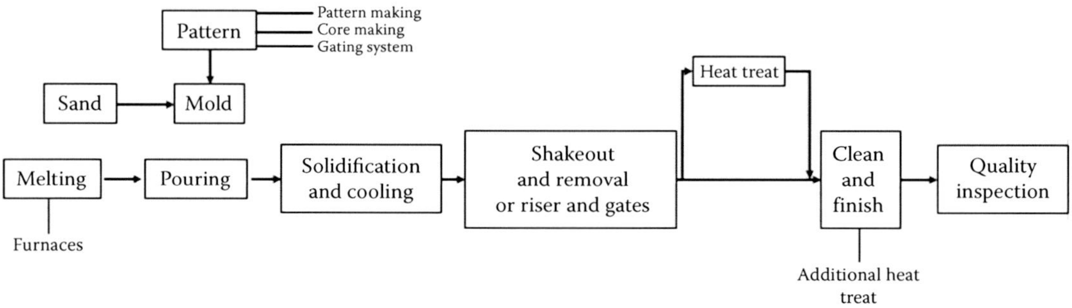

FIGURE 14.8 Outline of the steps involved in the production of a typical sand casting.

The sequence of steps involved in the production of a typical sand casting is shown in Figure 14.8. Most operations that are cast in sand molds use silica sand (SiO_2) as the mold material. Sand is the preferential choice for the mold material primarily because it is inexpensive coupled with its high-temperature characteristics and high melting point. The two general types of sand that are available are (1) naturally bonded (bank sand) and (2) synthetic sand. Since the chemical composition of the synthetic sand can be controlled with accuracy, it is the preferred choice in most foundries (Askeland et al. 2010; ASM Handbook 2010). However, there does exist a few factors that are important in the selection of sand for molds. Fine-grained sands having a round shape contribute to enhancing the strength of the mold. The fine grains also contribute to lowering the permeability of the mold. There exists a dire need for good permeability of molds and cores to allow both the gases and steam that is evolved during the casting process to easily escape. The molds should also have good collapsibility to facilitate the casting to shrink while it is cooling, thereby preventing the occurrence of defects in the cast product.

The sand molds are characterized by the types of sand that comprise them and by the methods that are used to produce them. An illustration of a sand mold and its key features is shown in Figure 14.9. The three basic types of sand molds that have been tried for purpose of casting are (1) green sand, (2) cold box, and (3) no-bake mold.

Green sand molding is by far the least expensive method for making molds. The sand is then recycled for subsequent use. For the case of skin-dried molds, the mold surfaces are dried prior to casting. The drying occurs by either storing the mold in ambient air or drying the mold using torches. Due to their high strength, these molds are generally preferred for the large castings. In the cold-box mold process, a choice of both organic and inorganic binders is blended into the sand to bond the grains chemically with the purpose of achieving greater strength. These molds are observably more dimensionally accurate than the green sand molds. However, they are expensive to produce. In the no-bake mold process, a synthetic liquid resin is mixed with the sand, and the resulting mixture can harden at room temperature (25°C).

The sand molds can be dried in an oven or baked prior to pouring the molten metal. These molds are much stronger than the green sand molds and impart better dimensional accuracy and surface finish to the casting. However, this method suffers from the following drawbacks: (1) distortion of the mold is often greater; (2) the castings are susceptible to hot tearing due to lower collapsibility of the mold; and (3) the production rate is generally lower because of the considerable time that is required for drying.

14.3.1.3.2 Shell Molding

This technique was first developed in the 1940s. Over the years, from 1940 up until now, it has shown significant growth both in technology and applications because it can be used to produce many different types of castings. These castings have close dimensional tolerances, have good surface finish, and are produced at a low cost. Essentially, the applications of shell molding include small

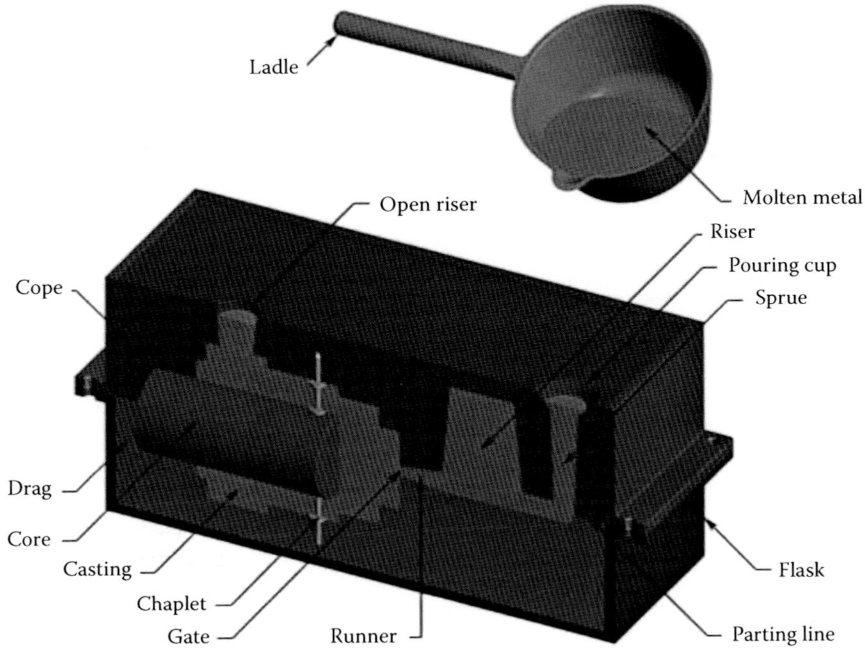

FIGURE 14.9 Schematic illustration of a sand mold showing the various features.

mechanical parts that require high precision, such as (1) cylinder heads, (2) connecting rods, and even (3) gear housings. The technique has also been used to produce high-precision molding cores.

In this process, the mounted pattern, made from either a ferrous or a nonferrous metal, such as aluminum, is initially heated to a range of 175°C to 350°C (350°F to 700°F). Subsequently, the pattern is coated with a parting agent, such as silicone, and then clamped to a box or chamber. The box contains fine sand that is mixed with 2.5–4.0% of a thermosetting resin binder, such as phenol-formaldehyde, which helps to coat the sand particles. Normally, the sand mixture is blown over the pattern, allowing it to form a coating on the pattern. The assembly is then placed in an oven for a short time to facilitate completion of curing of the resin. The oven consists of a metal box, with gas-fired burners that cover the shell mold to cure it. The shell hardens around the pattern and is removed from the pattern with the aid of the built-in ejector pins. Two half shells are made in this manner and then bonded or clamped together to form a mold. The thickness of the shell is determined by controlling the time the pattern is in contact with the mold.

The shell sand has a much smaller grain size and consequently a lower permeability than the sand used in green sand molding. Decomposition of the shell sand does produce a high volume of gas. The molds must be vented properly to ensure that the trapped air and gases escape and thus save the shell molds from having any defects. The high quality of the shell mold casting contributes to reducing the cleaning, machining, and other finishing costs that arise for the case of sand molds. Furthermore, shell molding can be and has been used to produce complex shapes with less labor, allowing the process to be receptive to automation. The shell molding process is shown in Figure 14.10.

14.3.1.3.3 Plaster Mold Casting

This process is also known as precision casting because of the high-dimensional accuracy and good surface finish that is obtained. A few of the parts that can be made using this technique are (1) gears, (2) valves, (3) fittings, (4) tooling, and even (5) ornaments. The castings are typically in the range of 12.5–250 g. In this process, the mold is made of plaster of Paris [gypsum or calcium sulfate ($CaSO_4$)] with the addition of talc and silica powder for improving strength and to concurrently control the time

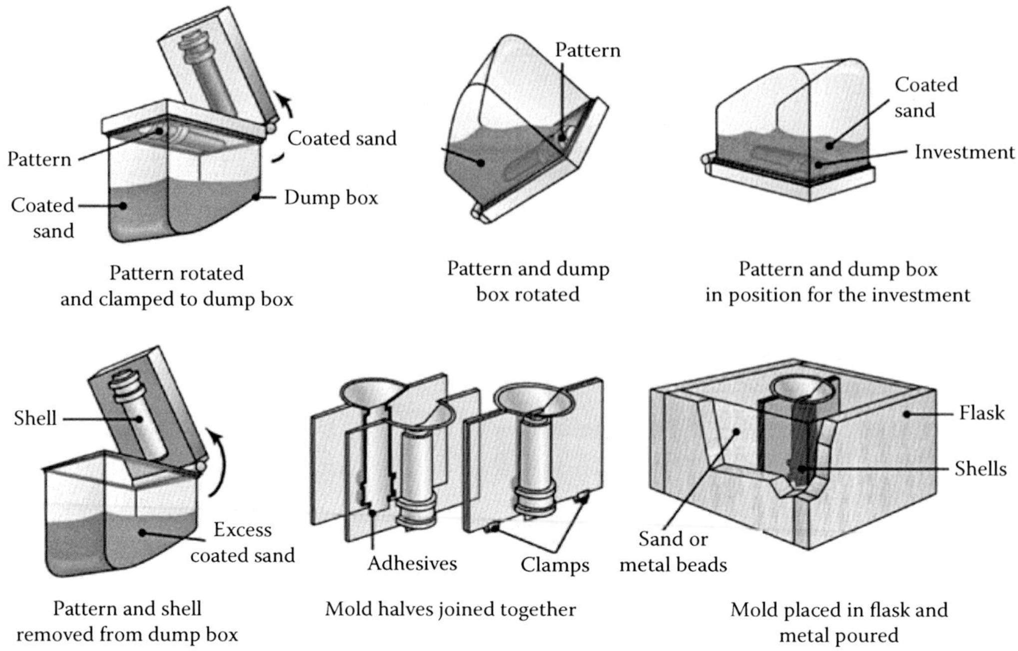

FIGURE 14.10 Shell molding process, also called the dump-box technique.

that is required for the plaster to set. These components are then mixed with water, and the resulting slurry is poured over the pattern. After the plaster has completely set or hardened, it is removed and the mold is dried in the temperature range of 125–250°C. Depending on the plaster chosen, a higher drying temperature can be used. The mold halves are then assembled to form the mold cavity and preheated to a temperature of 125°C. The molten metal is then poured into the mold.

Since the plaster molds have very low permeability, the gases that are evolved during solidification of the metal cannot escape. Consequently, the molten metal is poured either in vacuum or under pressure. The shell mold castings have good surface finish with fine details. Since the plaster molds have low thermal conductivity than other mold materials, the castings cool slowly and the resultant grain structure of the cast product is uniform.

14.3.1.3.4 Ceramic Mold Casting

The ceramic mold casting process is also referred to as the cope-and-drag investment casting process. The technique is quite like the plaster mold process, except that it uses refractory mold materials that can withstand exposure to high temperatures. Typical parts that have been produced using this technique are the following: (1) cutters for machining operations, (2) dies for metal working, (3) molds for making plastic and rubber components, and (4) impellers. Parts weighing as much as 700 kg have been successfully cast by this process. The slurry is a mixture of fine-grain zircon ($ZrSiO_4$), aluminum oxide (Al_2O_3), and fused silica (SiO_2), which are mixed with bonding agents and then poured over the pattern, which was placed in a flask.

The pattern can be made of either wood or metal. After setting, the molds (ceramic facing) are removed, dried, ignited to burn off the volatile matter, and then baked. The molds are then clamped firmly and used as an all ceramic mold. The high-temperature resistance of the refractory molding material chosen and used allows these molds to be used for the casting of (1) ferrous alloys, (2) other high-temperature alloys, (3) stainless steels, and (4) tool steels. An essential drawback of the process is the cost involved. Overall, the castings made using this process have good dimensional accuracy and surface finish over a wide range of sizes and shapes.

14.3.1.4 Expendable Mold, Expendable Pattern Casting Technique

The evaporative pattern and investment casting technique are referred to as expendable pattern casting process or expendable mold, expendable pattern process. They are unique in that a mold and pattern must be produced for each casting. Typical applications are engine blocks, cylinder heads, crankshafts, components of a brake, and bases of heavy machinery.

14.3.1.4.1 The Lost Form Process (Evaporative Pattern Casting)

This evaporative pattern process uses a polystyrene pattern that evaporates upon contact with the molten metal to form a cavity for the casting. This process is also known as the lost-foam casting. Through the years, it has become an important casting process for both the ferrous and nonferrous metals, particularly for those that are chosen for selection and used in the automobile industry.

In this process, polystyrene beads containing 5–8% pentane (a volatile hydrocarbon) are placed in a preheated die that is made of aluminum. Complex patterns are made by bonding various industrial pattern sections using a hot-melt adhesive. The polystyrene tends to expand and take on the shape of the die cavity. Additional heat is then applied to both fuse and bond the beads together. The die is then cooled and opened, and the polystyrene pattern removed. The pattern is then coated with a water-based refractory slurry, dried, and placed in a flask. The flask is then filled with loose fine sand, which surrounds and supports the pattern. Bonding agents are often used to provide additional strength. Without removing the polystyrene pattern, the sand is compacted periodically. The molten metal is then poured into the mold. The hot molten metal vaporizes the pattern and fills the mold cavity completely, and in the process replaces the space that was occupied by the polystyrene. This casting technique offers several advantages over other technically feasible casting methods. A few of these are the following (ASM Handbook 2010; Beeley 2002; Martin 2007):

a. The process is relatively simple because there are no parting lines, cores, or risers.
b. Inexpensive flasks can be used for the process.
c. Polystyrene is inexpensive and can be processed easily into patterns having complex shapes, various sizes, and fine surface details.
d. The casting process requires minimal cleaning and finishing operations.
e. The process can be automated and is economical for long production runs. The cost to produce the die coupled with the need for two sets of tooling are factors that must be considered.

14.3.1.4.2 Investment Casting

With specific reference to metals for use in engineering applications, the investment casting technique provides a unique opportunity to convert liquid metal to a solid shape of the precise form needed and in one integrating process. The term *investment casting* is used to denote the production of industrial metal components using a casting process that utilizes an expendable pattern. The basic technique, known as the lost-wax process, is one of the oldest, if not the oldest, metal-shaping processes known to mankind. Records pertaining to use of this process go back to the period from 4000 B.C. to 3000 B.C. (Green-Spikesley 1979; Investment Casting Handbook 1997; Sias 2006). In fact, use of this process by the early civilizations is recorded through both the Egyptian and Roman eras to the Bronze Age and medieval times in Europe. Use of this process to produce jewelry, status, and other related intricate works of art has been developed through centuries.

The industrial process is referred to as investment casting, which emerged in the 1940s in the United States and subsequently in the United Kingdom. It was essentially a combination of the skills of both the jeweler and the dentist and provided a route for the manufacture of complex components having a precise size and shape. The resultant product had good surface finish and could be made using alloys that are difficult to work using other methods. Over the years, the investment casting technique has seen sustained growth and development and is currently being used for a range of applications in a spectrum of industries. In this process, an expendable pattern, usually a proprietary

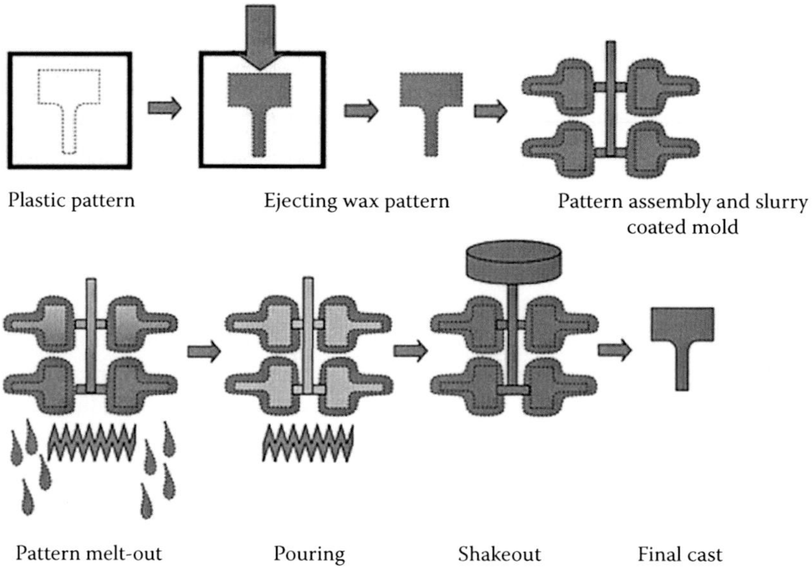

FIGURE 14.11 Schematic illustration of the investment casting or lost-wax process. Castings produced by this method can be made in very fine detail and from a variety of metals.

form of blended waxes, is produced using a permanent mold or pattern. The pattern is assembled onto a wax runner to form an assembly that is referred to as "cluster" or "tree." The completed assembly, which generally comprises several individual patterns in addition to the pouring cup, runners, risers, and in-gates, is filled with a refractory slurry followed by the application of a coating of dry refractory particles, which upon drying gives a thin refractory shell totally enclosing the assembly (Beeley 1982). For the ceramic shell process, the application of both slurry and stucco (refractory particles) is repeated, with drying between each successive coat, until a shell mold having sufficient thickness or the desired thickness is achieved. Following completion of the shell buildup, the expendable pattern material is removed by the use of heat and/or steam. The "dewaxed" mold is then fired at a temperature of the order of 1000°C to develop a strong ceramic bond. It is normal practice to cast the molds immediately following their removal from the firing furnace. The sequence of steps involved in investment casting or lost-wax process is shown in Figure 14.11. The resultant castings can be made in fine detail from a variety of metals.

Upon cooling to room temperature (25°C), the casting is separated from the runners and risers, and finished by a synergism of (1) shot blasting, (2) grinding to remove all metal strips, (3) heat treatment, if required, followed by (4) inspection and quality assurance testing.

14.3.1.4.2.1 Developments in the Process In concurrence with all other manufacturing processes, there is a continual updating of process techniques and the raw materials utilized. Traditionally a labor-intensive process, investment casting has in recent years seen a marked influx of specialized, highly automated production equipment coupled with the application of latest thinking in the domain specific to technological innovations. A few examples specific to the development process are (1) the application of computer graphics to pattern die design and production, (2) fully automatic wax injection machines and pattern dies, and (3) the use of robots and specialized ceramic shell building machines to produce molds (Beeley 1982).

14.3.1.4.2.2 Capabilities of the Process The use of an expendable pattern confers a unique advantage on the investment casting process in that the pattern is removed from the mold without any disturbance to the mold. This obviates the need for having a taper in the pattern, thereby enhancing dimensional accuracy over depth of the mold since it is unnecessary to split the mold once molding has been completed. Undercut sections, requiring additional process techniques such as (1) core assembly in sand casting and (2) using a permanent pattern and movable inserts in permanent mold or die casting, are a regular feature of the investment mold. The design capability of an investment casting will be determined by the foundry characteristics of the metal chosen rather than the molding technique used. This necessitates the need to adhere to basic design criteria to achieve both soundness and quality in the component.

14.3.1.4.2.3 Savings in Both Size and Weight A clear majority of investment castings in regular production will fall with the weight range of a few grams to a maximum of 5 kg. However, castings weighing up to a maximum of 100 kg have also been produced. In terms of linear dimensions, components of 1 m in length have been regularly produced, and much bigger and heavier castings have been produced by the investment process using specialized techniques involving pattern fabrication. Theoretically, there exists no finite limit on maximum size and weight of the components that can be produced by the process. Limitations, if any, are those imposed by the available plant and equipment.

14.3.1.4.2.4 Wall Thickness The investment casting technique allows one to produce thin wall thicknesses spread over a sizeable area. The limits generally used for minimum wall thickness are 1.5 mm for steels and 1.0 mm for the light alloys, subject to acceptability of the design. The production of thin wall thicknesses is assisted by (1) the elevated temperature of the mold during casting and (2) the use of the vacuum casting technique.

14.3.1.4.2.5 Surface Finish A smooth finish of the pattern surface, particularly wax, which is most widely used, coupled with the mold cavity surface being formed by a liquid slurry gives a smooth mold surface against which the molten metal must solidify. The smooth mold cavity surface is imparted to the finished casting, and an excellent as-cast surface finish is a positive attribute of the process along with close dimensional accuracy and freedom of design. Surface finish of the as-cast investment casting ranges between 1.5 and 3.0 μm.

14.3.1.5 Permanent Mold Casting Processes

14.3.1.5.1 Permanent Mold Casting

In permanent mold casting, two halves of a mold are made from materials that offer high resistance to both erosion and fatigue, such as cast iron, steels, bronzes, refractory metal alloys, and graphite (ASM Handbook 2010; Martin 2007). Typical parts made using this technique are (1) gear blanks, (2) cylinder heads, (3) connecting rods, (4) automobile pistons, and (5) kitchenware. Parts that can be made economically generally weigh less than 2500 g (2.5 kg). Special castings weighing a few hundred kilograms have been made using this process. In an industry-related environment, this process is also known as "hard-mold" casting.

Both the mold cavity and the gating system are machined into the mold and in the process become an integral part of it. To produce castings having internal cavities, cores made of metal or sand aggregate are placed in the mold prior to casting. Typical materials that have been chosen for the core are the following: (1) oil-bonded sand, (2) resin-bonded sand, (3) graphite, (4) gray iron, (5) hot-work die steel, and (6) low carbon steel. With an objective of increasing the life of permanent molds, the surfaces of the mold cavity are often coated with a refractory slurry, such as sodium silicate and clay, or sprayed with graphite after every few castings. The coatings serve both as parting agents and as thermal barriers, thereby contributing in a positive manner to controlling the rate of cooling of the casting. This casting technique has been used for the alloys of aluminum, copper, and magnesium, and gray iron due essentially to their lower melting points. The carbon steels and alloy steels can be cast using either graphite or heat-resistant metal molds. The permanent mold castings offer the advantages of good surface finish, close dimensional tolerances, uniform mechanical properties, and a high production rate.

14.3.1.5.2 Vacuum Casting

This technique is also known in industry circles as counter-gravity low-pressure (CL) process. A schematic of the process technique is shown in Figure 14.12.

This casting technique has proven itself to be a viable alternative to investment casting, shell-mold casting, and green sand casting. It is appropriate or suitable for the casting of thin-walled components having a complex shape. In this process, a mixture of fine sand and urethane is molded over metal dies and cured using amine vapor. The mold is then held on the arm of a robot and immersed into molten metal in an induction furnace. The vacuum reduces the air pressure inside the mold to about two-thirds of atmospheric pressure. This aids in drawing the molten metal into cavities in the mold through a gate that is positioned at the bottom of the mold. The process is often automated with production costs quite like that of green sand casting. Through the years, carbon steel, low alloy steel, and stainless-steel parts have been successfully vacuum-cast by this method.

14.3.1.5.3 Slush Casting

The thin-walled hollow castings can be made by permanent mold casting using a process called slush casting. The molten metal is poured into a metal mold. After the desired thickness of the solidified skin is obtained, the mold is inverted and the remaining liquid metal is poured out. The mold halves are then opened and the casting is removed. The slush casting technique is suitable for small production runs, and has been tried and successfully used for ornaments, decorative objects, and a few toys made from the low melting point metals, such as alloys of zinc, tin, and lead.

14.3.1.5.4 Pressure Casting

In pressure casting, also known or referred to in the industry environment as pressure pouring or low-pressure casting, the molten metal is forced upward using gas pressure into either a graphite mold or metal mold. The pressure is maintained until such time that the metal has solidified completely in the mold. The molten metal can also be forced upward into the mold cavity using a vacuum. This aids in the removal of dissolved gases while concurrently producing a casting that has low porosity levels.

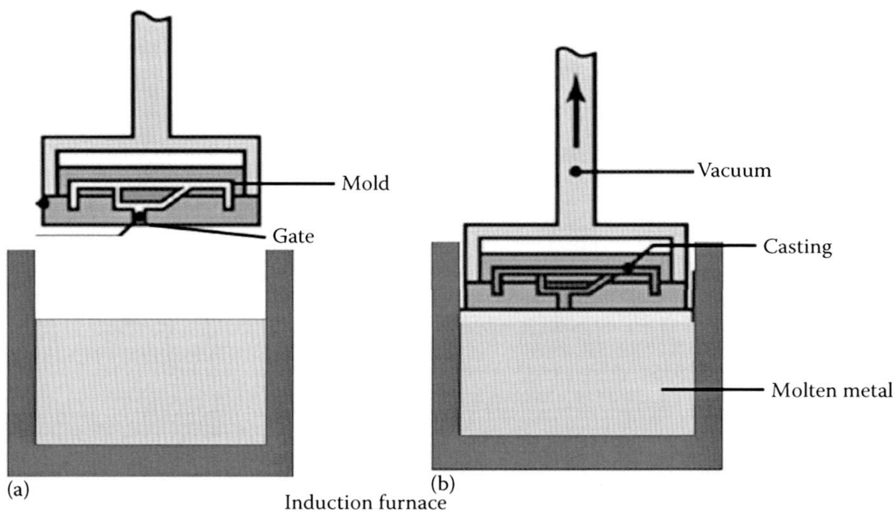

FIGURE 14.12 Schematic of the vacuum-casting process. The mold has a gate at the bottom. (a) Before immersion of the mold into the molten metal, and (b) after immersion of the mold in the molten metal.

In the early days of its development, the pressure casting technique was used for high-quality castings, such as steel wheels for the railroad car.

14.3.1.5.5 Die Casting

This is another example of the overall technique of permanent mold casting. This casting technique was developed in the early 1900s. Typical parts that have been made using the die casting technique are (1) housings for purpose of transmission, (2) business machines, (3) components of an appliance, (4) components of hand tools, and (5) toys (ASM Handbook 2010; Campbell 2003; Casting Defects 2002; Stefanescu 2008). The weight of most castings generally ranges from 100 g to about 25 kg. The cost of equipment, particularly the cost of dies, is high, but the labor cost is generally low primarily because the process is semiautomated or fully automated. The die casting technique was found to be both economical and affordable for large production runs.

In the *hot-chamber die casting* process (shown in Figure 14.13), a piston is used to force a specific volume of molten metal into the die. Pressures range up to 35 MPa, with an average pressure in the range of 15 MPa. The metal is held under pressure until such time that it solidifies in the die. To improve the life of the die and to concurrently aid in the rapid cooling of the metal thereby reducing cycle time, the dies are cooled by circulating either water or oil through various passageways in the die block. The low melting point metals, such as alloys of zinc, magnesium, tin, and lead, are commonly cast using this process.

In the *cold-chamber die casting* process, shown in Figure 14.14, molten metal is poured into the injection chamber. The chamber is not heated, hence the name cold chamber. The metal is forced into the die cavity at a sufficiently high pressure ranging from 20 to 70 MPa. High melting point alloys of

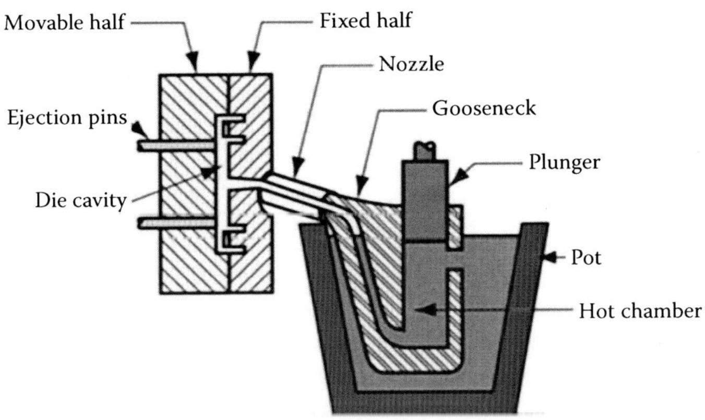

FIGURE 14.13 Schematic illustration of the hot-chamber die casting process.

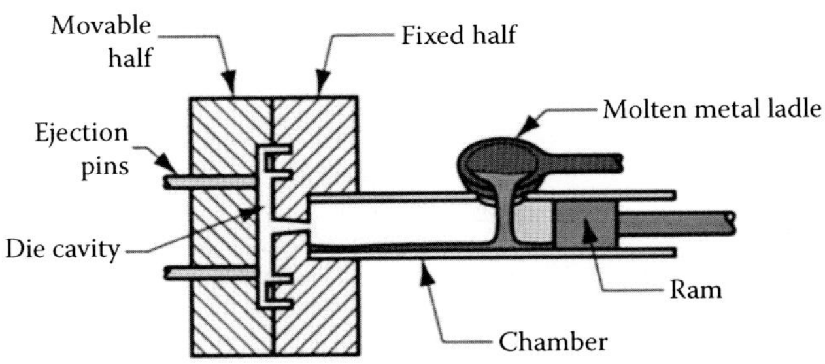

FIGURE 14.14 Schematic illustration of the cold-chamber die casting process. The machines are large when compared to the size of the casting. This is because high forces are required to keep two halves of the die closed.

aluminum, magnesium, and copper have been cast using this method. Certain ferrous alloys and other high-temperature metals can also be cast using this method. Temperatures of the molten metal begin at 600°C for the alloys of aluminum and some alloys of magnesium, and increase appreciably for the copper-base alloys and iron-base alloys.

14.3.1.5.6 Centrifugal Casting

The process of centrifugal casting puts to effective use the inertia that is caused by rotation to force the molten metal into cavities of the mold. The method or technique was initially put forth in the early 1800s. There are three types of centrifugal casting.

14.3.1.5.6.1 True Centrifugal Casting Hollow cylindrical parts, such as pipes, bushings, gun barrels, engine cylinder liners, bearing rings, and street lamp posts, have been successfully produced. In this process technique, molten metal is poured into a rotating mold with the axis of rotation being horizontal. The molds are made of steel, iron, or graphite and are often coated with a refractory lining with the intent of enhancing the useful life of the mold. The mold surfaces are often shaped so that pipes having various external designs can be cast with relative ease. Since the centrifugal force aids in ensuring uniform distribution of the molten metal, the inner surface of the casting is cylindrical. However, due essentially to differences in density, the lighter elements that include slag, inclusions, and pieces of refractory lining the mold tend to collect toward the inner surface of the casting or center of the casting. This results in variation in properties of the casting through the thickness. Normally, the pressure that is generated by the centrifugal force is high and is essential for the casting of parts having appreciable thickness. The unique advantage of this technique is that the castings produced are of good quality and maintain dimensional accuracy. A schematic of the centrifugal casting process is shown in Figure 14.15.

14.3.1.5.6.2 Semicentrifugal Casting This method has been used to produce cast parts that have rotational symmetry, such as wheel with spokes. A schematic illustration of this process technique is shown in Figure 14.16a.

14.3.1.5.6.3 Centrifuging This method is referred to in industry circles as centrifuge castings. The mold cavities are placed at a distance from the axis of rotation. The molten metal is then poured at the center and is forced into the mold cavity by use of centrifugal forces. The properties of the resultant casting can vary with distance from the axis of rotation, just as in the case of true centrifugal casting. A schematic of this casting process technique is shown in Figure 14.16b.

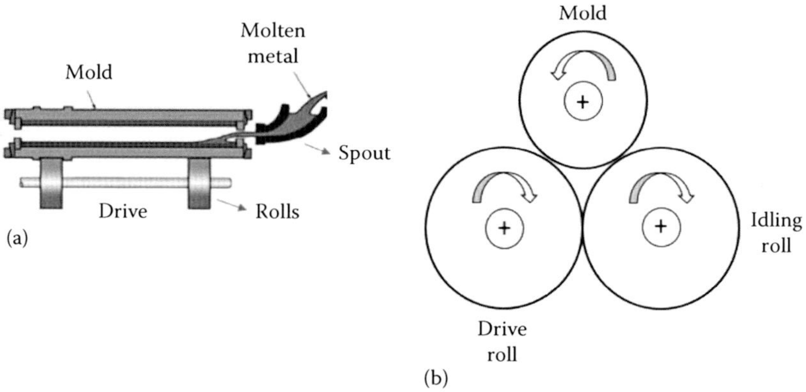

FIGURE 14.15 (a) Schematic illustration of the centrifugal casting processes. Can be used to cast pipes, cylinder lines, and similar shaded parts. (b) Side view of the machine.

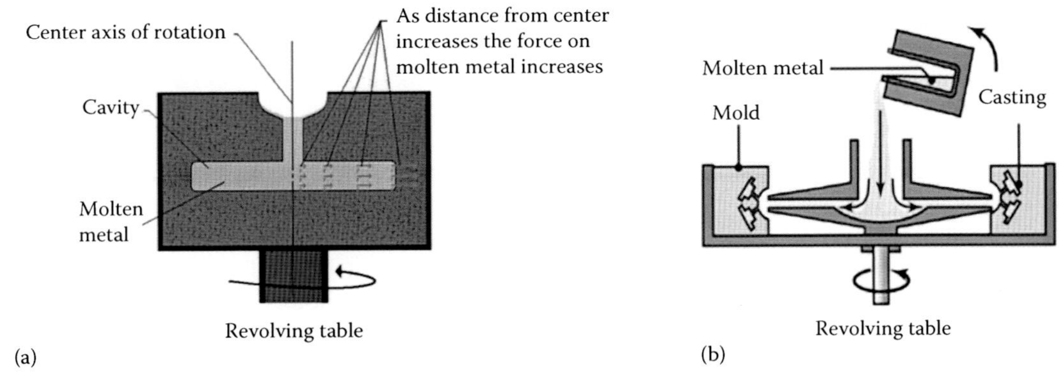

FIGURE 14.16 (a) Schematic illustration of the semicentrifugal casting process. It can be used to cast a wheel that has spokes. (b) Schematic illustration of the process of casting by centrifuging. The molds are placed at the periphery of the machine and molten metal is forced into the molds by centrifugal force.

14.3.1.5.7 Squeeze Casting

This casting technique is also referred to or known as the liquid–metal forging process. While invented in the 1930s, it saw actual development for viable industrial applications in the 1960s. Basically, the process involves solidification of the molten metal under high pressure. The sequence of operations in the squeeze casting process technique are shown in Figure 14.17. The products made by this process are mainly components for the automobile industry. The basic machinery includes the following: (1) a die, (2) a punch, and (3) an ejector pin. The pressure that is applied by the punch keeps the entrapped gases in solution. The occurrence of contact at the die–metal interface under conditions of high pressure facilitates ease in heat transfer. The resultant casting has a fine microstructure and good mechanical properties. The use of high pressure aids in overcoming all difficulty associated with feeding that can occur when the chosen metal that is being cast has a long freezing range. Both nonferrous and ferrous alloys can be used to make parts having a near net shape coupled with fine surface details or finish.

14.3.2 PM Processing

14.3.2.1 Overview

In this technique, the powder particles or the powder of a metal is compacted to the desired shape and at times complex shapes, and subsequently sintered (a technique or process that involves heating without melting) to form a solid piece. Early use of the PM technique was in Egypt around 3000 B.C. for making iron-based tools. A relatively modern use of this technique was in the early 1900s for

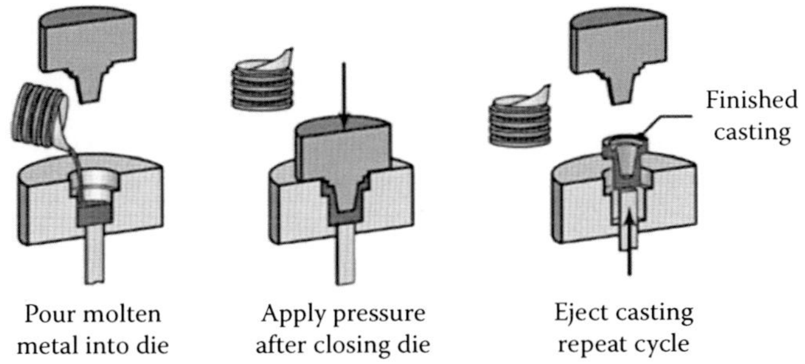

FIGURE 14.17 Sequence of operations in the squeeze casting process. This process technique combines the advantage of casting and forging.

making tungsten filaments for incandescent light bulbs. Through the years, the availability of a range of metal powder compositions coupled with an ability to produce parts to net dimensions through use of the net shape forming technique, in synergism with the overall economics of operation, gave the PM process technique far more than a few attractive and expanding applications (ASM Handbook 1998; Douvard 2009; German 2007).

A few of the parts and components currently being made using this technique are the following (Douvard 2009; German 2007): (1) automotive components, which in recent years constitute about 70% of the PM market, such as brake pads, connecting rods, cams, gears, bushings, and even piston rings; (2) cermets both as cutting tool and die material; (3) tool steel; (4) tungsten carbide; (5) graphite brushes impregnated with copper for use in electric motors; (6) magnetic materials; (7) metal filters; and (8) oil-impregnated bearings having controlled porosity. Noticeable advances in the technology relevant to PM have now made it possible to use this technique for the following purposes: (1) structural parts of an aircraft, such as landing gear components; (2) supports for engine mount; (3) engine disks; (4) impellers; and (5) engine frames. In fact, the PM processing technique has rapidly grown to become highly competitive with processes such as casting, forging, and machining. This is the case when relatively complex parts having intricate shapes, geometry, and size are to be made from a family of high-strength alloys. The most commonly chosen and used metals in the PM processing technique are iron, copper, tin, aluminum, nickel, titanium, and even the refractory metals. For parts that are made of brass, bronze, steels, and stainless steels, pre-alloyed powders are preferred and used. In this case, each powder particle is an alloy, just like the bulk component is (German 2007). The sources for the metals are generally bulk metals, their alloy counterparts, ores, salts, and chemical compounds (German 2006, 2007).

14.3.2.2 Key Operations in PM Processing

In a nutshell, the key operations in this technique are as follows:

1. Production of powders
2. Blending of the powder particles
3. Compaction of the powder particles
4. Sintering of the compact
5. Finishing operation

The sequence of processes and operations in producing PM parts is shown in Figure 14.18.

14.3.2.2.1 Production of Powder Particles

There are several methods currently available for producing powders. Most powders currently in use can often be produced by more than one method. Choice of the method is both governed and dictated by the requirements of the product. The resultant microstructure, surface properties, bulk properties, chemical purity, porosity, shape, and size distribution of the particles depend on the method chosen for producing the powders.

14.3.2.2.1.1 Atomization Atomization essentially involves a liquid–metal stream that is produced by injecting molten metal through an orifice. The stream of metal is often broken up further by jets of inert gas or air (Figure 14.19a), or water (Figure 14.19b), and is known as gas atomization or water atomization. The size and shape of the particles that are formed depend on the following: (1) temperature of the molten metal, (2) rate of flow, (3) size of the nozzle, and (4) characteristics of the jet.

The use of water often results in a slurry of metal powder and liquid at the bottom of the atomization chamber. The powders must be dried prior to use. The use of water facilitates in rapid cooling of the powder particles and thus a higher production rate. In gas atomization, the resultant particles are near spherical in shape. In centrifugal atomization, the stream of molten metal drops into

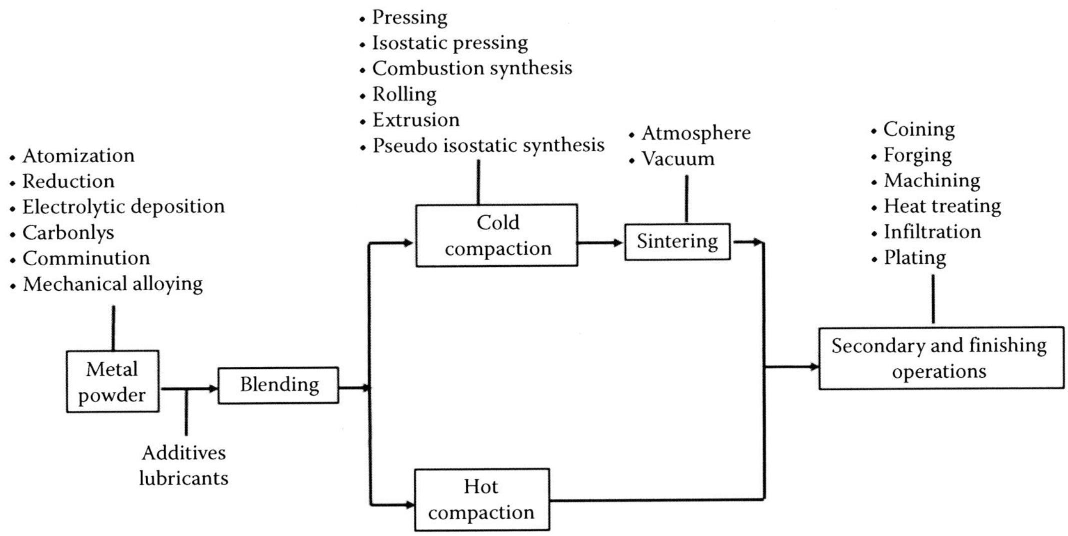

FIGURE 14.18 Outcome of processes and operations involved in producing powder metallurgy parts.

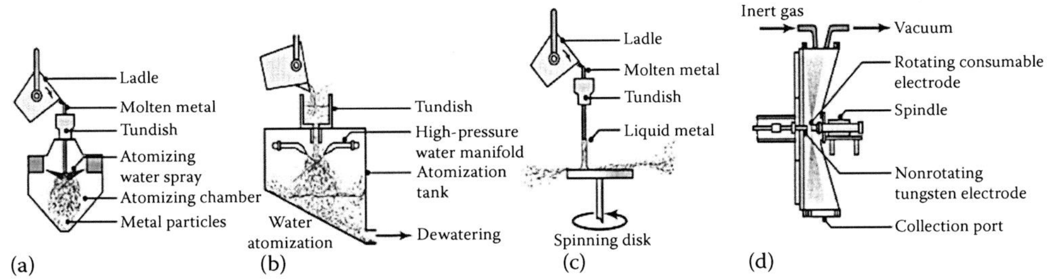

FIGURE 14.19 Methods of production of metal powders by the technique of atomization. (a) Gas atomization, (b) water atomization, (c) centrifugal atomization with a spinning disk, and (d) atomization with a rotating consumable electrode.

a rotating disk or cup (Figure 14.19c). The centrifugal forces aid in breaking up the molten metal stream and generates particles that are both fine in size and uniform in shape. A variation of this method is when a consumable electrode is rotated rapidly in a helium chamber. The centrifugal forces break up the molten tip of the electrode into metal particles. A schematic of this process technique is shown in Figure 14.19d.

14.3.2.2.1.2 Reduction Reduction, by way of removal of oxygen, of metal oxides makes use of gases such as hydrogen and carbon monoxide as reducing agents. In this method, the fine metallic oxides are reduced to the metal state. The resultant powders are both spongy and porous, and have either spherical or angular shapes of near-uniform size.

14.3.2.2.1.3 Electrolytic Deposition This method utilizes either aqueous solutions or fused salts. The resultant powders produced are among the purest kind and can be a wide variety to include several alloy compositions.

14.3.2.2.1.4 Comminution In the method of mechanical comminution, also known or referred to as pulverization, the process involves crushing, milling in a ball mill, or grinding brittle and even

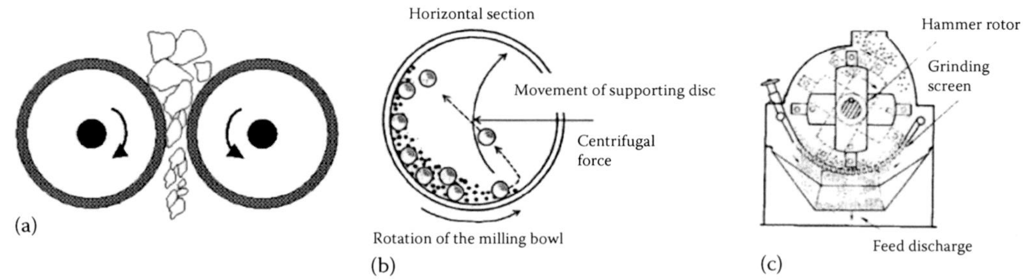

FIGURE 14.20 Methods of mechanical comminution to obtain fine particles: (a) roll crushing, (b) ball mill, and (c) hammer milling.

less ductile metals to get small particles. A ball mill is a machine with a rotating hollow cylinder that is often filled with mini-sized balls made from either steel or white cast iron (Figure 14.20b). The powders placed in the ball mill are impacted by the moving balls as the cylinder rotates. This causes the particles, essentially brittle in nature, to fracture easily resulting in smaller particles. Fracture of the fine particles also affects both their size and shape. For the case of very brittle materials, the particles that are produced will have an angular shape, while for the ductile and even less ductile metals, the particles produced can be flaky and not suitable or preferred for PM applications. The methods used commonly in both research and industrial environments for mechanical comminution are shown in Figure 14.20.

14.3.2.2.1.5 Mechanical Alloying In this method, powders of two or more pure metals are mixed in a ball mill. Upon being compacted by the hard balls, the powder particles tend to gradually fracture and bond together by the diffusion technique, entrapping the second phase and in the process forming alloy powder particles. The dispersed phase contributes to strengthening of the powder particles while even imparting improved electrical properties and magnetic properties to the powders.

14.3.2.2.1.6 Other Methods Other less commonly used or preferred methods that have been both tried and used in industry circles to produce powders are the following:

i. Precipitation from a chemical solution
ii. Production of fine chips of the concerned metal using the machining technique
iii. Vapor condensation

14.3.2.2.2 Blending of Metal Powders

The follow-on step in PM processing is blending. This is carried out to serve the following purposes:

a. Powders of different metals, alloys, and materials are blended or mixed to impart special physical and mechanical properties to the PM product. Mixtures of metals can also be produced by alloying the pure metal before producing the powder.
b. Even for a single metal, the powders can vary significantly both in size and shape. Hence, they are subject to blending with the prime objective of ensuring uniformity from part to part.
c. Lubricants are often mixed with the metal powders to either improve or enhance their flow characteristics. They tend to
    i. Reduce friction between the metal particles
    ii. Improve flow of the metal powder into the die, and in the process
    iii. Serve to improving life of the die

    The lubricants frequently preferred and chosen for use are stearic acid and zinc stearate.

d. Few additives, such as binders, are also used to impart sufficient green strength. Additives are also used to facilitate softening.

Mixing of the powders is often carried out under controlled conditions to avoid both contamination and deterioration. Deterioration of the chosen powders often results from excessive mixing, which tends to alter the shape of the powder particles and can cause the occurrence of work hardening at the fine microscopic level. The local work hardening experienced by the powder particles does make their subsequent compaction difficult to undertake.

14.3.2.2.3 Compaction of Metal Powders

This is the step that occurs when the blended powders are pressed into dies. The purpose of compaction is manifold in nature and essentially for the following:

i. Making the part sufficiently strong to facilitate or enable continued processing
ii. To obtain the required shape, density, and particle-to-particle contact

The metal powder particles are fed into a die by a feed shoe. The punch descends into the die. The presses used to compress the powder particles are actuated either mechanically or hydraulically, and the process is essentially done at room temperature (25°C). The sequence of steps involved in compaction are shown in Figure 14.21. The different stages in the compaction of powder particles, or powders, are shown in Figure 14.22.

Initially, the powder particles are loosely packed and there does exist a noticeable amount of porosity. With low applied pressure, the powder particles rearrange themselves, filling in the voids and eventually producing a much denser product. Continued compaction causes increased contact stress and even plastic deformation of the powder particles. This aids in increasing adhesion of the powders to each other. The pressed powder is known or referred to as "green compact" (German

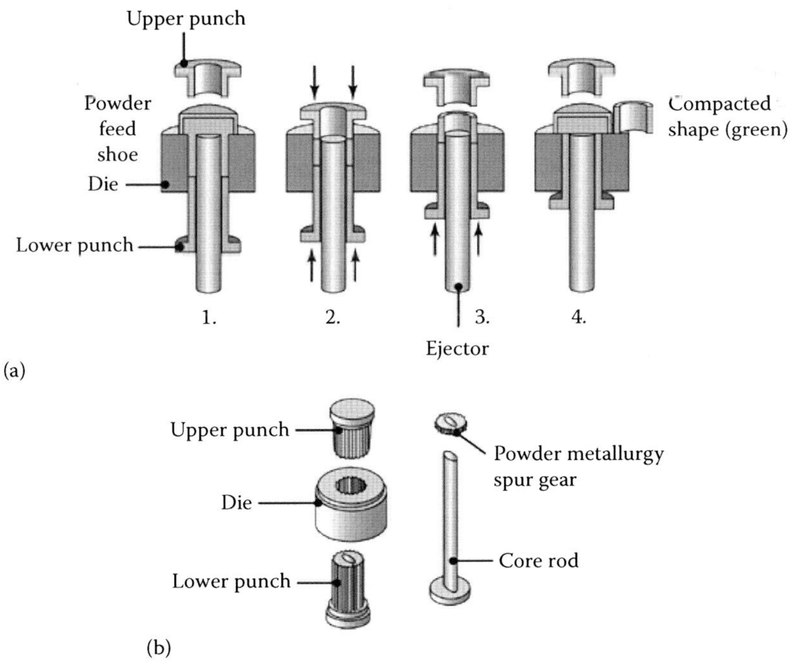

FIGURE 14.21 (a) Compaction of metal powders to form a bushing. The pressed powder part is called a green compact. (b) Typical tool and die setup for compacting a spur gear.

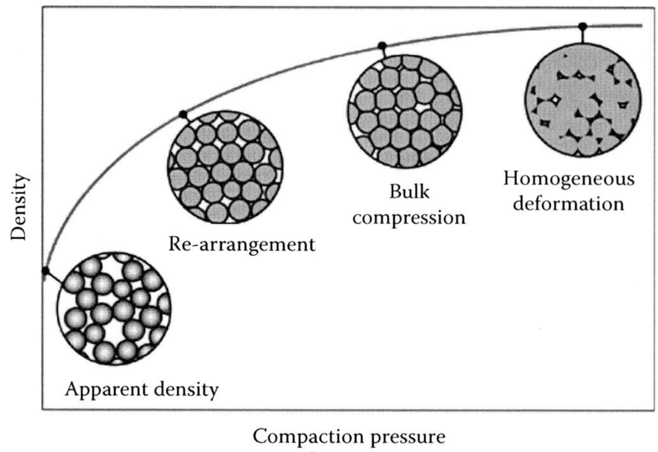

FIGURE 14.22 Influence of compaction pressure on density of metal powders. At low compaction pressures, the powders merely rearrange without deforming, resulting in noticeable increase in density. When the powders are closely packed, plastic deformation occurs at their interfaces, resulting in further increase in density of consolidated compact.

2006; Leander and West 2002). This is essentially because of its low strength. The green parts are usually fragile in nature and can easily crumble or become damaged. For the purpose of achieving high "green strength," the powders are fed properly into the die cavity, and adequate pressure must be developed throughout the consolidated part.

The density of the green compact produced depends upon the compacting pressure that was used (Figure 14.23). As the pressure is increased, the density of the compact approaches that of the chosen metal in its bulk form. An important factor in density is the size distribution of the particles. When all the powder particles are of the same size, there will exist some amount of porosity when they are packed together. The higher the density of the compacted part, the higher its strength and stiffness (elastic modulus). This is because the higher the density, the higher the amount of solid metal in the same volume, and hence the strength is higher. Due to the friction between the metal particles in the powder and surfaces of both the punch and die wall, the density within the compacted part can vary considerably. This variation can be minimized by proper design of both the punch and die while concurrently controlling friction.

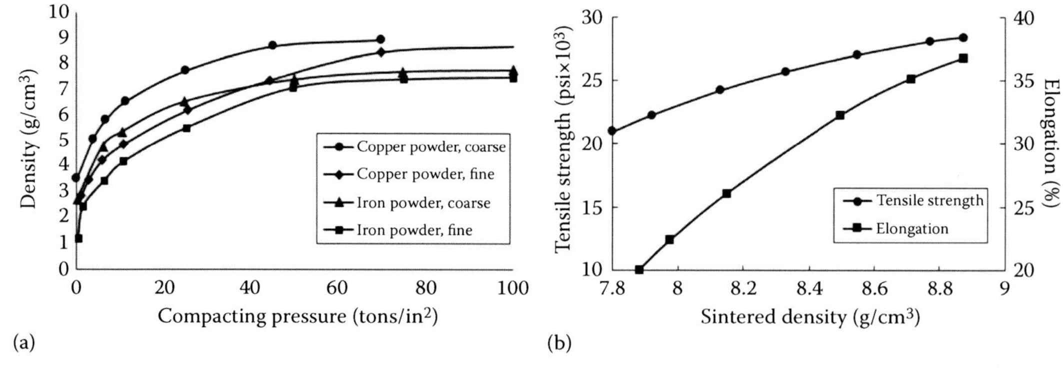

FIGURE 14.23 (a) Density of copper and iron powder compacts as a function of compacting pressure. (b) Effect of density of the consolidated compact on electrical conductivity, tensile strength, and elongation of copper powder compacts.

For a single action press, the pressure distribution along the length of the compact is given by the expression

$$P_s = P_o e^{-4\mu Kx/D} \tag{14.1}$$

where μ is the coefficient of friction between the powder particles and container, K is a factor that denotes interparticle friction during compaction, D is the diameter of the compact, and P is the pressure in the direction of compaction (x). The pressure prevailing at the bottom of the punch is P_o. Equation 14.1 also includes a variable to account for friction between the particles. This is given by

$$\sigma_r = KP_x \tag{14.2}$$

where σ_r is the stress in the radial direction. If there is no friction between the particles, then $K = 1$ and the powder behaves like a fluid, and $\sigma r = P_x$. The prevailing state is one of hydrostatic pressure. However, for the situation when there is high friction, then $K = 0$ and the pressure will be low near the punch. From Equation 14.1, it is seen that the pressure within the compact decays as the coefficient of friction (μ), the parameter K, and the length-to-diameter ratio increase.

14.3.2.2.3.1 Other Compacting and Shaping Processes In powder injection molding, also referred to as metal injection molding, very fine metal powders (less than 10 μm in size) are blended with a polymer or a wax-based binder. The resultant mixture undergoes a process similar to die-casting. The molded green parts are then placed in a low-temperature oven to burn off the plastic, or the binder (debinding) that was used is removed by use of solvent extraction. Often, a small amount of the binder is retained to provide sufficient green strength for purpose of transfer to a sintering furnace where the parts are sintered at a high temperature, as high as 1350°C.

The metals that are receptive to powder injection molding are those that melt at temperatures above 1000°C, such as carbon, stainless steels, copper, bronze, and even titanium. A few of the parts that are preferentially made using the powder injection molding technique are the following: (1) gun barrels, (2) components for watches, (3) impellers for sprinkler systems, (4) door hinges, and (5) surgical knives. A few of the advantages of powder injection molding over conventional compaction are the following:

1. Complex shapes, having small wall thickness of the order of a few millimeters, can be successfully molded, and then removed easily from the die.
2. Mechanical properties are the same as those for wrought products.
3. Dimensional tolerances are good.
4. High production rates can be achieved by using dies having multicavities.
5. Parts that are produced compete well with those made by investment casting, forging, and even complex machining.

The primary limitation with the powder injection molding technique is the high cost for small production runs coupled with the preferential need for fine powders.

14.3.2.2.3.1.1 Powder Forging In powder forging, the part that is produced following compaction and then sintered serves as a preform for a hot-forging operation. The forged product can be categorized to be dense, have a good surface finish, have good dimensional tolerances, and be a uniform and fine grain size. These superior properties obtained make the PM technique suitable for applications spanning (1) highly stressed automotive parts such as connecting rods and (2) jet engine components for use in the aerospace industry.

14.3.2.2.3.1.2 Powder Rolling This is also known as roll compaction. In this specific technique, the metal powder is fed into the roll gap of a two-high rolling mill and is compacted into a

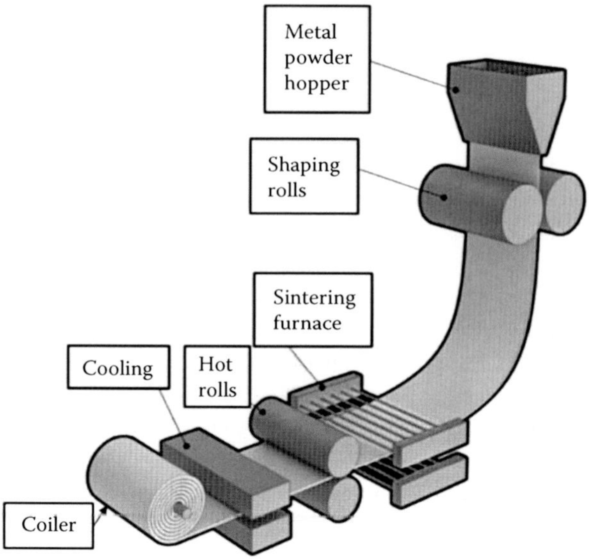

FIGURE 14.24 Illustration of metal powder rolling.

continuous strip. The rolling operation can be carried out either at room temperature (25°C) or at an elevated temperature. Sheet metal for use in electrical components and coins is made by this process. An illustration of the powder rolling process is shown in Figure 14.24.

14.3.2.2.3.1.3 Powder Extrusion A sizeable quantity of powders can be compacted by extrusion. The powders are encased in a metal container and then hot extruded. After sintering, the parts are often reheated and forged in a closed die to get the desired final shape.

14.3.2.2.3.1.4 Spray Deposition This is essentially a shape generation process and involves the following key components: (1) an atomizer, (2) a spray chamber with an inert atmosphere, and (3) a mold for producing preforms. The molds can be made in various shapes, such as billets, tubes, disks, and even cylinders. The best-known spray deposition process is the "Osprey" process. A schematic of the spray deposition process is shown in Figure 14.25. After the metal is atomized, it is deposited onto a cooled preform mold where it solidifies. The metal particles bond together giving a

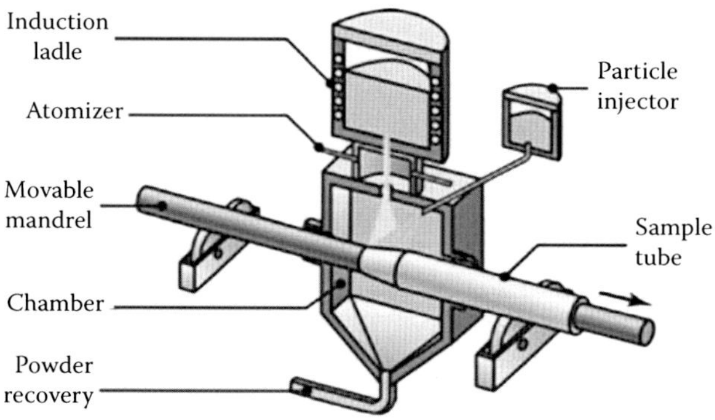

FIGURE 14.25 Spray deposition technique in which molten metal is sprayed over a rotating mandrel to produce seamless tubing and pipes.

density that is about 99% of the theoretical density of the solid metal. The spray-deposited form is then subjected to additional shaping and consolidation processes. The grain size of the resultant spray-deposited product is fine, and its mechanical properties are comparable with those of the wrought product made from the same alloy.

14.3.2.2.4 Sintering

The green compact can be safely categorized as being brittle and its strength is understandably low. During sintering, the green compact is heated in a furnace whose atmosphere is carefully controlled. The compact is heated to a temperature below the melting point of the chosen metal, but high enough to facilitate bonding or fusion of the individual particles to impart strength to the sintered part. Both the nature and strength of the bond between the particles, and eventually the sintered compact, involve a healthy synergism of the following mechanisms: (1) diffusion, (2) evaporation of volatile materials in the compact, (3) plastic flow, (4) recrystallization, (5) grain growth, and (6) shrinkage of the pores.

The key variables in sintering are temperature, time, and atmosphere in the furnace. Sintering temperature should be well within 70–90% of the melting point of the chosen metal or alloy. The sintering temperature of different metals is summarized in Table 14.2.

The time that is required for the sintering operation ranges from a minimum of 10 min for iron-base and copper-base alloys to as much as 8 h for tungsten and tantalum and their alloy counterparts. A furnace that aids, or facilitates, in continuous sintering generally consists of three chambers. These are categorized as

1. The burn-off chamber: this chamber aids in volatilizing the lubricants in the green compact to improve bond strength while concurrently minimizing or even preventing cracking.
2. A high-temperature chamber for sintering.
3. A cooling chamber.

To obtain an optimum combination of properties, a proper control of the furnace atmosphere is essential for successful sintering. Essentially, an oxygen-free atmosphere is important for controlling both the carburization and decarburization of iron and iron-based compacts, while concurrently preventing oxidation of the powders. For sintering of the refractory metal alloys and stainless steels, a vacuum environment is often preferred. The gases commonly chosen and used for sintering are hydrogen, dissociated ammonia, nitrogen, and even partially combusted hydrocarbon gas.

The mechanism governing sintering depends upon composition of the metal powder particles and the processing parameters chosen (German 2006, 2007; Leander and West 2002). The key and/or

TABLE 14.2
Sintering Temperatures and Time for Various Metals

Material	Temperature (°C)	Time (min)
Copper, brass, bronze	760–900	10–45
Iron and iron graphite	1000–1150	8–45
Nickel	1000–1150	30–45
Stainless steels	1100–1290	30–60
Alnico alloys (for permanent magnets)	1200–1300	120–150
Ferrites	1200–1500	10–600
Tungsten carbide	1430–1500	20–30
Molybdenum	2050	120
Tungsten	2350	480
Tantalum	2400	480

likely governing mechanisms will be (1) diffusion, (2) vapor phase transport, and (3) liquid-phase sintering (Cart 1989). With an increase in temperature, the adjacent powder particles begin to form a bond by the diffusion mechanism. This is shown in Figure 14.26.

The strength, density, ductility, thermal conductivity, and even electrical conductivity of the consolidated compact reveal an increase. Concurrently, the consolidated powder compact experiences shrinkage, and allowances must be provided to compensate for shrinkage, as are done with conventional castings.

One other sintering mechanism is vapor phase transport. In the vapor phase transport mechanism, which occurs when the material is heated close to its melting temperature, the metal atoms are gradually released to the vapor phase from the particles. At the interface of two particles, the melting temperature is higher and the vapor phase gradually resolidifies. Thus, with time, the interface grows and concurrently becomes stronger, while each particle tends to shrink.

When the two adjacent particles are from different metals, alloying tends to occur at the interface of the two particles. Should one of the particles have a lower melting point than the other particle, the particle will melt and due to surface tension effects tends to surround or envelope the particle that has not yet melted. This is shown in Figure 14.27.

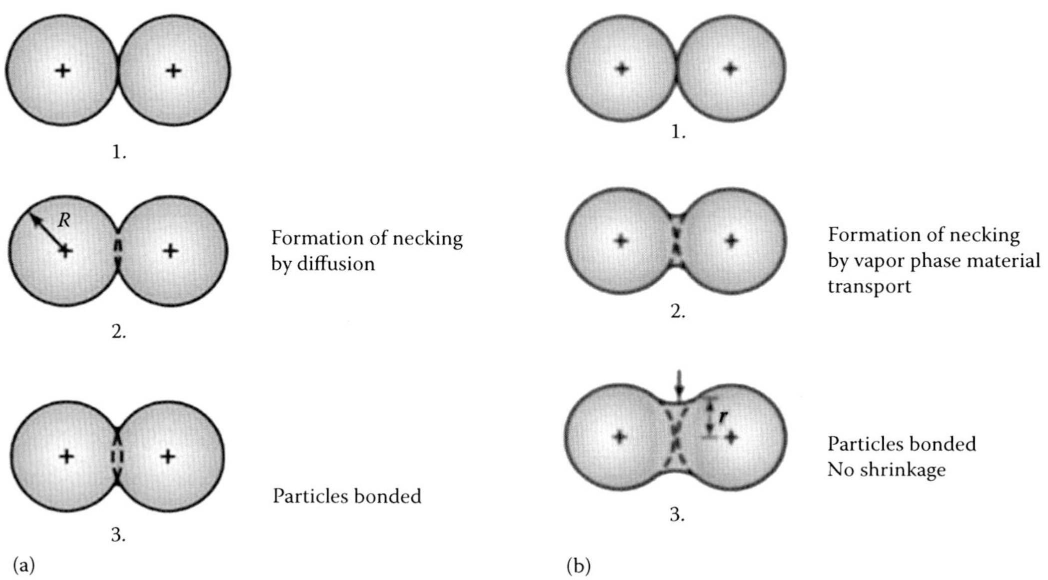

FIGURE 14.26 Schematic illustration of the two mechanisms used for sintering of metal powders. (a) Solid-state material transport, and (b) vapor phase material transport. Here R is the particle radius, r is the neck radius, and ρ is the radius of the neck profile.

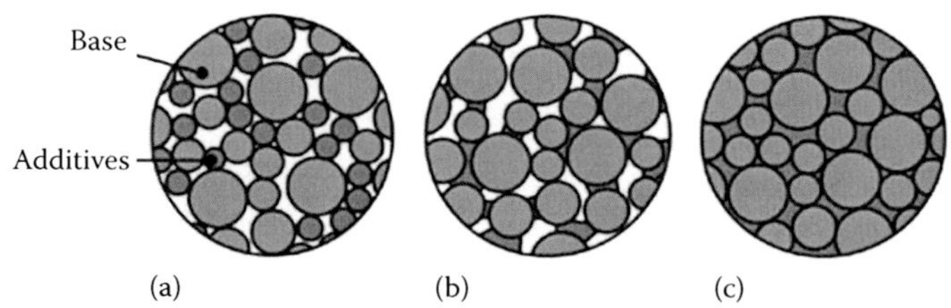

FIGURE 14.27 Schematic illustration of liquid phase sintering using a mixture of two powders. (a) Green compact of high melting point base metal and low melting point additive. (b) Liquid melting, wetting, and reprecipitation of surfaces. (c) Fully sintered solid material.

A good example of this mechanism, known as liquid phase sintering, is the presence of cobalt in tungsten–carbide tools and dies resulting in a stronger and dense part.

In spark sintering, loose metal powders are placed in a graphite mold, heated by electric current, subjected to a high-energy discharge, and then compacted, all in one step. The newer microwave sintering technique reduces the time that is required for sintering, thereby suppressing grain growth, which has a detrimental influence on strength (Mohan and Ramakrishnan 2002; Powder Metallurgy Design Manual 1995).

14.4 FORMING AND SHAPING OF METALS

14.4.1 Forging

14.4.1.1 Background

Forging can essentially be categorized to be a procedure in which the chosen workpiece is put to shape using compressive forces that are applied using various dies and tools. This technology was categorized to be one of the oldest and most important metal working operations, dating way back to 4000 B.C. At that time, the forging technique was used to make jewelry, coins, and various implements. This was made possible by hammering the chosen metal with tools that were made of stone. In the prevailing and ongoing period, parts that are forged include the following: (1) large rotors for turbines, (2) gears, (3) hand tools, (4) components of machinery, (5) railroad equipment, (6) transportation equipment, (7) aerospace equipment, and (8) delicate cutlery (Schey 2000).

Essentially during the forging operation, the flow of metal in a die and resultant grain structure of the chosen material can be controlled. The resultant forged parts have good strength and toughness and can be reliable for use in highly stressed applications, such as the drag line chain link, shown in Figure 14.28 (Byrer 1985; Dieter et al. 2003).

Often, simple forging operations are performed using a heavy hammer and an anvil, as has been traditionally practiced by blacksmiths. However, currently most forgings necessitate the need for a set of dies and powered hammers. When the forging operation is carried out at room temperature (25° C), it is referred to in industry circles as *cold forging* (Schey 2000). When carried out at an elevated temperature, it is known, or referred to, as *warm forging* or *hot forging*, depending on the homologous temperature. Due to high strength of the chosen workpiece material, the technique or operation of cold forging often necessitates the requirement for higher forces. Further, the chosen material must possess adequate ductility at room temperature (25°C) to be able to undergo the required deformation without cracking. The cold forged parts essentially have a good surface finish and dimensional accuracy to offer (Byrer 1985). In contrast, the operation of hot forging requires lower forces, but the dimensional accuracy and surface finish of the parts are often impaired (Byrer 1985; Dieter et al. 2003). Often, a forging must be subjected to additional finishing operations such as heat treatment to modify the properties and even machining for purpose of enabling accuracy in final dimension while concurrently ensuring a good surface finish. The above-mentioned finishing operations can be minimized by the precision forging technique, which is classified as a net-shape or near-net-shape forming process (Altan et al. 2005; Byrer 1985; Dieter et al. 2003).

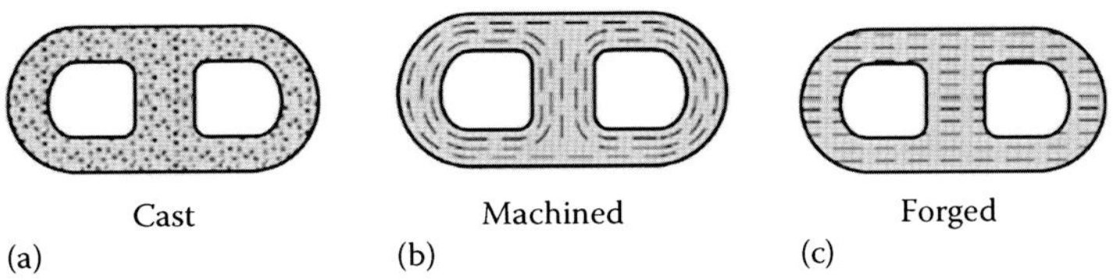

FIGURE 14.28 Schematic illustration of part of a drag-line chain link made by the processes of (a) casting, (b) machining, and (c) forging.

14.4.1.2 Open-Die Forging

This can be categorized to be a simple forging operation. Most open-die forgings generally weigh 15 to 500 kg. Forgings as heavy as a few tons have also been made. Part sizes may range from very small (e.g., pins, nuts, and bolts) to very large (e.g., shafts for the propellers of a ship). In this technique, the chosen metal workpiece is placed between two flat dies and reduced in height by compressing it. This is shown in Figure 14.29.

This process is also known as upset die forging or flat-die forging. Since constancy of volume is maintained, a reduction in height of the chosen workpiece material causes the diameter of the forged part to increase. In actual operations, there does exist friction at the die–workpiece interface, thus causing the product or part to take on a barrel shape. This mode of deformation is referred to as pancaking.

In barreling, the friction forces oppose outward flow of the workpiece material at the die interface(s). This is often minimized in industry operations using an effective lubricant. The tendency for the occurrence of barreling is favored while upsetting hot workpieces between cold dies. The material at the die surfaces tends to cool rapidly, while the rest of the workpiece material remains at a higher temperature. This then causes the material, both at the top and at the bottom of the workpiece, to experience greater resistance to deformation than the material at the center. Consequently, the central portion of the chosen workpiece material expands laterally than do the ends. Barreling due to thermal effects can be effectively reduced, or even eliminated, by using heated dies.

14.4.1.2.1 Cogging

This operation is often referred to in industry circles as drawing out. It is essentially an open-die forging operation in which the thickness of the chosen metal bar is reduced by successive steps that are performed at regular intervals. This is shown in Figure 14.30. The thickness of the chosen bars and rings can be reduced by similar open-die forging techniques as shown in Figure 14.30b and c. Since the contact area between the die and the chosen workpiece is small, a sizeable section of the chosen bar can be reduced in thickness without the need for large forces or heavy machinery. Cogging of a larger workpiece material is usually done using mechanized equipment and computer control. Through this, both the lateral and vertical movements of the die can be coordinated to produce the desired shape of the part.

14.4.1.3 Impression-Die Forging

In this technique, the workpiece takes the shape of the die cavity while being forged between two shaped dies. A schematic of this technique is shown in Figure 14.31. During deformation, some of the starting raw material easily flows outward to form a flash.

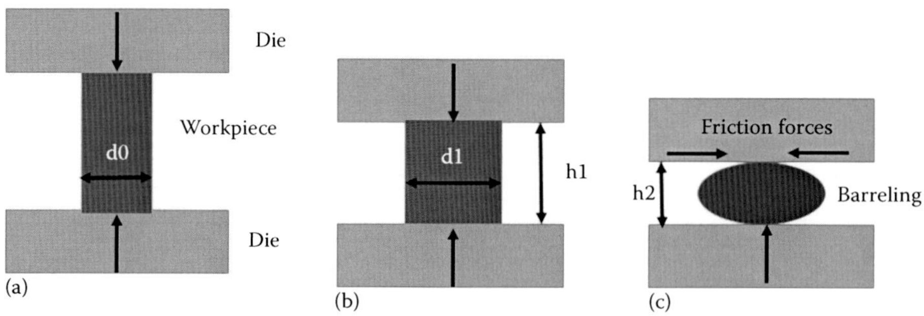

FIGURE 14.29 (a) Solid cylindrical billet upset between two flat dies. (b) Uniform deformation of the billet in the absence of friction. (c) Deformation and friction; with barreling of the billet caused by the presence of friction at the billet–die interface.

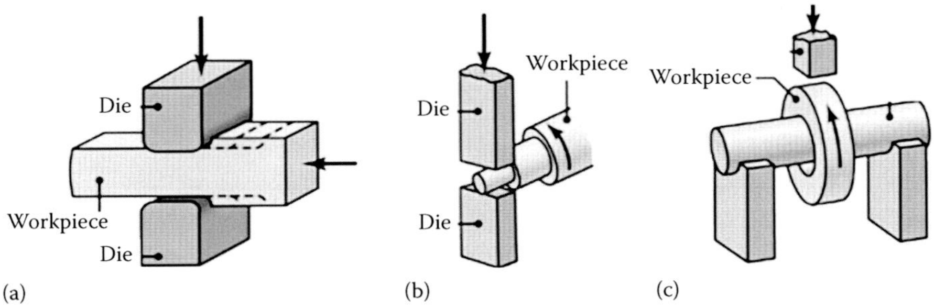

FIGURE 14.30 (a) Schematic illustration of a cogging operation on a rectangular blank. (b) Reducing diameter of bar by open die forging. (c) Thickness of a ring being reduced by open-die forging.

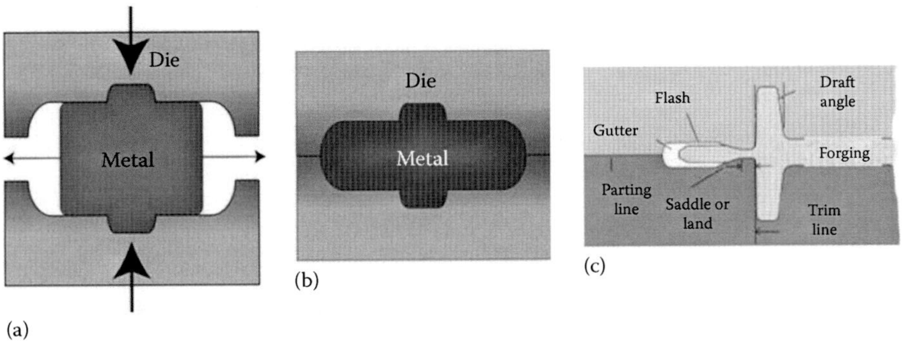

FIGURE 14.31 (a and b) Two stages in impression-die forging of a solid round billet. (c) Standard terminology for various features of a forging die.

The flash that is formed has an important role to play in impression-die forging. The high pressure and resultant high friction resistance in the flash present a severe constraint on radial outward flow of the chosen workpiece material in the die. Based on the principle of plastic deformation, which is an indication of favorable flow of the material in the direction of least resistance, the workpiece material tends to flow preferentially into the die cavity and in the process filling it completely.

The standard terminology for the various features of a hot forging die is shown in Figure 14.31c. With demands posed by technological advances, the dies are made in several pieces to even include the die inserts. The inserts can be easily replaced in case of either wear or failure in a region of the die. Often, the insert is made of a stronger and harder material than the die.

The blank that is to be forged is prepared initially by any one of the following methods:

i. Cropping from either an extruded or drawn bar stock of the chosen material
ii. Either casting or PM
iii. A preformed blank from a prior forging operation

The chosen blank is then placed on a lower die, and as the upper die descends, its shape is progressively altered.

14.4.1.4 Closed-Die Forging

In true closed-die forging, a flash does not form. This is also referred to in industry circles as flashless forging. The chosen workpiece is made to completely fill the die cavity. Actual control of the initial or starting volume of the chosen metal block along with die design is essential to produce a forging that has the required dimensional tolerances. An initial undersize blank would prevent complete filling of the chosen die cavity, while an oversized blank would generate excessive forces that are both favorable and conducive for causing premature failure of the die or the machine to jam.

14.4.1.5 Precision Forging

To reduce the number of finishing operations, and thereby enable savings in operational cost, there is a growing need to ensure precision in forging and the resultant forged product. A few of the parts made by precision forging are (1) gears, (2) connecting rods, and (3) turbine blades. This forging technique necessitates the need for (1) special and/or more complex dies; (2) precision control of the size, shape, and initial volume of the chosen metal blank; and importantly (3) an accurate positioning of the chosen blank in the die cavity. Further, this forging technique necessitates the need for the use of higher forces to obtain fine details on the forged part. This, in turn, has set the demand for a higher capacity in the equipment chosen and used. The two common metals chosen for purpose of precision forging through the years are the family of steels and alloys of titanium. With time and advances in technology, both aluminum alloys and magnesium alloys have also been found to be suitable for precision forging primarily because of the need for low forging loads coupled with low operating temperatures that these two lightweight metals require. This is shown in Figure 14.32.

14.4.1.6 Other Forging Operations

A few of the forging operations still finding use in industry environment are discussed next.

14.4.1.6.1 Coining

This is essentially a closed-die forging process operation that has been actively used for making coins, medallions, and even jewelry. It is also being used in recent years in a wide range of parts with high accuracy, such as gears, industrial seals, and useful medical devices. The chosen blank, also referred to as a slug, is coined in a complete closed-die cavity to produce or imprint the finer details. The pressure required can be as high as five to six times the strength of the chosen material. On some parts and with few materials, several coining operations may be required. Lubricants are often not used in the coining operation since they can often become entrapped in the die cavity or, because of their incompatibility, prevent full reproduction of die-surface details and surface finish.

14.4.1.6.2 Heading

This is also known in industry circles as "upset forging." It is essentially an upsetting operation that is performed on the end of either a metal rod or wire with the primary intent of increasing the cross section. Nails, bolt heads, screws, rivets, and few other fasteners are preferentially made using this technique. Often, the heading operation can be carried out in the cold, warm, or hot state. An important consideration to be given to this technique is the tendency for the rod or wire to buckle if

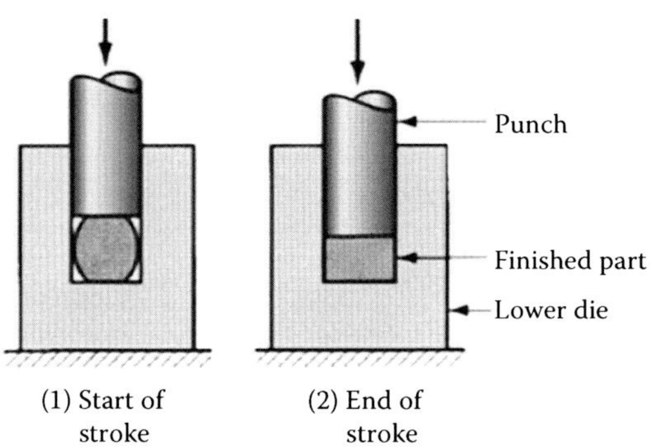

FIGURE 14.32 Schematic showing precision or flashless forging of a round billet.

the unsupported length-to-diameter ratio of the chosen rod or wire is a little too high. This ratio is limited to less than 3:1.

14.4.1.6.3 Piercing

This is a process that has been used for indenting the surface of a workpiece material using a punch in order to produce either a cavity or an impression. Often, the chosen workpiece material is confined in a container, such as a die cavity, or may even be unconstrained. The extent of surface deformation experienced by the workpiece material depends on how much it is constrained from flowing freely as the punch descends. A common example of piercing is the indentation of a hexagonal cavity on bolt heads. Piercing is normally followed by punching to produce a hole in a part.

14.4.1.6.4 Orbital Forging

In this process, the upper die moves on an orbital path and forms the required part incrementally. Typical components that can be forged by this process are disk-shaped and conical-shaped parts, such as bevel gears and gear blanks. The load that is required for forging is relatively small since at any instant the die contact is concentrated on a small area of the chosen workpiece material. The technique as an operation is relatively quiet, and parts can be easily forged within 10–25 cycles of the orbital die.

14.4.1.6.5 Incremental Forging

In this process, a tool forges the blank into the desired shape in several small steps. This operation technique is quite like cogging, in which a die is used to deform a blank to a different extent at different positions. Due to a small area of contact with the die, this process necessitates the requirement for much lower forces when compared one on one with both conventional forging and conventional impression die forging. The tools often chosen and used during this operation are noticeably simple and affordable in cost.

14.4.1.6.6 Isothermal Forging

This is also known as hot-die forging. The dies in this technique are heated to the same temperature as that of the chosen hot workpiece. The workpiece remains hot since no heat is lost to the dies. Consequently, its flow strength coupled with ductility are both maintained and put to effective use during forging. Also, the load that is required for forging is low, and material flow within the die cavity is noticeably improved. Complex parts can be isothermally forged, with good dimensional accuracy and to near-net shape using one stroke in a hydraulic press. The dies for hot forging are made of either nickel or molybdenum alloys due to their resistance to hot temperatures. For aluminum and its alloy counterparts, the dies made from steel are adequate and frequently chosen for use. This forging operation is expensive and the production rate can be observably slow. It can be considered economical provided that the quantity of forgings to be made is sufficiently high to justify high cost of the die.

14.4.1.6.7 Swaging

This process is also known in industrial circles as radial forging or rotary forging or simply as swaging. In this process, a solid rod or tube is often subjected to radial impact using a set of reciprocating dies. The dies are activated by means of a set of rollers within a cage. This is like a roller bearing. The workpiece remains stationary while the dies rotate (often moving radially in their slots), striking the workpiece at rates as high as 20 strokes per second. The swaging technique has been successfully used to assemble fittings over both cables and wires. The process has also been used for operations such as

i. Pointing (tapering the tip of a cylindrical part),
ii. Sizing (finalizing the dimensions of a specific part of interest)

TABLE 14.3
Forgeability of Metals (in Decreasing Order)

Metal or Alloy	Range of Hot Forging Temperatures (°C)
Aluminum alloys	400–550
Magnesium alloys	250–350
Copper alloys	600–900
Carbon steels and low alloy steels	850–1150
Martensitic stainless steels	1100–1250
Austenitic stainless steels	1100–1250
Titanium alloys	700–950
Iron-base super alloys	1050–1180
Cobalt base super alloys	1180–1250
Tantalum alloys	1050–1350
Molybdenum alloys	1150–1350
Nickel-base super alloys	1050–1200
Tungsten alloys	1200–1300

Swaging is limited to a maximum workpiece diameter of 150 mm and parts as small as 0.5 mm. This process is suitable for medium to high rates of production, with rates dependent on the overall complexity of the part. Overall, swaging is a versatile process and is limited in length and governed by the length of the bar that supports the mandrel.

The receptivity or ease of forgeability of the different metals is summarized in Table 14.3.

14.4.1.7 Typical Defects Induced during Forging Operations

In addition to noticeable evidence of cracking, both microscopic and macroscopic in nature, occurring on the surface, a few other defects develop during the operation of forging as a direct result of material flow pattern in the die. An insufficient volume of metal to fill the die cavity can often result in buckling of the workpiece during forging resulting in the formation of a lap. However, should the chosen workpiece be too thick, the excess material would resist to flow past the already formed portion of the forging resulting in the development of internal cracks. Internal defects are also favored to occur because of the following (Dieter et al. 2003; Schey 2000; Tschaetch 2007):

a. A nonuniform deformation of the chosen metal within the die cavity
b. Temperature gradients developed through the workpiece during forging
c. Microstructural changes caused by phase transformation

The grain flow pattern in the material during forging is also important. The grain boundaries of the end grains are exposed to the environment and can be attacked by it. This essentially results in the formation of a rough surface that acts as a potential stress raiser during subsequent use of the workpiece. Defects engineered or resulting from the forging operation are easily prone or susceptible to failure due to fatigue, corrosion, and wear during service life of the forging. Thus, it is important to inspect forgings prior to their placement in service.

14.4.2 Rolling of Metals to Engineer the Desired Product

14.4.2.1 Background

Rolling is a process that has been traditionally used to reduce the thickness and change the cross section of a chosen workpiece with the aid of compressive forces that are applied through a set of rolls. A neat depiction of the various flat rolling and sheet rolling processes are shown in Figure 14.33.

In fact, by all standards, rolling accounts for about 90% of all metals produced using metal working processes. The technique was first developed and tried around the 1500s. With time, progressive advances in steel making practices, coupled with the production of both ferrous alloys and nonferrous metals and their alloy counterparts, have now reached a stage of being able to integrate continuous casting with rolling processes. A combination of the two has shown to improve productivity while concurrently lowering production costs. Both metallic and nonmetallic materials are often subject to rolling to reduce their starting thickness while also being able to enhance their properties.

Hot rolling is rolling that is carried out at an elevated temperature (> 0.5 T_M). During the hot-rolling operation, a coarse grain, brittle, and porous structure of the ingot or piece of continuously cast metal is broken down into a wrought structure, having a finer grain size coupled with improved strength and hardness. Cold rolling is rolling that is performed at room temperature (25°C). During the cold-rolling process, the rolled sheet has higher strength and hardness and even better surface finish than the hot-rolled counterpart. However, cold rolling often results in a product that has anisotropic properties due to preferential orientations or mechanical fibering.

The rolled metal plates often have a thickness greater than 6 mm (0.25 in.) and find their selection for use in a plethora of structural applications spanning the hull of ships, boilers, bridges, and even heavy machinery. Plates can be even as thick as 300 mm (12 in.), and these are chosen for use as structural support. However, plates having a thickness of 150 mm find use for reactor vessels and plates having a thickness of 100–125 mm for frames of machinery.

When thickness of the product is less than 6 mm, it is often referred to as a sheet. Sheets are often provided to manufacturing facilities as either coils or flat sheets for further processing into a wide variety of sheet metal products. Sheets find use for applications ranging from (1) bodies of trucks,

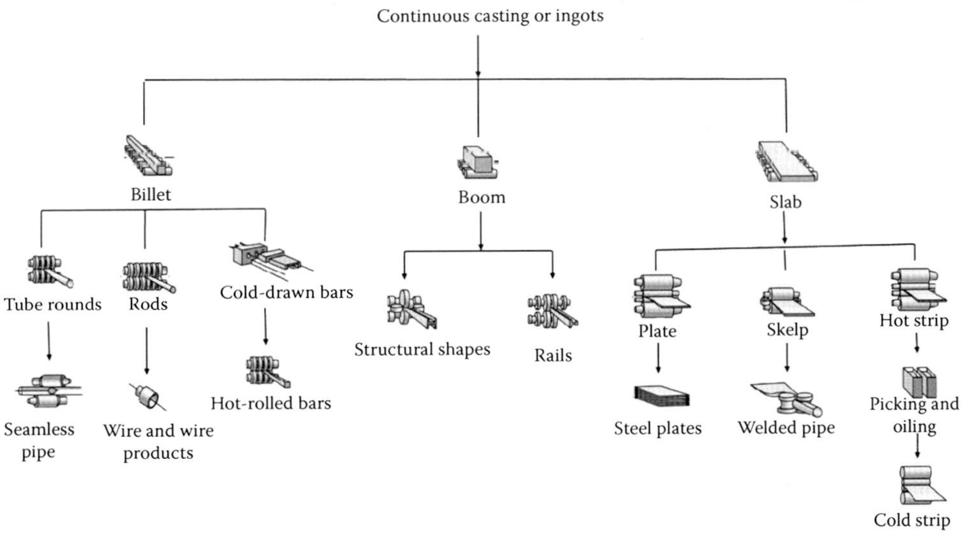

FIGURE 14.33 Schematic outline of various flat-rolling and sheet-rolling processes.

(2) wing and hull of an aircraft, (3) equipment for use in the kitchen, (4) equipment for use in the office, and (5) appliances, food containers, and beverage containers.

The sheets of steel chosen for use for the bodies of automobiles and appliances are typically around 0.7 mm thick. The beverage cans are made from aluminum sheet that is 0.28 mm thick. Aluminum foil that finds wide use in the kitchen has a thickness of 0.008 mm (0.0003 in.).

14.4.2.2 Flat Rolling

The flat rolling process is best depicted by the schematic shown in Figure 14.34. In this process, a metal strip having an initial thickness of h_o enters the roll gap and is reduced to a thickness of h_f by a pair of rolling rolls. The rolls are powered by an electric motor. The surface speed of the rolls is denoted as V_r. The velocity of the metal strip increases from its initial value at the point of entry (V_o) as it moves through the roll gap, and is the highest at the point of its exit from the roll, where it is denoted as V_f. The acceleration that the sheet metal experiences in the roll gap is very much like the flow of an incompressible fluid through a convergent channel.

There tends to occur relative sliding between the roll and the sheet metal strip along the contact length (L). This is essentially because the surface speed of the rigid roll is constant. At one point, the velocity of the sheet metal strip is the same as that of the roll. This is called, or referred to, as the *no-slip point* or *neutral point*. To the left of this point, the roll moves faster than the sheet metal strip. To the right of the no-slip point, the sheet metal strip moves faster than the roll indicating the presence and occurrence of friction forces on the sheet metal strip.

In the rolling operation, forward slip is defined in terms of the exit velocity of the sheet metal strip (V_f) and the surface speed of the roll (V_r).

$$\text{forward slip} = [(V_f - V_r)/V_r] \tag{14.3}$$

This then provides a direct measure of the relative velocity in the roll gap. By measuring the roll velocity and workpiece velocity in a rolling mill, the forward slip can be obtained and does give an indication of the location of the neutral point. Forward slip can also be used to correlate well with the surface finish of the rolled sheet metal strip. A low value of forward slip is preferred over a higher value. The roll pulls the material into the roll gap through a frictional force on the material. Thus, a frictional force must exist and be to the right. This also means that the frictional force to the left of the neutral point must be higher than the friction force to the right.

Draft is defined as the difference between the initial thickness and final thickness of the sheet metal strip, i.e., $[h_o - h_f]$. A large draft will tend to cause the roll to strip. Maximum draft is a function of roll radius (R) and the coefficient of friction (μ).

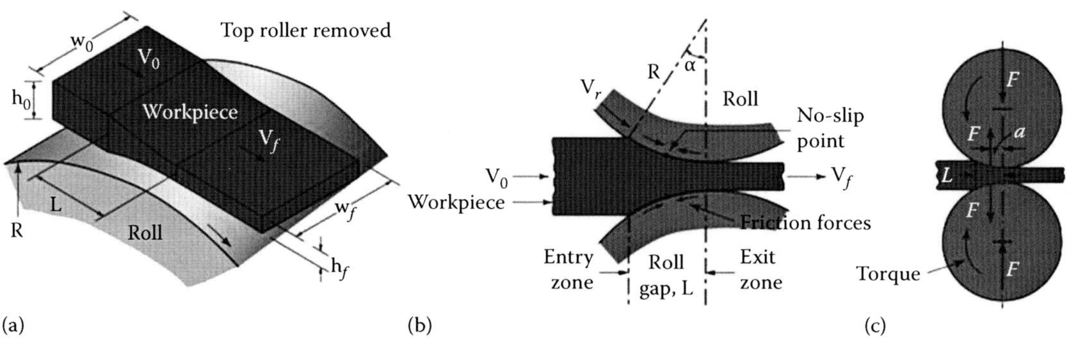

FIGURE 14.34 (a) Schematic illustration of the flat-rolling process. (b) Friction forces acting on the surfaces during rolling. (c) Roll force F and torque T acting on the rolls. The width of the metal strip (w) increases with rolling.

$$(h_o - h_f) = \mu^2[R] \tag{14.4}$$

The higher the friction (μ) and the larger the roll radius (*R*), the greater the maximum possible draft.

Force on the roll during the operation of flat rolling can be obtained using the following mathematical expression:

$$F = \sigma_{average}(Lw) \tag{14.5}$$

where *F* is the roll force, $\sigma_{average}$ is the average true stress of the sheet metal strip as it moves through the roll gap, *L* is the contact length of the roll strip, and *w* is the width of the sheet metal strip. This equation is for the case of frictionless condition. In the presence of friction, the actual roll force is obtained by increasing the calculated force by about 20% (Ginsberg and Balsam 2001; Hosford and Caddell 2011).

The presence of roll force can cause an observable amount of both deflection and flattening of the rolls. Such noticeable changes will tend to affect the rolling process, its overall efficiency, coupled with an ability to produce a uniform thickness in the rolled metal sheet. Under the influence of a high roll force, the columns of the roll stand would tend to deflect to an extent that is conducive for opening the roll gap. From a practical stand point, the magnitude of the roll force can be reduced by the following means (Ginsberg and Balsam 2001):

a. Using smaller diameter rolls to reduce the contact area
b. Ensuring smaller reduction in thickness for each pass, thereby minimizing the contact area

14.4.2.2.1 The Practice of Flat Rolling

Under ideal conditions, initial rolling of the chosen metal or material is done by hot rolling at a temperature above the recrystallization temperature of the chosen metal. The as-cast structure of the chosen metal is initially dendritic in nature, comprising of coarse and non-uniform grains. Consequently, the structure is brittle and can also be porous. Hot rolling successfully converts the as-cast microstructure to a wrought structure. Hot rolling is a viable method for reducing the grain size of a metal with resultant improvement in both strength and ductility. This is shown in Figure 14.35.

The wrought structure essentially consists of a healthy mixture of fine grains that provide a combination of good strength and enhanced ductility. This results as a direct consequence of breaking up the brittle grain boundaries coupled with closure of all the internal defects to include porosity. Typical temperature ranges for hot rolling are 450°C for aluminum alloys, up to 1250°C for alloy steels, and up to 1650°C for the refractory metal alloys (Ginsberg and Balsam 2001; Hosford and Caddell 2011; Lee 2004).

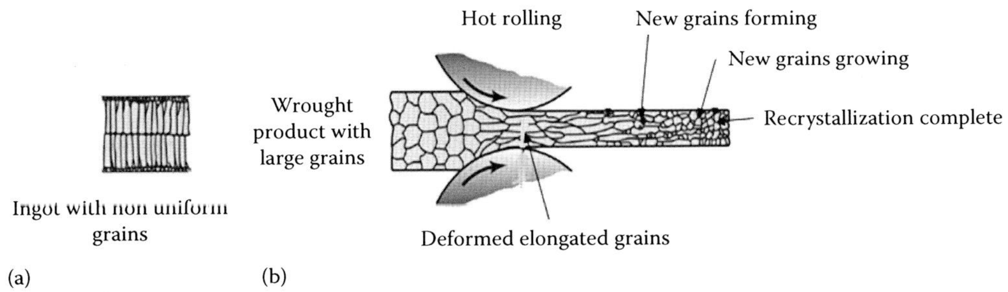

FIGURE 14.35 Change in grain structure experienced by the cast and large grained wrought metal during hot rolling. Hot rolling is an effective way to reduce grain size in metals while concurrently achieving improved strength and ductility.

The rolled product following the first hot-rolling operation is called bloom, slab, or billet. A bloom usually has a square cross section with a section size of 150 mm on one side. The slab is usually rectangular in cross section. The bloom is subject to further processing by shape rolling to get structural shapes, such as I-beams and railroad rails. The slabs are subject to additional rolling to get plates and sheets. The billets having a square cross section are smaller than the bloom. The billets are subject to additional rolling to get round rods and even bars, using shaped rolls. The hot-rolled round rods, denoted as wire rods, are used as the starting material for both rod and wire-drawing operations.

a. Cold rolling is performed at room temperature (25°C). It is often used to produce sheets and strips that have a much better surface finish, better dimensional tolerances, and, due to the presence of "local" strain hardening, better mechanical properties.
b. Pack rolling is a flat rolling operation in which two or more layers of the sheet metal are rolled together. This is done for purpose of increasing productivity. In aluminum foil pack-rolled in two layers, only the top surface and the bottom outer surface are in direct contact with the rolls. This essentially results in one side of the aluminum foil being shiny, while the other side is "matte." The foil-to-foil side has a "matte" and satin finish, whereas the side of the foil in contact with the roll is bright and shiny. This is because the polished surface is impressed onto the aluminum foil during rolling.

14.4.2.3 Defects Occurring in Rolled Plates and Sheets

The defects may be present on the surfaces of rolled plates and sheets. There can also and be a few too many internal structural defects. The presence of defects has an adverse influence on the overall appearance of the surface after rolling. Presence of defects does exert an influence on strength, ductility, formability, and other manufacturing characteristics of the rolled sheet metal. The following are surface defects that have been identified in sheet metals: (1) scales, (2) rust, (3) scratches, (4) gouges, (5) pits, and (6) cracks spanning both macroscopic and fine microscopic. Presence of these defects is often caused by the presence of both inclusions and impurities in the original cast metal structure. Typical defects resulting from flat rolling operation are shown in Figure 14.36.

The wavy edges (shown in Figure 14.36a) are the result of roll bending. The metal strip becomes observably thinner along its edges than at its center. Due to the need to maintain volume constancy during plastic deformation, the edges must elongate more than the material at the center. This causes the edges to buckle since they are constrained by the central region from expanding fully along the longitudinal direction or rolling direction.

Cracks span both fine microscopic and macroscopic, and can be either zipper cracks at the center of the structure (Figure 14.36b) or edge cracks (Figure 14.36c). This is due to the low ductility of the material at the rolling temperature. The quality of edges of the rolled sheet is important should additional forming operations be required. This then has necessitated the need for removal of the edge defects in rolled sheets using shearing and even slitting operations. Alligatoring (shown in Figure 14.36d) is caused because of non-uniform bulk deformation experienced by the billet during rolling or by the presence of a sizeable number of defects in the original cast structure.

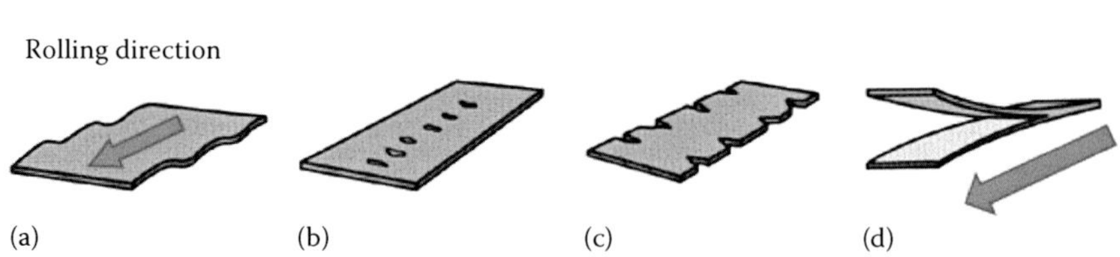

FIGURE 14.36 Schematic illustration of typical defects in flat rolling. (a) Wavy edges. (b) Zipper cracks in the center of the strip. (c) Edge cracks. (d) Alligatoring.

14.4.2.4 Other Industry-Relevant Rolling Processes

A few other rolling processes having significance in industry-relevant operations and the corresponding mills have been developed for producing shapes. Of the several processes, the most common ones are (1) shape rolling, (2) roll forging, (3) skew rolling, (4) ring rolling, and (5) thread rolling (Ginsberg and Balsam 2001; Hosford and Caddell 2011; Lee 2004).

14.4.2.4.1 Shape Rolling

Structural shapes, such as channels, I-beams, solid bars, and even railroad rails, are formed by the shape rolling technique. In this case, the heated metal stock is made to pass through specially designed rolls. Since the cross section of the starting material is reduced non-uniformly, a proper design of the roll sequence is essential to prevent the formation and presence of both external and internal defects, to retain dimensional tolerance to an acceptable level, and to concurrently minimize wear experienced by the roll.

14.4.2.4.2 Roll Forging

This is also known as cross-rolling; the cross section of a round bar is shaped by passing it through a pair of rolls having specially designed grooves. This process has been used to produce tapered shafts, leaf springs, hand tools, and even table knives. It is also a preliminary operation in forging prior to other forging processes.

14.4.2.4.3 Skew Rolling

This process is quite like roll forging and is widely used for making ball bearings. Either a round wire or a round rod is fed into the roll gap, and spherical blanks are continuously formed by the action of the rolling rolls. The as-formed balls often require further finishing. This is made possible by subjecting the balls to a ground and polishing treatment.

14.4.2.4.4 Ring Rolling

A ring-shaped blank is placed carefully between two rolls, one of which is driven while the other roll is idle. The thickness of the ring is gradually reduced by bringing the rolls closer together as they rotate. Since the volume of the chosen ring remains essentially constant during deformation, a reduction in the thickness of the ring results in an increase in its diameter thereby satisfying volume constancy. Ring rolling can be carried out either at room temperature or at an elevated temperature, depending essentially on size, strength, and ductility of the chosen metal or material. Typical applications of ring rolling are the following: (1) body casings for both rocket and jet engines, (2) ball-bearing race, (3) roller-bearing race, (4) flanges, and (5) reinforcing rings for pipes. The advantages of ring rolling upon comparison with other manufacturing processes are (1) short production times, (2) close dimensional tolerances, (3) low scrap, and (4) favorable grain flow in the product resulting in enhanced strength in the desired direction.

14.4.2.4.5 Thread Rolling

This is essentially a cold-forming process that is used to form both straight and twisted threads on both round rods and wires. The threads are formed on the chosen rod or wire using a pair of flat reciprocating dies. This is shown in Figure 14.37a. The threads can also be formed using two rolls and using a rotating die and a planetary die (Figure 14.37c). Parts that are often made using this process are screws, bolts, and similar threaded parts.

Based on the design of the die, the major diameter of a rolled thread may not be larger than a machined thread, as shown in Figure 14.38, and is about the same as the diameter of the chosen metal blank.

The process of thread rolling has the advantage of generating threads at a high production rate and without scrap. Also, the surface finish is smooth and the process aids in inducing compressive

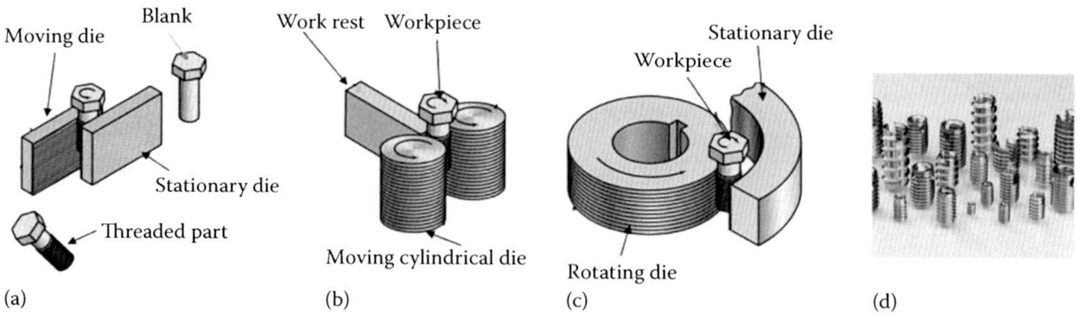

FIGURE 14.37 Thread-rolling process. (a) Reciprocating flat dies being used to produce a threaded fastener. (b) Two-roll dies. (c) Rotary or planetary die set. (d) Collection of the threaded parts.

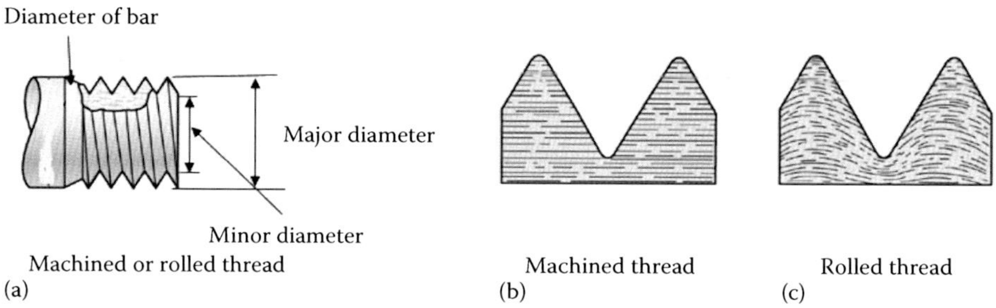

FIGURE 14.38 (a) Typical key features in a machined or rolled thread. (b) Grain flow in a machined thread. (c) Grain flow in a rolled thread.

residual stresses on the surface. The presence of residual stresses, which are compressive in nature, helps in improving fatigue life. Thread rolling is superior to the other methods of thread manufacture. This is because machining of the threads cuts through the grain flow lines of the material, while rolling of the threads results in a grain flow pattern that enhances the strength of the thread. Lubrication is both important and essential in thread rolling operation to obtain a good surface finish, to obtain integrity of the surface, and to minimize defects. Lubrication does influence the way the chosen metal deforms during deformation. This is an important consideration to minimize internal defects from being developed. The rolling dies are made of hardened steel and are expensive because of their complex shape.

14.4.3 Extrusion of Metals and Drawing

14.4.3.1 Background

Both extrusion and drawing operations offer numerous applications in the domain of manufacturing by way of discrete products from a variety of metals and their alloy counterparts. In extrusion, a cylindrical billet is forced through a die. A schematic of the direct extrusion process is shown in Figure 14.39.

A wide variety of both hollow and solid cross sections can be produced by the extrusion technique, which can be categorized to be semifinished products. A noticeable characteristic of the extrusion technique is that large deformation can take place without causing failure by fracture or rupture. This is primarily because the chosen metal or material is under the influence of high compressive stresses. The extruded metal products often have a constant cross section with the starting product of the same metal or material.

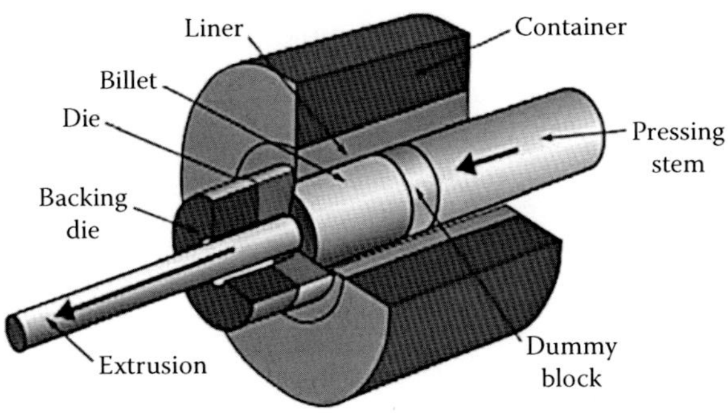

FIGURE 14.39 Schematic illustration of the direct extrusion process.

A few products made by the extrusion technique are (1) sliding doors, (2) window frames, (3) tubing, (4) ladder frames, (5) structural shapes, and (6) architectural shapes. An extrusion can be cut to the desired length, which is then used to make parts such as (1) coat hangers, (2) brackets, (3) gears, and (4) door handles. Commonly used metals for the extrusion process are aluminum, copper, lead, magnesium, and the family of steels. A few other metals and their alloy counterparts can also be extruded. Since each billet is often extruded individually, the extrusion technique can be categorized to be a batch or semicontinuous process. The process can be economical for both sheet as well as thicker products and for both long and short production runs. Tool costs are generally low, particularly for processing of both simple and solid cross sections. Depending on ductility of the chosen metal, the process is often carried out at room temperature (25°C) or at an elevated temperature. Extrusion that is carried out at room temperature (25°C) is referred to as cold extrusion. The cold extrusion process has been used for numerous applications such as (1) fasteners, (2) components for automobiles, (3) components of a bicycle, (4) components of a motorcycle, (5) heavy machinery, and (6) transportation equipment.

14.4.3.2 Process of Extrusion

There are essentially three basic types of extrusion. This is neatly depicted in Figure 14.40. In direct or forward extrusion, a billet is placed in a container and forced through a die (Figure 14.39). This is shown as well in Figure 14.40.

The die opening may be round, or it may have other shapes, depending on the shape requirement for the desired product. The purpose of the "dummy" block, shown in the figure, is to protect the tip of the pressing punch, particularly during hot extrusion.

In *indirect extrusion*, also known as *reverse*, *inverted*, or *backward* extrusion, the die moves toward the unextruded billet. Indirect extrusion has the advantage of having no billet–container friction, since there is no relative motion. This is shown in Figure 14.40a. This technique is often used on materials having very high friction, such as hot extrusion of high-strength steels and stainless steels.

In *hydrostatic extrusion*, the billet is often smaller in diameter than the container, which is filled with a fluid. The pressure is transmitted to the fluid by a ram, as shown in Figure 14.40b. The fluid pressure imparts triaxial compressive stresses acting on the chosen workpiece or metal resulting in improved formability. In this technique, there is less workpiece–container friction than in direct extrusion. Another less common type of extrusion that is used in industry is "lateral extrusion," also known as "side" extrusion. This is shown in Figure 14.40c.

The ratio of the cross section of the billet (A_o) to the cross-sectional area of the extruded part (A_f) is known as the extrusion ratio. Other variables that exert an influence on the extrusion process are (1) temperature of the billet, (2) ram speed, and (3) type of lubricant used. The force that is required for extrusion depends on the following: (1) strength of the chosen material, (2) extrusion ratio, and (3) friction between the billet, the container, and the surface of the die.

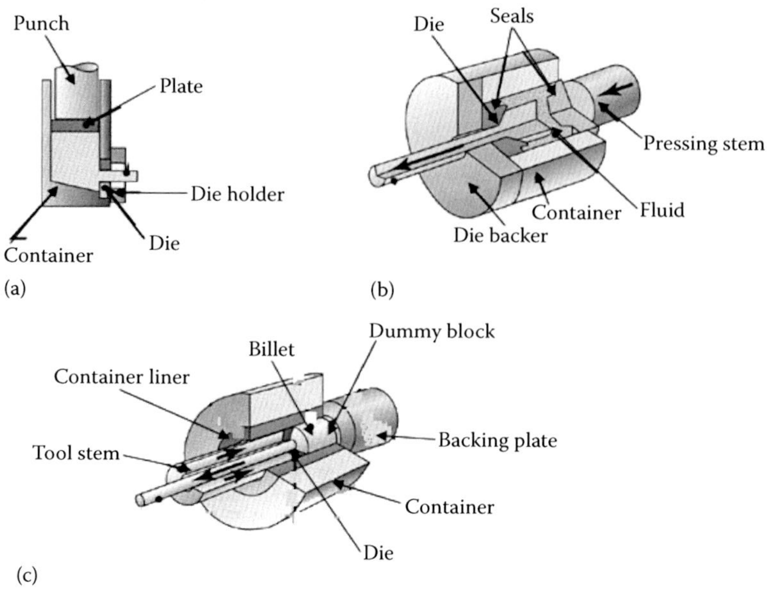

FIGURE 14.40 Types of extrusion: (a) indirect extrusion, (b) hydrostatic extrusion, and (c) lateral extrusion.

For a small value of the die angle (α), the extrusion pressure is obtained using the relationship (ASM Handbook 2005; Bauser et al. 2006)

$$P = \sigma_{ys}[1 + \tan\alpha/\mu][R^{\mu\cot(\alpha)} - 1] \tag{14.6}$$

In this expression σ_{ys} is the yield strength of the chosen metal, μ is the coefficient of friction, and R is the extrusion ratio. The extrusion force is obtained by multiplying this pressure with billet area and is expressed as

$$F = A_o K \ln[A_o/A_f] \tag{14.7}$$

In this expression, K is the extrusion constant that is determined experimentally. K is a measure of the strength of the material being extruded under prevailing conditions of friction.

The flow of the metal during extrusion, as in other forming processes, is important because of its influence on both the quality and properties of the extruded product. The material tends to flow longitudinally. This causes the extruded product to have an elongated grain structure or preferred orientation. Improper flow of the metal during extrusion results in the presence of defects in the extruded product.

14.4.3.3 Hot Extrusion

This process technique is preferred and used for both metals and their alloy counterparts that do not have sufficient ductility at room temperature (25°C). It is also used or preferred in situations that necessitate the need for reduced force during extrusion. As in all other elevated temperature operations, hot extrusion has special requirements that result from the high operating temperatures involved. In this technique, wear of the die can be excessive. Cooling of hot surfaces of the billet and the die can result in the workpiece billet experiencing highly non-uniform deformation. This necessitates the need for preheating of the extrusion dies as is done in the operation of hot forging.

Since the workpiece billet is at a high temperature, it tends to easily develop an oxide film on its surface, unless heating of the billet is done in an inert atmosphere. The oxide film can be abrasive and exert a profound influence on flow or flow pattern of the material being extruded. Presence of an oxide film also results in an extruded product that may be unacceptable when good surface finish is the requirement.

The die materials chosen for the purpose of hot extrusion are the hot work die steels. Coatings such as partially stabilized zirconia (PSZ) are often applied to the dies with the prime objective of enhancing their life. The dies made of PSZ are also used for the hot extrusion of both rods and tubes. However, the dies are not suitable for extruding complex shapes. This can be ascribed to the development and presence of stress gradients in the die that are conducive for favoring their premature failure.

Use of a lubricant, or lubrication, is both important and essential in hot extrusion. This is because of its influence on the following:

i. Material flow during extrusion
ii. Surface finish and integrity
iii. Quality of the product
iv. Forces required in extrusion

An excellent lubricant that is chosen and used for steels, stainless steels, high-temperature metals, and their alloy counterparts is glass. For those metals that are either prone or tend to stick or even weld themselves to the die and container, the billet is enclosed in a thin-walled container that is made of a softer and lower strength metal, such as copper or mild steel. This procedure is referred to in industry as *canning* or *jacketing*. The jacket, in addition to acting as a low-friction interface, prevents contamination of the billet by the surrounding environment.

14.4.3.4 Cold Extrusion

This is a general term that denotes a combination of operations, such as combination of direct extrusion and indirect extrusion and forging. A schematic of this technique is shown in Figure 14.41.

The cold extrusion technique is extensively used for components in automobiles, motorcycles, bicycles, appliances, and both transportation equipment and farm equipment. The force (F) in cold extrusion is estimated using the relationship

$$F - [1.7\sigma_{average} A_o \varepsilon] \tag{14.8}$$

where $\sigma_{average}$ is the average flow stress of the chosen metal, A_o is the initial cross-section area of the metal blank, and ε is the true strain that the workpiece experiences, based on its initial cross-sectional area and final cross-sectional area.

The advantages of the cold extrusion technique over hot extrusion are the following:

1. Improved mechanical properties resulting as a direct consequence of work hardening. This is true provided that the heat that is generated both by plastic deformation and friction does cause recrystallization of the microstructure of the extruded metal.

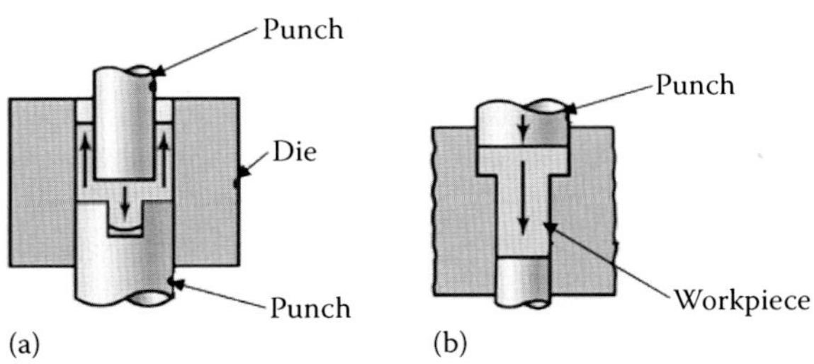

FIGURE 14.41 Two examples of cold extrusion. Arrows indicate the direction of metal flow during extrusion: (a) indirect extrusion, and (b) direct extrusion.

2. Good control of dimensional tolerances, thus reducing the need for subsequent machining and/or finishing operations.
3. Improved surface finish due to the absence of an oxide film and provided the appropriate lubricant was used.
4. The production rates and costs are competitive with those of other methods that are used for producing the same part.

Overall, the magnitude of stresses experienced by the tooling during extrusion is noticeably high, especially when the workpiece materials chosen are carbon steel and alloy steel. Lubrication becomes essential especially when working with steels. This is because of the possibility of sticking, or seizure, between the workpiece of interest and the tooling that can occur in the event of breakdown of the lubricant. The most effective method of lubrication is the application of phosphate-conversion coating on the workpiece, followed by a coating of soap or wax. Also, in cold extrusion, both tool design and the selection of the appropriate tool and die materials are essential for the overall success of the cold extrusion technique. Also important are selection and control of the workpiece material with specific regard to its quality and repeat accuracy of dimensions of the slug and its surface condition.

14.4.3.4.1 Impact Extrusion

This technique is quite like indirect extrusion. The process or technique is essentially included in the category of cold extrusion. The punch descends rapidly on the blank or slug. This causes the blank to extrude backwards. This is neatly shown in the schematic in Figure 14.42 (Ginsberg 2001; Gourley and Walker 2012).

Due to the need for maintaining volume constancy, the thickness of the tubular extruded region is a function of clearance between the punch and the die cavity. A few parts or products that are easily made using this process technique are the following: (1) collapsible tubes, (2) light fixtures, (3) automobile parts, and (4) small pressure vessels. Most nonferrous metals can be impact-extruded in vertical presses and at a high production rate.

14.4.3.4.2 Hydrostatic Extrusion

In hydrostatic extrusion, the pressure that is required in the chamber is supplied using a piston through an incompressible fluid medium that surrounds the billet. The high pressure in the chamber transmits some of the fluid to the die surfaces, where it contributes to reducing the friction. Hydrostatic extrusion is usually carried out at room temperature (25°C) after using vegetable oil as the fluid. Brittle materials can be extruded successfully by this method because the high hydrostatic pressure in synergism with low friction, the use of small die angles, and high extrusion ratios increase the ductility of the material. Hydrostatic extrusion has limited industrial application primarily because of the complex nature of tooling, the design of specialized equipment, and the need for long cycle times. A combination of these makes the technique uneconomical for most materials.

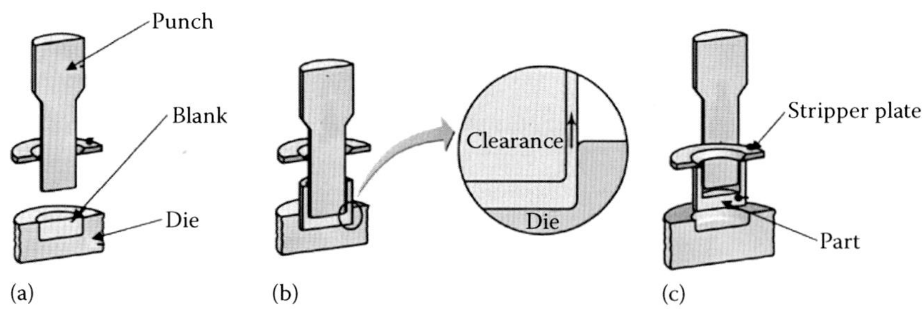

FIGURE 14.42 Schematic illustration of the impact extrusion process: (a) Punch descends on the die blank or slug. (b) Blank or slug is extruded backward. (c) The extruded part is stripped using a stripper plate.

14.4.3.5 Defects in Extrusion

Depending on the condition of the workpiece material and the process variables used, the extruded products can develop several types of defects that exert an influence on both their strength and quality of the product. Some of these defects are visible to the naked eye, while few other defects can be detected by the prudent use of nondestructive techniques. The three most common types of extrusion defects are the following: (1) surface cracking, (2) pipe cracking, and (3) internal cracking.

14.4.4 Drawing of Metals

14.4.4.1 Background

In drawing the cross section of a rod or wire, it is often reduced by pulling it through a die (hence the term "drawing"). The die used for this purpose is referred to as the draw die. A schematic of the drawing technique or process is shown in Figure 14.43.

The difference between drawing and extrusion is that in extrusion, the material is pushed through a die, whereas in drawing, it is pulled through a die. Drawn rod and wire products cover a wide range of applications that include the following: (1) shafts used for the transmission of power, (2) components of a machine, (3) components of load-bearing structures, (4) blanks for bolts and nuts, (5) electrical wiring, (6) cables, (7) welding electrodes, (8) springs, (9) paperclips, (10) spokes of a bicycle wheel, and (11) musical instruments containing strings. The major processing variables in drawing are quite like those in extrusion, that is, (1) reduction in the cross-sectional area, (2) die angle, (3) friction along the die–workpiece interface, and (4) drawing speed.

An expression for the force that is required for drawing under ideal and frictionless conditions is quite like the expression for extrusion and is given by the equation

$$F = \sigma_{\text{average}} A_f \ln[A_o/A_f] \quad (14.9)$$

In this equation, σ_{average} is the average true stress of the material in the die gap. Since work must be done to overcome friction, the force that is required for drawing often increases for the following: (1) a large reduction in the cross section area and (2) the presence of increasing friction. Further, due to non-uniform deformation within the die zone, additional energy, known as redundant work of deformation, is required.

Overall, achieving success in drawing requires a proper selection of the process parameters. During drawing, the reduction in the cross sectional area per pass ranges up to about 45%. The smaller the initial cross-section area, the smaller the reduction that can occur per pass. Fine wires are often drawn at 15–25% reduction per pass. Reduction in the cross-section area of higher than 45% can and does result in breakdown of the lubricant, resulting in rapid deterioration of the surface finish of the drawn component or product. Most drawing operations are done at room temperature (25°C).

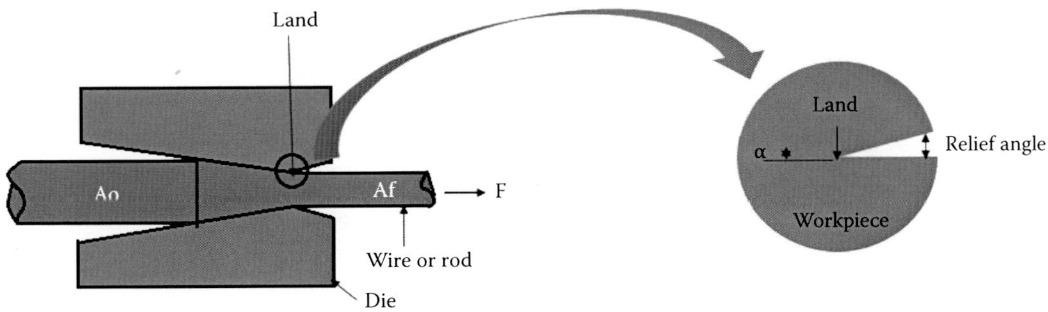

FIGURE 14.43 Key process variables in wire drawing. Reduction in cross-sectional area per pass, speed of drawing, temperature, and lubrication condition influence the drawing operation.

However, drawing of the large solid and even hollow sections is often done at an elevated temperature to reduce the force that is required for drawing (ASM Handbook 2005).

Use of a lubricant is essential in all drawing operations to improve both the die life and surface finish of the drawn product while concurrently reducing the drawing force and the temperature rise that can occur during the drawing operation. Lubrication becomes critical in the drawing of tubes. This is essentially because of the difficulty in maintaining a thick film of the lubricant at the tube–mandrel interface. For the drawing of rods, a common method of lubrication uses phosphate conversion coatings.

14.4.4.2 Defects in Drawing

Typical defects in a drawn rod or drawn wire are like those observed during extrusion. This includes center cracking. Seams are another major defect in drawing; these are longitudinal scratches or folds in the drawn product. A few other surface defects, such as scratches and die marks, can be attributed to the following: (1) an improper selection of the process parameters, (2) poor condition of the die, and (3) inadequate lubrication. Since they experience non-uniform deformation during drawing, cold-drawn products experience residual stresses. For light reductions of the order of a few percent, the longitudinal surface residual stresses are compressive in nature, while the bulk is in tension and fatigue life is significantly improved. However, heavier reductions tend to induce tensile residual stresses on the surface, while the bulk is in compression. The presence of residual stresses can cause the part to experience degradation due to stress corrosion cracking.

14.5 MACHINING OF METALS AND MATERIALS

14.5.1 Background

The process of cutting helps in the removal of material from various surfaces of a given workpiece by the production of chips. The commonly used machining operations, shown in Figure 14.44, are the following:

a. *Turning*. In this process, the workpiece is rotated and a cutting tool is used to remove a layer of material as the tool moves along its length.
b. *Cutting off*. In this process, the tool moves radially inward and in the process separates the piece on the right from the blank.
c. *Slab milling*. In this process, a rotating cutting tool removes a layer of material from the surface of the workpiece.
d. *End milling*. In this process, a rotating cutter travels along a certain depth in the workpiece to produce a cavity.

Essentially, in the turning process, the cutting tool is set at a certain depth of the cut required and gradually travels to the left as the workpiece rotates. The feed rate or feed is the distance the tool travels per unit revolution of the workpiece material (expressed in mm/revolution). It is this

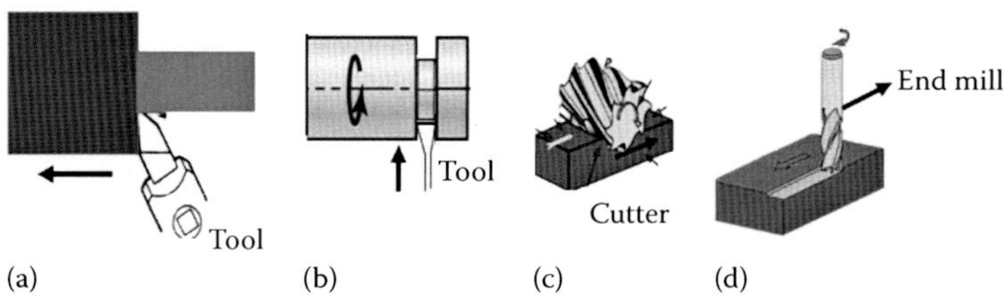

FIGURE 14.44 Few examples of commonly used machining operations. (a) Straight turning, (b) cutting off, (c) slab milling, and (d) end milling.

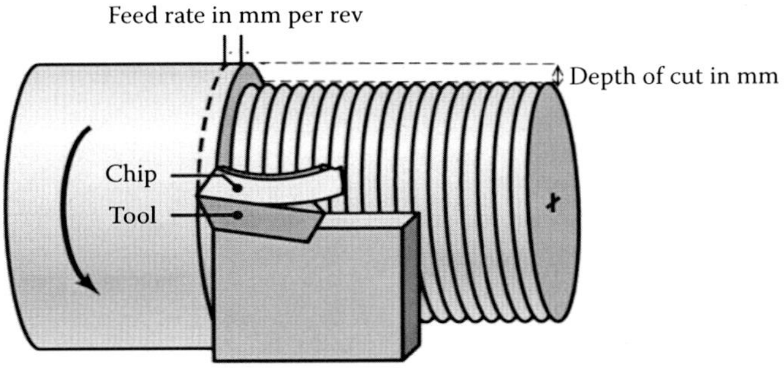

FIGURE 14.45 Schematic illustration of the turning operation, showing various features.

movement of the cutting tool that produces a chip, which moves up the face of the cutting tool. This is neatly depicted in Figure 14.45.

14.5.2 Mechanics of Cutting

The key variables influencing the basic cutting process are the following (ASM Handbook 1989; Boothroyd and Knight 2006):

i. Tool materials and nature of coatings present on the surface of the chosen workpiece, if any
ii. Shape of the cutting tool, surface finish, and sharpness of the chosen cutting tool
iii. Workpiece material and its processing history
iv. Speed, feed, and depth of cut during cutting
v. Use of cutting fluids
vi. Characteristics of the machine tool
vii. Type of work-holding device and fixtures used to hold the workpiece material and the cutting tool

The dependent variables in cutting are those that easily influenced by the changes made to the independent variables. A few of the dependent variables are as follows: (1) type of chip that is being produced; (2) force and energy that are dissipated during cutting; (3) rise in temperature experienced by the workpiece, tool, and chip; (4) wear of tool as a direct consequence of machining and nature of failure, if it occurs; and (5) surface finish imparted to the workpiece material and overall integrity of the surface.

Microscopic examination of the chips produced during machining has revealed that they are essentially produced by shearing. The shearing is confined to occur within a shear zone (i.e., a well-defined plane that is also referred to as the shear plane) and at an angle, which is referred to as the shear angle. Below the shear plane, the workpiece material remains undeformed, while above this plane, the chip that is formed is moving up the rake face of the tool. This is shown in Figure 14.46a. If the thickness of the chip is greater than the depth of the cut, then the chip velocity (V_C) must be lower than the cutting speed (V). To maintain mass continuity, we have

$$Vt_O = V_C t_C \tag{14.10}$$

or

$$V_C = V(r)$$

where $r = t_O/t_C$

$$V_C = V \sin\Phi/\cos(\Phi - \alpha) \tag{14.11}$$

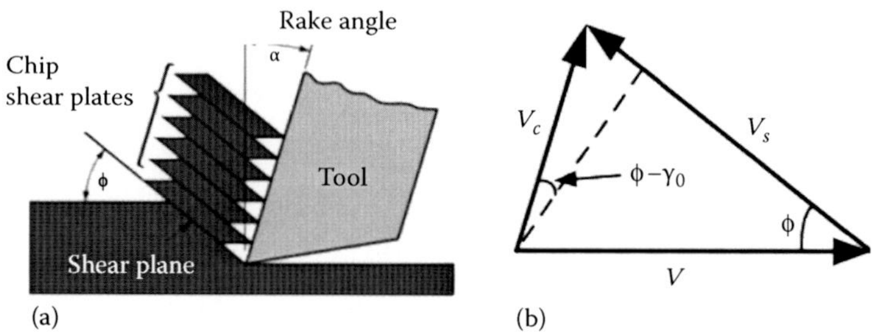

FIGURE 14.46 (a) Schematic illustration of basic mechanism of chip formation by shearing. (b) Velocity diagram showing angular relationships among the three speeds (V, V_S, V_C) in the cutting zone.

A velocity diagram can now be constructed, as shown in Figure 14.46b. From trigonometric relationships, we can show that

$$V/\cos(\Phi - \alpha) = V_S/\cos\alpha = V_C/\sin\Phi \tag{14.12}$$

where V_S is the velocity at which the shearing takes place on the shear plane.

14.5.3 Types of Chips Resulting from Machining by Metal Cutting

The four main types of chips resulting from metal cutting are (Boothroyd and Knight 2006; Cornier 2005; Shaw 2005)

1. Continuous chips
2. Built-up edge (BUE) chips
3. Serrated or segmented chips
4. Discontinuous chips

A chip has essentially two surfaces. The surface that is in contact with the rake face of the tool has a shiny appearance resulting because of sliding of the chip up the tool face. The other surface is from the exterior surface of the workpiece. It has a jagged and/or a rough appearance that is the result or outcome of shearing mechanism.

14.5.3.1 Continuous Chips

The continuous chips are a by-product of a ductile material that is machined at high cutting speeds and large rake angles. Deformation of the chosen workpiece material occurs along a narrow shear zone that is referred to as the primary shear zone. While the continuous chips produce a good surface finish, they are, however, not desirable as they tend to get entangled around the tool holder, fixturing, and even the workpiece material. Thus, they tend to interfere with the chip dispersal system. The problem is alleviated using a chip breaker or changing the process parameters, such as cutting speed, feed, and depth of the cut, or by using a cutting fluid during cutting.

14.5.3.2 Built-Up Edge

BUE chips consist of layers of material from the workpiece that get deposited at the tip of the cutting tool. This gives it the name built-up edge. As the BUE grows larger in size, it becomes unstable and eventually ruptures. While a portion of the BUE material is carried away by the tool side of the chip, a portion does get deposited onto the surface of the workpiece, thereby impairing the surface finish. A thin and stable BUE is regarded as being desirable since it tends to reduce tool wear by protecting its rake face. In general, the cold-worked metals have a lower tendency to form a BUE than an annealed counterpart. Due essentially to work hardening coupled with the deposition of successive layers of material, a BUE hardens with time. The tendency for the formation and presence of a BUE can be reduced by either one or a combination of these mechanisms: (1) an increase in cutting speed, (2) a decrease in the depth of the cut, (3) an increase in the rake angle, (4) the use of a cutting tool that has lower chemical affinity for the workpiece material, (5) the use of a sharp tool, and (6) the use of an effective cutting fluid.

14.5.3.3 Serrated Chips

These chips are also known or called *segmented* or *nonhomogeneous chips*. They are essentially semicontinuous chips having large zones of low shear strain and small zones of high shear strain. The chips have a sawtooth-like appearance. Metals that have low thermal conductivity and a strength that decreases sharply with rise in temperature, i.e., thermal softening, are prone to the formation of serrated chips. A typical example of material where this has been observed is titanium and its alloys.

14.5.3.4 Discontinuous Chips

These chips consist of segments that are attached either firmly or loosely to each other. These chips tend to form when the following conditions exist (ASM Handbook 1989; Trent and Wright 2000):

1. A brittle workpiece material that does not have the capacity to withstand the high shear strains that occur during machining
2. A workpiece material that contains a sizeable quantity of hard, brittle, and elastically deforming inclusions and impurities, or the presence and distribution of graphite flakes in gray cast iron
3. Very low or very high cutting speed (V)
4. Large depth of cut (d)
5. Low rake angle (α)
6. Absence of an ideal cutting fluid
7. An overall low stiffness of the machine tool or tool holder that is conducive for the occurrence of both the vibration and the chatter.

Due to the discontinuous nature of chip formation, the cutting forces vary continuously during machining. Thus, the stiffness of the cutting tool holder and other workpiece holding devices and the machine tool are key factors in machining with the formation of serrated or discontinuous chips. Vibration of the cutting tool can adversely affect the surface finish and dimensional accuracy of the machined part, and can cause premature wear or damage to the cutting tool (ASM Handbook 1989; Trent and Wright 2000; Smith G.T. 2008].

14.5.4 Temperatures during Cutting

The energy that is dissipated during cutting is often converted to heat, which contributes to raising the temperature both in the cutting zone and the surface of the workpiece. The rise in temperature is an important factor in machining because of its adverse influence on the following:

a. Lowers the strength, hardness, stiffness, and even wear resistance of the cutting tool. The cutting tool tends to soften and undergoes plastic deformation that can cause intrinsic changes in its shape.
b. The increased heat can cause dimensional changes in the part being machined depending on physical properties of the material that the part or component is made of. This makes it difficult to control dimensional accuracy and tolerance of the part being produced.
c. Excessive increase in temperature can cause thermal damage and even induce metallurgical changes on the machined surface, thereby affecting and/or influencing its properties.

The source of heat generation during machining are concentrated in

i. The primary shear zone
ii. Energy that is dissipated at the tool–chip interface
iii. Heat that is generated as the tool rubs against the machined surface, especially for the case of dull tools

Thus, one can expect severe temperature gradients to exist in the cutting zone. The rise in temperature in the cutting zone is determined using thermocouples embedded on both the tool and the workpiece. It is easier to determine the mean temperature with thermal electromotive force at the tool–chip interface, which acts as a hot junction between the two different materials, i.e., material of the tool and material of the chip.

14.5.5 Tool Life, Wear, and Failure of the Cutting Tool

Wear of the cutting tool is a major consideration in all machining operations, as are mold wear in casting operations and die wear in metal working processes. Tool wear has an adverse influence on the following: (1) tool life, (2) the quality of the machined surface, (3) dimensional accuracy of the machining process, and (4) overall economics of the cutting operation.

Wear is basically a gradual process. The rate of tool wear, defined as the volume worn per unit time, depends on the workpiece material, tool material, tool geometry, process parameters chosen, presence of coatings, if any, and characteristics of the machine tool. Tool wear and resultant changes in tool geometry can be classified as follows: (1) flank wear, (2) wear of the crater, (3) nose wear, (4) notching, (5) plastic deformation, (6) chipping, and (7) gross failure by fracture.

14.5.6 Machinability

The machinability of a material is defined in terms of the following four factors:

1. Surface finish and surface integrity of the machined part or component
2. Life of the tool
3. Requirements specific to both force and power
4. The level of difficulty associated with control of the chip after it has been produced

A good machinability is indicative of good surface finish coupled with an overall integrity of the surface, long tool life, and low force and power requirements. Due to the overall complex nature of the cutting operations, it is often difficult to establish specific relationships that quantitatively define the machinability of a material. In actual machining practice, both tool life and surface roughness are properties considered to be most important factors in governing machinability. The machinability ratings have been available for many years for each type of workpiece material and its condition. However, due to their qualitative nature, they are not particularly useful and reliable under all conditions of machining.

14.5.7 Materials for Cutting Tools

Selection of a cutting tool material for an application is an important factor in the operation of machining. Often in practice, a cutting tool is subject to a synergistic influence of high temperature, high contact stresses, and even rubbing along the tool–chip interface and the machined surface. Consequently, the cutting tool material must possess the following characteristics:

1. Hot hardness: To maintain hardness, strength, and wear resistance of the cutting tool at the temperatures encountered during machining. This property ensures that the cutting tool does not undergo any plastic deformation and retains both its shape and sharpness.

2. Toughness and impact strength so that the impact forces on the tool encountered repeatedly during interrupted cutting operations, such as milling and turning a shaft on a lathe, or forces that arise due to vibration and chatter during machining, do not chip or fracture the tool.
3. Thermal shock resistance to withstand the rapid temperature cycling, as encountered during interrupted cutting.
4. Wear resistance so that an acceptable tool life is obtained before replacement becomes essential.
5. Chemical stability and inertness with respect to the workpiece material to either avoid or minimize the occurrence of any adverse reactions, tool–chip diffusion, and adhesion that would contribute to tool wear.

To respond to these demanding requirements, a variety of cutting-tool materials having a wide range of mechanical, physical, and chemical properties have been developed over the years. The characteristics of the cutting tool material are summarized in Table 14.4. The tool materials that have been developed, put forth, and used are the following:

1. High speed steels
2. Cast cobalt alloys
3. Carbides
4. Coated tools
5. Alumina-based ceramics
6. Cubic boron nitride
7. Silicon–nitride based ceramics
8. Diamond
9. Whisker-reinforced materials
10. Nanomaterials

TABLE 14.4
General Operating Characteristics of Cutting Tool Materials

Tool Materials	General Characteristics	Modes of Tool Wear or Failure	Limitations
High-speed steels	High toughness, resistance to fracture, wide range of rough and finish cuts, good for interrupted cuts	Flank wear, crater wear	Low hot hardness, limited hardenability, limited wear resistance
Uncoated carbides	High hardness over a wide range of temperatures, toughness, wear resistance, versatile, wide range of applications	Flank wear and crater wear	Cannot use at low speeds because of cold welding of chips and microchipping
Coated carbides	Improved wear resistance over uncoated carbides, better friction and thermal properties	Flank wear and crater wear	Cannot use at low speed because of cold welding of chips and microchipping
Ceramics	High hardness at elevated temperatures, high abrasive wear resistance	Depth of cutline notching, microchipping, gross fracture	Low strength and low thermomechanical fatigue
Cubic born nitride (cBN; polycrystalline)	High hot hardness, toughness, cutting-edge strength	Depth-of-cut line notching, chipping, oxidation, graphitization	Low strength, low chemical stability than ceramics at higher temperatures
Diamond	High hardness and toughness, abrasive wear resistance	Chipping, oxidation, graphitization	Low strength and low chemical stability at high temperatures

Many of these materials also find use for dies and molds in the casting, forming, and shaping of both metallic and nonmetallic materials.

The family of carbon steels are by far the oldest tool materials that are used for the purpose of drills, taps, broaches, and even reamers. The low alloy steels and medium alloy steels were developed later for similar applications. Although inexpensive, and easily shaped and sharpened, these steels do not have sufficient hot hardness and wear resistance for the purpose of machining at high speeds, where the temperature rises significantly. Their use is essentially restricted to very low-speed cutting operations, such as wood working, and hence they are not of great importance in the machining of new and improved materials.

14.5.8 Cutting Fluids

The cutting fluids find extensive use in machining operations for the following reasons (Astakhov and Joksch 2012; Nachtman and Kalpakjian 1985; Smith 2008):

a. To reduce friction and wear, thereby improving both tool life and surface finish of the workpiece
b. To cool the cutting zone, thereby improving tool life while concurrently reducing both temperature and thermal distortion, if any, experienced by the workpiece
c. Reduce both force and energy consumption
d. To flush away the chips from the cutting zone and thereby prevent the chips from interfering with the cutting operation, especially during the operations of drilling and tapping operations
e. To protect the machined surfaces from environment-induced corrosion

Depending on the type of machining operation, the cutting fluid that is needed may be a coolant or a lubricant or both. The overall effectiveness of the cutting fluid depends on the following factors:

1. Type of machining operation
2. Tool and workpiece materials
3. Cutting speed
4. Method of application

Water is an excellent coolant and can effectively reduce the high temperatures developed in the cutting zone. Since water cannot be an effective lubricant, it cannot reduce friction while concurrently causing the occurrence of corrosion of both the workpiece and components of the machine tool.

The need and necessity for use of a cutting fluid largely depends on the severity of the specific machining operation. The degree of severity is established based on

a. The temperatures and forces encountered coupled with an inability of the tool material to withstand them
b. The tendency for formation of BUE
c. The ease with which the chips that are produced can be removed
d. The effectiveness with which the fluid can be supplied to the tip of the tool–workpiece interface

The relative severities of specific machining processes, in increasing order of severity, are sawing, turning, milling, drilling, gear cutting, thread cutting, tapping, and internal broaching (Nachtman and Kalpakjian 1985; Smith 2008).

The key characteristics and applications of four general types of cutting fluids that are commonly used in machining operations are as follows:

1. *Oils*. Also referred to as straight oils and include mineral, vegetable, and compounded and synthetic oils. These oils are typically used for low-speed operations where the rise in temperature is not significant.
2. *Emulsion*. Also called soluble oils, they are essentially a mixture of oil, water, and additives. They can be used for high-speed operations where the rise in temperature is significant. The presence of water does make an emulsion a highly effective coolant. The presence of oil either eliminates or reduces the tendency of water to cause oxidation of the workpiece surfaces.
3. *Semisynthesis*. These are essentially chemical emulsions containing little mineral oil that is diluted with water along with the presence of additives whose purpose is to reduce the size of the oil particles, thereby making them more effective.
4. *Synthetics*. These are chemicals with additives, diluted in water, and contain no oil.

The four basic methods used for the application of the cutting fluid are (1) flooding, (2) mist, (3) high-pressure systems, and (4) through the cutting tool. Selection and application of the cutting fluid should also consider the following factors: (1) nature of workpiece material, (2) machine tool components, (3) biological considerations, and (4) environment (Smith 2008).

14.5.9 Machining Processes

The machining processes are used to remove unwanted material from a workpiece to give it proper shape and size coupled with the required surface finish. This is often accomplished using various machine tools, such as lathe, shaper, milling cutters, etc. The lathe is the oldest and most commonly used machine tool. Its purpose is to essentially aid in the removal of material by rotating the workpiece against a single-point cutter. Several operations can be performed on a lathe. The operations performed on a lathe can be categorized into two groups:

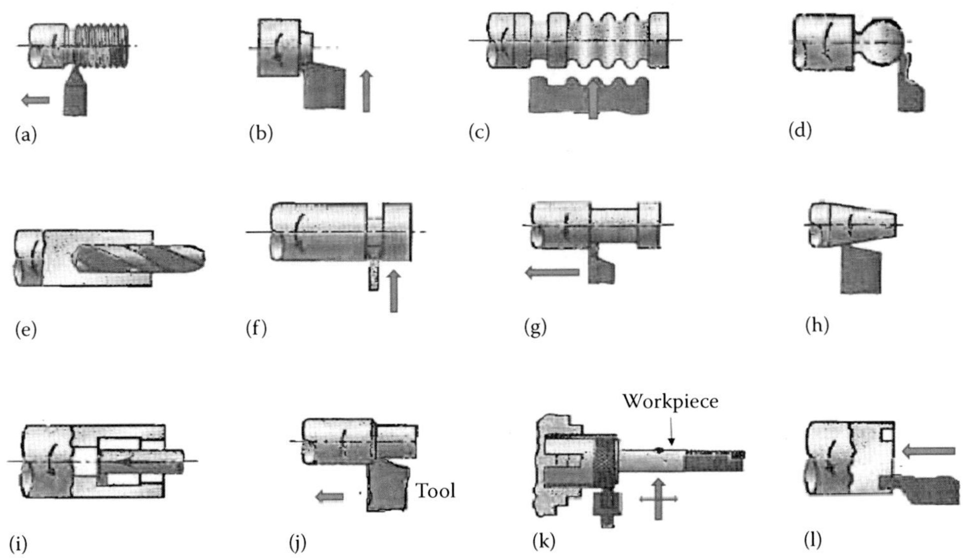

FIGURE 14.47 Few operations that can be performed on a lathe. (a) Threading, (b) facing, (c) cutting with a form of tool, (d) profiling, (e) drilling, (f) cutting off, (g) turning and external grooving, (h) taper turning, (i) boring and internal grooving, (j) straight turning, (k) knurling, and (l) face grooving.

a. Standard operations: these include (1) turning to include straight turning and taper turning, (2) facing, (3) drilling, (4) boring, (5) parting, (6) knurling, (7) reaming, (8) milling, and (9) tapping. Only a few of these are shown in Figure 14.47.
b. Special operations include (1) grinding, (2) milling, (3) copying or duplicating, (4) spinning, (5) taping, and (6) dyeing.

14.5.9.1 Turning

During turning, the workpiece rotates while it is being machined. This is essentially a straightforward metal-cutting method and the most widely used. It can be used to produce straight, conical, and even curved surfaces on the workpiece. It can be easily categorized to be a combination of two movements. The first is rotation of the workpiece and the second is feed movement of the tool. The feed movement of the tool is often along the axis of the workpiece, implying that the diameter of the workpiece will be reduced or turned down to a smaller size. Alternatively, the tool can be fed toward the center, at the end of the part, which results in the length of the workpiece being faced down. Situations in which the feed of the cutting tool is a combination of these two directions result in the production of a tapered surface or curved surface. The turning operation showing removal of a chip is illustrated in Figure 14.48.

14.5.9.2 Facing

This is a turning operation in which a flat surface is produced by feeding the cutting tool normal to the axis of rotation of the workpiece. The cutting speed continuously decreases as the axis of the workpiece is approached.

14.5.9.3 Drilling

This is essentially a process of producing a hole in the workpiece of interest by forcing a rotating drill against it. In fact, hole making is among the most common operations in manufacturing, and drilling is a major and commonly used hole-making process. One other process that is widely used for producing holes is punching. The drills typically have a high length-to-diameter ratio. Hence, they can produce deep holes in the workpiece of interest. The high length-to-diameter ratio makes the drill somewhat flexible but equally prone to fracture or produce inaccurate holes in the workpiece. During drilling, the metal chips that are produced within the hole do present an observable difficulty in their removal or disposal and thereby difficulty in ensuring effectiveness of the cutting fluid. The most common drill that is currently used is the conventional standard point twist drill. Other types of drill

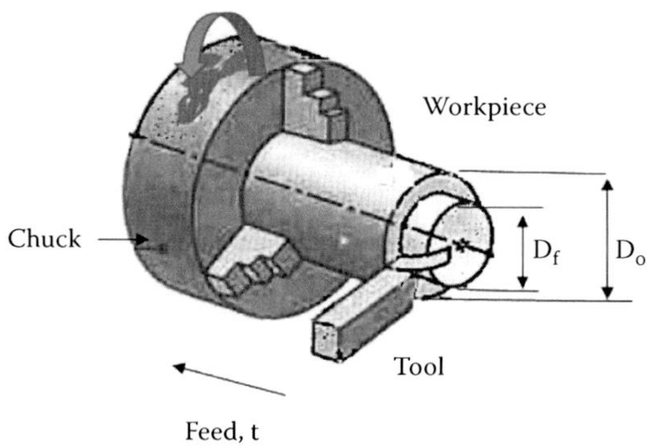

FIGURE 14.48 Illustration of the basic turning operation showing the depth of cut (d), feed (f), and spindle rotation speed (N) in revolutions/min. Cutting speed is the surface speed of the workpiece at the tip of the cutting tool.

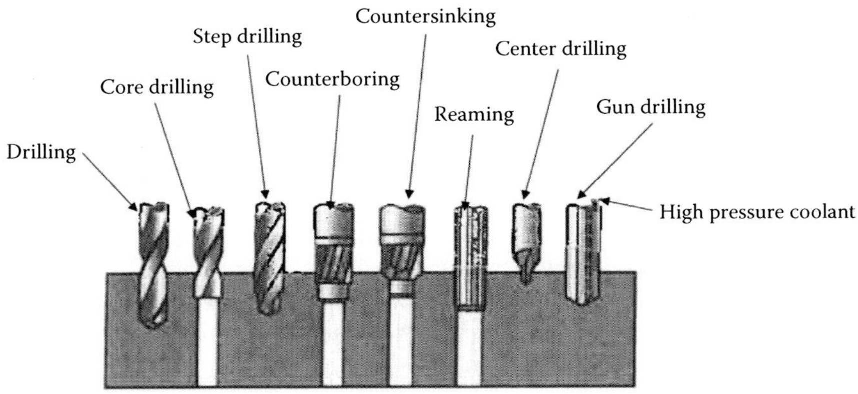

FIGURE 14.49 Various types of drills used in the operation of drilling.

that have been and can be used are (1) spade drill, (2) straight flute drill, (3) drill with carbide inserts, and (4) drill with a brazed carbide tip. A schematic of these drills is shown in Figure 14.49.

14.5.9.4 Boring

This is a process that is used for enlarging a hole that has already been drilled in the workpiece of interest. Thus, boring does not create the hole but merely machines or opens the hole up to the desired size. While performing boring on a lathe, the workpiece is usually held in the chuck or the faceplate. While the workpiece rotates, the tool is forced against the workpiece. Boring can also be done using a rotating tool with the workpiece remaining stationary. Boring, which is also referred to as internal turning, is done using cutting tools that are like those used in turning.

14.5.9.5 Parting

This is also called as cutting off. It is a process that is used to cut a piece from the end of a part. The tool is fed normal to the axis of rotation as in the case of facing. Parting essentially uses thin tools with considerable overhang. Hence, the accuracy of the process is difficult to achieve.

14.5.9.6 Knurling

This is a process that is used to produce regularly shaped roughness or indentations that are called knurls on a cylindrical surface. This is basically a chipless and cold-forming process that is performed by using a tool called the knurling tool on a lathe. The tool consists of a straight shank that is fitted with one or two knurling wheels in its front. The wheels are made of hardened tool steel and carry teeth on its outer surfaces. The tool is fed across the axis of the workpiece while being pressed against its surface. The straight, diagonal, and diamond-shaped knurlings are commonly used.

14.5.9.7 Reaming

This is a process of enlarging a machined hole to a proper size and having a smooth finish. It is done for two purposes: (1) to bring the hole to a more exact size and (2) to improve the internal finish of an existing hole. The cutting tool used in the process is called reamer. It is a multiple cutting-edge tool that has either straight or helically fluted edges. The operation of reaming is normally performed at a slow speed. The speed is usually two-thirds of the speed used for drilling the same material. For the purpose of close tolerances and a fine finish, the speed of operation should be slow. The feed is usually much higher than those used for drilling and depends on the material that is being reamed.

Several reamers are available for performing the operation. A few of the commonly used ones are (1) hand reamer, (2) machine (chucking) reamer, (3) shell reamer, and (4) adjustable reamer.

The hand and machine reamers are categorized to be basic types of reamers. Each reamer is used for a specific application. Reamers are usually made of high-speed steel or solid carbides, or have carbide tip cutting edges. No special machines are built for the operation of reaming. The drilling machine can be used for reaming just by changing the cutting tool.

14.5.9.8 Milling

This is a process that is used to cut away material by feeding a workpiece past a rotating cutter that has multiple tooth. At times, the workpiece remains stationary and the cutter is fed to the workpiece. The machined surface maybe flat, curved, angular, or simply a combination of these. Since several teeth are involved in the cutting operation, the rate of metal removal is high. The tool life is also extended. Milling is well suited and widely used for the purpose of mass production. The machine that is used for holding the workpiece, rotating the cutter, and concurrently feeding it to the workpiece is known as the milling machine.

There are several types of milling cutters. In many ways, these cutters can be used. These cutters are available in many standard and special types, forms, diameter, and width. The teeth on the cutter can be straight (i.e., parallel to the axis of rotation) or at a helix angle. The cutter may be right-handed (to turn clockwise) or left-handed (to turn counterclockwise). A few important milling operations that find extensive use in industry circles are (1) slab milling, (2) side milling, (3) end milling, (4) face milling, (5) up-milling, and (6) down-milling.

14.5.9.9 Tapping

This is a process that is used to produce internal threads in the case of a nut. The cutting tool, referred to as a tap, is a hardened screw made from tool steel. The lengthwise grooves on the tool are referred to as flutes. The tool is milled or ground across the threads. The flutes form a series of teeth that provides room for the chip along its entire length. A tap is used to cut threads in a plain hole that is drilled somewhat larger than the minor diameter of the thread. When the tap is turned into the hole, the teeth cut into the wall of the hole and remove the material to form threads having the same pitch as the threads on the tap. The most common types of taps used are (1) solid tap, (2) shell tap, (3) expansion tap, (4) inserted chaser tap, and (5) collapsible tap. It is the solid tap that is chosen and used in a large majority of tapping operations.

14.6 JOINING OF METALS

In a few to many situations, manufacturing of a part or component as a single unit may be impossible primarily because the part is made up of a few to several pieces of intricate shapes or many components. For such situations, different methods are often used by those in the industry spanning manufacturing to join the individual parts together with the primary objective of obtaining a complete unit. This can be made possible by joining. The processes of joining of parts or components to bring about a single unit can include the following: (1) welding, (2) soldering, (3) brazing, (4) adhesive bonding, and (5) conventional mechanical fastening. The choice of a specific joining process often depends on several competing factors, spanning (1) application; (2) joint design; (3) material involved; and (4) overall thickness, size, and shape of the components, that are to be held together or joined (Campbell 1999; Cart 1989; Chapman and Martin 1975).

14.6.1 The Process of Welding

14.6.1.1 Background

Basically, welding is a process by which two materials, often metals and their composite counterparts, are permanently held or joined together by the conjoint use of heat and pressure. A coalesce of the two materials, or metals, to be joined is made possible when atoms of the two chosen metals that are to be joined come close to each other such that there is only a boundary that separates the two of them. Cleaning of the surfaces prior to welding is very important. This is because the process of welding does promote oxidation of the metal to occur, which has a detrimental influence on joining of the two or more parts together to get a single unit. In solid-state welding, the layers that are contaminated are removed by either mechanical or chemical cleaning, or a combination of both prior to welding. In fusion welding, there exists a molten metal pool and the contaminants are removed using appropriate fluxing agents. With noticeable advances in technology and technology-related applications, many of the welding processes are often performed in a controlled environment or in an inert atmosphere (Campbell 1999).

14.6.1.2 Classification of Welding Processes

Through the years, commensurate with advances in technology, many welding processes have been developed. These processes tend to differ not only the way heat and pressure are applied but also how they are protected from oxidation and/or contamination during the welding process. An important method that has been used for classification is based on the composition of the joints. In accordance with this method, the joining processes are classified as follows:

a. Autogenous joining
b. Homogeneous joining
c. Heterogeneous joining

In the autogenous joining process, no filler material is used. All types of solid-phase welding and resistance welding come under this category. In the methods of homogeneous joining and heterogeneous joining, it is essential to use a filler material. The composition of the filler material that is used in homogeneous joining is like that of the parent material or metal. Typical examples of homogeneous joining include (1) arc welding, (2) gas welding, and (3) thermit welding. In the case of heterogeneous joining, composition of the filler material is different from that of the parent or original parts that are to be joined. Typical examples of heterogeneous joining are soldering and brazing processes.

Another method used in the manufacturing industry is based on the application of pressure. Accordingly, the processes are classified as follows:

a. *Plastic welding*. The two or more metals that are to be joined are heated to their plastic state and subsequently forged together by applying external pressure. The temperature to which the chosen metals are heated must be adequate to facilitate cohesion. This usually occurs at the subfusion temperature. Solid-state welding falls in this category.
b. *Fusion welding*. The metal at the joint is heated to its molten state and then allowed to solidify without the application of pressure. The molten metal flows easily to fill up the gap between the parts that are to be joined. This is very much like a casting process. The process may be either autogenous or homogeneous. Typical examples include arc welding and gas welding.

14.6.1.3 Welding and Weldability

Weldability of a material is defined as its ability to be easily welded. The weldability of a material often depends on various factors including the following: (1) metallurgical changes that occur during the process of welding, (2) hardness, (3) evolution and absorption of gas, and (4) extent and/or severity of oxidation. Among the family of metals, the low carbon steels have the best weldability. The weldability reveals an observable decrease as the carbon content in steel increases.

14.6.1.4 Selection Criteria for Welding

Welding has carved for itself the position of being the most dominant joining process in manufacturing, and there is hardly any of the commonly chosen and used metals that cannot be welded. There do exist several welding processes. However, all metals and materials cannot be easily welded by every process that is available. Therefore, selection of a specific welding process is important. For example, for sheet metal work in the industry spanning automobiles, resistance spot welding is chosen and used. Electron beam welding is chosen and used when dealing with reactive metals, and shielded metal arc welding (SMAW) is preferentially used for difficult-to-reach sections or for field welding.

The ultimate choice of a welding process often depends on several competing factors including the following: (1) application (i.e., automobile industry, aircraft industry, ship-building industry, pressure vessel fabrication, etc.); (2) joint design; (3) materials to be joined; and (4) thickness, size, and shape of the components to be joined. Other secondary considerations that must be examing are the following: (1) location of the joint (with specific reference to ease of accessibility), (2) welding position, and (3) economic considerations to include both equipment and labor cost. An overview of the appropriate welding process based on the metal or material to be joined is summarized in Table 14.5. Selection of appropriate welding process based on the joint desired is summarized in Table 14.6.

TABLE 14.5
Selection of Welding Processes Based on Metals to Be Joined

Materials to be Welded	Recommended Joining Processes
Plain carbon steel and low alloy steel	Arc welding, oxyacetylene welding, resistance welding, brazing, electron beam welding, adhesive binding
Cast iron	Oxyacetylene welding (best results), arc welding and adhesive bonding
Stainless steels	Arc welding, resistance welding, and brazing. Oxyacetylene welding, electron beam welding, soldering and adhesive bonding (commonly performed)
Aluminum and magnesium	Adhesive bonding (best results), arc welding, oxyacetylene welding, electron beam welding, resistance welding and brazing (commonly performed)
Copper and copper alloys	Brazing and soldering (best results), arc welding, oxyacetylene welding, electron beam welding resistance welding, and adhesive bonding (commonly preferred)
Thermoplastics and thermosets	Adhesive bonding (commonly preferred)
Ceramics	Adhesive bonding (best results), electron beam welding (commonly performed)
Dissimilar metals	Soldering and adhesive bonding (best results), electron beam welding (commonly performed)

TABLE 14.6
Selection of Welding Processes Based on Types of Joint

Number	Type of Joint	Recommended Joining Processes
1	Lap joint	Oxyacetylene welding, SMAW, gas tungsten arc welding (GTAW), gas metal arc welding (GMAW), resistance welding, spot welding
2	Butt joint (bars)	Resistance welding (butt and flash), friction welding, electron beam welding, thermit welding, diffusion bonding
3	Butt joint (tubes)	SMAW, GTAW, GMAW, resistance welding (butt and flash), friction welding, electron beam welding, diffusion bonding
4	T-joints or fillet	SMAW, submerged arc welding (SAW), GTAW, GMAW, electron bean welding, resistance welding, thermit welding, and diffusion bonding

14.6.1.5 Inspection to Certify Quality of the Weld

The primary objective of weld inspection is to assure the high quality of welded structures through a careful examination of the components at each stage of fabrication. The welds are inspected and tested to locate the presence of defects and flaws. The overall quality of a welded joint often depends on performance of the welding equipment, the welding procedures adopted, and the overall skill of the operator. The weld inspection methods are broadly classified into two groups:

a. Destructive testing
b. Nondestructive testing

The destructive tests are mostly mechanical in nature and used for the testing of soundness, integrity, strength, ductility, and toughness of the weld material. As the name suggests, the components following the test can no longer be used. Few of the most important destructive test methods extensively chosen and used are the following: (1) tension test, (2) tension-shear test, (3) bend test, (4) impact test, (5) hardness test, (6) fatigue test, (7) creep test, (8) corrosion test, (9) nick-break test, and (10) slug and shear test (Cart 1989; Chapman and Martin 1975).

14.6.1.6 Defects in Welding

A study of the presence of defects in welded joints is necessary to find out the reasons for failure arising from failure of the defective weldment and the resultant damage caused during the service life of the structure. Defects often reduce the strength of the welded joint. Defects can arise from any one or a combination of these factors: (1) use of substandard material, (2) defective welding equipment, (3) improper welding procedures, and (4) poor welding skills of the operator. The major defects or discontinuities that affect the overall quality of the weld are the following (DeGarmo 1997; Parmar 1997).

14.6.1.6.1 Porosity

Porosity is defined as the presence of voids, holes, or cavities usually of a spherical shape. It is essentially caused by gas that is entrapped in the weld metal during solidification. Additionally, the occurrence of chemical reactions during welding, the presence of contaminants, such as oil, grease, dirt, and rust on the electrode, filler metal and base metal, and insufficient shielding against gas may also be the reasons for the entrapment of gas and resultant porosity. The porosity in welds may be of the following types:

a. Uniformly scattered porosity
b. Cluster porosity
c. Starting porosity
d. Linear porosity

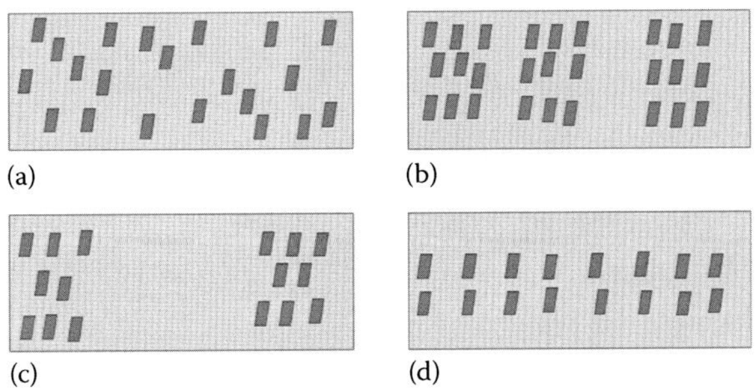

FIGURE 14.50 Various types of porosity in a weld metal. (a) Uniformly scattered, (b) cluster, (c) starting, and (d) linear.

These porosities are shown in Figure 14.50. The presence of blow holes in a welded component or at the region of the weld can be categorized to be an enlarged void. Under normal weld conditions, both porosity and blow holes would be scattered throughout the cross section of a weld. A few of the blow holes reach the surface and appear as fine pinholes. Some types of porosity are referred to as wormholes when they are long and continuous. When the pores are long in length and parallel to the root of the weld, it is referred to as *niping*. The methods that have been used in industry-relevant manufacturing operations to reduce porosity include the following: (1) proper selection of both the electrode and the filler metal, (2) proper cleaning of the contaminants and preventing them from entering the weld zone, and (3) slowing the weld speed to allow sufficient time for the gases that are generated to escape.

14.6.1.6.2 Cavities due to Shrinkage

Cavities due to shrinkage normally occur during the welding of thick plates by arc welding processes. This is especially true when dealing with a large amount of molten metal. The cavities due to shrinkage normally occur at the surface and may tend to extend inward during shrinkage effects experienced by the weld metal during solidification.

14.6.1.6.3 Cracks

Through the years, engineers working with welding have concluded that cracks can be easily considered to be the most dangerous of all weld defects. They tend to occur often in various directions and locations in the weld area, such as (1) base metal, (2) at the interface of the base and weld metal, (3) the surface of the weld metal, (4) under the weld bead, and (5) in the crater. The *transverse cracks, longitudinal cracks*, and *crater cracks* are defined as surface cracks primarily because they are formed on the surface of the weld. The crater cracks are generated in the crater due to an interruption in the welding process. These cracks are usually star in shape. The toe cracks tend to form in the adjacent base metal. The under-bead cracks are, by nature, internal cracks that occur at the subsurface. These cracks tend to occur in the high-temperature heat-affected zone. The various types of cracks are shown in Figure 14.51. The formation and presence of cracks can be due to the following reasons:

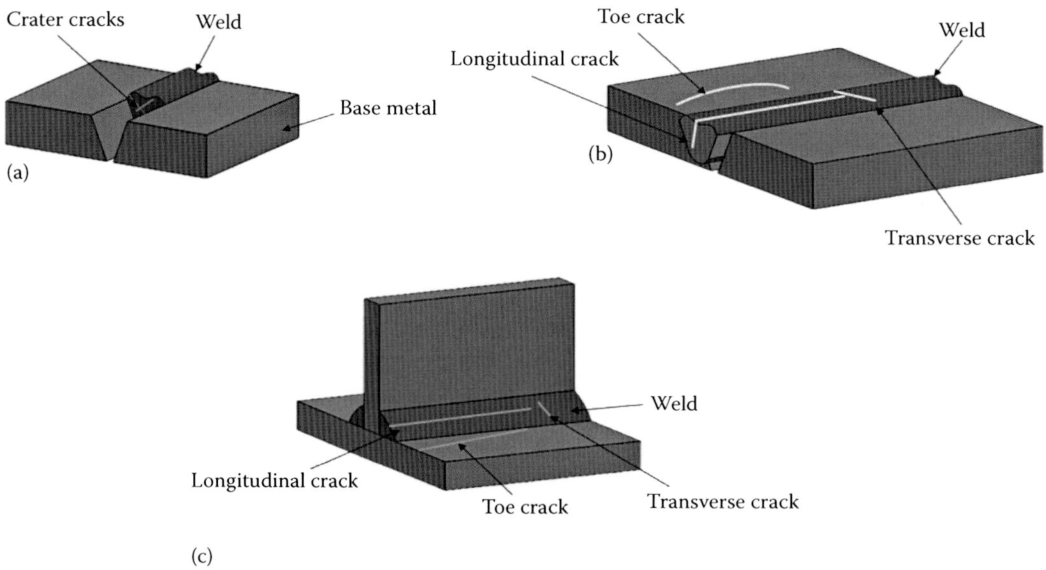

FIGURE 14.51 Schematic depicting the occurrence of cracks in welded joints. (a) Crater crack. (b) Longitudinal crack, transverse crack, and toe crack. (c) Longitudinal crack, transverse crack, and toe crack.

a. Differential thermal stresses generated in the weld zone due to temperature gradient.
b. An inability of the weld metal to contract during cooling resulting in the generation of localized stresses.
c. High sulfur and carbon content present in the base metal is conducive for the occurrence of embrittlement.

The formation and presence of cracks can be checked by changing the design of the joint to uniformly distribute the stresses within the weld metal. Additionally, the metals to be joined may be preheated and allowed to cool slowly once the process of welding is completed.

14.6.1.6.4 Presence of Slag Inclusions

Inclusions in the form of slag, such as oxides, sulfides, fluxes, and other foreign material, tend to get trapped in both the weld metal and the base metal. Consequently, they do not get a chance to escape and tend to float on the surface. The presence of inclusions tends to lower the strength of the joint. The most common cause for the presence of slag inclusions is inadequate cleaning of the area near the weld joint. The occurrence and presence of slag inclusions can be prevented or minimized by thorough cleaning of the surface prior to welding.

14.6.1.6.5 Distortion

Distortion tends to occur as a direct consequence of differential rate of heating and cooling in both the weld zone and adjacent metal resulting in the generation of stresses. After joining, the parts may not look properly aligned because of the distortion effect. Reducing the current used during welding, using a smaller diameter electrode, and proper clamping devices can reduce or prevent the occurrence of distortion.

14.6.1.6.6 Incomplete Fusion and Penetration

Incomplete fusion is an outcome of the following: (1) poor preparation of the joint, (2) wrong design of the joint, and (3) choice of incorrect weld parameters, such as welding current and welding speed, to name a few. Further, incomplete or poor penetration can also result from too little heat input due to a decrease in welding current. Both fusion and penetration are related to each other. If fusion is poor, then there occurs poor penetration. This problem can be overcome by improving the design of the joint, increasing the heat input, and lowering the welding speed. These two defects are shown in Figure 14.52.

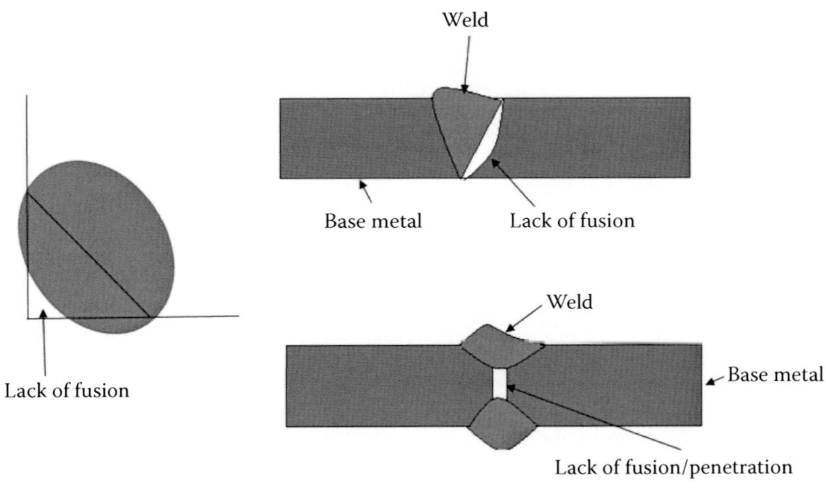

FIGURE 14.52 Lack of fusion and poor penetration in both fillet and butt welds.

14.6.1.6.7 Incorrect Profile of the Weld

The occurrence and presence of a wrong bead profile are an indication of lack of strength and bad appearance of the joint. Excessive reinforcement or excessive penetration often results as an outcome of bad weld profile and eventually becomes a source of stress concentration. Further, excessive reinforcement is an indication of wastage of filler metal, thereby increasing the cost of welding, without any noticeable gain in strength of the weld. This defect is shown in Figure 14.53.

14.6.1.6.8 Undercutting and Overlapping

In undercutting, a groove is formed in the base metal along the sides of the weld bead. Due to excessive welding, current, and arc voltage, excessive melting of the base metal tends to occur thereby facilitating in the occurrence of a sharp recess or notch.

Overlapping is favored to occur when excessive molten metal from the filler metal flows into the base metal and subsequently solidifies. This defect is often caused due to excessive welding currently and the wrong positioning of the electrode during welding. In fact, overlapping is just the opposite of undercutting. Both defects can be effectively reduced by (1) controlling the current and arc voltage during welding and (2) correct positioning of the electrode during the welding operation. This is shown in Figure 14.54.

14.6.1.6.9 Occurrence of Spatter

A spatter is a small metal particle that can be thrown randomly in any direction around the arc and is often deposited on the base metal. Presence of spatter tends to degrade the overall quality of the surface around the joint. Occurrence of excessive spatter is made possible because of (1) high welding current, (2) arc blow, and (3) wrong selection of electrodes.

14.6.1.7 Gas Welding Processes

Gas welding processes use the heat that is produced by burning gaseous fuels to join the parts. These gases are often burnt in the presence of oxygen and thereby generate a large amount of heat that can be used during welding. Three of the common gas welding techniques in use are (1) oxyfuel gas welding (OFW), (2) oxyacetylene welding, and (3) pressure gas welding (ASM Handbook 1993, 2011; Cary 2004).

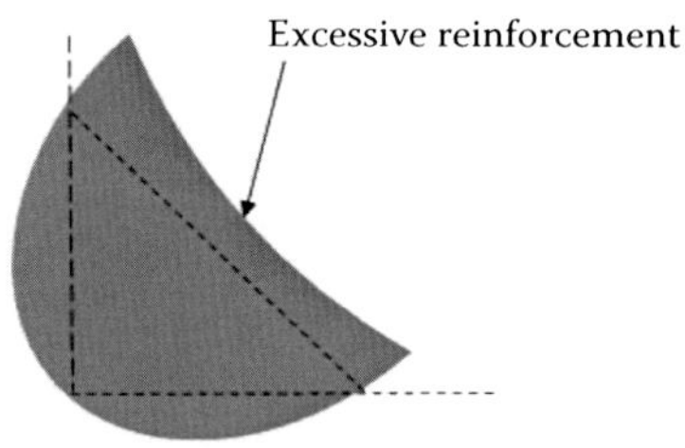

FIGURE 14.53 Incorrect profile of the weld.

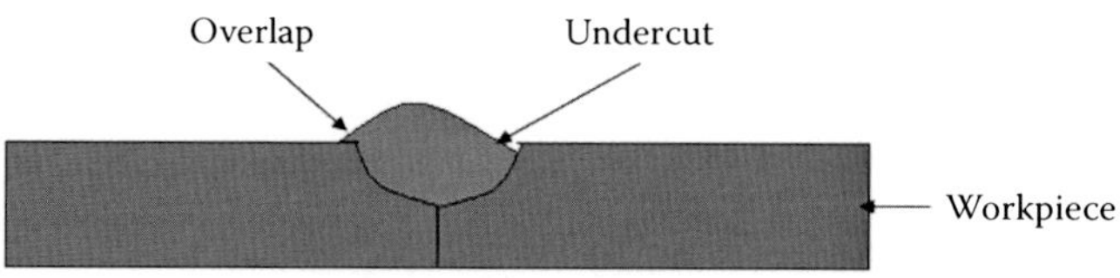

FIGURE 14.54 Schematic showing the outline of overlapping and undercutting.

14.6.1.7.1 OFW

This welding essentially includes all the processes in which a combination of gases is effectively used to produce a gas flame. The flame is used as the source of heat to melt the metals at the location of the joint. The most common fuel gases that have been chosen and used in industry are acetylene, propane, propylene, natural gas, and hydrogen. However, acetylene continues to be the most popular choice for purpose of gas welding. This is because the combination of acetylene and oxygen results in a high temperature when compared to other combinations.

14.6.1.7.2 Oxyacetylene Welding

Oxyacetylene welding essentially uses an acetylene–oxygen mixture for the generation of the required heat. The chemical reaction between acetylene and oxygen is exothermic in nature, and sufficient heat is generated by the reaction, which can reach a temperature as high as 3250°C. The heat is used to successfully melt the parts that are to be joined. The reaction takes place in two stages. In the first stage, both acetylene and oxygen react with each other to produce carbon monoxide (CO) and hydrogen (H_2). During this reaction, heat is liberated.

$$C_2H_2 + O_2 = 2CO + H_2 + \text{Heat}$$

In the second stage, the products that are formed during the first stage further react in the presence of excess oxygen to produce an adequate amount of heat that is required for welding.

$$4CO + 2H_2 + 3O_2 = 4CO_2 + 2H_2O + \text{Heat}$$

The acetylene gas that is used in the process is obtained when calcium carbide (CaC_2) chemically reacts with water (H_2O):

$$CaC_2 + H_2O = Ca(OH)_2 + C_2H_2$$

Since acetylene forms an explosive mixture in combination with air, it is not safe to store the gas at pressures greater than 0.1 MPa.

This technique has good use in sheet-metal fabrication and repair-related work. With the choice of appropriate variables, almost all the metals of interest in engineering can be welded. The overall quality of the weld does to a certain extent depend upon preparation of the surfaces prior to welding and the methods used to prevent contamination of the surface by the surrounding atmosphere. Overall, the advantages of oxyacetylene welding are the following:

1. Cost of equipment is low coupled with the need for little maintenance.
2. It is portable and can be used at any location.
3. The gas flame can be easily controlled.
4. The process can also be used for cutting.

For every advantage, the associated limitations arising from use of this welding technique are the following:

1. An inability to concentrate the arc flame causes a large area of the metal to become heated and resultant influence in promoting distortion.
2. In a few cases, there is noticeable loss in corrosion resistance of the welded product.
3. Due to its intrinsically slow nature, the technique has been largely replaced by arc welding.
4. The technique necessitates the need for proper training and skill level of the operator.

14.6.1.7.3 Pressure Gas Welding

The pressure gas welding technique has been used to make butt joints between the ends of objects, such as pipe and railroad rail. The ends are heated using an oxyacetylene flame to a semisolid state, and pressure is then applied. Further, in this technique, no filler metal is used. This technique is also known as *oxyacetylene pressure welding* or *hot pressure welding*. Since the parts that are to be welded are not melted, it can be classified as a solid-state welding process.

14.6.1.8 Specific Welding Processes: Arc Welding Process

Arc welding is a fusion process where the source of heat is an electric arc. The arc is produced between a metal electrode and the workpiece or between two electrodes and is basically used to heat the parts that are to be joined. This is shown in Figure 14.55. Overall, the arc acts as a concentrated source of heat. Thus, temperatures as high as 6000°C to 7000°C are easily produced. Welding can be performed using either a direct current (dc) or an alternating current (ac). The polarity of the chosen electrode can be either positive or negative. If a dc is used and the workpiece is connected to the positive terminal (anode) of the circuit, then the condition is known as *straight polarity*. When the workpiece is made negative and the electrode positive, then the condition is said to be *reverse polarity*. A few of the arc welding processes preferentially chosen and used in the industry environment are the following:

a. SMAW
b. SAW
c. GMAW
d. Flux-cored arc welding
e. GTAW

A few key highlights of each of these techniques are briefly presented.

14.6.1.8.1 SMAW

This is the most widely used arc welding process with the aid of a consumable electrode. It is one of the oldest, simplest, and most versatile joining processes. It is also referred to in industry circles as *stick welding* primarily because of the stick-like shape of the electrodes used in the process. The heat is produced by an arc between the flux-covered electrode and the workpiece. This is shown in Figure 14.56. The heat of the arc melts the tip of the electrode, the coating, and the workpiece that is just below the arc. The electrode coating, upon melting, forms a shielding gas that aids in protecting the molten metal from atmospheric contamination. The shielding gas also aids in stabilizing the arc. The flux that is used in the process combines with the impurities present in the molten metal and floats on the surface in the form of slag. The slag is then easily chipped away from the weld surface upon cooling The SMAW technique has been extensively used in many of the fabrication industries

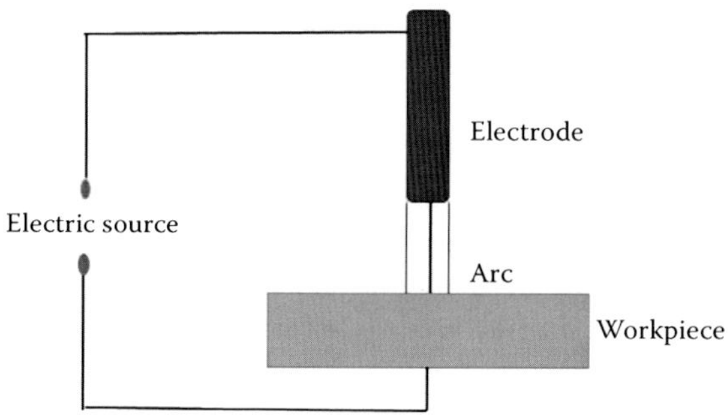

FIGURE 14.55 Schematic showing an arc welding process.

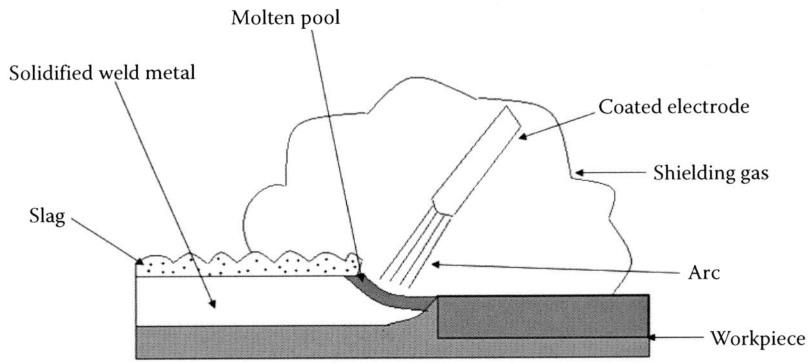

FIGURE 14.56 Schematic showing the profile of SMAW.

involved with (1) general construction, (2) ship building, (3) pipeline, and (4) pressure-vessel fabrication. It has also been used for operations involved with repair.

The notable advantages of this technique are that it is simple, portable, and less expensive when compared one on one with other arc welding processes. The welds can be made in all positions.

14.6.1.8.2 SAW

In this technique, an arc is produced between a consumable metal electrode and the workpiece. The arc is basically submerged or hidden under a heat layer of flux, hence the name *submerged arc welding*. The flux consists of calcium or magnesium silicate. A portion of the flux melts and combines with the impurities to form the slag that is finally chipped away upon cooling. The presence of residual slag in the joint can cause the occurrence of corrosion resulting in failure of the joint. Both ac and dc can be used in this process technique. Direct current reverse polarity (dcrp) helps in providing deeper penetration and a stronger joint. A dc straight polarity (dcsp) ensures a higher rate of deposition. The process of SAW is shown in Figure 14.57. A few of the advantages of this technique are the following:

i. Since the arc is completely hidden, there is no chance of any flash, spatter, or smoke visible from outside.
ii. Heavy currents can be used. This is often not preferred in other arc welding processes primarily because a heavier current causes higher infrared and ultraviolet radiation. Since the arc is covered by a flux, no such problem arises in the process of SAW.
iii. A higher current enables a high deposition rate coupled with deep penetration. The deposition rate can be much more than in SMAW.

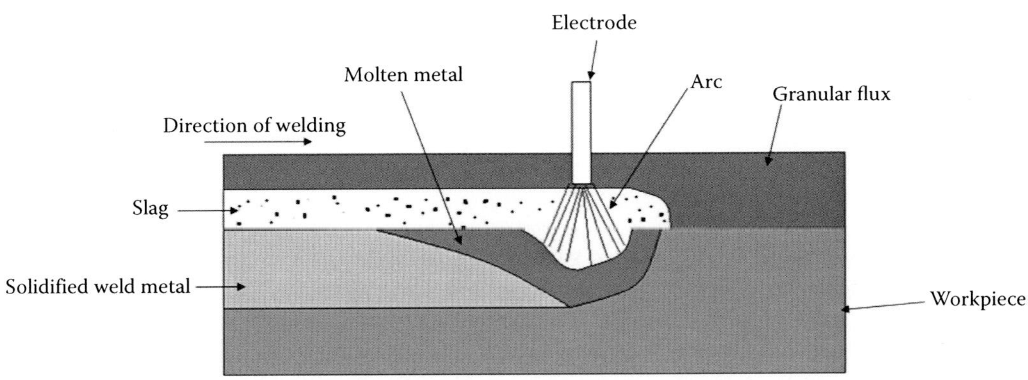

FIGURE 14.57 Schematic showing the SAW process.

iv. The welding speed is higher, which helps in joining thicker plates.
v. The overall weld is of high quality due to the high degree of cleanliness used through the entire welding operation.

14.6.1.8.3 GMAW

This is also known as metal inert gas welding. It has been used primarily for welding aluminum and stainless steels with an inert gas shielding. In this process, an arc is produced between a consumable electrode and the workpiece. The electrode is in the form of bare wire of small diameter, which provides the filler metal. Hence, no additional feed is required. The shielding gases that have been used in industry include argon, helium, carbon dioxide, and oxygen. The gases help in surrounding the arc and thereby protect the molten metal pool from contaminants present in the atmosphere, such as dust, dirt, and metal oxides. The shielding gases have been used either as a single gas or in combination with other gases. When carbon dioxide alone is used as the shielding gas, the process is termed as carbon dioxide welding. For the welding of nonferrous metals, argon or helium or mixtures of these two gases are preferred and used. However, carbon dioxide is the preferred choice and used for plain carbon steels and low-alloy steels. Carbon dioxide in combination with argon and helium is the preferred choice when welding high strength steels and stainless steels.

14.6.1.8.4 Flux-Cored Arc Welding

In this process, a continuous tubular electrode wire is used (when compared to a finite-length solid wire electrode used in SMAW) that is filled with flux, hence the name flux-cored arc welding. The flux that is contained in the tubular electrode performs the very same functions as a coating on a covered electrode. The flux acts as a deoxidizer, slag former, and arc stabilizer and can even provide alloying elements and the shielding gas. The diameter of the wire ranges from 0.5 to 4 mm with flux forming 5–25% by weight of the wire resulting in a deposition efficiency in the range of 85–95%. The noticeable advantages of this technique are the following:

i. The deposition rate is higher than GMAW. It is about twice that of SMAW for a comparable setup.
ii. It can be used for a wide range of thickness of the metal, as thin as 1.5 mm.
iii. The welds can be made in all positions using a smaller diameter wire.
iv. The process is easy to automate and is readily adaptable to flexible manufacturing systems and robotics.

It is used for the welding of a variety of joints in steels, stainless steels, and nickel-base alloys. Typical applications include bridges, high-rise buildings, ship building, and offshore drilling platforms. The few limitations of this technique are the following:

1. A large amount of smoke is produced during the operation because of the presence of flux on the electrode wire.
2. It can be successfully used to weld the ferrous metals, primarily steels.

14.6.1.8.5 GTAW

This was formerly known as tungsten inert gas welding. In this process, an arc is produced between a nonconsumable tungsten electrode and the workpiece. Inert gases such as argon or helium or a mixture of the two are employed to provide a protective shield around both the arc and the molten metal pool. All three welding currents may be used, i.e., dcsp, dcrp, and an ac. Selection of the current depends upon the type of material to be welded. DCSP gives deep penetration and faster welding of thicker workpieces and has been used for welding of steel, cast iron, copper alloys, and stainless steels. DCRP provides a weld pool with shallow penetration and is therefore suitable for only thin workpieces. This technique is rarely used because it tends to melt the tungsten electrode. AC is preferred for alloys of aluminum and magnesium because its cleaning action removes the oxides while concurrently improving the quality of the weld.

A filler metal is generally used when welding thicker pieces with well-prepared edges. The filler metal has a similar composition as that of the metal that is being welded. In applications where a close fit exists, the use of a filler metal may not be required. The noticeable advantages of this novel technique are the following:

i. It gives a clean weld joint due to the absence of use of flux in the process.
ii. There is less chance for weld spatter, thereby making the joint defect-free.
iii. It can be easily automated.
iv. Welding can be done in all positions.

Few other techniques that have been tried and progressively evolved for the purpose of welding metals and materials can be categorized under the category of solid-state welding processes (Little 1973; Sharma 1996). The processes can be performed both at room temperature (25°C) and at an elevated temperature. Filler metal is not used in several of these processes. The parts to be joined are initially cleaned to make them free of any oxide film, oil, and other contaminants present on the surface. These parts are then brought together to make proper contact with the application of high pressure. A few of these processes are

1. Forge welding
2. Cold welding
3. Friction welding
4. Inertia welding
5. Ultrasonic welding
6. Explosive welding
7. Diffusion welding

14.6.2 Soldering Technique

Soldering was categorized as being a process used for joining of metals and their alloy counterparts without melting them. This was made possible by using a fusable alloy, called solder. A typical solder melts below the temperature of 450°C and effectively acts as a filler metal that successfully reaches the gaps between the two parts that are to be joined. This is made possible through capillary action (ASM Handbook 1993; Parmar 1997; Sharma 1996).

The solders put to extensive use the alloys of tin–lead. These alloys have good strength at low temperatures. A 60Pb–40Sn solder has a low melting point and is a preferred candidate in several soldering-related operations. A few other solder compositions that have been tried and used are (1) tin–zinc, (2) lead–silver, (3) cadmium–silver, (4) zinc–aluminum, and (5) indium–tin. The tin–zinc and zinc–aluminum solders are the preferred choice for soldering of aluminum and its alloys. For the purpose of joining cryogenic parts and for the joining of glass with metal, a 50% indium and 50% tin solder has been the preferred choice.

The strength of a soldered joint is essentially due to the metallic bond that is formed, although adhesion and mechanical attachment make a small contribution toward strength. A good joint is characterized by a small amount of solder coupled with perfect adhesion, as opposed to the use of a large amount of the solder. To facilitate effective bonding, the surfaces that are to be joined must be thoroughly cleaned of all impurities, such as presence of an oxide layer on the surface and other impurities. A flux is used for this purpose, but it is highly essential that all oil, dirt, and even grease be removed from the surface prior to the application of a flux. The soldering flux normally chosen and used can be a solid, liquid, or a gaseous product (Gourley and Walker 2012; Humpston and Jacobson 2004).

Based on the mode of heating used, the important soldering methods used in industry-related circles are the following:

i. Iron soldering
ii. Torch soldering
iii. Dip soldering
iv. Wave soldering
v. Induction soldering
vi. Resistance soldering
vii. Ultrasonic soldering
viii. Furnace soldering

The soldering technique is extensively used in both electrical and electronic industries for the purpose of joining wires and light articles made of steel, copper, and brass. A few of the key advantages of the soldering technique are as follows:

a. The technique can be used to join various metals having a range of thicknesses.
b. It can be easily automated.
c. Automation aids in increasing the rate of production.
d. The process can be economical.

A few of the limitations associated with the soldering technique are the following:

1. It is not suitable for joining of parts that require adequate strength and parts that are subject to vibration and heat. This is essentially because the solder is weak and has a low melting point.
2. Cleaning of the surfaces prior to soldering is a mandatory requirement.
3. Due to small faying surfaces, a butt joint cannot be easily made.
4. Overall, the operation requires adequate amount of skill, besides being time consuming.
5. Both aluminum and stainless steels are difficult to solder because of the formation and presence of a strong, thin-oxide film (after cleaning). These metals require special soldering process such as *flux soldering*.

14.6.3 Brazing Technique

Quite like soldering, the brazing technique can be easily categorized to be a process for the joining of metals without melting the parent metal. In this process, the filler metal is placed in the gap between the parts that are to be joined and gradually heated to cause it to melt. The filler metal, now in a molten state, completely fills the gap by capillary action, and upon cooling and solidification, a strong joint is the outcome. Basically, the strength of the joint held together by brazing can be attributed to three sources (Gourley and Walker 2012):

1. Atomic forces between the metals at the interface
2. Alloying that occurs at elevated temperatures due to the diffusion of metals (both the base metal and filler metal)
3. Intergranular penetration

The filler metal chosen is usually a nonferrous alloy that has a melting point greater than 450°C, but lower than the melting point of the parent metal that is to be joined. Alloys of copper, silver, and aluminum are the most common filler metals used in the brazing technique. These filler metals are

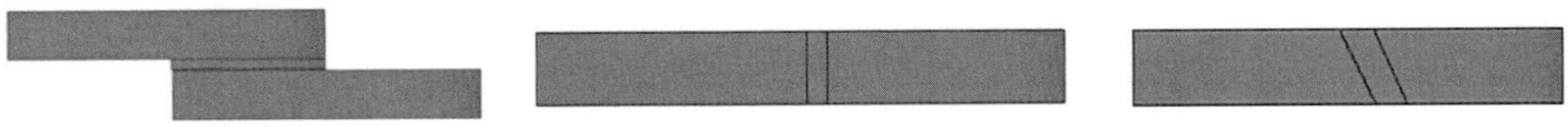

FIGURE 14.58 Schematic showing the joints common in brazing.

often used in many forms, such as ring, washer, rod, or other shapes to fit the joint that is being brazed. The basic types of joints common to the brazing technique are (1) lap, (2) butt, and (3) scarf. This is neatly depicted in Figure 14.58. Among these three, the lap joint is the strongest primarily because it offers a greater contact area. A butt joint proves a smooth joint having minimum thickness. The scarf joint is a compromise between the lap joint and the butt joint but suffers from the drawback of alignment.

The source of heat that is required for brazing is often obtained from several sources, and accordingly the brazing methods are classified as follows (ASM Handbook 1993; Sharma 1996):

1. Torch brazing
2. Furnace brazing
3. Induction brazing
4. Dip brazing
5. Infrared brazing
6. Resistance brazing

The brazing technique is suitable for the joining of dissimilar metals, such as joining of a ferrous metal to a nonferrous metal. It is also suited for the joining of metals having widely different melting points. This technique does find major application in the industries spanning electronics, utensil manufacturing, and maintenance. Few of the typical applications of the brazing technique include the following: (1) assembly of pipes to fittings, (2) carbide tips to cutting tools, (3) radiators, (4) electrical parts, (5) heat exchangers, and (6) repair of castings.

14.6.4 Adhesive Bonding Technique

This technique has been used for the joining of materials by means of an adhesive. Adhesion is facilitated due to molecular attraction between the adhesive and the workpiece. Since both metal and nonmetals can be bonded, adhesive bonding has grown significantly with the emergence of plastic and composite materials. After cleaning of the surfaces, the adhesive is applied between the faying surfaces by chemical or mechanical means. The strength of the joint is strongly governed by the absence of dirt, dust oil, and other contaminants. The presence of contaminants on the surfaces to be joined impairs the wetting ability of the adhesive and prevents its spreading evenly over the interface (Pizzi and Mittal 2003).

The adhesives of interest for use are normally available in different forms, such as liquid, paste, solid tape, and film. They can be applied at the required place by brushing, spraying, roller coating, and even dipping. The subsequent pressure that is applied to the joint must be carefully applied after placing the adhesive between the faying surfaces. The pressure that is applied should be sufficient to achieve a uniform spread of the adhesive, but not enough to squeeze out the liquid and thereby deprive the joint of any adhesive.

The adhesives chosen and used in industry-relevant applications can be classified as being of structural or nonstructural type. Structural adhesives give higher bond strength, and the resultant joint can be safely subjected to heavy loads for a long period. With advances in technology, the adhesives chosen and used are further classified as thermoplastics and thermosetting. Selection and use of an adhesive depends on the following: (1) materials being joined, (2) the joint design,

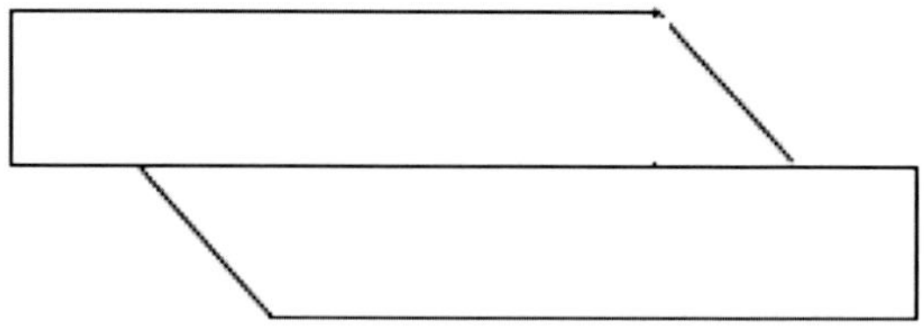

FIGURE 14.59 Schematic showing a bevel-lap joint.

and (3) operational requirements spanning strength, temperature fluctuation, and humidity level. Design for an adhesive bond should be such that the joint is shear rather than tension. An ideal joint is a bevel-lap joint, as shown in Figure 14.59. The presence of beveling reduces stress concentration at the bond edges. A few of the advantages of adhesive bonding are the following (Pizzi and Mittal 2003):

i. The process is versatile and offers the joining of several similar and dissimilar materials in different thicknesses, shapes, and sizes. Foils can be joined to each other or to thicker plates, and even very thin and fragile components can be bonded.
ii. Since the process is performed not at a high temperature, the heat-sensitive materials can be joined without causing any distortion due to thermal effects.
iii. There is an absence of heat-affected zone in adhesive bonding.
iv. The elastomeric nature of few adhesives does provide both shock and vibration protection.
v. Since the bond covers the entire area to be joined, the load to be carried is uniformly distributed, thereby minimizing local stress concentration.
vi. The technique or process is basically fast, and a large area can be bonded in a relatively short period.
vii. Since adhesives are basically bad conductors of electricity, there is an absence of electrochemical corrosion between the two dissimilar metals to be joined.
viii. An adhesive joint can withstand shear, compressive, and tensile forces, and their strength is comparable to a joint made possible through other methods of joining.
ix. Since an adhesive is essentially inexpensive, this technique is an inexpensive method of joining.

Few of the limitations of this joining technique are the following:

i. The joint should not be subjected to peeling forces. To increase the strength of the joint, additional fasteners at the stress points are essential.
ii. Adhesive joints are difficult to inspect once completed.
iii. Preparation of the surfaces is very important and essential. The surfaces to be bonded must be both mechanically and chemically cleaned to remove contaminants from the surface. Absence of these impurities and contaminants aids in increasing the strength of the joint.
iv. Adhesive joint has limited reliability. Its quality deteriorates with time and under hostile environmental conditions.
v. The joint cannot be subjected to elevated temperatures.
vi. Both joint design and bonding methods require care and skill.

14.7 CONCLUDING REMARKS: DEVELOPMENT, EMERGENCE, AND USE OF MANUFACTURING

In a nutshell, manufacturing can be aptly considered to be a human activity that pervades through many phases of our daily life. The outcome of manufacturing by way of products surrounds us on all fronts. Almost every item that we wear, we need to live in, and we need to travel, and most importantly what we eat, has been produced through some manufacturing process. The word *manufacturing* was derived from Latin (*manus* = hand; *factus* = made) and has been defined by dictionaries to be the making of goods and articles by hand, or using machinery, often on a large scale and with the involvement of labor. The growth of manufacturing is well recorded by the gradual developments, but it is the cumulative effects that have had an observable social impact and can be regarded as evolutionary. This, in turn, has led to the study of manufacturing by those pursuing the engineering curriculum. While several books and courses offered at both the undergraduate level and graduate level exist, this chapter attempted to present the various features, applications, advantages, and potential limitations on the traditionally used or referred to as conventional manufacturing processes and operations spanning casting, forming, machining, and joining processes. A basic knowledge of the fundamentals behind these conventional processes would provide the student, the practicing engineer, and the interested reader with the desired knowledge that would enable them to put the software and Internet-related facilities available to design products and processes. For each of the conventional manufacturing technique chosen, i.e., casting, forming, machining, and joining, several of the key techniques specific to each are neatly presented and appropriately discussed, trying to include the specific advantages, limitations, and viable applications of each technique. An attempt is made to include a section that highlights the key defects and their likely causes specific to each manufacturing technique presented and discussed. The use of the PM processing technique for manufacturing high-performance parts or products is also included and discussed.

It is hoped that by reading this chapter and understanding the fundamentals, the interested reader including the practicing engineer will have an appreciation for all the other chapters included in this bound volume. A knowledge of the fundamentals and gradual developments should also provide the interested reader with an appreciation of the critical role of manufacturing and related rapidly developing technologies in our daily lives and professional activities. In a nutshell, a thorough understanding of the intricacies included in this chapter should help the reader view manufacturing as an exciting, challenging, and important discipline in the domain enveloping engineering.

REFERENCES

Altan, T., Ngaile, G., and Shen, G., (editors), *Cold and Hot Forging: Fundamentals and Applications*, ASM International, Materials Park, Ohio, USA, 2005.

American Foundry Society, *Analysis of Casting Defects.*, New York, USA, 2002.

Ashby, M., Shercliff, H., and Cebin, D., *Materials Engineering Science, Processing and Design*, 2nd Edition, Butterworth-Heinemann, Oxford, United Kingdom, 2009.

Ashby, M.F., *Materials Selection in Mechanical Design*, 4th Edition, Butterworth Heinemann, 2010.

Ashby, M.F., and Jones, D.R.H., Engineering materials, volume I, *An Introduction to Properties, Applications and Microstructure*, 4th Edition, 2012; Volume 2: *An Introduction to Microstructure and Processing*, Butterworth-Heinemann, Oxford, United Kingdom, 2012.

Askeland, D.R., Fulat, P.P., and Wright, P.K., *The Science and Engineering of Materials*, 6th Edition, CL Engineering, Boston, Massachusett, 2010.

ASM Handbook, *Volume 16: Machining*, ASM International, Materials Park, Ohio, USA, 1989.

ASM Handbook, *Volume 6: Welding, Brazing and Soldering*, ASM International, Materials Park, Ohio, 1993.

ASM Handbook, *Volume 7: Powder Metal Technologies and Applications*, ASM International, Materials Park, Ohio, USA, 1998.

ASM Handbook, *Volume 14A: Metalworking: Bulk Forming*, ASM International, Materials Park, Ohio, USA, 2005.

ASM Handbook, *Volume 15: Casting*, ASM International, Materials Park, Ohio, 2010.

ASM Handbook, *Volume 6A: Welding Fundamentals and Processes*, ASM International, Materials Park, Ohio, 2011.

Astakhov, V.P., and Joksch, S., *Metalworking Fluids for Cutting and Grinding: Fundamentals and Recent Advances*, Woodhead Publishers, Cambridge, Massachusett, 2012.

Bauser, M., Sauer, G., and Siegert, K. (editors), *Extrusion*, 2nd Edition, ASM International, Materials Park, Ohio, USA, 2006.

Beeley, P., *Foundry Technology*, Butterwort-Heinemann, USA, 2002.

Beeley, P.R., Advances in the casting of metals, *Materials & Design*, Vol. 3, 1982, pp. 632–637.

Blair, M., Stevens, T.L. and Lindsey, B. (editors), *Steel Castings Handbook*, 6th Edition, ASM International, Materials Park, Ohio, USA, 1995.

Boothroyd, G., and Knight, W.A., *Fundamentals of Metal Machining and Machine Tools*, 3rd Edition, Marcel Dekker, Inc., New York, USA, 2006.

Byrer, T.G., (editor), *Forging Handbook*, Forging Industry Association, Chicago, 1985.

Campbell, J., *Castings*, 2nd Edition, Butterworth-Heinemann, 2003.

Campbell, J., *Complete Casting Handbook: Metal Casting Processes, Techniques and Design*, Butterworth Heinemann, 2011.

Campbell, J.S., *Principles of Manufacturing Materials and Processes*, Tata McGraw Hill Publishers, New Delhi, India, Oxford, United Kingdom, 1999.

Cart, H.B., *Modern Welding Technology*, 2nd Edition, Prentice-Hall, New Jersey, USA, 1989.

Cary, H.B., *Modern Welding Technology*, 6th Edition, Prentice-Hall, New York, USA, 2004.

Chapman, W.A.J., and Martin, S.J., *Workshop Technology*, Volume 1, Volume 2 *and* Volume 3, Arnold Publishers (India) Private Limited, New Delhi, India, 1975.

Cornier, D., *Machining and Metal Working Handbook*, McGraw-Hill, New York, USA, 2005.

DeGarmo, E.P., *Materials and Processes in Manufacturing*, 8th Edition, Prentice Hall of India Private Limited, New Delhi, India, 1997.

Dieter, G.E., Kuhn, H.A., and Semiatin, S.L. (editors), *Handbook of Workability and Process Design*, ASM International, Materials Park, Ohio, USA, 2003.

Douvard, D. (editor), *Powder Metallurgy*, ISTE Publishing, New York, USA, 2009.

German, R.M., *Powder Metallurgy and Particulate Materials Processing*, Metal Powder Industries Federation, USA, 2006.

German, R.M., *A-Z of Powder Metallurgy*, Elsevier Publishers, Cambridge, Massachusett, 2007.

Ginsberg, V.B., and Balsam R., *Flat Rolling Fundamentals*, CRC Press, New York, USA, 2001.

Gourley, R., and Walker, C. (editors), *Brazing and Soldering*, 2012, American Society for Metals, Materials Park, Ohio, USA, 2012.

Green-Spikesley, E., *Investment Casting in Materials in Engineering Applications*, Vol. 1, December 1979, pp. 328–334.

Hosford, W.F., and Caddell, R.M., *Metal Forming: Mechanics and Metallurgy*, 4th Edition, Prentice Hall, New York, USA, 2011.

Howard, W.F., and Caddell, R.M., *Metal Forming Mechanics and Metallurgy*, 4th Edition, Cambridge, United Kingdom, 2010.

Humpston, G., and Jacobson, D.M., *Principles of Soldering*, ASM International, Materials Park, Ohio, USA, 2004.

Investment Casting Handbook, Investment Casting Institute, Montvale, New Jersey, 1997.

Kalpakjian, S., and Schmid, S., *Manufacturing Engineering and Technology*, 7th Edition, Pearson Education Inc., New Jersey, USA, 2014.

Leander, F., and West, W.G., *Fundamentals of Powder Metallurgy*, Metal Powder Industries Federation, USA, 2002.

Lee, Y., *Rod and Bar Rolling: Theory and Applications*, CRC Press, New York, USA, 2004.

Little, R.L., *Welding and Welding Technology*, McGraw Hill, New York, USA, 1973.

Martin, A., *The Essential Guide to Mold Making and Slip Casting*, Lark Books, 2007.

Mohan, T.R.R. and Ramakrishnan, P. (editors), *Powder Metallurgy in Automotive Applications-II*, Science Publishers, Florida, USA, 2002.

Nachtman, E.S., and Kalpakjian, S., *Lubricants and Lubrication in Metalworking Operations*, Marcel Dekker, Inc., New York, USA, 1985.

Parmar, R.S., *Welding Processes and Technology*, 2nd Edition, Khanna Publishers, Delhi, India, 1997.

Pizzi, A., and Mittal, K.L., *Handbook of Adhesive Technology*, 2nd Edition, CRC Press, Florida, USA, 2003.

Powder Metallurgy Design Manual, 2nd Edition, Metal Powder Industries Federation, Princeton, New Jersey, 1995.

Schey, J.A., *Introduction to Manufacturing Processes*, 3rd Edition, McGraw Hill Publishers, Boston, 2000.

Sharma, P.C., *A Textbook of Production Technology*, S. Chand and Company Limited, New Delhi, India, 1996.

Shaw, M.C., *Metal Cutting Principles*, 2nd Edition, Oxford University Press, United Kingdom, 2005.

Sias, F.R., *Lost-Wax Casting*, Woodsmere Press, South Carolina, USA, 2006.

Smith, G.T., *Cutting Tool Technology: Industrial Handbook*, Springer, New York, USA, 2008.

Stefanescu, D.M., *Science and Engineering of Casting Solidification*, 2nd Edition, Springer, USA, 2008.

Trent, E.M., and Wright, P.K., *Metal Cutting*, 4th Edition, Butterworth-Heinemann, Oxford, United Kingdom, 2000.

Tschaetch, H., *Metal Forming Practice: Processes, Machines, Tools*, Spitzer, New York, USA, 2007.

15 Advances in Understanding the Kinetics of Solidification

Modeling Microstructural Evolution

Mohsen Eshraghi and Sergio D. Felicelli

CONTENTS

15.1 INTRODUCTION

Solidification is a crucial step for many manufacturing processes, such as casting, welding, and additive manufacturing. While solidification happens during processing of all types of materials, solidification of metallic alloys has been of importance to scientists and engineers. The importance comes from the fact that the solidification microstructure has a significant influence on the properties of the solidified materials. The kinetics of solidification also determines the distribution of solute atoms, which eventually leads to microsegregation, secondary phases, and formation of various defects like freckles and pores, to name just a few, which exert enormous influence on mechanical properties.

Computational materials engineering has received attention due to its potential to shorten product and process development time, while lowering cost and improving outcome. By combining the bedrock computational physics and informatics with systematic experiments and advanced manufacturing, we can reduce the cost, risk, and cycle time for new product development. Such an approach drastically reduces the time-consuming cycle of experimentation and testing needed to move new materials to the marketplace [1]. Computational modeling of solidification microstructure can help scientists to gain a better understanding of the kinetics governing the microscopic features of the solidification process. From an industrial point of view, modeling microstructural evolution enables engineers to predict the properties of the material and subsequently modify the process parameters in order to produce materials of higher quality. Direct numerical simulation of the solidification microstructure formed during the manufacturing processes will provide a relation between macroscopically observable variables like cooling rate or temperature gradient and difficult-to-measure dynamic microscopic features like solute redistribution and dendrite arm spacing. Although much observation has been done to examine the static microstructures during different stages of solidification, it has never been possible to capture and examine the dynamic response of these features in an evolving mushy zone. In addition to providing information to better understand the process, this can help in assessing, validating, and improving numerical tools for the study of mushy zone in macroscale. Nevertheless, several physical phenomena are involved during the solidification processes that in turn make the simulation complex.

In this manuscript, some of the recent numerical studies on modeling kinetics of solidification and microstructural evolution are presented and discussed. The focus will be on lattice Boltzmann method (LBM) as a novel numerical technique for solidification modeling. The two-dimensional (2D) and three-dimensional (3D) simulation results of dendrite growth are presented and discussed. Large-scale simulations of dendritic solidification, using the advantages of supercomputing, are presented as well. In addition, application of LBM for the purposes of modeling convection effects, and formation of defects, such as, freckles, and microporosities, are presented.

15.2 MODELING DENDRITE GROWTH

Dendrites are tree-like microstructures that form during solidification. They form when molten metal or alloy freezes from liquid state, which may happen during many industrial manufacturing processes. The morphology, size, and spacing between dendritic arms have a significant influence on material properties.

Analytical and numerical models have been developed to describe various aspects of dendritic solidification, such as (a) cell-to-dendrite transition, (b) dendrite tip radius, and (c) primary dendrite spacing [2–13]. Most of these models assume the primary dendrite to be an axisymmetric, branchless, and a 2D feature growing under diffusive solutal and thermal transport. Given that mechanical properties depend upon the spacing and distribution of dendrites, numerous experiments have been carried out to measure primary spacing as a function of process parameters and the physical properties of the alloy and compared with model predictions [14–34].

The majority of models of dendritic growth at the microscopic scale can be categorized into three main types: (i) those based on the phase-field (PF) method [35–39], (ii) models based on the level set (LS) method [40–43], and (iii) models that perform a direct interface tracking (DIT) [44–46]. Of these methods, PF is probably the most powerful, because it can deal directly with morphological complexity by introducing a field variable that eliminates the need to explicitly find the interface. However, it is computationally taxing even when combined with adaptive meshing methods. An additional disadvantage is that the solid–liquid interface can be tracked to first-order accuracy. The LS method also interchanges the interface with a field variable (the LS) but requires knowledge of the direction in which the solid front is advancing, its velocity, and calculation of the vector normal to the interface, which makes it less effective than the PF method in complex 3D geometries. DIT is the simplest and computationally the most efficient of the methods, but it is also the less powerful because it requires the calculation of the temperature gradients at the interface in addition to the normal velocity and curvature of the interface. The technique has been used efficiently to reproduce the complex dendritic structure during crystal growth in undercooled melts [47,48]. These techniques enforce the freezing interface temperature (Gibbs–Thomson) relation and energy balance (Stefan) condition. However, the complexity for handling interfaces in all possible solidification conditions, such as advancing/receding, merging/splitting, and interface unit normal vector/curvature computations, limits the applicability of the DIT method to single solid phase systems.

Compared with the extensive work that has been done in two dimensions, fewer 3D calculations of dendritic solidification at the microscale have been reported [49–64]. Some of this work included convection, as well as binary [57,61] and multicomponent [62] alloys. However, in three dimensions, the microscopic solvability theory is not definitively established, as it is for two dimensions [50]. This stems from the form of the Gibbs–Thompson relation and the surface energy relation, and simplifications must be used to obtain an axisymmetric form of the interface [65]. Therefore, validation of the 3D results is increasingly difficult. Further, regardless of the method of choice, the computational effort is significantly larger than for the simple 2D case. This is due in part to the fact that growth velocities are higher in three dimensions, due to the presence of capillary length scale, that cannot be fully resolved, and solution of the energy equation forces the need to use computational meshes that are much smaller than is required for solving the thermal diffusion length scale in order to obtain reasonably well-converged results.

In addition, methods based on the cellular automata (CA) have been developed and offered an excellent alternative [58,62,64,66–74], with the potential to allow for reasonably accurate calculations in regions on the order of a cubic centimeter. The CA models are often characterized as being simple in their construction while concurrently being able to produce very complicated behavior. This property of CA models has been exploited to produce computer simulations of various aspects of microstructural evolution that occurs during solidification. Results of a series of 3D simulations of nonisothermal "free" dendritic growth were presented by Brown and Bruce [75]. In their study, the changes in dendrite morphology for different conditions were both quantified and discussed. A modification of their model was also developed to examine the effects of composition on microstructural development for a simple eutectic system. As the composition moves toward the eutectic, the simulated microstructure changes from combined dendritic/lamellar to completely lamellar.

One of the most elaborate early CA models for the purpose of calculating grain growth, combining stochastic nucleation, diffusional growth, and macroscopic heat conduction, was proposed by Gandin and Rappaz [76] as an extension to the stand-alone CA model. The combined CA–finite element (FE) model, referred to as the CAFE model, applies to a nonuniform temperature field. The temperature at each cell location and time-step is interpolated using the values at the neighboring nodal points of the FE mesh, which, in turn, is used for calculating the evolution of microstructure using a CA-based code. The corrected solid fraction is fed back to the FE code to compute the new temperature field.

Guillemot and coworkers [77] developed a 2D CAFE model for the prediction of grain structure formation during solidification. While the FE method solves the macroscopic conservation equations for heat and mass transfer, the CA method is used at a mesoscopic scale to simulate growth of the mushy zone domain. The limit between this mushy zone domain and the undercooled liquid melt in which it develops defines the envelope of a single grain.

CA–finite difference models have been applied for the purpose of simulating dendrite growth and formation of equiaxed grains in directionally solidified nickel-based superalloys [69,78]. Realistic dendritic structures and complex solute concentration profiles at the growth front were simulated. It was observed that solute interaction between the primary dendrites occurred well below their tips, while strong solute interaction occurred between the diffusion fields of secondary and tertiary arms. The influence of thermal gradient and growth velocity on columnar-to-equiaxed transition (CET) was investigated and the results were combined on a CET map. Results revealed that a decrease in thermal gradient and an increase in growth rate favor a CET.

A 2D model for the simulation of a binary dendritic growth with convection was developed by Li and coworkers [79] in order to investigate the effects of convection on the morphology of the dendrite. The model is based on a CA technique for the calculation of evolution of solid/liquid interface. The dynamics of an interface controlled by temperature, solute diffusion and Gibbs–Thomson effects, is coupled with the continuum model for energy, solute, and momentum transfer with liquid convection.

A 3D modified CA model (3D MCA) was developed by Zhu and Hong [72] in order to simulate the evolution of microstructures during solidification of alloys. Different from the classical CA in which only the temperature field was calculated, this model included the solute redistribution both in the liquid and solid during solidification. The relationship between growth velocity of a dendrite tip and local undercooling, which consists of thermal, constitutional, and curvature undercooling terms, was calculated using the KGT (Kurz–Giovanola–Trivedi) and LKT (Lipton–Kurz–Trivedi) models. The finite volume method, coupled with the CA model, was used to compute both the temperature and solute fields in the calculation domain. The 3D modified CA model was used to predict different microstructures, such as (a) free dendritic growth from an undercooled melt and (b) the competitive dendritic growth in practical casting solidification.

Sanchez and Stefanescu [80] and Zhu and Stefanescu [73] proposed a model based on the CA technique for the simulation of dendritic growth controlled by solute effects in the low Peclet number regime. One of the innovative aspects of this model is that it does not use an analytical solution (like KGT or LKT) to determine the velocity of the solid–liquid interface as is common in other models but solves the solute conservation equation that is subject to the boundary conditions at the interface. This methodology has been extended to 3D models [58].

Yin and Felicelli [81] developed a coupled CA–FE model to calculate dendrite growth during solidification of metals having a hexagonal close packed crystal structure. The model solved the conservation equations of mass, energy, and solutes in order to calculate (a) the temperature field, (b) solute concentration, and (c) the morphology of dendritic growth. The dendrite growth for Mg–8.9 wt% Al alloy, which is similar in composition to alloy AZ91, was simulated. Figure 15.1a–c shows the simulated evolution of equiaxed dendrite growth for different simulation times of 0.0212 s, 0.0424 s, and 0.0636 s, respectively. It is seen that during the early stages of solidification, dendrites develop primary arms, which follow the crystallographic orientation, as shown in Figure 15.1a.

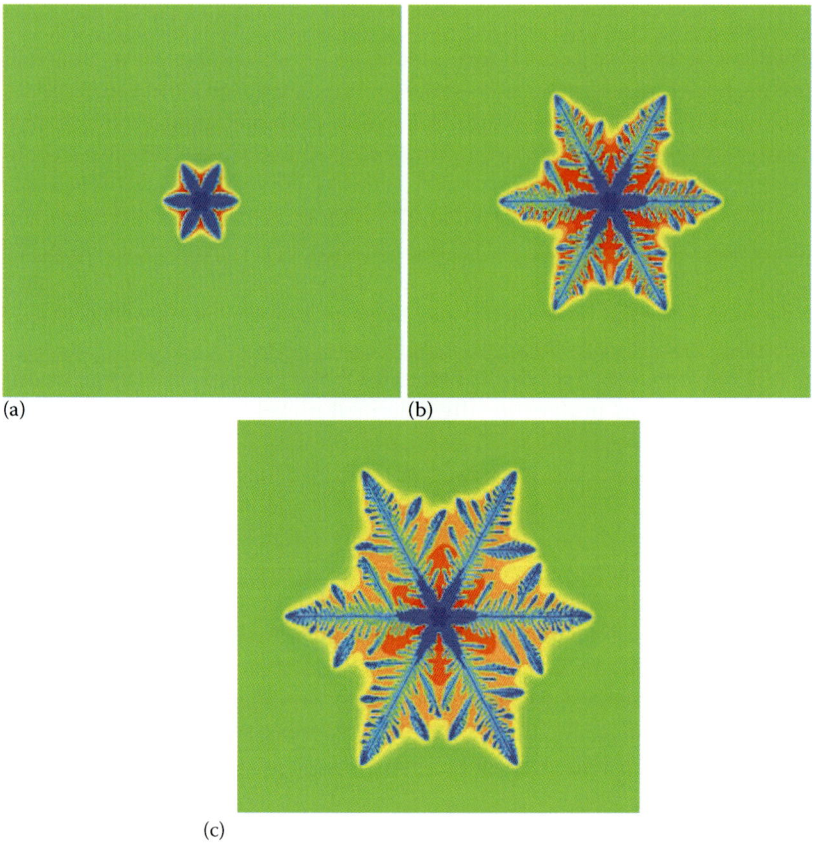

FIGURE 15.1 Solute distribution around Mg–8.9 wt% Al alloy dendrites at different simulation times: (a) 0.0212 s, (b) 0.0424 s, and (c) 0.0636 s. (From Yin, H., Felicelli, S.D., *Model. Simul. Mater. Sci. Eng.*, 17, 075011, 2009. With permission.)

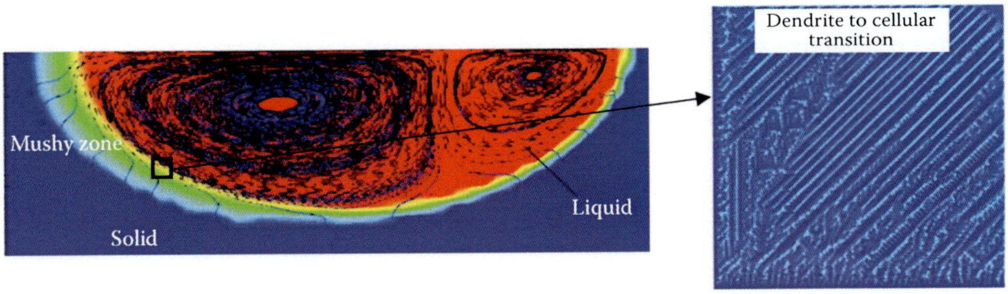

FIGURE 15.2 (a) Multiscale model in the LENS pool, showing dendrite-to-cellular transition (b). (From Yin, H., Felicelli S.D., *Acta Mater.*, 58, 1455–1465, 2010. With permission.)

As solidification proceeds, the primary arms become larger and the secondary arms begin to appear (Figure 15.1b). With further solidification, some tertiary dendritic arms form on the secondary arms (Figure 15.1c). The simulated structure compares qualitatively well with the measured microstructure of an AZ91 dendrite reported by Fu and coworkers [82].

The CA–FE model was also used for multiscale simulation of dendrite growth in a laser engineered net shaping (LENS) molten pool, as shown in Figure 15.2 [83]. The continuum transport equations were solved at the pool scale using FE, while the dendrite growth was solved at the microscale using the CA (FE–CA scheme). The case shown is for a fast-moving laser beam; hence, the cooling rate is high and dendrite-to-cellular transition occurs. Although Marangoni convection was simulated as occurring in the pool to illustrate the real physics, simulation of dendrite growth at the microscale was actually done without convection. A simulation of the whole pool under convection with dendrite resolution is computationally unaffordable for a FE–CA scheme.

In spite of these progresses, the modeling of 3D dendrites remains a formidable task because of (a) difficulties associated with the presence of a complex interface, (b) its interaction with fluid flow and solutal boundary layers, and (c) the need for interface tracking. As a result, the simulations could not go beyond the microscale domain, most of them dealing with a few dendrites with no convection coupled with scarce fluid flow. However, the advent of new massively parallel supercomputers of "petaflop" performance has enabled the modeling of 3D dendrite to the next scale level. Large-scale simulations allow for direct numerical simulation of a 3D mushy zone involving hundreds or thousands of dendrites, growing under either forced or natural convection. Taking advantage of local-type nature of CA and LBMs, it is possible to achieve an efficient parallelization and scalability, enabling the simulation of cm-size domains having microscale resolution.

15.3 LATTICE BOLTZMANN METHOD RESEARCH

The LBM is emerging as a powerful technique for the simulation of single and multiphase flows in complex geometries. Proponents of the LBM consider it to have the potential to become a versatile computational fluid dynamics platform that is superior to the existing continuum-based computational fluid dynamics methods [84]. Unlike conventional computational fluid dynamics methods that directly simulate the continuum-based governing equations (for example, the Navier-Stokes equations), the LBM is based on kinetic equations that incorporate microscopic physics in a mesoscopic approach. In the flow context, flows are considered to be composed of a collection of pseudo-particles that are represented by a velocity distribution function. The fluid portions reside and interact on the nodes of a grid. System dynamics and complexity emerge as a result of the repeated application of local rules for the motion, collision, and redistribution of these coarse-grained droplets [85]. The LBM is an ideal approach for scale-bridging simulations. Because of its microscopic origin, the LBM has many advantages over conventional methods of computational fluid dynamics. The advantages include, among others, (i) a clear physical meaning, (ii) simple local-type calculations,

(iii) easy handling of complex geometry and boundary conditions, (iv) good accuracy and numerical stability, (v) constitutive versatility, and (vi) efficient parallelization.

Motivated by the good features of the LBM, many researchers are using the method beyond the purely hydrodynamic arena. Several works have been reported in the published literature using the LBM in problems involving heat transfer, phase change, reaction systems, and other material processing applications. Ho and coworkers [86] used the LBM to study non-Fourier heat conduction in a planar layer. Jiaung and coworkers [87] derived an LBM approach to solve an enthalpy formulation of heat conduction with phase change in a mushy zone. Raj and coworkers [84] combined the LBM with the discrete transfer method to study solidification with radiation in a semitransparent material. Medvedev and coworkers [88] and Miller and coworkers [89] used the LBM in combination with the PF method to simulate dendritic growth and alloy solidification in the presence of fluid flow. Brewster [90] demonstrated the computational efficiency of the LBM solving large 3D fluid flow calculations using a desktop computer. In one of the few known applications combining LBM with the classical CA, Chopard and Dupuis [91] simulated the erosion and deposition process in water channels. Applications to dendrite growth in two [92,93] and three dimensions [94,95] have recently appeared.

In spite of the recognized advantages of the LBM, the method is not free of limitations. Lattice spacing, time spacing, fluid viscosity, and other parameters are not independent, and care must be taken to constrain their values to obtain both stable and accurate results. An application of the method outside the hydrodynamic field is currently emerging. Simulations involving heat transfer, alloy solidification, and dendritic growth are recent and have been limited to simple cases. Although results are encouraging, there is still much ground to break.

15.4 LATTICE BOLTZMANN–CELLULAR AUTOMATA MODEL FOR DENDRITIC SOLIDIFICATION

The authors developed an all-local technique in which the microscopic CA is used to calculate solidification and interface position while the 3D lattice Boltzmann (LB) method is used to compute the solutal and thermal fields (and the velocity in convective problems). The basic structure of the LBM is very similar to a CA, and a combination of both methods seems a natural approach. Given their local-type nature and their excellent scalability for many processors [96], a dendritic growth model based on these methods is an attractive choice for exploitation of large-scale parallelization in three dimensions. A new paradigm was created that brought LBM to the frontline of massively parallel simulations of the solidification processes. As is usually done in these simulations, we did not focus on the problem of nucleation but assumed a distribution of solid seeds having preferred crystallographic orientations. The velocity of the interface was not calculated from analytical models but determined from the solution of the transport equations and boundary conditions at the interface [62,73].

A CA algorithm was used to capture new interface cells. Every cell was characterized by (a) temperature, (b) solute concentration, (c) crystallographic orientation, and (d) fraction of solid. The state of each cell, in each time step, was determined from the state of itself and its neighbors in previous time step. However, an exact solid/liquid interface was implicitly calculated by the fraction of solid in each cell. During solidification, a solute partition occurs between the solid and the liquid and the solute is rejected to the liquid at the interface.

Multidendrite growth was simulated using the LB–CA model [93]. A uniform initial temperature and composition was taken for all simulations. Seven nuclei with random preferential growth orientation ranging from 0 to 90 degrees, with respect to the horizontal direction, were randomly distributed in the calculation domain. Figure 15.3 shows the calculated evolution of multiequiaxed dendrite growth in an Al–3 wt% Cu alloy. The dendrite morphologies calculated with solute transport only by diffusion are shown in Figure 15.3a. The dendrites develop the main primary arms along their crystallographic orientation, but without secondary arms. The growth of some main arms is constrained by the nearby dendrites. In Figure 15.3b, a heat flux is prescribed on the four

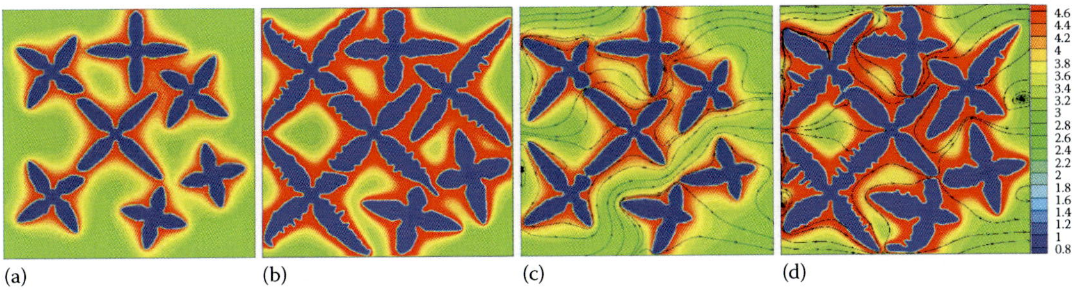

FIGURE 15.3 Multiequiaxed dendrite growth in Al–3.0 wt% Cu alloy. (a) Solute diffusion only; (b) heat flux at boundaries; (c) inflow velocity on the left side; (d) Inflow velocity and heat flux at boundaries. (From Yin, H., Felicelli, S.D., Wang, L., *Acta Mater.*, 59, 3124–3136, 2011. With permission.)

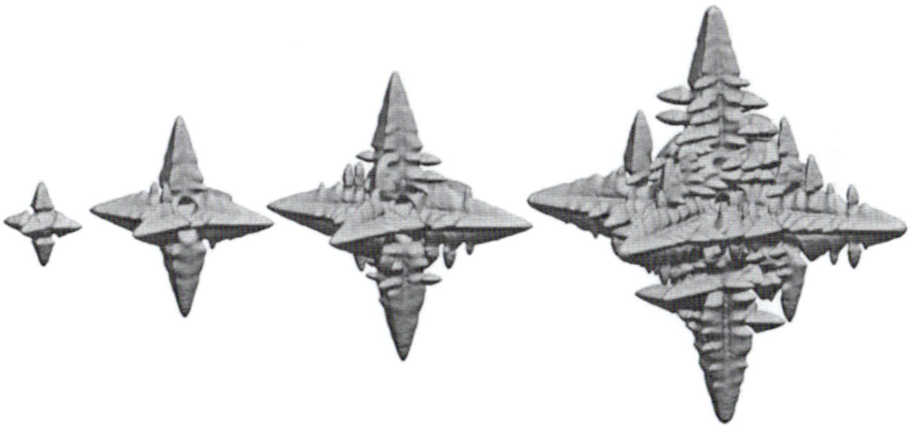

FIGURE 15.4 Simulated dendrite morphologies for Al–3 wt% Cu alloy. From left to right, after 3, 7, 10, and 15 ms. (From Eshraghi, M., Felicelli, S.D., Jelinek, B. *J. Cryst. Growth*, 354, 129–134, 2012. With permission.)

boundaries, causing the cooling rate to enhance the side branching of dendrites. An inflow from the left boundary is added to the simulation shown in Figure 15.3c. It is observed that the primary dendrite arms are coarser and longer in the upstream direction than those in the downstream direction. The dendrite morphologies with heat and mass transport by both diffusion and convection due to inflow imposed at the left side are shown in Figure 15.3d. The liquid flow washes away the interdendrite composition and promotes dendrite growth as well as merging between dendrites.

The LB–CA model was extended to three dimensions to enable 3D simulation of dendrite growth [94]. A cubic domain with a uniform mesh size and a constant undercooling was used to simulate the dendrite growth. The morphology of the dendrite growing in a 120 μm × 120 μm × 120 μm domain at different time steps is shown in Figure 15.4. The simulations are conducted using a 4.5°C undercooled melt. It can be seen that during the initial stages of solidification, the primary arms grow along their crystallographic orientation without any secondary arms. As solidification proceeds, the primary arms grow and coarsen, the secondary arms start to grow perpendicular to the primary arms, and the tertiary arms form perpendicular to the secondary arms thereafter.

15.5 LARGE-SCALE SIMULATION OF DENDRITIC MICROSTRUCTURES

PetaFLOPS computing capability for unclassified, open science research is already a reality. The current path to petaflop and beyond (exaflops) computers involves distributed memory architectures with multicore processors and 3D interconnect between the nodes/processors. The number of processors to be connected to achieve the peak operation count mandates a fast and multiconnected network between the processors; practical limitations still make this network a limiting factor of petaflop and exaflop systems [97]. Accordingly, the algorithms that take advantage of the locality of reference and long, multiple streams of data perform well on these computer architectures.

Locality of CA and LB methods refers to the characteristic that the update of a computational cell is based solely on the state of its neighborhood. This means that parallel algorithms based on domain decomposition will, for the most part, compute the local state based on values in the local memory and communicate between the local processors in a fixed pattern. High-quality domain decomposition algorithms are available for initial domain to processor assignment and subsequent partitioning updates are performed using local redistributions. Hierarchical domain decomposition methods have demonstrated their suitability for hierarchical computer architectures [98].

Compared to linear equation solvers, the CA and LB algorithms usually involve limited global synchronization operations and their performance exhibits close to ideal scaling. With an increasing number of processors and cores, this goal becomes more challenging and requires a closer interaction with hardware and software details of the target machines. Nevertheless, CA and LBM have been consistently achieving peak performance on teraflop computers [96].

To parallelize the LB codes for 3D problems, the lattice is partitioned onto a 3D processor grid, and message passing interface is used for communication. Ghost cells are used to hold copies of the data from neighboring processors. Computational implementations of the LBM have been extensively studied and optimized both for a single processor and for many, up to 2048, processors [96,99,100].

Facilitated by the simplicity of the LB core algorithm, abundant research has been done on parallel computing studies using LB codes. Many optimization techniques have been developed. For example, cache blocking, which aids in taking advantage of memory hierarchy so as to minimize number of cache misses loop unrolling, which replicates the original loop body multiple times, adjusts the loop termination code and eliminates (a) redundant branch instructions [100]; (b) optimal data structure in the context of both single threaded and distributed memory parallel execution [96,101], etc. Because of the local nature of the LBM stencil, there is a weak coupling between different domains in the message passing interface domain decomposition partitioning. Therefore, the LBM has proved to scale excellently with a number of processors [102,103].

We have developed 2D and 3D LB–CA models [53,93,94,104,105] for dendrite growth (with and without convection) that are able to simulate the growth of several equiaxed or columnar dendrites in binary alloys. Through our allocation at the Extreme Science and Engineering Discovery Environment (XSEDE) national supercomputer network, we parallelized the 2D LB–CA model. This model solves the full set of conservation equations. The results show an impressive efficiency of the model, with a linear speed-up to 5000 computing cores and a near perfect scale-up of up to 41,000 computing cores. With this capability, we have been able to simulate 2D domains of 17 cm × 8 cm in grids of 165 billion nodes containing 11.3 million dendrites [53]. This has never been reported before for the modeling of dendrite growth. A subset of simulated domain containing several thousand dendrites is shown in Figure 15.5. The domain presents a flow of the Al–3 wt% Cu melt between solidifying dendrites. The computational domain of 2.4 mm × 1.8 mm size is meshed on a regular 8000 × 6000 lattice having 3264 dendrite nucleation sites.

The result of 3D simulation of columnar dendritic growth in the melt of Al–3 wt% Cu alloy with 4.5°C undercooling is shown in Figure 15.6a [104]. The domain mesh is 720 × 720 × 720, which is equivalent to 216 × 216 × 216 $(\mu m)^3$. Columnar dendrites growing in different orientations and having well-developed side branches can be observed in the simulation snapshots. This is similar to the morphology observed in the experimental micrographs. Dendrites compete with each other, and the ones having orientations other than 90 degrees are blocked by the perpendicular dendrites. Therefore, the dendrites that survive to grow to the top are all parallel to each other. The flat tips or sides observed in some cases are due to the dendrites touching the domain boundaries.

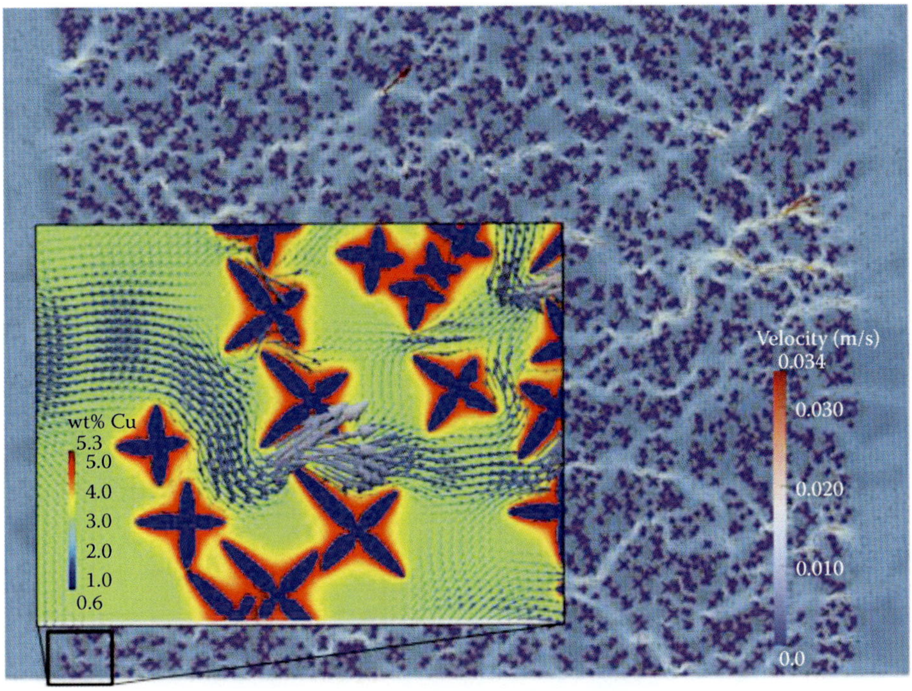

FIGURE 15.5 Two-dimensional simulation of dendrite growth in a 2.4 mm × 1.8 mm size domain containing 3264 dendrites [53]. The arrows represent the velocity vectors of the melt. The colors of the background and dendrites represent solute concentration, while the color and size of the arrows represent the magnitude of the velocity.

At the beginning of the simulation, a number of solid seeds having random positions and crystallographic orientation were placed at the bottom of a domain of the undercooled molten alloy. This condition is similar to what is observed in many industrial applications, where alloys are directionally solidified in conditions that produce a complex array of columnar dendrites.

The solute distribution around the 3D columnar dendrites is shown in Figure 15.6b. The legend shows the weight percent of copper. During solidification, since the solubility of copper in solid is less than its solubility in liquid, the extra solute is rejected to the interface, resulting in a higher copper concentration between the dendritic arms. At a later stage of solidification, the high-concentration region can cause microsegregation or form eutectic phases.

The low computational cost and great scalability of the LB–CA model enabled us to perform large-scale 3D simulations in macrosize domains. A 3D domain discretized with 3300 × 3300 × 3300 grid cells, around 36 billion grid points in total, is shown in Figure 15.7 [104]. With a mesh size of Δx = 0.3 μm, this domain represents a volume close to 1 mm^3. About 4000 seeds with random positions and crystallographic orientations were initially distributed at the bottom of the domain. The dendrites grow in the undercooled melt and then develop side branches. Again, the tilted dendrites are blocked by the dendrites growing in perpendicular directions.

15.6 INFLUENCE OF IMPOSED CONVECTION ON DENDRITE MORPHOLOGY

Many experiments have been carried out to determine the primary dendrite tip radius in metallic and transparent organic alloys [33,34,106–110]. Invariably, the discrepancy between experimental observations and model predictions has been attributed to convection in the melt. Convection has been treated by a pseudo-analytical approach of defining an overall mushy-zone Rayleigh Number for the growth conditions [111–124], and correlating an increasing propensity for convection with

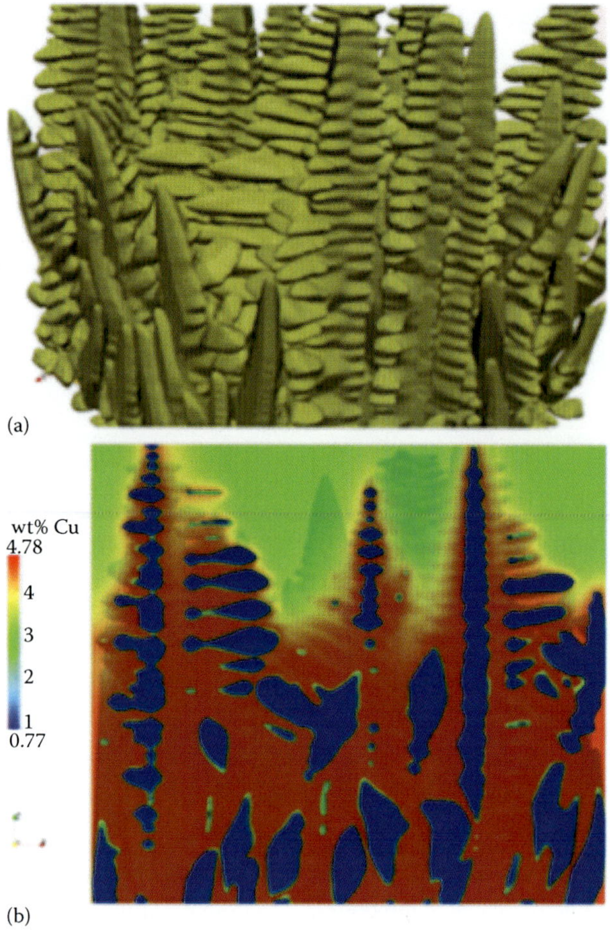

FIGURE 15.6 (a) columnar dendrites growing in an undercooled melt of Al–3 wt% Cu alloy and (b) sectional cut showing solute distribution around 3D columnar dendrites. (From Eshraghi, M., Jelinek, B., Felicelli, S.D., *J. Metals*, 67, 1786–1792, 2015. With permission.)

observed reduction in mean spacing in terrestrial grown samples when compared with those solidified in microgravity [112,114,115]. However, such a treatment ignores the fact that "dendrites" are not branchless features and dendrite array is not made up of dendrites having identical morphology. It ignores the fact that the orientation and intensity of convection produces radial and axial solutal inhomogeneities throughout the mushy zone, which affects the morphology of the dendrite at a local level. The primary dendrites in terrestrial grown samples do not have identical lengths, identical side-branching patterns, and uniform ordering across entire mushy zone. Driven by the availability of primary dendrite models, such as Hunt-Lu [12] and a desire to validate them, the emphasis in the published literature has been on measuring the average behavior [14–34]. Limited effort has been devoted to measuring the distribution and ordering of primary dendrites in an array [21–26] and examining its dependence on interdendritic thermosolutal convection. The nonavailability of analytical, or numerical, tools to predict such interdependence in a 3D array of dendrites is, to a large extent, responsible for the lack of interest shown by experimentalists. Yet, from the point of view of component mechanical property, it is not only important to predict and control the mean primary

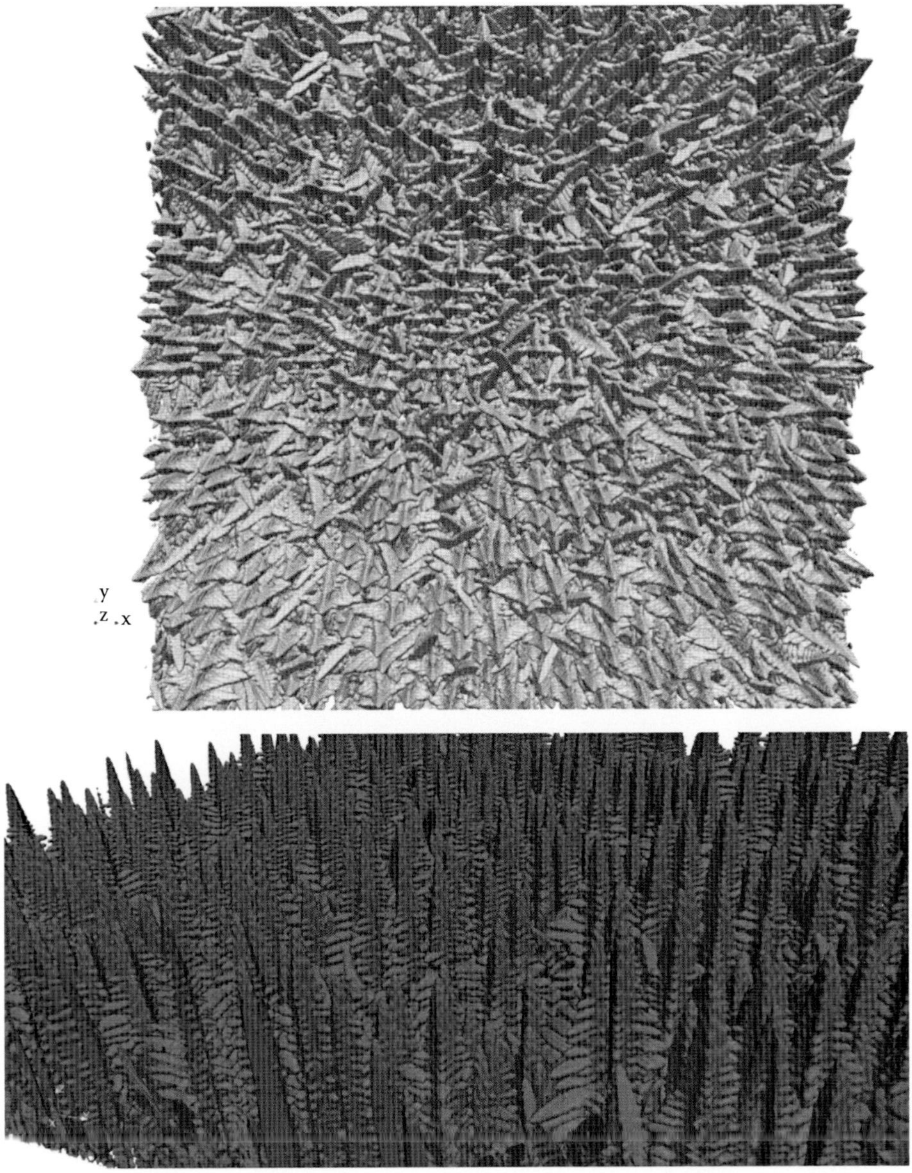

FIGURE 15.7 Large-scale simulation of the growth of 3D columnar dendritic microstructure in a 1 mm^3 domain with approximately 36 billion grid points and 4000 initial seeds. (The top image is from Esraghi, M., Jelinek, B., Felicelli, S.D., *J. Metals*, 67, 1786–1792, 2015. With permission.)

spacing on any cross-section, it is also equally, if not more, important to predict and, where possible, control the extent of nonhomogeneity in the distribution of primary dendrites in a casting [125].

Imposed shear flows on dendritic and cellular arrays have been used during directional solidification (DS) of transparent organic "alloys" in 100–300 μm gaps between the glass slides. These experiments showed the cells to bend toward the downstream of shear flow [126] while the dendrites bend towards the upstream [126–129]. Side branching is enhanced upstream of the primary dendrite arms and suppressed downstream [126–129]. Also, the primary spacing is significantly larger in the presence of shear flow [128,129]. This is contrary to an observed reduction in primary spacing due to the thermosolutal convection that is described previously, indicating that changes in dendrite morphology caused by an imposed shear flow do not represent those that are associated with thermosolutal convection during DS [125].

So, it is known that melt flow can significantly alter the growth kinetics by affecting solutal gradient around the dendrites. But how should the simulations be handled? It is now well recognized that growth of the dendrites in a 3D simulation is different from that in a 2D simulation. Two dimensional models cannot capture true dendrite morphology (side arms etc.), especially when fluid flow is involved because it does significantly alter the local growth kinetics by influencing solutal gradient around the dendrites. While melt convection is blocked by dendrite arms in 2D simulations, flow can go around the arms in 3D simulations thereby yielding different morphology and arm spacing. An accurate representation of morphological evolution of dendrite and its side branches in the presence of convection does require simulation in three dimensions.

The LB–CA model described in the preceding section was extended to solve for fluid flow and count in the convection effects. The evolution of dendritic morphologies after 2, 4, 6, and 8 ms under the effect of melt convection is shown in Figure 15.8 [130]. A uniform flow with an inlet velocity of 7 mm/s enters the domain from the left face. The fluid convection affects solute distribution around the dendrites and consequently alters the kinetics of dendritic growth. The cubic simulation domain contains 2883 cubic

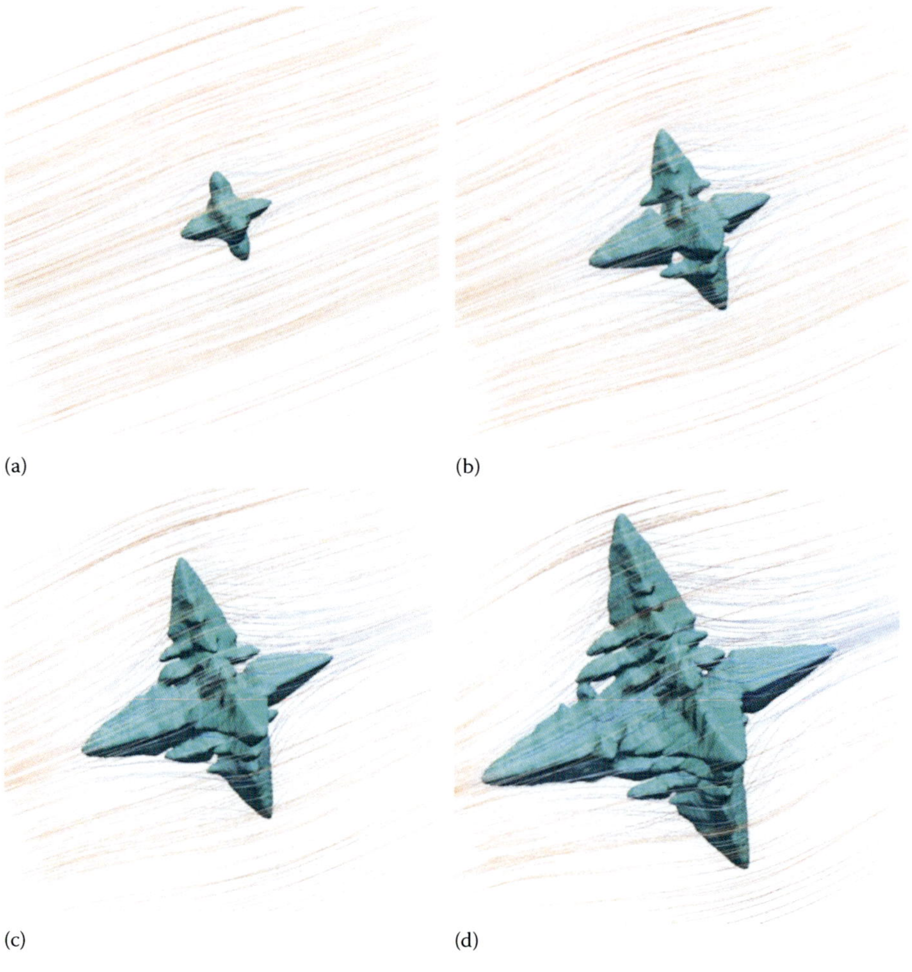

FIGURE 15.8 Evolution of dendritic structures under melt convection after (a) 2, (b) 4, (c) 6, and (d) 8 ms. (From M. Eshraghi et al., *Metals*, 7(11), 474, 2017)

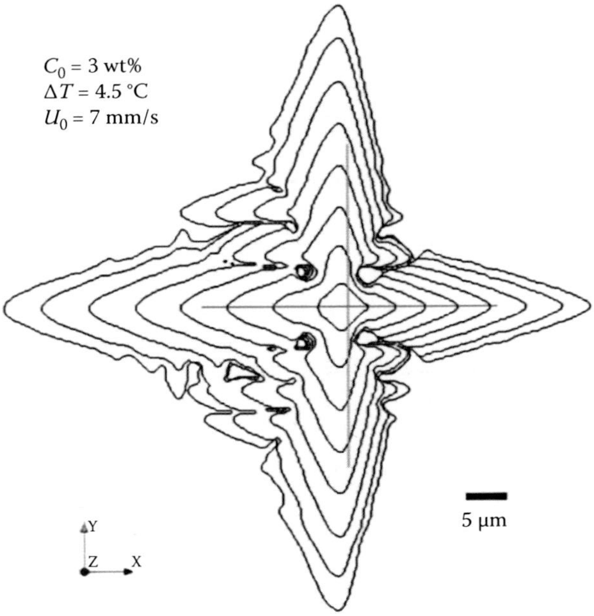

FIGURE 15.9 2D sections showing the morphological changes of the 3D dendrite growing under melt convection. (From M. Eshraghi et al., *Metals*, 7(11), 474, 2017.)

cells with $\Delta x = 0.3$ μm. The domain is uniformly cooled down up to 4.5°C below the melting point ($\Delta T = 4.5$°C). The streamlines depict how flow travels around the growing dendrite. Moreover, the morphology of the dendrite in 2D sections perpendicular to the z-direction and passing through the dendrite center is shown in Figure 15.9 [130]. The wireframes show the morphologies, from inside to outside, following 0.75 ms, 2.00 ms, 3.25 ms, 4.50 ms, 5.75 ms, 7.00 ms, 8.25 ms, and 9.00 ms. As solidification proceeds, the primary arms grow and coarsen, and subsequently, the secondary arms start to grow perpendicular to the primary arms. Melt convection washes the solute from the upstream primary and secondary arms and transports it downstream. This leads to a lower concentration in the upstream area and a higher concentration downstream, resulting in a higher growth rate upstream and a lower growth rate downstream. This matches the findings of previous studies [52,55,56,61,81,92]. Although the transverse arms (the ones in the y and z directions) are not significantly affected by convection, the secondary branches grow faster on the upstream side of the transverse arms.

The melt flow around a dendrite is shown in Figure 15.10 [130]. Velocity values are higher in the areas far from the dendrite and lower in the vicinity of a dendrite. The bottom part of Figure 15.7 shows solute concentration on two perpendicular planes passing through the center of the dendrite. As solidification advances, since solubility of solute in the solid is less than its solubility in the liquid, the extra solute is rejected to the interface. At later times, when solidification is close to completion, there are regions containing the liquid encompassed by the solid phase, holding a high concentration. Those regions may end up causing microsegregation or forming eutectic phases.

Columnar dendrite growth was simulated in the presence of forced convection and the results are presented in Figure 15.11 [131]. At the beginning, arrays of seeds with crystallographic orientation perpendicular to the bottom face were placed in the domain. This is to help show the effect of fluid flow on the morphology of the dendrite. A uniform flow of 7 mm/s velocity enters from the left side, perpendicular to the growth direction. The melt convection washes the solute from the interface on the left face of the dendrite and accelerates growth kinetics in that direction. Similar to what was observed in the equiaxed dendrites, the dendrite arms grow faster in the upstream direction. The secondary arms perpendicular to the flow direction also grow faster when convection is present.

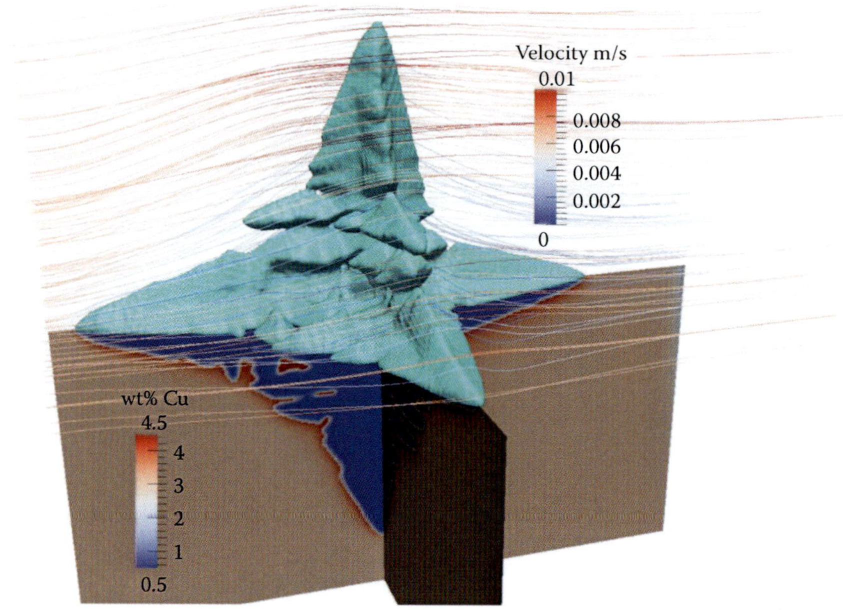

FIGURE 15.10 Solute distribution and slow flied around a dendrite growing under forced melt convection. (From M. Eshraghi et al., *Metals*, 7(11), 474, 2017.)

FIGURE 15.11 3D columnar dendrites growing in an undercooled melt of Al–3 wt% Cu. The domain contains around 173 million 3D lattice points that corresponds to a 180 × 180 × 144 (μm)^3 physical domain. (From Eshraghi, M., PhD Dissertation, Mississippi State University, 2013. With permission.)

15.7 EFFECT OF NATURAL CONVECTION AND FORMATION OF MICROSEGREGATION DEFECTS

Segregation defects formed during the solidification of metallic alloys are of significant importance for industries in the domain of casting, as they tend to degrade the mechanical properties of the casting. Among these defects are the channel-like macrosegregation defects, also known as freckles, observed during DS of some metallic alloys. The channels form in the mushy zone between the dendritic arms, affecting the mechanical properties and causing subsequent rejection of the cast product. Complex thermosolutal convection induces the formation of these defects. Studying the flow pattern during solidification can provide a more comprehensive picture of this phenomenon. Flemings and coworkers [132–134] considered the interdendritic fluid flow through a fixed dendritic solid network and defined a "local solute redistribution equation (LSRE)" to be a mathematical model to explain the behavior of flow during solidification. Mehrabian and coworkers [135] suggested Darcy's law to calculate interdendritic flow velocity, which induces microsegregation. They assumed the interdendritic spacing between dendrites to behave as a porous medium. Fujii and coworkers [136] extended the work of Mehrabian and Flemings [137] to ternary alloys to study macrosegregation in multicomponent low-alloy steels. They solved a coupled set of equations given by Darcy's law and LSRE. However, they made use of experimentally measured temperature for their analysis. Bennon and Incropera [138] and Beckermann and Viskanta [139] derived continuum or volume-averaged models for macrosegregation during alloy solidification. Wang and Beckermann [140] used a unified solute diffusion model for both columnar and equiaxed dendrite growth, in which nucleation, growth kinetics, and dendrite morphology were taken into account. Warren and Boettinger [141] obtained a 2D model for solving the energy and mass conservation equations during solidification, thereby avoiding tracking of the liquid–solid interface. Also, Nastac [142] developed a comprehensive stochastic model to include time-dependent calculations for the purpose of temperature distribution, solute redistribution in the liquid and solid phases, curvature, and growth anisotropy without further assumptions on nucleation and growth of dendritic crystals. Ramirez and Beckermann [143] suggested a criterion to predict the formation of freckles in lead–tin and nickel-based superalloys based on a maximum value of the Rayleigh number.

Shevchenko and coworkers [144] investigated the formation of channel-like defects during DS of gallium–25 wt% indium alloy within a Hele-Shaw cell under the influence of thermosolutal convection by means of X-ray radioscopy. They found instability in density stratification as the reason for the emergence of convective plumes carrying solute-rich liquid upward at the mushy zone. In one other study [145], they demonstrated the effect of fluctuations generated by electromagnetic force, for both the local and global flow pattern. They mentioned that these fluctuations can block the chimney development. Wang and coworkers [146] studied the segregation and density profiles in the mushy zone of CMSX-4 superalloys solidified during both downward and upward directional casting process. The experimental results of density calculations suggest that the interdendritic melt becomes lighter with an increasing depth in the mushy zone of the upward solidified samples, which indicates the occurrence of density inversion. Density inversion did not emerge during downward solidification, which suggests that downward solidification can avoid the formation of freckles.

The numerical results obtained in these studies also confirm the existence of complex thermosolutal convection patterns in the melt. Felicelli and coworkers [147] explained the emergence and survival of channels by a 2D mathematical model for lead–10 wt% tin alloy. Also, Felicelli and coworkers [148] developed a 3D FE model using a thermodynamic function to express the solidification path. Recently, Yuan and Lee [149] developed a 3D microscale model for freckling in Pb–Sn alloys. They found instability in density stratification as the reason for the emergence of convective plumes carrying solute-rich liquids up the mushy zone. Nonetheless, in their study, the motion of flow was constricted to a thin slab, limiting the 3D features specific to the motion of the fluid.

A 3D model was recently developed by the authors to study the onset of freckling and motion of the molten alloy during dendritic solidification [105]. Pb–10 wt% Sn alloy was considered for the simulations. The gradient of solute concentration leads to density inversion, moving the liquid having higher concentration upward. Since Pb–Sn alloy has a relatively high partition coefficient, the solute rejection is significant, resulting in a higher concentration gradient and a stronger buoyancy force. That is the reason that this alloy forms more freckles during the casting processes.

The morphology of dendrites while growing under the influence of natural convection is shown in Figure 15.12a. The height of the domain was 0.512 mm, which was sufficient to allow for the formation of rising plumes and competition of dendrites. Sixteen primary nuclei were placed at the bottom of the domain with a primary arm spacing equal to 12 μm. The initial temperature of the

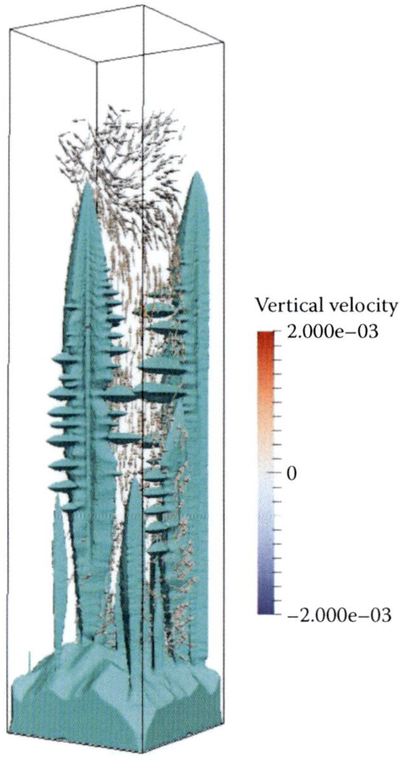

FIGURE 15.12 Simulated motion of rising plume flowing around the dendrite under the effect of thermosolutal convection. (From Hashemi, M., Master's Thesis, The University of Akron, 2016. With permission.)

domain was uniform with an initial undercooling of 3.5 K, and then a constant cooling rate equal to 2.994 K/s was applied to the domain. Only a part of the domain is shown in Figure 15.12, the purpose of enlarging the image to show more details. The solute concentration around the dendrites increases with time and the high partition coefficient of the Pb–Sn alloy creates a strong buoyancy force during solidification and a rising plume.

There is an active competition between the dendrites; only few dendrites survive to fill the domain. The dendrites that are closer to one another can influence each other's growth more significantly. The longest dendrite located at the corner in Figure 15.12 has a faster growth with more side branching. It is observed that the branching of this dendrite blocks the growth of other dendrites. However, the dendrites that are far from the corner dendrite grow fast and show branching as well. Finally, only two dendrites survive, resulting in larger primary arm spacing. Subsequently, during solidification, the two survived dendrites start to generate new branches and fill out the whole main domain. This side branching is caused by the solute-depleted liquid flowing from outside of the channel area. The flow has a very low velocity when it passes over the secondary and tertiary arms, as shown in Figure 15.13. In this case, the velocity is one or two orders of magnitude smaller than reality in the bulk channel velocity found in the channel area with less branching and, hence, more permeability. When solidification is completed, the solute-rich liquid in the channel areas will generally form microsegregation defects known as "freckles." The competition between casting speed and the peak velocity of bulk liquid in the mushy zone does play a significant role in the mechanism of freckle formation. It has been suggested that a faster velocity of the liquid in the mushy zone, in comparison with growth velocity of the dendrite, is a probable reason for the formation of microsegregated channels [149–151]. Branching can retard the motion of liquid and block the

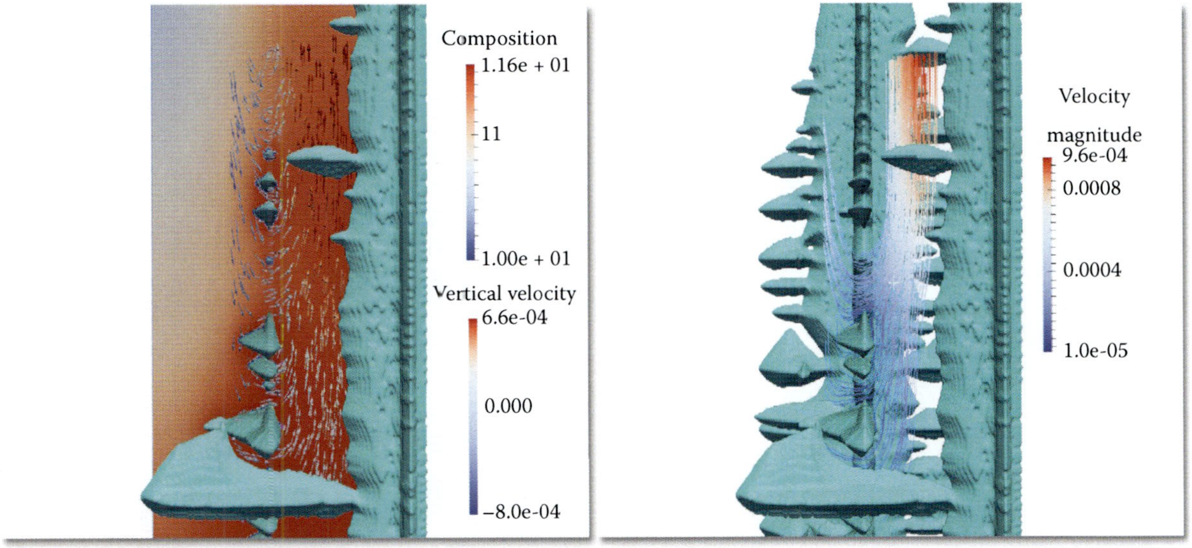

FIGURE 15.13 Motion of the molten alloy in the channel area (a) contour plot of solute concentration and velocity vectors and (b) streamlines of flow around the secondary arms. (From Hashemi, M., Master's Thesis, The University of Akron, 2016. With permission.)

growth of dendrites. Therefore, it plays an important role in the formation of microsegregated channels. The variation of primary arm spacing due to a competition between dendrites can be another factor for the formation of freckles. A microstructural level simulation is needed to observe the effect of branching and variation of primary arm spacing. In addition, it should be noted again that 2D models cannot explain the actual motion of liquid phase and the kinetics of dendritic growth is different in a 3D simulation in comparison with a 2D simulation.

15.8 MODELING MICROPOROSITY FORMATION AND BUBBLE–DENDRITE INTERACTION

The formation of shrinkage porosity and bubbles during solidification disturbs the dendritic array network and degrades the mechanical properties of castings, whether these are large commercial castings of aluminum alloys or steel alloys or directionally solidified single crystal turbine blades. Since metals are opaque, early studies regarding the formation, migration, and growth of gaseous bubbles were based on postsolidification microstructural characterization of cast samples. Lee and Hunt [152] were the first to use X-rays to examine the formation of hydrogen pores in thin slabs of Al–Cu alloys in situ during DS. They observed the number of bubbles to increase and their size decreased with an increase in growth speed and hydrogen content, and the pores developed a "worm-hole-like" shape. However, due to the poor spatial resolution of this technique, research has focused on transparent organic alloys for in situ observation of bubble formation and its interaction with the planar, cellular, and dendritic solid–liquid interface [153–161]. It has been reported that the shape of the bubble changes from drop, to elongated, to irregular with an increase in growth speed in succinonitrile alloy [157]. The bubbles can be elongated, expanded, shrunk, rippled, or closed, depending on the bubble growth rate-to-solidification rate ratio [158]. Han [159] observed the bubbles to nucleate near the base of primary dendrites because of decreased gas solubility at lower temperature. They remain adhered to the solid and grow in size or they can get detached and migrate, often in sudden jumps at speeds as high as 14 cm s^{-1} [159]. Bubble engulfment in the growing solid has been treated [160–161] in a manner akin to particle-pushing and the engulfment process [162–164]. However, it is important to remember that since the experiments on the transparent alloy invariably used narrow spaced (~250 μm) rectangular cross-section glass crucibles, all their observations are essentially between bubble and a 2D array of primary dendrites and are affected by "wall" effects. Contrary to earlier belief, it is now recognized that the basic premise of such experiments, i.e., 2D dendrites represent morphology of a 3D array, is false [165,166]. Understanding pore–mushy zone interaction in real castings requires both experimental observations and also theoretical/numerical modelling with 3D array of dendrites.

The formation of microporosity has been simulated by several researchers by solving fluid flow, solute diffusion, and PF equations. Some of the models used Darcy's law coupled with continuity and conservation equations [167,168]. Others developed a diffusion controlled model for bubble growth [169,170] or continuum stochastic models by combining finite difference and CA methods [171,172]. In one other effort [173], an enthalpy-based method was used along with LS to simulate the evolution of hydrogen bubbles during solidification of aluminum alloys in two dimensions. Growth, movement, and engulfment of bubbles were simulated, but microstructures were not considered in the model. Tiedje and coworkers [174] developed a multizone model for microporosity formation during solidification of Al–Si castings. The effect of cooling rate and porosity distribution was studied, but the effect of solid phase evolution was not considered in the model. Recently, a PF model was proposed to simulate the formation and growth of porosities in a confined solid structure, but the effect of dendrite growth was not considered [175]. Although these models provided a better understanding about the mechanism governing microporosity formation and dynamics, they did not describe the effect of dendritic structures and their evolution on the dynamics of bubble formation. As a matter of fact, experimental observations have proven that the evolution of a solid network and dendrite growth has a significant influence of the morphology and distribution on microporosities [157,159,161].

Due to its special characteristics, the LBM has been employed in recent years to simulate bubble dynamics. However, limited literature is available about using LB modeling to study bubble formation in the context of solidification and dendrite growth. Wu and coworkers [176] developed a 2D LB–CA model for simulation of bubble dynamics during dendritic solidification of alloys. The effect of dendrite growth on nucleation, growth, and movement of bubbles was considered. They simulated the evolution of dendrites and bubble formation in Al–4 wt% Cu alloy during DS. They considered a 250 μm × 800 μm domain and assumed nonwetting conditions to exist between the gas and melt. The initial gas content was taken as 4.5%. As shown in Figure 15.14, subsequent initial competition, three dendrites oriented in the direction of thermal gradient overgrow the others. As solidification progresses, because gas solubility in solid is less than gas solubility in liquid, gas gets rejected to the interface. Once liquid is supersaturated by the gas, bubbles nucleate. This starts to happen when there is about 14% solid in the domain. The bubble nucleation is more likely to happen between the secondary arms, where gas concentration is the highest. The bubbles then keep growing by absorbing gas elements from the surrounding liquid. As the dendrites grow and secondary branches develop, the bubbles get squeezed and elongated between dendritic arms, increasing the pressure inside the bubbles. As the internal pressure of bubbles increases, they tend to escape the interdendritic areas and move forward to regions with more liquid available. In addition, the nearby bubbles tend to coalesce together to form a larger bubble. The three bubbles shown inside the circle in Figure 15.14c coalesced to form a larger bubble shown in Figure 15.14d. The bubble then moves forward and coalesced with the neighboring bubbles to form a larger worm-like bubble as depicted in Figure 15.14e and f. The same group extended their work to a pseudo-3D model and studied both bubble nucleation and bubble–dendrite interactions [177]. However, they considered an artificial binary alloy in a very thin domain with 2.5 μm thickness. We are currently working to extend the models described in the previous sections to simulate bubble–dendrite interactions in large 3D domains with real physical properties.

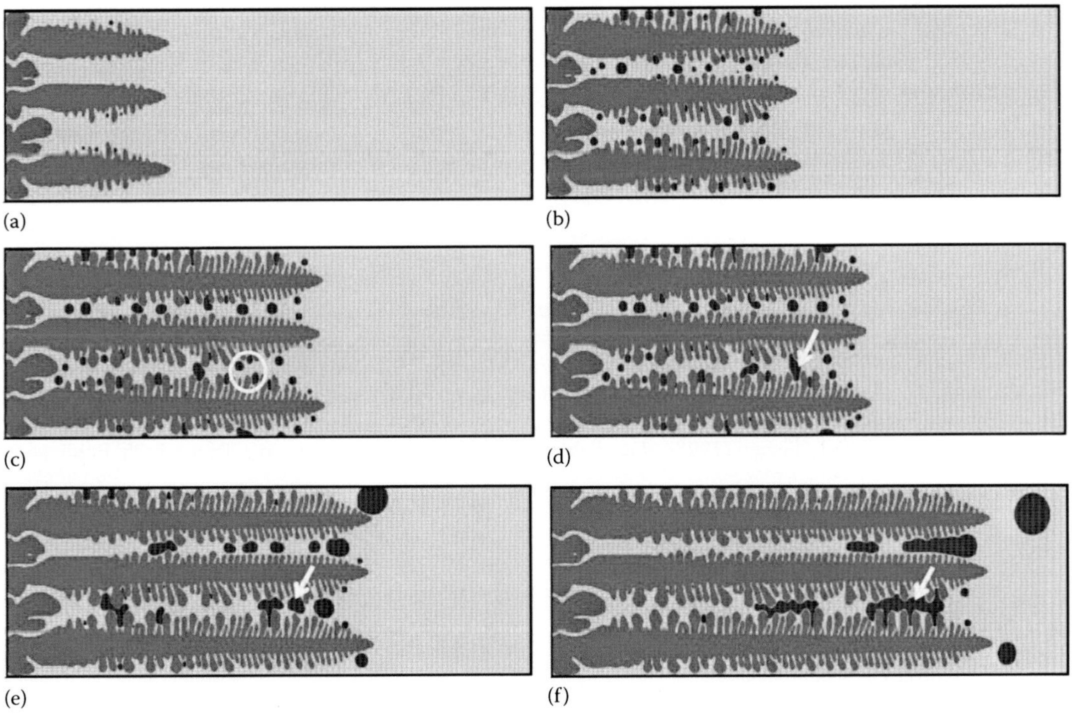

FIGURE 15.14 Simulation of dendrite growth and bubble formation during DS of a binary alloy at different solid fractions: (a) 15%, (b) 28%, (c) 36%, (d) 37%, (e) 43%, and (f) 54%. (From Wu, W., Zhu, M.F., Sun, D.K., Dai, T., Han, Q.Y., Raabe, D., *Mater. Sci. Eng.*, 33, 012103, 2012. With permission.)

15.9 CONCLUDING REMARKS AND FUTURE VISION

The simulation of solidification microstructures involves a complex combination of physical phenomena at different scales. The bridging of multiple scales through filtering and averaging procedures necessarily introduces errors and uncertainty that lead to a loss of important lower-scale features. An ideal solution would be to directly capture these features in macroscale simulations. The incredible advance of supercomputing power, currently at petaflop capability and moving fast toward the exaflop range, is making it possible to start thinking of macroscale simulations with microscale resolution as a viable alternative. The works described in this chapter represent small steps toward achieving this objective. Since the computing power comes from an increasingly large number of processors rather than individual processor speed, the ability to succeed depends on the development of new methods and algorithms for solidification modeling that fully utilize the nominal power in current and coming supercomputers. The microscale simulations presented in this review explore the capabilities of local-type methods like LB and CA with the objective of developing efficient and scalable algorithms capable of simulating scales approaching actual part size. As the power of modern supercomputers continues to increase, we seek to develop models that exhibit the potential to perform a direct numerical simulation of microstructure formed during manufacturing processes like casting and welding. In a similar way that we can nowadays take digital pictures with our cameras at an increasingly larger number of megapixels and then zoom in to visualize small details of the photograph, we envision that in a not so distant future, we will be able to simulate manufacturing processes of actual parts at full-scale and then zoom in at a particular location of our model to observe the resulting microstructure.

ACKNOWLEDGMENTS

The authors acknowledge The University of Akron; California State University, Los Angeles; Center for Advanced Vehicular Systems (CAVS) in Mississippi State University; National Aeronautics and Space Administration (NASA) through Grant Number NNX16AT75G; and National Science Foundation through Grant Numbers CBET-0931801 and PREM-1523588 for their sponsorship and the Extreme Science and Engineering Discovery Environment (XSEDE) for providing the computational resources.

REFERENCES

1. Horstemeyer, M.F. (2009). "Multiscale modeling: A review," in *Practical Aspects of Computational Chemistry*, ed. Leszczynski J. and M.K. Shukla, Springer Science+Business Media, Netherlands, pp. 87–135.
2. Flemings, M.C. (1974). *Solidification Processing*. McGraw Hill, New York.
3. Biloni, H. "Solidification," in *Physical Metallurgy I*, ed. R.W. Kahn and P. Haasen, North-Holland Physics Publishing, Amsterdam, The Netherlands, pp. 477–579.
4. Kurz, W. and D.J. Fisher (1998). *Fundamentals of Solidification*. 4th ed., Trans Tech Publication, Zurich.
5. Dantzig, J.A. and M. Rappaz (2009). *Solidification*. EPFL Press, Lausanne.
6. Kurz, W. and D.J. Fisher (1981). "Dendrite growth at the limit of stability: Tip radius and spacing." *Acta Metallurgica* 29: 11–20.
7. Burden, M.H. and J.D. Hunt (1974). "Cellular and dendritic growth." *Journal of Crystal Growth* 22: 99–108.
8. Kurz, W. and D.J. Fisher (1979). "Dendrite growth in eutectic alloys: The couple zone." *International Metals Review* 5(6): 177–204.
9. Trivedi, R. (1980). "Theory of dendritic growth during the directional solidification of binary alloys." *Journal of Crystal Growth* 49: 219–232.
10. Hunt, J.D. (1979). *Solidification and Casting of Metals*. The Metals Society, London, p. 3.
11. Trivedi, R. (1984). "Interdendritic spacing: Part II. A comparison between theory and experiments." *Metallurgical Transactions A* 15A: 977–984.

12. Lu, S.Z. and J.D. Hunt (1992). "A numerical analysis of dendrite and cellular array growth: The spacing adjustment mechanism." *Journal of Crystal Growth* 123: 17–34.
13. Hunt, D. and S.Z. Lu (1996). "Numerical modeling of cellular/dendritic array growth: Spacing and structure predictions." *Metallurgical and Material Transactions A* 27A: 611–23.
14. Klaren, C.M., J.D. Verhoeven and R. Trivedi (1980). "Primary dendrite spacing of lead dendrites in Pb–Sn and Pb–Au alloys." *Metallurgical Transactions A* 11A: 1853–1861.
15. McCartney, D.G. and J.D. Hunt (1981). "Measurement of cell and primary dendrite arm spacings in directionally solidified aluminum alloys." *Acta Metallurgica* 29: 1851–1863.
16. Mason, J.T., J.D. Verhoeven and R. Trivedi (1982). "Primary dendrite spacing I. Experimental studies." *Journal of Crystal Growth* 59: 516–524.
17. Mason, J.T., J.D. Verhoeven and R. Trivedi (1984). "Primary dendrite spacing: Part II. Experimental studies of Pb–Pd and Pb–Au alloys." *Metallurgical Transactions A* 15A: 1665–1676.
18. Somboonsuk, K., J.T. Mason and R. Trivedi (1984). "Interdendritic spacing: Part I experimental studies." *Metallurgical Transactions A* 15A: 967–975.
19. Miyata, Y. and T. Suzuki (1985). "Cellular and dendritic growth: Part II. Theory." *Metallurgical Transactions A* 16A: 1807–1814.
20. Chopra, M.A. and S.N. Tewari (1991). "Growth speed dependence of primary arm spacing in directionally solidified Pb-10 wtpct tin." *Metallurgical Transactions A* 22A: 2467–2474.
21. Billia, B., H. Jamgotchian and H. N. Thi Stastical (1991). "Analysis of the disorder of two-dimensional cellular arrays in directional solidification." *Metallurgical Transactions A* 22A: 3041–3050.
22. Noel, N., H. Jamgotchian and B. Billia (1997). "In situ and real time observation of the formation and dynamics of a cellular interface in a succinonitrile–0.5 wt% acetone alloy directionally solidified in a cylinder." *Journal of Crystal Growth* 181: 117–132.
23. Odell, S.P., G.L. Ding and S.N. Tewari (1999). "Cell/dendrite distribution in directionally solidified hypoeutectic Pb–Sb alloys." *Metallurgical and Material Transactions A* 30A: 2159–2165.
24. Hui, J., R. Chen, R. Tiwari, X. Wu, S.N. Tewari and R. Trivedi (2002). "Primary dendrite distribution and disorder during directional solidification of Pb–Sb alloys." *Metallurgical and Material Transactions A* 33A: 3499–3510.
25. Trivedi, R., P. Mazumder and S.N. Tewari (2002). "The effect of convection on disorder in primary cellular and dendritic arrays." *Metallurgical and Material Transactions A* 33A: 3763–3775.
26. Tschopp, M.A., J.D. Miller, A.L. Oppedal and K.N. Solanki (2014). "Characterizing the local primary dendrite arm spacing in directionally solidified dendritic microstructures." *Metallurgical and Material Transactions A* 45A: 426–437.
27. Chopra, M.A. and S.N. Tewari (1991). "Growth speed dependence of primary arm spacings in directionally solidified Pb-10 wt% Sn." *Metallurgical and Material Transactions A* 22A: 2467–2474.
28. Han, S.H. and R. Trivedi (1994). "Primary spacing selection in directionally solidified alloys." *Acta Metallurgical Material* 42: 25–41.
29. Tewari, S.N., Y.H. Weng, G.L. Ding and R. Trivedi (2002). "Cellular array morphology during directional solidification." *Metallurgical and Material Transactions A* 33A: 1229–1243.
30. Gunduz, M. and E. Cardili (2002). "Directional solidification of Al–Cu alloys." *Material Science and Engineering A* 3127: 167–185.
31. Tewari, S.N. and V. Laxmanan (1987a). "A critical examination of the dendrite growth models: Comparison of theory with experimental data." *Acta Metallurgical* 35: 175–183.
32. Tewari, S.N. and V. Laxmanan (1987b). "Cellular dendritic transition in directionally solidified binary alloys." *Metallurgical Transactions A* 18A 167–170.
33. Trivedi, R. and K. Somboonsuk (1984). "Constrained growth and spacing." *Materials Science and Engineering* 65: 65–74.
34. Esaka, H. and W. Kurz Columner (1984). "Dendritic growth: A comparison of theory." *Journal of Crystal Growth* 69: 362–366.
35. Badillo, A. and C. Beckermann (2006). "Phase field simulation of the columnar-to-equiaxed transition in alloy solidification." *Acta Materialia* 54: 2015–2026.
36. Boettinger, W.J., J.A. Warren, C. Beckermann and A. Karma (2002). "Phase-field simulations of solidification." *Annual Reviews of Materials Research* 32: 132–194.
37. Hoyt, J.J., M. Asta and A. Karma (2003). "Atomistic and continuum modeling of dendritic solidification." *Material Science and Engineering R* 41: 121–163.
38. Karma, A. and W.J. Rappel (1998). "Quantitative phase-field modeling of dendritic growth in two and three dimensions." *Physical Review E* 57: 4323–4349.

39. Provatas, N., N. Goldenfeld and J.A. Dantzig (1999). "Adaptive mesh refinement computation of solidification microstructures using dynamic data structures." *Journal of Computational Physics* 148: 265–290.
40. Gibou, F., R. Fedkiw, R. Caflisch and S. Osher (2003). "A level set approach for the numerical simulation of dendritic growth." *Journal of Scientific Computing* 19: 183–199.
41. Gibou, F., R.P. Fedkiw, L.T. Cheng and M. Kang (2002). "A second-order-accurate symmetric discretization of the Poisson equation in irregular domains." *Journal of Computational Physics* 176: 205–227.
42. Kelly, S.M., S.L. Kampe and C.R. Crowe (2000). "Microstructural study of laser formed Ti–6Al–4V." *Materials Research Society Symposium on Proceedings* 625: 3–8.
43. Osher, S. and R.P. Fedkiw (2001). "Level set methods: An overview and some recent results." *Journal of Computational Physics* 169: 463–502.
44. Merle, R. and J. Dolbow (2002). "Solving thermal and phase change problems with extended finite element method." *Computational Mechanics Journal* 28: 339–350.
45. Schmidt, A. (1998). "Approximation of crystalline dendritic growth in two space dimensions." *Acta Mathematica Universitatis Comenianae* 67: 57–68.
46. Udaykumar, H.S., R. Mittal and W. Shyy (1999). "Computation of solid-liquid phase fronts in the sharp interface limit on fixed grids." *Journal of Computational Physics* 153: 535–574.
47. Zhao, P. and Heinrich J.C. (2001). "Front-tracking finite element method for dendritic solidification." *Journal of Computational Physics*. 173: 765–796.
48. Al-Rawahi, N. and G. Tryggvason (2004). "Numerical simulation of dendritic solidification with convection: Three-dimensional flow." *Journal of Computational Physics* 194: 677–696.
49. Guo, Z., J. Mi, P. Grant (2012). "An implicit parallel multigrid computing scheme to solve coupled thermal-solute phase-field equations of dendrite evolution." *Journal of Computational Physics* 231: 1781–1796.
50. Karma, A. and W.J. Rappel (1997). "Phase-field simulation of three-dimensional dendrites: Is solvability theory correct?" *Journal of Crystal Growth* 174: 54–64.
51. Karma, A. and W.J. Rappel (1998). "Quantitative phase-field modeling of dendritic growth in two and three dimensions." *Physical Review E* 57: 4323–4349.
52. Jeong, J.H., N. Goldenfeld and J.A Dantzig (2001). "Phase field model for three-dimensional dendritic growth with fluid flow." *Physical Review E* 65: 041602.
53. Jelinek, B., M. Eshraghi, S.D. Felicelli and J.F. Peters (2014). "Large-scale parallel lattice Boltzmann—Cellular automaton model of two-dimensional dendritic growth." *Computer Physics Communications* 185: 939–947.
54. Kobayashi, R. (1994). "A numerical approach to three-dimensional dendritic solidification." *Experimental Mathematics* 3: 59–81.
55. Lu, Y., C. Beckermann and A. Karma (2002). "Convective effects in three-dimensional dendritic." *Materials Research Society Symposium on Proceedings* 701: T2.2.1–T2.2.10.
56. Lu, Y., C. Beckermann and J.C. Ramirez (2005). "Three-dimensional phase-field simulations of the effect of convection on free dendritic growth." *Journal of Crystal Growth* 280: 320–334.
57. Narski, J. and M. Picasso (2006). "Adaptive 3-D finite elements with high aspect ratio for dendritic growth of a binary alloy including fluid flow induced by shrinkage." *FDMP* 1: 1–13.
58. Pan, S. and M.F. Zhu (2010). "A three-dimensional sharp interface model for the quantitative simulation of solutal dendrite growth." *Acta Materialia* 58: 340–352.
59. Schmidt, A. (1996). "Computation of three dimensional dendrites with finite elements." *Journal of Computational Physics* 125: 293–312.
60. Yamanaka, A., T. Aoki, S. Ogawa and T. Takaki (2011). "GPU-accelerated phase-field simulation of dendritic solidification in binary alloy." *Journal of Crystal Growth* 318: 40–45.
61. Yuan, L. and P.D. Lee (2010). "Dendritic solidification under natural and forced convection in binary alloys." *2D versus 3D Simulation, Modelling and Simulation in Material Science and Engineering* 18: 055008.
62. Zhang, X., J. Zhao, H. Jiang and M.F. Zhu (2012). "A three-dimensional cellular automaton model for dendritic growth in multi-component alloys." *Acta Materialia* 60: 2249–2257.
63. Zhao, P., J.C. Heinrich and D.R. Poirier (2007). "Numerical simulation of crystal growth in three dimensions using a sharp-interface finite element method." *International Journal for Numerical Methods in Engineering* 71: 25–46.

64. Zhu, M.F., S. Pan, D. Sun and H. Zhao (2010). "Numerical simulation of microstructure evolution during alloy solidification by using cellular automaton method." *ISI Journal International* 50: 1851–1858.
65. Barbieri, A. and J.S. Langer (1989). "Prediction of dendritic growth rates in the linearized solvability theory." *Physical Reviews A* 39: 5314–5325.
66. Choudhury, A., K. Ruther, E. Wesner, A. August, B. Nestler and M. Rettenmayr (2012). "Comparison of phase-field and cellular automaton models for dendritic solidification in Al–Cu alloy." *Computational Material Science* 55: 263–268.
67. Gandin, C.A., J.L. Desbiolles, M. Rappaz and P.L. Thevoz (1999). "A three-dimensional cellular automaton–finite element model for the prediction of solidification grain structures." *Metallurgical and Material Transactions A* 30: 3153–3165.
68. Guillemot, G., C.A. Gandin, H. Combeau, R. Heringer (2004). "A new cellular automaton–finite element coupling scheme for alloy solidification." *Modelling and Simulation in Material Science and Engineering* 12: 545–556.
69. Wang, W., P.D. Lee and M. McLean (2003). "A model of solidification microstructures in nickel-based superalloys: Predicting primary dendrite spacing selection." *Acta Materialia* 51: 2971–2987.
70. Wei, L., X. Lin, M. Wang and W. Huang (2012). "A cellular automaton model for a pure substance solidification with interface reconstruction method." *Computational Material Science* 54: 66–74.
71. Wei, L., X. Lin, M. Wang and W. Huang (2012). "Orientation selection of equiaxed dendritic growth by three-dimensional cellular automaton model." *Physica B* 407: 2471–2475.
72. Zhu, M.F. and C.P. Hong (2002). "A three dimensional modified cellular automaton model for the prediction of solidification microstructures." *ISI Journal International* 42: 520–526.
73. Zhu, M.F. and D.M. Stefanescu (2007). "Virtual front tracking model for the quantitative modeling of dendritic growth in solidification of alloys." *Acta Materialia* 55: 1741–1755.
74. Zhu, M.F., S.Y. Lee and C.P. Hong (2004). "Modified cellular automaton model for the prediction of dendritic growth with melt convection." *Physical Review E* 69: 061610.
75. Brown, S.G.R. and N.B. Bruce (1995). "3-Dimensional cellular-automaton models of microstructural evolution during solidification." *Journal of Materials Science* 30: 1144–1150.
76. Gandin, C.A. and M. Rappaz (1994). "Coupled finite-element cellular-automaton model for the prediction of dendritic grain structures in solidification processes." *Acta Metallurgica et Materialia* 42: 2233–2246.
77. Guillemot, G., C.A. Gandin and M. Bellet (2007). "Interaction between single grain solidification and macrosegregation: Application of a cellular automaton–finite element model." *Journal of Crystal Growth* 303: 58–68.
78. Dong, H.B., X.L. Yang, P.D. Lee and W. Wang (2004). "Simulation of equiaxed growth ahead of an advancing columnar front in directionally solidified Ni-based superalloys." *Journal of Materials Science* 39: 7207–7212.
79. Li, D., R. Li and P. Zhang (2007). "A cellular automaton technique for modelling of a binary dendritic growth with convection." *Applied Mathematical Modelling* 31: 971–998.
80. Sanchez, L.B. and D.M. Stefanescu (2003). "Growth of solutal dendrites: A cellular automaton model and its quantitative capabilities." *Metallurgical and Materials Transactions A* 34A: 367–382.
81. Yin, H. and S.D. Felicelli (2009). "A cellular automaton model for dendrite growth in magnesium alloy AZ91" *Modeling and Simulation in Materials Science and Engineering* 17: 075011.
82. Fu, Z.H., Q. Xu and S.H. Xiong (2007). "Microstructure simulation of magnesium alloy." *Material Science Forum* 546–549: 133–137.
83. Yin, H. and S.D. Felicelli (2010). "Dendrite growth simulation during solidification in the LENS process." *Acta Materialia* 58: 1455–1465.
84. Raj, R., A. Prasad, P.R. Parida and S.C. Mishra (2006). "Analysis of solidification of a semitransparent planar layer using the lattice Boltzmann method and the discrete transfer method." *Numerical Heat Transfer Part A* 49: 279–299.
85. Raabe, D. (2004). "Overview of the lattice Boltzmann method for nano- and microscale fluid dynamics in materials science and engineering." *Modeling and Simulation in Materials Science and Engineering* 12: R13–R46.
86. Ho, J.R., C.P. Kuo and W.S. Jiaung (2003). "Study of heat transfer in multilayered structure within the framework of dual-phase-lag heat conduction model using lattice Boltzmann method." *International Journal Heat Mass Transfer* 46: 55–69.

87. Jiaung, W.S., J.R. Ho and C.P. Kuo (2001). "Lattice Boltzmann method for the heat conduction problem with phase change." *Numerical Heat Transfer Part B* 39: 167–187.
88. Medvedev, D. and K. Kassner (2005). "Lattice Boltzmann scheme for dendritic growth in presence of convection." *Journal of Crystal Growth* 275: e1495–e1500.
89. Miller, W., I., Rasin and S. Succi (2006). "Lattice Boltzmann phase-field modeling of binary-alloy solidification." *Physica A* 362: 78–83.
90. Brewster, J.D. (2007). "Lattice Boltzmann simulations of three-dimensional fluid flow on a desktop computer." *Analytical Chemistry* 79: 2965–2971.
91. Chopard, B. and A. Dupuis (2002). "Lattice Boltzmann models: An efficient and simple approach to complex flow problems." *Computer Physics Communications* 147: 509–515.
92. Sun, D., M.F. Zhu, S. Pan and D. Raabe (2009). "Lattice Boltzmann modeling of dendritic growth in a forced melt convection." *Acta Materialia* 57: 1755–1767.
93. Yin, H., S.D. Felicelli and L. Wang (2011). "Simulation of dendritic microstructure with lattice Boltzmann and cellular automaton methods." *Acta Materialia* 59: 3124–3136.
94. Eshraghi, M., S.D. Felicelli and B. Jelinek (2012). "Three-dimensional simulation of solutal dendrite growth using lattice Boltzmann and cellular automaton methods." *Journal of Crystal Growth* 354: 129–134.
95. Selzer, M., M. Jainta and B. Nestler (2009). "A lattice-Boltzmann model to simulate the growth of dendritic and eutectic microstructure under the influence of fluid flow." *Physica Status Solidi B* 246(6): 1197–1205.
96. Wellein, G., P. Lammers, G, Hager, S. Donath and T. Zeiser (2005). "Towards optimal performance for lattice Boltzmann applications on terascale computers," in *Parallel Computational Fluid Dynamics: Theory and Applications*, eds. A. Deane, G. Brenner, A.E. et al., Proceedings of the 2005 International Conference on Parallel Computational Fluid Dynamics, May 24–27, College Park, MD, pp. 31–40.
97. HPC Wire (June 2012). Exascale computing: The view from Argonne. http://www.hpcwire.com
98. Grinberg, L. and G.E. Karniadakis (2010). "A new domain decomposition method with overlapping patches for ultrascale simulations: Application to biological flows." *Journal of Computer Physics* 229: 5541–5563.
99. Aidun, C.K. and J.R. Clausen (2010). "Lattice-Boltzmann method for complex flows." *Annual Review of Fluid Mechanics* 42: 439–472.
100. Wu, X. (2008). *Performance Analysis and Optimization of Parallel Scientific Applications on CMP Cluster Systems*. First International Workshop on Simulation and Modelling in Emergent Computational Systems, Portland, Oregon.
101. Pohl, T. (2003). "Optimization and Profiling of the Cache Performance of Parallel Lattice Boltzmann Codes." *Parallel Processing Letter* 13: 549–560.
102. Oliker, L. (2008). "Scientific application performance on leading scalar and vector supercomputering platforms." *International Journal of High Performance Computing Applications* 22: 5–20.
103. Wittmann, M., T. Zeiser, G. Hager and G. Wellein (2012). "Domain decomposition and locality optimization for large-scale lattice Boltzmann simulations." *Computers and Fluids* 10: 1016.
104. Eshraghi, M., B. Jelinek and S.D. Felicelli (2015). "Large-scale three-dimensional simulation of dendritic solidification using lattice Boltzmann method." *Journal of Metals* 67(8): 1786–1792.
105. Hashemi, M. (2016). "Lattice Boltzmann simulation of natural convection during dendritic growth." Master's Thesis, The University of Akron.
106. Yu, L., G.L. Ding, J. Reye, S.N. Ojha and S.N. Tewari (2000). "Mushy zone morphology during directional solidification of Pb-5.8Sb alloy." *Metallurgical and Material Transactions* 31A 2275–2285.
107. Tewari, S.N. (1986). "Dendrite characteristics in directionally solidified Pb 8% Au and Pb 3% Pd alloys." *Metallurgical Transactions* 17A 2279–2290.
108. Tewari, S.N. and R. Trivedi (1991). "Shape selection criterion for cellular array during constrained growth of binary alloys—Need for low gravity experiment." *Microgravity Science and Technology* 4: 240–244.
109. Song, H. and S.N. Tewari (1996). "Time dependence of tip morphology during cellular/dendritic arrayed growth." *Metallurgical and Material Transactions A* 27A 1111–1119.
110. Melendez, A.J. and C. Beckermann (2012). "Measurements of dendrtite tip growth and sidebranching in succinonitrile–acetone alloys." *Journal Crystal Growth* 340: 175–189.
111. Dupouy, M.D., D. Camel and J.J. Favier (1989). "Natural convection in directional dendritic solidification of metallic alloys—I. Macroscopic effects." *Acta Metallurgica* 37(4): 1143–1157.

112. Dupouy, D., D. Camel and J.J. Favier (1992). "Natural convective effects in directional dendritic solidification of binary metallic alloys: Dendrite array primary spacing." *Acta Metallurgica et Materialia* 40(7): 1791–1801.
113. Dupov, M.D. and D. Camel (1998). "Effects of gravity on columnar dendritic growth of metallic alloys: Flow pattern and mass transfer." *Journal of Crystal Growth* 183: 469–489.
114. Thi, H.N., Y. Dabo, B. Drevet, M.D. Dupouy, D. Camel, B. Billia, J.D. Hunt and A. Chilton (2005). "Directional solidification of Al–1.5 wt% Ni alloys under diffusion transport in space fluid-flow localization on earth." *Journal of Crystal Growth* 281: 654–668.
115. Drevet, B., H.N. Thi, D. Camel, B. Billia and M.D. Dupouy (2000). "Solidification of aluminum–lithium alloys near the cell/dendrite transition-influence of solutal convection." *Journal of Crystal Growth* 218: 419–433.
116. Tewari, S.N. and R. Shah (1996). "Macrosegregation during dendritic arrayed growth: Influence of primary arm spacings and mushy zone length." *Metallurgical Transactions A* 27A 1353–1361.
117. Beckermann, C., J.P. Gu and W.J. Boettinger (2000). *Metallurgical and Material Transactions A* 31A: 2545–2552.
118. Song, H., S.N. Tewari and H. C. de Groh III (1996). "Natural convection during directional solidification of Pb–Sn alloys." *Metallurgical and Material Transactions A* 27A: 1095–1110.
119. Tewari, S.N., R. Tiwari and G. Magadi (2004). "Mushy zone Rayleigh number to describe macrosegregation and channel formation in directionally solidified metallic alloys." *Metallurgical and Materials Transactions A* 35A: 2927–2934.
120. Tewari, S.N. and R. Tiwari (2003). "A mushy zone Rayleigh number to describe interdendritic convection during directional solidification of hypoeutectic Pb–Sb and Pb–Sn alloys." *Metallurgical and Materials Transactions A* 34A: 2365–2376.
121. Ojha, S.N., G. Ding, Y. Lu, J. Rye and S.N. Tewari (1999). "Macrosegregation caused by thermosolutal convection during directional solidification of Pb–Sb alloys." *Metallurgical and Material Transactions A* 30A: 2167–2171.
122. Tewari, S.N., R. Shah and M.A. Chopra (1993). "Thermosolutal convection and macrosegregation caused by solute rejection at cell/dendrite tips." *Metallurgical and Material Transactions A* 24A: 1661–1669.
123. Ojha, S.N. and S.N. Tewari (2004). "Mushy zone characteristics and macrosegregation during directional solidification." *Transactions of the Indian Institute of Metals* 57(5): 475–483.
124. Ojha, S.N. and S.N. Tewari (2000). "Effect of growth rate on macrosegregation during directional solidification of a Pb–Sb alloy." *International Journal of Cast Metals Research* 13207–213.
125. Tewari, S. (2016). Personal Communication.
126. Huang, T., S. Liu, Y. Yang, D. Lu and Y. Zhou (1993). "Coupling of couette flow and crystal morphologies in directional freezing." *Journal of Crystal Growth* 128: 167.
127. Huang, T., L. Deyang and Z. Yaoche (1988). "Diffusion-convection effects on constrained dendritic growth in dilute alloys." *Acta Astronautica* 12(9): 997.
128. Esaka, H., T. Taenaka, H. Ohishi, S. Mizoguchi and H. Kajioka (1988). "A model experiment for solidification–deflection mechanism of primary dendrite due to fluid flow." *International Symposium on Scale Modeling* July 18–22, 1988, Tokyo, The Japan Society of Mechanical Engineers, pp. 287–293.
129. Okamoto, T., K. Kishitake and 1. Bessho (1975). "Dendritic structure in unidirectionally solidified cyclohexanol." *Journal of Crystal Growth* 29: 131–136.
130. Eshraghi, M., M. Hashemi, B. Jelinek, and S.D. Felicelli. (2017). "Three-dimensional lattice Boltzmann modeling of dendritic solidification under forced and natural convection." Metals, 7(11), 474.
131. Eshraghi, M. (2013). "Modeling dendritic solidification using lattice Boltzmann and cellular automaton methods." PhD Dissertation, Mississippi State University.
132. Flemings, M.C. and G.E. Nereo (1967). "Macrosegregation: Part I." *Transactions of the Metallurgical Society of AIME* 239: 1449–1461.
133. Flemings, M.C., R. Mehrabian and G.E. Nereo (1968). "Macrosegregation: Part II." *Transactions of the Metallurgical Society of AIME* 242. 41–49.
134. Flemings, M.C. and G.E. Nereo (1968). "Macrosegregation: Part III." *Transactions of the Metallurgical Society of AIME* 242: 50–55.
135. Mehrabian, R., M. Keane and M.C. Flemings (1970). "Interdendritic fluid flow and macrosegregation; influence of gravity." *Metallurgical and Materials Transactions B* 1: 1209–1220.
136. Fujii, T., D.R., Poirier and M.C. Flemings (1979). "Macrosegregation in a multicomponent low alloy steel." *Metallurgical Transactions B* 10: 331–339.

137. Mehrabian, R. and M.C. Flemings (1970). "Macrosegregation in ternary alloys." *Metallurgical and Materials Transactions B* 1: 455–464.
138. Bennon, W.D. and F.P. Incropera (1987). "A continuum model for momentum, heat and species transport in binary solid-liquid phase change systems. I. Model formulation." *International Journal of Heat and Mass Transfer* 30: 2161–2170.
139. Beckermann, C. and R. Viskanta (1988). "Double-diffusive convection during dendritic solidification of a binary mixture." *PhysicoChemical Hydrodynamics* 10: 195–213.
140. Wang, C.Y. and C. Beckermann (1993). "A unified solute diffusion model for columnar and equiaxed dendritic alloy solidification." *Materials Science and Engineering A* 171(1): 199–211.
141. Warren, J.A. and W.J. Boettinger (1995). "Prediction of dendritic growth and microsegregation patterns in a binary alloy using the phase-field method." *Acta Metallurgica et Materialia* 43(2): 689–703.
142. Nastac, L. "Numerical modeling of solidification morphologies and segregation patterns in cast dendritic alloys." *Acta Materialia* 47(17): 4253–4262.
143. Ramirez, J.C. and C. Beckermann (2003). "Evaluation of a Rayleigh-number-based freckle criterion for Pb–Sn alloys and Ni-base superalloys." *Metallurgical and Materials Transactions A* 34(7): 1525–1536.
144. Shevchenko, N., S. Boden, G. Gerbeth, and S. Eckert (2013). "Chimney formation in solidifying Ga-25wt pctIn alloys under the influence of thermosolutal melt convection." *Metallurgical and Materials Transactions A* 44(8): 3797–3808.
145. Shevchenko, N., S. Eckert, S. Boden and G. Gerbeth (2012). "In situ X-ray monitoring of convection effects on segregation freckle formation." *Materials Science and Engineering* 33(1): 012035.
146. Wang, F., D. Ma, J. Zhang and A. Bührig-Polaczek (2015). "Investigation of segregation and density profiles in the mushy zone of CMSX-4 superalloys solidified during downward and upward directional solidification processes." *Journal of Alloys and Compounds* 620: 24–30.
147. Felicelli, S.D., J.C. Heinrich and D.R. Poirier (1991). "Simulation of freckles during vertical solidification of binary alloys." *Metallurgical Transactions B* 22(6) 847–859.
148. Felicelli, S.D., D.R. Poirier and J.C. Heinrich (1998). "Modeling freckle formation in three dimensions during solidification of multicomponent alloys." *Metallurgical and Materials Transactions B* 29(4): 847–855.
149. Yuan, L. and P.D. Lee (2012). "A new mechanism for freckle initiation based on microstructural level simulation." *Acta Materialia* 60(12): 4917–4926.
150. Frueh, C., D.R. Poirier and S.D. Felicelli (2002). "Predicting freckle-defects in directionally solidified Pb–Sn alloys." *Materials Science and Engineering A* 328(1): 245–255.
151. Flemings, M.C. (1974). "Solidification processing." *Metallurgical Transactions* 5(10): 2121–2134.
152. Lee, P.D. and J.D. Hunt (1997). "Hydrogen porosity in directionally solidified aluminum–copper alloys: In situ observation." *Acta Materialia* 45: 4155–4169.
153. Jamgotchian, H., R. Trivedi and B. Billia (1993). "Interface dynamics and coupled growth in directional solidification in presence of bubbles." *Journal of Crystal Growth* 134: 181.
154. Akamatsu, S. and G. Faivre (1996). "Residual-impurity effects in directional solidification: Long-lasting recoil of the front and nucleation-growth of gas bubbles." *Journal de Physique I France* 8: 503.
155. Xing, H., J.Y. Wang, C.L. Chen, K.X. Jin and Z.F. Shen (2010). "Morphological evolution of the interface microstructure in the presence of bubbles during directional solidification." *Scripta Materialia* 63: 1228.
156. Wei, P.S., C.C. Huang and K.W. Lee (2003). "Nucleation of bubbles on a solidification front—experiment and analysis." *Metallurgical and Materials Transactions B* 34: 321–332.
157. Xing, H., J.Y. Wang, C.L. Chen, K.X. Jin and Z.F. Shen (2010). "Morphological evolution of the interface microstructure in the presence of bubbles during directional solidification." *Scripta Materialia* 63: 1228–1231.
158. Wei, P.S. and S.Y. Hsiao (2012). "Pore shape development from a bubble captured by a solidification front." *International Journal of Heat and Mass Transfer* 55: 8129–8138.
159. Han, Q. (2006). "Motion of bubbles in the mushy zone." *Scripta Materialia* 55: 871–874.
160. Park, M.S., A.A. Golovin and S.H. Davis (2006). "The encapsulation of particles and bubbles by an advancing solidification front." *Journal of Fluid Mechanics* 560: 415–436.
161. Xing, H., J.Y. Wang, C.L. Chen, Z.F. Shen and C.W. Zhao (2012). "Bubbles engulfment and entrapment by cellular and dendritic interfaces during directional solidification." *Journal of Crystal Growth* 338: 256–261.
162. Stefanescu, D.M. and A.V. Catalina (1998). *ISI Journal International* 38: 503–506.

163. Juretzko, F.R., B.K. Dhindaw, D.M. Stefanescu, S. Sen and P. Curreri (1998). "Particle engulfment and pushing by solidifying interfaces: Part 1. Ground experiments." *Metallurgical and Materials Transactions A* 29: 1691–1696.
164. Kao, J.C.T., A.A. Golovin and S.H. Davis (2009). "Particle capture in binary solidification." *Journal of Fluid Mechanics* 625: 299–320.
165. Bergman, N., R. Trivedi, B. Billia, B. Echbarria, A. Karma, S. Liu, C. Weiss and N. Mangelinck (2005). "Necessity of investigating microstructure formation during directional solidification of transparent alloys in 3D." *Advances in Space Research* 36: 80–85.
166. Gurevich, S., A. Karma, M. Plapp and R. Trivedi (2010). "Phase field study of three-dimensional steady-state growth shapes in directional solidification." *Physical Reviews E* 81: 011603.
167. Pequet, C. and M. Rappaz (2000). "Modeling of porosity formation during the solidification of aluminium alloys using a mushy zone refinement method." *Modeling of Casting, Welding and Advanced Solidification Processes* 9: 71–79.
168. Pequet, C., M. Gremaud and M. Rappaz (2002). "Deformation of liquid capsules enclosed by elastic membranes in simple shear flow: Large deformations and the effect of fluid viscosities." *Metallurgical and Materials Transactions A* 33: 2095–2106.
169. Atwood, R.C., S. Sridhar, W. Zhang and P.D. Lee (2000). "Diffusion controlled growth of hydrogen pores in aluminium–silicon casting: In situ observation and modeling." *Acta Materialia* 48: 405–417.
170. Carlson, K.D., Z. Lin, C. Beckermann (2007). "Modeling the effect of finite-rate hydrogen diffusion on porosity formation in aluminum alloys." *Metallurgical and Materials Transactions B* 38: 541–555.
171. Lee, P.D. and R.C Atwood (2004). "Simulation of the three-dimensional morphology of solidification porosity in an aluminium–silicon alloy." *Acta Materialia* 51: 5447–5466.
172. Dong, S.Y., S.M. Xiong and B.C. Liu (2004). "Numerical simulation of microporosity evolution of aluminum alloy castings." *Journal of Materials Science and Technology–Shenyang* 20: 23–26.
173. Karagadde, S., S. Sundarraj and P. Dutta (2012). "A model for growth and engulfment of gas microporosity during aluminum alloy solidification process." *Computational Materials Science* 65: 383–394.
174. Tiedje, N.S., J.A. Taylor and M.A. Easton (2013). "A new multi-zone model for porosity distribution in Al–Si alloy castings." *Acta Materialia* 61: 2037–3049.
175. Meidani, H., A. Jacot and M. Rappaz (2015). "Multiphase-field modeling of micropore formation in metallic alloys." *Metallurgical and Materials Transactions A* 46: 23–26.
176. Wu, W., M.F. Zhu, D.K. Sun, T. Dai, Q.Y. Han and D. Raabe (2012). "Modelling of dendritic growth and bubble formation." *Materials Science and Engineering* 33: 012103.
177. Sun, D., M.F. Zhu, J. Wang and B. Sun (2016). "Lattice Boltzmann modeling of bubble formation and dendritic growth in solidification of binary alloys." *International Journal of Heat and Mass Transfer* 94: 474–487.

16 Analysis of Solidification Kinetics in Mold during Continuous Casting Process

Ambrish Maurya and Pradeep K. Jha

CONTENTS

16.1 INTRODUCTION

Continuous casting process is the most preferred technique to solidify molten steel to produce semifinished steel in the form of slab, billet, and bloom. With the development in technologies and modernization of living brought about by industrial revolution, demand for steel is increasing continuously along with improved quality at a reduced cost. To improve the quality of steel, several techniques are being practiced at each step during the steel making process, i.e., from extraction of iron ore to final product development. Every step has its valuable contribution in both quality improvement and cost saving. However, the solidification phase is considered to be most sophisticated step with regard to achieving metallurgical standards and smooth functioning of the process. In the continuous casting process, solidification of molten steel starts from the water-cooled copper mold and ends in the secondary cooling zone (SCZ). The desired rate of solidification is achieved by controlled heat extraction from the mold as well as from SCZ, in accordance with the process parameters, such as (a) mold dimensions, (b) casting speed, etc. A schematic of the continuous casting process with the cooling zones and heat flux removed at different sections is shown in Figure 16.1.

In a continuous casting mold, heat is continuously removed through circulating water in channels equipped within the copper mold. The rate of heat extraction is controlled by controlling the flow rate of the water and water temperature. Heat transfer from the meniscus is suppressed by providing a slag/flux powder layer of having lower density than the molten steel. The prime reason to control the rate of heat transfer from the mold is to ensure a smooth running of the process at high casting speed coupled with good quality of the casting. Yu and coworkers [1] reported that during initial solidification, cracks form and propagate in the solidifying shell surface at the lower part or just below the mold. The cracks typically occur in the longitudinal or transverse direction to the casting direction [2,3]. Other types of defects that result due to solidification and fluid flow in the mold are (a) oscillation marks, (b) bleeder marks, (c) depressions, (d) nonmetallic inclusions, (e) bulging, (f) breakout, and (g) centerline segregation [4,5].

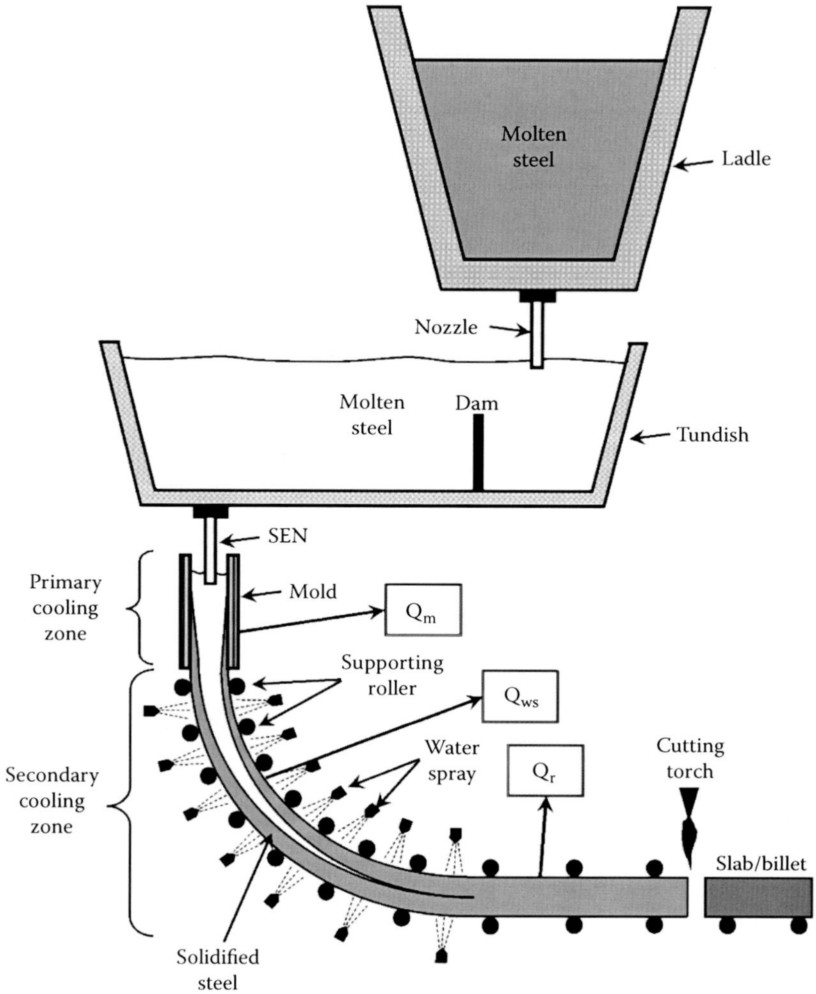

FIGURE 16.1 Schematic of continuous casting process.

16.2 MECHANISM OF SOLIDIFICATION IN A MOLD

The mechanism of solidification or heat transfer in the mold is a complex phenomenon, where about 15%–30% of the total heat is extracted during the process. Figure 16.2 shows the schematic diagram of heat transfer during the solidification process. Samarasekera and Brimacombe [6] have reported that heat is released during the solidification process as (a) conduction through the solidifying shell, (b) radiation across the steel/mold gap (gap is created due to solidification shrinkage), (c) conduction through the copper mold, and (d) convection to the mold cooling water. The net heat flux from the solidification front to the mold cooling water can be given as [6]

$$\dot{q} = \frac{1}{R_T}(T_s - T_w), \tag{16.1}$$

where, T_s is the solidus temperature, T_w is the mold cooling water temperature, and R_T is the total resistance to heat flow. The total resistance to heat flow can be calculated as

$$R_T = \frac{\delta_s}{k_s} + R_g + \frac{\delta_w}{k_w} + \frac{1}{h_c}. \tag{16.2}$$

Complexity in solidification, or heat transfer, increases with the addition of mold slag/flux, mold oscillation, and continuous movement of liquid steel and solidified shell down the mold. In a continuous casting mold, superheated liquid steel enters the mold through the submerged entry nozzle. As the mold is water cooled, heat is extracted from the molten steel at the mold wall. Both the superheat and latent heat of the steel are removed, primarily during temperature change prior to solidification and later during phase change. Thus, a continuous transition takes place from liquid to take solid state. When the steel reaches below the liquidus temperature, nucleation of solid cells is initiated due to local undercooling. During the initial stage, solid cell grows independently and forms dendrites, which tend to agglomerate on further decrease in temperature. Liquid steel accommodates in the dendritic cluster to form a semisolid body called mush. Quantitatively, the region or volume of these mushes depends upon both composition of the steel and cooling rate. Later, with a further decrease in temperature, the mushy zone converts to form a solid shell near the mold wall. The thickness of the solid shell increases down the mold. The solidified shell is continuously pulled by the rolls at the exit of the mold. The partially solidified shell having a higher solid fraction is also set to move within the solid shell. This induces a pressure gradient and stresses in the mush that often lead to the formation of defects, such as (a) macrosegregation, (b) porosity,

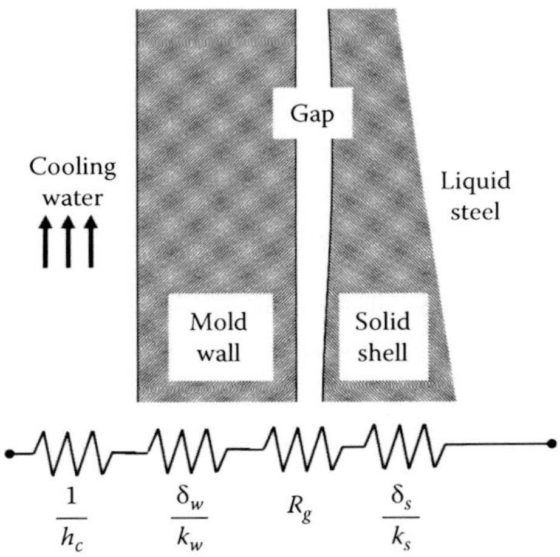

FIGURE 16.2 Schematic diagram of heat transfer from the steel to the mold cooling water.

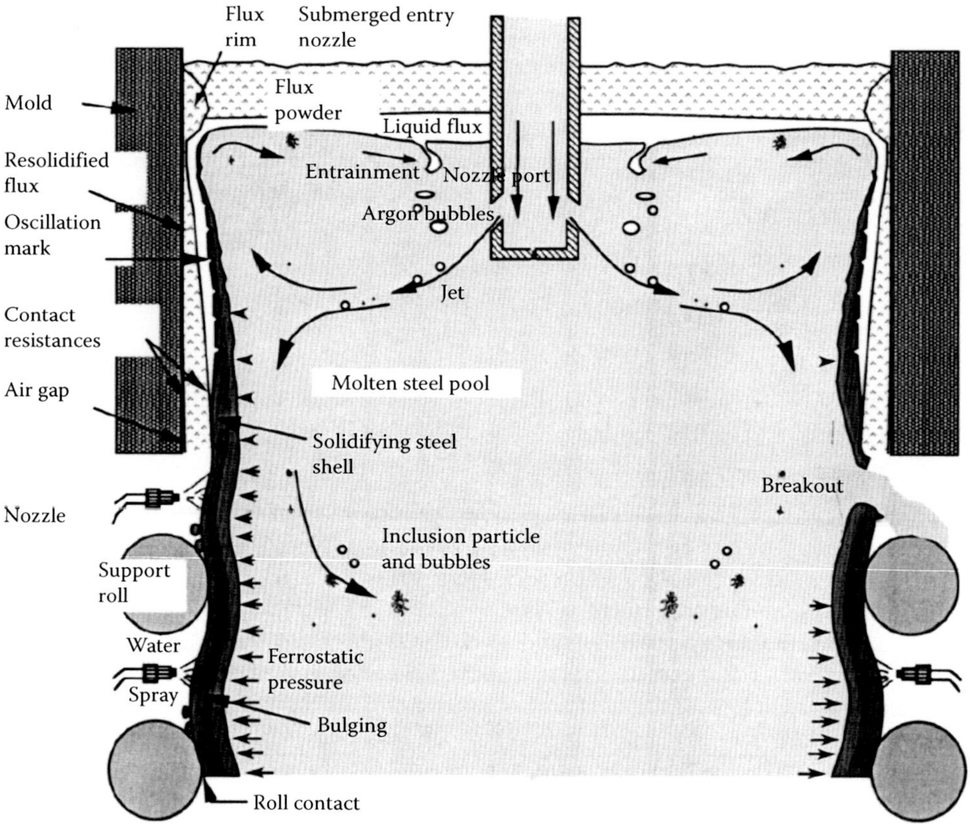

FIGURE 16.3 Schematic of phenomena occurs in continuous casting slab mold. (From Meng, Y.A., Thomas, B.G., *Metall. Mater. Trans. B*, 34, 685–705, 2003. With permission.)

and (c) hot tear. The thickness of the solidified shell [7] with respect to time can be calculated using the following relationship:

$$d = k \,.\, \sqrt{t}, \tag{16.3}$$

where d is thickness of the solidified shell, t is solidification time, and k is solidification constant.

While the steel solidifies, solid shell shrinks and creates an air gap between the mold wall and the solidified shell. Shrinkage of the shell increases with an increase in shell thickness down the mold. The air gap between the solid shell and mold wall decreases the heat transfer rate or interrupts the solidification process, which is generally avoided by the use of a taper mold. Due to a high temperature difference between the mold wall and the solid shell, sticking may tend to occur between the two. Thus, flux powder is added from the top of a mold, which moves into the thin gap between the solid shell and mold wall and acts as a lubricant. Mold is also oscillated to increase the inflow rate of flux powder and to concurrently avoid sticking. Several other phenomena also occur in the mold that have to be considered during the solidification process of steel in a continuous casting mold. A complete study of solidification behavior in a continuous casting mold may include following considerations:

- Heat transfer between steel–mold, mold–water, mold–environment, and steel–slag–mold
- Flow turbulence and nature of solidification for specified steel composition

- Inlet and boundary conditions
- Mold oscillation (frequency, stroke and profile of cyclic motion)
- Flow rate of the flux powder
- Melting and resolidification of the flux powder
- Solidification shrinkage and mold taper design
- Solid shell friction and stickiness with the mold wall

Figure 16.3 shows a schematic of a slab caster mold showing the occurrence of various phenomena in the mold. Studying these phenomena all together makes the analysis complex, and hence, only few of them are considered at a time. Experimental examination of each phenomena is almost impossible because of high temperature, opaque condition, and continuous steel flow. Therefore, mathematical modeling is a faster and more trustworthy tool for the analysis of a continuous casting mold. Thomas and coworkers [8] discussed and examined the accuracy of models with respect to reference reproducibility and a comparison with the real process. The authors have further validated their model for fluid flow, heat transfer, and solidification against the experimental measurement taken at AK Steel–Mansfield caster. Meng and Thomas [9] have developed a numerical model to predict temperature distribution, heat-flux profile, shell thickness, thickness of the resolidified and liquid powder layer, rise of mold water temperature, and ideal taper of the mold walls. These researchers have also reported the complexities that often occur during solidification and heat transfer in the mold.

16.3 MODELING OF SOLIDIFICATION IN A MOLD

The main challenges in continuous casting process are observation, control, and optimization of the process. Looking into the mold, very little can be seen as to what is going on inside the mold because of the high-temperature environment and also the top surface being covered with slag/flux. Several sensors and instruments were used to observe and control various attributes during an industrial run. Because of high-temperature melt flow, continuous running of the process, and high instrument cost, the observations and controls were restricted to only few aspects, such as (a) strand surface and mold temperature, (b) water flow rate, (c) temperature in the mold, (d) flux powder flow rate, etc. These restrictions demand for alternative ways to both analyze and optimize the phenomena occurring in the mold for purpose of smooth and controlled functioning of process. Physical water modeling on the basis of similar kinematic viscosity of liquid steel and water and use of the particle image velocimetry method comes into the picture for studying the fluid flow behavior. However, heat transfer, solidification, and other related aspects cannot be studied using physical modeling. Pin-shooting test [10] is a preferred method to measure the length of the liquid core with high accuracy coupled with ease of operation. However, it has a high operational cost.

Mathematical modeling is one of the most practiced tools for researchers to study the continuous casting process. It saves resources, time, and labor without affecting the process. Researchers frequently use various mathematical modeling tools to study: (a) fluid flow, (b) heat transfer, (c) solidification, (d) slag flow, (e) mold level fluctuation, (f) inclusion removal, (g) stress analysis, (h) crack prediction, and (i) grain structure prediction, to name a few. Over the years, few to several research articles investigating solidification in a mold have been reported and available on web of science. However, due to the complexity of the continuous casting process, the reproducibility of real caster by mathematical modeling is always a big question. Thomas and coworkers [8] recommended that the researcher should establish an ability to reproduce and document the evidence of accuracy rather than focusing on end results. Modeling of solidification includes coupling of the solidification model with fluid flow and heat transfer. Thus, it is necessary to simultaneously validate both fluid flow and solidification, which has rarely been reported.

16.3.1 Generic Assumptions and Boundary Conditions

Modeling of solidification continues with a number of assumptions. These assumptions are essentially due to the complexity of the process because of the simultaneous occurrence of numerous phenomena. The limitation of computational resources, coupled with limited capability of computational tools, restricts the consideration of most phenomena simultaneously. In general, solidification modeling is carried out with fluid flow only. However, literature considering mold oscillation, solidification shrinkage, slag entrapment, thermal stresses, and even inclusion tracking can also be found. The assumptions considered during modeling increases with an increase in complexity or number of parameters. The general assumptions and boundary conditions considered while modeling solidification during continuous casting in a mold are as follows:

- Due to complexity in modeling heat transfer due to water circulation in mold, convective heat transfer through the mold is considered with appropriate heat transfer coefficient calculated using an empirical formula.
- Temperature gradient along the transverse direction is noticeably high when compared to the longitudinal direction. Thus, heat conduction along the longitudinal direction is neglected to reduce calculation.
- Heat loss from the meniscus is assumed to be adiabatic since the presence of flux powder at the meniscus minimizes the loss of heat to a negligible level.
- Often, half or quarter symmetry of mold cross-section is assumed to reduce the computational time.
- Mold oscillation is neglected, and to avoid sticking of mold wall with solidifying shell, the mold walls are assumed to move at casting velocity with the solidifying shell.
- Any shrinkage due to solidification, movement of nonmetallic inclusions, and wall shear stresses is neglected.
- Mold level fluctuation is neglected while calculating turbulence in the molten steel.
- Covering of meniscus by the flux powder and its entrapment during steel flow inside the mold are ignored.
- The molten steel is considered to be Newtonian incompressible liquid.
- Solidified steel shell is set to move with a pull velocity equal to the casting speed and along the casting direction.
- Velocity inlet and velocity outlet boundary conditions are generally applied at both the inlet and outlet of the mold, to maintain the flow rate in the mold.

The mentioned assumptions and boundary conditions are not true for all cases because different researchers incorporate parameters and boundary conditions differently for analyzing the effect and to improve the overall level of accuracy. Thus, a few of the mentioned assumptions and boundary conditions are changed depending upon the modeling technique used, boundary conditions chosen, type of analysis, and the number of parameters considered during modeling.

16.3.2 Mathematical Model

On the basis of assumptions and boundary conditions, a mathematical model is developed. The model consists of fluid flow equations (mass continuity equation, momentum conservation Navier Stokes equation, and turbulence model equations) and a heat transfer equation involving solidification. A generic solidification model has a phase change from solid to liquid and an intermediate state, referred to as the mushy region (liquid coexist in solid dendrites). There are few other techniques available to model phase change during solidification. Brent and coworkers [11] broadly classified the solidification modeling methodologies into three groups as (a) empirical, (b) classical, and (c) enthalpy. An alternative to the empirical and classical approaches outlined earlier is the enthalpy formulation, which allows for a fixed-grid solution to be undertaken. The enthalpy approach removes the requirement of explicit conditions for purpose of energy conservation at the liquid–solid interface. Thus, standard solution procedures were followed for fluid flow and energy equations. The equations to be solved while computing using the enthalpy approach are described in the following section.

16.3.2.1 Solidification Equations

The energy conservation equation for solidification can be defined as

$$\rho \frac{\partial H}{\partial t} + \rho \nabla . \left(\vec{U} H\right) = \nabla . \left(k_{eff} \nabla T\right) + Q_L, \tag{16.4}$$

where, Q_L is the source term and H is the enthalpy of the material, which can be computed as the sum of sensible heat (h) and latent heat content (ΔH), and k_{eff} is effective conductivity.

Sensible enthalpy and latent heat content at the reference temperature (T_{ref}) with the reference enthalpy (h_{ref}) and latent heat of material (L) are defined as

$$h = h_{ref} + \int_{T_{ref}}^{T} c_p dT \tag{16.5}$$

and

$$\Delta H_o = L\beta, \tag{16.6}$$

where β is the liquid fraction at temperature T.

The source term (Q_L) has two terms in it: (i) an explicit latent heat term, which describes the liquid/solid transformation, and (ii) a convective term, which considers movement of the mushy zone. Source term can be expressed as

$$Q_L = \rho L \frac{\partial(1 - \beta)}{\partial t} + \rho L \vec{U}_{pull} . \nabla(1 - \beta). \tag{16.7}$$

In Equation 16.7, latent heat L has been used because the source term is associated with the phase change and describes the rate of latent heat evolution during the phase change or liquid/solid transformation. The region having a solid fraction value of 1 will move along the casting direction of the casting speed. It is to be noted that the term $(1 - \beta)$ shows the solid fraction of the material. The liquid fraction can be calculated by determining the temperature [12,13] as

$$\beta = \begin{cases} 0 & if\ T < T_{solidus} \\ \dfrac{T - T_{solidus}}{T_{liquidus} - T_{solidus}} & if\ T_{solidus} < T < T_{liquidus}\ . \\ 1 & if\ T > T_{liquidus} \end{cases} \tag{16.8}$$

16.3.2.2 Fluid Flow Equations

The continuity equation and transient momentum conservation Navier-Stokes equation can be expressed as follows:

$$\nabla . \vec{U} = 0, \tag{16.9}$$

$$\frac{\partial}{\partial t}\left(\rho \vec{U}\right) + \rho \nabla \left(\vec{U}\vec{U}\right) = -\nabla P + \nabla . \left\{\mu_{eff}\left(\nabla . \vec{U}\right)\right\} + \rho \vec{g} + S. \tag{16.10}$$

Effective viscosity (μ_{eff}) is the sum of dynamic viscosity (μ_l) and turbulent viscosity (μ_t). For solidification, enthalpy–porosity technique is being used where the mushy zone is treated as a pseudo porous medium and porosity in each cell will be equal to the liquid fraction in that cell. Porosity of the porous medium indicates the liquid content in solid or liquid dispered through the solid dendrites. When porosity of the cell equals to zero, the cell will be treated as completely solidified and velocity in this zone extinguishes as there is no liquid. To move the solidified zone in the casting direction, the momentum sink term "S" is added to the right-hand side of the Navier-Stokes equation (Equation 16.10). This moves the newly solidified material at a constant pull velocity, which is set to be equal to the casting velocity. The momentum sink term is expressed as

$$S = \frac{(1-\beta)^2}{(\beta^3+\xi)} A_{mush}\left(\vec{U} - \vec{U}_{pull}\right). \tag{16.11}$$

A small positive constant ξ is provided in the denominator of Equation 16.11 to avoid zero in the denominator. A_{mush} denotes the mushy zone constant and is influenced by grain morphology at the solidification front [14]. So, care must be taken while assigning a value to it. The value of mushy zone constant is taken to be between 10^4 to 10^8. In the momentum sink term, the relative velocity between the molten liquid and solid is used rather than an absolute velocity of the liquid.

Several authors [15,16] have reported the realizable k-ε turbulence model as most suitable for solidification modeling. Realizable k-ε turbulence model avoids the singularity at low value of turbulence [16,17], which can be found in the mushy zone. The partial differential equations for turbulent kinetic energy (k) and dissipation rate (ε) are given by

$$\rho\frac{\partial k}{\partial t} + \nabla\,.\left(\rho k\vec{U}\right) = \nabla\,.\left[(\mu_l + \alpha_k\mu_t)\nabla k\right] + G + \rho\varepsilon + S_k, \tag{16.12}$$

$$\rho\frac{\partial \varepsilon}{\partial t} + \nabla\,.\left(\rho\varepsilon\vec{U}\right) = \nabla\,.\left[(\mu_l + \alpha_\varepsilon\mu_t)\nabla\varepsilon\right] + C_{1\varepsilon}\frac{\varepsilon}{k}G - C_{2\varepsilon}\rho\frac{\varepsilon^2}{k+\sqrt{v\varepsilon}} + S_\varepsilon, \tag{16.13}$$

where, α_k and α_ε are the inverse effective Prandtl numbers, $C_{1\varepsilon} = 1.44$ and $C_{2\varepsilon} = 1.92$ [18,19] are the model parameters, G is the generation of the turbulence kinetic energy due to mean velocity gradient. Sinks S_k and S_ε are added to the turbulence kinetic and dissipation equations, in the mushy zone and solidified zone to account for the presence of solid matter.

16.4 INFLUENCE OF FLUID FLOW ON SOLIDIFICATION

Flow behavior of molten steel inside the mold plays a major role in guiding the solidification behavior and initial grain formation in the mold. Meng and Thomas [8] stated that turbulence in fluid flow inside the mold does affect the superheat delivery to the liquid–solid interface at the solidification front. Redistribution of solute is dependent on convection flow of the molten steel. Chen and coworkers [20] numerically simulated the fluid flow, heat transfer, and solidification in a continuous casting slab caster mold. Fluid flow distribution, shown in Figure 16.4a, had four recirculation zones (two above nozzle inlet and two below nozzle) and similar recirculation zones can be seen in the temperature distribution plot shown in Figure 16.4b. The thickness of solidified shell, shown in Figure 16.4a, has been found to vary according to fluid flow, as shell thickness increases suddenly after the bottom recirculation zone. Comparing the two, researchers have advocated that fluid flow highly affects the temperature distribution in the mold and the solidification behavior. Several other researchers have also advocated an alteration to the fluid flow behavior in order to have the desired solidification profile inside the continuous casting mold.

Over the years, various technologies and change in design parameters have both emerged to alter fluid flow behavior inside the mold. For example, in a slab caster mold, nozzle design [21] and nozzle

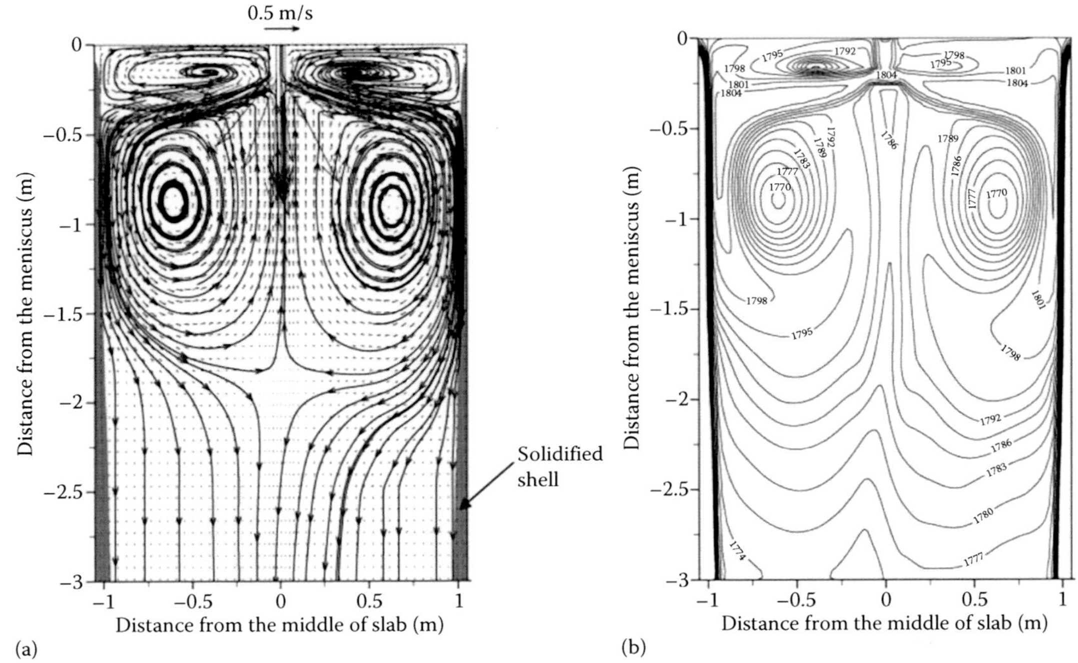

FIGURE 16.4 (a) Fluid flow and (b) temperature distribution in middle–longitudinal section of slab. (From Chen, R., Shen, H.F., Liu, B.C., *Ironmaking Steelmaking*, 38, 546–551, 2011. With permission.)

immersion depth [22] are changed, argon gases are injected, and electromagnetic braking system is applied [23,24], while in a billet and bloom caster mold, an electromagnetic stirring technique is very frequently used [25,26]. Electromagnetic braking and stirring have been reported to be the most favorable technique to alter fluid flow. This is because it not only changes the solidification behavior but also redefines the microstructures of the solid shell formed.

16.4.1 Electromagnetic Braking

Electromagnetic braking is a localized MHD technique, used to brake the flow of liquid steel coming directly from the inlet nozzle. Electromagnetic braking technique is generally used in a slab caster mold having a bifurcated submerged entry nozzle. Electromagnetic braking greatly suppresses the flow turbulence and biased vorticity which reduces the surface defects and internal defects, stabilizes the flow, and improves the of inclusions [23]. The main motive of using electromagnetic braking system is to have metallurgical assistance during casting by minimizing the penetration of the oxide inclusions in molten steel pool and stabilizing flow at the inlet. These finally reduce the entrapment of inclusions and liquid slag in the mushy zone or solid shell formed in the mold. When the molten flux moves in the thin gap between the solidified shell and mold wall, it starts solidifying due to the high cooling rate at the mold wall. The solidified flux stuck over the mold at meniscus and forms a flux rim (refer to Figure 16.3). Electromagnetic braking increases the meniscus temperature and helps in remelting the flux [24]. Literature reports on the physical effect of electromagnetic braking system on solidification during continuous casting in a mold are scarcely available. Few researchers have numerically proven the utility of electromagnetic braking towards altering the solidification behavior. Ha and coworkers [27] and Kim and coworkers [24] simulated the fluid flow, heat transfer, and solidification in a mold with electromagnetic brake and reported the use of electromagnetic braking to decrease the convective heat transfer and increase the temperature gradient near the solidification front, which leads to increase in diffusion heat flux at the shell surface. Surface

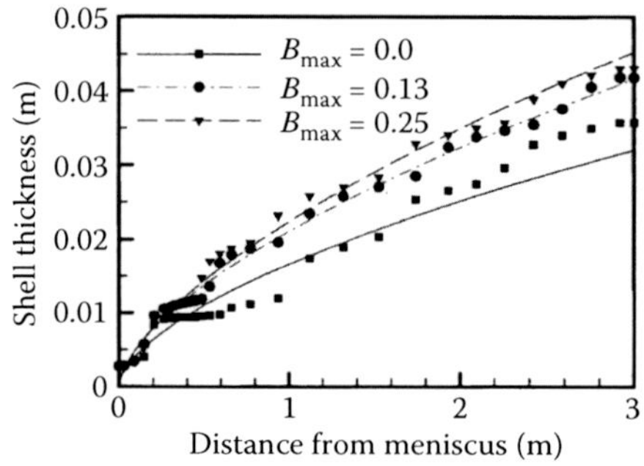

FIGURE 16.5 Solid shell thickness at narrow face for different flux densities of electromagnetic braking system. (From Kim, D., Kim, W., Cho, K., *ISIJ Int.*, 40, 670–676, 2000. With permission.)

temperature of the shell along the narrow face decreases, while temperature along wide face hardly changes. Figure 16.5 shows an increase in the thickness of the solid shell by application of electromagnetic brake and with an increase of magnetic flux density, as predicted by Kim and coworkers [24]. Optimum value of magnetic flux density applied with electromagnetic braking system is the main concern in having a positive outcome, and an extremely high value of magnetic flux density may result in inappropriate steel flow that can result in defects.

16.4.2 Electromagnetic Stirring

Stirring of molten metal during the solidification process has various metallurgical advantages, along with the fulfillment of few other process requirements. Electromagnetic stirring technique is being widely used in a continuous casting mold for casting square or round billets and blooms, resulting in high-quality steel cast products. Electromagnetic stirring controls the fluid flow, heat transfer, and solidification of steel in the mold and results in (a) increased cleanliness of the cast, (b) increased columnar to equiaxed grain transition, (c) decreased central segregation and porosity, and (d) reduction in other surface and subsurface defects. The mechanism of electromagnetic stirring is based on the fact that when a time variable magnetic field is applied over a conductor, it induces eddy current in it. These together generate a force called Lorentz force, which depends upon the intensity and other characteristics of the applied magnetic field.

The stirring of molten steel generates centrifugal forces pushing the high-temperature melt towards the low-temperature solidified shell at the mold wall. It increases the temperature near the solid front that leads to remelting of the solidified shell and grain manipulation. Electromagnetic stirring has maximum stirring effect at its lateral central plane, while it also has some effect below the stirrer position, which gradually decreases moving down the mold. Spitzer and coworkers [28] reported the onset of secondary fluid flow because of stirring. In secondary flow, the steel moves in an upward and downward direction adjacent to the mold wall, and further, it returns back to the center of the stirrer (shown in Figure 16.6). The secondary flow defines the total stirring zone beyond the stirrer. Literature supports the fact that by increasing the stirring intensity and length of stirring zone, the quality of steel cast can be improved coupled with reduced defects. However, too high a stirring intensity can lead to white band formation on the strand surface.

Abundance of literature can be found advocating benefits of electromagnetic stirring by implementing stirrer over industrial mold and finally finding microstructures and macrostructures of the cast. This cannot be justified without using the optimum stirring conditions. Although several researchers have performed fluid flow analysis inside the mold with the application of electromagnetic field, modeling of fluid flow and electromagnetic stirring with the consideration of

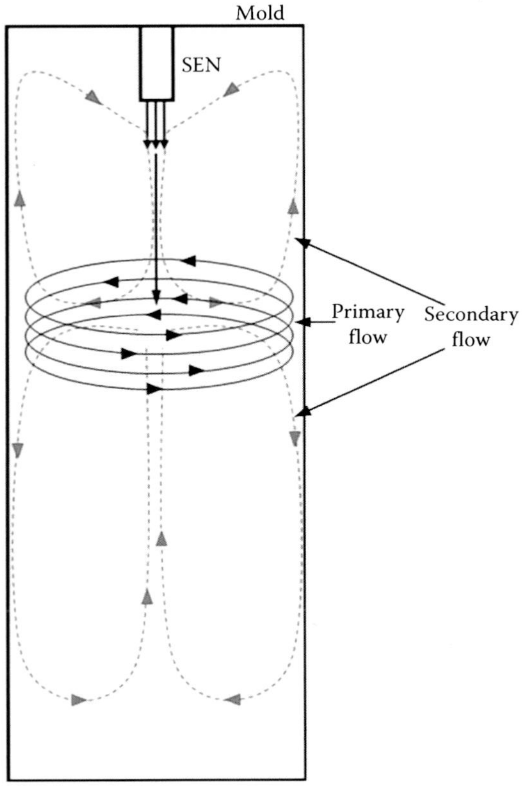

FIGURE 16.6 Schematic of stirring in round mold.

solidification have rarely been found. Ren and coworkers [29] numerically predicted the effect of electromagnetic stirring on solidification behavior in continuous casting mold. They reported that superheat of the molten metal was quickly removed by the application of electromagnetic stirring. The inclusion of solidification model with electromagnetic stirring results in a rapid decrease in swirling flow. Maurya and Jha [30] have also advocated the decrease in swirling flow due to solidification and predicted the break in solid shell at a high stirring intensity (shown in Figure 16.7), although shell thickness at the mold exit was reported to be approximately the same.

16.5 CONSEQUENCES OF INAPPROPRIATE SOLIDIFICATION IN THE MOLD

From initiation of solidification until such time the formed solid shell exits the mold, numerous complex phenomena occur and have to be simultaneously considered for proper functioning of the process and to have a defect-free casting. These phenomena often tend to affect the solidification process inside the mold, complicate grain manipulation, and cause defects in the final cast product or disrupt the casting process. A considerable amount of research has been conducted and technologies developed with the intent of controlling these phenomena that occur during the process. Heat transfer and fluid flow inside the mold significantly influences the process and is somehow interconnected with the other phenomenon that occurs inside the mold. Cho and coworkers [31] reported that appropriate heat transfer analysis should be performed to control heat flux in the mold to produce a defect-free casting. Researchers have proposed several techniques as well as suggested measures, which are to be taken care with respect to solidification within the mold that occur during the casting process. The problem arises due to inappropriate solidification in a continuous casting mold and the remedies are discussed in the following.

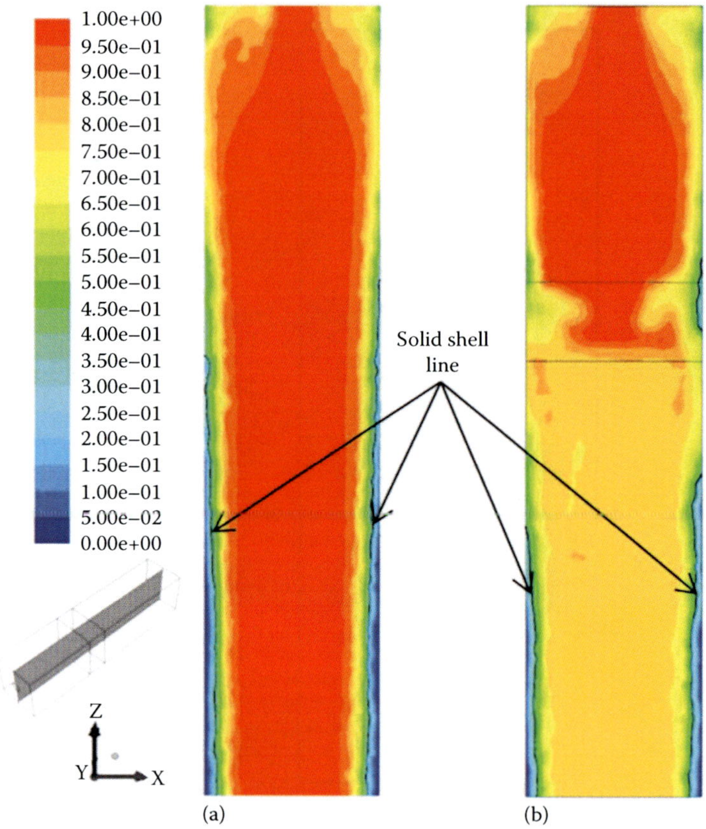

FIGURE 16.7 Liquid fraction contour of the steel showing solid shell growth in the mold (a) without using electromagnetic stirring and (b) after using electromagnetic stirring. (From Maurya, A., Jha, P. K., *Arch. Metall. Mater.*, 2016, Accepted for publication. With permission.)

16.5.1 Breakout

Breakout is sudden failure of the continuous casting process that occurs due to rapture of solidified steel shell beneath the mold. The process suddenly stops after breakout, which not only results in an economic loss due to drop in production but also is dangerous to the operator due to scattering of hot liquid steel. The reasons found for the cause of breakout are the following: (i) inadequate solid shell thickness at the mold exit [32], (ii) physical irregularities in the solid shell [33], and (iii) sticking of solid shell with the mold wall [34].

The developing solid shell continuously attains the ferrostatic pressure of hot liquid steel core, which is shown in Figure 16.3. Within the mold, the solid shell gets continuous support from the mold wall while, beneath the mold exit, simple beam-type supports are present with the help of rollers. Thus, the thickness of the solidified shell at the mold exit should be appropriate enough to sustain the ferrostatic pressure of the liquid core inside. If the thickness of solid shell at the mold exit is inadequate to sustain the ferrostatic pressure, then breakout can happen and the process gets stopped [32]. To avoid this from occurring, the mold should have an adequate heat transfer rate according to the casting process parameters so that the liquid steel can release sufficient heat to get enough shell thickness at the mold exit with the purpose of sustaining ferrostatic pressure. On the other hand, physical irregularities in the solid shell are caused by an improper solidification due to solidification shrinkage, uneven flow, entrapment of slag, and excessive turbulence in the mold. The slag between the solidified shell and the mold wall acts as a lubricating agent against mold

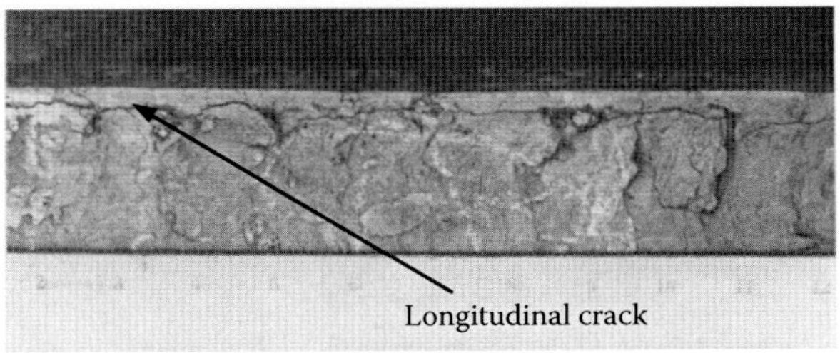

FIGURE 16.8 Longitudinal corner cracks in a slab. (From Brimacombe, J.K., Sorimachi, K., In, S., Casting, C., *Metall. Trans. B*, 8, 489–505, 1977. With permission.)

oscillation. The uneven, or less flow of slag can cause sticking of the solid shell to the mold wall. This can tear the solidified shell during mold oscillation or withdrawal of the solid shell from the mold and finally result in breakout beneath the mold exit. The lack of slag flow occurs due to high casting speed coupled with inappropriate heat transfer rate and solidification. Ghosh and coworkers [35] reported that cracks developed due to thermal stresses can also lead to breakout. Brimacombe and coworkers [5] revealed the picture of a longitudinal crack in a casting as shown in Figure 16.8. They reported that such longitudinal cracks do initiate high up in the mold and often cause breakout below the mold.

16.5.2 Bulging

Below the mold, supporting/guiding rolls are often provided to pull the solidified shell and guide it to move on a predefined path. The rolls apply pressure on the strand from the outside for pulling, whereas from the inside, strand experiences ferrostatic pressure of the liquid steel core. An imbalance of these pressures causes deformation to occur in the strand, mainly between the two rolls, which is also known as bulging. Bulging in the slab strand beneath the mold is shown in Figure 16.9. Excessive bulging can cause (a) centerline segregation, (b) internal cracks because of excessive tensile strains along the solidification front [36], (c) subsurface off-corner longitudinal hot tear cracks, and (d) breakout in extreme cases. The basic reasons behind bulging are a premature solidified shell thickness and uneven solidification arising as a consequence of cooling conditions. Ha and coworkers [37] reported that both casting speed and roll alignment have an important contribution in causing the bulging defect. After the mold exit, a mixture of water and air spray in between two rolls can be increased to enhance the cooling and an increase in thickness of the solid shell.

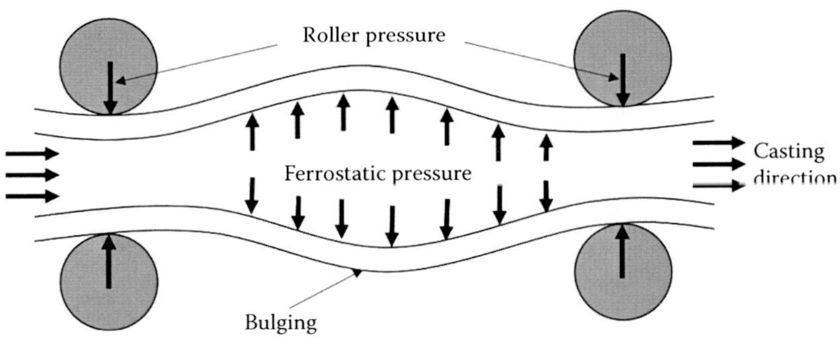

FIGURE 16.9 Bulging in a slab strand.

16.5.3 Grain Structure

The quality of steel cast depends on the grain structure formed (columnar or equiaxed) and its size after final solidification. In continuous casting of steel, dendritic crystals of equiaxed or columnar microstructures are formed. Columnar dendritic structures are formed near the mold wall due to a higher degree of undercooling and high heat transfer rate from the mold. Moving toward the mold center, the tip of the columnar grain has a lower degree of undercooling [38]. This results in the formation of fine equiaxed grains, growing independently in all directions. Hence, columnar to equiaxed grain transition takes place at the center of the mold. Increased heat transfer rate experienced by the mold increases the temperature gradient at the solidification front, which correspondingly decreases the columnar-to-equiaxed grain transition. The equiaxed grains are preferred in a continuously cast product primarily because of its metallurgical benefits. M'Hamdi and coworkers [38] reported the equiaxed grain structure of a cast product to have less segregation. To increase the columnar-to-equiaxed transition, researchers have suggested the following: (i) a decrease in the degree of superheat given to the molten steel at the mold inlet, (ii) decreased cooling rate, (iii) increase in strand thickness, and (iv) an increase of fluid flow turbulence in the mold. However, change in strand thickness and fluid flow turbulence is not possible without a change in casting setup. Lowering of the superheat to a large extent can cause other surface defects. Thus, too high cooling rate from mold wall should be avoided and a low degree of superheat should be preferred.

Electromagnetic stirring has been proposed as one of the most effective techniques for continuous casting of an equiaxed structure billet/bloom cast. It can be seen in Figure 16.10 that when there is no stirring in the mold, the area covered by columnar grains is much more when compared to area of the equiaxed grains at the center. However, after electromagnetic stirring of the molten steel during solidification in the mold, the zone of columnar grains is significantly reduced while the central equiaxed grain zone increases. After applying electromagnetic stirring to the mold, Stransky and coworkers [39] found that the columnar dendritic grain structure was limited to about 1/4 to 1/3 of the billet width.

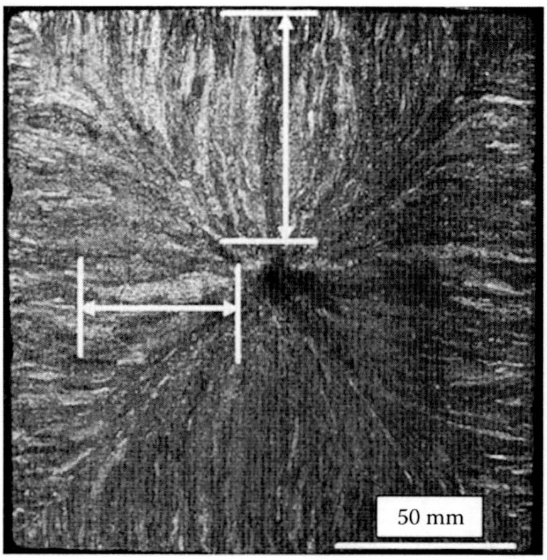

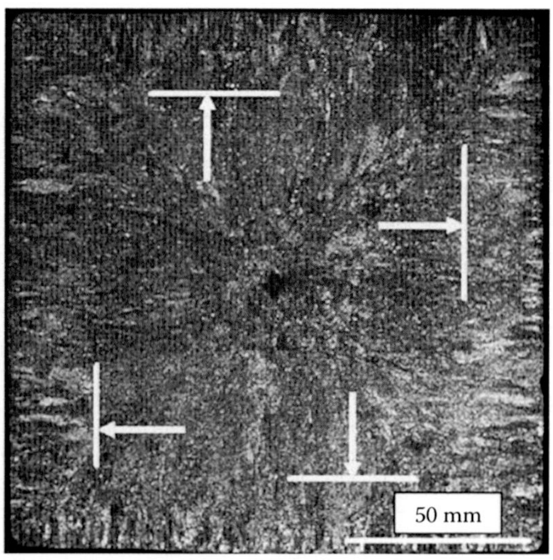

FIGURE 16.10 Growth of dendrites in billet casted in mold (a) without using electromagnetic stirring and (b) after electromagnetic stirring. (From Stransky, K., Kavicka, F., Sekanina, B., Stetina, J., Gontarev, V., Dobrovska, J., *Mater. Technol.*, 45, 163–166, 2011. With permission.)

16.5.4 Stresses

Stresses (thermal or mechanical) are the prime cause for crack initiation and subsequent crack propagation during the continuous casting process. The strand inside the mold experiences thermal stresses, generated due to both high-temperature gradient and phase transformation, while mechanical stresses due to friction between the mold wall and solidified shell. Allazadeh and Garcia [40] numerically predicted the residual stress distribution during continuously casting of steel and reported that too high a cooling rate leaves the residual stresses on the strand surface, which increases the probability for crack formation and intensifies the growth of existing cracks in the microstructure. Brimacombe et al. [41] suggested that temperature at the mold exit be maintained at 1100°C–1050°C and to avoid high cooling in the spray zone. A high-temperature brittleness curve of the steel reveals the highly elastic steel to fall in the temperature range of 900°C–1100°C, embrittlement at 700°C–900°C, and least elastic at temperature below 700°C. Subsequent to exit from the mold, a strand has to pass through the rolls for bending and straightening process. It has been proposed by Ma and coworkers [42] that during the bending/straightening zone, the strand surface temperature should be kept either above or in between the embrittlement range, to avoid excessive stresses, which can cause the occurrence of transverse corner surface cracks. Thus, to have a strand with least stresses, heat transfer from the mold should be optimized and the mold exit temperature should be kept in the specified range.

16.5.5 Shrinkage

Shrinkage because of solidification is a general phenomenon and cannot be avoided completely. During the continuous casting process, solidification shrinkage creates a gap between the solidified shell and mold wall. It is categorized as a restrictive condition since during the process, heat flow is restricted from the steel through the mold wall and leads to a decrease in solid shell growth and forms thin and weak solid shell at the mold exit. This can cause breakouts, bulging, longitudinal off-corner gutter, and longitudinal cracks. To overcome this, few researchers [43–45] have suggested the use of a tapered mold such that the mold wall remains in contact with the solidifying steel shell. The mold taper has to be calculated in advance to adapt to the shrinkage for the specific steel composition, casting speed, cooling rate, and mold dimension.

Shrinkage increases with a decrease in temperature or an increase in solid shell thickness. In a continuous casting mold, heat flux near the meniscus is high, while moving down the mold, the heat flux decreases. In accordance with the decrease in heat flux down the mold, shrinkage due to solidification has a parabolic profile, as shown in Figure 16.11. Zhu and Kumar [45] have predicted the solidification shrinkage and suggest a parabolic taper profile for the mold. Mold taper should be

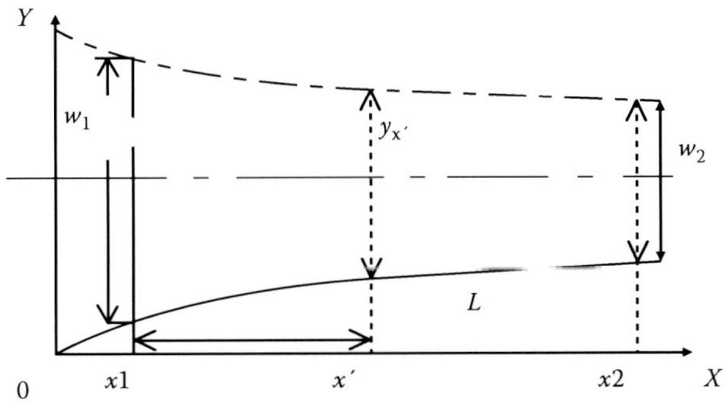

FIGURE 16.11 Schematic of parabolic mold taper. (From Zhu, L.G., Kumar, R.V., *Ironmaking Steelmaking*, 34, 76–82, 2007. With permission.)

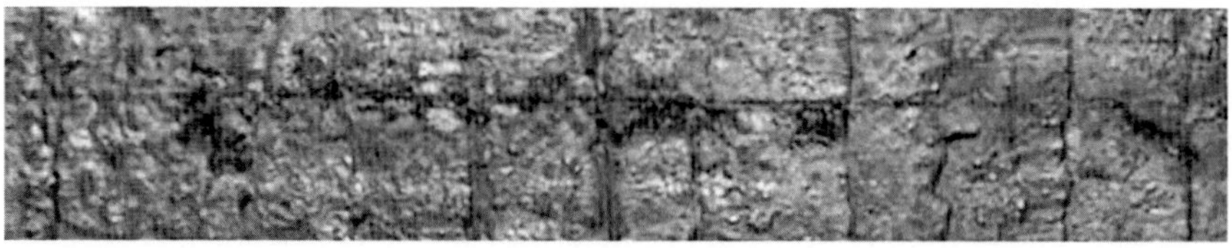

FIGURE 16.12 Oscillation marks on the slab strand. (From Lundkvist, P., Bergquist, B. *Ironmaking Steelmaking*, 41, 304–309, 2013. With permission.)

ideally designed according to the shrinkage observed, or calculated numerically. Thomas and Ojeda [44] reported that extra taper can increase the following: (a) strand mold friction, (b) shell rapture, (c) mold wear, (d) breakout, and (e) transverse cracks.

16.5.6 Friction and Sticking

Because of high cooling rate and continuous movement of the solidified shell with respect to the mold wall, friction is produced between the shell and mold wall. On occasions, friction between them increases, resulting in sticking of the solid shell with the mold. Sticking can cause rupture of the solidified shell or breakout at the mold exit. In order to reduce the friction and avoid sticking between the mold wall and solid shell, slag powder is supplied from the top of the mold top. The slag powder fills the thin gap between the solidified shell and mold wall and acts as a lubricating agent. In 1930, Siegfreid Junghans proposed the idea of mold oscillation to accelerate the flow of slag to avoid sticking phenomena. Mold oscillation variables, such as (i) frequency of oscillation, (ii) stroke, and (iii) profile of cyclic motion, are optimized on the basis of casting speed, cooling rate, mold slag flow, and mold geometry. A repetitive reciprocation of mold generates dynamic pressure on the solidified shell that results in bending of the shell. Thus, this leaves dents on the strand surface in the form of oscillation mark, which can cause formation of transverse cracks.

Figure 16.12 shows the oscillation marks on the slab strand, observed by Lundkvist and Bergquist [46]. Toh and coworkers [47] reported that oscillation marks can be suppressed by lowering the generation of dynamic pressure. Low mold flux viscosity, high oscillation frequency, and short oscillation stroke are the simplest countermeasures that decrease the generation of dynamic pressure. However, excessive changes in countermeasures are not suggested because it can cause other casting complications or defects in the final casting. Another consequence of the mold oscillation is the mold level fluctuation. The increase in mold or meniscus level fluctuation above a critical limit causes the slag entrapment in the liquid steel, which results in poor quality of the cast product or can even cause breakout. Thus, sticking and friction between the mold and the solidified shell can be reduced by both supplying slag for the lubrication and mold oscillation. Conversely, the mold oscillation can result in oscillation marks on the strand, which can be minimized by controlling the mold oscillation parameters. Luk'yanov and coworkers [34] have developed an intelligent system to predict the breakout below the mold and suggested that an increase in the temperature of the mold avoids sticking of the solid shell with the mold.

16.6 CONCLUDING REMARKS

A continuous casting process is a means of fast, economical, and quality production of steel slab/billet/bloom. The initial stage of solidification during continuous casting in the mold is very crucial in attaining desired metallurgical prospective and smooth running of the process. Solidification in the mold is a complex phenomenon due to its dependence on several other processes running simultaneously inside the mold. The salient aspects of the study are summarized in the following:

- The steel releases heat to the mold cooling water by several paths, through conduction, convection, and radiation process.
- The heat transfer and the solidification behavior are influenced by the flow field of liquid steel in the mold.
- Due to high temperature, opaque condition, and a continuous flow of steel, experimental investigations are restricted to a few aspects only. Numerical modeling is a faster and trustworthy method for analysis, with few assumptions.
- Advance mold design and modern techniques are practiced for the alteration of solidification behavior and to achieve desired quality of the cast. Electromagnetic braking in the slab and electromagnetic stirring in the billet/bloom have emerged as prominent techniques to alter fluid flow and solidification behavior.
- Breakout is the most severe consequence of inadequate solidification, which leads to sudden stoppage of the process and creates hazardous working environment.
- Other consequences of an inappropriate heat flow and solidification condition are bulging beneath the mold exit, undesired grain structure of the casting, friction and sticking between the solidified steel shell and mold wall, and stresses leading to various defects in the cast.
- In a nutshell, solidification during continuous casting in the mold should be optimized, numerically or experimentally, to have desired property of the cast.

REFERENCES

1. Yu, C. H., Suzukl, M., Shibata, H., and Emi, T. 1996. Simulation of crack formation on solidifying steel shell in continuous casting mold. *ISIJ International*, 36:159–62.
2. Nakato, H., Ozawa, M., Kinoshita, K., Habu, Y., and Emi, T. 1984. Factors affecting the formation of shell and longitudinal cracks in mold during high speed continuous casting of slabs. *Transactions ISIJ*, 24:957–65.
3. Harada, S., Tanaka, S., Misumi, H., Mizoguch, S., and Horiguchi, H. 1990. A formation mechanism of transverse cracks on CC slab surface. *ISIJ International*, 30:310–16.
4. Camisani-Calzolari, F. R., Craig, I. K., and Pistorius, P. C. 2003. Defects and mold variable prediction in continuous casting. In: *Application of Computers and Operations Research in the Minerals Industries*, South African Institute of Mining and Metallurgy, Cape Town, South Africa, 253–60.
5. Brimacombe, J. K., Sorimachi, K., In, S., and Casting, C. 1977. Crack formation in the continuous casting of steel. *Metallurgical Transactions B*, 8:489–505.
6. Samarasekera, I. V., and Brimacombe, J. K. 1988. Heat extraction capability of continuous-casting billet moulds. In: W. O. Philbrook Memorial Symposium Conference Proceedings, Toronto, Ontario 157–71.
7. Milkowska-Piszczek, K., Dziarmagowski, M., Buczek, A., and Pioro, J. 2012. The methods of calculating the solidifying strand shell thickness in a continuous casting machine. *Archives of Materials Sciences and Engineering*, 57:75–9.
8. Thomas, B. G., Malley, R. O., Shi, T., Meng, Y., Creech, D., and Stone, D. 2000. Validation of fluid flow and solidification simulation of a continuous thin-slab caster. In: *Modeling of Casting, Welding and Advanced Solidification Processes IX*, (eds. Peter R.S., Preben N.H. and James G.C.) Shaker Verlag GmbH, Aachen, Germany, 20:769–76.
9. Meng, Y. A. and Thomas, B. G. 2003. Heat-transfer and solidification model of continuous slab casting: CON1D. *Metallurgical and Materials Transactions B*, 34:685–705.
10. Zhang, T., Li, J., Yang, H., Song, F., and Huang, T. 2013. Numerical simulating study on the solidification process of continuous casting billet. In: EPD Congress 2013 (eds M. L. Free and A. H. Siegmund), John Wiley & Sons, Inc., Hoboken, NJ, USA, 117–24.
11. Brent, A. D., Voller, V. R., and Reid, K. J. 1988. Enthalpy–porosity technique for modeling convestion–diffusion phase change: Application to the melting of a pure metal. *Numerical Heat Transfer*, 13:297–318.
12. ANSYS, 2013. Academic Research. Release 14.0, Help System, Fluent, ANSYS, Inc.
13. Maurya, A. and Jha, P. K. 2017. Mathematical modelling of solidification in a curved strand during continuous casting of steel. *Journal of the Institution of Engineers (India): Series C*, 98:45–52.
14. Jiang, D. and Zhu, M. 2015. Flow and solidification in billet continuous casting machine with dual electromagnetic stirrings of mold and the final solidification. *Steel Research International*, 86:993–1003.

15. Shih, T. H., Liou, W. W., Shabbir, A., Yang, Z., and Zhu, J. 1995. A new k-ε eddy-viscosity model for high Reynolds number turbulent flows—Model development and validation. *Computers Fluids*, 24:227–38.
16. Shamsi, M. R. R. I. and Ajmani, S. K. 2007. Three dimensional turbulent fluid flow and heat transfer mathematical model for the analysis of a continuous slab caster. *ISIJ International*, 47:433–42.
17. Bielnicki, M., Jowsa, J., and Cwudziński, A. 2015. Multiphase numerical model of molten steel and slag behavior in the continuous casting mould. *Archives of Metallurgy and Materials*, 60:257–62.
18. Jha, P. K., Dash, S. K., and Kumar, S. 2001. Fluid flow and mixing in a six strand billet caster tundish: A parametric study. *ISIJ International*, 41:1437–46.
19. Jha, P. K. and Dash, S. K. 2004. Employment of different turbulence models to the design of optimum steel flows in a tundish. *International Journal of Numerical Methods for Heat & Fluid Flow*, 14:953–79.
20. Chen, R., Shen, H. F., and Liu, B. C. 2011. Numerical simulation of fluid flow and solidification in continuous slab casting mould based on inverse heat transfer calculation. *Ironmaking & Steelmaking*, 38:546–51.
21. Zare, M. H., Meysami, A. H., Mahmoudi, S., Hajisafari, M., and Mazar Atabaki, M. 2013. Simulation of flow field and steel/slag interface in the mold region of a thin slab steel continuous caster with tetra-furcated nozzle. *Journal of Manufacturing Processes*, 15:307–17.
22. Ramos, I. C., Morales, R. D., Garcia-Hernandez, S., and Ceballos-Huerta, A. 2014. Effects of immersion depth on flow turbulence of liquid steel in a slab mold using a nozzle with upward angle rectangular ports. *ISIJ International*, 54:1797–806.
23. Chaudhary, R., Thomas, B. G., and Vanka, S. P. 2012. Effect of electromagnetic ruler braking (EMBr) on transient turbulent flow in continuous slab casting using large eddy simulations. *Metallurgical and Materials Transactions B*, 43:532–53.
24. Kim, D., Kim, W., and Cho, K. 2000. Numerical simulation of the coupled turbulent flow and macroscopic solidification in continuous casting with electromagnetic brake. *ISIJ International*, 40:670–76.
25. Davidson, P. A. and Boysan, F. 1987. The importance of secondary flow in the rotary electromagnetic stirring of steel during continuous casting. *Applied Scientific Research*, 44:241–59.
26. Maurya, A. and Jha, P. K. 2017. Influence of electromagnetic stirrer position on fluid flow and solidification in continuous casting mold. *Applied Mathematical Modelling*, 48:736–48.
27. Ha, M. Y., Lee, H. G., and Seong, S. H. 2003. Numerical simulation of three-dimensional flow, heat transfer, and solidification of steel in continuous casting mold with electromagnetic brake. *Journal of Materials Processing Technology*, 133:322–39.
28. Spitzer, K. H., Dubke, M., and Schwerdtfeger, K. 1986. Rotational electromagnetic stirring in continuous casting of round strands. *Metallurgical Transactions B*, 17:119–31.
29. Ren, B., Chen, D., Wang, H., and Long, M. 2015. Numerical fluid flow and solidification in bloom continuous casting mould with electromagnetic stirring. *Ironmaking & Steelmaking*, 42:401–08.
30. Maurya, A. and Jha, P. K. 2016. Study of fluid flow and solidification in billet caster continuous casting mold with electromagnetic stirring. *Archives of Metallurgy and Materials*, (In press).
31. Cho, J. W., Emi, T., Shibata, H., and Suziki, M. 1998. Heat transfer across mold flux film in mold during initial solidification in continuous casting of steel. *ISIJ International*, 38:834–42.
32. Janik, M. and Dyja, H., 2004. Modelling of three-dimensional temperature field inside the mould during continuous casting of steel. *Journal of Materials Processing Technology*, 157–158:177–82.
33. Lopez, P. E. R., Sjostrom, U., Jonsson, T., Lee, P. D., Mills, K. C., Petäjäjärvi, M., and Pirinen, J. 2012. Industrial application of a numerical model to simulate lubrication, mould oscillation, solidification and defect formation during continuous casting. *Materials Science and Engineering*, 33:1–10.
34. Luk'yanov, S. I., Suspitsyn, E. S., Krasilnikov, S. S., and Shvidchenko, D. V. 2015. Intelligent system for prediction of liquid metal breakouts under a mold of slab continuous casting machines. *The International Journal of Advanced Manufacturing Technology*, 79:1861–68.
35. Ghosh, S., Mitra, K., Basu, B., and Jategaonkar, Y. A. 2004. Control of meniscus-level fluctuation by optimization of spray cooling in an industrial thin slab casting machine using a genetic algorithm. *Materials and Manufacturing Processes*, 19:549–62.
36. Bellet, M. and Heinrich, A. 2004. A Two-dimensional finite element thermomechanical approach to a global stress-strain analysis of steel continuous casting. *ISIJ International*, 44:1686–95.
37. Ha, J. S., Cho, J. R., Lee, B. Y., and Ha, M. Y. 2001. Numerical analysis of secondary cooling and bulging in the continuous casting of slabs. *Journal of Materials Processing Technology*, 113:257–61.
38. M'Hamdi, M., Combeau, H., and Lesoult, G. 1999. Modelling of heat transfer coupled with columnar dendritic growth in continuous casting of steel. *International Journal of Numerical Methods for Heat & Fluid Flow*, 9:296–317.

39. Stransky, K., Kavicka, F., Sekanina, B., Stetina, J., Gontarev, V., and Dobrovska, J. 2011. The effect of electromagnetic stirring on the crystallization of concast billets. *Materials and Technology*, 45:163–66.
40. Allazadeh, M. R. and Garcia, C. I. 2011. FEM technique to study residual stresses developed in continuously cast steel during solid–solid phase transformation. *Ironmaking & Steelmaking*, 38:566–76.
41. Brimacombe, J. K. 1999. The challenge of quality in continuous casting processes. *Metallurgical and Materials Transactions B*, 30:553–66.
42. Ma, F. J., Wen, G. H., Tang, P., Yu, X., Li, J. Y., Xu, G. D., and Mei, F. 2010. Causes of transverse corner cracks in microalloyed steel in vertical bending continuous slab casters. *Ironmaking & Steelmaking*, 37:73–79.
43. Dippenaar, R. J., Samarasekera, I. V., and Brimacombe, J. K. 1986. Mold taper in continuous casting billet machines. *ISS Transactions*, 7:33–43.
44. Thomas, B. G. and Ojeda, C. 2003. Ideal taper prediction for slab casting. In: *2003 ISSTech Steelmaking Conference*, Indianapolis, USA, 295–308.
45. Zhu, L. G. and Kumar, R. V. 2007. Modelling of steel shrinkage and optimisation of mould taper for high speed continuous casting. *Ironmaking & Steelmaking*, 34:76–82.
46. Lundkvist, P. and Bergquist, B. 2013. Experimental study of oscillation mark depth in continuous casting of steel. *Ironmaking & Steelmaking*, 41:304–9.
47. Toh, T., Takeuchi, E., Hojo, M., Kawai, H., and Matsumura, S. 1997. Electromagnetic control of initial solidification in continuous casting of steel by low frequency alternating magnetic field. *ISIJ International*, 37:1112–19.

17 Use of Fused Deposition Modeling Assisted Investment Casting for Developing Functionally Graded Material

Rupinder Singh

CONTENTS

17.1 INTRODUCTION

In the recent past, the rapid development of functionally graded materials via metal matrix composites (MMCs) has been noticed especially in the areas of aerospace and automotive industries (Ralph et al., 1997, Singh et al., 2016a). Synergistically, the composite leverages and maximizes the characteristics of each of the constituents, resulting in a (functional, structural, or multifunctional) material with properties not achievable by the constituents individually (Pech-Canul and Valdez, 2015). MMCs have superior properties such as high strength, light weight and high stiffness, increased wear resistance, enhanced high-temperature performance, and better thermal and mechanical fatigue and creep resistance over conventional alloys (Sajjadi et al., 2010, Singh and Singh, 2016a). MMCs in general consist of at least two components, namely, matrix and the reinforcement (Surappa, 2003, Singh, 2014). Matrix is mostly an alloy and the reinforcement is usually ceramic. The difference in MMCs from an alloy is that, in an alloy, the second phase forms as a result of a eutectic or eutectoid reaction where in MMCs, the second phase and the matrix alloy are mixed together (Kumar et al., 2013, Radhika et al., 2015). Because of the different physical properties of the reinforcement, a few advantages of MMCs over conventional alloys are the combination of high strength, high elastic modulus, high toughness and impact resistance, low sensitivity to changes in temperature or thermal shock, high surface durability, low sensitivity to surface flaws, high electric and thermal conductivity, minimum exposure to the potential problem of moisture absorption resulting in environmental degradation, and improved fabric ability with conventional working equipment (Sajjadi et al., 2012, Singh et al., 2012). Early studies on MMCs were concentrated on continuously reinforced MMCs (Kumar et al., 2011, Singh and Singh, 2013a). However, despite encouraging results, the high cost of the reinforcement and the highly labor-intensive manufacturing process restricted the effective use of these materials to military and highly specialized applications (Mishra et al., 2012, Singh and Singh, 2015). On the other hand, in recent years, particulate reinforced composites attracted attention because of the availability of a range of reinforcements at competitive costs and successful development of manufacturing processes that utilize the production of MMCs with reproducible microstructure and properties (Surappa and Rohatgi, 1981, Degischer, 1997, Harrigan, 1998). Furthermore, use of discontinuous reinforcements has the advantage over fiber reinforced composites that it minimizes the problems associated with the latter, such as fiber damage, microstructural heterogeneity, fiber mismatch, and interfacial reactions. For some applications that are subjected to severe loadings or extreme thermal fluctuations such as automotive components, discontinuously reinforced composites proved to show near isotropic properties with substantial increase in strength and stiffness compared to monolithic alloys (Mazahery et al., 2009, Kumar and Singh, 2015a, Singh and Singh, 2017). However, discontinuously reinforced composites are not homogeneous and material properties are sensitive to the properties of the constituent, the interfacial properties, and the geometric shape of the reinforcement (*Metals Handbook: Casting*, 1988, Previtali et al., 2008, Singh et al., 2016b, 2017).

Aluminum and its alloys have multipurpose applications in automobile and aerospace industries due to their unique characteristics such as low density, high strength-to-weight ratio, high thermal and conductivity, and good corrosion resistance. However, poor surface-dependent properties are of serious concern for prolonged use (Panda et al., 2016, Vatankhah Barenji et al., 2016). Currently, aluminum matrix composites (AMCs), which result from the incorporation of hard ceramic particles into aluminum alloys, have been widely used to improve the mechanical behavior of the aluminum alloys (Urena et al., 2004, Naher et al., 2007). A clear interface with no porosity or reaction products, as well as homogeneous distribution of the particles, is required to obtain improved properties (Bienia et al., 2003).

The properties of MMCs depend on many factors, such as different properties of matrix material, size, shape, hardness, volume fraction and distribution of the reinforcement, etc. (Fang et al., 1999). The thermal conductivity and microstructure of the aluminum oxide particulate reinforced aluminum composites were investigated (Tatar and Ozdemir, 2007). This chapter highlights a novel way to prepare functionally graded materials by using fused deposition modeling and investment casting.

17.1.1 Metallic Matrix Types

There is a wide range of metals one can choose to use as the matrix of MMCs (Singh, 2014, Kumar and Singh, 2015b). Some important metallic matrixes and their properties are given in the following.

17.1.1.1 Aluminum Alloys

Aluminum alloys are used mainly in the aerospace industry because of their low density and high strength, toughness, and resistance to corrosion. To improve the strength of aluminum and its alloys without a loss in ductility, ceramic particles can be added, forming a composite material.

17.1.1.2 Titanium Alloys

Titanium alloys are also used in the aerospace industry due to their low density. They retain strength at high temperatures and they have good oxidation and corrosion resistance. However, it is an expensive material.

17.1.1.3 Magnesium Alloys

Magnesium and its alloys are also light materials. Among other commercially used alloys, its density is the lowest, which is 1.74 g/cm^3. Magnesium castings are used in aircraft gearbox housings, chainsaw housings, and electronic equipment.

17.1.1.4 Copper Alloys

Copper is used as an electrical conductor and has good thermal conductivity. It can be easily cast and worked on. One major application in which copper is used as the matrix material is the production of niobium-based super conductors.

17.1.1.5 Intermetallic Compounds

Intermetallic compounds have extremely low ductility. This can be overcome by some other methods, but it is the best to use them as matrix materials to produce a composite to improve toughness.

17.1.1.6 Reinforcements

Particulate reinforced MMCs are being used for wide range of industrial applications. The most commonly used particulate are silicon carbide, aluminum oxide, titanium diboride, boron carbide, silicon dioxide, titanium carbide, tungsten carbide, boron nitride, zirconium dioxide, etc. During processing, chemical reaction can occur between the matrix and the particulate. For example, silicon carbide is reactive with aluminum alloys, whereas without the presence of magnesium, aluminum oxide is more stable in aluminum alloys. However, when the matrix is magnesium, this stability is reversed as magnesium has high affinity for oxygen. When fibers are considered, generally, coating is applied on the surface, but particulate reinforcement is generally introduced into the matrix in virgin state. For structural applications, reinforcement particle size diameter is in the range of 10–30 μm and about 10–30 vol.% of particles is used. Nevertheless, MMCs in which these values are outside these ranges are also used and available commercially (for example, particularly fine particles and high volume contents). For electronic substrate applications, higher particle volume percentages are used, such as around 70 vol.%.

17.2 COMMERCIAL APPLICATIONS

Despite the improvement in mechanical and thermal properties of MMCs, they are used in specific applications. Shortcomings such as complex processing routes and the high cost of the final product are the greatest barriers for the large-scale industrial production of MMCs (Singh and Singh, 2013a,b). So, improvements in reinforcement fabrication and processing techniques are essential to increase the commercial availability. Some examples of currently produced parts (especially particulate reinforced MMCs) are brake discs, drums, calipers, or back plate applications that are found in automotive braking systems.

17.3 INVESTMENT CASTING

Investment casting is also known as the lost wax process (Sabudin, 2004). This process is one of the oldest manufacturing processes (Hunt, 1980). The Egyptians used it to make gold jewelry (hence the name investment) some 5000 years ago. Intricate shapes can be made with high accuracy (Taylor, 1983, Chhabra and Singh, 2011). In addition, metals that are hard to machine or fabricate are good candidates for this process. It can be used to make parts that cannot be produced by normal manufacturing techniques, such as turbine blades that have complex shapes or airplane parts that have to withstand high temperatures (Singh and Singh, 2016b). The mold is made by making a pattern using wax or some other material that can be melted away (Jones and Marquis, 1995). This wax pattern is dipped in refractory slurry, which coats the wax pattern and forms a skin. This is dried and the process of dipping in the slurry and drying is repeated until a robust thickness is achieved (Barnett, 1988). After this, the entire pattern is placed in an oven and the wax is melted away. This leads to a mold that can be filled with the molten metal. Because the mold is formed around a one-piece pattern (which does not have to be pulled out from the mold as in a traditional sand casting process), very intricate parts and undercuts can be made (Jones and Yuan, 2003, Garg et al., 2015). The wax pattern itself is made by duplicating using a stereo lithography or similar model, which has been fabricated using a computer solid model master. The materials used for the slurry are a mixture of plaster of paris, a binder and powdered silica, a refractory, for low-temperature melts (Wu, 1997). For higher temperature melts, alumina-silicate is used as a refractory and silica is used as a binder. Depending on the fineness of the finish desired, additional coatings of ethyl silicate may be applied. The mold thus produced can be used directly for light castings or be reinforced by placing it in a larger container and reinforcing it more slurry. Just before the pour, the mold is preheated to about 1000°C to remove any residues of wax and harden the binder. The pour in the preheated mold also ensures that the mold will fill completely. Pouring can be done using gravity, pressure, or vacuum conditions (Csáky et al., 2015). Attention must be paid to mold permeability when using pressure, to allow the air to escape as the pour is done. Tolerances of 0.5% of length are routinely possible, and as low as 0.15% is possible for small dimensions. Castings can weigh from a few grams to 35 kg, although the normal size ranges from 200 g to about 8 kg. Normal minimum wall thicknesses are about 1 mm to about 0.5 mm (0.040–0.020 in) for alloys that can be cast easily. The types of materials that can be cast are aluminum alloys, bronzes, tool steels, stainless steels, satellite, and precious metals (Rzadkosz et al., 2015). The parts made with investment castings often do not require any further machining, because of the close tolerances that can be achieved (Cheah et al., 2005, Singh et al., 2016c).

17.3.1 Benefits of Investment Casting Over Other Casting Process

i. *Tolerances*: Investment casting routinely holds to a tolerance of ±0.005. This is not possible with other types of casting processes.
ii. *Amortization lowers tooling cost*: The initial wax injection mold to produce the patterns, averaged over the entire production quantity, is often lower than other casting tooling costs. Quantity tooling produces a quality part and will be more cost efficient in the long run.
iii. *Better for the environment*: An investment casting is usually produced from wax patterns, which in most cases can be reclaimed. The pattern is a great way to see your part before it is cast, thus eliminating expensive revisions and reducing metal scrap. More importantly, the investment casting process produces parts to net or near net shapes, which significantly

reduces or eliminates the amount of secondary machining. Most scrap from secondary services like machining can be reused as well.

iv. *Design and casting versatility*: Investment casting works over 100 different ferrous and nonferrous casting alloys. This allows our process to be used in a variety of industries as it produces a wide range of cast and casting-based assemblies. The casting process provides the maximum design flexibility for manufacturing complex, multipart products in many cases.

v. *Intricate design*: When using investment casting, design engineers can easily incorporate features such as logos, product IDs/numbers, and letters into their components. Through holes, slots, blind holes, and external and internal splines, gears and thread profiles can often be cast to reduce secondary machining time and total part cost.

The investment casting process is well known for casting small and complicated shapes (Eddy et al., 1974, Barnett, 1988, Clegg, 1991, Beeley and Smart, 1999). First of all, a part replica is prepared with the help of wax or any other favorable material. This replica is surrounded with ceramic material (wax replica is dewaxed later) and the metal is poured into the cavity of the mold. Figure 17.1 shows a schematic of the investment casting process.

The various steps involved in the investment casting process are described in the following:

i. *Pattern*: The wax patterns or replicas of the part are made by using dies.
ii. *Tree formation*: In this step, a number of part replicas (patterns) are attached to a riser. The size of the pattern defines how many patterns can attach to a single riser.
iii. *Slurry coating*: The patterns are dipped into the slurry (clay). Clay is the mixture of zircone flour, colloidal silica, ethanol, and distilled water.
iv. *Stucco coating*: Stucco coating is basically the coating of pattern trees with refractory grains. The number of coatings basically depends upon the required thickness of the mold. Coating is started with the finest refractory and then proceeds toward the coarse one.
v. *Mold baking*: The sand molds are baked at a temperature of about 1100°C in diesel furnace for 10–30 minutes. Baking increases the strength of the mold and also evaporation of pattern material and moisture, etc.

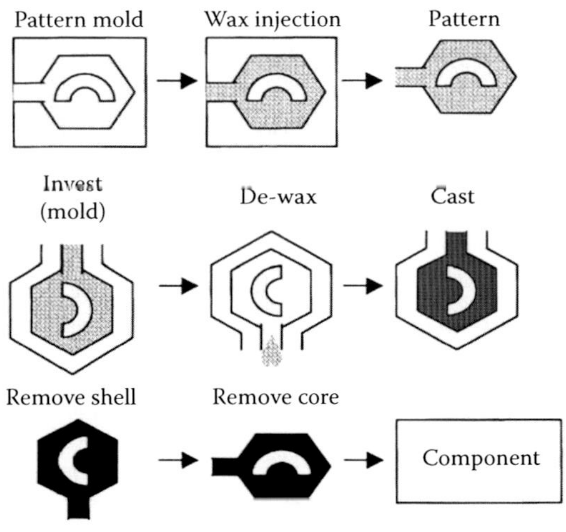

FIGURE 17.1 Schematic of investment casting. (From Clegg, A.J., *Precision Casting Processes*, Pergamon Press, Oxford, 1991.)

vi. *Pouring*: Pouring is processed immediately after the baking. The molten metal is poured into the mold cavities. Then it is allowed to cool at room temperature.
vii. *Shakeout*: After the metal get solidifies, the sand is chipped out on a pneumatic vibration machine (mechanical vibrations) or by chemical cleaning.
viii. *Final product*: At last, the riser is separated from the casted products. After a minor postmachining, castings are ready to use.

17.3.2 Advantages of Investment Casting Process

i. In general, investment casting produces high accuracy and fine surface finish. Tolerances close to ±0.1 mm can be achieved on small components.
ii. Extremely smooth surfaces are easily produced in investment-cast parts.
iii. Investment casting is adaptable to more complex and smaller designs in both limited and production lots.
iv. Investment casting is adaptable to all metallic alloys.

17.3.3 Disadvantages of Investment Casting Process

i. The limitation of size and weight by physical and economic considerations renders the process best applicable to castings weighing from a few grams to 5 kg. Thus, large objects are impractical owing to equipment size limits.
ii. The raw materials, special tooling, equipment, and technology required are expensive in comparison to sand casting.
iii. There is a slower production cycle that makes the process less applicable where the smooth surfaces and fine details are not essential.

17.3.4 Applications of Investment Casting Process

Investment casting has many areas of applications, such as the following:

i. Dental tools
ii. Electrical equipments
iii. Electronics
iv. Guns
v. Hand tools
vi. Jewelry
vii. Machine tools
viii. Materials handling equipments
ix. Agriculture equipments
x. Pumps, etc

17.4 FUSED DEPOSITION MODELING

In order to reduce product development time and cost of manufacturing, fused deposition modeling is one of the cost-effective solutions, which offers the potential to completely revolutionize the process of manufacturing. In this technology, the shape of the physical part is generated by adding the material layer by layer (Singh, 2013, Garg et al., 2015). Although rapid prototyping (RP) techniques are still in their development phase, they are considered a major breakthrough in production engineering. Different types of commercial wire materials are being used as feed stock filament for fused deposition modeling process. Some of these commercial filament materials are listed in the following.

17.4.1 Acrylonitrile Butadiene Styrene (ABS) PLUS Thermoplastic

- Environmentally stable—no appreciable warp-age, shrinkage, or moisture absorption
- 40% stronger than standard ABS material
- ABS thermoplastic
- 25%–30% stronger than standard ABS material
- Greater tensile, impact, and flexural strength
- Layer bonding is significantly stronger for a more durable part
- Versatile material: good for form, fit, and moderate functional applications

17.4.2 ABS Thermoplastic (Polycarbonate Acrylonitrile Butadiene Styrene)

- Most desirable properties of both Polycarbonate (PC) and ABS materials
- Superior mechanical properties and heat resistance of PC
- High impact strength

17.4.3 PC-ISO (International Standards Organization) Thermoplastic

- Biocompatible material
- Ideal material for medical and food packaging industries
- Sterilizable using gamma radiation or ethylene oxide sterilization method
- Highest heat and chemical resistance of all fused deposition modeling materials
- Mechanically superior material, greater strength
- Sterilizable via steam autoclave, chemical, and addition sterilization

Fused deposition modeling is a three-dimensional (3D) printing technique based on the deposition of successive layers of thermoplastic materials following their softening/melting. Such a technique holds huge potential for the manufacturing of pharmaceutical products and is currently under extensive investigation (Beeley and Smart, 1995, Vaezi et al., 2013). The uniqueness of fused deposition modeling technique is more economical for small- to medium-sized parts in the shortest lead time (Haq, 2015). Now, it has progressed one step further from fabricating a prototype to a fully functional product, and one of the key industries that shows a huge impact in the emergence of this technology is the medical industry (Espalin et al., 2010). Fused deposition modeling works on an "additive" principle by laying down material in layers. A plastic filament or metal wire is unwound from a coil and then this is transferred to an extrusion nozzle (Cheah et al., 2005). The nozzle is heated to melt the polymeric materials and can be moved in both horizontal and vertical directions by a numerically controlled mechanism, directly controlled by a computer-aided manufacturing software package (Cunico, 2013). The model or part is produced by extruding small beads of thermoplastic material to form layers as the material hardens immediately after extrusion from the nozzle. Stepper motors or servo motors are typically employed to move the extrusion head. Fused deposition modeling begins with a software process, which processes an STL file, mathematically slicing and orienting the model for the build process. If required, support structures are automatically generated.

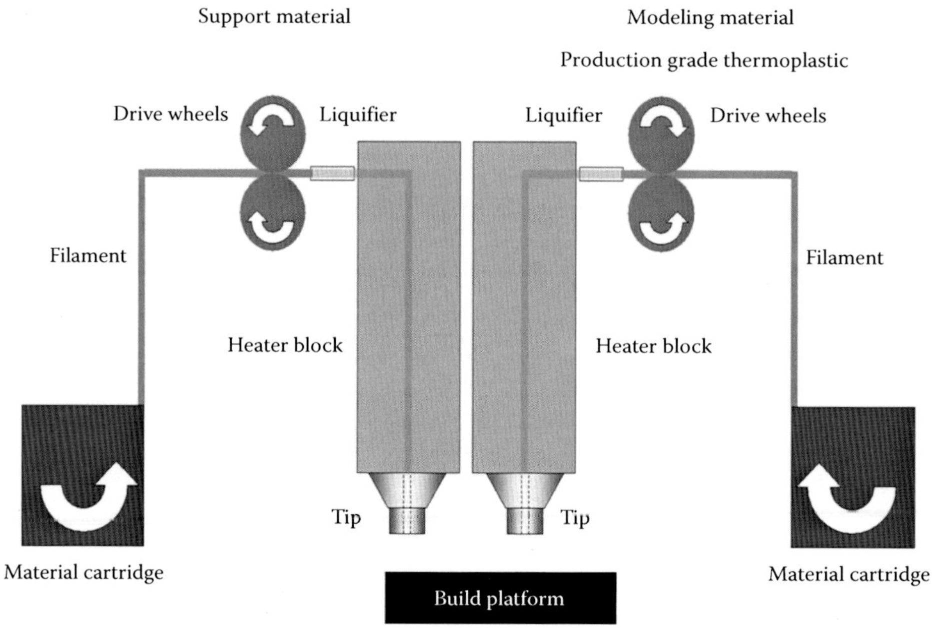

FIGURE 17.2 Schematic of fused deposition modeling. (From Singh, J., Investigation of metal matrix composite prepared by investment casting using reinforced fused deposition modeling pattern, M.Tech Thesis, Guru Nanak Dev Engineering College, Ludhiana, India, 2014.)

The machine dispenses two different types of materials—one for the model and one for a disposable support structure. The thermoplastics are liquefied and deposited by an extrusion head, which follows a tool-path defined by a computer-aided design (CAD) file. The materials are deposited in layers as fine as 0.254 mm thick, and the part is built from the bottom–up one layer at a time (Everhart et al., 2013, Jian and Kuthe, 2013, Kumar et al., 2013). Figure 17.2 shows a schematic of fused deposition modeling.

17.4.4 Advantages of Fused Deposition Modeling

1. Low maintenance costs
2. Thin parts produced fast
3. Tolerance of ±0.005 overall
4. No supervision required
5. No toxic materials
6. Very compact size

17.4.5 Disadvantages of Fused Deposition Modeling

1. There are seam lines between layers.
2. The extrusion head must continue moving, or else material bumps up.
3. Supports may be required.

17.5 SCREW EXTRUDER MACHINE

In the extrusion of plastics, raw thermoplastic material in the form of small beads (of 2–5 mm), often called resin in the industry, is gravity fed from a top mounted hopper into the barrel of the extruder. The material enters through the feed throat (an opening near the rear of the barrel) and comes into contact with the screw. The rotating screw (normally turning at up to 120 rpm) forces the plastic beads forward into the barrel, which is heated to the desired melt temperature of the molten plastic, which can range from 200°C to 275°C. Depending on the type of polymer, three or more independent controlled heater zones gradually increase the temperature of the barrel from the rear (where the plastic enters) to the front, as shown in Figure 17.3.

This allows the plastic beads to melt gradually as they are pushed through the barrel and lowers the risk of overheating, which may cause degradation in the polymer. Most of the screws have three zones as under the following:

Feed zone: Also called the solids conveying zone. This zone feeds the resin into the extruder, and the channel depth is usually the same throughout the zone.

Melting zone: Also called the transition or compression zone. Most of the resin is melted in this section, and the channel depth gets progressively smaller.

Metering zone: Also called the melt conveying zone. In this zone, the channel depth is again same throughout. This zone melts the matter and mixes it at uniform temperature and composition.

17.5.1 Advantages of Extrusion Process

1. There is high production of volume materials in this process.
2. The cost is less as compared to other modeling processes.
3. There is efficient melting of material.
4. Many types of raw materials can be used in extrusion process.
5. There is good mixing of the materials.

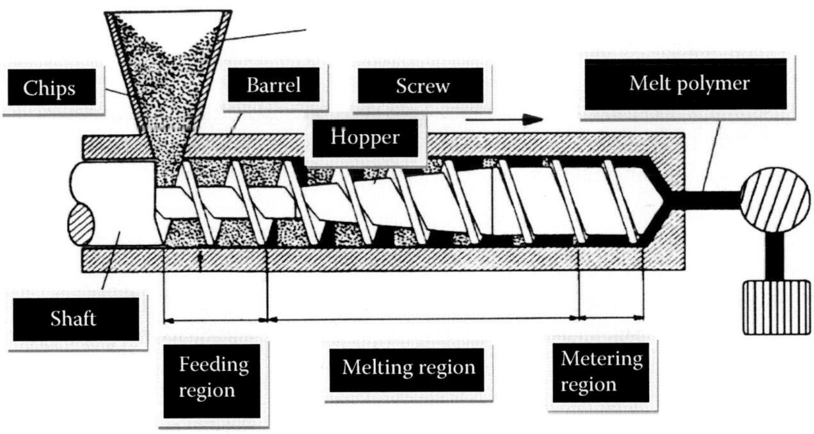

FIGURE 17.3 Screw extruder machine. (From Singh, J., Investigation of metal matrix composite prepared by investment casting using reinforced fused deposition modeling pattern, M.Tech Thesis, Guru Nanak Dev Engineering College, Ludhiana, India, 2014.)

17.5.2 Disadvantage of Extrusion Process

1. There is limited complexity of the parts. Usually, different contours and sharp edges are difficult to generate.
2. It is for uniform cross-sectional shapes only.

17.5.3 Applications of Extrusion Process

1. Polymer compounding
2. Nanomaterials research
3. Industrial hybrid polymer optimization
4. Continuous granulation and hot melt extrusion for pharmaceuticals
5. Processing high-performance and engineering polymers
6. Extruding and mixing ceramic compounds

Pandey et al. (2003) presented a semiempirical model for evaluation of the surface roughness of a layered manufactured part by fused deposition modeling. The increase in surface roughness value due to staircase effect in fused deposition modeling parts is one of the major problems. An attempt was made to address this problem using simple material removal method, namely, hot cutter machining. Analysis of variance was used to find out the significance index for process variables and confidence level for the statistical model. It was concluded that the proposed machining method is able to produce surface finish of 0.3 μm. Lee et al. (2004) highlighted the use of rapid prototyping to reduce the costs associated with single-part or small-quantity production as they can be applied to the fabrication of sacrificial investment casting patterns containing complex and intricate designs with significant cost and lead-time savings. In this effort, a benchmark model was designed to assess the fused deposition modelling process for creating sacrificial investment casting patterns. In addition, an indirect approach toward producing wax patterns via silicone rubber molding was investigated. Cost and lead time comparisons between the two investment casting pattern production methods were carried out and highlighted. Masood and Song (2004) developed new composite material for rapid tooling applications in injection molding process. The tensile properties of various combinations of composite materials were investigated experimentally. The research showed a reduction in cost and cycle time of new parts development. A new metal polymer composite material for fused deposition modelling machine for direct rapid tooling applications was developed in order to reduce time and cost (Masood and Song, 2005). In this study, Fe powder (as filler) and nylon P301 (as matrix material) were selected and thermal properties were studied and compared with other composite used in rapid tooling. The feed stock filament was used successively without any modification in fused deposition modelling machine and thermal properties were acceptable for RP tooling applications for injection molding.

Nikzad et al. (2011) investigated the thermomechanical properties of a new metal–polymer composite consisting of ABS and 10% fine Fe particles by volume. Thermal properties tested include glass transition temperature using dynamic thermal analysis and heat capacity using differential scanning calorimeter. The tensile strength and dynamic mechanical properties were also tested. It has been observed that the addition of 10% Fe powder improves thermal properties and storage modulus of the fused deposition modeling grade ABS, resulting in more thermally stable prototypes producible on Stratasys fused deposition modeling 300 machine, while tensile strength drops significantly. Previtali et al. (2008) studied the application of traditional investment casting process to obtain components in discontinuously particle reinforced AMCs. Further, Pattnaik et al. (2012) reviewed various investigations made by researchers in different stages of investment casting. Singh and Singh (2013) investigated the effect of investment casting process variable (like the

TABLE 17.1
Input Parameters

Parameter 1	Parameter 2	Parameter 3
Size of pattern (mm)	Density of pattern	Number of layers

TABLE 17.2
Output Parameters

Parameter 1	Parameter 2	Parameter 3
Surface roughness	Dimensional accuracy	Wear

volume/surface area ratio of components, slurry layer's combination, and type of metal at three levels by Taguchi's parametric approach) to identify the main factors controlling surface hardness, surface roughness, and dimensional accuracy. Saude et al. (2013) investigated the dynamic mechanical properties of a copper–ABS composite material for possible fused deposition modeling feedstock. The material used consists of copper powder filled in ABS in different compounding ratio. It was observed that increments by vol.% of copper filler in ABS matrix affects the storage modules (E') and tan δ.

The literature review reveals that a lot of research work has been reported on MMCs, fused deposition modelling, and investment casting but much lesser work has been reported on the preparation of functionally graded materials by combining fused deposition modeling and investment casting. The casted functionally graded materials will be subjected to mechanical and metallurgical testing (like dimensional accuracy, surface roughness, and microstructure analysis). The input and output parameters of the process are given in Tables 17.1 and 17.2, respectively.

17.6 CASE STUDY

17.6.1 Objective of Study

- Development of reinforced feed stock wire for fused deposition modeling using single screw extruder
- Development of functionally graded materials by using fused a deposition-modeling-based pattern in investment casting
- Mechanical and tribological analysis of functionally graded materials prepared by the proposed route

Based on Table 17.1, the study was conducted by using fused deposition modeling and investment casting process (as per Table 17.3).

The methodology involved in the preparation of functionally graded materials is shown in Figure 17.4. The first step in the present study is to select proportions of filament ingredient (nylon and aluminum oxide) using a melt flow indexer. The number of tests was performed by taking different proportions of nylon and aluminum oxide. Finally, out of these proportions, the particular proportion whose melt flow index (MFI) matches with ABS is shown in Table 17.4. It was found that the 60% nylon, 30% aluminum, and 10% aluminum oxide proportion matches with the MFI of ABS material (commercially used in fused deposition modeling machines).

TABLE 17.3
Input Parameters and Their Levels

Parameter 1			Parameter 2			Parameter 3		
Size of Pattern/Linear Dimension of Cubical Surface (mm)			Density of Pattern[a]			Number of Layers		
L1	L2	L3	L1	L2	L3	L1	L2	L3
26	30	34	Low	High	Solid	7	8	9

[a] Low density: 5.21×10^{-4} g/mm^3; high density: 7.787×10^{-4} g/mm^3; solid density: 9.35×10^{-4} g/mm^3.

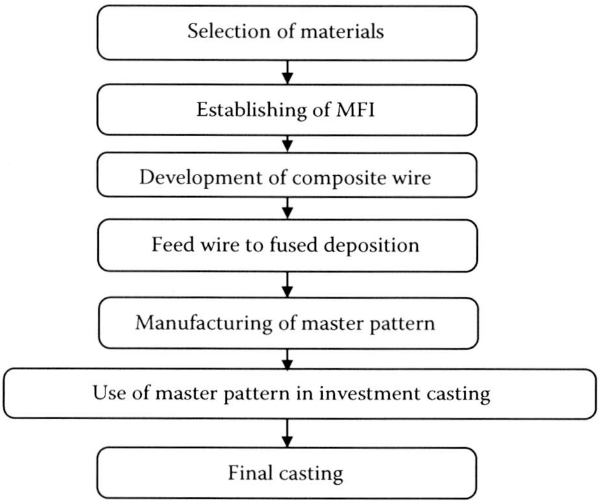

FIGURE 17.4 Process flow diagram.

TABLE 17.4
MFI of Different Materials

S. No.	Composition/Proportion	Values (g/2min) 1	2	3	4	5	6	7	8	9	10	Avg. Value
1	ABS	0.548	0.568	0.508	0.477	0.46	0.4	0.423	0.453	0.49	0.491	0.481
2	Pure nylon 6	1.933	1.985	2.411	1.92	2.16	2.174	2.196	2.04	2.075	2.321	2.122
3	40% Aluminum oxide + 60% nylon 6	0.385	0.428	0.393	0.422	0.70	0.463	0.442	0.481	0.697	0.67	0.509
4	10% Aluminum oxide + 30% Al + 60% nylon 6	0.476	0.491	0.500	0.450	0.466	0.510	0.385	0.568	0.390	0.481	0.480

Finally, for preparing wire using a single screw extruder (see Figure 17.5), the material proportion as 10% aluminum oxide + 30% aluminum + 60% nylon 6 was used. The wire prepared was in the diameter range of 1.65–1.80 mm for its successful running in commercial fused deposition modelling setup.

The cubical benchmark/master patterns were designed using CAD software package (see Figure 17.6). The CAD model was then converted into stereolithography file format (STL file). Figure 17.7 shows a 3D view of the reinforced cubical patterns.

FIGURE 17.5 Filament wire preparation using single screw extruder. (From Singh, J., Investigation of metal matrix composite prepared by investment casting using reinforced fused deposition modeling pattern, M.Tech Thesis, Guru Nanak Dev Engineering College, Ludhiana, India, 2014.)

FIGURE 17.6 2D drawing of component.

After the preparation of reinforced fused deposition modelling patterns, the assembly of these patterns is made on an ABS riser (see Figure 17.8). The final shape appears like a tree. It should be noted that the selection of how many patterns can be attached to a single tree depends upon the size of the patterns and metal to pour. As a commercial practice, normally, the size of the cast component (based on weight constraint) is kept below 5 kg.

The patterns tree was dipped into a slurry (clay) as shown in Figure 17.9. Clay was a mixture of zircone flour, colloidal silica, and distilled water. The mixture was rotated in a drum for about 72 hours and further needed 12 hours for adjusting its viscosity. The patterns were coated with a slurry layer, which acts as a binder and holds the mold sand.

After this step, stucco coating is applied. Stucco coating is basically coating of the pattern with silica sand (see Figure 17.10). This step plays an important role in the formation of sand mold. Starting from the fine sand, coating proceeds toward the coarse sand. As there was no established standard for limiting the total number of layers, in this research work, the total number of layers was varied to seven, eight, and nine (see Table 17.5).

FIGURE 17.7 3D view of reinforced fused deposition modeling patterns.

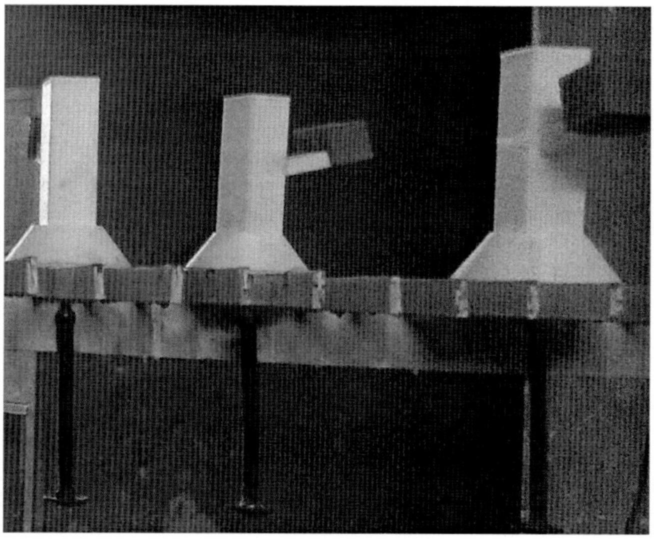

FIGURE 17.8 Pattern assembly in the form of a tree.

The next step was baking the sand molds at 1150°C in diesel furnace for 30 minutes (see Figure 17.11). The importance of baking is that it increases the strength of the mold and evaporates the nylon from the mold cavity.

Pouring was processed immediately after baking. The metal was melted at their respective pouring material. The molds were supported with sand. Figure 17.12 shows the pouring of molten metal from the bucket into the cavity of the mold for present work AL 6063 was used as matrix.

After pouring of molten metal in the cavities, it was allowed to get cooled. The cast tree was clamped in a pneumatic vibrating machine and the air pressure produces vibrations imparted to the cast tree. After cutting the components from the riser, the end products, which were cube in shape and of different sizes, are shown in Figure 17.13.

After conducting the pilot experimentation, the design of the final experiments was made according to Taguchi's L9 orthogonal array, as shown in Table 17.6.

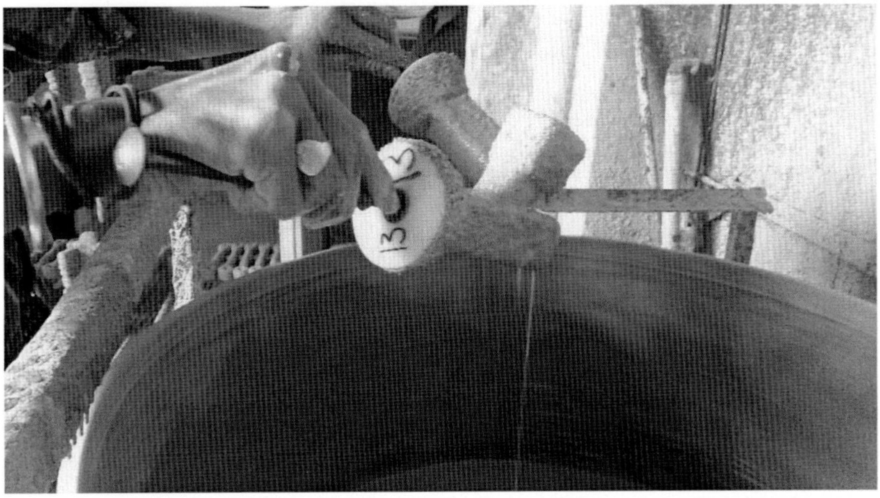

FIGURE 17.9 Slurry coating.

FIGURE 17.10 Stucco coating.

Table 17.7 shows the observation of final experimentation for surface roughness according to the design of the L9 orthogonal array.

Table 17.8 shows the observation of final experimentation for dimensional accuracy according to design of the L9 orthogonal array.

Table 17.9 shows the final experiment observation for wear according to the design of the L9 orthogonal array.

Surface roughness was measured with the help of a Mitutoyo Surface roughness tester (Model SJ-210). For measuring the roughness of a surface, a diamond stylus was placed on the flat surface and then it was allowed to slide along the surface. The roughness over the surface caused displacement of the probe, which results in generation of an analog signal (see Figure 17.14). Any up and down movement of the stylus was recorded and displayed by a digital means.

S/N ratio and means for surface roughness were plotted using Minitab 16 software for "smaller the better" type condition as surface roughness required to be minimum for casted specimens.

TABLE 17.5
Coating Description

Experiment Number	Zircone Sand/Adhesive Coatings	50–80 Grit Sand Coatings	30–80 Grit Sand Coatings	16–30 Grit Sand Coatings
1	1	1	3	3
2	1	1	4	3
3	1	1	4	4
4	1	1	4	3
5	1	1	4	4
6	1	1	3	3
7	1	1	4	4
8	1	1	3	3
9	1	1	4	3

FIGURE 17.11 Mold baking.

Figure 17.15a and b shows the main effects plots for S/N ratio and means, respectively. In case of parameter A, it has been observed that the surface roughness of cast component was best when 26-mm cube size was used. Further, as the size of the pattern increased from 26 mm, surface roughness became higher. It has been found that with an increase in the volume of the cube, there was an increase in the R_a value of aluminum MMC, perhaps due to fused deposition modeling staircase affect. This may be due to the fact that the slicing step in fused deposition modeling is based upon approximation of the original model depending upon geometry and produces a physical object. The larger the volume of geometry, the more will be the impact of staircase affect and, hence, the roughness of the fused deposition modeling parts. Further, in case of parameter B, it has been seen that an increase in density affects the roughness of the castings. Generally, pattern density was an option available in fused deposition modelling system that affects the part weight. This means that patterns fabricated at solid density have maximum quantity of aluminum oxide particles;

FIGURE 17.12 Pouring of molten metal.

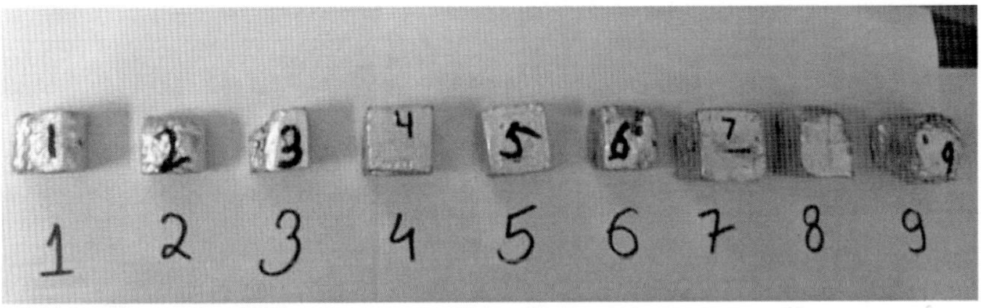

FIGURE 17.13 Cast products.

TABLE 17.6
Control Log of Experimentation

Exp. number	Variable 1 (Size)	Variable 2 (Density)	Variable 3 (Number of Layers)
1	26 mm	Low	7
2	26 mm	High	8
3	26 mm	Solid	9
4	30 mm	Low	8
5	30 mm	High	9
6	30 mm	Solid	7
7	34 mm	Low	9
8	34 mm	High	7
9	34 mm	Solid	8

TABLE 17.7
Observation of Final Experimentation for Surface Roughness

Serial Number	Component Size	Density	Number of Layers	R_a (μm)		
				L1	L2	L3
1	26 mm	Low	7	4.426	5.303	5.632
2	26 mm	High	8	5.676	4.136	4.86
3	26 mm	Solid	9	4.954	5.148	5.279
4	30 mm	Low	8	4.71	4.64	5.028
5	30 mm	High	9	4.985	5.176	5.149
6	30 mm	Solid	7	5.249	5.875	4.968
7	34 mm	Low	9	4.929	4.865	4.662
8	34 mm	High	7	5.124	5.291	6.104
9	34 mm	Solid	8	5.702	5.414	4.989

Note: L1, L2, and L3 are three repetitions in order to control experimental errors.

TABLE 17.8
Observations of Final Experimentation for Dimensional Accuracy

Serial Number	Component Size	Density	Number of Layers	Δd (mm)		
				L1	L2	L3
1	26 mm	Low	7	0.03	0.04	0.02
2	26 mm	High	8	0.02	0.04	0.03
3	26 mm	Solid	9	0.04	0.03	0.05
4	30 mm	Low	8	0.06	0.05	0.07
5	30 mm	High	9	0.17	0.15	0.16
6	30 mm	Solid	7	0.07	0.06	0.08
7	34 mm	Low	9	0.12	0.11	0.13
8	34 mm	High	7	0.03	0.04	0.02
9	34 mm	Solid	8	0.05	0.04	0.03

Note: L1, L2, and L3 are three repetitions in order to control experimental errors.

hence, rough casting surfaces were produced. Similarly, in the case of number of investment casting slurry layers, it has been found that the surface roughness of the castings was produced best at nine layers. As the number of layers was increased to nine, the heat transfer rate of the composite matrix was decreased and smaller-sized grains were bound to form comparative to seven and eight layers.

The best setting of input parameters for surface roughness is shown in Table 17.10.

The percentage deviation is basically the quantitative value by which the metal component shrinks or changes its dimensions after casting. The side of the cube is the most critical dimension in the present study. So the dimensions of master patterns and that of the final casted products were measured by using coordinate measuring machine. The difference in both sizes gives the deviation (Δd) (see Table 17.8). The main effect plot for S/N ratio and means for dimensional accuracy were plotted by using Minitab 16 at "smaller the better" condition as deviation (Δd) required to be

TABLE 17.9
Observation of Final Experimentation for Wear (μm)

Serial Number	Component Size	Density	Number of Layers	Wear (μm)		
				L1	L2	L3
1	26 mm	Low	7	120	131	115
2	26 mm	High	8	143	140	151
3	26 mm	Solid	9	148	140	145
4	30 mm	Low	8	232	225	241
5	30 mm	High	9	139	134	147
6	30 mm	Solid	7	102	94	105
7	34 mm	Low	9	136	135	127
8	34 mm	High	7	87	104	85
9	34 mm	Solid	8	199	186	195

Note: L1, L2, and L3 are three repetitions in order to control experimental errors.

FIGURE 17.14 Surface roughness tester (make: Mitutoyo SJ-210).

minimum for casted specimens (see Figure 17.16a and b). In the case of parameter A, it has been observed that the deviation of casted functionally graded materials was better when 26-mm cube size was used. Further, as the size of the pattern increased from 26 mm, it was found that deviation became poor. It was found that with an increase in the volume of the cube, there was an increase in the Δd of functionally graded materials, perhaps due to fused deposition modeling staircase effect. Further, in the case of parameter B, it was seen that increase in density improves the deviation of the castings. Generally, pattern density was an option available in fused deposition modeling system that affects the part weight. This means that patterns fabricated at solid density have a maximum quantity of aluminum oxide particles; hence, fewer deviations were produced. Similarly, in case of number of investment casting slurry layers, it has been found that deviations of the castings were produced best at seven layers. At seven layers, the heat transfer rate of the matrix metal was increased, so shrinkage occurs that may be filled by reserved metal and minimized deviation.

The best setting of input parameters for dimensional accuracy is shown in Table 17.11.

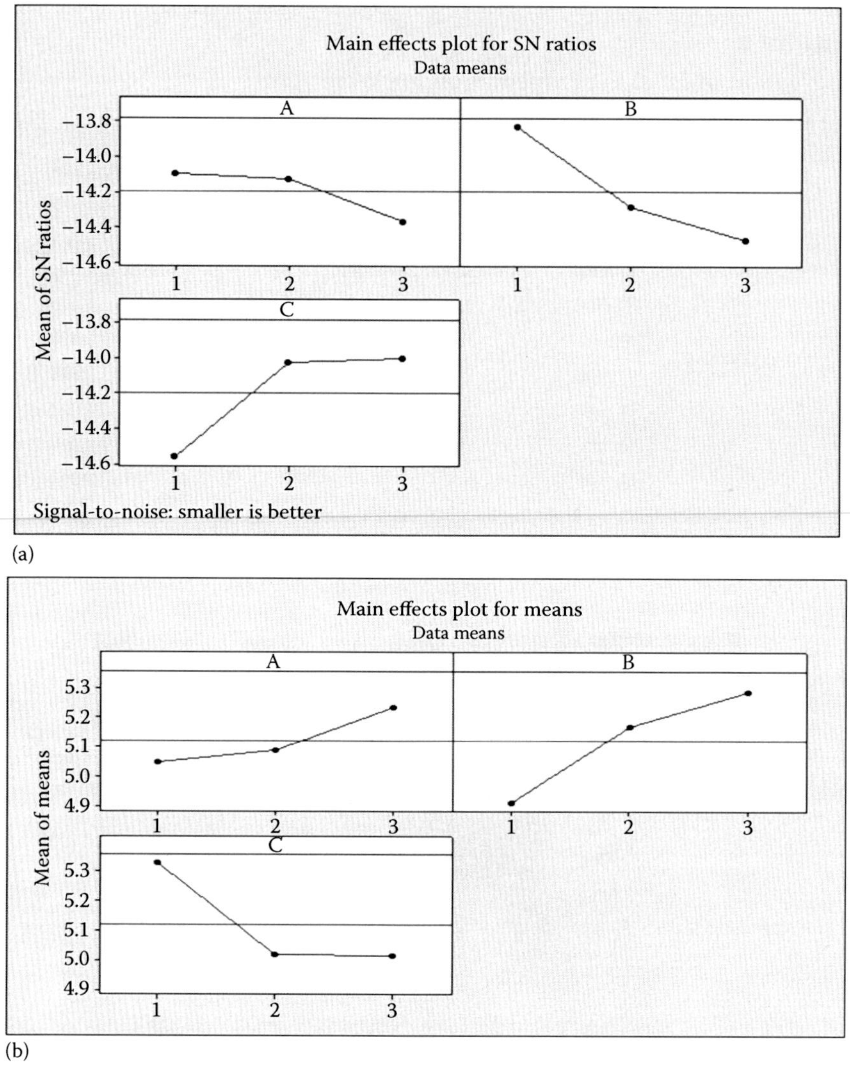

FIGURE 17.15 (a) S/N ratio and (b) means for surface roughness. A: Size of the pattern; B: part layer density; C: number of layers.

TABLE 17.10
Best Input Parameters Setting for Surface Roughness

Size of Pattern	Density of Pattern	Number of Layers	Surface Roughness (μm)
26	Low	9	4.520

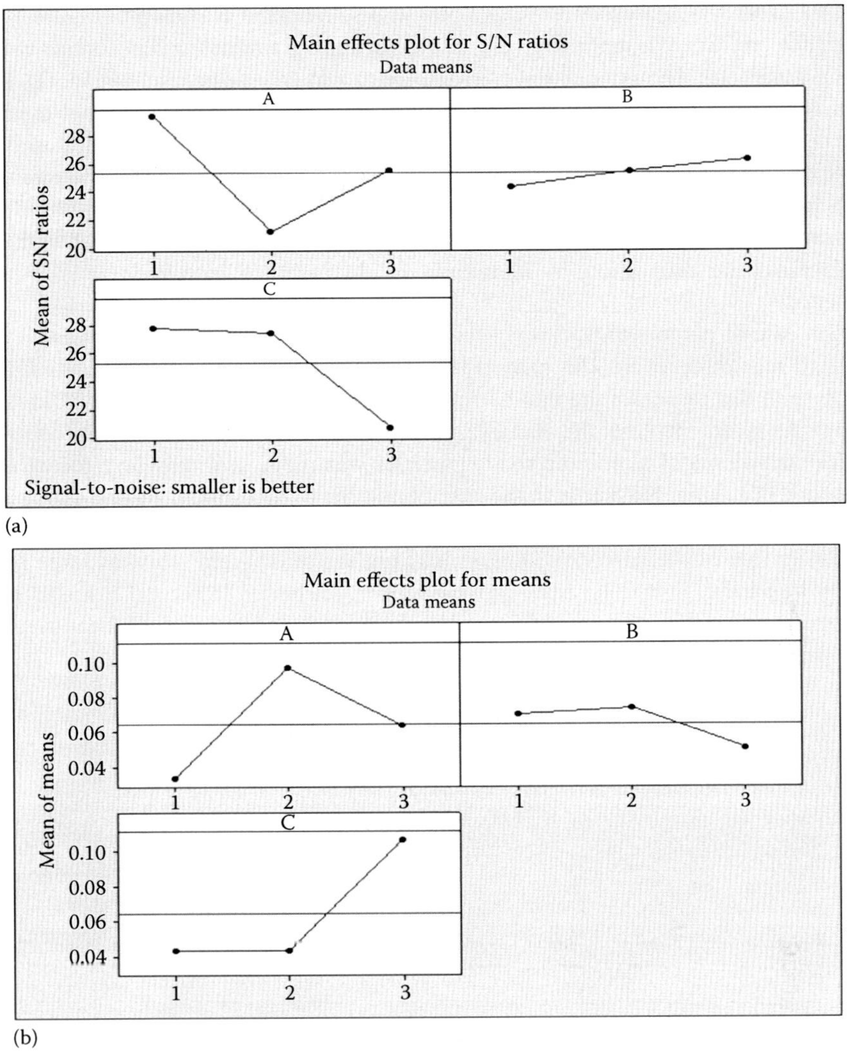

FIGURE 17.16 (a) S/N ratio and (b) means for deviation (Δd).

TABLE 17.11
Best Setting of Input Parameters for Deviation

Size of Pattern (mm)	Density of Pattern	Number of Layers	Deviation (mm)
26	Solid	7	0.05

Further, the wear was analyzed by using a pin-on-disc machine as shown in Figure 17.17. A pin-on-disc machine consists of a stationary "pin" under an applied normal load in contact with a rotating disc of EN-31 material. The pin can have any shape to simulate a specific contact (Ø-10 mm), but spherical tips are often used to simplify the contact geometry. The S/N ratio and means for wear were plotted using Minitab 16 at a "smaller the better" condition as wear required to be minimum for casted specimens. Table 17.9 and Figure 17.18a and b, respectively, show the main effect plots for the S/N ratio and mean. It was observed that the wear of casted functionally graded materials was minimum when 34-mm cube size was used. This may be due to the fact that as the size of the pattern increases, the proportion of aluminum oxide particles increases; thus, wear property becomes better.

Further, in case of parameter B, it was seen that an increase in density improves the wear of the castings up to a middle level. The pattern density was an option available in fused deposition modelling system that increases the number of layers in pattern development (low, high, and solid, respectively). As layers increase, the abrasive particle also increases. Thus, wear property improves. But in solid, the volume of aluminum oxide particles was more so it may be rejected out from the mold at the time of baking. Similarly, in the case of number of investment casting slurry layers, it has been found that the wear property of the castings was best at seven layers. In the case of seven layers, the heat transfer rate of the composite matrix was higher; thus, grain size was large and their orientation may be parallel in the first case, so wear property is better, In second case, the grain may be relatively smaller from the previous case, so wear increased, and in last case, heat transfer rate was very slow; perhaps fine grains were formed so wear property changes.

The best setting of input parameters for wear is shown in Table 17.12.

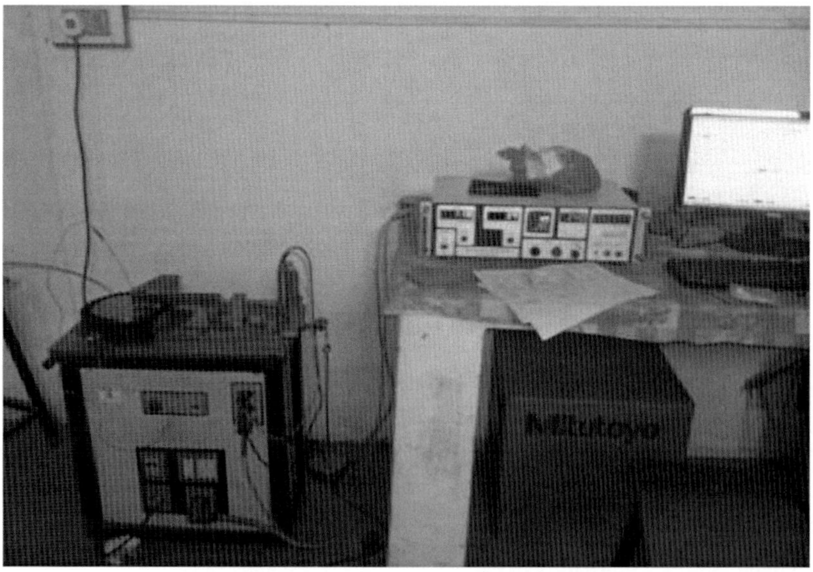

FIGURE 17.17 Pin-on-disc machine (make: Ducom Material Characterization).

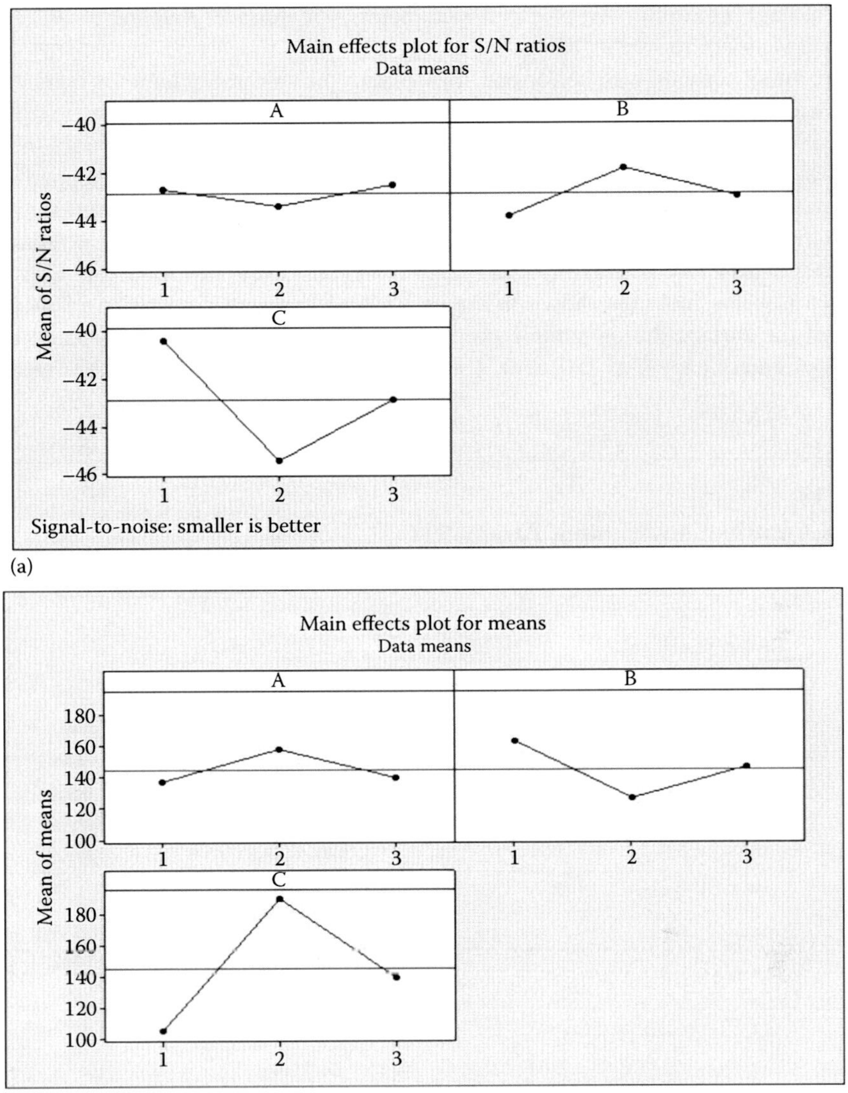

FIGURE 17.18 (a) S/N ratio and (b) means for wear.

TABLE 17.12
Best Setting of Input Parameters for Wear

Size of Pattern (mm)	Density of Pattern	Number of Layers	Wear (μm)
34	High	7	138

The percentage contributions of input parameters for surface roughness (R_a), dimensional accuracy, and wear are shown in Table 17.13.

Microstructural analysis was performed to compare the microstructures of composite castings produced using reinforced fused deposition modelling pattern in investment casting process under the optical microscope to determine the reinforcement pattern in the casting structure. The microstructure of cast specimens was studied using a Qualitech optical microscope equipped with a digital camera. A sample was cut from the bottom of the castings as it was supposed that most of the aluminum oxide particles fell down and gathered at the lower edge, as shown in Figure 17.19.

The corresponding face (containing aluminum oxide abrasives) of the sample was shielded and mounted in a Bakelite mold (as shown in Figure 17.20) to prevent the erosion of aluminum oxide particle from the surface during sample preparation. The specimens were prepared using silicon carbide coated emery papers of 150, 300, 400, 600, 1000, 2000, 3000, and 4000 grades of emery

TABLE 17.13
Percentage Contributions of Input Parameters

Serial Number	Parameters	% Contribution for R_a	% Contribution for Δd	% Contribution for Wear
1	Size of pattern	9.51	51.10	2.79
2	Density of pattern	52.79	2.49	14.74
3	Number of layers	23.38	39.12	75.75
4	Error	14.32	7.29	6.72

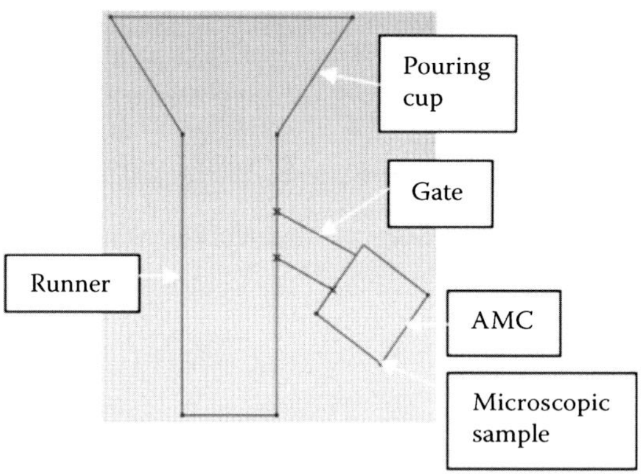

FIGURE 17.19 Description for sample selection. (From Singh, R., Singh, S., *Ref. Module Mater. Sci. Mater. Eng.*, 1–54, 2016.)

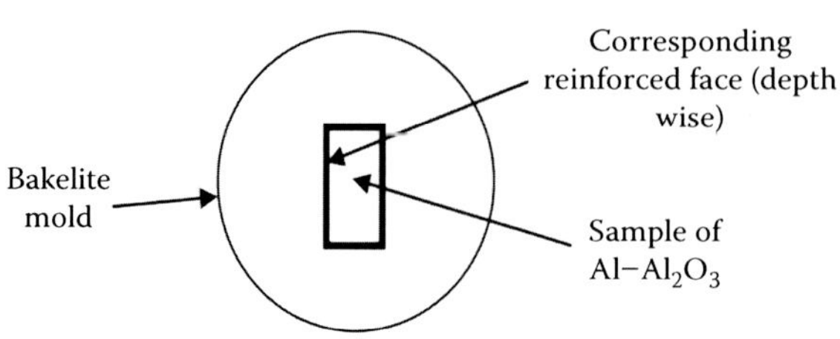

FIGURE 17.20 Mounting of samples in Bakelite.

paper. Before optical observation, the samples were mechanically polished using alumina powder and etched by hydrofluoric (HF) solution (HF 48%: 1 ml; and H_2O: 200 ml) to obtain better contrast. The samples were examined by optical microscope and then photographed.

Figure 17.21 shows the optical micrographs of castings obtained at 100x magnification. The bright area, geometrical shapes, and dark black regions (in the above given micrographs) are Al matrix, aluminum oxide particles, and Bakelite material, respectively.

From microstructure analysis, it has been observed that most of the aluminum oxide particles were presented at the specimen–Bakelite interface, which means that the shielded surface of casting is reinforced. Further, it has been found that the number of particle increases with an increase in the density of fused deposition modeling reinforced pattern and by increasing the aluminum oxide proportion in fused deposition modeling filament. It can be seen that the distribution of aluminum oxide particles is of nonuniform nature. This may be due to the use of a single-screw extruder for the development of alternative fused deposition modeling filaments, which were nonhomogeneous. Further, the results of tensile strength of the filament (tested along the same length) were dissimilar. However, with the use of a twin-screw extruder, homogeneous filaments can be prepared. For counter verification of results, SEM for all the samples was performed. Figure 17.22 shows a SEM

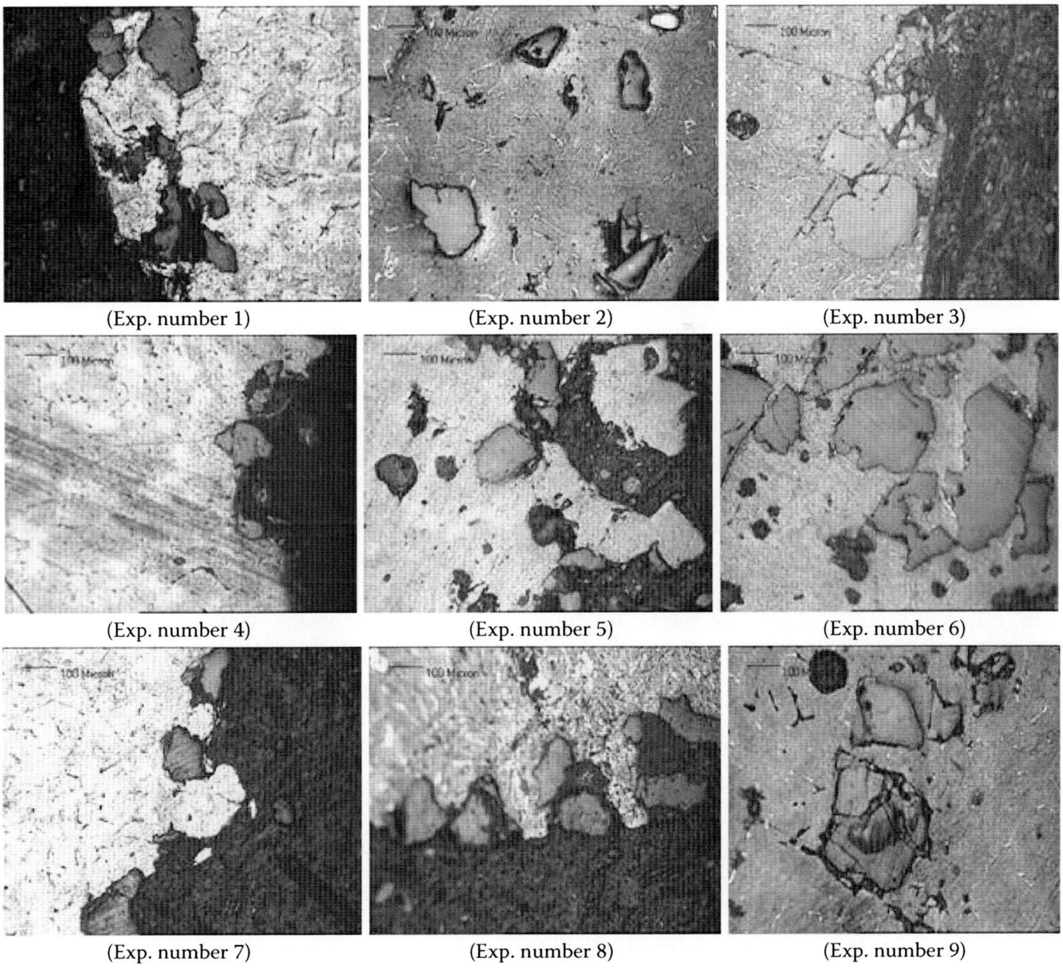

FIGURE 17.21 Photomicrographs of aluminum–aluminum oxide composite (as per Table 17.6).

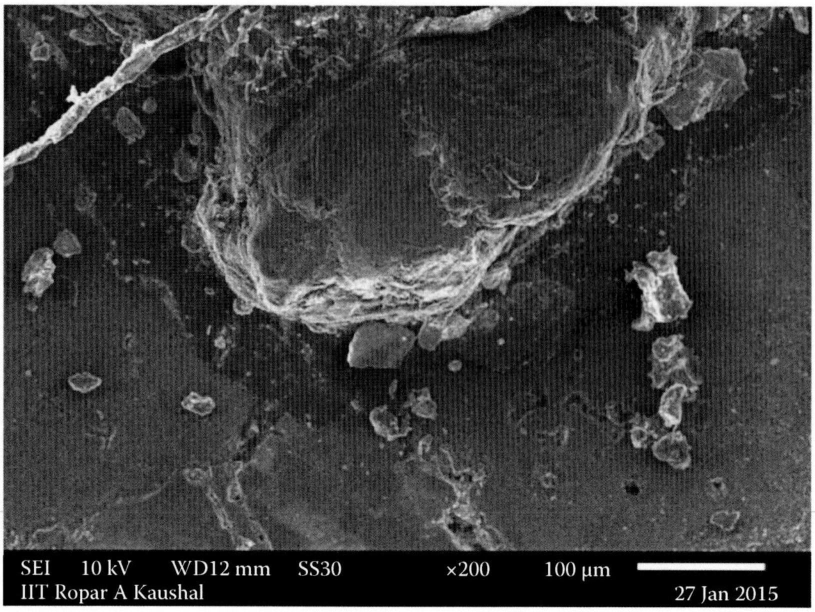

FIGURE 17.22 SEM graph of experiment number 1.

graph for experiment 1 as per Table 17.6, ensuring the presence of aluminum oxide particles on the bearing surface, and hence, the material can be regarded as functionally graded (as it ensures variation of abrasive particles along the depth of cast components). It should be noted that the proportion of aluminum oxide particles can be adjusted as per customer applications.

17.7 CONCLUSIONS

The present chapter describes an alternative route for developing a functionally graded material through the use of fused deposition modelling assisted investment casting process. The material so developed will offer a high surface wear resistant property due to the presence of abrasive reinforcements, whereas at the same time, the core of the material is un-reinforced and thus will contribute toward higher damping and shock absorbing strength. From the present case study, it has been found that the various studied quality characteristics of the developed materials are influenced by input process parameters. For surface roughness, dimensional accuracy, and wear resistant properties, different input variables are found to contribute, as discussed in the result and discussion sections. It often becomes difficult for engineers or scientists to give priority to a particular output response while sacrificing the other, where the response of input parameters varies. Further, from the microstructural analysis, it has been found that the distribution of the reinforcement particles is not uniform, which may be due to two reasons: (1) the feedstock filament developed by using single screw extruders does, by itself, not have uniform distribution of aluminum oxide particles and affects the distribution of the same in the final cast parts or (2) when the reinforced patterns coated with clay were unable to hold the aluminum oxide particles on the surface and, as a result, most of the particles settled down due to gravitation.

Future studies should be focused on the development of a uniformly distributed feedstock filament through twin screw extruder for producing comparatively better castings. Moreover, the mold design should also be taken into consideration in order to control mainly the streamline flow in various units of the mold.

REFERENCES

ASM international handbook committee. *Metals Handbook: Casting* (1988), Vol. 15, ASM International, Materials Park, Ohio, USA.

Barnett, S.O. (1988). "Investment casting—The multi-process technology", *Foundry Trade Journal*, 11(3), 33–37.

Beeley, P.R. and R.F. Smart (1995). "Investment casting (material science)", *Maney Material Science*, 511.

Bienia, J., M. Walczak, B. Surowska, and J. Sobczaka (2003). "Microstructure and corrosion behaviour of aluminum fly ash composites", *Journal of Optoelectronics and Advanced Materials*, 5(2), 493–502.

Cheah, C.M., K.C. Chua, W.C. Lee, C. Feng, and K. Totong (2005). "Rapid prototyping and tooling techniques: A review of applications for rapid investment casting", *International Journal of Advance Manufacturing Technology*, 25(3–4), 308–320.

Chhabra, M. and R. Singh (2011). "Rapid casting solutions: A review", *Rapid Prototyping Journal*, 17(5), 328–350.

Clegg, A.J. (1991). *Precision Casting Processes*, Pergamon Press, Oxford.

Csáky, V., R.J. Neto, T.P. Duarte, and J.L. Alves (2015). "A framework for custom design and fabrication of cranio-maxillofacial prostheses using investment casting", *Engineering Optimization IV* (Editor-Rodrigues et al.), Taylor & Francis Group, London, 941–945, ISBN 978-1-138-02725-1.

Cunico, M.W.M. (2013). "Study and optimization of fused deposition modeling process parameters for support-material-free deposition of filaments and increased layer adherence", *Virtual and Physical Prototyping*, 2(2), 127–134.

Degischer, H.P. (1997). "Innovative light metals: Metal matrix composites and foamed aluminium", *Materials & Design*, 18(4–6), 221–226.

Eddy, W.P., R.J. Barbero, W.I. Dieters, B.J. Esarey, L. Frey, and J.R. Gros (1974). "Investment casting", *American Society for Metals* (Editor—Lyman, T.), Ohio, 237-261.

Espalin, D., K. Arcaute, D. Rodriguez, F. Medina, M. Posner, and R.B. Wicker (2010). "Fused deposition modeling of patient-specific polymethylmethacrylate implants", *Rapid Prototyping Journal* 16, 164–173.

Everhart, W., S. Lekakh, V. Richards, J. Chen, H. Li, and K. Chandrashekhara (2013). "Corner strength of investment casting shells", *International Journal of Metal Casting*, Winter, 21–27.

Fang, C.K., R.L. Fang, W.P. Weng, and T.H. Chuang (1999). "Applicability of ultrasonic testing for the determination of volume fraction of particulates in alumina-reinforced aluminum matrix composites", *Materials Characterization*, 43(4), 217–226.

Garg, P., R. Singh, and I.P.S. Ahuja (2015). "Investigations on dimensional accuracy of the components prepared by hybrid investment casting", *Journal of Manufacturing Processes*, 20, 525 533.

Haq, R.H.B.A. (2015). Characterization and development of polycaprolactone (PCL)/montmorillonite (MMT)/hydroxapaptite (HA) nanocomposites for fused deposition modelling (FDM) process, Doctorate thesis, Faculty of Mechanical and Manufacturing Engineering, Universiti Tun Hussein Onn Malaysia.

Harrigan, W.C. (1998). "Commercial processing of metal matrix composites", *Materials Science and Engineering A*, 244, 75–79.

Hunt, L.B. (1980). "The long history of lost wax casting", *Golden Bulletin*, 13(2), 63–79.

Jian, P. and A.M. Kuthe (2013). "Feasibility study of manufacturing using rapid prototyping: Fused deposition modeling approach", *Procedia Engineering*, 63, 4–11.

Jones, S. and P.M. Marquis (1995). "Role of silica binders in investment casting", *British Ceramics Transaction*, 94(2), 94–96.

Jones, S. and C. Yuan (2003). "Advances in shell moulding for investment casting", *Journal of Material Processing Technology*, 135, 258–265.

Kumar, G.B.V., C.S.P. Rao, and N. Selvaraj (2011). "Mechanical and tribological behaviour of particulate reinforced aluminum metal matrix composites—A review", *Journal of Minerals & Materials Characterisation & Engineering*, 10(1), 59–91.

Kumar, K. and S. Sultan (2015a). "Effect of electromagnetic field and mechanical milling in the synthesis of metal matrix nano composite", In: Proceedings of International Conference of Advance Research and Innovation, ISBN 978-93-5156-328-0, pp. 84–95.

Kumar, S. and R. Singh (2015b). "Mathematical modeling for surface hardness in FDM assisted vacuum moulding of Al–SiC metal matrix composite", *Additive Manufacturing Journal*, 1, 3–10.

Kumar, S., R. Singh, and R. Singh (2013). Development of metal matrix composites by combining stir casting and ABS replica based investment casting, Proceedings of International Conference on Smart Technology for Mechanical Engineering, DTU, New Delhi.

Lee, C.W., C.K. Chua, C.M. Cheah, L.H. Tan, and C. Feng (2004). "Rapid investment casting: Direct and indirect approaches via fused deposition modeling", *International Journal of Advanced Manufacturing Technology*, 23, 93–101.

Masood, S.H. and W.Q. Song (2004). "Development of new metal/polymer materials for rapid tooling using fused deposition modeling", *Materials & Design*, 25, 587–594.

Masood, S.H. and W.Q. Song (2005). "Thermal characteristics of a new metal/polymer material for fused deposition modeling rapid prototyping process", *Assembly Automation*, 25(4), 309–315.

Mazahery, A., H. Abdizadeh, and R. Baharvandi (2009). "Development of high-performance A356/nano-Al_2O_3 composites", *Materials Science and Engineering A*, 518, 61–64.

Mishra, A.K., R. Sheokand, and R.K. Srivastava (2012). "Tribological behaviour of Al-6061/SiC metal matrix composite by Taguchi`s techniques" *International Journal of Scientific and Research Publications*, 2 (10), 1–9.

Naher, S., D. Brabazon, and L. Looney (2007). "Computational and experimental analysis of particulate distribution during Al–SiC MMC fabrication", *Composites Part A: Applied Science and Manufacturing*, 38 (3), 719–729.

Nikzad, M., S.H. Masood, and I. Sbarski (2011). "Thermo-mechanical properties of a highly filled polymeric composites for fused deposition modeling", *Material & Design*, 32(6), 3448–3456.

Panda, B., A. Garg, Z. Jian, A. Heidarzadeh, and L. Gao (2016). "Characterization of the tensile properties of friction stir welded aluminum alloy joints based on axial force, traverse speed, and rotational speed", *Frontiers of Mechanical Engineering*, 11(3), 289–298.

Pandey, P.M., N.V. Reddy, and S.G. Dhande (2003). "Real time adaptive slicing for fused deposition modeling", *International Journal of Machine Tools and Manufacture*, 43(1), 61–71.

Pattnaik, D.S., B. Karunakar, and P.K. Jha (2012). "Developments in investment casting process: A review", *Journal of Materials Processing Technology*, 212, 2332–2348.

Pech-Canul, M.I. and S. Valdez (2015). *Contemporary Concepts and Applications in the Field of Composites Materials, SAMPE.* Society for the Advancement of Material and Process Engineering, Baltimore, MD.

Previtali, B., D. Pocci, and C. Taccardo (2008). "Application of traditional investment casting process to aluminum matrix composites", *Composites Part (A)*, 39, 1606–1617.

Radhika, N., K.T. Vijaykarthik, and P. Shivaram (2015). "Adhesive wear behaviour of aluminium hybrid metal matrix composites using genetic algorithm", *Journal of Engineering Science and Technology*, 10, 258–268.

Ralph, B., H.C. Yuen, and W.B. Lee (1997). "The processing of metal matrix composites-an overview", *Journal of Material Processing Technology*, 63, 339–353.

Rzadkosz, S., J. Zych, A. Garbacz-Klempka, M. Kranc, J. Kozana, M. Piękoś, J. Kolczyk, L. Jamrozowicz, and T. Stolarczyk (2015). "Copper alloys in investment casting technology", *METABK*, 54(1), 293–296.

Sabudin, S.B. (2004). Defect analysis on investment casting: A case study, Master's Thesis, Kolej Universiti Teknologi Tun Hussein Onn.

Sajjadi, S.A., H.R. Ezatpour, and H. Beygi (2010). Microstructure and mechanical properties of Al–Al_2O_3 micro and nano composites fabricated by stir casting, In: Proceedings of 14th National Conference on Materials Science and Engineering, Tehran, Iran, pp. 325–332.

Sajjadi, S.A., H.R. Ezatpour, and M.T. Parizi (2012). "Comparison of microstructure and mechanical properties of A356 aluminum alloy/Al_2O_3 composites fabricated by stir and compo-casting processes", *Materials & Design*, 34, 106–111.

Saude, N., S.H. Masood, M. Nikzad, M. Ibrahim, and M.H.I. Ibrahim (2013). "Dynamic mechanical properties of Copper-ABS composites for fused deposition modeling feedstock", *International Journal of Engineering Research and Applications*, 3(3), 1257–1263.

Singh, R. India Patent File No. 2847/DEL/2013, September 26, 2013.

Singh, J. (2014). Investigation of metal matrix composite prepared by investment casting using reinforced fused deposition modeling pattern, M.Tech Thesis, Guru Nanak Dev Engineering College, Ludhiana, India.

Singh, J. and R. Singh (2013a). "Macro-model for development of Al–SiC metal matrix composite with vacuum moulding: Designed experiments", *Materials Science Forum*, 751, 21–26.

Singh, R., P. Bedi, F. Fraternali, and I.P.S. Ahuja (2016a). "Effect of single particle size, double particle size and triple particle size Al_2O_3 in nylon-6 matrix on mechanical properties of feed stock filament for FDM", *Composites Part B: Engineering*, 106, 20–27.

Singh, R., J. Singh, and J. Singh (2012). "Macro-model for development of Al–Al2O3 metal matrix composite with vacuum moulding: Designed experiments", *Journal of the Institution of Engineers (India): Series C*, 93(4), 325–330.

Singh R., J. Singh, and S. Singh (2016b). "Investigation for dimensional accuracy of aluminum metal matrix composite prepared by fused deposition modeling assisted investment casting using nylon-6 waste based reinforced filament", *Measurement*, 78, 253–259.

Singh, R., J. Singh, and S. Singh (2017). "Investigation of dry sliding wear properties of aluminium matrix composite prepared through FDMAIC process using reinforced pattern", *Proceedings of the National Academy of Sciences, India Section A: Physical Sciences*, In press.

Singh, R. and N. Singh (2017). "Effect of hybrid reinforcement of SiC and Al_2O_3 in Nylon-6 matrix on mechanical properties of feed stock filament for FDM", *Advances in Materials and Processing Technologies*, In press.

Singh, R. and S. Singh (2013b). "Effect of process parameters on surface hardness, dimensional accuracy and surface roughness of investment cast components", *Journal of Mechanical Science and Technology*, 27 (1), 191–197.

Singh, R. and S. Singh (2016a). "Development of functionally graded material through the fused deposition modeling assisted investment casting process: A case study", *Reference Module in Materials Science and Materials Engineering*, 1–54.

Singh, R., S. Singh, and P. Kapoor (2016c). "Investigating the surface roughness of implants prepared by combining fused deposition modeling and investment casting", *Proceedings of IMechE Part-E (Journal of Process Mechanical Engineering)*, 230(5), 403–410.

Singh, S. and R. Singh (2015). "Development of aluminium matrix composite using hybrid FDM pattern for investment casting applications", *International Journal of Mechanical Engineering and Material Science*, 8(1), 1–6.

Singh, S. and R. Singh (2016b). "Development of functionally graded material by fused deposition modeling assisted investment casting", *Journal of Manufacturing Processes*, 24, 38–45.

Surappa, M.K. (2003). "Aluminium matrix composites challenges and opportunities", *Sadhana*, 28, 319–334.

Surappa, M.K. and P.K. Rohatgi (1981). "Preparation and properties of aluminium alloy ceramic particle composites", *Journal of Material Science*, 16, 983–993.

Tatar, C. and N. Ozdemir (2010). "Investigation of thermal conductivity and microstructure of the α-Al2O3 particulate reinforced aluminium composites (Al/Al2O3-MMC) by powder metallurgy method", *Physica B: Condensed Matter*, 405(3), 896–899.

Taylor, P.R. (1983). An illustrated history of lost wax casting, In: Proceeding of the 17th Annual BICTA Conference.

Urena, A., E.E. Martınez, P. Rodrigo, and L. Gil. (2004). "Oxidation treatments for SiC particles used as reinforcement in aluminium matrix composites", *Composites Science and Technology*, 64(12), 1843–1854.

Vaczi, M., H. Seitz, and S. Yang (2013). "A review on 3D micro additive manufacturing technologies", *The International Journal of Advanced Manufacturing Technology*, 67, 1721–1754.

Vatankhah Barenji, R., V.M. Khojastehnezhad, H.H. Pourasl, and A. Rabiezadeh, A. (2016). "Wear properties of Al–Al_2O_3/TiB_2 surface hybrid composite layer prepared by friction stir process", *Journal of Composite Materials*, 50(11), 1457–1466.

Wu, X. (1997). Study on ceramic mold investment casting based on ice patterns made by rapid prototyping method, Bachelor degree thesis, Tsinghua University, Beijing, China.

18 Influence of Nanofragmentation and Microstructural Optimization during Hot Working of Metals and Alloys

G. S. Avadhani, K. R. Y. Simha, and Y. V. R. K. Prasad

CONTENTS

18.1 INTRODUCTION

Hot working of metals and alloys such as maraging steels are vital for applications where high strength, fracture toughness, and weldability are critical. Special steels derive their superior mechanical properties from heat treatment where complex intermetallic precipitates strengthen the iron and nickel martensite phase. These steels are preferred for pressure vessels, rocket motor casings, landing gears for aircraft, marine, and cryogenic applications. The manufacture of these components involves processes of hot working like rolling, forging, or ring rolling. An interesting aspect of high-speed and high-temperature rolling and forging processes concerns microstructural evolution. One also often observes that extremely fine and fragmented grains are produced under a combined loading of pressure, shear stress, strain rate, and temperature. Conventional processing maps provide useful information under uniaxial compression at various temperatures and strain rates. However, shear stresses are almost always present in common forging and rolling operations due to biaxial flow, heterogeneities, and other geometrical factors. The effects of shear stresses are also evident near rolls and platens on account of friction and adhesion. Lubrication at high temperatures is also unpredictable. This shear stress is a chief cause for grain fragmentation.

Nowadays, bulk metalworking of steels by processes like rolling is carried out at elevated temperatures where deformation of austenite takes place. In general, very large strains are imposed when performing these operations, where simultaneous softening processes occur during deformation to prevent the occurrence of fracture. An understanding of the hot deformation characteristics of the austenite phase is valuable both from the point of view of workability optimization as well as microstructural control. The aim of the present chapter is to characterize the constitutive response for optimizing the hot working process parameters. As an example, maraging steel, which is used for the manufacturing of rocket casings using ring rolling process, is discussed as a potential candidate for high pressure torsion (HPT). This chapter, organized in two parts, presents the potential of combining conventional uniaxial test data with HPT, in order to achieve both nanofragmentation as well as microstructural optimization. Part I deals with a brief introduction to the current status of processing maps pioneered by Prasad and coworkers. Part II highlights the complexity of extending such maps to address the combined effects of compression, pressure, and torsion.

18.2 PART I: PROCESSING MAPS

Processing maps assume considerable importance in optimizing the processing of strategic materials for nuclear and space applications. Some important examples include the development of magnesium alloy plates for missiles, aluminum alloys and shape memory alloy for aircraft applications, zircaloy and newer zirconium–niobium–copper alloys for nuclear applications, austenitic stainless steels for fast breeder reactors, maraging steels for rocket casings, and magnesium–aluminum alloys for satellite applications.

Mechanical processing is an essential step in shaping materials into engineering components, which require dimensional accuracy as well as specified microstructure and mechanical properties. Steels, for example, are primarily processed [1] by bulk metalworking methods using rolling, forging, or extrusion, which are generally conducted at an elevated temperature (in the austenitic range) so that large strains may be imposed in a single step of operation without the onset of fracture. Among all mechanical processing methods, the bulk metalworking stage is considered to be of primary importance for two reasons. First, during this stage major microstructural changes occur and these have a profound influence on the subsequent processing steps. Second, in view of the large tonnage of material being processed by bulk metalworking, any improvement in processing techniques has a multiplying effect on overall productivity in manufacturing. Thus, considerable effort has gone into developing techniques for the design and optimization of bulk metalworking processes. The ultimate objective is to manufacture merchant products, or components, with controlled microstructure and properties, without macrostructural or microstructural defects, on a repeatable basis in a manufacturing environment. Hitherto, processes were developed using trial-and-error techniques that are expensive as well as time-consuming and may not always lead to a successful solution or optimization, particularly for advanced materials like super-alloys, intermetallics, and high strength materials like maraging steel. In recent years, however, the trial-and-error techniques have been replaced by modeling techniques developed on the basis of science-based principles.

18.2.1 Workability

The engineering parameter of importance in mechanical processing is commonly termed "workability," which refers to the relative ease with which a material can be shaped by plastic flow without the onset of fracture. An evaluation of workability involves both measurement of the resistance to deformation (strength) and the extent of possible plastic deformation before fracture (ductility). Dieter [2] presents a detailed review of all the parameters influencing workability. It is clear that the workability is influenced not only by the microstructure of the material, applied temperature, strain rate, and strain but also by the stress-state in the deformation zone. It is therefore convenient to consider workability to consist of two independent parts: state-of-stress (SOS) workability and intrinsic workability.

18.2.2 State-of-Stress Workability

State-of-stress workability (SOS) depends upon the geometry of the deformation zone, which occurs when a workpiece is subjected to a three-dimensional stress state. This is represented as a stress tensor with nine components or six independent components of which the three shear stress components contribute to the plastic flow of the material while the normal components are responsible for the occurrence of internal fracture. For good SOS workability, the normal components should be essentially compressive since any tensile component will open up weak interfaces in the microstructure and enhance cracking. The SOS is controlled by the nature of applied stress and geometry of deformation zone, both of which are different for different metalworking processes. The SOS workability is thus specific to the mechanical working process and is independent of the material behavior. For example, it may be optimized in rolling by roll pass design, in forging by preform (blocking die) design and in extrusion by the design of the geometry of the die cavity. Optimization of SOS workability is best achieved using simulation techniques for which commercial packages like DEFORM [3] are capable of accepting experimental constitutive equations for a given material to make the simulation realistic.

18.2.3 Intrinsic Workability

Intrinsic workability depends upon the initial microstructure, as decided by the alloy chemistry, and prior processing history and its response to applied temperature, strain rate, and strain in processing. This response is embedded implicitly in flow stress variation with temperature, strain rate and strain, and is represented mathematically as a constitutive equation. However, as a part of the explicit response of the material to imposed process parameters, certain microstructural changes (mechanisms) that occur within the material need to be characterized. For example, under certain conditions, response may be in terms of microstructural damage processes like wedge cracking and unstable or localized flow. Alternately, occurrence of mechanisms like dynamic recrystallization (DRX) may cause favorable reconstitution of the microstructure and enhance intrinsic hot workability. For obtaining good intrinsic workability, it is essential to choose processing conditions corresponding to DRX and avoid microstructural damage processes and instabilities.

18.2.4 Workability Tests

The standard workability tests are discussed in detail by Dieter [2] in terms of their advantages and limitations. The classic methods involve determining flow stress by a uniform compression test or by a torsion test at temperatures and strain rates of interest. However, the widely known tensile test is not suitable since large strains cannot be achieved in this test due to the formation of a neck. On the other hand, tensile tests are useful for obtaining hot ductility. In a torsion test, large strains can be attained without the onset of geometric instability. However, disadvantages in this test are (1) strain and strain rate gradients exist from center to the surface of a solid specimen, and (2) the imposed large rotational strain results in a fiber structure. Due to its simplicity, the compression test is considered to be a standard bulk workability test in view of the following advantages:

1. The average stress-state is similar to that of many bulk deformation processes without problems of necking or material reorientation.
2. Constant true strain rate can be achieved in a relatively simple fashion.
3. The specimen geometry is simple.

18.2.5 Modeling Hot Deformation

Intrinsic workability of a material is decided by its constitutive behavior and is implicitly embedded in the constitutive equations relating flow stress to temperature, strain rate, and strain. For evaluating the mechanisms of hot deformation explicitly from such relations, material modeling techniques have been developed and include (1) analysis of shapes of stress–strain curves, (2) evaluation of kinetic parameters, and (3) developing processing maps. These have been reviewed recently by Prasad and Seshacharyulu [4].

18.2.6 Stress–Strain Behavior

The shape of a stress–strain curve is considered to contain some information related to the mechanisms of hot deformation [5]. For example, flow-softening type of stress–strain behavior with an initial peak stress or oscillations is often considered as a manifestation of DRX. However, similar stress–strain behavior can also be due to flow instability. Continuous flow softening is typically observed when flow localization occurs [6] and also during globularization of lamellar microstructures in titanium alloys [7]. Further, DRX may occur in cases where the behavior is steady state [8]. It is therefore not possible to conclude unequivocally on the deformation mechanism from the shape of the stress–strain curves alone. However, the information may be considered as supporting evidence.

18.2.7 Kinetic Analysis

One of the early attempts [5] to evaluate the mechanisms of hot working was to represent the steady state stress as a function of temperature and strain rate using a kinetic rate equation basically of the type:

$$\dot{\epsilon} = A\dot{\sigma}^n \exp[-Q/RT] \tag{18.1}$$

where $\dot{\epsilon}$ is the strain rate, σ is the steady-state flow stress, n is the stress exponent, Q is the activation energy for hot deformation, R is the gas constant, T is the temperature in Kelvin, and A is a constant. The mechanism of hot deformation may be evaluated on the basis of the stress exponent and the apparent activation energy. For example, if the apparent activation energy is close to that for diffusion, the mechanism is diffusion controlled. Also, if the stress exponent $n = 2.0$, the mechanism is considered to be superplastic. Furthermore, it is customary in kinetic analysis to evaluate the temperature compensated strain rate parameter, popularly known as Zener–Hollomon parameter, defined as

$$Z = \dot{\epsilon} \exp(Q/RT) \tag{18.2}$$

Extensive microstructural correlations have been obtained [5,9] in a wide variety of materials using the Z parameter approach. However, in commercial alloys having complex microstructures, the apparent activation energy estimates have not resulted in unambiguous interpretations since they were vastly different from those for diffusional processes and the stress exponent itself is strain rate dependent. Furthermore, this approach does not provide information for an optimization of intrinsic workability or for predicting flow instabilities.

18.2.8 Atomistic Models

Frost and Ashby [10] were the first to represent the materials' response in the form of a deformation mechanism map. These are plots of normalized stress versus homologous temperature (expressed as a fraction of the melting point) showing the area of dominance of each flow mechanism, calculated using fundamental parameters. The emphasis in these maps has been essentially on creep mechanisms, applicable to lower strain rates ($<10^{-3}$ s^{-1}), and the maps are very useful for alloy design. However, mechanical processing is done at strain rates that are orders of magnitude higher than those encountered in creep deformation and may involve different microstructural regimes.

With a view to develop a map that is directly relevant to processing, Raj [11] extended the deformation mapping concept to represent limiting conditions for the occurrence of microstructural damage mechanisms so that a safe regime for processing may be identified. The Raj map represents the limiting conditions for two damage mechanisms: (1) cavity formation at hard particles in a soft matrix occurring at lower temperatures and higher strain rates, and (2) wedge cracking at grain boundary triple junctions occurring at higher temperatures and lower strain rates. At very high strain rates, another unsafe regime representing adiabatic heating was also identified. He concluded that in principle there is always a region that may be termed safe for processing where neither of the two damage mechanisms nor adiabatic shearing occurs. In this regime, both dynamic recovery and DRX processes occur. Using an atomistic approach and the basis of fundamental parameters, Raj developed processing maps for pure metals as well as dilute alloys, which are extremely useful for the metalworking industry in the selection of optimum processing windows. However, the Raj maps suffer from the following drawbacks: (1) the required atomistic data for commercial alloys are inadequate to make calculations; (2) a prior knowledge of the deformation processes occurring in the materials is essential; (3) only some selected damage mechanisms were considered disregarding many more (e.g., intercrystalline cracking, porosity, etc.) in materials having complex microstructures; and (4) this model is based on an atomistic approach and cannot be integrated with continuum models that relate microscopic and macroscopic aspects of large plastic deformation. However, this is a pioneering approach in materials modeling applicable to processing.

18.2.9 Processing Maps

The difficulties encountered in the Raj maps were addressed by a dynamic material model (DMM) developed by Prasad et al. [12]. A processing map is an outcome of this model, which is an explicit representation of the response of a material, in terms of microstructural mechanisms, to the imposed process parameters. DMM is essentially a continuum model that is developed integrating the concepts of systems engineering [13], extremum principles of irreversible thermodynamics as applied to continuum mechanics of large plastic flow [14] and those describing the stability and self-organization of chaotic systems [15]. DMM considers mechanical processing as a system, an example of which is shown in Figure 18.1, with reference to the forging process. The system consists of various subsystems such as source of power (e.g., a hydraulic power pack), stores of power (tools like anvil, ram, and die), and dissipater of power (workpiece). Energy is generated by the source, transmitted to the tools to store the power and transfer it to the workpiece through an

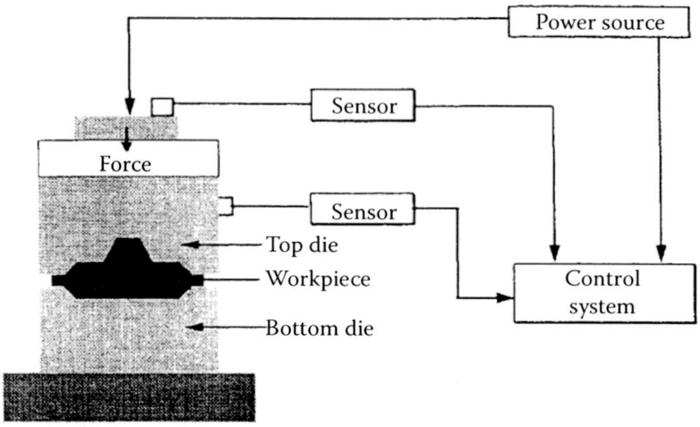

FIGURE 18.1 Material processing system elements with forging as an example.

interface (lubricant). The workpiece itself dissipates the energy while it undergoes plastic flow to take the shape imposed by the deformation zone. In this system, it is important to note that power or energy per second is to be considered and not energy per se, since the dissipation response of the system depends on how fast or slow the energy is input, bringing in time as an independent variable to make the system dynamic. While an integration of all the system elements has not been achieved so far, characteristics of the dissipater element (workpiece) are considered to be most important in designing the control system [16]. The workpiece dissipates the input power in a nonlinear fashion and the constitutive equation of the workpiece describes the manner in which energy is converted at any instant. The large irreversible plastic deformation imposed on the workpiece drives the system away from equilibrium.

On the basis of the description above for the work-piece characteristics, the instantaneous response of the work-piece material to the applied strain rate to impose a given plastic strain, at a constant temperature (T), and for a given history (M), is given by the dynamic constitutive equation:

$$\sigma = K\dot{\epsilon}^{m} \text{ at constant } T, M(\text{history}) \tag{18.3}$$

where σ is the effective stress, ($\dot{\epsilon}$) is the effective strain rate, and K and m are constants. The instantaneous total power dissipated is represented by a rectangle of area ($\sigma \, . \, \dot{\epsilon}$) as shown in Figure 18.2a. The constitutive equation describes the path chosen by the system to reach the applied strain rate (limiting condition). A different path may be chosen by the material for a different strain rate and hence will have different K and m values. The total power (P) consists of two complementary functions: G content and J co-content:

$$P = \sigma\,\dot{\epsilon} = G + J = \int_0^{\dot{\epsilon}} \sigma \, d\dot{\epsilon} = \int_0^{\sigma} \dot{\epsilon} \, d\sigma \tag{18.4}$$

This power split-up is shown schematically in Figure 18.2b. For finding the instantaneous (dynamic) values of G and J for deformation at a given temperature and strain rate, m is a constant as chosen by the system, while for different and widely varying strain rates, m may be dependent on the strain rate. The physical interpretations of G and J follow from the thermodynamic principles discussed by Malvern [17]. The total power dissipated is related to the rate of entropy production [14] as

$$P = \sigma\,\dot{\epsilon} = \frac{dS}{dT} T \geq 0 \tag{18.5}$$

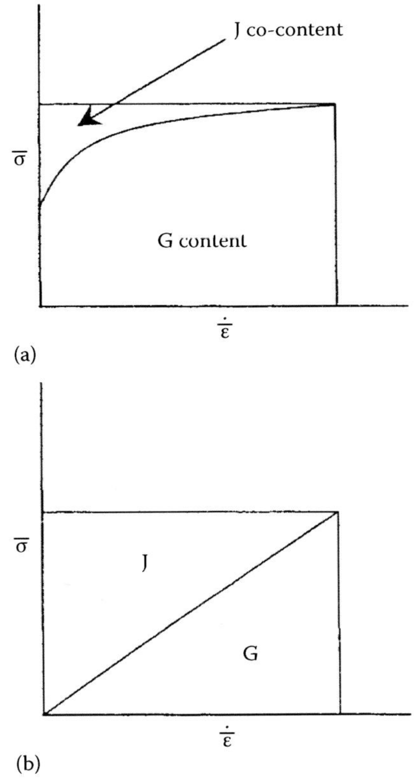

FIGURE 18.2 Schematic representation of the constitutive equation of (a) nonlinear power dissipater and (b) an ideal linear dissipater.

where T is the temperature, dS/dt is the rate of entropy production, and the inequality sign applies for irreversible deformation. Malvern [17] has shown that during an irreversible process, the total rate of entropy production consists of two separable parts, namely, internal entropy rate and thermal entropy rate. These two correspond to the rate of entropy production due to metallurgical changes and due to heat conduction, respectively. It is obvious that in hot deformation most (90–95%) of the dissipation is due to a temperature rise and only a small amount of energy is dissipated through microstructural changes. The power partitioning between G and J is controlled by the constitutive flow behavior of the material and is decided by the strain rate sensitivity (m) of flow stress since

$$\frac{dJ}{dG} = \frac{\dot{\varepsilon} d\sigma}{\sigma d\dot{\varepsilon}} = \frac{\dot{\varepsilon}\sigma \, \partial \ln \sigma}{\sigma \dot{\varepsilon} \, \partial \ln \dot{\varepsilon}} \approx \frac{\Delta \log \sigma}{\Delta \log \dot{\varepsilon}} = m \tag{18.6}$$

and thus m is a power partitioning factor. For an evaluation of an instantaneous value of J co-content, at each strain rate and temperature, m has a constant value corresponding to the limiting strain rate. Hence J is evaluated from the integral

$$J = \int_0^{\sigma} \dot{\varepsilon} \, d\sigma = \int_0^{\sigma} K' \, \sigma^{1/m} d\sigma \tag{18.7}$$

where K' $(1/K)$ is another constant. By combining Equation 18.7 with Equation 18.3, we can obtain J as

$$J = \frac{m\sigma\,\dot{\epsilon}}{m+1} \tag{18.8}$$

The metallurgical dissipation process may be characterized by the variation of J co-content with temperature and strain rate but normalization with the input power ($\sigma.\epsilon^{\cdot}$) sharpens the variation. Thus, a comparison with the linear dissipater (m = 1) (Figure 18.2b) in which maximum possible dissipation occurs through J co-content [Jmax = $\sigma.\epsilon^{\cdot}/2$] leads to the definition of a dimensionless parameter called efficiency of power dissipation (η), given by

$$\eta = \frac{J}{J_{max}} = \frac{2m}{m+1} \tag{18.9}$$

This parameter may be plotted as a function of temperature and strain rate at a particular strain to obtain a power dissipation map. In view of the nonlinear variation of flow stress with strain rate and temperature, the map exhibits hills and valleys. An easier method of representation is in the form of a 2-D iso-efficiency contour map, obtained by sectioning the 3-D map at constant efficiency levels. The contour map exhibits several efficiency domains with successively increasing iso-efficiency contours. The domains are separated by valleys, which are referred to as bifurcations. This behavior of the dissipative system is similar to the self-organization of chaotic systems suggested by Prigogine [15] and the efficiency of power dissipation during hot deformation represents the relative rate of entropy production, the highest being that for a linear dissipater. Each of the domains in the processing map represents a specific microstructural mechanism that contributes to power dissipation and is deterministic in the sense that kinetic laws are obeyed. Processing under conditions of a deterministic domain will ensure microstructural control and reproducibility. It may be noted that in dissipative systems, the strain rate and initial conditions determine the motion of the microsystem. The strain components only define the frame of the macrosystem and the entropy production inside the microsystem is a function of strain rate alone [14]. Hence, power dissipation maps at the different strains are not significantly different.

The parameter η, which is the efficiency of power dissipation, serves as the most useful index for characterizing dynamic material behavior for the following reasons:

1. It defines unique temperature and strain rate combinations for safe and optimum processing with microstructural control.
2. In its evaluation, specific atomistic rate-controlling mechanisms need not be evaluated or assumed. This aspect is advantageous, particularly when more than one mechanism is operating during hot deformation.
3. It is a continuum parameter and can be integrated with the finite element method.

18.2.10 Instability Map

A continuum instability criterion is developed [18,19] on the basis of extremum principles of irreversible thermodynamics [14]. According to this, unstable flow will occur if the differential quotient satisfies the inequality

$$\frac{dD}{d\dot{R}} < \frac{D}{\dot{R}} \tag{18.10}$$

where $\dot{R} = (\dot{\epsilon}\,.\,\dot{\epsilon})^{1/2}$ and D is the dissipative function, which represents constitutive behavior of the material. Since J determines the dissipation through metallurgical processes, the dissipation function related to metallurgical stability is given by J. By substituting $D = J$ in Equation 18.10, the condition for microstructural instability at a constant temperature can be obtained in terms of a dimensionless parameter) $\xi(\dot{\epsilon})$ given by

$$\xi(\dot{\epsilon}) = \frac{\partial \ln (m/(m+1))}{\partial \ln \dot{\epsilon}} + \mathrm{m} < 0 \tag{18.11}$$

The $\xi(\dot{\epsilon})$ parameter may be evaluated as a function of temperature and strain rate and plotted to obtain an instability map. According to Equation 18.11, unstable flow is expected when the value of *m* is negative (e.g., dynamic strain ageing) and/or the variation of *m* with strain rate is negative. The well-known manifestations of flow instabilities are adiabatic shear bands, flow localization, dynamic strain ageing (Luder's bands), kink bands, mechanical twinning, and flow rotations. This criterion may be physically interpreted in the following simple terms: if the material does not produce entropy constitutively at a rate that matches the rate of input entropy (through imposed process parameters), the flow becomes unstable and leads to flow localization.

The important advantages of continuum instability criterion are the following:

1. It can be superimposed onto the power dissipation map to delineate the regimes of unstable flow. Such a combined map is referred to as a Processing Map.
2. It can be conveniently used as a nonholonomic constraint (go/no-go decision) in finite element analysis to avoid undesirable microstructures in the component.
3. It is very useful from an industry viewpoint because processing can be performed at the fastest possible rates without entering into the unstable regimes.

The above concept of processing map has been applied to over 160 materials [20] including metals of different purity, conventional materials like aluminum alloys, copper alloys, and steels, advanced materials like super-alloys and zirconium alloys, and emerging generation of materials like titanium aluminides and metal matrix composites. These results have been confirmed by extensive microstructural investigations and a thorough understanding of the effects of several metallurgical variables on hot deformation mechanisms such as stacking fault energy, impurity content, second phase, hard particles, reinforcement, and change in initial microstructure on the hot deformation mechanisms has been obtained. Another variation of DMM is known as polar reciprocity model [21]. In this model, the reciprocal of efficiency of power dissipation is used. However, it does not add any further to the understanding of hot deformation.

18.2.11 Hot Deformation Mechanisms

During hot deformation, several microstructural mechanisms occur; some of them are safe but others cause microstructural damage. The safe mechanisms include dynamic recovery and dynamic recrystallization while others include wedge cracking, intercrystalline cracking, ductile fracture and flow localizations, and adiabatic shear bands. Superplasticity is also a safe mechanism as long as the strains involved are not too large since large strains can cause porosity in the component [22].

18.2.12 Dynamic Recovery

Dynamic recovery (DRY) involves mechanisms such as cross-slip, climb, and node unpinning, which occur generally in the homologous temperature range of 0.4–0.6Tm (Tm is the melting point temperature in degrees kelvin). DRY permits dislocations to unravel from hardening networks and annihilate each other. The process that occurs during DRY is the diffusion of rate controlling atomic species. The dynamically recovered microstructure contains well-defined subgrains with relatively dislocation-free interiors. DRY causes work hardening but the rate is much lower than that obtained in cold working. McQueen and Jonas [23] have reviewed the process of DRY in terms of flow curves, effect of strain rate and temperature, and microstructural development. Often, during postdeformation cooling, the occurrence of dynamic recovery results in static recrystallization, referred to as meta-DRX [24], which results in considerable grain refinement.

18.2.13 Dynamic Recrystallization

DRX is the major softening process that occurs during hot deformation (>0.7 Tm). This process refers to the occurrence of simultaneous recrystallization during deformation by nucleation and growth process [24,25]. DRX is different from static recrystallization in the sense that DRX characteristics are decided by the rate of nucleation versus rate of growth under given imposed conditions of temperature and strain rate. In static recrystallization on the other hand, a fixed amount of stored energy (dependent on cold work) is released by thermally activated dislocation recovery and grain boundary migration and may be termed as a kinetic process. DRX is a beneficial process in hot deformation since it not only gives stable flow and good workability to the material by simultaneously softening it but also reconstitutes the microstructure. For example, DRX breaks down the as-cast microstructure to produce wrought microstructure, globularizes the acicular preform microstructure (e.g., titanium and zirconium materials), redistributes the prior particle boundary defects in P/M (powder metallurgy) compacts to facilitate further processing or eliminates discrete particle effects by transferring mechanical energy across the interfaces of a hard particle to refine them. In the last three decades, numbers of models have been put forth by researchers to explain the mechanism of DRX and these are discussed below.

DRX is often manifested as oscillations in the flow curves in tests at constant strain rate. With increasing strain rate or decreasing temperature, it has been reported that the flow curve systematically changes from the multiple peaks to single peak type. Luton and Sellars [25] proposed a model to explain this behavior by considering the critical strain(ε_c) for the onset of recrystallization process and the strain or time required for completion of the process (ε_X). These two quantities exhibit different stress dependencies and when $\varepsilon_c < \varepsilon_X$, one recrystallization cycle is completed before the next one begins resulting in an oscillatory curve. When $\varepsilon_c > \varepsilon_X$, the second wave of recrystallization starts before the completion of first cycle and results in a single peak behavior. This model could not predict the gradual dampening of oscillations at high strains as observed in experimental curves. In order to overcome some of the shortcomings of this model, Sah and coworkers [26] have proposed a more complex model by considering the influence of grain size changes on the recrystallization kinetics and explained the gradual dampening of the oscillations at high strains. Wray [27] proposed a model based on a parameter termed as rate of recrystallization, determined from the time interval between the un-recrystallized and the recrystallized states to explain the oscillatory flow curve. Stuwe and Ortner [28] proposed an impingement model in which the new grains stop growing when they meet other grains traveling in the opposite direction, and they assumed that the dynamically recrystallized grains nucleate successively at the centers of the original grains.

A more realistic approach was put forward by Sandstrom and Langneborg [29]. They consider the mean dislocation density within the subgrains (ρ), to determine flow stress and the dislocation density in the subgrain boundaries (ρ_d) for providing the migration force for grain boundaries. By employing the postulated volume distributions of (ρ) and (ρ_d) they produced both single and multiple peak curves. However, this model overlooks the important role of grain boundaries in providing the nucleation sites for recrystallization.

Sakai and Jonas [9] have shown that the transition in flow curve shape is closely associated with a change in the type of recrystallization taking place during deformation and they reported that multiple peak flow appears when grain coarsening occurs and single peak flow appears when grain refinement occurs. Based on the metallographic studies, it was shown that the transition from single to multiple peak behavior depends on the relationship between the initial grain size and that established under steady-state conditions.

Monte-Carlo techniques have recently attracted much attention to simulate DRX. Peczak and Luton [30] have simulated DRX using different nucleation models and have reported many of the essential features of DRX. Derby and Ashby [31] have developed a model for DRX in which the nucleation rate is equated to a simple growth law such that a constant grain size is maintained. Ravichandran and Prasad [8] proposed a model on the basis of the rate of nucleation and rate of grain boundary migration. In this model,

DRX is considered to consist of two competing processes: formation of interfaces (nucleation) and migration of interfaces (growth). Nucleation consists of the formation of a grain boundary due to dislocation generation, simultaneous recovery, and rearrangement. Under hot working conditions the material acts essentially as a good dissipater of power and therefore the driving force for the migration of grain boundaries is the reduction of total interface energy. When nucleation and growth occur simultaneously, the slower of these two will be controlling DRX. Under conditions of constant true strain rate, the rate of formation of interfaces will compete with the rate of migration in such a way as to maintain a constant true strain rate with strain and the relative values of these two rates will decide the shape of the stress–strain curve. If these two rates result in comparable changes in the interface area, steady-state stress–strain curves are expected. If the rate of interface formation is slower than the rate of migration, a certain strain will have to elapse before a critical interface configuration is achieved for the purpose of migration. At the critical strain, large number of interfaces migrate leading to flow softening. Mathematical details of rates of interface formation and migration are explained in ref. [32,33]. On the basis of this model, it can be shown that [8], in low SFE materials like copper, the DRX is controlled by interface formation whereas in high SFE materials (like Al), interface migration controls the DRX process.

18.2.14 Flow Instability Processes

The microstructural manifestations of flow instabilities are many but the most common process is the occurrence of adiabatic shear bands. At high strain rates, heat generated due to the local temperature rise by plastic deformation is not conducted away to the cooler regions of the body since the time available for this is insufficient. The flow stress in the deformation band will be lowered and further plastic flow will be localized. The band gets intensified and nearly satisfies adiabatic conditions. Such bands are called adiabatic shear bands and tend to exhibit cracking, recrystallization, or phase transformation along macroscopic shear planes [34] and hence have SOS dependent manifestation. Their intensity depends on physical properties of the material like specific heat and conductivity in addition to the deformation characteristics.

Flow localization is another most common manifestation of flow instability in the microstructure [6]. This is less intense than adiabatic shear band formation and gives rise to microstructural inhomogeneity where localized shear bands may be curved or wavy.

18.2.15 Hot Working of Steels

Bulk metalworking of steels is generally carried out using hot rolling process at temperatures well above the critical temperature and involves deformation of austenite phase. Austenite has an FCC structure and is very ductile at elevated temperature thereby permitting the application of large plastic flow without the onset of fracture processes. In the early investigations on pure iron, Reynolds and Tegart [35] and Robbins et al. [36] have shown that the hot ductility of γ-iron increases with an increase in temperature. For plain carbon steels deformed in the austenitic range of temperature, DRX occurs [37–40], which is responsible for imparting excellent ductility as well as microstructural change. Mild steel in the austenitic range exhibits flow softening and oscillations, a classic picture as obtained by Rossard and Blain [37]. Such features suggest that γ-iron has low stacking fault energy and its constitutive behavior is not significantly affected by alloying dilutely with carbon, manganese, and nickel [39–42]. However, in concentrated alloys of austenite like austenitic stainless steels [41,43,44] or steels containing alloying additions that form carbides (e.g., molybdenum, niobium) [45,46] or oxides and nitrides [40], the hot working behavior changes considerably since the DRX process is slowed down by lowering of stacking fault energy or by the pinning down of grain boundaries by carbides, oxides, or nitrides, respectively. In such cases, it is observed that the strain rates for hot working are lower and the optimum hot working temperatures are higher. The influence of microstructural factors like grain size, precipitates of carbides, nitrides or oxides as well as the chemical composition on the hot ductility of steel was reviewed by Mintz and coworkers [47].

In general, a ductility trough was noticed for all steels in the temperature range 650–800°C. However, at temperatures above the austenitizing temperature, the ductility increases steeply with temperature and strain rate [47].

18.2.16 Maraging Steels

Maraging steels are characterized by two unique properties: ability to form martensite without quenching and to exhibit age hardening characteristics. These steels have emerged as alternative materials to conventional quenched and tempered steels for such advanced technologies as aerospace, nuclear, and gas turbine to name only a few. They possess the following advantages:

1. Excellent combination of strength and toughness
2. Good weldability
3. Minimum distortion during hardening

18.2.17 Mechanical Properties and Applications

The tensile strength of maraging steels lies in the range of 1400–2800 MPa and their resistance to brittle-fracture is excellent. This depends on the plasticity, toughness, and fineness of the martensitic matrix structure, morphology of the precipitates, and the impurity content (carbon, sulfur, phosphorus, and nitrogen). Maraging steels have good toughness at ultrahigh yield strengths (KIC = 100 MPa$\sqrt{m}$). Toughness of maraging steels can be improved by keeping carbon and nitrogen at low levels [48].

Maraging steels possess superior strength/toughness property coupled with good weldability and high service reliability because of high resistance to brittle fracture. A combination of these properties provides a distinct advantage over other high strength alloys like thermo-mechanically treated En-24 steels. Maraging steels find extensive application in the aerospace industry. Solid fuel rocket casing, engine housing, high-pressure tank, rocket fastener, landing gears, gear drives, and jet propellers are made of maraging steels. They are also used in the tools and machine building industry for such applications as dies and punches for hot extrusion, molds for pressure castings, tools for upsetting, operating shaft, microtools, and valve springs.

18.2.18 Processing Map of M250 Maraging Steel

The processing map obtained for M250 grade Maraging Steel [49] is shown in Figure 18.3 at a strain of 0.3. The general features of the maps for other strains are similar indicating that the influence of strain is not significant. Upset forging in the form of hot compression tests was used for generating flow stress data as a function of temperature, strain rate, and strain in view of its specific advantages discussed earlier. The processing maps are interpreted on the basis of microstructural observations, apparent activation energy analysis, and hot ductility measurements.

The processing map exhibits a single domain in the temperature range 1000–1200°C and strain rate range 0.001–0.01 s^{-1} with a peak efficiency of 43% occurring at 1125°C and 0.001 s^{-1}. Typical microstructures recorded on the specimens deformed at 1050, 1150, and 1200°C, at 0.001 s^{-1} are shown in Figure 18.4a–c. Since the material undergoes martensitic transformation on air-cooling, the features of austenite deformation are masked. However, it is possible to distinguish the prior austenitic boundaries in all the microstructures. The microstructures clearly show that a significant change of prior austenitic grain size occurs as a function of temperature in this domain. The variation of prior austenitic grain size as a function of temperature at a strain rate of 0.001 s^{-1} is shown in Figure 18.5a and compared with that of the efficiency of power dissipation in Figure 18.5b. The temperature of peak efficiency results in a significant change in grain size. Such features are indicative of DRX, which imparts very good workability to the material during hot working. The

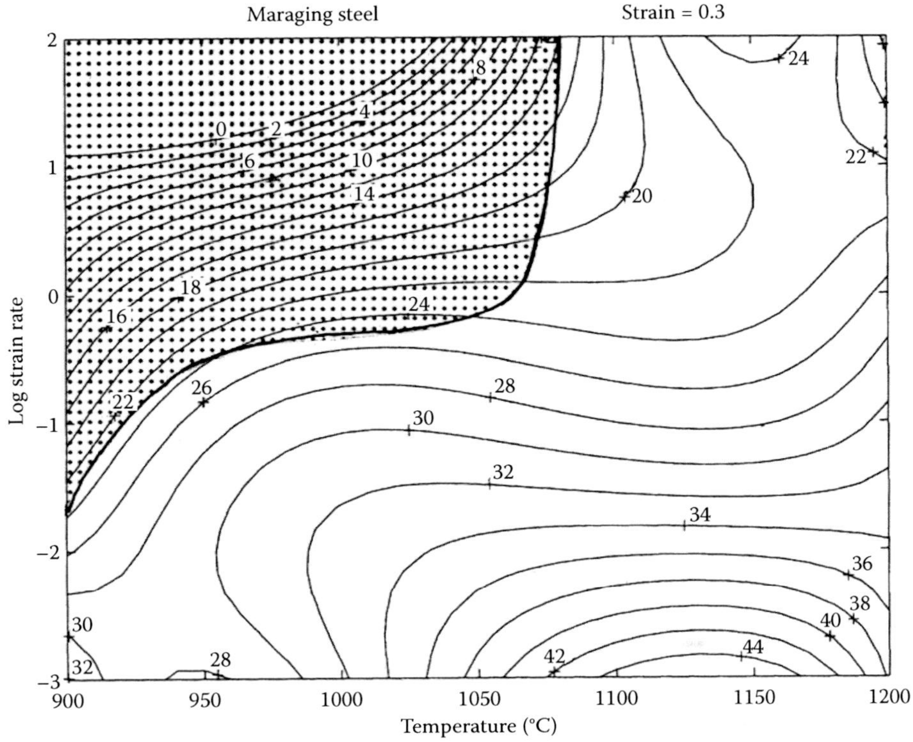

FIGURE 18.3 Processing map obtained for maraging steel at a strain of 0.3. Contour numbers represent percent efficiency of power dissipation. Shaded region corresponds to flow instability. (From G.S. Avadhani, *JMEPEG* **12**, (2003), 609.)

variation of tensile ductility of the material with temperature at a nominal strain rate of 0.001 s^{-1} is shown in Figure 18.5c. The ductility maximum has occurred at the temperature for the peak efficiency, as is commonly observed in materials undergoing DRX. Beyond this temperature, the occurrence of extensive oxidation has resulted in a drastic drop in the tensile ductility. On the basis of the microstructural observations, grain size variation, and the ductility variation, the domain is interpreted to represent the DRX process.

The variation of average grain diameter with Z (Zener–Holloman parameter introduced earlier) is shown in Figure 18.6, which exhibits a linear relation. The variation of grain size with Z may be expressed in the form of the equation

$$\log(d) = 4.9 - 0.27 \log(Z) \tag{18.12}$$

18.2.19 Instability Manifestation

The processing map of maraging steel (Figure 18.3) exhibits a regime of flow instability up to 1075° C and at higher strain rates (>1 s^{-1}). At lower temperatures (e.g., 900°C), the instability regime extends up to strain rates of 0.05 s^{-1}. Typical macrostructure recorded on a specimen deformed at 950°C/100 s^{-1} is shown in Figure 18.7a, which exhibits localized shear bands. The microstructures recorded at locations inside and outside the band at higher magnification are shown in Figure 18.7b and c, which reveal the occurrence of martensitic transformation in these regions during air cooling. The microstructure is not uniform over the entire specimen as expected from the occurrence of flow localization.

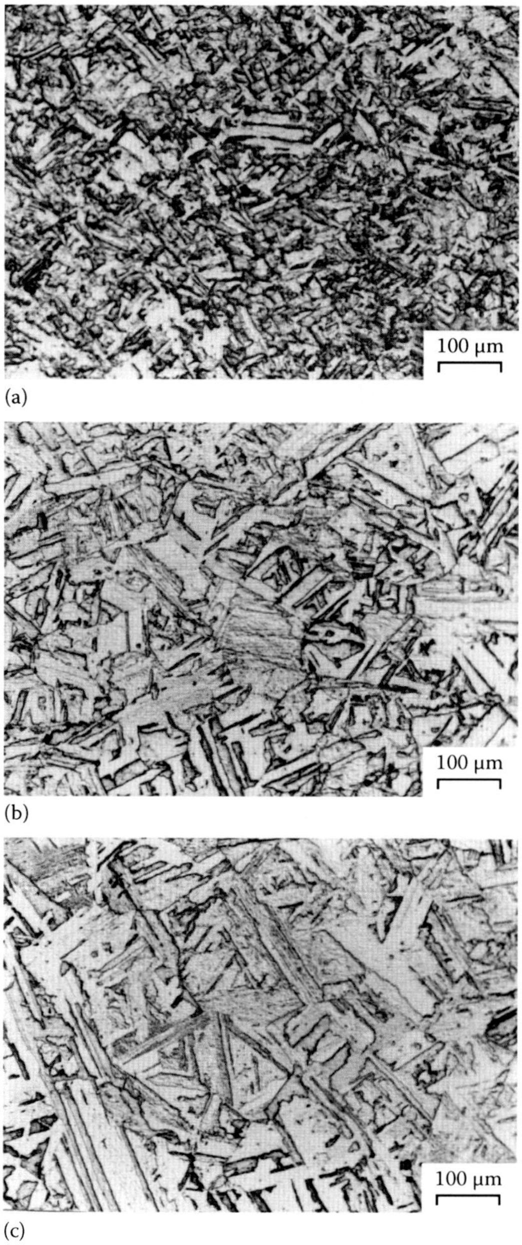

FIGURE 18.4 Microstructures recorded on maraging steel specimens deformed at (a) 1050°C, (b) 1150°C and (c) 1200°C, at 0.001 s^{-1} in the dynamic recrystallization domain. (From G.S. Avadhani, *JMEPEG* **12**, (2003), 609.)

18.2.20 Conclusion

The following conclusions are drawn from a study carried out on the hot deformation behavior of maraging steel material, using a processing map approach:

1. Maraging steel undergoes DRX with a peak efficiency of 44% at the optimum conditions of 1125°C and 0.001 s^{-1}.

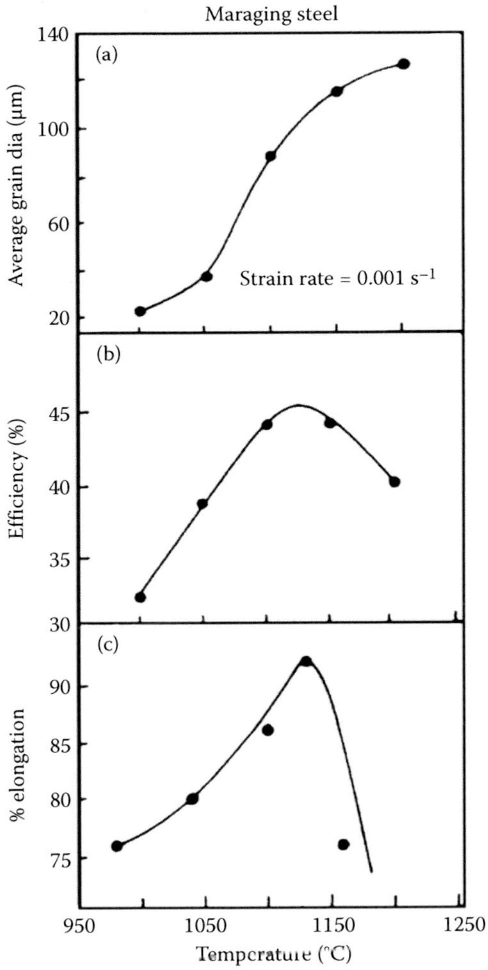

FIGURE 18.5 Maraging steel: (a) variation of average grain diameter, (b) efficiency of power dissipation, and (c) tensile ductility with temperature at 0.001 s^{-1} (dynamic recrystallization domain). (From G.S. Avadhani, *JMEPEG* **12**, (2003), 609.)

2. The apparent activation energy for DRX in this steel is estimated to be equal to 355 kJ/mol.
3. Unstable flow manifested as flow localization, occurs when maraging steel is deformed at higher strain rates (>1 s^{-1}), below 1075°C. This regime should be avoided during processing.

18.3 PART II: NANOFRAGMENTATION

Reverting back to the HPT technique, which has enjoyed vast success and attention ever since the Nobel Laureate Bridgman first introduced it in 1935 at Harvard University [50]. HPT requires a modified servo hydraulic testing facility and also a modified cylindrical coin specimen of 6–10 mm diameter and 1–2 mm thickness, as shown in Figure 18.8. In this figure, separate actuators are required for applying pressure and torsion independently. However, if needed, these actuators may be coupled using a control system as shown. It is necessary to understand the fact that the final grain growth and fragmentation depend on the effective stress and effective strain as discussed in the following section.

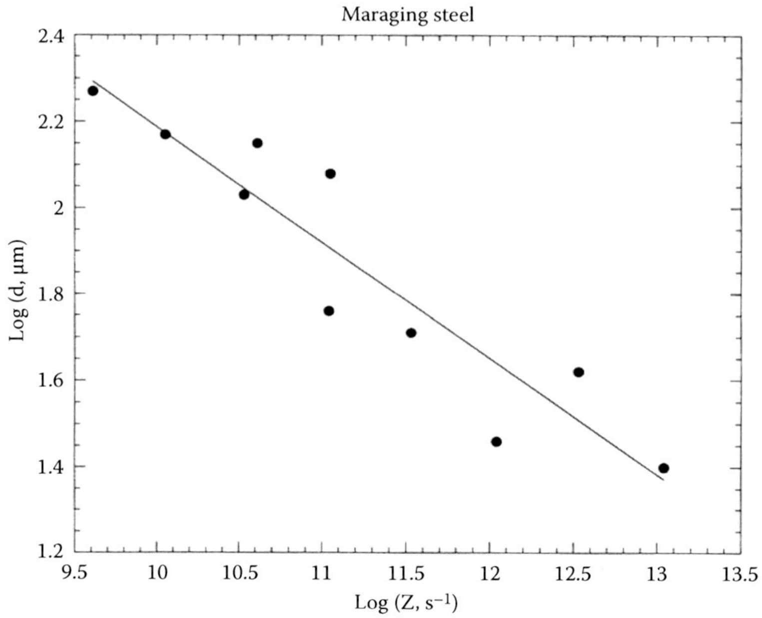

FIGURE 18.6 Variation of average grain diameter (d) with the Zener–Hollomon parameter (Z) in the dynamic recrystallization domain for maraging steel. (From G.S. Avadhani, *JMEPEG* **12**, (2003), 609.)

18.3.1 Effective Stress and Effective Strain

An interesting aspect of high-speed and high-temperature rolling and forging processes is concerned with the microstructural evolution. It is also often observed that extremely fine and fragmented grains are produced under a combined loading of pressure, shear stress, strain rate, and temperature. As stated earlier, conventional processing maps provide useful information under uniaxial compression at various temperatures and strain rates. However, shear stresses are almost always present in common forging and rolling operations due to biaxial flow, heterogeneities, and other geometrical factors. The effects of shear stresses are also evident near the platens on account of friction and adhesion. Lubrication at high temperatures is also unpredictable. This shear stress is a chief cause for grain fragmentation.

A series of recent HPT investigations [51–59] on various materials have highlighted the vast unexplored potential of achieving nanomaterials in bulk for the engineering industry. Nanoscience and technology initiatives have largely focused on advanced manufacturing strategies such as additive manufacturing (AM) and 3-D printing for small volume parts and components, contributing to less than one percent of the total global turnover!

A considerable amount of work has been undertaken to show the advantage of HPT for producing different nanomaterials, including intermetallic compounds [60]. The principle of HPT is explained in great detail by Edalati and Horita [50]. Also, attempts have been made for materials selection in the nanoscale range using new Ashby type plots [61].

But the need is there to generate a processing window in the nanorange from the data obtained from HPT. This will enable us to get process parameters to produce bulk nanomaterials without any trial and error.

18.3.2 State-of-Stress (SOS): Octahedral Plane

Work and power in hot working depend upon the SOS as well as the state of deformation (SOD). Further, the latter (SOD), includes both the state of plastic strain (SOPS) and the rate of deformation

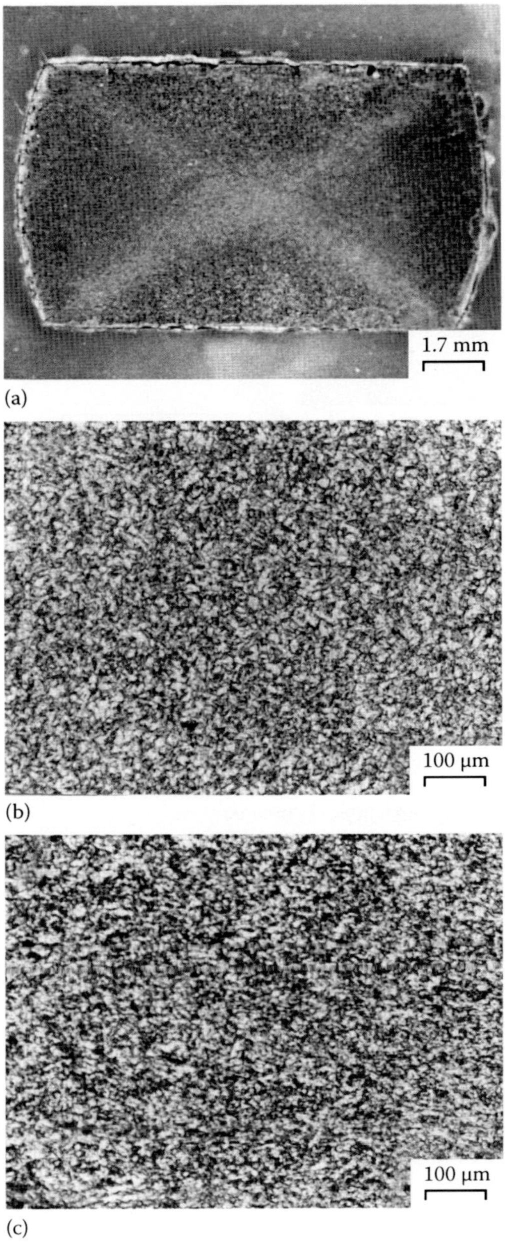

FIGURE 18.7 Macrostructure recorded on maraging steel specimen deformed at 950°C/100 s^{-1}, showing (a) localized shear bands and (b) microstructure inside and (c) outside the band. (The compression axis is vertical.) (From G.S. Avadhani, *JMEPEG* **12**, (2003), 609.)

(ROD). In the general case of three dimensional metalworking processes, it becomes necessary to define a unique effective measure for stress, strain, and rate of strain.

It is precisely in this context that octahedral planes assume significance. Referring stresses and strains to these privileged planes, we can define the effective stress to be

$$\sigma_e = \tau_{oct} = \frac{1}{3}\sqrt{(\sigma_x - \sigma_y)^2 + (\sigma_y - \sigma_z)^2 + (\sigma_z - \sigma_x)^2 + 6\left(\tau_{xy}^2 + \tau_{yz}^2 + \tau_{zx}^2\right)} \tag{18.13}$$

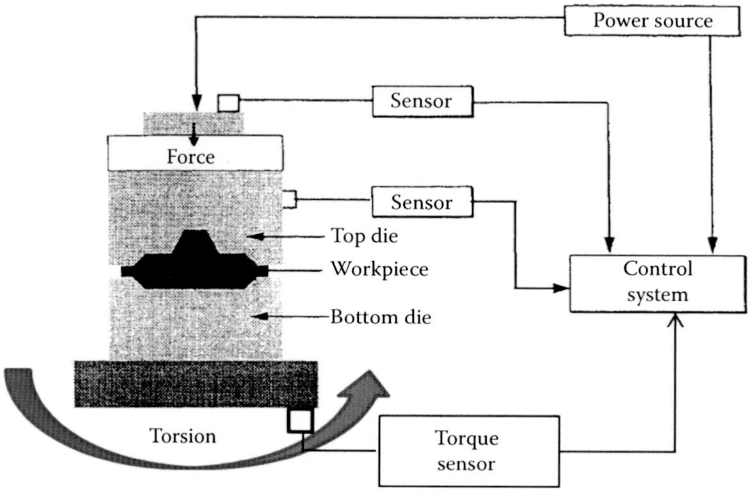

FIGURE 18.8 HPT processing system elements.

It may be noted that σ_e^2 is related to second deviatoric stress invariant J2 (Dieter [2], p. 47, SI metric edition):

$$J_2 = \frac{3}{2}\sigma_e^2 \tag{18.14}$$

18.3.3 HPT and High Compression Torsion

For the case of HPT,

$$\sigma_e = \sqrt{\frac{2}{3}}\tau \tag{18.15}$$

For the case of high compression torsion (HCT),

$$\sigma_e = \frac{1}{3}\sqrt{\sigma_x^2 + 6\tau^2} \tag{18.16}$$

Similar expressions are available for defining the effective strain and the effective rate of strain as follows:

$$\epsilon_e = \frac{\sqrt{2}}{3}\sqrt{(\epsilon_x - \epsilon_y)^2 + (\epsilon_y - \epsilon_z)^2 + (\epsilon_z - \epsilon_x)^2 + \frac{3}{2}\left(\gamma_{xy}^2 + \gamma_{yz}^2 + \gamma_{zx}^2\right)} \tag{18.17}$$

The effective strain rate is

$$\dot{\epsilon}_e = \frac{\sqrt{2}}{3}\sqrt{(\dot{\epsilon}_x - \dot{\epsilon}_y)^2 + (\dot{\epsilon}_y - \dot{\epsilon}_z)^2 + (\dot{\epsilon}_z - \dot{\epsilon}_x)^2 + \frac{3}{2}\left(\dot{\gamma}_{xy}^2 + \dot{\gamma}_{yz}^2 + \dot{\gamma}_{zx}^2\right)} \tag{18.18}$$

Sometimes, the effective shear strain and the effective shear strain rate are used in place of ϵ_e and $\dot{\epsilon}_e$. These quantities denoted by γ_e and $\dot{\gamma}_e$ are given by

$$\gamma_e = \sqrt{3}\epsilon_e \tag{18.19}$$

and

$$\dot{\gamma}_e = \sqrt{3}\dot{\epsilon}_e \tag{18.20}$$

From the above HPT and HCT equations, it is important to distinguish the two different SOS scenarios of HPT and HCT. HPT implies a 3-D hydrostatic state of stress. HCT, on the other hand, applies to uniaxial compression. In reality, however, the SOS will be tri-axial with the largest compressive stress acting along the loading axis of the machine. This situation is somewhat similar to a uniaxial strain state experienced under shock loading. The corresponding state of stress can be represented by the following equation as

$$\tau_{ij} = \begin{bmatrix} \sigma_x & 0 & \tau \\ 0 & \sigma_y & 0 \\ \tau & 0 & \sigma_z \end{bmatrix} \tag{18.21}$$

The main compressive stress exerted by the machine is along the loading axis is σ_Z. The pressure in the transverse direction (σ_X or σ_Y) is roughly equal to $\sigma_z/2$. In Equation 18.21, τ is the torsional shear stress applied independently.

Both HPT and HCT demand a deeper understanding of the influence of SOS and processing temperature. At present, there is a considerable amount of data on hot working under compression. To date, the data gathered on HPT is restricted to a small sample of metals and alloys. Combining all these data could pave the way for initiating HCT investigations. The first and foremost requirement in this endeavor concerns the design and development of a testing machine conceptualized in Figure 18.8. Furthermore, there is new evidence that three-dimensional effects associated with specific SOS can be captured using a single triaxiality parameter defined as the ratio of mean stress to the von-Mises stress [62,63].

18.3.4 Conclusion

Thus, HPT requires a more elaborate procedure for determining the optimum processing parameters. Nevertheless, if optimization of its deformation behavior is achieved, it would be very beneficial for the production of bulk nanomaterials in a predictable manner such as with the processing maps approach.

18.4 SUMMARY

It has been shown clearly that the development of processing maps of materials is a very important science-based method to decide on the optimum hot-working parameters (temperature and strain rate or speed) for the repeatable manufacturing of a particular component without costly and time-consuming trial-and-error methods. For example, the processing map of maraging steel presented reveals optimum regions for guiding the manufacturing operations. Prasad and co-workers have recently revised the book, "Hot working Guide", in which processing maps for more than 300 materials are documented in detail [64].

It would indeed be very beneficial to use the technique of referring to a processing map, before considering for the manufacture of any engineering component, which would be cost effective, saving both time and energy. Further grain refinement, including nanofragmentation, is possible by combining torsion with high pressure/compression. However, designing this combined SOS-testing machine demands a careful appreciation of effective stress, effective strain, and effective strain rates. These quantities should be carefully controlled and monitored—besides temperature and the starting microstructure. The starting microstructure also becomes crucial for generating HPT processing maps.

Nanofragmentation is largely determined by the presence of defects, precipitates, and dislocations topologies. Further, low angle grain boundaries and increasing grain boundary volume add further complexity. It should also be noted here that energy and power consumption in HPT processing will be at least one order of magnitude higher than in conventional metal forming machinery. This implies scaling up components and processes to handle much larger tonnage, reaching values up to several GN (giga newtons) for bulk operations such as rolling and forging. Another important aspect of nanostructured materials pertains to thickness of the component. Stable nanostructures, which are reasonably homogeneous, are limited to thicknesses of few hundreds of microns. Therefore, achieving a homogeneous microstructure for millimeter-thick components becomes quite challenging for future development. This chapter has covered the salient aspects of advanced hot working concepts involving HPT and HCT for bulk manufacturing.

REFERENCES

1. H.E. McGannon (Ed.), *The Making Shaping and Treating of Steel*, United States Steel Corp., Pittsburgh, PA (1964).
2. G.E. Dieter, in *Metals Handbook*, 9th edn. Vol.14, American Society for Metals, Metals Park, OH, pp. 363–387 (1988).
3. A.N. Bramley and D.J. Mynors, The use of forging simulation tools. *Mater. Des.* **21**, (2000), 279.
4. Y.V.R.K. Prasad and T. Seshachryulu, Modelling of hot deformation for microstructural control, *Intl. Mat. Rev.* **43**, (1998), 6.
5. J.J. Jonas, C.M. Sellars and W.J.M.G. Tegart, Strength and structure under hot-working conditions, *Metall. Rev.* **14**, (1969), 1.
6. S.L. Semiatin and J.J. Jonas, *Formability and Workability of Metals: Plastic Instability and Flow Localization*, American Society for Metals, Metals Park, OH (1984).
7. Y.V.R.K. Prasad, T. Seshacharyulu, C. Medeiros, W.G. Frazier and J.C. Malas III, Hot deformation mechanisms in Ti–6Al–4V with transformed β starting microstructure: Commercial v. extra low interstitial grade, *Mater. Sci. Tech.* **16**, (2000), 1029.
8. N. Ravichandran and Y.V.R.K. Prasad, Dynamic recrystallization during hot deformation of aluminum: A study using processing maps, *Metall. Trans.* **22A**, (1991), 2339.
9. T. Sakai and J.J. Jonas, Dynamic recrystallization: Mechanical and micro-structural considerations, *Acta Metall.* **17**, (1984), 189.
10. H.J. Frost and M.F. Ashby, *Deformation Mechanism Maps*, Pergamon Press, Oxford, London (1982).
11. R. Raj, Development of a processing map for use in warm-forming and hot-forming processes, *Metall. Trans.* **12A**, (1981), 1089.
12. Y.V.R.K. Prasad, H.L. Gegel, S.M. Doraivelu, J.C. Malas III, J.T. Morgan, K.A. Lark and B.R. Barker, Modelling of dynamic material behavior in hot deformation: Forging of Ti-6242, *Metall. Trans.* **15A**, (1984), 1883.
13. P.E. Wellstead, *Introduction to Physical Systems Modeling*, Academic Press, London (1979).
14. H. Ziegler, in *Progress in Solid Mechanics*, I.N. Sneddon and R. Hill (Eds.). John-Wiley and Sons, New York, vol. 4, p. 93 (1963).
15. I. Prigogine, Self-organization in non-equilibrium systems, *Science*, **201**, (1978), 77.
16. J.C. Malas III, W.G. Frazier, S. Venugopal, E.A. Medina, S. Mederios, R. Srinivasan, R.D. Irvin, W.M. Mullins and A. Choudary, Optimization of microstructure development during hot working using control theory, *Metall. Mater. Trans.* **28A**, (1997), 1921.
17. L.E. Malvern, *Introduction to the Mechanics of Continuum Medium*, Prentice-Hall, Inc. Englewood Cliffs, New Jersey, p. 197 (1969).
18. A.K.S. Kalyan Kumar, Criteria for Predicting Metallurgical Instabilities in Processing M.Sc. (Eng.) dissertation, Indian Institute of Science, Bangalore, (1987).
19. Y.V.R.K. Prasad, Recent advances in the science of mechanical processing, *Indian J. Technol.* **28**, (1990), 435.
20. Y.V.R.K. Prasad and S. Sasidhara, (Eds.). *Hot Working Guide: A Compendium of Processing Maps*, ASM International, Materials Park, OH (1997).
21. T. Rajagopalachary and V.V. Kutumbarao, Intrinsic hot workability map for a titanium alloy IMI685, *Scr. Mater.* **35**, (1996), 311.

22. T.G. Nieh, J. Wadsworth and O.D. Sherby, *Superplasticity of Metals and Ceramics*, Cambridge University Press, UK (1997).
23. H.J. McQueen and J.J. Jonas, in *Treatise in Materials Science and Technology*, R.J. Arsenault (Ed.), Academic Press, New York, p. 6, 393 (1975).
24. W. Roberts, in *Deformation Processing and Structure*, G. Krauss (Ed.). ASM, Metals Park, Ohio, p. 109 (1984).
25. M.J. Luton and C.M. Sellars, Dynamic recrystallization in nickel and nickel–iron alloys during high temperature deformation, *Acta Metall.* **17**, (1969), 1033.
26. J.P. Sah, G. Richardson and C.M. Sellars, Grain-size effects during dynamic recrystallization of nickel, *Met. Sci.* **8**, (1974), 325.
27. P.J. Wray, Effect of composition and initial grain size on the dynamic recrystallization of austenite in plain carbon steels, *Metall. Trans.* **15A**, (1984), 2009.
28. H.P. Stuwe and B. Ortner, Recrystallization in hot working and creep, *Met. Sci.* **8**, (1974), 161.
29. R. Sandstrom and R. Lagneberg, A model for hot working occurring by recrystallization, *Acta Metall.* **23**, (1975), 387.
30. P. Peczak and M.J. Luton, The effect of nucleation models on dynamic recrystallization I. Homogeneous stored energy distribution, *Phil. Mag.* **B68**, (1993), 115.
31. B. Derby and M.F. Ashby, On dynamic re-crystallization, *Scr. Metall.* **21**, (1987), 879.
32. J. Friedel, *Dislocations*, Ed. R. Smoluchowski, N. Kurti, Pergamon Press, Oxford, p. 164 (1964).
33. D.A. Smith, C.M.F. Rae and C.R.M. Grovenor, *Grain Boundary Structure and Kinetics*, R.W. Balluffi (Ed.), ASM, Metals Park, OH, p. 347 (1980).
34. S.P. Timothy, The structure of adiabatic shear bands in metals: A critical review, *Acta Metall.* **35**, (1982), 301.
35. R.A. Reynolds and W.J.M.G. Tegart, Torsion testing to assess bulk workability, *J. Iron Steel Inst.* **200**, (1962), 1044.
36. J.L. Robbins, D.C. Shepard and O.D. Sherby, Bulk workability tests, *J. Iron Steel Inst.* **199**, (1961), 175.
37. C. Rossard and P. Blain, Torsion testing for evaluation of plastic behaviour of materials, *Mem. Sci. Rev. Met.* **56**, (1959), 285.
38. J.J. Jonas and T. Sakai in *Deformation, Processing and Structure*, George Krauss (Ed.), ASM, Metals Park, OH, p. 185 (1984).
39. B. Mintz, R. Abushnosha and J.J. Jonas, Influence of dynamic re-crystallization on the tensile ductility of steels in the temperature range 700–1150°C, *ISIJ Int.* **32**, (1992), 241.
40. L.E. Cepeda, J.M. Rodrigues Ibabe and J.J. Urcola, Influence of composition and thermal history on the dynamic recrystallisation and subsequent hot ductility of mild steels, *ISIJ Int.* **33**, (1993) 799.
41. K.E. Hughes, K.D. Nair and C.M. Sellars, Temperature and flow stress during the hot extrusion of steel. Metals technology, *Metals Tech.* **1**, (1974), 161.
42. N.S. Mishra, *Hot Working Guide: A Compendium of Processing Maps*, Y.V.R.K. Prasad and S. Sashidhara (Eds.), ASM International, Materials Park, OH, p. 337 (1997).
43. S. Venugopal, S.L. Mannan and Y.V.R.K. Prasad, Optimization of hot workability in stainless steel, *Metall. Trans.* **23A**, (1992), 3093.
44. S. Venugopal, S.L. Mannan and Y.V.R.K. Prasad, Processing maps for hot working of commercial grade wrought stainless steel type AISI 304, *Mater. Sci. Eng.* **A177**, (1994) 143,
45. S. Venugopal, S.L. Mannan and Y.V.R.K. Prasad, Processing map for hot working of stainless steel type AISI 316L, *Mater. Sci. Tech.* **9**, (1993), 899.
46. P. Sricharoenchai, C. Nagasaki and J. Kihara, Hot ductility of high purity steels containing niobium, *ISIJ Int.* **32**, (1992), 1102.
47. B. Mintz, S. Yue and J.J. Jonas, Hot ductility of steels and its relationship to the problem of transverse cracking during continuous casting, *Int. Mater. Rev.* **36**, (1991), 187.
48. S.I. Krasnikova and S.I. Kuzmenko, On toughness studies in maraging steels, *Steel USSR* **12**, (1982), 63.
49. G.S. Avadhani, Optimization of process parameters for the manufacturing of rocket casings: A study using processing maps, *JMEPEG* **12**, (2003), 609.
50. K. Edalati and Z. Horita, A review on high-pressure torsion (HPT) from 1935 to 1988, *Mater. Sci. Eng.* **A 652**, (2016), 325.
51. R.B. Figueiredoa and T.G. Langdon, Using severe plastic deformation for the processing of advanced engineering materials, *Mater. Trans.* **50**, (2009), 1613.
52. H. Iwaoka and Z. Horita, High-pressure torsion of thick Cu and Al–Mg–Sc ring samples *J. Mater. Sci.* **50**, (2015), 4888.

53. R.Z. Valiev, Y. Estrin, Z. Horita, T.G. Angdon, M.J. Zehetbauer and Y. Zhu, Producing bulk ultrafine-grained materials by severe plastic deformation, *JOM*, **68**, (2016), 1216.
54. S.A. Alsubaie, P. Bazarnik, M. Lewandowska, Y. Huang and T.G. Langdon, Evolution of microstructure and hardness in an AZ80 magnesium alloy processed by high-pressure torsion, *J. Mater. Res. Technol.* **5**, (2016), 152.
55. X. Yang, J. Yi, S. Ni, Y. Du and M. Song, Microstructural evolution and structure-hardness relationship in an AL-4wt.%mg alloy processed by high-pressure torsion, *JMEPEG* **25**, (2016), 1909.
56. T. Santos Pinheiroa, J. Gallegob, C. Bolfarinia, C. Shyinti Kiminamia, A. Moreira Jorge Jr. and W.J. Bottaa, Microstructural evolution of Ti–6Al–7Nb alloy during high pressure torsion, *Mater. Res.* **15**, (2012), 792.
57. X.H. An, Q.Y. Lin, G. Sha, M.X. Huang, S.P. Ringer, Y.T. Zhu, and X.Z. Liao, Microstructural evolution and phase transformation in twinning-induced plasticity steel induced by high-pressure torsion, *Acta Mater.* **109**, (2016), 300.
58. T. Raghu, I. Balasundar and M. Sudhakara Rao, Isothermal and near isothermal processing of titanium alloys, *Def. Sci. J.* **61**, (2011), 72.
59. P. Sonnek and J. Petruželka, The use of Processing Maps for Prediction of Metal flow Stability in Hot Forming, *METAL* 15.–17.5., Ostrava, Czech Republic (2001).
60. A. Alhamidi, K. Edalati and Z. Horita, Production of nanograined intermetallics using high-pressure torsion, *Mater. Res.* **16**, (2013), 672.
61. Y. Zou, Materials Selection in Micro-or Nano-Mechanical Design: Towards New Ashby Plots for Small –Sized Materials, *MSE-A* **680** (2017) 421–425.
62. Y. Bai and T. Wierzbicki, A new model of metal plasticity and fracture with pressure and Lode dependence, *Int. J. Plast.* **24**, (2008), 1071.
63. M. Dunand, Ductile Fractureat Intermediate Stress Triaxialities: Experimental Investigations and Micro-mechanical Modeling, ScD Thesis, MIT, May 2013.
64. Y.V.R.K. Prasad, K.P. Rao and S. Sashidhara, *Hot Working Guide: A Compendium of Processing Maps*, 2nd Edition, ASM International, Materials Park, OH (2015).

19 Techniques of Powder Processing Alnico Magnets
Innovations and Applications

Emma M. White, Aaron G. Kassen, Iver E. Anderson, and Steve Constantinides

CONTENTS

19.1 INTRODUCTION

19.1.1 PERMANENT MAGNETS

Traction drive motors for electric vehicles and generators for wind turbines are clean propulsion and green energy technologies with a critical need for permanent magnets. The majority of these applications currently use Nd–Fe–B-based permanent magnets due to their high maximum energy product (~55 MGOe), the highest among all types of magnets [1–4]. Neodymium has been identified as a critical material (Figure 19.1) due to supply constraints as a rare earth element, and equipment manufacturers are averse to rely on rare earths because of market and price uncertainty, stemming from a near monopoly position (by China) for production of such critical elements as Dy, Nd, and Pr [5–7]. Additionally with increasing temperature, and especially for operation above 120°C, the energy product of Nd–Fe–B falls rapidly and requires added dysprosium, which is a critical material with an even greater cost and supply risk [8].

These factors have led to the renewed search for rare earth permanent magnet material alternatives with sufficient energy density for compact, high performance electric motors and generators.

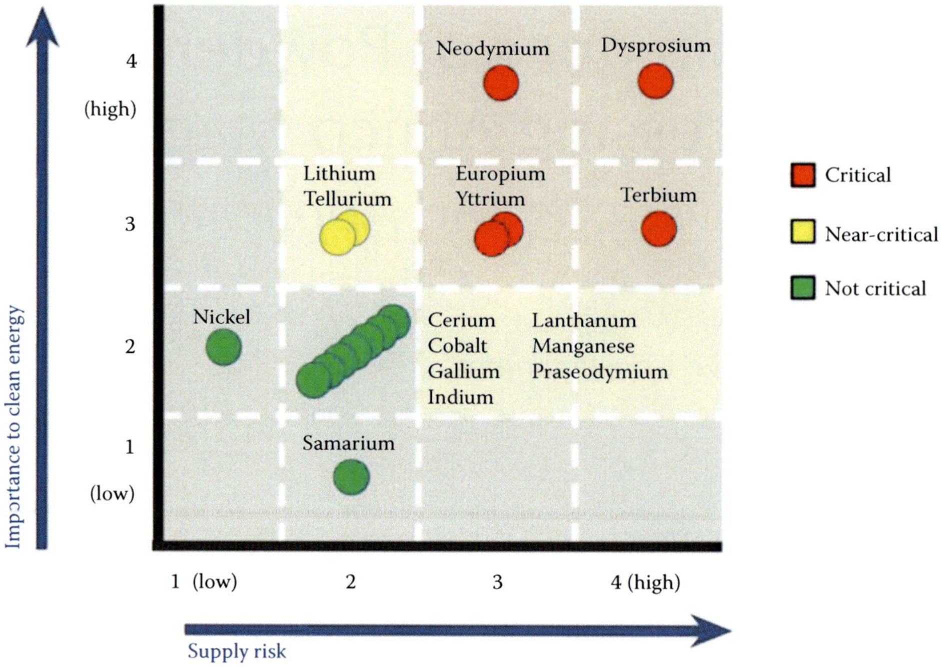

FIGURE 19.1 Critical materials according to importance and supply as identified by the United States Department of Energy. (From DOE *Critical Materials Strategy Report* 2011.)

The development of different magnet materials relative to their room temperature magnetic energy product is shown in Figure 19.2 [9]. Rare earth-containing permanent magnets have the highest energy products, followed by alnico permanent magnets. Alnico has potential for further magnetic property improvement due to the advent of modern processing methods and the increased understanding available from many high-resolution characterization tools that can be directed to "tune" the nanostructure of these complex magnetic alloys. Alnico is a promising near-term non-rare earth permanent magnet because of its impressive saturation magnetization and remanence, good corrosion resistance, improved mechanical properties, and excellent thermal stability, featuring a nearly flat temperature dependence of magnetic properties up to 400°C [6].

19.1.2 Sintered Alnico: History and Commercialization

Some inventions, upon commercial issue, are immediately disruptive but many require (1) a period of acceptance followed by (2) general adoption and, for some, (3) eventual market domination. Most inventions are adopted over an extended period of time during which the technology is continually improved. Alnico was such an invention—it represented a quantum improvement over existing permanent magnet materials: high carbon steel, cobalt steel (MK Steel, KS Steel), chrome steel, and similar products—but its development spanned several decades. The first alnico material became known as Alni, reported by Mishima, Honda, and Matsumoto between 1931 and 1933 [10–12]. Development followed rapidly in Japan, United Kingdom, the United States, and Russia [13]. According to Betteridge "The original Mishima patents were wide in scope, and included alloys of iron, nickel and aluminum with or without additions of several other metals" [14]. Continuing development of alnico over a period of 45 years resulted in numerous grades of cast and sintered alnico with improved energy product and coercivity. Material developments included the following substantial examples and are summarized in Table 19.1 [17].

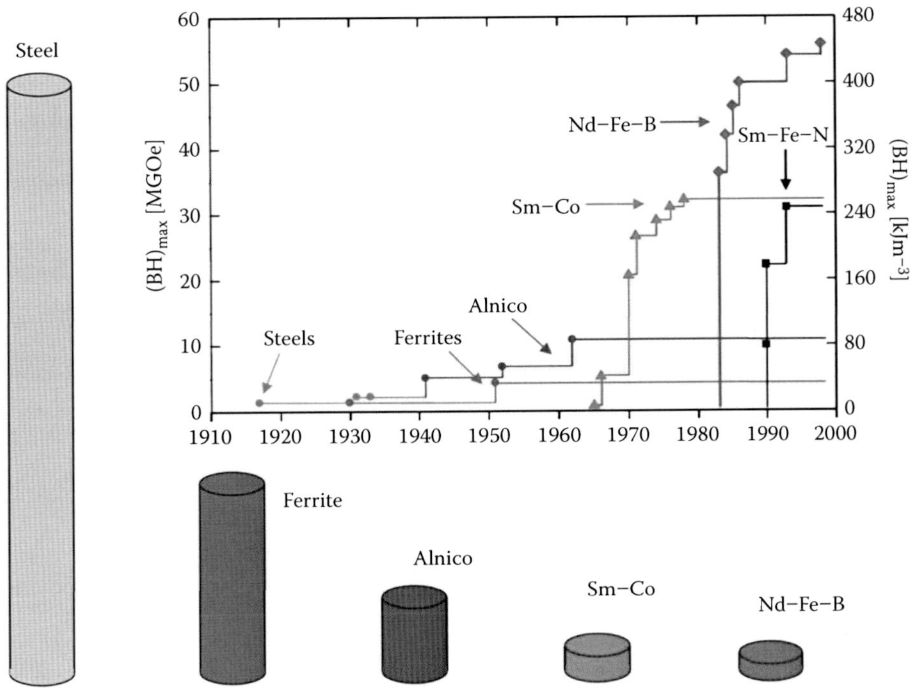

FIGURE 19.2 Development of different hard magnet materials and energy density at room temperature over time and comparable volumes of different materials required to achieve the same field. (From O. Gutfleisch et al., *Adv. Mater.*, 23, 821–842, 2011.)

1. Alni: cobalt-free, isotropic. Marginally improved from the best alternate steels of the time.
2. (Re-)introduction of cobalt with copper to enhance both residual induction and coercivity (Alnico 1) [15].
3. Increased titanium content for higher coercivity (Alnico 4) [16].
4. Spinodal decomposition in a magnetic field to enhance magnetic properties (1938; Alnico 2, 5).
5. Directional grain casting (casting onto a chilled surface).
6. Casting into a hot mold located on a chill surface (Alnico 5–7).
7. Increased percentage of refractory metals for increased coercivity (Mo, W, Nb, Zr, Ti) (Alnico 8).
8. Finally, in Alnico 9, optimized composition regarding cobalt with titanium plus casting into a hot mold on a chill plate (to achieve directional grain structure); followed by spinodal decomposition in a high magnetic field (3,000 Oe, 0.3 tesla); final "draw" thermal treatment at 650°C followed by 580°C.

This list shows that an evolutionary improvement in properties required ever more complex composition and processing. Alnico is functionally a self-organized nanostructured alloy due to the spinodal decomposition of the high temperature solid solution into a magnetic Fe–Co–rich ($\alpha 1$) phase and an Al–Ni–rich ($\alpha 2$) phase during cooling [18,19]. As mentioned, this class of permanent magnets was developed in the late 1930s and the compositions, complex processing, and heat treatments were optimized empirically over the following decades into the 1960s [20–26]. The alnico spinodal nanoscale structure elongates along the applied field direction after thermal annealing in a magnetic field, resulting in coercivity from shape anisotropy [18,19]. Modern examination of several alnico grades has shown how changes in chemistry and processing produce noticeably different

TABLE 19.1
Main Classes of Commercial (Cast) Alnico Alloys from Gould

			Composition, wt.% (bal. Fe)					Magnetic Properties					
								B_r		$(BH)_{max}$		H_e	
Class	Characteristic	Typical Brand Names	Al	Ni	Co	Cu	Others	kG	T	MGOe	kJ/m^3	Oe	kA/m
1	Isotropic, Co free	Alni Alnico 3 Nial Nipermag	12–14	24–30	0	0–3	0–1Ti	5–6	0.5–0.6	~1.3	~10	680–450	54–40
2	Isotropic, Low Co	AlNiCo 120 Alnico 1 Alnico 4	11–13	21–28	3–5	2–4	0–1Ti	5.5–7.5	0.55–0.75	1.4–1.5	11–12	700–450	56–36
3	Isotropic, Medium Co	Alnico Alnico 2 AlNiCo160 Nialco Reco 160	9–11	16–20	12–14	3–6	0–1Ti	6.5–8.0	0.65–0.8	1.6–1.75	13–14	620–500	50–40
4	Isotropic, Extra high H_e	AlNiCo 190 Hynico II Alnico 12	8–10	18–21	17–20	2–4	4–8Ti 0–1Nb	6–7	0.6–0.7	1.75–2.0	14–16	900–750	72–60
5	Field Treated random grain	Alcomax II Alcomax III Alnico 5 AlNiCo 500 Ticonal G Ticonal 600 Coalnimax	7.8–8.5	12–15	23–25	2–4	0–0.5Ti 0–1Nb	12–13	1.2–1.3	5–5.5	40–44	650–580	52–46

(Continued)

TABLE 19.1 (CONTINUED)
Main Classes of Commercial (Cast) Alnico Alloys from Gould

			Composition, wt.% (bal. Fe)					Magnetic Properties					
								B_r		$(BH)_{max}$		H_e	
Class	Characteristic	Typical Brand Names	Al	Ni	Co	Cu	Others	kG	T	MGOe	kJ/m^3	Oe	kA/m
6	Field treated random grain, higher H_e	Alcomax IV Alnico 6 Ticonal E Ticonal 800 Coercimax I	7.8–8.5	13–16	23–25	2–4	1–1.5Ti 1–2Nb	10.5–11.5	1.05–1.15	3.8–4.5	30–36	780–700	62–56
7	Field treated random grain, high H_e	Alnico 8 AlNiCo 450 Hycomax III Ticonal X Ticonal 1500 Coercimax II	7–8	14–16	32–36	4	4–6Ti 0–1Nb	8–9	0.8–0.9	5–5.6	40–45	1750–1400	140–110
8	Field treated random grain, extra–high H_e	Sermalloy A_1 Hycomax IV Ticonal 2000	7–8	14–15	37–40	3	7–8Ti	7.4–7.8	0.74–0.78	5.5–6.0	44–48	2100–1900	170–150
9	Field treated directed–grain form of Class 5	Alnico 5–7 AlNiCo 700 Columax Ticonal GX Ugimax Ansemax	7.8–8.5	13–15	24–25	2–4	0–1Nb	13–14	1.3–1.4	7–8	56–64	780–700	62–56
10	Field treated directed–grain form of Class 7	Alnico 9 Columnar Hycomax Ticonal XX	7–8	14–16	32–36	4	4–6Ti 0–1Nb ~0.3S	10–11	1.0–1.1	7.5–9.5	60–75	1750–1400	140–110

Source: J.E. Gould; *Cobalt Alloy Permanent Magnets*; Centre d'Information du Cobalt s.a, Brussels (1971).
Note: Development of sintered grades followed for each grade shown here

micro- and nanoscale structures, and result in substantial differences in magnetic properties, especially remanence and coercivity [18].

Commercial grades of alnico are composed of six elements plus one or more additives. The naming convention for alnico products has been sequential and follows the improvements in properties. The first grade introduced commercially was called Alnico 1, the second product called Alnico 2, etc. After Alnico 6, it was learned that dramatic improvements could be made in the properties of the Alnico 5 via thermal processing in a magnetic field without changes in composition. Therefore, this newest product became Alnico 5–7. The last two alnico grade improvements have been Alnico 8 and Alnico 9, both of which are titanium-containing grades. The Alnico 8 composition and processing was optimized to achieve high coercivity while Alnico 9 compromised some coercivity to maximize energy product. The highest maximum energy product achieved for Alnico 9 was 11.5 MGOe (92 kJ/m^3), though in large-scale production values between 9.5 and 10 MGOe (76 to 80 kJ/m^3) are more common.

Alnico 5–7 is a grain-aligned alloy and displays the highest remanence (13.5 kG) of any of the commercial alnico compositions [3]. Alnico 8 is isotropic and exhibits the highest coercivity (1.9 kOe) of any of the commercially available alnico grades [3]. However, all these grades have coercivities and energy products more or less two or three times below the theoretically possible value from the spinodally decomposed nanostructure [3]. Further processing and composition optimization can theoretically produce alnico permanent magnets that attain an energy product of ~20 MGOe, which at elevated temperatures (150–180°C) is comparable to both Dy enriched Nd-Fe-B and Sm-Co-based permanent magnets [3].

Consistent with magnetic material production over the preceding 100 years, alnico started as a melt and cast type product. The composition and physical properties that made for acceptable permanent magnetic characteristics also resulted in materials that were brittle (crystalline and non-malleable). Traditional metallurgy via melt-cast into sand molds was acceptable for large parts cast to near net shape. The presence of sprues, runners, and a large melt pool (to accommodate shrinkage during cooling) could be tolerated as a relatively small percentage of the volume of large parts. As demand grew for smaller magnets, a more appropriate processing method was sought that was more material efficient and capable of dimensional accuracy on small parts, achieved with little or no machining [17]. Such a process existed in powder metallurgy, a growing field of commercial endeavor. As reported by Gould [17]:

> All the alnico alloys, as outlined in Section 5.1 for isotropic and in Section 5.2 for anisotropic materials, can also be manufactured by mixing together appropriate fine metal powders, pressing a compact to the desired shape with some allowance for shrinkage, sintering for alloy homogenization and compact densification at a temperature in the range 1250–1330°C, followed by the heat treatment appropriate to the alloy. When correctly carried out the resultant magnetic properties are similar to those of the cast alloy, subject only to the modifications caused by residual porosity of the sintered compact. The mechanical properties are usually somewhat better than for the cast material because of the finer grain size produced by sintering. Production by sintering rather than casting is only economically advantageous for the manufacture of very large quantities of small magnets because of the high cost of pressing tools and of suitable powdered raw materials.

Patents for sintered alnico in the United States were issued starting in 1940 (e.g., 2192743 and 2192744, assigned to General Electric) with additional patents issues at least to 1951–2 with U.S. patents 2546047 and 2617723 [27].

In summary, the advantages of powder metallurgy processing include [28]

1. Near net shapes—some magnets can be used without any finish machining.
2. Manufactured without the waste associated with sprues, runners, and a melt pool.
3. Permit manufacture of very small parts (less than 1 g) and even moderate size parts (15 g).

A typical flow chart for the manufacture of alnico is presented in Figure 19.3. Alloy raw material in suitable form and purity is melt and cast or melt and atomized. For the cast alloy, crushing and milling are required to obtain fine particles of an average of 50 μm. Atomized alloy requires screening for particle size control [29]. The powder is blended with a die lubricant and minor additives. The blended powder can then be compacted in a die cavity at 550 to 800 MPa to obtain adequate strength for handling during loading into de-lube and sintering processes [30].

Dimensional shrinkage of magnets during sintering is adequately predictable and controllable to often permit use of magnets without finish grind. For a medium smooth surface, a light sandblast may be used to clean the surface, and applications requiring a smooth surface with precision dimensions can be finish ground on the pole faces and on the magnet perimeter when required.

By its nature, the powder metallurgy process prevents incorporation of the directional grain, chill cast, or hot mold-chill cast methods. Thus, product produced using powder metallurgy lacks the oriented macrocrystalline structure and grain orientation of cast alloy. The sintering process results in a fine grain structure. Minor ingredients can effect development of the beneficial alnico microstructure thus limiting minor constituent content [31–34]. Further, included porosity results in lower residual

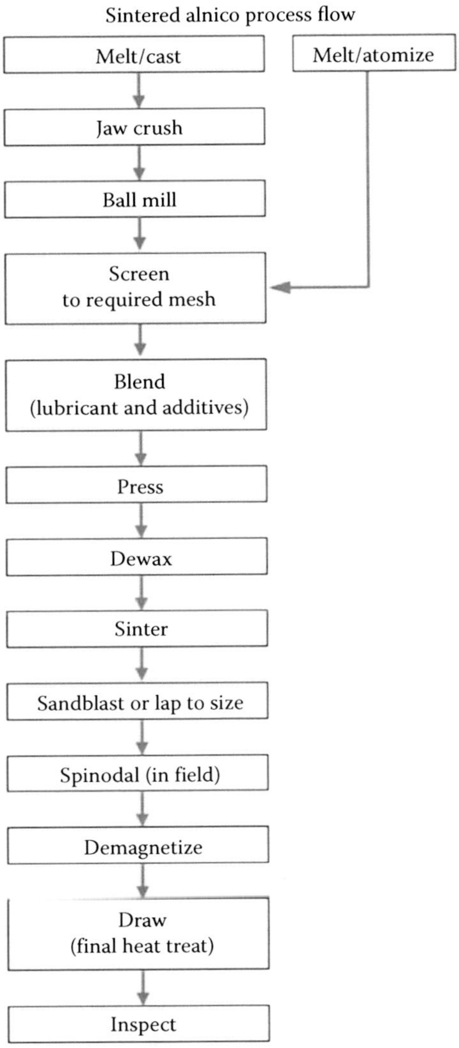

FIGURE 19.3 Typical process flow for the manufacture of sintered alnico.

induction (Br) and maximum energy product, (BH)max. Application of a magnetic field during spinodal decomposition somewhat improves the magnetic properties through enhancement of grains most closely aligned with the field direction, sacrificing those grains perpendicular to the field direction. The magnets exhibit partially oriented properties—a cross between isotropic and anisotropic. An inherent benefit of the absence of macrocrystals and presence of the "woven matte-like" microstructure of powder metallurgy processed alnico is significantly greater strength. Tensile strength and transverse modulus of rupture of cast and sintered alnico are compared in Table 19.2 with physical strength of sintered magnets being from two to ten times greater than that of cast alloy magnets.

The two most commonly manufactured grades of powder metallurgy (sintered) alnico are 5 and 8. Each of these grades may consist of several subgrades—materials that have been tailored to a customer's specific requirements and then became a general commercial product. Typical variants are "H" with somewhat higher coercivity, "B" with higher remanence, and "HE" with highest energy product. There is no naming convention, so each manufacturer's literature will require careful reading. A collection of both cast and sintered alnico magnets from a commercial manufacturer is included in Figure 19.4.

With growth in production of ferrite magnets starting in 1956, and the introduction of SmCo 1:5 (1967) and SmCo 2:17 (1974), confirmed research into alnico came to a virtual halt and was only resurrected in 2010 due to problems associated with pricing and availability of rare earths (for NdFeB magnets). Efforts for improving alnico have focused on increasing Hcb and "squaring" of the hysteresis loop.

Current standard alnico magnet production involves directionally solidified casting to achieve the best microstructural orientation and therefore second quadrant squareness and energy product [35]. Directional solidification is a complex, expensive processing method that leads to large quantities of excess scrap material from extensive machining of each casting and high production costs. Fortunately, materials processing and characterization advances present opportunities for dramatically enhanced alnico magnets, allowing preconceived notions about alnico alloys and processing to be challenged. For example, alnico 8 is currently understood to derive its good intrinsic coercivity (compared to alnico 9) primarily from shape anisotropy, resulting from elongation of the nanostructured Fe–Co spinodal rods. However, if the reduced remnant magnetization of Alnico 8 with its isotropic microstructure [18] could be increased by a solid state method, the microstructure could be aligned without reducing its coercivity, greatly increasing the energy product. Encouragement for this pathway comes from early single crystal and texturing experiments performed by

TABLE 19.2
Physical Strength of Alnico Magnets Made by Melt-Cast and by Powder Metallurgy

Alnico Grade	Tensile Strength, MPa		Transverse Modulus of Rupture, MPa	
	Cast	Sintered	Cast	Sintered
2	20	450	50	480
3	80	n/a	160	n/a
4	60	n/a	165	n/a
5	40	350	65	415
6	160	380	310	760
5–7	35	n/a	55	n/a
8	65	360	205	360
9	50	n/a	55	n/a

FIGURE 19.4 Collection of cast alnico (larger magnets near top) and sintered alnico (smaller magnets near the bottom). (Courtesy of HPMG.)

Durrand-Charre and Liu who showed, when preferred microstructural orientation is promoted, improved magnetic energy product and remanent magnetization are possible [36,37]. Specifically, Durrand-Charre showed, if preferential orientation could be achieved in alnico within approximately 15° of the optimal [100] alignment, enhanced properties could be achieved over typical isotropic magnets, in agreement with Chikazumi and Bunge [36,38,39].

Optimizing the composition and processing of sintered alnico magnets could achieve high coercivities and energy products, vital to motor applications, while lowering production costs through near-net shape processing and simplified manufacturing methods through compression molding and sintering [40]. If an aligned powder-processed sintered alnico permanent magnet with high remanence and an optimized composition for high coercivity could be produced, it would be a competitive alternative to rare earth permanent magnets for middle-energy-product applications, as well as high-temperature applications.

19.1.3 High-Pressure Gas Atomization Powder Production

Any powder processing method begins with consideration of the powder properties and characteristics, including size, shape, and composition, which affect subsequent molding, sintering, and densification operations. These powder properties are directly related to the method used to produce the powder. Atomization is widely considered the best method for high quantity production of metal alloy powders due to the flexibility and control inherent to the process [41]. High-pressure gas atomization, using inert gases, is used for spherical powders of reactive and/or specialty alloy powders [41]. In the high-pressure gas atomization process solid metal powders are formed from impingement of high-velocity gas jets on a molten metal stream which causes the stream to break up into fine droplets and rapidly solidify [41]. Figure 19.5 details the components and setup of an experimental bottom pour high-pressure gas atomization system. Prealloyed, fine, spherical powders within the alnico 8 family have been produced using a similar setup to that shown in Figure 19.5. Kassen and coworkers and Anderson and coworkers further describe the gas atomization processing and resulting powder characteristics [42–44].

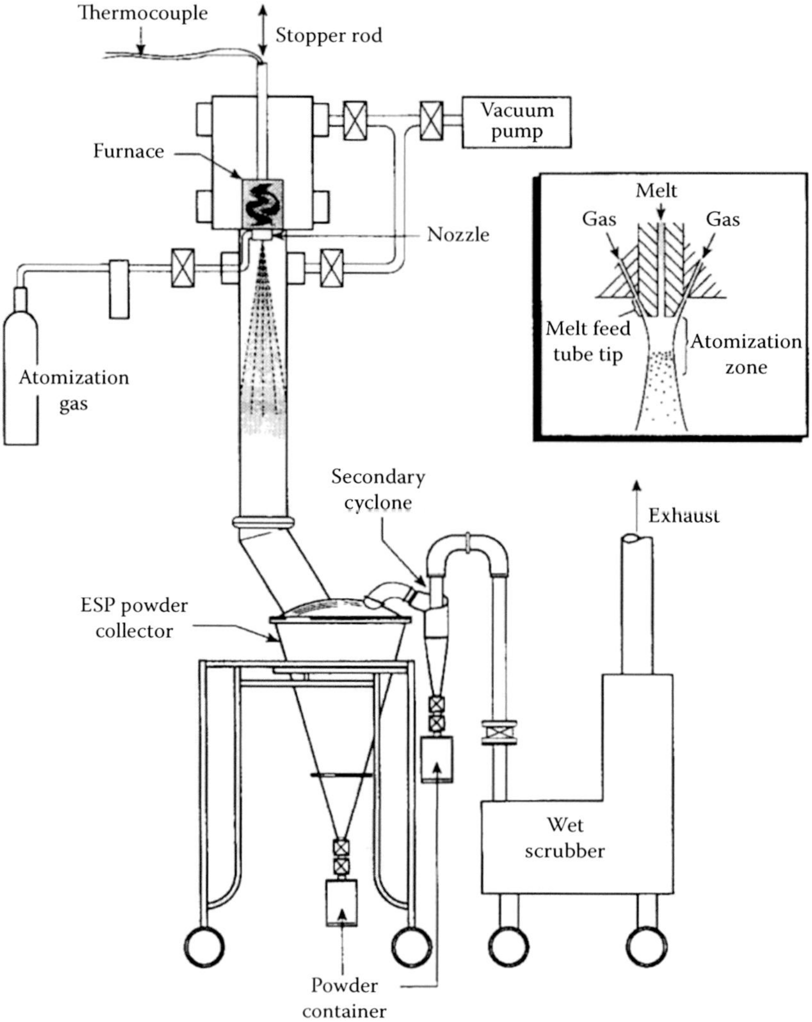

FIGURE 19.5 Experimental bottom pour high-pressure gas atomization system.

19.2 CONSTANT UNIAXIAL STRESS APPROACH FOR TEXTURED ABNORMAL GRAIN GROWTH

19.2.1 Dead Weight Driven Case

19.2.1.1 Experimental Methods

High-pressure gas atomized alnico 8H prealloyed powder was produced at Ames Laboratory and size classified as detailed by Anderson and coworkers [43,45]. A sample in the form of 90 wt.% 32–38 μm diameter powders + 10 wt.% 3–15 μm diameter powders was compounded with a 6 wt.% solution of a polypropylene carbonate binder (QPAC 40®), which has extremely clean burnout characteristics [43]. These compounded powder samples were compression-molded into two cylindrical dies: 0.9525 cm (dia.) at 156 MPa and a scaled up 2.54 cm (dia.) at 175 MPa, to produced green bodies. The binder was burnt out as detailed by Anderson and coworkers and then the samples were vacuum sintered between 1–12 h at 1250°C (below the solidus of 1281°C from differential scanning calorimetry), to produce samples with density >99% (cast density = 7.3 g/cm^3) [43].

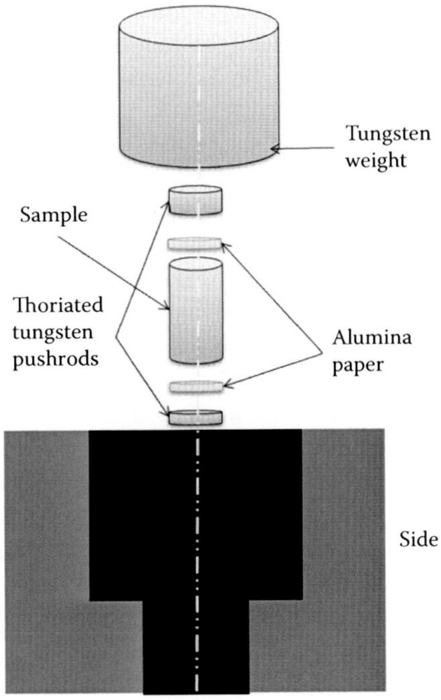

FIGURE 19.6 Schematic of uniaxial loading apparatus for alnico specimen texturing.

Experimental specimens were cut from the larger bulk specimens to 3.25 mm (dia.) and centerless ground to 3 mm (dia.) and faced off to 8 mm length with parallel, flat ends.

The sintered specimens underwent two different processing paths for direct comparison. Control groups of samples were vacuum sintered for 1 to 12 h without loading to measure densification rate and baseline magnetic properties. The other set of samples received a primary (4 h) sintering step to obtain >99% density, followed by a novel grain growth step under an applied uniaxial stress during sintering for an additional 4 h. The stressed specimens were placed in a uniaxial loading apparatus (Figure 19.6) for the second 4 h sintering step in vacuum ($<5 \times 10^{-6}$ torr) at 1250°C. The applied stress varied up to 1248 kPa (900 g dead weight) to determine the effect of the different loadings on the development of the microstructure. The higher pressures were excursion-limited by the apparatus to approximately 10% axial strain, which resulted in a reduction of applied stress at the limit, while the pressures lower than 345 kPa (250 g) allowed continuous strain to occur during the experiment under continuous pressure. After the stress-biased grain growth process was completed, centerless grinding was performed a second time to ensure sample uniformity for magnetic property measurements and compensate for any creep. All finished samples received a final solutionizing (1250°C, 30 min) heat treatment with an oil quench, followed by magnetic annealing (~1.0 T, 840°C, 10 min) and draw annealing (cycles at 650°C for 5 h and 580°C for 15 h). The magnetic properties were measured using a closed loop Laboratorio Elettrofisico AMH-500 hysteresisgraph at a maximum applied field of 12 kOe [43].

Analysis of the longitudinal microstructures and grain orientation were performed using an AMRAY 1845 scanning electron microscope (SEM), JEOL JAMP-7830F auger electron spectroscopy (AES) system, and an FEI Teneo SEM, each fitted with an electron backscatter detector (EBSD) to measure final grain size and orientation to correlate with magnetic properties.

19.2.1.2 Results and Discussion—Compression Molding & Sintering

Chemical analysis of the as-atomized (dia. <45 μm) powder (NSL Analytical, Cleveland, OH) verified that the match between the desired and final compositions was nearly perfect. Inert gas fusion analysis of contamination of the atomized powder (dia.<45 μm) showed the bulk oxygen level to be 190 ppmw, less than half of the aerospace specification for powders. The binder decomposition produced no significant carbon increase in the final sintered magnets as carbon determination showed a maximum of 209 ppmw (8 h sintered sample), below industry values of 300 ppmw [46].

This compression molding process formed near-final shape magnets as detailed by Anderson and coworkers, with initial green bodies of measured average relative densities of 70% (geometric) of cast density [43]. Shrinkage was consistently predictable after debinding and final sintering, with approximately 9% shrinkage for samples sintered for 4 h and a total 10% lineal shrinkage occurring for samples sintered for 8 h. A densification curve was established, showing rapid densification in the first 0.5 h and over 99% density by 4 h [43].

Archimedes density values correlate with the residual porosity evident in the micrographs of Figure 19.7. Despite the sample in Figure 19.7c appearing to show higher than expected porosity along prior particle boundaries, further examination and energy-dispersive x-ray spectroscopy mapping confirmed that the microscale features were spheroidized alumina particles that had coarsened during extended sintering from the original thin passivating oxide layers on the high-pressure gas atomized powders.

Grain contrast imaging allowed the grain size to be measured in each specimen, indicating average grain sizes of 23 μm, 30 μm, and 1.3 mm, after sintering for 1 h, 4 h, and 8 h, respectively. Most of the spheroidized alumina particles (Figure 19.7c) from the prior particle boundaries became intragranular, showing that they were not effective at pinning the grain boundaries after extended sintering at 1250°C.

19.2.1.3 Results and Discussion—Uniaxial Stress Application during Extended Sintering

The 4 h sintered samples were loaded with a uniaxial compressive stress for an additional 4 h in the previously included set up. A uniaxial stress condition was ensured by the fixture in each 3 mm (dia.) × 8 mm (height) specimen subjected to a deadweight stress, ranging up to 1248 kPa (900 g) longitudinally applied to the cylindrical specimens. The sintering temperature of 1250°C and a time of 4 h was used since it was known to promote abnormal grain growth in the occasional sample, as discussed earlier and detailed by Anderson and coworkers [43].

The growth of the diameter and shrinkage of the height of each specimen was measured at the bottom, middle, and top to determine any resulting plastic deformation or shear/creep that was promoted by each level of stress as shown in Figure 19.8. To illustrate the results of the most reliable experiments this plot extends only up to samples stressed to 345 kPa (250 g). The fixture "excursion limited" some of the higher stress samples (1248 kPa or 900 g) by the fixture dimensions, where the posttest plastic strain averaged approximately +8% in the diameter and −11% in height.

Electron backscatter diffraction analysis of the longitudinal sections of the resultant microstructures showed samples subject to the highest loads of 1248 kPa were observed to have a primarily <111> final texture in the normal direction and near <110> type texture along the axial direction (see Figure 19.9a) after plastic deformation at 1250°C. In the body-centered cubic (BCC) structure, slip systems under uniaxial compression are activated, where the slip plane {110} aligns normal to the compressive axis and the slip direction <111> normal will rotate away from the compressive axis [47]. The texture that evolved during the apparent shear/creep and grain growth within the samples is indicative of these classic BCC slip systems from pencil-glide and grain rotation at these high temperatures [47].

In alnico, optimal alignment would be the <001> easy direction along the magnet sample axial direction whereas these samples show a suboptimal final orientation of the normal to the slip planes from the grain rotation mechanism (Figure 19.9a). The reduction in magnetic properties for misoriented microstructures can be approximated with a Schmidt style law approach that involves two rotations (theta-phi) to align the crystal and axial coordinate systems to any <001> type direction [48]. However, a simple approximation can be derived by considering a simple tilt angle of approximately 45° between the axial direction and resulting alignment plane normal and the desired

FIGURE 19.7 Scanning electron microscopy (backscattered electron) images of 1250°C sintered samples: (a) 1 h; Archimedes 98.8% dense; (b) 4 h sinter; 99.1% dense; and (c) 8 h sinter; 99.6% dense sintered samples showing grain contrast and residual porosity. Also observed in the 8 h sinter micrograph is coarsened spheroidized alumina particles decorating prior particle boundaries, e.g., indicated by the arrow in (c) verified by energy-dispersive x-ray spectroscopy mapping.

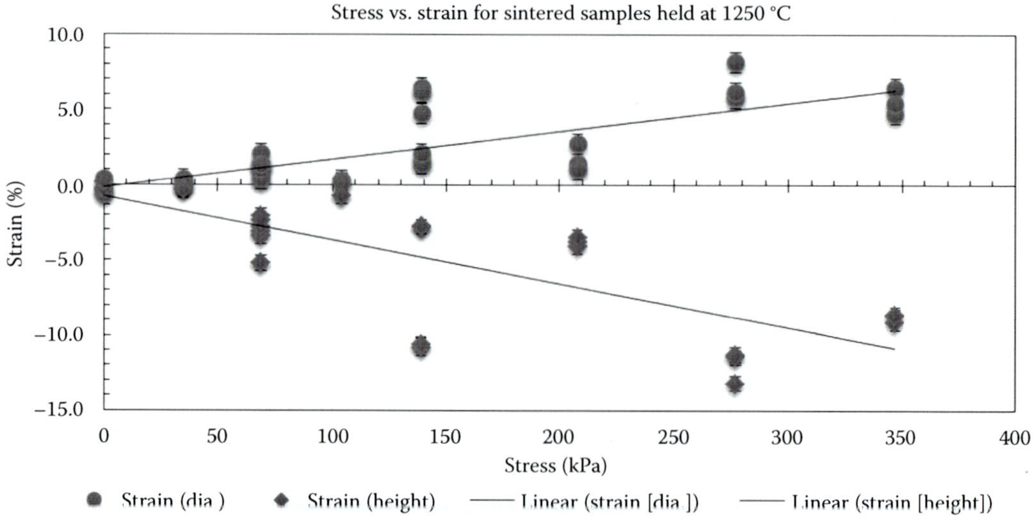

FIGURE 19.8 Summary of results for applied stress vs. strain plot for sintered alnico held for an additional 4 h at 1250°C.

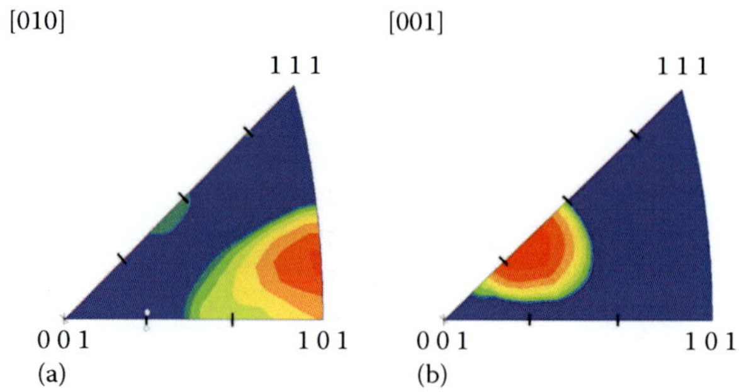

FIGURE 19.9 Orientation summary showing(a) inverse pole figure along the axial direction for 1248 kPa (900 g) loading (with 15° tick marks) and (b) inverse pole figure along the axial direction for 75 g loading (with 15° tick marks).

<001> orientation. The ~45° misalignment for cubic symmetry approaches a minimum for magnetic properties as indicated in studies by both Higuchi and Zhou, who show the angular relationship between crystal orientation and magnetic properties of alnico [48,49]. The resulting magnetic properties of the 1248 kPa (900 g) loading sample corresponded to an overall reduction in texture related parameters with a remanence of 8.41 kG, remanence ratio of 0.71, and energy product of 4.34 MGOe.

Conversely, samples loaded below ~250 g showed a clear improvement in texture (Figure 19.9b) as compared to the high stress/strain situation (Figure 19.9a). The low loaded sample of Figure 19.9b demonstrated enhanced orientation of nearly the [001] direction with improvements in all texture related properties: remanence of 9.01 kG, remanence ratio of 0.76, and an energy product of 5.02 MGOe. Figure 19.9b displays the measured tendency to align near the <115> direction, resulting in a texture close to the ideal orientation of [001]. A sufficiently low uniaxial stress causes a slightly anisotropic grain boundary energy that prefers anisotropic grain boundary mobility. This produces grain growth bias in this alnico cubic system with nearly isotropic grain boundary energy. Successive experiments have advanced these parameters even further to a remanence ratio of 0.79 and energy product of 5.65 MGOe.

From this work it is evident that two distinct textures can be achieved depending on the loading employed, which activates two different mechanisms: either grain rotation and growth that would require postprocess shaping/machining to utilize its preferred orientation, or grain boundary energy biased growth that could be used directly as a near-final shape magnet. The lower level of stress may be more applicable to simplified mass production methods. Follow-up work should be able to fully optimize the texture and should result in further increase of remanence ratio values, creating an improved energy product and square hysteresis loop shape as required by motor manufacturers.

19.2.1.4 Conclusions

Gas atomized prealloyed alnico powder was used to create a highly dense fine-grained sintered alnico microstructure as a starting condition for controlled abnormal grain growth processing, which can convert weak isotropic magnets into high energy permanent alnico magnets with a large grained-aligned microstructure. The described novel processing method utilizes uniaxial stressing of sintered alnico under a static load in a solutionized (BCC) phase field at very high-temperatures (0.97 $T_{solidus}$). Under these conditions, two rapid grain growth mechanisms are available to create different microstructural textures: either plastic deformation-induced grain rotation at high loads or, more ideally, grain boundary energy biased grain growth at low loads. An alnico alloy with a large-grained, clearly textured microstructure, approaching the 15° beneficial threshold of Durand-Charre, was demonstrated using this solid-state approach at low loads, with an improvement of 30% in energy product over the alternative highly stressed grain rotation mechanism.

19.2.2 Thermal Gradient Driven Case

The following example is offered as an alternative method to generate a well-aligned large-grained magnet from alnico magnet alloys that continues to use the "constant uniaxial stress approach for textured abnormal grain growth"; however, the constant uniaxial stress is provided by a temperature gradient due to the difference in thermal expansion coefficient, rather than a dead weight loading situation. Past work by Steinort and coworkers [12], Makino and coworkers [50], and Luborsky and coworkers [51] has shown that achieving a certain thermal gradient across a sample within a critical secondary recrystallization temperature regime can result in enough strain to result in abnormal grain growth.

19.2.2.1 Experimental Methods

The gas atomized, fine, spherical alnico alloy powders produced at Ames Laboratory were size-classified and then compression-molded using a QPAC®40 poly(propylene carbonate) binder (from Empower Materials Inc., New Castle, Delaware) into a 9.5 mm diameter, ~1 cm tall cylinder. The samples were debound and vacuum-sintered for 4 h to obtain >99% density. Kassen and coworkers and Anderson and coworkers include further details about the gas atomization process, resulting powder characteristics, and sintering procedures [42–44]. Each sintered sample then was placed in a tube furnace set to 1260°C and on a water-cooled, nickel superalloy cold finger to generate a >20°C/cm gradient through the height of the cylinder as shown in Figure 19.10. The furnace was purged with high purity nitrogen or argon gas to minimize oxidation, and two type K thermocouples monitored the temperature across the height of the sample. A polished disk of oriented alnico 9, provided by Arnold Magnetic Technologies, was placed in contact with the hotter end of the sample in the furnace to epitaxially seed the preferred grain growth direction. Electron backscatter diffraction analysis of the alnico 9 disk, verifying the texture, is included in Figure 19.11. A second sample was prediffusion bonded, using a molybdenum die with internal alumina insulation for 4 h at 1225°C, to an alnico 9 disk to (more effectively) epitaxially seed the abnormal grain growth grain orientation of the sintered alnico powders. Whereas the furnace control was not quite optimized for the first experiment, the second diffusion-bonded sample achieved 1250°C at the cold finger end with 1280°C at the furnace end of the sample, resulting in a 30°C/cm gradient within the critical recrystallization range.

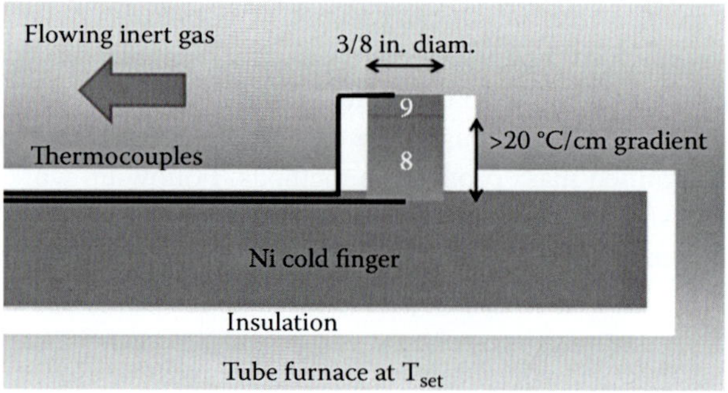

FIGURE 19.10 Cold finger experimental setup to achieve a thermal gradient.

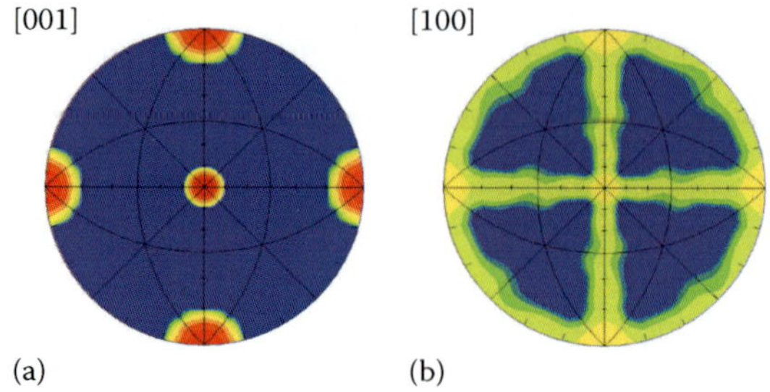

FIGURE 19.11 (a and b) Inverse pole figures along the [001] and [100] directions for the alnico 9 seed disk.

Cylinders of 3 mm (dia.) × 8 mm (height) were machined from the larger samples and underwent solutionizing, magnetic annealing and draw annealing heat treatments. The rods were heated to 1250°C for 30 min in an Oxy-Gon quench furnace (from OXY-GON Industries, Inc., Epsom, New Hampshire) with a vacuum of 10^{-6} to 10^{-7} torr, followed by an oil quench to solutionize the microstructure, to eliminate γ-phase precipitation, and to minimize spinodal decomposition of the high-temperature solid solution phase. After being wrapped in tantalum foil and sealed in a quartz vial, the rods then were magnetically annealed under a ~1T field at 840°C for 10 min. Finally, draw anneals were performed at 650°C for 5 h and 580°C for 15 h to achieve a full heat treatment. The magnetic properties were measured on a closed-loop Laboratorio Elettrofisico AMH-500 hysteresisgraph under a maximum applied field of 12 kOe. Scanning electron microscopy and electron backscatter diffraction analysis were performed on polished surfaces using an FEI Quanta field emission gun 250 scanning electron microscope, FEI Tecnai scanning electron microscope, and a JEOL JAMP-7830F field emission auger electron spectroscopy instrument, respectively.

19.2.2.2 Results and Discussion—Grain Growth

Scanning electron microscopy analysis of the polished transverse cross-sectional end surfaces of the initial experimental sample showed significant differences between the hot/alnico 9 seeded side and the end in contact with the cold finger. Figure 19.12a shows a montage scanning electron micrograph of the alnico 9 (seeded) end, while Figure 19.12b shows the cold finger end. The primary goal of this work was to achieve grain growth within the sample. It is apparent this was achieved, at least at the hot end (see Figure 19.12a) that was seeded with alnico 9, as evidenced by the ~2 mm size grains in the sample. The center of the hot end retained a fine grain size, indicating further time and/or higher temperatures might be necessary in order to achieve complete recrystallization. The cold finger end (see Figure 19.12b) showed mostly a fine-grained structure with two larger grains (see arrows).

Orientations of the recrystallized grains in Figure 19.12 were analyzed using electron backscatter diffraction and are shown in Figures 19.13 and 19.14 with overlays on the montage scanning electron

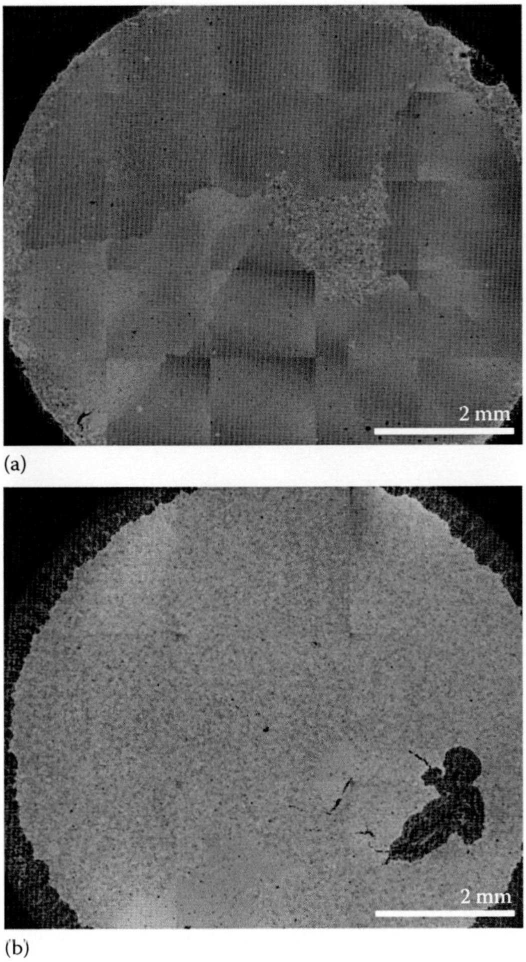

FIGURE 19.12 (a) Scanning electron montage micrograph of hot end that was seeded with an alnico 9 disk; (b) montage micrograph of cold end that was in contact with the cold finger.

micrographs. The fine-grain sections are clearly randomly oriented (see Figure 19.14), while a large portion of the recrystallized region of the alnico 9 (seeded) end has tended towards the preferred [001] orientation, as shown by the section labeled "d" in Figure 19.13.

The sectioned samples for magnetic property measurements were chosen from the areas shown in Figure 19.15 (corresponding to the sample shown in Figure 19.13), with three longitudinal cross sections from the labeled locations. Figure 19.16 shows a scanning electron micrograph of the polished "Long. 1" (longitudinal) cross section of the cylindrical sample where the grain size transition occurs over halfway through the height of the sample. Figure 19.17 includes electron backscatter diffraction overlays on the longitudinal cross section with the orientations of the grains correlating well with the transverse cross section.

The other longitudinal cross sections were examined and showed similar results and good correspondence with the transverse cross section. It is evident from these results that over 50% of the sample height and over 90% of the sample area within that height exhibited abnormal grain growth. Processing defects, such as the crack extending across the middle of the sample from the left side in Figure 19.16, seemed to be a barrier to continued grain growth, possibly due to both oxygen contamination and grain pinning effects.

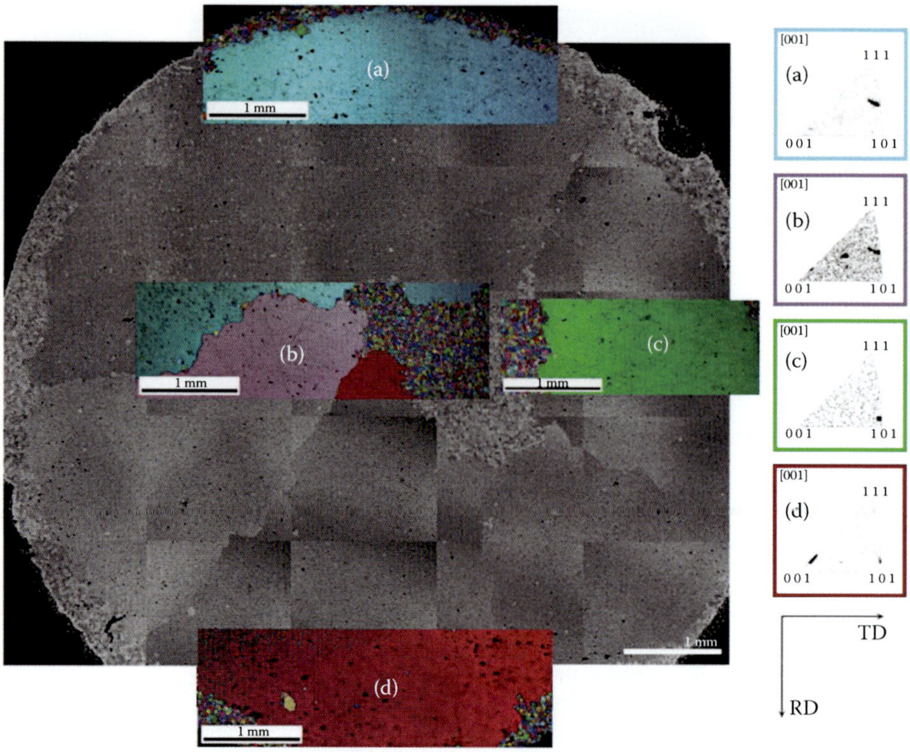

FIGURE 19.13 Electron backscatter diffraction grain texture overlaid on the montage micrograph with corresponding inverse pole figures showing orientation.

A fine-grain microstructure was evident around the circumference of the entire sample. Further examination of the edge of the sample using energy-dispersive x-ray spectroscopy mapping, included in Figure 19.18, showed significant oxidation of both Ti and Al at the edge of the sample. Oxygen contamination promotes γ-phase formation at the grain boundaries and, apparently, stabilization of the fine-grained microstructure.

Magnetic property measurements of the three cylinders machined from the bulk sample are included in Table 19.3. Comparison with Arnold Magnetic Technologies sintered alnico 8H and cast alnico 9 showed our samples' properties to lie between those of the two magnet types, as expected for the initial results of a sintered microstructure with some grain alignment. The values for B_r, H_c, H_{ci}, and BH_{max} of the pink, green, and teal cylinders bridge the gap between the Arnold sintered alnico 8H and cast alnico 9. This initial work showed the potential for improvement in B_r, H_c, H_{ci}, and BH_{max} of the sintered samples by further grain growth, texture, and compositional adjustments through improving the furnace atmosphere, optimizing the temperature, and the time to achieve full grain growth, as well as the preferred texture within the sintered alnico magnet.

Scanning electron micrographs (Figure 19.19) of the second sample showed complete abnormal grain growth throughout the width and height of the sample. Properties for these specimens started to show improved values over typical isotropic alnico, indicating that enhanced texturing likely occurred in these specimens, as shown in Table 19.4. The second set of samples specifically showed remanence ratios as high as 0.76. These values are clearly higher than typical values for specimens of isotropic sintered alnico (0.72). It also is true that the initial samples had good squareness and remanence/saturation ratio values, compared to their sintered counterparts, showing the beginnings of developing an underlying texture. An even greater remanence/saturation ratio would be expected as the grain orientation of the sample approaches the ideal <001> texture; fully realizing this

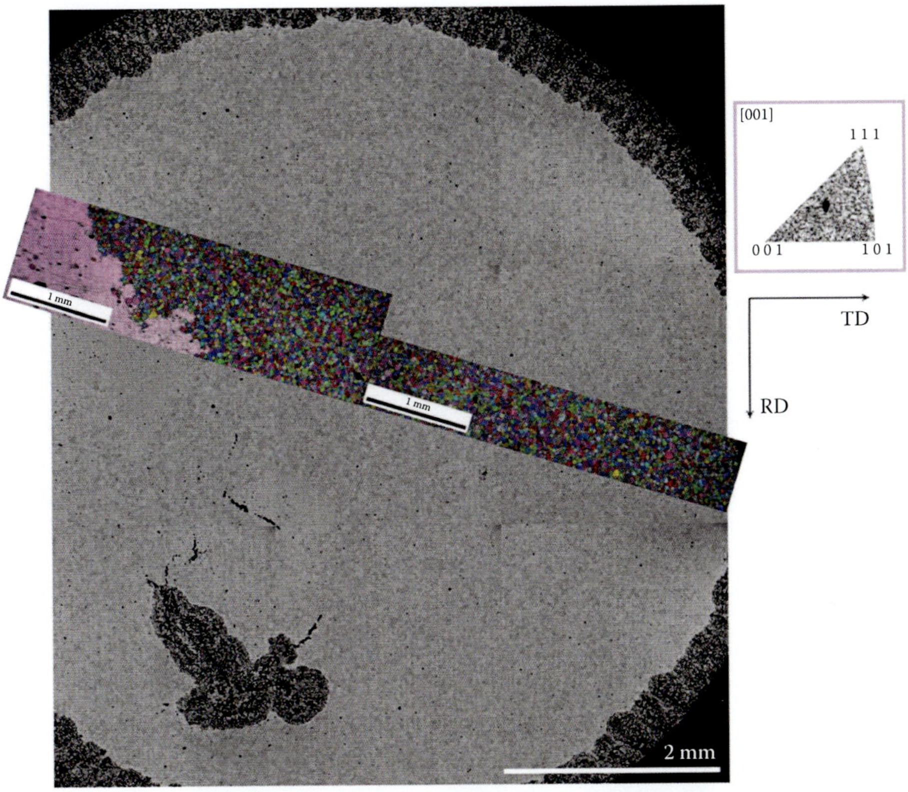

FIGURE 19.14 Electron backscatter diffraction grain texture overlaid on the montage micrograph with corresponding inverse pole figure showing orientation of the lower left rectangular area.

improved texture should grant an even further improvement in energy product without impacting (decreasing) coercivity.

19.2.2.3 Conclusions

Grain growth and some texturing control have been demonstrated through solid-state thermal processing of a fully dense powder processed sample of the modified alnico 8 composition. Diffusion bonding of directionally grown alnico 9 shows promise for being able to control grain orientation and epitaxially seed the [001] direction for grain growth. The γ-phase appears to stabilize the grain size and random orientation and must be minimized along with oxidation of the sample. These thermal gradient samples show improved energy products over other sintered commercially available alnico 8 magnets.

19.3 SUMMARY AND CONCLUSIONS

In summary, great progress has been made in improving alnico permanent magnet properties (both magnetic and mechanical) from their initial conception and development from the 1930s into the 1960s. Modern characterization methods have elucidated many features of the nanostructured alloy, which have enabled greater understanding of the alnico alloys and further processing and composition optimization. Additionally, the advent of advanced processing techniques, including solid-state stress-driven abnormal grain growth with a controlled orientation, shows promise for improving the magnetic properties even further towards their theoretical maximums, with the potential for broadened applications, especially at high temperatures, i.e., traction drive motors. Alnico alloys are

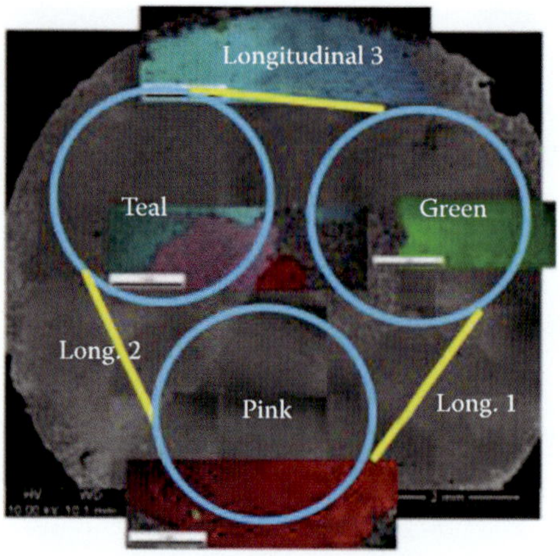

FIGURE 19.15 Rough locations of the three cylinders that were cut for magnetic property measurements, as well as the locations for the longitudinal cross sections that were examined microstructurally.

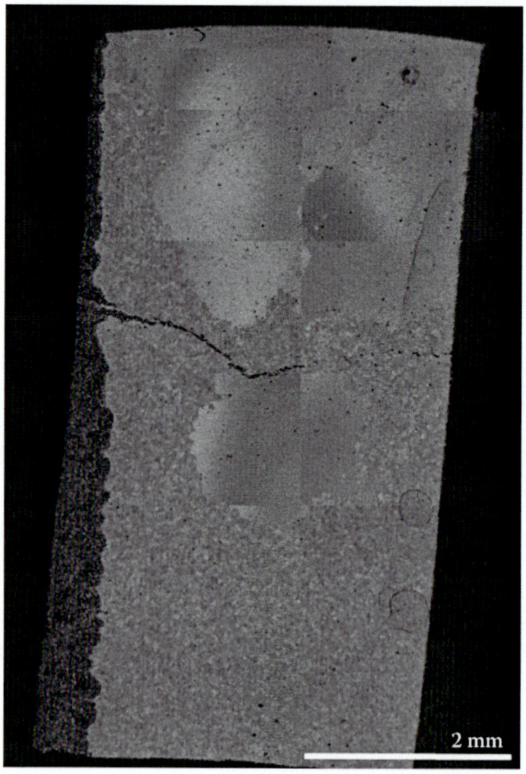

FIGURE 19.16 Scanning electron montage micrograph of the "Long. 1" (longitudinal) cross section with the hot end at the top of the image and the cold finger end at the bottom. Note that ~1 mm has already been cut off both ends for the transverse cross sections.

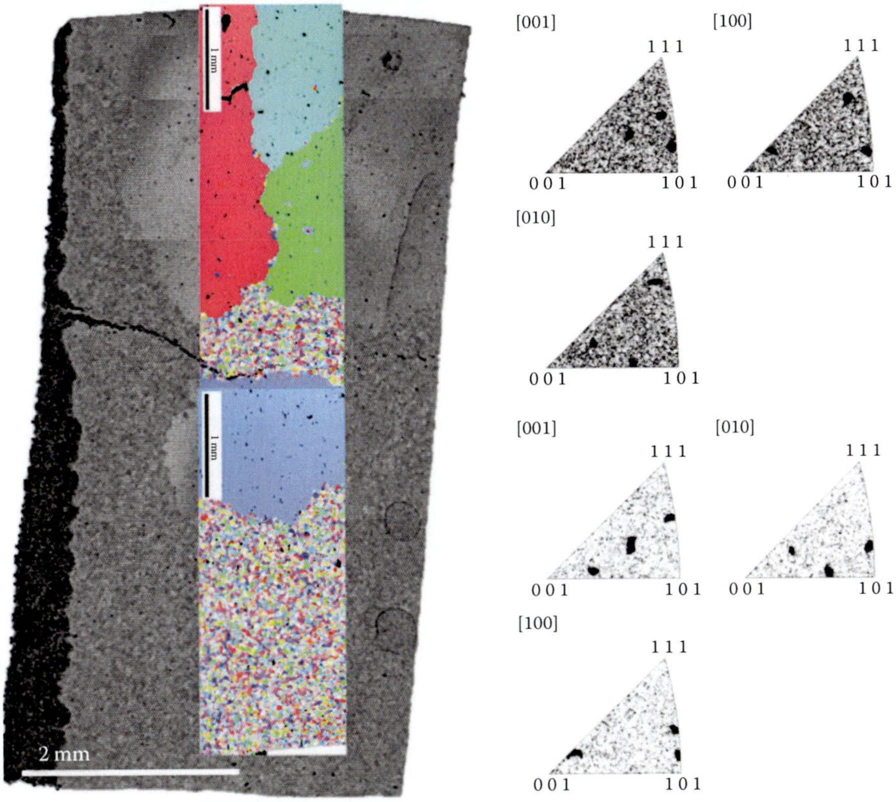

FIGURE 19.17 Electron backscatter diffraction grain texture overlaid on the montage micrograph of longitudinal section 1 ("Long. 1") with corresponding inverse pole figures showing orientations of large grains and randomly oriented fine grains.

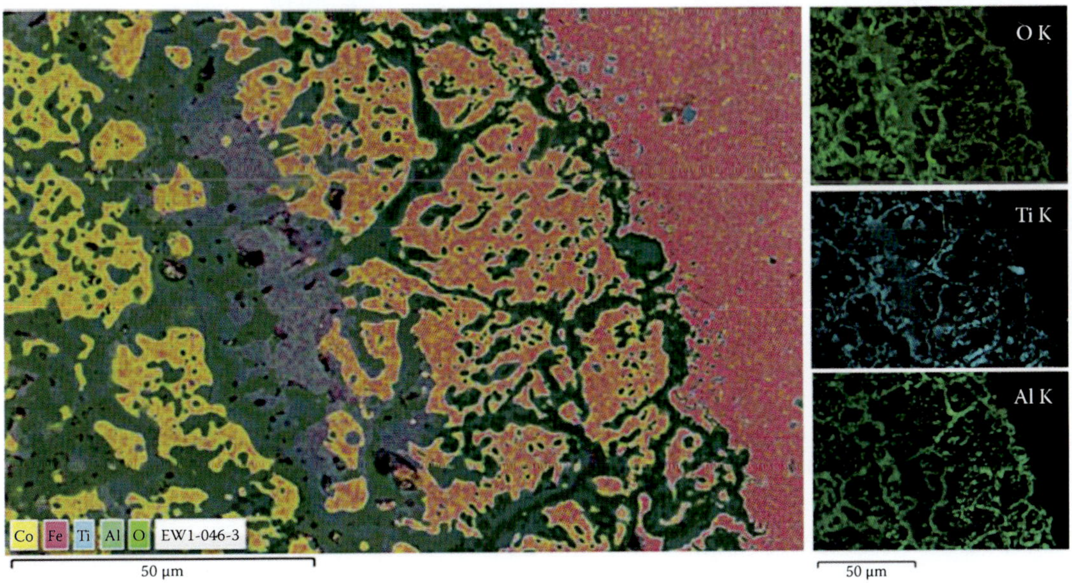

FIGURE 19.18 Energy-dispersive x-ray spectroscopy map of select elements along the edge of longitudinal section 1 ("Long. 1").

TABLE 19.3
Magnetic Property Measurements of Thermal Gradient Grain Growth Samples Compared to Arnold Magnetic Technologies (AMT) Commercial Magnet Properties

Sample ID	B_r (kG)	H_c (Oe)	H_k (Oe)	H_{ci} (Oe)	BH_{max} (MGOe)	Squareness
Pink	8.8	1580	435	1701	4.89	0.26
Green	8.6	1565	420	1705	4.67	0.25
Teal	8.0	1527	364	1672	4.15	0.22
AMT sintered alnico 8H	6.7	1800	–	2020	4.5	–
AMT cast alnico 9	10.6	1500	–	1500	9.0	–

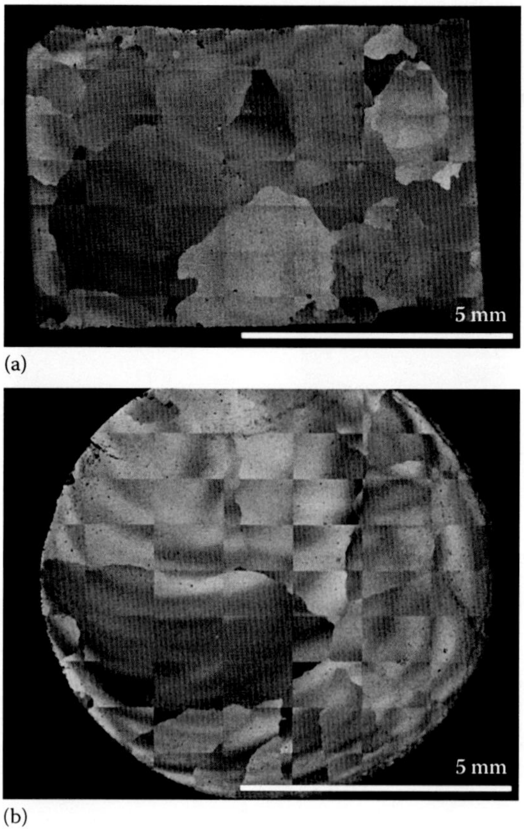

FIGURE 19.19 Scanning electron micrographs of the second thermal gradient sample: (a) longitudinal cross section and (b) transverse cross section.

one part of an important complex puzzle to solve the issue of limited supplies and higher costs of rare earth permanent magnets for critical technologies.

ACKNOWLEDGMENTS

Steve Constantinides is gratefully acknowledged for providing the significant text for this chapter as well as the alnico 9 samples (from Arnold Magnetic Technologies) for these experiments and helpful discussions throughout the course of this research. The authors would like to thank Rick Schmidt

TABLE 19.4
Magnetic Property Measurements of Thermal Gradient Grain Growth Samples Compared to Magnetic Materials Producers Association Sintered Alnico 8

Sample	M_r (kG)	BH_{max}	H_{ci} (Oe)	M_r/M_s
First-1	8.8	4.89	1701	0.73
First-2	8.6	4.67	1705	0.71
First-3	8.0	4.15	1672	0.69
Second-1	8.9	5.3	1580	0.76
Second-2	8.3	4.5	1580	0.72
Sintered alnico 8	6.7	4.5	2020	0.72

for assistance with the compression molding and Wei Tang for composition and processing assistance. Dave Byrd and Ross Anderson performed the gas atomization experiments and provided powder-screening expertise. Jim Anderegg provided operation of the auger electron spectroscopy instrument and some electron backscatter diffraction data. Matt Kramer, Lin Zhou, and Liangfa Hu afforded valuable informative discussions during the course of this research. This work was supported by the U.S. Department of Energy (DOE), Energy Efficiency and Renewable Energy Office (EERE), Vehicle Technologies (VT), Electric Drive Train (EDT) program. This research was performed at Ames Laboratory, which is operated for the U.S. DOE by Iowa State University under contract no. DE-AC02-07CH11358.

REFERENCES

1. D. Brown, B.-M. Ma, and Z. Chen, Developments in the processing and properties of NdFeb-type permanent magnets, *J. Magn. Magn. Mater.*, 248, 432–440 (2002).
2. G. Bai, R.W. Gao, Y. Sun, G.B. Han, and B. Wang, Study of high-coercivity sintered NdFeB magnets, *J. Magn. Magn. Mater.*, 308, 20–23 (2007).
3. O. Gutfleisch, A. Bollero, A. Handstein, D. Hinz, A. Kirchner, A. Yan, K.-H. Muller, and L. Schultz, Nanocrystalline high performance permanent magnets, *J. Magn. Magn. Mater.*, 242–245, 1277–1283 (2002).
4. D.W. Scott, B.M. Ma, Y.L. Liang, and C.O. Bounds, Microstructural control of NdFeB cast ingots for achieving 50 MGOe sintered magnets, *J. Appl. Phys.*, 79, 4830–4832 (1996).
5. *DOE Critical Materials Strategy Report* 2011.
6. M.J. Kramer, R.W. McCallum, I.E. Anderson, and S. Constantinides, Prospects for Non-Rare Earth Permanent Magnets for Traction Motors and Generators, *JOM*, 64, 752–763 (2012).
7. R. Skomski, P. Manchanda, P. Kumar, B. Balamurugan, A. Kashyap, and D. Sellmyer, Predicting the Future of Permanent-Magnet Materials, *J. IEEE Trans. Magn.*, 49, 3215–3220 (2013).
8. M.S. Walmer, C.H. Chen, and M.H. Walmer, A new class of Sm-TM magnets for operating temperatures up to 550/spl deg/C, *IEEE Trans. Magn.*, 36, 3376–3381 (2000).
9. O. Gutfleisch, M.A. Willard, E. Bruck, C.H. Chen, S.G. Sankar, and J. Ping Liu, Magnetic materials and devices for the 21st century: stronger, lighter, and more energy efficient, *Adv. Mater.*, 23, 821–842 (2011).
10. T. Mishima, Nickel-aluminum steel for permanent magnets, *Ohm*, 19, 353 (1932).
11. T. Mishima, British Patent 392661 (Appl. 1931), U.S. Patents 2027996 and 2027994 (1936).
12. E. Steinort, E.R. Cronk, S.J. Garvin, and H. Tiderman, Formation of monocrystalline alnico magnets by secondary recrystallization methods, *J. Appl. Phys*, suppl. to 33, 1, 1310–1313 (March 1962).
13. B.G. Livshits, V.S. L'vov, *Vysokokoertsitivnye splavy na zhelezonikel'aliuminievoi asmove* (Trans.: High-coercivity alloys on a nickel–iron–aluminum base), Moscow (1960).
14. W. Betteridge, Nickel–aluminium permanent magnet alloys, *J. Iron Steel Inst.* 139, 187–208 (1939).
15. C. Heck, *Magnetic Materials and Their Applications*, Crane, Russak & Company, Inc. (1974); pp. 271–272.

16. W. Zumbusch, Die Entwicklung der kupfer- und/oder titanhaltigen Alnico-Magnete zum heutigen Gütestand (Trans.: Development of copper and/or titanium-containing AlNiCo magnets to their present quality), *Z. Metall.*, 49, 1, 1–8 (1958).
17. J.E. Gould, *Cobalt Alloy Permanent Magnets*, *Centre d'Information du Cobalt s.a*, Brussels (1971).
18. L. Zhou, M.K. Miller, P. Lu, L. Ke, R. Skomski, H. Dillon, Q. Xing, A. Palasyuk, M.R. McCartney, D.J. Smith, S. Constantinides, R.W. McCallum, I.E. Anderson, V. Antropov, and M.J. Kramer, Architecture and magnetism of alnico, *Acta Mater.*, 74, 224–233 (2014).
19. L. Zhou, W. Tang, W. Guo, J.D. Poplawsky, I.E. Anderson, and M. Kramer, Spinodal Decomposition in an Alnico Alloy, *Microsc. Microanal.*, 22 (Suppl. 3), 670–671 (2016).
20. C. Kittel, E.A. Nesbitt, and W. Schokley, Theory of Magnetic Properties and Nucleation in Alnico V, *Phys. Rev.*, 77, 839–840 (1950).
21. R.B. Campbell and C.A. Julien, Structure of Alnico V, *J. Appl. Phys.*, 32, 192S–194S (1961).
22. C.A. Julien and F.G. Jones, Alpha-Sub-Gamma Phase in Alnico 8 Alloys, *J. Appl. Phys.*, 36, 1173–1174 (1965).
23. E.A. Nesbitt and H.J. Williams, Mechanism of Magnetization in Alnico V, *Phys. Rev.*, 80, 112–114 (1950).
24. R.A. McCurrie, *Ferromagnetic Materials 3*, North Holland, Amsterdam (1982).
25. E. Cronk, Recent Developments in High-Energy Alnico Alloys, *J. Appl. Phys.*, 37, 1097–1100 (1966).
26. V. Sergeyev and T.Y. Bulygina, Magnetic properties of alnico 5 and alnico 8 phases at the sequential stages of heat treatment in a field, *IEEE Trans. Magn.*, mag-6, 194–198 (1970).
27. R.J. Studders and D.G. Eberling, Patent US2617723, *Sintered High Energy Permanent Magnets* (November 1952).
28. W. Hotop, The special position of powder-metallurgy in the production of permanent-magnets, *Metall.*, 7, 1–9 (1953).
29. C. Song, B. Han, and Y. Li, Study on alnico permanent magnet powders prepared by atomization, *J. Mater. Sci. Technol.*, 20, 3, 347–349 (2004).
30. H. Böhnert, Transport of material by sintering alnico magnets, *Physics*, 80B, 177–198 (1975).
31. D.A. Oliver and J. Shedden, Cooling of Permanent Magnet Alloys in a Constant Magnetic Field, *Nature*, 142, 209 (1938).
32. Q. Xing, M.K. Miller, L. Zhou, H.M. Dillon, R.w. McCallum, I.E. Anderson, S. Constantinides, and M.J. Kramer, Phase and elemental distributions in alnico magnetic materials, *IEEE Trans. Magn.*, 49, 3314–3317 (2013).
33. H. Fahlenbracht, Beitraag zu den magnetischen Eigenschaften und der Verwendbarkeit pulvermetallurgisch hergestellter Dauermagnete (Trans: Contribution to the magnetic properties and the availability of powder-metallurgically produced permanent magnets), *Arch. Eisenhüttenwerk*, 20, 301–304 (1949).
34. E. Planchard and C. Bronner, Improvement to the preparation of sintered Al–Ni–Co alloys, *IEEE Trans. Mag.*, 6, 304 (1970).
35. Y.L. Sun, J.T. Zhao, Z. Liu, W.X. Xia, S.M. Zhu, D. Lee, and A.R. Yan, The phase and microstructure analysis of Alnico magnets with high coercivity, *J. Magn. Magn. Mater.*, 379, 58–62 (2015).
36. M. Durand-Charre, C. Bronner, and J.-P. Lagarde, Relation between magnetic properties and crystallographic texture of columnar Alnico 8 permanent magnets, *IEEE Trans. Mag.*, 14 797–799 (1978).
37. T. Liu, W. Li, M. Zhu, Z. Guo, and Y. Li, Effect of Co on the thermal stability and magnetic properties of AlNiCo 8 alloys, *J. Appl. Phys.*, 115, 17A751 (2014).
38. S. Chikazumi, *Physics of Ferromagnetism*, Sōshin Chikazumi. English edition prepared with the assistance of C.D. Graham, Jr. 2nd ed., Oxford University Press, Inc. (1999).
39. H.J. Bunge, Texture and Magnetic Properties, *Textures Microstruct.*, 11, 75–91 (1989).
40. H.M. Dillon, Effects of heat treatment and processing modifications on microstructure in alnico 8H permanent magnet alloys for high temperature applications, M.S. thesis, Iowa State University, Ames, IA (2014).
41. H.A. Kuhn, A. Lawley, *Powder Metallurgy Processing*, Academic Press, Inc., New York (1978), pp. 13–15.
42. A.G. Kassen, E.M.H. White, A. Palasyuk, L. Zhou, R.W. McCallum, and I.E. Anderson, Compression Molding of High-Pressure Gas Atomized Powders to Form Near-Net Shape Alnico-Based Permanent Magnets, *Adv. Powder Met. Particulate Mater.*, 07–13-07–27 (2015).
43. I.E. Anderson, A.G. Kassen, E.M.H. White, L. Zhou, W. Tang, A. Palasyuk, K.W. Dennis, R.W. McCallum, and M.J. Kramer, Novel pre-alloyed powder processing of modified alnico 8: Correlation of microstructure and magnetic properties, *J. Appl. Phys.*, 117, 17D138-(1–4) (2015).

44. I.E. Anderson, E.M.H. White, M.J. Kramer, A.G. Kassen, and K.W. Dennis, U.S. Utility Patent 62/ 390,513. March 31, 2016.
45. I.E. Anderson, D. Byrd, and J. Meyer, Highly tuned gas atomization for controlled preparation of coarse powder, *Materialwiss. Werkstofftech.*, 41, 504–512 (2010).
46. S. Constantinides, Aaron Kassen (ed.) (2014). Personal Communication.
47. W.F. Hosford, *Mechanical Behavior of Materials*. Cambridge University Press, New York (2009), pp. 120–125.
48. A. Higuchi and T. Miyamoto, Some relationships between crystal textures and magnetic properties of alnico 8, *IEEE Trans. Mag.*, MAG6, 218–220 (1970).
49. L. Zhou, M.K. Miller, H. Dillon, A. Palasyuk, S. Constantinides, R.W. McCallum, I.E. Anderson, and M.J. Kramer, Role of the Applied Magnetic Field on the Microstructural Evolution in Alnico 8 Alloys, *Metall. Mater. Trans. E*, 1, 27–35 (2014).
50. N. Makino and Y. Kimura, Techniques to Achieve Texture in Permanent Magnet Alloy Systems, *J. Appl. Phys.*, 36, 1185–1190 (1965).
51. F.E. Luborsky and K.T. Grain Growth in Some Alnico Alloys, Aust. *Trans. Met. Soc. AIME*, 227, 791–793 (1963).

20 Techniques, Trends, and Advances in Conventional Machining Practices for Metals and Composite Materials

Ramanathan Arunachalam, Sathish Kannan, and Sayyad Zahid Qamar

CONTENTS

20.1 INTRODUCTION

With advancements in additive manufacturing technology like 3-D printing, machining being a subtractive process has been dealt a serious blow. For example, recently the Federal Aviation Authority approved 3-D printed parts for the Boeing 787 aircraft (Moon 2017). Although subtractive processes cannot be wiped out completely from the manufacturing industry, they need to be competitive with additive or other net shape manufacturing techniques. Even additive manufacturing parts, depending on printing speed, may require a postprocessing operation like machining. Similarly, net shape manufactured or additive manufacturing components need holes for assembly of components, and drilling is the most economical machining technique that can be easily employed to produce them. This is the reason why drilling is considered to be an important machining process and accounts for a major fraction of all the machining processes (Mathew and Vijayaraghavan 2016). The market distribution for milling, turning, drilling, and other machining processes are 39%, 30%, 17% and 14%, respectively, but at the expected growth rate drilling is expected to become the most common (Bobzin 2016). Conventional machining processes can be competitive in the near future only if advances or the latest technologies are implemented. One of the enabling technologies that can render conventional machining competitive with additive manufacturing is high speed machining using multi-axis (such as 5-axis) machining centers. Recently, high speed machining technology has been extended to prototypes in aluminium, titanium and copper for the aerospace, electric/electronic, biomedical, and defense industries. They have found that high-speed machining technology can compete favorably with additive manufacturing techniques for the production of hardened steel molds/dies for net-shape manufacture of forgings, plastic injection moldings, and even die-castings. Not only for prototypes but also several critical components for the Airbus A320 aircraft and the new Airbus A380 aircraft have been successfully produced by machining of solid titanium blocks using high-speed machining centers equipped with special tooling systems (Grzesik 2017). With developments in machine tool technology, the prices of such machining centers are easily affordable. For example, the Haas UMC-750 vertical universal Computer Numerical Control (CNC) machine, is capable of 8000 rpm, is available at a price starting from $163,995 (Haas Automation 2017a). Sebastian and coworkers (Sebastian et al. 2016) compared additive manufacturing (using powder bed fusion) and high-speed machining processes in terms of cost for eight parts. Based on their studies, they concluded that additive manufacturing is more expensive than high-speed machining especially when the additive manufacturing parts need postprocessing. Parts produced using additive manufacturing can be justified only if the required shape is complex. The machine tools market in the United States has been forecast to grow at a compound annual growth rate of 4.34% during the 2017–2021 period according to Wise Guy Reports (MaschinenMarkt International 2017). This is a good indicator of the future business potential for the machine tool industry. With this scenario, it is quintessential to advance conventional machining processes to become highly competitive.

With this in mind, this chapter focuses on the advances that are promising to make the conventional machining techniques a viable alternative to the manufacturing industries. For this to happen, machine shops need to take advantage of the advanced technologies presently available, although they may be initially expensive. A study by Makino (Clark 2014) revealed how investing in a best technical solution, like high-performance machining centers, can yield a high return on investment instead of the lower-cost options.

As mentioned earlier, postprocessing serves to increase the production cost and needs to be minimized or eliminated. In conventional machining, especially with high-speed machining technology, it is possible to minimize postprocessing. Hard part machining offers this advantage by making use of the latest cutting tools like cubic boron nitride, poly crystalline diamond and coated carbides to avoid subsequent grinding processes (Bhemuni et al. 2015). The main factor that affects the machining processes is machinability of the material employed. The various factors that contribute to the machinability of materials are summarized in Figure 20.1.

20.2 CONVENTIONAL MACHINING OF ENGINEERING AND ENGINEERED MATERIALS

This section reviews the machining issues with some of the common materials, both engineering and engineered, used in two major industrial sectors, such as aerospace and automotive. These materials represent the major engineering materials and engineered materials chosen for use in industrial components.

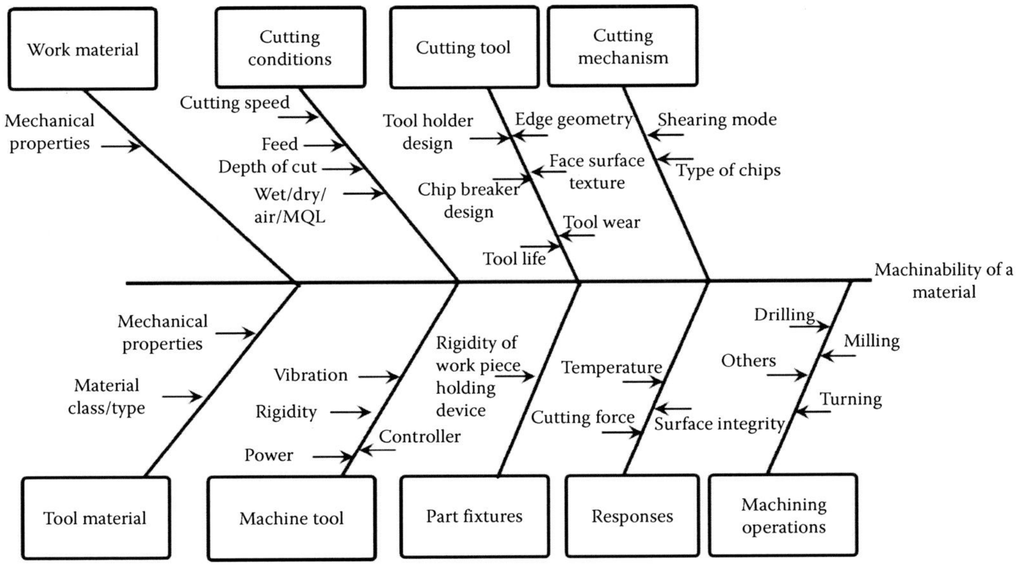

FIGURE 20.1 Factors affecting machinability of materials.

20.2.1 Aerospace

Over the past 55 years, the aeronautic industry has successfully risen to become a leading industry in the world. During this time, the share of this market has increased steadily and demand for air travel has grown significantly. Currently, there are thousands of aeronautical companies in Europe alone supporting approximately 450,000 skilled jobs that generate a turnover in excess of €110 billion (European Commission 2011). The Advisory Council for Aviation Research in Europe has published their "Flightpath 2050" long-term strategic objectives to be as follows (Flightpath 2050, European Commission 2011): (a) 75% reduction in carbon dioxide per passenger kilometer, (b) 90% reduction in greenhouse gas emissions, and (c) 65% reduction in noise. This will require major cost-efficient measures in the following four areas: (1) advanced manufacturing technologies for airframes and landing gear (lightweight designs and materials e.g., carbon fiber composites, etc.), (2) systems (lean burn, higher engine bypass ratio) and equipment, (3) lightweight aeroengine structures (large static casings, rotor discs, rotor blades and vanes), and (4) advanced aeroengine materials (ceramic matrix composites, advanced nickel based, and titanium-based super alloys, etc.).

For an improved aerodynamic performance, the airframe to include the wings, tail section, and aeroengine components result in complex geometries that need to be machined. This necessitates the need for design of advanced component profile shapes, which creates challenges. In the last few years, research activities have focused on advanced lightweight metallic alloys, high-temperature alloys, and even composite materials addressing weight and manufacturing challenges (e.g., free-form machining), to achieve the desired aerodynamic performance in the area of fuel efficiency.

In this context, the present chapter provides an overview of developments in the area of machining of advanced aerospace alloys (e.g., nickel, titanium, and aluminium) and novel composites such as (a) carbon reinforced polymer composites, (b) ceramic matrix composites, and (c) carbon/titanium stacks, etc. used for airframe structures and aeroengine components. A critical assessment of machining behavior, tool wear, and surface integrity is presented herein as well. Further, advances in high-performance machining technologies are reviewed in detail. Some of the key engineering materials used in aircraft frame, aircraft engine, and related components such as landing gear and their machinability issues are summarized in Table 20.1.

TABLE 20.1
Commonly Used Materials in Aerospace Industries and Their Key Machinability

Alloys and Composites	Aerospace Components	Commonly Used Materials	Machining Improvement Areas	Key References
Nickel based super alloys	Turbine disc, compressor disc, blisk-integrally bladed rotors, shafts, turbine and compressor blades and vanes, large static casing and structures	Astroloy, Hastelloy®, B/C/C-276/X, INCONEL® 601/617/625/700/706/718,IN100,Incoloy® 901, RR-X (Rolls-Royce), FGH95, INCONEL® 718, Allvac™718 plus1 Aramid, CMSX-x, MAR-M200, Nimonic®, Rene 41, Udimet®, Waspaloy®	Depth of cut notch, built-up edge, flank wear, crater wear, torn workpiece surface finish, workpiece glazing, surface quality, pick up, smearing, laps, crack and micro voids, residual stress relaxation, geometrical accuracy, tool chipping, white layer, form accuracy, bore accuracy, swash, run out.	Turning [Polvorosa et al. (2017), Sharman et al. (2015), Cantero et al. (2013), Thakur et al. (2016), S.L. Soo et al. (2016)] Millig [Shokrani (2012), Aramcharoen (2014)] Drilling [Soo et al. (2011), Herbert et al. (2012)] Broaching [Klocke et al. (2013), Chen et al. (2014)] Tearing surface [Rahman et al. (2013), Razak et al. (2014)]
Cobalt-based, heat-resistant alloy (150–425 HB) (≤45 HRC)	Gas turbine hot end components, combustor parts, aerofoils-vanes, blades, static casings and structures	Wrought: AiResist 213, Haynes 25 (L605), Haynes 188, J-1570, Stellite Cast: AiResist 13, Haynes 21, Mar-M302, Mar-M509, Nasa C0-W-Re, WI- 52		Surface cavities/crack/micro voids [Chang et al. (2015), Zhou et al. (2012)] Grooves [Axinte et el. (2006), Zhou et al. (2011)]
Iron-based, heat-resistant alloys (135–320 HB) (≤34 HRC)	Turbine disc, blades, shafts, rings, and static casing and structures for gas turbines, flap tracks	Wrought: A-286, Discaloy, Incoloy® 801, N-155, 16-25-6, 19-9 DL Cast: ASTM A297, A351, A608, A567		White layer [Zhou et al. (2011), Bushlya et al. (2011)] Residual stresses [Ezilarasan (2013)] Plucking [Rahman et al. (2013), Razak et al. (2014)] Laps/redeposited material [Axinte et al. (2006))

(Continued)

TABLE 20.1 (CONTINUED)
Commonly Used Materials in Aerospace Industries and Their Key Machinability

Alloys and Composites	Aerospace Components	Commonly Used Materials	Machining Improvement Areas	Key References
Titanium and titanium alloys (110–450 HB) (≤48 HRC)	Compressor disc, blisk-integrally bladed rotors, compressor blades and vanes, fan casing, underwing fittings, center wing box main frames, sub-structure for anti-crash frames, landing gear application, engine pylons, engine mounts	Pure: Ti98.8, Ti99.9, Ti-5553, TC21, Ti-6-4, g-TiAl alloy (Ti–48A1–2Cr–2Nb), Ti-6242s, Ti-6246, Ti-834, Ti-4.5-2-2 XD, Ti-4.5A1–4.5Mn, Ti-6-6-2, TA-48, Ti–6A1–7nb, Ti3A18V6Cr4Mo4Zr, Ti4A12Sn4Zr2Mo, Ti10V2Fe 3A1, Ti13V11Cr3A1, Ti5A15Mo5V3Cr Ti5A12.5Sn, Ti8A11Mo1V, Ti10V2F23A1 Gamma-titanium Aluminide	High-chemical reactivity causes the chip to weld to the tool, low thermal conductivity, high tool tip temperature and excessive plastic deformation wear, higher cutting forces, depth-of-cut notch, premature failure of the cutting tool due to high stresses, low Modulus of Elasticity (Young's Modulus), greater workpiece spring back and deflection of thin-walled structures, vibration, chatter and poor surface finish, produces abrasive, tough, and stringy chips.	Tool wear, Surface Quality & integrity issues Corduan et al. (2003), Hughes et al. (2004), Ezugwu et al. (2005), Ginting et al. (2009), Ugarte et al. (2012) Da Silva et al. (2013), R.M'Saoubi et al. (2015) O.Hatt et al. (2017)
Aluminium based alloys	Honey comb structures, wing spars/ribs, hydraulic and pneumatics parts, upper-wing structure and fuselage skin	7055-T7751 plate, 7055-T77511, 7150-T7751, Alclad 2524-T3 sheets, 2524-T351 plates, 7010/7050-T7651 and A12024-T351, 6061, AIRWARETM 2050-T84, 7075, Aluminium-Lithium (A1-Li) alloys such as Alcoa 2099, Scalmalloy-1	Metal removal rates, productivity improvement, tool run out, high cutting loads, thin walled cross section, tearing or compression of the material.	Wang et al. (2016), Nouari et al. (2003), Abdelhafeez et al. (2016), Giasin et al. (2016), Teimouri et al. (2017)
Composites-composite metal stakcs, fiber metal laminates	Large static casing and structures-compressor/turbine/fan casing and structures. turbine blades, combustor parts	Polymers: Epoxy, Phenolic, Polyimide, Polyetheretherketone (PEEK) Fibers: Carbon fiber/Graphite fiber (high strength or high modulus) – Glass fibers – Ceramic fibers – Polymer fibers (Kevlar, Polyethylene) – Tungsten fibers, CFRP/carbon-fiber reinforced polymers, FRP/Ti stakcs, CFRC-SiC-SiC ceramic fiber composites, copper mesh/CFRP/woven ply, Ti6A14V/U.D CFRP/A17050-T651, CFRP/A17010-T7451, TI6A14V/U.D-CFRP/A17050,GLARE 3-3/2-0.3, Glass aluminium reinforced epoxy (GLARE) Titanium fiber metal composites	Matrix cratering, delamination, fuzzing micro crack, fiber/matrix debonding, spalling, fiber pullout, fiber breaking, resin loss, surface cavities, thermal alteration, discoloration ring, damage ring, delamination, hole size error, roundness error, position error, surface drag, burr, cracking, feed marks, tearing surface, debris of microchips, surface plucking, deformed grains, surface cavities, toll edge rounding, flank wear, micro chipping.	Soussia et al. (2014), Xu et al. (2016), Ramulu et al. (2001), Park et al. (2011), Sreejith et al. (2000), Sharif & Rahim (2007), A. Faraz et al. (2009) R. M'Saoubi et al. (2015)

20.2.2 Automotive

The global sales of passenger cars is expected to hit 78 million units in 2017 (Statistica 2017). In addition to this several thousands of buses, trucks, and vans are being produced. The automotive industry has registered a growth of almost 30% over the period 1995–2005 (International Organization of Motor Vehicle Manufacturers 2017). In the United States, the automobile industry is responsible for 3% of the Gross Domestic Product (International Organization of Motor Vehicle Manufacturers 2017). Hence, the automotive industry is considered to be one of the major industrial sectors which contributes significantly to the high growth of individual countries, and the global economy.

In any automobile, the engine and the power train constitute heart of the system and contribute to vehicle weight. Most of the engine components, such as engine blocks, cylinder liners, connecting rods, crankshafts, camshafts, etc., are now increasingly made with engineering materials, such as aluminum, magnesium, titanium, and their alloy counterparts. Engineered materials such as metal matrix composites as well as fiber-reinforced plastics, are also finding increasing use in the automobile industry. Currently, among the several engineering and engineered materials used in automobiles, the major components, especially engine and transmission systems, are made of high-strength aluminium alloys including aluminum-based metal–matrix composites (McCune and Weber 2001) and other lightweight materials, such as magnesium and titanium, to a lesser degree. Engine and transmission system parts involve finish machining and drilling subsequent to the primary manufacturing process. High-performance plastic materials, such as polypropylene, polyurethane, and polyvinyl chloride are also being increasingly chosen for applications in interiors, bodies, and exteriors systems, such as bumpers, dashboard, body panels, interior and exterior trims, and tanks for oil, fuel and others (Patil et al. 2017). Glass or carbon fiber reinforced plastics are also being increasingly chosen for use for purpose of light-weighting of high performance cars.

Most of these components are manufactured through net-shape manufacturing processes, such as casting, forging, molding, and others. With net-shape manufacturing processes gaining importance in the context of lower manufacturing costs and scrap, it appears that conventional machining might have a lessened scope in the future. However, this is not the case and even components produced using net-shape processes do require some finish machining as well as drilling of holes or slots for purpose of assembly. Most of the advancements in conventional machining have focused on improving the efficiency of the machining processes. Some of the key materials chosen for use in the automotive industry to produce various components, and their key machinability issues are summarized in Table 20.2. Materials, such as titanium, although not extensively chosen for use in automotive applications have been covered in the aerospace alloys section. However, the aluminium alloys and composite materials, which have been discussed in the aerospace section, are repeated here but from an automotive perspective.

20.3 TECHNIQUES AND ADVANCES IN CONVENTIONAL MACHINING OF ENGINEERING AND ENGINEERED MATERIALS

Advances in machining processes can be broadly classified into two categories, based on how advancements have been achieved: (i) machine tools (high-speed machining, hybrid machining), and (ii) machining operation or process (high-performance cooling strategies, high performance or high efficiency, multitasking, hard part machining, cutting tools technology, and tool materials). The following sections discuss these categories in detail.

20.3.1 Advances in Conventional Machining Techniques

20.3.1.1 High-Speed Machining (HSM)

High-speed machining or high-speed cutting has been of interest to manufacturers since three decades (Grzesik 2017). However, only in the last decade have the developments in machine tool

TABLE 20.2
Commonly Used Materials in Automotive Industries and their Key Machinability

Alloys and Composites	Automotive Components	Commonly Used Materials	Machining Improvement Areas	References
Aluminium based high strength alloys	Wheels, panels, and structures	6061, 7050-T7451	Higher machining forces due to the ductile nature, poor surface finish, chip control difficulties and high wear rate of tool materials; energy consumption during machining processes—a sustainability issue; material removal mechanism and finished surface quality in Ultra HSM	Santos et al. 2016, Sekhar and Singh 2015, Wang et al. 2014, Wang and Liu 2016
Al based MMCs	Engine cylinder bores, pistons, brake disks and drums, drive shafts and other parts	Matrix: LM25 alloy reinforcements: SiC, Al_2O_3 and graphite	Tool wear, cutting forces and surface roughness	Dabade et al. 2010, Singh et al. 2015
Plastics and fiber reinforced plastic (FRP) composites	Bumpers, dashboard, body panels, frames, interior and exterior trim and tanks for oil, fuel and other structural components; leaf springs, drive shafts, and various chassis parts	Polypropylene, polyurethane and PVC; glass–polyester composites; carbon–polymer composites	Delamination, fiber/matrix pull-out or debonding, matrix cracking and poor surface roughness of the hole wall are common issues faced with FRP composites; severe tool wear and significant subsurface damage also occurs	Nassar et al. 2016, Xu et al. 2016
Magnesium alloys	Wheels, instrument panel, transfer cases, steering wheel, seat and inner door frames, intake manifold, cylinder head cover and side mirror brackets	AZ91D, AZ31	Brittle nature and risk of inflammability (high affinity to oxygen) and reactivity; built up edge (BUE)	Blawert et al. 2004, Carou et al. 2014, Sunil et al. 2016
Intermetallic TiAl-based alloys	Valves, roller leavers and turbochargers for combustion engines	Titanium Aluminide	Poor productivity and surface integrity	Mathew and Vijayaraghavan 2016

technology, as well as the control programs, enabled this technology to be realized in machine shops on a large scale. Until now, the definition of high-speed machining has been a tricky one because—depending on the workpiece material, the cutting speeds that can be classified under high-speed machining vary (Arunachalam and Mannan 2000; Grzesik 2017). High-speed machining is capable of competing with additive manufacturing, especially in producing monolithic components, for not only the automotive but also for the aerospace industry, as well as the die and mold industry. Although the initial investment is higher for high-speed machining, the benefits, such as fivefold increases in material removal rates, up to 70% reduction in machining times, 25–50% reduction in machining cost, lower cutting forces, better surface finish, and exponential increase in productivity (Grzesik 2017) can help offsetting the higher initial investment for high speed machining center. The only problem is tool life, which is reduced and can be compensated with use of advanced cutting tool materials now being developed.

In order to be both competitive and reliable, high-speed machining centers should have the minimum kinematical and functional requirements, such as drives, spindle speeds, axes, and traverse rates, as shown in Figure 20.2. The technology that has recently revolutionized high-speed machining is linear motor drives. In addition, the development of direct drive motors has also improved precision and accuracy. High-speeds do not make sense if they are not complemented by high accelerations of the axes and rapid traverse rates. The advancements in high-speed machining have been superimposed in Figure 20.2 that illustrates the factors that affect the performance of high-speed machining. A typical advance high-speed machining center from one of the largest machine tool manufacturers in the world, which is capable of meeting those requirements, is shown in Figure 20.3. This machine uses a linear drive in all axes and can achieve accelerations of up to 2 g and a rapid traverse of up to 90 m/min. This can increase productivity by up to 20 percent. The linear direct drive technology has been implemented in the rotary/tilt axes in order to catch up with both the feed rates and acceleration of the linear axes thereby enabling true high-speed machining simultaneously in all of the five axes. These kinds of rates are required for the manufacturing of (a) turbo machinery parts, (b) intricate molds and dies, as well as (c) complex and challenging shapes of medical parts (Sudhakar 2017). This is a true advancement in recent years in the area of high-speed machining. The symmetrical bridge-type construction provides high stability while the direct measuring system provides ultimate precision. All these features render these types of high-speed machining centers versatile for several industrial applications, such as (a) aerospace, (b) energy, (c) die and mold, (e) precision, and others.

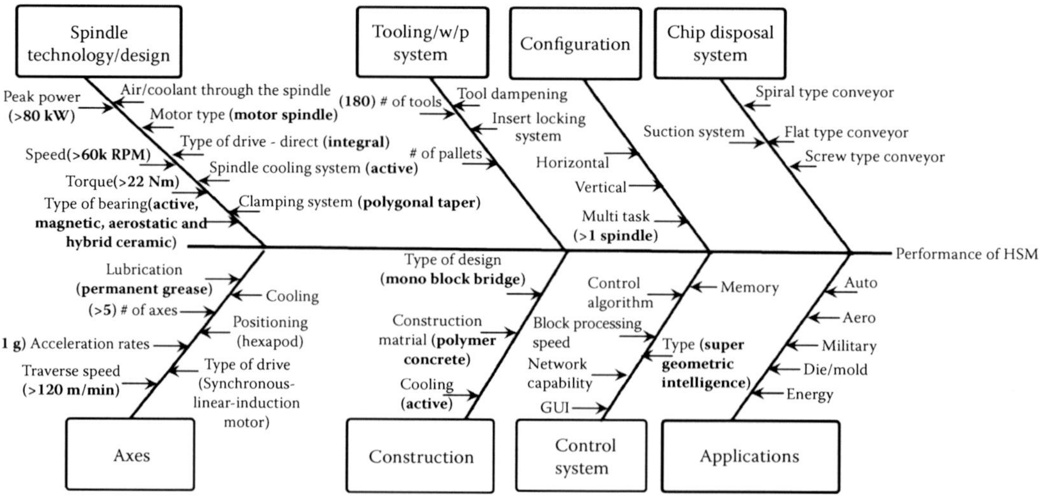

FIGURE 20.2 Factors affecting the performance of HSM and advances in those factors (indicated in bold) in improving the performance of HSM.

FIGURE 20.3 HSC 75 linear (DMG Mori 2017) Five-axis high-speed precision center with linear drives. Technical data: Rotational spindle; max. speed 40,000 rpm; rapid traverses rate along X/Y/Z axes: 90/90/90 m/min; feed rate 90 m/min; max. acceleration: >2g; tool magazine for 180 tools; CNC control system: DMG MORI ERGOline® Control with Siemens 840D solutionline/DMG MORI ERGOline® Control with Heidenhain iTNC 530; clamping taper: HSK-E50. (Courtesy of DMG MORI.)

Notable advances in the hardware system are not enough unless the software portion can also meet up with the requirements of the high-speed machining centers. Newer machines have the capability of an expert control system where the operator can optimize performance settings of the machine through simple prioritization of three target values: (a) accuracy, (b) speed, and (c) surface finish. Many performance settings are also predefined in the factory and user experience can be easily transferred into additional machine performance settings (Sudhakar 2017). Makino recently introduced the SGI.5 (Makino 2017), the latest version of Super Geometric Intelligence software for the high-feed rate, tight-tolerance machining of complex 3-D contoured shapes, which can shorten machining cycle time while maintaining precise surface finish.

High-speed machining is applicable for machining components for many industries such as (a) aerospace, (b) automotive, (c) energy, (d) die and mold, and (e) others. In the aerospace industry, high-speed machining is competing with additive manufacturing in producing components made of titanium. With advancements in cutting tool materials, machine tools, and cooling strategies, it is possible to eliminate secondary processes, such as grinding. Large structural aerospace components can be machined from a solid block providing high structural integrity and avoiding joining or fastening processes.

Inconel 718 has been machined under quite high speed (300 m/min) using cubic boron nitride cutting tools that were textured on both the rake and flank face resulting in improved tool life (Sugihara et al. 2017). High-speed machining of aluminum alloys used in aerospace, automotive, and marine industries have been improved by using polycrystalline diamond and chemical vapor deposition-coated diamond tools. This was possible because of their low coefficient of friction, low chemical affinity, high thermal conductivity and high-temperature inertness (Kalyan and Samuel 2015). Polycrystalline diamond tools are also suitable for high-speed machining of titanium and its alloys but quite expensive (Hosseini and Kishawy 2014). However, with the advantages of high-speed machining, the higher cost of polycrystalline diamond tools can be compensated.

20.3.1.2 Hybrid Machining Techniques

Hybrid machining can be best described as the concurrent and controlled utilization of multiple machining processes, energy sources, tools, or stimulation mechanisms to achieve better performance in material removal. The performance behavior of a hybrid machining process will be different from that of machining processes performed individually (Kozak and Rajurkar 2000). An integration of these machining processes to get a single hybrid machining process can be done either by applying these processes independently on a single machine or by applying them simultaneously in an approach called "assisted machining" (Aspinwall et al. 2001).

Assisted machining uses an external source to assist the primary machining process in material removal. The objective of "assisting" is to attain more effective machining by overcoming the limitations of the basic machining process. The external assisting source may include vibration, heat, abrasive particles, magnetic field, and other media (Zhu et al. 2013). The main machining process may be of conventional type or nonconventional type. Here, we focus on assisted "conventional" machining process as per the scope of this chapter.

Some common benefits of hybrid machining processes are improved machinability of difficult-to-machine materials, improved tool life, higher geometrical accuracy, better surface integrity, and higher machining efficiency. Typical applications of hybrid machining processes include precise machining of composites, ceramics, less-machinable metals, and alloys, etc. (Zhu et al. 2013). As assisted hybrid machining process form the major core of hybrid processes, their different classes are discussed below.

20.3.1.2.1 Vibration-Assisted Machining (VAM)

Vibration-assisted machining superimposes vibration on the motion of a cutting tool by oscillating the tool at a high frequency in a linear (one dimensional) or elliptical (two dimensional) path of small amplitude. Major benefits of vibration-assisted machining are extended tool life, improved surface finish and form accuracy, ductile regime machining of brittle materials, reduced forces and stresses, reduced cutting temperatures, and suppression of burr formation (Brehl and Dow 2008; Kumar et al. 2014).

Cutting systems in vibration-assisted machining are mainly of three types. One-dimensional (linear) resonant systems use discrete high frequencies greater than 20 kHz, and displacement amplitudes of less than 6 μm. The tool works parallel to the surface of the workpiece and collinear with the direction of cutting force. In a 2-D resonant system, the tool supporting structure is made to vibrate at resonant frequencies in two dimensions, thereby creating an elliptical tool path whose major axis is in line with the cutting force and the minor axis in line with the thrust force. When piezoelectric actuator stacks are excited by sinusoidal voltage signals, and a mechanical linkage is used to convert the linear expansion and contraction of the stack into an elliptical tool path, the vibration-assisted machining system is of 2-D nonresonant type.

Vibration-assisted drilling works efficiently on materials like hard alloys, brittle materials, and composites that are difficult to machine by conventional drilling (Kumar et al. 2014). Thrust force is reduced besides providing accurate and precise material removal and longer tool life (Chang and Bone 2009; Kadivar et al. 2014). One of the latest techniques used to drill composite laminates is vibration-assisted twist drilling. Here, vibrations of small amplitude with high frequency (>1000 Hz) or low frequency (<1000 Hz) are superimposed on a twist drill bit along the direction of feed to provide pulsed intermittent cutting. It has been found to reduce the damage due to delamination, which occurs in conventional drilling (Aoki et al. 2005; Mehbudia et al. 2013). The principle of vibration-assisted drilling is shown in Figure 20.4.

Vibration-assisted milling differs from conventional milling by facilitating the formation of thinner chip through successive overlapping of toolpaths (Kumar et al. 2014). This leads to reduced heat generation and increased tool life. Use of ultrasonic vibrations was recently introduced, the technique being named Ultrasonic Vibration-Assisted Milling. The process improves the machining

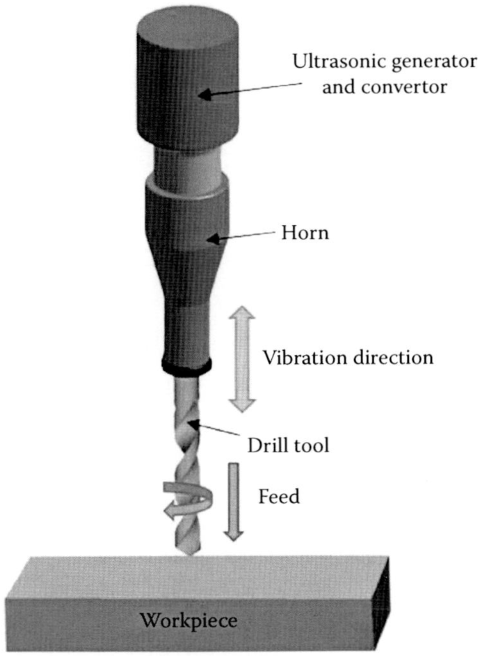

FIGURE 20.4 Schematic diagram for vibration-assisted drilling.

process of difficult-to-machine materials and allows for a larger applicable area. Motion of the tool-tip during vibration-assisted milling is illustrated in Figure 20.5.

In one kind of vibration-assisted turning, ultrasonic vibrations are used to assist the turning process, and the process is called Ultrasonic-Assisted Turning. The process tends to change the material response during machining, resulting in a reduction of cutting force coupled with improved surface roughness. Compared to conventional machining, the residual stresses induced and the heat generated are lower (Adnan and Subbiah 2010; Guo and Ehmann 2013; Jamshidi and Nategh 2013; Muhammad et al. 2012; Patil et al. 2014). Ultrasonic-assisted turning is used in machining of aerospace super-alloys, high-strength alloys like nickel and titanium alloys, and some ferrous and brittle materials. Figure 20.6 shows ultrasonic vibrations in the radial, tangential, and feed directions during vibration-assisted turning.

In vibration-assisted grinding, ultrasonic vibrations can improve delivery of the over tool-workpiece contact zone through periodic separation, thereby facilitating efficient heat removal (Kumar et al. 2014; Qu et al. 2000; Xu et al. 2016). Due to the imposed vibration, cutting can be done with more than one edge, and the grinding wheel also gets self-sharpened (Liang et al. 2010; Liu et al. 2012). Subclasses of ultrasonically-assisted grinding are one dimensional axial, one dimensional vertical, and two-dimensional elliptical. A new grinding technique called vibration-assisted ball centerless grinding has been recently introduced in order to reduce processing time during spherical surface grinding (Wu et al. 2011; Xu et al. 2012). The ball is energized with two kinds of direction-adjustable ultrasonic vibrations and is controlled to rotate in two directions for achieving spherical surface grinding. The vibration-assisted grinding process is shown in Figure 20.7, with coolant supply, and directions of ultrasonic vibration, tool, and feed.

20.3.1.2.2 Thermally Assisted Machining (TAM)

Thermally assisted machining uses an external heat source to heat the workpiece locally near the cutting tool, thereby softening the workpiece and changing its microstructure. The reduced hardness results in lower forces and reduced wear of the cutting tool. Normally used heat sources are plasma

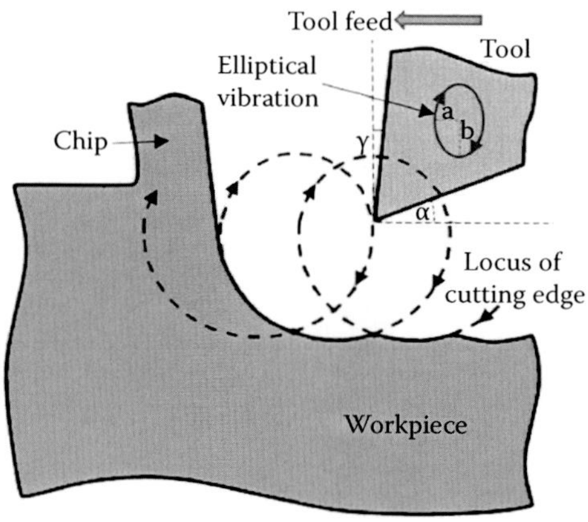

FIGURE 20.5 Schematic illustration of the motion of the tool-tip in vibration-assisted milling.

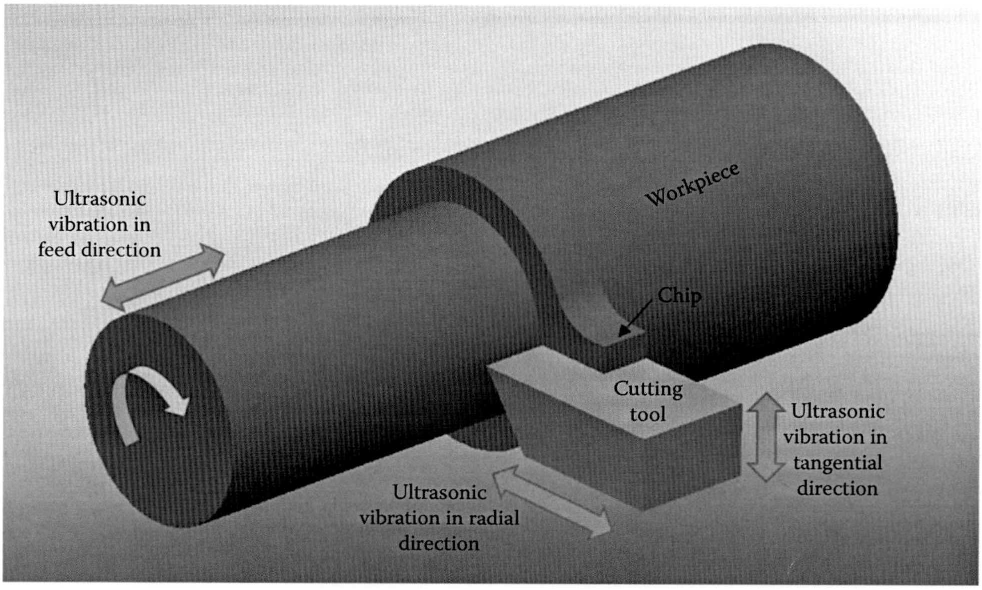

FIGURE 20.6 Vibration-assisted turning process, showing ultrasonic vibrations in the radial, tangential, and feed directions.

and laser, laser-assisted machining being the most widely used (Zhu et al. 2013). A comparison of different thermally assisted hybrid techniques is given in Table 20.3 (Jeon and Lee 2012). As laser and plasma are the most common thermal sources in hybrid processes, they are discussed in detail below.

20.3.1.2.3 Laser-Assisted Machining (LAM)

Due to the ease in control of the beam and laser beam delivery, laser-assisted machining is the most common thermally assisted machining hybrid process. A laser beam is focused on the workpiece to heat and soften the material. This leads to ductile deformation while cutting, resulting in force reduction, lower tool wear, and an improved surface finish (Jeon and Lee 2012; Zhao et al. 2010).

In laser-assisted drilling, the laser beam must be focused on the workpiece near the tip of the drill bit. Another method for delivering the heat is by using an absorbent liquid like liquid blackbody (dispersed carbon colloidal solution), which absorbs the laser beam. The heated absorbent goes inside the drilled hole and heats the workpiece. The laser beam is sometimes delivered from behind the workpiece to avoid absorption by the cutting tool (Ito et al. 2017). A schematic layout of

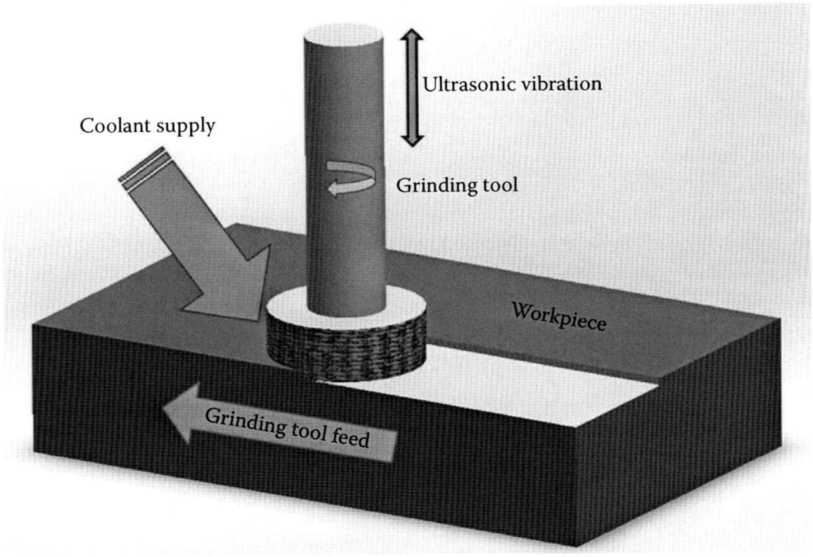

FIGURE 20.7 Vibration-assisted grinding process, showing coolant supply, and directions of ultrasonic vibration, tool, and feed.

a laser-assisted drilling system is shown in Figure 20.8; laser beam is delivered from behind, and a liquid blackbody is used as an absorbent.

There are two common laser delivery configurations in laser-assisted milling. In the first one, a laser beam is placed near the milling tool, while the second one integrates the laser on the tool spindle. A new technique guides the laser beam through the spindle directly to the spot of chip removal on the workpiece near the cutting edge. In this way, the laser beam is exposed for a very short time, giving flexible adjustments for the heat input, resulting in efficient plasticizing of the material. Other advantages include reduction in flaking, surface cracks, and material distortion (Hoffmeister and Wenda 2000; Ito et al. 2017; Zhu et al. 2013). In Figure 20.9 is shown the laser-assisted milling process; the laser beam is inclined to the workpiece surface to avoid its absorption by the cutting tool.

For milling of glass materials, a different approach called selective laser-assisted milling is used. The glass can absorb two types of laser beams; one with a wavelength in the micrometer range and the other with an ultra-short pulse laser. Both of them are ineffective in producing a precise surface for the purpose of machining; the first one gives a larger area, while the second gives a smaller area.

TABLE 20.3
Advantages and Disadvantages of Some Thermally Assisted Hybrid Machining Methods

Heat Source	Advantage	Disadvantage
Electricity	Simple equipment; even heat distribution	Precise control difficult
Induction coil	Easy to use; high-capacity preheating	Limited tool mobility; high-concentration preheating difficult
Gas flame	Low initial investment cost	High-concentration preheating difficult
Plasma	High degree of heat concentration; direct contact not needed	Precise control difficult
Laser	High degree of heat concentration; easy control of heat source	Costly equipment; absorption rate different for different material

Source: Jeon, Y. and Lee, C.M., *Int J. Precis. Eng. Manuf.*, 13: 311, 2012. https://doi.org/10.1007/s12541-012-0040-4.

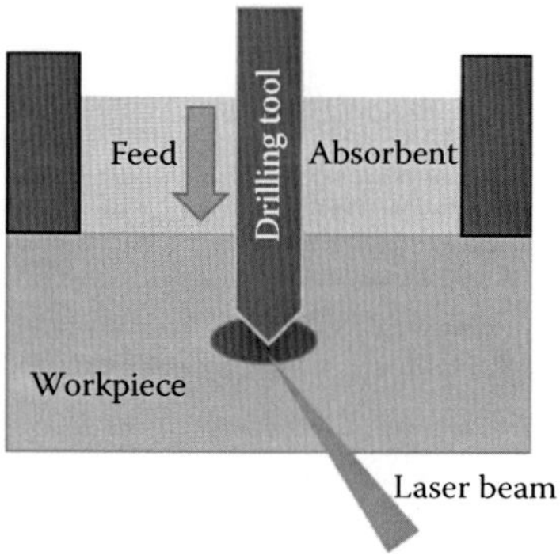

FIGURE 20.8 Schematic layout of a laser-assisted drilling system, where the laser beam is delivered from the backside, a liquid blackbody being used as an absorbent.

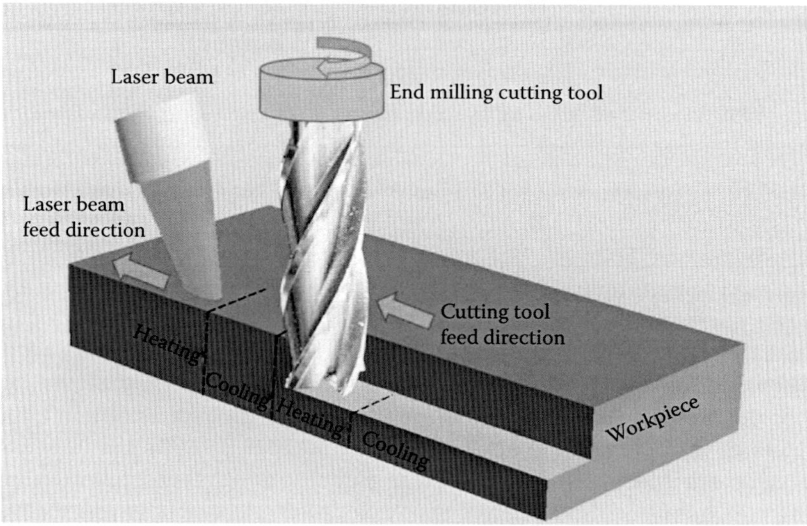

FIGURE 20.9 Laser-assisted milling process, where the laser beam has an incline to the workpiece surface to avoid its absorption into the cutting tool.

A fiber laser having a wavelength out of the absorption band for glass is made to absorb into a small area that is selected by coating it with black body coating. This enables selective heating of the area prior to material removal with the cutting tool (Ito et al. 2017).

Since the cutting tool does not rotate while turning, laser-assisted turning is considered to be the most favorable laser integration due to the ease in incorporating the laser beam (Sun et al. 2010; Sundaram et al. 2008). It is effective in machining materials like nickel and ceramics. Laser-assisted turn-mill can machine workpieces other than round shapes by performing turning and milling simultaneously (Cha et al. 2015). As shown in Figure 20.10, a laser beam provides intense localized heating of the workpiece ahead of the cutting region in laser-assisted turning.

Even when a two-step approach was used in conventional grinding with the idea of minimizing surface microcracks and subsurface damages in ceramics, failure at the microscale level was reported (Gabler and Pleger 2010; Hoffmeister and Wenda 2000). A new approach, called laser-assisted grinding, was developed where the ceramic is induced with thermal cracks by laser irradiation. By adjusting the laser power, speed, and spot size, volume of the crack can be controlled. This cracked material is then removed by microgrinding (Kumar et al. 2011). On a macroscale, thermal softening

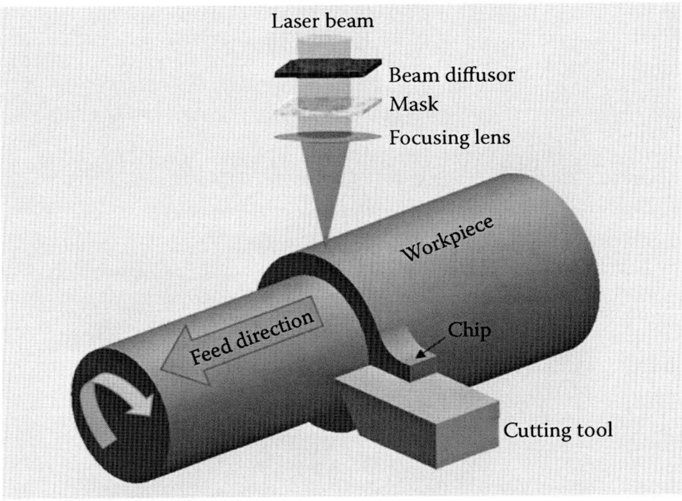

FIGURE 20.10 Laser beam provides intense localized heating to the workpiece ahead of the cutting region in laser-assisted turning.

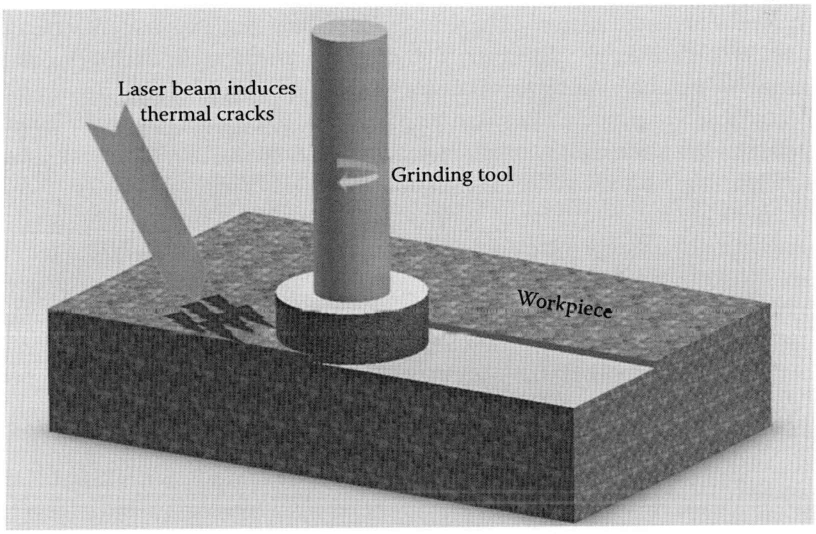

FIGURE 20.11 In laser-assisted grinding, localized thermal stresses weaken the surface material (ceramic), which can then be removed at a higher material removal rate.

is achieved by laser irradiation to plasticize the region for the purpose of finish grinding. The principle of laser-assisted grinding is shown in Figure 20.11. Localized thermal stresses tend to weaken the surface material (ceramic), which can then be removed at a higher material removal rate.

In laser-assisted shearing, the underside of the sheet metal where shearing is to be done, is heated by the laser beam. The cutting stamp is then punched from the topside. This results in reduced forces and warping of the edges (Brecher and Emonts 2010).

20.3.1.2.4 Plasma-Enhanced Machining (PEM)

High-temperature alloys find extensive application in the aerospace sector. However, conventional machining of these alloys is difficult because of the twin problems of high wear and bad surface finish. Thermally enhanced machining, such as laser-assisted machining and plasma-enhanced

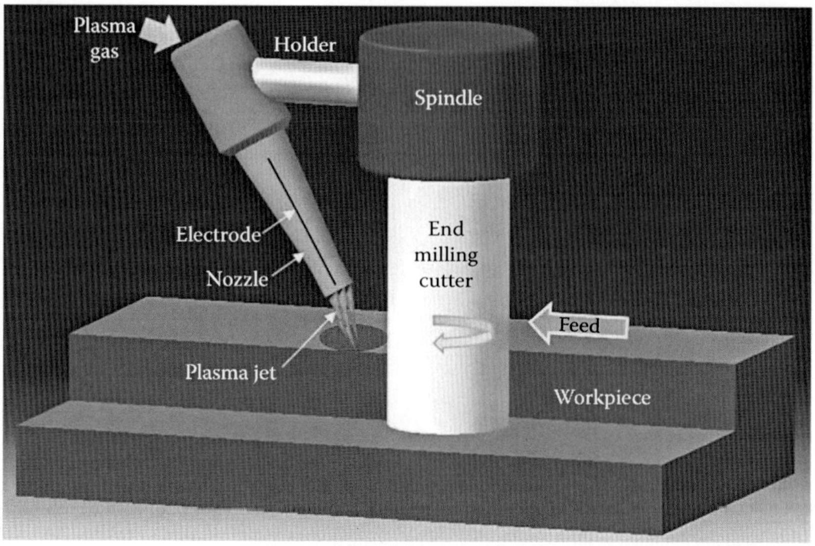

FIGURE 20.12 Schematic illustration of the plasma-enhanced machining process for end milling.

machining, can be viable alternatives. These techniques offer higher material removal rate, longer tool life, and better surface finish. However, high power lasers are quite expensive, preventing wide usage of laser-assisted machining by the industry. Plasma-enhanced machining can provide heating rates equivalent to lasers at a much lower cost. In plasma-enhanced machining, a plasma jet is used to heat and soften the workpiece while doing conventional machining like milling or turning (Leshock et al. 2001). A schematic illustration of the plasma-enhanced machining process for end milling is shown in Figure 20.12.

The characteristics of plasma-enhanced machining must be clearly understood for generating and transferring an optimum level of localized heating to the workpiece. Maximum temperatures should be known in order to mitigate thermal damage of the workpiece material. Temperature at the tip of the cutting tool should also be known for better process control. Plasma-enhanced machining is especially well suited for the machining of electrically conducting superalloys, as high localized energy (peak temperature reaching 16,000 K) can be generated for low gas flow rates (2017; Kitagawa et al. 1988; Komanduri et al. 1985; Konig et al. 1990; Novak et al. 1997; Yao 2000).

20.3.1.2.5 Abrasion/Magnetic Field–Assisted Machining (A/MFAM)

In finishing/polishing processes, the use of conventional machining is at times impractical, or there is rapid tool wear or tool chatter/vibration (Najariana et al. 2014). In such cases, magnetically assisted machining is now widely used. The working element is a homogeneous mixture of magnetic particles/fluid and abrasive particles/grains. This magnetic abrasive fluid is forced by the imposed magnetic field to remove material from the workpiece (Gandarias 2017; Zou and Shinmura 2007). The workpiece is made to rotate on a spindle to create an oscillating magnetic field which can induce vibrations in the fluid (Paswan 2017). The fluid can flow into difficult-to-reach areas and improve the finishing process. Major benefits of the process include the following: (a) smooth surface finish, (b) high accuracy, (c) little surface damage (due to low machining force generated by the magnetic field), (d) high-speed operation, and (e) easy adaptation on existing machines (Thomas 2016). Rough finishing, deburring, and precision finishing can be done concurrently. Typical applications include optics, microelectromechanical systems, medical components, electronic components, fluid components, to name a few (Kashyap 2016).

Some new variations of magnetic field-assisted machining have been recently developed, using permanent magnets in the system. These techniques can achieve a high degree of efficiency and

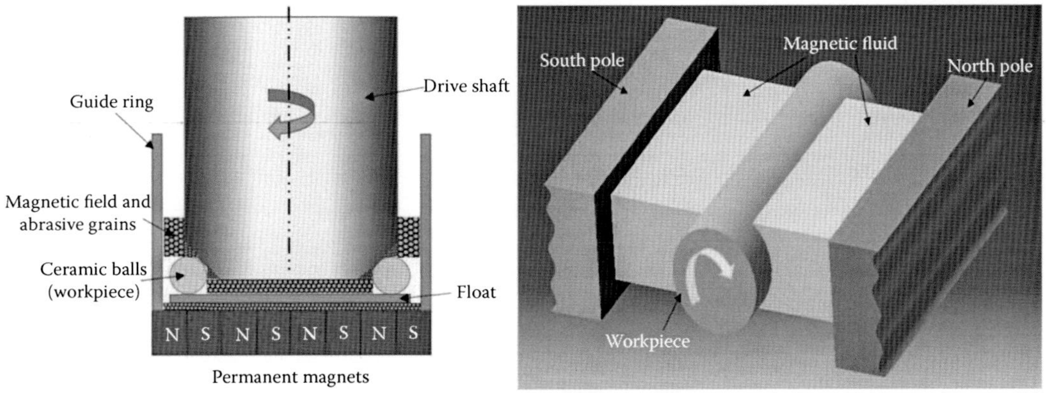

FIGURE 20.13 Magnetic field–assisted polishing (using an abrasive fluid) of balls (left) and rollers (right).

accuracy in internal finishing and deburring of the tubes. Combination of a smooth finishing action coupled with a high machining force can increase the internal roundness of the tube, and also yield a better surface finish (Zou and Shinmura 2007).

Another new type of magnetic-assisted finishing uses abrasive slurries (abrasive particles and exceptionally fine ferromagnetic grains in a liquid medium, such as water or kerosene) supported by magnetic field for polishing of ceramic balls and bearing rollers. A very fine polish is achieved in an economical manner, generating an almost defect-free surface (El-Hofy 2005). The use of an abrasive fluid together with magnetic field assisted polishing of balls and rollers is shown in Figure 20.13.

20.3.1.2.6 Media-Assisted (Superimposed) Machining (MAM)

Overheating can lead to severe wear at the tool-workpiece interface. Media-assisted machining tries to address this problem by using high-pressure or ultra-low temperature cutting fluids. By choosing an optimum combination of cooling lubricant and pressure, chip breakage, chip control, cutting speed, material removal, and tool life can be improved. This helps in reducing the machining time while concurrently increasing productivity. The details of these processes are discussed in Section 20.3.2.1. In addition to those techniques, there are two further advanced techniques. In laser-assisted and cryogenic machining, a CO_2 laser is used to change workpiece material (titanium alloys) properties, and CNC machining is used with liquid nitrogen as the coolant (Dandekar et al. 2010). Plasma-enhanced cryogenic machining consists of a cooling chamber that supplies liquid nitrogen to cool the cutter during plasma-enhanced machining process (Wang et al. 2003).

20.3.2 Advance Strategies in Conventional Machining Processes

20.3.2.1 High-Performance Cooling Strategies

The cutting fluid used during metal cutting processes, plays a vital role by cooling the surface of the work piece and the cutting tool, removing chips from the cutting zone and by lubricating the tool work piece interface. For centuries, cutting fluids has been the natural conventional choice in the manufacturing industry to reduce heat generated during machining. Due to the poor machinability of various engineering materials, such as nickel alloys, titanium alloys as well as fiber reinforced composite materials, several research studies have been conducted to meet the need of improving the machinability of these materials through high-performance cooling technologies.

Selecting the most suitable cooling strategy for each machining task depends mainly on the workpiece material and the machining parameters. Figures 20.14 and 20.15 show the classification of cutting fluids and types of cooling techniques. Conventional wet machining with emulsion is still the most applied system based on a wide range of field investigations. Innovative cooling strategies

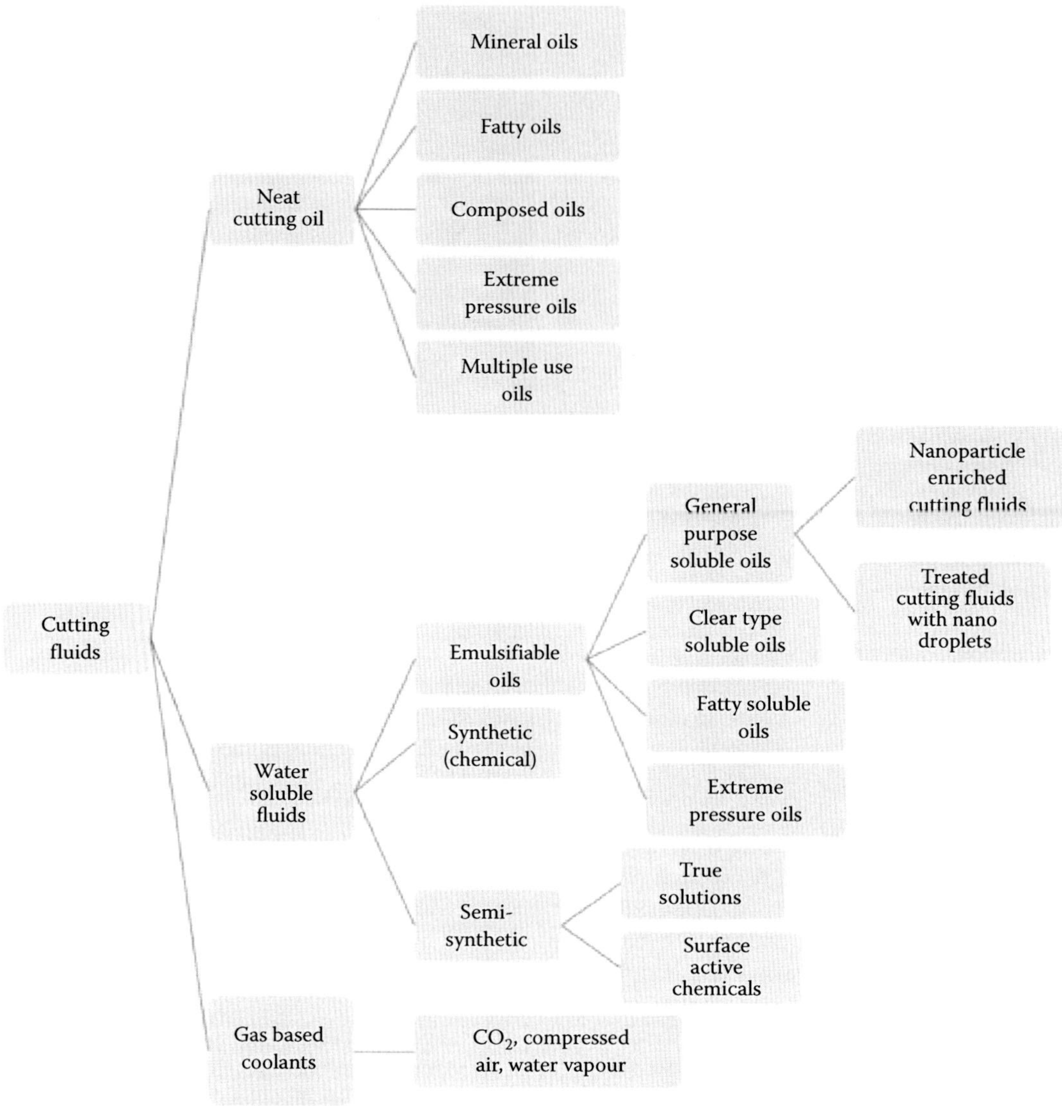

FIGURE 20.14 Cutting fluids classification.

especially in machining of engineering materials are necessary to achieve more efficient cutting conditions in high-performance cutting and high-speed cutting. High-pressure cooling helps in facilitating chip control, a stable cutting process, and acceptable tool life. Due to the high-energy consumption of the machine, as well as cost intensive maintenance and disposal, this strategy shows its benefits in roughing operations where high material removal rates tend to reduce specific energy consumption.

The effectiveness and application of various cooling and lubricating strategies are shown in Table 20.4. In recent years, near-dry machining also known as Minimum Quantity Lubrication (MQL) has emerged as a progressive and more sustainable solution for significantly reducing the amount of coolant/lubricant in machining applications. The new minimum quantity lubrication system delivers a steady flow of compressed air and small quantity of cutting oil directly to the cutting tool or tap, which reduces heat, removes chips, and concurrently provides lubrication (Figure 20.16). It's not

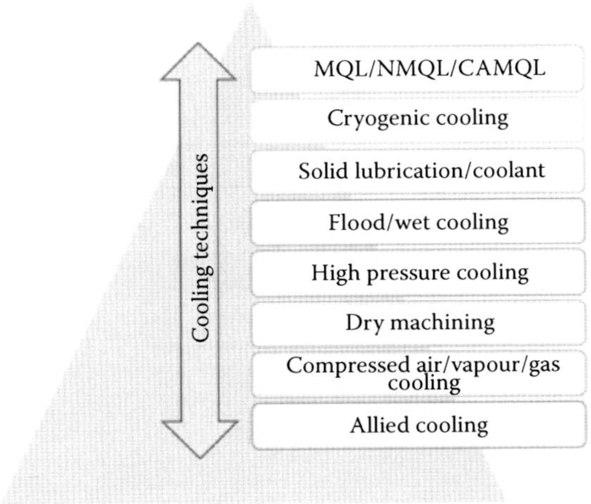

FIGURE 20.15 Different cooling techniques.

much oil, but it is just the right amount to be effective. This extends tool life and, in some cases, eliminates the requirement for coolant during milling operations.

Solid lubricants comprising of graphite and molybdenum disulfide dispersed in oil and delivered through minimum quantity lubrication during turning of Inconel 718 showed improvements in tool life with no significant microstructural changes or influence on surface roughness and residual stresses (Marques et al. 2016). During turning of Ti–6Al–4V alloy, a novel technique of internal oils on water cooling method, with an appropriate amount of water, or external oils on water cooling from a proper location yields a lower surface roughness and slower flank wear rate, and bare areas on the cutting tool substrate are absent because of better lubrication (Lin et al. 2015). High-pressure air in

TABLE 20.4
Effectiveness and Application of Various Cooling and Lubricating Strategies

Effects of the Cooling and Lubricating Strategy	Flood Cooling	Dry Cutting Using Compressed Air	Minimum Quantity Lubrication Using Oil	Cryogenic Cooling Using Liquid Nitrogen	Hybrid Cooling Using Liquid Nitrogen and Minimum Quantity Lubrication
Cooling effectiveness	↑	↓	→	↑↑	↑↑
Lubrication effectiveness	↑↑	↓	↑↑	→	↑↑
Removal of chips	↑	↑	→	↑	↑
Cooling of machine tool	↑	↓	↓	→	→
Workpiece cooling	↑	↓	↓	↑	↑
Dust and particle control	↑	↓	→	→	↑
Product quality (surface integrity)	↑	↓	→	↑↑	↑↑
Legend	↑↑ Very good	↑ Good	→ Average	↓ Poor	

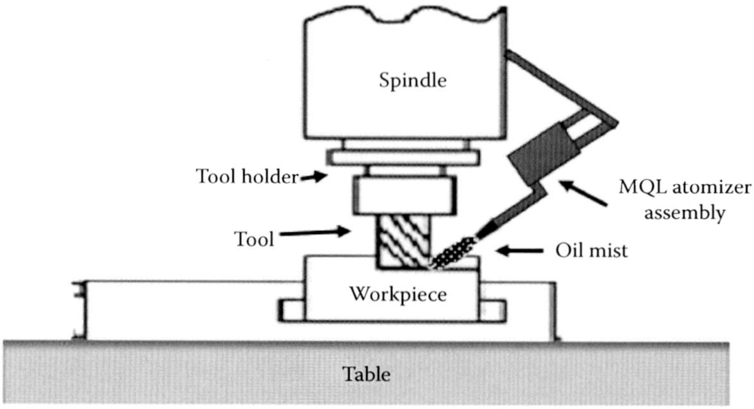

FIGURE 20.16 MQL system.

minimum quantity lubrication during milling the Ti–6Al–4V alloy shows drastic reduction in friction at higher cutting speeds. However, at lower speeds there is no major difference between dry cutting and high-pressure air (Liu et al. 2015). Novel lubrication medium such as use of ethanol-blended metal removal fluids reduced cutting force by over 65% compared to dry machining of the Ti–6Al–4V alloy. Ethanol-blended metal removal fluids facilitated adsorption of hydroxy groups from ethanol by the carbon at the surfaces of the cutting tool resulting in negligible adhesion to the cutting tool (Krishnamurthy et al. 2017). Similar improvement in tool life can also be achieved using an atomization-based cutting fluid spray system. The minimum quantity lubrication condition assists in decreasing notch wear during finish turning of pure iron material compared to wet cutting, dry cutting, and hand operation to brush rapeseed oil on unmachined workpiece surfaces (Kong et al. 2016). Due to smaller size of the droplet when using vegetable-based cutting fluid, it has better chance to reach the tool–workpiece contact zone in minimum quantity lubrication than in flood lubrication. Among the different vegetable-based cutting fluids such as palm oil, coconut oil, sunflower oil, and soya bean oil, the Palm oil suits better in terms of minimum cutting force requirement and minimum vibration during milling of 7075 hybrid aluminium metal matrix composite (Shankar et al. 2017). Drilling of A356/15 silicon carbide metal matrix composites using minimum quantity lubrication showed an improved surface roughness with increasing cutting speed and decreasing feed rate. The surface roughness is lower than other cutting environments with an optimal condition identified at a cutting speed of 15 m/min, and a feed rate of 0.1 mm/rev (Kilickap 2016). Minimum quantity lubrication cutting fluids have also been shown to be effective while form tapping A306 cast aluminum alloy. The minimum quantity lubrication system has high potential to reduce the value of torque while minimizing friction and avoiding tool breakage. Furthermore, form tapping has been found to be a suitable alternative, since it does not generate chips, which reduces cost associated with recycling, storage, and waste disposal (Filho et al. 2017). Grinding of nickel-based alloy GH4169 shows minimum quantity lubrication grinding using vegetable oil generates a lower friction coefficient (decreased by 50.1%) and specific grinding energy (decreased by 49.4%) compared with flood grinding. A similar reduction in specific energy has been noticed during minimum quantity lubrication grinding of alumina engineering ceramic using synthetic oil poly-alpha-olefin and is suitable for rough grinding. Among the grinding fluids, castor oil achieves the best lubrication property and best surface quality of the workpiece (Y. Wang et al. 2016). The minimum quantity lubrication plus water (1:1,1:3, and 1:5 parts of oil per parts of water) technique is a potential alternative for cooling-lubrication when applied properly. Grinding performance for the minimum quantity lubrication plus water in the proportion of 1:5, with a wheel cleaning system (at a 30° inclination angle of the air nozzle) is the best, when compared to minimum quantity lubrication without water, and close to the conventional flood coolant (Ruzzi et al. 2017).

Recently, an air-assisted approach using air as carrier medium has been proven to be effective when grinding AISI 1045 steel with boric acid as the lubricant. Coarse particles and concentrated mixture gives favorable results in general, and lubricant/air mixture ratio is a significant factor for microhardness and residual stresses. Minimum quantity lubrication also cut down tangential forces by 50% (Saleem et al. 2017). Milling carbon fiber-reinforced polymer using a solid carbide cutting tool results in higher abrasive wear of the tool under dry machining compared to the chilled air cooling method. This demonstrates the latter technique to be sustainable for the machining of carbon fiber-reinforced polymer (Khairusshima et al. 2016). The vibration signals generated during intermittent turning and their dependence on the process parameters show dependency of the vibrations on flow rate of the minimum quantity lubrication system, its interaction with feed rate, and no dependency on the type of interruption. The influence of the minimum quantity lubrication system was found to be greater when machining at the lower feed rate. In general, under dry conditions, the higher the vibrations, the higher the surface roughness, while the opposite occurs when the minimum quantity lubrication system is used (Carou et al. 2016).

The economics of using cutting fluids have changed dramatically over the past two decades. In the early 1980s, buying, managing, and disposing of cutting fluids accounted for less than 5% of the cost of most machining jobs. Today, fluids—including their management and disposal—account for 17% of the cost of the average job. Because cutting tools account for only about 5% of the total cost of a machining project, accepting a slightly shorter tool life for the chance to eliminate the cost and difficulties of maintaining cutting fluids could be a less expensive choice. As a result, dry cutting is applied to avoid the problems associated with cutting fluid such as (a) contamination, (b) disposal, and (c) hazardous components. Dry machining does not cause pollution of air or water resources. Hence, disposal cost of cutting fluids is reduced. Dry machining of duplex stainless steel using different carbide tools has shown to induce an almost three-fold growth of tool life in comparison to that obtained during cutting with fluids (Krolczyka et al. 2017). Dry milling is also shown to be a viable option using a solid carbide tool for aluminium titanium carbide metal matrix composites (Das et al. 2016).

The application of cryogenic media offers the advantage of dry machining in combination with rapid cooling due to environmental and health effects of mist application. The two gases, carbon dioxide and liquid nitrogen, have to be distinguished regarding the mechanisms for generating low temperatures. A new aerosol strategy using small quantities of oil particles, in minimum quantity lubrication form, combined with cryogenic carbon dioxide or liquid nitrogen has also been developed to provide lubrication of the cutting zone. Aerosol dry lubrication with carbon dioxide or the carbon dioxide cryogenic cooling during the turning of titanium shows good promise to reduce cutting-edge temperature and limit tool wear (Busch et al. 2016). Similar observations were made concerning the reduction in tool temperature by an average of 30% during machining of Ti–6Al–4V alloy, Inconel 718, and AZ31 magnesium alloy using cryogenic cooling (Danish et al. 2017; Kramer et al. 2014). However, Iturbe et al. (2016), during finish turning of Inconel 718, have reported conventional cooling to be a better option, from both a machinability and surface integrity point of view, when compared to liquid nitrogen cooling and minimum quantity lubrication using KLUBERTCUT CO 6-150 oil. Cryogenic cooling proved to be the most promising lubrication strategy for external turning of gamma titanium aluminides as they could successfully counteract the huge thermal load on hard cutting edges resulting in improved surface quality and productivity (Klocke et al. 2013). Regardless of the as-delivered condition of the alloy (wrought and additive manufactured), the adoption of cryogenic cooling during machining significantly affects the surface properties of Ti–6Al–4V alloy. Usage of an atomized-cryogenic-liquid spray technique and liquid nitrogen feeding system significantly enhances the machinability of aluminium and 5% titanium carbide particulate composites in comparison with cryogenically chilled argon, wet and the dry machining conditions, respectively. (Josyula and Narala 2017; Josyula et al. 2016). A significant improvement in surface integrity can be achieved through cryogenic machining (liquid nitrogen) of Ti–6Al–7Nb alloy by up to 35% and 6.6% respectively, compared with a dry and flood-cooled machining. Hardness of the cryogenically machined surface layer increased with the formation of a severe plastic deformation layer along with the ultrafine grain layer. (Sun et al. 2016). The same effect is also seen in the case of NiTi shape memory alloys, which had substantial effects on phase transformation behavior of machined specimens. Magnesium AZ31B and Co-Cr-Mo alloys are commonly used in biomedical applications, which require machining. Cryogenic processing of these materials is sustainable and a nontoxic alternative to conventional flood-cooled, minimum quantity lubrication, and dry machining processes.

Ti–6Al–4V titanium alloy, when cryogenic milled using coated solid carbide cutters, produced 18% and 21% lower surface roughness than dry machining and wet machining, irrespective of the cutting parameters (Shokrani et al. 2016). Diamond coated carbide indexable inserts used for drilling of aluminium-silicon carbide metal matrix composite with an internal supply of carbon dioxide resulted in higher tool life while generating the best precision and surface finish compared to the other cooling strategies (Sadik and Grenmyr 2016).

Fiber metal laminates are used in the upper fuselage of the Airbus A380 aircraft. Machining fiber metal laminates causes various forms of damage around the hole either in the composite layers or the metallic sheets, which deteriorates the surface quality of the metal/composite. Microstructural studies of the holes drilled on Glass-reinforced aluminium 2B 11/10-0.4 fiber metal laminates using cryogenic and minimum quantity lubrication system shows an increase in deterioration of the borehole surface with an increase in spindle speed and feed rate near the edge of the hole. This is because of the damage modes such as (a) erosion, (b) surface delamination, and (c) interlayer burr that develop during the drilling process (Giasin et al. 2016, 2017). End milling of multidirectional composites under pressured cryogenic nitrogen gas showed that surface quality at high temperatures to contain severe subsurface damage due to bent fibers leading to fracture (Jia et al. 2016). In contrast, the hardened matrix at low temperatures offered fibers the better support, and the composites were easily squeezed to rupture, which led to a superior cutting performance. Cryogenic machining of the Aramid fiber composites decreases the cutting force and temperature, changes the chip breaking mechanism with reduced defects (Wang et al. 2016). Compression end mill or a tool with a big helix angle reduces the cutting force and produces a good surface quality (Siqi et al. 2016). Fewer defects, reduction in grinding forces (by 13%) and zone temperature (by 38%) is also noted during grinding of stainless steel 316 under cryogenic cooling compared to other methods (Manimaran et al. 2014). The G ratio also indicates low wear rates for cryogenic grinding. As a drawback, the cryogenic grinding, the spindle power increases when liquid nitrogen is used (Reddy and Ghosh 2014).

A comparison of dry environment and high-pressure coolant jet during hard turning of EN 24T steel using coated carbide cutting tools revealed a reduced surface roughness. However, higher hardness of the EN 24T material induces higher average surface roughness due to a higher restraining force against tool imposed cutting force (Mia et al. 2016). High-pressure water jet assistance stabilizes the evolution of flank wear of a WC/Co (H13A) uncoated tungsten carbide tool insert during rough and finish turning of a Ti17 titanium alloy (Ayed 2013). However, the nature of critical degradation in this case was notch wear leading to sudden rupture of the cutting edge. An in-house high-pressure-assisted machining setup developed by Behera et al. (2016) to optimize coolant pressure during turning of Inconel 718 produced a better surface finish, marginally lower machining force, and lower tool wear for a jet pressure of 80 bar. A schematic illustration of a typical through-tool air blast option that supplies high-pressure/high-flow air through the tool and to the cutting edge is shown in Figure 20.17. This option is ideal for applications that require dry cutting and where a coolant cannot be used. Also other option such as the high-pressure coolant system, shown schematically in Figure 20.18 adds value through a faster feed rate and an optimized drilling operation, leading to reduced cycle time and high-quality parts.

20.3.2.1.1 Nanofluids

The last few years has witnessed significant research on the new types of nanofluids. Nanofluid is a fluid that contains nanometer-sized solid particles. As result of its remarkable improvement in

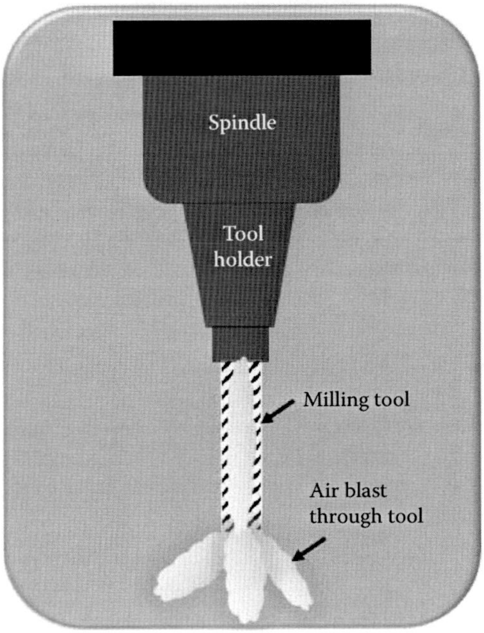

FIGURE 20.17 Air blast system.

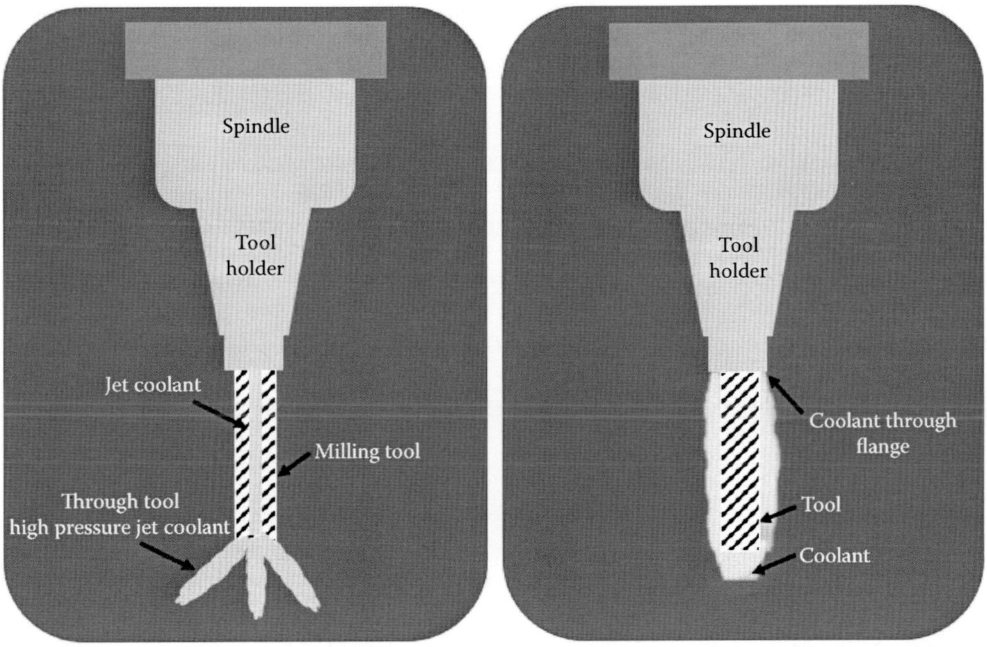

FIGURE 20.18 Through-tool high pressure coolant.

thermophysical and heat transfer capabilities, nanoparticles are being introduced in conventional lubricants. This enables a reduction in coefficient of friction and wear which enhances the efficiency and reliability of machine tools. The presence of nanoparticles in the base fluid contributes to better flow of mixing and a higher thermal conductivity when compared to pure fluid. Some of the key information on machining using nanofluids is summarized in Table 20.5.

TABLE 20.5
Summary of Nano Cutting Fluids

Material	Nano Particles	Method	Machining Process	Remarks	References
Nimonic 90 alloy	Al_2O_3 and colloidal solution of silver (Ag) nanoparticles mixed with water	Minimum quantity lubrication	Turning	Reduced cutting forces, tool wear and chip curling during machining, good surface finish, reduced abrasion wear	Chetan et al. (2016)
AISI 1045 steel	Graphite-LB2000 nanofluid, vegetable and ester based oil	Dry, minimum quantity lubrication	Turning	Reduced cutting force and temperature at high cutting speeds	Su et al. (2016)
AISI D2 steel	5 g of multi wall nano carbon tubes into the SAE20W40 oil base fluid (250 ml)	Minimum quantity lubrication; nano fluid minimum quantity lubrication	Turning	Reduced cutting zone temperature, and surface roughness	Sharma et al. (2015)
Hardened AISI52100	Oil-based nanofluids with MoS_2 nanoparticles	Minimum quantity lubrication	Vibration assisted grinding	Ultrasonic vibration significantly decreases the grinding normal force	Molaie et al. (2016)
AISI 304, AISI 1040	Ethylene-glycol-based TiO_2	Minimum quantity lubrication; flood	Milling, turning	Reduction in workpiece temperature by 30% and reduces tool wear; attrition and oxidation at the cutting edge	Yogeswaran et al. (2015), Sharma et al. (2016), Muthusamy et al. (2016)
AISI 1040	Vegetable oils-nanomolybdenum disulfide ($nMoS_2$) in coconut (CC), sesame (SS) and canola (CAN) oils	Minimum quantity lubrication	Turning	Cutting forces, temperatures, tool wear and surface roughness are approximately reduced by 37, 21, 44, and 39% respectively by using CC+ $nMoS_2$ at 0.5%	Padmini et al. (2016)
Nickel-based alloy GH4170	0.5 vol% to 4.0 vol% Al_2O_3 nanoparticles to palm oil	Minimum quantity lubrication	Grinding	Smallest force ratio of 0.281 is obtained when the volume concentration of the nanoparticles is 1.5%, minimum roughness of Ra = 0.301 μm, slender grinding debris, and maximum drop wetting area are realized when the volume concentration is 2.0%	Wang et al. (2017)
Nickel-based alloy	Carbon nanotube (CNT) nanoparticles with Palm oil as the base oil	Minimum quantity lubrication	Grinding	Volume fraction of 2% nanofluid achieved 21.93 N grinding force, the lowest grinding temperature of 109.8°C, and the lowest proportionality coefficient of 42.7%	Li et al. (2017)

20.3.2.1.2 *Future Trends in High-Performance Cooling Strategies*

It has been shown that the use of petroleum-based cutting fluids causes significant environmental pollution. Manufacturing cost data shows that cutting fluids are approximately 16% of the total costs. This has caused a shift in focus from biodegradability to renewability in order to protect the environment. Several issues need to be addressed in terms of an environmentally compatible cutting fluid which is biodegradable, less toxic, renewable, and bioaccountable. Having the least negative environmental impact, vegetable-based lubricants are expected to increase around 58% in 2018 compared to 2011. Keeping this in focus, some of the drawbacks related to bio-based fluids such as (a) low thermal and oxidative stability, (b) lubrication application techniques, and (c) recycling methods needs to be overcome. Nanolubricant is a novel engineering material consisting of nanosized particles dispersed in base oil which could be an effective method for reducing friction during machining. Nanoparticles have shown enormous potential to be an innovative, effective alternative to flood lubrication. Therefore, future research should concentrate on an optimum utilization of the high performance of nanobased lubricants in the machining industry. In summary, future research should focus on composition, selection, application techniques, quantity optimization, and recycling of the vegetable-based cutting fluids for green machining.

20.3.2.2 Hard Part Machining

The demand for parts made of extremely hard and tough steels has led to the development of hard part machining. It is advancement in the machining of parts harder than 45R_C. This has several advantages since in most cases the slow grinding processes can be eliminated thereby improving productivity coupled with an elimination of secondary process, setup time. Other advantages include better product quality and accuracy, process flexibility, and most importantly, cost reductions. Most often cubic boron nitride and, to a lesser extent, ceramic cutting tool materials are used for the hard part machining (Koepfer 2010). Recently Chinchanikar and Choudhury, (2014) developed coated carbide tools using high power impulse magnetron sputtering for hard turning of AISI 4340 low alloy steel having a hardness of 55 (54–57) R_C. They obtained better tool life, especially under conditions of minimum quantity lubrication. Coated carbide tools could be an alternative to the expensive cubic boron nitride tools used in hard part machining. Hard machining is one of the potential processes for the machining of hardened steels in the die and mold industries. Hard machining can be easily carried out in high-speed machining centers.

20.3.2.3 Cutting Tool Technology

The tool holder also has an important function of properly holding the insert in the right position. Generally, during any machining operation, forces are generated and can be significant depending on the material and tool geometry. These forces may alter the position of the insert resulting in vibration, which affects both the surface finish and accuracy. Sandvik Coromant recently have developed iLock™ (Sandvik Coromant 2017b), a secure locking interface between the insert and the tool holder, which prevents the cutting forces from affecting the tool position resulting in better performance and dimensional accuracy. Another new technology (Sandvik Coromant 2017d), that Sandvik Coromant has developed for reducing vibration during machining is "Silent tools," which has a dampening system inside the tool body. This allows for higher overhang tools as well as other advantages, such as higher cutting parameters, which can increase the material removal rates. Additional benefits claimed are more secure and vibration-free process with close tolerances and good surface finish. All these benefits result in lowering the manufacturing cost of the component.

Another advancement in recent times is multitask machining that is capable of multiple operations, such as milling, turning, drilling, boring, and others, in a single setup without changing the workpiece setup. This reduces the setup time, improves productivity as well as accuracy ultimately resulting in lower manufacturing costs for the components. This does not require any special machines but most existing machining centers would be able to do this by adopting cutting tools designed for multitasking. For example CoroPlex MT system (Sandvik Coromant 2017a) of is capable of multifunctional milling and turning. Similarly, Haas automation has a dual spindle lathe (Haas Automation 2017b), which is capable of producing complex shapes in a single setup; programming is simple with just mirroring of the program generated for a single spindle machine. Other advancements in tool design, such as (a) through-tool air blast, and (b) high precision/pressure coolant, have been discussed in Section 20.3.2.1.

20.3.2.4 Trends and Advances in Cutting Tool Materials

The performance of cutting tool materials depend on several factors as illustrated in Figure 20.19. Most of the advancements in cutting tools address one of these factors. With the development of newer advanced materials that exhibit excellent mechanical properties, it has become increasingly difficult to machine them to produce the components. In truth, the use of these materials has been limited by the available cutting tool materials. Cutting tool manufacturers are trying to meet these demands through newer cutting tool materials mostly in the form of coatings on the most prominent carbide cutting tools. There has been no recent addition to the existing main classes of tool materials, such as (a) high-speed steel, (b) tungsten carbides, (c) cermets, (d) ceramics (alumina and silicon nitride), and (e) super-/ultra-hard materials, such as polycrystalline cubic boron nitride (PCBN) and polycrystalline diamond (PCD). Practically, it is not necessary to develop newer grades or classes of tool material but focus on coatings that can enhance the performance as per the requirement of the machining process. That is why even in 2005, the total market share of chemical vapor deposited and physical vapor deposited coated carbide tools was around 53% (MaschinenMarkt International 2017). With the advancement in surface coating technologies, especially hybrid coating technologies, such as (a) pulsed magnetron sputter ion plating physical vapor deposition (PVD) technology, (b) high power impulse magnetron sputtering, (c) unbalanced magnetron sputtering, and (d) laser chemical vapor deposition (CVD) (Bouzakis et al. 2012), it is possible to coat any of the cutting tool materials, not necessarily carbides, in order to incorporate the required properties for machining advanced materials. Most of these technologies are capable of producing nanostructured coatings, which are excellent candidates for tooling applications. The newer trends in coating technology are multilayered ones that can combine the mechanical properties of ideal cutting tool materials into one cutting tool.

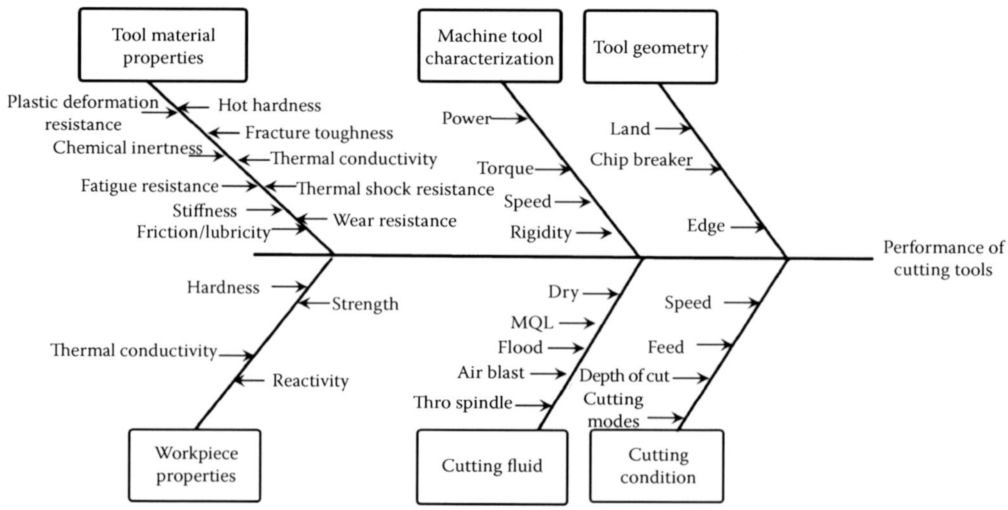

FIGURE 20.19 Factors affecting the performance of cutting tools.

Nanolayered physical vapor deposited titanium aluminium nitride coatings with a titanium/aluminium ratio of 46/54 were deposited on carbide tools using high power impulse magnetron sputtering (Skordaris et al. 2016). The coating structure consisted of successive nanolayers of titanium aluminium nitride (24 nm) and titanium nitride layers (3 nm) amounting to a total thickness of 2, 4, and 8 μm. The nanolayered coated tools exhibited improved wear resistance compared to a single layer of the same thickness when milling hardened steels at high cutting speeds. The highest thickness (8 μm) performed the best, yielding a tool life of almost 365,000 cuts. The better performance of nanolayer coatings when compared to multilayer and single layer coatings is due to the large number of interfaces, which act as inhibitors of crack propagation (Caliskan et al. 2013). Such coatings exhibit good mechanical properties, such as (a) hardness, (b) toughness, and (c) adhesion. The performance of nanolayer-aluminium titanium nitride/titanium nitride and multilayer nanocomposite titanium aluminium silicon nitride/titanium silicon nitride/titanium aluminium nitride coated carbide tools during high-speed machining of hardened steel was evaluated by Halil Caliskan and coworkers (Caliskan et al. 2013). The nanolayer coated tool performed the best among the compared tools and its tool life was 25, 77 and 2300% of that of the nanocomposite, commercial titanium nitride/titanium aluminium nitride coated and uncoated tools, respectively. The high-temperature hardness exhibited by nanostructured aluminium$_{0.8}$ titanium$_{0.2}$ nitride coating enabled an increase in productivity of 400% when machining steels and cast irons (Köpf et al. 2017).

Two novel nanocomposite coatings of aluminium titanium nitride/silicon nitride and aluminium chromium nitride/silicon nitride have been developed using physical vapor deposition technique and used in high-speed machining of titanium alloys under both dry and minimum quantity lubrication conditions (Liu et al. 2013). The advanced coatings exhibited high hardness, high toughness, and low friction allowing them to be used at high cutting speeds of 120 m/min and feed rate of 0.1 mm/rev. The aluminium titanium nitride/silicon nitride coated tool having a lower thickness of 1–4 μm exhibited better performance (in terms of tool life) than the aluminum chromium nitride/silicon nitride (thickness varying form 1–7 μm) under conditions of minimum quantity lubrication mainly because of the higher hardness.

The advantage of coatings is that even an inexpensive substrate material could be developed into a high-performance cutting tool. High-speed steel cutters have been deposited with aluminum chromium nitride coating and aluminum chromium silicon nitride multilayer nanocomposite coating (Wu et al. 2015). When machining alloys steels, the aluminum chromium silicon nitride coating showed better machining efficiency when compared to aluminum chromium nitride.

Recently, chemical vapor deposition technology is being used to develop diamond-coated cutting tools that are gaining increasing importance for machining of carbon fiber-reinforced plastic, graphite, nonferrous metals, and ceramics (Bobzin 2016). These diamond-coated tools may replace the expensive polycrystalline diamond tools in the near future once the issues of adhesion as well as producing sharp edges and a small cutting edge-included angle are resolved (Ahmad 2009). With the increased usage of carbon fiber-reinforced plastic in newer aircrafts, such as the Boeing 787 (25%) and the Airbus A350 (53%), diamond-coated tools are the potential candidates that can produce quality holes. Unlike engineering materials in which chips are formed, predominantly by plastic deformation, in fiber-reinforced plastics, the chip formation is by compression shearing and fracture of the fiber reinforcement and matrix. This necessitates proper cutting tool material and geometry. In fiber-reinforced plastics, because of the abrasive nature of the fibers and inhomogeneity of the material, microchipping occurs (Ahmad 2009). In addition to polycrystalline diamond and diamond-coated tools, nanolayer or nanocomposite coated carbide tools can also meet the demanding requirements for machining fiber-reinforced plastics because of their higher hardness and toughness.

Similar to fiber-reinforced plastics, metal matrix composites are also another class of engineered materials that are finding increasing applications, especially for lightweight and high-strength applications in automotive and aerospace industries. In metal matrix composites, the higher hardness of the reinforcement phase causes abrasive wear and the inhomogeneity causes microchipping, as in fiber-reinforced plastics. Among the tool materials, polycrystalline diamond is recommended for machining of metal matrix composites (Nicholls et al. 2016; Singh et al. 2015). However, because of the higher cost, diamond-coated and other nanolayer and nanocomposite-coated carbides (CC), along with high-performance cooling strategies, will improve the machinability of metal matrix composites in the near future.

Sandvik Coromant has come up with a technical breakthrough in coating alumina layer on carbide inserts using chemical vapor deposition technology. Using this Inveio™ (Sandvik Coromant 2017c) technology they are able to tightly pack, as well as align the direction of crystal growth, in one direction instead of being random. The crystals are lined up in the same direction with the strongest part facing the top surface as shown in the microscopic image in Figure 20.20. Each crystal direction in the microscopic image is given a unique color to illustrate. This arrangement exhibits significant improvement in both crater wear and flank wear resistance and also swiftly takes the heat away from the cutting zone and extends tool life.

Similar to the Inveio™ technology, Sandvik has also developed another new technology for PVD coatings called as Zertivo™ (Sandvik Coromant 2017e), which improves the adhesion between the substrate and the coating coupled with an optimized cutting edge integrity, resulting in longer tool life and secure machining, especially when machining steels. The new physical vapor deposition coating is designed to create a perfect balance between protection against abrupt chipping and continuous wear resistance. This is an essential prerequisite when high-speed machining of complex parts, dies, or molds that involve complex tool paths or deep cavities.

It is not necessary to always come up with expensive coating technologies to improve the performance of cutting tools in machining materials. For example, a simple microtextures (dimple or channel) on the rake face of uncoated cutting tool (without chip breakers) can modify the formation of built-up-edge resulting in better surface finish or reduced wear when machining plain carbon steel (Kummel et al. 2015). Sugihara et al. (2017) also used surface texture on cubic boron nitride cutting tool for high-speed machining of Inconel 718 and obtained a higher tool life. Further research is required to extend this to machining of other materials.

It is quite evident from the descriptions above that the presence of advanced coatings on cutting tools can meet the demands of minimum quantity lubrication, high-speed machining, high performance, and hard machining. Development of coatings from both academia and cutting tool manufacturers have been discussed and it is apparent that a good collaboration between both parties could benefit aerospace, automotive, energy, and die and mold industries to a great extent.

FIGURE 20.20 Microscope image of CVD Alumina coating from Sandvik Coromant using Inveio™ technology exhibiting crystals lined up in the same direction, towards the top surface (Sandvik Coromant 2017c). (Courtesy of Sandvik Coromant.)

It is not enough to focus on high-performance coatings for cutting tools, but also the geometry of the cutting edge. In the case of ceramic cutting tools that lack sufficient toughness, it is important to increase the strength of the cutting edge since it is critical to tool life and surface integrity in ceramic tool applications. Honing the cutting edge and/or adding a negative land guard the edge from chipping and breakage by changing the shear forces at the edge to compressive forces (Arunachalam and Mannan 2000). Cutting edge preparation and cutting tool edge geometry can also significantly affect heat distribution in the cutting tool (Bhemuni et al. 2015). Cutting edge preparation is not only important for ceramic cutting tools but even in advanced cutting tools, such as nanostructured hard coatings on cemented carbide cutting tools. Also chamfer, with the angle of 32°, helps to strengthen the cutting edge (Caliskan et al. 2013).

20.4 CONCLUDING REMARKS

The trends of techniques and advances in conventional machining processes have been discussed in this chapter, with a focus on two major industrial sectors, namely aerospace and automotive. Advancements that can help in improving machinability when using conventional machining processes have been summarized in Figure 20.21. From these discussions, it is clear that conventional machining processes can be competitive with additive manufacturing only if advancements in machining technology are implemented. In fact additive manufacturing when combined with conventional machining could complement each other to create a hybrid process capable of producing advanced components for the aerospace and automotive industries. High-speed machining is one enabling technology capable of meeting almost all the requirements of the machining processes, such as high productivity/performance/efficiency, high-performance cooling, hard machining, and multitasking. With the present emphasis on sustainability, it is important that conventional machining processes are sustainable. One such step in that direction is to make use of high-performance cooling strategies, as reviewed in this chapter. In addition to the conventional machining processes, hybrid processes can help in improving the manufacturing processes, especially in machining of both engineering and engineered materials.

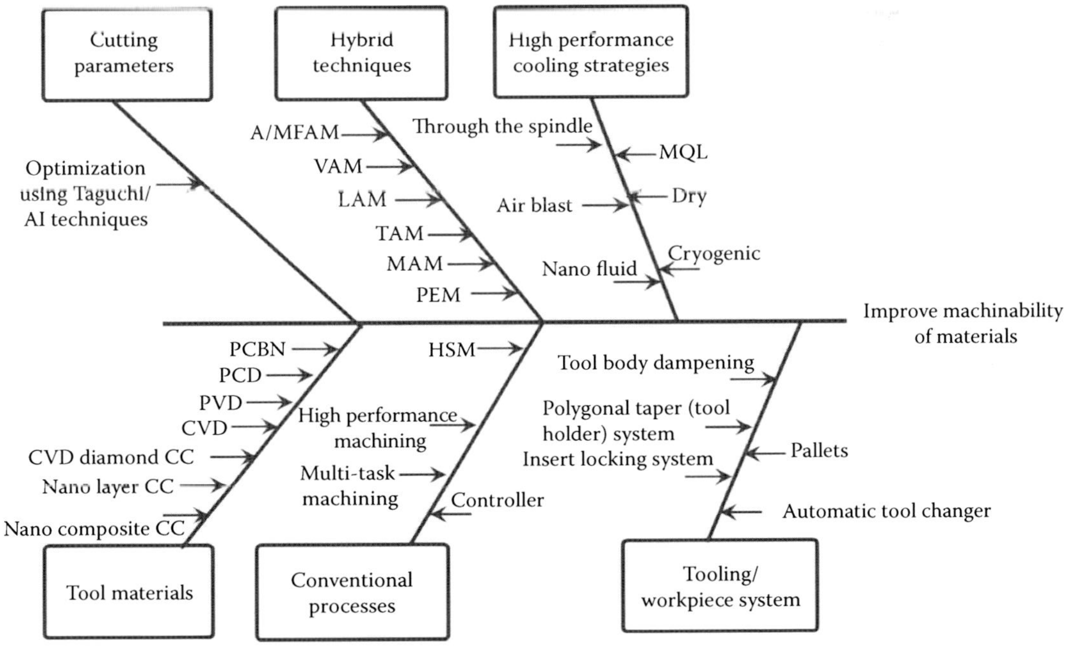

FIGURE 20.21 Advances in factors that improve the machinability of materials.

ACKNOWLEDGMENTS

Sayyad Zahid Qamar would like to thank Mr. Josiah Cherian Chekotu (M.Sc. student in the Department of Mechanical & Industrial Engineering at Sultan Qaboos University) for his contribution in preparing the manuscript, especially the schematic illustrations. Similarly, Ramanathan Arunachalam would like to thank Pradeep Kumar Krishnan (Caledonian College of Engineering, Sultanate of Oman) for his contribution in preparing the cause and effect diagrams used in this chapter.

REFERENCES

(2017). AML Engineering: Cross processes. Accessed on May 16, 2017, from http://aml.engineering.columbia.edu/ntm/CrossProcess/CrossProcessSect5.htm.

Abdelhafeez, A., S. Soo, D. Aspinwall, A. Dowson and A. Arnold (2016). A coupled Eulerian Lagrangian finite element model of drilling titanium and aluminium alloys. *SAE International Journal of Aerospace* **9**(1): 198–207.

Adnan, A. S. and S. Subbiah (2010). Experimental investigation of transverse vibration-assisted orthogonal cutting of Al-2024. *International Journal of Machine Tools and Manufacture* **50**: 294–302.

Ahmad, J. (2009). *Machining of Polymer Matrix Composites*, Springer, USA.

Aoki, S., S. Hirai and T. Nishimura (2005). Prevention from delamination of composite material during drilling using ultrasonic vibration. *Key Engineering Materials* **291–292**: 465–470.

Aramcharoen, A. and S. K. Chuan (2014). An experimental investigation on cryogenic milling of Inconel 718 and its sustainability assessment. *Procedia CIRP* **14**: 529–534.

Arunachalam, R. and M. A. Mannan (2000). Machinability of nickel-based high temperature alloys. *Machining Science and Technology* **4**(1): 127–168.

Aspinwall, D. K., R. C. Dewes, J. M. Burrows and M. A. Paul (2001). Hybrid high speed machining (HSM): System design and experimental results for grinding/HSM and EDM/HSM. *CIRP Annals—Manufacturing Technology* **50**(1): 145–148.

Axinte, D. A., P. Andrews, W. Li, N. Gindy and P. J. Withers (2006). Turning of advanced Ni based alloys obtained via powder metallurgy route. *CIRP Annals—Manufacturing Technology* **55**(1): 117–120.

Axinte, D. A., N. K. Gindy, I. Fox, and I. Unanue (2004). Process monitoring to assist the workpiece surface quality in machining. *International Journal of Machine Tools and Manufacture* **44**(10): 1091–1108.

Ayed, Y., G. Germain, A. Ammar and B. Furet (2013). Degradation modes and tool wear mechanisms in finish and rough machining of Ti17 Titanium alloy under high-pressure water jet assistance. *Wear* **305**: 228–237.

Behera, B. C., S. Ghosh and P. V. Rao (2016). Use of high pressure jet during turning of Inconel 718 super alloy and optimization of jet pressure. Proceedings of 6th International and 27th All India Manufacturing Technology, Design and Research Conference (AIMTDR-2016), Pune, India.

Bhemuni, V., S. R. Chalamalasetti, P. K. Konchada and V. V. Pragada (2015). Analysis of hard turning process: Thermal aspects. *Advances in Manufacturing* **3**(4): 323–330.

Blawert, C., N. Hort and K. U. Kainer (2004). Automotive applications of magnesium and its alloys. *Transactions of the Indian Institute of Metals* **57**(4): 397–408.

Bobzin, K. (2016). High-performance coatings for cutting tools. *CIRP Journal of Manufacturing Science and Technology* **18**: 1–9; https://doi.org/10.1016/j.cirpj.2016.11.004.

Bouzakis, K. D., N. Michailidis, G. Skordaris, E. Bouzakis, D. Biermann and R. M'Saoubi (2012). Cutting with coated tools: Coating technologies, characterization methods and performance optimization. *CIRP Annals—Manufacturing Technology* **61**(2): 703–723.

Brecher, C. and M. Emonts (2010). Laser-assisted shearing: New process developments for the sheet metal industry. *International Journal of Advanced Manufacturing Technology* **48**(1–4): 133–141.

Brehl, D. E. and T. A. Dow (2008). Review of vibration-assisted machining. *Precision Engineering* **32**(3): 153–172.

Bushlya, V., J. M. Zhou, F. Lenrick, P. Avdovic and J. E. Stahl (2011). Characterization of white layer generated when turning aged Inconel 718. *Procedia Engineering* **19**: 60–66.

Caliskan, H., C. Kurbanoglu, P. Panjan, M. Cekada, and D. Kramar (2013). Wear behavior and cutting performance of nanostructured hard coatings on cemented carbide cutting tools in hard milling. *Tribology International* **62**: 215–222.

Cantero, J. L., J. Dıaz, A. lvarez, M. H. Miguelez, and N. C. Marin (2013). Analysis of tool wear patterns in finishing turning of Inconel 718. *Wear* **297**: 885–894.
Carou, D., E. M. Rubio, C. H. Lauro and J. P. Davim (2014). Experimental investigation on finish intermittent turning of UNS M11917 magnesium alloy under dry machining. *International Journal of Advanced Manufacturing Technology* **75**(9–12): 1417–1429.
Carou, D., E. M. Rubio, C. H. Lauro and J. P. Davim (2016). The effect of minimum quantity lubrication in the intermittent turning of magnesium based on vibration signals. *Measurement* **94**: 338–343.
Cha, N. H., W. S. Woo and C. M. Lee (2015). A study on the optimum machining conditions for laser-assisted turn-mill. *International Journal of Precision Engineering and Manufacturing* **16**(11): 2327–2332.
Chang, L., R. Chengzu, W. Guofeng, Y. Yinwei and Z. Lu (2015). Study on surface defects in milling Inconel 718 superalloy. *Journal of Mechanical Sciences and Technology* **29**(4): 1723–1730.
Chang, S. S. F. and G. M. Bone (2009). Thrust force model for vibration assisted drilling of aluminum 6061-T6. *International Journal of Machine Tools and Manufacture* **49**: 1070–1076.
Chen, Z., R. L. Peng, P. Avdovic, J. Moverare, F. Karlsson, J. M. Zhou and S. Johansson (2014). Analysis of thermal effect on the residual stresses of broached Inconel 718. *Advanced Material Research* **966**: 574–579.
Chetan, H., B. C. Behera, S. Ghosh and P.V. Rao (2016). Application of nanofluids during minimum quantity lubrication: A case study in turning process. *Tribology International* **101**: 234–246.
Chinchanikar, S. and S. K. Choudhury (2014). Hard turning using HiPIMS-coated carbide tools: Wear behavior under dry and minimum quantity lubrication (MQL). *Measurement: Journal of the International Measurement Confederation* **55**: 536–548.
Clark, T. and T. Scherpenberg (2014). High-performance machining center ROI: How to determine a machine's true value. Accessed from https://www.makino.com/Resources/Whitepapers/
Corduan, N., T. Himbert, G. Poulachon, M. Dessoly, M. Lambertin, J. Vigneau and B. Payoux (2003). Wear mechanisms of new tool materials for Ti–6Al–4V high performance machining. *CIRP Annals—Manufacturing Technology* **52**(1): 73–76.
Da Silva, R. B., A. R. Machado, E. O. Ezugwu, J. Bonney and W. F. Sales (2013). Tool life and wear mechanisms in high speed machining of Ti–6Al–4V alloy with PCD tools under various coolant pressures. *Journal of Materials Processing Technology* **213**: 1459–1464.
Dabade, U. A., H. A. Sonawane and S. S. Joshi (2010). Cutting forces and surface roughness in machining Al/SiCp composites of varying composition. *Machining Science and Technology* **14**(2): 258–279.
Dandekar, C. R., Y. C. Shin, J. Barnes (2010). Machinability improvement of titanium alloy (Ti–6Al–4V) via lam and hybrid machining. *International Journal of Machine Tools and Manufacture* **50**(2): 174–182.
Danish, M., T. L. Ginta, K. Habib, D. Carou, A. M. A. Rani and B. R. Saha (2017). Thermal analysis during turning of AZ31 magnesium alloy under dry and cryogenic conditions. *International Journal of Advanced Manufacturing Technology* **91**(5–8): 2855–2868; https://doi.org/10.1007/s00170-016-9893-5.
Das, B., S. Roy, R. N. Rai and S. C. Saha (2016). Study on machinability of in situ Al–4.5%Cu–TiC metal matrix composite-surface finish, cutting force prediction using ANN. *CIRP Journal of Manufacturing Science and Technology* **12**: 67–78.
DMG Mori (2017). HSC 75 Linear. Accessed May 17, 2017, from http://us.dmgmori.com/products/milling-machines/high-speed-precision-cutting-centers/hsc/hsc-75-linear#Intro.
Duchosal, A., S. Werda, R. Serra, C. Courbon and R. Leroy (2016). Experimental method to analyze the oil mist impingement over an insert used in MQL milling process. *Measurement* **86**: 283–292.
El-Hofy, H. (2005). *Advanced Machining Processes: Nontraditional and Hybrid Machining Processes*. McGraw-Hill, Egypt.
Kilickap, E. (2016). Effect of cutting environment and heat treatment on the surface roughness of drilled Al/SiC MMC. *Materials Testing* **58**(4): 357–361.
European Commission (2011). Flightpath 2050—Europe's vision for aviation. Report of the high-level group on aviation research., Publications Office of the European Union, Luxembourg.
Ezilarasan, C. and A. Velayudham (2013). Effect of machining parameters on surface integrity in machining NimonicC-263 superalloy using whisker-reinforced ceramic insert. *Journal of Materials Engineering and Performance* **22**(6): 1619–1628.
Ezugwu, E. O., R. B. Da Silva, J. Bonney and A. R. Machado (2005). Evaluation of the performance of CBN tools when turning Ti–6Al–4V alloy with high pressure coolant supplies. *International Journal of Machine Tools and Manufacture* **45**: 1009–1014.
Faraz, A., D. Biermann and K. Weinert (2009). Cutting edge rounding: An innovative tool wear criterion in drilling CFRP composite laminates. *International Journal of Machine Tools and Manufacture* **49**: 1185–1196.

Filho, S. L. M. R., J. T. Vieira, J. A. D. Oliveira, E. M. Arruda and L. C. Brandao (2017). Comparison among different vegetable fluids used in minimum quantity lubrication systems in the tapping process of cast aluminum alloy. *Journal of Cleaner Production* **140**(3): 1255–1262.

Gabler, J. and S. Pleger (2010). Precision and micro CVD diamond-coated grinding tools. *International Journal of Machine Tools and Manufacture* **50**: 420–424.

Gandarias, E (2017). Abrasive technologies. Accessed May 5, 2017, from https://www.slideshare.net/endika55/abrasive-technologies-71766273.

Giasin, K. and S. Ayvar-Soberanis (2017). Microstructural investigation of drilling induced damage in fibre metal laminates constituents. Composites Part A: *Applied Science and Manufacturing* **97**: 166–178.

Giasin, K., A. Alma Hodzic, V. Phadnis and S. Ayvar-Soberanis (2016). Assessment of cutting forces and hole quality in drilling Al2024 aluminium alloy: Experimental and finite element study. *International Journal of Advanced Manufacturing Technology* **87**(5): 2041–2061.

Giasin, K., S. Ayvar-Soberanis and A. Hodzic (2016). Evaluation of cryogenic cooling and minimum quantity lubrication effects on machining GLARE laminates using design of experiments. *Journal of Cleaner Production* **135**: 533–548.

Ginting. A. and M. Nouari (2009). Surface integrity of dry machined titanium alloys. *International Journal of Machine Tools and Manufacture* **49**: 325–332.

Grzesik, W. (2017). *Advanced Machining Processes of Metallic Materials* (second edition). Elsevier B. V, Amsterdam.

Guo, P. and K. F. Ehmann (2013). An analysis of the surface generation mechanics of the elliptical vibration texturing process. *International Journal of Machine Tools and Manufacture* **64**: 85–95.

Haas Automation (2017a). Haas UMC-750. Accessed from http://int.haascnc.com/mt_spec1.asp?intLanguageCode=1033&id=UMC-750&webID=UNIVERSAL.

Haas Automation (2017b). What's new. Accessed May 17, 2017, from http://int.haascnc.com/whatsnew.asp?intLanguageCode=1033.

Herbert, C. R. J., D. A. Axinte, M. C. Hardy and P. D. Brown (2012). Investigation in to the characteristics of white layers produced in a nickel-based superalloy from drilling operations. *International Journal of Machining Science and Technology* **16**(4): 40–52.

Hoffmeister, H. W. and A. Wenda (2000). Novel grinding tools for machining precision micro parts of hard and brittle materials. Proceedings of 15th ASPE Annual Meeting, Scottsdale, Arizona.

Hosseini, A. and H. A. Kishawy (2014). Cutting tool materials and tool wear. *Machining of Titanium Alloys*, Springer-Verlag, Berlin, pp. 31–56.

Hughes, J. I., A. R. C. Sharman and K. Ridgway (2004). The effect of tool edge preparation on tool life and workpiece surface integrity. *Proceedings of the Institute of Mechanical Engineers, Part B: Journal of Engineering Manufacture* **218**: 1113–1122.

International Organization of Motor Vehicle Manufacturers (2017). Economic contributions OICA. Accessed May 4, 2017, from http://www.oica.net/category/economic-contributions/.

Ito, Y., T. Kizaki, R. Shinomoto, M. Ueki, N. Sugita and M. Mitsuishi (2017). High-efficiency and precision cutting of glass by selective laser-assisted milling. *Precision Engineering* **47**: 498–507.

Ito, Y., M. Ueki, T. Kizaki, N. Sugita and M. Mitsuishi (2017). Precision cutting of glass by laser-assisted machining. International Conference on Sustainable Materials Processing and Manufacturing SMPM, Kruger National Park, South Africa.

Iturbe, A., E. Hormaetxe, A. Garay and P. J. Arrazola (2016). Surface integrity analysis when machining Inconel 718 with conventional and cryogenic cooling. *Procedia CIRP* **45**: 67–70.

Jamshidi, H. and M. J. Nategh (2013). Theoretical and experimental investigation of the frictional behavior of the tool–chip interface in ultrasonic vibration assisted turning. *International Journal of Machine Tools and Manufacture* **65**: 1–7.

Jawahir, I. S., H. Attia, D. Biermann, J. Duflou, F. Klocke, D. Meyer, S. T. Newman, F. Pusavec, M. Putz, J. Rech, V. Schulze, D. Jeon, Y. and C. M. Lee (2012). Current research trend on laser assisted machining. *International Journal of Precision Engineering and Manufacturing* **13**(2): 311–317.

Jeon, Y. and Lee, C.M. (2012). Current research trend on laser assisted machining. *International Journal of Precision Engineering Manufacturing*. **13**: 311; https://doi.org/10.1007/s12541-012-0040-4.

Jia Z. Z., R. Fu, F. Wang, B. Qian, C. He (2016). Temperature effects in end milling carbon fiber reinforced polymer composites. *Polymer Composites* 1–11, doi:10.1002/pc.23954.

Josyula, S. K. and S. K. R. Narala (2017). Machinability enhancement of stir cast Al-TiCp composites under cryogenic condition. *Journal of Materials and Manufacturing Processes* **32**(15): 1764–1774; http://dx.doi.org/10.1080/10426914.2017.1303151.

Josyula, S. K., S. K. R. Narala, E. G. Charan, H. A. Kishawy (2016). Sustainable machining of metal matrix composites using liquid nitrogen. *Procedia CIRP* **40**: 568–573.

Kadivar, M. A., J. Akbari, R. Yousefi, A. Rahi, M. G. Nick (2014). Investigating the effects of vibration method on ultrasonic-assisted drilling of Al/SiCp metal matrix composites. *Robotics and Computer-Integrated Manufacturing* **30**: 344–350.

Kalyan, C. and G. L. Samuel (2015). Cutting mode analysis in high speed finish turning of AlMgSi alloy using edge chamfered PCD tools. *Journal of Materials Processing Technology* **216**: 146–159.

Kashyap, S (2016). Nanofinishing. Accessed May 6, 2017, from https://www.slideshare.net/EmeraldPolytechnic/nanofinishing-61211431.

Katja, B., H. Carsten, P. Bernhard, S. Andrea and W. Rafael (2016). Investigation of cooling and lubrication strategies for machining high-temperature alloys. *Procedia CIRP* **41**: 835–840.

Khairussaleh, N. K. M., C. H. C. Haron and J. Ghani (2016). Study on wear mechanism of solid carbide cutting tool in milling CFRP. *Journal of Materials Research* **31**(13): 1893–1899.

Kitagawa, T., K. Katsuhiro and A. Kubo (1988). Plasma hot machining for high hardness metals. *Japan Society of Precision Engineering* **22**(2): 145–151.

Klocke, F., L. Settineri, D. Lung, P. Claudio Priarone and M. Arft (2013). High performance cutting of gamma titanium aluminides: Influence of lubricoolant strategy on tool wear and surface integrity. *Wear* **302**: 1136–1144.

Klocke, F., P. Vogtel, S. Gierlings, D. Lung and D. Veselovac (2013). Broaching of Inconel 718 with cemented carbide. *Production Engineering* **7**: 593–600.

Koepfer, C. (2010), Hard turning as an alternative to grinding. Accessed May 17, 2017, from http://www.productionmachining.com/articles/hard-turning-as-an-alternative-to-grinding.

Komanduri, R., D. G. Flom and M. Lee (1985). Highlights of the DARPA advanced machining research program. *Transactions of the ASME. Journal of Engineering for Industry* **107**: 325–335.

Kong, J., Z. Xia, D. Xu and N. He (2016). Investigation on notch wear mechanism in finish turning pure iron material with uncoated carbide tools under different cooling/lubrication conditions. *International Journal of Advanced Manufacturing Technology* **86**: 97–105.

Konig, W., L. Cronjager, G. Spur, H. K. Tonshoff, M. Vigneau, W. J. Zdeblick (1990). *Annals CIRP*, **39**(2): 673–681.

Köpf, A., J. Keckes, J. Todt, R. Pitonak and R. Weissenbacher (2017). Nanostructured coatings for tooling applications. *International Journal of Refractory Metals and Hard Materials* **62**: 219–224.

Kozak, J. and Rajurkar, K. P (2000). Hybrid machining process evaluation and development. Proceedings of 2nd International Conference on Machining and Measurements of Sculptured Surfaces, Krakow, Poland.

Kramer, A., F. Klocke, H. Sangermann and D. Lung (2014). Influence of the lubricoolant strategy on thermo mechanical tool load. *CIRP Journal of Manufacturing Science and Technology* **7**: 40–47.

Krishnamurthy, G., S. Bhowmick, S. W. Altenhof and A. T. Alpas (2017). Increasing efficiency of Ti-alloy machining by cryogenic cooling and using ethanol in MRF. *CIRP Journal of Manufacturing Science and Technology* **18**: 159–172.

Krolczyk, G. M., P. Nieslony, R. W. Maruda and S. Wojciechowski (2017). Dry cutting effect in turning of a duplex stainless steel as a key factor in clean production. *Journal of Cleaner Production* **142**(4): 3343–3354.

Kumar, M., S. Melkote and G. Lahoti (2011). Laser-assisted microgrinding of ceramics. *CIRP Annals—Manufacturing Technology* **60**: 367–370.

Kumar, M. N., S. K. Subbu, P. V. Krishna and A. Venugopal (2014). Vibration assisted conventional and advanced machining: A review. *Procedia Engineering* **97**: 1577–1586.

Kummel, J., D. Braun, J. Gibmeier, J. Schneider, C. Greiner, V. Schulze and A. Wanner (2015). Study on micro texturing of uncoated cemented carbide cutting tools for wear improvement and built-up edge stabilisation. *Journal of Materials Processing Technology* **215**: 62–70.

Leshock, C. E., J. N. Kim and Y. C. Shin (2001). Plasma enhanced machining of Inconel 718: Modeling of workpiece temperature with plasma heating and experimental results. *International Journal of Machine Tools and Manufacture* **41**(6). 877–897.

Li, B., C. Li, Y. Zhang, Y. Wang, M. Yang, D. Jia, N. Zhang and Q. Wu (2017). Effect of the physical properties of different vegetable oil-based nanofluids on MQLC grinding temperature of Ni-based alloy. *International Journal of Advanced Manufacturing Technology* **89**: 3459.

Liang, Z., Y. Wu and X. Wang (2010). A new two-dimensional ultrasonic assisted grinding (2D-UAG) method and its fundamental performance in mono crystal silicon machining. *International Journal of Machine Tools and Manufacture* **50**: 728–736.

Lin, H., C. Wang, Y. Yuan, Z. Chen, Q. Wang and W. Xiong (2015). Tool wear in Ti–6Al–4V alloy turning under oils on water cooling comparing with cryogenic air mixed with minimal quantity lubrication. *International Journal of Advanced Manufacturing Technology* **81**: 87–101.

Liu, J., D. Zhang, L. Qin and L. Yan (2012). Feasibility study of the rotary ultrasonic elliptical machining of carbon fiber reinforced plastics (CFRP). *International Journal of Machine Tools and Manufacture* **53**: 141–150.

Liu, Z. Q., M. Chen and Q. L. An (2015). Investigation of friction in end-milling of Ti–6Al–4V under different green cutting conditions. *International Journal of Advanced Manufacturing Technology* **78**(5): 1181–1192.

Liu, Z., Q. An, J. Xu, M. Chen and S. Han (2013). Wear performance of (Nc-AlTiN)/(a-Si_3N_4) coating and (Nc-AlCrN)/(a-Si_3N_4) coating in high-speed machining of titanium alloys under dry and minimum quantity lubrication (MQL) conditions. *Wear* **305**(1–2): 249–259.

Makino (2017). Makino SGI.5 shortens machining cycle times while maintaining precise surface finishes. Accessed May 16, 2017, from https://www.makino.com/about/news/Makino-SGI5-Shortens-Machining-Cycle-Times-While-Maintaining-Precise-Surface-Finishes/678/.

Manimaran, G., M. P. Kumar and R. Venkatasamy (2014). Influence of cryogenic cooling on surface grinding of stainless steel 316. *Cryogenics* **59**: 76–83.

Marques, A., C. Guimarães, R. Batista da Silva, M. C. Fonseca, W. F. Sales and A. R. Machado (2016). Surface integrity analysis of Inconel 718 after turning with different solid lubricants dispersed in neat oil delivered by MQL. *Procedia Manufacturing* **5**: 609–620.

MaschinenMarkt International (2017). USA: Development of machine tools market. Accessed May 18. 2017, from http://www.maschinenmarkt.international/index.cfm?pid=14044&pk=604860&print=true&printtype=article.

Mathew, N. T. and L. Vijayaraghavan (2016). High-throughput dry drilling of titanium aluminide. *Materials and Manufacturing Processes* **32**(2): 199–208.

McCune, R. C. and G. A. Weber (2001). Automotive engine materials. *Encyclopedia of Materials—Science and Technology*, **C**: 426–434.

Mehbudia, P., V. Baghlania, J. Akbaria, A. R. Bushroab and N. A. Mardib (2013). Applying ultrasonic vibration to decrease drilling-induced delamination in GFRP laminates. *Procedia CIRP* **6**: 577–582.

Mia, M., M. A. Khan and N. R. Dhar (2016). High-pressure coolant on flank and rake surfaces of tool in turning of Ti–6Al–4V: Investigations on surface roughness and tool wear. *International Journal of Advanced Manufacturing Technology* **90**: 1825–1834; doi:10.1007/s00170-016-9512-5.

Molaie, M. M., J. Akbari and M. R. Movahhedy (2016). Ultrasonic assisted grinding process with minimum quantity lubrication using oil-based nanofluids. *Journal of Cleaner Production* **129**: 212–222.

Moon, M. (2017). Boeing uses first FAA-approved 3D-printed parts for the 787. Accessed from https://www.engadget.com/2017/04/11/boeing-faa-approved-3d-printed-metals-787/.

M'Saoubi, R., D. Axinte, S. L. Soo, C. Nobel, A. Attia, G. Kappmeyer, S. Engin and W. M. Sim (2015). High performance cutting of advanced aerospace alloys and composite materials. *CIRP Annals—Manufacturing Technology* **64**: 557–580.

Muhammad, R., N. Ahmed, A. Roya and V. V. Silberschmidta (2012). Numerical modeling of vibration-assisted turning of Ti-15333. *Procedia CIRP* **1**: 377–382.

Muthusamy, Y., K. Kadirgama, M. M. Rahman, D. Ramasamy and K. V. Sharma (2016). Wear analysis when machining AISI 304 with ethylene glycol/TiO_2 nanoparticle-based coolant. *International Journal of Advanced Manufacturing Technology* **82**(1–4): 327–340.

Najariana, F., M. Y. Noordin and D. Kurniawan (2014). Magnetic field assisted machining. Proceedings of the 2014 International Conference on Industrial Engineering and Operations Management, Bali, Indonesia.

Nassar, M. M. A., R. Arunachalam and K. I. Alzebdeh (2016). Machinability of natural fiber reinforced composites: A review. *International Journal of Advanced Manufacturing Technology* **88**(9): 2985–3004.

Nicholls, C. J., B. Boswell, I. J. Davies and M. N. Islam (2016). Review of machining metal matrix composites. *International Journal of Advanced Manufacturing Technology* **90**(9): 2429–2441.

Nouari, M., G. List, F. Girot and D. Coupard (2003). Experimental analysis and optimization of tool wear in dry machining of aluminium alloys. *Wear* **255**:1359–1368.

Novak, J. W., Y. C. Shin, and F. P. Incropera (1997). Assessment of plasma enhanced machining of improved machinability of Inconel 718—Transactions of ASME. *Journal of Manufacturing Science and Engineering* **119**: 125–129.

Padmini, R., P. V. Krishna and G. K. M. Rao (2016). Effectiveness of vegetable oil based nanofluids as potential cutting fluids in turning AISI 1040 steel. *Tribology International* **94**: 490–501.

Park, K. H., A. Beal, D. Kim, P. Kwon and J. Lantrip (2011). Tool wear in drilling of composite/titanium stacks using carbide and polycrystalline diamond tools. *Wear* **271**(11–12): 2826–2835.

Paswan, S. (2017). Abrasive machining and finishing. Accessed May 5, 2017, from https://www.slideshare.net/SunilPaswan4/m-processes-notes8-1.

Patil, A., A. Patel and R. Purohit (2017). An overview of polymeric materials for automotive applications. *Proceedings—Materials Today* **4**(2): 3807–3815.

Patil, S., S. Joshi, A. Tewari and S. S. Joshi (2014). Modelling and simulation of effect of ultrasonic vibrations on machining of Ti6Al4V. *Ultrasonics* **54**(2): 694–705.

Polvorosa, R., A. Suárez, L.N. López de Lacalle, I. Cerrillo, A. Wretland and F. Veiga (2017). Tool wear on nickel alloys with different coolant pressures: Comparison of Alloy 718 and Waspaloy. *Journal of Manufacturing Processes* **26**: 44–56.

Qu, W., K. Wang, M. H. Miller, Y. Huang and A. Chandra (2000). Using vibration-assisted grinding to reduce subsurface damage. *Journal of the International Societies for Precision Engineering and Nanotechnology* **24**: 329–337.

Rahman, M. M., N. H. Razak and K. Kadirgama (2013). Experiment study on surface integrity in end milling of Hastelloy C-2000 super alloy. Proceedings of International Conference on Mechanical Engineering Research (ICMER), Pahang, Malaysia.

Ramulu, M., T. Branson and D. Kim (2001). A study on the drilling of composite and titanium stacks. *Composite Structures* **54**(1): 67–77.

Razak, N. H., M. M. Rahman and K. Kadirgama (2014). Experimental study on surface integrity in end milling of Hastelloy C-2000 superalloy. *International Journal of Automotive and Mechanical Engineering* **9**: 1578–1587.

Reddy, P. and A. Gosh (2014). Effect of cryogenic cooling on spindle power and G ratio in grinding of hardened bearing steel. *Procedia Materials Science* **5**: 2622–2628.

Ruzzi, R., R. Belentani, H. J. de Mello, R. C. Canarim, D. M. D'Addona, A. E. Diniz, P. Roberto de Aguiar and E. C. Bianchi (2017). MQL with water in cylindrical plunge grinding of hardened steels using CBN wheels, with and without wheel cleaning by compressed air. *International Journal of Advanced Manufacturing Technology*, **90**(1): 329.

Sadik, M. I. and G. Gustav (2016). Application of different cooling strategies in drilling of metal matrix composite (MMC). *Materials Science Forum* **836/837**: 3–12.

Saleem, M. Q., A. H. Ahmad, A. Raza and M. A. M. Qureshi (2017). Air-assisted boric acid solid powder lubrication in surface grinding: An investigation into the effects of lubrication parameters on surface integrity of AISI 1045. *International Journal of Advanced Manufacturing Technology* **91**: 3561–3572; doi:10.1007/s00170 016 9982 5.

Sandvik Coromant (2017a). CoroPlex MT. Accessed May 17, 2017, from http://www.sandvik.coromant.com/en-gb/products/coroplex_mt/Pages/default.aspx.

Sandvik Coromant (2017b). iLock—The ingenious locking interface. Accessed May 16, 2017, from http://www.sandvik.coromant.com/en-gb/knowledge/technologies/ilock/pages/default.aspx

Sandvik Coromant (2017c). Inveio™—Uni-Directional Crystal Orientation. Accessed Aug 23, 2017, from http://www.sandvik.coromant.com/en-gb/knowledge/technologies/inveio/Pages/default.aspx.

Sandvik Coromant (2017d). Silent tools. Accessed May 16, 2017, from http://www.sandvik.coromant.com/en-gb/knowledge/technologies/silent-tools/pages/default.aspx.

Sandvik Coromant (2017e). ZertivoTM technology. Accessed May 16, 2017, from http://www.sandvik.coromant.com/en-gb/knowledge/technologies/zertivo technology/pages/default.aspx.

Santos, M. C., A. R. Machado, W. F. Sales, M. A. S. Barrozo and E. O. Ezugwu (2016). Machining of aluminum alloys: A review. *International Journal of Advanced Manufacturing Technology* **86**(9–12): 3067–3080.

Sebastian, H., L. Pejryd and J. Ekengren (2016). Additive manufacturing and high speed machining—Cost comparison of short lead time manufacturing methods. *Procedia CIRP* **50**: 384–389.

Sekhar, R. and T. P. Singh (2015). Mechanisms in turning of metal matrix composites: A review. *Journal of Materials Research and Technology* **4**(2): 197–207.

Shankar, S., T. Mohanraj and K. Ponappa (2017). Influence of vegetable based cutting fluids on cutting force and vibration signature during milling of aluminium metal matrix composites. *Malaysian Tribology Society* **12**: 1–17.

Sharif, S. and E. A. Rahim (2007). Performance of coated and uncoated carbide tools when drilling titanium alloy Ti–6Al–4V. *Journal of Materials Processing Technology* **185**(1–3): 72–76.

Sharma, A. K., A. K. Tiwari, R. K. Singh and A. R. Dixit (2016). Tribological investigation of TiO_2 nanoparticle based cutting fluid in machining under minimum quantity lubrication (MQL). *Materials Today: Proceedings* **3**(6): 2155–2162.

Sharma, P., B. S. Sidhu and J. Sharma (2015). Investigation of effects of nanofluids on turning of AISI D2 steel using minimum quantity lubrication. *Journal of Cleaner Production* **108**: 72–79.

Sharman, A. R. C., J. J. Hughes and K. Ridgway (2015). The effect of tool nose radius on surface integrity and residual stresses when turning Inconel 718. *Journal of Materials Processing Technology* 123–132.

Shokrani, A., V. Dhokia and S. T. Newman (2016). Investigation of the effects of cryogenic machining on surface integrity in CNC end milling of Ti–6Al–4V titanium alloy. *Journal of Manufacturing Processes* **21**: 172–179.

Shokrani, A., V. Dhokia, S. T. Newman and R. I. Asrai (2012). An initial study of the effect of using liquid nitrogen coolant on the surface roughness of Inconel 718 nickel-based alloy in CNC milling. *Procedia CIRP* **3**: 121–125.

Singh, P., S. S. Sidhu and H. S. Payal (2015). Fabrication and machining of metal matrix composites: A review. *Materials and Manufacturing Processes* **31**(5): 553–573.

Siqi, L., C. Yan, F. Yucan and H. Andong (2016). Study on the cutting force and machined surface quality of milling AFRP. *Materials Science Forum* **836/837**: 155–160.

Skordaris, G., K. D. Bouzakis, T. Kotsanis, P. Charalampous, E. Bouzakis, O. Lemmer and S. Bolz (2016). Film thickness effect on mechanical properties and milling performance of nano-structured multilayer PVD coated tools. *Surface and Coatings Technology* **307**: 452–460.

Soo, S. L., R. Hood, D. K. Aspinwall, W. E. Voice and C. Sage (2011). Machinability and surface integrity of RR1000 nickel-based superalloy. *CIRP Annals—Manufacturing Technology* **60**(1): 89–92.

Soussia, A. B., A. Mkaddem and M. El Mansori (2014). Rigorous treatment of dry cutting of FRP—Interface consumption concept: A review. *International Journal of Mechanical Sciences*, **83**: 1–29.

Sreejith, P. S., R. Krishnamurthy, S. K. Malhotra and K. Narayanasamy (2000). Evaluation of PCD tool performance during machining of carbon/phenolic ablative composites. *Journal of Materials Processing Technology*, **104**(1): 53–58.

Statistica (2017). Automotive industry—Statistics & facts. Accessed May 4, 2017, from https://www.statista.com/topics/1487/automotive-industry/.

Su, Y., L. Gong, B. Li, Z. Liu and D. Chen (2016). Performance evaluation of nanofluid MQL with vegetable-based oil and ester oil as base fluids in turning. *International Journal of Advanced Manufacturing Technology* **83**(9): 2083–2089.

Sudhakar, M. A. L. (2017). New developments in high speed machining technology. Accessed from http://www.moldmakingtechnology.com/articles/new-developments-in-high-speed-machining-technology.

Sugihara, T., Y. Nishimoto and T. Enomoto (2017). Development of a novel cubic boron nitride cutting tool with a textured flank face for high-speed machining of Inconel 718. *Precision Engineering* **48**: 75–82.

Sun, S., M. Brandt and M. S. Dargusch (2010). Thermally enhanced machining of hard-to-machine materials—A review. *International Journal of Machine Tools and Manufacture* **50**(8): 663–680.

Sun, Y., B. Huang, A. Puleo, J. Schoop and I. S. Jawahir (2016). Improved surface integrity from cryogenic machining of Ti–6Al–7Nb alloy for biomedical applications. *Procedia CIRP* **45**: 63–66.

Sundaram, M. M., G. B. Pavalarajan and K. P. Rajurkar (2008). A study on process parameters of ultrasonic assisted micro EDM based on Taguchi method. *Journal of Materials Engineering and Performance* **17**(2): 210–215.

Sunil, B. R., K. V. Ganesh, P. Pavan, G. Vadapalli, C. Swarnalatha, P. Swapna, P. Bindukumar and G. P. K. Reddy (2016). Effect of aluminum content on machining characteristics of AZ31 and AZ91 magnesium alloys during drilling. *Journal of Magnesium and Alloys* **4**(1): 15–21.

Teimouri, R., S. Amini and N. Mohagheghian (2017). Experimental study and empirical analysis on effect of ultrasonic vibration during rotary turning of aluminum 7075 aerospace alloy. *Journal of Manufacturing Processes* **26**: 1–12.

Thakur, A. and S. Gangopadhyay (2016). State-of-the-art in surface integrity in machining of nickel-based super alloys. *International Journal of Machine Tools and Manufacture* **100**: 25–54.

Thomas, M. (2016). Technological advances in fine abrasive processes. Accessed May 6, 2017, from http://slideplayer.com/slide/10297263/

Ugarte, A., R. M'Saoubi, A. Garay and P. J. Arrazola (2012). Machining behaviour of Ti–6Al–4V and Ti-5553 Alloys in interrupted cutting with PVD coated cemented carbide. Proceedings from the Fifth CIRP Conference on High Performance Cutting Procedia CIRP, Zurich, Switzerland.

Ulutan, D. and T. Ozel (2011). Machining induced surface integrity in titanium and nickel alloys: A review. *International Journal of Machine Tools & Manufacture* **51**: 250–280.

Umbrello (2016). Cryogenic manufacturing processes. *CIRP Annals—Manufacturing Technology* **65**: 713–736.

Wang, B., Z. Liu, Q. Song, Y. Wan and Z. Shi (2016). Proper selection of cutting parameters and cutting tool angle to lower the specific cutting energy during high speed machining of 7050-T7451 aluminum alloy. *Journal of Cleaner Production* **129**: 292–304.

Wang, B. and Z. Liu (2016). Investigations on deformation and fracture behavior of workpiece material during high speed machining of 7050-T7451 aluminum alloy. *CIRP Journal of Manufacturing Science and Technology* **14**: 43–54.

Wang, B., Z. Liu, Q. Song, Y. Wan and Z. Shi (2014). Proper selection of cutting parameters and cutting tool angle to lower the specific cutting energy during high speed machining of 7050-T7451 aluminum alloy. *Journal of Cleaner Production* **129**: 292–304.

Wang, F., Y. Wang, B. Hou, J. Zhang and Y. Li (2016). Effect of cryogenic conditions on the milling performance of aramid fiber. *International Journal of Advanced Manufacturing Technology* **83**(1): 429.

Wang, Y., C. Li, Y. Zhang, M. Yang, B. Li, D. Jia, Y. Hou and C. Mao, (2016). Experimental evaluation of the lubrication properties of the wheel/workpiece interface in minimum quantity lubrication (MQL) grinding using different types of vegetable oils. *Journal of Cleaner Production* **127**: 487–499.

Wang, Y., C. Li, Y. Zhang, M. Yang, X. Zhang, N. Zhang and J. Dai (2017). Experimental evaluation on tribological performance of the wheel/workpiece interface in minimum quantity lubrication grinding with different concentrations of Al_2O_3 nanofluids. *Journal of Cleaner Production* **142**(4): 3571–3583.

Wang, Z. Y., K. P. Rajurkar, J. Fan, S. Lei, Y. C. Shin and G. Petreseu (2003). Hybrid machining of Inconel 718. *International Journal of Machine Tools and Manufacture* **43**(13): 1391–1396.

Wu, W., W. Chen, S. Yang, Y. Lin, S. Zhang, T. Y. Cho, G. H. Lee and S. C. Kwon (2015). Design of AlCrSiN multilayers and nanocomposite coating for HSS cutting tools. *Applied Surface Science* **351**: 803–810.

Wu, Y., W. Xu, M. Fujimoto and T. Tachibana (2011). Ceramic balls machining by centerless grinding using a surface grinder. *Advancements in Abrasion Technology XIV* **325**: 103–109.

Xu, J., A. Mkaddem and M. E. Mansori (2016). Recent advances in drilling hybrid FRP/Ti composite: A state-of-the-art review. *Composite Structures* **135**: 316–338.

Xu, W., D. Cui, Y. Wu (2016). Sphere forming mechanisms in vibration-assisted ball centerless grinding. *International Journal of Machine Tools and Manufacture* **108**: 83–94.

Xu, W., Y. Wu, M. Fujimoto, and T. Tachibana (2012). A new ball machining method by centerless grinding using a surface grinder. *International Journal of Abrasion Technology* **5**: 107–118.

Xu, W., L. Zhang and Y. Wu (2016). Effect of tool vibration on chip formation and cutting forces in the machining of fiber-reinforced polymer composites. *Machining Science and Technology* **20**(2): 312–329.

Yao, Y.L. (2000). Cross process innovations. Accessed from http://aml.engineering.columbia.edu/ntm/CrossProcess/CrossProcessSect5.htm.

Yogeswaran, M., K. Kadirgama, M. M. Rahman and R. Devarajan (2015). Temperature analysis when using ethylene-glycol-based TiO_2 as a new coolant for milling. *International Journal of Automotive and Mechanical Engineering* **11**: 2272.

Zhao, F., W. Z. Bernstein, G. Naik and G. J. Cheng (2010). Environmental assessment of laser assisted manufacturing: Case studies on laser shock peening and laser assisted turning. *Journal of Cleaner Production* **18**: 1311–1319.

Zhou, J. M., V. Bushlya, R. L. Peng, S. Johansson, P. Avdovic and J. E. Stahl (2011). Effects of tool wear on sub surface deformation of nickel-based super alloy. *Procedia Engineering* **19**: 407–413.

Zhou, J. M., V. Bushlya and J. E. Stahl (2012). An investigation of surface damage in the high speed turning of Inconel 718 with use of whisker reinforced ceramic tools. *Journal of Materials Processing Technology* **212**: 372–381.

Zhu, Z., V. G. Dhokia, A. Nassehi and S. T. Newman (2013). A review of hybrid manufacturing processes—State of the art and future perspectives. *International Journal of Computer Integrated Manufacturing* **26**(7): 596–615.

Zou, Y. and T. Shinmura (2007). Development of effective magnetic deburring method for a drilled hole on the inside of tubing using a magnetic machining jig. *Key Engineering Materials* **329**: 243–248.

21 Advances and Applications of Nontraditional Machining Practices for Metals and Composite Materials

Ramanathan Arunachalam, Rajasekaran Thanigaivelan, and Sivasrinivasu Devadula

CONTENTS

21.1 INTRODUCTION

Additive manufacturing is growing at a rapid pace and is perceived as a potential disruptive technology, especially in the area of manufacturing complex-shaped structural metallic components. Although additive manufacturing exhibits several advantages, it has limitations in meeting the component quality requirements of major industries such as aerospace and automotive. Although conventional machining can compete with present developments, it still cannot catch up with the developments in additive manufacturing. In addition to the complex shapes and quality requirements of components (surface integrity), rapid developments in the area of engineering and engineered materials have resulted in materials that exhibit very high strength, toughness, hardness, temperature resistance, and other physical properties. The higher hardness of advanced materials has made cutting tool–based conventional machining processes almost impossible since the workpiece materials are harder than the cutting tool materials. Also, the demand for lower product cost has also put excess pressure on the manufacturing engineers to produce components at a lower price. In this situation, nontraditional machining practices seem to be a viable alternative manufacturing route to produce complex-shaped components not only in metallic but also in composite materials. Nontraditional machining practices can also be integrated with either conventional machining processes or additive manufacturing to improve the manufacturing processes. Hence, the need to investigate the advances and applications of nontraditional machining practices arises.

In fact, nontraditional machining practices are sometimes also referred to as advanced machining processes because they have been in common usage only in the last few decades (although they were introduced in 1950s). Nontraditional machining practices can be classified based on the principal energy used for removal of the material as shown in Figure 21.1. Figure 21.1 also shows some of the well-known machining processes under each classification. In the case of hybrid processes, there are many developments taking place, but the more popular ones have been listed and discussed to some degree. In the following sections, the principles, applications, issues, and advances in these nontraditional machining practices are discussed, but the focus is on the advances taking place in order to address the issues with these processes. Very briefly, the principles, advantages, and disadvantages of these processes are discussed here in the context of advances. We recommend that the reader refer to the numerous literatures already available on these topics for further details of the processes.

The major advantage of nontraditional machining practices is that energy in its direct form is utilized for removal of material and it is a noncontact process. Hence, these processes are more efficient and cutting forces and vibration are eliminated, which is a major issue with conventional machining processes. This enables it to be applied for even brittle or fragile materials. For example, the abrasive jet machining process could be used to produce an intricate shape even in a brittle material such as an eggshell. Most nontraditional machining practices such as electric discharge machining, electrochemical machining, and ultrasonic machining are capable of producing complex full-form machining as opposed to single-point machining by conventional machining processes. There are several more advantages such as the ability to reach inaccessible areas, high–aspect ratio holes, burr-free surfaces, and lower or zero tool wear (depending on the nontraditional machining practices and micro- or nanofeatures).

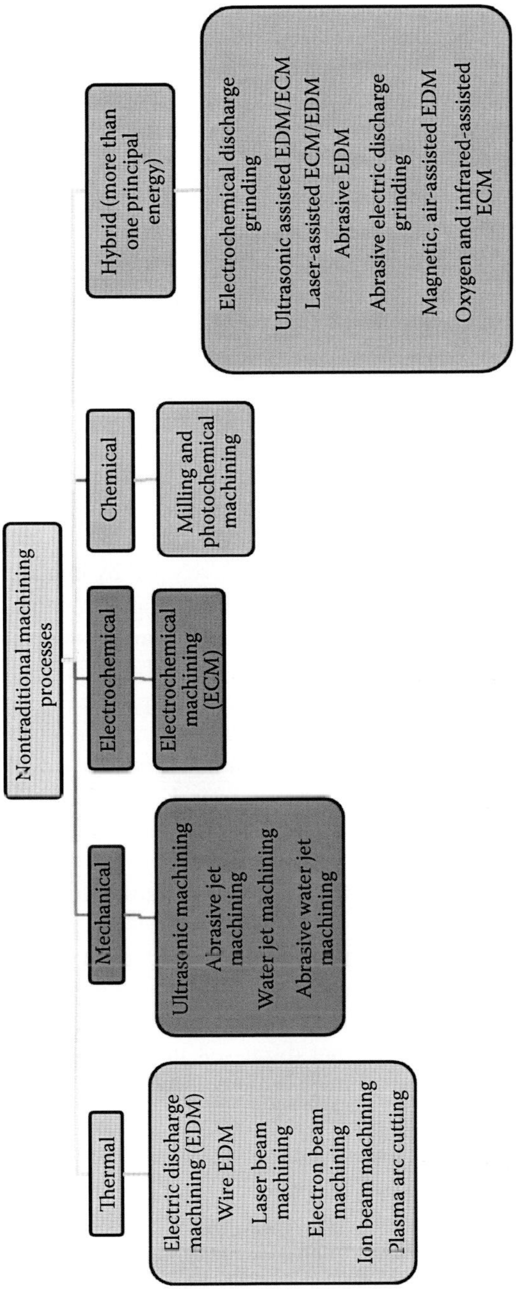

FIGURE 21.1 Classification of nontraditional machining processes.

TABLE 21.1
Relative Economic Criteria Comparison for Some of the Common Machining Processes

	Criterion				
Machining Process	**Capital Investment**	**Tooling/Fixtures**	**Power Requirements**	**MRR Efficiency**	**Tool Wear**
EDM	Medium	High	Low	High	High
ECM	Very high	Medium	Medium	Low	Very low
CHEM	Medium	Low	High	Medium	Very low
LBM	Medium	Low	Very low	Very high	Very low
PAM	Very low	Low	Very low	Very low	Very low
USM	Low	Low	Low	High	Medium
Conventional	Low	Low	Low	Very low	Low
HSM	High	Medium	Medium	High	Low

Source: Samant, A. (2011a). Laser machining. In: *Laser Machining of Advanced Materials*, CRC Press.

Each of the machining processes, whether traditional or nontraditional, has its own advantages and disadvantages. With so many competing nontraditional machining practices that could be employed for producing the same component, the user is in a dilemma to select an appropriate machining process in order to produce a specific component. Recently, a decision guidance framework for nontraditional machining practices selection has been developed by Prasad and Chakraborty (2015). The expert system is capable of choosing suitable nontraditional machining practices for a particular work material and shape combination, and for the selected nontraditional machining practices, it can also recommend the optimum process parameters. Table 21.1 shows a relative comparison of some of the common conventional as well as nontraditional machining practices in terms of important criteria that influence the economic aspect. This table could also be a starting point in selecting an appropriate machining process based on the component to be manufactured.

Although most nontraditional machining practices can be applied at all levels, macro-domain, micro-domain (micro-machining), and nano-domain (nano-machining), it is not within the scope of this chapter to discuss the developments in micro-domain and nano-domain. In fact, it is from macro-domain that many of the material removal principles are transferred to micro- or nano-domain. In this chapter, since the focus is on the advances and applications of those principles, the domain is not of great importance. For this reason, for most of the processes, a schematic illustration of the basic material removal principle is not provided and discussed in great detail since it is readily available in the published literature. In addition, all hybrid machining processes that involve a physical tool coming in contact with the workpiece are not discussed in this chapter since it is discussed to a greater extent in the chapter on advances in conventional machining processes.

21.2 APPLICATIONS OF NONTRADITIONAL MACHINING PRACTICES FOR METALS AND COMPOSITE MATERIALS

With the developments taking place in nontraditional machining practices, the applications for these processes are widening and much newer applications are being explored. Table 21.2 summarizes the applications of the major nontraditional machining practices along with the materials on which these processes could be applied as well as the principal mechanism of material removal. In addition, Table 21.2 also summarizes the key issues in those nontraditional machining practices. The table is self-explanatory and so is not discussed here. Section 21.3 discusses how some of these key issues have been addressed by the advances or developments in nontraditional machining practices.

TABLE 21.2
Summary of Applications, Materials, Material Removal Mechanism, and Key Issues of Various Nontraditional Machining Practices

NTMP	Applications (Components Produced)	Materials (Metals and Composites)	Major Mechanism of Material Removal	Key Issues	References
Electric discharge machining	Thermal and pneumatic valves, orthopedic implants, orthodontics, stents, gas turbine blades, cutting tools, space vehicle, screws, machinery parts, food-handling equipment, automobile components, aircraft components, frames of eyeglasses, and surgical tools	Nitinol, high-speed steel, stainless steel, pyrolytic carbon, ceramic matrix composites, MMCs, titanium alloys, Cu–W, Cu, Ag–W, Monel metals, Inconel 718, plastic mold steel	Fusion and vaporization	Machining of advanced engineering materials, tool wear, material removal rates, surface quality, flushing of debris, electrical parameters, cost of the setup, bidirectional material migration, tool tip concavity formation, work hardening, re-solidified layer, microcracks and microvoids, hazardous dielectric and sustainability of electric discharge machining process	Gao and Liu 2003; Jafferson et al. 2014; Jothimurugan and Amirthagadeswaran 2014; Huang et al. 2015; Liew et al. 2013; Ng et al. 2016; Prihandana et al. 2009; Shervani-Tabar et al. 2013; Shen et al. 2016; Sidhu et al. 2014; Talla et al. 2015; Tang and Du 2014; Yan et al. 2005
Wire electric discharge machining	Gas turbine engines for components such as turbine disk, blades, combustors, aircraft turbines, rocket engines, power generation turbines, nuclear plants, chemical treatment plants, surgical instruments, and watches	Nitinol, high-speed steel, stainless steel, ceramic matrix composites, metal matrix composites, titanium alloys, Inconel 718,708 and SKD11	Fusion and vaporization	Stable machining condition, surface integrity, machining of difficult to cut materials, dimensional accuracy and shape, wire tension, speed of wire moving, type of wire	Atzeni et al. 2015; Li et al. 2014; Liu et al. 2014; Maher et al. 2015a,b; Nani 2016; Sharma et al. 2015; Yan et al. 2015; Zhang et al. 2013a
Laser beam machining	Cut thick plates for large structures such as ships, pressure vessels, and bridges, cutting of battery electrodes; drilling of watch jewels, diamond drawing dies; cooling holes in nozzle guide veins and gas turbines, fuel injection nozzles; metal mold	Metals, alloys, ceramics, and composites—mild steel, Al alloy (Al6061), stainless steel, SiC ceramics, biomaterials, intermetallics	Fusion and vaporization	Formation of striations on cut surface, resulting in poor surface quality and higher roughness; burr formation; recast layer; heat-affected zone; kerf width; reflectivity of the workpiece surface, hole wall geometry can be irregular	Chun et al. 2014; Ferraris et al. 2016; Lee et al. 2013; Lutey et al. 2017; Morimoto et al. 2015; Samant 2011a,b,c,d; Swift and Booker 2013; Wei et al. 2008

(Continued)

TABLE 21.2 (CONTINUED)
Summary of Applications, Materials, Material Removal Mechanism, and Key Issues of Various Nontraditional Machining Practices

NTMP	Applications (Components Produced)	Materials (Metals and Composites)	Major Mechanism of Material Removal	Key Issues	References
	for microscale surface patterning/ micropin arrays; metal aerospace filters; oil–water separation filters				
Electron beam machining	Multiple small-diameter holes in very thin and thick materials, injector nozzle holes, small extrusion die holes; filters and screens; irregular-shaped holes and slots, engraving, features in silicon wafers for the electronics industry	Practically any material regardless of its type, electrical conductivity and hardness—ceramics, tungsten	Fusion and vaporization	Thermal stresses result in very small heat-affected zone, small recast layers and low distortion of thin parts possible, vacuum integrity, high aspect ratio holes may slightly become tapered; requires auxiliary backing material	Swift and Booker 2013; Youssef and El-Hofy 2008
Ion beam machining	Machining and finishing a wide range of materials, smoothing of silicon surfaces, removing single crystal diamond turning tool marks from nickel parts and molds	Carbides, optical glasses and group III–V semiconductors such as indium antimonide and indium arsenide	Mechanical (ion impact)	Requirement of vacuum environment for operation, surface roughening and heating; nonuniform ion beam profile and material redeposition, ion beam–induced damage	Allen et al. 2009; Kannegulla and Cheng 2016
Plasma arc cutting	Profile cutting of metals, cutting nonconductive materials	Stainless steel, Al, and other nonferrous metals; ceramics; nonconductive materials such as textiles, nylon, and polypropylene	Fusion and vaporization	HAZ, reduced accuracy and poor surface quality, high power requirement, toxic fumes, possible eye injuries	Krajcarz 2014; Youssef and El-Hofy 2008
Ultrasonic machining	Automobile components, aircraft components, orthopedic prostheses, as articulating bearing surfaces, hip joint balls or knee condyles	Pure titanium and alloys, borosilicate glass, biomass, soda lime glass, silicon carbide matrix, alumina zirconia ceramic composite, zirconia bioceramics,	Mechanical erosion	Power rating of the machine, static load, slurry concentration and abrasive grit size. The effect of local contact deflections, machining of biomass, surface deformation and residual stress,	Agarwal 2015; Ahluwalia et al. 2014; Asami and Miura 2015a,b; Baek et al. 2013; Bhosale et al. 2014; Das et al. 2016a,b; Feucht et al. 2014;

(Continued)

TABLE 21.2 (CONTINUED)
Summary of Applications, Materials, Material Removal Mechanism, and Key Issues of Various Nontraditional Machining Practices

NTMP	Applications (Components Produced)	Materials (Metals and Composites)	Major Mechanism of Material Removal	Key Issues	References
		carbon fiber–reinforced silicon carbide composites, nickel-base alloys (e.g., Inconel) and titanium-aluminum alloys, tungsten carbide–cobalt composite, alumina ceramics		type of vibration, precision hole machining, material removal rate, tool wear rate and surface topography of composites, different types of complex shape fabrication, inner surface machining of deep cavity component, machining of advanced materials, hole quality, profile accuracy of noncircular holes	Kataria et al. 2015; Lalchhuanvela et al. 2013; Patil et al. 2014; Teimouri and Baseri 2013; Wang et al. 2016b; Wu et al. 2015
Waterjet machining	Gaskets for automotive, marine, motorcycle and aircrafts. It is also used to produce high-precision parts in aerospace, automobile, electronics, food industry, munition demilitarization, and textiles. Deburring, descaling, peening, surface treatment	Rubber, polymer matrix composites, glass, foam	Mechanical erosion	Only limited materials can be processed economically. Difficulty to maintain dimensional accuracy for thicker materials due to striation development and kerf taper	Miller 2012; Srinivasu and Axinte 2014b
Abrasive waterjet machining	Various parts in aerospace, automotive, energy, oil and gas, transportation, agriculture, demilitarization, architecture industry	Metals, alloys, ceramics, composites, glass, foams	Mechanical erosion	Only limited materials can be processed economically. Difficulty to maintain dimensional accuracy for thicker materials due to striation development and kerf taper	Miller 2012; Srinivasu and Axinte 2011, 2014a; Babashov and Mammadova 2015
Abrasive jet machining	Microfluidics, microdrilling, cutting microfabrication technologies. It is also used for cutting slots,	Brittle and heat-sensitive materials such as quartz, glass, sapphire, ceramics	Mechanical erosion	Dusty environment and difficult-to-control jet divergence. Masking adds up the production cost	Ahmadzadeh et al. 2017; Ghobeity et al. 2009b; Jafar et al. 2016

(Continued)

TABLE 21.2 (CONTINUED)
Summary of Applications, Materials, Material Removal Mechanism, and Key Issues of Various Nontraditional Machining Practices

NTMP	Applications (Components Produced)	Materials (Metals and Composites)	Major Mechanism of Material Removal	Key Issues	References
	cleaning hard surfaces, deburring, and polishing				
Electrochemical machining	Blisk, turbine blades, surgical instruments, watches, bearing components, stethoscope, healing abutments, implants, prefabricated abutments, high-temperature fasteners, chemical processing and pressure vessels, heat exchanger tubing	SS-304, nickel-based superalloy, Ti40 titanium alloy, TB6 Ti alloy, Gcr15 bearing steel, aluminum alloy, Inconel 718, aluminum matrix composites	Chemical dissolution	Proper flushing of machining area, machining localization, shape of the workpiece, difficult-to-machine convex parts and concave parts, high-quality holes, MRR, overcut, surface quality, uniform current density and generation of spikes, finishing of particle-reinforced aluminum matrix composites, types of electrolyte	Ghosal and Bhattacharyya 2013; Liu et al. 2016; 2017; Paczkowski and Zdrojewski, 2017; Wang et al 2017a; Zhengyang et al. 2015; Zhu et al. 2015
Chemical machining	Plates enveloping walls of rockets and airplanes like wing skins; helicopter vent screens, instrument panels, flat springs, artwork; printed circuit boards, fine screens, honeycomb structures, decorative panels; applications also include removal of recast layer after electric discharge machining, removal of burrs after conventional machining	Aluminum alloys, magnesium alloys, copper alloys, titanium alloys and steel alloys; ferrous, nickel, titanium, magnesium and copper alloys, and silicon	Ablative reaction (etching)	Safety and environmental aspects when dealing with highly toxic chemicals used as maskants/etchants. Etchant attack on areas of porosity in the material, adhesion of masking, surface roughness may be quite high	Benhabib 2003; Swift and Booker 2013

21.3 ADVANCES IN NONTRADITIONAL MACHINING PROCESSES

21.3.1 Electric Discharge Machining

In electric discharge machining, the material removal occurs due to melting and evaporation of work material between the electrode and workpiece. Material removal in electric discharge machining is influenced by various machining parameters and techniques adapted in the hope of improving the material removal rate, tool wear rate, and surface quality discussed in this section. Advances in electric discharge machining are broadly classified into two categories, namely, modification in tool and workpiece mechanism and dielectric medium. The following sections briefly discuss the advanced strategies involved in the electric discharge machining process shown in Figure 21.2.

21.3.1.1 Ultrasonic Vibration of Tool and Workpiece

In electric discharge machining, the performance of the material removal process mainly depends on discharge energy. Hence, increasing this energy leads to larger craters and worsens the surface quality. The major reason for worsening of the surface quality is improper renewal of dielectric in the interelectrode gap. During machining, the debris formed due to electric discharge is accumulated in between the tool and the workpiece. This accumulation of machined product reduces the resistance between the tool and the workpiece, leading to irregular discharge, which contributes to significant tool wear and a lower material removal rate. In the view of improving dielectric circulation, researchers since 1989 (Kremer et al. 1989) have incorporated various methods such as suction of dielectric through tool or workpiece and use of orbital motion electrodes. The latter methods were not successful because no machining occurred in front of the hole. The vibrating electrode method is a commonly adapted technique for scavenging the debris from the machining zone and also yielded significant results.

Electrode vibration is classified into tool vibration, workpiece vibration, and simultaneous vibration of electrode. The schematic representation of the vibrating electrode is shown in Figure 21.3.

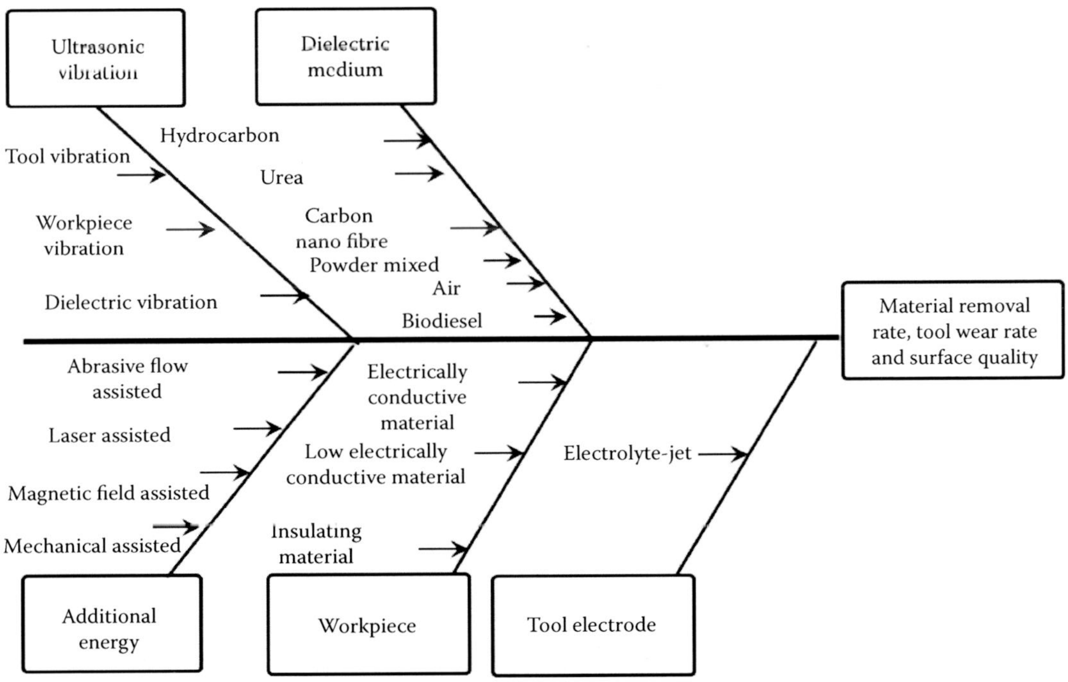

FIGURE 21.2 Advanced strategies adopted in electric discharge machining.

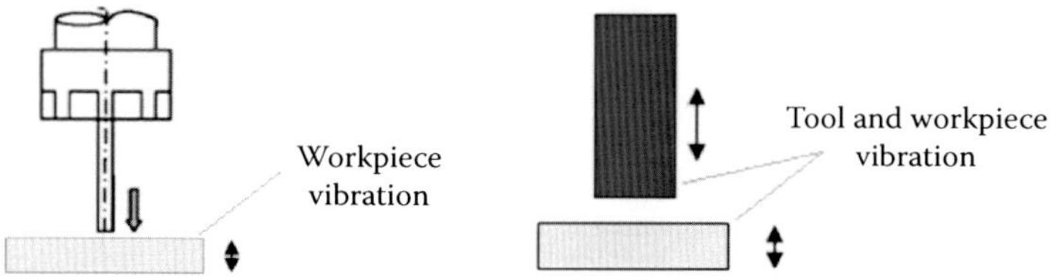

FIGURE 21.3 Schematic representation of the vibrating workpiece and tool.

Gao and Liu (2003) have used the ultrasonically vibrated workpiece and improved the electric discharge machining performance. They used tungsten electrode on a stainless steel workpiece with high-frequency vibration. This high-frequency vibration causes large variation in pressure between the tool and the workpiece, leading to a higher material removal rate. The use of tungsten electrode with high-frequency electrode vibration contributes to machining high–aspect ratio holes. A typical ultrasonic-assisted electric discharge machine is shown in Figure 21.4.

The vibrating electrode plays a significant role in machining advanced engineering materials such as nickel, alloys, titanium, and metal matrix composites. The surface quality improved from 2.543 to 2.050 μm and improved material removal rate by three times. The influence of a simultaneous vibrating electrode and workpiece in an electric discharge machine contributes to longer life and larger growth of the bubble, which, in turn, reduced the inside pressure of the bubble. This rapid decrease in pressure ejects the molten materials from the crater, resulting in an increase in the material removal rate. This type of technique will be more suitable for micro-electric discharge machines for machining high–aspect ratio holes. Hence, vibrating the tool, dielectric, and workpiece incurs additional setup, and the expenses involved significantly increases the machine cost (Shervani-Tabar et al. 2013). The cost analysis is performed by incorporating the ultrasonic vibration

FIGURE 21.4 Tool vibrated electric discharge machining (Pi Tech Blog 2017). (Courtesy of PI-USA.)

and magnetic fields individually in an electric discharge machine. The experimental results were compared; the magnetic field–assisted electric discharge machine provides better results, and the cost of the setup is 10 times cheaper than the vibration setup. The ultrasonic vibrator cost was approximately US$315, whereas the magnetic field–assisted setup costs US$30 (Jafferson et al. 2014). Thus, the use of ultrasonic vibrations develops a high-frequency pumping action, by evacuating the debris away and renewing new fresh dielectric owing to a higher discharge energy, resulting in a higher removal rate. Moreover, this high-frequency pumping action creates more turbulence and cavitations, resulting in better ejection of molten metal. This, of course, accelerates the material removal and also lessens the recast layers and micro-cracks, thereby increasing the fatigue life of the components.

21.3.1.2 Dielectric

The basic functions of a dielectric are to insulate the workpiece, for ionization, and to cool the electrode and workpiece. Normally, the dielectrics used in an electric discharge machine are deionized water, mineral oil, and kerosene. The advent of new materials and composites forces the researchers to identify the suitable dielectric in an electric discharge machine with the hope of improving the material removal rate, tool wear rate, and surface quality. Table 21.3 summarizes the new dielectric medium used for machining advanced materials.

21.3.1.3 Insulating Materials

Electric discharge machining is a well-established process for machining conductive materials; at present, electric discharge machines are also applied for machining ceramics through electrode-assisted methods. In an electrode-assisted method, two types of discharge phenomenon occur: first,

TABLE 21.3
Summary of the New Dielectric Mediums Used for Machining Advanced Materials

Advanced New Dielectric	Workpiece Material	Benefits	References
Urea with distilled water	Pure titanium metals	Improving the surface roughness and tool wear rate	Yan et al. 2005
Carbon nanofiber mixed dielectric	Reaction bonded silicon carbide	Prevent tool tip concavity, bidirectional material migration	Liew et al. 2013
Copper and graphite powder suspended dielectric medium	Metal matrix composites	Higher material removal rates and surface quality	Sidhu et al. 2014
Micro molybdenum disulfide suspended dielectric medium	Copper–tungsten, copper–silver–tungsten	Increase in frequency of discharge	Prihandana et al. 2009
Aluminum suspended dielectric medium	Metal matrix composites	Higher material removal rate and surface quality	Talla et al. 2015
Servotherm oils	Monel metal	High machining efficiency coupled with superior surface finish	Jothimurugan and Amirthagadeswaran 2014
Tap water air and biodiesel.	Titanium (Ti–6Al–4V)	High material removal rate, decrease the machining cost environmental friendly	Tang and Du 2014; Ng et al. 2016
Air dielectric (nitrogen, acetylene, argon gas)	Inconel 718, TiNi-based shape memory alloys, high-chromium high-carbon die	Reduces the resolidified layer, microcracks and microvoids	Shen et al. 2016; Huang et al. 2015

discharge creates carbonized products on the surface of the workpiece, which act as a positive electrode, and second, discharge results in material removal. Addition of electrically conductive materials to a ceramic matrix makes the insulator a low conductive ceramic. Generally, ceramics are insulators, and some ceramics such as silicon carbide and boron carbide possess sufficient electrical conductivity ($>10^{-2}$ S/cm) as a result of being machined in an electric discharge machine. Hence, the nonconductive ceramics can also be made conductive by the addition of borons, nitrides, and carbides of transition metals in the insulating matrix (Ferraris et al. 2016). Material removal is in the higher side for the electrode-assisted method and low conductive composites due to joule heating. Major shape consistency is achieved through the electrode-assisted method of electric discharge machining.

21.3.1.4 Electrolyte Jet Electric Discharge Machine

In electric discharge machines, tool or electrode cost is a major consumable expenditure because of the high tool wear. Several researchers have carried out research to reduce tool wear but they have not been successful. Zhang et al. (2016) investigated a novel electrolyte jet electric discharge machine in which the tool electrode is replaced by an electrolyte jet. A three-dimensional model of the setup is illustrated in Figure 21.5. The electrolyte jet makes use of the electrostatic field to create the tool electrode, which is in the form of a fine pulse jet. Because it uses an electrolyte (sodium chloride) that is not expensive and can be easily replaced, there is no tool wear and no associated issues such as tool compensation problem. Although this advancement has been made in the microdomain, this could be applied at the macro level as well.

21.3.2 Wire Electric Discharge Machine

In a wire-cut electric discharge machine, the workpiece is machined with a sequence of electrical sparks that is produced between the workpiece and the wire electrode. The wire electrode discharges high-frequency pulses to the workpiece through a very small gap with a dielectric fluid. Numerous sparks can be seen instantaneously at the machining zone, that is, 100,000 times per second. A Sodick AP250L Wire EDM (Sodick 2017) with linear drives and advance corner control is shown in Figure 21.6. The key issues of wire-cut electric discharge machines are machining of difficult-to-cut materials, surface integrity, and dimensional accuracy, and these issues are addressed with new strategies. The use of a non-round electrode can expand the machining gap, and with a high electrode rotation speed, the machined debris can be ejected out of the machining area successfully, and hence

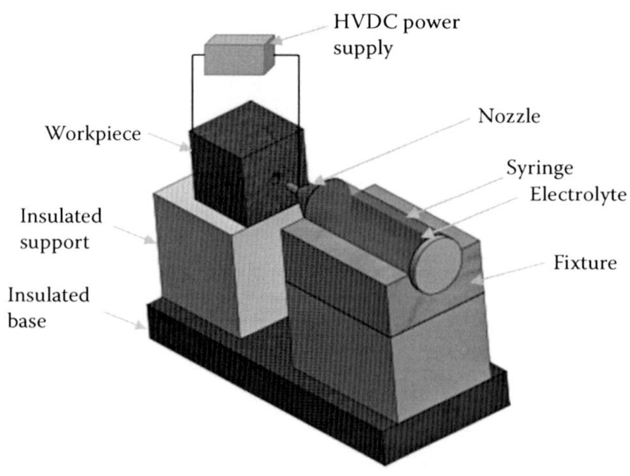

FIGURE 21.5 3D model of the electrolyte-jet electric discharge machining setup.

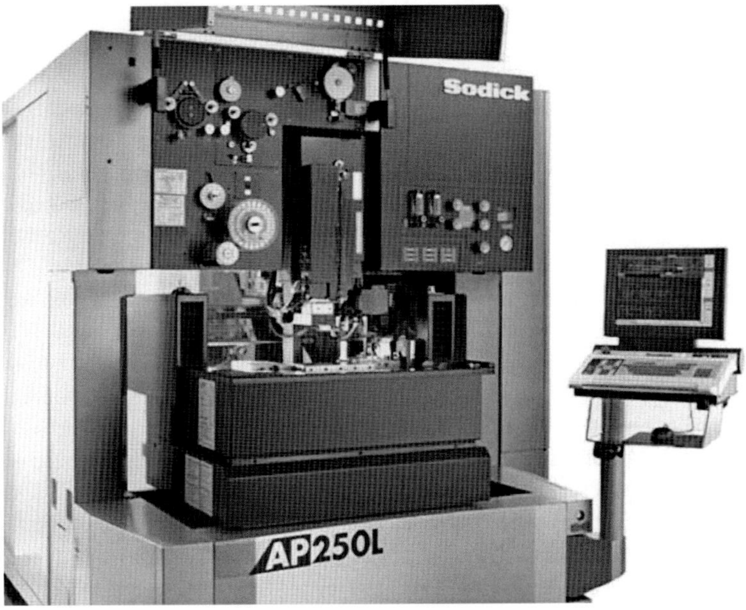

FIGURE 21.6 Precision wire EDM (Sodick 2017). Salient features: Linear motor drives (X, Y, U, and V axes), glass scale feedback (X, Y, U, and V), super jet annealing high-speed AWT, motion controller (K-SMC), and advanced corner control. (Courtesy of Sodick Incorporated.)

a stable machining condition can be obtained (Liu et al. 2016). The major challenges faced when machining nickel-based alloys are low thermal conductivity, work-hardening behavior, and high adhesion characteristic onto the tool face; hence, the surface integrity of the end products can be substandard. To overcome the disadvantages, the wire-cut electric discharge machine is operated at different energy modes and average roughness can be significantly reduced at low energy mode. To improve dimensional and shape accuracy, the wire electrode is activated by the ultrasonic vibration, which leads to an increase in intensity for the breakdown of the dielectric liquid. Moreover, conversion of electro-erosive processes to tensile forces is relatively lower in a wire electrode. At the same time, the wire electrode has not been entitled to mechanical stresses close to the breaking limit. Because of this, the frequent breakages of the wire electrode were circumvented, contributing to the dimensional accuracy of the workpieces. In addition, proper design of a mechatronics system for the wire transport system also contributes to straightness accuracy. The speed of the moving and coated wire plays a prominent role in material removal and surface quality (Atzeni et al. 2015; Zhang et al. 2013a). The higher value of wire speed contributes to a higher material removal rate and surface roughness. The coating of the wire electrode with zinc ensures high precision machining of Inconel alloy.

21.3.3 Laser Beam Machining

Laser beam machining works on the principle of light absorption of the workpiece materials, resulting in melting, dissociation/decomposition, evaporation, and material expulsion from the desired area. Among the beam-based processes, laser beam machining is one of the most widely used thermal energy–based nontraditional machining practices. Unlike electric discharge machining, laser beam machining can be applied to not only conductive materials, but also for a wide variety of materials. Laser beam machining is mostly used in cutting applications, and such machines are popularly called laser cutting machines. Among the several innovations in laser beam machining is the fiber laser. Fiber lasers exhibit high efficiency and excellent beam quality, because of which it is applied in several processes such as welding, cladding, and cutting (Marimuthu et al. 2017). They are also easier to deliver, and thus, laser cutting machines using fiber laser have become popular for drilling and cutting (profiling) several materials including superalloys and nuclear and polymeric materials. Figure 21.7 shows a photo of a new model of the Cincinnati Incorporated fiber laser cutting machine (Cincinnati Inc. 2017) that has innovative features such as linear motor drives that can achieve traverse speeds as high as 300 m/min and accelerations up to 3G. The machine tool is equipped with an air assist laser cutting system that uses clean, filtered air instead of the expensive nitrogen gas. Another feature is the automatic nozzle changer, which increases the productivity by allowing nozzles for the laser head to be automatically changed and cleaned.

Morimoto et al. (2015) studied the effect of orbital and vertical oscillations of the laser focus position on striations (affects surface roughness) when cutting a thick steel plate. The experimental results showed that the formation of striations can be controlled by adjusting oscillation conditions such as radius and frequency. Orbital oscillation resulted in smaller pitch and shallower striations while the effectiveness on the quality of surface roughness was smaller for the vertical oscillations.

With demand on productivity, it is important that the machining processes are capable of achieving it. Research on increasing the drilling speed in trepanning 5-mm-thick nimonic alloy by fiber laser has been carried out by Marimuthu et al. (2017). The results showed that the high average power of the quasi-continuous wave fiber lasers can simultaneously achieve higher speeds as well as holes with lower (taper, recast layer, oxide layer) and better hole surface morphology. They also used compressed air as an assist gas and obtained recast layer thickness that was comparable to oxygen assist gas, but the only issue was higher surface roughness. Cutting speeds of 3–5 m/s were achieved while cutting lithium cobalt oxide Li-ion battery electrodes using a 500-W average power nanosecond pulsed laser source (Lutey et al. 2017). Inconel 625 superalloy was machined using laser by submerging the workpiece in water (Roy et al. 2017). This had a significant impact (completely or partially removed) on the taper, heat-affected zone, recast layer, debris removal, and surface morphology. The advanced strategies adopted by many researchers and laser machine tool system manufacturers on parameters that affect the quality of the machined surface are summarized in Figure 21.8.

With the expanding applications for advanced structural ceramic materials, the main bottleneck is economically manufacturing components. Laser beam machining is one of the promising candidates that can produce three-dimensional parts at lower cost and higher productivity. Wang and Zeng (2007)

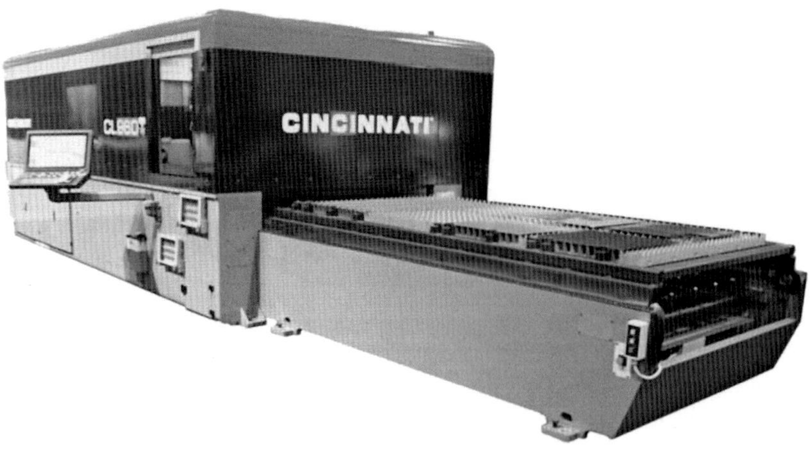

FIGURE 21.7 CL-900 Series Fiber Laser Cutting System (Cincinnati Inc. 2017). Salient features: Fiber technology; linear motor drive system for speed, accuracy, and reliability; flexible glass fiber beam delivery; easy-to-use human–machine interface; up to 6000 W power; interfaces with automation. (Courtesy of Cincinnati Incorporated.)

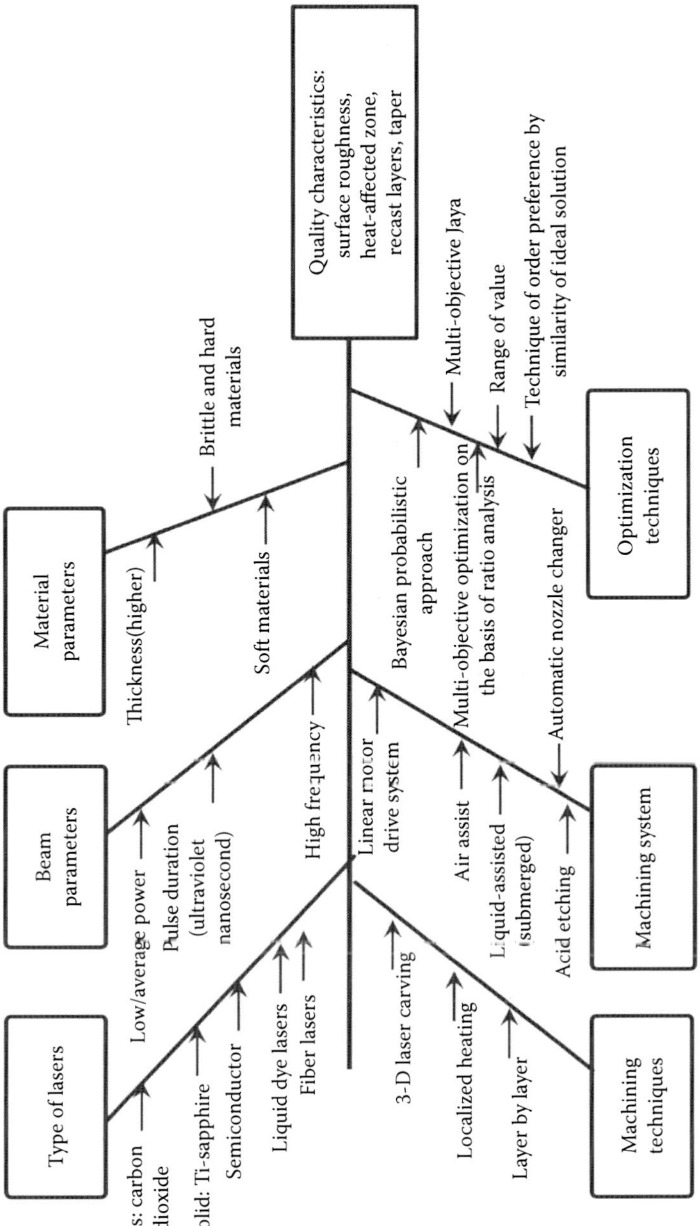

FIGURE 21.8 Advanced strategies in laser beam machining.

used a three-dimensional laser carving technique to fabricate ceramic components. Review of the published literature by Wang et al. (2017b) has confirmed the use of laser beam machining as a tool for producing high-quality holes in ceramics. In order to reduce the level of recast layer, localized heating is recommended. Another effective approach during drilling of the holes is liquid assistance, which reduces the recast layer and also removes molten slag because of the cooling and evaporating effect. Similarly, to remove laser irradiation zone from the material, acid etching is a potential approach. A layer-by-layer strategy was utilized by Hu and Xu (2016) when pocket milling of carbon fiber–reinforced plastic laminates. They used a 532-nm nanosecond pulsed laser using layer-by-layer strategy and obtained good quality surface with lower heat-affected zone and carbonization at an optimized process parameter.

21.3.4 Electron Beam Machining

Electron beam machining is a relatively new nontraditional machining practice when compared to laser beam machining, but the principle of material removal is similar: high-speed electrons bombard the workpiece material, resulting in melting and vaporization. Also, like laser beam machining, electron beam machining could be applied in several manufacturing processes such as lithography, drilling, welding, and hardening. However, among the energy beam–based nontraditional machining practices, electron beam machining is considered superior because of its controllability and rapid energy transfer gradient (Kim et al. 2016). Although electron beam machining is superior in controllability, the high capital cost limits its availability to a few installations. Issues such as focusing of beam sources and increasing the energy density need attention. Also, in order to process advanced materials such as composites and polymers, electron interactions with nonmetallic materials need to be investigated so that, in the future, these materials can be processed by electron beam machining (Kim et al. 2016). Not much advancement has been made in the area of electron beam machining, although recently, electron beam melting (an additive manufacturing process) is being investigated to produce titanium aluminides that may replace the heavy nickel-based alloys in aerospace industries (Tebaldo and Faga 2017).

21.3.5 Ion Beam Machining

In ion beam machining, a stream of charged atoms (ions) of an inert gas, such as argon, is accelerated and directed toward the workpiece. The kinetic energy of the beam knocks atoms from the workpiece, thereby removing material at atomic scale (Allen et al. 2009). This is the reason why ion beam machining and focused ion beam machining are being increasingly used to generate nanostructures. Recently, an ion beam system has been used for surface texturing as well as smoothing applications. As discussed in Section 21.3.4, ion beam machining also requires special facilities and higher capital investment, because of which, not much research or advancements have been made in this process (Allen et al. 2009).

21.3.6 Plasma Arc Cutting

Plasma arc cutting works on the principle of high energy stream of ionized gas (plasma) as a source of heat to melt/vaporize materials. Plasma arcs are extremely hot, and temperatures can be in the range of 25,000°C. This high temperature can melt or vaporize any material, and thus, plasma arc cutting finds wide application in cutting of materials, with metals being the most common. Among the beam energy–based nontraditional machining practices, plasma arc cutting is much cheaper, and portable systems that have their own power source that could be used in remote locations have been developed.

An advanced multitasking machine from Kaltenbach (Kaltenbach 2017) capable of plasma arc cutting is shown in Figure 21.9. The machine is designed for high cutting performance and optimum cutting quality. Other features include common cut and cross-cut technology that significantly

FIGURE 21.9 Kaltenbach KF 1614 Plate Cutting and Drilling Centre (Kaltenbach 2017). Salient features: diverse manufacturing technologies such as drilling, milling, counter sinking, thread-cutting, contour marking, and oxyfuel and plasma cutting bundled on a single machine. (Courtesy of Kaltenbach.)

reduces waste and a dust removal system that automatically cleans the filter. The Lincoln Electric Company (Lincoln Electric 2017) has come up with a wider range of portable plasma cutters that are capable of efficiently cutting most common materials such as mild steel, stainless steel, aluminum, brass, and copper.

Most of the developments in academic setup are mostly on optimizing the process parameters in plasma arc cutting. Response surface methodology and grey relational analysis with principal component analysis were reasonably accurate and can be used for prediction within the limits of the factors investigated. Response surface methodology and grey relational analysis with principal component analysis have been used with reasonable accuracy and so are recommended for the prediction of optimum process parameters such as feed rate, current, voltage, and torch height (Maity and Bagal 2015).

21.3.7 Ultrasonic Machining

In ultrasonic machining, the material removal process is through mechanical force where the workpiece profile generated by it on hard and brittle materials such as ceramics, semiconductors, and glass depends on the optimal control of the process parameters. The material removal mechanism is due to the abrasive action of grit-suspended liquid slurry circulating between the workpiece and a tool that is vibrating perpendicular to the workpiece at a frequency above the audible range. The current application of ultrasonic machining is for ascertaining the cost-effective machining solutions for comparatively tough and ductile metals such as titanium, nickel alloys, ceramic metal matrix composites, and metal matrix composites. The recent advancement in ultrasonic machining is presented in Figure 21.10.

21.3.7.1 Type of Abrasive and Force

The material removal process in ultrasonic machining is due to micro-brittle fracture on the surface of the brittle material. Further, the formation of radial cracks due to the repeated impact of abrasive particle which is attributed for material removal in ultrasonic machining. In order to understand the effect of processing velocity, static load, and feed rate on residual stress, plastic deformation, and work hardening, experiments are performed by Wu et al. (2015), who found out that a higher static load produces a deeper work hardening region, higher compressive residual stress, and larger surface deformation. The size of abrasive particles, such as fine, medium, and coarse particles, has a significant influence on hole oversize, roundness, and conicity. Coarse particles produced larger

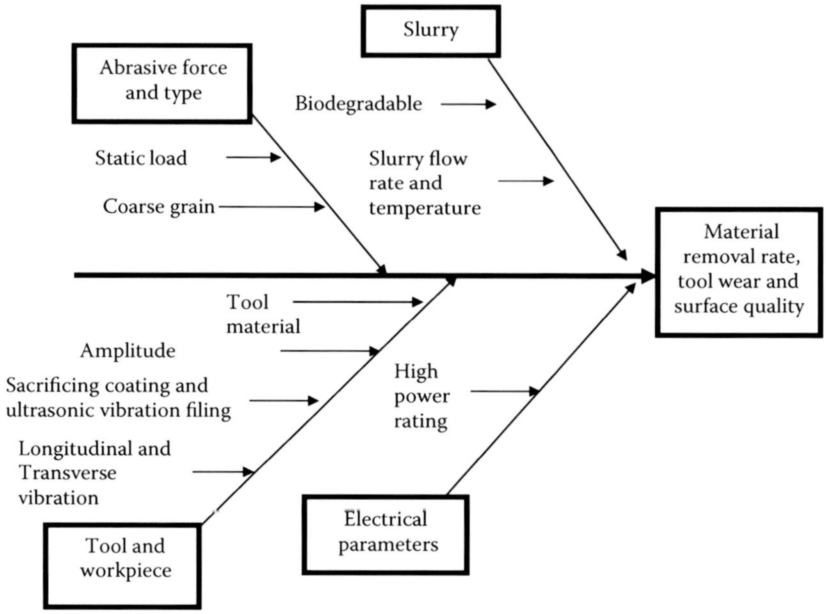

FIGURE 21.10 Advanced strategies adopted in ultrasonic machining.

oversize than fine particles because larger particles erode large chunks of material from the work surface, resulting in higher oversize.

21.3.7.2 Tool and Workpiece Mechanism

In the ultrasonic machining process, the ultrasonic vibration of the tool horn and abrasive slurry plays a prominent role in material removal. A new ultrasonic machining method using ultrasonic complex vibration caused by the longitudinal and torsional vibration was proposed by Asami and Miura (2015a) to improve the machining speed and the machining accuracy. A new fabrication method to achieve precision hole machining through sacrificing the coating on the substrate is proposed by Baek et al. (2013). In this method, a hard wax is deposited on the glass substrate, and holes are precisely fabricated in the coated glass using ultrasonic machining. Last, a wax coating is removed by cleaning. The presence of wax coating protects the surface of the glass by accepting the cracks rather than on the surface. Hence, the surface cracks and roundness error of the machined holes are generated in the sacrificed coating attributes to improve the performance of ultrasonic machining. For machining a deeper cavity, a new method known as ultrasonic vibration filing is considered for inner surface machining of a deep-cavity component made of carbon fiber–reinforced silicon carbide composites. The carbon fiber with ultrasonic vibration creates high-frequency scratching at short times. These small fragments produced from scratching detach the big block of material and improve surface integrity and surface quality (Wang et al. 2016b). Other important factors such as tool vibrating amplitude, tool material (stainless steel, silver steel, nimonic), and tool profile (solid and hollow tool) also play a significant role in improving the output parameters.

21.3.7.3 Slurry

The slurry concentration is considered as an important parameter because the increase in concentration contributes to higher tool wear. Although slurry flow rate is not making any significant impact on hole accuracy under normal working conditions, when comparing lower flow rate to higher flow rate, better accuracy has been obtained in the latter. At higher slurry flow rates, angular deviation and dimensional deviation are low and a good quality of the cut profile is obtained. Kumar (2014) has mentioned the temperature of the slurry as one of the parameters for generating optimal surface quality, and a comparative temperature study is not yet reported. Ultrasonic machining is applied in the biomedical field to treat blood clots, which are a health risk involving the circulatory system. The biodegradable starch slurry plays a significant role in enhancing the biomass machining efficiency. Hard clots (fibrin-rich) prepared from rabbit blood were exposed in vitro concomitantly to ultrasound (1 MHz) and starch slurry. Starch slurry particles have a 200% increase in material removal efficiency (Ahluwalia et al. 2014).

21.3.7.4 Electrical Parameters

The maximum material removal rate has been achieved with a high power rating value. Low values of power rating are recommended to achieve minimum values of overcuts for both large-diameter and small-diameter holes.

21.3.8 Waterjet Machining

Waterjet machining is an advanced machining process in which kinetic energy associated with high-velocity waterjet is used for removing material. It can be used for various applications (peening, cleaning, drilling, laser guiding, surface treatment, and cutting soft materials) and for processing a wide variety of materials in the aerospace industry, mining industry, manufacturing industry, medical industry, and so on. Waterjet machining is a preferred process when (i) the material being cut is sensitive to high temperature generated by other machining processes and (ii) there is a demand for lower material contamination and residual stresses.

In waterjet machining, high-pressure water from the pump is passed through an accumulator to reduce pressure fluctuations. The pressurized water passes through a small-diameter orifice (e.g., 0.25 mm) that changes pressure energy to kinetic energy and is capable of eroding material. Figures 21.11 and 21.12 show the schematic and photograph of a waterjet machine tool, respectively.

Although water is considered to be a carrier and an acceleration medium in abrasive waterjet, it has a considerable role in both waterjets and abrasive waterjets. There have been few investigations that explain the material removal mechanism using the high-pressure waterjets (Kong et al. 2010). The cutting mechanism may involve plastic deformations of the material, formation of microcracks, propagation of microcracks, and water hammer pressure. Furthermore, very limited information is available for the notoriously difficult-to-machine gamma titanium aluminides used by high-added-value (e.g., aerospace, automotive) manufacturing industries (Kong et al. 2010). A very recent study has reported a four-stage material removal mechanism that includes plastic deformation and crack initiation, stress wave propagation, micropits due to joint of crack lines, and intergranular cracking fracture, triple split, and translamellar and interlamellar fracture (Kong et al. 2010).

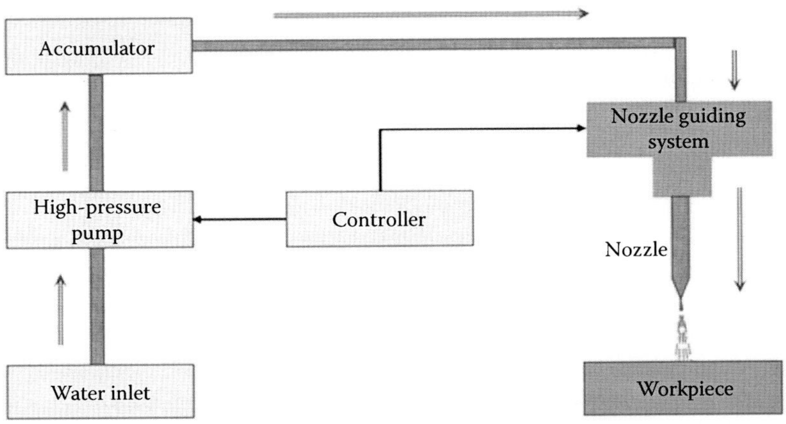

FIGURE 21.11 Schematic illustration of the waterjet machining process.

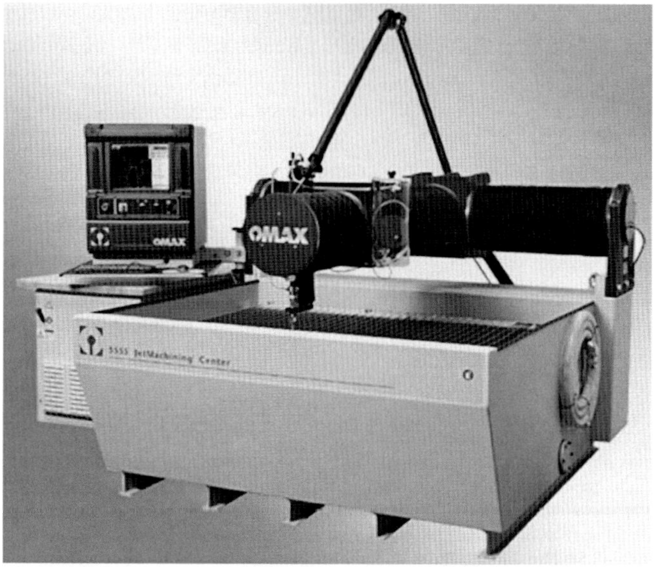

FIGURE 21.12 OMAX 5555 Precision Jetmachining Center (Omax 2017a). (Courtesy of OMAX Corporation.)

21.3.8.1 Advances with Waterjet Processing

21.3.8.1.1 Waterjet-Guided Lasers

Waterjet-guided laser processing is a new kind of laser machining technology (Li et al. 2011). A pulsed laser will be focused into the waterjet nozzle and the waterjet acts as an optical fiber, leading the laser beam onto the workpiece. Using the water-guided laser for cutting titanium sheets results in an increase in machining quality, and the cut surface is smooth and flatter. The top edge of the kerf does not display any jagged pattern, and the dimensional accuracy of the kerf improved. This results in the wider use of the waterjet-guided laser for micromachining applications.

21.3.8.1.2 Pulsed Waterjet

The pulsed waterjet is used to improve the erosion of material layers by interrupting the flow. In this method, the pressure fluctuations are developed by the principle of self-resonance. The interrupted flow has a larger impact stress due to the water hammer pressure, which enhances the local erosive intensity. The greater ratio of impacted area per volume of water coming from the jet, thus exposing larger areas of the surface to the water hammer pressure, results in higher material removal (Foldyna et al. 2012). The high amplitudes of velocity at the output of the nozzle will produce well-developed pulses at a certain standoff distance from the nozzle, which results in a higher impact pressure, and at low amplitude, a poorly developed waterjet is produced, which has very low impact pressure on the cut surface, resulting in lower cutting efficiency.

21.3.8.1.3 Pure Waterjet Milling

Pure waterjet milling of gamma titanium aluminide, which is a difficult-to-cut material, is demonstrated, with the benefits of reduced running costs (no cost of abrasives) and the elimination of grit embedment (Kong et al. 2010). The capability of the pure waterjet milling process is demonstrated by examining the geometrical accuracy and the resulting pocket surface quality. Results have shown that the threshold water hammer pressure for eroding the target material and for achieving uniform erosion was in the vicinity of 800 MPa and greater than 1 GPa. The knowledge generated during the investigation of the material removal mechanism and surface morphology enables the successful generation of 3D pure waterjet milled features.

21.3.9 Abrasive Waterjet Machining

Abrasive waterjet machining is a nontraditional advanced method of processing materials in which a high-velocity waterjet entrained with abrasives is used for eroding materials. In this process, a high-velocity waterjet is entrained with abrasives to improve its cutting capability. Figures 21.13 and 21.14 show a schematic illustration of the process and the machine tool, respectively, presenting various components of an abrasive waterjet system. Abrasive waterjet machining can also be viewed as an improved version of waterjet machining, which is capable of machining only softer materials. Abrasive waterjet machining, on the other hand, can machine almost any material that ranges from hard to soft, brittle to ductile, reflective to dull surfaces, and single-layered to multilayered material. In this process, high-pressure water is allowed to pass through an orifice and is transformed into a high-velocity waterjet. Abrasives are allowed to mix with the high-velocity jet and pass through a focusing tube in order to obtain the necessary acceleration, velocity, and jet coherency.

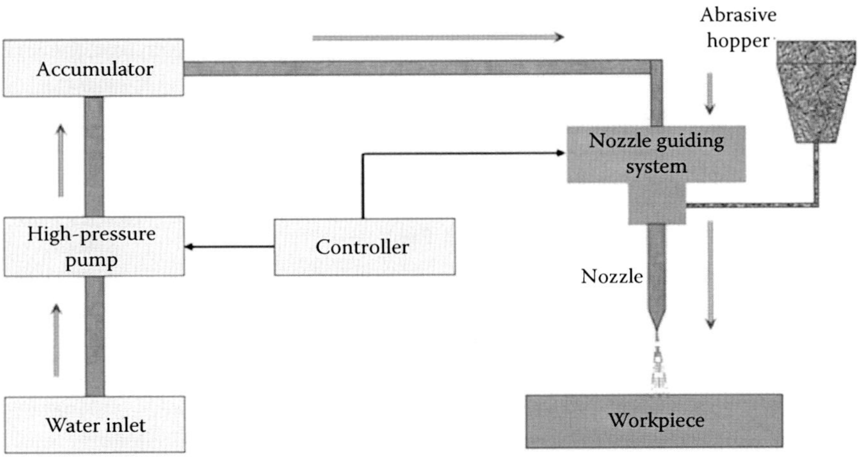

FIGURE 21.13 Schematic illustration of the abrasive waterjet machining process.

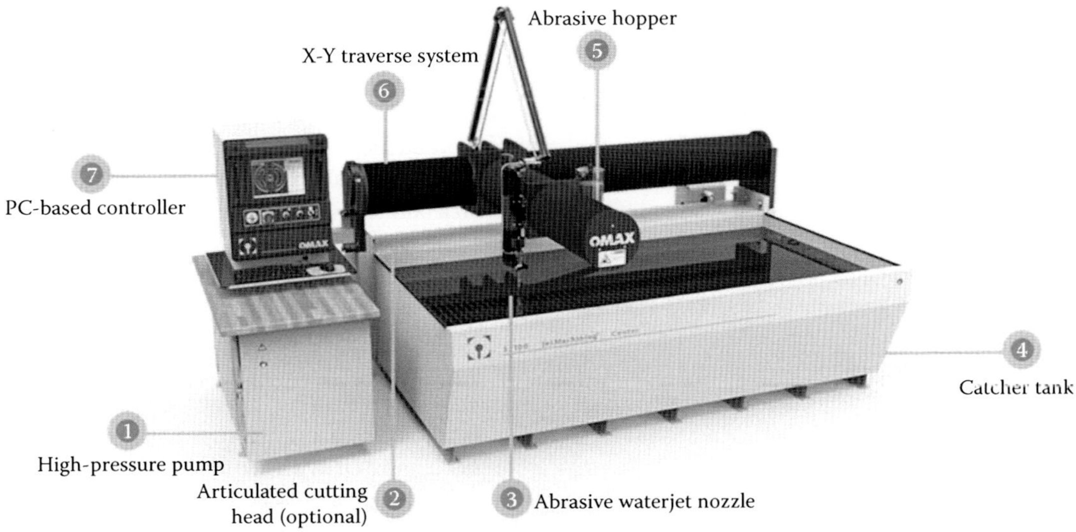

FIGURE 21.14 OMAX 55100 JetMachining Center (Omax 2017b). (Courtesy of OMAX Corporation.)

21.3.9.1 Process Modeling

The jet footprint is the building block of any surface generation with abrasive waterjets. Hence, the control over the shape and dimensional features of the jet footprints is of utmost importance. To this end, in the recent past, efforts have been made by employing experimental, analytical, and finite element approaches. A comprehensive experimental study is performed by varying the jet impingement angle, and the shape of the jet footprint is analyzed and the physical phenomenon involved in the formation of the shape of the jet footprint at various jet impingement angles is explained. Furthermore, the variation in the characteristics of the jet footprint is analyzed in the presence of the change in the jet traverse rate (Srinivasu et al. 2009).

A geometrical model based on a nonlinear partial differential equation for the jet footprint generated at the normal jet impingement, which is the building block for the controlled depth milling, is proposed (Axinte et al. 2010). A modeling approach is reported to predict the jet footprint that is capable of taking into consideration the nonorthogonal jet impingements when the jet path relative to the target surface is arbitrary direction (Kong et al. 2012). The model is powerful and applicable for any target material, and the model needs a single shallow jet footprint generated at a jet traverse rate. The material influence function (i.e., etching rate) is then obtained, and from this, the jet footprint of the arbitrary moving jet at any jet traverse rate can be obtained.

Furthermore, an enhanced model that can predict the surface geometry for overlapped jet footprints and an analysis of error propagation for predicting the material removed in successive layers are successfully developed and demonstrated (Billingham et al. 2013).

21.3.9.2 Advances with Abrasive Waterjet Processing

21.3.9.2.1 Gear Manufacturing

Production of gears for the automotive industry during 2008 is estimated to have been between 2000 and 2500 trillion, from which 1000 to 1400 trillion pieces were high-quality gears (Karpuschewski et al. 2008). Increasing standards on environmental impacts associated with products force the modern manufacturing industry to take a critical approach on making processes environmental friendly. Gear form cutting (broaching and milling) and gear generating (shaping and hobbing) are two of the main gear manufacturing methods. Usually, production involves three stages: (i) soft machining, (ii) heat treatment, and (iii) hard machining. Limitations of the conventional machining methods (difficult to cut materials, high cost of specialized cutting tools, high cutting tool failures, high cost of machine tools, stresses generated, and cutting force) affect the component life and cost and require alternative ways of material removal that do not affect the mechanical properties while delivering high-quality products in a cost-effective way, which is the goal of the industry. A large volume of material is removed during conventional gear production, resulting in higher lead time, greater use of cutting fluids (and its disposal), chip handling (and its disposal), and dust. In this context, abrasive waterjet cutting is considered to be a good addition to the current production system due to the low amount of applied force, negligible heat generated during machining, minimal change in material properties, versatility, lower initial investment, and environmental friendliness (no chip generation and no need for cutting fluids). To demonstrate the capability of the proposed approach in manufacturing spur gears and helical gears, forged gear blanks typically used in the automobile industry were used for initial tests and gears were produced and individual teeth have been separated from each gear and tested from different perspectives—metrological, surface integrity, productivity, and production cost comparisons (Babashov and Mammadova 2015). While improvements were achieved from environment and surface integrity perspectives, abrasive waterjet cutting cannot replace the soft machining stage because of increased lead time and production cost. A novel hybrid gear manufacturing method was proposed from this research and compared with the conventional method employed in the automotive industry. The proposed hybrid approach is composed of an abrasive waterjet cutting process as the major material removal method and a conventional five-axis machining method for final finishing or maintenance of tight tolerances and, at the same time, for decreasing initial investment and adding further flexibility to the production system.

For precision gears with a module below 1 mm, the time limitations and costs associated with the design of the cutting tool can be eliminated by using abrasive waterjets. The design of a hybrid manufacturing system configured by the use of abrasive waterjets and finishing processes using conventional machining methods are proposed (Gotia et al. 2016). The technical feasibility is analyzed to produce high-precision ring gears using a five-axis abrasive waterjet machining system to achieve Deutsches Institut für Normung (DIN) standards quality levels. For this purpose, a gear with a module of 0.55 mm, 199 teeth, and 110 mm in the outer diameter and 130 teeth and 72 mm in the inner diameter with a thickness of 6 mm is studied; the selected material is Armox T500, a high-strength steel. The results indicate high potential of producing ISO (International Organization for Standardization) quality standard gears. Certain quality characteristics defined in DIN and ISO (for instance, surface roughness—values as low as R_a = 0.8 μm) can be accurately achieved by using abrasive waterjet machining. Other quality features such as profile deviation are related to parameters such as cutting power, feed rate, abrasive feed rate, and so on. The displayed values ranged from Q10 to Q11 according to DIN-3967, which allows for use of further finishing operations such as grinding.

21.3.9.2.2 Patterning of Grinding Wheels

Grinding of difficult-to-machine superalloys requires coolant at the cutting face to reduce the heat-affected zone on the workpiece. As greater material removal rates are required, the depth of cut is increased, which leads to coolant delivery becoming problematic. A novel utilization of abrasive waterjets (i.e., to mill macroscopic patterns) is proposed, which can hold lubricant in the grinding zone and on the surface of a porous ceramic grinding wheel (Axinte et al. 2009). Furthermore, the process utilizes available grinding wheels without developing them from scratch and smaller amounts of abrasives, which makes the proposed approach environmentally friendly. The feasibility of this approach is evaluated by generating complex patterns, such as (a) standard zigzag patterns, (b) parallel zigzag patterns, (c) two-crossed parallel zigzag patterns, and (d) three-crossed zigzag patterns.

21.3.9.2.3 Honeycomb Structures Machining

Machining of hollow core honeycomb sandwich materials is a challenging job because the mechanical properties of various layers (skin and core) are quite dissimilar, and the conventional solid cutting tool experiences a sudden change of conditions while machining that might lead to damage on the tool/part. Although abrasive waterjets are highly efficient in the machining of advanced composite materials because of their unique characteristics, efficient machining of sandwich structures by abrasive waterjets needs to address many challenges. Various issues are observed in abrasive waterjet machining of carbon- and glass-fiber skin-based, aluminum sandwich structure composite materials, and the limitations of abrasive waterjet in processing these exotic materials are highlighted from an extensive experimental investigation (Srinivasu and Mihai 2013). Finally, possible strategies for efficient machining of honeycomb structures were proposed for future investigation.

21.3.9.2.4 Modeling of Kerf Characteristics in Multilayered Materials

Understanding and controlling the kerf profile in multilayered structures are a difficult task because various layers made of different materials respond to erosion in a different manner and result in a completely different kerf shape due to the material removal mechanism dependency on the material property of the specific layers, jet divergence, and position of the specific layers. Therefore, it is important to understand and develop predictive models of resulting kerf profile in multilayered structures so that they can be used in controlling the accuracy of the resulting kerf, which, in turn, dictates the final part accuracy. For the first time, an analytical model for predicting the kerf profile generated in multilayered structures machining with abrasive waterjets is proposed (Ngangkham et al. 2016). A discretized form of the Hashish model is used to determine depth of cut. The effect of jet divergence from the experimentally obtained values, upon passing through the upper layer, has been considered. The developed predictive model was validated by the kerf shapes obtained from the experimental trials on metal–adhesive–rubber multilayered structures. Kerf profiles obtained from the simulations have captured the resulting convergent–divergent profile while effectively cutting metal–rubber laminate composites. Furthermore, the effectiveness of the proposed analytical model was demonstrated by generating the various kerf shapes generated at various jet traverse rates.

Fabrication of complex three-dimensional freeform (flat pockets, ramp surface, concave surface, and convex surface) parts in advanced engineered materials (ceramics, heat-sensitive alloys) is a difficult task with conventional machining approaches because of high hardness, inhomogeneity, brittleness of advanced engineered materials, and heat generated in machining. To this end, for the first time, in the field of abrasive waterjet machining, five-axis milling was attempted to generate three-dimensional shapes, such as flat pockets, ramp surface, concave surface, and convex surface, in difficult-to-cut advanced engineered materials, by exploiting unique characteristics of abrasive waterjets, such as no heat-affected zone, minimum cutting force, and so on (Axinte and Srinivasu 2009). Novel tool path strategies are developed by considering issues with the jet (undesired secondary machining by deflected jet, divergence of jet plume, time dependence of erosion), as the conventional tool path strategies available in the computer-aided design/computer-aided manufacturing packages cannot be employed for abrasive waterjet milling. Proposed five-axis abrasive waterjet milling strategies are demonstrated on ceramics (silicon carbide and alumina), composites (glass fiber, aramid fiber, and carbon fiber), heat-sensitive alloys (titanium alloys), and shape memory alloys. Finally, theoretical and analytical modeling studies to understand jet footprint at various jet impingement angles are performed to get a hold on the control issues of the machined surface. All the above activities are realized in a project called Freeform-JET.

Controlled abrasive waterjet milling is employed for this purpose and has successfully demonstrated surface generation. The main idea behind controlling jet penetration is to control the jet traverse rate and to keep the pump pressure and mass flow of the abrasives constant. Jet penetration for any jet feed speed needs to be determined by means of analytical or lookup tables. An excellent and fully integrated system, with analytical prediction of jet penetration and computer-aided design/computer-aided manufacturing package for controlled depth milling, is reported.

21.3.9.2.5 Diamond Machining and Diamond Composite Nozzle

Up to now, there is scarce information on the use of abrasive waterjet in cutting superhard materials (e.g., diamond-based materials). A preliminary study of the capability of abrasive waterjet cutting of polycrystalline diamond using abrasives of different hardness (alumina, silicon carbide, and diamond) is presented (Axinte et al. 2009). The jet traverse rate has been adjusted to enable full jet penetration. It was found that not only the material removal rates but also the nozzle wear ratios, along with kerf quality, vary significantly with the types of abrasives. Further, more in-depth studies of the cut surfaces helped reveal the material removal mechanism when different types of abrasives are employed. The experimental results showed that alumina and silicon carbide yield modest material removal rates and that the use of diamond abrasives can greatly increase (4200 times) the productivity. Despite some limitations (e.g., cost of diamond abrasives, extensive nozzle wear rates) that can be overcome through further developments, it is believed that this preliminary research gives an indication of the capability of the abrasive waterjet to profile diamond-based structures for high-value engineering applications where conventional methods cannot be applied or are not productive enough.

21.3.9.2.6 Tool Path Strategies

For the first time, tool path strategies for milling of various three-dimensionnal shapes, such as convex, concave, and ramp surfaces, are proposed by considering the uniquely positive characteristics of high-energy abrasive waterjets as well as the limitations due to the aggressiveness of these jets, such as reflection of secondary jets, which has considerable erosive capability (Axinte and Srinivasu 2009). A method is proposed to generate scanning paths to be used in automated abrasive waterjet polishing of free-form surfaces (Khakpour et al. 2015). This method is able to produce trajectories with constant offset distance between curves on surfaces with holes and complex boundaries without reconfiguration of their triangular mesh model. For this, the particular requirements of this polishing technique to be kept along the path are investigated. Next, a reference curve is obtained, and using geodesic distances in specific directions, the adjacent offset curves are found. Finally, if needed, the main trajectory is divided into a set of continuous subtrajectories. By defining two indices, the effect of the shape of the surface and the configuration of the generated path on the uniformity of the distribution of the waterjet is evaluated. Through several examples, it is shown that the method can effectively generate scanning paths adapted to the requirements of this technique.

21.3.10 Abrasive Jet Machining

In abrasive jet machining, material removal takes place by the application of a high-speed stream of abrasive particles carried out by a gaseous medium from a nozzle. Gas can also work as a coolant and hence the surface finish is good and can achieve tight tolerances. Abrasive jet machining is not only used to cut hard and brittle materials that are sensitive to heat and easily chip away but also is employed for deburring and cleaning applications. Abrasive jet machining is free from the heat-affected zone, chatter, shock, and vibrations. Abrasive jet machining is widely used to manufacture electronic devices, liquid crystal displays, tibo-elements, and semiconductors. In the recent past, micro-abrasive jet machining has been widely used for various precision manufacturing applications. Figure 21.15 shows the schematic of the abrasive jet machine, which consists of an air compressor, a dehumidifier to control the humidity in the system, and an abrasive tank to supply the abrasives needed for the process. It is integrated with a kinematic system to provide necessary motion to the nozzle, which accelerates the abrasives needed for eroding the material.

21.3.10.1 Advances in Abrasive Jet Machining Process

21.3.10.1.1 Dusty Environment

Abrasive jet machining results in a dusty environment due to airborne, rebounding abrasive particles that eventually settle. In an effort to improve the cleanness of the process, a method in which the target is covered with a layer of liquid is proposed (Jafar et al. 2016). For this, a variety of workpiece/target

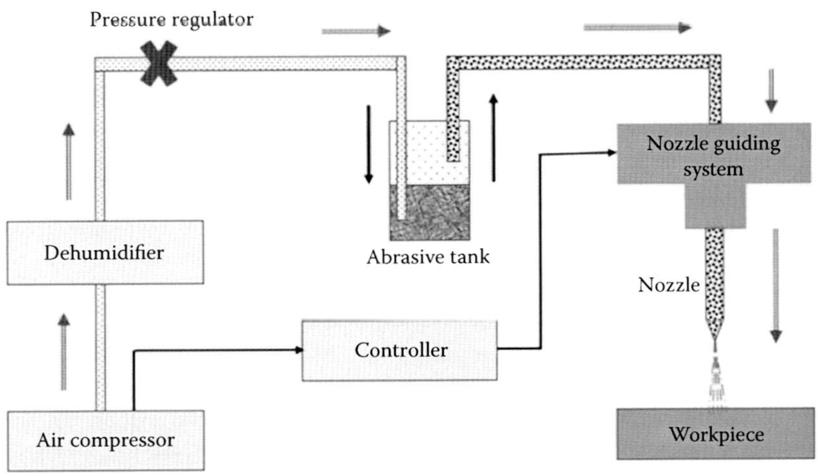

FIGURE 21.15 Schematic illustration of the abrasive jet machining process.

surface-covering mediums, such as a film of water, glycerin, and polymer solutions, are used. As a result, it is found that dust particles are reduced by up to 61% and that glycerin (61%) absorbed more particles than water (36%) and polymer (42%) solutions.

21.3.10.1.2 Diverged Jet and Stray Particles

Because the high-speed air particles coming out of the nozzle are divergent in nature, erosion-resistant masks need to be used for machining to suppress the effect of stray particles in order to get clean cuts and also to maintain the sharp edges with tight dimensional tolerances. For this application, 50- and 100-μm-thick ultraviolet light curing self-adhesive masks are commercially available and very useful in the abrasive jet machining process (Ahmadzadeh et al. 2017). In the ultraviolet masking process, first, the desired pattern is printed over a clear acetate transparency sheet and then the sheet is placed over the unexposed mask material and the material is exposed in the appropriate ultraviolet light; the dark surface remains uncured and erosion resistant, and the surface under light becomes brittle and erodes easily. By using ultraviolet light, with energy densities between 516–774 and 387–516 mJ/cm^2, the desired design pattern can be achieved.

21.3.10.1.3 Geometrical Accuracy of Machined Features

Geometrical accuracy is very important for assembly of various parts. Usually, abrasive jet machined microholes are tapered, resulting in poor cylindricity. In conventional micro-abrasive jet machining during machining time, the position of the nozzle and workpiece remains stationary and the distance between the nozzle and the workpiece (nozzle standoff distance) is set at a certain value at the beginning of the experiment. However, from the beginning of the experiment, the material starts to erode and, hence, the nozzle standoff distance continues to change in a dynamic manner. The exit and entrance diameters of the microhole are very much affected by the nozzle standoff distance value. By making the feed rate to the nozzle equal to the average rate of change in workpiece thickness, the nozzle standoff distance value can remain constant and can control the taper angle of the hole to improve the cylindricity of the microholes (Kumar et al. 2016). To maintain the nozzle standoff distance value, the material is kept constant and a certain feed is given to the nozzle that is controlled by the open loop nozzle feed system. The entrance diameter of the hole was reduced by approximately 29%, whereas the taper angle is reduced by approximately 58%, which results in improved cylindricity.

21.3.10.1.4 Surface Profile Waviness

With abrasive jet machining, very flat masked and unmasked planar areas of constant elevation can be machined by a technique in which the target oscillates relative to the nozzle in a direction perpendicular to the scan direction in order to generate a uniform erosive flux. In abrasive jet machining, the planar area and transitional slopes in glass are machined with minimum surface waviness using the oscillation technique, in which the target oscillates transversely in the overall direction (Ghobeity et al. 2009b). If the oscillation velocity is sufficiently greater than the scanning velocity, the target receives an approximately uniform energy flux, resulting in a high degree of flatness for both masked and unmasked planar areas micro-machined in glass. The scan velocity was in the 0.01–1 mm/s range and the oscillation speed was in the 80–110 mm/s range. In the rapid target oscillation technique, a uniform erosive energy is available to the target surface. This produces planar areas that have a relatively high degree of flatness.

21.3.10.1.5 Reduction of Surface Roughness by Post-Blasting

Abrasive jet machining is used to machine micro-features in brittle and ductile materials. Surface roughness is greater in the abrasive jet machining process compared to other machining processes. After the post-blast process, surface roughness was reduced by 60% (Jafar et al. 2013). A particle jet of low kinetic energy is used in post blast technique. By three methods, the kinetic energy of the particle in the abrasive jet machining process can be reduced: lowering the pressure and the impact angle, and using smaller particles.

21.3.10.1.6 *Highly Divergent Nature of Air-Driven Jet*

Since the air-driven jet is highly divergent in nature, a mask is needed, which is attached to the surface of the target material. However, it is time consuming and adds additional cost. Shadow masking is an alternate technique that can be applied over the surface, thus allowing direct writing of features on the surface (Nouhia et al. 2015). This eliminates the additional cost and time of fabricating and attaching traditional masks and allows a straightforward adjustment of the mask opening.

21.3.10.1.7 *A Rotating Mask System for Sculpting of Three-Dimensional Features Using Abrasive Jet Micro-Machining*

Channels with nonconventional shaped cross sections have applications especially in microfluidics, whereas most microfabrication technologies are limited in the range of shapes they can produce. Abrasive jet machining has directional etching capabilities, which makes it suitable for producing a wide variety of micro-features with three-dimensional topographies. Traditionally, for such a complex pattern, elastomeric masks have been used in the past, which are directly attached to the target, allowing only the abrasive jet machining of shallow features without any control over their cross-sectional shape. The use of a rotating mask apparatus is a new technique that allows instantaneous control over abrasive jet machining erosive footprint size and shape (Sookhak Lari et al. 2016). To achieve this, the rotating mask apparatus uses different types of holes with ultraviolet light–cured elastomeric thin films that are suspended across slots in a rotating disk held under the abrasive jet and over the target material. The maximum error between the desired and produced topographies was found to be less than 9.1%.

21.3.10.1.8 *Cryogenic Abrasive Jet Machining*

A novel cryogenic abrasive jet machining apparatus is used to investigate the temperature dependence of polydimethylsiloxane solid particles by using aluminum oxide particles with temperatures between −178°C and 17°C for different angles of attack (Gradeen et al. 2012). Between temperatures −82°C and −127°C, the erosion mechanism is ductile in nature, and at −178°C, it is brittle in nature.

21.3.10.1.9 *Flexible Magnetic Abrasive in Abrasive Jet Machining*

The magnetic abrasive is made from thermoplastic polymer (as base material) and thermally coated with silicon carbide particles and magnetic materials. The flexible magnetic abrasive is used in abrasive jet machining not only to restrain the direction of the abrasive particles but also for a more uniform direction to enhance the processing area and improve the material removal rate; it also has the slip-scratch effect to obtain better surface roughness than traditional machining (Ke et al. 2012).

21.3.10.1.10 *Gradient Ceramic Nozzles*

In abrasive jet machining, the nozzle is the most critical part. Ceramics have a great potential as nozzle materials because of their high wear resistance. Compared with conventional ceramics, the gradient ceramic nozzles exhibited high erosion wear resistance (Deng et al. 2007). The mechanism responsible for nozzle erosion is found to be tensile stress at entry point, which is greatly reduced in gradient ceramic nozzles compared with conventional ceramic nozzles. A gradient structure is produced by hot-pressing. The aim is to reduce the tensile stress at the entrance area of the nozzle.

21.3.10.1.11 *Abrasive Jet Polishing*

Abrasive jet machining principles can be used for the metal surface polishing process. Abrasive jet polishing is a new nonconventional method for polishing small and complex surfaces (Chen et al. 2017). Abrasive jet polishing has the following advantages: high precision, easy to control, small machining force, good flexibility, and absence of thermal distortion compared to conventional polishing techniques. In recent years, abrasive jet polishing has been rapidly developed for polishing hard-to-polish materials because it has unique advantages and flexibilities over conventional methods of polishing.

21.3.10.1.12 *Boron Carbide Ceramic Nozzles*

Boron carbide is one of the most widely used ceramics. Boron carbide nozzles are manufactured by hot-pressing. Because of its improved mechanical property, such as high hardness, high melting point, better wear resistance, high Young's modulus, good chemical inertness, and high thermal conductivity, boron carbide is a widely used material in modern engineering applications and is a promising candidate for wear resistance components (Deng 2005). The third hardest material after diamond and cubic boron nitride is boron carbide. Abrasive jet machining nozzles are made of dense sintered boron carbide (Deng 2005); these nozzles are extremely hard and wear resistant and have a long life. Because of their low wear rate, boron carbide nozzles can maintain the internal geometry and minimum compressor requirement and deliver maximum blasting effectiveness as well as save nozzle replacement downtime.

21.3.10.1.13 *Machining of Large Hollow Components*

The new machining setup is introduced in which abrasive particles are placed inside the sealed zone and then a compressed air jet is provided, which is mixed with abrasive particles (Zakharov et al. 2016). In this system, with the help of two circular plunge venturi nozzles that are symmetrical within the working zone, compressed air is supplied through plugs, rods, and hogs. Wear-resistant materials are used to make cleaning elements, which continue to function during the machining process. The nozzles are inclined at a small angle with the surface to create a vertical flux for maximum machining efficiency.

21.3.10.1.14 *Using Abrasive with Hot Air Jet*

For machining glass and other abrasive materials, abrasive jet machining is one of the most useful machining techniques. For abrasive hot air jet machining, the abrasive is combined with hot air jet (Jagannatha et al. 2012). Abrasive hot air jet machining can perform many operations such as surface etching, grooving, drilling, and micro-finish on glass and its composites. Above 100°C, the effect of hot air temperature is more significant on material removal rate than at a lower temperature. It is observed that the material removal rate is 1.4 to 1.7 times higher at high temperatures than at room temperature. It is highest at 310°C. Surface roughness is also reduced by up to 4%–10% in hot air machining.

21.3.11 Electrochemical Machining

The growing demand for microproducts with better surface integrity and accuracy on different applications, namely, micro-fluidics systems and nondefective stress-free drilled holes in aerospace and automobile manufacturing systems, attracts research in electrochemical machining. Electrochemical machining is an unconventional machining process in which material removal takes place as a result of anodic dissociation in an electrolyte under the influence of an electric field. This chapter presents different advanced strategies of electrochemical machining with the aim of improving the material removal rate, surface quality, and accuracy. A photograph of an electrochemical machining machine with a special three-coordinate electrochemical machine for the production of grooves, holes, and complicated bowls in cylindrical soft shells with a big diameter is shown in Figure 21.16.

The requirement for high machining efficiency and quality constantly motivates researchers and scientists to consider new strategies in electrochemical machining. The three important parameters that the researchers can play upon and control are the tool electrode (cathode), electrolyte, and electrical parameters and additional energy (hybrid process). The fishbone diagram in Figure 21.17 shows the various strategies that can be adopted to improve the performance of the electrochemical machining process. Proper control of electrochemical machining variables helps optimize the machining efficiency and quality to a greater extent.

FIGURE 21.16 sET6090-3D: Three-coordinate special electrochemical machine (Indec 2017). (Courtesy of ECM, LLC.)

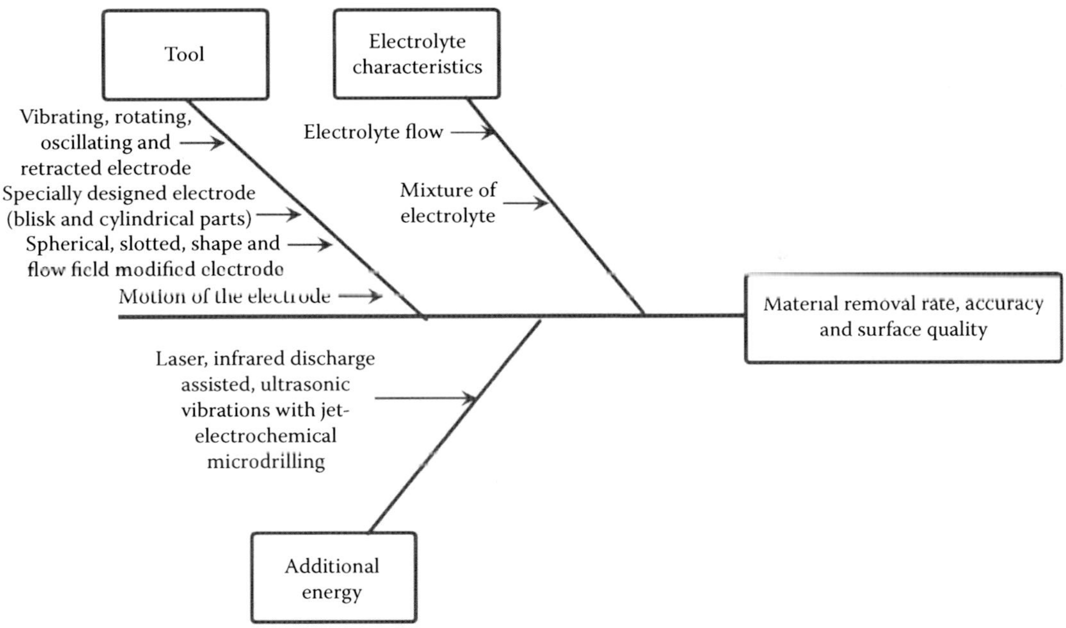

FIGURE 21.17 Advanced electrochemical machining strategies to improve the performance.

21.3.11.1 Design of Tool Electrode and Mechanism

In electrochemical machining, renewal of electrolyte and scavenging of machined products (debris) from the machining zone are important for efficient machining, good surface quality, and accuracy. Proper flushing of machining area is ensured using an ultrasonic-assisted vibrating electrode (Ghosal and Bhattacharyya 2013; Paczkowski and Zdrojewski 2017) and with the proper cathode design. The use of a vibrating tool electrode improves machining localization. In order to improve mass transfer, the oscillating cathode electrode motion (shown in Figure 21.18) is used, which will oscillate

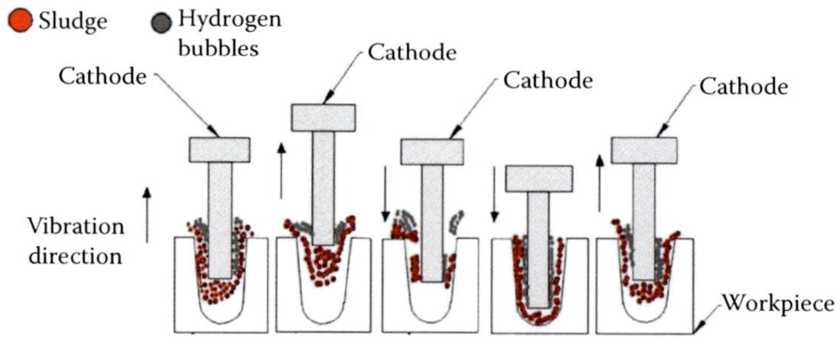

FIGURE 21.18 Oscillating cathode electrode motion.

according to the sine law to change the machining gap periodically. This mechanism helps flush out the machined products. Therefore, the use of periodic cathode vibration helps evacuate electrolysis products in the machining gap and improves the mass transfer of the electrolyte, resulting in stable machining of narrow slit.

One of the main advantages of electrochemical machining is machining of complex shapes on superalloys such as compressor blades and key components of gas turbines. Cathode tool design also plays a major role in machining different workpiece shapes. For example, in order to machine a cylindrical workpiece, Wang et al. (2017a) developed a tool attachment with rotating cylindrical electrodes with different speeds and dimensions as shown in Figure 21.19. Both the anode and cathode cylindrical electrodes were kept on two relatively rotating shafts at the same rotational speed. The rotational speed of cylindrical electrodes in sodium nitrate solution shows the significant influence on material removal rate. Apart from using a new cathode design, researchers now focus on new methods of using auxiliary electrodes. The use of an auxiliary anode improves machining localization, leading to high-accuracy dimples, and is also useful for machining convex shapes.

For machining integrally bladed rotors, different types of tool electrodes, such as equal-thickness cathode and extended cathode, decrease trailing edge by 31.6%, and the surface roughness R_a is

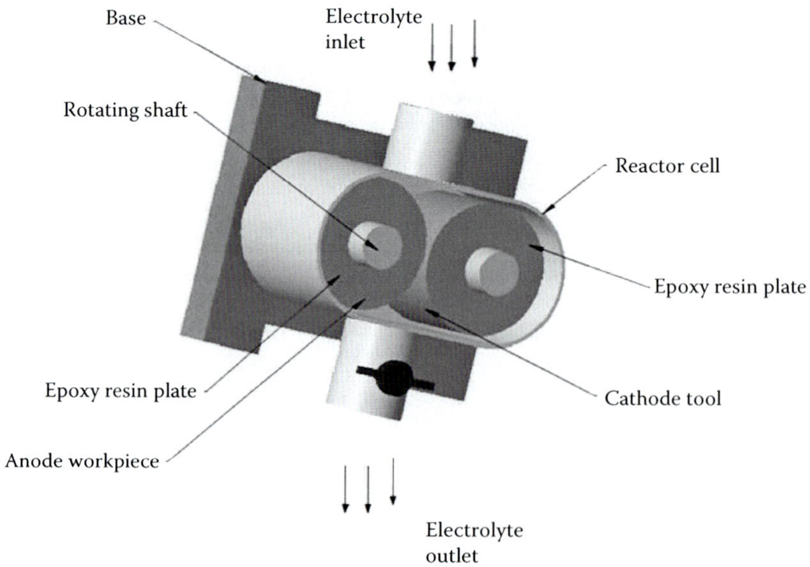

FIGURE 21.19 Rotating cylindrical electrodes with different speeds.

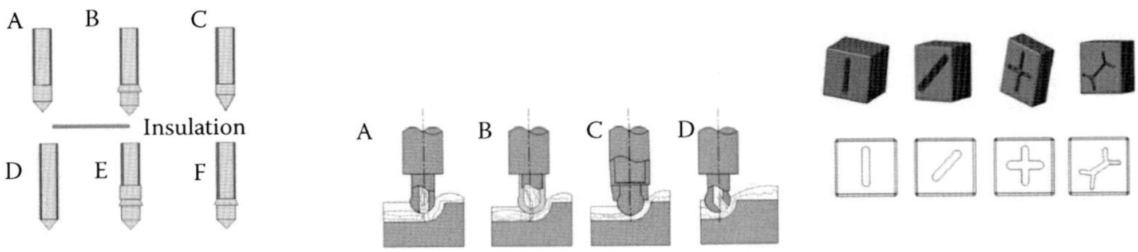

FIGURE 21.20 Different tool electrode shapes.

decreased by 29.6%. In the case of machining blades in electrochemical machining, the cathode and anode feed rate plays a prominent role and the optimal combination of cathode feed and anode position plays a vital role in machining accuracy. This type of feeding arrangement is useful for minimizing the errors during machining of convex and concave parts (Qu and Xu 2013). Moreover, different tool electrode shapes (Figure 21.20), such as pointed cathodes and spherical cathodes with different drain passages, outlet slot patterns, and the turning profile of the cathode tool can provide adequate electrolyte to the machining gap, thus leading to a better flow stabilization and electrolyte flow marks.

21.3.11.2 Electrolyte Characteristics

Electrolyte characteristics play an important role in electrochemical machining and have a great influence on material removal rate and accuracy. Researchers have developed new combinations of electrolytes to improve the output parameters and optimize electrochemical machining performances. High material removal rate and good surface quality are achieved in stainless steel and cemented carbide workpiece with the use of different electrolyte mixtures such as ammonia with sodium nitrate, acidified sodium nitrate, and sodium chlorate. (Thanigaivelan et al. 2013). The mix of nano copper powder in electrolyte improves the electrical conductivity of the electrolyte and breaks the gas layer that causes the uniform current density and prevents the generation of spikes, resulting in high material removal rate and good surface quality. Apart from the concentration, electrolyte flow characteristics play a significant role in electrochemical machining. Different types of flow patterns such as continuous electrolytic free jet, progressive-pressure electrolyte, pulsating flow, and Π-shaped electrolyte flow mode are incorporated in electrochemical machining (Hackert-Oschatzchen et al. 2016; Qu et al. 2013, 2014; Xu et al. 2014). Each flow pattern has a significant contribution toward deeper calottes, improving efficiency and accuracy, heat transfer, material removal rate, surface profile, and electrolyte impact to the cathode.

21.3.12 Chemical Machining

Chemical machining is one of the oldest nontraditional machining practices and is more suitable for producing shallow intricate cavities in materials independent of the strength or hardness. The process uses a simple principle of controlled dissolution of the workpiece material when it comes in contact with an etchant. Chemical milling and photochemical machining are two main variations of the chemical machining process. The main application of chemical machining is in the aerospace industry (for light-weighting of airframe structures produced by other manufacturing processes). However, the main concern with this process is the safety and environmental issues, especially because of the toxicity of both etchant and masking materials. In recent years, with environmental issues becoming very serious, advancements in chemical machining have been made in relation to this. Many initiatives have been taken in the "greening" of the photochemical machining process (Allen 2005). Aqueous cleaning solutions are being used to replace solvents used in metal cleaning.

Similarly, a liquid photoresist is coated from aqueous solution rather than organic solvents. On-site etchant regeneration has been applied to solve the issue of waste etchant disposal. Photographic processing has been eliminated by using laser direct imaging of photoresists and high-resolution inkjet printing of resists. With the developments in sensors and automation systems, the efficiency of the process has improved.

21.3.13 Hybrid Nontraditional Machining Practices

Hybrid machining processes use a combination of the above two or three techniques for machining the material. They have been developed mainly to address the shortcomings of one technique so that it could be used for various applications. These techniques could even be conventional (contact of physical tool), but in this chapter, processes that involve only nonconventional (contact-less) or nontraditional techniques are discussed. Hybrid techniques could be classified in general into two: assisted and mixed processes (Lauwers et al. 2014). Figure 21.21 shows the classification of hybrid techniques relevant to nontraditional machining practices with examples. In the assisted type, the principal material removal would be through primary energy. For example, in a hybrid process of electric discharge machining with ultrasonic vibration and assisted magnetic force, thermal energy is the primary source for material removal and ultrasonic vibration and magnetic force assist (removes debris from the interelectrode gap) the process and thereby improves the performance (Lin et al. 2014). In the case of mixed processes, two or more techniques are applied more or less simultaneously. For example, in the electrochemical discharge machining process, two techniques, namely, electrochemical reaction and electrodischarge action, are combined at the same time to enhance machining performance. In the following paragraphs, some of the recent hybrid machining processes are discussed briefly. Some of the hybrid processes (such as assisted ones) in waterjet machining (waterjet-guided lasers), electric discharge machining (ultrasonic assisted), and abrasive jet machining (cryogenic, flexible magnetic abrasive, and hot air jet) have been discussed in their respective subsections. Processes that involve a physical tool coming in contact with the workpiece are covered in the chapter that discusses advances in conventional machining processes.

21.3.13.1 Electric Discharge Machining–Pulse Electrochemical Machining

Electric discharge machining–pulse electrochemical machining is a mixed hybrid nontraditional machining practice that uses pulse electrochemical machining (basically an electrochemical

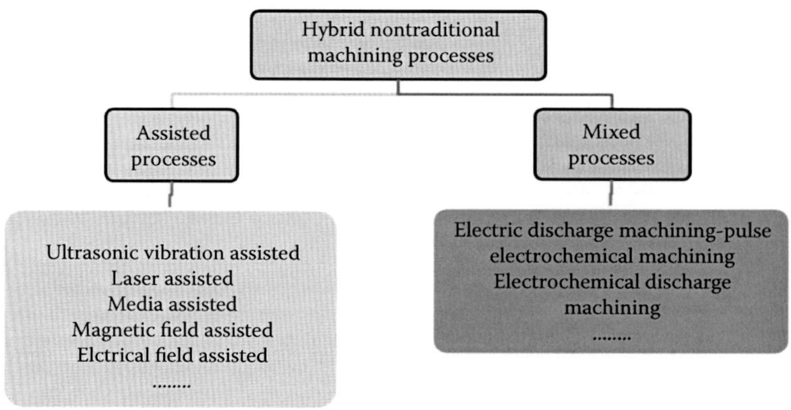

FIGURE 21.21 Classification of hybrid nontraditional machining processes.

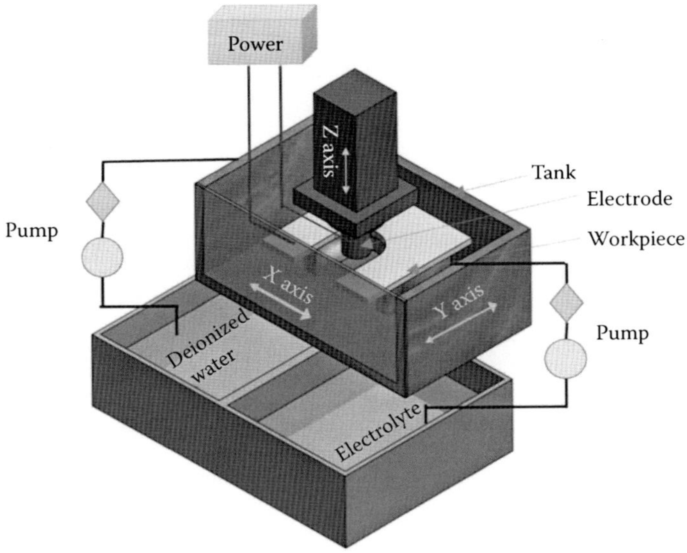

FIGURE 21.22 3D model of the combined electric discharge machining and pulse electrochemical machining setup.

machining process using a pulsed current) along with electric discharge machining. Although electric discharge machining is one of the most widely used nontraditional machining practices, it has the inherent disadvantage of recast layer, which inhibits its usage for producing components for the aerospace industry. To overcome this difficulty, Ma et al. (2015) used electric discharge machining and pulse electrochemical machining to improve the hole quality when drilling a superalloy (GH4169). In this research, electric discharge machining was initially used to drill holes and then later pulse electrochemical machining was used to remove the recast layer, resulting in high shape accuracy and surface quality. Figure 21.22 shows the combined electric discharge machining and pulse electrochemical machining setup.

21.3.13.2 Rotary Magnetic Field–Assisted Dry Electric Discharge Machining with Ultrasonic Vibration of Workpiece

Dry electric discharge machining is gaining popularity because of its sustainable nature. However, the dielectric function of flushing out debris needs to be compensated by other systems and thus researchers have ventured into solving this issue by assisted hybrid processes—ultrasonic vibration, magnetic field, and others. Teimouri and Baseri (2013) experimentally investigated the effect of rotary magnetic field–assisted dry electric discharge machining with ultrasonic vibration of workpiece (SPK cold work steel). The used setup is schematically illustrated in Figure 21.23. The created magnetic field helped expel the debris from the machining gap and thus made the process stable. It also helped in achieving higher material removal rates and a smoother surface finish. Similar results were obtained with ultrasonic vibration of the workpiece, which acted as a pump and helped in driving away the debris from the machining gap.

21.3.13.3 Electric Discharge Machining with Ultrasonic Vibration and Assisted Magnetic Force

This process is similar to the hybrid process described in Section 21.3.13.2, but machining takes place under wet conditions using a dielectric medium. The principle used in electric discharge

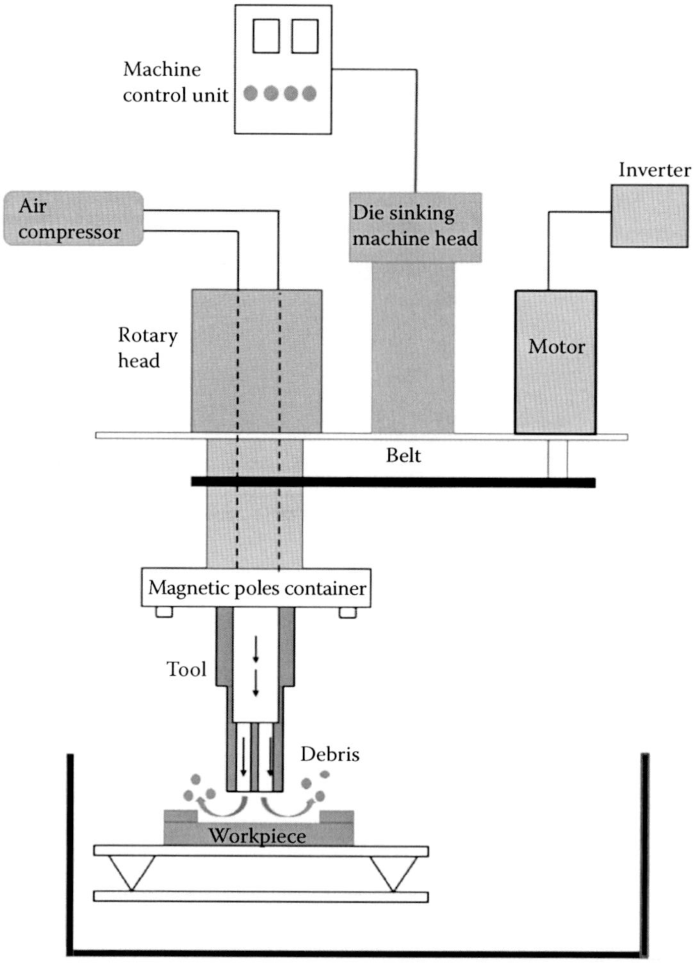

FIGURE 21.23 Schematic illustration of the rotary magnetic field–assisted dry electric discharge machining with ultrasonic vibration of workpiece.

machining with ultrasonic vibration and assisted magnetic force setup used by Lin et al. (2014) is illustrated in Figure 21.24. SKD 61 steel, another commonly used die and mold material, was machined using this hybrid process. As highlighted earlier, ultrasonic vibration and the magnetic force assisted in removing the debris from the machining gap, resulting in higher material removal rates and a finer surface integrity. Compared to the dry electric discharge machining process, this could be applied for machining larger areas.

21.3.13.4 Laser-Assisted Jet Electrochemical Machining

Laser-assisted jet electrochemical machining is an assisted hybrid process in which the laser is a secondary energy source that is used for locally heating the workpiece and thereby enhancing the kinetics of electrochemical reactions (De Silva et al. 2011). The working principle is illustrated in Figure 21.25. De Silva et al. used laser-assisted jet electrochemical machining to machine several aspects and studied the thermal effects since the use of laser can affect the surface. Because of the lower power density of the laser, the workpiece did not suffer from any thermal issues such as heat damage or structural alterations. With developments of portable laser systems, laser-assisted jet electrochemical machining could be a potential hybrid nontraditional machining practice that could find applications in the aerospace industry.

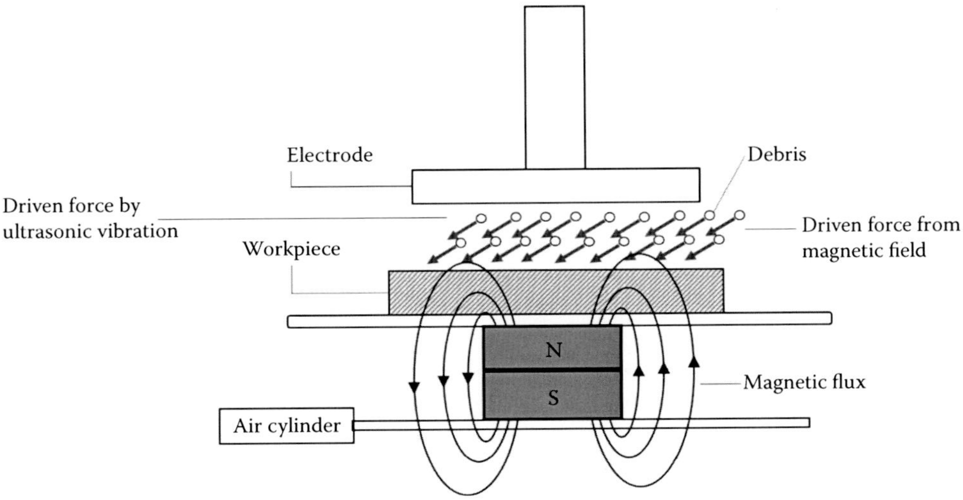

FIGURE 21.24 Principle of hybrid electric discharge machining with ultrasonic vibration and assisted magnetic force.

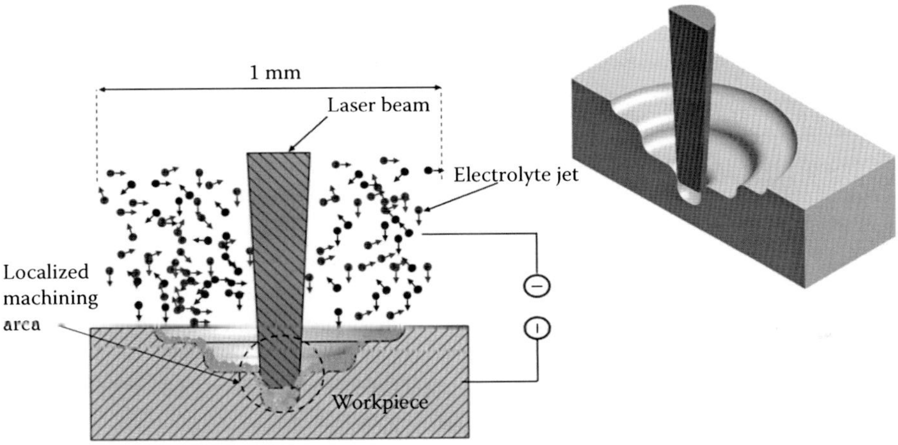

FIGURE 21.25 Illustration of the laser-assisted jet electrochemical machining process.

21.4 CONCLUDING REMARKS

It is quite evident from the discussion that nontraditional machining practices can overcome the difficulties of conventional machining processes, especially for materials that are very difficult to machine. It is not recommended to eliminate conventional machining processes; instead, it is suggested to supplement them with nontraditional machining practices. The newer development in machine tools has allowed integrating the nontraditional machining practices into conventional machine tools, and this has enabled single-setup production of components that could compete with additive manufacturing. These advances have enabled nontraditional machining to be applied in many industries such as aerospace, automotive, energy/power, biomedical, and even in the nuclear field for producing several components. This chapter provides an overview of the various applications, key issues, and developments that have taken place in the last decade and would serve as a guide for manufacturing engineers in choosing appropriate nontraditional machining practices based on the requirement. The following are some of the key highlights on the developments in nontraditional machining practices:

1. The use of ultrasonic vibrations in electric discharge machining contributes to higher discharge energy, resulting in higher removal rates irrespective of the material strength.
2. The electrode-assisted method is identified as a key method for machining insulating materials in electric discharge machining.
3. The electrolyte jet electric discharge machining method can be applied in macro-domain to overcome tool wear issues.
4. The higher value of wire speed in wire electric discharge machining contributes to higher material removal rates and surface roughness.
5. Advanced laser beam machine tools are capable of producing high-quality holes as well as three-dimensional features in engineered materials such as ceramics at lower costs and higher productivity.
6. Recent developments in plasma arc cutting systems have made them more efficient and environment friendly and thus are finding increasing applications for cutting most engineering materials.
7. A new ultrasonic machining method using ultrasonic complex vibration caused by the longitudinal and torsional vibration, sacrificial coating on the substrate, and ultrasonic vibration nano-filing is considered to improve machining speed, machining accuracy, surface integrity, and for a deeper cavity.
8. Waterjet machining can effectively generate clean parts without the issue of abrasive contamination, which is a strong requirement from the medical field as well as for industries where component failure is of extreme concern.
9. Controlled abrasive waterjets can be employed in generating three-dimensional complex parts in advanced engineering materials, such as ceramics, composites, superalloys, and heat-sensitive alloys.
10. Tool path strategy and jet footprint are the critical elements in complex part machining by high-energy jets, and further research is needed in this area to employ them as a standard shop floor process.
11. By adding additional elements (setups, hot air) to the conventional abrasive jet machining approaches, complex and high-precision parts with fine surface quality can be manufactured. This will take the conventional abrasive jet machining one step ahead in terms of its capabilities.
12. In electrochemical machining, the different tool electrode shapes and motions such as slotted and tool oscillation, electrolyte characteristics such as powder mixed electrolytes, and different flow patterns contribute significantly to the material removal rate and surface profile.
13. The chemical machining process is becoming "greener" with developments in aqueous cleaning solutions and etchant regeneration techniques and thus will have wider applications for light-weighting aerospace components.
14. The recently developed hybrid nontraditional machining practices possess features (such as removal of recast layer/debris) that are capable of overcoming the difficulties associated with nontraditional machining practices and hence are suitable to machine most of the recently developed advanced difficult-to-cut materials.
15. With the trend in miniaturization of components, the various hybrid nontraditional machining processes could be scaled down to produce microcomponents.

ACKNOWLEDGMENTS

The first author would like to thank Pradeep Kumar Krishnan (Caledonian College of Engineering, Sultanate of Oman) for his contribution in preparing the schematic for the hybrid machining processes; the second author would like to thank K.G. Saravanan (Sona College of Technology, Salem,

India) for his contribution in preparing the schematic for electrochemical machining/electric discharge machining processes; and the third author would like to thank Mr. Rajesh Ranjan Ravi and Mr. Ngangkham Peter Singh (Indian Institute of Technology Madras, India) for their support.

REFERENCES

Agarwal, S. (2015). On the mechanism and mechanics of material removal in ultrasonic machining. *International Journal of Machine Tools and Manufacture* **96**: 1–14.

Ahluwalia, D., Borrelli, M.J., Smithson, K., Rajurkar, K.P., and Malshe, A.P. (2014). Ultrasonic machining of biomass using biodegradable slurry. *CIRP Annals—Manufacturing Technology* **63**(1): 217–220.

Ahmadzadeh, F., Tsai, S.S.H., and Papini, M. (2017). Effect of curing parameters and configuration on the efficacy of ultra violet light curing self-adhesive masks used for abrasive jet micro-machining. *Precision Engineering* **49**: 354–364.

Allen, D.M. (2005). Photochemical machining: From manufacturing. *CIRP Annals—Manufacturing Technology* **53**(2): 559–572.

Allen, D.M., Shore, P., Evans, R.W., Fanara, C., O'Brien, W., Marson, S., and O'Neill, W. (2009). Ion beam, focused ion beam, and plasma discharge machining. *CIRP Annals—Manufacturing Technology* **58**(2): 647–662.

Asami, T. and Miura, H. (2015a). Study of ultrasonic machining by longitudinal-torsional vibration for processing brittle materials—Observation of machining marks. *Physics Procedia* **70**:118–121.

Asami, T., and Miura, H. (2015b). Ultrasonic welding of dissimilar metals by vibration with planar locus. *Acoustical Science and Technology* **36**(3): 232–239.

Atzeni, E., Bassoli, E., Gatto, A., Iuliano, L., Minetola, P., and Salmi, A. (2015). Surface and sub surface evaluation in coated-wire electrical discharge machining (WEDM) of INCONEL® alloy 718. *Procedia CIRP* **33**: 388–393.

Axinte, D.A., and Srinivasu, D.S. (2009). Freeform jet, Nottingham Innovative Manufacturing Research Centre—Final report, University of Nottingham, UK.

Axinte, D.A., Srinivasu, D.S., Billingham, J., and Cooper, M. (2010). Geometrical modelling of abrasive waterjet footprints: A study for 90° jet impact angle. *CIRP Annals—Manufacturing Technology* **59**(1): 341–346.

Axinte, D.A., Srinivasu, D.S., Kong, M.C., and Butler-Smith, P.W. (2009). Abrasive waterjet cutting of polycrystalline diamond: A preliminary investigation. *International Journal of Machine Tools and Manufacture* **49**(10): 797–803.

Babashov, V., and Mammadova, G. (2015). An investigation on the suitability of abrasive waterjet cutting for hybrid manufacturing of gears in automotive industry. Master of Science Thesis, Department of Production Engineering, KTH—Royal Institute of Technology, Sweden.

Baek, D.K., Ko, T.J., and Yang, S.H. (2013). Enhancement of surface quality in ultrasonic machining of glass using a sacrificing coating. *Journal of Materials Processing Technology* **213**(4): 553–559.

Benhabib, B. (2003). *Manufacturing: Design, Production, Automation, and Integration*. CRC Press.

Bhosale, S.B., Pawade, R.S., and Brahmankar, P.K. (2014). Effect of process parameters on MRR, TWR and surface topography in ultrasonic machining of alumina–zirconia ceramic composite. *Ceramics International* **40**(8): 12831–12836.

Billingham, J., Miron, C.B., Axinte, D.A., and Kong, M.C. (2013). Mathematical modelling of abrasive water jet foot prints for arbitrarily moving jets: Part II—Overlapped single and multiple straight paths. *International Journal of Machine Tools and Manufacture* **68**: 30–39.

Chen, F. Miao, X., Tang, Y., and Yin, S. (2017). A review on recent advances in machining methods based on abrasive jet polishing. *The International Journal of Advanced Manufacturing Technology* **90**(1): 785–799.

Chun, D.M., Davaasuren, G., Ngo, C.V., Kim, C.S., Lee, G.Y., and Ahn, S.H. (2014). Fabrication of transparent super hydrophobic surface on thermoplastic polymer using laser beam machining and compression molding for mass production. *CIRP Annals—Manufacturing Technology* **63**(1): 525–528.

Cincinnati Inc. (2017). CL-900—Series fiber laser cutting system. Accessed May 29. https://www.e-ci.com/cl-900-series-fiber-laser-cutting-system/

Das, A.K., Kumar, P., Sethi, A., Singh, P.K., and Hussain, M. (2016a). Influence of process parameters on the surface integrity of micro-holes of SS304 obtained by micro-EDM. *Journal of the Brazilian Society of Mechanical Sciences and Engineering* **38**(7): 2029–2037.

Das, S., Doloi., B, and Bhattacharyya, B. (2016b). Fabrication of stepped hole on zirconia bioceramics by ultrasonic machining. *Machining Science and Technology* **20**(4): 681–700.

De Silva, A.K.M., Pajak, P.T., McGeough, J.A., and Harrison, D.K. (2011). Thermal effects in laser assisted jet electrochemical machining. *CIRP Annals—Manufacturing Technology* **60**(1): 243–246.

Deng, J. (2005). Erosion wear of boron carbide ceramic nozzles by abrasive air-jets. *Materials Science and Engineering* **408**(1–2): 227–233.

Deng, J., Wu, F., and Zhao, J. (2007). Wear mechanisms of gradient ceramic nozzles in abrasive air-jet machining. *International Journal of Machine Tools & Manufacture* **47**(12–13): 2031–2039.

Ferraris. E, Vleugels, J., Guo, Y., Bourell, D., Kruth, J.P., and Lauwers, B. (2016). Shaping of engineering ceramics by electro, chemical and physical. *CIRP Annals—Manufacturing Technology* **65**: 761–784.

Feucht, F., Ketelaer, J., Wolff, A., Mori, M., and Fujishima, M. (2014). Latest machining technologies of hard to cut materials by ultrasonic machine tool. *Procedia CIRP* **14**: 148–152.

Foldyna, J., Klich, J., Hlavacek, P., Zelenak, M., and Scucka, J. (2012). Erosion of metals by pulsating water jet. *Technical Gazette* **19**(2): 381–386.

Gao, C., and Liu, Z. (2003). A study of ultrasonically aided micro-electrical-discharge machining by the application of workpiece vibration. *Journal of Materials Processing Technology* **139**: 226–228.

Ghobeity, A., Ciampini, D., and Papini, M. (2009a). An analytical model of the effect of particle size distribution on the surface profile evolution in abrasive jet micromachining. *Journal of Materials Processing Technology* **209**(20): 6067–6077.

Ghobeity, A., Papini, M., and Spelt, J.K. (2009b). Abrasive jet micro-machining of planar areas and transitional slopes in glass using target oscillation. *Journal of Materials Processing Technology* **209**(11): 5123–5132.

Ghosal, B., and Bhattacharyya, B. (2013). Influence of vibration on micro-tool fabrication by electrochemical machining. *International Journal of Machine Tools & Manufacture* **64**: 49–59.

Gotia, B., and Loya Mucino, J. (2016). Advanced hybrid manufacturing process for high precision ring of a planetary gear—Main focus on abrasive waterjet machining. Department of Production Engineering, KTH—Royal Institute of Technology, Sweden.

Gradeen, A.G., Spelt, J.K., and Papini, M. (2012). Cryogenic abrasive jet machining of polydimethylsiloxane at different temperatures. *Wear* **274–275**(27): 335–344.

Hackert-Oschatzchen, M., Lehnert, N., Martin, A., and Schubert, A. (2016). Jet electrochemical machining of particle reinforced aluminum matrix composites with different neutral electrolytes. IOP Conf. Series: Materials Science and Engineering **118**(1).

Hu, J., and Xu, H. (2016). Pocket milling of carbon fiber-reinforced plastics using 532-nm nanosecond pulsed laser: An experimental investigation. *Journal of Composite Materials* **50**(20): 2861–2869.

Huang, T.S., Hsieh, S.-F., Chen, S.-L., Lin, M.-H., Ou, S.-F., and Chang, W.-T. (2015). The effect of acetylene as a dielectric on modification of TiNi-based shape memory alloys by dry EDM. *Journal of Material Research* **30**(22): 3484–3492.

Indec (2017). sET6090-3D—Three-coordinate special electrochemical machine. http://www.indec-ecm.com/en/equipment/xet/ET6090-3D/

Jafar, R.H.M., Papini, M., Spelt, J.K. (2013). Simulation of erosive smoothing in the abrasive jet micro-machining of glass. *Journal of Materials Processing Technology* **213**(12): 2254–2261.

Jafar, R.H.M., Hadavi, V., Spelt, J.K., and Papini, M. (2016). Dust reduction in abrasive jet micro-machining using liquid films. *Powder Technology* **301**: 1270–1274.

Jafferson, J.M., Hariharan, P., and Ram Kumar, J. (2014). Effects of ultrasonic vibration and magnetic field in micro-EDM milling of nonmagnetic material. *Materials and Manufacturing Processes* **29**(3): 357–363.

Jagannatha, N., Hiremath Somashekhar, S., Sadashivappa, K., and Arun, K.V. (2012). Machining of soda lime glass using abrasive hot air jet: An experimental study. *Machining Science and Technology* **16**(3): 459–472.

Jothimurugan, R., and Amirthagadeswaran, K.S. (2014). Performance of additive mixed kerosene-servotherm in electrical discharge machining of monel 400^{TM}. *Materials and Manufacturing Processes* **31**(4): 432–438.

Kaltenbach (2017). *KF 1614—Plate cutting and drilling centre*. Accessed May 29. https://www.kaltenbach.com/en/products/machines/plate-cutting-and-drilling-centre/kf-1614/#/?playlistId=0&videoId=0/

Kannegulla, A., and Cheng, L.-J. (2016). Metal assisted focused-ion beam nanopatterning. *Nanotechnology* **27**(36): 36LT01.

Karpuschewski, B., Knoche, H.J., and Hipke, M. (2008). Gear finishing by abrasive processes. *CIRP Annals—Manufacturing Technology* **57**: 621–640.

Kataria, R., Kumar, J., and Pabla, B.S. (2015). Experimental investigation into the hole quality in ultrasonic machining of WC-Co composite. *Materials and Manufacturing Processes* **30**(7): 921–933.

Ke, J.-H., Tsai, F.-C., Hung, J.-C., and Yan, B.-H. (2012). Characteristics study of flexible magnetic abrasive in abrasive jet machining. *Procedia CIRP* **1**: 679–680.

Khakpour, H., Birglen, L., and Tahan, S.-A. (2015). Uniform scanning path generation for abrasive waterjet polishing of free-form surfaces modeled by triangulated meshes. *International Journal of Machine Tools & Manufacture* **77**: 1167–1176.

Kim, J., Lee, W.J., and Park, H.W. (2016). The state of the art in the electron beam manufacturing processes. *International Journal of Precision Engineering and Manufacturing* **17**(11): 1575–1585.

Kong, M.C., Axinte, D.A., and Voice, W. (2010). Aspects of material removal mechanism in plain waterjet milling on gamma titanium aluminide. *Journal of Materials Processing Technology* **210**(3): 573–584.

Kong, M.C., Anwar, S., Billingham, J., and Axinte, D.A. (2012). Mathematical modelling of abrasive waterjet footprints for arbitrarily moving jets. *International Journal of Machine Tools and Manufacture* **53** (1): 58–68.

Krajcarz, D. (2014). Comparison metal water jet cutting with laser and plasma cutting. *Procedia Engineering* **69**: 838–843.

Kremer, D., Lebrun, J.L., Hosari, B., and Moisan, A. (1989). Effects of ultrasonic vibrations on the performance in EDM. *Annals of CIRP* **38**(1): 199–202.

Kumar, J. (2014). Investigations into the surface quality and micro-hardness in the ultrasonic machining of titanium (ASTM GRADE-1). *Journal of Brazilian Society of Mechanical Sciences and Engineering* **36** (4): 807–823.

Kumar, A., and Hiremath, S.S. (2016). Improvement of geometrical accuracy of micro holes machined through micro abrasive jet machining. *Procedia CIRP* **46**: 47–50.

Lalchhuanvela, H., Doloi, B., and Bhattacharyya, B. (2013). Analysis on profile accuracy for ultrasonic machining of alumina ceramics. *The International Journal of Advanced Manufacturing Technology* **67** (5): 1683–1691.

Lauwers, B., Klocke, F., Klink, A., Erman Tekkaya, A., Neugebauer, R., and McIntosh, D. (2014). Hybrid processes in manufacturing. *CIRP Annals—Manufacturing Technology* **63**(2): 561–583.

Lee, S.W., Shin, H.S., and Chu, C.N. (2013). Fabrication of micro-pin array with high aspect ratio on stainless steel using nanosecond laser beam machining. *Applied Surface Science* **264**: 653–663.

Li, C., Yang, L., and Wang, Y. (2011). A research on surface morphology of cutting of titanium sheet with water-jet guided laser and conventional laser. *Applied Mechanics and Materials* **120**(10): 366–370.

Li, L., Wei, X.T., Guo, Y.B., Li. W., and Liu, J.F (2014). Surface integrity of Inconel 718 by wire-EDM at different energy modes. *Journal of Materials Engineering and Performance* **23**(8): 3051–3057.

Liew, P.J., Yan, J., and Kuriyagawa, T. (2013). Carbon nanofiber assisted micro electro discharge machining of reaction-bonded silicon carbide. *Journal of Materials Processing Technology* **213**: 1076–1087.

Lin, Y.C., Chuang, F.P., Wang, A.C., and Chow, H.M. (2014). Machining characteristics of hybrid EDM with ultrasonic vibration and assisted magnetic force. *International Journal of Precision Engineering and Manufacturing* **15**(6): 1143–1149.

Lincoln Electric (2017). Tomahawk® 375 Plasma Cutter—K2806-1. Accessed May 29. http://www.lincolnelectric.com/en

Liu, J.F., Li, L., and Guo, Y.B. (2014). Surface integrity evolution from main cut to finish trim cut in W-EDM of shape memory alloy. *Procedia CIRP* **13**: 137–142.

Liu, W., Ao, S., Li, Y., Liu, Z., Wang, Z., Luo, Z., Wang, Z., and Song, R. (2017). Jet electrochemical machining of TB6 titanium alloy. *The International Journal of Advanced Manufacturing Technology* **90** (5): 2397–2409.

Liu, W., Ao, S., Li, Y., Liu, Z., Zhang, H., Luo, Z., and Yu, H. (2016). Investigation on the profile of microhole generated by electrochemical micromachining using retracted tip tool. *The International Journal of Advanced Manufacturing Technology* **87**(1–4): 877–889.

Lutey, A.H.A., Fortunato, A., Carmignato, S., and Fiorini, M. (2017). High speed pulsed laser cutting of Li-ion battery electrodes. *Optics & Laser Technology* **94**: 90–96.

Ma, N., Yang, X., Gao, M., Song, J., Liu, G., and Xu, W. (2015). A Study of electro discharge machining pulse electrochemical machining combined machining for holes with high surface quality on super alloy. *Advances in Machining Engineering*.

Maher, I., Sarhan, A.A., Barzani, M.M., and Hamdi, M. (2015a). Increasing the productivity of the wire-cut electrical discharge machine associated with sustainable production. *Journal of Cleaner Production* **108**: 247–255.

Maher, I., Sarhan, A.A., and Hamdi, M. (2015b). Review of improvements in wire electrode properties for longer working time and utilization in wire EDM machining. *The International Journal of Advanced Manufacturing Technology* **76**(1–4): 329–351.

Maity, K.P., and Bagal, D.K. (2015). Effect of process parameters on cut quality of stainless steel of plasma arc cutting using hybrid approach. *The International Journal of Advanced Manufacturing Technology* **78** (1–4): 161–175.

Marimuthu, S., Antar, M., Dunleavey, J., Chantzis, D., Darlington, W., and Hayward, P. (2017). An experimental study on quasi-CW fibre laser drilling of nickel superalloy. *Optics & Laser Technology* **94**: 119–127.

Miller, P.L. (2012). Abrasive waterjets—A nontraditional process for the safe and environmentally friendly demilitarization of underwater high-explosive munitions. *Marine Technology Society Journal* **9**: 83–91.

Morimoto, Y., He, D., Hijikata, W., Shinshi, T., Nakai, T., and Nakamura, N. (2015). Effect of high-frequency orbital and vertical oscillations of the laser focus position on the quality of the cut surface in a thick plate by laser beam machining. *Precision Engineering* **40**: 112–123.

Nani, V.-M. (2016). The ultrasound effect on technological parameters for increase in performances of W-EDM machines. *The International Journal of Advanced Manufacturing Technology* **88**(1): 519–528.

Ng, P.S, Kong, A. and Yeo, S.H. (2016). Investigation of biodiesel dielectric in sustainable electrical discharge machining. *The International Journal of Advanced Manufacturing Technology* **90** (9): 2549–2556.

Ngangkham, P.S., Srinivasu, D.S., and Ramesh Babu, N. (2016). Modeling of kerf profile generated in multi-layered composites with abrasive waterjet. *Material Science Forum* **874**: 219–224.

Nouhi, A., Sookhak Lari, M.R., Spelt, J.K., and Papini, M. (2015). Implementation of a shadow mask for direct writing in abrasive jet micro-machining. *Journal of Materials Processing Technology* **223**: 232–239.

Omax (2017a). OMAX 5555 JetMachining Center. https://www.omax.com/omax-machine/5555

Omax (2017b). OMAX 55100 JetMachining Center. https://www.omax.com/omax-machine/55100

Paczkowski, T., and Zdrojewski, J. (2017). Monitoring and control of the electrochemical machining process under the conditions of a vibrating tool electrode. *Journal of Materials Processing Technology* **244**: 204–214.

Patil, S., Joshi, S., Tewari, A., and Joshi, S.S. (2014). Modelling and simulation of effect of ultrasonic vibrations on machining of Ti6Al4V. *Ultrasonics* **54**(2): 694–705.

Pi Tech Blog (2017). Sonodrive 300 serial-production vibratory spindle. http://www.pi-usa.us/blog/accelerated-edm-microstructuring-aided-by-electroceramic-transducers/

Prasad, K., and Chakraborty, S. (2015). A decision guidance framework for non-traditional machining processes selection. *Ain Shams Engineering Journal*. https://doi.org/10.1016/j.asej.2015.10.013.

Prihandana, G.S., Mahardika, M., Hamdi, M., and Mitsui, K. (2009). The current methods for improving electrical discharge machining processes. *Recent Patent in Mechanical Engineering* **2**(1): 61–68.

Qu, N.S., Hu, Y., Zhu, D., and Xu, Z.Y. (2014). Electrochemical machining of blisk channels with progressive-pressure electrolyte flow. *Materials and Manufacturing Processes*, **29**(5): 572–578.

Qu, N.S., Fang, X.L., Zhang, Y.D., and Zhu, D. (2013). Enhancement of surface roughness in electrochemical machining of Ti6Al4V by pulsating electrolyte. *The International Journal of Advanced Manufacturing Technology* **69**(9–12): 2703–2709.

Qu, N.S., and Xu, Z.Y (2013). Improving machining accuracy of electrochemical machining blade by optimization of cathode feeding directions. *The International Journal of Advanced Manufacturing Technology* **68**(5–8): 1565–1572.

Roy, N., Kuar, A.S., and Mitra, S. (2017). Underwater pulsed laser beam cutting with a case study. *Microfabrication and Precision Engineering*: 189–212.

Samant, A. (2011a). Laser machining. In: *Laser Machining of Advanced Materials*, CRC Press.

Samant, A. (2011b). Laser machining of biomaterials. In: *Laser Machining of Advanced Materials*, CRC Press: 109–130.

Samant, A. (2011c). Laser machining of intermetallics. In: *Laser Machining of Advanced Materials*, CRC Press: 177–194.

Samant, A. (2011d). Laser machining of structural ceramics. In: *Laser Machining of Advanced Materials*, CRC Press: 41–107.

Sharma, P., Chakradhar, D., and Narendranath, S. (2015). Evaluation of WEDM performance characteristics of Inconel 706 for turbine disk application. *Materials & Design* **88**: 558–566.

Shen, Y., Liu, Y., Dong, H., Zhang, K., Lv, L., Zhang, X., Wu, X., Zheng, C., and Ji, R. (2016). Surface integrity of Inconel 718 in high-speed electrical discharge machining milling using air dielectric. *The International Journal of Advanced Manufacturing Technology* **90**(1): 691–698.

Shervani-Tabar, M.T., Maghsoudi, K., and Shabgard, M.R. (2013). Effects of simultaneous ultrasonic vibration of the tool and the workpiece in ultrasonic assisted EDM. *International Journal for Computational Methods in Engineering Science and Mechanics* **14**: 1–9.

Sidhu, S.S., Batish, A. and Kumar, S. (2014). Study of surface properties in particulate-reinforced metal matrix composites (MMCs) using powder-mixed electrical discharge machining (EDM). *Materials and Manufacturing Processes* **29**(1): 46–52.

Sodick. (2017). Sodick's linear motor technology—AP250L Wire EDM. https://www.sodick.com/products/wire-edm/ap250l

Sookhak Lari, M.R., Ghazavi, A., and Papini, M. (2016). A rotating mask system for sculpting of three-dimensional features using abrasive jet micro-machining. *Journal of Materials Processing Technology* **243**: 62–74.

Srinivasu, D.S., and Axinte, D.A. (2011). An analytical model for top width of jet footprint in abrasive waterjet milling: A case study on SiC ceramics. *Proceedings of the Institution of Mechanical Engineers, Part B: Journal of Engineering Manufacture* **225**(3): 319–335.

Srinivasu, D.S., and Axinte, D.A. (2014a). Mask less pocket milling in composites by abrasive waterjets: An experimental investigation. *ASME—Journal of Manufacturing Science and Engineering* **136**(4): 1144–1157.

Srinivasu, D.S., and Axinte, D.A. (2014b). Surface integrity analysis of plain waterjet milled advanced engineering composite materials. *Procedia CIRP* **13**: 371–376.

Srinivasu, D.S., Axinte, D.A., Shipway, P.H., and Folkes, J. (2009). Influence of kinematic operating parameters on kerf geometry in abrasive waterjet machining of silicon carbide ceramics. *International Journal of Machine Tools and Manufacture* **49**(14): 1077–1088.

Srinivasu, D.S., and Nicolescu, M. (2013). Issues in the machining of hollow core honeycomb sandwich structures by abrasive waterjet machining. *CIRP Sponsored Conference on Supervising and Diagnostics of Machining Systems*, Karpacz, Poland.

Swift, K.G., and Booker, J.D. (2013). Non-traditional machining processes. In: *Manufacturing Process Selection Handbook*: 205–226.

Talla, G., Sahoo, D.K., Gangopadhyay, S., and Biswas, C.K. (2015). Modeling and multi-objective optimization of powder mixed electric discharge machining process of aluminum/alumina metal matrix composite. *Engineering Science and Technology, an International Journal* **18**: 369–373.

Tang, L., and Du, Y.T. (2014). Multi-objective optimization of green electrical discharge machining Ti–6Al–4V in tap water via Grey-Taguchi method. *Materials and Manufacturing Processes* **29**(5): 507–513.

Tebaldo, V., and Faga, M.G. (2017). Influence of the heat treatment on the microstructure and machinability of titanium aluminides produced by electron beam melting. *Journal of Materials Processing Technology* **244**. 289–303.

Teimouri, R., and Baseri, H. (2013). Experimental study of rotary magnetic field-assisted dry EDM with ultrasonic vibration of workpiece. *International Journal of Advanced Manufacturing Technology* **67** (5–8): 1371–1384.

Thanigaivelan, R., Arunachalam, R.M., Karthikeyan, B., and Loganathan, P. (2013). Electrochemical micromachining of stainless steel with acidified sodium nitrate electrolyte. *Procedia CIRP* **6**: 351–355.

Wang, C., and Zeng, X. (2007). Study of laser carving three-dimensional structures on ceramics: Quality controlling and mechanisms. *Optics and Laser Technology* **39**(7): 1400–1405.

Wang, D., Zhu, Z., Zhu, D., He, B., and Ge, Y. (2017a). Reduction of stray currents in counter-rotating electrochemical machining by using a flexible auxiliary electrode mechanism. *Journal of Materials Processing Technology* **239**: 66–74.

Wang, H., Lin, H., Wang, C., Zheng, L., and Hu, X. (2017b). Laser drilling of structural ceramics: A review. *Journal of the European Ceramic Society* **37**(4): 1157–1173.

Wang, Y., Gong, H., Fang, F.Z., and Ni, H. (2016a). Kinematic view of the cutting mechanism of rotary ultrasonic machining by using spiral cutting tools. *The International Journal of Advanced Manufacturing Technology* **83**(1–4): 461–474.

Wang, Y., Sarin, V.K., Lin, B., Li, H., and Gillard, S. (2016b). Feasibility study of the ultrasonic vibration filing of carbon fiber reinforced silicon carbide composites. *International Journal of Machine Tools & Manufacture* **101**: 10–17.

Wei, W., Zhu, D., Allen, D.M., and Almond, H.J.A. (2008). Non-traditional machining techniques for fabricating metal aerospace filters. *Chinese Journal of Aeronautics* **21**(5): 441–447.

Wu, B., Zhang, L., Zhang, J., Murakami, R.I., and Pyoun, Y.S. (2015). An investigation of ultrasonic nanocrystal surface modification machining process by numerical simulation. *Advances in Engineering Software* **83**: 59–69.

Xu, Z., Sun, L., Hu, Y., and Zhang, J. (2014). Flow field design and experimental investigation of electrochemical machining on blisk cascade passage. *The International Journal of Advanced Manufacturing Technology* **71**(1–4): 459–469.

Xu, Z., Liu, J., Xu, Q., Gong, T., Zhu, D., Qu, N. (2015). The tool design and experiment on electrochemical machining of a blisk using multiple tube electrodes. *International Journal of Advanced Manufacturing Technology* **79**(1–4): 531–539.

Yan, B.H., Tsai, H.C., and Huang, F.Y. (2005). The effect in EDM of a dielectric of a urea solution in water on modifying the surface of titanium. *International Journal of Machine Tools & Manufacture* **45**: 194–200.

Yan, M.-T., Wang, P.-W., and Lai, J.-C. (2015). Improvement of part straightness accuracy in rough cutting of wire EDM through a mechatronic system design. *The International Journal of Advanced Manufacturing Technology* **84**(9): 2623–2635.

Youssef, H.A., and El-Hofy, H. (2008). Nontraditional machine tools and operations. *Machine Technology—Machine Tools and Operations*: 391–493.

Zakharov, O.V., Khudobin, L.V., Vetkasov, N.I. Klyarov, I.A., and Kochetkov, A.V. (2016). Abrasive-jet machining of large hollow components. *Russian Engineering Research* **36**: 469.

Zhang, G., Zhang, Z., Guo, J., Ming, W., Li, M., and Huang, Y. (2013a). Modeling and optimization of medium-speed WEDM process parameters for machining SKD11. *Materials and Manufacturing Processes* **28**(10): 1124–1132.

Zhang, Y., Wilt, K., Kang, X., Han, N., and Zhao, W. (2016). The experimental investigation of electrostatic field-induced electrolyte jet (E-Jet) micro-electrical machining mechanism. *The International Journal of Advanced Manufacturing Technology* **87**: 1–10.

Zhang, Z., Peng, H., and Yan, J. (2013b). Micro-cutting characteristics of EDM fabricated high-precision polycrystalline diamond tools. *International Journal of Machine Tools and Manufacture* **65**: 99–106.

Zhu, Z. Wang, D., Bao, J., Wang, N., and Zhu, D. (2015). Cathode design and experimental study on the rotate-print electrochemical machining of revolving parts. *International Journal of Advanced Manufacturing Technology* **80**(9–12): 1957–1963.

22 Recent Trends and Advances in Friction Stir Welding and Friction Stir Processing of Metals

Puthuveettil Sreedharan Robi, Sukhomay Pal, and Biswajit Parida

CONTENTS

22.1 INTRODUCTION

Friction stir welding is an emerging technique for welding similar and dissimilar materials with superior weld properties. It has proven to be very effective in replacing the fusion welding process and is being practiced by a number of industries in sectors, such as aerospace, shipbuilding, railway, automobile, and electrical. The reason for the recent focus of attention to this process is its energy efficiency, versatility, and environment friendliness in addition to the fact that it can be used to weld almost all difficult-to-weld materials to include low–melting point materials like aluminum (Al) and magnesium (Mg). Friction stir welding has the ability to produce high-quality similar joints like magnesium, copper, and iron alloys and dissimilar joints in aluminum–copper, aluminum–magnesium, and aluminum–steel, to name a few, without melting the parent material and thereby eliminating the defects that are likely to form during the fusion welding processes. Friction stir welding also has the ability to join plates of different thicknesses with superior weld quality, which is difficult to accomplish using other conventional welding process.

Friction stir welding is an upgraded version of the conventional friction welding process, invented by The Welding Institute, United Kingdom (Thomas et al. 1991). It is a solid-state joining process in which the joined material is plasticized by frictional heat that is generated between the surface of the plates to be welded and the contact surface of a special rotating tool. The rotating consumable tool that is inserted into the two materials to be welded mainly consists of three parts: shank, shoulder, and pin. The shank fits into the machine spindle through a collet. The shoulder is responsible for the generation of frictional heat due to the rubbing action between the contact surface of the workpiece and retaining the plasticized material in the weld zone. The pin penetrates into the workpiece and mixes the materials of the components to be welded. A small amount of heat is generated between the pin surface and the workpiece. Heat is also generated as a result of the plastic deformation of the workpiece. Though tools designed for different applications may have slightly different shapes for the tool pin and shoulder, all tools maintain this three-element design.

The mechanics behind friction stir welding is complicated and requires a balance of thermal and mechanical interactions as well as plastic flow of the material. The localized heating softens the material around the pin while a combination of tool rotation and translation along the weld line leads to movement of the plasticized material from the front of the pin to the back of the pin. These two motions cause a difference in relative velocity between both sides of the weld, leading to a higher relative velocity of the material on the advancing side compared to the retarding side (Mishra and Ma 2005). The material movement around the pin can also be quite complex depending on the geometrical features of the tool (Mishra and Ma 2005). In friction stir welding, the material experiences intense plastic deformation at an elevated temperature that promotes fine, equiaxed, and recrystallized grains (Rhodes et al. 1997) and results in improved mechanical properties for the weld. There has been a substantial increase in research publications after 2005 in this area. The bulk of the published data is related to aluminum alloys. However, recent reports are currently available for other metals and alloy systems. Comprehensive literature on friction stir welding of aluminum alloys (1xxx to 7xxx series, aluminum die casting alloys, high-strength age-hardened, aluminum–silicon–based alloys, etc.), magnesium alloys (AZ and AM series), copper, brass, steel, and titanium are available.

Despite several advantages, welding of the high–melting point alloys remains a challenge. Friction stir welding of materials that have a high melting point requires sufficiently high frictional heat generation to plasticize the workpiece material near the weld zone. This can be achieved with high welding forces and spindle speeds, which imposes a higher force and torque on the tool, often resulting in poor performance and failure of the tool. An assisted or hybrid friction stir welding system has been developed to overcome this drawback. Hybrid friction stir welding introduces additional localized heating, without melting of the material, immediately ahead of the weld zone, thereby softening the material. This results in less mechanical energy being delivered through the tool, thereby reducing the force on the tool and deflections in the machine and fixtures, and enables a higher welding speed to be attained.

Several variants of assisted friction stir welding have been developed over the past few years. They have been developed in combination with laser, electric arc, plasma arc, resistance heating, induction heating, and ultrasonic vibration (Padhy et al. 2015). The easiest method to introduce additional heat source is to attach an underpower fusion welding torch in front of the friction stir welding tool. Location of the heat source with respect to the rotating tool depends on the type of welding, that is, similar or dissimilar material, to be joined. A schematic diagram of the laser-assisted friction stir welding is shown in Figure 22.1. Song et al. (2009) investigated the effect of laser preheating on the microstructure and mechanical properties of the friction stir welded Inconel 600 alloy. They found the laser-assisted friction stir welding to be 1.5 times faster than the normal friction

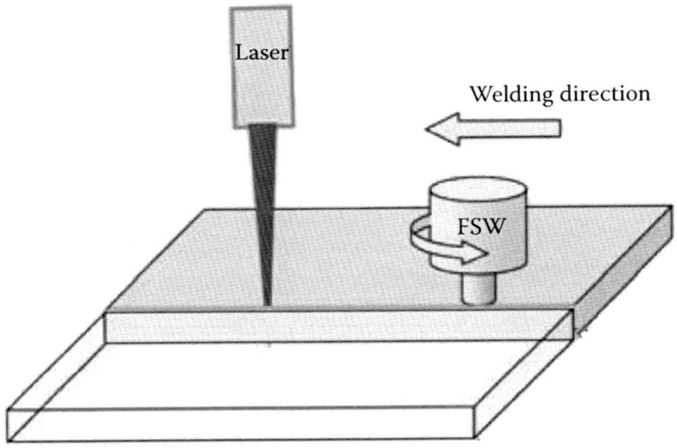

FIGURE 22.1 Schematic illustration of the principle of laser-assisted friction stir welding.

stir welding process, with a 10% improvement in tensile strength. Chang et al. (2011) carried out laser-assisted friction stir welding of dissimilar Al6061-T6/AZ31 alloy plates using a third material foil between the faying edges with the purpose of increasing the weld strength. A defect-free joint with a reduction in brittle nickel-based intermetallic phases was achieved by heating using a laser power source of 2 kW. The tensile strength of the joint was increased by 66% of the magnesium base metal. Merklein and Giera (2008) also conducted laser-assisted friction stir welding of steel and aluminum sheets, resulting in an increase in welding speed and a reduction in tool wear rate. Bang and Bang (2010) and Bang et al. (2013) investigated laser and tungsten inert gas (TIG) welding torch–assisted friction stir welding for the purpose of joining Al6061-T6 aluminum alloy to SS400 low-carbon steel. A joint strength of ~90%–93% with reference to the aluminum alloy compared to ~60%–78% without the use of external heating was achieved. It was also shown that the tool life increased and welding force decreased with the aid of an external heat source.

Electric arc is another efficient and cheaper preheating source. Joo (2013) used TIG-assisted friction stir welding for joining magnesium alloy (AZ31B) to mild steel. The tensile strength of the hybrid welds was approximately 91% that of the base magnesium alloy, which is higher than that of the normal friction stir welds. This is due to the enhanced plastic flow of the material and the partial annealing effect in both the magnesium alloy and mild steel material that is achieved by preheating of the mild steel side using a TIG heat source. Yaduwanshi et al. (2016) compared the performance of plasma-assisted friction stir welding with that of conventional friction stir welding in joining similar aluminum alloy and dissimilar aluminum–copper material. Preheating enhanced softening and smooth material flow during stirring, leading to an increase in weld strength. Optimizing offsetting of the preheating source toward the copper side ensured both a defect-free and good-quality welds. Ferrando (2012) and Long and Khanna (2005) used a resistance heating source to develop hybrid friction stir welding. Lai et al. (2014) applied ultrasonic vibrations directly on the friction stir welding tool to improve weld performance during friction stir welding of the 2024 aluminum alloy. This is shown in Figure 22.2a. The implementation of ultrasonic vibration during friction stir welding indicates that the flow stress of the material decreases significantly without appreciable heating. The transfer of vibration through the tool also affects the physical properties of the tool material. Shi et al. (2015) and Liu et al. (2015) developed a hybrid system. This is shown in Figure 22.2b, for transmitting ultrasonic vibration directly to the workpiece material ahead of a rotating tool using a sonotrode. Since the ultrasonic tool is close to the weld zone, this method is most efficient for transmitting the ultrasonic energy and producing a good weld joint having improved mechanical properties using wider process parameters than the friction stir welding process. A detailed review of heat-assisted friction stir welding can be found in the article by Padhy et al. (2015).

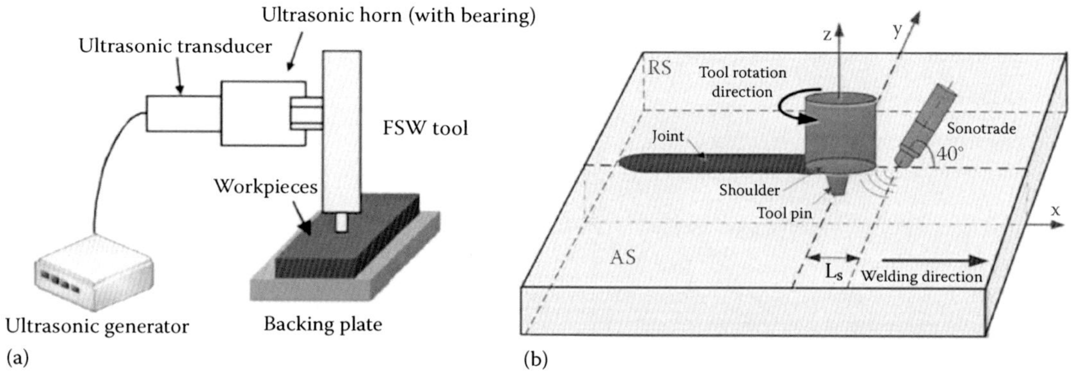

FIGURE 22.2 Different methods to apply ultrasonic vibration during friction stir welding process. (a) Ultrasonic vibrations applied directly on the friction stir welding tool (From Lai, R. L., He, D. Q., Liu, L., Ye, S. Y. and Yang, K. (2014), *Int. J. Adv. Manuf. Technol.*: A study of the temperature field during ultrasonic-assisted friction-stir welding, Vol. 73, pp. 321–327.); (b) Ultrasonic vibration applied using a sonotrode (From Shi, L., Wu, C. S. and Liu, X. C. (2015), *J. Mater. Process. Technol.*: Modeling the effects of ultrasonic vibration on friction stir welding, Vol. 222, pp. 91–102.)

22.2 INFLUENCE OF PROCESS PARAMETERS

The parameters involved in a typical friction stir welding process are shown in Figure 22.3. These are categorized into preset and variable parameters. The preset parameters are those that can be controlled during the welding setup period and include tool geometry (i.e., geometry and diameter of shoulder and pin and pin length) and welding medium. Once the setup is complete and the welding begins, these parameters remain constant. The tool rotational speed, welding speed, axial load, dwell time, and tilt angle can be varied during welding, and these are collectively known or referred to as the welding parameters. The tool hardness, base plate characteristics (composition and surface condition), and joint configurations have not been included as inputs since their influence has not been studied in detail. Of these, a few important parameters are tool rotating speed, welding speed, plunging depth, tool shoulder diameter, and tool pin profile. Both temperature distribution and material flow pattern are influenced by the welding parameters, tool geometry, and joint design. This, in turn, influences the evolution of microstructure during the welding process. The important parameters and their effects on mechanical and metallurgical properties are discussed in following subsections.

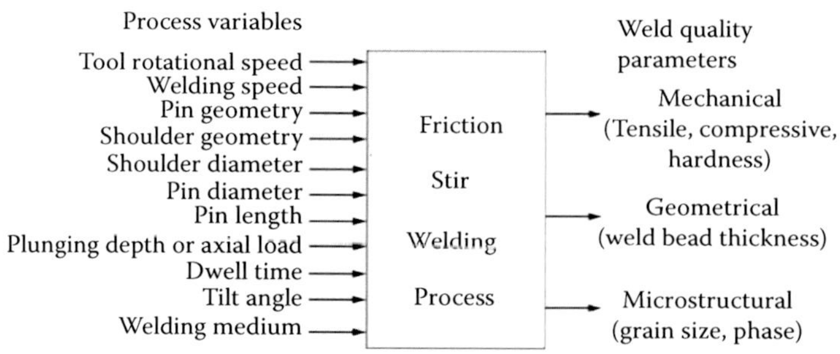

FIGURE 22.3 Typical parameters involved in friction stir welding.

22.2.1 Influence of Tool Geometry

The tool geometry plays a vital role in material flow, which in turn governs the following: welding speed, heat generation, power required, the uniformity of microstructure, and mechanical properties. The depth of insertion of the pin into the workpiece is important for producing a sound weld. The initial heating during the tool plunge stage is due to friction between the pin and the workpiece. Subsequently, the friction between the workpiece and shoulder is the primary source of heat generation. The function of the tool is to "stir" and "move" the material while the shoulder provides a confinement for the heated volume of material. Another important process parameter is the tilt angle, that is, the angle the tool makes with the vertical axis. A suitable tilt of spindle toward the trailing direction ensures that the shoulder holds the stirred material and moves it efficiently from the front to the back of the pin.

The tool dimension is critical to produce a sound weld. Scialpi et al. (2007) studied the influence of shoulder geometry on the microstructure and mechanical properties of friction stir welded 6082 aluminum alloy. They found that the tool having fillet and cavity crown was the best in terms of crown quality compared to the tool with fillet and scroll. The fillet and cavity crown showed a higher strength and elongation. Fujii et al. (2006) reported that for 1050-H24 material, a columnar tool without threads was the best in terms of mechanical properties. The tool shape was not significant for 6061-T6. The triangular prism tool and the tool having a column with threads were suitable for welding at both high and medium rotational speeds. The shape of the tool metal was not significant for lower rotational speeds. Boz and Kurt (2004) concluded that the screw-type tool with a higher pitch acted like a drill rather than like a stirrer and forced the weld metal outward. As a result of this, weld metal accumulated toward the shoulder of the stirrer, thereby not favoring the welding process. Though the best bonding was achieved with lower pitched stirrers, it had no effect on both microstructure and mechanical properties. Use of a square cross-section stirrer reveals a combination of poor metallographic features and mechanical properties. Elangovan and other researchers (Elangovan and Balasubramanian 2007, 2008a,b) studied the effects of tool pin profile on the formation of friction stir processing/friction stir welding zone in both the 6061 and 2219 aluminum alloy. The joints produced by the square pin tool resulted in a defect-free joint having superior tensile properties. In friction stir spot welds of the 6061 aluminum alloy, the tensile shear strength increased with an increasing probe length (Tozaki et al. 2007), while the cross-tension strength was not appreciably affected by the length of the probe. Kumar and Kailas (2008b) observed that the pin transfers the material layer by layer, while the shoulder transfers the material by bulk. They also observed an onion ring pattern in the weld zone. This pattern is formed as a result of the combined effect of pin-driven material flow and a vertical movement of the material due to shoulder interaction. The friction stir welding of T-joints requires a large amount of heat input due to the necessary forging action that is required to fulfill the joint fillets (Acerra et al. 2010). A slight increase in diameter of the tool shoulder results in sound welds even for a large tool feed rate. D'Urso and Giardini (2010) observed the threaded tool to be effective compared to the plane tool for lap joining of the AA6022-T4 aluminum alloy. Tools with a conical cavity shoulder required a higher spindle torque than the tool having a flat shoulder (Leal et al. 2011) and had a lesser influence on microstructure, hardness, and formation of defects. Use of tools having a conical cavity shoulder resulted in excellent surface appearance with some reduction in thickness (Leal et al. 2008), whereas a scrolled tool shoulder produced a less smooth weld without appreciable reduction in thickness. For the scrolled tool shoulder, the amount of material flow from the advancing side to the retreating side of the tool was higher than that obtained for a tool having a conical cavity. A bigger tool provided higher joint resistance (Donati et al. 2009), though it often resulted in high temperatures during friction stir welding of aluminum T-joints.

Investigations using different tool pin geometries revealed that the weakest joints are obtained while using straight cylindrical tools compared to other tool geometries (Elangovan and Balasubramanian

2007; Shanmuga Sundaram and Murugan 2010). Apart from the basic straight cylindrical tool pin, other types of pin geometry investigated were (a) tapered cylindrical (Elangovan and Balasubramanian 2007, 2008b; Imam et al. 2013; Liu et al. 2012b; Sharma et al. 2012), (b) threaded (Boz and Kurt 2004; Elangovan and Balasubramanian 2007, 2008b; Hao et al. 2013; Liang et al. 2012; Rajakumar et al. 2011b; Silva et al. 2011), (c) conical threaded (Zhang et al. 2012a), (d) square (Elangovan and Balasubramanian 2007, 2008b; Heidarzadeh et al. 2012; Imam et al. 2013), (e) triangular (Elangovan and Balasubramanian 2007, 2008b), (f) tapered cylinder with grooves (Shanmuga Sundaram and Murugan 2010), (g) tapered square (Imam et al. 2013; Shanmuga Sundaram and Murugan 2010), (h) tapered hexagon (Shanmuga Sundaram and Murugan 2010), and (i) paddle-shaped tool (Shanmuga Sundaram and Murugan 2010). The straight cylindrical tools are most widely used for friction stir welding owing to low fabrication cost, best self-cleaning phenomena, better pin profile effectiveness, and very long tool life.

22.2.2 Influence of Tool Rotational Speed

The tool rotational speed significantly affects the strength of the weld. Kim et al. (2006) investigated the effect of welding speed and rotation speed on the microstructure of the stir zone by measuring the distribution of silicon particles in aluminum die casting alloys. They found fine recrystallized grains in the nugget zone. Increasing the tool rotational speed resulted in an increase in tensile shear strength and a decrease in cross-tension strength of the friction stir welded 6061 aluminum alloy joints (Tozaki et al. 2007). Empirical relationships between tool rotational speed and base metal properties such as yield strength, elongation, and hardness were also developed (Balasubramanian 2008). These correlations were very effective in deciding the friction stir welding process parameters for fabricating defect-free joints from the known base metal properties. Buffa et al. (2009) found an operative parameter range of 300 to 500 rev/min of rotational speed to weld AA2024-T4 with AA7175-T73511 for both T-joint and lap joint using the friction stir welding process. Kwon et al. (2009) found defect-free welds at speeds of 500–3000 rev/min during friction stir welding of aluminum plates. The grain size decreased with an increase in tool rotation speed. The average hardness of the nugget zone was 33% higher than that of the base metal at 500 rev/min.

Investigation of the friction stir welding process by several authors reveals an improvement in the tensile properties with an increase in tool rotational speed up to the maximum value (Hao et al. 2013; Heidarzadeh et al. 2012; Sharma et al. 2012; Simoncini and Forcellese 2012; Zhang et al. 2012b). Further increase in tool rotational speed resulted in a decrease in properties due to the formation of tunnel- and void-type defects. Thermal stresses were reduced at lower tool rotational speeds during welding of dissimilar materials, namely, A356 and 6061 aluminum alloy for tool rotational speeds in the range 1000 to 1400 rev/min (Ghosh et al. 2010). Poor surface finish was observed in commercially pure copper–aluminum friction stir welded joint while increasing the tool rotational speeds (Xue et al. 2011). The tensile properties were low at lower tool rotation speeds. The joints showed sound bending properties at 600 rev/min. The friction stir welded joints of AA7075-T6 showed higher strength (Rajakumar et al. 2011b) at 1400 tool rev/min due to the fine-grained microstructure of weld nugget and uniformly distributed finer $MgZn_2$ particles in the weld nugget. The welding power increases with an increase in rotational speed (El-Hafez 2011). Though the measured power at lower rotation speed agrees with corresponding values calculated using the model proposed by Heurtier et al. (2006), the theoretical values exceed the experimental results at higher speeds and are attributed to a higher slip between the tool and the metal. Lertora and Gambaro (2010) suggested higher welding ratios (rotational speed/welding speed) during friction stir welding of the AA8090 aluminum–lithium alloy to obtain coarse equiaxed grains with fewer precipitates at the grain boundaries. Zhou et al. (2010) found the stir zone temperature to be below the β-transus and bimodal microstructure for a rotation speed of 400 rev/min. When rotation speeds of 500 and 600 rev/min were used, the stir zone temperature was above the β-transus temperature, resulting in full lamellar microstructure characterized by basket-weave $\alpha+\beta$ lamellae.

Spindle speed is one of the most significant factors in the friction stir welding process since the maximum temperature that can be generated is at the tool shoulder (Record et al. 2007). The joint strength for friction stir lap welded joints of Al-5083 and St-12 alloy sheets was improved by increasing the tool rotation speed (Movahedi et al. 2012). Toktas and Toktas (2012) observed an increase in hardness values with tool rotation speed. However, at low welding speed, the yield strength and the ultimate tensile increased with an increase in tool rotation speed. A decrease in tool rotation speed decreased the average grain size of α-aluminum at the weld nugget, resulting in a shift in the minimum hardness region from the heat-affected zone to the weld nugget zone (Aydin et al. 2012; Sharma et al. 2012). The corrosion resistance property of 7475 aluminum joints increased (Gupta et al. 2012) with increasing rotation speed. The stir zone exhibited better corrosion resistance compared to thermomechanically affected zone. Koilraj et al. (2012) reported an optimum rotational speed of 700 rev/min for carrying out friction stir welding of AA2219-T87 and AA5083-H321. Higher spindle torque was required for tools having a conical cavity shoulder (Leal et al. 2011). Aval et al. (2011c) found weaker welds with a coarse grain size in the weld nugget with an increase in rotational speed during friction stir welding of AA5086–AA6061 aluminum alloys. Fu et al. (2012a) observed porosities and insufficient penetration in the clockwise rotation of tools during welding of AZ31B magnesium alloys. However, the counterclockwise rotation was found beneficial for improving the mechanical property of the joints.

22.2.3 Influence of Welding Speed

The welding speed (i.e., the tool traverse speed) had a strong effect on microhardness and tensile strength of aluminum alloys (Adamowski and Szkodo 2007; Deqing and Shuhua 2004; D'Urso and Giardini 2010; Liu et al. 2012b). Sakthivel et al. (2009) reported fine equiaxed and more homogeneous grains in the nugget zone at a lower welding speed. As the weld speed increased, weld zone size became narrow, accompanied by an increase in hardness and a decrease in ultimate tensile strength. The best mechanical properties were obtained at a lower traverse speed owing to the presence of homogeneous grains and a higher heat input. The resistance of the material increased with an increase in travel speed during friction stir welding (Adamowski and Szkodo 2007).

During friction stir welding of ADC12 (aluminum–silicon–copper alloy), silicon particle size decreased with an increase in welding speed (Kim et al. 2006). Studies on the friction stir welding of similar materials such as AA6082 Al (Cavaliere et al. 2008) and A356 and 6061 aluminum alloys (Ghosh et al. 2010) and dissimilar welding of AA5086–AA6061 aluminum alloys (Fu et al. 2012) reveal that at lower weld speeds, the grain size at the nugget zone decreased to a minimum value with an increase in weld speed. With an increase in welding speed, the minimum hardness region was shifted from the heat-affected zone to the nugget zone (Sharma et al. 2012). A further increase in weld speed did not show any further variation in grain size. Weaker welds and coarser grain were found in the weld nugget with decreasing welding speed during dissimilar welding of AA5086–AA6061 aluminum alloys (Fu et al. 2012b).

Investigation of the dependency of weld speed on properties of different friction welded joints revealed an improvement in mechanical properties with an increase in speed up to a certain value (Cavaliere et al. 2008). The optimum welding speed depends on the composition and initial microstructure of the alloy plates (Elangovan and Balasubramanian 2008b; Ghosh et al. 2010; Zhang et al. 2012b). In underwater friction stir welding of 2219 aluminum alloy (Liu et al. 2011), the tensile strength increased with an increase in speed. At a low welding speed of 50 mm/min, the joint fractured in the heat-affected zone adjacent to the thermomechanically affected zone on the retarding side. Fracture occurred at the thermomechanically affected zone adjacent to the nugget zone on the advancing side when the speed was increased to 100 and 150 mm/min.

A welding speed of 60 mm/min yielded higher strength properties (Rajakumar et al. 2011b) of AA7075-T6 alloy weld. Aydin et al. (2012) and El-Hafez (2011) found an increase in hardness of the heat-affected zone with an increase in welding speed for aluminum alloys. Gupta et al. (2012) found maximum tensile strength for a traverse speed of 50 mm/min for 7475 aluminum joints. Both the axial thrust and horizontal force were significantly affected by the welding speed (Arora et al. 2010a). The ultimate tensile strength, percentage elongation, and joint efficiency increased with a decrease in welding speed (Fu et al. 2012a; Sharma et al. 2012). Steuwer et al. (2011a) reported a decrease in width of the tensile section of the residual stress profile with an increase in the traverse speed for W-Re tool material. The mean hardness and ultimate tensile strength of the nugget zone increased with increasing traverse speed (El-Rayesa and El-Danaf 2012).

22.2.4 Influence of Axial Load or Plunging Depth

The axial load significantly affects the frictional heat generation during friction stir welding. Record et al. (2007) reported plunge depth as one of the most significant factors of the friction stir welding process, which, in turn, depends on the plunging force. The temperature generated at the shoulder also depends on the axial force. An axial force of 7 kN showed superior tensile properties in friction stir welded joints of the AA6061 Al alloy (Elangovan et al. 2008). Kumar and Kailas (2008a) reported defect-free 7020-T6 Al alloy welds above 8.1 kN, and the maximum strength was produced at an axial load of 8.8 kN load. The axial thrust was affected significantly by shoulder diameter and welding speed (Steuwer et al. 2011a).

22.2.5 Influence of Tool Tilt Angle

The weld nugget shape is influenced by the tool tilt angle (Barlas and Ozsarac 2012). These researchers found the best result for a tool rotation speed of 1100 rev/min and a tool tilt angle of 2° while joining the 5754 aluminum alloy. The joint strength was 86% of the base metal. The optimal tool tilt angle for friction stir welding of the AA6061 Al alloy and SS400 low-carbon steel was 1° (Chen 2009). Weld trials revealed a narrower basin-shaped weld nugget with an increase in the tilt angle from 0° to 2°. Other investigations indicate the contribution of the tilt angle to be the least (5.96%) compared to tool rotational speed and welding speed (Bozkurt 2012; Robson et al. 2010). The thickness of the weld zone during friction stir welding of polyethylene (Arici and Selale 2007; Squeo et al. 2007) decreased with an increase in tool tilt angle due to excess flashing and reduced the tensile strength of the joint. The fatigue strength of the friction stir welded A6005-T5 alloy improved with an increase in tilt angle (Kim et al. 2010). It was also observed that the use of scrolled shoulder facilitated the friction stir welding process for a zero tilt angle (Mishra and Mahoney 2007; Yan et al. 2010) due to direct deformation of the material from the edge of the shoulder to the pin.

22.2.6 Influence of Number of Weld Passes

Leal and Loureiro (2008) reported that the voids produced by a single pass or two overlapping passes during friction stir welding can be eliminated by three or four overlapping passes. Several studies (Robson et al. 2010; Steuwer et al. 2011a) revealed a reduction in weld strength with an increase in the number of passes. The strength reduction was greatest after the first pass and the reduction in strength on subsequent passes became increasingly small. El-Rayesa and El-Danaf (2012) observed an increase in the stir zone grain size, fragmentation, dissolution, and re-precipitation of second-phase particles with an increase in the number of passes.

22.3 MICROSTRUCTURAL STUDIES

The microstructure of the friction stir welding joint depends on the thermophysical properties of the base metal and heat generated during the process, which, in turn, depends on both process parameters and tool geometry. The heat generated during friction stir welding results in significant microstructural evolution within and around the three zones (i.e., nugget zone, thermomechanically affected zone, and heat-affected zone). A typical cross section of the friction stir welding joint with number of zones is shown in Figure 22.4. Friction stir welding of the 60656 AA alloy revealed fine equiaxed grains in the nugget zone, fine and highly elongated grains in the retreating side of the thermomechanically affected zone, and slightly elongated coarse grains in the heat-affected zone (Cabibbo et al. 2007).

Wei and Nelson (2011) used a convex scrolled shoulder step spiral tool. They found a linear relationship of heat input with ferrite grain size and bainite lath size. The differences in tool geometry and welding parameters induced significant changes in the material flow path as well as in the nugget zone microstructure of AA6016-T4 welds (Rodrigues et al. 2009). Yeni et al. (2008) reported a recrystallized fine-grain structured nugget zone. The grain sizes increased from the weld region to the base metal. The formation of fine grains in FSP 7050 aluminum alloy was initiated by recrystallization (Rhodes et al. 2003).

The fracture location in friction stir welding 2519 aluminum joints was at the boundary between the thermomechanically affected zone and the heat-affected zone characterized by low hardness (Fonda and Bingert 2004). The softening at the thermomechanically affected zone/heat-affected zone boundary was due to coarsening of the grains with simultaneous transformation of the strengthening precipitates during the welding process. The comminuting effect induced on the ceramic reinforcement during the friction stir welding process led to a significant reduction in particle size and area as well as rounding of the particles (Ceschini et al. 2007). This also reduced the hardness from the base material to the middle line of the nugget zone. Analysis of the microstructure of friction stir welding AA5251 aluminum alloy (Attallah et al. 2007) revealed that the thermomechanically affected zone/nugget zone strength was primarily controlled by grain boundary strengthening. Studies reveal fine and equiaxed grains at the nugget zone due to dynamic recrystallization during the friction stir welding process (Sakthivel and Mukhopadhyay 2007; Xie et al. 2007), resulting in higher hardness at this region compared to the thermomechanically affected zone, heat-affected zone, and base metal.

Analysis of the microstructure and properties of similar and dissimilar material joints by friction stir welding was carried out by several investigators. The microstructure revealed fine equiaxed grains at the nugget zone of similar materials (Azimzadegan and Serajzadeh 2010; Cho et al. 2011; Miyano et al. 2011; Zahmatkesh et al. 2010) and dissimilar material joint (Scialpi et al. 2008); presence of intermetallic compounds of Al_4Cu_9, AlCu, and Al_2Cu near the Al/Cu interface in Al–Cu alloy joints (Zadeh et al. 2008); and dissolution of the precipitates and formation of coarse precipitates in the heat-affected zone in the 6065 alloy (Simar et al. 2008) and AA 2219 (Arora et al. 2010b).

Study of material flow with threaded and unthreaded pin during the friction stir welding of the 7020 T6 aluminum alloy (Lorrain et al. 2010) showed the same features. The microstructure for friction stir welding copper plates (Xue et al. 2011) revealed faster cooling by water quenching and retained the initial microstructure of the base material with a higher dislocation density compared to cooling in air. The amounts of both heat input to the welded material and plastic deformation increased with an increase in plunging depth (Cho et al. 2011), and the fraction of low angle boundaries in the stir zone was significantly increased as compared to that in the base material.

FIGURE 22.4 Friction stir welding joint with four distinct zones. BM: unaffected/based material; HAZ: heat affected zone; NZ: weld nugget/nugget zone; TMZ: thermo-mechanically affected zone.

TABLE 22.1
Studies of Microstructural and Mechanical Properties

Workpiece Material	Tool Material	Operating Parameters[a] REV/MIN, WS (mm/min), AL (kN), PD (mm), TA (°), Others	Remarks	Reference
6061 Al alloy	High-carbon steel	REV/MIN: 1200 WS: 75 AL: 7	Square pin profiled tool produced good joints; shoulder diameter of 18 mm showed superior tensile properties.	Liu and Ma 2008
			FSW exhibited strength enhancement of 34% compared to GMAW and 15% compared to GTAW; fine, equiaxed grains in the nugget zone.	Lakshminarayanan et al. 2009
	Highly wear-resistant steel	–	Proof strength and ultimate tensile strength decreased to 43% and 28%, respectively, with respect to the base material.	Ceschini et al. 2007
	–	REV/MIN: 400–1600 WS: 100–400	Joints welded at 400 mm/min exhibited higher strength with 45° shear fracture; WS was the dominant factor in determining the tensile properties and fracture mode.	Ren et al. 2007
	High-carbon steel	REV/MIN: 862–1337 WS: 32.43–127.5 AL: 5.62–10.37	Highest hardness value of 121 HV obtained at 1100 rev/min.	Rajakumar et al. 2011a
	H13 steel	REV/MIN: 400–600 WS: 50 PD: 0.1 TA: 2.5	Lowest hardness in the heat-affected zone on advancing side; tensile properties along the skin higher than those along the stringer.	Cui et al. 2012
2024-T3 and 7075-T7351	–	REV/MIN: 215 WS: 113	YS reduced and exhibited rate sensitivity.	Chao et al. 2001
2024T3 and 6082T6 Al alloy	56NiCrMoV7 tool steel	REV/MIN: 2085–1800 WS: 460–762	Stress decreased with increasing strain rate; modifications in grain size.	Cerri and Leo 2009
63% Cu–37% Zn brass	Hot work steel, X32CrMo33	REV/MIN: 1250–1600 WS: 100–225 PD: 2.8	Grain refinement in the nugget zone; strength and ductility performance of 105% and 84%, respectively, at 1250 (rev/min) and 125(WS).	Cam et al. 2009
6056 Al alloy	–	REV/MIN: 1800 WS: 900 TA: 3	Strain rate and temperature gradients are steeper in advancing side than retreating; ultimate tensile strength and YS were 90% and 66%, respectively, of the base alloy.	Cabibbo et al. 2007
Al alloy	–	REV/MIN: 1000 WS: 50–175	Fine equiaxed grains in the nugget zone; ultimate tensile strength increased with decreasing WS.	Sakthivel et al. 2009
7075	–	REV/MIN: 350 WS: 120	Microstructure depends on the tool design, processing parameters and cooling rate.	Su et al. 2005

(Continued)

TABLE 22.1 (CONTINUED)
Studies of Microstructural and Mechanical Properties

Workpiece Material	Tool Material	Operating Parameters[a] REV/MIN, WS (mm/min), AL (kN), PD (mm), TA (°), Others	Remarks	Reference
	H13 tool steel	REV/MIN: 100–1400 WS: 40–80 TA: 3	REV/MIN of 1300; the WS of 40 mm/min produced defect-free weld.	Azimzadegan and Serajzadeh 2010
2024 and 7075 Al alloy	H13 steel quenched at 1020 °C	REV/MIN: 1040 WS: 104 PD: 2.9 TA: 2	Recrystallization phenomena occurred; inhomogeneous decrease of the mechanical characteristics.	Barcellona et al. 2006
	SKD61	REV/MIN: 1200 WS: 42–198	Maximum tensile strength of the joints of 423MPa at 1.7 mm/s (WS) when 2024 alloy plate on advancing side.	Khodir and Shibayanagi 2008
	–	REV/MIN: 400–2000 WS: 254 AL: 9.8–13.4 TA: 3	Weld efficiency in terms of tensile strength at 1000 rev/min was 96%; minimum hardness value of naturally aged samples was 88% of the base metal (BM).	Silva et al. 2011
Al & Cu	Tool steel	REV/MIN: 151–1400 WS: 57–330	Different microhardness levels from 136 to 760HV0.2 were produced in the nugget zone.	Ouyang et al. 2006
	SPK quenched and tempered tool steel	REV/MIN: 750–1500 WS: 30–375 TA: 3	Shear load decreased with increasing rev/min and decreasing WS.	Zadeh et al. 2008
2095 Al alloy	–	REV/MIN: 1000 WS: 252	Fine equiaxed grain structure with high grain boundary misorientation angles.	Salem 2003
6016-T4 Al alloy	–	REV/MIN: 180–320 WS: 1800–1120 PD: 2.5 TA: 0	Reduction in elongation of 30% and 70%, respectively, for the conical and scrolled shoulder welds.	Rodrigues et al. 2009
Mg Zn Y Zr alloy		REV/MIN: 800 WS: 100 TA: 2.5	Joining efficiency was 95% of BM.	Xie et al. 2007
01420 Al–Li alloy	1Cr18Ni9Ti stainless steel	REV/MIN: 400–1960 WS: 23.5–85.7 AL: 1–7	Ultimate tensile strength of the joints was 86% of BM; bending angle of the joints can reach 180°.	Wei et al. 2007
Cu	–	REV/MIN: 1000 WS: 30	Higher nugget zone hardness; joint efficiency of 85%.	Sakthivel and Mukhopadhyay 2007
	–	REV/MIN: 400–800 WS: 50 PD: 0.2 TA: 2.5	YS exhibited a linear correlation with the lowest hardness values in the heat-affected zone.	Xue et al. 2011b
Steel HSLA 65	PCBN	REV/MIN: 300–600 WS: 51–203 TA: 0.5	Ferrite grain size and bainite lath size increased 150% with increasing heat input of 2.27 kJ/mm.	Wei and Nelson 2011

(Continued)

TABLE 22.1 (CONTINUED)
Studies of Microstructural and Mechanical Properties

Workpiece Material	Tool Material	Operating Parameters[a] REV/MIN, WS (mm/min), AL (kN), PD (mm), TA (°), Others	Remarks	Reference
	WC-based material	REV/MIN: 600 WS: 50–250 AL: 14 PD: 1.7 TA: 3	Nugget zone consisted of fine equiaxed grains of α and γ phases; tensile strength and hardness increased with the increasing WS.	Saeid et al. 2008
	Si_3N_4-based material	REV/MIN: 400 WS: 100–300 TA: 3	WS of 100 mm/min and a rev/min of 400 were the most preferable conditions.	Miyano et al. 2011
	PCBN	REV/MIN: 600–1100 WS: 7.62–254 PD: 1.65 TA: 0	Fine-grained microstructure in the nugget zone; increase in PD decreased grain size and increased the hardness of the nugget zone.	Cho et al. 2011
	PCBN	REV/MIN: 324–1200 WS: 150–198 AL: 9–16	Different weld properties were obtained depending on the welding parameters used.	Miles et al. 2009
	PCBN and W-Re tool	REV/MIN: 400–600 WS: 50–500 AL: 12–55 End force (kN): 0.3–15	Nugget zone microstructure was found to be a mixture of martensite, bainite, and proeutectoid ferrite.	Barnes et al. 2012
2024 and 6082 Al alloy	56NiCrMoV7-KU material	REV/MIN: 1810–2085 WS: 460–762	Grain dimension of less than 3 μm in the nugget zone; fatigue limits in the range 40–75 MPa.	Scialpi et al. 2008
7050-T7451 Al alloy	H13 tool steel	REV/MIN: 540 WS: 408 AL: 28 No. of passes: 5	Residual stress and *x*-force reduced with number of passes.	Brown et al. 2009
6061-T6 and 6082-T6 Al alloy	–	REV/MIN: 1120 WS: 224 TA: 2.5	Failures occurred near the weld edge line.	Moreira et al. 2009
2219 Al alloy	H13 steel	REV/MIN: 400 WS: 180 PD: 10.6	Joint efficiency of 72%; corrosion resistance of the nugget zone was better than BM.	Srinivasan et al. 2010
	–	REV/MIN: 800 WS: 100 AL: 4.6 Welding medium (air and water)	Improved tensile strength and decreased plasticity due to underwater friction stir welding.	Hui-jie et al. 2010
	H13 tool steel	PD: 7–35 TA: 2	40% of Al_2Cu precipitates dissolved in the nugget zone; hardness decreased from BM to the joint line; reduced tensile strength.	Arora et al. 2010b
7020 Al alloy	H13 steel	REV/MIN: 1120–1800 WS: 20–80 TA: 3	Average size of the α-Al primary phase and ductility increased with increasing rev/min.	Gaafer et al. 2010

(*Continued*)

TABLE 22.1 (CONTINUED)
Studies of Microstructural and Mechanical Properties

Workpiece Material	Tool Material	Operating Parameters[a] REV/MIN, WS (mm/min), AL (kN), PD (mm), TA (°), Others	Remarks	Reference
2198 T3 Al alloy	–	REV/MIN: 500–900 WS: 150–300 Specimen position; run-in side, center and outside	Ultimate tensile strength increased with decreasing rev/min; higher YS at higher WS.	Bitondo et al. 2010
Ti–6Al–4V	W-1 pct La_2O_3 tool	REV/MIN: 120–800 WS: 50.8–203.2	25% of the heat transferred out of the workpiece and into the tool; fine, interpenetrating α platelets and small α colonies in the nugget zone.	Pilchak et al. 2011
2199-T8E74 Al alloy	–	REV/MIN: 800 WS: 400	Hardness distribution exhibited a U-shaped profile.	Steuwer et al. 2011b
5052 Al alloy	High-speed steel tool	REV/MIN: 1120–1400 WS: 60–100 TA: 3	Equiaxed grains with strong in-grain misorientation and presence of grain-interior dislocation structure in the nugget zone.	Kumbhar et al. 2011
Mg alloy	H13 tool steel	REV/MIN: 2000–3000 WS: 30–70 TA: 2	Flow stress values of joints were lower than BM.	Forcellese et al. 2012
	SKD61 tool steel	REV/MIN: 110 WS: 200–400 AL: 8 TA: 3	Equiaxed grains in the nugget zone; tensile strength and microhardness lowered.	Chen et al. 2012
	High-carbon steel	REV/MIN: 1000–2000 WS: 22.2–135 AL: 2–4	Rev/min of 1600, WS of 40.2 mm/min, and AL of 3 kN showed higher tensile properties.	Padmanaban and Balasubramanian 2010
	High-carbon steel	REV/MIN: 1200 WS: 30–150 AL: 5	WS has significant influences on the formation of defects in the nugget zone, grain size, and hardness of the nugget zone.	Rose et al. 2012
	High-carbon steel	Tool shoulder diameter:15, 18, 21 D/d ratio: 2.5, 3, and 3.5	Friction stir processing produced uniformly refined and equiaxed homogeneous microstructure with fine dynamically recrystallized grains and improved mechanical properties.	Venkateswarlu et al. 2013
	High-carbon steel	REV/MIN: 1200 WS: 90 AL: 5	The tensile strength and yield strength are enhanced by approximately 12% and 18% as compared to pulsed gas tungsten arc welding joints.	Rajakumar et al. 2013a
	High-carbon steel	REV/MIN: 800–1600 WS: 30–150 AL: 3–7	Joint efficiency of 82% has been found; 1194 rev/min, WS of 92.19 mm/min and AL of 5.05 kN	Rajakumar et al. 2013b

(Continued)

TABLE 22.1 (CONTINUED)
Studies of Microstructural and Mechanical Properties

Workpiece Material	Tool Material	Operating Parameters[a] REV/MIN, WS (mm/min), AL (kN), PD (mm), TA (°), Others	Remarks	Reference
2017A-T451 Al alloy	–	REV/MIN: 300 WS: 120 TA: 2	Weld efficiency of 90%; partial recovery of hardness in the nugget zone.	Ahmed et al. 2012
2519-T87 Al alloy	–	REV/MIN: 250 WS: 30	The θ′ precipitates are larger in size and lower in density in the heat-affected zone than the thermomechanically affected zone.	Liang et al. 2012
6061/ZrB_2 composites	High-carbon, high-Cr steel	REV/MIN: 1150 WS: 50 AL: 6	Parallel band-like distribution of ZrB2 particles and rotated elongated grains in the thermomechanically affected zone; reduced ductility and wear rate.	Dinaharan and Murugan 2012
Al and Mg alloy	H13 steel	REV/MIN: 600–3000 WS: 30–300 TA: 2	Grain refinement in the nugget zone.	Simoncini and Forcellese 2012
	–	REV/MIN: 780 WS: 30 TA: 2°	The hardness of the nugget zone is higher than that of the thermomechanically affected zone and the heat-affected zone.	Li et al. 2012
	High-carbon, high-Cr steel	REV/MIN: 1900 DT: 2 s	The plastic deformation and high-temperature exposure induced the grain boundary and the interfacial diffusion causing local melting.	Suhuddin et al. 2013
Al–Mg–Sc alloy	M2 steel	REV/MIN: 650 WS: 158 AL: 23	Fine, fragmented, and dynamically recrystallized grains in the nugget zone.	Subbaiah et al. 2012
1100 Al alloy	–	REV/MIN: 562–1037 WS: 40.54–159.5 AL: 3.62–8.37	Maximum tensile strength of 105 MPa, hardness value of 67 HV, and minimum corrosion rate of 0.699×10^{-4} in the nugget zone with 893 (rev/min), 100 mm/min (WS), 6.5 kN (AL), shoulder and pin diameter of 14.8 and 4.9 mm, and 45.4 HRc (tool hardness)	Rajakumar and Balasubramanian 2012
High-density polyethylene	SAE 1050 steel	REV/MIN: 1500–3000 WS: 45–115 TA: 1–3	Ultimate tensile strength and joint efficiency improved by 112% and 105% of BM; optimum welding parameters for ultimate tensile strength 3000 (rev/min), 115 mm/min (WS), and 3° (TA).	Bozkurt 2012

[a] AL, axial load; DT, dwell time; PCBN, polycrystalline cubic boron nitride; PD, plunging depth; PS, plunging speed; REV/MIN, tool rotational speed; TA, tilt angle; WS, welding speed; YS, yield strength.

Bakavos et al. (2011) observed a complex material flow in the weld zone of friction stir spot weld with the pinless tool and high depth of penetration of the deformation zone with the profile tool. Zadpoor et al. (2010) studied the global and local mechanical properties and microstructure for friction stir welding of 2024-T3 and 7075-T6 alloys, revealing a heterogeneous texture in the nugget zone with large intermetallic particles in the heat-affected zone and nugget zone. The yield strength and plasticity parameters vary around the weld centerline.

Kumbhar et al. (2011) observed equiaxed grains in the nugget zone with strong in-grain misorientation and presence of dislocation at grain interior in friction stir welding 5052 Al alloy. Li et al. (2011) reported material-loss defects in friction stir welds due to poor material flow or insufficient heat generation at the tool/workpiece interface. Studies reveal that grain size varied from the top to the bottom of the nugget zone of AA2017A-T451 welded plates (Ahmed et al. 2012). Liang et al. (2012) welded 2519-T87 aluminum alloy and found dynamically recrystallized nugget zone where the grain size increased from the center of the nugget zone to thermo-mechanically affected zone.

Investigation of the type, distribution, dissolution, and precipitation behaviors in different regions of the top surface of friction stir welded AA2524-T3 aluminum joints revealed the presence of a circular Al_2Cu phase, a rod-shaped Al_2CuMg phase, and an Fe-containing impurity phase (Fu et al. 2012b). The temperature of the nugget zone was below the β-transus temperature when friction stir welding was applied to Ti–6Al4V, resulting in a mixture of dynamic recrystallized (DRX) α and transformed β (i.e., acicular martensite α′ and lamellar α+β) at the nugget zone (Liu et al. 2009).

Abnormal grain growth was observed during friction stir welding of AA5083-H18 sheets (Chen et al. 2011). During the fabrication of a lap-butt joint using dissimilar aluminum alloys by friction stir welding (Li and Shen 2012), three micro flow patterns were observed in SZ, namely, circumfluence, laminar flow, and turbulent flow. Friction stir welding of dissimilar 6082/5083 alloys was characterized by steady flow behavior in the 5083 plate, whereas 6082 was sensitive to flow softening (Leitao et al. 2012). Homogeneous distribution of ZrB_2 particles was observed in the nugget zone and parallel band-like distribution in the thermomechanically affected zone of friction stir welding aluminum matrix composites (Dinaharan and Murugan 2012). A fully recrystallized microstructure was observed for friction stir welding of dissimilar alloys (aluminum and magnesium) (Venkateswaran and Reynolds 2012). Equiaxed grains were observed in the nugget zone of friction stir welded AMX602 magnesium alloys (Chen et al. 2012).

Formation and distribution of brittle structures during friction stir welding of Al and Cu by varying the tool shoulder geometry (Galvao et al. 2012) revealed different morphology and intermetallic content at the weld nugget. Liu et al. (2012a) introduced external nonrotational shoulder for friction stir welding of the 2219-T6 aluminum alloy. They found fine and equiaxed grains in the nugget zone.

Several studies were carried out to understand the microstructural evolution in different weld zones during friction stir welding operation of different alloys. A summary of the findings of these investigations is presented in Table 22.1.

22.4 MECHANICAL PROPERTIES

Heat generation in the friction stir welding process leads to substantial changes in post-weld mechanical properties such as strength, ductility, fatigue, and fracture toughness. Aluminum alloys are classified into heat-treatable alloys and non–heat-treatable (solid solution–hardened) alloys. Friction stir welding creates a softened region around the weld center in a number of precipitation-hardened aluminum alloys. The change in hardness of the friction stir welds is different for precipitation-hardened and solid solution–hardened aluminum alloys. Friction stir welding joints of aluminum alloy exhibited higher fatigue strength compared to tungsten arc welding, electron beam welding processes (Malarvizhi and Balasubramanian 2011), and gas metal arc welding (Kim et al. 2010).

The friction stir welding process reduced the yield stress of both AA2024-T3 and AA7075-T7351 under both high strain rate and quasi-static loading conditions and found similar strain hardening at various strain rates (Chao et al. 2001). Cam et al. (2009) proposed higher rotational and traverse speeds for higher strength of the weld, whereas higher ductility can be achieved at lower rotational and traverse speeds.

The mechanical properties of the friction stir welding joints are influenced by various factors such as (a) distribution of various phases, (b) heat generated during stirring, (c) grain refinement due to dynamic recrystallization at high strain rate and elevated temperatures, (d) welding speed, (e) tool rotation speed, and (f) tool geometry. The traverse speed is the most influencing factor affecting hardness and strength of the nugget zone (Sarsilmaz and Caydas 2009). The heat input increases at higher welding speed, thereby obtaining homogeneous grains having low residual stresses and improved ductility and strength (Barcellona et al. 2006; Ouyang et al. 2006; Sakthivel et al. 2009; Salem 2003; Su et al. 2005). However, sound joints were obtained in stainless steels at a lower welding speed (Saeid et al. 2008). At a higher speed of 250 mm/min, groove-like defects were observed in the same material due to insufficient heat input.

Tool rotation speed and change in diameter of both the shoulder and pin changed the position and inclination of the lower hardness zones but did not affect the hardness values along these zones (Liu and Ma 2008). The hardness along the lower hardness zones increased with an increase in welding speed, due to higher heating and cooling rates. However, no noticeable effect was observed on both heating and cooling rates with a change in diameter of shoulder and pin. The tensile strength of joints increased with an increase in welding speed and was independent of (a) the dimension of the shoulder and pin and (b) the rotation rate. The tensile strength and hardness of the nugget zone of AA7020-O aluminum alloy joints increased with an increase in welding speed (Gaafer et al. 2010).

The joint efficiency decreases with an increase in plate thickness since energy per unit length increases due to a decrease in the advancing speed and an increase in the plunge force (Eberl et al. (2010). The strength and elongation was found to be lower than the parent metal during friction stir welding of 6-mm-thick magnesium–zinc–yttrium–zirconium plates (Xie et al. 2007). Variation in the welding speed affects the hardness profile at the weld region (Ren et al. 2007). At higher speeds, the lowest hardness profile distribution was found to be inclined at 45° with the butting surface, whereas at lower speeds, this was found to be nearly vertical. The hardness at the nugget zone is lower for underwater friction stir welding of the 2219 aluminum alloy compared to normal friction stir welding (Hui-jie et al. 2010).

Plunge force affects quality of the friction stir welds. The quality of the friction stir welded joints deteriorates due to depressions and formation of waving burrs when the plunging forces exceed a certain value (Wei et al. 2007). At a low pressure, both tunneling and groove-type defects appear. Depending on material and plate thickness, optimization of the plunging force is essential for obtaining defect-free joints. Investigation of microstructural evolution and mechanical behavior of 6082T6-6082T6, 2024T3-2024T3, and 6082T6-2024T3 friction stir welded joints by Cerri et al. (2010) revealed a decrease in flow stress with an increase in temperature and decreasing strain rate. However, the ductility was poor and independent of temperature and strain rate.

The effect of tool profile also plays an important role in achieving sound weld quality. Pinless tool resulted in higher tensile strength and ductility when compared to the tool with a pin (Simoncini and Forcellese 2012) when joining AA5754 and AZ31 thin sheets. The clamping procedure influenced the tensile residual stress and fatigue life of a 6005 aluminum alloy weld (Shahri and Sandstrom 2012). Forcellese et al. (2012) used pin and pinless tool configurations to join AZ31 thin sheets by the friction stir welding process. They concluded that both ultimate tensile strength and ductility increased at a higher rotational speed/welding speed ratio until a peak value was reached and then decreased. Higher strength and ductility were found for the case of pin tool configuration, and a more homogeneous microstructure was obtained for the pinless tool configuration. Miles et al. (2009) focused on the properties and microstructures of friction stir welded high-strength steel sheets.

Khodir and Shibayanagi (2008) studied the microstructure and mechanical properties of dissimilar joints of 2024-T3 to 7075-T6 Al alloys. They found that a rise in welding speed caused the formation of a kissing bond and pores especially when the 2024 aluminum alloy plate was located on the retreating side. Minimum hardness was observed in the heat-affected zone of both sides, and their values increased with welding speed. In case of dissimilar and similar micro-friction stir welding butt welds (Scialpi et al. 2008), tensile failure occurred in the nugget zone as a result of irregularities in thickness rather than the presence of defects. Excellent fatigue properties were also obtained. Heat-affected zone hardness and transverse tensile strength were reduced with an increase in the number of passes for the 7050-T7451 aluminum alloy (Brown et al. 2009).

Moreira et al. (2009) welded two dissimilar aluminum alloys. They concluded that the welded joints of AA6082-T6 material revealed lower yield and ultimate stresses and the dissimilar joints displayed intermediate properties. The hardness profile of the dissimilar joint was lower on the AA6082-T6 alloy plate side. Srinivasan et al. (2010) found the weakest region in the nugget zone/thermomechanically affected zone interface of AA2219-T87 aluminum alloy joints.

22.5 FRICTION STIR PROCESSING

Friction stir processing technology is an extension of friction stir welding for materials processing. In friction stir processing, a rotating tool is plunged partially or fully into the material for local microstructure changes. The intense plastic deformation, material mixing, and thermal exposure during friction stir processing cause the creation of fine, uniform, and pore-free structures, thereby resulting in significant microstructural refinement, densification, and homogeneity of the processed zone. By controlling the depth of the process zone, modification of the microstructure and mechanical properties of the surfaces and subsurfaces can be achieved. The properties of certain alloys can be tailored by friction stir processing to achieve special characteristics like development of superplastic materials and thermal barrier coating. Few recent developments in these areas are highlighted below.

22.5.1 Microstructural Modification by Friction Stir Processing

Microstructure of cast components can be modified with concomitant improvement in the mechanical properties. Extensively used cast alloys like aluminum–silicon and aluminum–silicon–magnesium alloys are generally brittle due to the presence of needle-shaped silicon phases in the aluminum matrix. Addition of elements such as calcium, sodium, strontium, antimony, and so on to the molten aluminum alloy is found to modify the morphology of the silicon phase, resulting in an improvement in mechanical properties. Properties of the cast alloys can be further improved to a great extent by friction stir processing, during which the brittle silicon phases can be broken down to very fine particles and become uniformly distributed in the alloy matrix. The microstructure of the A356 alloy was modified by friction stir processing, resulting in improved yield strength and ultimate tensile strength of the cast alloy (Ma et al. 2006). The microstructure of the friction stir processing alloy revealed an absence of porosity, a very fine grained structure, and a uniform distribution of fine silicon particles (Figure 22.5). The size of the silicon particles in the matrix reduced from ~16 μm in the as-cast condition to ~2 μm with an increase in tensile elongation from 2% to as high as 30% after friction stir processing.

Application of friction stir processing to a magnesium–aluminum–zinc alloy resulted in significant grain refinement and dissolution/breakup of the coarse eutectic β-$Mg_{17}Al_{12}$ phase in the matrix (Mathis et al. 2005; Tsujikawa et al. 2007). The yield strength, ultimate tensile strength, hardness, and percentage elongation of the cast alloy improved significantly via friction stir processing (Feng and Ma 2007). The surface properties of cast nickel–aluminum bronzes, which are used for marine applications, were improved by friction stir processing. Selective modification of the near-surface layers of cast NiAl bronze was achieved by friction stir processing, where the as-cast microstructure was converted to that of the wrought condition (Ferrara and Canton 1982; Oh-Ishi and McNelley 2004). Friction stir processing resulted in grain refinement and microstructural homogenization and improved the tensile properties and fatigue life of the cast alloy. The friction stir processing of as-cast nickel–aluminum bronze materials provides microstructure refinement and homogenization as well as closure of porosity and improved mechanical properties.

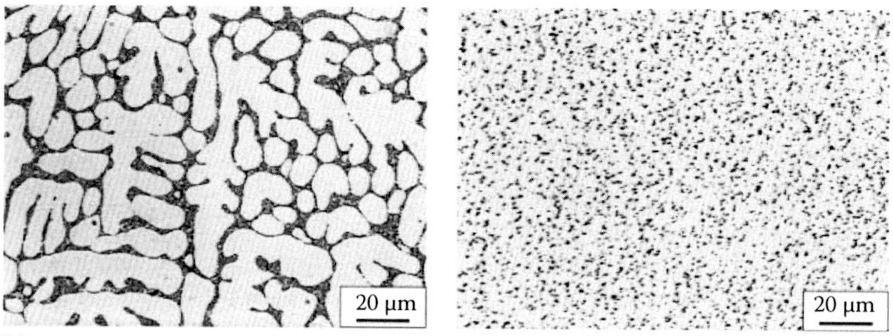

(a) (b)

FIGURE 22.5 Micrographs of (a) cast A356 and (b) friction stir processed A356. (From Ma, Z. Y., Sharma, S. R. and Mishra, R. S. (2006), *Metallurg. Mater. Trans. A*: Microstructural modification of as-cast Al–Si–Mg alloy by friction stir processing, Vol. 37A, pp. 3323–3336.)

22.5.2 Surface Composites by Friction Stir Processing

The intense plastic deformation associated with the movement of material from the front to the back of the rotating pin results in a very fine grained microstructure. During friction stir processing, the true plastic strain may reach a value as high as 40 (Heartier et al. 2002). It is possible to manufacture composite materials on the surface of metallic substrates by incorporating ceramic particles during the severe plastic deformation and mixing by friction stir processing. Mishra et al. (2003) first used friction stir processing to fabricate aluminum–silicon carbide particle surface composite by applying a thin layer of silicon carbide powder particles on the surface of an aluminum alloy plate. A composite layer of ~100 μm with a uniform distribution of silicon carbide particles with good bonding (Figure 22.6) was obtained at the surface of the aluminum alloy plate. The surface/bulk composites can also be processed through friction stir processing by reinforcement predeposition technique. In this technique, grooves were cut on AZ31 plates along the friction stir processing direction and filled with reinforcement particles like silicon carbide and multiwalled carbon nanotubes (Morisada et al. 2006a,b). A thick surface composite having very fine grains was obtained by conducting friction stir processing along the grooves. Several authors have reported successful attempts at processing bulk composites by friction stir processing technique, namely, nanosized SiO_2 or ZrO_2 on AZ61 plates (Chang et al. 2006; Lee et al. 2006) and nitinol particles on 1100 Al matrix (Dixit et al. 2007).

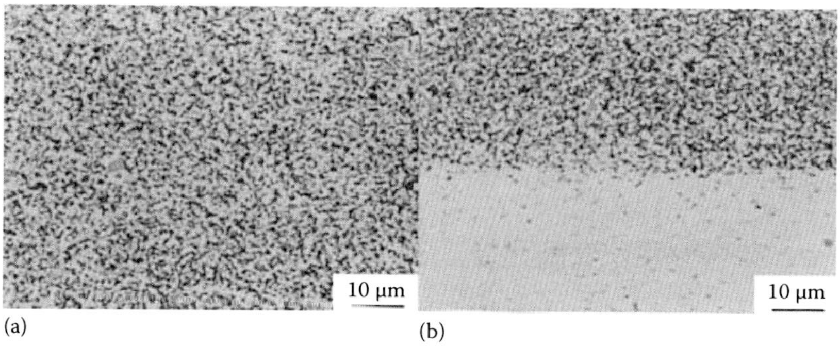

(a) (b)

FIGURE 22.6 Optical micrograph showing (a) uniform distribution SiC particles in aluminum matrix, and (b) good bonding between surface composite and aluminum substrate. (From Mishra, R. S., Ma, Z. Y. and Charit, I. (2003), *Mater. Sci. Eng. A*: Friction stir processing: A novel technique for fabrication of surface composite, Vol. A341, pp. 307–310.)

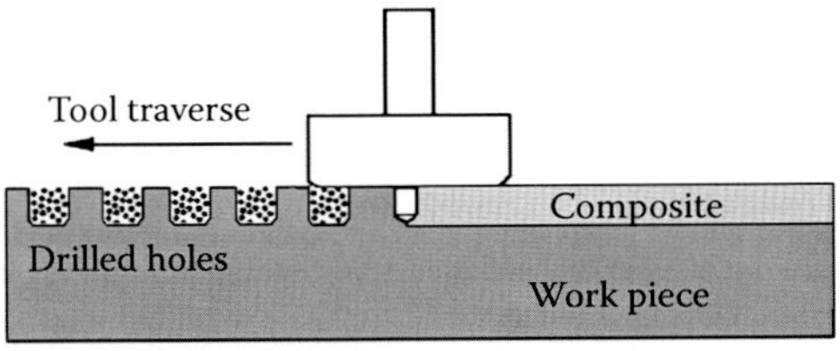

FIGURE 22.7 Method of placing reinforcement particles by the drilled hole technique in the fabrication of surface composites.

Surface composite fabrication can be carried out using the drilled hole method (Li et al. 2013). A schematic of this process is shown in Figure 22.7. In this technique, blind holes were drilled on the workpiece and filled with reinforcement particles and subjected to friction stir processing. The intense plastic deformation coupled with mixing resulted in a surface composite with reduced agglomeration and a good distribution of the reinforcements in addition to obtaining a fine grain size. In these approaches, the distribution of particles was limited by the pin size, and the reinforced area is limited to the stir zone.

Several approaches have been developed to fabricate surface composites. The workpiece was coated with reinforcement particles by either powder coating or plasma spray coating and subjected to friction stir processing. A composite layer of around 600 μm with a uniform distribution of aluminum oxide (Al_2O_3) in AA 2024 alloy plate was processed by this technique (Zahmatkesh and Enayati 2010). Surface composites were also fabricated by high-velocity oxy-fuel spraying of an Al_2O_3/A356 composite powder mixture onto the surface of grit blasted A356 plates and followed by friction stir processing (Mazaheri et al. 2011). Hodder et al. (2012) obtained a 48 wt% Al_2O_3 composite layer on the surface of an AA6061 alloy by a combination of cold spraying and friction stir processing. Friction stir processing resulted in an increase in surface hardness from 85 H_V to 137 H_V.

22.5.3 Friction Stir Processing for Development of Superplastic Materials

Materials having a fine-grained microstructure to assist superplastic deformation can be obtained by friction stir processing. It is generally known that superplastic forming can be achieved for materials having a high strain-rate sensitivity. For this to occur, the chosen material should have (a) two phases (a continuous ductile phase and a discontinuous brittle phase), (b) a strain rate sensitivity exponent in the range 0.3–1.0, (c) a fine grain size of the order of <5 μm, and (d) a processing temperature above 40% of melting temperature. Friction stir processing can be effectively used for processing composite materials with hard second-phase particles well distributed in a fine grain–sized matrix. This concept can be applied for the development of materials, which can be subjected to superplastic deformation. Dieguez et al. (2012) investigated the tensile deformation behavior of friction stir processed 7075-T651 aluminum alloy. Friction stir processing produced a refined region with an average grain size of 4.65 μm. Elongation of the order ~900% was achieved for the material when deformed at 400°C and 0.001 s^{-1} strain rate for a maximum deformation stress of 9 MPa.

Friction stir processed A356 alloy exhibited superplasticity during tensile deformation at strain rates in the range 0.003–0.1 s^{-1} and temperatures 470–570°C (Ma et al. 2004). Friction stir processing resulted in severe breakage of the silicon particles, which were uniformly distributed in the metal matrix during mixing. An average grain size of ~3 μm was obtained when subjected to friction stir processing. An elongation of 650% was achieved as well as a maximum strain rate sensitivity of 0.45. High strain rate superplasticity was also reported for friction stir processed aluminum alloys such as (a) 2024 aluminum (Charit and Mishra 2003) and (b) aluminum–magnesium–zirconium alloy (Ma et al. 2003).

22.6 MODELING OF FRICTION STIR WELDING PROCESS

Modeling of the friction stir welding process provides an understanding of the mechanical and thermal processes during joining. Heat is generated by friction between the tool and the workpiece, resulting in intense plastic deformation and a temperature rise within and around the stir zone. Because of this, significant microstructural evolution such as (a) grain refinement, (b) dissolution and coarsening of precipitates, (c) breakup and redistribution of dispersoids, and (d) texturing takes place. Various models developed by researchers in this area are highlighted in the subsequent subsection.

22.6.1 Statistical Models

The temperature distribution during friction stir welding of the 6061 aluminum alloy predicted, using a second-order polynomial equation (Hwang et al. 2008), a uniform temperature distribution inside the pin, and heat transfer was from the rim of the pin to the edge of the workpiece. The advancing side temperature was slightly higher than the retreating side. Both tensile strength and hardness at the thermomechanically affected zone were about half of the base metal. The ultimate tensile strength and tensile elongation of the dissimilar friction stir welded joints of aluminum alloys were predicted within a ±10% error using regression models (Shanmuga Sundaram and Murugan 2010). It was found that a straight cylindrical tool produced the lowest tensile strength and tensile elongation, whereas a tapered hexagon tool produced a higher pulsating effect and a smooth material flow, which resulted in the highest tensile strength and tensile elongation of the dissimilar friction stir welded joints. Tensile elongation decreased with an increase in tool revolution per minute and increased with an increase in welding speed. Different empirical models were developed to predict downward force, yield strength, tensile strengths, and hardness of friction stir welding joints using various process parameters (Bitondo et al. 2011; Palanivel and Mathews 2012; Rajakumar and Balasubramanian 2012; Rajakumar et al. 2010).

The contribution of various parameters on tensile strength of friction stir welded joints evaluated using the Taguchi technique (Lakshminarayanan and Balasubramanian 2008) revealed that a maximum contribution was for tool rotational speed. This was followed by traverse speed and axial force. Other researchers have also developed mathematical models to establish an empirical relationship between the different process parameters and ultimate tensile strength (Blignault et al. 2012; Dinaharan and Murugan 2012; Karthikeyan and Kumar 2011) during friction stir welding of different materials. These studies indicate the axial force to be the most affected by tool diameter, rotational speed, and welding speed, whereas the horizontal force is influenced by welding speed, pin diameter, and an interaction of tool diameter with the rotational speed.

22.6.2 Thermomechanical Modeling

A mathematical model for heat generation during various stages of friction stir welding process was developed by Durdanovic et al. (2009). The precise amount of heat generated was complicated due to various uncertainties and assumptions used for simplification of the model. Fratini et al. (2009) combined neural network with a finite element model to predict the average grain size during the friction stir welding process using local numerical values of strain, strain rate, temperature, and Zener–Hollomon parameter as the inputs. A good agreement between the predicted values and experimental results was obtained. Arora et al. (2011) developed a three-dimensional heat transfer and visco-plastic flow model for an optimum diameter of tool shoulder. Superior tensile properties were obtained by optimizing both the shoulder diameter and tool rotational speed.

A finite element model was developed to analyze the friction stir welding of corner fillet geometries (Buffa et al. 2010). The temperature distributions predicted by the model showed a good agreement with experimental data, and useful information was derived from an analysis of deformation as well as forces on the welding tool. Biswas and Mandal (2011) developed a three-dimensional finite element transient thermal model to study the effect of tool geometry on thermal history of the workpiece and observed a lower temperature rise while using a concave shoulder. Conical tool pins exhibited a lower peak temperature compared to a cylindrical pin. The result indicated that with a 100% increase in tool rotational speed, the increase in peak temperature and reduction in plunging force was 13% and 40%, respectively.

An improved continuum finite element model was developed for successful prediction and simulation of material flow and distribution and other relevant field variables using temperature, strain, and strain rate–dependent flow stress (Fratini et al. 2010). The effect of shoulder size on temperature distribution and material deformation during friction stir welding of an aluminum alloy was studied using a fully coupled thermomechanical model (Zhang et al. 2009). The study indicated that increasing the shoulder size increased the maximum temperature and resulted in a uniform distribution of temperature under the tool shoulder. Using a three-dimensional finite element model, the welding heat input (Aval et al. 2011a) was predicted. The frictional power had a major effect on both heat generation and temperature distribution within the metal. The study also revealed that the temperature during friction stir welding was asymmetrically distributed and the peak temperatures were higher on the advancing side when compared to the retreating side. Investigation of the thermomechanical and microstructural issues during dissimilar friction stir welding of AA5086–AA6061 by Aval et al. (2011b) revealed finer grains and a large thermally affected region on the AA6061 side compared to the AA5086 side. For the AA5086, the microhardness was found to be higher than the base metal due to the presence of refined grains as a result of recrystallization in the weld nugget. On the AA6061 side, softening by dissolution of hardening phases in the thermo-mechanically affected zone was observed.

22.6.3 Soft Computational Models

During friction stir welding, heat is generated by friction between the tool and the workpiece, which results in intense plastic deformation and temperature increase within and around the stirring zone. Because of this, a significant microstructural evolution, including grain size, grain boundary character, dissolution and coarsening of precipitates, breakup and redistribution of dispersoids, and texture, takes place. Different researchers used different modeling and optimization techniques to study and optimize the friction stir welding process. Among them, the regression, thermomechanical, and mathematical models are frequently used for the analysis of both thermal and mechanical properties with respect to different process parameters. The main drawbacks of these models are that they are time consuming and they lack in reliability. The soft computation approaches to model the friction stir weld and to optimize the process parameters were beneficial considering the time and resources saved in several experimental tests. While soft computational modeling and optimization techniques are few in application for the friction stir welding process, some of the relevant reviews are briefly discussed.

Artificial neural network is one intelligent system, which is useful in a variety of real-world applications since it can deal with complex and highly interactive processes. For modeling of the weld quality, different types of neural network can be used. These are (a) multilayer neural network, (b) radial basis function, and (c) self-organizing map. These techniques can very efficiently map the nonlinear relationships during the friction stir welding process. Popularly, a multilayer neural network trained with the back-propagation error algorithm was used to model weld quality (Ghetiya and Patel 2014; Kim et al. 2003; Lightfoot et al. 2005; Lim and Gweon 1999; Okuyucu et al. 2007; Pal et al. 2007). Kim et al. (2003) developed an intelligent system for automatic determination of optimal welding parameters for each pass and welding position. Lakshminarayanan and Balasubramanian (2009) found better prediction capability of the neural network model compared to the response surface model. The feed forward multilayer neural network model was more robust and accurate in estimating the values of tensile strength. Pal et al. (2007) developed six different types of radial basis function neural network models to predict distortion during welding. Different research works (Buffa et al. 2012; Fratini et al. 2009) linked the finite element method with a neural network to predict the mechanical and microstructural properties of friction stir welded joints. Generalized feed-forward and modular feed-forward neural networks along with a multilayer neural network were developed by Lightfoot et al. (2005) to model weld-induced deformation of a ship plate. Lim and Gweon (1999) proposed a neural network model to estimate joint strength of spot welds during pulsed laser spot welding. Okuyucu et al. (2007) developed a neural network model to correlate the input parameters (welding speed and tool rotation speed) of the friction stir welding process with the mechanical properties of aluminum butt joints.

Tansel et al. (2010) developed a genetically optimized neural network system in order to obtain the optimal process parameters. The average estimation error of an artificial neural network was less than 0.5%. A multilevel adaptive fuzzy control was used to improve the quality of the weld (Davis et al. 2012). Multilevel adaptive fuzzy control was able to maintain nearly constant spindle power for a wide range of process parameters even when the process was subjected to significant variations and external disturbances. Liao and Daftardar (2009) constructed two surrogate models, one linear and one nonlinear, to relate friction stir welding process parameters with maximum temperature at a selected location, using the simulation data generated by a thermal model. They implemented five population-based metaheuristics, namely, (a) ant colony optimization, (b) differential evolution, (c) genetic algorithm, (d) harmony search, and (e) particle swarm optimization. For a loose temperature range, a linear model produced better optimization results, whereas a nonlinear model produced better results for a tight temperature range.

22.6.4 Other Models

Lammlein et al. (2009) developed a computational fluid dynamics process model to analyze thermal, tensile, and macro-section forces. It was found that the 90° tool was better than the 60°, 80°, or 120° tools as it retained a reasonable percentage of strength of the parent material (50%–60%). A small cone inclusive angle required a higher spindle speed and created more flash while a large angle produced a larger heat-affected zone. McNelley et al. (2008) developed two models to interpret microstructure and micro-texture data of welded joint and concluded that both recrystallization and grain refinement occur during the friction stir welding process. A microstructure model to track the evolution and distribution of precipitate phases during the weld thermal cycle (Robson et al. 2010) revealed dissolution of the precipitates in the nugget during each pass. A post-weld aging treatment was found to be ineffective in recovering the weld properties of the heat-affected zone after multiple weld passes. An approximate analytical technique for the calculation of three-dimensional material flow during friction stir welding by considering the motion of an incompressible fluid induced by a solid rotating disk was developed by Arora et al. (2011). A mathematical model was developed to predict the ultimate tensile strength and tensile elongation of the joints for 95% confidence level in friction stir welded AA6061-T4 aluminum alloy joints (Heidarzadeh et al. 2012). The same study revealed that the ultimate tensile strength of the joint increases with an increase in tool rotational speed, welding speed, and tool axial force, up to a maximum value, and then decrease. The tensile elongation of the joints increased with an increase in rotational speed and axial force but decreased with an increase in welding speed. Arora et al. (2009) modeled torque, power requirement, and geometry of the stir zone during friction stir welding of the AA2524 aluminum alloy by solving the equations of conservation of mass, momentum, and energy. It was concluded that because of a higher heat generation rate and high temperature, the torque required for welding decreased with an increase in tool rotation speed. The size of the thermomechanically affected zone revealed a marginal increase with an increase in tool rotational speed.

22.7 DEFECTS IN FRICTION STIR WELDING

Friction stir welding is characterized by high weld quality, but defects still exist for unsuitable joining parameters (Vijay and Murugan 2010). In the present review, the research status of welding defects has been introduced from the aspects of characterization of defects, affecting factors and testing methods used. Defect characterization and the affecting factors are based on microstructural analysis and mechanical property. A "characteristic defect" is a microstructural or geometric anomaly peculiar to friction stir welding that adversely affects form, fit, and function. Defects in friction stir welding are related to processing temperature, material flow pattern, and joint geometry, which, in turn, are functions of process parameters and geometry.

The nondestructive ultrasonic method (Ghidini et al. 2005) was used as a promising technique to detect and characterize defects. The impact test was used to identify the presence of a root defect and zone of weakness (Thomas et al. 2005). Unintentional offsetting of the pin can cause several types of discontinuities, or defects, including lack of penetration or a displaced bond line developing a "stuck" or "weak" bond having no strength. Presently, the best methods for nondestructive evaluation of friction stir welds include the following: (a) ultrasonic inspection, (b) radiographic inspection, and (c) eddy current inspection depending on individual needs. Further advanced nondestructive evaluation processes using a slight variation of the PAUT called Matrix Phased Array Ultrasonic Testing have been developed (Lamarre and Moles 2000). It uses a series of conventional ultrasonic elements that are grouped together to function as a unit and can be directed at various angles as they pass through the part that is being inspected. This results in giving a complete map of the joint area.

Santos et al. (2011) introduced an eddy current probe to test AA2024 welded joints. They were able to both detect and size micro root defects in friction stir welding joints with a depth of 60 μm. Liu et al. (2008) evaluated the defects in friction stir welded aluminum alloy using a new ultrasonic method, which is based on multiple-incident angle reflection at the joints by using a spot-focused beam. Lombard et al. (2006) found pseudo-bond defects in the form of planar regions on the fracture surface and in the form of onion-skin defects in the friction stir welded 5083-H321 alloy. Occurrences of these defects were found to be a function of tensile strength and tool speed. Santos et al. (2007) inspected an AA5083-H111 welded sample using an online nondestructive integrated inspection system that used a data fusion algorithm with fuzzy logic and fuzzy inference functions.

22.8 MONITORING OF FRICTION STIR WELDING PROCESS

Monitoring can be defined as the procedure or method of observation of any physical process or phenomenon and acquiring valuable information through which performance of the process can be evaluated either continuously or periodically. This process involves two crucial steps: (a) acquisition of valuable information through observation and (b) an estimation or prediction of the output from the acquired information. Monitoring in the case of the friction stir welding process can be implemented for the prediction of process output. The output from the process can be represented in terms of quality of the weld. The quality of the welds relates to (a) strength of the welds, (b) hardness distribution of the welds when compared to the base material properties, and (c) absence of superficial or internal defects. To have better control over the outcome of the process, monitoring should be effective to record every possible change during the process. Process signals can be an effective option for detecting changes that occur during the friction stir welding process over time. Monitoring for the purpose of predicting weld quality of friction stir welding can be achieved with signals acquired during the process. Moreover, acquired signals give a more direct representation of the physical process. Since signals are sampled over time, information of the signal can play an important role in detecting any unusual event that occurs during the process, which will undoubtedly enhance the monitoring process. The following paragraphs outline different research efforts carried out by researchers for developing methodologies for effective monitoring of the friction stir welding process.

Longhurst et al. (2016) presented a methodology to monitor the friction stir welding process for identification of defects in the welded samples. Frequency information of the current signal acquired from the main spindle motor of a friction stir welding system is a key feature for monitoring the occurrence of defects in the welded samples. A Hall effect current sensor was used for acquisition of current signal from the main spindle motor. The researchers claimed that with the inclusion of a void in the welded sample, during the welding process, the frequency of the current signals follow an increase of 1–4 Hz. Although the increase is not remarkably high to be accepted as a feature for the purpose of defect identification, the proposed research opened up a new avenue to monitor the friction stir welding process using an inexpensive current sensor. However, the researchers did not put more emphasis on exploring the proposed method over a range of various process parameters. Applicability of infrared thermography for online monitoring of the friction stir welding process was presented by Serio et al. (2016). As a concluding remark, it was stated that temperatures of the joints reveal an asymmetric behavior about the joint line and the process of friction stir welding is a nonstationary welding process. A new indicator, termed "maximum heating slope value," was proposed to correlate tool rotational speed and welding speed during the process for the purpose of monitoring behavior of the joining process. Imam et al. (2013) developed a methodology, based on temperature distribution of friction stir welded joints, to monitor the process. The temperatures of various significant zones in the friction stir welded samples were acquired and processed to correlate with process behavior. It was commented that a weld nugget zone temperature below 350°C can result in the formation of tunnel defects in joining the 6063-T4 aluminum alloy using the friction stir welding process. It was also reported that better ductility of the welded samples was observed at nugget zone temperatures above 450°C. Image processing–based monitoring of first mode metal transfer during the friction stir welding process was studied by Sinha et al. (2008). Images of welds were captured to obtain image features in terms of gray-level distribution, texture, pattern, and contours. The effect of pin failure and pin depth was investigated with the computed image features. These researchers reported the proposed methodology to be effective and could be extended toward prediction of process parameters for unknown materials and would be efficient for online condition monitoring of the friction stir welding process. Monitoring of weld quality in terms of ultimate tensile strength of the joints was presented by Jene et al. (2008) with features computed from force signals acquired during the friction stir welding process. Short time Fourier transform (STFT) has been adopted as the signal processing tool. Frequency spectra of the force signals were reported to have significant information pertaining to process behavior. It was reported that high frequency amplitudes in the frequency spectra were observed for welds without any notable imperfections or defects.

Monitoring of the friction stir welding process using acoustic emission signals acquired during the process was reported in the research study presented by Soundararajan et al. (2006). Fast Fourier transform (FFT), short time Fourier transform, and wavelet transform were applied for the processing of acoustic signals. As a salient finding from this research, it was reported that high-frequency bands in the frequency spectra disappeared when the tool lost contact with the work-piece. From wavelet analysis of signals, it was commented that a good weld often results in high fluctuations in the approximation part of the decomposed signal compared to a defective weld. The research work claimed that among the three signal processing methods adopted, wavelet transform results in more information than FFT and STFT. Processing of the acoustic emission signal with a wavelet transform to monitor the friction stir welding process was also reported by Chen et al. (2003). It was reported that the defects formed during the friction stir welding process have different signal band energy characteristics and the same can be an effective indicator for the purpose of identification of defects in the welded samples. However, research work has still not been tested for the characterization of defects, such as (a) localization of defects and (b) size of the defects. In-process gap detection was attempted by Yang et al. (2008). Force signals acquired during the friction stir welding process were processed with FFT to compute the power spectral density of the signals. It was commented that gaps present in the welded plates can be detected by observing the power spectral density of the signals. A statistical feature–based classification scheme was proposed by Fleming et al. (2008). The acquired force signals were processed using FFT algorithm for the detection of gaps in the welded joints. Linear discriminant analysis and principal component analysis were adopted to develop methodologies for the identification of welds having gaps. Mehta et al. (2013) proposed a novel method to measure torque during friction stir welding with input electrical signatures of the driving motor. The developed method was tested, with results obtained from a well-accepted mathematical model for the computation of torque and force during the friction stir welding process. The developed method can be an effective solution to monitor the friction stir welding process through measurement of torque during the friction stir welding process.

Das et al. (2017a) proposed a strategy to monitor weld quality in the friction stir welding process with a main spindle motor current signal and a welding motor current signal. Wavelet packet transform (WPT) has been adopted as the tool for processing acquired signals. Wavelet packet coefficients extracted from signals are presented as features of the signals to monitor the ultimate tensile strength and yield strength of the joints. The research also proposed a methodology for the effective selection of suitable mother wavelet function for wavelet packet decomposition. Artificial neural network models were developed for the prediction of joint qualities with the computed signal features. The developed artificial neural network models show appreciable accuracy for the prediction of joint qualities with features of the signal.

Defect identification in friction stir welded samples was attempted by Das et al. (2016a) through processing vertical (or plunge) force acquired during the friction stir welding process. A new strategy has been developed combining Hilbert–Huang transform (HHT) and WPT. The developed HHT–WPT method eliminates the limitation faced by HHT of incurring high frequency band in the first IMF (1998). The developed strategy is reproduced in Figure 22.8a. From the analysis, an instantaneous phase and instantaneous frequency information were extracted as well as a correlation with the formation of defects in the welded samples. It was reported that features of the signal are effective in discriminating defective welds from defect-free welds with reliability and accuracy. The signal features computed for defect identification were reproduced in Figure 22.8b and c. In Figure 22.8b, it can be seen that the computed instantaneous phase from the vertical force signals shows a different trend against defective and defect-free welded samples. It is observed that the instantaneous phase against defective samples follows negative phase angles whereas signals against defect-free welded samples follow a positive phase angle. The variation in phase angle is presented as the key indicator for the purpose of identification of defects in the friction stir welding process. Apart from the phase angle representation, the proposed method also provides an indicator for the identification of defects in friction stir welded samples in terms of computed instantaneous frequency of the vertical force signal. From Figure 22.8c, it is observed that areas under the frequency spectrum for defective cases are higher than that under the frequency spectrum of defect-free welding cases. Moreover, a jump in the frequency spectrum of defective cases was observed compared to the frequency spectrum for defect-free welding cases. The proposed methodology can be effective in the identification of defective welds from real-time data and offers preliminary safeguard in monitoring of the friction stir welding process.

Monitoring of the friction stir welding process through an easy identification of defects in the welded samples is reported by Das et al. (2017b). In the research presented, torque signals acquired during the friction stir welding process were analyzed with discrete wavelet transform (DWT). Effective signal features were computed from the approximation part of the signals and a new indicator, DI, has been proposed for defect identification during the friction stir welding process. The proposed indicator is free from process conditions and can be effectively implemented for classification of defective welds from defect-free welds with ease. The research work also presented support vector regression models for weld quality monitoring during the friction stir welding process. The developed data-driven model yields an average accuracy of 0.53% for monitoring ultimate tensile strength of the welded samples.

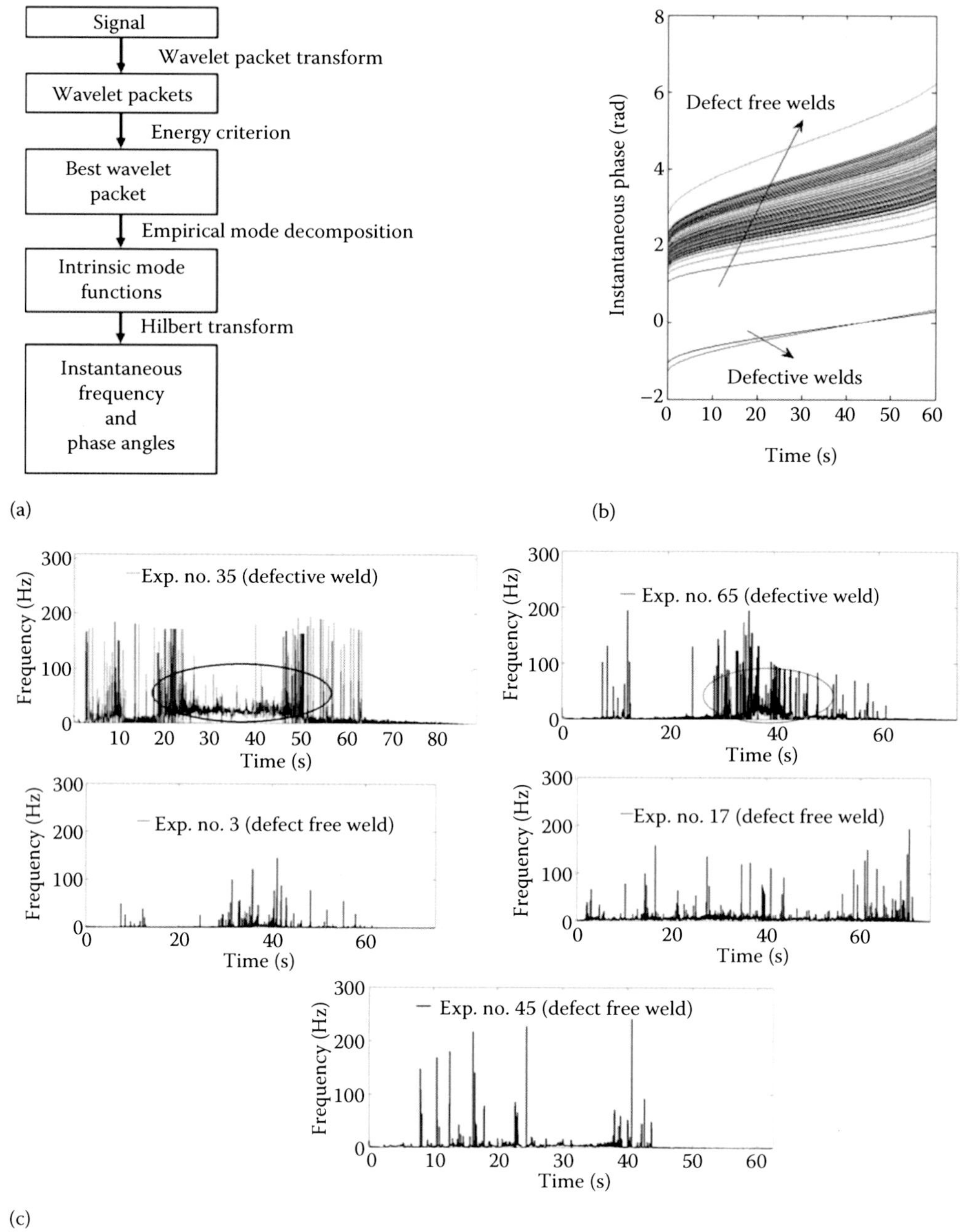

FIGURE 22.8 (a) Methodology proposed by Das et al. (2016a) for identification of defects in friction stir welded samples with proposed features, (b) instantaneous phase, and (c) instantaneous frequency.

An image information–based monitoring scheme for the friction stir welding process has been developed by Das et al. (2016b). Top surface images of the welded samples were captured, and features were extracted by implementing fractal theory and wavelet transform. Fractal dimensions were computed from the images and correlated to the ultimate tensile strength of the joints. Compared to process parameters, computed fractal dimensions show a more definite trend with the ultimate tensile strength of the joints. The ultimate tensile strength of the joints follows a decreasing trend with an increase in fractal dimension. A new indicator, CI, has been developed to monitor weld quality in the friction stir welding process through processing of images with wavelet transform. The proposed indicator also reveals a decreasing tend of ultimate tensile strength with increase in CI.

Implementation of fractal theory in processing in-process tool rotational speed signal acquired during the friction stir welding process for the purpose of defect identification was proposed by Das et al. (2016c). Higuchi's and Katz's algorithm was considered for an estimation of the fractal dimensions of speed signal. The computed fractal dimensions were presented as an independent indicator for the purpose of identification of defects during the friction stir welding process. It was reported that occurrence of defects in the welded samples results in low fractal dimension compared to the defect-free welded samples.

In another study presented by Das et al. (2017c), a strategy was developed to monitor weld quality during the friction stir welding process. Vertical and transverse force signals acquired during the friction stir welding process were analyzed in the time frequency domain for extraction of signal features. The computed signal features were then combined to obtain a new indicator termed as CP to monitor the ultimate tensile strength of the friction stir welded samples. It was reported that with an increase in CP, the ultimate tensile strength of the joints follows an increasing trend. Later support vector regression models were developed to model joint quality in terms of the ultimate tensile strength of the joints.

22.9 SUMMARY AND FUTURE TRENDS IN FRICTION STIR WELDING AND FRICTION STIR PROCESSING

Current developments in process models, joint microstructure, and mechanical properties and monitoring of welding process have been addressed in this review. The main highlights are as follows:

1. Friction stir welding is an effective technique for welding almost all materials including dissimilar joints and components of different thicknesses.
2. Local heat-assisted friction stir welding is proven very effective for joining high–melting point metals and alloy systems. It increases welding speed, tool life, and weld quality.
3. Tool geometry, rotational speed, welding speed, and axial load are the most significant parameters in joining of similar materials. The above parameters, along with tool offset distance, are very critical for the friction stir welding of dissimilar materials.
4. The microstructure of the weld zone reveals fine equiaxed grains in the nugget zone, highly elongated grains in the retreating side of TMAZ, and slightly elongated coarse grains in the HAZ. In general, grain sizes increases from the center of the weld region to the base metal.
5. The wavelet transform is found to be an effective signal processing technique to monitor weld quality in friction stir welding.
6. The features extracted from top surface images of the welds using fractal dimensions and 2D WPT also show a noteworthy trend with the ultimate tensile strength of the joints.

Despite considerable interest in this technology during the past decade, a basic physical understanding of the process is still lacking. Some important areas, including material flow, tool geometry design, tool wear, microstructural stability, and welding of dissimilar alloys and metals, require further understanding. Tools can be standardized with a better quantitative understanding of the principles of heat transfer, material flow, tool–workpiece contact conditions, and effects of various process parameters. Cost-effective and longer-life tools need to be developed to implement friction stir welding in commercial applications of high–melting point materials. Current friction stir welding process submodels are complex, time consuming, and difficult to use in real time. Further, they suffer from lack of reliability of the predicted results. This is because the physics is highly complex and current phenomenological models do not contain any model component designed to ensure good agreement with experimental results. Keeping in view of increasing day-to-day application in aerospace and automobile industries, friction stir welding of dissimilar, lightweight materials such as aluminum to magnesium alloys should be studied. Computerization and automation of the friction stir welding process can lead to its application in reconfigurable manufacturing systems where dissimilar components can be joined with minimum distortion and complexity. Further studies should be focused on defect formation mechanisms and testing techniques in order to find out the mutual action of various factors and for effective control of defect during friction stir welding. A theoretical understanding of the multiphysics behind the process is needed. With further research efforts and an increased understanding of the process, an increasing number of applications will be found for the fabrication and processing of materials.

Friction stir processing is a versatile solid-state technique for welding of similar or nonsimilar materials, improving the microstructure and mechanical properties of cast alloys, enhancing surface modification, producing composite materials and surface composites, and so on. The unique advantages of this technique are the breaking up and, to a great extent, dissolution of some of the constituent brittle phases in an as-cast alloy matrix along with considerable grain size refinement. Friction stir processing is an alternative technique for the fabrication of bulk and surface nanocomposites exhibiting high hardness. Looking at the advantages, the application of this technique can be extended to the fabrication of thermal barrier coatings for high-temperature superalloy turbine blade applications. The lengthy and cumbersome process of thermal barrier coating for turbine blade materials can be overcome by friction stir processing. In addition, the two interface layers, namely, the metal–bond coat and the bond coat–ceramic coat, can be eliminated by friction stir processing. The distinctive interface layers with wide variation in the thermal expansion properties results in debonding and subsequent failure of the coating materials. This can be avoided by the incorporating the coating materials by friction stir processing, where the surface can be coated by ceramic materials or high entropy materials.

REFERENCES

Acerra, F., Buffa, G., Fratini, L. and Troiano, G. (2010), *Int. J. Adv. Manuf. Technol.*: On the FSW of AA2024-T4 and AA7075-T6 T-joints: An industrial case study, Vol. 48, pp. 1149–1157.

Adamowski, J. and Szkodo, M. (2007), *J. Achieve. Mater. Manuf. Eng.*: Friction stir welds (FSW) of aluminum alloy AW6082-T6, Vol. 20, pp. 403–406.

Ahmed, M. M. Z., Wynne, B. P., Rainforth, W. M. and Threadgill, P. L. (2012), *Mater. Charact.*: Microstructure, crystallographic texture and mechanical properties of friction stir welded AA2017A, Vol. 64, pp. 107–117.

Arici, A. and Selale, S. (2007), *Sci. Technol. Weld. Joining*: Effects of tool tilt angle on tensile strength and fracture locations of friction stir welding of polyethylene, Vol. 6, pp. 536–539.

Arora, A., DebRoy, T. and Bhadeshia, H. K. D. H. (2011), *Acta Mater.*: Back-of-the-envelope calculations in friction stir welding—Velocities, peak temperature, torque, and hardness, Vol. 59, pp. 2020–2028.

Arora, A., Nandan, R., Reynolds, A. P. and DebRoy, T. (2009), *Scripta Mater.*: Torque, power requirement and stir zone geometry in friction stir welding through modeling and experiments, Vol. 60, pp. 13–16.

Arora, K. S., Pandey, S., Schaper, M. and Kumar, R. (2010a), *Int. J. Adv. Manuf. Technol.*: Effect of process parameters on friction stir welding of aluminum alloy 2219-T87, Vol. 50, pp. 941–952.

Arora, K. S., Pandey, S., Schaper, M. and Kumar, R. (2010b), *J. Mater. Sci. Technol.*: Microstructure evolution during friction stir welding of aluminum alloy AA2219, Vol. 26, pp. 747–753.

Attallah, M. M., Claire, L. D. and Strangwood, M. (2007), *J. Mater. Sci.*: Microstructure–microhardness relationships in friction stir welded AA525, Vol. 42, pp. 7299–7306.

Aval, H. J., Serajzadeh, S. and Kokabi, A. H. (2011a), *Int. J. Adv. Manuf. Technol.*: Theoretical and experimental investigation into friction stir welding of AA 5086, Vol. 52, pp. 531–544.

Aval, H. J., Serajzadeh, S. and Kokabi, A. H. (2011b), *J. Mater. Sci.*: Thermo-mechanical and microstructural issues in dissimilar friction stir welding of AA5086–AA6061, Vol. 46, pp. 3258–3268.

Aval, K. H. J., Serajzadeh, S., Kokabi, A. H. and Loureiro, A. (2011c), *Sci. Technol. Weld. Joining*: Effect of tool geometry on mechanical and microstructural behaviors in dissimilar friction stir welding of AA 5086-AA 6061, Vol. 16, pp. 597–604.

Aydin, H., Tutar, M., Durmus, A., Bayram, A. and Sayaca, T. (2012), *Trans. Indian Inst. Metals*: Effect of welding parameters on tensile properties and fatigue behavior of friction stir welded 2014-T6 aluminum alloy, Vol. 65, No. 1, pp. 21–30.

Azimzadegan, T. and Serajzadeh, S. (2010), *J. Mater. Eng. Perform.*: An investigation into microstructures and mechanical properties of aa7075-t6 during friction stir welding at relatively high rotational speeds, Vol. 19, No. 9, pp. 1256–1263.

Bakavos, D., Chen, Y., Babout, L. and Prangnell, P. (2011), *Metall. Mater. Trans. A*: Material interactions in a novel pinless tool approach to friction stir spot welding thin aluminum sheet, Vol. 42, No. 5, pp. 1266–1282.

Balasubramanian, V. (2008), *Mater. Sci. Eng. A*: Relationship between base metal properties and friction stir welding process parameters, Vol. 480, pp. 397–403.

Bang, H. S. and Bang, H. S. (2010), *J. Korean Weld. Joining Soc.*: A study on the weldability and mechanical characteristics of dissimilar materials butt joints by laser assisted friction stir welding, Vol. 28, pp. 678–683.

Bang, H. S., Bang, H. S., Song, H. and Joo, S. (2013), *Mater. Design*: Joint properties of dissimilar Al6061-T6 aluminum alloy/Ti–6%Al–4%V titanium alloy by gas tungsten arc welding assisted hybrid friction stir welding, Vol. 51, pp. 544–551.

Barcellona, A., Buffa, G., Fratini, L. and Palmeri, D. (2006), *J. Mater. Process. Technol.*: On microstructural phenomena occurring in friction stir welding of aluminium alloys, Vol. 177, pp. 340–343.

Barlas, Z. and Ozsarac, U. (2012), *Weld. J.*: Effects of FSW parameters on joint properties of AlMg3 Alloy, Vol. 91, pp. 16–22.

Barnes, S. J., Bhatti, A. R., Steuwer, A., Johnson, R., Altenkirch, J. and Withers, P. J. (2012), *Metall. Mater. Trans. A*: Friction stir welding in HSLA-65 steel: Part I. Influence of weld speed and tool material on microstructural development, Vol. 43, No. 7, pp. 2342–2355.

Biswas, P. and Mandal, N. R. (2011), *Weld. J.*: Effect of tool geometries on thermal history of FSW of AA1100, Vol. 90, pp. 129–135.

Bitondo, C., Prisco, U., Squilace, A., Buonadonna, P. and Dionoro, G. (2011), *Int. J. Adv. Manuf. Technol.*: Friction-stir welding of AA 2198 butt joints: Mechanical characterization of the process and of the welds through DOE analysis, Vol. 53, pp. 505–516.

Bitondo, C., Prisco, U., Squillace, A., Giorleo, G., Buonadonna, P., Dionoro, G. and Campanile, G. (2010), *Int. J. Mater. Form.*: Friction stir welding of AA 2198-T3 butt joints for aeronautical applications, Vol. 3, pp. 1079–1082.

Blignault, C., Hattingh, D. G. and James, M. N. (2012), *J. Mater. Eng. Perform.*: Optimizing friction stir welding via statistical design of tool geometry and process parameters, Vol. 21, pp. 927–935.

Boz, M. and Kurt, A. (2004), *Mater. Design*: The influence of stirrer geometry on bonding and mechanical properties in friction stir welding process, Vol. 25, pp. 343–347.

Bozkurt, Y. (2012), *Mater. Design*: The optimization of friction stir welding process parameters to achieve maximum tensile strength in polyethylene sheets, Vol. 35, pp. 440–445.

Brown, R., Tang, W. and Reynolds, A. P. (2009), *Mater. Sci. Eng. A*: Multi-pass friction stir welding in alloy 7050-T7451: Effects on weld response variables and on weld properties, Vol. 513–514, pp. 115–121.

Buffa, G., Fratini, L., Arregi, B. and Penalva, M. (2010), *Int. J. Mater. Form.*: A new friction stir welding based technique for corner fillet joints: Experimental and numerical study, Vol. 3, pp. 1039–1042.

Buffa, G., Fratini, L. and Micari, F. (2012), *J. Manuf. Process.*: Mechanical and microstructural properties prediction by artificial neural networks in FSW processes of dual phase titanium alloys, Vol. 14, pp. 289–296.

Buffa, G., Fratini, L. and Ruisi, V. (2009), *Int. J. Mater. Form.*: Friction stir welding of tailored joints for industrial applications, Vol. 2, pp. 311–314.

Cabibbo, M., McQueen, H. J., Evangelist, E., Spigarelli, S., Paol, M. D. and Falchero, A. (2007), *J. Mater. Sci. Eng. A*: Microstructure and mechanical property studies of AA6056 friction stir welded plate, Vol. 460–461, pp. 86–94.

Cam, G., Mistikoglu, S. and Pakdil, M. (2009), *Weld. J.*: Microstructural and mechanical characterization of friction stir butt joint welded 63% Cu–37% Zn brass plate, Vol. 11, pp. 225–232.

Cavaliere, P., Squillace, A. and Panell, F. (2008), *J. Mater. Process. Technol.*: Effect of welding parameters on mechanical and microstructural properties of AA6082 joints produced by friction stir welding, Vol. 200, pp. 364–372.

Cerri, E. and Leo, P. (2009), *J. Mater. Design*: Warm and room temperature deformation of friction stir welded thin aluminum sheets, Vol. 31, pp. 1392–1402.

Cerri, E., Leo, P., Wang, X. and Embury, J. D. (2010), *Metall. Mater. Trans. A*: Mechanical properties and microstructural evolution of friction-stir-welded thin sheet aluminum alloys, Vol. 42(5), pp. 1283–1295.

Ceschini, L., Boromei, I., Minak, G., Morri, A. and Tarterini, F. (2007), *Compos. Part A: Appl. Sci. Manuf.*: Microstructure, tensile and fatigue properties of AA6061/20 vol.% Al_2O_3p friction stir welded joints, Vol. A 38(4), pp. 1200–1210.

Chang, C. I., Wang, U. N., Pei, H. R., Lee, C. J. and Huang, J. C. (2006), *Mater. Trans.*: On the hardening of friction stir processed Mg-AZ31 based composites with 5–20% nano-ZrO2 and nano-SiO2 particles, Vol. 47, pp. 2942–2949.

Chang, W. S., Rajesh, S. R., Chun, C. K. and Kim, H. J. (2011) *J. Mater. Sci. Technol.*: Microstructure and mechanical properties of hybrid laser-friction stir welding between AA6061-T6 Al alloy and AZ31 Mg alloy, Vol. 27, No. 3, pp. 199–204.

Chao, Y. J., Wang, Y. and Miller, K. W. (2001), *Weld. J.*: Effect of friction stir welding on dynamic properties of AA2024-T3 and AA7075-T7351, pp. 196–200.

Charit, I. and Mishra, R. S. (2003), *Mater. Sci. Eng. A*: High strain rate superplasticity in a commercial 2024 Al alloy via friction stir processing, Vol. 359, pp. 290–296.

Chen, C., Kovacevic, R. and Jandgric, D. (2003), *Int. J. Mach. Tool Manuf.*: Wavelet transform analysis of acoustic emission in monitoring friction stir welding of 6061 aluminum, Vol. 43, pp. 1383–1390.

Chen, J., Fujii, H., Sun, Y., Morisada, Y., Kondoh, K. and Hashimoto, K. (2012), *Mater. Sci. Eng. A*: Effect of grain size on the microstructure and mechanical properties of friction stir welded non-combustive magnesium alloys, Vol. 549, pp. 176–184.

Chen, K. E., Gan, W., Okamoto, K., Chung, K. and Wagoner, R. H. (2011), *Metall. Mater. Trans. A*: The mechanism of grain coarsening in friction-stir-welded AA5083 after heat treatment, Vol. 42, No. 2, pp. 488–507.

Chen, T. (2009), *J. Mater. Sci.*: Process parameters study on FSW joint of dissimilar metals for aluminum–steel, Vol. 44, pp. 2573–2580.

Cho, H. H., Han, H. N., Hong, S. T., Park, J. H., Kwon, Y. J., Kim, S. H. and Steel, R. J. (2011), *Mater. Sci. Eng. A*: Microstructural analysis of friction stir welded ferritic stainless, Vol. 528, pp. 2889–2894.

Cui, L., Yang, X., Zhou, G., Xu, X. and Shen, Z. (2012), *Mater. Sci. Eng. A*: Characteristics of defects and tensile behaviors on friction stir welded AA6061-T4 T-joints, Vol. 543, pp. 58–68.

D'Urso, G. and Giardini, C. (2010), *Int. J. Mater. Form.*: The influence of process parameters and tool geometry on mechanical properties of friction stir welded aluminum lap joints, Vol. 3, pp. 1011–1014.

Das, B., Pal, S. and Bag, S. (2016a), *J. Manuf. Process.*: A combined wavelet packet and Hilbert–Huang transform for defect detection and modelling of weld strength in friction stir welding process, Vol. 22, pp. 260–268.

Das, B., Pal, S. and Bag, S. (2016b), *Sci. Technol. Weld. Joining*: Monitoring of friction stir welding process using weld image information, Vol. 21, No. 4, pp. 317–324.

Das, B., Bag, S. and Pal, S. (2016c), *Manuf. Lett.*: Defect detection in friction stir welding process through characterization of signals by fractal dimension, Vol. 7, pp. 6–10.

Das, B., Pal, S. and Bag, S. (2017a), *Int. J. Adv. Manuf. Technol.*: Weld quality prediction in friction stir welding using wavelet analysis, Vol. 89, pp. 711–725.

Das, B., Pal, S. and Bag, S. (2017b), *J. Manuf. Process.*: Torque based defect detection and weld quality modelling in friction stir welding process, Vol. 27, pp. 8–17.

Das, B., Pal, S. and Bag, S. (2017c), *Measurement*: Design and development of force and torque measurement setup for real time monitoring of friction stir welding process, Vol. 103, pp. 186–198.

Davis, T. A., Ngo, P. D. and Shin, Y. C. (2012), *Int. J. Adv. Manuf. Technol.*: Multi-level fuzzy control of friction stir welding power, Vol. 59, pp. 559–567.

Deqing, W. and Shuhua, L. (2004), *J. Mater. Sci.*: Study of friction stir welding of aluminum, Vol. 39, pp. 1689–1693.

Dieguez, T., Burgueño, A. and Svoboda, H. (2012), *Procedia Materials Science*: Superplasticity of a friction stir processed 7075–T651 aluminum alloy, Vol. 1, pp. 110–117.

Dinaharan, I. and Murugan, N. (2012), *Mater. Sci. Eng. A*: Effect of friction stir welding on microstructure, mechanical and wear properties of AA6061/ZrB2 in situ cast composites, Vol. 543, pp. 257–266.

Dixit, M., Newkirk, J. W. and Mishra, R. S. (2007), *Scripta Mater.*: Properties of friction stir-processed Al 1100–NiTi composite, Vol. 56, pp. 541–544.

Donati, L., Tomesani, L. and Morri, A. (2009), *Int. J. Mater. Form.*: Structural T-joint produced by means of friction stir welding (FSW) with filling material, Vol. 2, pp. 295–298.

Durdanovic, M. B., Mijajloric, M. M., Milcic, D. S. and Stamenkovic, D. S. (2009), *Tribol. Ind.*: Heat generation during friction-stir-welding (FSW) process, Vol. 31, pp. 8–14.

Eberl, I., Hantrais, C., Ehrtsrom, J. C. and Nardin, C. (2010), *Sci. Technol. Weld. Joining.*: Friction stir welding dissimilar alloys for tailoring properties of aerospace parts, Vol. 15, pp. 699–705.

Elangovan, K. and Balasubramanian, V. (2007), *Mater. Sci. Eng. A*: Influences of pin profile and rotational speed of the tool on the formation of friction stir processing zone in AA2219 aluminium alloy, Vol. 459, pp. 7–18.

Elangovan, K. and Balasubramanian, V. (2008a), *Mater. Design*: Influences of tool pin profile and tool shoulder diameter on the formation of friction stir processing zone in AA6061 aluminium alloy, Vol. 29, pp. 362–373.

Elangovan, K. and Balasubramanian, V. (2008b), *Int. J. Mater. Process. Technol.*: Influences of tool pin profile and welding speed on the formation of friction stir processing zone in AA2219 aluminium alloy, Vol. 200, pp. 163–175.

Elangovan, K., Balasubramanian, V., and Valliappan, M. (2008), *Int. J. Adv. Manuf. Tech.*: Influences of tool pin profile and axial force on the formation of friction stir processing zone in AA6061 aluminium alloy, Vol. 38, pp. 285–295.

El-Hafez, H. (2011), *J. Mater. Eng. Perform.*: Mechanical properties and welding power of friction stirred AA2024-T35 joints, Vol. 20, No. 6, pp. 839–845.

El-Rayesa, M. M. and El-Danaf, E. A. (2012), *J. Mater. Process. Technol.*: The influence of multi-pass friction stir processing on the microstructural and mechanical properties of Aluminum Alloy 6082, Vol. 212, pp. 1157–1168.

Feng, A. H. and Ma, Z. Y. (2007), *Scripta Mater.*: Enhanced mechanical properties of Mg–Al–Zn cast alloy via friction stir processing, Vol. 56, pp. 397–400.

Ferrando, W. A. (2012), Electrically assisted friction stir welding. The United States of America as represented by the Secretary of the Navy, Washington, DC, USA, US patent no. 8 164 021 B1, published 21 April 2012.

Ferrara, R. J. and Caton, T. E. (1982), *Mater. Performance*: Review: Dealloying of cast aluminum bronze and nickel-aluminum bronze alloys in seawater service, Vol. 21, pp. 30–34.

Fleming, P., Lammlein, D., Wilkes, D., Fleming, K., Bloodworth, T., Cook, G., Strauss, A., DeLapp, D., Leinert, T., Bement, M. and Prater, T. (2008), *Sensor Rev.*: In-process gap detection in friction stir welding, Vol. 28, No. 1, pp. 62–67.

Fonda, R. W. and Bingert, J. F. (2004), *Metall. Mater. Trans. A*: Micro-structural evolution in the heat-affected zone of a friction stir weld, Vol. 35A, pp. 1487–1499.

Forcellese, A., Gabrielli, F. and Simoncini, M. (2012), *Mater. Design*: Mechanical properties and microstructure of joints in AZ31 thin sheets obtained by friction stir welding using "pin" and "pinless" tool configurations, Vol. 34, pp. 219–229.

Fratini, L., Buffa, G. and Monaco, L. L. (2010), *Sci. Technol. Weld. Joining*: Improved FE model for simulation of friction stir welding of different materials, Vol. 15, pp. 199–207.

Fratini, L., Buffa, G. and Palmeri, D. (2009), *Comput. Struct.*: Using a neural network for predicting the average grain size in friction stir welding processes, Vol. 87, pp. 1166–1174.

Fu, R. D., Ji, H. S., Li, Y. J. and Liu, L. (2012a), *Sci. Technol. Weld. Joining*: Effect of weld conditions on microstructures and mechanical properties of friction stir welded joints on AZ31B magnesium alloys, Vol. 17, pp. 174–179.

Fu, R., Xu, H., Luan, G., Dong, C., Zhang, F. and Li, G. (2012b), *Mater. Charact.*: Top surface microstructure of friction stir welded AA2524-T3 aluminum alloy joints, Vol. 65, pp. 48–54.

Fujii, H., Cui, L., Maeda, M. and Nogi, K. (2006), *Mater. Sci. Eng. A*: Effect of tool shape on mechanical properties and microstructure of friction stir welded aluminum alloys, Vol. 419, pp. 25–31.

Gaafer, A. M., Mahmoud, T. S. and Mansour, E. H. (2010), *Mater. Sci. Eng. A*: Microstructural and mechanical characteristics of AA7020-O Al plates joined by friction stir welding, Vol. 527, pp. 7424–7429.

Galvao, I., Oliveira, J. C., Loureiro, A. and Rodrigues, D. M. (2012), *Intermetallics*: Formation and distribution of brittle structures in friction stir welding of aluminium and copper: Influence of shoulder geometry, Vol. 22, pp. 122–128.

Ghetiya, N. D. and Patel, K. M. (2014), *2nd International Conference on Innovations in Automation and Mechatronics Engineering, Procedia Technology*: Prediction of tensile strength in friction stir welded aluminium alloy using artificial neural network, Vol. 14, pp. 274–281.

Ghidini, T., Vugrin, T. and Dalle Donne, C. (2005), *Weld. Int.*: Residual stresses, defects and non-destructive evaluation of FSW joints, Vol. 19, pp. 783–790.

Ghosh, M., Kumar, K., Kailas, S. V. and Ray, A. K. (2010), *Mater. Design*: Optimization of friction stir welding parameters for dissimilar aluminum alloys, Vol. 31, pp. 3033–3037.

Gupta, R. K., Das, H. and Pal, T. K. (2012), *J. Mater. Eng. Perform.*: Influence of processing parameters on induced energy, mechanical and corrosion properties of FSW butt joint of 7475 AA, Vol. 21, No. 8, pp. 1645–1654.

Hao, H. L., Ni, D. R., Huang, H., Wang, D., Xiao, B. L., Nie, Z. R. and Ma, Z. Y. (2013), *Mater. Sci. Eng. A*: Effect of welding parameters on microstructure and mechanical properties of friction stir welded Al–Mg–Er alloy, Vol. 559, pp. 889–896.

Heartier, P., Desrayaud, C. and Montheillt, F. (2002), *Mater. Sci. Forum*: A thermomechanical analysis of the friction stir welding process, pp. 1537–1542.

Heidarzadeh, A., Khodaverdizadeh, H., Mahmoudi, A. and Nazar, E. (2012), *Mater. Design*: Tensile behavior of friction stir welded AA 6061-T4 aluminum alloy joints, Vol. 37, pp. 166–173.

Hodder, K. J., Izadi, H., McDonald, A. G. and Gerlich, A. P. (2012), *Mat. Sci. Eng. A*: Fabrication of aluminum–alumina metal matrix composites via cold gas dynamic spraying at low pressure followed by friction stir processing, Vol. A556, pp. 114–121.

Heurtier, O. P., Jones, M. J., Desrayaud, C., Driver, J. H., Montheillet, F. and Allehaux, D. (2006), *J. Mater. Process. Technol.*: Mechanical and thermal modeling of friction stir welding, Vol. 171, pp. 348–357.

Hui-jie, L., Hui-jie, Z., Yong-xian, H. and Lei, Y. (2010), *Trans. Nonferrous Met. Soc. China*: Mechanical properties of underwater friction stir welded 2219 aluminum alloy, Vol. 20, pp. 1387–1391.

Hwang, Y. M., Kang, Z. W., Chiou, Y. C. and Hsu, H. H. (2008), *Int. J. Mach. Tool. Manuf.*: Experimental study on temperature distributions within the workpiece during friction stir welding of aluminum alloys, Vol. 48, pp. 778–787.

Imam, M., Biswas, K. and Racherla, V. (2013), *Mater. Design*: Effect of weld morphology on mechanical response and failure of friction stir welds in a naturally aged aluminium alloy, Vol. 44, pp. 23–34.

Jene, T., Dobmann, G., Wagner, G. and Eifler, D. (2008), *Weld. World*: Monitoring of the friction stir welding process to describe parameter effects on joint quality, Vol. 52, pp. 47–53.

Joo, S. M. (2013), *Metals Mater. Int.*: Joining of dissimilar AZ31B magnesium alloy and SS400 mild steel by hybrid gas tungsten arc friction stir welding, Vol. 19, No. 6, pp. 1251–1257.

Karthikeyan, L. and Senthil Kumar, V. S. (2011), *Mater. Design*: Relationship between process parameters and mechanical properties of friction stir processed AA6063-T6 aluminum alloy, Vol. 32, pp. 3085–3091.

Khodir, S. A. and Shibayanagi, T. (2008), *Mater. Sci. Eng. B*: Friction stir welding of dissimilar AA2024 and AA7075 aluminum alloys, Vol. 148, pp. 82–87.

Kim, I. S., Jeong, Y. J., Lee, C. W. and Yarlagadda, P. K. D. V. (2003), *Int. J. Adv. Manuf. Technol.*: Prediction of welding parameters for pipeline welding using an intelligent system, Vol. 22, pp. 713–719.

Kim, W. K., Won, S. T. and Goo, B. C. (2010), *Int. J. Precis. Eng. Manuf.*: A study on mechanical characteristics of the friction stir welded A6005-T5 extrusion, Vol. 11, pp. 931–936.

Kim, Y. G., Fujii, H., Tsumur, T., Komazaki, T. and Nakata, K. (2006), *Mater. Lett.*: Effect of welding parameters on microstructure in the stir zone of FSW joints of aluminum die casting alloy, Vol. 60, pp. 3830–3837.

Koilraj, M., Sundareswaran, V., Vijayan, S. and Koteswara Rao, S. R. (2012), *Mater. Design*: Friction stir welding of dissimilar aluminum alloys AA2219 to AA5083—Optimization of process parameters using Taguchi technique, Vol. 42, pp. 1–7.

Kumar, K. and Kailas, S. V. (2008a), *Mater. Design*: On the role of axial load and the effect of interface position on the tensile strength of a friction stir welded aluminium alloy, Vol. 29, pp. 791–797.

Kumar, K. and Kailas, S. V. (2008b), *Mater. Sci. Eng. A*: The role of friction stir welding tool on material flow and weld formation, Vol. 485, pp. 367–374.

Kumbhar, N. T., Sahoo, S. K., Samajdar, I., Dey, G. K. and Bhanumurthy, K. (2011), *Mater. Design*: Microstructure and microtextural studies of friction stir welded aluminium alloy 5052, Vol. 32, pp. 1657–1666.

Kwon, Y. J., Shim, S. B. and Park, D. H. (2009), *Trans. Nonferrous Met. Soc. China*: Friction stir welding of 5052 aluminum alloy plates, Vol. 19, pp. 23–27.

Lai, R. L., He, D. Q., Liu, L., Ye, S. Y. and Yang, K. (2014), *Int. J. Adv. Manuf. Technol.*: A study of the temperature field during ultrasonic-assisted friction-stir welding, Vol. 73, pp. 321–327.

Lakshminarayanan, A. K. and Balasubramanian, V. (2008), *Trans. Nonferrous Met. Soc. China*: Process parameters optimization for friction stir welding of RDE-40 aluminium alloy using Taguchi technique, Vol. 18, pp. 548–554.

Lakshminarayanan, A. K. and Balasubramanian, V. (2009), *Trans. Nonferrous Met. Soc. China*: Comparison of RSM with ANN in predicting tensile strength of friction stir welded AA7039 aluminium alloy joints, Vol. 19, pp. 9–18.

Lakshminarayanan, K., Balasubramanian, V. and Elangovan, K. (2009), *Int. J. Adv. Manuf. Technol.*: Effect of welding processes on tensile properties of AA6061 aluminium alloy joints, Vol. 40, pp. 286–296.

Lamarre, A. and Moles, M. (2000), 15th World Conference on Non-destructive Testing Roma (Italy): Ultrasound phased array inspection technology for the evaluation of friction stir welds, pp. 15–21.

Lammlein, D. H., DeLapp, D. R., Fleming, P. A., Strauss, A. M. and Cook, G. E. (2009), *Mater. Design*: The application of shoulderless conical tools in friction stir welding: An experimental and theoretical study, Vol. 30, pp. 4012–4022.

Leal, R. M., Leitao, C., Loureiro, A., Rodrigues, D. M. and Vila, P. (2008), *Mater. Sci. Eng. A*: Material flow in heterogeneous friction stir welding of thin aluminium sheets: Effect of shoulder geometry, Vol. 498, pp. 384–391.

Leal, R. M. and Loureiro, A. (2008), *Mater. Design*: Effect of overlapping friction stir welding passes in the quality of welds of aluminium alloys, Vol. 29, pp. 982–991.

Leal, R. M., Sakharova, N., Vilac, P., Rodrigues, D. M. and Loureiro, A. (2011), *Sci. Technol. Weld. Joining*: Effect of shoulder cavity and welding parameters on friction stir welding of thin copper sheets, Vol. 16, pp. 146–152.

Lee, C. L., Huang, J. C. and Hsisn, P. J. (2006), *Scripta Mater.*: Mg based nano-composites fabricated by friction stir processing, Vol. 54, pp. 1415–1420.

Leitao, C., Louro, R. and Rodrigues, D. M. (2012), *Mater. Design*: Analysis of high temperature plastic behaviour and its relation with weldability in friction stir welding for aluminium alloys AA5083-H111 and AA6082-T6, Vol. 37, pp. 402–409.

Lertora, E. and Gambaro, C. (2010), *Int. J. Mater. Form.*: AA8090 Al–Li alloy fsw parameters to minimize defects and increase fatigue life, Vol. 3, pp. 1003–1006.

Li, B. and Shen, Y. (2012), *Mater. Design*: A feasibility research on friction stir welding of a new-typed lap-butt joint of dissimilar Al alloys, Vol. 34, pp. 725–731.

Li, B., Shen, Y. and Hu, W. (2011), *Mater. Design*: The study on defects in aluminum 2219-T6 thick butt friction stir welds with the application of multiple non-destructive testing methods, Vol. 32, pp. 2073–2084.

Li, B., Shen, Y., Luo, L., and Hu, W. (2013), *Mater. Sci. Eng. A*: Fabrication of TiC p/Ti–6Al–4V surface composite via friction stir processing (FSP): Process optimization, particle dispersion-refinement behavior and hardening mechanism, Vol. 574, pp. 75–85.

Li, D., Cui, Z., Yang, Q., Sun, B. and Sun, M. (2012), *J. Shanghai Jiaotong Univ. (Sci.)*: Microstructure and property of friction stir welding joint of 7075Al and AZ31BMg, Vol. 17, pp. 679–683.

Liang, X., Li, H., Li, Z., Hong, T., Ma, B., Liu, S. and Liu, Y. (2012), *Mater. Design*: Study on the microstructure in a friction stir welded 2519-T87 Al alloy, Vol. 35, pp. 603–608.

Liao, T. W. and Daftardar, S. (2009), *Sci. Technol. Weld. Joining*: Model based optimization of friction stir welding processes, Vol. 14, No. 5, pp. 426–435.

Lightfoot, M. P., Bruce, G. J., Mcpherson, N. A. and Woods, K. (2005), *Weld. J.*: The application of artificial neural networks to weld-induced deformation in ship plate, Vol. 84, No. 2, pp. 23–30.

Lim, D. C. and Gweon, D. G. (1999), *J. Manuf. Process.*: In-process joint strength estimation in pulsed laser spot welding using artificial neural networks, Vol. 18, No. 2, pp. 31–42.

Liu, F., Lui, S., Guo, E. and Li, L. (2008), *17th World Conference on Nondestructive Testing, 25–28 Oct 2008, Shanghai, China*: Ultrasonic Evaluation of Friction Stir Welding, pp. 1–7.

Liu, F. C. and Ma, Z. Y. (2008), *Metall. Mater. Trans. A*: Influence of tool dimension and welding parameters on microstructure and mechanical properties of friction-stir-welded 6061-T651 aluminum alloy, Vol. 29, pp. 362–373.

Liu, H. J., Li, J. Q. and Duan, W. J. (2012a), *Int. J. Adv. Manuf. Technol.*: Friction stir welding characteristics of 2219-T6 aluminum alloy assisted by external non-rotational shoulder, Vol. 64, pp. 1685–1694.

Liu, H. J., Zhou, L. and Liu, Q. W. (2009), *Scripta Mater.*: Microstructural evolution mechanism of hydrogenated Ti–6Al–4V in the friction stir welding and post-weld dehydrogenation process, Vol. 61, pp. 1008–1011.

Liu, H. J., Zhang, H., Pan, Q. and Yu, L. (2012b), *Int. J. Mater. Form.*: Effect of friction stir welding parameters on microstructural characteristics and mechanical properties of 2219-T6 aluminum alloy joints, Vol. 5, pp. 235–241.

Liu, H. J., Zhang, H. J. and Yu, L. (2011), *J. Mater. Eng. Perform.*: Homogeneity of mechanical properties of underwater friction stir welded 2219-T6 aluminum alloy, Vol. 20, No. 8, pp. 1419–1422.

Liu, X. C., Wu, C. S. and Padhy, G. K. (2015), *Sci. Technol. Weld. Joining*: Improved weld macrosection, microstructure and mechanical properties of 2024Al-T4 butt joints in ultrasonic vibration enhanced friction stir welding, Vol. 20, pp. 345–352.

Lombard, H., Hattingh, D. G., Steuwer, A. and James, M. N. (2006), Proceedings of Crack Paths, Parma Italy: Relationships among FSW process parameters, defects, crack paths and fatigue strength in 5083-H321 aluminium alloy.

Long, X. and Khanna, S. K., (2005), *Sci. Technol. Weld. Joining*: Modelling of electrically enhanced friction stir welding process using finite element method, Vol. 10, pp. 482–487.

Longhurst, W. R., Wilbur, I. C., Osborne, B. E. and Gaither, B. W. (2016), *Proc. Inst. Mech. Eng. Part B: J. Eng. Manuf.*: Process monitoring of friction stir welding via the frequency of the spindle motor current, DOI: https://doi.org/10.1177/0954405416654089.

Lorrain, O., Favier, V., Zahrouni, H. and Lawrjaniec, D. (2010), *Int. J. Mater. Form.*: Friction stir welding using unthreaded tools: Analysis of the flow, Vol. 3, pp. 1043–1046.

Ma, Z. Y., Mishra, R. S. and Mahoney, M. W. (2004), *Scripta Mater.*: Superplasticity in cast A356 induced via friction stir processing, Vol. 50, pp. 931–935.

Ma, Z. Y., Mishra, R. S., Mahomey, M. W. and Crimes, R. (2003), *Mater Sci. Eng A.*: High strain rate superplasticity in friction stir processed Al–Mg–Zr alloy, Vol. 351, pp. 148–153.

Ma, Z. Y., Sharma, S. R. and Mishra, R. S. (2006), *Metallurg. Mater. Trans. A*: Microstructural modification of as-cast Al–Si–Mg alloy by friction stir processing, Vol. 37A, pp. 3323–3336.

Malarvizhi, S. and Balasubramanian, V. (2011), *J. Mater. Eng. Perform.*: Effects of welding processes and post-weld aging treatment on fatigue behavior of AA2219 aluminium alloy joints, Vol. 20, pp. 359–367.

Mathis, K., Gubicza, J. and Nam, N. H. (2005), *J. Alloys Compounds*: Microstructure and mechanical behavior of AZ91 Mg alloy processed by equal channel angular pressing, Vol. 394, pp. 194–199.

Mazaheri, Y., Karimzadeh, F. and Enayati, M. H. (2011), *J. Mater. Process. Technol.*: A novel technique for development of A356/Al_2O_3 surface nanocomposite by friction stir processing, Vol. 211, pp. 1614–1619.

McNelley, T. R., Swaminathan, S. and Su, J. Q. (2008), *Scripta Mater.*: Recrystallization mechanisms during friction stir welding/processing of aluminum alloys, Vol. 58, pp. 349–354.

Mehta, M., Chatterjee, K. and De, A. (2013), *Sci. Tech. Weld. Joining*: Monitoring torque and traverse force in friction stir welding from input electrical signatures of driving motors, Vol. 18, No. 3, pp. 191–197.

Merklein, M. and Giera, A. (2008), *Int. J. Mater. Forming*: Laser assisted friction stir welding of drawable steel-aluminium tailored hybrids, Vol. 1, pp. 1299–1302.

Miles, M. P., Nelson, T. W., Steel, R., Olsen, E. and Gallagher, M. (2009), *Sci. Technol. Weld. Joining*: Effect of friction stir welding conditions on properties and microstructures of high strength automotive steel, Vol. 14, pp. 228–232.

Mishra, R. S. and Ma, Z. Y. (2005), *Mater. Sci. Eng. R*: Friction stir welding and processing, Vol. 50, pp. 1–78.

Mishra, R. S., Ma, Z. Y. and Charit I. (2003), *Mater. Sci. Eng. A*: Friction stir processing: A novel technique for fabrication of surface composite, Vol. A341, pp. 307–310.

Mishra, R. S. and Mahoney M. W. (2007), *Friction Stir Welding and Processing*, USA, ASM International.

Miyano, Y., Fujii, H., Sun, Y., Katada, Y., Kuroda, S. and Kamiya, O. (2011), *Mater. Sci. Eng. A*: Mechanical properties of friction stir butt welds of high nitrogen-containing austenitic stainless steel, Vol. 528, pp. 2917–2921.

Moreira, P. M. G. P., Santos, T., Tavares, S. M. O., Richter-Trummer, V., Vilaça, P. and Castro, P. M. S. T. (2009), *Mater. Design*: Mechanical and metallurgical characterization of friction stir welding joints of AA6061-T6 with AA6082-T6, Vol. 30, pp. 180–187.

Morisada, Y., Fujii, H., Nagaoka, T. and Fukusumi, M. (2006a), *Mater. Sci. Eng. A*: MWCNTs/AZ31 surface composites fabricated by friction stir processing, Vol. A419, pp. 344–348.

Morisada, Y., Fujii, H., Nagaoka, T. and Fukusumi, M. (2006b), *Mater. Sci. Eng. A*: Effect of friction stir processing with SiC particles on microstructure and hardness of AZ31, Vol. A433, pp. 50–54.

Movahedi, M., Kokabi, A. H., Seyed Reihani, S. M. and Najafi, H. (2012), *Sci. Technol. Weld. Joining*: Effect of tool travel and rotation speeds on weld zone defects and joint strength of aluminium steel lap joints made by friction stir welding, Vol. 17, No. 2, pp. 162–167.

Oh-Ishi, K. and McNelley, T. R. (2004), *Metall. Mater. Trans. A*: Microstructural modification of as-cast NiAl bronze by friction stir processing, Vol. 35A, pp. 2951–2961.

Okuyucu, H., Kurt, A. and Arcaklioglu, E. (2007), *Mater. Design*: Artificial neural network application to the friction stir welding of aluminum plates, Vol. 28, pp. 78–84.

Ouyang, J., Yarrapareddy, E. and Kovacevic, R. (2006), *J. Mater. Process. Technol.*: Microstructural evolution in the friction stir welded 6061 aluminum alloy (T6-temper condition) to copper, Vol. 172, pp. 110–122.

Padmanaban, G. and Balasubramanian, V. (2010), *Int. J. Adv. Manuf. Technol.*: An experimental investigation on friction stir welding of AZ31B magnesium alloy, Vol. 49, pp. 111–121.

Padhy, G. K., Wu C. S. and Gao, S. (2015), *Sci. Technol. Weld. Joining*: Auxiliary energy assisted friction stir welding—Status review, Vol. 20, No. 8, pp. 631–649.

Pal, S., Pal, S. K. and Samantaray, A. K. (2007), *Sci. Technol. Weld. Joining*: Radial basis function neural network model based prediction of weld plate distortion due to pulsed metal inert gas welding, Vol. 12, No. 8, pp. 725–731.

Palanivel, R. and Koshy Mathews, P. (2012), *Journal of Central South University*: Prediction and optimization of process parameter of friction stir welded AA5083-H111 aluminum alloy using response surface methodology, Vol. 19, pp. 1–8.

Pilchak, A. L., Tang, W., Sahiner, H., Reynolds, A. P. and Williams, J. C. (2011), *Metall. Mater. Trans. A*: Microstructure evolution during friction stir welding of mill-annealed Ti–6Al–4V, Vol. 42A, pp. 745–762.

Rajakumar, S. and Balasubramanian, V. (2012), *Mater. Design*: Establishing relationships between mechanical properties of aluminium alloys and optimized friction stir welding process parameters, Vol. 40, pp. 17–35.

Rajakumar, S., Balasubramanian, V. and Razalrose, A. (2013a), *Mater. Design*: Friction stir and pulsed current gas metal arc welding of AZ61A magnesium alloy: A comparative study, Vol. 49, pp. 267–278.

Rajakumar, S., Muralidharan, C. and Balasubramanian, V. (2010), *Trans. Nonferrous Met. Soc. China*: Establishing empirical relationships to predict grain size and tensile strength of friction stir welded AA 6061-T6 aluminium alloy joints, Vol. 20, pp. 1863–1872.

Rajakumar, S., Muralidharan, C. and Balasubramanian, V. (2011a), *Mater. Design*: Predicting tensile strength, hardness and corrosion rate of friction stir welded AA6061-T6 aluminium alloy joints, Vol. 32, pp. 2878–2890.

Rajakumar, S., Muralidharan, C., Balasubramanian, V. (2011b), *Mater. Design*: Influence of friction stir welding process and tool parameters on strength properties of AA7075-T6 aluminium alloy joints, Vol. 32, pp. 535–549.

Rajakumar, S., Razalrose, A. and Balasubramanian, V. (2013b), *Int. J. Adv. Manuf. Technol.*: Friction stir welding of AZ61A magnesium alloy A parametric study, Vol. 68, pp. 277–292.

Record, J. H., Covington, J. L., Nelson, T. W., Sorensen, C. D. and Webb, B. W. (2007), *Weld. J.*: A look at the statistical identification of critical process parameters in friction stir welding, Vol. 4, pp. 97–103.

Ren, S. R., Ma, Z. Y. and Chen, L. Q. (2007), *Scripta Mater.*: Effect of welding parameters on tensile properties and fracture behavior of friction stir welded Al–Mg–Si alloy, Vol. 56, pp. 69–72.

Rhodes, C. G., Mahoney, M. W., Bingel, W. H. and Calabrese, M. (2003), *Scripta Mater.*: Fine-grain evolution in friction-stir processed 7050 aluminum, Vol. 48, pp. 1451–1455.

Rhodes, C. G., Mahoney, M. W., Bingel, W. H., Spurling, R. A. and Bampton, C. C. (1997), *Scripta Mater.*: Effects of friction stir welding on microstructure of 7075 aluminum, Vol. 36, pp. 69–75.

Robson, J. D., Upadhyay, P. and Reynolds, A. P. (2010), *Sci. Technol. Weld. Joining*: Modelling microstructural evolution during multiple pass friction stir welding, Vol. 15, pp. 613–618.

Rodrigues, D. M., Loureiro, A., Leitao, C., Leal, R. M., Chaparro, B. M. and Vilaca, P. (2009), *J. Mater. Design*: Influence of friction stir welding parameters on the micro-structural and mechanical properties of AA 6016-T4 thin welds, Vol. 30, pp. 1913–1921.

Rose, R., Manisekar, K. and Balasubramanian, V. (2012), *J. Mater. Eng. Perform.*: Influences of welding speed on tensile properties of friction stir welded AZ61A magnesium alloy, Vol. 21, pp. 257–265.

Saeid, T., Abdollah-zadeh, A., Assadi, H. and Ghaini, F. M. (2008), *Mater. Sci. Eng. A*: Effect of friction stir welding speed on the microstructure and mechanical properties of a duplex stainless steel, Vol. 496, pp. 262–268.

Sakthivel, T. and Mukhopadhyay, J. (2007), *J. Mater. Sci.*: Microstructure and mechanical properties of friction stir welded copper, Vol. 42, pp. 8126–8129.

Sakthivel, T., Senegar, G. S. and Mukhopadhyay, J. (2009), *Int. J. Adv. Manuf. Technol.*: Effect of welding speed on micro-structure and mechanical properties of friction-stir welded aluminum, Vol. 43, pp. 468–473.

Salem, H. G. (2003), *Scripta Mater.*: Friction stir weld evolution of dynamically recrystallized AA2095 weldments, Vol. 49, pp. 1103–1110.

Santos, T. G., Vilaça, P. and Miranda, R. M. (2011), *J. Mater. Process. Technol.*: Electrical conductivity field analysis for evaluation of FSW joints in AA6013 and AA7075 alloys, Vol. 211, pp. 174–180.

Santos, T., Vilaça, P. and Quintino, L. (2007), Developments in NDT focusing defects in FSW of aluminium alloys. IIW Doc III-1426-07.

Sarsilmaz, F. and Caydas, U. (2009), *Int. J. Adv. Manuf. Tech.*: Statistical analysis on mechanical properties of friction stir welded AA1050/AA5083 couples, Vol. 43, pp. 248–255.

Scialpi, A., De Filippis, L. A. C. and Cavaliere, P. (2007), *J. Mater. Design*: Influence of shoulder geometry on microstructure and mechanical properties of friction stir welded 6082 aluminum alloy, Vol. 28, pp. 1124–1129.

Scialpi, A., Giorgi, M. D., Filippis, L. A. C. D., Nobile, R. and Panella, F. W. (2008), *Mater. Design*: Mechanical analysis of ultra-thin friction stir welding joined sheets with dissimilar and similar materials, Vol. 29, pp. 928–936.

Serio, L. M., Palumbo, D., Galietti, U., Filippis, L. A. C. and Ludovico, A. D. (2016), Nondestructive testing and evaluation: Monitoring of the friction stir welding process by means of thermography, Vol. 31, No. 4, pp. 371–383.

Shahri, M. M. and Sandstrom, R. (2012), *J. Mater. Process. Technol.*: Influence of fabrication stresses on fatigue life of friction stir welded aluminium profiles, Vol. 212, pp. 1488–1494.

Shanmuga Sundaram, N. and Murugan, N. (2010), *Mater. Design*: Tensile behavior of dissimilar friction stir welded joints of aluminium alloys, Vol. 31, pp. 4184–4193.

Sharma, C., Dwivedi, D. K. and Kumar, P. (2012), *Mater. Design*: Effect of welding parameters on microstructure and mechanical properties of friction stir welded joints of AA7039 aluminum alloy, Vol. 36, pp. 379–390.

Shi, L., Wu, C. S. and Liu, X. C. (2015), *J. Mater. Process. Technol.*: Modeling the effects of ultrasonic vibration on friction stir welding, Vol. 222, pp. 91–102.

Silva, A. A. M., Arruti, E., Janeiro, G., Aldanondo, E., Alvarez, P. and Echeverria, A. (2011), *Mater. Design*: Material flow and mechanical behaviour of dissimilar AA2024-T3 and AA7075-T6 aluminium alloys friction stir welds, Vol. 32, pp. 2021–2027.

Simar, A., Brechet, Y., Meester, B., Denquin, A. and Pardoen, T. (2008), *Mater. Sci. Eng. A*: Microstructure, local and global mechanical properties of friction stir welds in aluminium alloy 6005A-T6, Vol. 486, pp. 85–95.

Simoncini, M. and Forcellese, A. (2012), *Mater. Design*: Effect of the welding parameters and tool configuration on micro- and macro-mechanical properties of similar and dissimilar FSWed joints in AA5754 and AZ31 thin sheets, Vol. 41, pp. 50–60.

Sinha, P., Muthukumaran, S., Sivakumar, R. and Mukherjee, S. K. (2008), *Int. J. Adv. Manuf. Technol.*: Condition monitoring of first mode metal transfer in friction stir welding by image processing techniques, Vol. 36, pp. 484–489.

Song, K. H., Tsumura, T. and Nakata, K. (2009), *Mater. Trans.*: Development of microstructure and mechanical properties in laser-FSW hybrid welded Inconel 600, Vol. 50, No. 7, pp. 1832–1837.

Soundararajan, V., Atharifar, H. and Kovacevic, R. (2006), *Proc. Inst. Mech. Eng. Part B: J. Eng. Manuf.*: Monitoring and processing the acoustic emission signals from the friction stir welding process, Vol. 220, pp. 1673–1685.

Squeo, E. A., Bruno, G., Guglielmotti, A. and Quadrini, F. (2007), The Annals of "Dunărea de jos" University of Galaţi Fascicle V: Friction stir welding of polyethylene sheets, pp. 241–246.

Srinivasan, P. B., Arora, K. S., Dietzel, W., Pandey, S. and Schaper, M. K. (2010), *J. Alloy. Compd.*: Characterization of microstructure, mechanical properties and corrosion behaviour of an AA2219 friction stir weldment, Vol. 492, pp. 631–637.

Steuwer, A., Barnes, S. J., Altenkirch, J., Johnson, R. and Withers, P. J. (2011a), *Metall. Mater. Trans. A*: Friction stir welding of HSLA-65 Steel: Part II. The influence of weld speed and tool material on the residual stress distribution and tool wear, Vol. 43, No. 7, pp. 2356–2365.

Steuwer, A., Dumont, M., Altenkirch, J., Birosca, S., Deschamps, A., Prangnell, P. B. and Withers, P. J. (2011b), *Acta Mater.*: A combined approach to microstructure mapping of an Al–Li AA2199 friction stir weld, Vol. 59, pp. 3002–3011.

Su, J. Q., Nelson, T. W. and Sterling, C. J. (2005), *Mater. Sci. Eng. A*: Microstructure evolution during FSW/FSP of high strength aluminum alloys, Vol. 405, pp. 277–286.

Subbaiah, K., Geetha, M., Govindaraju, M. and Rao, S. R. K. (2012), *T. Indian Inst. Metals*: Mechanical properties of friction stir welded cast Al–Mg–Sc alloys, Vol. 65, pp. 155–158.

Suhuddin, U. F. H., Fischer, V. and dos Santos, J. F. (2013), *Scripta Mater.*: The thermal cycle during the dissimilar friction spot welding of aluminum and magnesium alloy, Vol. 68, pp. 87–90.

Tansel, I. N., Demetgul, M., Okuyucu, H. and Yapici, A. (2010), *Int. J. Adv. Manuf. Technol.*: Optimizations of friction stir welding of aluminum alloy by using genetically optimized neural network, Vol. 48, pp. 95–101.

Thomas, W. M., Nicholas, E. D., Needhman, J. C., Murch, M. G., Temple-Smith, P., Dawes, C. J. (1991), International patent application PCT/GB92/02203 and GB patent application 9125978.8, UK Patent office, London.

Thomas, W. M., Norris, I. M., Staines, D. J. and Lucas, W. (2005), The evaluation of root defects in FSW by 'through-hole' impact testing-preliminary studies. http://www.twi.co.uk/technical-knowledge/published-papers/the-evaluation-of-root-defects-in-fsw-by-through-hole-impact-testing-preliminary-studies-july-2005/, date accessed: 13/05/2013.

Toktas, A. and Toktas, G. (2012), *J. Mater. Eng. Perform.*: Effect of welding parameters and aging process on the mechanical properties of friction stir-welded 6063-T4 Al alloy, Vol. 21, pp. 936–945.

Tozaki, Y., Uematsu, Y. and Tokaji, K. (2007), *Int. J. Mach. Tool Manuf.*: Effect of tool geometry on microstructure and static strength in friction stir spot welded aluminium alloys, Vol. 47, pp. 2230–2236.

Tsujikawa, M., Chung, S. W., Morishige, T., Chiang, L. F. and Takigawas, Y. (2007), *Mater. Trans.*: Microstructural evolution of friction stir processed cast Mg–5.9 mass%Y–2.6 mass%Zn alloy in high temperature deformation, Vol. 48, pp. 618–621.

Venkateswaran, P. and Reynolds, A. P. (2012), *Mater. Sci. Eng. A*: Factors affecting the properties of friction stir welds between aluminum and magnesium alloys, Vol. 545, pp. 26–37.

Venkateswarlu, G., Devaraju, D., Davidson, M. J., Kotiveerachari, B. and Tagore, G. R. N. (2013), *Mater. Design*: Effect of overlapping ratio on mechanical properties and formability of friction stir processed Mg AZ31B alloy, Vol. 45, pp. 480–486.

Vijay, S. J. and Murugan, N. (2010), *Mater. Design*: Influence of tool pin profile on the metallurgical and mechanical properties of friction stir welded Al–10 wt.% TiB2 metal matrix composite, Vol. 31, pp. 3585–3589.

Wei, L. Y. and Nelson, T. W. (2011), *Weld. J.*: Correlation of microstructures and process variables in FSW HSLA-65 steel, Vol. 5, pp. 95–101.

Wei, S., Hao, C. and Chen, J. (2007), *Mater. Sci. Eng. A*: Study of friction stir welding of 01420 aluminum-lithium alloy, Vol. 452–453, pp. 170–177.

Xie, G. M., Ma, Z. Y., Geng, L. and Chen, R. S. (2007), *Mater. Sci. Eng. A*: Microstructural evolution and mechanical properties of friction stir welded Mg–Zn–Y–Zr alloy, Vol. 471, pp. 63–68.

Xue, P., Ni, D. R., Wang, D., Xiao, B. L. and Ma, Z. Y. (2011a), *Mater. Sci. Eng. A*: Effect of friction stir welding parameters on the microstructure and mechanical properties of the dissimilar Al–Cu joints, Vol. 528, pp. 4683–4689.

Xue, P., Xiao, B. L., Zhang, Q. and Ma, Z. Y. (2011b), *Scripta Mater.*: Achieving friction stir welded pure copper joints with nearly equal strength to the parent metal via additional rapid cooling, Vol. 64, pp. 1051–1054.

Yaduwanshi, D. K., Bag, S. and Pal, S. (2016), *Mater. Design*: Numerical modeling and experimental investigation on plasma-assisted hybrid friction stir welding of dissimilar materials, Vol. 92, pp. 166–183.

Yan, D. P., Chen, Z. W. and Littlefair, G. (2010), *FSW of Thick Section 6061 Al Alloys Using Scroll Shoulder Tool: Tool Design, Weld Zone Forming Mechanism and Weld Quality Control*, LAP Lambert Academic Publishing.

Yang, Y., Kalya, P., Landers, R. G. and Krishnamurthy, K. (2008), *Int. J. Mach. Tool Manuf.*: Automatic gap detection in friction stir butt welding operations, Vol. 48, pp. 1161–1169.

Yeni, C., Sayer, S., Ertugrul, O. and Pakdil, M. (2008), *Arch. Mater. Sci. Eng.*: Effect of post weld-aging on the mechanical and micro-structural properties of friction-stir welded aluminum alloy 7075, Vol. 34, pp. 105–109.

Zadeh, A., Saeid, T. and Sazgari, B. (2008), *J. Alloy. Compd.*: Microstructural and mechanical properties of friction stir welded aluminum/copper lap joints, Vol. 460, pp. 535–538.

Zadpoor, A. A., Sinke, J. and Benedictus, R. (2010), *Metall. Mater. Trans. A*: Global and local mechanical properties and microstructure of friction stir welds with dissimilar materials and/or thicknesses. Vol. 41, No. 13, pp. 3365–3378.

Zahmatkesh, B. and Enayati, M. H. (2010), *Mater. Sci. Eng. A*: A novel approach for development of surface nanocomposite by friction stir processing, Vol. 527, pp. 6734–6740.

Zahmatkesh, B., Enayati, M. H. and Karimzadeh, F. (2010), *Mater. Design*: Tribological and microstructural evaluation of friction stir processed Al2024 alloy, Vol. 31, pp. 4891–4896.

Zhang, H. J., Liu, H. J. and Yu, L. (2012a), *J. Mater. Eng. Perform.*: Effect of water cooling on the performances of friction stir welding heat-affected zone, Vol. 21, No. 7, pp. 1182–1187.

Zhang, Z., Liu, Y. L. and Chen, J. T. (2009), *Int. J. Adv. Manuf. Technol.*: Effect of shoulder size on the temperature rise and the material deformation in friction stir welding, Vol. 45, pp. 889–895.

Zhang, Z., Xiao, B. L. and Ma, Z. Y. (2012b), *J. Mater. Sci.*: Effect of welding parameters on microstructure and mechanical properties of friction stir welded 2219Al-T6 joints, Vol. 47, pp. 4075–4086.

Zhou, L., Liu, H. J. and Liu, Q. W. (2010), *J. Mater. Sci.*: Effect of process parameters on stir zone microstructure in Ti–6Al–4V friction stir welds, Vol. 45, pp. 39–45.

23 Laser Welding of NiTinol Shape Memory Alloy

An Overview

Thangaraju Deepan Bharathi Kannan, Paulraj Sathiya, and Thillaigovindan Ramesh

CONTENTS

23.1 INTRODUCTION

Shape memory alloys belong to a class of smart materials that have the ability to respond to the change in temperature and mechanical load. Early in 1951, Chang and Read first noted the shape memory behavior in AuCd alloys. Cu-based alloys such as Cu–Zn–Al, Cu–Al–Ni, FePt, and FePd are a few of the recently developed shape memory alloys along with NiTinol. NiTinol was discovered in 1965 by Buehler and Wiley at the Naval Ordnance Laboratory, USA (Deepan Bharathi Kannan et al. 2016; Elahinia et al. 2012; Mohd Jani et al. 2014; Van der Wijst 1992). NiTinol stands for its constituents nickel, titanium, and Naval Ordnance Laboratory, and it has three unique properties, namely, shape memory effect, superelasticity, and a very good biocompatibility. Because of the excellent corrosion resistance, NiTinol is the only shape memory alloy used as implant in the human body. NiTinol exists in two phases: austenite and martensite. Austenite is a high-temperature stable phase and has body-centered cubic crystal structures. Martensite is stable at low temperature, exists in monoclinic crystal structure, and is softer than the austenite phase (Mihalcz 2001; Sun and Huang 2009). The transformation of martensite from austenite involves diffusionless, ordered movement of atoms and it involves lattice distortion. The relative position of the atoms does not change during martensite transformation. The martensite transformation occurring in NiTinol alloys differs in a wide range of aspects from the martensite transformation occurring in ferrous-based alloys. In the martensite phase of NiTinol, the nature of alloying is substitutional in nature and the hardness is comparatively lower than the parent austenite phase.

The four important temperatures with respect to NiTinol are as follows:

i. Austenite start temperature (A_s)—It is the temperature at which the austenite phase starts to form.
ii. Austenite finish temperature (A_f)—It is the temperature at which austenite formation is complete and the NiTinol exists fully in the austenite phase.
iii. Martensite start temperature (M_s)—It is the temperature at which the martensite phase starts to form on cooling.
iv. Martensite finish temperature (M_f)—It is the temperature at which martensite phase formation is complete and the NiTinol exists fully in the martensite phase (Chan et al. 2011).
v. In addition to the martensite and austenite phases, NiTinol shows an intermediate phase called R-phase. While cooling NiTinol from austenite, instead of converting into martensite, sometimes R-phase is formed. The lattice distortion taking place during austenite to R-phase is very small compared with lattice distortion associated with martensite transformation. The crystal structure and lattice parameters of phases present in NiTinol are given in Table 23.1.

The shape memory effect and superelasticity behavior of NiTinol are mainly attributed to the phase transformations occurring between austenite and martensite. In NiTinol, it is generally possible to have three types of shape memory effect, namely, one-way memory effect (OWME), two-way memory effect (TWME), and all-round memory effect (ARME). Shape memory effect is the ability of the material to revert to its original shape on the application of heat. The pictorial representation of the OWME is shown in Figure 23.1.

It can be seen from Figure 23.1 that the material in the initial position has twinned martensite, and on application of load, the material deforms by transforming into detwinned martensite. On the application of heat (i.e., the temperature should be at least equal to austenite finish temperature), the material gets back to its original shape as a phase change occurs from detwinned martensite to austenite phase. The material again transforms into the martensite phase on cooling to martensite finish temperature. Similarly, the NiTinol can be trained by some thermomechanical treatment to remember its shape in both the martensite phase and austenite phase, and it is possible through the introduction of irreversible strain in the unstable martensite phase. This phenomenon is called TWME (Elahinia et al. 2012). The shape change in both the phases can be done by cooling and heating and without the need to apply external load. Training for TWME in NiTinol can be done easily if the grain size is coarser, and with a decrease in grain size, training becomes a more tedious task (Elahinia et al. 2004). The pictorial representation of TWME is shown in Figure 23.2.

ARME is observed in Ni-rich NiTinol and it is achieved through a diffusion-based training process. The pictorial representation of the ARME is shown in Figure 23.3. ARME involves the following steps:

i. Aging (at 800°C) followed by quenching.
ii. Putting the sample in a stainless tube and heating at a temperature between 400°C and 500°C.

TABLE 23.1
Crystallographic Information of NiTinol Phases

Sl. No.	Phase	Crystal Structure	Lattice Parameters
1	Austenite	Cubic B2	a = 3.015 Å
2	Martensite	Monoclinic	a = 2.98 Å, b = 4.108 Å, c = 4.646 Å, β = 97.78°
3	R-phase	Trigonal	–

Source: Adharapurapu, R. R. 2007. Phase transformation in nickel–rich nickel–titanium alloys: Influence of strain rate, temperature, thermomechanical treatment and nickel composition on the shape memory and superelastic characteristics. University of California, San Diego.

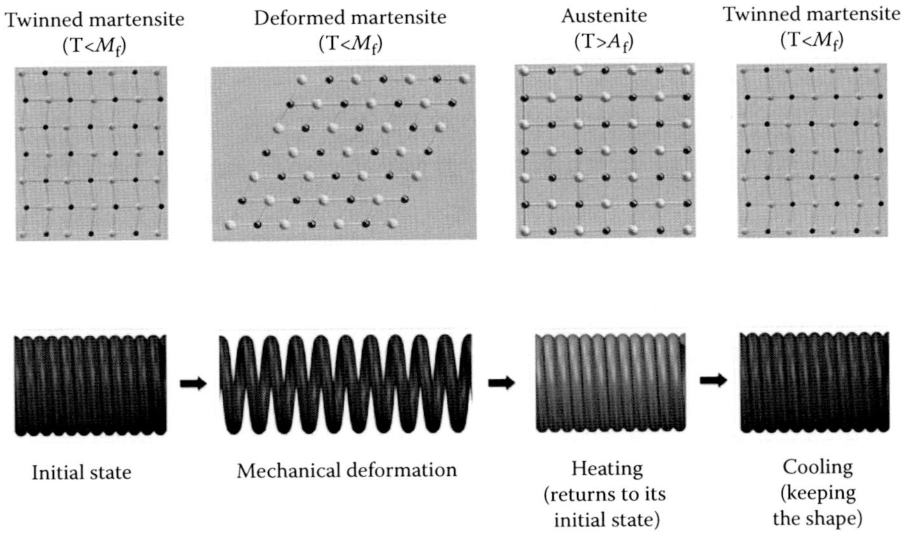

FIGURE 23.1 Pictorial representation of NiTinol's shape memory behavior.

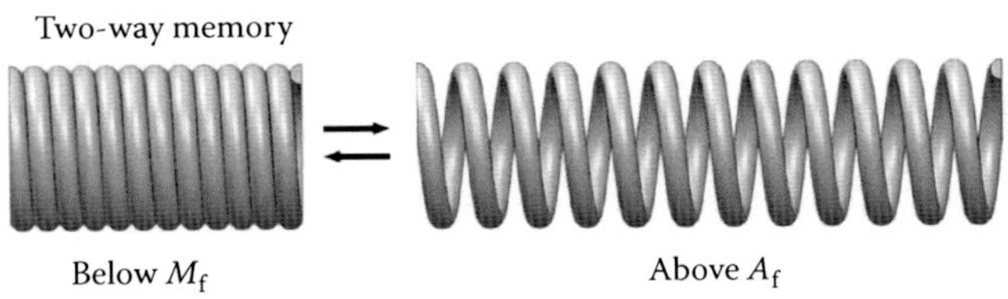

FIGURE 23.2 Two-way shape memory effect.

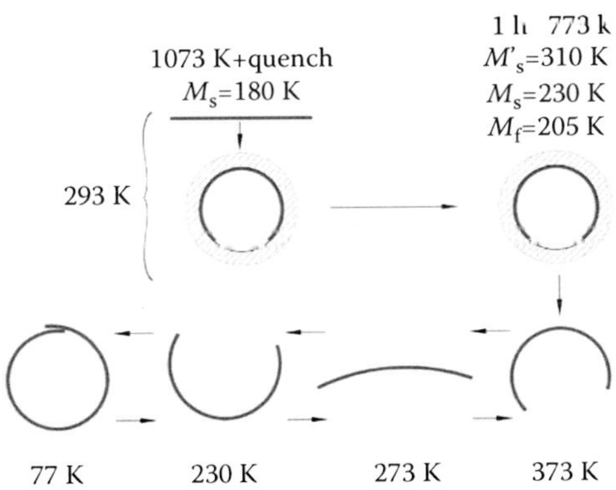

FIGURE 23.3 All-round memory effect.

iii. Cooling the sample at a temperature greater than the R-phase starting temperature. This step results in the sample having a curved shape.
iv. Cooling between 30°C and −70°C makes the sample curl in the opposite direction.
v. Heating above 30°C, the sample retracts back to the shape obtained after putting it in the stainless tube, and this transformation between the shapes can be repeated (Mirshekari et al. 2016).

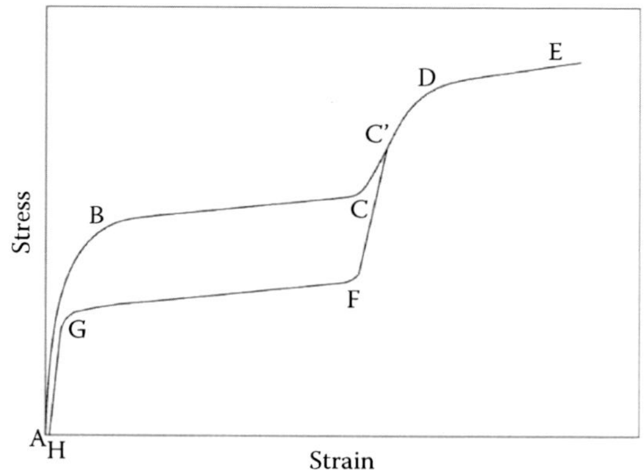

FIGURE 23.4 Pseudoelasticity behavior of NiTinol.

Another important property of NiTinol is pseudoelasticity or superelasticity. At temperatures above A_f, NiTinol can be stress induced to form the martensite phase, and it can be reversed back to the austenite phase upon unloading. This phenomenon is called pseudoelasticity, and it is possible to recover a strain rate of approximately 10%, which is significantly more than all the known metals or alloys. The maximum temperature above which martensite can no longer be stress induced is called the martensite deformation limit temperature (M_d); NiTinol behaves like an ordinary material above this temperature (Duerig and Pelton 1994). The pictorial representation of pseudoelasticity is shown in Figure 23.4.

From Figure 23.4, it can be understood that the portion AB denotes the elastic deformation of the austenite phase. On reaching B, martensite starts to form, and on reaching point C, martensite formation will be complete. If the load is continuously applied beyond C, elastic deformation continues, and on reaching point D, plastic deformation starts and it starts to fracture at point E. If the load is released, the material starts to revert, and at point F, few traces of austenite start to form and it is completed on reaching point G. The GH region denotes the parent phase elastic unloading.

23.1.1 Applications of Shape Memory Alloys

Owing to the shape memory effect, superelasticity, and a very good biocompatibility, the applications of NiTinol extend to many fields such as automobile, aerospace, medical, fashion, and robotics industries. With the need to develop automobiles with better fuel efficiencies, many researchers are concentrating on replacing the conventional materials with new modern materials such that it can reduce the overall weight. The shape memory behavior makes NiTinol one of the ideal candidates to be used as a sensor and actuator in many automobile parts. The mechanical compactness and simplicity of NiTinol shape memory alloys not only reduce the overall weight of the automobiles but also help in achieving better performance (Butera 2008; Butera et al. 2007; Stoeckel 1990; Stoeckel and Tinschert 1991). In almost all automobile applications, NiTinol is used as a linear actuator, but in recent times, efforts are also being made to utilize NiTinol in other areas such as aerodynamic and aesthetics applications. During the mid-1990s, General Motors and Chevrolet were constantly utilizing NiTinol for manufacturing many automobile components. Chevrolet Corvette is the first vehicle to come out in market with a shape memory alloy actuator for easier trunk lid closing (GM 2013; Suzuki 1986; Zychowicz 1992). The major problems in extending NiTinol's application in automobile industries are transformation temperature and compatibility with batteries used in the automobiles (Hodgson et al. 1990). Rear view mirror folding, lock/latch control, and climate control flap adjustment are some of the other examples where NiTinol is preferred in automobile industries.

In aerospace applications, NiTinol is used as actuators, manipulators, and inflatable structures, and in addition, it is also used in low shock release mechanisms, rotor blade tip morphing, and vibration damper; even a small unmanned aircraft, which is completely driven by devices made of NiTinol, is created (Carpenter and Lyons 2001; Cleveland 2008; Godard et al. 2003; Humbeeck 1999; Lortz and Tang 1998; McDonald Schetky 1991; Singh and Chopra 2002). This unmanned aircraft has the capability to withstand any aerodynamic pressure under any weather conditions (Huettl and Willey 2000; Long and Vezain 1998).

NiTinol's robotic applications are biologically inspired. NiTinol is widely used in such applications mainly because of its size and shape, which, in turn, help in achieving more degrees of freedom. NiTinol is mainly used as micro-actuators and artificial muscles, and its usage was drastically increased after 1990. Micro-grippers are being developed with the help of NiTinol, which can activate wirelessly and without internal power source (Caldwell and Taylor 1988; Fujita 1989; Kuribayashi 1989). BATMAV and Bat Robot are some of the flying robots developed with the help of NiTinol shape memory alloys. A NiTinol actuator wire with a thickness of just 1/100 inch has the capability to lift 2 pounds (Dahlgren and Gelbart 2009; Honma et al. 1985; Mohamed Ali and Takahata 2010; Sreekumar et al. 2007).

NiTinol's contribution to the biomedical field is immense and is mainly due to its excellent biocompatibility. It has better properties than other biomaterials such as stainless steel. In stainless steel, the maximum recoverable strain is 1%, whereas in NiTinol, it is about 8%. A comparison of the physical and mechanical properties of NiTinol and stainless steel is shown in Table 23.2.

NiTinol deformation behavior is similar to that of human bone, which makes it an ideal candidate for the manufacture of bone implants. The stress–strain curve for NiTinol, stainless steel, and human bone is shown in Figure 23.5.

Another advantage of NiTinol is that the shape of the implant can be changed even after implantation with the help of body temperature. NiTinol is used in almost all medical fields, including orthodontic (braces, palatal arches, files), orthopedic (head, bone, spine, legs), and vascular (aorta, arteries, vena cava filter, vessels, valves) areas, to name a few. Biomedical instruments such as catheters, scopes, suture, stents, and so on are also made with the help of NiTinol shape memory alloys (Lim et al. 1996; Pfeifer et al. 2013; Terzo 2006). Plastic, reconstructive, and aesthetic surgery equipments are also manufactured using NiTinol shape memory alloys (Suzuki 1986; Zychowicz 1992). Recently, research is being carried out to utilize NiTinol for solving penile erectile dysfunction problems and work is being done to effectively control the erection at the required time

TABLE 23.2
Properties of NiTinol and Stainless Steel

Sl. No.	Property	NiTi	Stainless Steel
1	Recovered elongation	8%	0.8%
2	Biocompatibility	Excellent	Fair
3	Effective modulus	Approx. 48 GPa	193 GPa
4	Torqueability	Excellent	Poor
5	Density	6.45 g/cm^3	8.03 g/cm^3
6	Magnetic	No	Yes
7	Ultimate tensile strength (UTS)	Approx. 1240 MPa	760 MPa
8	Coefficient of thermal expansion Martensite Austenite	6.6 × 10^{-6} cm/cm/°C 11.0 × 10^{-6} cm/cm/°C	17.3 × 10^{-6} cm/cm/°C

Source: Elahinia, M., Hashemi, M., Tabesh, M., and Bhaduri, S. B. 2012. Manufacturing and processing of NiTi implants: A review. *Progress in Materials Science* 57: 911–946.

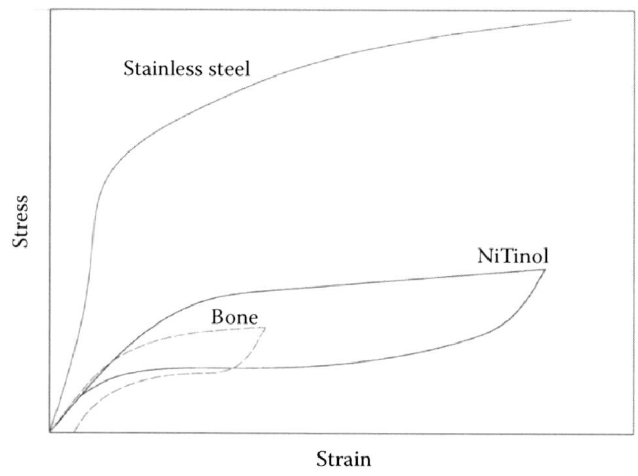

FIGURE 23.5 Comparison of stress–strain curve of NiTinol, stainless steel, and human bone.

(www.Sciencealert.com). The contribution of NiTinol to the fashion industry is getting bigger day by day. In Japan, women's innerwear is being manufactured with the help of NiTinol wire as it gives exceptional comfort and good durability. NiTinol is also utilized to manufacture special types of jackets that shrink upon reaching 45°C (this temperature can be changed as per the requirement). These jackets, which automatically shrink, are specifically designed for summer (www.edition.cnn.com.)

23.2 MANUFACTURING PROCESSES INVOLVED WITH NiTinol

Successful utilization of any material to a particular application depends on its processing capabilities, and NiTinol is no exception. Melting, casting, followed by primary and secondary metal working and heat treatment are the general manufacturing sequences followed with respect to manufacturing of NiTinol shape memory alloys. The major problems associated with manufacturing of NiTinol are controlling the Ni/Ti ratio and interstitial contamination by oxygen and carbon elements. Carbon presence in the molten NiTi leads to the formation of TiC, which might increase the Ni content and thus alter the transformation temperatures. Till now, some of the major manufacturing processes used for shaping NiTinol are casting (vacuum arc remelting, vacuum induction melting, and electron beam melting) and powder metallurgy through conventional methods, such as conventional sintering, spark plasma sintering, self-propagating high-temperature synthesis, and so on. Since Ti is more reactive with oxygen, it is recommended to use vacuum induction melting for primary alloy preparation followed by vacuum arc melting, which helps achieve microstructural homogenization. With the development of additive manufacturing techniques such as selective laser sintering, selective laser melting, laser-engineered net shaping, and so on, manufacturers prefer additive manufacturing over traditional casting processes. At present, a lot of work is being done in additive manufacturing of NiTinol, because it has the capability to produce porous NiTi components and intricate components required for biomedical applications. Machining of NiTinol shape memory alloys is a tedious task because of its higher hardness, which drastically increases the tool wear. Hence, nonconventional methods such as laser cutting, water jet machining, and plasma cutting are preferred for machining NiTinol shape memory alloys (Elahinia et al. 2012). Welding of NiTinol is comparatively a difficult task as NiTinol is easily susceptible to property variation when subjected to higher temperatures, and in most cases, mechanical options such as swaging and crimping are preferred for joining NiTinol. To date, NiTinol has been successfully joined using tungsten inert gas welding (Song et al. 2008), plasma arc welding (Schlossmacher et al. 1997), percussive arc welding

(Schlossmacher et al. 1994), resistance welding (Vieira et al. 2011), and laser welding processes. In some cases, heat treatment processes are also used to fine-tune the properties as per the requirement. Annealing of NiTinol will help increase the austenite transformation temperatures.

In this chapter, an attempt is made to discuss laser welding of NiTinol and the problems related to laser joining of NiTinol, methodologies followed in analyzing the quality of laser welded NiTinol samples, and future scope in laser joining of NiTinol.

The reasons for choosing laser welding for joining NiTinol are as follows:

i. Lesser distortion
ii. Higher power density
iii. Higher welding speed
iv. Lesser heat-affected zone

Laser welding is under the category of fusion welding processes, where the coalescence between the materials to be joined is done with the help of a coherent, monochromatic, and highly directional laser beam. After its discovery in 1960, laser has been widely used in almost all fields and has played a major role in the welding industries. Laser welding involves the following three steps: melting the workpiece to form the weld pool, allowing the weld pool to grow to the required size, and stabilizing the weld pool until solidification. Of all the various welding processes available, laser welding has the highest power density of the order 109 W/cm^2 (Sokolov and Salminen 2014). The main advantage of laser is its ability to concentrate a higher energy density in a small area. All the three types of laser (solid state, gas, and semiconductor) are employed in the joining process, and specifically, gas and solid-state lasers are widely used. Solid-state lasers make use of crystals such as ruby, sapphire, and some artificially doped crystals such as neodymium doped yttrium aluminium garnet (Nd:YAG) rods and ytterbium:yttrium aluminium garnet (Yb:YAG). Both Nd:YAG (1093 nm) and Yb:YAG (1030 nm) lasers are widely preferred in joining processes as they have a smaller wavelength, which, in turn, helps increase absorption. Laser welding is generally carried out in two modes, namely, continuous wave and pulsed mode. Continuous wave mode is either on or off, whereas pulsed lasers use individual pulses for creating weld. Pulsed laser mode is comparatively economical than continuous mode as it consumes lesser energy and it produces weld with a smaller heat-affected zone. In pulsed mode laser welding, the weld qualities are controlled by pulse energy (J), pulse width (ms), and spot size (mm), whereas in continuous wave mode, the weld qualities are controlled by laser power (W), welding speed (mm/min), and focal position. In both continuous wave and pulsed laser mode, shielding gases are used to protect the weld from oxidation. The selection of shielding gases depends on the materials used, and in most cases, argon or helium is preferred. In some works, the mixture of shielding gases, such as He + Ar, Ar + N_2, and so on, in different ratios is used in order to get the desired depth of penetration, bead width, and microstructure. The differences in weld quality with respect to different shielding gases used are mainly attributed to the different ionization potential, thermal conductivity, and density of the shielding gases. Shielding gas flow rate is expressed in liters per minute, and optimized flow rate has to be selected depending on the requirements. Now, importance is given for selecting the optimized shielding gas blown distance and shielding gas blown angle in order to avoid the formation of porosities. These porosities in turn affect the stability of the keyhole. The overall properties of the weld will be drastically reduced and hence care should be taken in selecting the proper shielding gas blown distance and blown angle.

Because of its higher power density, laser welding is carried out in keyhole mode. Welding processes such as gas tungsten arc welding and gas metal arc welding produce weld in conduction mode as a result of its lower power density. Keyhole mode welding helps improve the depth of penetration, and even thicker plates can be welded with a single pass. When laser strikes the surface to be welded, the surface not only melts but also forms a cavity that extends through the entire thickness of the workpiece and forms a keyhole. Figure 23.6 gives a clear idea of the shape of the keyhole made and the conduction mode obtained during laser joining of NiTinol.

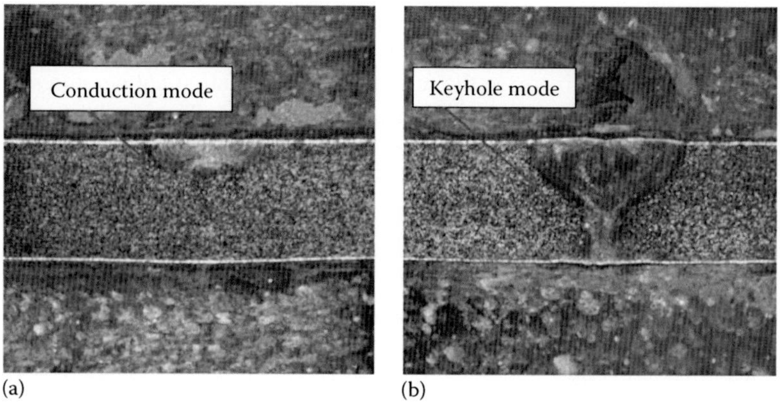

FIGURE 23.6 (a) Weld bead in conduction mode. (b) Weld bead obtained in keyhole mode.

Figure 23.6 shows the bead on plate welding of NiTinol body type NiTinol shape memory alloys having a thickness of 1 mm, and it can be seen that full penetration was not achieved in conduction mode, whereas in the keyhole mode, full penetration was achieved. The reason for the full penetration might be attributed to the higher density and good absorption of the laser. The major problem in laser welding is the reflection of the laser by the metal to be joined. Absorption is generally higher in gases followed by liquids and solids. In laser welding, once the laser strikes the surface, it vaporizes the surface. These vapors present in the top surface helps absorb more amounts of lasers and prevents reflection. In addition, the keyhole shape should be made stable, and it can be achieved by balancing the pressures acting to close and open the keyhole. The pressures that are acting in the keyhole are as follows:

i. Ablation pressure (P_{ap})—It will try to open the keyhole.
ii. Metallostatic/hydrostatic pressure (P_h)—It will try to close the keyhole.
iii. Capillary pressure (P_s/P_{cp})—It will try to close the keyhole.
iv. Convection-induced pressure (P_c)—It will try to close the keyhole.

In order to make the keyhole stable, the sum of pressures trying to open the keyhole should be equal to the sum of the pressures trying to close the keyhole. Of the four pressures acting in the keyhole, metallostatic pressure and convection-induced pressure values are very small and, hence, can be neglected. Thus, the ablation pressure should be equal to the capillary pressure. The graph explaining the conditions in order to obtain a stable keyhole is shown in Figures 23.7 and 23.8. From the graph, it can be seen that at two points A and B, the ablation pressure and capillary pressure are the same. At point A, once the keyhole size exceeds a_A, the keyhole collapses as the ablation pressure exceeds the capillary pressure, whereas in point B, the ablation pressure is equal to the capillary pressure, and even beyond point B, a stable keyhole is obtained as the ablation pressure is less than

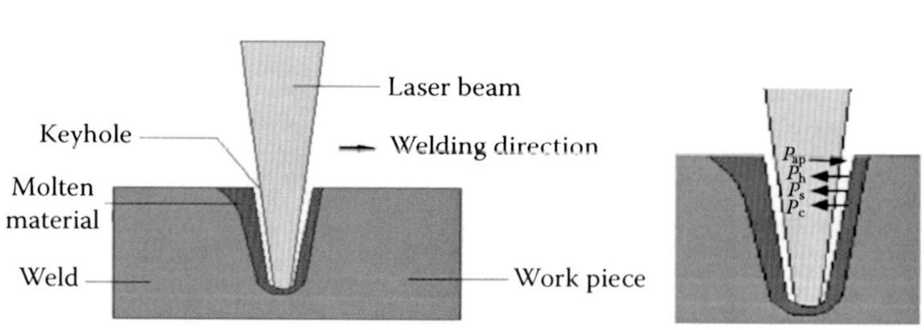

FIGURE 23.7 Schematic of laser welding processes with various pressures acting on the keyhole.

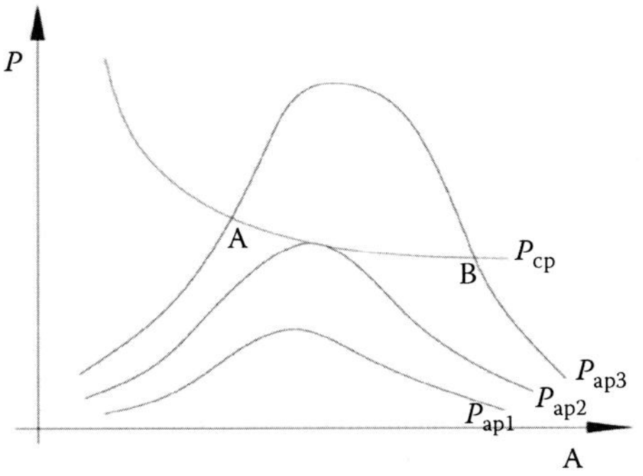

FIGURE 23.8 Conditions for keyhole stability in laser welding.

the capillary pressure. Generally, the keyhole will be stable when its size is 70% more than the heat source size. If the keyhole is not stable, pores will form in the weld, which, in turn, affects the mechanical and corrosion properties of the weld.

23.3 METALLURGICAL ASPECTS RELATED TO LASER WELDING OF NiTinol

In this section of the chapter, the microstructural variations of NiTinol with respect to the laser welding are discussed. A good understanding of the material property relation with the microstructure will help fine-tune the properties as per the requirement. The phase diagram of NiTinol is presented in Figure 23.9.

NiTinol exists as a single phase at equiatomic compositions. In Ni-rich solid solution, a eutectic reaction forming $TiNi_3$ was observed. Solubility on the Ni-rich side was found to be decreasing with a decrease in temperature. Ti_2Ni_3 is one of the intermetallic phases observed in the NiTinol phase diagram and many authors observed a peritectoid reaction at 625°C (Adharapurapu 2007). $TiNi_3$ and

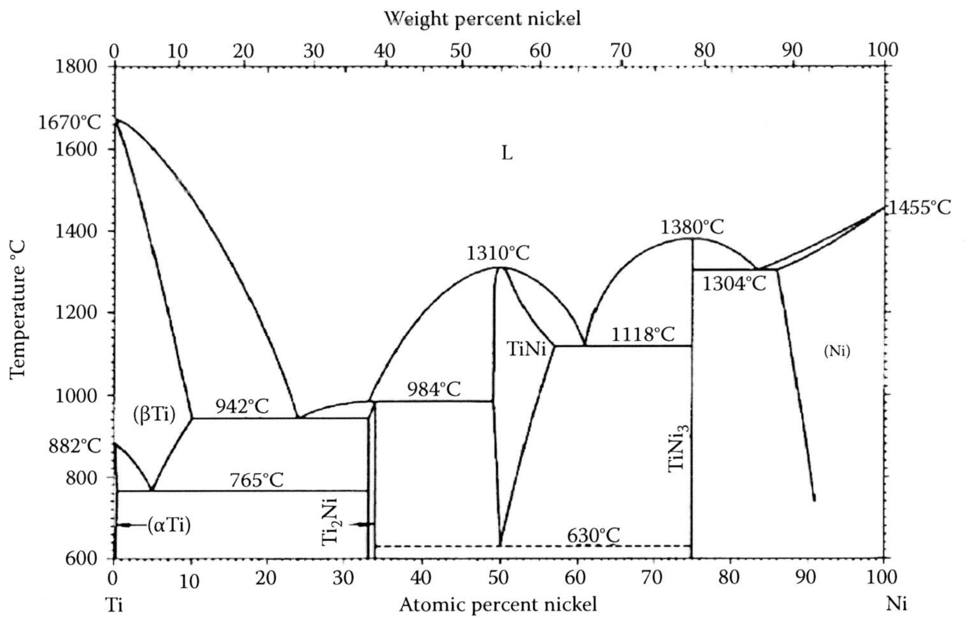

FIGURE 23.9 Phase diagram of NiTinol.

Ti_3Ni_4 are the two important phases found in the phase diagram of NiTi due to different phase reactions, namely, eutectoid and peritectoid reactions. The formation of different intermetallic phases such as Ti_3Ni_4, Ti_2Ni_3, and $TiNi_3$ in Ni-rich NiTinol (Ni -55% to 65%) by solution treatment, aging, and cold and hot working helped in tailoring the shape memory effect and superelasticity. Different precipitates are formed in NiTinol shape memory depending on aging temperature and time. Ti_3Ni_4 is formed at a small aging temperature and at a short period of time, Ti_2Ni_3 is formed at an intermediate aging temperature and at an intermediate period of time, whereas $TiNi_3$ is formed at a high aging temperature and at a long period of time. In order to prepare the samples for microstructural analysis, the weld beads have to be cut along the weld cross section and the samples have to be mounted based on cold mounting or hot mounting. The samples have to be polished with Sic sand papers of different grit size starting from 100, 200, 400, 600, 800, 1000, and 1200 (Deepan Bharathi Kannan et al. 2016) followed by cloth polishing using alumina powder in order to get a mirror finish. The etchant plays a vital role in revealing the microstructure and hence the right combination of etchant has to be selected. For NiTinol shape memory alloys, two combinations of etchants are used, namely, water + nitric acid + hydrofluoric acid in the ratio of 5:4:1 and acetic acid + nitric acid + hydrofluoric acid in the ratio of 2:5:1 (Deepan Bharathi Kannan et al. 2016). Sample preparation plays a vital role in getting clear and required microstructure features in NiTinol welded samples. The oxide layers present in NiTinol have the tendency to form cracks during solidification. Tam et al. (2011) recommended the cleaning of NiTinol sheet using acid solution in order to get clear microstructure images from the optical microscope and scanning electron microscope. The samples immersed in acid solution for an optimum time clearly revealed the microstructure as compared with samples that are not immersed in acid solution. The optimum time suggested for acid immersion was 20 s as prolonged immersion will deteriorate the thickness or diameter of the NiTinol samples, and thus, the immersion in acid solution has to be done with utmost care. The major parameters in laser welding that influenced the microstructure formation were laser power and welding speed. If the laser power was taken at a lower value, required penetration may not be achieved, and if the laser power was taken at a higher value, it might lead to vaporization of the NiTinol material. Similarly, the welding speed had to be selected at an optimum value as higher welding speeds may cause incomplete penetration and vaporization of NiTinol at lower welding speeds. The general features that were observed from the NiTinol's weld microstructure were shape, size of the grains, presence of intermetallic phases, and change in percentage composition of the elements. In most of the works related to laser welding of NiTinol, a transformation from planar to cellular to dendritic grains was observed during the transition from the weld interface to the weld center. The microstructures revealing different grain shapes and sizes in different weld zones of the laser welded NiTinol sample are shown in Figure 23.10.

This grain shape variation was mainly influenced by the ratio of temperature gradient (G) and growth rate (R). The variation of grain shape with respect to the G/R ratio is given in Figure 23.10, and it can be observed that when the G/R ratio was high, it led to planar mode, and when the G/R ratio was low, it led to the formation for dendritic grains. Similarly, the product of temperature gradient and growth rate, which is nothing but the cooling rate, controls the grain size in different zones of the weld. The GR value was generally low at the weld interface and it increased moving toward the weld center. Thus, the grains in the weld center were comparatively finer to those away from the weld centerline (Kou 2003). A pictorial representation of grain shape variation with respect to G and R is given in Figure 23.11.

Zoeram et al. (2015) studied the influence of welding speed on the microstructure of NiTinol and found that with the increase in welding speed, the amount of brittle intermetallic phases being formed in the weld was increasing. It was observed that the heat input to the weld decreased with an increase in welding speed. This decreased heat input to the weld increased the cooling rate leading to fine-grain formation. With an increase in welding speed, the Ti/Ni ratio increased, which, in turn, affected the properties of the weld. Song et al. (2008) welded NiTinol sheet in pulsed mode and found that current and impulse width had a major role in controlling the microstructures. Laser welding carried

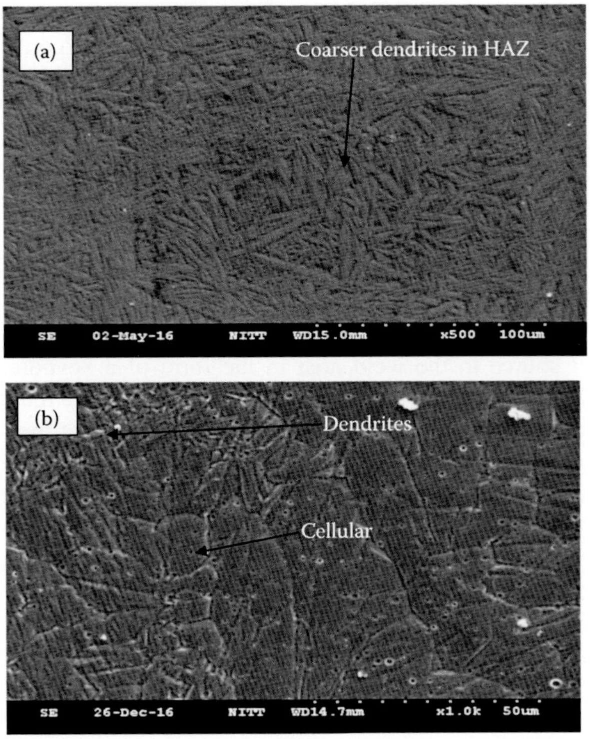

FIGURE 23.10 (a) Microstructure of laser welded NiTinol containing dendrites. (b) Microstructure of laser welded NiTinol showing different grain shapes.

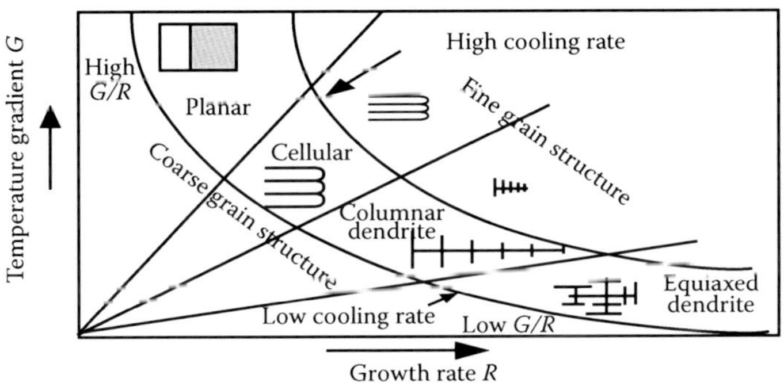

FIGURE 23.11 Effect of G and R on grain shape and size.

out at higher impulse width resulted in the formation of cracks. Chan et al. (2011) studied the influence of laser welding parameters and found that, comparatively, welding speed had a greater impact on the weld microstructure than the laser power. A finer grain size was obtained in the weld, when the welding speed was at a higher level. Selection of correct shielding gas for laser welding plays a vital role in protecting the weld from oxidation. In addition, flow rate has to be selected at an optimum value, and in most of the works related to laser welding of NiTinol, argon was used as shielding gas because it has low thermal conductivity compared with helium and nitrogen. The optimum argon gas flow rates recommended are between 12 and 14 L/min. The weld bead profiles in laser welding are highly influenced by material thermal capacity, power intensity, and total heat input and heat loss. Chan and Man (2010) joined NiTinol with Nd:YAG laser welding with two different shielding gases, namely, argon and helium. Welding carried out with argon shielding gas produced a higher aspect ratio compared to the weld aspect ratio obtained using helium as shielding gas. The reason might be attributed to the lower thermal conductivity of argon gas. Deepan Bharathi Kannan et al. (2017) studied the influence of shielding gas blown distance in laser welding of NiTinol and found that when the shielding gas blown distance was at 16 mm, the porosity formation was completely avoided and small pores were seen in the weld when the shielding gas blown distance was less than 16 mm. They recommended studying the influence of shielding gas blown angle on weld microstructure and mechanical properties. Chan et al. (2011) studied the bead profile change with respect to heat input to the weld and observed that bead profile shape changed from hemispherical to triangular to rectangular with increase in heat input, and they suggested transferring the energy from the power source to the weld area in the form of a keyhole shape. The amount of intermetallic phases present in the weld varied drastically with respect to the change in welding speed and power. Both speed and welding on the whole controlled the heat input supplied to the weld. When the heat input to the weld was lower, the amount of brittle intermetallic phases present in the weld was lower and vice versa.

Chan and Man (2010) observed that the weld center contained higher amounts of fine dendritic grains, and this was attributed to the small spot size and the laser's higher power energy density. With the extension of NiTinol's usage in a wide range of applications such as atomic energy, ocean development, medical device, and so on, the need to join NiTinol with other materials leads to further complications for designers and manufacturing engineers. Stainless steel is one of the main materials to which NiTinol is joined for the purpose of biomedical applications. Joining NiTinol with stainless steel by the traditional laser welding method produces a lot of defects due to the difference in thermal and mechanical properties. Joining NiTinol to stainless steel results in the formation of many intermetallic phases such as $TiNi_3$, $NiTi_2$, $TiNi_3$, and some brittle intermetallic phases such as TiFe and $TiFe_2$. The presence of brittle phases in the weld leads to crack formation, which, in turn, drastically affects the tensile strength. Ti present in NiTinol is chemically very active and readily reacts with other elements present in the weld during solidification. These problems can be overcome by the usage of some filler materials, and the filler materials should be selected in such a way that it helps retain the shape memory, biocompatibility, tensile strength, and ductility equal to the base metal. Rare earth elements are an ideal option for filler materials because they help solidify branch grains and retain the strength and toughness of stainless steel. Cerium is one of the rare earth metals. Zhao et al. (2008) welded NiTinol shape memory alloys using cerium as a filler material and observed that, except in fusion boundary, chemical compositions of the cerium welded NiTinol were almost the same as that of the weld obtained without the cerium filler material. Cerium welded NiTinol had some white and gray granule particles that were dispersed in the grain boundaries, whereas the weld obtained without cerium contained only some dark particles (Ni_3Ti) in addition to the parent austenite phase. Li et al. (2012) used nickel as a filler material to join NiTinol with stainless steel, and it was concluded that the thickness of the filler material used in laser welding had a greater impact on the microstructures. A trend of a decrease in the amount of brittle intermetallic phases was observed with an increase in thickness of the Ni filler material used. From the energy dispersion spectroscopy analysis, it was found that with an increase in Ni thickness, the amount of Ni content in the base metal increased, whereas Ti, Fe, and Cr content decreased. Li et al. (2013) recommended cobalt as a filler material for laser welding of NiTinol as it produced favorable microstructures. The thickness of the cobalt filler material should be higher in order to reduce the brittle intermetallic phase formation. In few of the works, the authors used filler materials for joining NiTinol and observed changes in microstructures that were advantageous in improving the properties. Zhao et al. (2010) compared the microstructural features of the laser welded NiTinol obtained using cerium, niobium, and nonadditive weld and observed that grains in the nonadditive weld were comparatively coarser than the grains in the additive weld, which indicated that both Ce and Nb helped reduce the formation of coarser grains. The percentage of Ce and Nb in the weld was approximately 3%, and they existed in the eutectic of Nb+NiTi and Ce+NiTi. The x-ray pattern of the additive and nonadditive weld was almost the same except for some oxide peaks that formed because of cerium. Post-weld heat treatment of laser welded NiTinol shape memory alloys increased the intensity of the brittle intermetallic phases and reduced the grain size. The XRD graphs showing the difference in intensities of the different phases present in the weld are presented in Figure 23.12a and b.

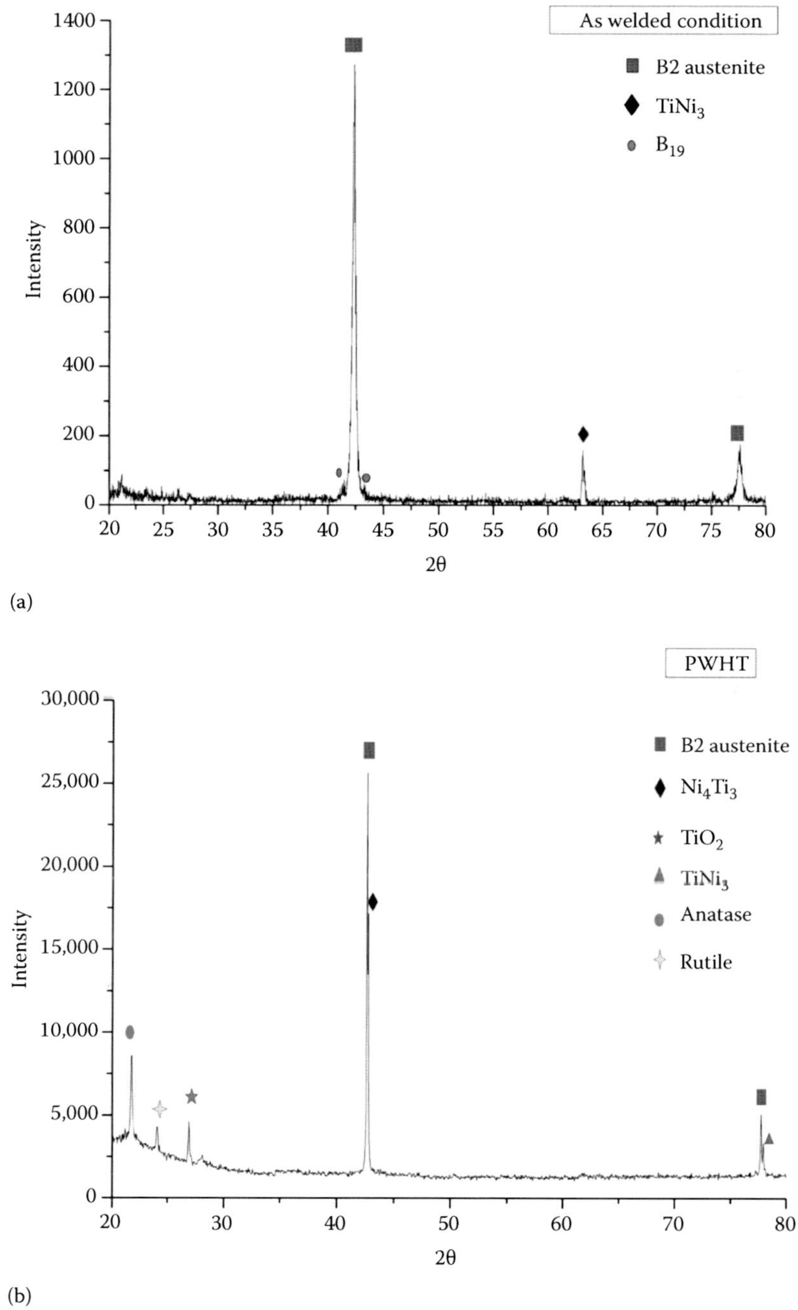

FIGURE 23.12 (a) XRD analysis on laser weld. (b) XRD analysis on PWHT NiTinol.

On the whole, from the above literatures, it is understood that laser welding is suitable for joining NiTinol shape memory alloys and it is possible to get the required microstructure by selecting the right combination of input parameters. In order to join NiTinol with other materials, it is recommended to use rare earth metals as filler materials. Welding speed and power are the two most important factors in laser welding that must be given utmost importance as they control microstructural features to a greater extent.

23.4 LASER WELDING EFFECT ON PHASE TRANSFORMATION TEMPERATURES

Differential scanning calorimetry (DSC) analysis is done on the NiTinol welded sample to get an idea regarding the transformation temperature of the two phases, namely, austenite and martensite start and finish temperatures, respectively. In addition to that, a clear idea regarding the presence of R-phase can be found from the DSC analysis. The presence of R-phase makes it difficult to identify martensite start temperatures. Shape memory effect also depends on the transformation temperatures of martensite and austenite phases. For example, in staple fixation of the human body, upon reaching human body temperature, the staple changes its shape by transforming into austenite. If importance is not given in laser welding of NiTinol for controlling transformation temperatures, it will not meet its intended purpose. Actuator efficiencies are also greatly affected by the modified phase transformation temperatures obtained in the welded NiTinol samples. During the initial production process of NiTinol, the following factors mainly control the phase transformation temperatures: Ni concentration, ternary alloying elements, precipitation in Ni-rich alloys, aging process, and presence of defects. The DSC analysis is done by using 20 g of the welded sample either in powdered form or in solid form. The temperature range for laser welded DSC analysis depends on the base metal, and it is generally varied at a rate of 5°C/min. From the graph obtained using DSC analysis, the phase transformation temperatures were calculated by the intersection of the tangent to the peak slope with the baseline. R-phase formation in the NiTinol samples can be avoided by annealing the samples. R-phase transformation occurs as a result of the formation of Ni-rich precipitates, the influence of cold work leading to high dislocation densities. The dislocations present in the laser welded NiTinol act as an obstacle for the direct transformation from austenite to martensite. Two-stage transformations are generally observed in Ni-rich NiTinol shape memory alloys, whereas multiphase transformation is observed in equiatomic NiTinol alloys. Phase transformation temperature in laser welded NiTinol specimens were mainly influenced by the following:

i. Grain size
ii. Presence of defects
iii. Change in Ni content
iv. Poor shielding of the weld

When the heat input to the weld was maximum, it led to a lower cooling rate, which, in turn, resulted in the formation of coarser grains. The presence of coarser grains reduced the number of nucleation sites for martensite formation, leading to a change in martensite transformation temperatures. The presence of defects and brittle intermetallic phases caused dislocation and creation of residual stress in the laser welded NiTinol and caused significant variations in the phase transformation temperatures. When the shielding of the weld was not done properly, the oxygen and hydrogen content in the weld increased significantly, altering the phase transformation temperatures by the phase reorientation phenomenon. Hence, importance should be given in selecting a suitable shielding gas, an optimum shielding gas flow rate, blown distance, and blown angle. Formation of phases like Ni_3Ti_4 during solidification of the weld led to a change in the nickel distribution, and the Ni/Ti ratio was altered. This change in Ni/Ti ratio significantly influenced the phase transformation temperatures. The low coherence stress field present around the Ni_3Ti_4 precipitates results in an austenite to R-phase to martensite transformation. Song et al. (2008) joined a NiTinol wire of two different compositions at different welding parameters and observed that the weld's phase transformation temperature was completely different from the base metals. Falvo et al. (2005) observed a significant variation in the weld's phase transformation temperatures and concluded that the reason for the change in phase transformation temperature after welding was that it had reset the influence of heat treatment and cold working during the production process. Annealing of NiTinol shape memory alloys helped prevent the formation of R-phase (Khan and Zhou 2010). Tuissi et al. (1999) chose two different state NiTinol shape memory alloys, namely, partially annealed and fully recrystallized. Laser weld obtained with a fully recrystallized state did not show the formation of R-phase, whereas the laser weld obtained under partially annealed conditions contained R-phase, and it was observed after cooling from the austenite phase. The appearance of increased heat flux before the martensite start temperature indicates that a lot of nucleation points are available for the formation of the martensite phase (Gugel et al. 2008). Gong et al. (2011) successfully applied laser welding for joining NiTinol, and the DSC analysis results showed that the annealed base metal and weld metal had similar phase transformation temperatures. The presence of Ti_3Ni_4 in the weld did not significantly affect the phase transformation temperatures. Post-weld heat treatment of the laser welded NiTinol had a greater impact on the phase transformation temperatures. The DSC curve obtained in the weld and post-weld heat-treated sample is shown in Figure 23.13a and b.

From Figure 23.13a and b, it can be seen that, after post-weld heat treatment, there was a slight variation in austenite start and finish temperatures, whereas, martensite start and finish temperatures were drastically changed, which might be attributed to the change in Ni content after post-weld heat treatment. Post-weld heat treatment of the NiTinol resulted in the formation of Ni_3Ti phases, which might influence the change in phase transformation temperatures. On the whole, it is understood that the phase transformation temperature played a vital role in designing a component. It is observed that annealing helped prevent the formation of R-phase. If the morphology and morphometry of the base metal and precipitates formed during laser welding were the same, then the R-phase did not form in the weld. The microstructural inhomogeneity in the weld due to the presence of the precipitates alone is not a sufficient condition for multistage transformation. In addition, the degree of supersaturation during solidification has to be low for multistage transformation to occur in NiTinol shape memory alloys. There was no single particular parameter that had a greater influence on the phase transformation temperatures, but as a whole, heat treatment significantly affected the phase transformation temperatures.

In a few works (Li et al. 2006), the shape memory effect of the laser welded NiTinol samples are analyzed with the help of a bend test. The weld joint is bent to 90° by applying load and load is released after 500 s and the corresponding angle is noted as θ1. The welded sample is then heated at 100°C, which will make the sample return to its original position. This angle is noted as θ2. With these two angles, the shape memory effect is calculated using the following formula (Deepan Bharathi Kannan et al. 2016):

$$\lambda = (\theta 1 - \theta 2/\theta 1) \times 100\%.$$

The pictorial representation of the bend test is shown in Figure 23.14.

23.5 LASER WELDING EFFECT ON MECHANICAL PROPERTIES

23.5.1 Tensile Strength

In this section, mechanical properties such as tensile strength and hardness of laser welded NiTinol are discussed in detail. Tensile test on NiTinol welded sample is done based on the standard ASTM E8. The standard tensile test specimen is shown in Figure 23.15.

In a few of the works, substandard sizes were also used for tensile test because NiTinol is quite expensive. In laser welding of NiTinol, tensile strength is mainly controlled by two parameters,

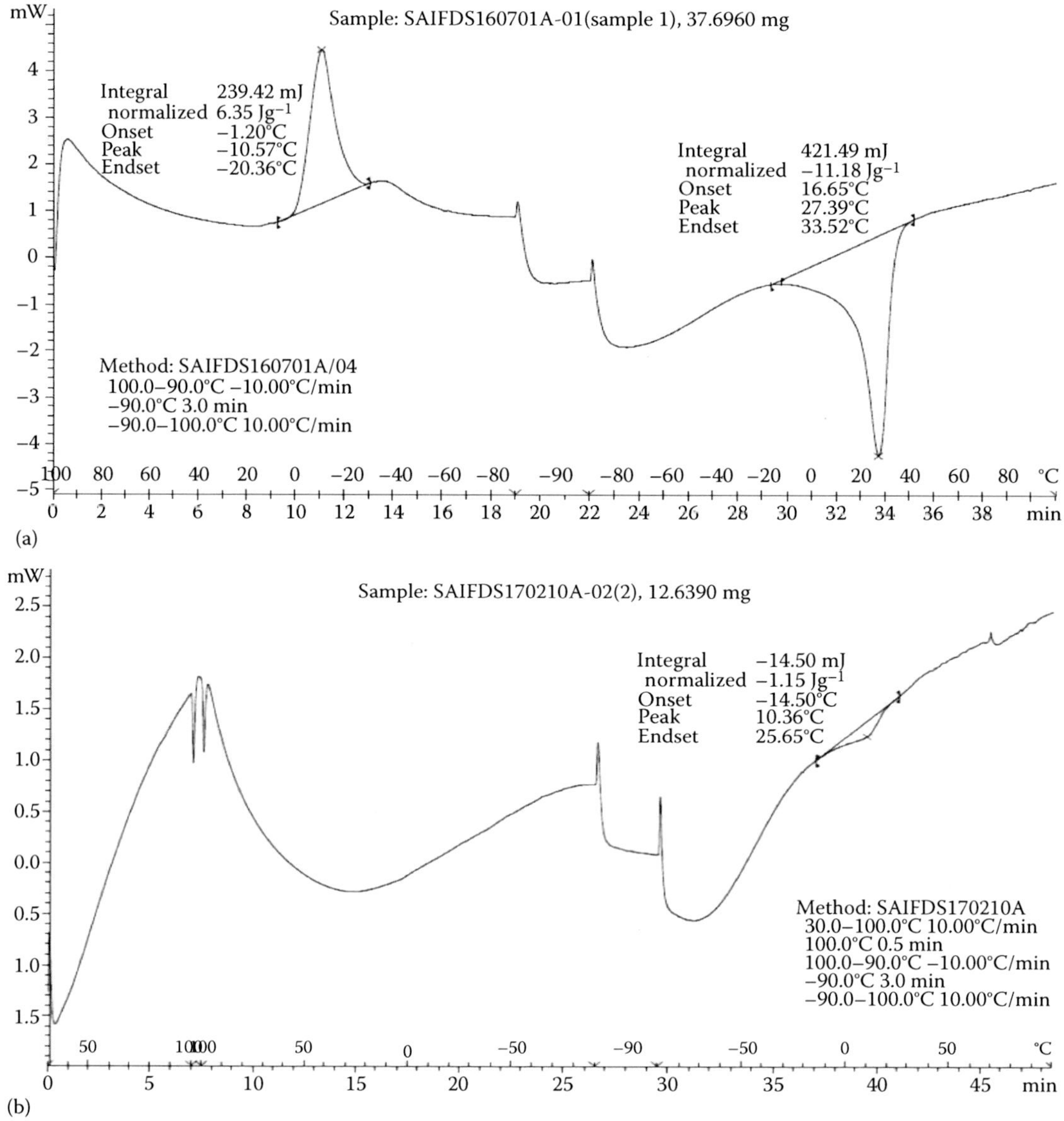

FIGURE 23.13 (a) DSC graph for laser welded sample. (b) DSC graph for PWHT sample.

namely, welding speed and laser power. As discussed in Section 23.3, heat input to the weld is mainly controlled by these two factors. When the heat input to the weld is maximum, it led to the formation of more brittle intermetallic phases. These brittle intermetallic phases reduced the tensile strength by a greater extent. These brittle intermetallic phases are generally accumulated in the grain boundaries during solidification and helps induce crack formation. Laser welding should be done either by using lower laser power or higher welding speed as it reduces the total heat input to the weld, which, in turn, reduces the brittle intermetallic phases. Ni_3Ti and Ti_2Ni were the important brittle intermetallic phases that were formed in larger proportion in Ni-rich and Ti rich NiTinol shape memory alloys, respectively. Similarly, lower power and higher welding speed in laser welding increase the cooling rate, which, in turn, helps in achieving finer grains. These fine grains also help in increasing the tensile strength. In the case of pulsed mode laser welding, impulse width and current values played a vital role in controlling the tensile strength. The laser weld obtained at lower value of

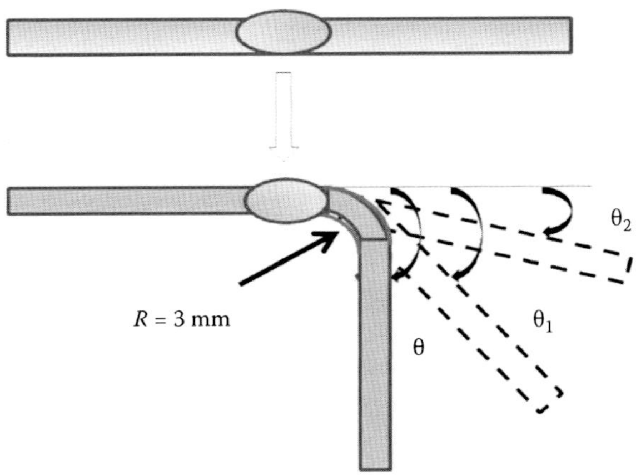

FIGURE 23.14 Schematic of bend test.

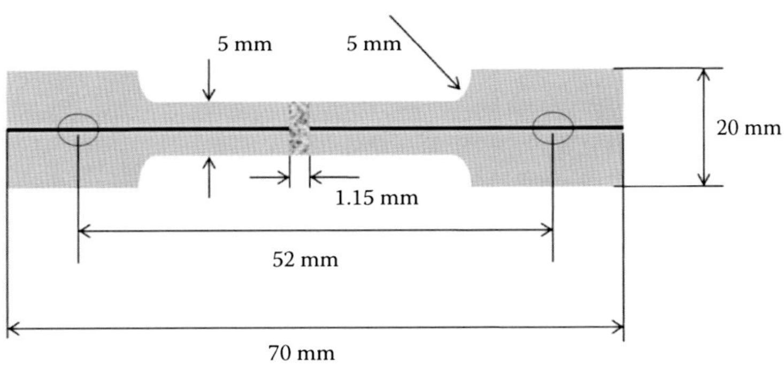

FIGURE 23.15 Schematic of tensile test specimen.

impulse width resulted in maximum tensile strength and elongation percentage. Gugel et al. (2008) concluded that the presence of Ti-rich precipitates on the grain boundaries reduced the tensile strength and ductility by a greater extent. Samples were broken at the weld center and the tensile strength of the weld was reduced by 30%. In some works, laser welding of NiTinol was done by using filler materials Nb and Ce. Nonadditive laser welded samples broke at the weld center, whereas Nb- and Ce-added weld specimens broke at the fusion line. Comparatively, Ce weld specimens possessed greater tensile strength than Nb-added and nonadditive weld specimens. Nonadditive weld revealed brittle mode of fracture, whereas Ce- and Nb-added weld shown mixed mode of fracture. The size of the cup and cones in the Ce-added laser weld was finer than the Nb-added laser weld, justifying the improved strength in the Ce-added weld. Importance should be given for sample preparation, as it helps increase the tensile strength of the laser welded samples. Immersion of NiTinol samples in a solution containing nitric acid and hydrochloric acid helped remove the oxide layer present in the sample, which helped improve the weld strength by a significant value. In case of dissimilar welding of NiTinol with stainless steel, both tensile strength and elongation percentages were greatly influenced by the filler materials and their thickness. In laser welding of NiTinol with stainless steel, the dissimilar weld tensile strength was lower than the base metal and it improved significantly when welded with filler materials like Ni, Co, Cu, and Ag. These filler materials played a significant role in the formation of brittle intermetallic phases, and the intensity of the brittle intermetallic phases varied and was greatly influenced by the filler material thickness. When Ni was used as a filler material, tensile strength and elongation percentage improved as it reduced the formation of $TiFe_2$ and $TiCr_2$ and increased the content of gamma Fe phases. At higher thickness, despite the $TiFe_2$ and $TiCr_2$ content, the formation of $TiNi_3$ phase, gas pores, and shrinkage cavities reduced the tensile strength and elongation percentage. The same phenomenon of increasing tensile strength and elongation percentage was observed at lower thickness of Co, Cu, and Ag and again started to decrease at higher thickness. However, the intermetallic phase formation varied with different filler materials. For example, in the case of copper as filler material, the good plasticity and toughness of the Cu solid solution and the lower brittleness of Cu–Ti intermetallic phases helped improve tensile strength at lower thickness, whereas at higher thickness of the Cu filler material, $TiNi_3$ phase formation in the weld reduced the tensile strength and elongation percentage. Similarly, in the case of Co as a filler material, Ti–Co–based intermetallic formation affected the strength. The other major reason for the failure of the NiTi–SS laser welded joints was higher stress concentration at the toes of the weld. Mirshekari et al. (2013) compared the laser welded NiTi–NiTi joints with laser welded NiTi–SS and found that similar joints possessed better tensile strength and elongation percentage than dissimilar joints. The formation of brittle intermetallic phases like $TiFe_2$, $TiCr_2$, TiFe, FeNi, $TiNi_3$, and Ti_2Ni was cited as the reason for the reduced tensile strength. Similar joints made between NiTi and NiTi showed ductile mode of failure, whereas the dissimilar joints made between NiTi and SS showed brittle mode of failure. Fractography obtained in laser welded NiTinol is shown in Figure 23.16a, and the river markings seen in the fractography reveal brittle mode of fracture. The crack formed in the weld due to higher heat input is shown in Figure 23.16b. In a few works, the effect of post-weld treatment on tensile strength was studied, and it was found that brittle intermetallic phase intensity increased after post-weld heat treatments, which reduced the overall strength. In other works, the pseudoelastic behavior of the laser welded joints was analyzed at different temperatures, and it was observed that the laser weld and base metal pseudoelastic behavior was almost similar (Alberty Vieria et al. 2011).

On the whole, it is observed that welding speed, laser power, and pulse width are the parameters that have to be given importance to obtain the required tensile strength in laser welding of NiTinol. The presence of brittle intermetallic phases and defects such as cracks and porosities played a vital role in controlling the tensile strength. Post-weld heat treatment is one of the options for modifying tensile strength and elongation percentage. However, importance should be given to soaking time and temperature range as they directly affect the formation of intermetallic phases.

23.5.2 Microhardness

In laser welding of NiTinol, microhardness values are measured along the transverse direction at five different points, and its average value is taken for analysis. In addition to that, from the microhardness analysis, the width of the different zones, namely, the weld zone and heat-affected zone, can be measured. Vickor's microhardness tester is used for measuring the microhardness value, and the measurement is done based on the standard ASTM E384. With respect to parameters, as discussed in previous sections, welding speed and laser power had a significant control over the microhardness values in different zones of the weld. Grain size and intermetallic phase presence in the different zones of the weld influenced the microhardness value. The microhardness value distribution is generally nonuniform throughout the different zones of the weld, which was mainly due to the varying cooling rates in different weld zones. In a few works related to laser welding of NiTinol, a trend of increasing microhardness value was observed from the weld zone to the base metal, and the exact opposite trend (i.e., decreasing hardness value from the weld metal to the base metal) was observed in few works (Deepan Bharathi Kannan et al. 2017). Falvo et al. (2005) observed no significant variation in different zones of the weld, and the weld had a microhardness value almost similar to that of base metal. Initial production process of the base metal such as cold working had a greater influence on the microhardness values. Mirshekari et al. (2013) compared the microhardness values of similar NiTi–NiTi joints with dissimilar NiTi–SS joints and found that

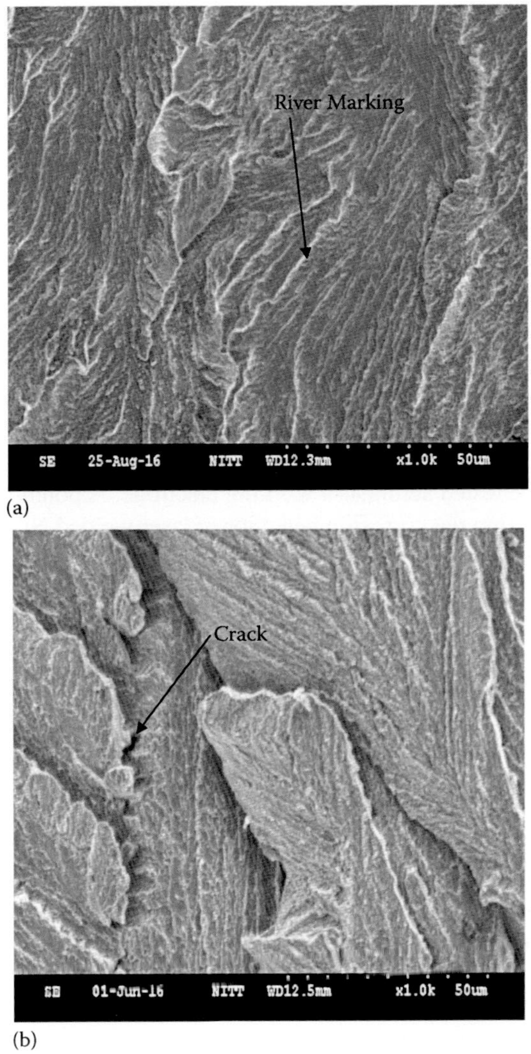

FIGURE 23.16 (a) Fractography of laser weld. (b) Fractography revealing cracks.

dissimilar joints had better microhardness values than similar joints. Increased microhardness value in dissimilar joints was mainly due to the presence of higher amounts of intermetallic phases. In the case of dissimilar joining of NiTinol with stainless steel, the TiNi side was hugely influenced by laser welding because it has lower thermal conductivity compared with stainless steel. Increased power and time resulted in coarser grain formation in the heat-affected zone, which, in turn, led to lower hardness values (Li et al. 2006). Tuissi et al. (1999) recommended using fully recrystallized base metal for laser welding as it produced weld with better hardness than the base metal. The filler material thickness used in laser welding had a significant influence on the weld hardness values. When the filler material such as nickel, cobalt, and copper was less than 50 μm, the microhardness value was higher due to the presence of brittle intermetallic phases such as $TiNi_3$, $TiFe_2$, and $TiCr_2$. With the increase in the size and amount of these brittle intermetallic phases, microhardness values changed accordingly. When the filler material thickness exceeded 50 μm, the microhardness values were greatly reduced due to the presence of increased gamma Fe phase and softer Cu–Ti and Co–Ti intermetallic phases. When laser welding was done by using filler material for joining NiTinol and stainless steel, the weld had higher hardness values than the base metal, and the stainless steel side had a comparatively lower hardness value than the TiNi side. Only very few works are reported in the literature with respect to post-weld heat treatment effect on the microhardness values, and it was found that with respect to the different heat treatment temperature and soaking time, the intermetallic phase formation varied, which, in turn, affected the hardness values.

23.6 LASER WELDING EFFECT ON CORROSION BEHAVIOR

The corrosion behavior of the laser welded NiTinol should be given utmost importance, especially because it is used for biomedical applications. When NiTinol welded samples are used as bio-implants, blood, saliva, and other body fluids get in contact with the implants, leading to deterioration of the implant with time. Steps should be taken to prevent the excess release of nickel ions into the human body as it very harmful and causes kidney and cardiovascular problems. A corrosion test is generally carried out with the help of electrochemical systems. An electrochemical system contains three electrodes, namely, a counterelectrode made of graphite, a saturated calomel reference electrode, and the sample to be tested acting as a working electrode. A potentiodynamic polarization test is mostly carried out in testing the corrosion behavior of laser welded NiTinol shape memory alloys. In the potentiodynamic polarization test, the corrosion test is carried out by varying the voltage with respect to time and the corresponding current is measured. When any sample is immersed in a solution, an electrochemical reaction takes place between the sample and the solution, resulting in the corrosion of the sample. This reaction is measured in terms of a potential called corrosion potential (ECORR) or open circuit potential. The corrosion potential value does not change with number of experiments as it is a characteristic reaction taking place between a particular sample and solution. The corrosion potential value should be closer to the positive value or should be greater than the base metal value, in order to have better corrosion resistance. Corrosion current density is another factor that is measured in potentiodynamic polarization experiments, which gives information regarding the corrosion rate. If the corrosion current density value for the given sample is minimum, then the corrosion rate is minimum. Corrosion current density (A/cm^2) is calculated from a potentiodynamic curve or tafel plot by superimposing a straight line along the linear portion of the anodic or cathodic curve obtained in the tafel plot and extrapolating it through ECORR.

The standard for carrying out corrosion test on NiTinol welded samples is ASTM G61-86. Electrochemical tests related to NiTinol were carried out in 0.9% NaCl solution, which is an equivalent of human body solution. Scan rate should be selected in such a way that the corrosion test is carried out for a long time because it results in higher accuracy. In most of the experiments related to laser welded NiTinol, the scan rate was set at 0.8 mV/s and the voltage was varied between −1 V and 1 V.

The corrosion behavior of laser welded NiTinol samples is generally affected by the surface finish, grain size, and presence of intermetallic phases. Good surface finish is the most important criterion in order to obtain good corrosion resistance. Hence, it is recommended to clean the surface using acetone, hydrochloric acid, and nitric acid to remove the oxide layers and dirt present in the surface. The corrosion resistance of the laser welded NiTinol samples depends on the stability of the TiO_2 passive layer formed over the surface. Fine grains and higher grain densities help create more nucleation sites to form a stable passive layer leading to higher corrosion resistance. Hence, it is recommended to use a small beam diameter, a higher welding speed, and a lower power to obtain better corrosion resistance in laser joining of NiTinol shape memory alloys. Yan et al. (2007) studied the corrosion behavior of laser welded NiTinol shape memory alloys and found that the TiO_2 layer formed on the weld surface helps create a chemical and physical barrier against Ni oxidation by altering the Ni oxidation path. The authors suggested that a defect-free, smoother carbide absence will help increase corrosion resistance. The authors were able to produce weld with better corrosion resistance than the base metal. Similarly, Yan et al. (2006) were able to produce weld with better corrosion resistance using the laser welding process and concluded that the absence of carbides in the weld helped improve corrosion resistance. Chan et al. (2012b) studied the stress corrosion cracking behavior of the laser welded NiTinol shape memory alloys and found that the weld was more susceptible to stress corrosion cracking than the base metal. The reason might be attributed to the presence of coarse grains in the heat-affected zone. When laser welded NiTinol samples were subjected to stress, the pits formed in the heat-affected zone aggravated the crack formation.

NiTinol is widely used as orthodontic wires because of their good biocompatibility and corrosion resistance but the problem arises when it is joined with other materials like stainless steel. Toothpaste contains fluoride, which causes brittle failure of orthodontic wires through hydrogen absorption. Mechanical movements of teeth also cause instability in the passive layer of orthodontic wires, which protects against the corrosive oral environment. The feasibility of using a composite arch wire made of NiTinol and stainless steel in dental applications was studied by Zhang et al. (2014). They were able to successfully join NiTinol with stainless steel, and they observed that with an increase in fluoride concentration, the corrosion resistance of the composite arch wire decreased. The authors also observed that Cu ions were released due to the fluoride environment exposure, and the amount of Cu ions released was well within the clinical requirement and does not cause any harm to the human body. Sevilla et al. (2008) studied the feasibility of using NiTi wires for dental applications. The authors also compared the corrosion resistance of the NiTi–NiTi laser weld with that of the CuNiTi–CuTiNi laser weld and found that NiTi–NiTi comparatively had better corrosion resistance than the Cu-added NiTinol samples. The amount of ions released into the artificial saliva medium was further studied, and it was found that it was well within the allowed level. Hence, the authors recommended the usage of laser welding of NiTinol for dental applications.

In some works, because of changes induced by thermal effect in laser welding, weld corrosion resistance was slightly decreased. In order to overcome these effects, researchers have tried post-weld treatment to improve corrosion behavior. Chan et al. (2012a) studied the effect of post-weld heat treatment on laser welded NiTinol alloy and found that, after post-weld heat treatment, laser weld has shown a significant increase in corrosion resistance. Laser welded samples had poor corrosion resistance owing to the presence of large dendritic grains and microstructural heterogeneities, which, in turn, led to the formation of an instable passive layer over the surface. In post-weld heat treatment, the temperature and soaking time played a vital role. Post weld heat treatment of laser welded samples at 350°C (1 h soaking time) showed better resistance to corrosion than the samples post-weld heat treated at 450°C (1 h soaking time). The reason for the increased corrosion resistance for post-weld heat treatment at 350°C was the formation of Ni4Ti3, which helped form a stable TiO_2 passive layer on the surface. Similarly, Chan et al. (2011) compared the corrosion behavior of as-received NiTinol with that of laser welded NiTinol, post-weld heat-treated samples at different temperatures, namely, 573 K, 773 K, 973 K, and 1173 K. The authors observed that post-weld heat treatment at temperatures above recrystallization temperature resulted in reduced corrosion resistance due to the presence of coarser grains. The authors recommended post-weld heat treatment at 573 K because it had better corrosion resistance than the as-received base metal, laser weld, and post-weld heat-treated samples at 773 K, 973 K, and 1173 K. The reason for the increased corrosion resistance in post-weld heat-treated samples at 573 K was the homogenization of the microstructure, a finer grain size, and an increased Ti/Ni ratio. The microstructure showing the corroded sample is shown in Figure 23.17.

From Figure 23.17, it can be seen that at the locations where intermetallic phases are present, the corrosion effect is severe. The reason for the severe attack in the intermetallic phases might be the poor bonding between the atoms in intermetallic phases. Even with increased formation of intermetallic compounds like Ni_4Ti_3 in the post-weld heat-treated samples, the corrosion resistance was far better than the base metal, which might be attributed to the good surface finish, smaller grain size, and higher Ti/Ni ratio. Increased Ti/Ni ratio in the post-weld heat-treated samples helped form more stable passive films, which, in turn, prevented the sample's corrosion upon exposure to the human

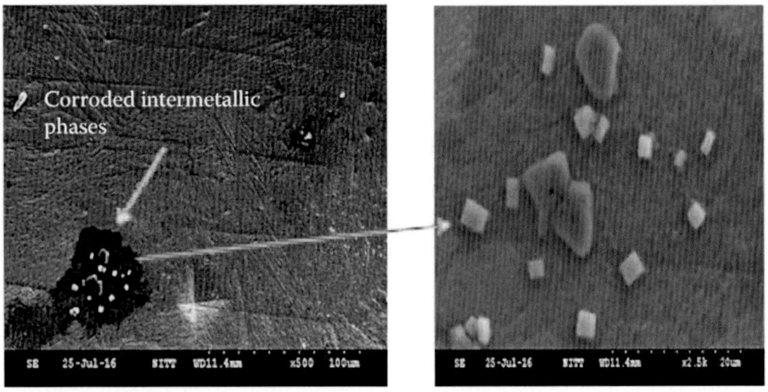

FIGURE 23.17 Microstructures of corroded samples.

body-simulated solution. On the whole, it is observed that laser welding is suitable for joining NiTinol shape memory alloys because it produced weld joints with corrosion resistance required for clinical applications. However, some novel approaches have to be found in laser welding, in order to improve the resistance against stress corrosion cracking.

23.7 CONCLUSION

In this chapter, an effort is made to review the various aspects related to laser welding of NiTinol. Laser welding is found to be a suitable process to effectively join NiTinol shape memory alloys. The major parameters that influence the microstructural, mechanical, and corrosion behavior were described in a detailed manner. The problem of reflection in laser welding can be overcome by using a suitable surface roughness on the material to be joined. Laser welding of NiTinol can be done by using a gas alternator, which has the capability to send different shielding gases alternatively and prevent the formation of porosities by effective shielding. Laser welding of NiTinol should be tried with dual beams by splitting a single laser beam, which will help improve the weld quality by preventing the weld defects formed due to higher cooling rate and by improving the keyhole stability. Laser welding should be expanded for use to other shape memory alloys such as AuCd, NiTiCr, and so on. Heat treatment of the weld has to be explored with the help of a laser source. In the future, other welding processes such as plasma arc welding and friction stir processing have to be explored for joining NiTinol shape memory alloys.

REFERENCES

Adharapurapu, R. R. 2007. Phase transformation in nickel-rich nickel–titanium alloys: Influence of strain rate, temperature, thermomechanical treatment and nickel composition on the shape memory and superelastic characteristics. University of California, San Diego.

Alberty Vieria, L., Braz Fernandes, F. M., Miranda, R. M., Silva, R. J. C., Quintino, L., Cuesta, A., and Ocana, J. L. 2011. Mechanical behaviour of Nd:YAG laser welded superelastic NiTi. *Material Science and Engineering A*. 528: 5560–5565.

Butera, F. 2008. Shape memory actuators for automotive applications. *Advanced Materials Processes* 166: 37–53.

Butera, F., Coda, A., and Vergani, G. 2007. Shape memory actuators for automotive applications. In: *Nanotec IT Newsletter*. Roma: AIRI/nanotec IT. pp. 12–16.

Caldwell, D. G., and Taylor, P. M. 1988. Artificial muscles as robotic actuators. In: *IFAC Robot Control Conference (Syroc 88)*. Karlsruhe, Germany. pp. 401–406.

Carpenter, B., and Lyons, J. 2001. EO-1 technology validation report. In: *Lightweight Flexible Solar Array Experiment*. NASA/GSFC Last updated: August 2001. p. 8.

Chan, C. W., and Man, H. C. 2010. Laser welding of thin foil nickel–titanium shape memory alloy. *Optics and Lasers in Engineering* 49: 121–126.
Chan, C. H., Man, H. C., and Yue, T. M. 2011. Effects of process parameters upon the shape memory alloy and pseudo-elastic behaviors of laser welded NiTi thin foil. *Metallurgical Materials Transaction A* 42: 2264–2270.
Chan, C. W., Man, H. C., and Yue, T. M. 2012a. Effect of post weld heat treatment on the oxide film and corrosion behaviour of laser welded shape memory NiTi wires. *Corrosion Science* 56: 158–167.
Chan, C. W., Man, H. C., and Yue, T. M. 2012b. Susceptibility to stress corrosion cracking of NiTi laser weldment in Hanks solution. *Corrosion Science* 57: 260–269.
Cleveland, M. A. 2008. Apparatus and method for releasably joining elements. US Patent 7367738B2. The Boeing Co.
Dahlgren, J. M., and Gelbart, D. 2009. System for mechanical adjustment of medical implants. US Patent 2009/0076597A12009.
Deepan Bharathi Kannan, T., Ramesh, T., and Sathiya, P. 2016. A review of similar and dissimilar micro-joining of NiTinol. *The Minerals, Metals, & Materials Society* 68: 1227–1245.
Deepan Bharathi Kannan, T., Sathiya, P., and Ramesh, T. 2017. Experimental investigation and characterization of laser welded NiTinol shape memory alloys. *Journal of Manufacturing Processes* 25: 253–261.
Duerig, T. W., and Pelton, A. R. 1994. Ti–Ni Shape memory alloys. In *Materials Properties Handbook, Titanium Alloys*. Materials Park, OH: American Society for Metals. pp. 1035–1048.
Elahinia, M., Hashemi, M., Tabesh, M., and Bhaduri, S. B. 2012. Manufacturing and processing of NiTi implants: A review. *Progress in Materials Science* 57: 911–946.
Elahinia, M. H., Seigler, T., Leo, D. J., and Ahmadian, M. 2004. Nonlinear stress-based control of a rotary SMA-actuated manipulator. *Journal of Intelligent Materials System Structure* 15: 495–508.
Falvo, A., Furgiuele, F. M., and Maletta, C. 2005. Laser welding of a NiTi alloy: Mechanical and shape memory behaviour. *Materials and Engineering A* 412: 235–240.
Fujita, H. 1989. Studies of micro actuators in Japan. In: *IEEE International Conference on Robotic Automation*. Institute of Industrial Science, Tokyo University. pp. 1559–1564.
GM. 2013. Chevrolet debuts lightweight 'smart material' on Corvette. General Motors News.
Godard, O. J., Lagoudas, M. Z., and Lagoudas, D. C. 2003. Design of space systems using shape memory alloys. In: *Smart Structures and Materials*. International Society for Optics and Photonics. pp. 545–558.
Gong, W.-H., Chen, Y.-H., and Ke, L.-M. 2011. Microstructure and properties of laser micro welded joint of TiNi shape memory alloy. *Transaction of Nonferrous Metals Society of China* 21: 2044–2048.
Gugel, H., Schuermann, A., and Theisen, W. 2008. Laser welding of NiTi wires. *Material Science and Engineering A* 481–482: 668–671.
Hodgson, D. E, Wu, M. H., and Biermann, R. J. 1990. Shape memory alloys. In *ASM Handbook*. ASM International. pp. 897–902.
Honma, D., Miwa, Y., and Iguchi. 1985. Micro robots and micro mechanisms using shape memory alloy to robotic actuators. *Robotic Systems* 2: 3–25.
Huettl, B., and Willey, C. 2000. Design and development of miniature mechanisms for small spacecraft. In: *14th AIAA/USU Small Satellite Conference*. North Logan, UT, USA: Utah State University Research Foundation. pp. 1–14.
Humbeeck, J. V. 1999. Non-medical applications of shape memory alloys. *Materials Science Engineering A*, 134–148.
Khan, M. I., and Zhou, Y. 2010. Effects of local phase conversion on the tensile loading of pulsed Nd:YAG laser processes NiTinol. *Materials Science and Engineering A* 527: 6235–6238.
Kou, S. 2003. *Welding Metallurgy*. Second Edition,Wiley Publication.
Kuribayashi, K. 1989. Millimeter size joint actuator using shape memory alloy. In: *Micro Electro Mechanical Systems, Proceedings, an Investigation of Microstructures, Sensors, Actuators, Machines and Robots*. IEEE. pp. 139–144.
Li, H., Sun, D., Cai, X., Deng, P., and Gu, X. 2013. Laser welding of TiNi shape memory alloy and stainless steel using Co filler metal. *Optics and Laser Technology* 45: 453–460.
Li, H. M., Sun, D. Q., Cai, X. L., Dong, P., and Wang, W. Q. 2012. Laser welding of TiNi shape memory alloy and stainless using Ni interlayer. *Materials & Design* 39: 285–293.
Li, M. G., Sun, D. Q., Qiu, X. M., Sun, D. X., and Yin, S. Q. 2006. Effects of laser brazing parameters on microstructure and properties of TiNi shape memory alloy and stainless steel joint. *Material Science and Engineering A* 424: 17–22.

Lim, G., Park, K., Sugihara, M., Minami, K., and Esashi, M. 1996. Future of active catheters. *Sensors and Actuators A* 56: 113–121.

Long, C. F. L., and Vezain, G. A. P. 1998. Single actuation pushing device driven by a material with form memory. US Patent 5829253: Societe Nationale Industrielle et Aerospatiale, Paris Cedex, France. p. 12.

Lortz, B. K., and Tang, A. 1998. *Separation device using a shape memory alloy retainer*. US Patent 5722709. Hughes Electronics, Los Angeles, CA, USA.

McDonald Schetky, L. 1991. Shape memory alloy applications in space systems. *Materials & Design* 12: 29–32.

Mihalcz, I. 2001. Fundamental characteristics and design method for nickel–titanium shape memory alloy. *Periodica Polytechnica Mechanical Engineering* 45: 75–86.

Mirshekari, G. R., Saatchi, A., Kermanpur, A., and Sadrnezhaad, S. K. 2013. Laser welding of NiTi shape memory alloy: Comparison of the similar and dissimilar joints to AISI 304 stainless steel. *Optics and Laser Technology* 54: 151–158.

Mirshekari, G. R., Saatchi, A., Kermanpur, A., and Sadrnezhaad, S. K. 2016. Effect of post weld heat treatment on mechanical and corrosion behaviour of NiTi and stainless steel laser welded wires. *Journal of Materials Engineering and Performance Engineering* 25: 2395–2402.

Mohamed Ali, M. S., and Takahata, K. 2010. Frequency-controlled wireless shape-memory alloy micro-actuators integrated using an electroplating bonding process. *Sensors and Actuators A* 163: 363–372.

Mohd Jani, J., Leary, M., Subic, A., and Gibson, M. 2014. A review of shape memory alloy research, applications and opportunities. *Materials & Design* 56: 1078–1113.

Pfeifer, R., Müller, C. W., Hurschler, C., Kaierle, S., Wesling, V., and Haferkamp, H. 2013. Adaptable orthopedic shape memory implants. *Procedia CIRP* 5: 253–258.

Schlossmacher, P., Haas, T., and Schussler, A. 1994. Proceedings of the First International Conference on Shape Memory and Superelastic Technologies (Pacific Grove, California). p. 85.

Schlossmacher, P., Haas, T., and Schussler, A. 1997. Laser-welding of a Ni-rich TiNi shape memory alloy: Mechanical behavior. *J. Phys. IV* 7: 251.

Sevilla, P., Martorell, F., Libenson, C., Planell, J. A., and Gil, F. J. 2008. Laser welding of NiTi orthodontic archwires for selective force application. *Journal of Material Science: Materials in Medicine* 19: 525–529.

Singh, K., and Chopra, L. 2002. Design of an improved shape memory alloy actuator for rotor blade tracking. In: *Smart Structures and Materials*. SPIE. pp. 244–266.

Sokolov, M., and Salminen, A. 2014. Improving laser beam welding efficiency. *Engineering* 6: 559–571.

Song, Y. G., Li, W. S., Li, L., and Zheng, Y. F. 2008. The influence of laser welding parameters on the microstructure and mechanical property of the as-jointed NiTi alloy wires. *Materials Letters* 62: 2325–2328.

Sreekumar, M., Nagarajan, T., Singaperumal, M., Zoppi, M., and Molfino, R. 2007. Critical review of current trends in shape memory alloy actuators for intelligent robots. *Industrial Robot: An International Journal* 34: 285–294.

Stoeckel, D. 1990. Shape memory actuators for automotive applications. *Materials & Design* 11: 302–307.

Stoeckel, D., and Tinschert, F. 1991. Temperature compensation with thermovariable rate springs in automatic transmissions. SAE Technical Paper Series: SAE.

Sun, L., and Huang, W. M. 2009. Nature of the multistage transformation in shape memory alloys upon heating. *Material Science Heat Treatment* 51: 573–578.

Suzuki, M. 1986. Rotatable door mirror for a motor vehicle. US Patent 4626085, G02B 7/18 ed. Kabushiki Kaisha Tokai Rika Denki Seisakusho, Aichi, Japan. p. 9.

Tam, B., Khan, M. I., and Zhou, Y. 2011. Mechanical and functional properties of laser welded Ti–55.8 wt pct Ni NiTinol wires. *The Mineral, Metals and Materials Society and ASM International* 42A: 2166–2175.

Terzo, G. 2006. Taking the pulse of the stent market. In: *Investment Dealers' Digest*. p. 12.

Tuissi, A., Besseghini, S., Ranucci, T., Squatrito, F., and Pozzi, M. 1999. Effect of Nd-YAG laser welding on the functional properties of the Ni-49.6 at% Ti. *Materials Science and Engineering A* 273–275: 813–817.

Van der Wijst, M. W. M. 1992. Shape memory alloys featuring NiTinol. DCT rapporten; Vol. 1992.085. Eindhoven: Technische Universiteit Eindhoven.

Vieira, L. A., Fernandes, F. M. B., Miranda, R. M., Silva, R. J. C., Quintino, L., Cuesta, A., and Ocana, J.L. 2011. Mechanical behaviour of Nd:YAG laser welded superelastic NiTi. *Materials Science and Engineering A* 528: 5560.

Yan, X.-J., Yang, D.-Z., and Liu, X. P. 2006. Electrochemical behaviour of YAG laser welded NiTi shape memory alloy. *Transaction of Nonferrous Metals Society of China* 16: 572–576.

Yan, X.-J., Yang, D.-Z., and Liu, X. P. 2007. Corrosion behavior of a laser welded NiTi shape memory alloy. *Materials Characterization* 58: 623–628.

Zhang, C., Zhao, S., Sun, X., Sun, D., and Sun, X. 2014. Corrosion of laser welded NiTi shape memory alloy and stainless steel composite wires with a copper interlayer upon exposure to fluoride and mechanical stress. *Corrosion Science* 82: 404–409.

Zhao, X., Lan, L., Sun, H., Huang, J., and Zhang, H. 2010. Mechanical properties of additive laser-welded NiTi alloy. *Materials Letters* 64: 628–631.

Zhao, X., Wang, W., Chen, L., Liu, F., and Huang, J. 2008. Microstructures of cerium added laser weld of a TiNi alloy. *Materials Letters* 62: 1551–1553.

Zoeram, A. S., Mousavi, A. A., and Mohsenifar, F. 2015. Effects of welding speed on mechanical and fracture behaviour of NiTinol laser-joints. *Weld World* 60: 11–19.

Zychowicz, R. 1992. Exterior view mirror for a motor vehicle. US Patent 5166832. Britax (GECO) SA. p. 5.

Index

A

B

C

D

E

F

G

H

M

N

O

P

Q

R

S

T

U

V

W

X

Z